APPLICATIONS TO GEOMETRY

Application	Section	Page	Mathematics that is used
Rotation Matrices	1.2, 2.1, 4.4, 6.6	22–23, 98–99, 269–270, 427–431	Matrix-vector products and matrix products
Area/Volume of a Parallelepiped	3.1	206–208	Determinants
Equations of Conic Sections	4.4	269–270	Change of [illegible]
Orthogonal Projections and Reflections	5.4, 6.1, 6.3	328–329, 366, 393–397	Dot produc[illegible]
Diagonals of a Rhombus	6.1	367	Dot products
Orthogonal Operators on the Euclidean Plane	6.5	416–419	Orthogonal matrices
Rigid Motions	6.5	419–421	Orthogonal matrices
Conic Sections	6.6	427–431	Diagonalization of symmetric matrices
Changing Shapes	6.7	443–444	Singular values

T4-AKY-382

APPLICATIONS TO OTHER AREAS OF MATHEMATICS

Application	Section	Page	Mathematics that is used
LU Decomposition	2.6	152–164	Elementary row operations
Partitioned Matrices and Block Multiplication	2.5	146–150	Matrix multiplication
Cramer's Rule	3.2	217–218	Determinants
QR Factorization	6.2	381–384	Orthogonal matrices
Singular Value Decomposition	6.7	438–452	Orthonormal bases
Pseudoinverse of a Matrix	6.7	451–452	Orthonormal bases
Spectral Decomposition	6.6	431–433	Orthonormal bases of eigenvectors
Lagrange Interpolating Polynomials	7.3	516–518	Vector spaces and dimension
Systems of Differential Equations	5.5	339–343	Diagonalization
Difference Equations	5.5	344–348	Diagonalization
Approximation by Trigonometric Polynomials	7.5	539–543	Spanning sets and dimension
Approximation by Legendre Polynomials	7.5	537–538	Gram-Schmidt process
Frobenius Norm and Frobenius Inner Product	6.6, 6.7	533–534	Trace of a matrix
Stochastic and Transition Matrices	1.2, 5.5	21–22, 334–339	Matrix arithmetic and eigenvectors
Quadratic Forms	6.6	427–431	Diagonalization

ELEMENTARY LINEAR ALGEBRA

A MATRIX APPROACH

ELEMENTARY LINEAR ALGEBRA

A MATRIX APPROACH

SECOND EDITION

LAWRENCE E. SPENCE

ARNOLD J. INSEL

STEPHEN H. FRIEDBERG

Illinois State University

Upper Saddle River, New Jersey 07458

Library of Congress Cataloging-in-Publication Data on File.

Editorial Director, Computer Science, Engineering, and Advanced Mathematics: Marcia J. Horton
Senior Editor: Holly Stark
Editorial Assistant: Jennifer Lonschein
Senior Managing Editor/Production Editor: Scott Disanno
Art Director: John Christiana
Cover Designer: Tamara Newnan
Art Editor: Thomas Benfatti
Manufacturing Manager: Alexis Heydt-Long
Manufacturing Buyer: Lisa McDowell
Senior Marketing Manager: Tim Galligan

© 2008 Pearson Education, Inc.
Pearson Education, Inc.
Upper Saddle River, NJ 07458

NOTICE:
This work is protected by U.S. copyright laws and is provided solely for the use of college instructors in reviewing course materials for classroom use. Dissemination or sale of this work, or any part (including on the World Wide Web), is not permitted.

All rights reserved. No part of this book may be reproduced in any form or by any means, without permission in writing from the publisher.

Pearson Prentice Hall™ is a trademark of Pearson Education, Inc.

MATLAB is a registered trademark of The Mathworks, Inc., 3 Apple Hill Drive, Natick, MA 01760-2098.

The author and publisher of this book have used their best efforts in preparing this book. These efforts include the development, research, and testing of the theories and programs to determine their effectiveness. The author and publisher make no warranty of any kind, expressed or implied, with regard to these programs or the documentation contained in this book. The author and publisher shall not be liable in any event for incidental or consequential damages in connection with, or arising out of, the furnishing, performance, or use of these programs.

Printed in the United States of America

10 9 8 7 6 5 4 3 2 1

ISBN: 0-13-600110-6

Pearson Education Ltd., *London*
Pearson Education Australia Pty. Ltd., *Sydney*
Pearson Education Singapore, Pte. Ltd.
Pearson Education North Asia Ltd., *Hong Kong*
Pearson Education Canada, Inc., *Toronto*
Pearson Educación de Mexico, S.A. de C.V.
Pearson Education—Japan, *Tokyo*
Pearson Education Malaysia, Pte. Ltd.
Pearson Education, Inc., *Upper Saddle River, New Jersey*

To our families:
Linda, Stephen, and Alison
Barbara, Thomas, and Sara
Ruth Ann, Rachel, Jessica, and Jeremy

CONTENTS

PREFACE

In response to concern that the first course in linear algebra was not meeting the needs of the students who take it, the Linear Algebra Curriculum Study Group was formed in January 1990. With the support of the National Science Foundation, it sponsored a workshop on the undergraduate linear algebra curriculum at the College of William and Mary in August of that year. Participants at the workshop included representatives from mathematics departments and other disciplines whose students study linear algebra.

Its recommendations,[1] published in January 1993, have inspired a number of textbooks, including this one. In its core syllabus, abstract vector spaces are regarded as a supplementary topic. What distinguishes this book from many others is its complete development of the concepts of linear algebra in $\mathcal{R}^n$ before the introduction of abstract vector spaces. We agree with the following statement by the Linear Algebra Curriculum Study Group:

> *Furthermore, an overemphasis on abstraction may overwhelm beginning students to the point where they leave the course with little understanding or mastery of the basic concepts they may need in later courses and their careers.*

Although we believe that the first linear algebra course is a natural one in which to introduce mathematical theory and proofs, this can be accomplished without discussion of abstract vector spaces. In addition, if abstract vector spaces are included in the first linear algebra course, we believe that they can be taught more effectively after a thorough presentation of the essential topics in the more familiar context of $\mathcal{R}^n$. Nevertheless, we have written this book so that it is possible to teach concepts in the context of abstract vector spaces immediately after the corresponding topics are discussed in $\mathcal{R}^n$. (Suggestions for doing so are included in the sample course descriptions on page xiv.)

Although there is no use of calculus until the introduction of vector spaces in Chapter 7, the material is aimed at students who have the mathematical maturity obtained by having taken one year of calculus. The core topics can be comfortably covered within one semester, but there is adequate material for a two-quarter course.

PEDAGOGICAL APPROACH

This text is written for a matrix-oriented course, as recommended by the Linear Algebra Curriculum Study Group. In our experience, such a course results in greater understanding of the concepts of linear algebra and serves the needs of students in many disciplines. It begins with the study of matrices, vectors, and systems of linear equations and gradually leads to more complicated concepts and general principles, such as linear independence, subspaces, and bases. As mentioned, this text develops all the core content of linear algebra in $\mathcal{R}^n$ before introducing abstract vector spaces. This provides students additional opportunities to visualize concepts in the familiar context of the Euclidean plane and 3-space before encountering the abstraction of vector spaces.

Our approach is based on an early introduction of the rank of a matrix. This concept is then encountered in other contexts throughout the book. For example, the

[1] David Carlson, Charles R. Johnson, David C. Lay, and A. Duane Porter. "The Linear Algebra Curriculum Study Group Recommendations for the First Course in Linear Algebra." *The College Mathematics Journal*, 24 (1993), pp. 41–46.

rank of a matrix is used initially to check if solutions of a system of linear equations exist and are unique. Later, it is used to test if sets are linearly independent or are spanning sets for $\mathcal{R}^n$. Then, in Chapter 2, it is used to determine whether linear transformations are one-to-one or onto.

Even though a course taught from this book may devote less time to the study of abstract mathematics than a more traditional course, we have found that it s till serves as an excellent prerequisite to abstract algebra and an abstract second course in linear algebra, such as one using our text *Linear Algebra*.

TECHNOLOGY

The Linear Algebra Study Group recommends that technology be used in the first course in linear algebra. In our experience, the use of technology, whether through computer software or supercalculators, greatly enhances a course taught from this book by freeing students from tedious computations and enabling them to concentrate on conceptual understanding.

Most sections include exercises designed to be worked by means of MATLAB or similar technology. Additional technology exercises are found at the end of each chapter. For MATLAB users, our website contains data files and M-files that can be downloaded.

For the convenience of those wishing to use technology, we have added an appendix (Appendix D) with an introduction to MATLAB. It provides sufficient background to prepare students to perform the calculations required for this book and to work the technology exercises.

EXAMPLES AND PRACTICE PROBLEMS

Our examples motivate and illustrate definitions and theoretical results. These are written to be understood by students so that an instructor need not discuss each example in class. In fact, in our own teaching, we almost never discuss examples from the text, but rather present similar examples, leaving the text examples to be read by students. Many examples are accompanied by similar practice problems that enable students to test their understanding of the material in the text. Complete solutions of these practice problems are included within the text in order to help prepare students to work the exercises in each section.

EXERCISES

We felt that the first edition of this book had an ample number of computational exercises. Yet some of our conscientious students asked us for more. Therefore we have significantly increased the number of computational exercises in this edition. As in the first edition, except for the specially marked exercises that use technology, all our computational exercises are designed so that the calculations involve "nice" numbers. This permits students to concentrate on the linear algebra concepts rather than the computations.

Most sections include approximately 20 true/false exercises that are designed to test a student's understanding of the conceptual ideas in each section. In response to reviewer suggestions, we have moved these to follow the straightforward computational exercises so that students can gain confidence by working computational exercises before encountering the true/false exercises. Answers to all true/false questions are given in the book, so students will know if they have misunderstood some concept.

For a proof-oriented course, we have included a significant number of accessible exercises requiring proofs. These are ordered according to difficulty.

Finally, each chapter ends with a set of review exercises that provide practice with all the main topics in each chapter. The exercises are designed for students to use as preparation for a chapter examination. As noted, there are also technology exercises at the end of each chapter and in the exercise sets of most sections.

For a list of suggested exercises for each section, see our website.

APPLICATIONS

This book includes a wide variety of applications, and they are given when the necessary prerequisites have been introduced. There are applications to economics (the Leontief input–output model), electrical networks (current flow through an electrical network), population change (the Leslie matrix), traffic flow, scheduling ($(0, 1)$-matrices), anthropology, Google searches, counting problems (difference equations), predator–prey models and harmonic motion (systems of differential equations), least-squares approximation, computer graphics, principal component analysis, and music (applying trigonometric polynomials).

These applications are entirely independent and so may be taught according to the interests of the instructor and students. Although we do not assume that any particular applications will be covered, we believe that the core material is greatly enhanced when its use is demonstrated through applications.

ORGANIZATION OF THE TEXT

In Chapter 1, students are introduced to matrices, vectors, and systems of linear equations. The study of linear combinations of vectors leads naturally to the span of a set of vectors and the concepts of linear dependence and independence. The rank and nullity of a matrix are introduced early and used to determine when a system of equations has solutions, and how many solutions there are.

In Chapter 2, we introduce matrix multiplication, along with matrix inverses and linear transformations. The relationship between linear transformations and matrices is emphasized so that matrix techniques can be used to answer questions such as whether a linear transformation is one-to-one or onto.

Because we use determinants mainly in the context of eigenvalues, we provide a short but complete treatment in Chapter 3.

The important topics of subspaces, bases, and dimension are covered in Chapter 4. We follow these concepts with a discussion of coordinate systems and matrix representations of linear transformations.

Although Chapters 3 and 4 can be interchanged, we prefer to cover determinants before subspaces so that there is no delay between the discussions of change of coordinates and diagonalization in Chapter 5. Chapter 5 contains the important results about eigenvalues, eigenvectors, and diagonalization of matrices and linear operators.

With the introduction of orthogonality in Chapter 6, we are able to emphasize the geometry of vectors, matrices, and linear transformations. The important applications of orthogonal projections and least-squares lines illustrate the usefulness of these ideas. We continue with a discussion of orthogonal matrices and operators and the study of symmetric matrices. The chapter concludes with the singular value decomposition of a matrix and applications to principal component analysis and computer graphics.

Chapter 7 introduces abstract vector spaces. Because of the careful foundation that has been developed in Euclidean spaces, most of the concepts in earlier chapters—such as span, linear independence, and subspace—are easily generalized.

We focus mainly on function and matrix spaces. For example, a nice application in this context arises when Lagrange interpolating polynomials are used to find a basis for a polynomial space. Differential operators and integrals are now explored as special cases of linear transformations. Finally, inner product spaces are introduced, and the Gram–Schmidt process is applied to produce the Legendre polynomials. Least-squares theory and trigonometric polynomials are used to explore periodic motion in the setting of musical notes.

Although we prefer to introduce abstract vector spaces only after the complete development of all the core topics in $\mathcal{R}^n$, the book is written to allow abstract vector spaces to be introduced earlier, alongside the corresponding topics in $\mathcal{R}^n$. For details, see the sample syllabi that follow.

We have included a number of optional topics, for example, the LU, QR, and singular value decomposition, of a matrix, the spectral decomposition of a symmetric matrix, Lagrange interpolating polynomials, block multiplication, the Moore–Penrose generalized inverse, and quadratic forms. In addition, there are many applications of the subject matter throughout the book.

SAMPLE SYLLABI

The following list shows the sections of this book that cover the Linear Algebra Curriculum Study Group's core syllabus:

Linear Algebra Curriculum Study Group's Core Syllabus
Matrix Addition and Multiplication (4 days: Sections 1.1, 1.2, 2.1, and 2.5)
Systems of Linear Equations (5 days: Sections 1.3, 1.4, 2.3, 2.4, 2.6)
Determinants (2 days: Sections 3.1 and 3.2)
Properties of $\mathcal{R}^n$ (10 days: Sections 1.2, 1.6, 1.7, 4.1, 4.2, 4.3, 2.7, 2.8, 6.1, and 6.2)
Eigenvalues and Eigenvectors (4 days: Sections 5.1, 5.2, 5.3, and 6.5)
More on Orthogonality (3 days: Sections 6.2, 6.3, and 6.4)

Suggested Syllabus for a One-Semester 3-Credit Course
Chapter 1: Sections 1.1, 1.2, 1.3, 1.4, 1.6, 1.7
Chapter 2: Sections 2.1, 2.3, 2.4, 2.7, 2.8
Chapter 3: Sections 3.1, 3.2 (omitting optional material)
Chapter 4: Sections 4.1, 4.2, 4.3, 4.4
Chapter 5: Sections 5.1, 5.2, 5.3
Chapter 6: Sections 6.1, 6.2, 6.3, 6.4, 6.5
Supplementary material selected from Sections 1.5, 2.2, 2.6, 5.5, 6.6, 6.7, 7.1, 7.2, 7.3, 7.5

Suggested Syllabus for a One-Semester 3-Credit Course that Integrates Abstract Vector Spaces with $\mathcal{R}^n$
Chapter 1: Sections 1.1, 1.2, 1.3, 1.4, 1.6, 1.7
Chapter 2: Sections 2.1, 2.3, 2.4, 2.7, 2.8
Chapter 3: Sections 3.1, 3.2 (omitting optional material)
Chapter 4: Sections 4.1, 7.1, 7.2, 4.2, 4.3, 4.4, 7.3
Chapter 5: Sections 5.1, 5.2, 7.4, 5.3
Chapter 6: Sections 6.1, 6.2, 6.3, 6.4, 7.5, 6.5

Suggested Syllabus for a Two-Quarter Sequence of 3-Credit Courses
Chapter 1: Sections 1.1, 1.2, 1.3, 1.4, 1.5, 1.6, 1.7
Chapter 2: Sections 2.1, 2.3, 2.4, 2.5, 2.6, 2.7, 2.8

Chapter 3: Sections 3.1, 3.2
Chapter 4: Sections 4.1, 4.2, 4.3, 4.4, 4.5
Chapter 5: Sections 5.1, 5.2, 5.3, 5.5
Chapter 6: Sections 6.1, 6.2, 6.3, 6.4, 6.5, 6.6
Supplementary material selected from Sections 2.2, 6.6, 6.7, 7.1, 7.2, 7.3, 7.4, 7.5

SUPPLEMENTS

A *Student Solutions Manual* (ISBN 0-13-239734-X) is available for purchase by students. It includes solutions of all the true/false exercises and over half of the odd-numbered exercises.

An *Instructor's Solutions Manual* that includes solutions of all the exercises is available from the publisher.

ACKNOWLEDGMENTS

This second edition has been greatly improved by the comments of the following reviewers: Adam Avilez (Mesa Community College), Roe Goodman (Rutgers University), Chungwu Ho (Evergreen Valley College), Steve Kaliszewski (Arizona State University), Noah Rhee (University of Missouri-Kansas City), and Edward Soares (College of the Holy Cross). We especially appreciate the many detailed comments provided by Jane Day (San Jose State University).

In addition, we are indebted to Development Editor Paul Trow, who provided significant editorial and mathematical suggestions about several chapters of our manuscript and to W.R. Winfrey (Concord College) for writing the introductory application for each chapter. Finally, we would like to express our appreciation to Senior Editor Holly Stark and Senior Managing Editor Scott Disanno of Pearson Prentice Hall for their help during the writing and production of this book.

To find the latest information about this book, consult our home page on the World Wide Web.

We encourage comments, which can be sent to us by e-mail or ordinary post. Our home page and e-mail addresses are as follows:

home page: **http://www.cas.ilstu.edu/math/matrix**
e-mail: **matrix@math.ilstu.edu**

LAWRENCE E. SPENCE
ARNOLD J. INSEL
STEPHEN H. FRIEDBERG

TO THE STUDENT

Linear algebra is concerned with *vectors* and *matrices,* and with special functions called *linear transformations* that are defined on vectors. Since vectors arise in a wide variety of settings, linear algebra can be applied to many disciplines. In fact, it is one of the most important tools of applied mathematics. In this book, we present applications of linear algebra to economics, physics, biology, statistics, computer graphics, and other fields.

Like most areas of mathematics, linear algebra has its own terminology and notation. To be successful in your study of linear algebra, you must be able to solve problems and communicate ideas with its language and symbolism. Developing these abilities requires more than just attending lectures or casually reading this book—it requires active involvement with the subject matter. One of the best ways to be involved with mathematics is by working exercises. Most sections include practice problems, which are very similar to the examples they follow. These are intended for you to work as you read through the section to give you practice with certain basic computations and concepts before you begin the exercises. Complete solutions to the practice problems are given at the end of the section. The exercise sets in this book begin with problems that offer practice in basic computations and true/false questions that test your understanding of important ideas in the section. Then come exercises that ask for conjectures, explanations, or justifications. Finally, most exercise sets end with questions that involve technology. These different types of exercises help you learn different aspects of linear algebra.

Not only do the exercises help you to check your understanding of important concepts, but they also provide you an opportunity to practice the vocabulary and symbolism that you are learning. For this reason, regular work on exercises is essential for success.

In mathematics courses prior to linear algebra, the emphasis is often on performing calculations. In linear algebra, however, you are expected to understand the concepts and facts that are the basis for computations. In order to learn these concepts, you must be able to use the terminology and notation of linear algebra; so begin by learning the definitions of important terms and illustrate them with examples. The key results in the text are usually found in theorems or statements enclosed in a box. Pay particular attention to these, being certain that you understand thoroughly what is being said. Then try to express the result in your own words. If you can't communicate an idea in writing, then you probably don't understand it well.

In addition, we offer four specific suggestions that will enable you to get the most from your study of linear algebra.

- **Carefully read each section *before* the classroom discussion occurs.**

  Some students use a textbook only as a source of examples when working exercises. This approach does not allow them to get the full benefit from either the textbook or the classroom discussions. By reading the text before a discussion occurs, you get an overview of the material to be discussed and know what you understand and what you don't. This enables you to be a more active participant in the classroom discussion. Although some learning occurs when you replicate examples from the text, this by itself is not sufficient for you to understand the material. Reading the text before the classroom discussion helps you to learn the material more quickly and prepares you to do the exercises.

- **Prepare regularly for each class.**

  You cannot expect to learn to play the piano merely by attending lessons once a week—long and careful practice between lessons is necessary. So it is with linear algebra. At the least, you should study the material presented in class. Often, however, the current lesson builds on previous material that must be reviewed. Usually, there are exercises to work that deal with the material presented in class. In addition, you should prepare for the next class by reading ahead.

  Each new lesson usually introduces several important concepts or definitions that must be learned in order for subsequent sections to be understood. As a result, falling behind in your study by even a single day prevents you from understanding the new material that follows. Many of the concepts of linear algebra are deep; to fully understand these ideas requires time. It is simply not possible to absorb an entire chapter's worth of material in a single day. To be successful, you must learn the new material as it arises and not wait to study until you are less busy or until an exam is imminent.

- **Ask questions of yourself and others.**

  Mathematics is not a spectator sport. It can be learned only by the interaction of study and questioning. Certain natural questions arise when a new topic is introduced: What purpose does it serve? How is the new topic related to previous ones? What examples illustrate the topic? For a new theorem, one might also ask: Why is it true? How does it relate to previous theorems? Why is it useful? Be sure you don't accept something as true unless you believe it. If you are not convinced that a statement is correct, you should ask for further details.

- **Review often.**

  As you attempt to understand new material, you may become aware that you have forgotten some previous material or that you haven't understood it well enough. By relearning such material, you not only gain a deeper understanding of previous topics, but you also enable yourself to learn new ideas more quickly and more deeply. When a new concept is introduced, search for related concepts or results and write them on paper. This enables you to see more easily the connections between new ideas and previous ones. Moreover, expressing ideas in writing helps you to learn, because you must think carefully about the material as you write. A good test of your understanding of a section of the textbook is to ask yourself if, with the book closed, you can explain in writing what the section is about. If not, you will benefit from reading the section again.

We hope that your study of linear algebra is successful and that you take from the subject concepts and techniques that are useful in future courses and in your career.

1 INTRODUCTION

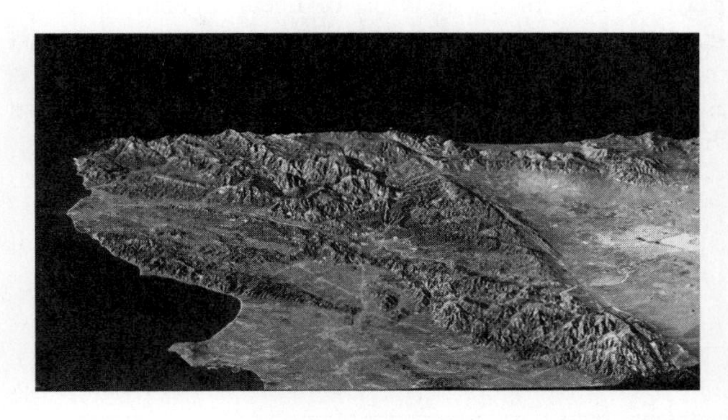

For computers to process digital images, whether satellite photos or x-rays, there is a need to recognize the edges of objects. Image edges, which are rapid changes or discontinuities in image intensity, reflect a boundary between dissimilar regions in an image and thus are important basic characteristics of an image. They often indicate the physical extent of objects in the image or a boundary between light and shadow on a single surface or other regions of interest.

The figures on the left indicate changes in image intensity of the figures above, when moving from right to left. We see that real intensities can change rapidly, but not instantaneously. In principle, the edge may be found by looking for very large changes over small distances.

However, a digital image is discrete rather than continuous: it is a matrix of nonnegative entries that provide numerical descriptions of the shades of gray for the pixels in the image, where the entries vary from 0 for a white pixel to 1 for a black pixel. An analysis must be done using the discrete analog of the derivative to measure the rate of change of image intensity in two directions.

The *Sobel* matrices, $S_1 = \begin{bmatrix} -1 & 0 & 1 \\ -2 & 0 & 2 \\ -1 & 0 & 1 \end{bmatrix}$ and $S_2 = \begin{bmatrix} 1 & 2 & 1 \\ 0 & 0 & 0 \\ -1 & -2 & -1 \end{bmatrix}$ provide a method for measuring these intensity changes. Apply the Sobel matrices S_1 and S_2 in turn to the 3x3 subimage centered on each pixel in the original image. The results are the changes of intensity near the pixel in the horizontal and the vertical directions, respectively. The ordered pair of numbers that are obtained is a vector in the plane that provides the direction and magnitude of the intensity change at the pixel. This vector may be thought of as the discrete analog of the gradient vector of a function of two variables studied in calculus.

Replace each of the original pixel values by the lengths of these vectors, and choose an appropriate threshold value. The final image, called the *thresholded image*, is obtained by changing to black every pixel for which the length of the vector is greater than the threshold value, and changing to white all the other pixels. (See the images below.)

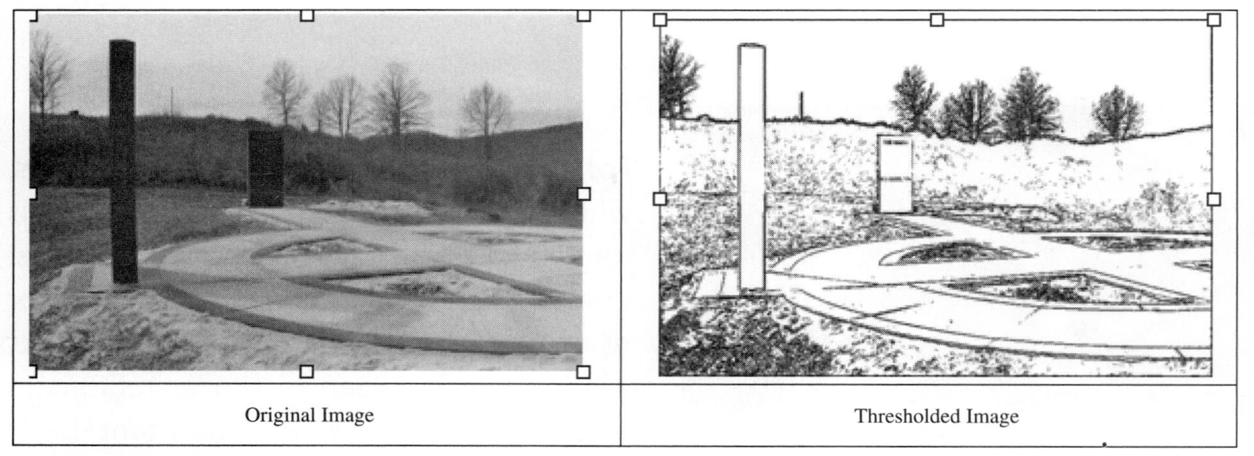

Original Image | Thresholded Image

Notice how the edges are emphasized in the thresholded image. In regions where image intensity is constant, these vectors have length zero, and hence the corresponding regions appear white in the thresholded image. Likewise, a rapid change in image intensity, which occurs at an edge of an object, results in a relatively dark colored boundary in the thresholded image.

CHAPTER 1 MATRICES, VECTORS, AND SYSTEMS OF LINEAR EQUATIONS

The most common use of linear algebra is to solve systems of linear equations, which arise in applications to such diverse disciplines as physics, biology, economics, engineering, and sociology. In this chapter, we describe the most efficient algorithm for solving systems of linear equations, *Gaussian elimination.* This algorithm, or some variation of it, is used by most mathematics software (such as MATLAB).

We can write systems of linear equations compactly, using arrays called *matrices* and *vectors*. More importantly, the arithmetic properties of these arrays enable us to compute solutions of such systems or to determine if no solutions exist. This chapter begins by developing the basic properties of matrices and vectors. In Sections 1.3 and 1.4, we begin our study of systems of linear equations. In Sections 1.6 and 1.7, we introduce two other important concepts of vectors, namely, generating sets and linear independence, which provide information about the existence and uniqueness of solutions of a system of linear equations.

1.1 MATRICES AND VECTORS

Many types of numerical data are best displayed in two-dimensional arrays, such as tables.

For example, suppose that a company owns two bookstores, each of which sells newspapers, magazines, and books. Assume that the sales (in hundreds of dollars) of the two bookstores for the months of July and August are represented by the following tables:

	July	
Store	1	2
Newspapers	6	8
Magazines	15	20
Books	45	64

and

	August	
Store	1	2
Newspapers	7	9
Magazines	18	31
Books	52	68

The first column of the July table shows that store 1 sold \$1500 worth of magazines and \$4500 worth of books during July. We can represent the information on July sales more simply as

$$\begin{bmatrix} 6 & 8 \\ 15 & 20 \\ 45 & 64 \end{bmatrix}.$$

Such a rectangular array of real numbers is called a *matrix*.[1] It is customary to refer to real numbers as **scalars** (originally from the word *scale*) when working with a matrix. We denote the set of real numbers by $\mathcal{R}$.

Definitions A **matrix** (*plural, matrices*) is a rectangular array of scalars. If the matrix has m rows and n columns, we say that the **size** of the matrix is $\boldsymbol{m}$ **by** $\boldsymbol{n}$, written $m \times n$. The matrix is **square** if $m = n$. The scalar in the ith row and jth column is called the **(i, j)-entry** of the matrix.

If A is a matrix, we denote its (i,j)-entry by a_{ij}. We say that two matrices A and B are **equal** if they have the same size and have equal corresponding entries; that is, $a_{ij} = b_{ij}$ for all i and j. Symbolically, we write $A = B$.

In our bookstore example, the July and August sales are contained in the matrices

$$B = \begin{bmatrix} 6 & 8 \\ 15 & 20 \\ 45 & 64 \end{bmatrix} \quad \text{and} \quad C = \begin{bmatrix} 7 & 9 \\ 18 & 31 \\ 52 & 68 \end{bmatrix}.$$

Note that $b_{12} = 8$ and $c_{12} = 9$, so $B \neq C$. Both B and C are 3×2 matrices. Because of the context in which these matrices arise, they are called *inventory matrices.*

Other examples of matrices are

$$\begin{bmatrix} \frac{2}{3} & -4 & 0 \\ \pi & 1 & 6 \end{bmatrix}, \quad \begin{bmatrix} 3 \\ 8 \\ 4 \end{bmatrix}, \quad \text{and} \quad \begin{bmatrix} -2 & 0 & 1 & 1 \end{bmatrix}.$$

The first matrix has size 2×3, the second has size 3×1, and the third has size 1×4.

Practice Problem 1 ▶ Let $A = \begin{bmatrix} 4 & 2 \\ 1 & 3 \end{bmatrix}$.

(a) What is the $(1, 2)$-entry of A?

(b) What is a_{22}? ◀

Sometimes we are interested in only a part of the information contained in a matrix. For example, suppose that we are interested in only magazine and book sales in July. Then the relevant information is contained in the last two rows of B; that is, in the matrix E defined by

$$E = \begin{bmatrix} 15 & 20 \\ 45 & 64 \end{bmatrix}.$$

E is called a *submatrix* of B. In general, a **submatrix** of a matrix M is obtained by deleting from M entire rows, entire columns, or both. It is permissible, when forming a submatrix of M, to delete none of the rows or none of the columns of M. As another example, if we delete the first row and the second column of B, we obtain the submatrix

$$\begin{bmatrix} 15 \\ 45 \end{bmatrix}.$$

---

[1] James Joseph Sylvester (1814–1897) coined the term *matrix* in the 1850s.

MATRIX SUMS AND SCALAR MULTIPLICATION

Matrices are more than convenient devices for storing information. Their usefulness lies in their *arithmetic*. As an example, suppose that we want to know the total numbers of newspapers, magazines, and books sold by both stores during July and August. It is natural to form one matrix whose entries are the sum of the corresponding entries of the matrices B and C, namely,

$$\begin{array}{r} \text{Store} \\ \text{Newspapers} \\ \text{Magazines} \\ \text{Books} \end{array} \begin{array}{c} \begin{array}{cc} 1 & \;\;2 \end{array} \\ \begin{bmatrix} 13 & 17 \\ 33 & 51 \\ 97 & 132 \end{bmatrix} \end{array}.$$

If A and B are $m \times n$ matrices, the **sum** of A and B, denoted by $A + B$, is the $m \times n$ matrix obtained by adding the corresponding entries of A and B; that is, $A + B$ is the $m \times n$ matrix whose (i,j)-entry is $a_{ij} + b_{ij}$. Notice that the matrices A and B must have the same size for their sum to be defined.

Suppose that in our bookstore example, July sales were to double in all categories. Then the new matrix of July sales would be

$$\begin{bmatrix} 12 & 16 \\ 30 & 40 \\ 90 & 128 \end{bmatrix}.$$

We denote this matrix by $2B$.

Let A be an $m \times n$ matrix and c be a scalar. The **scalar multiple** cA is the $m \times n$ matrix whose entries are c times the corresponding entries of A; that is, cA is the $m \times n$ matrix whose (i,j)-entry is ca_{ij}. Note that $1A = A$. We denote the matrix $(-1)A$ by $-A$ and the matrix $0A$ by O. We call the $m \times n$ matrix O in which each entry is 0 the $m \times n$ **zero matrix**.

Example 1 Compute the matrices $A + B$, $3A$, $-A$, and $3A + 4B$, where

$$A = \begin{bmatrix} 3 & 4 & 2 \\ 2 & -3 & 0 \end{bmatrix} \quad \text{and} \quad B = \begin{bmatrix} -4 & 1 & 0 \\ 5 & -6 & 1 \end{bmatrix}.$$

Solution We have

$$A + B = \begin{bmatrix} -1 & 5 & 2 \\ 7 & -9 & 1 \end{bmatrix}, \quad 3A = \begin{bmatrix} 9 & 12 & 6 \\ 6 & -9 & 0 \end{bmatrix}, \quad -A = \begin{bmatrix} -3 & -4 & -2 \\ -2 & 3 & 0 \end{bmatrix},$$

and

$$3A + 4B = \begin{bmatrix} 9 & 12 & 6 \\ 6 & -9 & 0 \end{bmatrix} + \begin{bmatrix} -16 & 4 & 0 \\ 20 & -24 & 4 \end{bmatrix} = \begin{bmatrix} -7 & 16 & 6 \\ 26 & -33 & 4 \end{bmatrix}.$$

Just as we have defined addition of matrices, we can also define **subtraction**. For any matrices A and B of the same size, we define $A - B$ to be the matrix obtained by subtracting each entry of B from the corresponding entry of A. Thus the (i,j)-entry of $A - B$ is $a_{ij} - b_{ij}$. Notice that $A - A = O$ for all matrices A.

If, as in Example 1, we have

$$A = \begin{bmatrix} 3 & 4 & 2 \\ 2 & -3 & 0 \end{bmatrix}, \quad B = \begin{bmatrix} -4 & 1 & 0 \\ 5 & -6 & 1 \end{bmatrix}, \quad \text{and} \quad O = \begin{bmatrix} 0 & 0 & 0 \\ 0 & 0 & 0 \end{bmatrix},$$

then

$$-B = \begin{bmatrix} 4 & -1 & 0 \\ -5 & 6 & -1 \end{bmatrix}, \quad A - B = \begin{bmatrix} 7 & 3 & 2 \\ -3 & 3 & -1 \end{bmatrix}, \quad \text{and} \quad A - O = \begin{bmatrix} 3 & 4 & 2 \\ 2 & -3 & 0 \end{bmatrix}.$$

Practice Problem 2 ▶ Let $A = \begin{bmatrix} 2 & -1 & 1 \\ 3 & 0 & -2 \end{bmatrix}$ and $B = \begin{bmatrix} 1 & 3 & 0 \\ 2 & -1 & 4 \end{bmatrix}$. Compute the following matrices:

(a) $A - B$
(b) $2A$
(c) $A + 3B$ ◀

We have now defined the operations of matrix addition and scalar multiplication. The power of linear algebra lies in the natural relations between these operations, which are described in our first theorem.

THEOREM 1.1

(Properties of Matrix Addition and Scalar Multiplication) Let A, B, and C be $m \times n$ matrices, and let s and t be any scalars. Then

(a) $A + B = B + A$. (commutative law of matrix addition)
(b) $(A + B) + C = A + (B + C)$. (associative law of matrix addition)
(c) $A + O = A$.
(d) $A + (-A) = O$.
(e) $(st)A = s(tA)$.
(f) $s(A + B) = sA + sB$.
(g) $(s + t)A = sA + tA$.

PROOF We prove parts (b) and (f). The rest are left as exercises.

(b) The matrices on each side of the equation are $m \times n$ matrices. We must show that each entry of $(A + B) + C$ is the same as the corresponding entry of $A + (B + C)$. Consider the (i,j)-entries. Because of the definition of matrix addition, the (i,j)-entry of $(A + B) + C$ is the sum of the (i,j)-entry of $A + B$, which is $a_{ij} + b_{ij}$, and the (i,j)-entry of C, which is c_{ij}. Therefore this sum equals $(a_{ij} + b_{ij}) + c_{ij}$. Similarly, the (i,j)-entry of $A + (B + C)$ is $a_{ij} + (b_{ij} + c_{ij})$. Because the associative law holds for addition of scalars, $(a_{ij} + b_{ij}) + c_{ij} = a_{ij} + (b_{ij} + c_{ij})$. Therefore the (i,j)-entry of $(A + B) + C$ equals the (i,j)-entry of $A + (B + C)$, proving (b).

(f) The matrices on each side of the equation are $m \times n$ matrices. As in the proof of (b), we consider the (i,j)-entries of each matrix. The (i,j)-entry of $s(A + B)$ is defined to be the product of s and the (i,j)-entry of $A + B$, which is $a_{ij} + b_{ij}$. This product equals $s(a_{ij} + b_{ij})$. The (i,j)-entry of $sA + sB$ is the sum of the (i,j)-entry of sA, which is sa_{ij}, and the (i,j)-entry of sB, which is sb_{ij}. This sum is $sa_{ij} + sb_{ij}$. Since $s(a_{ij} + b_{ij}) = sa_{ij} + sb_{ij}$, (f) is proved. ■

Because of the associative law of matrix addition, sums of three or more matrices can be written unambiguously without parentheses. Thus we may write $A + B + C$ instead of either $(A + B) + C$ or $A + (B + C)$.

MATRIX TRANSPOSES

In the bookstore example, we could have recorded the information about July sales in the following form:

Store	Newspapers	Magazines	Books
1	6	15	45
2	8	20	64

This representation produces the matrix

$$\begin{bmatrix} 6 & 15 & 45 \\ 8 & 20 & 64 \end{bmatrix}.$$

Compare this with

$$B = \begin{bmatrix} 6 & 8 \\ 15 & 20 \\ 45 & 64 \end{bmatrix}.$$

The rows of the first matrix are the columns of B, and the columns of the first matrix are the rows of B. This new matrix is called the *transpose* of B. In general, the **transpose** of an $m \times n$ matrix A is the $n \times m$ matrix denoted by A^T whose (i,j)-entry is the (j,i)-entry of A.

The matrix C in our bookstore example and its transpose are

$$C = \begin{bmatrix} 7 & 9 \\ 18 & 31 \\ 52 & 68 \end{bmatrix} \quad \text{and} \quad C^T = \begin{bmatrix} 7 & 18 & 52 \\ 9 & 31 & 68 \end{bmatrix}.$$

Practice Problem 3 ▶ Let $A = \begin{bmatrix} 2 & -1 & 1 \\ 3 & 0 & -2 \end{bmatrix}$ and $B = \begin{bmatrix} 1 & 3 & 0 \\ 2 & -1 & 4 \end{bmatrix}$. Compute the following matrices:

(a) A^T

(b) $(3B)^T$

(c) $(A+B)^T$ ◀

The following theorem shows that the transpose preserves the operations of matrix addition and scalar multiplication:

THEOREM 1.2

(Properties of the Transpose) Let A and B be $m \times n$ matrices, and let s be any scalar. Then

(a) $(A+B)^T = A^T + B^T$.

(b) $(sA)^T = sA^T$.

(c) $(A^T)^T = A$.

PROOF We prove part (a). The rest are left as exercises.

(a) The matrices on each side of the equation are $n \times m$ matrices. So we show that the (i,j)-entry of $(A+B)^T$ equals the (i,j)-entry of $A^T + B^T$. By the definition of transpose, the (i,j)-entry of $(A+B)^T$ equals the (j,i)-entry of $A+B$, which is $a_{ji} + b_{ji}$. On the other hand, the (i,j)-entry of $A^T + B^T$ equals the sum of the (i,j)-entry of A^T and the (i,j)-entry of B^T, that is, $a_{ji} + b_{ji}$. Because the (i,j)-entries of $(A+B)^T$ and $A^T + B^T$ are equal, (a) is proved. ■

VECTORS

A matrix that has exactly one row is called a **row vector**, and a matrix that has exactly one column is called a **column vector**. The term *vector* is used to refer to either a row vector or a column vector. The entries of a vector are called **components**. In this book, we normally work with column vectors, and we denote the set of all column vectors with n components by $\mathcal{R}^n$.

We write vectors as boldface lower case letters such as **u** and **v**, and denote the ith component of the vector **u** by u_i. For example, if $\mathbf{u} = \begin{bmatrix} 2 \\ -4 \\ 7 \end{bmatrix}$, then $u_2 = -4$. Occasionally, we identify a vector **u** in $\mathcal{R}^n$ with an n-tuple, $(u_1, u_2, \ldots, u_n)$.

Because vectors are special types of matrices, we can add them and multiply them by scalars. In this context, we call the two arithmetic operations on vectors **vector addition** and **scalar multiplication**. These operations satisfy the properties listed in Theorem 1.1. In particular, the vector in $\mathcal{R}^n$ with all zero components is denoted by $\mathbf{0}$ and is called the **zero vector**. It satisfies $\mathbf{u} + \mathbf{0} = \mathbf{u}$ and $0\mathbf{u} = \mathbf{0}$ for every **u** in $\mathcal{R}^n$.

Example 2

Let $\mathbf{u} = \begin{bmatrix} 2 \\ -4 \\ 7 \end{bmatrix}$ and $\mathbf{v} = \begin{bmatrix} 5 \\ 3 \\ 0 \end{bmatrix}$. Then

$$\mathbf{u} + \mathbf{v} = \begin{bmatrix} 7 \\ -1 \\ 7 \end{bmatrix}, \qquad \mathbf{u} - \mathbf{v} = \begin{bmatrix} -3 \\ -7 \\ 7 \end{bmatrix}, \qquad \text{and} \qquad 5\mathbf{v} = \begin{bmatrix} 25 \\ 15 \\ 0 \end{bmatrix}.$$

For a given matrix, it is often advantageous to consider its rows and columns as vectors. For example, for the matrix $\begin{bmatrix} 2 & 4 & 3 \\ 0 & 1 & -2 \end{bmatrix}$, the **rows** are $\begin{bmatrix} 2 & 4 & 3 \end{bmatrix}$ and $\begin{bmatrix} 0 & 1 & -2 \end{bmatrix}$, and the **columns** are $\begin{bmatrix} 2 \\ 0 \end{bmatrix}$, $\begin{bmatrix} 4 \\ 1 \end{bmatrix}$, and $\begin{bmatrix} 3 \\ -2 \end{bmatrix}$.

Because the columns of a matrix play a more important role than the rows, we introduce a special notation. When a capital letter denotes a matrix, we use the corresponding lower case letter in boldface with a subscript j to represent the jth column of that matrix. So if A is an $m \times n$ matrix, its jth column is

$$\mathbf{a}_j = \begin{bmatrix} a_{1j} \\ a_{2j} \\ \vdots \\ a_{mj} \end{bmatrix}.$$

GEOMETRY OF VECTORS

For many applications,[2] it is useful to represent vectors geometrically as directed line segments, or arrows. For example, if $\mathbf{v} = \begin{bmatrix} a \\ b \end{bmatrix}$ is a vector in $\mathcal{R}^2$, we can represent **v** as an arrow from the origin to the point (a, b) in the xy-plane, as shown in Figure 1.1.

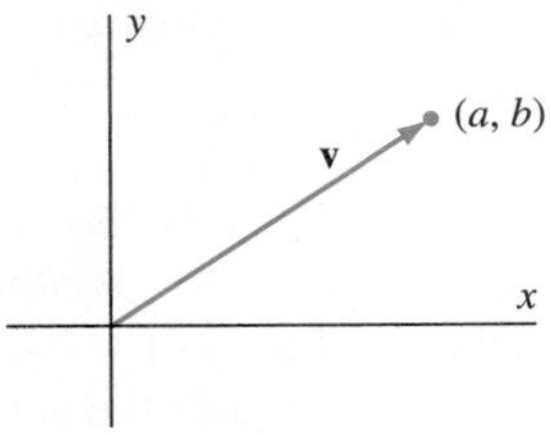

y
v
(a, b)
x

Figure 1.1 A vector in $\mathcal{R}^2$

[2] The importance of vectors in physics was recognized late in the nineteenth century. The algebra of vectors, developed by Oliver Heaviside (1850–1925) and Josiah Willard Gibbs (1839–1903), won out over the algebra of quaternions to become the language of physicists.

Example 3

Velocity Vectors A boat cruises in still water toward the northeast at 20 miles per hour. The velocity $\mathbf{u}$ of the boat is a vector that points in the direction of the boat's motion, and whose length is 20, the boat's speed. If the positive y-axis represents north and the positive x-axis represents east, the boat's direction makes an angle of 45° with the x-axis. (See Figure 1.2.) We can compute the components of $\mathbf{u} = \begin{bmatrix} u_1 \\ u_2 \end{bmatrix}$ by using trigonometry:

$$u_1 = 20\cos 45^\circ = 10\sqrt{2} \qquad \text{and} \qquad u_2 = 20\sin 45^\circ = 10\sqrt{2}.$$

Therefore, $\mathbf{u} = \begin{bmatrix} 10\sqrt{2} \\ 10\sqrt{2} \end{bmatrix}$, where the units are in miles per hour.

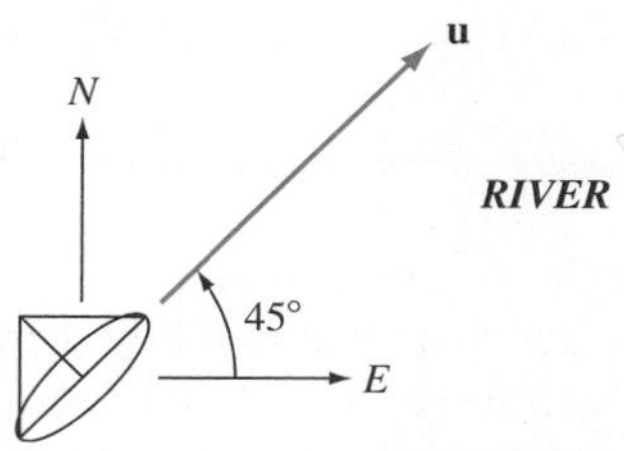

Figure 1.2

VECTOR ADDITION AND THE PARALLELOGRAM LAW

We can represent vector addition graphically, using arrows, by a result called the *parallelogram law*.[3] To add nonzero vectors $\mathbf{u}$ and $\mathbf{v}$, first form a parallelogram with adjacent sides $\mathbf{u}$ and $\mathbf{v}$. Then the sum $\mathbf{u}+\mathbf{v}$ is the arrow along the diagonal of the parallelogram as shown in Figure 1.3.

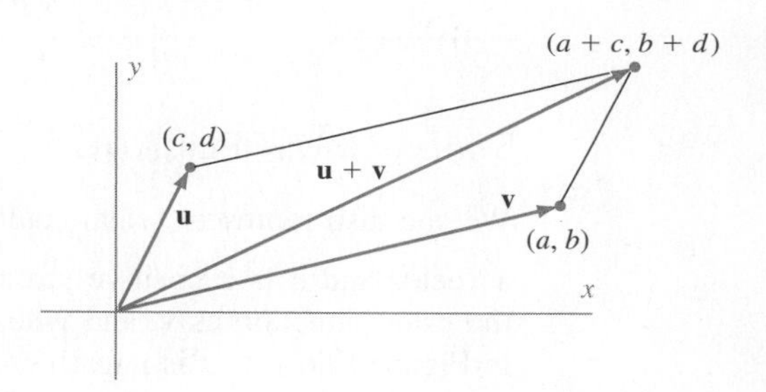

Figure 1.3 The parallelogram law of vector addition

Velocities can be combined by adding vectors that represent them.

Example 4

Imagine that the boat from the previous example is now cruising on a river, which flows to the east at 7 miles per hour. As before, the bow of the boat points toward the northeast, and its speed relative to the water is 20 miles per hour. In this case, the vector $\mathbf{u} = \begin{bmatrix} 10\sqrt{2} \\ 10\sqrt{2} \end{bmatrix}$, which we calculated in the previous example, represents the boat's velocity (in miles per hour) relative to the river. To find the velocity of the boat relative to the shore, we must add a vector $\mathbf{v}$, representing the velocity of the river, to the vector $\mathbf{u}$. Since the river flows toward the east at 7 miles per hour, its velocity vector is $\mathbf{v} = \begin{bmatrix} 7 \\ 0 \end{bmatrix}$. We can represent the sum of the vectors $\mathbf{u}$ and $\mathbf{v}$ by using the parallelogram law, as shown in Figure 1.4. The velocity of the boat relative to the shore (in miles per hour) is the vector

$$\mathbf{u}+\mathbf{v} = \begin{bmatrix} 10\sqrt{2}+7 \\ 10\sqrt{2} \end{bmatrix}.$$

[3] A justification of the parallelogram law by Heron of Alexandria (first century C.E.) appears in his *Mechanics*.

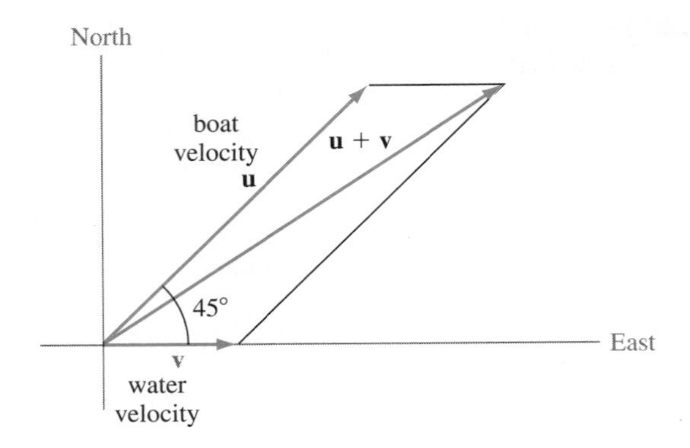

Figure 1.4

To find the speed of the boat, we use the Pythagorean theorem, which tells us that the length of a vector with endpoint (p, q) is $\sqrt{p^2 + q^2}$. Using the fact that the components of $\mathbf{u} + \mathbf{v}$ are $p = 10\sqrt{2} + 7$ and $q = 10\sqrt{2}$, respectively, it follows that the speed of the boat is

$$\sqrt{p^2 + q^2} \approx 25.44 \text{ mph}.$$

SCALAR MULTIPLICATION

We can also represent scalar multiplication graphically, using arrows. If $\mathbf{v} = \begin{bmatrix} a \\ b \end{bmatrix}$ is a vector and c is a positive scalar, the scalar multiple $c\mathbf{v}$ is a vector that points in the same direction as $\mathbf{v}$, and whose length is c times the length of $\mathbf{v}$. This is shown in Figure 1.5(a). If c is negative, $c\mathbf{v}$ points in the opposite direction from $\mathbf{v}$, and has length $|c|$ times the length of $\mathbf{v}$. This is shown in Figure 1.5(b). We call two vectors **parallel** if one of them is a scalar multiple of the other.

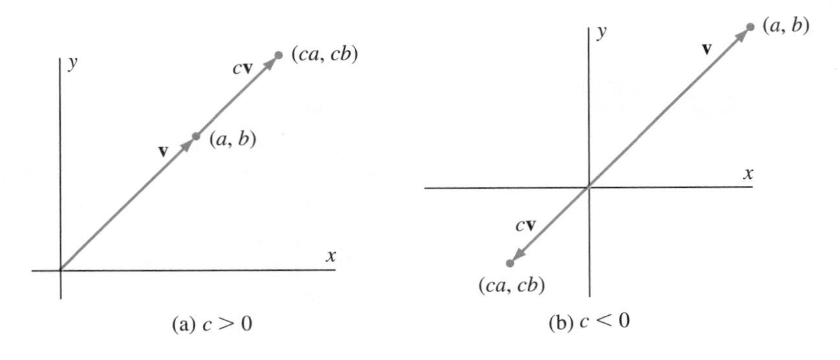

Figure 1.5 Scalar multiplication of vectors

VECTORS IN $\mathcal{R}^3$

If we identify $\mathcal{R}^3$ as the set of all ordered triples, then the same geometric ideas that hold in $\mathcal{R}^2$ are also true in $\mathcal{R}^3$. We may depict a vector $\mathbf{v} = \begin{bmatrix} a \\ b \\ c \end{bmatrix}$ in $\mathcal{R}^3$ as an arrow emanating from the origin of the xyz-coordinate system, with the point (a, b, c) as its

Figure 1.6 Vectors in $\mathcal{R}^3$

endpoint. (See Figure 1.6(a).) As is the case in $\mathcal{R}^2$, we can view two nonzero vectors in $\mathcal{R}^3$ as adjacent sides of a parallelogram, and we can represent their addition by using the parallelogram law. (See Figure 1.6(b).) In real life, motion takes place in 3-dimensional space, and we can depict quantities such as velocities and forces as vectors in $\mathcal{R}^3$.

EXERCISES

In Exercises 1–12, compute the indicated matrices, where

$$A = \begin{bmatrix} 2 & -1 & 5 \\ 3 & 4 & 1 \end{bmatrix} \quad \text{and} \quad B = \begin{bmatrix} 1 & 0 & -2 \\ 2 & 3 & 4 \end{bmatrix}.$$

1. $4A$
2. $-A$
3. $4A - 2B$
4. $3A + 2B$
5. $(2B)^T$
6. $A^T + 2B^T$
7. $A + B$
8. $(A + 2B)^T$
9. A^T
10. $A - B$
11. $-(B^T)$
12. $(-B)^T$

In Exercises 13–24, compute the indicated matrices, if possible, where

$$A = \begin{bmatrix} 3 & -1 & 2 & 4 \\ 1 & 5 & -6 & -2 \end{bmatrix} \quad \text{and} \quad B = \begin{bmatrix} -4 & 0 \\ 2 & 5 \\ -1 & -3 \\ 0 & 2 \end{bmatrix}.$$

13. $-A$
14. $3B$
15. $(-2)A$
16. $(2B)^T$
17. $A - B$
18. $A - B^T$
19. $A^T - B$
20. $3A + 2B^T$
21. $(A + B)^T$
22. $(4A)^T$
23. $B - A^T$
24. $(B^T - A)^T$

In Exercises 25–28, assume that $A = \begin{bmatrix} 3 & -2 \\ 0 & 1.6 \\ 2\pi & 5 \end{bmatrix}$.

25. Determine a_{12}.
26. Determine a_{21}.
27. Determine $\mathbf{a}_1$.
28. Determine $\mathbf{a}_2$.

In Exercises 29–32, assume that $C = \begin{bmatrix} 2 & -3 & 0.4 \\ 2e & 12 & 0 \end{bmatrix}$.

29. Determine $\mathbf{c}_1$.
30. Determine $\mathbf{c}_3$.
31. Determine the first row of C.
32. Determine the second row of C.

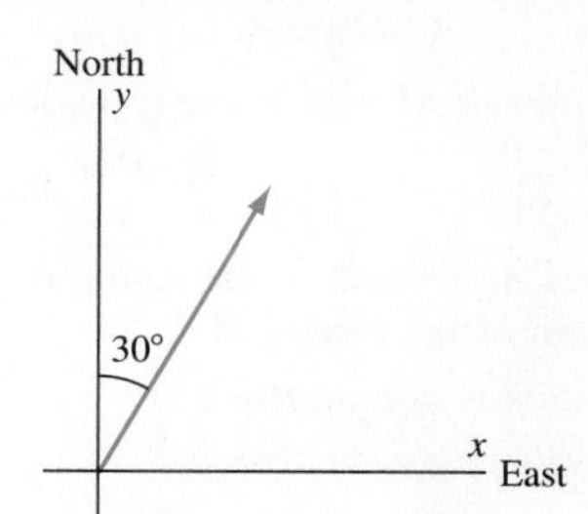

Figure 1.7 A view of the airplane from above

33. An airplane is flying with a ground speed of 300 mph at an angle of 30° east of due north. (See Figure 1.7.) In addition, the airplane is climbing at a rate of 10 mph. Determine the vector in $\mathcal{R}^3$ that represents the velocity (in mph) of the airplane.
34. A swimmer is swimming northeast at 2 mph in still water.
    (a) Give the velocity of the swimmer. Include a sketch.
    (b) A current in a northerly direction at 1 mph affects the velocity of the swimmer. Give the new velocity and speed of the swimmer. Include a sketch.
35. A pilot keeps her airplane pointed in a northeastward direction while maintaining an airspeed (speed relative to the surrounding air) of 300 mph. A wind from the west blows eastward at 50 mph.

(a) Find the velocity (in mph) of the airplane relative to the ground.

(b) What is the speed (in mph) of the airplane relative to the ground?

36. Suppose that in a medical study of 20 people, for each i, $1 \leq i \leq 20$, the 3×1 vector $\mathbf{u}_i$ is defined so that its components respectively represent the blood pressure, pulse rate, and cholesterol reading of the ith person. Provide an interpretation of the vector $\frac{1}{20}(\mathbf{u}_1 + \mathbf{u}_2 + \cdots + \mathbf{u}_{20})$.

In Exercises 37–56, determine whether the statements are true or false.

37. Matrices must be of the same size for their sum to be defined.
38. The transpose of a sum of two matrices is the sum of the transposed matrices.
39. Every vector is a matrix.
40. A scalar multiple of the zero matrix is the zero scalar.
41. The transpose of a matrix is a matrix of the same size.
42. A submatrix of a matrix may be a vector.
43. If B is a 3×4 matrix, then its rows are 4×1 vectors.
44. The $(3, 4)$-entry of a matrix lies in column 3 and row 4.
45. In a zero matrix, every entry is 0.
46. An $m \times n$ matrix has $m + n$ entries.
47. If $\mathbf{v}$ and $\mathbf{w}$ are vectors such that $\mathbf{v} = -3\mathbf{w}$, then $\mathbf{v}$ and $\mathbf{w}$ are parallel.
48. If A and B are any $m \times n$ matrices, then
$$A - B = A + (-1)B.$$
49. The (i,j)-entry of A^T equals the (j,i)-entry of A.
50. If $A = \begin{bmatrix} 1 & 2 \\ 3 & 4 \end{bmatrix}$ and $B = \begin{bmatrix} 1 & 2 & 0 \\ 3 & 4 & 0 \end{bmatrix}$, then $A = B$.
51. In any matrix A, the sum of the entries of $3A$ equals three times the sum of the entries of A.
52. Matrix addition is commutative.
53. Matrix addition is associative.
54. For any $m \times n$ matrices A and B and any scalars c and d, $(cA + dB)^T = cA^T + dB^T$.
55. If A is a matrix, then cA is the same size as A for every scalar c.
56. If A is a matrix for which the sum $A + A^T$ is defined, then A is a square matrix.

---

57. Let A and B be matrices of the same size.
    (a) Prove that the jth column of $A + B$ is $\mathbf{a}_j + \mathbf{b}_j$.
    (b) Prove that for any scalar c, the jth column of cA is $c\mathbf{a}_j$.
58. For any $m \times n$ matrix A, prove that $0A = O$, the $m \times n$ zero matrix.
59. For any $m \times n$ matrix A, prove that $1A = A$.
60. Prove Theorem 1.1(a).
61. Prove Theorem 1.1(c).
62. Prove Theorem 1.1(d).
63. Prove Theorem 1.1(e).
64. Prove Theorem 1.1(g).
65. Prove Theorem 1.2(b).
66. Prove Theorem 1.2(c).

A square matrix A is called a **diagonal matrix** *if* $a_{ij} = 0$ *whenever* $i \neq j$. *Exercises 67–70 are concerned with diagonal matrices.*

67. Prove that a square zero matrix is a diagonal matrix.
68. Prove that if B is a diagonal matrix, then cB is a diagonal matrix for any scalar c.
69. Prove that if B is a diagonal matrix, then B^T is a diagonal matrix.
70. Prove that if B and C are diagonal matrices of the same size, then $B + C$ is a diagonal matrix.

A (square) matrix A is said to be **symmetric** *if* $A = A^T$. *Exercises 71–78 are concerned with symmetric matrices.*

71. Give examples of 2×2 and 3×3 symmetric matrices.
72. Prove that the (i,j)-entry of a symmetric matrix equals the (j,i)-entry.
73. Prove that a square zero matrix is symmetric.
74. Prove that if B is a symmetric matrix, then so is cB for any scalar c.
75. Prove that if B is a square matrix, then $B + B^T$ is symmetric.
76. Prove that if B and C are $n \times n$ symmetric matrices, then so is $B + C$.
77. Is a square submatrix of a symmetric matrix necessarily a symmetric matrix? Justify your answer.
78. Prove that a diagonal matrix is symmetric.

A (square) matrix A is called **skew-symmetric** *if* $A^T = -A$. *Exercises 79–81 are concerned with skew-symmetric matrices.*

79. What must be true about the (i,i)-entries of a skew-symmetric matrix? Justify your answer.
80. Give an example of a nonzero 2×2 skew-symmetric matrix B. Now show that every 2×2 skew-symmetric matrix is a scalar multiple of B.
81. Show that every 3×3 matrix can be written as the sum of a symmetric matrix and a skew-symmetric matrix.
82.[4] The **trace** of an $n \times n$ matrix A, written trace(A), is defined to be the sum
$$\text{trace}(A) = a_{11} + a_{22} + \cdots + a_{nn}.$$
Prove that, for any $n \times n$ matrices A and B and scalar c, the following statements are true:
    (a) $\text{trace}(A + B) = \text{trace}(A) + \text{trace}(B)$.
    (b) $\text{trace}(cA) = c \cdot \text{trace}(A)$.
    (c) $\text{trace}(A^T) = \text{trace}(A)$.
83. *Probability vectors* are vectors whose components are nonnegative and have a sum of 1. Show that if $\mathbf{p}$ and $\mathbf{q}$ are probability vectors and a and b are nonnegative scalars with $a + b = 1$, then $a\mathbf{p} + b\mathbf{q}$ is a probability vector.

[4] This exercise is used in Sections 2.2, 7.1, and 7.5 (on pages 115, 495, and 533, respectively).

In the following exercise, use either a calculator with matrix capabilities or computer software such as MATLAB to solve the problem:

84. Consider the matrices

$$A = \begin{bmatrix} 1.3 & 2.1 & -3.3 & 6.0 \\ 5.2 & 2.3 & -1.1 & 3.4 \\ 3.2 & -2.6 & 1.1 & -4.0 \\ 0.8 & -1.3 & -12.1 & 5.7 \\ -1.4 & 3.2 & 0.7 & 4.4 \end{bmatrix}$$

and

$$B = \begin{bmatrix} 2.6 & -1.3 & 0.7 & -4.4 \\ 2.2 & -2.6 & 1.3 & -3.2 \\ 7.1 & 1.5 & -8.3 & 4.6 \\ -0.9 & -1.2 & 2.4 & 5.9 \\ 3.3 & -0.9 & 1.4 & 6.2 \end{bmatrix}.$$

(a) Compute $A + 2B$.
(b) Compute $A - B$.
(c) Compute $A^T + B^T$.

SOLUTIONS TO THE PRACTICE PROBLEMS

1. (a) The $(1, 2)$-entry of A is 2.
   (b) The $(2, 2)$-entry of A is 3.

2. (a) $$A - B = \begin{bmatrix} 2 & -1 & 1 \\ 3 & 0 & -2 \end{bmatrix} - \begin{bmatrix} 1 & 3 & 0 \\ 2 & -1 & 4 \end{bmatrix} = \begin{bmatrix} 1 & 4 & -1 \\ 1 & 1 & -6 \end{bmatrix}$$

   (b) $$2A = 2\begin{bmatrix} 2 & -1 & 1 \\ 3 & 0 & -2 \end{bmatrix} = \begin{bmatrix} 4 & -2 & 2 \\ 6 & 0 & -4 \end{bmatrix}$$

   (c) $$A + 3B = \begin{bmatrix} 2 & -1 & 1 \\ 3 & 0 & -2 \end{bmatrix} + 3\begin{bmatrix} 1 & 3 & 0 \\ 2 & -1 & 4 \end{bmatrix}$$
   $$= \begin{bmatrix} 2 & -1 & 1 \\ 3 & 0 & -2 \end{bmatrix} + \begin{bmatrix} 3 & 9 & 0 \\ 6 & -3 & 12 \end{bmatrix}$$
   $$= \begin{bmatrix} 5 & 8 & 1 \\ 9 & -3 & 10 \end{bmatrix}$$

3. (a) $$A^T = \begin{bmatrix} 2 & 3 \\ -1 & 0 \\ 1 & -2 \end{bmatrix}$$

   (b) $$(3B)^T = \begin{bmatrix} 3 & 9 & 0 \\ 6 & -3 & 12 \end{bmatrix}^T = \begin{bmatrix} 3 & 6 \\ 9 & -3 \\ 0 & 12 \end{bmatrix}$$

   (c) $$(A + B)^T = \begin{bmatrix} 3 & 2 & 1 \\ 5 & -1 & 2 \end{bmatrix}^T = \begin{bmatrix} 3 & 5 \\ 2 & -1 \\ 1 & 2 \end{bmatrix}$$

1.2 LINEAR COMBINATIONS, MATRIX–VECTOR PRODUCTS, AND SPECIAL MATRICES

In this section, we explore some applications involving matrix operations and introduce the product of a matrix and a vector.

Suppose that 20 students are enrolled in a linear algebra course, in which two tests, a quiz, and a final exam are given. Let $\mathbf{u} = \begin{bmatrix} u_1 \\ u_2 \\ \vdots \\ u_{20} \end{bmatrix}$, where u_i denotes the score of the ith student on the first test. Likewise, define vectors $\mathbf{v}$, $\mathbf{w}$, and $\mathbf{z}$ similarly for the second test, quiz, and final exam, respectively. Assume that the instructor computes a student's course average by counting each test score twice as much as a quiz score, and the final exam score three times as much as a test score. Thus the *weights* for the tests, quiz, and final exam score are, respectively, 2/11, 2/11, 1/11, 6/11 (the weights must sum to one). Now consider the vector

$$\mathbf{y} = \frac{2}{11}\mathbf{u} + \frac{2}{11}\mathbf{v} + \frac{1}{11}\mathbf{w} + \frac{6}{11}\mathbf{z}.$$

The first component y_1 represents the first student's course average, the second component y_2 represents the second student's course average, and so on. Notice that $\mathbf{y}$ is a sum of scalar multiples of $\mathbf{u}$, $\mathbf{v}$, $\mathbf{w}$, and $\mathbf{z}$. This form of vector sum is so important that it merits its own definition.

Definitions A **linear combination** of vectors $\mathbf{u}_1, \mathbf{u}_2, \ldots, \mathbf{u}_k$ is a vector of the form

$$c_1\mathbf{u}_1 + c_2\mathbf{u}_2 + \cdots + c_k\mathbf{u}_k,$$

where $c_1, c_2, \ldots, c_k$ are scalars. These scalars are called the **coefficients** of the linear combination.

Note that a linear combination of one vector is simply a scalar multiple of that vector.

In the previous example, the vector $\mathbf{y}$ of the students' course averages is a linear combination of the vectors $\mathbf{u}$, $\mathbf{v}$, $\mathbf{w}$, and $\mathbf{z}$. The coefficients are the weights. Indeed, any weighted average produces a linear combination of the scores.

Notice that

$$\begin{bmatrix} 2 \\ 8 \end{bmatrix} = (-3)\begin{bmatrix} 1 \\ 1 \end{bmatrix} + 4\begin{bmatrix} 1 \\ 3 \end{bmatrix} + 1\begin{bmatrix} 1 \\ -1 \end{bmatrix}.$$

Thus $\begin{bmatrix} 2 \\ 8 \end{bmatrix}$ is a linear combination of $\begin{bmatrix} 1 \\ 1 \end{bmatrix}$, $\begin{bmatrix} 1 \\ 3 \end{bmatrix}$, and $\begin{bmatrix} 1 \\ -1 \end{bmatrix}$, with coefficients -3, 4, and 1. We can also write

$$\begin{bmatrix} 2 \\ 8 \end{bmatrix} = \begin{bmatrix} 1 \\ 1 \end{bmatrix} + 2\begin{bmatrix} 1 \\ 3 \end{bmatrix} - 1\begin{bmatrix} 1 \\ -1 \end{bmatrix}.$$

This equation also expresses $\begin{bmatrix} 2 \\ 8 \end{bmatrix}$ as a linear combination of $\begin{bmatrix} 1 \\ 1 \end{bmatrix}$, $\begin{bmatrix} 1 \\ 3 \end{bmatrix}$, and $\begin{bmatrix} 1 \\ -1 \end{bmatrix}$, but now the coefficients are 1, 2, and -1. So the set of coefficients that express one vector as a linear combination of the others need not be unique.

Example 1

(a) Determine whether $\begin{bmatrix} 4 \\ -1 \end{bmatrix}$ is a linear combination of $\begin{bmatrix} 2 \\ 3 \end{bmatrix}$ and $\begin{bmatrix} 3 \\ 1 \end{bmatrix}$.

(b) Determine whether $\begin{bmatrix} -4 \\ -2 \end{bmatrix}$ is a linear combination of $\begin{bmatrix} 6 \\ 3 \end{bmatrix}$ and $\begin{bmatrix} 2 \\ 1 \end{bmatrix}$.

(c) Determine whether $\begin{bmatrix} 3 \\ 4 \end{bmatrix}$ is a linear combination of $\begin{bmatrix} 3 \\ 2 \end{bmatrix}$ and $\begin{bmatrix} 6 \\ 4 \end{bmatrix}$.

Solution (a) We seek scalars x_1 and x_2 such that

$$\begin{bmatrix} 4 \\ -1 \end{bmatrix} = x_1\begin{bmatrix} 2 \\ 3 \end{bmatrix} + x_2\begin{bmatrix} 3 \\ 1 \end{bmatrix} = \begin{bmatrix} 2x_1 \\ 3x_1 \end{bmatrix} + \begin{bmatrix} 3x_2 \\ 1x_2 \end{bmatrix} = \begin{bmatrix} 2x_1 + 3x_2 \\ 3x_1 + x_2 \end{bmatrix}.$$

That is, we seek a solution of the system of equations

$$\begin{aligned} 2x_1 + 3x_2 &= 4 \\ 3x_1 + x_2 &= -1. \end{aligned}$$

Because these equations represent nonparallel lines in the plane, there is exactly one solution, namely, $x_1 = -1$ and $x_2 = 2$. Therefore $\begin{bmatrix} 4 \\ -1 \end{bmatrix}$ is a (unique) linear combination of the vectors $\begin{bmatrix}2\\3\end{bmatrix}$ and $\begin{bmatrix}3\\1\end{bmatrix}$, namely,

$$\begin{bmatrix}4\\-1\end{bmatrix} = (-1)\begin{bmatrix}2\\3\end{bmatrix} + 2\begin{bmatrix}3\\1\end{bmatrix}.$$

(See Figure 1.8.)

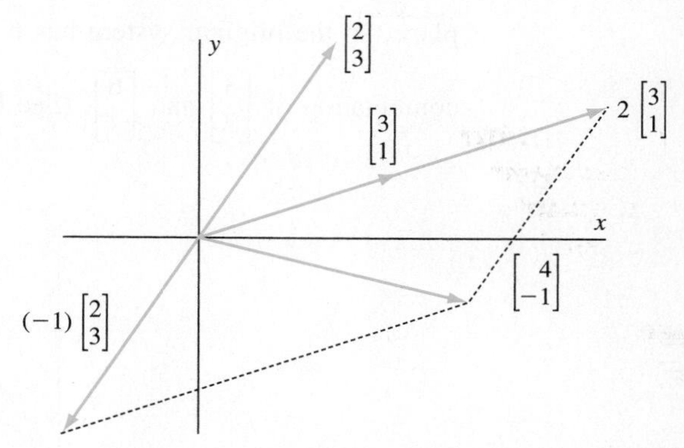

Figure 1.8 The vector $\begin{bmatrix}4\\-1\end{bmatrix}$ is a linear combination of $\begin{bmatrix}2\\3\end{bmatrix}$ and $\begin{bmatrix}3\\1\end{bmatrix}$.

(b) To determine whether $\begin{bmatrix}-4\\-2\end{bmatrix}$ is a linear combination of $\begin{bmatrix}6\\3\end{bmatrix}$ and $\begin{bmatrix}2\\1\end{bmatrix}$, we perform a similar computation and produce the set of equations

$$\begin{aligned} 6x_1 + 2x_2 &= -4 \\ 3x_1 + x_2 &= -2. \end{aligned}$$

Since the first equation is twice the second, we need only solve $3x_1 + x_2 = -2$. This equation represents a line in the plane, and the coordinates of any point on the line give a solution. For example, we can let $x_1 = -2$ and $x_2 = 4$. In this case, we have

$$\begin{bmatrix}-4\\-2\end{bmatrix} = (-2)\begin{bmatrix}6\\3\end{bmatrix} + 4\begin{bmatrix}2\\1\end{bmatrix}.$$

There are infinitely many solutions. (See Figure 1.9.)

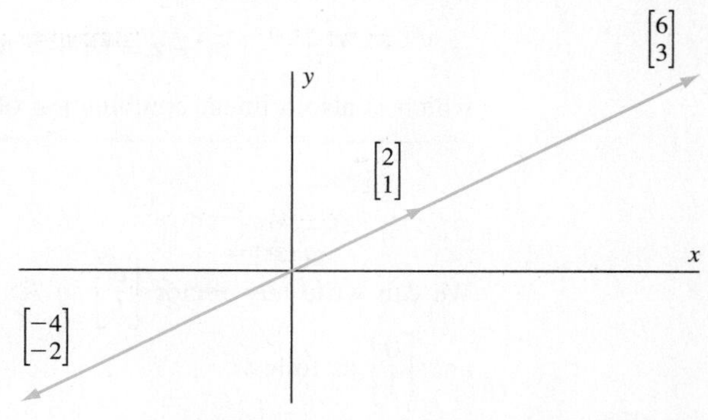

Figure 1.9 The vector $\begin{bmatrix}-4\\-2\end{bmatrix}$ is a linear combination of $\begin{bmatrix}6\\3\end{bmatrix}$ and $\begin{bmatrix}2\\1\end{bmatrix}$.

(c) To determine if $\begin{bmatrix} 3 \\ 4 \end{bmatrix}$ is a linear combination of $\begin{bmatrix} 3 \\ 2 \end{bmatrix}$ and $\begin{bmatrix} 6 \\ 4 \end{bmatrix}$, we must solve the system of equations

$$\begin{aligned} 3x_1 + 6x_2 &= 3 \\ 2x_1 + 4x_2 &= 4. \end{aligned}$$

If we add $-\frac{2}{3}$ times the first equation to the second, we obtain $0 = 2$, an equation with no solutions. Indeed, the two original equations represent parallel lines in the plane, so the original system has no solutions. We conclude that $\begin{bmatrix} 3 \\ 4 \end{bmatrix}$ is not a linear combination of $\begin{bmatrix} 3 \\ 2 \end{bmatrix}$ and $\begin{bmatrix} 6 \\ 4 \end{bmatrix}$. (See Figure 1.10.)

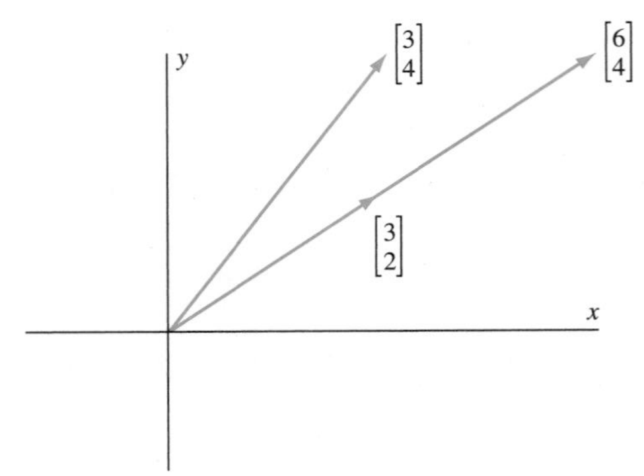

Figure 1.10 The vector $\begin{bmatrix} 3 \\ 4 \end{bmatrix}$ is *not* a linear combination of $\begin{bmatrix} 3 \\ 2 \end{bmatrix}$ and $\begin{bmatrix} 6 \\ 4 \end{bmatrix}$.

Example 2 Given vectors $\mathbf{u}_1$, $\mathbf{u}_2$, and $\mathbf{u}_3$, show that the sum of any two linear combinations of these vectors is also a linear combination of these vectors.

Solution Suppose that $\mathbf{w}$ and $\mathbf{z}$ are linear combinations of $\mathbf{u}_1$, $\mathbf{u}_2$, and $\mathbf{u}_3$. Then we may write

$$\mathbf{w} = a\mathbf{u}_1 + b\mathbf{u}_2 + c\mathbf{u}_3 \qquad \text{and} \qquad \mathbf{z} = a'\mathbf{u}_1 + b'\mathbf{u}_2 + c'\mathbf{u}_3,$$

where a, b, c, a', b', c' are scalars. So

$$\mathbf{w} + \mathbf{z} = (a + a')\mathbf{u}_1 + (b + b')\mathbf{u}_2 + (c + c')\mathbf{u}_3,$$

which is also a linear combination of $\mathbf{u}_1$, $\mathbf{u}_2$, and $\mathbf{u}_3$.

STANDARD VECTORS

We can write any vector $\begin{bmatrix} a \\ b \end{bmatrix}$ in $\mathcal{R}^2$ as a linear combination of the two vectors $\begin{bmatrix} 1 \\ 0 \end{bmatrix}$ and $\begin{bmatrix} 0 \\ 1 \end{bmatrix}$ as follows:

$$\begin{bmatrix} a \\ b \end{bmatrix} = a\begin{bmatrix} 1 \\ 0 \end{bmatrix} + b\begin{bmatrix} 0 \\ 1 \end{bmatrix}$$

The vectors $\begin{bmatrix} 1 \\ 0 \end{bmatrix}$ and $\begin{bmatrix} 0 \\ 1 \end{bmatrix}$ are called the *standard vectors* of $\mathcal{R}^2$. Similarly, we can write any vector $\begin{bmatrix} a \\ b \\ c \end{bmatrix}$ in $\mathcal{R}^3$ as a linear combination of the vectors $\begin{bmatrix} 1 \\ 0 \\ 0 \end{bmatrix}$, $\begin{bmatrix} 0 \\ 1 \\ 0 \end{bmatrix}$, and $\begin{bmatrix} 0 \\ 0 \\ 1 \end{bmatrix}$ as follows:

$$\begin{bmatrix} a \\ b \\ c \end{bmatrix} = a \begin{bmatrix} 1 \\ 0 \\ 0 \end{bmatrix} + b \begin{bmatrix} 0 \\ 1 \\ 0 \end{bmatrix} + c \begin{bmatrix} 0 \\ 0 \\ 1 \end{bmatrix}$$

The vectors $\begin{bmatrix} 1 \\ 0 \\ 0 \end{bmatrix}$, $\begin{bmatrix} 0 \\ 1 \\ 0 \end{bmatrix}$, and $\begin{bmatrix} 0 \\ 0 \\ 1 \end{bmatrix}$ are called the *standard vectors* of $\mathcal{R}^3$.

In general, we define the **standard vectors** of $\mathcal{R}^n$ by

$$\mathbf{e}_1 = \begin{bmatrix} 1 \\ 0 \\ \vdots \\ 0 \end{bmatrix}, \quad \mathbf{e}_2 = \begin{bmatrix} 0 \\ 1 \\ \vdots \\ 0 \end{bmatrix}, \quad \ldots, \quad \mathbf{e}_n = \begin{bmatrix} 0 \\ 0 \\ \vdots \\ 1 \end{bmatrix}.$$

(See Figure 1.11.)

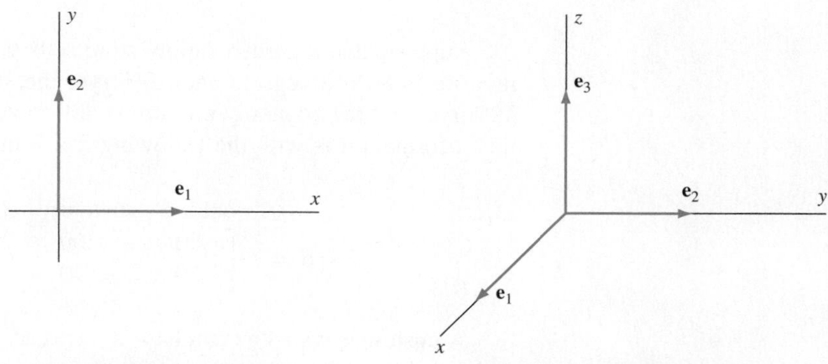

Figure 1.11

From the preceding equations, it is easy to see that every vector in $\mathcal{R}^n$ is a linear combination of the standard vectors of $\mathcal{R}^n$. In fact, for any vector $\mathbf{v}$ in $\mathcal{R}^n$,

$$\mathbf{v} = v_1\mathbf{e}_1 + v_2\mathbf{e}_2 + \cdots + v_n\mathbf{e}_n.$$

(See Figure 1.13.)

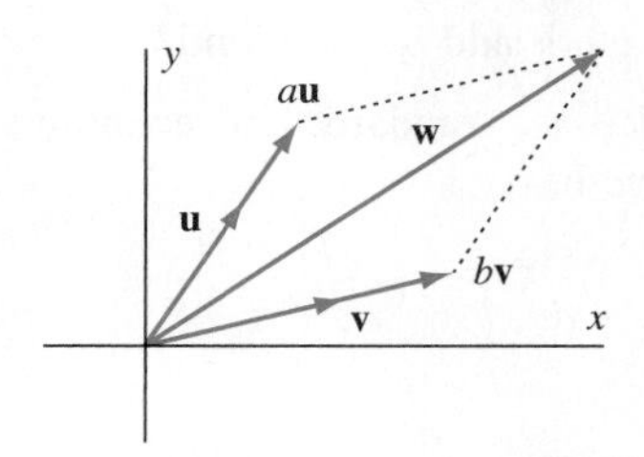

Figure 1.12 The vector **w** is a linear combination of the nonparallel vectors **u** and **v**.

Now let **u** and **v** be nonparallel vectors, and let **w** be any vector in $\mathcal{R}^2$. Begin with the endpoint of **w** and create a parallelogram with sides $a\mathbf{u}$ and $b\mathbf{v}$, so that **w** is its diagonal. It follows that $\mathbf{w} = a\mathbf{u} + b\mathbf{v}$; that is, **w** is a linear combination of the vectors **u** and **v**. (See Figure 1.12.) More generally, the following statement is true:

> If **u** and **v** are any nonparallel vectors in $\mathcal{R}^2$, then every vector in $\mathcal{R}^2$ is a linear combination of **u** and **v**.

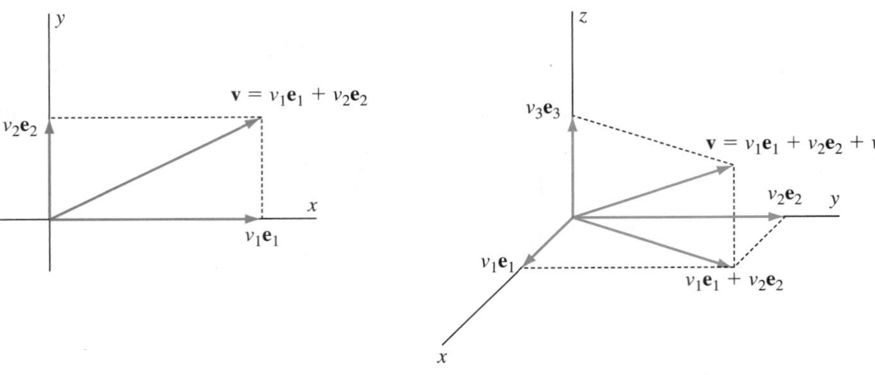

The vector $\mathbf{v}$ is a linear combination of standard vectors in $\mathcal{R}^2$.

The vector $\mathbf{v}$ is a linear combination of standard vectors in $\mathcal{R}^3$.

Figure 1.13

Practice Problem 1 ▶ Let $\mathbf{w} = \begin{bmatrix} -1 \\ 10 \end{bmatrix}$ and $\mathcal{S} = \left\{ \begin{bmatrix} 2 \\ 1 \end{bmatrix}, \begin{bmatrix} 3 \\ -2 \end{bmatrix} \right\}$.

(a) Without doing any calculations, explain why $\mathbf{w}$ can be written as a linear combination of the vectors in $\mathcal{S}$.

(b) Express $\mathbf{w}$ as a linear combination of the vectors in $\mathcal{S}$. ◀

Suppose that a garden supply store sells three mixtures of grass seed. The deluxe mixture is 80% bluegrass and 20% rye, the standard mixture is 60% bluegrass and 40% rye, and the economy mixture is 40% bluegrass and 60% rye. One way to record this information is with the following 2×3 matrix:

$$B = \begin{array}{c} \text{deluxe} \quad \text{standard} \quad \text{economy} \\ \begin{bmatrix} .80 & .60 & .40 \\ .20 & .40 & .60 \end{bmatrix} \end{array} \begin{array}{l} \text{bluegrass} \\ \text{rye} \end{array}$$

A customer wants to purchase a blend of grass seed containing 5 lb of bluegrass and 3 lb of rye. There are two natural questions that arise:

1. Is it possible to combine the three mixtures of seed into a blend that has exactly the desired amounts of bluegrass and rye, with no surplus of either?
2. If so, how much of each mixture should the store clerk add to the blend?

Let x_1, x_2, and x_3 denote the number of pounds of deluxe, standard, and economy mixtures, respectively, to be used in the blend. Then we have

$$\begin{aligned} .80x_1 + .60x_2 + .40x_3 &= 5 \\ .20x_1 + .40x_2 + .60x_3 &= 3. \end{aligned}$$

This is a *system of two linear equations in three unknowns.* Finding a solution of this system is equivalent to answering our second question. The technique for solving general systems is explored in great detail in Sections 1.3 and 1.4.

Using matrix notation, we may rewrite these equations in the form

$$\begin{bmatrix} .80x_1 + .60x_2 + .40x_3 \\ .20x_1 + .40x_2 + .60x_3 \end{bmatrix} = \begin{bmatrix} 5 \\ 3 \end{bmatrix}.$$

Now we use matrix operations to rewrite this matrix equation, using the columns of B as

$$x_1 \begin{bmatrix} .80 \\ .20 \end{bmatrix} + x_2 \begin{bmatrix} .60 \\ .40 \end{bmatrix} + x_3 \begin{bmatrix} .40 \\ .60 \end{bmatrix} = \begin{bmatrix} 5 \\ 3 \end{bmatrix}.$$

Thus we can rephrase the first question as follows: Is $\begin{bmatrix} 5 \\ 3 \end{bmatrix}$ a linear combination of the columns $\begin{bmatrix} .80 \\ .20 \end{bmatrix}$, $\begin{bmatrix} .60 \\ .40 \end{bmatrix}$, and $\begin{bmatrix} .40 \\ .60 \end{bmatrix}$ of B? The result in the box on page 17 provides an affirmative answer. Because no two of the three vectors are parallel, $\begin{bmatrix} 5 \\ 3 \end{bmatrix}$ is a linear combination of any pair of these vectors.

MATRIX–VECTOR PRODUCTS

A convenient way to represent systems of linear equations is by *matrix–vector products*. For the preceding example, we represent the variables by the vector $\mathbf{x} = \begin{bmatrix} x_1 \\ x_2 \\ x_3 \end{bmatrix}$ and define the *matrix–vector product* $B\mathbf{x}$ to be the linear combination

$$B\mathbf{x} = \begin{bmatrix} .80 & .60 & .40 \\ .20 & .40 & .60 \end{bmatrix} \begin{bmatrix} x_1 \\ x_2 \\ x_3 \end{bmatrix} = x_1 \begin{bmatrix} .80 \\ .20 \end{bmatrix} + x_2 \begin{bmatrix} .60 \\ .40 \end{bmatrix} + x_3 \begin{bmatrix} .40 \\ .60 \end{bmatrix}.$$

This definition provides another way to state the first question in the preceding example: Does the vector $\begin{bmatrix} 5 \\ 3 \end{bmatrix}$ equal $B\mathbf{x}$ for some vector $\mathbf{x}$? Notice that for the matrix–vector product to make sense, the number of columns of B must equal the number of components in $\mathbf{x}$. The general definition of a matrix–vector product is given next.

Definition Let A be an $m \times n$ matrix and $\mathbf{v}$ be an $n \times 1$ vector. We define the **matrix–vector product** of A and $\mathbf{v}$, denoted by $A\mathbf{v}$, to be the linear combination of the columns of A whose coefficients are the corresponding components of $\mathbf{v}$. That is,

$$A\mathbf{v} = v_1\mathbf{a}_1 + v_2\mathbf{a}_2 + \cdots + v_n\mathbf{a}_n.$$

As we have noted, for $A\mathbf{v}$ to exist, the number of columns of A must equal the number of components of $\mathbf{v}$. For example, suppose that

$$A = \begin{bmatrix} 1 & 2 \\ 3 & 4 \\ 5 & 6 \end{bmatrix} \quad \text{and} \quad \mathbf{v} = \begin{bmatrix} 7 \\ 8 \end{bmatrix}.$$

Notice that A has two columns and $\mathbf{v}$ has two components. Then

$$A\mathbf{v} = \begin{bmatrix} 1 & 2 \\ 3 & 4 \\ 5 & 6 \end{bmatrix} \begin{bmatrix} 7 \\ 8 \end{bmatrix} = 7 \begin{bmatrix} 1 \\ 3 \\ 5 \end{bmatrix} + 8 \begin{bmatrix} 2 \\ 4 \\ 6 \end{bmatrix} = \begin{bmatrix} 7 \\ 21 \\ 35 \end{bmatrix} + \begin{bmatrix} 16 \\ 32 \\ 48 \end{bmatrix} = \begin{bmatrix} 23 \\ 53 \\ 83 \end{bmatrix}.$$

Returning to the preceding garden supply store example, suppose that the store has 140 lb of seed in stock: 60 lb of the deluxe mixture, 50 lb of the standard mixture, and 30 lb of the economy mixture. We let $\mathbf{v} = \begin{bmatrix} 60 \\ 50 \\ 30 \end{bmatrix}$ represent this information. Now the matrix–vector product

$$B\mathbf{v} = \begin{bmatrix} .80 & .60 & .40 \\ .20 & .40 & .60 \end{bmatrix} \begin{bmatrix} 60 \\ 50 \\ 30 \end{bmatrix}$$

$$= 60 \begin{bmatrix} .80 \\ .20 \end{bmatrix} + 50 \begin{bmatrix} .60 \\ .40 \end{bmatrix} + 30 \begin{bmatrix} .40 \\ .60 \end{bmatrix}$$

$$= \begin{array}{c} \text{seed (lb)} \\ \begin{bmatrix} 90 \\ 50 \end{bmatrix} \end{array} \begin{array}{c} \\ \text{bluegrass} \\ \text{rye} \end{array}$$

gives the number of pounds of each type of seed contained in the 140 pounds of seed that the garden supply store has in stock. For example, there are 90 pounds of bluegrass because $90 = .80(60) + .60(50) + .40(30)$.

There is another approach to computing the matrix–vector product that relies more on the entries of A than on its columns. Consider the following example:

$$A\mathbf{v} = \begin{bmatrix} a_{11} & a_{12} & a_{13} \\ a_{21} & a_{22} & a_{23} \end{bmatrix} \begin{bmatrix} v_1 \\ v_2 \\ v_3 \end{bmatrix}$$

$$= v_1 \begin{bmatrix} a_{11} \\ a_{21} \end{bmatrix} + v_2 \begin{bmatrix} a_{12} \\ a_{22} \end{bmatrix} + v_3 \begin{bmatrix} a_{13} \\ a_{23} \end{bmatrix}$$

$$= \begin{bmatrix} a_{11}v_1 + a_{12}v_2 + a_{13}v_3 \\ a_{21}v_1 + a_{22}v_2 + a_{23}v_3 \end{bmatrix}$$

Notice that the first component of the vector $A\mathbf{v}$ is the sum of products of the corresponding entries of the first row of A and the components of $\mathbf{v}$. Likewise, the second component of $A\mathbf{v}$ is the sum of products of the corresponding entries of the second row of A and the components of $\mathbf{v}$. With this approach to computing a matrix–vector product, we can omit the intermediate step in the preceding illustration. For example, suppose

$$A = \begin{bmatrix} 2 & 3 & 1 \\ 1 & -2 & 3 \end{bmatrix} \quad \text{and} \quad \mathbf{v} = \begin{bmatrix} -1 \\ 1 \\ 3 \end{bmatrix}.$$

Then

$$A\mathbf{v} = \begin{bmatrix} 2 & 3 & 1 \\ 1 & -2 & 3 \end{bmatrix} \begin{bmatrix} -1 \\ 1 \\ 3 \end{bmatrix} = \begin{bmatrix} (2)(-1) + (3)(1) + (1)(3) \\ (1)(-1) + (-2)(1) + (3)(3) \end{bmatrix} = \begin{bmatrix} 4 \\ 6 \end{bmatrix}.$$

In general, you can use this technique to compute $A\mathbf{v}$ when A is an $m \times n$ matrix and $\mathbf{v}$ is a vector in $\mathcal{R}^n$. In this case, the ith component of $A\mathbf{v}$ is

$$[a_{i1}\ \ a_{i2}\ \ \dots\ \ a_{in}]\begin{bmatrix} v_1 \\ v_2 \\ \vdots \\ v_n \end{bmatrix} = a_{i1}v_1 + a_{i2}v_2 + \cdots + a_{in}v_n,$$

which is the matrix–vector product of the ith row of A and $\mathbf{v}$. The computation of all the components of the matrix–vector product $A\mathbf{v}$ is given by

$$A\mathbf{v} = \begin{bmatrix} a_{11} & a_{12} & \dots & a_{1n} \\ a_{21} & a_{22} & \dots & a_{2n} \\ \vdots & & & \vdots \\ a_{m1} & a_{m2} & \dots & a_{mn} \end{bmatrix}\begin{bmatrix} v_1 \\ v_2 \\ \vdots \\ v_n \end{bmatrix} = \begin{bmatrix} a_{11}v_1 + a_{12}v_2 + \cdots + a_{1n}v_n \\ a_{21}v_1 + a_{22}v_2 + \cdots + a_{2n}v_n \\ \vdots \\ a_{m1}v_1 + a_{m2}v_2 + \cdots + a_{mn}v_n \end{bmatrix}.$$

Practice Problem 2 ▶ Let $A = \begin{bmatrix} 2 & -1 & 1 \\ 3 & 0 & -2 \end{bmatrix}$ and $\mathbf{v} = \begin{bmatrix} 3 \\ 1 \\ -1 \end{bmatrix}$. Compute the following vectors:

(a) $A\mathbf{v}$

(b) $(A\mathbf{v})^T$ ◀

Example 3 A sociologist is interested in studying the population changes within a metropolitan area as people move between the city and suburbs. From empirical evidence, she has discovered that in any given year, 15% of those living in the city will move to the suburbs and 3% of those living in the suburbs will move to the city. For simplicity, we assume that the metropolitan population remains stable. This information may be represented by the following matrix:

$$\begin{array}{cc} & \quad\quad\quad\text{From} \\ & \quad\quad\ \text{City}\quad\text{Suburbs} \\ \text{To} & \begin{array}{l}\text{City}\\ \text{Suburbs}\end{array}\begin{bmatrix} .85 & .03 \\ .15 & .97 \end{bmatrix} = A \end{array}$$

Notice that the entries of A are nonnegative and that the entries of each column sum to 1. Such a matrix is called a **stochastic matrix**. Suppose that there are now 500 thousand people living in the city and 700 thousand people living in the suburbs. The sociologist would like to know how many people will be living in each of the two areas next year. Figure 1.14 describes the changes of population from one year to the next. It follows that the number of people (in thousands) who will be living in the city next year is $(.85)(500) + (.03)(700) = 446$ thousand, and the number of people living in the suburbs is $(.15)(500) + (.97)(700) = 754$ thousand.

If we let $\mathbf{p}$ represent the vector of current populations of the city and suburbs, we have

$$\mathbf{p} = \begin{bmatrix} 500 \\ 700 \end{bmatrix}.$$

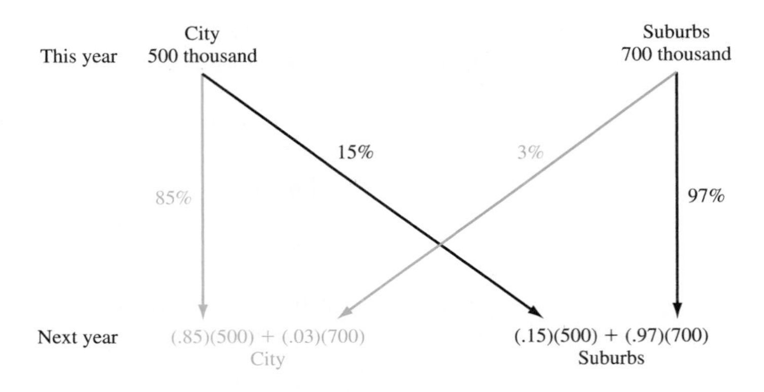

Figure 1.14 Movement between the city and suburbs

We can find the populations in the next year by computing the matrix–vector product:

$$A\mathbf{p} = \begin{bmatrix} .85 & .03 \\ .15 & .97 \end{bmatrix} \begin{bmatrix} 500 \\ 700 \end{bmatrix} = \begin{bmatrix} (.85)(500) + (.03)(700) \\ (.15)(500) + (.97)(700) \end{bmatrix} = \begin{bmatrix} 446 \\ 754 \end{bmatrix}$$

In other words, $A\mathbf{p}$ is the vector of populations in the next year. If we want to determine the populations in two years, we can repeat this procedure by multiplying A by the vector $A\mathbf{p}$. That is, in two years, the vector of populations is $A(A\mathbf{p})$.

IDENTITY MATRICES

Suppose we let $I_2 = \begin{bmatrix} 1 & 0 \\ 0 & 1 \end{bmatrix}$ and $\mathbf{v}$ be any vector in $\mathcal{R}^2$. Then

$$I_2\mathbf{v} = \begin{bmatrix} 1 & 0 \\ 0 & 1 \end{bmatrix} \begin{bmatrix} v_1 \\ v_2 \end{bmatrix} = v_1 \begin{bmatrix} 1 \\ 0 \end{bmatrix} + v_2 \begin{bmatrix} 0 \\ 1 \end{bmatrix} = \begin{bmatrix} v_1 \\ v_2 \end{bmatrix} = \mathbf{v}.$$

So multiplication by I_2 leaves every vector $\mathbf{v}$ in $\mathcal{R}^2$ unchanged. The same property holds in a more general context.

Definition For each positive integer n, the $n \times n$ **identity matrix** I_n is the $n \times n$ matrix whose respective columns are the standard vectors $\mathbf{e}_1, \mathbf{e}_2, \ldots, \mathbf{e}_n$ in $\mathcal{R}^n$.

For example,

$$I_2 = \begin{bmatrix} 1 & 0 \\ 0 & 1 \end{bmatrix} \quad \text{and} \quad I_3 = \begin{bmatrix} 1 & 0 & 0 \\ 0 & 1 & 0 \\ 0 & 0 & 1 \end{bmatrix}.$$

Because the columns of I_n are the standard vectors of $\mathcal{R}^n$, it follows easily that $I_n\mathbf{v} = \mathbf{v}$ for any $\mathbf{v}$ in $\mathcal{R}^n$.

ROTATION MATRICES

Consider a point $P_0 = (x_0, y_0)$ in $\mathcal{R}^2$ with polar coordinates (r, α), where $r \geq 0$ and α is the angle between the segment $\overline{OP_0}$ and the positive x-axis. (See Figure 1.15.) Then $x_0 = r\cos\alpha$ and $y_0 = r\sin\alpha$. Suppose that $\overline{OP_0}$ is rotated by an angle θ to the

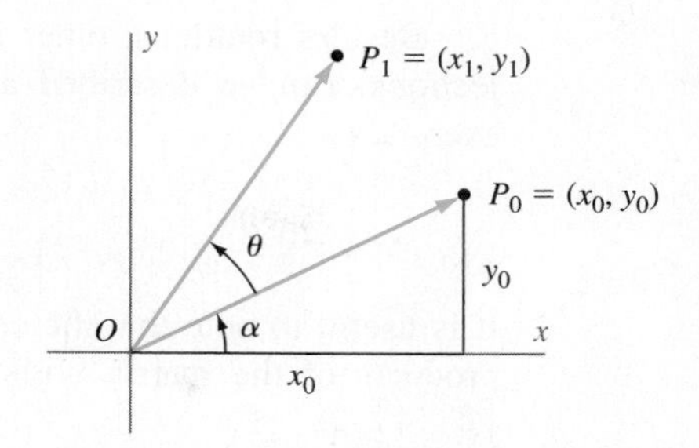

Figure 1.15 Rotation of a vector through the angle θ.

segment $\overline{OP_1}$, where $P_1 = (x_1, y_1)$. Then $(r, \alpha + \theta)$ represents the polar coordinates for P_1, and hence

$$\begin{aligned} x_1 &= r\cos(\alpha + \theta) \\ &= r(\cos\alpha\cos\theta - \sin\alpha\sin\theta) \\ &= (r\cos\alpha)\cos\theta - (r\sin\alpha)\sin\theta \\ &= x_0\cos\theta - y_0\sin\theta. \end{aligned}$$

Similarly, $y_1 = x_0 \sin\theta + y_0 \cos\theta$. We can express these equations as a matrix equation by using a matrix–vector product. If we define A_θ by

$$A_\theta = \begin{bmatrix} \cos\theta & -\sin\theta \\ \sin\theta & \cos\theta \end{bmatrix},$$

then

$$A_\theta \begin{bmatrix} x_0 \\ y_0 \end{bmatrix} = \begin{bmatrix} \cos\theta & -\sin\theta \\ \sin\theta & \cos\theta \end{bmatrix} \begin{bmatrix} x_0 \\ y_0 \end{bmatrix} = \begin{bmatrix} x_0\cos\theta - y_0\sin\theta \\ x_0\sin\theta + y_0\cos\theta \end{bmatrix} = \begin{bmatrix} x_1 \\ y_1 \end{bmatrix}.$$

We call A_θ the **θ-rotation matrix**, or more simply, a **rotation matrix**. For any vector $\mathbf{u}$, the vector $A_\theta\mathbf{u}$ is the vector obtained by rotating $\mathbf{u}$ by an angle θ, where the rotation is counterclockwise if $\theta > 0$ and clockwise if $\theta < 0$.

Example 4

To rotate the vector $\begin{bmatrix} 3 \\ 4 \end{bmatrix}$ by 30°, we compute $A_{30^\circ}\begin{bmatrix} 3 \\ 4 \end{bmatrix}$; that is,

$$\begin{bmatrix} \cos 30^\circ & -\sin 30^\circ \\ \sin 30^\circ & \cos 30^\circ \end{bmatrix} \begin{bmatrix} 3 \\ 4 \end{bmatrix} = \begin{bmatrix} \frac{\sqrt{3}}{2} & -\frac{1}{2} \\ \frac{1}{2} & \frac{\sqrt{3}}{2} \end{bmatrix} \begin{bmatrix} 3 \\ 4 \end{bmatrix} = \begin{bmatrix} \frac{3\sqrt{3}}{2} - \frac{4}{2} \\ \frac{3}{2} + \frac{4\sqrt{3}}{2} \end{bmatrix} = \frac{1}{2}\begin{bmatrix} 3\sqrt{3} - 4 \\ 3 + 4\sqrt{3} \end{bmatrix}.$$

Thus when $\begin{bmatrix} 3 \\ 4 \end{bmatrix}$ is rotated by 30°, the resulting vector is $\frac{1}{2}\begin{bmatrix} 3\sqrt{3} - 4 \\ 3 + 4\sqrt{3} \end{bmatrix}$.

It is interesting to observe that the 0°-rotation matrix A_{0°, which leaves a vector unchanged, is given by $A_{0^\circ} = I_2$. This is quite reasonable because multiplication by I_2 also leaves vectors unchanged.

Besides rotations, other geometric transformations (such as reflections and projections) can be described as matrix–vector products. Examples are found in the exercises.

PROPERTIES OF MATRIX–VECTOR PRODUCTS

It is useful to note that the columns of a matrix can be represented as matrix–vector products of the matrix with the standard vectors. Suppose, for example, that $A = \begin{bmatrix} 2 & 4 \\ 3 & 6 \end{bmatrix}$. Then

$$A\mathbf{e}_1 = \begin{bmatrix} 2 & 4 \\ 3 & 6 \end{bmatrix}\begin{bmatrix} 1 \\ 0 \end{bmatrix} = \begin{bmatrix} 2 \\ 3 \end{bmatrix} \quad \text{and} \quad A\mathbf{e}_2 = \begin{bmatrix} 2 & 4 \\ 3 & 6 \end{bmatrix}\begin{bmatrix} 0 \\ 1 \end{bmatrix} = \begin{bmatrix} 4 \\ 6 \end{bmatrix}.$$

The general result is stated as (d) of Theorem 1.3.

For any $m \times n$ matrix A, $A\mathbf{0} = \mathbf{0}'$, where $\mathbf{0}$ is the $n \times 1$ zero vector and $\mathbf{0}'$ is the $m \times 1$ zero vector. This is easily seen since the matrix–vector product $A\mathbf{0}$ is a sum of products of columns of A and zeros. Similarly, for the $m \times n$ zero matrix O, $O\mathbf{v} = \mathbf{0}'$ for any $n \times 1$ vector $\mathbf{v}$. (See (f) and (g) of Theorem 1.3.)

THEOREM 1.3

(Properties of Matrix–Vector Products) Let A and B be $m \times n$ matrices, and let $\mathbf{u}$ and $\mathbf{v}$ be vectors in $\mathcal{R}^n$. Then

(a) $A(\mathbf{u} + \mathbf{v}) = A\mathbf{u} + A\mathbf{v}$.

(b) $A(c\mathbf{u}) = c(A\mathbf{u}) = (cA)\mathbf{u}$ for every scalar c.

(c) $(A + B)\mathbf{u} = A\mathbf{u} + B\mathbf{u}$.

(d) $A\mathbf{e}_j = \mathbf{a}_j$ for $j = 1, 2, \ldots, n$, where $\mathbf{e}_j$ is the jth standard vector in $\mathcal{R}^n$.

(e) If B is an $m \times n$ matrix such that $B\mathbf{w} = A\mathbf{w}$ for all $\mathbf{w}$ in $\mathcal{R}^n$, then $B = A$.

(f) $A\mathbf{0}$ is the $m \times 1$ zero vector.

(g) If O is the $m \times n$ zero matrix, then $O\mathbf{v}$ is the $m \times 1$ zero vector.

(h) $I_n\mathbf{v} = \mathbf{v}$.

PROOF We prove part (a) and leave the rest for the exercises.

(a) Because the ith component of $\mathbf{u} + \mathbf{v}$ is $u_i + v_i$, we have

$$\begin{aligned} A(\mathbf{u} + \mathbf{v}) &= (u_1 + v_1)\mathbf{a}_1 + (u_2 + v_2)\mathbf{a}_2 + \cdots + (u_n + v_n)\mathbf{a}_n \\ &= (u_1\mathbf{a}_1 + u_2\mathbf{a}_2 + \cdots + u_n\mathbf{a}_n) + (v_1\mathbf{a}_1 + v_2\mathbf{a}_2 + \cdots + v_n\mathbf{a}_n) \\ &= A\mathbf{u} + A\mathbf{v}. \end{aligned}$$

■

It follows by repeated applications of Theorem 1.3(a) and (b) that the matrix–vector product of A and a linear combination of $\mathbf{u}_1, \mathbf{u}_2, \ldots, \mathbf{u}_k$ yields a linear combination of the vectors $A\mathbf{u}_1, A\mathbf{u}_2, \ldots, A\mathbf{u}_k$. That is,

For any $m \times n$ matrix A, any scalars $c_1, c_2, \ldots, c_k$, and any vectors $\mathbf{u}_1, \mathbf{u}_2, \ldots, \mathbf{u}_k$ in $\mathcal{R}^n$,

$$A(c_1\mathbf{u}_1 + c_2\mathbf{u}_2 + \cdots + c_k\mathbf{u}_k) = c_1A\mathbf{u}_1 + c_2A\mathbf{u}_2 + \cdots + c_kA\mathbf{u}_k.$$

EXERCISES

In Exercises 1–16, compute the matrix–vector products.

1. $\begin{bmatrix} 3 & -2 & 1 \\ 4 & 0 & 2 \end{bmatrix} \begin{bmatrix} 1 \\ -2 \\ 5 \end{bmatrix}$

2. $\begin{bmatrix} 1 & -3 \\ 0 & 2 \\ -1 & 4 \end{bmatrix} \begin{bmatrix} 1 \\ 2 \end{bmatrix}$

3. $\begin{bmatrix} 2 & -1 & 3 \\ 1 & 0 & -1 \\ 0 & 2 & 4 \end{bmatrix} \begin{bmatrix} 2 \\ 1 \\ 2 \end{bmatrix}$

4. $\begin{bmatrix} 4 & 2 \\ 7 & -3 \end{bmatrix} \begin{bmatrix} 5 \\ 1 \end{bmatrix}$

5. $\begin{bmatrix} 1 & 0 \\ 0 & 1 \end{bmatrix} \begin{bmatrix} a \\ b \end{bmatrix}$

6. $\begin{bmatrix} 2 & 1 & 3 \end{bmatrix} \begin{bmatrix} -2 \\ 4 \\ 6 \end{bmatrix}$

7. $\begin{bmatrix} 3 & 0 \\ 2 & 1 \end{bmatrix}^T \begin{bmatrix} 4 \\ 5 \end{bmatrix}$

8. $\begin{bmatrix} 1 & 0 & 0 \\ 0 & 1 & 0 \\ 0 & 0 & 1 \end{bmatrix} \begin{bmatrix} a \\ b \\ c \end{bmatrix}$

9. $\begin{bmatrix} s & 0 & 0 \\ 0 & t & 0 \\ 0 & 0 & u \end{bmatrix} \begin{bmatrix} a \\ b \\ c \end{bmatrix}$

10. $\begin{bmatrix} 4 \\ 2 \\ -3 \end{bmatrix}^T \begin{bmatrix} 2 \\ -1 \\ 0 \end{bmatrix}$

11. $\begin{bmatrix} 2 & -3 \\ -4 & 5 \\ 3 & -1 \end{bmatrix} \begin{bmatrix} 4 \\ 2 \end{bmatrix}$

12. $\begin{bmatrix} 3 & -3 \\ -2 & 4 \\ 1 & 2 \end{bmatrix} \begin{bmatrix} 0 \\ 1 \end{bmatrix}$

13. $\begin{bmatrix} 3 & -1 & 4 \\ -2 & 6 & -1 \end{bmatrix} \begin{bmatrix} 0 \\ 1 \\ 0 \end{bmatrix}$

14. $\begin{bmatrix} 2 & -3 & 4 \\ -4 & 5 & -2 \\ 3 & -1 & 0 \end{bmatrix} \begin{bmatrix} 1 \\ 1 \\ 1 \end{bmatrix}$

15. $\left(\begin{bmatrix} 3 & 0 \\ -2 & 4 \end{bmatrix}^T + \begin{bmatrix} 1 & 2 \\ 3 & -3 \end{bmatrix}^T \right) \begin{bmatrix} 4 \\ 5 \end{bmatrix}$

16. $\left(\begin{bmatrix} 3 & 0 \\ -2 & 4 \end{bmatrix} + \begin{bmatrix} 1 & 2 \\ 3 & -3 \end{bmatrix} \right) \begin{bmatrix} 4 \\ 5 \end{bmatrix}$

In Exercises 17–28, an angle θ and a vector $\mathbf{u}$ are given. Write the corresponding rotation matrix, and compute the vector found by rotating $\mathbf{u}$ by the angle θ. Draw a sketch and simplify your answers.

17. $\theta = 45^\circ$, $\mathbf{u} = \mathbf{e}_2$

18. $\theta = 0^\circ$, $\mathbf{u} = \mathbf{e}_1$

19. $\theta = 60^\circ$, $\mathbf{u} = \begin{bmatrix} 3 \\ 1 \end{bmatrix}$

20. $\theta = 30^\circ$, $\mathbf{u} = \begin{bmatrix} 1 \\ 2 \end{bmatrix}$

21. $\theta = 210^\circ$, $\mathbf{u} = \begin{bmatrix} -1 \\ -3 \end{bmatrix}$

22. $\theta = 135^\circ$, $\mathbf{u} = \begin{bmatrix} 2 \\ -1 \end{bmatrix}$

23. $\theta = 270^\circ$, $\mathbf{u} = \begin{bmatrix} -2 \\ 3 \end{bmatrix}$

24. $\theta = 330^\circ$, $\mathbf{u} = \begin{bmatrix} 4 \\ 1 \end{bmatrix}$

25. $\theta = 240^\circ$, $\mathbf{u} = \begin{bmatrix} -3 \\ -1 \end{bmatrix}$

26. $\theta = 150^\circ$, $\mathbf{u} = \begin{bmatrix} 5 \\ -2 \end{bmatrix}$

27. $\theta = 300^\circ$, $\mathbf{u} = \begin{bmatrix} 3 \\ 0 \end{bmatrix}$

28. $\theta = 120^\circ$, $\mathbf{u} = \begin{bmatrix} 0 \\ -2 \end{bmatrix}$

In Exercises 29–44, a vector $\mathbf{u}$ and a set $\mathcal{S}$ are given. If possible, write $\mathbf{u}$ as a linear combination of the vectors in $\mathcal{S}$.

29. $\mathbf{u} = \begin{bmatrix} 1 \\ 1 \end{bmatrix}$, $\mathcal{S} = \left\{ \begin{bmatrix} 1 \\ 0 \end{bmatrix}, \begin{bmatrix} 0 \\ 1 \end{bmatrix} \right\}$

30. $\mathbf{u} = \begin{bmatrix} 1 \\ -1 \end{bmatrix}$, $\mathcal{S} = \left\{ \begin{bmatrix} 4 \\ -4 \end{bmatrix} \right\}$

31. $\mathbf{u} = \begin{bmatrix} 1 \\ -1 \end{bmatrix}$, $\mathcal{S} = \left\{ \begin{bmatrix} 4 \\ 4 \end{bmatrix} \right\}$

32. $\mathbf{u} = \begin{bmatrix} 1 \\ 1 \end{bmatrix}$, $\mathcal{S} = \left\{ \begin{bmatrix} 1 \\ 0 \end{bmatrix}, \begin{bmatrix} 0 \\ -1 \end{bmatrix} \right\}$

33. $\mathbf{u} = \begin{bmatrix} 1 \\ 1 \\ 2 \end{bmatrix}$, $\mathcal{S} = \left\{ \begin{bmatrix} 1 \\ 0 \\ 1 \end{bmatrix}, \begin{bmatrix} 1 \\ 0 \\ -1 \end{bmatrix} \right\}$

34. $\mathbf{u} = \begin{bmatrix} 1 \\ 1 \end{bmatrix}$, $\mathcal{S} = \left\{ \begin{bmatrix} 1 \\ 0 \end{bmatrix}, \begin{bmatrix} 0 \\ -1 \end{bmatrix}, \begin{bmatrix} 0 \\ 0 \end{bmatrix} \right\}$

35. $\mathbf{u} = \begin{bmatrix} -1 \\ 11 \end{bmatrix}$, $\mathcal{S} = \left\{ \begin{bmatrix} 1 \\ 3 \end{bmatrix}, \begin{bmatrix} 2 \\ -1 \end{bmatrix} \right\}$

36. $\mathbf{u} = \begin{bmatrix} 1 \\ 1 \end{bmatrix}$, $\mathcal{S} = \left\{ \begin{bmatrix} 1 \\ 0 \end{bmatrix}, \begin{bmatrix} 0 \\ -1 \end{bmatrix}, \begin{bmatrix} 1 \\ 1 \end{bmatrix} \right\}$

37. $\mathbf{u} = \begin{bmatrix} 3 \\ 8 \end{bmatrix}$, $\mathcal{S} = \left\{ \begin{bmatrix} 1 \\ 2 \end{bmatrix}, \begin{bmatrix} 2 \\ 3 \end{bmatrix}, \begin{bmatrix} -2 \\ -5 \end{bmatrix} \right\}$

38. $\mathbf{u} = \begin{bmatrix} a \\ b \end{bmatrix}$, $\mathcal{S} = \left\{ \begin{bmatrix} 1 \\ 1 \end{bmatrix}, \begin{bmatrix} 2 \\ -1 \end{bmatrix} \right\}$

39. $\mathbf{u} = \begin{bmatrix} 3 \\ 5 \\ -5 \end{bmatrix}$, $\mathcal{S} = \left\{ \begin{bmatrix} 2 \\ 0 \\ -1 \end{bmatrix}, \begin{bmatrix} -1 \\ 1 \\ 0 \end{bmatrix} \right\}$

40. $\mathbf{u} = \begin{bmatrix} 2 \\ -2 \\ 8 \end{bmatrix}$, $\mathcal{S} = \left\{ \begin{bmatrix} 0 \\ 1 \\ 2 \end{bmatrix}, \begin{bmatrix} -1 \\ 3 \\ 0 \end{bmatrix} \right\}$

41. $\mathbf{u} = \begin{bmatrix} 3 \\ -2 \\ 1 \end{bmatrix}$, $\mathcal{S} = \left\{ \begin{bmatrix} 2 \\ -1 \\ 2 \end{bmatrix}, \begin{bmatrix} 3 \\ -2 \\ 1 \end{bmatrix}, \begin{bmatrix} -4 \\ 1 \\ 3 \end{bmatrix} \right\}$

42. $\mathbf{u} = \begin{bmatrix} 5 \\ 6 \\ 7 \end{bmatrix}$, $\mathcal{S} = \left\{ \begin{bmatrix} 1 \\ 0 \\ 0 \end{bmatrix}, \begin{bmatrix} 0 \\ 1 \\ 0 \end{bmatrix}, \begin{bmatrix} 0 \\ 0 \\ 1 \end{bmatrix} \right\}$

43. $\mathbf{u} = \begin{bmatrix} -4 \\ -5 \\ -6 \end{bmatrix}$, $\mathcal{S} = \left\{ \begin{bmatrix} 1 \\ 0 \\ 0 \end{bmatrix}, \begin{bmatrix} 0 \\ 1 \\ 0 \end{bmatrix}, \begin{bmatrix} 0 \\ 0 \\ 1 \end{bmatrix} \right\}$

44. $\mathbf{u} = \begin{bmatrix} -1 \\ 3 \\ 2 \end{bmatrix}$, $\mathcal{S} = \left\{ \begin{bmatrix} 1 \\ -1 \\ 1 \end{bmatrix}, \begin{bmatrix} 0 \\ -2 \\ 3 \end{bmatrix}, \begin{bmatrix} -1 \\ 3 \\ 2 \end{bmatrix} \right\}$

T&F *In Exercises 45–64, determine whether the statements are true or false.*

45. A linear combination of vectors is a sum of scalar multiples of the vectors.

46. The coefficients in a linear combination can always be chosen to be positive scalars.

47. Every vector in $\mathcal{R}^2$ can be written as a linear combination of the standard vectors of $\mathcal{R}^2$.
48. Every vector in $\mathcal{R}^2$ is a linear combination of any two nonparallel vectors.
49. The zero vector is a linear combination of any nonempty set of vectors.
50. The matrix–vector product of a 2×3 matrix and a 3×1 vector is a 3×1 vector.
51. The matrix–vector product of a 2×3 matrix and a 3×1 vector equals a linear combination of the rows of the matrix.
52. The product of a matrix and a standard vector equals a standard vector.
53. The rotation matrix A_{180° equals $-I_2$.
54. The matrix–vector product of an $m \times n$ matrix and a vector yields a vector in $\mathcal{R}^n$.
55. Every vector in $\mathcal{R}^2$ is a linear combination of two parallel vectors.
56. Every vector $\mathbf{v}$ in $\mathcal{R}^n$ can be written as a linear combination of the standard vectors, using the components of $\mathbf{v}$ as the coefficients of the linear combination.
57. A vector with exactly one nonzero component is called a standard vector.
58. If A is an $m \times n$ matrix, $\mathbf{u}$ is a vector in $\mathcal{R}^n$, and c is a scalar, then $A(c\mathbf{u}) = c(A\mathbf{u})$.
59. If A is an $m \times n$ matrix, then the only vector $\mathbf{u}$ in $\mathcal{R}^n$ such that $A\mathbf{u} = \mathbf{0}$ is $\mathbf{u} = \mathbf{0}$.
60. For any vector $\mathbf{u}$ in $\mathcal{R}^2$, $A_\theta \mathbf{u}$ is the vector obtained by rotating $\mathbf{u}$ by the angle θ.
61. If $\theta > 0$, then $A_\theta \mathbf{u}$ is the vector obtained by rotating $\mathbf{u}$ by a clockwise rotation of the angle θ.
62. If A is an $m \times n$ matrix and $\mathbf{u}$ and $\mathbf{v}$ are vectors in $\mathcal{R}^n$ such that $A\mathbf{u} = A\mathbf{v}$, then $\mathbf{u} = \mathbf{v}$.
63. The matrix vector product of an $m \times n$ matrix A and a vector $\mathbf{u}$ in $\mathcal{R}^n$ equals $u_1\mathbf{a}_1 + u_2\mathbf{a}_2 + \cdots + u_n\mathbf{a}_n$.
64. A matrix having nonnegative entries such that the sum of the entries in each column is 1 is called a stochastic matrix.

---

65. Use a matrix–vector product to show that if $\theta = 0^\circ$, then $A_\theta \mathbf{v} = \mathbf{v}$ for all $\mathbf{v}$ in $\mathcal{R}^2$.
66. Use a matrix–vector product to show that if $\theta = 180^\circ$, then $A_\theta \mathbf{v} = -\mathbf{v}$ for all $\mathbf{v}$ in $\mathcal{R}^2$.
67. Use matrix–vector products to show that, for any angles θ and β and any vector $\mathbf{v}$ in $\mathcal{R}^2$, $A_\theta(A_\beta \mathbf{v}) = A_{\theta+\beta}\mathbf{v}$.
68. Compute $A_\theta^T(A_\theta \mathbf{u})$ and $A_\theta(A_\theta^T \mathbf{u})$ for any vector $\mathbf{u}$ in $\mathcal{R}^2$ and any angle θ.
69. Suppose that in a metropolitan area there are 400 thousand people living in the city and 300 thousand people living in the suburbs. Use the stochastic matrix in Example 3 to determine
    (a) the number of people living in the city and suburbs after one year;
    (b) the number of people living in the city and suburbs after two years.
70. Let $A = \begin{bmatrix} 1 & 2 & 3 \\ 4 & 5 & 6 \\ 7 & 8 & 9 \end{bmatrix}$ and $\mathbf{u} = \begin{bmatrix} a \\ b \\ c \end{bmatrix}$. Represent $A\mathbf{u}$ as a linear combination of the columns of A.

In Exercises 71–74, let $A = \begin{bmatrix} -1 & 0 \\ 0 & 1 \end{bmatrix}$ *and* $\mathbf{u} = \begin{bmatrix} a \\ b \end{bmatrix}$.

71. Show that $A\mathbf{u}$ is the reflection of $\mathbf{u}$ about the y-axis.
72. Prove that $A(A\mathbf{u}) = \mathbf{u}$.
73. Modify the matrix A to obtain a matrix B so that $B\mathbf{u}$ is the reflection of $\mathbf{u}$ about the x-axis.
74. Let C denote the rotation matrix that corresponds to $\theta = 180^\circ$.
    (a) Find C.
    (b) Use the matrix B in Exercise 73 to show that
    $$A(C\mathbf{u}) = C(A\mathbf{u}) = B\mathbf{u} \quad \text{and}$$
    $$B(C\mathbf{u}) = C(B\mathbf{u}) = A\mathbf{u}.$$
    (c) Interpret these equations in terms of reflections and rotations.

In Exercises 75–79, let $A = \begin{bmatrix} 1 & 0 \\ 0 & 0 \end{bmatrix}$ *and* $\mathbf{u} = \begin{bmatrix} a \\ b \end{bmatrix}$.

75. Show that $A\mathbf{u}$ is the projection of $\mathbf{u}$ on the x-axis.
76. Prove that $A(A\mathbf{u}) = A\mathbf{u}$.
77. Show that if $\mathbf{v}$ is any vector whose endpoint lies on the x-axis, then $A\mathbf{v} = \mathbf{v}$.
78. Modify the matrix A to obtain a matrix B so that $B\mathbf{u}$ is the projection of $\mathbf{u}$ on the y-axis.
79. Let C denote the rotation matrix that corresponds to $\theta = 180^\circ$. (See Exercise 74(a).)
    (a) Prove that $A(C\mathbf{u}) = C(A\mathbf{u})$.
    (b) Interpret the result in (a) geometrically.
80. Let $\mathbf{u}_1$ and $\mathbf{u}_2$ be vectors in $\mathcal{R}^n$. Prove that the sum of two linear combinations of these vectors is also a linear combination of these vectors.
81. Let $\mathbf{u}_1$ and $\mathbf{u}_2$ be vectors in $\mathcal{R}^n$. Let $\mathbf{v}$ and $\mathbf{w}$ be linear combinations of $\mathbf{u}_1$ and $\mathbf{u}_2$. Prove that any linear combination of $\mathbf{v}$ and $\mathbf{w}$ is also a linear combination of $\mathbf{u}_1$ and $\mathbf{u}_2$.
82. Let $\mathbf{u}_1$ and $\mathbf{u}_2$ be vectors in $\mathcal{R}^n$. Prove that a scalar multiple of a linear combination of these vectors is also a linear combination of these vectors.
83. Prove (b) of Theorem 1.3.
84. Prove (c) of Theorem 1.3.
85. Prove (d) of Theorem 1.3.
86. Prove (e) of Theorem 1.3.
87. Prove (f) of Theorem 1.3.
88. Prove (g) of Theorem 1.3.
89. Prove (h) of Theorem 1.3.

In Exercises 90 and 91, use either a calculator with matrix capabilities or computer software such as MATLAB to solve each problem.

90. In reference to Exercise 69, determine the number of people living in the city and suburbs after 10 years.

91. For the matrices

$$A = \begin{bmatrix} 2.1 & 1.3 & -0.1 & 6.0 \\ 1.3 & -9.9 & 4.5 & 6.2 \\ 4.4 & -2.2 & 5.7 & 2.0 \\ 0.2 & 9.8 & 1.1 & -8.5 \end{bmatrix}$$

and

$$B = \begin{bmatrix} 4.4 & 1.1 & 3.0 & 9.9 \\ -1.2 & 4.8 & 2.4 & 6.0 \\ 1.3 & 2.4 & -5.8 & 2.8 \\ 6.0 & -2.1 & -5.3 & 8.2 \end{bmatrix}$$

and the vectors

$$\mathbf{u} = \begin{bmatrix} 1 \\ -1 \\ 2 \\ 4 \end{bmatrix} \quad \text{and} \quad \mathbf{v} = \begin{bmatrix} 7 \\ -1 \\ 2 \\ 5 \end{bmatrix},$$

(a) compute $A\mathbf{u}$;
(b) compute $B(\mathbf{u}+\mathbf{v})$;
(c) compute $(A+B)\mathbf{v}$;
(d) compute $A(B\mathbf{v})$.

SOLUTIONS TO THE PRACTICE PROBLEMS

1. (a) The vectors in $\mathcal{S}$ are nonparallel vectors in $\mathcal{R}^2$.
   (b) To express $\mathbf{w}$ as a linear combination of the vectors in $\mathcal{S}$, we must find scalars x_1 and x_2 such that

$$\begin{bmatrix} -1 \\ 10 \end{bmatrix} = x_1 \begin{bmatrix} 2 \\ 1 \end{bmatrix} + x_2 \begin{bmatrix} 3 \\ -2 \end{bmatrix} = \begin{bmatrix} 2x_1 + 3x_2 \\ x_1 - 2x_2 \end{bmatrix}.$$

   That is, we must solve the following system:

$$\begin{aligned} 2x_1 + 3x_2 &= -1 \\ x_1 - 2x_2 &= 10 \end{aligned}$$

   Using elementary algebra, we see that $x_1 = 4$ and $x_2 = -3$. So

$$\begin{bmatrix} -1 \\ 10 \end{bmatrix} = 4 \begin{bmatrix} 2 \\ 1 \end{bmatrix} - 3 \begin{bmatrix} 3 \\ -2 \end{bmatrix}.$$

2. (a) $A\mathbf{v} = \begin{bmatrix} 2 & -1 & 1 \\ 3 & 0 & -2 \end{bmatrix} \begin{bmatrix} 3 \\ 1 \\ -1 \end{bmatrix} = \begin{bmatrix} 4 \\ 11 \end{bmatrix}$

   (b) $(A\mathbf{v})^T = \begin{bmatrix} 4 \\ 11 \end{bmatrix}^T = \begin{bmatrix} 4 & 11 \end{bmatrix}$

1.3 SYSTEMS OF LINEAR EQUATIONS

A **linear equation** in the variables (unknowns) $x_1, x_2, \ldots, x_n$ is an equation that can be written in the form

$$a_1x_1 + a_2x_2 + \cdots + a_nx_n = b,$$

where $a_1, a_2, \ldots, a_n$, and b are real numbers. The scalars $a_1, a_2, \ldots, a_n$ are called the **coefficients**, and b is called the **constant term** of the equation. For example, $3x_1 - 7x_2 + x_3 = 19$ is a linear equation in the variables x_1, x_2, and x_3, with coefficients 3, -7, and 1, and constant term 19. The equation $8x_2 - 12x_5 = 4x_1 - 9x_3 + 6$ is also a linear equation because it can be written as

$$-4x_1 + 8x_2 + 9x_3 + 0x_4 - 12x_5 = 6.$$

On the other hand, the equations

$$x_1 + 5x_2x_3 = 7, \qquad 2x_1 - 7x_2 + x_3^2 = -3, \qquad \text{and} \qquad 4\sqrt{x_1} - 3x_2 = 15$$

are *not* linear equations because they contain terms involving a product of variables, a square of a variable, or a square root of a variable.

A **system of linear equations** is a set of m linear equations in the same n variables, where m and n are positive integers. We can write such a system in the form

$$\begin{aligned} a_{11}x_1 + a_{12}x_2 + \cdots + a_{1n}x_n &= b_1 \\ a_{21}x_1 + a_{22}x_2 + \cdots + a_{2n}x_n &= b_2 \\ &\vdots \\ a_{m1}x_1 + a_{m2}x_2 + \cdots + a_{mn}x_n &= b_m, \end{aligned}$$

where a_{ij} denotes the coefficient of x_j in equation i.

For example, on page 18 we obtained the following system of 2 linear equations in the variables x_1, x_2, and x_3:

$$\begin{aligned} .80x_1 + .60x_2 + .40x_3 &= 5 \\ .20x_1 + .40x_2 + .60x_3 &= 3 \end{aligned} \tag{1}$$

A **solution** of a system of linear equations in the variables $x_1, x_2, \ldots, x_n$ is a vector $\begin{bmatrix} s_1 \\ s_2 \\ \vdots \\ s_n \end{bmatrix}$ in $\mathcal{R}^n$ such that every equation in the system is satisfied when each x_i is replaced by s_i. For example, $\begin{bmatrix} 2 \\ 5 \\ 1 \end{bmatrix}$ is a solution of system (1) because

$$.80(2) + .60(5) + .40(1) = 5 \quad \text{and} \quad .20(2) + .40(5) + .60(1) = 3.$$

The set of all solutions of a system of linear equations is called the **solution set** of that system.

Practice Problem 1 ▶ Determine whether (a) $\mathbf{u} = \begin{bmatrix} -2 \\ 3 \\ 2 \\ 1 \end{bmatrix}$ and (b) $\mathbf{v} = \begin{bmatrix} 5 \\ 8 \\ 1 \\ 3 \end{bmatrix}$ are solutions of the system of linear equations

$$\begin{aligned} x_1 + \phantom{x_2 +{}} 5x_3 - x_4 &= 7 \\ 2x_1 - x_2 + 6x_3 \phantom{{}- x_4} &= 8. \end{aligned}$$ ◀

SYSTEMS OF 2 LINEAR EQUATIONS IN 2 VARIABLES

A linear equation in two variables x and y has the form $ax + by = c$. When at least one of a and b is nonzero, this is the equation of a line in the xy-plane. Thus a system of 2 linear equations in the variables x and y consists of a pair of equations, each of which describes a line in the plane.

$$a_1x + b_1y = c_1 \quad \text{is the equation of line } \mathcal{L}_1.$$

$$a_2x + b_2y = c_2 \quad \text{is the equation of line } \mathcal{L}_2.$$

Geometrically, a solution of such a system corresponds to a point lying on both of the lines $\mathcal{L}_1$ and $\mathcal{L}_2$. There are three different situations that can arise.

If the lines are different and parallel, then they have no point in common. In this case, the system of equations has no solution. (See Figure 1.16.)

If the lines are different but not parallel, then the two lines have a unique point of intersection. In this case, the system of equations has exactly one solution. (See Figure 1.17.)

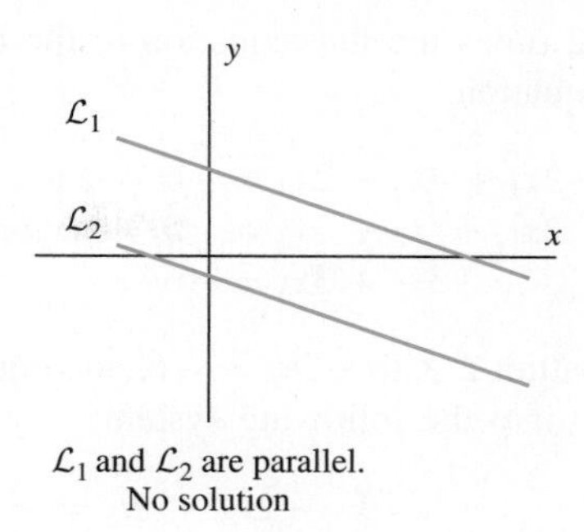

$\mathcal{L}_1$ and $\mathcal{L}_2$ are parallel.
No solution

Figure 1.16

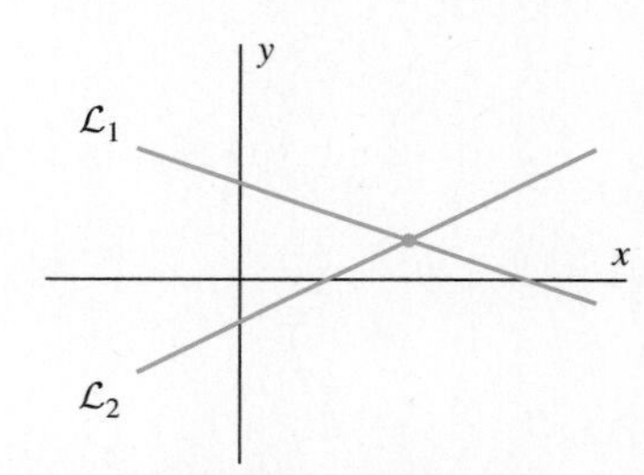

$\mathcal{L}_1$ and $\mathcal{L}_2$ are different but not parallel.
Exactly one solution

Figure 1.17

Finally, if the two lines coincide, then every point on $\mathcal{L}_1$ and $\mathcal{L}_2$ satisfies both of the equations in the system, and so every point on $\mathcal{L}_1$ and $\mathcal{L}_2$ is a solution of the system. In this case, there are infinitely many solutions. (See Figure 1.18.)

As we will soon see, no matter how many equations and variables a system has, there are exactly three possibilities for its solution set.

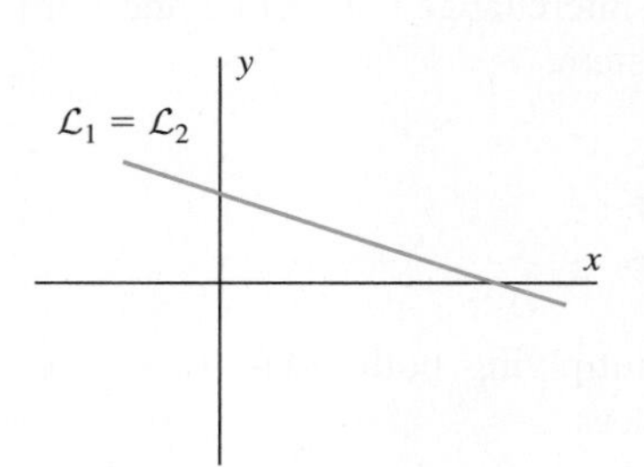

$\mathcal{L}_1$ and $\mathcal{L}_2$ are the same.
Infinitely many solutions

Figure 1.18

> Every system of linear equations has no solution, exactly one solution, or infinitely many solutions.

A system of linear equations that has one or more solutions is called **consistent**; otherwise, the system is called **inconsistent**. Figures 1.17 and 1.18 show consistent systems, while Figure 1.16 shows an inconsistent system.

ELEMENTARY ROW OPERATIONS

To find the solution set of a system of linear equations or determine that the system is inconsistent, we replace it by one with the same solutions that is more easily solved. Two systems of linear equations that have exactly the same solutions are called **equivalent**.

Now we present a procedure for creating a simpler, equivalent system. It is based on an important technique for solving a system of linear equations taught in high school algebra classes. To illustrate this procedure, we solve the following system of three linear equations in the variables x_1, x_2, and x_3:

$$\begin{aligned} x_1 - 2x_2 - x_3 &= 3 \\ 3x_1 - 6x_2 - 5x_3 &= 3 \\ 2x_1 - x_2 + x_3 &= 0 \end{aligned} \tag{2}$$

We begin the simplification by eliminating x_1 from every equation but the first. To do so, we add appropriate multiples of the first equation to the second and third equations so that the coefficient of x_1 becomes 0 in these equations. Adding -3 times the first equation to the second makes the coefficient of x_1 equal 0 in the result.

$$\begin{array}{rl} -3x_1 + 6x_2 + 3x_3 = -9 & (-3 \text{ times equation 1}) \\ 3x_1 - 6x_2 - 5x_3 = 3 & (\text{equation 2}) \\ \hline -2x_3 = -6 & \end{array}$$

Likewise, adding -2 times the first equation to the third makes the coefficient of x_1 0 in the new third equation.

$$\begin{array}{rcl} -2x_1 + 4x_2 + 2x_3 &=& -6 \quad (-2 \text{ times equation 1}) \\ 2x_1 - x_2 + x_3 &=& 0 \quad (\text{equation 3}) \\ \hline 3x_2 + 3x_3 &=& -6 \end{array}$$

We now replace equation 2 with $-2x_3 = -6$, and equation 3 with $3x_2 + 3x_3 = -6$ to transform system (2) into the following system:

$$\begin{aligned} x_1 - 2x_2 - x_3 &= 3 \\ -2x_3 &= -6 \\ 3x_2 + 3x_3 &= -6 \end{aligned}$$

In this case, the calculation that makes the coefficient of x_1 equal 0 in the new second equation also makes the coefficient of x_2 equal 0. (This does not always happen, as you can see from the new third equation.) If we now interchange the second and third equations in this system, we obtain the following system:

$$\begin{aligned} x_1 - 2x_2 - x_3 &= 3 \\ 3x_2 + 3x_3 &= -6 \\ -2x_3 &= -6 \end{aligned} \tag{3}$$

We can now solve the third equation for x_3 by multiplying both sides by $-\frac{1}{2}$ (or equivalently, dividing both sides by -2). This produces

$$\begin{aligned} x_1 - 2x_2 - x_3 &= 3 \\ 3x_2 + 3x_3 &= -6 \\ x_3 &= 3. \end{aligned}$$

By adding appropriate multiples of the third equation to the first and second, we can eliminate x_3 from every equation but the third. If we add the third equation to the first and add -3 times the third equation to the second, we obtain

$$\begin{aligned} x_1 - 2x_2 &= 6 \\ 3x_2 &= -15 \\ x_3 &= 3. \end{aligned}$$

Now solve for x_2 by multiplying the second equation by $\frac{1}{3}$. The result is

$$\begin{aligned} x_1 - 2x_2 &= 6 \\ x_2 &= -5 \\ x_3 &= 3. \end{aligned}$$

Finally, adding 2 times the second equation to the first produces the very simple system

$$\begin{aligned} x_1 &= -4 \\ x_2 &= -5 \\ x_3 &= 3, \end{aligned} \tag{4}$$

whose solution is obvious. You should check that replacing x_1 by -4, x_2 by -5, and x_3 by 3 makes each equation in system (2) true, so that $\begin{bmatrix} -4 \\ -5 \\ 3 \end{bmatrix}$ is a solution of system (2). Indeed, it is the only solution, as we soon will show.

In each step just presented, the names of the variables played no essential role. All of the operations that we performed on the system of equations can also be performed on matrices. In fact, we can express the original system

$$\begin{aligned} x_1 - 2x_2 - x_3 &= 3 \\ 3x_1 - 6x_2 - 5x_3 &= 3 \\ 2x_1 - x_2 + x_3 &= 0 \end{aligned} \tag{2}$$

as the matrix equation $A\mathbf{x} = \mathbf{b}$, where

$$A = \begin{bmatrix} 1 & -2 & -1 \\ 3 & -6 & -5 \\ 2 & -1 & 1 \end{bmatrix}, \qquad \mathbf{x} = \begin{bmatrix} x_1 \\ x_2 \\ x_3 \end{bmatrix}, \qquad \text{and} \qquad \mathbf{b} = \begin{bmatrix} 3 \\ 3 \\ 0 \end{bmatrix}.$$

Note that the columns of A contain the coefficients of x_1, x_2, and x_3 from system (2). For this reason, A is called the **coefficient matrix** (or the **matrix of coefficients**) of system (2). All the information that is needed to find the solution set of this system is contained in the matrix

$$\begin{bmatrix} 1 & -2 & -1 & 3 \\ 3 & -6 & -5 & 3 \\ 2 & -1 & 1 & 0 \end{bmatrix},$$

which is called the **augmented matrix** of the system. This matrix is formed by augmenting the coefficient matrix A to include the vector $\mathbf{b}$. We denote the augmented matrix by $[A\ \ \mathbf{b}]$.

If A is an $m \times n$ matrix, then a vector $\mathbf{u}$ in $\mathcal{R}^n$ is a solution of $A\mathbf{x} = \mathbf{b}$ if and only if $A\mathbf{u} = \mathbf{b}$. Thus $\begin{bmatrix} -4 \\ -5 \\ 3 \end{bmatrix}$ is a solution of system (2) because

$$A\mathbf{u} = \begin{bmatrix} 1 & -2 & -1 \\ 3 & -6 & -5 \\ 2 & -1 & 1 \end{bmatrix} \begin{bmatrix} -4 \\ -5 \\ 3 \end{bmatrix} = \begin{bmatrix} 3 \\ 3 \\ 0 \end{bmatrix} = \mathbf{b}.$$

Example 1

For the system of linear equations

$$\begin{aligned} x_1 + 5x_3 - x_4 &= 7 \\ 2x_1 - x_2 + 6x_3 &= -8, \end{aligned}$$

the coefficient matrix and the augmented matrix are

$$\begin{bmatrix} 1 & 0 & 5 & -1 \\ 2 & -1 & 6 & 0 \end{bmatrix} \qquad \text{and} \qquad \begin{bmatrix} 1 & 0 & 5 & -1 & 7 \\ 2 & -1 & 6 & 0 & -8 \end{bmatrix},$$

respectively. Note that the variable x_2 is missing from the first equation and x_4 is missing from the second equation in the system (that is, the coefficients of x_2 in the first equation and x_4 in the second equation are 0). As a result, the $(1,2)$- and $(2,4)$-entries of the coefficient and augmented matrices of the system are 0.

In solving system (2), we performed three types of operations: interchanging the position of two equations in a system, multiplying an equation in the system by a nonzero scalar, and adding a multiple of one equation in the system to another. The analogous operations that can be performed on the augmented matrix of the system are given in the following definition.

Definition Any one of the following three operations performed on a matrix is called an **elementary row operation**:

1. Interchange any two rows of the matrix. **(interchange operation)**
2. Multiply every entry of some row of the matrix by the same nonzero scalar. **(scaling operation)**
3. Add a multiple of one row of the matrix to another row. **(row addition operation)**

To denote how an elementary row operation changes a matrix A into a matrix B, we use the following notation:

1. $A \xrightarrow{\mathbf{r}_i \leftrightarrow \mathbf{r}_j} B$ indicates that row i and row j are interchanged.
2. $A \xrightarrow{c\mathbf{r}_i \to \mathbf{r}_i} B$ indicates that the entries of row i are multiplied by the scalar c.
3. $A \xrightarrow{c\mathbf{r}_i + \mathbf{r}_j \to \mathbf{r}_j} B$ indicates that c times row i is added to row j.

Example 2 Let

$$A = \begin{bmatrix} 2 & 1 & -1 & 3 \\ 1 & 2 & 1 & 3 \\ 3 & 1 & 0 & 2 \end{bmatrix} \quad \text{and} \quad B = \begin{bmatrix} 1 & 2 & 1 & 3 \\ 0 & 1 & 1 & 1 \\ 0 & -5 & -3 & -7 \end{bmatrix}.$$

The following sequence of elementary row operations transforms A into B:

$$A = \begin{bmatrix} 2 & 1 & -1 & 3 \\ 1 & 2 & 1 & 3 \\ 3 & 1 & 0 & 2 \end{bmatrix} \xrightarrow{\mathbf{r}_1 \leftrightarrow \mathbf{r}_2} \begin{bmatrix} 1 & 2 & 1 & 3 \\ 2 & 1 & -1 & 3 \\ 3 & 1 & 0 & 2 \end{bmatrix}$$

$$\xrightarrow{-2\mathbf{r}_1 + \mathbf{r}_2 \to \mathbf{r}_2} \begin{bmatrix} 1 & 2 & 1 & 3 \\ 0 & -3 & -3 & -3 \\ 3 & 1 & 0 & 2 \end{bmatrix}$$

$$\xrightarrow{-3\mathbf{r}_1 + \mathbf{r}_3 \to \mathbf{r}_3} \begin{bmatrix} 1 & 2 & 1 & 3 \\ 0 & -3 & -3 & -3 \\ 0 & -5 & -3 & -7 \end{bmatrix} \xrightarrow{-\frac{1}{3}\mathbf{r}_2 \to \mathbf{r}_2} \begin{bmatrix} 1 & 2 & 1 & 3 \\ 0 & 1 & 1 & 1 \\ 0 & -5 & -3 & -7 \end{bmatrix} = B.$$

We may perform several elementary row operations in succession, indicating the operations by stacking the individual labels above a single arrow. These operations are performed in top-to-bottom order. In the previous example, we could indicate how to transform the second matrix into the fourth matrix of the example by using the following notation:

$$\begin{bmatrix} 1 & 2 & 1 & 3 \\ 2 & 1 & -1 & 3 \\ 3 & 1 & 0 & 2 \end{bmatrix} \xrightarrow[\;]{\substack{-2\mathbf{r}_1 + \mathbf{r}_2 \to \mathbf{r}_2 \\ -3\mathbf{r}_1 + \mathbf{r}_3 \to \mathbf{r}_3}} \begin{bmatrix} 1 & 2 & 1 & 3 \\ 0 & -3 & -3 & -3 \\ 0 & -5 & -3 & -7 \end{bmatrix}$$

Every elementary row operation can be reversed. That is, if we perform an elementary row operation on a matrix A to produce a new matrix B, then we can perform an elementary row operation of the same kind on B to obtain A. If, for example, we obtain B by interchanging two rows of A, then interchanging the same rows of B yields A. Also, if we obtain B by multiplying some row of A by the nonzero constant c, then multiplying the same row of B by $\frac{1}{c}$ yields A. Finally, if we obtain B by adding c times row i of A to row j, then adding $-c$ times row i of B to row j results in A.

Suppose that we perform an elementary row operation on an augmented matrix $[A\ \mathbf{b}]$ to obtain a new matrix $[A'\ \mathbf{b}']$. The reversibility of the elementary row operations assures us that the solutions of $A\mathbf{x} = \mathbf{b}$ are the same as those of $A'\mathbf{x} = \mathbf{b}'$. *Thus performing an elementary row operation on the augmented matrix of a system of linear equations does not change the solution set. That is, each elementary row operation produces the augmented matrix of an equivalent system of linear equations.* We assume this result throughout the rest of Chapter 1; it is proved in Section 2.3. Thus, because the system of linear equations (2) is equivalent to system (4), there is only one solution of system (2).

REDUCED ROW ECHELON FORM

We can use elementary row operations to simplify any system of linear equations until it is easy to see what the solution is. First, we represent the system by its augmented matrix, and then use elementary row operations to transform the augmented matrix into a matrix having a special form, which we call a *reduced row echelon form*. The system of linear equations whose augmented matrix has this form is equivalent to the original system and is easily solved.

We now define this special form of matrix. In the following discussion, we call a row of a matrix a **zero row** if all its entries are 0 and a **nonzero row** otherwise. We call the leftmost nonzero entry of a nonzero row its **leading entry**.

Definitions A matrix is said to be in **row echelon form** if it satisfies the following three conditions:

1. Each nonzero row lies above every zero row.
2. The leading entry of a nonzero row lies in a column to the right of the column containing the leading entry of any preceding row.
3. If a column contains the leading entry of some row, then all entries of that column below the leading entry are 0.[5]

If a matrix also satisfies the following two additional conditions, we say that it is in **reduced row echelon form**.[6]

4. If a column contains the leading entry of some row, then all the other entries of that column are 0.
5. The leading entry of each nonzero row is 1.

---

[5] Condition 3 is a direct consequence of condition 2. We include it in this definition for emphasis, as is usually done when defining the row echelon form.

[6] Inexpensive calculators are available that can compute the reduced row echelon form of a matrix. On such a calculator, or in computer software, the reduced row echelon form is usually obtained by using the command `rref`.

A matrix having either of the forms that follow is in reduced row echelon form. In these diagrams, a $*$ denotes an arbitrary entry (that may or may not be 0).

$$\begin{bmatrix} 1 & * & 0 & 0 & * \\ 0 & 0 & 1 & 0 & * \\ 0 & 0 & 0 & 1 & * \\ 0 & 0 & 0 & 0 & 0 \end{bmatrix} \qquad \begin{bmatrix} 1 & 0 & * & * & 0 & 0 & * \\ 0 & 1 & * & * & 0 & 0 & * \\ 0 & 0 & 0 & 0 & 1 & 0 & * \\ 0 & 0 & 0 & 0 & 0 & 1 & * \\ 0 & 0 & 0 & 0 & 0 & 0 & 0 \end{bmatrix}$$

Notice that the leading entries (which must be 1's by condition 5) form a pattern suggestive of a flight of stairs. Moreover, these leading entries of 1 are the only nonzero entries in their columns. Also, each nonzero row precedes all of the zero rows.

Example 3 The following matrices are *not* in reduced row echelon form:

$$A = \begin{bmatrix} 1 & 0 & 0 & 6 & 3 & 0 \\ 0 & 0 & 1 & 5 & 7 & 0 \\ 0 & 1 & 0 & 2 & 4 & 0 \\ 0 & 0 & 0 & 0 & 0 & 1 \end{bmatrix} \qquad B = \begin{bmatrix} 1 & 7 & 2 & -3 & 9 & 4 \\ 0 & 0 & 1 & 4 & 6 & 8 \\ 0 & 0 & 0 & 2 & 3 & 5 \\ 0 & 0 & 0 & 0 & 0 & 0 \\ 0 & 0 & 0 & 0 & 0 & 0 \end{bmatrix}$$

Matrix A fails to be in reduced row echelon form because the leading entry of the third row does not lie to the right of the leading entry of the second row. Notice, however, that the matrix obtained by interchanging the second and third rows of A is in reduced row echelon form.

Matrix B is not in reduced row echelon form for two reasons. The leading entry of the third row is not 1, and the leading entries in the second and third rows are not the only nonzero entries in their columns. That is, the third column of B contains the first nonzero entry in row 2, but the $(2, 3)$-entry of B is not the only nonzero entry in column 3. Notice, however, that although B is not in *reduced* row echelon form, B is in row echelon form.

A system of linear equations can be easily solved if its augmented matrix is in reduced row echelon form. For example, the system

$$\begin{aligned} x_1 \qquad\qquad &= -4 \\ x_2 \qquad &= -5 \\ x_3 &= 3 \end{aligned}$$

has a solution that is immediately evident.

If a system of equations has infinitely many solutions, then obtaining the solution is somewhat more complicated. Consider, for example, the system of linear equations

$$\begin{aligned} x_1 - 3x_2 \quad + 2x_4 \quad &= 7 \\ x_3 + 6x_4 \quad &= 9 \\ x_5 &= 2 \\ 0 &= 0. \end{aligned} \tag{5}$$

The augmented matrix of this system is

$$\begin{bmatrix} 1 & -3 & 0 & 2 & 0 & 7 \\ 0 & 0 & 1 & 6 & 0 & 9 \\ 0 & 0 & 0 & 0 & 1 & 2 \\ 0 & 0 & 0 & 0 & 0 & 0 \end{bmatrix},$$

which is in reduced row echelon form.

Since the equation $0 = 0$ in system (5) provides no useful information, we can disregard it. System (5) is consistent, but it is not possible to find a unique value for each variable because the system has infinitely many solutions. Instead, we can solve for some of the variables, called **basic variables**, in terms of the others, called the **free variables**. The basic variables correspond to the leading entries of the augmented matrix. In system (5), for example, the basic variables are x_1 , x_3, and x_5 because the leading entries of the augmented matrix are in columns 1, 3, and 5, respectively. The free variables are x_2 and x_4. We can easily solve for the basic variables in terms of the free variables by moving the free variables and their coefficients from the left side of each equation to the right.

The resulting equations

$$\begin{aligned} x_1 &= 7 + 3x_2 - 2x_4 \\ x_2 &\quad \text{free} \\ x_3 &= 9 - 6x_4 \\ x_4 &\quad \text{free} \\ x_5 &= 2 \end{aligned}$$

provide a **general solution** of system (5). This means that for every choice of values of the free variables, these equations give the corresponding values of x_1, x_3, and x_5 in one solution of the system, and furthermore, every solution of the system has this form for some values of the free variables. For example, choosing $x_2 = 0$ and $x_4 = 0$ gives the solution $\begin{bmatrix} 7 \\ 0 \\ 9 \\ 0 \\ 2 \end{bmatrix}$, and choosing $x_2 = -2$ and $x_4 = 1$ yields the solution $\begin{bmatrix} -1 \\ -2 \\ 3 \\ 1 \\ 2 \end{bmatrix}$.

The general solution can also be written in *vector form* as

$$\begin{bmatrix} x_1 \\ x_2 \\ x_3 \\ x_4 \\ x_5 \end{bmatrix} = \begin{bmatrix} 7 + 3x_2 - 2x_4 \\ x_2 \\ 9 - 6x_4 \\ x_4 \\ 2 \end{bmatrix} = \begin{bmatrix} 7 \\ 0 \\ 9 \\ 0 \\ 2 \end{bmatrix} + x_2 \begin{bmatrix} 3 \\ 1 \\ 0 \\ 0 \\ 0 \end{bmatrix} + x_4 \begin{bmatrix} -2 \\ 0 \\ -6 \\ 1 \\ 0 \end{bmatrix}.$$

In vector form, it is apparent that every solution of the system is the sum of $\begin{bmatrix} 7 \\ 0 \\ 9 \\ 0 \\ 2 \end{bmatrix}$ and an arbitrary linear combination of the vectors $\begin{bmatrix} 3 \\ 1 \\ 0 \\ 0 \\ 0 \end{bmatrix}$ and $\begin{bmatrix} -2 \\ 0 \\ -6 \\ 1 \\ 0 \end{bmatrix}$, with the coefficients being the free variables x_2 and x_4, respectively.

Example 4 Find a general solution of the system of linear equations

$$\begin{aligned} x_1 \qquad\qquad + 2x_4 &= 7 \\ x_2 \qquad - 3x_4 &= 8 \\ x_3 + 6x_4 &= 9. \end{aligned}$$

Solution Since the augmented matrix of this system is in reduced row echelon form, we can obtain the general solution by solving for the basic variables in terms of the other variables. In this case, the basic variables are x_1, x_2, and x_3, and so we solve for x_1, x_2, and x_3 in terms of x_4. The resulting general solution is

$$\begin{aligned} x_1 &= 7 - 2x_4 \\ x_2 &= 8 + 3x_4 \\ x_3 &= 9 - 6x_4 \\ x_4 &\quad \text{free.} \end{aligned}$$

We can write the general solution in vector form as

$$\begin{bmatrix} x_1 \\ x_2 \\ x_3 \\ x_4 \end{bmatrix} = \begin{bmatrix} 7 \\ 8 \\ 9 \\ 0 \end{bmatrix} + x_4 \begin{bmatrix} -2 \\ 3 \\ -6 \\ 1 \end{bmatrix}.$$

There is one other case to consider. Suppose that the augmented matrix of a system of linear equations contains a row in which the only nonzero entry is in the last column, for example,

$$\begin{bmatrix} 1 & 0 & -3 & 5 \\ 0 & 1 & 2 & 4 \\ 0 & 0 & 0 & 1 \\ 0 & 0 & 0 & 0 \end{bmatrix}.$$

The system of linear equations corresponding to this matrix is

$$\begin{aligned} x_1 \qquad\quad - 3x_3 &= 5 \\ x_2 + 2x_3 &= 4 \\ 0x_1 + 0x_2 + 0x_3 &= 1 \\ 0x_1 + 0x_2 + 0x_3 &= 0. \end{aligned}$$

Clearly, there are no values of the variables that satisfy the third equation. Because a solution of the system must satisfy every equation in the system, it follows that this system of equations is inconsistent. More generally, the following statement is true:

> Whenever an augmented matrix contains a row in which the only nonzero entry lies in the last column, the corresponding system of linear equations has no solution.

It is not usually obvious whether or not a system of linear equations is consistent. However, this is apparent after calculating the reduced row echelon form of its augmented matrix.

Practice Problem 2 ▶ The augmented matrix of a system of linear equations has

$$\begin{bmatrix} 0 & 1 & -4 & 0 & 3 & 0 \\ 0 & 0 & 0 & 1 & -2 & 0 \\ 0 & 0 & 0 & 0 & 0 & 1 \end{bmatrix}$$

as its reduced row echelon form. Determine whether this system of linear equations is consistent and, if so, find its general solution. ◀

SOLVING SYSTEMS OF LINEAR EQUATIONS

So far, we have learned the following facts:

1. A system of linear equations can be represented by its augmented matrix, and any elementary row operations performed on that matrix do not change the solutions of the system.
2. A system of linear equations whose augmented matrix is in reduced row echelon form is easily solved.

Two questions remain. Is it always possible to transform the augmented matrix of a system of linear equations into a reduced row echelon form by a sequence of elementary row operations? Is that form unique? The first question is answered in Section 1.4, where an algorithm is given that transforms any matrix into one in reduced row echelon form. The second question is also important. If there were different reduced row echelon forms of the same matrix (depending on what sequence of elementary row operations is used), then there could be different solutions of the same system of linear equations. Fortunately, the following important theorem assures us that there is only one reduced row echelon form for any matrix. It is proved in Appendix E.

THEOREM 1.4

Every matrix can be transformed into one and only one matrix in reduced row echelon form by means of a sequence of elementary row operations.

In fact, Section 1.4 describes an explicit procedure for performing this transformation. If there is a sequence of elementary row operations that transforms a matrix A into a matrix R in reduced row echelon form, then we call R the **reduced row echelon form of A**. Using the reduced row echelon form of the augmented matrix of a system of linear equations $A\mathbf{x} = \mathbf{b}$, we can solve the system as follows:

Procedure for Solving a System of Linear Equations

1. Write the augmented matrix $[A \;\; \mathbf{b}]$ of the system.
2. Find the reduced row echelon form $[R \;\; \mathbf{c}]$ of $[A \;\; \mathbf{b}]$.
3. If $[R \;\; \mathbf{c}]$ contains a row in which the only nonzero entry lies in the last column, then $A\mathbf{x} = \mathbf{b}$ has no solution. Otherwise, the system has at least one solution. Write the system of linear equations corresponding to the matrix $[R \;\; \mathbf{c}]$, and solve this system for the basic variables in terms of the free variables to obtain a general solution of $A\mathbf{x} = \mathbf{b}$.

Example 5 Solve the following system of linear equations:

$$\begin{aligned} x_1 + 2x_2 - x_3 + 2x_4 + x_5 &= 2 \\ -x_1 - 2x_2 + x_3 + 2x_4 + 3x_5 &= 6 \\ 2x_1 + 4x_2 - 3x_3 + 2x_4 &= 3 \\ -3x_1 - 6x_2 + 2x_3 + 3x_5 &= 9 \end{aligned}$$

Solution The augmented matrix of this system is

$$\begin{bmatrix} 1 & 2 & -1 & 2 & 1 & 2 \\ -1 & -2 & 1 & 2 & 3 & 6 \\ 2 & 4 & -3 & 2 & 0 & 3 \\ -3 & -6 & 2 & 0 & 3 & 9 \end{bmatrix}.$$

In Section 1.4, we show that the reduced row echelon form of this matrix is

$$\begin{bmatrix} 1 & 2 & 0 & 0 & -1 & -5 \\ 0 & 0 & 1 & 0 & 0 & -3 \\ 0 & 0 & 0 & 1 & 1 & 2 \\ 0 & 0 & 0 & 0 & 0 & 0 \end{bmatrix}.$$

Because there is no row in this matrix in which the only nonzero entry lies in the last column, the original system is consistent. This matrix corresponds to the system of linear equations

$$\begin{aligned} x_1 + 2x_2 \qquad\qquad - x_5 &= -5 \\ x_3 \qquad\quad &= -3 \\ x_4 + x_5 &= 2. \end{aligned}$$

In this system, the basic variables are x_1, x_3, and x_4, and the free variables are x_2 and x_5. When we solve for the basic variables in terms of the free variables, we obtain the following general solution:

$$\begin{aligned} x_1 &= -5 - 2x_2 + x_5 \\ x_2 &\quad \text{free} \\ x_3 &= -3 \\ x_4 &= 2 \qquad\quad - x_5 \\ x_5 &\quad \text{free} \end{aligned}$$

This is the general solution of the original system of linear equations.

Practice Problem 3 ▶ The augmented matrix of a system of linear equations has

$$\begin{bmatrix} 0 & 1 & -3 & 0 & 2 & 4 \\ 0 & 0 & 0 & 1 & -1 & 5 \\ 0 & 0 & 0 & 0 & 0 & 0 \end{bmatrix}$$

as its reduced row echelon form. Write the corresponding system of linear equations, and determine if it is consistent. If so, find its general solution, and write the general solution in vector form. ◀

EXERCISES

In Exercises 1–6, write (a) the coefficient matrix and (b) the augmented matrix of the given system.

1. $\begin{aligned} -x_2 + 2x_3 &= 0 \\ x_1 + 3x_2 \qquad &= -1 \end{aligned}$

2. $2x_1 - x_2 + 3x_3 = 4$

3. $\begin{aligned} x_1 + 2x_2 &= 3 \\ -x_1 + 3x_2 &= 2 \\ -3x_1 + 4x_2 &= 1 \end{aligned}$

4. $\begin{aligned} x_1 \qquad + 2x_3 - x_4 &= 3 \\ 2x_1 - x_2 + \qquad\quad x_4 &= 0 \end{aligned}$

5. $\begin{aligned} 2x_2 - 3x_3 &= 4 \\ -x_1 + x_2 + 2x_3 &= -6 \\ 2x_1 + \qquad\quad x_3 &= 0 \end{aligned}$

6. $\begin{aligned} x_1 - 2x_2 + x_4 + 7x_5 &= 5 \\ x_1 - 2x_2 \qquad\quad + 10x_5 &= 3 \\ 2x_1 - 4x_2 + 4x_4 + 8x_5 &= 7 \end{aligned}$

In Exercises 7–14, perform the indicated elementary row operation on

$$\begin{bmatrix} 1 & -1 & 0 & 2 & -3 \\ -2 & 6 & 3 & -1 & 1 \\ 0 & 2 & -4 & 4 & 2 \end{bmatrix}.$$

7. Interchange rows 1 and 3.
8. Multiply row 1 by -3.
9. Add 2 times row 1 to row 2.
10. Interchange rows 1 and 2.

11. Multiply row 3 by $\frac{1}{2}$.
12. Add -3 times row 3 to row 2.
13. Add 4 times row 2 to row 3.
14. Add 2 times row 1 to row 3.

In Exercises 15–22, perform the indicated elementary row operation on

$$\begin{bmatrix} 1 & -2 & 0 \\ -1 & 1 & -1 \\ 2 & -4 & 6 \\ -3 & 2 & 1 \end{bmatrix}.$$

15. Multiply row 1 by -2.
16. Multiply row 2 by $\frac{1}{2}$.
17. Add -2 times row 1 to row 3.
18. Add 3 times row 1 to row 4.
19. Interchange rows 2 and 3.
20. Interchange rows 2 and 4.
21. Add -2 times row 2 to row 4.
22. Add 2 times row 2 to row 1.

In Exercises 23–30, determine whether the given vector is a solution of the system

$$\begin{aligned} x_1 - 4x_2 \quad\quad + 3x_4 &= 6 \\ x_3 - 2x_4 &= -3. \end{aligned}$$

23. $\begin{bmatrix} 1 \\ -2 \\ -5 \\ -1 \end{bmatrix}$ 24. $\begin{bmatrix} 2 \\ 0 \\ -1 \\ 1 \end{bmatrix}$ 25. $\begin{bmatrix} 3 \\ 0 \\ 2 \\ 1 \end{bmatrix}$ 26. $\begin{bmatrix} 4 \\ 1 \\ 1 \\ 2 \end{bmatrix}$

27. $\begin{bmatrix} 6 \\ -3 \\ 0 \\ 0 \end{bmatrix}$ 28. $\begin{bmatrix} 6 \\ 0 \\ -3 \\ 0 \end{bmatrix}$ 29. $\begin{bmatrix} 9 \\ 0 \\ -5 \\ -1 \end{bmatrix}$ 30. $\begin{bmatrix} -1 \\ -1 \\ -1 \\ 1 \end{bmatrix}$

In Exercises 31–38, determine whether the given vector is a solution of the system

$$\begin{aligned} x_1 - 2x_2 + x_3 + x_4 + 7x_5 &= 1 \\ x_1 - 2x_2 + 2x_3 \quad\quad + 10x_5 &= 2 \\ 2x_1 - 4x_2 \quad\quad + 4x_4 + 8x_5 &= 0. \end{aligned}$$

31. $\begin{bmatrix} 0 \\ 0 \\ 1 \\ 0 \\ 0 \end{bmatrix}$ 32. $\begin{bmatrix} 0 \\ 1 \\ 0 \\ 0 \\ 0 \end{bmatrix}$ 33. $\begin{bmatrix} 2 \\ 1 \\ 1 \\ 0 \\ 0 \end{bmatrix}$ 34. $\begin{bmatrix} 0 \\ -2 \\ 4 \\ 0 \\ -1 \end{bmatrix}$

35. $\begin{bmatrix} 1 \\ 0 \\ 1 \\ 1 \\ 0 \end{bmatrix}$ 36. $\begin{bmatrix} 0 \\ -1 \\ 0 \\ -1 \\ 0 \end{bmatrix}$ 37. $\begin{bmatrix} 0 \\ 3 \\ -1 \\ 3 \\ 0 \end{bmatrix}$ 38. $\begin{bmatrix} 0 \\ 1 \\ 0 \\ 1 \\ 0 \end{bmatrix}$

In Exercises 39–54, the reduced row echelon form of the augmented matrix of a system of linear equations is given. Determine whether this system of linear equations is consistent and, if so, find its general solution. In addition, in Exercises 47–54, write the solution in vector form.

39. $\begin{bmatrix} 1 & -1 & 2 \end{bmatrix}$ 40. $\begin{bmatrix} 1 & 0 & -4 \\ 0 & 1 & 5 \end{bmatrix}$

41. $\begin{bmatrix} 1 & -2 & 6 \\ 0 & 0 & 0 \end{bmatrix}$ 42. $\begin{bmatrix} 1 & -4 & 5 \\ 0 & 0 & 0 \\ 0 & 0 & 0 \end{bmatrix}$

43. $\begin{bmatrix} 1 & -3 & 0 \\ 0 & 0 & 1 \\ 0 & 0 & 0 \end{bmatrix}$ 44. $\begin{bmatrix} 1 & 0 & -6 \\ 0 & 1 & 3 \\ 0 & 0 & 0 \end{bmatrix}$

45. $\begin{bmatrix} 1 & -2 & 0 & 4 \\ 0 & 0 & 1 & 3 \\ 0 & 0 & 0 & 0 \end{bmatrix}$ 46. $\begin{bmatrix} 1 & -2 & 0 & 0 \\ 0 & 0 & 1 & 0 \\ 0 & 0 & 0 & 1 \end{bmatrix}$

47. $\begin{bmatrix} 1 & 0 & 0 & -3 & 0 \\ 0 & 1 & 0 & -4 & 0 \\ 0 & 0 & 1 & 5 & 0 \end{bmatrix}$ 48. $\begin{bmatrix} 1 & 0 & -1 & 3 & 9 \\ 0 & 1 & 2 & -5 & 8 \\ 0 & 0 & 0 & 0 & 0 \end{bmatrix}$

49. $\begin{bmatrix} 0 & 1 & 0 & 0 & -3 \\ 0 & 0 & 1 & 0 & -4 \\ 0 & 0 & 0 & 1 & 5 \end{bmatrix}$ 50. $\begin{bmatrix} 1 & -2 & 0 & 0 & -3 \\ 0 & 0 & 1 & 0 & -4 \\ 0 & 0 & 0 & 1 & 5 \end{bmatrix}$

51. $\begin{bmatrix} 1 & 3 & 0 & -2 & 6 \\ 0 & 0 & 1 & 4 & 7 \\ 0 & 0 & 0 & 0 & 0 \end{bmatrix}$ 52. $\begin{bmatrix} 0 & 1 & 0 & 3 & -4 \\ 0 & 0 & 1 & 2 & 9 \\ 0 & 0 & 0 & 0 & 0 \end{bmatrix}$

53. $\begin{bmatrix} 1 & -3 & 2 & 0 & 4 & 0 \\ 0 & 0 & 0 & 0 & 0 & 1 \\ 0 & 0 & 0 & 0 & 0 & 0 \end{bmatrix}$

54. $\begin{bmatrix} 0 & 0 & 1 & -3 & 0 & 2 & 0 \\ 0 & 0 & 0 & 0 & 1 & -1 & 0 \\ 0 & 0 & 0 & 0 & 0 & 0 & 0 \end{bmatrix}$

55. Suppose that the general solution of a system of m linear equations in n variables contains k free variables. How many basic variables does it have? Explain your answer.
56. Suppose that R is a matrix in reduced row echelon form. If row 4 of R is nonzero and has its leading entry in column 5, describe column 5.

T&F *In Exercises 57–76, determine whether the following statements are true or false.*

57. Every system of linear equations has at least one solution.
58. Some systems of linear equations have exactly two solutions.
59. If a matrix A can be transformed into a matrix B by an elementary row operation, then B can be transformed into A by an elementary row operation.
60. If a matrix is in row echelon form, then the leading entry of each nonzero row must be 1.
61. If a matrix is in reduced row echelon form, then the leading entry of each nonzero row is 1.
62. Every matrix can be transformed into one in reduced row echelon form by a sequence of elementary row operations.
63. Every matrix can be transformed into a unique matrix in row echelon form by a sequence of elementary row operations.

64. Every matrix can be transformed into a unique matrix in reduced row echelon form by a sequence of elementary row operations.
65. Performing an elementary row operation on the augmented matrix of a system of linear equations produces the augmented matrix of an equivalent system of linear equations.
66. If the reduced row echelon form of the augmented matrix of a system of linear equations contains a zero row, then the system is consistent.
67. If the only nonzero entry in some row of an augmented matrix of a system of linear equations lies in the last column, then the system is inconsistent.
68. A system of linear equations is called consistent if it has one or more solutions.
69. If A is the coefficient matrix of a system of m linear equations in n variables, then A is an $n \times m$ matrix.
70. The augmented matrix of a system of linear equations contains one more column than the coefficient matrix.
71. If the reduced row echelon form of the augmented matrix of a consistent system of m linear equations in n variables contains k nonzero rows, then its general solution contains k basic variables.
72. A system of linear equations $A\mathbf{x} = \mathbf{b}$ has the same solutions as the system of linear equations $R\mathbf{x} = \mathbf{c}$, where $[R\ \ \mathbf{c}]$ is the reduced row echelon form of $[A\ \ \mathbf{b}]$.
73. Multiplying every entry of some row of a matrix by a scalar is an elementary row operation.
74. Every solution of a consistent system of linear equations can be obtained by substituting appropriate values for the free variables in its general solution.
75. If a system of linear equations has more variables than equations, then it must have infinitely many solutions.
76. If A is an $m \times n$ matrix, then a solution of the system $A\mathbf{x} = \mathbf{b}$ is a vector $\mathbf{u}$ in $\mathcal{R}^n$ such that $A\mathbf{u} = \mathbf{b}$.

77.[7] Let $[A\ \ \mathbf{b}]$ be the augmented matrix of a system of linear equations. Prove that if its reduced row echelon form is $[R\ \ \mathbf{c}]$, then R is the reduced row echelon form of A.

78. Prove that if R is the reduced row echelon form of a matrix A, then $[R\ \ \mathbf{0}]$ is the reduced row echelon form of $[A\ \ \mathbf{0}]$.
79. Prove that for any $m \times n$ matrix A, the equation $A\mathbf{x} = \mathbf{0}$ is consistent, where $\mathbf{0}$ is the zero vector in $\mathcal{R}^m$.
80. Let A be an $m \times n$ matrix whose reduced row echelon form contains no zero rows. Prove that $A\mathbf{x} = \mathbf{b}$ is consistent for every $\mathbf{b}$ in $\mathcal{R}^m$.
81. In a matrix in reduced row echelon form, there are three types of entries: The leading entries of nonzero rows are required to be 1s, certain other entries are required to be 0s, and the remaining entries are arbitrary. Suppose that these arbitrary entries are denoted by asterisks. For example,

$$\begin{bmatrix} 0 & 1 & * & 0 & * & 0 & * \\ 0 & 0 & 0 & 1 & * & 0 & * \\ 0 & 0 & 0 & 0 & 0 & 1 & * \end{bmatrix}$$

is a possible reduced row echelon form for a 3×7 matrix. How many different such forms for a reduced row echelon matrix are possible if the matrix is 2×3?
82. Repeat Exercise 81 for a 2×4 matrix.
83. Suppose that B is obtained by one elementary row operation performed on matrix A. Prove that the same type of elementary operation (namely, an interchange, scaling, or row addition operation) that transforms A into B also transforms B into A.
84. Show that if an equation in a system of linear equations is multiplied by 0, the resulting system need not be equivalent to the original one.
85. Let S denote the following system of linear equations:

$$\begin{aligned} a_{11}x_1 + a_{12}x_2 + a_{13}x_3 &= b_1 \\ a_{21}x_1 + a_{22}x_2 + a_{23}x_3 &= b_2 \\ a_{31}x_1 + a_{32}x_2 + a_{33}x_3 &= b_3 \end{aligned}$$

Show that if the second equation of S is multiplied by a nonzero scalar c, then the resulting system is equivalent to S.
86. Let S be the system of linear equations in Exercise 85. Show that if k times the first equation of S is added to the third equation, then the resulting system is equivalent to S.

SOLUTIONS TO THE PRACTICE PROBLEMS

1. (a) Since $2(-2) - 3 + 6(2) = 5$, $\mathbf{u}$ is not a solution of the second equation in the given system of equations. Therefore $\mathbf{u}$ is not a solution of the system. Another method for solving this problem is to represent the given system as a matrix equation $A\mathbf{x} = \mathbf{b}$, where

$$A = \begin{bmatrix} 1 & 0 & 5 & -1 \\ 2 & -1 & 6 & 0 \end{bmatrix} \quad \text{and} \quad \mathbf{b} = \begin{bmatrix} 7 \\ 8 \end{bmatrix}.$$

Because

$$\mathbf{u} = \begin{bmatrix} 1 & 0 & 5 & -1 \\ 2 & -1 & 6 & 0 \end{bmatrix} \begin{bmatrix} -2 \\ 3 \\ 2 \\ 1 \end{bmatrix} = \begin{bmatrix} 7 \\ 5 \end{bmatrix} \neq \mathbf{b},$$

$\mathbf{u}$ is not a solution of the given system.

[7] This exercise is used in Section 1.6 (on page 70).

(b) Since $5+5(1)-3=7$ and $2(5)-8+6(1)=8$, $\mathbf{v}$ satisfies both of the equations in the given system. Hence $\mathbf{v}$ is a solution of the system. Alternatively, using the matrix equation $A\mathbf{x}=\mathbf{b}$, we see that $\mathbf{v}$ is a solution because

$$A\mathbf{v} = \begin{bmatrix} 1 & 0 & 5 & -1 \\ 2 & -1 & 6 & 0 \end{bmatrix} \begin{bmatrix} 5 \\ 8 \\ 1 \\ 3 \end{bmatrix} = \begin{bmatrix} 7 \\ 8 \end{bmatrix} = \mathbf{b}.$$

2. In the given matrix, the only nonzero entry in the third row lies in the last column. Hence the system of linear equations corresponding to this matrix is not consistent.
3. The corresponding system of linear equations is

$$\begin{aligned} x_2 - 3x_3 \quad + 2x_5 &= 4 \\ x_4 - \quad x_5 &= 5. \end{aligned}$$

Since the given matrix contains no row whose only nonzero entry lies in the last column, this system is consistent. The general solution of this system is

$$\begin{aligned} &x_1 \quad \text{free} \\ &x_2 = 4 + 3x_3 - 2x_5 \\ &x_3 \quad \text{free} \\ &x_4 = 5 + x_5 \\ &x_5 \quad \text{free.} \end{aligned}$$

Note that x_1, which is not a basic variable, is therefore a free variable.

The general solution in vector form is

$$\begin{bmatrix} x_1 \\ x_2 \\ x_3 \\ x_4 \\ x_5 \end{bmatrix} = \begin{bmatrix} 0 \\ 4 \\ 0 \\ 5 \\ 0 \end{bmatrix} + x_1 \begin{bmatrix} 1 \\ 0 \\ 0 \\ 0 \\ 0 \end{bmatrix} + x_3 \begin{bmatrix} 0 \\ 3 \\ 1 \\ 0 \\ 0 \end{bmatrix} + x_5 \begin{bmatrix} 0 \\ -2 \\ 0 \\ 1 \\ 1 \end{bmatrix}.$$

1.4 GAUSSIAN ELIMINATION

In Section 1.3, we learned how to solve a system of linear equations for which the augmented matrix is in reduced row echelon form. In this section, we describe a procedure that can be used to transform any matrix into this form.

Suppose that R is the reduced row echelon form of a matrix A. Recall that the first nonzero entry in a nonzero row of R is called the *leading entry* of that row. The positions that contain the leading entries of the nonzero rows of R are called the **pivot positions** of A, and a column of A that contains some pivot position of A is called a **pivot column** of A. For example, later in this section we show that the reduced row echelon form of

$$A = \begin{bmatrix} 1 & 2 & -1 & 2 & 1 & 2 \\ -1 & -2 & 1 & 2 & 3 & 6 \\ 2 & 4 & -3 & 2 & 0 & 3 \\ -3 & -6 & 2 & 0 & 3 & 9 \end{bmatrix}$$

is

$$R = \begin{bmatrix} 1 & 2 & 0 & 0 & -1 & -5 \\ 0 & 0 & 1 & 0 & 0 & -3 \\ 0 & 0 & 0 & 1 & 1 & 2 \\ 0 & 0 & 0 & 0 & 0 & 0 \end{bmatrix}.$$

Here the first three rows of R are its nonzero rows, and so A has three pivot positions. The first pivot position is row 1, column 1 because the leading entry in the first row of R lies in column 1. The second pivot position is row 2, column 3 because the leading entry in the second row of R lies in column 3. Finally, the third pivot position is row 3, column 4 because the leading entry in the third row of R lies in column 4. Hence the pivot columns of A are columns 1, 3, and 4. (See Figure 1.19.)

The pivot positions and pivot columns are easily determined from the reduced row echelon form of a matrix. However, we need a method to locate the pivot positions so we can compute the reduced row echelon form. The algorithm that we use to obtain the reduced row echelon form of a matrix is called **Gaussian elimination**.[8] This

[8] This method is named after Carl Friedrich Gauss (1777–1855), whom many consider to be the greatest mathematician of all time. Gauss described this procedure in a paper that presented his calculations to determine the orbit of the asteroid Pallas. However, a similar method for solving systems of linear equations was known to the Chinese around 250 B.C.

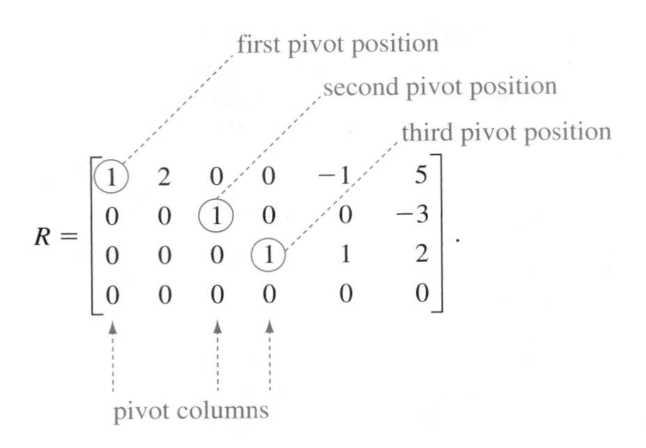

Figure 1.19 The pivot positions of the matrix R

algorithm locates the pivot positions and then makes certain entries of the matrix 0 by means of elementary row operations. We assume that the matrix is nonzero because the reduced row echelon form of a zero matrix is the same zero matrix. Our procedure can be used to find the reduced row echelon form of any nonzero matrix. To illustrate the algorithm, we find the reduced row echelon form of the matrix

$$A = \begin{bmatrix} 0 & 0 & 2 & -4 & -5 & 2 & 5 \\ 0 & 1 & -1 & 1 & 3 & 1 & -1 \\ 0 & 6 & 0 & -6 & 5 & 16 & 7 \end{bmatrix}.$$

Step 1. Determine the leftmost nonzero column. This is a pivot column, and the topmost position in this column is a pivot position.

Since the second column of A is the leftmost nonzero column, it is the first pivot column. The topmost position in this column lies in row 1, and so the first pivot position is the row 1, column 2 position.

pivot position

$$A = \begin{bmatrix} 0 & \textcircled{0} & 2 & -4 & -5 & 2 & 5 \\ 0 & 1 & -1 & 1 & 3 & 1 & -1 \\ 0 & 6 & 0 & -6 & 5 & 16 & 7 \end{bmatrix}$$

pivot columns

Step 2. In the pivot column, choose any nonzero[9] entry in a row that is not above the pivot row, and perform the appropriate row interchange to bring this entry into the pivot position.

Because the entry in the pivot position is 0, we must perform a row interchange. We must select a nonzero entry in the pivot column. Suppose that we select the entry 1. By interchanging rows 1 and 2, we bring this entry into the pivot position.

$$\xrightarrow{\mathbf{r}_1 \leftrightarrow \mathbf{r}_2} \begin{bmatrix} 0 & 1 & -1 & 1 & 3 & 1 & -1 \\ 0 & 0 & 2 & -4 & -5 & 2 & 5 \\ 0 & 6 & 0 & -6 & 5 & 16 & 7 \end{bmatrix}$$

[9] When performing calculations by hand, it may be advantageous to choose an entry of the pivot column that is ± 1, if possible, in order to simplify subsequent calculations.

Step 3. Add an appropriate multiple of the row containing the pivot position to each lower row in order to change each entry below the pivot position into 0.

In step 3, we must add multiples of row 1 of the matrix produced in step 2 to rows 2 and 3 so that the pivot column entries in rows 2 and 3 are changed to 0. In this case, the entry in row 2 is already 0, so we need only change the row 3 entry. Thus we add -6 times row 1 to row 3. This calculation is usually done mentally, but we show it here for the sake of completeness.

$$\begin{array}{rrrrrrrl} 0 & -6 & 6 & -6 & -18 & -6 & 6 & (-6 \text{ times row } 1) \\ 0 & 6 & 0 & -6 & 5 & 16 & 7 & (\text{row } 3) \\ \hline 0 & 0 & 6 & -12 & -13 & 10 & 13 & \end{array}$$

The effect of this row operation is to transform the previous matrix into the one shown at the right.

$$\xrightarrow{-6\mathbf{r}_1+\mathbf{r}_3\to\mathbf{r}_3} \begin{bmatrix} 0 & 1 & -1 & 1 & 3 & 1 & -1 \\ 0 & 0 & 2 & -4 & -5 & 2 & 5 \\ 0 & 0 & 6 & -12 & -13 & 10 & 13 \end{bmatrix}$$

During steps 1–4 of the algorithm, we can ignore certain rows of the matrix. We depict such rows by shading them. At the beginning of the algorithm, no rows are ignored.

Step 4. Ignore the row containing the pivot position and all rows above it. If there is a nonzero row that is not ignored, repeat steps 1–4 on the submatrix that remains.

We are now finished with row 1, so we repeat steps 1–4 on the submatrix below row 1.

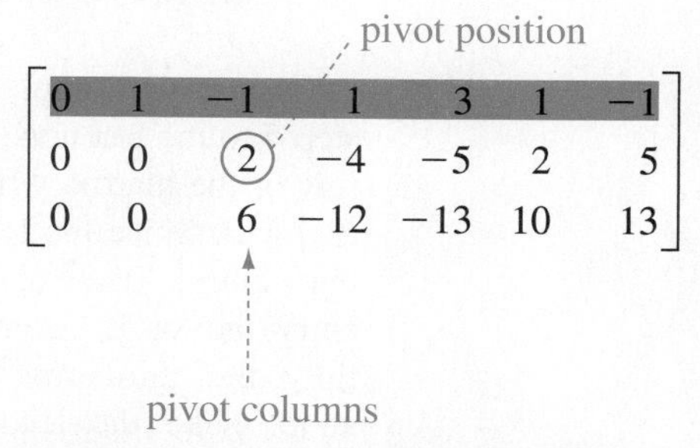

The leftmost nonzero column of this submatrix is column 3, so column 3 becomes the second pivot column. Since the topmost position in the second column of the submatrix lies in row 2 of the entire matrix, the second pivot position is the row 2, column 3 position.

Because the entry in the current pivot position is nonzero, no row interchange is required in step 2. So we continue to step 3, where we must add an appropriate multiple of row 2 to row 3 in order to create a 0 in row 3, column 3. The addition of -3 times row 2 to row 3 is shown as follows:

$$\begin{array}{rrrrrrrl} 0 & 0 & -6 & 12 & 15 & -6 & -15 & (-3 \text{ times row } 2) \\ 0 & 0 & 6 & -12 & -13 & 10 & 13 & (\text{row } 3) \\ \hline 0 & 0 & 0 & 0 & 2 & 4 & -2 & \end{array}$$

The new matrix is shown at the right.

$$\xrightarrow{-3\mathbf{r}_2+\mathbf{r}_3\to\mathbf{r}_3} \begin{bmatrix} 0 & 1 & -1 & 1 & 3 & 1 & -1 \\ 0 & 0 & 2 & -4 & -5 & 2 & 5 \\ 0 & 0 & 0 & 0 & 2 & 4 & -2 \end{bmatrix}$$

We are now finished with row 2, so we repeat steps 1–4 on the submatrix below row 2, which consists of a single row.

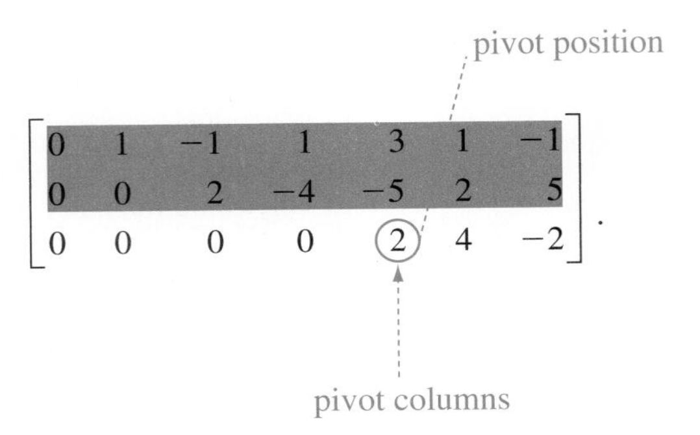

At this stage, column 5 is the leftmost nonzero column of the submatrix. So column 5 is the third pivot column, and the row 3, column 5 position is the next pivot position. Because the entry in the pivot position is nonzero, no row interchange is needed in step 2. Moreover, because there are no rows below the row containing the present pivot position, no operations are required in step 3. Since there are no nonzero rows below row 3, steps 1–4 are now complete and the matrix is in row echelon form.

The next two steps transform a matrix in row echelon form into a matrix in *reduced* row echelon form. Unlike steps 1–4, which started at the top of the matrix and worked down, steps 5 and 6 start at the last nonzero row of the matrix and work up.

Step 5. If the leading entry of the row is not 1, perform the appropriate scaling operation to make it 1. Then add an appropriate multiple of this row to every preceding row to change each entry above the pivot position into 0.

We start by applying step 5 to the last nonzero row of the matrix, which is row 3. Since the leading entry (the $(3, 5)$-entry) is not 1, we multiply the third row by $\frac{1}{2}$ to make the leading entry 1. This produces the matrix at the right.

$$\xrightarrow{\frac{1}{2}\mathbf{r}_3 \to \mathbf{r}_3} \begin{bmatrix} 0 & 1 & -1 & 1 & 3 & 1 & -1 \\ 0 & 0 & 2 & -4 & -5 & 2 & 5 \\ 0 & 0 & 0 & 0 & 1 & 2 & -1 \end{bmatrix}$$

Now we add appropriate multiples of the third row to every preceding row to change each entry above the leading entry into 0. The resulting matrix is shown at the right.

$$\xrightarrow[-3\mathbf{r}_3 + \mathbf{r}_1 \to \mathbf{r}_1]{5\mathbf{r}_3 + \mathbf{r}_2 \to \mathbf{r}_2} \begin{bmatrix} 0 & 1 & -1 & 1 & 0 & -5 & 2 \\ 0 & 0 & 2 & -4 & 0 & 12 & 0 \\ 0 & 0 & 0 & 0 & 1 & 2 & -1 \end{bmatrix}$$

Step 6. If step 5 was performed on the first row, stop. Otherwise, repeat step 5 on the preceding row.

Since we just performed step 5 using the third row, we now repeat step 5 using the second row. To make the leading entry in this row a 1, we must multiply row 2 by $\frac{1}{2}$.

$$\xrightarrow{\frac{1}{2}\mathbf{r}_2\to\mathbf{r}_2}\begin{bmatrix} 0 & 1 & -1 & 1 & 0 & -5 & 2 \\ 0 & 0 & 1 & -2 & 0 & 6 & 0 \\ 0 & 0 & 0 & 0 & 1 & 2 & -1 \end{bmatrix}$$

Now we must change the entry above the leading entry in row 2 to 0. The resulting matrix is shown at the right.

$$\xrightarrow{\mathbf{r}_2+\mathbf{r}_1\to\mathbf{r}_1}\begin{bmatrix} 0 & 1 & 0 & -1 & 0 & 1 & 2 \\ 0 & 0 & 1 & -2 & 0 & 6 & 0 \\ 0 & 0 & 0 & 0 & 1 & 2 & -1 \end{bmatrix}$$

We just performed step 5 on the second row, so we repeat it on the first row. This time the leading entry in the row is already 1, so no scaling operation is needed in step 5. Moreover, because there are no rows above this row, no other operations are needed. We see that the preceding matrix is in reduced row echelon form.

Steps 1–4 of the preceding algorithm are called the *forward pass*. The forward pass transforms the original matrix into a matrix in row echelon form. Steps 5 and 6 of the algorithm are called the *backward pass*. The backward pass further transforms the matrix into reduced row echelon form.

Example 1 Solve the following system of linear equations:

$$\begin{aligned} x_1 + 2x_2 - x_3 + 2x_4 + x_5 &= 2 \\ -x_1 - 2x_2 + x_3 + 2x_4 + 3x_5 &= 6 \\ 2x_1 + 4x_2 - 3x_3 + 2x_4 &= 3 \\ -3x_1 - 6x_2 + 2x_3 + 3x_5 &= 9 \end{aligned}$$

Solution The augmented matrix of this system is

$$\begin{bmatrix} 1 & 2 & -1 & 2 & 1 & 2 \\ -1 & -2 & 1 & 2 & 3 & 6 \\ 2 & 4 & -3 & 2 & 0 & 3 \\ -3 & -6 & 2 & 0 & 3 & 9 \end{bmatrix}.$$

We apply the Gaussian elimination algorithm to transform this matrix into one in reduced row echelon form.

Operations / **Resulting Matrix**

The first pivot position is in row 1, column 1. Since this entry is nonzero, we add appropriate multiples of row 1 to the other rows to change the entries below the pivot position to 0.

$$\xrightarrow[\substack{-2\mathbf{r}_1+\mathbf{r}_3\to\mathbf{r}_3 \\ 3\mathbf{r}_1+\mathbf{r}_4\to\mathbf{r}_4}]{\mathbf{r}_1+\mathbf{r}_2\to\mathbf{r}_2}\begin{bmatrix} 1 & 2 & -1 & 2 & 1 & 2 \\ 0 & 0 & 0 & 4 & 4 & 8 \\ 0 & 0 & -1 & -2 & -2 & -1 \\ 0 & 0 & -1 & 6 & 6 & 15 \end{bmatrix}$$

The second pivot position is row 2, column 3. Since the entry in this position is presently 0, we interchange rows 2 and 3.

$$\xrightarrow{\mathbf{r}_2\leftrightarrow\mathbf{r}_3}\begin{bmatrix} 1 & 2 & -1 & 2 & 1 & 2 \\ 0 & 0 & -1 & -2 & -2 & -1 \\ 0 & 0 & 0 & 4 & 4 & 8 \\ 0 & 0 & -1 & 6 & 6 & 15 \end{bmatrix}$$

Next we add -1 times row 2 to row 4.

$$\xrightarrow{-\mathbf{r}_2+\mathbf{r}_4\to\mathbf{r}_4}\begin{bmatrix}1&2&-1&2&1&2\\0&0&-1&-2&-2&-1\\0&0&0&4&4&8\\0&0&0&8&8&16\end{bmatrix}$$

The third pivot position is in row 3, column 4. Since this entry is nonzero, we add -2 times row 3 to row 4.

$$\xrightarrow{-2\mathbf{r}_3+\mathbf{r}_4\to\mathbf{r}_4}\begin{bmatrix}1&2&-1&2&1&2\\0&0&-1&-2&-2&-1\\0&0&0&4&4&8\\0&0&0&0&0&0\end{bmatrix}$$

At this stage, step 4 is complete, so we continue by performing step 5 on the third row. First we multiply row 3 by $\frac{1}{4}$.

$$\xrightarrow{\frac{1}{4}\mathbf{r}_3\to\mathbf{r}_3}\begin{bmatrix}1&2&-1&2&1&2\\0&0&-1&-2&-2&-1\\0&0&0&1&1&2\\0&0&0&0&0&0\end{bmatrix}$$

Then we add 2 times row 3 to row 2 and -2 times row 3 to row 1.

$$\xrightarrow[-2\mathbf{r}_3+\mathbf{r}_1\to\mathbf{r}_1]{2\mathbf{r}_3+\mathbf{r}_2\to\mathbf{r}_2}\begin{bmatrix}1&2&-1&0&-1&-2\\0&0&-1&0&0&3\\0&0&0&1&1&2\\0&0&0&0&0&0\end{bmatrix}$$

Now we must perform step 5 using row 2. This requires that we multiply row 2 by -1.

$$\xrightarrow{-\mathbf{r}_2\to\mathbf{r}_2}\begin{bmatrix}1&2&-1&0&-1&-2\\0&0&1&0&0&-3\\0&0&0&1&1&2\\0&0&0&0&0&0\end{bmatrix}$$

Then we add row 2 to row 1.

$$\xrightarrow{\mathbf{r}_2+\mathbf{r}_1\to\mathbf{r}_1}\begin{bmatrix}1&2&0&0&-1&-5\\0&0&1&0&0&-3\\0&0&0&1&1&2\\0&0&0&0&0&0\end{bmatrix}$$

Performing step 5 with row 1 produces no changes, so this matrix is the reduced row echelon form of the augmented matrix of the given system. This matrix corresponds to the system of linear equations

$$\begin{aligned}x_1 + 2x_2 \qquad\qquad - x_5 &= -5\\ x_3 \qquad\qquad &= -3\\ x_4 + x_5 &= 2.\end{aligned} \tag{6}$$

As we saw in Example 5 of Section 1.3, the general solution of this system is

$$\begin{aligned}x_1 &= -5 - 2x_2 + x_5\\ x_2 &\quad \text{free}\\ x_3 &= -3\\ x_4 &= 2 \qquad\quad - x_5\\ x_5 &\quad \text{free.}\end{aligned}$$

Practice Problem 1 ▶ Find the general solution of

$$\begin{aligned}x_1 - x_2 - 3x_3 + x_4 - x_5 &= -2\\ -2x_1 + 2x_2 + 6x_3 \qquad - 6x_5 &= -6\\ 3x_1 - 2x_2 - 8x_3 + 3x_4 - 5x_5 &= -7.\end{aligned}$$

◀

THE RANK AND NULLITY OF A MATRIX

We now associate with a matrix two important numbers that are easy to determine from its reduced row echelon form. If the matrix is the augmented matrix of a system of linear equations, these numbers provide significant information about the solutions of the corresponding system of linear equations.

Definitions The **rank** of an $m \times n$ matrix A, denoted by $\operatorname{rank} A$, is defined to be the number of nonzero rows in the reduced row echelon form of A. The **nullity** of A, denoted by $\operatorname{nullity} A$, is defined to be $n - \operatorname{rank} A$.

Example 2 For the matrix

$$\begin{bmatrix} 1 & 2 & -1 & 2 & 1 & 2 \\ -1 & -2 & 1 & 2 & 3 & 6 \\ 2 & 4 & -3 & 2 & 0 & 3 \\ -3 & -6 & 2 & 0 & 3 & 9 \end{bmatrix}$$

in Example 1, the reduced row echelon form is

$$\begin{bmatrix} 1 & 2 & 0 & 0 & -1 & -5 \\ 0 & 0 & 1 & 0 & 0 & -3 \\ 0 & 0 & 0 & 1 & 1 & 2 \\ 0 & 0 & 0 & 0 & 0 & 0 \end{bmatrix}.$$

Since the reduced row echelon form has three nonzero rows, the rank of the matrix is 3. The nullity of the matrix, found by subtracting its rank from the number of columns, is $6 - 3 = 3$.

Example 3 The reduced row echelon form of the matrix

$$B = \begin{bmatrix} 2 & 3 & 1 & 5 & 2 \\ 0 & 1 & 1 & 3 & 2 \\ 4 & 5 & 1 & 7 & 2 \\ 2 & 1 & -1 & -1 & -2 \end{bmatrix}$$

is

$$\begin{bmatrix} 1 & 0 & -1 & -2 & -2 \\ 0 & 1 & 1 & 3 & 2 \\ 0 & 0 & 0 & 0 & 0 \\ 0 & 0 & 0 & 0 & 0 \end{bmatrix}.$$

Since the latter matrix has two nonzero rows, the rank of B is 2. The nullity of B is $5 - 2 = 3$.

Practice Problem 2 ▶ Find the rank and nullity of the matrix

$$A = \begin{bmatrix} 0 & 1 & 0 & -1 & 0 & 1 & 2 \\ 0 & 0 & 1 & -2 & 0 & 6 & 0 \\ 0 & 0 & 0 & 0 & 1 & 2 & -1 \\ 0 & 0 & 0 & 0 & 0 & 0 & 0 \\ 0 & 0 & 0 & 0 & 0 & 0 & 0 \end{bmatrix}.$$

◀

In Examples 2 and 3, note that each nonzero row in the reduced row echelon form of the given matrix contains exactly one pivot position, so the number of nonzero rows equals the number of pivot positions. Consequently, we can restate the definition of rank as follows:

> The rank of a matrix equals the number of pivot columns in the matrix.
> The nullity of a matrix equals the number of nonpivot columns in the matrix.

It follows that in a matrix in reduced row echelon form with rank k, the standard vectors $\mathbf{e}_1, \mathbf{e}_2, \ldots, \mathbf{e}_k$ must appear, in order, among the columns of the matrix. In Example 2, for instance, the rank of the matrix is 3, and columns 1, 3, and 4 of the reduced row echelon form are

$$\mathbf{e}_1 = \begin{bmatrix} 1 \\ 0 \\ 0 \\ 0 \end{bmatrix}, \quad \mathbf{e}_2 = \begin{bmatrix} 0 \\ 1 \\ 0 \\ 0 \end{bmatrix}, \quad \text{and} \quad \mathbf{e}_3 = \begin{bmatrix} 0 \\ 0 \\ 1 \\ 0 \end{bmatrix},$$

respectively. Thus, if an $m \times n$ matrix has rank n, then its reduced row echelon form must be precisely $\begin{bmatrix} \mathbf{e}_1 & \mathbf{e}_2 & \ldots & \mathbf{e}_n \end{bmatrix}$. In the special case of an $n \times n$ (square) matrix,

$$\begin{bmatrix} \mathbf{e}_1 & \mathbf{e}_2 & \ldots & \mathbf{e}_n \end{bmatrix} = \begin{bmatrix} 1 & 0 & \cdots & 0 \\ 0 & 1 & \cdots & 0 \\ \vdots & \vdots & \ddots & \vdots \\ 0 & 0 & \cdots & 1 \end{bmatrix} = I_n,$$

the $n \times n$ identity matrix; therefore we have the following useful result:

> If an $n \times n$ matrix has rank n, then its reduced row echelon form is I_n.

Consider the system of linear equations in Example 1 as a matrix equation $A\mathbf{x} = \mathbf{b}$. Our method for solving the system is to find the reduced row echelon form $[R \;\; \mathbf{c}]$ of the augmented matrix $[A \;\; \mathbf{b}]$. The system $R\mathbf{x} = \mathbf{c}$ is then equivalent to $A\mathbf{x} = \mathbf{b}$, and we can easily solve $R\mathbf{x} = \mathbf{c}$ because $[R \;\; \mathbf{c}]$ is in reduced row echelon form. Note that each basic variable of the system $A\mathbf{x} = \mathbf{b}$ corresponds to the leading entry of exactly one nonzero row of $[R \;\; \mathbf{c}]$, so the number of basic variables equals the number of nonzero rows, which is the rank of A. Also, if n is the number of columns of A, then the number of free variables of $A\mathbf{x} = \mathbf{b}$ equals n minus the number of basic variables. By the previous remark, the number of free variables equals $n - \operatorname{rank} A$, which is the nullity of A. In general,

> If $A\mathbf{x} = \mathbf{b}$ is the matrix form of a consistent system of linear equations, then
>
> (a) the number of basic variables in a general solution of the system equals the rank of A;
>
> (b) the number of free variables in a general solution of the system equals the nullity of A.
>
> Thus a consistent system of linear equations has a unique solution if and only if the nullity of its coefficient matrix equals 0. Equivalently, a consistent system of linear equations has infinitely many solutions if and only if the nullity of its coefficient matrix is positive.

The original system of linear equations in Example 1 is a system of 4 equations in 5 variables. However, it is equivalent to system (6), the system of 3 equations in 5 variables corresponding to the reduced row echelon form of its augmented matrix. In Example 1, the fourth equation in the original system is *redundant* because it is a linear combination of the first three equations. Specifically, it is the sum of -3 times the first equation, 2 times the second equation, and the third equation. The other three equations are nonredundant. In general, the rank of the augmented matrix $[A\ \mathbf{b}]$ tells us the number of nonredundant equations in the system $A\mathbf{x} = \mathbf{b}$.

Example 4 Consider the following system of linear equations:

$$\begin{aligned} x_1 + x_2 + x_3 &= 1 \\ x_1 \phantom{{}+x_2} + 3x_3 &= -2 + s \\ x_1 - x_2 + rx_3 &= 3 \end{aligned}$$

(a) For what values of r and s is this system of linear equations inconsistent?

(b) For what values of r and s does this system of linear equations have infinitely many solutions?

(c) For what values of r and s does this system of linear equations have a unique solution?

Solution Apply the Gaussian elimination algorithm to the augmented matrix of the given system to transform the matrix into one in row echelon form:

$$\begin{bmatrix} 1 & 1 & 1 & 1 \\ 1 & 0 & 3 & -2+s \\ 1 & -1 & r & 3 \end{bmatrix} \xrightarrow[-\mathbf{r}_1+\mathbf{r}_3 \to \mathbf{r}_3]{-\mathbf{r}_1+\mathbf{r}_2 \to \mathbf{r}_2} \begin{bmatrix} 1 & 1 & 1 & 1 \\ 0 & -1 & 2 & -3+s \\ 0 & -2 & r-1 & 2 \end{bmatrix}$$

$$\xrightarrow{-2\mathbf{r}_2+\mathbf{r}_3 \to \mathbf{r}_3} \begin{bmatrix} 1 & 1 & 1 & 1 \\ 0 & -1 & 2 & -3+s \\ 0 & 0 & r-5 & 8-2s \end{bmatrix}$$

(a) The original system is inconsistent whenever there is a row whose only nonzero entry lies in the last column. Only the third row could have this form. Thus the original system is inconsistent whenever $r - 5 = 0$ and $8 - 2s \neq 0$; that is, when $r = 5$ and $s \neq 4$.

(b) The original system has infinitely many solutions whenever the system is consistent and there is a free variable in the general solution. In order to have a free variable, we must have $r - 5 = 0$, and in order for the system also to be consistent, we must have $8 - 2s = 0$. Thus the original system has infinitely many solutions if $r = 5$ and $s = 4$.

(c) Let A denote the 3×3 coefficient matrix of the system. For the system to have a unique solution, there must be three basic variables, and so the rank of A must be 3. Since deleting the last column of the immediately preceding matrix gives a row echelon form of A, the rank of A is 3 precisely when $r - 5 \neq 0$; that is, $r \neq 5$.

Practice Problem 3 ▶ Consider the following system of linear equations:

$$\begin{aligned} x_1 + 3x_2 &= 1 + s \\ x_1 + rx_2 &= 5 \end{aligned}$$

(a) For what values of r and s is this system of linear equations inconsistent?

(b) For what values of r and s does this system of linear equations have infinitely many solutions?

(c) For what values of r and s does this system of linear equations have a unique solution? ◀

The following theorem provides several conditions that are equivalent[10] to the existence of solutions for a system of linear equations.

THEOREM 1.5

(Test for Consistency) The following conditions are equivalent:

(a) The matrix equation $A\mathbf{x} = \mathbf{b}$ is consistent.

(b) The vector $\mathbf{b}$ is a linear combination of the columns of A.

(c) The reduced row echelon form[11] of the augmented matrix $[A\ \ \mathbf{b}]$ has no row of the form $[0\ \ 0\ \ \cdots 0\ \ d]$, where $d \neq 0$.

PROOF Let A be an $m \times n$ matrix, and let $\mathbf{b}$ be in $\mathcal{R}^m$. By the definition of a matrix–vector product, there exists a vector

$$\mathbf{v} = \begin{bmatrix} v_1 \\ v_2 \\ \vdots \\ v_n \end{bmatrix}$$

in $\mathcal{R}^n$ such that $A\mathbf{v} = \mathbf{b}$ if and only if

$$v_1\mathbf{a}_1 + v_2\mathbf{a}_2 + \cdots + v_n\mathbf{a}_n = \mathbf{b}.$$

Thus $A\mathbf{x} = \mathbf{b}$ is consistent if and only if $\mathbf{b}$ is a linear combination of the columns of A. So (a) is equivalent to (b).

Finally, we prove that (a) is equivalent to (c). Let $[R\ \ \mathbf{c}]$ be the reduced row echelon form of the augmented matrix $[A\ \ \mathbf{b}]$. If statement (c) is false, then the system of linear equations corresponding to $R\mathbf{x} = \mathbf{c}$ contains the equation

$$0x_1 + 0x_2 + \cdots + 0x_n = d,$$

where $d \neq 0$. Since this equation has no solutions, $R\mathbf{x} = \mathbf{c}$ is inconsistent. On the other hand, if statement (c) is true, then we can solve every equation in the system of linear equations corresponding to $R\mathbf{x} = \mathbf{c}$ for some basic variable. This gives a solution of $R\mathbf{x} = \mathbf{c}$, which is also a solution of $A\mathbf{x} = \mathbf{b}$. ■

TECHNOLOGICAL CONSIDERATIONS*

Gaussian elimination is the most efficient procedure for reducing a matrix to its reduced row echelon form. Nevertheless, it requires many tedious computations. In fact, the number of arithmetic operations required to obtain the reduced row echelon form of an $n \times (n+1)$ matrix is typically on the order of $\frac{2}{3}n^3 + \frac{1}{2}n^2 - \frac{7}{6}n$. We can

[10] Statements are called *equivalent* (or *logically equivalent*) if, under every circumstance, they are all true or they are all false. Whether any one of the statements in Theorem 1.5 is true or false depends on the particular matrix A and vector $\mathbf{b}$ being considered.

[11] Theorem 1.5 remains true if (c) is changed as follows: Every row echelon form of $[A\ \ \mathbf{b}]$ has no row in which the only nonzero entry lies in the last column.

* The remainder of this section may be omitted without loss of continuity.

easily program this algorithm on a computer or programmable calculator and thus obtain the reduced row echelon form of a matrix. However, computers or calculators can store only a finite number of decimal places, so they can introduce small errors called *roundoff errors* in their calculations. Usually, these errors are insignificant. But when we use an algorithm with many steps, such as Gaussian elimination, the errors can accumulate and significantly affect the result. The following example illustrates the potential pitfalls of roundoff error. Although the calculations in the example are performed on the TI-85 calculator, the same types of issues arise when any computer or calculator is used to solve a system of linear equations.

For the matrix

$$\begin{bmatrix} 1 & -1 & 2 & 3 & 1 & -1 \\ 3 & -1 & 2 & 4 & 1 & 2 \\ 7 & -2 & 4 & 8 & 1 & 6 \end{bmatrix},$$

the TI-85 calculator gives the reduced row echelon form as

$$\begin{bmatrix} 1 & 0 & -1\text{E}{-14} & 0 & -.999999999999 & 2 \\ 0 & 1 & -2 & 0 & 4 & 1.3\text{E}{-12} \\ 0 & 0 & 0 & 1 & 2 & -1 \end{bmatrix}.$$

(The notation $a\text{E}b$ represents $a \times 10^b$.) However, by exact hand calculations, we find that the third, fifth, and sixth columns should be

$$\begin{bmatrix} 0 \\ -2 \\ 0 \end{bmatrix}, \quad \begin{bmatrix} -1 \\ 4 \\ 2 \end{bmatrix}, \quad \text{and} \quad \begin{bmatrix} 2 \\ 0 \\ -1 \end{bmatrix},$$

respectively. On the TI-85 calculator, numbers are stored with 14 digits. Thus a number containing more than 14 significant digits is not stored exactly in the calculator. In subsequent calculations with that number, roundoff errors can accumulate to such a degree that the final result of the calculation is highly inaccurate. In our calculation of the reduced row echelon form of the matrix A, roundoff errors have affected the $(1, 3)$-entry, the $(1, 5)$-entry, and the $(2, 6)$-entry. In this instance, none of the affected entries is greatly changed, and it is reasonable to expect that the true entries in these positions should be 0, -1, and 0, respectively. But can we be absolutely sure? (We will learn a way of checking if these entries are 0, -1, and 0 in Section 2.3.)

It is not always so obvious that roundoff errors have occurred. Consider the system of linear equations

$$\begin{aligned} kx_1 + (k-1)x_2 &= 1 \\ (k+1)x_1 + \qquad kx_2 &= 2. \end{aligned}$$

By subtracting the first equation from the second, this system is easily solved, and the solution is $x_1 = 2 - k$ and $x_2 = k - 1$. But for sufficiently large values of k, roundoff errors can cause problems. For example, with $k = 4{,}935{,}937$, the TI-85 calculator gives the reduced row echelon form of the augmented matrix

$$\begin{bmatrix} 4935937 & 4935936 & 1 \\ 4935938 & 4935937 & 2 \end{bmatrix}$$

as

$$\begin{bmatrix} 1 & .999999797404 & 0 \\ 0 & 0 & 1 \end{bmatrix}.$$

Since the last row of the reduced row echelon form has its only nonzero entry in the last column, we would incorrectly deduce from this that the original system is inconsistent!

The analysis of roundoff errors and related matters is a serious mathematical subject that is inappropriate for this book. (It is studied in the branch of mathematics called *numerical analysis*.) We encourage the use of technology whenever possible to perform the tedious calculations associated with matrices (such as those required to obtain the reduced row echelon form of a matrix). Nevertheless, a certain amount of skepticism is healthy when technology is used. Just because the calculations are performed with a calculator or computer is no guarantee that the result is correct. In this book, however, the examples and exercises usually involve simple numbers (often one- or two-digit integers) and small matrices; so there is little chance of serious errors resulting from the use of technology.

EXERCISES

In Exercises 1–16, determine whether the given system is consistent, and if so, find its general solution.

1. $2x_1 + 6x_2 = -4$

2. $\begin{aligned} x_1 - x_2 &= 3 \\ -2x_1 + 2x_2 &= -6 \end{aligned}$

3. $\begin{aligned} x_1 - 2x_2 &= -6 \\ -2x_1 + 3x_2 &= 7 \end{aligned}$

4. $\begin{aligned} x_1 - x_2 - 3x_3 &= 3 \\ 2x_1 + x_2 - 3x_3 &= 0 \end{aligned}$

5. $\begin{aligned} 2x_1 - 2x_2 + 4x_3 &= 1 \\ -4x_1 + 4x_2 - 8x_3 &= -3 \end{aligned}$

6. $\begin{aligned} x_1 - 2x_2 - x_3 &= 3 \\ -2x_1 + 4x_2 + 2x_3 &= -6 \\ 3x_1 - 6x_2 - 3x_3 &= 9 \end{aligned}$

7. $\begin{aligned} x_1 - 2x_2 - x_3 &= -3 \\ 2x_1 - 4x_2 + 2x_3 &= 2 \end{aligned}$

8. $\begin{aligned} x_1 + x_2 - x_3 - x_4 &= -2 \\ 2x_2 - 3x_3 - 12x_4 &= -3 \\ x_1 + x_3 + 6x_4 &= 0 \end{aligned}$

9. $\begin{aligned} x_1 - x_2 - 3x_3 + x_4 &= 0 \\ -2x_1 + x_2 + 5x_3 &= -4 \\ 4x_1 - 2x_2 - 10x_3 + x_4 &= 5 \end{aligned}$

10. $\begin{aligned} x_1 - 3x_2 + x_3 + x_4 &= 0 \\ -3x_1 + 9x_2 - 2x_3 - 5x_4 &= 1 \\ 2x_1 - 6x_2 - x_3 + 8x_4 &= -2 \end{aligned}$

11. $\begin{aligned} x_1 + 3x_2 + x_3 + x_4 &= -1 \\ -2x_1 - 6x_2 - x_3 &= 5 \\ x_1 + 3x_2 + 2x_3 + 3x_4 &= 2 \end{aligned}$

12. $\begin{aligned} x_1 + x_2 + x_3 &= -1 \\ 2x_1 + x_2 - x_3 &= 2 \\ x_1 - 2x_3 &= 3 \\ -3x_1 - 2x_2 &= -1 \end{aligned}$

13. $\begin{aligned} x_1 + 2x_2 + x_3 &= 1 \\ -2x_1 - 4x_2 - x_3 &= 0 \\ 5x_1 + 10x_2 + 3x_3 &= 2 \\ 3x_1 + 6x_2 + 3x_3 &= 4 \end{aligned}$

14. $\begin{aligned} x_1 - x_2 + x_3 &= 7 \\ x_1 - 2x_2 - x_3 &= 8 \\ 2x_1 - x_3 &= 10 \\ -x_1 - 4x_2 - x_3 &= 2 \end{aligned}$

15. $\begin{aligned} x_1 - x_3 - 2x_4 - 8x_5 &= -3 \\ -2x_1 + x_3 + 2x_4 + 9x_5 &= 5 \\ 3x_1 - 2x_3 - 3x_4 - 15x_5 &= -9 \end{aligned}$

16. $\begin{aligned} x_1 - x_2 + x_4 &= -4 \\ x_1 - x_2 + 2x_4 + 2x_5 &= -5 \\ 3x_1 - 3x_2 + 2x_4 - 2x_5 &= -11 \end{aligned}$

In Exercises 17–26, determine the values of r, if any, for which the given system of linear equations is inconsistent.

17. $\begin{aligned} -x_1 + 4x_2 &= 3 \\ 3x_1 + rx_2 &= 2 \end{aligned}$

18. $\begin{aligned} 3x_1 + rx_2 &= -2 \\ -x_1 + 4x_2 &= 6 \end{aligned}$

19. $\begin{aligned} x_1 - 2x_2 &= 0 \\ 4x_1 - 8x_2 &= r \end{aligned}$

20. $\begin{aligned} x_1 + rx_2 &= -3 \\ 2x_1 &= -6 \end{aligned}$

21. $\begin{aligned} x_1 - 3x_2 &= -2 \\ 2x_1 + rx_2 &= -4 \end{aligned}$

22. $\begin{aligned} -2x_1 + x_2 &= 5 \\ rx_1 + 4x_2 &= 3 \end{aligned}$

23. $\begin{aligned} -x_1 + rx_2 &= 2 \\ rx_1 - 9x_2 &= 6 \end{aligned}$

24. $\begin{aligned} x_1 + rx_2 &= 2 \\ rx_1 + 16x_2 &= 8 \end{aligned}$

25. $\begin{aligned} x_1 - x_2 + 2x_3 &= 4 \\ 3x_1 + rx_2 - x_3 &= 2 \end{aligned}$

26. $\begin{aligned} x_1 + 2x_2 - 4x_3 &= 1 \\ -2x_1 - 4x_2 + rx_3 &= 3 \end{aligned}$

In Exercises 27–34, determine the values of r and s for which the given system of linear equations has (a) no solutions, (b) exactly one solution, and (c) infinitely many solutions.

27. $\begin{aligned} x_1 + rx_2 &= 5 \\ 3x_1 + 6x_2 &= s \end{aligned}$

28. $\begin{aligned} -x_1 + 4x_2 &= s \\ 2x_1 + rx_2 &= 6 \end{aligned}$

29. $\begin{aligned} x_1 + 2x_2 &= s \\ -4x_1 + rx_2 &= 8 \end{aligned}$

30. $\begin{aligned} -x_1 + 3x_2 &= s \\ 4x_1 + rx_2 &= -8 \end{aligned}$

31. $\begin{aligned} x_1 + rx_2 &= -3 \\ 2x_1 + 5x_2 &= s \end{aligned}$

32. $\begin{aligned} x_1 + rx_2 &= 5 \\ -3x_1 + 6x_2 &= s \end{aligned}$

33. $\begin{aligned} -x_1 + rx_2 &= s \\ 3x_1 - 9x_2 &= -2 \end{aligned}$

34. $\begin{aligned} 2x_1 - x_2 &= 3 \\ 4x_1 + rx_2 &= s \end{aligned}$

In Exercises 35–42, find the rank and nullity of the given matrix.

35. $\begin{bmatrix} 1 & -1 & -1 & 0 \\ 2 & -1 & -2 & 1 \\ 1 & -2 & -2 & 2 \\ -4 & 2 & 3 & 1 \\ 1 & -1 & -2 & 3 \end{bmatrix}$

36. $\begin{bmatrix} 1 & -3 & -1 & 2 \\ -2 & 6 & 2 & -4 \\ 3 & -9 & 2 & 1 \\ 1 & -3 & 4 & -3 \\ -1 & 3 & -9 & 8 \end{bmatrix}$

37. $\begin{bmatrix} -2 & 2 & 1 & 1 & -2 \\ 1 & -1 & -1 & -3 & 3 \\ -1 & 1 & -1 & -7 & 5 \end{bmatrix}$

38. $\begin{bmatrix} 1 & 0 & -2 & -1 & 0 & -1 \\ 2 & -1 & -6 & -2 & 0 & -4 \\ 0 & 1 & 2 & 1 & 1 & 1 \\ -1 & 2 & 6 & 3 & 1 & 2 \end{bmatrix}$

39. $\begin{bmatrix} 1 & 1 & 1 & 1 \\ 1 & 2 & 4 & 2 \\ 2 & 0 & -4 & 1 \end{bmatrix}$

40. $\begin{bmatrix} 1 & 0 & 1 & -1 & 6 \\ 2 & -1 & 5 & -1 & 7 \\ -1 & 1 & -4 & 1 & -3 \\ 0 & 1 & -3 & 1 & 1 \end{bmatrix}$

41. $\begin{bmatrix} 1 & -2 & 0 & -3 & 1 \\ 2 & -4 & -1 & -8 & 8 \\ -1 & 2 & 1 & 5 & -7 \\ 0 & 0 & 1 & 2 & -6 \end{bmatrix}$

42. $\begin{bmatrix} 1 & -2 & -1 & 0 & 3 & -2 \\ 2 & -4 & -2 & -1 & 5 & 9 \\ -1 & 2 & 1 & 1 & -2 & 7 \\ 0 & 0 & 0 & 1 & 1 & 5 \end{bmatrix}$

43. A mining company operates three mines that each produce three grades of ore. The daily yield of each mine is shown in the following table:

	Daily Yield		
	Mine 1	Mine 2	Mine 3
High-grade ore	1 ton	1 ton	2 tons
Medium-grade ore	1 ton	2 tons	2 tons
Low-grade ore	2 tons	1 ton	0 tons

(a) Can the company supply exactly 80 tons of high-grade, 100 tons of medium-grade, and 40 tons of low-grade ore? If so, how many days should each mine operate to fill this order?

(b) Can the company supply exactly 40 tons of high-grade, 100 tons of medium-grade, and 80 tons of low-grade ore? If so, how many days should each mine operate to fill this order?

44. A company makes three types of fertilizer. The first type contains 10% nitrogen and 3% phosphates by weight, the second contains 8% nitrogen and 6% phosphates, and the third contains 6% nitrogen and 1% phosphates.

(a) Can the company mix these three types of fertilizers to supply exactly 600 pounds of fertilizer containing 7.5% nitrogen and 5% phosphates? If so, how?

(b) Can the company mix these three types of fertilizers to supply exactly 600 pounds of fertilizer containing 9% nitrogen and 3.5% phosphates? If so, how?

45. A patient needs to consume exactly 660 mg of magnesium, 820 IU of vitamin D, and 750 mcg of folate per day. Three food supplements can be mixed to provide these nutrients. The amounts of the three nutrients provided by each of the supplements is given in the following table:

	Food Supplement		
	1	2	3
Magnesium (mg)	10	15	36
Vitamin D (IU)	10	20	44
Folate (mcg)	15	15	42

(a) What is the maximum amount of supplement 3 that can be used to provide exactly the required amounts of the three nutrients?

(b) Can the three supplements be mixed to provide exactly 720 mg of magnesium, 800 IU of vitamin D, and 750 mcg of folate? If so, how?

46. Three grades of crude oil are to be blended to obtain 100 barrels of oil costing \$35 per barrel and containing 50 gm of sulfur per barrel. The cost and sulfur content of the three grades of oil are given in the following table:

	Grade		
	A	B	C
Cost per barrel	\$40	\$32	\$24
Sulfur per barrel	30 gm	62 gm	94 gm

(a) Find the amounts of each grade to be blended that use the least oil of grade C.

(b) Find the amounts of each grade to be blended that use the most oil of grade C.

47. Find a polynomial function $f(x) = ax^2 + bx + c$ whose graph passes through the points $(-1, 14)$, $(1, 4)$, and $(3, 10)$.

48. Find a polynomial function $f(x) = ax^2 + bx + c$ whose graph passes through the points $(-2, -33)$, $(2, -1)$, and $(3, -8)$.

49. Find a polynomial function $f(x) = ax^3 + bx^2 + cx + d$ whose graph passes through the points $(-2, 32)$, $(-1, 13)$, $(2, 4)$, and $(3, 17)$.

50. Find a polynomial function $f(x) = ax^3 + bx^2 + cx + d$ whose graph passes through the points $(-2, 12)$, $(-1, -9)$, $(1, -3)$, and $(3, 27)$.
51. If the third pivot position of a matrix A is in column j, what can be said about column j of the reduced row echelon form of A? Explain your answer.
52. Suppose that the fourth pivot position of a matrix is in row i and column j. Say as much as possible about i and j. Explain your answer.

T&F *In Exercises 53–72, determine whether the statements are true or false.*

53. A column of a matrix A is a pivot column if the corresponding column in the reduced row echelon form of A contains the leading entry of some nonzero row.
54. There is a unique sequence of elementary row operations that transforms a matrix into its reduced row echelon form.
55. When the forward pass of Gaussian elimination is complete, the original matrix has been transformed into one in row echelon form.
56. No scaling operations are required in the forward pass of Gaussian elimination.
57. The rank of a matrix equals the number of pivot columns in the matrix.
58. If $A\mathbf{x} = \mathbf{b}$ is consistent, then the nullity of A equals the number of free variables in the general solution of $A\mathbf{x} = \mathbf{b}$.
59. There exists a 5×8 matrix with rank 3 and nullity 2.
60. If a system of m linear equations in n variables is equivalent to a system of p linear equations in q variables, then $m = p$.
61. If a system of m linear equations in n variables is equivalent to a system of p linear equations in q variables, then $n = q$.
62. The equation $A\mathbf{x} = \mathbf{b}$ is consistent if and only if $\mathbf{b}$ is a linear combination of the columns of A.
63. If the equation $A\mathbf{x} = \mathbf{b}$ is inconsistent, then the rank of $[A\ \mathbf{b}]$ is greater than the rank of A.
64. If the reduced row echelon form of $[A\ \mathbf{b}]$ contains a zero row, then $A\mathbf{x} = \mathbf{b}$ must have infinitely many solutions.
65. If the reduced row echelon form of $[A\ \mathbf{b}]$ contains a zero row, then $A\mathbf{x} = \mathbf{b}$ must be consistent.
66. If some column of matrix A is a pivot column, then the corresponding column in the reduced row echelon form of A is a standard vector.
67. If A is a matrix with rank k, then the vectors $\mathbf{e}_1, \mathbf{e}_2, \ldots, \mathbf{e}_k$ appear as columns of the reduced row echelon form of A.
68. The sum of the rank and nullity of a matrix equals the number of rows in the matrix.
69. Suppose that the pivot rows of a matrix A are rows $1, 2, \ldots, k$, and row $k+1$ becomes zero when applying the Gaussian elimination algorithm. Then row $k+1$ must equal some linear combination of rows $1, 2, \ldots, k$.
70. The third pivot position in a matrix lies in row 3.
71. The third pivot position in a matrix lies in column 3.
72. If R is an $n \times n$ matrix in reduced row echelon form that has rank n, then $R = I_n$.

---

73. Describe an $m \times n$ matrix with rank 0.
74. What is the smallest possible rank of a 4×7 matrix? Explain your answer.
75. What is the largest possible rank of a 4×7 matrix? Explain your answer.
76. What is the largest possible rank of a 7×4 matrix? Explain your answer.
77. What is the smallest possible nullity of a 4×7 matrix? Explain your answer.
78. What is the smallest possible nullity of a 7×4 matrix? Explain your answer.
79. What is the largest possible rank of an $m \times n$ matrix? Explain your answer.
80. What is the smallest possible nullity of an $m \times n$ matrix? Explain your answer.
81. Let A be a 4×3 matrix. Is it possible that $A\mathbf{x} = \mathbf{b}$ is consistent for every $\mathbf{b}$ in $\mathcal{R}^4$? Explain your answer.
82. Let A be an $m \times n$ matrix and $\mathbf{b}$ be a vector in $\mathcal{R}^m$. What must be true about the rank of A if $A\mathbf{x} = \mathbf{b}$ has a unique solution? Justify your answer.
83. A system of linear equations is called *underdetermined* if it has fewer equations than variables. What can be said about the number of solutions of an underdetermined system?
84. A system of linear equations is called *overdetermined* if it has more equations than variables. Give examples of overdetermined systems that have
    (a) no solutions,
    (b) exactly one solution, and
    (c) infinitely many solutions.
85. Prove that if A is an $m \times n$ matrix with rank m, then $A\mathbf{x} = \mathbf{b}$ is consistent for every $\mathbf{b}$ in $\mathcal{R}^m$.
86. Prove that a matrix equation $A\mathbf{x} = \mathbf{b}$ is consistent if and only if the ranks of A and $[A\ \mathbf{b}]$ are equal.
87. Let $\mathbf{u}$ be a solution of $A\mathbf{x} = \mathbf{0}$, where A is an $m \times n$ matrix. Must $c\mathbf{u}$ be a solution of $A\mathbf{x} = \mathbf{0}$ for every scalar c? Justify your answer.
88. Let $\mathbf{u}$ and $\mathbf{v}$ be solutions of $A\mathbf{x} = \mathbf{0}$, where A is an $m \times n$ matrix. Must $\mathbf{u} + \mathbf{v}$ be a solution of $A\mathbf{x} = \mathbf{0}$? Justify your answer.
89. Let $\mathbf{u}$ and $\mathbf{v}$ be solutions of $A\mathbf{x} = \mathbf{b}$, where A is an $m \times n$ matrix and $\mathbf{b}$ is a vector in $\mathcal{R}^m$. Prove that $\mathbf{u} - \mathbf{v}$ is a solution of $A\mathbf{x} = \mathbf{0}$.
90. Let $\mathbf{u}$ be a solution of $A\mathbf{x} = \mathbf{b}$ and $\mathbf{v}$ be a solution of $A\mathbf{x} = \mathbf{0}$, where A is an $m \times n$ matrix and $\mathbf{b}$ is a vector in $\mathcal{R}^m$. Prove that $\mathbf{u} + \mathbf{v}$ is a solution of $A\mathbf{x} = \mathbf{b}$.
91. Let A be an $m \times n$ matrix and $\mathbf{b}$ be a vector in $\mathcal{R}^m$ such that $A\mathbf{x} = \mathbf{b}$ is consistent. Prove that $A\mathbf{x} = c\mathbf{b}$ is consistent for every scalar c.

92. Let A be an $m \times n$ matrix and $\mathbf{b}_1$ and $\mathbf{b}_2$ be vectors in $\mathcal{R}^m$ such that both $A\mathbf{x} = \mathbf{b}_1$ and $A\mathbf{x} = \mathbf{b}_2$ are consistent. Prove that $A\mathbf{x} = \mathbf{b}_1 + \mathbf{b}_2$ is consistent.

93. Let $\mathbf{u}$ and $\mathbf{v}$ be solutions of $A\mathbf{x} = \mathbf{b}$, where A is an $m \times n$ matrix and $\mathbf{b}$ is a vector in $\mathcal{R}^m$. Must $\mathbf{u} + \mathbf{v}$ be a solution of $A\mathbf{x} = \mathbf{b}$? Justify your answer.

In Exercises 94–99, use either a calculator with matrix capabilities or computer software such as MATLAB to solve each problem.

In Exercises 94–96, use Gaussian elimination on the augmented matrix of the system of linear equations to test for consistency, and to find the general solution.

94. $$\begin{aligned} 1.3x_1 + 0.5x_2 - 1.1x_3 + 2.7x_4 - 2.1x_5 &= 12.9 \\ 2.2x_1 - 4.5x_2 + 3.1x_3 - 5.1x_4 + 3.2x_5 &= -29.2 \\ 1.4x_1 - 2.1x_2 + 1.5x_3 - 3.1x_4 - 2.5x_5 &= -11.9 \end{aligned}$$

95. $$\begin{aligned} x_1 - x_2 + 3x_3 - x_4 + 2x_5 &= 5 \\ 2x_1 + x_2 + 4x_3 + x_4 - x_5 &= 7 \\ 3x_1 - x_2 + 2x_3 - 2x_4 + 2x_5 &= 3 \\ 2x_1 - 4x_2 - x_3 - 4x_4 + 5x_5 &= 6 \end{aligned}$$

96. $$\begin{aligned} 4x_1 - x_2 + 5x_3 - 2x_4 + x_5 &= 0 \\ 7x_1 - 6x_2 + 3x_4 + 8x_5 &= 15 \\ 9x_1 - 5x_2 + 4x_3 - 7x_4 + x_5 &= 6 \\ 6x_1 + 9x_3 - 12x_4 - 6x_5 &= 11 \end{aligned}$$

In Exercises 97–99, find the rank and the nullity of the matrix.

97. $$\begin{bmatrix} 1.2 & 2.3 & -1.1 & 1.0 & 2.1 \\ 3.1 & 1.2 & -2.1 & 1.4 & 2.4 \\ -2.1 & 4.1 & 2.3 & -1.2 & 0.5 \\ 3.4 & 9.9 & -2.0 & 2.2 & 7.1 \end{bmatrix}$$

98. $$\begin{bmatrix} 2.7 & 1.3 & 1.6 & 1.5 & -1.0 \\ 1.7 & 2.3 & -1.2 & 2.1 & 2.2 \\ 3.1 & -1.8 & 4.2 & 3.1 & 1.4 \\ 4.1 & -1.1 & 2.1 & 1.2 & 0.0 \\ 6.2 & -1.7 & 3.4 & 1.5 & 2.0 \end{bmatrix}$$

99. $$\begin{bmatrix} 3 & -11 & 2 & 4 & -8 \\ 5 & 1 & 0 & 8 & 5 \\ 11 & 2 & -9 & 3 & -4 \\ 3 & 14 & -11 & 7 & 9 \\ 0 & 2 & 0 & 16 & 10 \end{bmatrix}$$

SOLUTIONS TO THE PRACTICE PROBLEMS

1. The augmented matrix of the given system is

$$\begin{bmatrix} 1 & -1 & -3 & 1 & -1 & -2 \\ -2 & 2 & 6 & 0 & -6 & -6 \\ 3 & -2 & -8 & 3 & -5 & -7 \end{bmatrix}.$$

Apply the Gaussian elimination algorithm to the augmented matrix of the given system to transform the matrix into one in row echelon form:

$$\begin{bmatrix} 1 & -1 & -3 & 1 & -1 & -2 \\ -2 & 2 & 6 & 0 & -6 & -6 \\ 3 & -2 & -8 & 3 & -5 & -7 \end{bmatrix}$$

$$\xrightarrow[-3\mathbf{r}_1 + \mathbf{r}_3 \to \mathbf{r}_3]{2\mathbf{r}_1 + \mathbf{r}_2 \to \mathbf{r}_2} \begin{bmatrix} 1 & -1 & -3 & 1 & -1 & -2 \\ 0 & 0 & 0 & 2 & -8 & -10 \\ 0 & 1 & 1 & 0 & -2 & -1 \end{bmatrix}$$

$$\xrightarrow{\mathbf{r}_2 \leftrightarrow \mathbf{r}_3} \begin{bmatrix} 1 & -1 & -3 & 1 & -1 & -2 \\ 0 & 1 & 1 & 0 & -2 & -1 \\ 0 & 0 & 0 & 2 & -8 & -10 \end{bmatrix}$$

$$\xrightarrow{\frac{1}{2}\mathbf{r}_3 \to \mathbf{r}_3} \begin{bmatrix} 1 & -1 & -3 & 1 & -1 & -2 \\ 0 & 1 & 1 & 0 & -2 & -1 \\ 0 & 0 & 0 & 1 & -4 & -5 \end{bmatrix}$$

$$\xrightarrow{-\mathbf{r}_3 + \mathbf{r}_1 \to \mathbf{r}_1} \begin{bmatrix} 1 & -1 & -3 & 0 & 3 & 3 \\ 0 & 1 & 1 & 0 & -2 & -1 \\ 0 & 0 & 0 & 1 & -4 & -5 \end{bmatrix}$$

$$\xrightarrow{\mathbf{r}_2 + \mathbf{r}_1 \to \mathbf{r}_1} \begin{bmatrix} 1 & 0 & -2 & 0 & 1 & 2 \\ 0 & 1 & 1 & 0 & -2 & -1 \\ 0 & 0 & 0 & 1 & -4 & -5 \end{bmatrix}$$

The final matrix corresponds to the following system of linear equations:

$$\begin{aligned} x_1 - 2x_3 + x_5 &= 2 \\ x_2 + x_3 - 2x_5 &= -1 \\ x_4 - 4x_5 &= -5 \end{aligned}$$

The general solution of this system is

$$\begin{aligned} x_1 &= 2 + 2x_3 - x_5 \\ x_2 &= -1 - x_3 + 2x_5 \\ x_3 & \quad \text{free} \\ x_4 &= -5 + 4x_5 \\ x_5 & \quad \text{free.} \end{aligned}$$

2. The matrix A is in reduced row echelon form. Furthermore, it has 3 nonzero rows, and hence the rank of A is 3. Since A has 7 columns, its nullity is $7 - 3 = 4$.

3. Apply the Gaussian elimination algorithm to the augmented matrix of the given system to transform the matrix into one in row echelon form:

$$\begin{bmatrix} 1 & 3 & 1+s \\ 1 & r & 5 \end{bmatrix} \xrightarrow{-\mathbf{r}_1 + \mathbf{r}_2 \to \mathbf{r}_2} \begin{bmatrix} 1 & 3 & 1+s \\ 0 & r-3 & 4-s \end{bmatrix}$$

(a) The original system is inconsistent whenever there is a row whose only nonzero entry lies in the last column. Only the second row could have this form. Thus the original system is inconsistent whenever $r - 3 = 0$ and $4 - s \neq 0$; that is, when $r = 3$ and $s \neq 4$.

(b) The original system has infinitely many solutions whenever the system is consistent and there is a free variable in the general solution. In order to have a free variable, we must have $r - 3 = 0$, and in order for the system also to be consistent, we must have $4 - s = 0$. Thus the original system has infinitely many solutions if $r = 3$ and $s = 4$.

(c) Let A denote the coefficient matrix of the system. For the system to have a unique solution, there must be two basic variables, so the rank of A must be 2. Since deleting the last column of the preceding matrix gives a row echelon form of A, the rank of A is 2 precisely when $r - 3 \neq 0$; that is, when $r \neq 3$.

1.5* APPLICATIONS OF SYSTEMS OF LINEAR EQUATIONS

Systems of linear equations arise in many applications of mathematics. In this section, we present two such applications.

THE LEONTIEF INPUT–OUTPUT MODEL

In a modern industrialized country, there are hundreds of different industries that supply goods and services needed for production. These industries are often mutually dependent.

The agricultural industry, for instance, requires farm machinery to plant and harvest crops, whereas the makers of farm machinery need food produced by the agricultural industry. Because of this interdependency, events in one industry, such as a strike by factory workers, can significantly affect many other industries. To better understand these complex interactions, economic planners use mathematical models of the economy, the most important of which was developed by the Russian-born economist Wassily Leontief.

While a student in Berlin in the 1920s, Leontief developed a mathematical model, called the *input–output model*, for analyzing an economy. After arriving in the United States in 1931 to be a professor of economics at Harvard University, Leontief began to collect the data that would enable him to implement his ideas. Finally, after the end of World War II, he succeeded in extracting from government statistics the data necessary to create a model of the U.S. economy. This model proved to be highly accurate in predicting the behavior of the postwar U.S. economy and earned Leontief the 1973 Nobel Prize for Economics.

Leontief's model of the U.S. economy combined approximately 500 industries into 42 sectors that provide products and services, such as the electrical machinery sector. To illustrate Leontief's theory, which can be applied to the economy of any country or region, we next show how to construct a general input–output model. Suppose that an economy is divided into n sectors and that sector i produces some commodity or service S_i $(i = 1, 2, \ldots, n)$. Usually, we measure amounts of commodities and services in common monetary units and hold costs fixed so that we can compare diverse sectors. For example, the output of the steel industry could be measured in millions of dollars worth of steel produced.

For each i and j, let c_{ij} denote the amount of S_i needed to produce one unit of S_j. Then the $n \times n$ matrix C whose (i, j)-entry is c_{ij} is called the **input–output matrix** (or the **consumption matrix**) for the economy.

To illustrate these ideas with a very simple example, consider an economy that is divided into three sectors: agriculture, manufacturing, and services. (Of course, a model of any real economy, such as Leontief's original model, will involve many more sectors and much larger matrices.) Suppose that each dollar's worth of agricultural output requires inputs of \$0.10 from the agricultural sector, \$0.20 from the

* This section can be omitted without loss of continuity.

manufacturing sector, and \$0.30 from the services sector; each dollar's worth of manufacturing output requires inputs of \$0.20 from the agricultural sector, \$0.40 from the manufacturing sector, and \$0.10 from the services sector; and each dollar's worth of services output requires inputs of \$0.10 from the agricultural sector, \$0.20 from the manufacturing sector, and \$0.10 from the services sector.

From this information, we can form the following input–output matrix:

$$C = \begin{array}{c} \begin{array}{ccc} \text{Ag.} & \text{Man.} & \text{Svcs.} \end{array} \\ \begin{bmatrix} .1 & .2 & .1 \\ .2 & .4 & .2 \\ .3 & .1 & .1 \end{bmatrix} \end{array} \begin{array}{l} \\ \text{Agriculture} \\ \text{Manufacturing} \\ \text{Services} \end{array}$$

Note that the (i, j)-entry of the matrix represents the amount of input from sector i needed to produce a dollar's worth of output from sector j. Now let x_1, x_2, and x_3 denote the total output of the agriculture, manufacturing, and services sectors, respectively. Since x_1 dollar's worth of agricultural products are being produced, the first column of the input-output matrix shows that an input of $.1x_1$ is required from the agriculture sector, an input of $.2x_1$ is required from the manufacturing sector, and an input of $.3x_1$ is required from the services sector. Similar statements apply to the manufacturing and services sectors. Figure 1.20 shows the total amount of money flowing among the three sectors.

Note that in Figure 1.20 the three arcs leaving the agriculture sector give the total amount of agricultural output that is used as inputs for all three sectors. The sum of the labels on the three arcs, $.1x_1 + .2x_2 + .1x_3$, represents the amount of agricultural output that is consumed during the production process. Similar statements apply to the other two sectors. So the vector

$$\begin{bmatrix} .1x_1 + .2x_2 + .1x_3 \\ .2x_1 + .4x_2 + .2x_3 \\ .3x_1 + .1x_2 + .1x_3 \end{bmatrix}$$

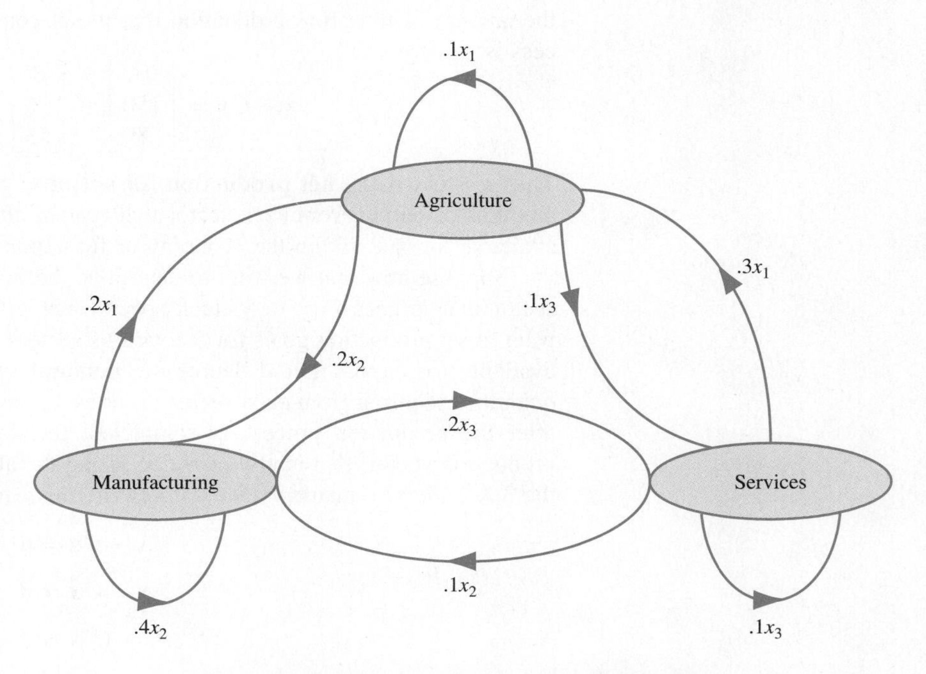

Figure 1.20 The flow of money among the sectors

gives the amount of the total output of the economy that is consumed during the production process. This vector is just the matrix–vector product $C\mathbf{x}$, where $\mathbf{x}$ is the **gross production vector**

$$\mathbf{x} = \begin{bmatrix} x_1 \\ x_2 \\ x_3 \end{bmatrix}.$$

> For an economy with input–output matrix C and gross production vector $\mathbf{x}$, the total output of the economy that is consumed during the production process is $C\mathbf{x}$.

Example 1

Suppose that in the economy previously described, the total outputs of the agriculture, manufacturing, and services sectors are \$100 million, \$150 million, and \$80 million, respectively. Then

$$C\mathbf{x} = \begin{bmatrix} .1 & .2 & .1 \\ .2 & .4 & .2 \\ .3 & .1 & .1 \end{bmatrix} \begin{bmatrix} 100 \\ 150 \\ 80 \end{bmatrix} = \begin{bmatrix} 48 \\ 96 \\ 53 \end{bmatrix},$$

and so the portion of the gross production that is consumed during the production process is \$48 million of agriculture, \$96 million of manufacturing, and \$53 million of services.

Since, in Example 1, the amount of the gross production consumed during the production process is

$$C\mathbf{x} = \begin{bmatrix} 48 \\ 96 \\ 53 \end{bmatrix},$$

the amount of the gross production that is not consumed during the production process is

$$\mathbf{x} - C\mathbf{x} = \begin{bmatrix} 100 \\ 150 \\ 80 \end{bmatrix} - \begin{bmatrix} 48 \\ 96 \\ 53 \end{bmatrix} = \begin{bmatrix} 52 \\ 54 \\ 27 \end{bmatrix}.$$

Thus $\mathbf{x} - C\mathbf{x}$ is the **net production** (or **surplus**) vector; its components indicate the amounts of output from each sector that remain after production. These amounts are available for sale within the economy or for export outside the economy.

Suppose now that we want to determine the amount of gross production for each sector that is necessary to yield a specific net production. For example, we might want to set production goals for the various sectors so that we have specific quantities available for export. Let $\mathbf{d}$ denote the **demand vector**, whose components are the quantities required from each sector. In order to have exactly these amounts available after the production process is completed, the demand vector must equal the net production vector; that is, $\mathbf{d} = \mathbf{x} - C\mathbf{x}$. Using the algebra of matrices and vectors and the 3×3 identity matrix I_3, we can rewrite this equation as follows:

$$\mathbf{x} - C\mathbf{x} = \mathbf{d}$$

$$I_3\mathbf{x} - C\mathbf{x} = \mathbf{d}$$

$$(I_3 - C)\mathbf{x} = \mathbf{d}$$

Thus the required gross production is a solution of the equation $(I_3 - C)\mathbf{x} = \mathbf{d}$.

For an economy with $n \times n$ input–output matrix C, the gross production necessary to satisfy exactly a demand $\mathbf{d}$ is a solution of $(I_n - C)\mathbf{x} = \mathbf{d}$.

Example 2 For the economy in Example 1, determine the gross production needed to meet a consumer demand for \$90 million of agriculture, \$80 million of manufacturing, and \$60 million of services.

Solution We must solve the matrix equation $(I_3 - C)\mathbf{x} = \mathbf{d}$, where C is the input–output matrix and

$$\mathbf{d} = \begin{bmatrix} 90 \\ 80 \\ 60 \end{bmatrix}$$

is the demand vector. Since

$$I_3 - C = \begin{bmatrix} 1 & 0 & 0 \\ 0 & 1 & 0 \\ 0 & 0 & 1 \end{bmatrix} - \begin{bmatrix} .1 & .2 & .1 \\ .2 & .4 & .2 \\ .3 & .1 & .1 \end{bmatrix} = \begin{bmatrix} .9 & -.2 & -.1 \\ -.2 & .6 & -.2 \\ -.3 & -.1 & .9 \end{bmatrix},$$

the augmented matrix of the system to be solved is

$$\begin{bmatrix} .9 & -.2 & -.1 & 90 \\ -.2 & .6 & -.2 & 80 \\ -.3 & -.1 & .9 & 60 \end{bmatrix}.$$

Thus the solution of $(I_3 - C)\mathbf{x} = \mathbf{d}$ is

$$\begin{bmatrix} 170 \\ 240 \\ 150 \end{bmatrix},$$

so the gross production needed to meet the demand is \$170 million of agriculture, \$240 million of manufacturing, and \$150 million of services.

Practice Problem 1 ▶ An island's economy is divided into three sectors—tourism, transportation, and services. Suppose that each dollar's worth of tourism output requires inputs of \$0.30 from the tourism sector, \$0.10 from the transportation sector, and \$0.30 from the services sector; each dollar's worth of transportation output requires inputs of \$0.20 from the tourism sector, \$0.40 from the transportation sector, and \$0.20 from the services sector; and each dollar's worth of services output requires inputs of \$0.05 from the tourism sector, \$0.05 from the transportation sector, and \$0.15 from the services sector.

(a) Write the input–output matrix for this economy.

(b) If the gross production for this economy is \$10 million of tourism, \$15 million of transportation, and \$20 million of services, how much input from the tourism sector is required by the services sector?

(c) If the gross production for this economy is \$10 million of tourism, \$15 million of transportation, and \$20 million of services, what is the total value of the inputs consumed by each sector during the production process?

(d) If the total outputs of the tourism, transportation, and services sectors are \$70 million, \$50 million, and \$60 million, respectively, what is the net production of each sector?

(e) What gross production is required to satisfy exactly a demand for \$30 million of tourism, \$50 million of transportation, and \$40 million of services? ◀

CURRENT FLOW IN ELECTRICAL CIRCUITS

When a battery is connected in an electrical circuit, a current flows through the circuit. If the current passes through a resistor (a device that creates resistance to the flow of electricity), a drop in voltage occurs. These voltage drops obey *Ohm's law*,[12] which states that

$$V = RI,$$

where V is the voltage drop across the resistor (measured in volts), R is the resistance (measured in ohms), and I is the current (measured in amperes).

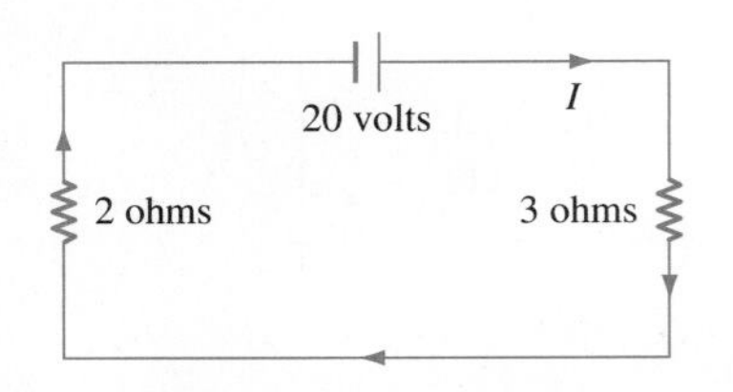

Figure 1.21 A simple electrical circuit

Figure 1.21 shows a simple electrical circuit consisting of a 20-volt battery (indicated by ⊣⊢) and two resistors (indicated by ⩘⩘) with resistances of 3 ohms and 2 ohms. The current flows in the direction of the arrows. If the value of I is positive, then the flow is from the positive terminal of the battery (indicated by the longer side of the battery) to the negative terminal (indicated by the shorter side). In order to determine the value of I, we must utilize *Kirchhoff's*[13] *voltage law.*

> **Kirchhoff's Voltage Law**
>
> In a closed path within an electrical circuit, the sum of the voltage drops in any one direction equals the sum of the voltage sources in the same direction.

In the circuit shown in Figure 1.21, there are two voltage drops, one of $3I$ and the other of $2I$. Their sum equals 20, the voltage supplied by the single battery. Hence

$$\begin{aligned} 3I + 2I &= 20 \\ 5I &= 20 \\ I &= 4. \end{aligned}$$

Thus the current flow through the network is 4 amperes.

Practice Problem 2 ▶ Determine the current through the following electrical circuit:

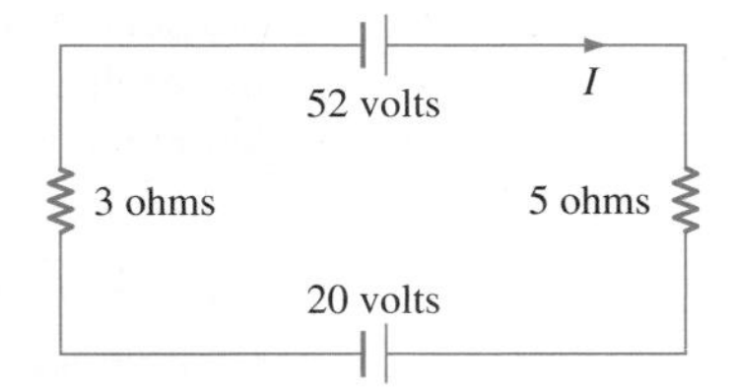

◀

---

[12] Georg Simon Ohm (1787–1854) was a German physicist whose pamphlet *Die galvanische Kette mathematisch bearbeitet* greatly influenced the development of the theory of electricity.

[13] Gustav Robert Kirchhoff (1824–1887) was a German physicist who made significant contributions to the fields of electricity and electromagnetic radiation.

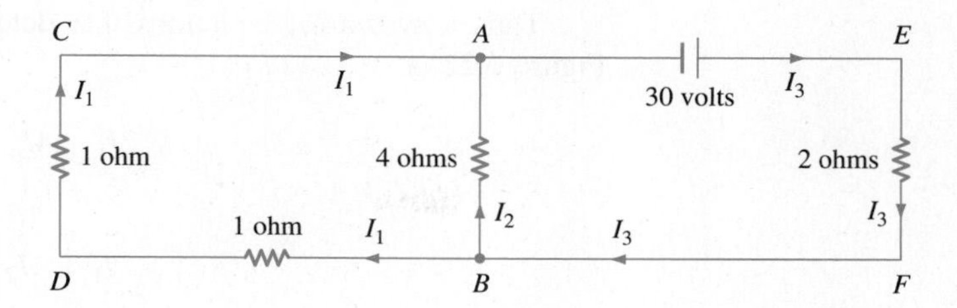

Figure 1.22 An electrical circuit

A more complicated circuit is shown in Figure 1.22. Here the junctions at A and B (indicated by the dots) create three branches in the circuit, each with its own current flow. Starting at B and applying Kirchhoff's voltage law to the closed path $BDCAB$, we obtain $1I_1 + 1I_1 - 4I_2 = 0$; that is,

$$2I_1 - 4I_2 = 0. \tag{7}$$

Note that, since we are proceeding around the closed path in a clockwise direction, the flow from A to B is opposite to the direction indicated for I_2. Thus the voltage drop at the 4-ohm resistor is $4(-I_2)$. Moreover, because there is no voltage source in this closed path, the sum of the three voltage drops is 0.

Similarly, from the closed path $BAEFB$, we obtain the equation

$$4I_2 + 2I_3 = 30, \tag{8}$$

and from the closed path $BDCAEFB$, we obtain the equation

$$2I_1 + 2I_3 = 30. \tag{9}$$

Note that, in this case, equation (9) is the sum of equations (7) and (8). Since equation (9) provides no information not already given by equations (7) and (8), we can discard it. A similar situation occurs in all of the networks that we consider, so we may ignore any closed paths that contain only currents that are accounted for in other equations obtained from Kirchoff's voltage law.

At this point, we have two equations (7) and (8) in three variables, so another equation is required if we are to obtain a unique solution for I_1, I_2, and I_3. This equation is provided by another of Kirchhoff's laws.

> **Kirchoff's Current Law**
>
> The current flow into any junction equals the current flow out of the junction.

In the context of Figure 1.22, Kirchhoff's current law states that the flow into junction A, which is $I_1 + I_2$, equals the current flow out of A, which is I_3. Hence we obtain the equation $I_1 + I_2 = I_3$, or

$$I_1 + I_2 - I_3 = 0. \tag{10}$$

Notice that the current law also applies at junction B, where it yields the equation $I_3 = I_1 + I_2$. However, since this equation is equivalent to equation (10), we can ignore it. In general, if the current law is applied at each junction, then any one of the resulting equations is redundant and can be ignored.

Thus a system of equations that determines the current flows in the circuit in Figure 1.22 is

$$\begin{aligned} 2I_1 - 4I_2 \qquad &= 0 \qquad (7)\\ 4I_2 + 2I_3 &= 30 \qquad (8)\\ I_1 + I_2 - I_3 &= 0. \qquad (10) \end{aligned}$$

Solving this system by Gaussian elimination, we see that $I_1 = 6$, $I_2 = 3$, and $I_3 = 9$, and thus the branch currents are 6 amperes, 3 amperes, and 9 amperes, respectively.

Practice Problem 3 ▶ Determine the currents in each branch of the following electrical circuit:

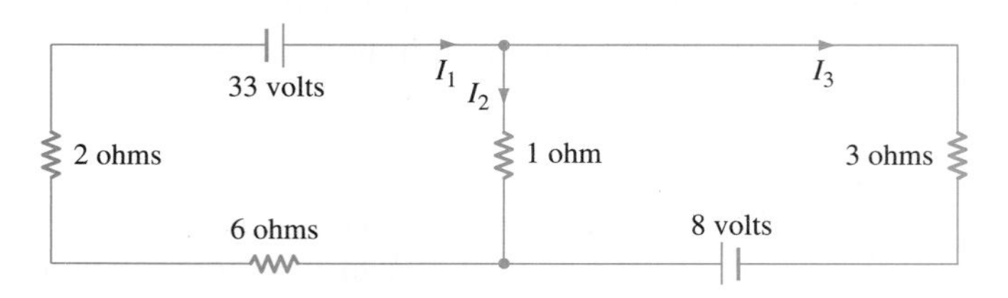

◀

EXERCISES

In Exercises 1–6, determine whether the statements are true or false.

1. The (i,j)-entry of the input–output matrix represents the amount of input from sector i needed to produce one unit of output from sector j.
2. For an economy with $n \times n$ input–output matrix C, the gross production necessary to satisfy exactly a demand $\mathbf{d}$ is a solution of $(I_n - C)\mathbf{x} = \mathbf{d}$.
3. If C is the input–output matrix for an economy with gross production vector $\mathbf{x}$, then $C\mathbf{x}$ is the net production vector.
4. In any closed path within an electrical circuit, the algebraic sum of all the voltage drops in the same direction equals 0.
5. At every junction in an electrical circuit, the current flow into the junction equals the current flow out of the junction.
6. The voltage drop at each resistor in an electrical network equals the product of the resistance and the amount of current through the resistor.

In Exercises 7–16, suppose that an economy is divided into four sectors (agriculture, manufacturing, services, and entertainment) with the following input–output matrix:

$$C = \begin{array}{c} \begin{array}{cccc} \text{Ag.} & \text{Man.} & \text{Svcs.} & \text{Ent.} \end{array} \\ \begin{bmatrix} .12 & .11 & .15 & .18 \\ .20 & .08 & .24 & .07 \\ .18 & .16 & .06 & .22 \\ .09 & .07 & .12 & .05 \end{bmatrix} \end{array} \begin{array}{l} \\ \text{Agriculture} \\ \text{Manufacturing} \\ \text{Services} \\ \text{Entertainment} \end{array}$$

7. What amount of input from the services sector is needed for a gross production of \$50 million by the entertainment sector?
8. What amount of input from the manufacturing sector is needed for a gross production of \$100 million by the agriculture sector?
9. Which sector is least dependent on services?
10. Which sector is most dependent on services?
11. On which sector is agriculture least dependent?
12. On which sector is agriculture most dependent?
13. If the gross production for this economy is \$30 million of agriculture, \$40 million of manufacturing, \$30 million of services, and \$20 million of entertainment, what is the total value of the inputs from each sector consumed during the production process?
14. If the gross production for this economy is \$20 million of agriculture, \$30 million of manufacturing, \$20 million of services, and \$10 million of entertainment, what is the total value of the inputs from each sector consumed during the production process?
15. If the gross production for this economy is \$30 million of agriculture, \$40 million of manufacturing, \$30 million of services, and \$20 million of entertainment, what is the net production of each sector?
16. If the gross production for this economy is \$20 million of agriculture, \$30 million of manufacturing, \$20 million of services, and \$10 million of entertainment, what is the net production of each sector?

17. The input–output matrix for an economy producing transportation, food, and oil follows:

$$\begin{array}{c} \text{Tran.} \quad \text{Food} \quad \text{Oil} \\ \begin{bmatrix} .2 & .20 & .3 \\ .4 & .30 & .1 \\ .2 & .25 & .3 \end{bmatrix} \begin{array}{l} \text{Transportation} \\ \text{Food} \\ \text{Oil} \end{array} \end{array}$$

   (a) What is the net production corresponding to a gross production of \$40 million of transportation, \$30 million of food, and \$35 million of oil?

   (b) What gross production is required to satisfy exactly a demand for \$32 million of transportation, \$48 million of food, and \$24 million of oil?

18. The input–output matrix for an economy with sectors of metals, nonmetals, and services follows:

$$\begin{array}{c} \text{Met.} \quad \text{Nonm.} \quad \text{Svcs.} \\ \begin{bmatrix} .2 & .2 & .1 \\ .4 & .4 & .2 \\ .2 & .2 & .1 \end{bmatrix} \begin{array}{l} \text{Metals} \\ \text{Nonmetals} \\ \text{Services} \end{array} \end{array}$$

   (a) What is the net production corresponding to a gross production of \$50 million of metals, \$60 million of nonmetals, and \$40 million of services?

   (b) What gross production is required to satisfy exactly a demand for \$120 million of metals, \$180 million of nonmetals, and \$150 million of services?

19. Suppose that a nation's energy production is divided into two sectors: electricity and oil. Each dollar's worth of electricity output requires \$0.10 of electricity input and \$0.30 of oil input, and each dollar's worth of oil output requires \$0.40 of electricity input and \$0.20 of oil input.

   (a) Write the input–output matrix for this economy.

   (b) What is the net production corresponding to a gross production of \$60 million of electricity and \$50 million of oil?

   (c) What gross production is needed to satisfy exactly a demand for \$60 million of electricity and \$72 million of oil?

20. Suppose that an economy is divided into two sectors: nongovernment and government. Each dollar's worth of nongovernment output requires \$0.10 in nongovernment input and \$0.10 in government input, and each dollar's worth of government output requires \$0.20 in nongovernment input and \$0.70 in government input.

   (a) Write the input–output matrix for this economy.

   (b) What is the net production corresponding to a gross production of \$20 million in nongovernment and \$30 million in government?

   (c) What gross production is needed to satisfy exactly a demand for \$45 million in nongovernment and \$50 million in government?

21. Consider an economy that is divided into three sectors: finance, goods, and services. Suppose that each dollar's worth of financial output requires inputs of \$0.10 from the finance sector, \$0.20 from the goods sector, and \$0.20 from the services sector; each dollar's worth of goods requires inputs of \$0.10 from the finance sector, \$0.40 from the goods sector, and \$0.20 from the services sector; and each dollar's worth of services requires inputs of \$0.15 from the finance sector, \$0.10 from the goods sector, and \$0.30 from the services sector.

   (a) What is the net production corresponding to a gross production of \$70 million of finance, \$50 million of goods, and \$60 million of services?

   (b) What is the gross production corresponding to a net production of \$40 million of finance, \$50 million of goods, and \$30 million of services?

   (c) What gross production is needed to satisfy exactly a demand for \$40 million of finance, \$36 million of goods, and \$44 million of services?

22. Consider an economy that is divided into three sectors: agriculture, manufacturing, and services. Suppose that each dollar's worth of agricultural output requires inputs of \$0.10 from the agricultural sector, \$0.15 from the manufacturing sector, and \$0.30 from the services sector; each dollar's worth of manufacturing output requires inputs of \$0.20 from the agricultural sector, \$0.25 from the manufacturing sector, and \$0.10 from the services sector; and each dollar's worth of services output requires inputs of \$0.20 from the agricultural sector, \$0.35 from the manufacturing sector, and \$0.10 from the services sector.

   (a) What is the net production corresponding to a gross production of \$40 million of agriculture, \$50 million of manufacturing, and \$30 million of services?

   (b) What gross production is needed to satisfy exactly a demand for \$90 million of agriculture, \$72 million of manufacturing, and \$96 million of services?

23. Let C be the input–output matrix for an economy, $\mathbf{x}$ be the gross production vector, $\mathbf{d}$ be the demand vector, and $\mathbf{p}$ be the vector whose components are the unit prices of the products or services produced by each sector. Economists call the vector $\mathbf{v} = \mathbf{p} - C^T\mathbf{p}$ the *value-added vector*. Show that $\mathbf{p}^T\mathbf{d} = \mathbf{v}^T\mathbf{x}$. (The single entry in the 1×1 matrix $\mathbf{p}^T\mathbf{d}$ represents the gross domestic product of the economy.) *Hint:* Compute $\mathbf{p}^T\mathbf{x}$ in two different ways. First, replace $\mathbf{p}$ by $\mathbf{v} + C^T\mathbf{p}$, and then replace $\mathbf{x}$ by $C\mathbf{x} + \mathbf{d}$.

24. Suppose that the columns of the input–output matrix

$$\begin{array}{c} \qquad\; \text{Ag.} \quad \text{Min.} \quad \text{Tex.} \\ C = \begin{bmatrix} .1 & .2 & .1 \\ .2 & .4 & .2 \\ .3 & .1 & .1 \end{bmatrix} \begin{array}{l} \text{Agriculture} \\ \text{Minerals} \\ \text{Textiles} \end{array} \end{array}$$

measure the amount (in tons) of each input needed to produce one ton of output from each sector. Let p_1, p_2, and p_3 denote the prices per ton of agricultural products, minerals, and textiles, respectively.

(a) Interpret the vector $C^T\mathbf{p}$, where $\mathbf{p} = \begin{bmatrix} p_1 \\ p_2 \\ p_3 \end{bmatrix}$.

(b) Interpret $\mathbf{p} - C^T\mathbf{p}$.

In Exercises 25–29, determine the currents in each branch of the given circuit.

25.

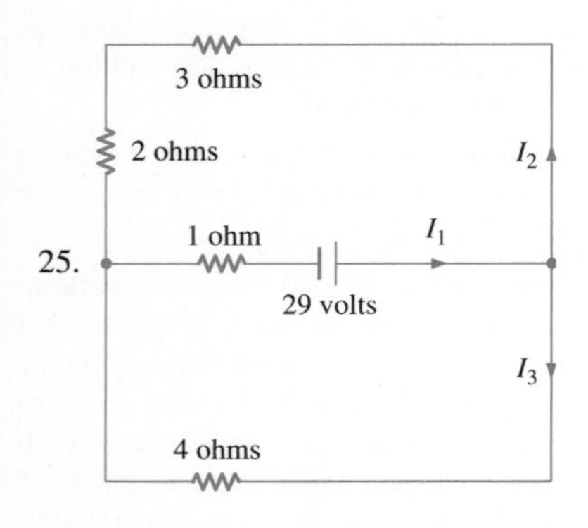

26.

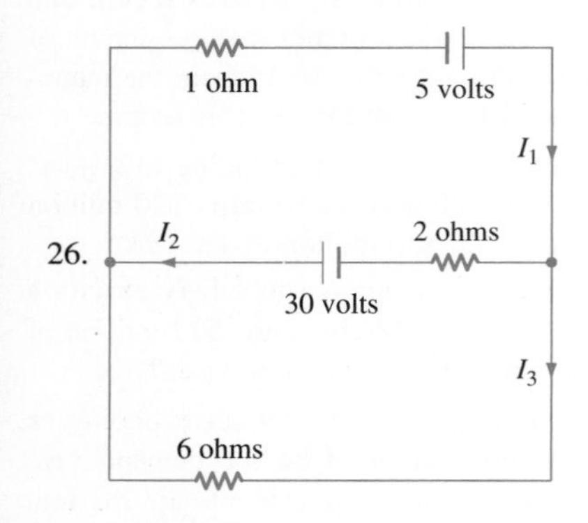

27.

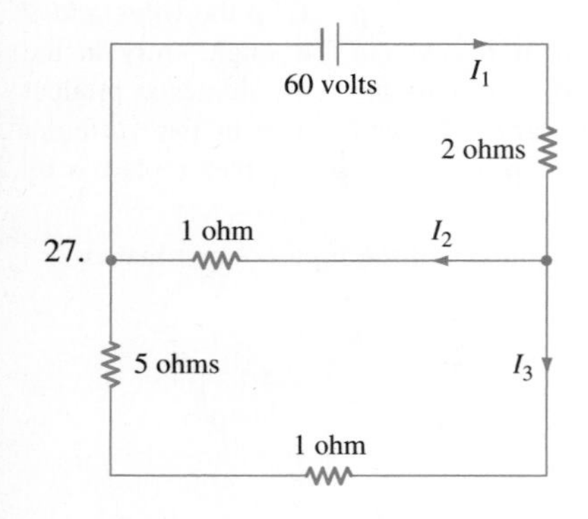

28.

29.

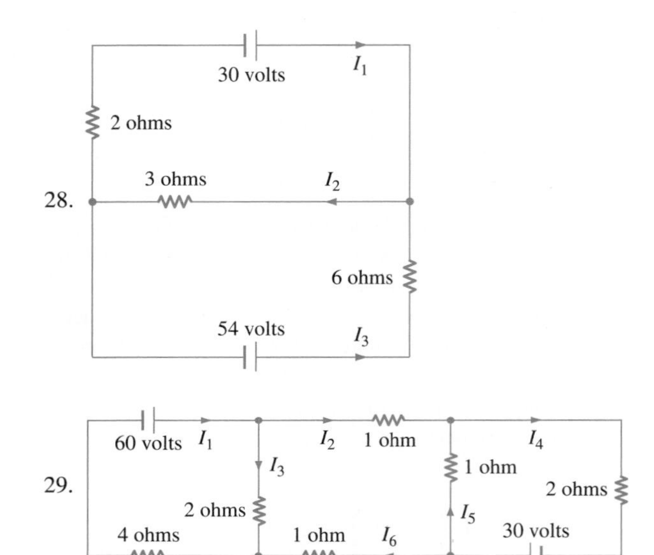

30. In the following electrical network, determine the value of v that makes $I_2 = 0$:

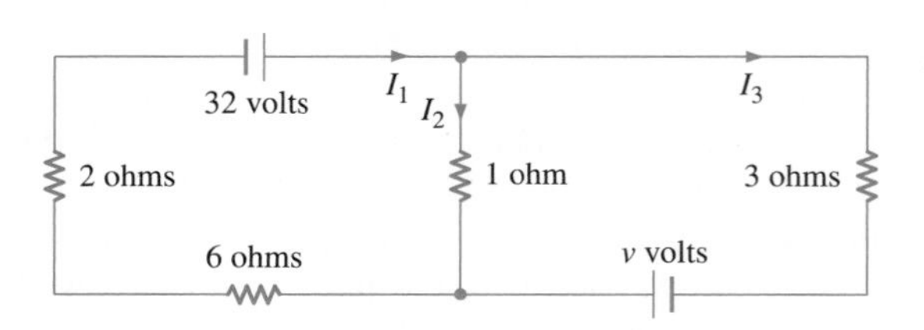

In the following exercise, use either a calculator with matrix capabilities or computer software such as MATLAB to solve the problem:

31. Let

$$C = \begin{bmatrix} .12 & .03 & .20 & .10 & .05 & .09 \\ .21 & .11 & .06 & .11 & .07 & .07 \\ .05 & .21 & .11 & .15 & .11 & .06 \\ .11 & .18 & .13 & .22 & .03 & .18 \\ .16 & .15 & .07 & .12 & .19 & .14 \\ .07 & .23 & .06 & .05 & .15 & .19 \end{bmatrix}$$

and

$$\mathbf{d} = \begin{bmatrix} 100 \\ 150 \\ 200 \\ 125 \\ 300 \\ 180 \end{bmatrix},$$

where C is the input–output matrix for an economy that has been divided into six sectors, and $\mathbf{d}$ is the net production for this economy (where units are in millions of dollars). Find the gross production vector required to produce $\mathbf{d}$.

SOLUTIONS TO THE PRACTICE PROBLEMS

1. (a) The input-output matrix is as follows:

$$C = \begin{array}{c} \text{Tour.} \quad \text{Tran.} \quad \text{Svcs.} \\ \begin{bmatrix} .3 & .2 & .05 \\ .1 & .4 & .05 \\ .3 & .2 & .15 \end{bmatrix} \end{array} \begin{array}{l} \text{Tourism} \\ \text{Transportation} \\ \text{Services} \end{array}$$

(b) Each dollar's worth of output from the services sector requires an input of \$0.05 from the tourism sector. Hence a gross output of \$20 million from the services sector requires an input of 20(\$0.05) = \$1 million from the tourism sector.

(c) The total value of the inputs consumed by each sector during the production process is given by

$$C\begin{bmatrix} 10 \\ 15 \\ 20 \end{bmatrix} = \begin{bmatrix} .3 & .2 & .05 \\ .1 & .4 & .05 \\ .3 & .2 & .15 \end{bmatrix}\begin{bmatrix} 10 \\ 15 \\ 20 \end{bmatrix} = \begin{bmatrix} 7 \\ 8 \\ 9 \end{bmatrix}.$$

Hence, during the production process, \$7 million in inputs is consumed by the tourism sector, \$8 million by the transportation sector, and \$9 million by the services sector.

(d) The gross production vector is

$$\mathbf{x} = \begin{bmatrix} 70 \\ 50 \\ 60 \end{bmatrix},$$

and so the net production vector is

$$\mathbf{x} - C\mathbf{x} = \begin{bmatrix} 70 \\ 50 \\ 60 \end{bmatrix} - \begin{bmatrix} .3 & .2 & .05 \\ .1 & .4 & .05 \\ .3 & .2 & .15 \end{bmatrix}\begin{bmatrix} 70 \\ 50 \\ 60 \end{bmatrix}$$

$$= \begin{bmatrix} 70 \\ 50 \\ 60 \end{bmatrix} - \begin{bmatrix} 34 \\ 30 \\ 40 \end{bmatrix} = \begin{bmatrix} 36 \\ 20 \\ 20 \end{bmatrix}.$$

Hence the net productions of the tourism, transportation, and services sectors are \$36 million, \$20 million, and \$20 million, respectively.

(e) To meet a demand

$$\mathbf{d} = \begin{bmatrix} 30 \\ 50 \\ 40 \end{bmatrix},$$

the gross production vector must be a solution of the equation $(I_3 - C)\mathbf{x} = \mathbf{d}$. The augmented matrix of the system is

$$\begin{bmatrix} .7 & -.2 & -.05 & 30 \\ -.1 & .6 & -.05 & 50 \\ -.3 & -.2 & .85 & 40 \end{bmatrix},$$

and so the gross production vector is

$$\begin{bmatrix} 80 \\ 105 \\ 100 \end{bmatrix}.$$

Thus the gross productions of the tourism, transportation, and services sectors are \$80 million, \$105 million, and \$100 million, respectively.

2. The algebraic sum of the voltage drops around the circuit is $5I + 3I = 8I$. Since the current flow from the 20-volt battery is in the direction opposite to I, the algebraic sum of the voltage sources around the circuit is $52 + (-20) = 32$. Hence we have the equation $8I = 32$, so $I = 4$ amperes.

3. There are two junctions in the given circuit, namely, A and B in the following figure:

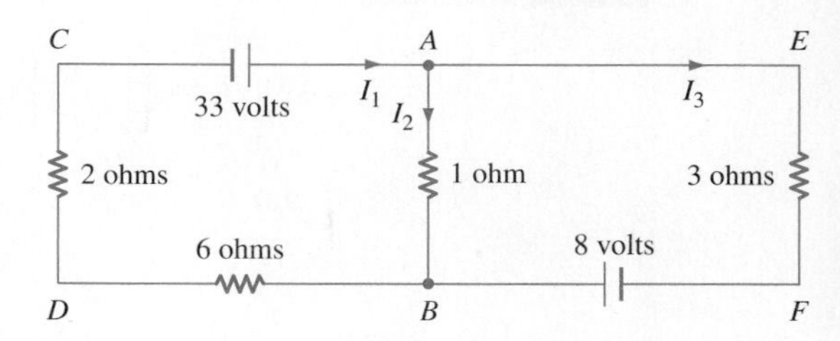

Applying Kirchoff's current law to junction A or junction B gives $I_1 = I_2 + I_3$. Applying Kirchoff's voltage law to the closed path $ABDCA$ yields

$$1I_2 + 6I_1 + 2I_1 = 33.$$

Similarly, from the closed path $AEFBA$, we obtain

$$3I_3 + 1(-I_2) = 8.$$

Hence the system of equations describing the current flows is

$$\begin{aligned} I_1 - I_2 - I_3 &= 0 \\ 8I_1 + I_2 &= 33 \\ -I_2 + 3I_3 &= 8. \end{aligned}$$

Solving this system gives $I_1 = 4$, $I_2 = 1$, and $I_3 = 3$. Hence the branch currents are 4 amperes, 1 ampere, and 3 amperes, respectively.

1.6 THE SPAN OF A SET OF VECTORS

In Section 1.2, we defined a linear combination of vectors $\mathbf{u}_1, \mathbf{u}_2, \ldots, \mathbf{u}_k$ in $\mathcal{R}^n$ to be a vector of the form $c_1\mathbf{u}_1 + c_2\mathbf{u}_2 + \cdots + c_k\mathbf{u}_k$, where $c_1, c_2, \ldots, c_k$ are scalars. For a given set $\mathcal{S} = \{\mathbf{u}_1, \mathbf{u}_2, \ldots, \mathbf{u}_k\}$ of vectors from $\mathcal{R}^n$, we often need to find the set of all the linear combinations of $\mathbf{u}_1, \mathbf{u}_2, \ldots, \mathbf{u}_k$. For example, if A is an $n \times p$ matrix, then the set of vectors $\mathbf{v}$ in $\mathcal{R}^n$ such that $A\mathbf{x} = \mathbf{v}$ is consistent is precisely the set of all the linear combinations of the columns of A. We now define a term for such a set of linear combinations.

Definition For a nonempty set $\mathcal{S} = \{\mathbf{u}_1, \mathbf{u}_2, \ldots, \mathbf{u}_k\}$ of vectors in $\mathcal{R}^n$, we define the **span of** $\mathcal{S}$ to be the set of all linear combinations of $\mathbf{u}_1, \mathbf{u}_2, \ldots, \mathbf{u}_k$ in $\mathcal{R}^n$. This set is denoted by Span $\mathcal{S}$ or Span $\{\mathbf{u}_1, \mathbf{u}_2, \ldots, \mathbf{u}_k\}$.

A linear combination of a single vector is just a multiple of that vector. So if $\mathbf{u}$ is in $\mathcal{S}$, then every multiple of $\mathbf{u}$ is in Span $\mathcal{S}$. Thus the span of $\{\mathbf{u}\}$ consists of all multiples of $\mathbf{u}$. In particular, the span of $\{\mathbf{0}\}$ is $\{\mathbf{0}\}$. Note, however, that if $\mathcal{S}$ contains even one nonzero vector, then Span $\mathcal{S}$ contains infinitely many vectors. Other examples of the span of a set follow.

Example 1 Describe the spans of the following subsets of $\mathcal{R}^2$:

$$\mathcal{S}_1 = \left\{\begin{bmatrix} 1 \\ -1 \end{bmatrix}\right\}, \quad \mathcal{S}_2 = \left\{\begin{bmatrix} 1 \\ -1 \end{bmatrix}, \begin{bmatrix} -2 \\ 2 \end{bmatrix}\right\}, \quad \mathcal{S}_3 = \left\{\begin{bmatrix} 1 \\ -1 \end{bmatrix}, \begin{bmatrix} -2 \\ 2 \end{bmatrix}, \begin{bmatrix} 2 \\ 1 \end{bmatrix}\right\},$$

and

$$\mathcal{S}_4 = \left\{\begin{bmatrix} 1 \\ -1 \end{bmatrix}, \begin{bmatrix} -2 \\ 2 \end{bmatrix}, \begin{bmatrix} 2 \\ 1 \end{bmatrix}, \begin{bmatrix} -1 \\ 3 \end{bmatrix}\right\}$$

Solution The span of $\mathcal{S}_1$ consists of all linear combinations of the vectors in $\mathcal{S}_1$. Since a linear combination of a single vector is just a multiple of that vector, the span of $\mathcal{S}_1$ consists of all multiples of $\begin{bmatrix} 1 \\ -1 \end{bmatrix}$—that is, all vectors of the form $\begin{bmatrix} c \\ -c \end{bmatrix}$ for some scalar c. These vectors all lie along the line with equation $y = -x$, as pictured in Figure 1.23.

The span of $\mathcal{S}_2$ consists of all linear combinations of the vectors $\begin{bmatrix} 1 \\ -1 \end{bmatrix}$ and $\begin{bmatrix} -2 \\ 2 \end{bmatrix}$. Such vectors have the form

$$a\begin{bmatrix} 1 \\ -1 \end{bmatrix} + b\begin{bmatrix} -2 \\ 2 \end{bmatrix} = a\begin{bmatrix} 1 \\ -1 \end{bmatrix} - 2b\begin{bmatrix} 1 \\ -1 \end{bmatrix} = (a - 2b)\begin{bmatrix} 1 \\ -1 \end{bmatrix},$$

where a and b are arbitrary scalars. Taking $c = a - 2b$, we see that these are the same vectors as those in the span of $\mathcal{S}_1$. Hence Span $\mathcal{S}_2$ = Span $\mathcal{S}_1$. (See Figure 1.24.)

The span of $\mathcal{S}_3$ consists of all linear combinations of the vectors $\begin{bmatrix} 1 \\ -1 \end{bmatrix}$, $\begin{bmatrix} -2 \\ 2 \end{bmatrix}$, and $\begin{bmatrix} 2 \\ 1 \end{bmatrix}$. Note that the vectors $\begin{bmatrix} 1 \\ -1 \end{bmatrix}$ and $\begin{bmatrix} 2 \\ 1 \end{bmatrix}$ are not parallel. Hence an arbitrary vector

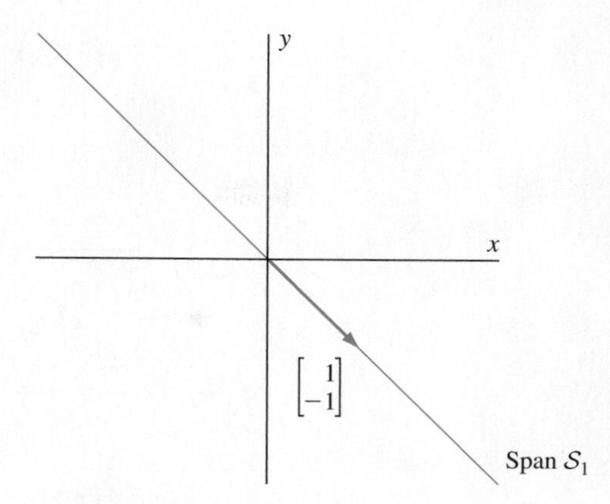

Figure 1.23 The span of $\mathcal{S}_1$

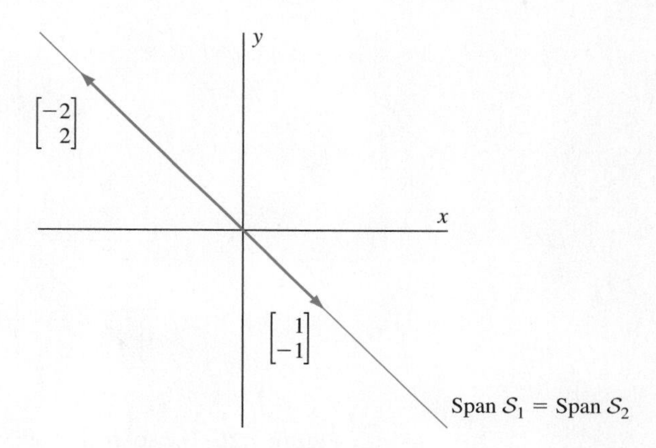

Figure 1.24 The span of $\mathcal{S}_2$

$\mathbf{v}$ in $\mathcal{R}^2$ is a linear combination of these two vectors, as we learned in Section 1.2. Suppose that $\mathbf{v} = a\begin{bmatrix}1\\-1\end{bmatrix} + b\begin{bmatrix}2\\1\end{bmatrix}$ for some scalars a and b. Then

$$\mathbf{v} = a\begin{bmatrix}1\\-1\end{bmatrix} + 0\begin{bmatrix}-2\\2\end{bmatrix} + b\begin{bmatrix}2\\1\end{bmatrix},$$

so every vector in $\mathcal{R}^2$ is a linear combination of the vectors in $\mathcal{S}_3$. It follows that the span of $\mathcal{S}_3$ is $\mathcal{R}^2$.

Finally, since every vector in $\mathcal{R}^2$ is a linear combination of the nonparallel vectors $\begin{bmatrix}1\\-1\end{bmatrix}$ and $\begin{bmatrix}2\\1\end{bmatrix}$, every vector in $\mathcal{R}^2$ is also a linear combination of the vectors in $\mathcal{S}_4$. Therefore the span of $\mathcal{S}_4$ is again $\mathcal{R}^2$.

Example 2 For the standard vectors

$$\mathbf{e}_1 = \begin{bmatrix}1\\0\\0\end{bmatrix}, \quad \mathbf{e}_2 = \begin{bmatrix}0\\1\\0\end{bmatrix}, \quad \text{and} \quad \mathbf{e}_3 = \begin{bmatrix}0\\0\\1\end{bmatrix}$$

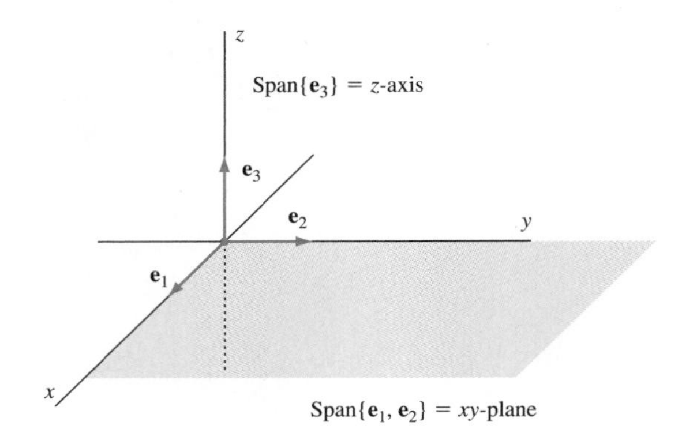

Figure 1.25 The span of $\{\mathbf{e}_1, \mathbf{e}_2\}$ in $\mathcal{R}^3$

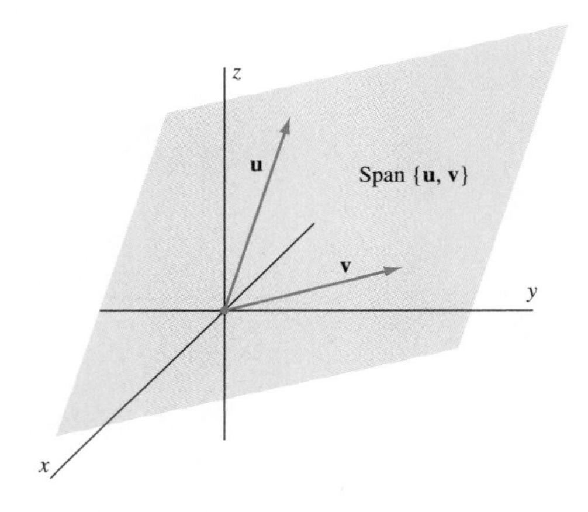

Figure 1.26 The span of $\{\mathbf{u}, \mathbf{v}\}$, where $\mathbf{u}$ and $\mathbf{v}$ are nonparallel vectors in $\mathcal{R}^3$

in $\mathcal{R}^3$, we see that the span of $\{\mathbf{e}_1, \mathbf{e}_2\}$ is the set of vectors of the form

$$a\mathbf{e}_1 + b\mathbf{e}_2 = a\begin{bmatrix}1\\0\\0\end{bmatrix} + b\begin{bmatrix}0\\1\\0\end{bmatrix} = \begin{bmatrix}a\\b\\0\end{bmatrix}.$$

Thus Span $\{\mathbf{e}_1, \mathbf{e}_2\}$ is the set of vectors in the xy-plane of $\mathcal{R}^3$. (See Figure 1.25.) More generally, if $\mathbf{u}$ and $\mathbf{v}$ are nonparallel vectors in $\mathcal{R}^3$, then the span of $\{\mathbf{u}, \mathbf{v}\}$ is a plane through the origin. (See Figure 1.26.)

Furthermore, Span $\{\mathbf{e}_3\}$ is the set of vectors that lie along the z-axis in $\mathcal{R}^3$. (See Figure 1.25.)

From the preceding examples, we see that saying "$\mathbf{v}$ belongs to the span of $\mathcal{S} = \{\mathbf{u}_1, \mathbf{u}_2, \ldots, \mathbf{u}_k\}$" means exactly the same as saying "$\mathbf{v}$ equals some linear combination of the vectors $\mathbf{u}_1, \mathbf{u}_2, \ldots, \mathbf{u}_k$." So our comment at the beginning of the section can be rephrased as follows:

> Let $\mathcal{S} = \{\mathbf{u}_1, \mathbf{u}_2, \ldots, \mathbf{u}_k\}$ be a set of vectors from $\mathcal{R}^n$, and let A be the matrix whose columns are $\mathbf{u}_1, \mathbf{u}_2, \ldots, \mathbf{u}_k$. Then a vector $\mathbf{v}$ from $\mathcal{R}^n$ is in the span of $\mathcal{S}$ (that is, $\mathbf{v}$ is a linear combination of $\mathbf{u}_1, \mathbf{u}_2, \ldots, \mathbf{u}_k$) if and only if the equation $A\mathbf{x} = \mathbf{v}$ is consistent.

Example 3 Is

$$\mathbf{v} = \begin{bmatrix} 3 \\ 0 \\ 5 \\ -1 \end{bmatrix} \quad \text{or} \quad \mathbf{w} = \begin{bmatrix} 2 \\ 1 \\ 3 \\ -1 \end{bmatrix}$$

a vector in the span of

$$\mathcal{S} = \left\{ \begin{bmatrix} 1 \\ 2 \\ 1 \\ 1 \end{bmatrix}, \begin{bmatrix} -1 \\ 1 \\ -2 \\ 1 \end{bmatrix}, \begin{bmatrix} 1 \\ 8 \\ -1 \\ 5 \end{bmatrix} \right\}?$$

If so, express it as a linear combination of the vectors in $\mathcal{S}$.

Solution Let A be the matrix whose columns are the vectors in $\mathcal{S}$. The vector $\mathbf{v}$ belongs to the span of $\mathcal{S}$ if and only if $A\mathbf{x} = \mathbf{v}$ is consistent. Since the reduced row echelon form of $[A \;\; \mathbf{v}]$ is

$$\begin{bmatrix} 1 & 0 & 3 & 1 \\ 0 & 1 & 2 & -2 \\ 0 & 0 & 0 & 0 \\ 0 & 0 & 0 & 0 \end{bmatrix},$$

$A\mathbf{x} = \mathbf{v}$ is consistent by Theorem 1.5. Hence $\mathbf{v}$ belongs to the span of $\mathcal{S}$.

To express $\mathbf{v}$ as a linear combination of the vectors in $\mathcal{S}$, we need to find the actual solution of $A\mathbf{x} = \mathbf{v}$. Using the reduced row echelon form of $[A \;\; \mathbf{v}]$, we see that the general solution of this equation is

$$\begin{aligned} x_1 &= 1 - 3x_3 \\ x_2 &= -2 - 2x_3 \\ x_3 & \quad \text{free.} \end{aligned}$$

For example, by taking $x_3 = 0$, we find that

$$1 \begin{bmatrix} 1 \\ 2 \\ 1 \\ 1 \end{bmatrix} - 2 \begin{bmatrix} -1 \\ 1 \\ -2 \\ 1 \end{bmatrix} + 0 \begin{bmatrix} 1 \\ 8 \\ -1 \\ 5 \end{bmatrix} = \begin{bmatrix} 3 \\ 0 \\ 5 \\ -1 \end{bmatrix} = \mathbf{v}.$$

In the same manner, $\mathbf{w}$ belongs to the span of $\mathcal{S}$ if and only if $A\mathbf{x} = \mathbf{w}$ is consistent. Because the reduced row echelon form of $[A \;\; \mathbf{w}]$ is

$$\begin{bmatrix} 1 & 0 & 3 & 0 \\ 0 & 1 & 2 & 0 \\ 0 & 0 & 0 & 1 \\ 0 & 0 & 0 & 0 \end{bmatrix},$$

Theorem 1.5 shows that $A\mathbf{x} = \mathbf{w}$ is not consistent. Thus $\mathbf{w}$ does not belong to the span of $\mathcal{S}$.

Practice Problem 1 ▶ Are $\mathbf{u} = \begin{bmatrix} -1 \\ 3 \\ 1 \end{bmatrix}$ and $\mathbf{v} = \begin{bmatrix} 1 \\ 1 \\ 2 \end{bmatrix}$ in the span of $\mathcal{S} = \left\{ \begin{bmatrix} 2 \\ -1 \\ 1 \end{bmatrix}, \begin{bmatrix} -1 \\ 1 \\ 0 \end{bmatrix} \right\}$? ◀

In our examples so far, we started with a subset $\mathcal{S}$ of $\mathcal{R}^n$ and described the set $V = \text{Span}\,\mathcal{S}$. In other problems, we might need to do the opposite: Start with a set V and find a set of vectors $\mathcal{S}$ for which $\text{Span}\,\mathcal{S} = V$. If V is a set of vectors from $\mathcal{R}^n$ and $\text{Span}\,\mathcal{S} = V$, then we say that $\mathcal{S}$ is a **generating set** for V or that $\mathcal{S}$ **generates** V. Because every set is contained in its span, a generating set for V is necessarily contained in V.

Example 4 Let

$$\mathcal{S} = \left\{ \begin{bmatrix} 1 \\ 0 \\ 0 \end{bmatrix}, \begin{bmatrix} 1 \\ 1 \\ 0 \end{bmatrix}, \begin{bmatrix} 1 \\ 1 \\ 1 \end{bmatrix}, \begin{bmatrix} 1 \\ -2 \\ -1 \end{bmatrix} \right\}.$$

Show that $\text{Span}\,\mathcal{S} = \mathcal{R}^3$.

Solution Because $\mathcal{S}$ is contained in $\mathcal{R}^3$, it follows that $\text{Span}\,\mathcal{S}$ is contained in $\mathcal{R}^3$. Thus, in order to show that $\text{Span}\,\mathcal{S} = \mathcal{R}^3$, we need only show that an arbitrary vector $\mathbf{v}$ in $\mathcal{R}^3$ belongs to $\text{Span}\,\mathcal{S}$. Thus we must show that $A\mathbf{x} = \mathbf{v}$ is consistent for every $\mathbf{v}$, where

$$A = \begin{bmatrix} 1 & 1 & 1 & 1 \\ 0 & 1 & 1 & -2 \\ 0 & 0 & 1 & -1 \end{bmatrix}.$$

Let $[R\ \ \mathbf{c}]$ be the reduced row echelon form of $[A\ \ \mathbf{v}]$. No matter what $\mathbf{v}$ is, R is the reduced row echelon form of A by Exercise 77 of Section 1.3. Since

$$R = \begin{bmatrix} 1 & 0 & 0 & 3 \\ 0 & 1 & 0 & -1 \\ 0 & 0 & 1 & -1 \end{bmatrix}$$

has no zero row, there can be no row in $[R\ \ \mathbf{c}]$ in which the only nonzero entry lies in the last column. Thus $A\mathbf{x} = \mathbf{v}$ is consistent by Theorem 1.5, and so $\mathbf{v}$ belongs to $\text{Span}\,\mathcal{S}$. Since $\mathbf{v}$ is an arbitrary vector in $\mathcal{R}^3$, it follows that $\text{Span}\,\mathcal{S} = \mathcal{R}^3$.

The following theorem guarantees that the technique used in Example 4 can be applied to test whether or not any subset of $\mathcal{R}^m$ is a generating set for $\mathcal{R}^m$:

THEOREM 1.6

The following statements about an $m \times n$ matrix A are equivalent:

(a) The span of the columns of A is $\mathcal{R}^m$.

(b) The equation $A\mathbf{x} = \mathbf{b}$ has at least one solution (that is, $A\mathbf{x} = \mathbf{b}$ is consistent) for each $\mathbf{b}$ in $\mathcal{R}^m$.

(c) The rank of A is m, the number of rows of A.

(d) The reduced row echelon form of A has no zero rows.

(e) There is a pivot position in each row of A.

PROOF Since, by Theorem 1.5, the equation $A\mathbf{x} = \mathbf{b}$ is consistent precisely when $\mathbf{b}$ equals a linear combination of the columns of A, statements (a) and

(b) are equivalent. Also, because A is an $m \times n$ matrix, statements (c) and (d) are equivalent. The details of these arguments are left to the reader.

We now prove that statements (b) and (c) are equivalent. First, let R denote the reduced row echelon form of A and $\mathbf{e}_m$ be the standard vector

$$\begin{bmatrix} 0 \\ \vdots \\ 0 \\ 1 \end{bmatrix}$$

in $\mathcal{R}^m$. There is a sequence of elementary row operations that transforms A into R. Since each of these elementary row operations is reversible, there is also a sequence of elementary row operations that transforms R into A. Apply the latter sequence of operations to $[R\ \ \mathbf{e}_m]$ to obtain a matrix $[A\ \ \mathbf{d}]$ for some $\mathbf{d}$ in $\mathcal{R}^m$. Then the system $A\mathbf{x} = \mathbf{d}$ is equivalent to the system $R\mathbf{x} = \mathbf{e}_m$.

If (b) is true, then $A\mathbf{x} = \mathbf{d}$, and hence $R\mathbf{x} = \mathbf{e}_m$, must be consistent. But then Theorem 1.5 implies that the last row of R cannot be a zero row, for otherwise $[R\ \ \mathbf{e}_m]$ would have a row in which the only nonzero entry lies in the last column. Because R is in reduced row echelon form, R must have no nonzero rows. It follows that the rank of A is m, establishing (c).

Conversely, assume that (c) is true, and let $[R\ \ \mathbf{c}]$ denote the reduced row echelon form of $[A\ \ \mathbf{b}]$. Since A has rank m, R has no nonzero rows. Hence $[R\ \ \mathbf{c}]$ has no row whose only nonzero entry is in the last column. Therefore, by Theorem 1.5, $A\mathbf{x} = \mathbf{b}$ is consistent for every $\mathbf{b}$. ■

Practice Problem 2 ▶ Is $\mathcal{S} = \left\{ \begin{bmatrix} 1 \\ 0 \\ 1 \end{bmatrix}, \begin{bmatrix} -1 \\ 1 \\ 2 \end{bmatrix}, \begin{bmatrix} 1 \\ 3 \\ 9 \end{bmatrix}, \begin{bmatrix} 2 \\ -1 \\ 1 \end{bmatrix} \right\}$ a generating set for $\mathcal{R}^3$? ◀

MAKING A GENERATING SET SMALLER

In Example 3, we found that $\mathbf{v}$ is in the span of $\mathcal{S}$ by solving $A\mathbf{x} = \mathbf{v}$, a system of 4 equations in 3 variables. If $\mathcal{S}$ contained only 2 vectors, the corresponding system would consist of 4 equations in 2 variables. In general, it is easier to check if a vector is in the span of a set with fewer vectors than it is to check if the vector is in the span of a set with more vectors. The following theorem establishes a useful property that enables us to reduce the size of a generating set in certain cases.

THEOREM 1.7

Let $\mathcal{S} = \{\mathbf{u}_1, \mathbf{u}_2, \ldots, \mathbf{u}_k\}$ be a set of vectors from $\mathcal{R}^n$, and let $\mathbf{v}$ be a vector in $\mathcal{R}^n$. Then Span $\{\mathbf{u}_1, \mathbf{u}_2, \ldots, \mathbf{u}_k, \mathbf{v}\} =$ Span $\{\mathbf{u}_1, \mathbf{u}_2, \ldots, \mathbf{u}_k\}$ if and only if $\mathbf{v}$ belongs to the span of $\mathcal{S}$.

PROOF Suppose that $\mathbf{v}$ is in the span of $\mathcal{S}$. Then $\mathbf{v} = a_1\mathbf{u}_1 + a_2\mathbf{u}_2 + \cdots + a_k\mathbf{u}_k$ for some scalars $a_1, a_2, \ldots, a_k$. If $\mathbf{w}$ is in Span $\{\mathbf{u}_1, \mathbf{u}_2, \ldots, \mathbf{u}_k, \mathbf{v}\}$, then $\mathbf{w}$ can be written $\mathbf{w} = c_1\mathbf{u}_1 + c_2\mathbf{u}_2 + \cdots + c_k\mathbf{u}_k + b\mathbf{v}$ for some scalars $c_1, c_2, \ldots, c_k, b$. By substituting $a_1\mathbf{u}_1 + a_2\mathbf{u}_2 + \cdots + a_k\mathbf{u}_k$ for $\mathbf{v}$ in the preceding equation, we can write $\mathbf{w}$ as a linear combination of the vectors $\mathbf{u}_1, \mathbf{u}_2, \ldots, \mathbf{u}_k$. So the span of $\{\mathbf{u}_1, \mathbf{u}_2, \ldots, \mathbf{u}_k, \mathbf{v}\}$ is contained in the span of $\{\mathbf{u}_1, \mathbf{u}_2, \ldots, \mathbf{u}_k\}$. On the other hand, any vector in Span $\{\mathbf{u}_1, \mathbf{u}_2, \ldots, \mathbf{u}_k\}$ can be written as a linear combination of the vectors $\mathbf{u}_1, \mathbf{u}_2, \ldots, \mathbf{u}_k, \mathbf{v}$ in which the coefficient of $\mathbf{v}$ is 0; so the span of $\{\mathbf{u}_1, \mathbf{u}_2, \ldots, \mathbf{u}_k\}$ is also contained in the span of $\{\mathbf{u}_1, \mathbf{u}_2, \ldots, \mathbf{u}_k, \mathbf{v}\}$. It follows that the two spans are equal.

Conversely, suppose that $\mathbf{v}$ does not belong to the span of S. Note that $\mathbf{v}$ is in the span of $\{\mathbf{u}_1, \mathbf{u}_2, \ldots, \mathbf{u}_k, \mathbf{v}\}$ because $\mathbf{v} = 0\mathbf{u}_1 + 0\mathbf{u}_2 + \cdots + 0\mathbf{u}_k + 1\mathbf{v}$. Hence Span $\{\mathbf{u}_1, \mathbf{u}_2, \ldots, \mathbf{u}_k\} \neq$ Span $\{\mathbf{u}_1, \mathbf{u}_2, \ldots, \mathbf{u}_k, \mathbf{v}\}$ because the second set contains $\mathbf{v}$, but the first does not. ■

Theorem 1.7 provides a method for reducing the size of a generating set. If one of the vectors in $\mathcal{S}$ is a linear combination of the others, it can be removed from $\mathcal{S}$ to obtain a smaller set having the same span as $\mathcal{S}$. For instance, for the set $\mathcal{S}$ of three vectors in Example 3, we have

$$\begin{bmatrix} 1 \\ 8 \\ -1 \\ 5 \end{bmatrix} = 3 \begin{bmatrix} 1 \\ 2 \\ 1 \\ 1 \end{bmatrix} + 2 \begin{bmatrix} -1 \\ 1 \\ -2 \\ 1 \end{bmatrix}.$$

Hence the span of $\mathcal{S}$ is the same as the span of the smaller set

$$\left\{ \begin{bmatrix} 1 \\ 2 \\ 1 \\ 1 \end{bmatrix}, \begin{bmatrix} -1 \\ 1 \\ -2 \\ 1 \end{bmatrix} \right\}.$$

Practice Problem 3 ▶ For the set $\mathcal{S}$ in Practice Problem 2, find a subset with the fewest vectors such that Span $\mathcal{S} = \mathcal{R}^3$. ◀

EXERCISES

In Exercises 1–8, determine whether the given vector is in

$$\text{Span} \left\{ \begin{bmatrix} 1 \\ 0 \\ 1 \end{bmatrix}, \begin{bmatrix} -1 \\ 1 \\ 1 \end{bmatrix}, \begin{bmatrix} 1 \\ 1 \\ 3 \end{bmatrix} \right\}.$$

1. $\begin{bmatrix} -1 \\ 4 \\ 7 \end{bmatrix}$
2. $\begin{bmatrix} 0 \\ 0 \\ 1 \end{bmatrix}$
3. $\begin{bmatrix} 0 \\ 5 \\ 2 \end{bmatrix}$
4. $\begin{bmatrix} 2 \\ -1 \\ 3 \end{bmatrix}$
5. $\begin{bmatrix} -1 \\ 1 \\ 1 \end{bmatrix}$
6. $\begin{bmatrix} -3 \\ 2 \\ 1 \end{bmatrix}$
7. $\begin{bmatrix} 1 \\ 1 \\ -1 \end{bmatrix}$
8. $\begin{bmatrix} -5 \\ 3 \\ 1 \end{bmatrix}$

In Exercises 9–16 determine whether the given vector is in

$$\text{Span} \left\{ \begin{bmatrix} 1 \\ 2 \\ -1 \\ 1 \end{bmatrix}, \begin{bmatrix} 2 \\ -1 \\ 1 \\ 0 \end{bmatrix}, \begin{bmatrix} -1 \\ 2 \\ 0 \\ 3 \end{bmatrix} \right\}.$$

9. $\begin{bmatrix} 0 \\ -1 \\ 1 \\ 0 \end{bmatrix}$
10. $\begin{bmatrix} 9 \\ 6 \\ 1 \\ 9 \end{bmatrix}$
11. $\begin{bmatrix} -8 \\ 9 \\ -7 \\ 2 \end{bmatrix}$
12. $\begin{bmatrix} 2 \\ 0 \\ -3 \\ 4 \end{bmatrix}$
13. $\begin{bmatrix} 4 \\ 0 \\ 5 \\ 8 \end{bmatrix}$
14. $\begin{bmatrix} -1 \\ 0 \\ 2 \\ 3 \end{bmatrix}$
15. $\begin{bmatrix} -1 \\ 2 \\ -3 \\ 5 \end{bmatrix}$
16. $\begin{bmatrix} 5 \\ -2 \\ -2 \\ -7 \end{bmatrix}$

In Exercises 17–20, determine the values of r for which $\mathbf{v}$ *is in the span of* $\mathcal{S}$.

17. $\mathcal{S} = \left\{ \begin{bmatrix} 1 \\ 0 \\ -1 \end{bmatrix}, \begin{bmatrix} -1 \\ 3 \\ 2 \end{bmatrix} \right\}, \mathbf{v} = \begin{bmatrix} 2 \\ r \\ -1 \end{bmatrix}$
18. $\mathcal{S} = \left\{ \begin{bmatrix} 1 \\ 2 \\ -1 \end{bmatrix}, \begin{bmatrix} -1 \\ -2 \\ 2 \end{bmatrix} \right\}, \mathbf{v} = \begin{bmatrix} 1 \\ r \\ 2 \end{bmatrix}$
19. $\mathcal{S} = \left\{ \begin{bmatrix} -1 \\ 2 \\ 2 \end{bmatrix}, \begin{bmatrix} 1 \\ -1 \\ 0 \end{bmatrix} \right\}, \mathbf{v} = \begin{bmatrix} 2 \\ r \\ -8 \end{bmatrix}$
20. $\mathcal{S} = \left\{ \begin{bmatrix} -1 \\ 1 \\ 1 \end{bmatrix}, \begin{bmatrix} 2 \\ -3 \\ 1 \end{bmatrix} \right\}, \mathbf{v} = \begin{bmatrix} r \\ 4 \\ 0 \end{bmatrix}$

In Exercises 21–28, a set of vectors in $\mathcal{R}^n$ is given. Determine whether this set is a generating set for $\mathcal{R}^n$.

21. $\left\{\begin{bmatrix}1\\-1\end{bmatrix}, \begin{bmatrix}-2\\2\end{bmatrix}\right\}$

22. $\left\{\begin{bmatrix}1\\-2\end{bmatrix}, \begin{bmatrix}-2\\1\end{bmatrix}\right\}$

23. $\left\{\begin{bmatrix}1\\-4\end{bmatrix}, \begin{bmatrix}3\\2\end{bmatrix}, \begin{bmatrix}-2\\8\end{bmatrix}\right\}$

24. $\left\{\begin{bmatrix}-2\\4\end{bmatrix}, \begin{bmatrix}1\\-2\end{bmatrix}, \begin{bmatrix}-3\\6\end{bmatrix}\right\}$

25. $\left\{\begin{bmatrix}1\\0\\-2\end{bmatrix}, \begin{bmatrix}-1\\1\\4\end{bmatrix}, \begin{bmatrix}1\\2\\-2\end{bmatrix}\right\}$

26. $\left\{\begin{bmatrix}-1\\2\\1\end{bmatrix}, \begin{bmatrix}-1\\1\\3\end{bmatrix}, \begin{bmatrix}1\\-3\\1\end{bmatrix}\right\}$

27. $\left\{\begin{bmatrix}-1\\1\\2\end{bmatrix}, \begin{bmatrix}0\\-1\\2\end{bmatrix}, \begin{bmatrix}3\\-7\\2\end{bmatrix}, \begin{bmatrix}-5\\7\\6\end{bmatrix}\right\}$

28. $\left\{\begin{bmatrix}-1\\3\\0\end{bmatrix}, \begin{bmatrix}0\\1\\1\end{bmatrix}, \begin{bmatrix}2\\-1\\5\end{bmatrix}, \begin{bmatrix}2\\-1\\1\end{bmatrix}\right\}$

In Exercises 29–36, an $m \times n$ matrix A is given. Determine whether the equation $A\mathbf{x} = \mathbf{b}$ is consistent for every $\mathbf{b}$ in $\mathcal{R}^m$.

29. $\begin{bmatrix}1 & 0\\-2 & 1\end{bmatrix}$

30. $\begin{bmatrix}1 & -2\\2 & -4\end{bmatrix}$

31. $\begin{bmatrix}1 & 0 & -3\\-1 & 0 & 3\end{bmatrix}$

32. $\begin{bmatrix}1 & 1 & 2\\-1 & -3 & 4\end{bmatrix}$

33. $\begin{bmatrix}1 & -1\\0 & 1\\-2 & 2\end{bmatrix}$

34. $\begin{bmatrix}1 & 0 & -1\\2 & -1 & 1\\0 & 3 & -2\\1 & 1 & -3\end{bmatrix}$

35. $\begin{bmatrix}1 & 2 & 3\\2 & 3 & 4\\3 & 4 & 6\end{bmatrix}$

36. $\begin{bmatrix}1 & 0 & 2 & 1\\2 & 1 & 3 & 2\\3 & 4 & 4 & 5\end{bmatrix}$

In Exercises 37–44, a set $\mathcal{S}$ of vectors in $\mathcal{R}^n$ is given. Find a subset of $\mathcal{S}$ with the same span as $\mathcal{S}$ that is as small as possible.

37. $\left\{\begin{bmatrix}1\\3\end{bmatrix}, \begin{bmatrix}0\\1\end{bmatrix}\right\}$

38. $\left\{\begin{bmatrix}-1\\1\end{bmatrix}, \begin{bmatrix}2\\-2\end{bmatrix}, \begin{bmatrix}1\\0\end{bmatrix}\right\}$

39. $\left\{\begin{bmatrix}1\\0\\-1\end{bmatrix}, \begin{bmatrix}-2\\0\\2\end{bmatrix}, \begin{bmatrix}0\\1\\0\end{bmatrix}\right\}$

40. $\left\{\begin{bmatrix}1\\-1\\2\end{bmatrix}, \begin{bmatrix}2\\-3\\0\end{bmatrix}, \begin{bmatrix}0\\0\\0\end{bmatrix}\right\}$

41. $\left\{\begin{bmatrix}1\\-2\\1\end{bmatrix}, \begin{bmatrix}-2\\4\\-2\end{bmatrix}, \begin{bmatrix}0\\0\\0\end{bmatrix}\right\}$

42. $\left\{\begin{bmatrix}1\\0\\1\end{bmatrix}, \begin{bmatrix}1\\1\\0\end{bmatrix}, \begin{bmatrix}0\\1\\1\end{bmatrix}\right\}$

43. $\left\{\begin{bmatrix}-1\\0\\1\end{bmatrix}, \begin{bmatrix}0\\1\\2\end{bmatrix}, \begin{bmatrix}1\\2\\3\end{bmatrix}\right\}$

44. $\left\{\begin{bmatrix}1\\0\\0\end{bmatrix}, \begin{bmatrix}1\\1\\0\end{bmatrix}, \begin{bmatrix}1\\1\\1\end{bmatrix}, \begin{bmatrix}0\\0\\1\end{bmatrix}\right\}$

T&F *In Exercises 45–64, determine whether the statements are true or false.*

45. Let $\mathcal{S} = \{\mathbf{u}_1, \mathbf{u}_2, \ldots, \mathbf{u}_k\}$ be a nonempty set of vectors in $\mathcal{R}^n$. A vector $\mathbf{v}$ belongs to the span of $\mathcal{S}$ if and only if $\mathbf{v} = c_1\mathbf{u}_1 + c_2\mathbf{u}_2 + \cdots + c_k\mathbf{u}_k$ for some scalars $c_1, c_2, \ldots, c_k$.
46. The span of $\{\mathbf{0}\}$ is $\{\mathbf{0}\}$.
47. If $A = [\mathbf{u}_1\ \mathbf{u}_2\ \ldots\ \mathbf{u}_k]$ and the matrix equation $A\mathbf{x} = \mathbf{v}$ is inconsistent, then $\mathbf{v}$ does not belong to the span of $\{\mathbf{u}_1, \mathbf{u}_2, \ldots, \mathbf{u}_k\}$.
48. If A is an $m \times n$ matrix, then $A\mathbf{x} = \mathbf{b}$ is consistent for every $\mathbf{b}$ in $\mathcal{R}^m$ if and only if the rank of A is n.
49. Let $\mathcal{S} = \{\mathbf{u}_1, \mathbf{u}_2, \ldots, \mathbf{u}_k\}$ be a subset of $\mathcal{R}^n$. Then the span of $\mathcal{S}$ is $\mathcal{R}^n$ if and only if the rank of $[\mathbf{u}_1\ \mathbf{u}_2\ \ldots\ \mathbf{u}_k]$ is n.
50. Every finite subset of $\mathcal{R}^n$ is contained in its span.
51. If $\mathcal{S}_1$ and $\mathcal{S}_2$ are finite subsets of $\mathcal{R}^n$ such that $\mathcal{S}_1$ is contained in Span $\mathcal{S}_2$, then Span $\mathcal{S}_1$ is contained in Span $\mathcal{S}_2$.
52. If $\mathcal{S}_1$ and $\mathcal{S}_2$ are finite subsets of $\mathcal{R}^n$ having equal spans, then $\mathcal{S}_1 = \mathcal{S}_2$.
53. If $\mathcal{S}_1$ and $\mathcal{S}_2$ are finite subsets of $\mathcal{R}^n$ having equal spans, then $\mathcal{S}_1$ and $\mathcal{S}_2$ contain the same number of vectors.
54. Let $\mathcal{S}$ be a nonempty set of vectors in $\mathcal{R}^n$, and let $\mathbf{v}$ be in $\mathcal{R}^n$. The spans of $\mathcal{S}$ and $\mathcal{S} \cup \{\mathbf{v}\}$ are equal if and only if $\mathbf{v}$ is in $\mathcal{S}$.
55. The span of a set of two nonparallel vectors in $\mathcal{R}^2$ is $\mathcal{R}^2$.
56. The span of any finite nonempty subset of $\mathcal{R}^n$ contains the zero vector.
57. If $\mathbf{v}$ belongs to the span of S, so does $c\mathbf{v}$ for every scalar c.
58. If $\mathbf{u}$ and $\mathbf{v}$ belong to the span of S, so does $\mathbf{u} + \mathbf{v}$.
59. The span of $\{\mathbf{v}\}$ consists of every multiple of $\mathbf{v}$.
60. If S is a generating set for $\mathcal{R}^m$ that contains k vectors, then $k \geq m$.
61. If A is an $m \times n$ matrix whose reduced row echelon form contains no zero rows, then the columns of A form a generating set for $\mathcal{R}^m$.
62. If the columns of an $n \times n$ matrix A form a generating set for $\mathcal{R}^n$, then the reduced row echelon form of A is I_n.
63. If A is an $m \times n$ matrix such that $A\mathbf{x} = \mathbf{b}$ is inconsistent for some $\mathbf{b}$ in $\mathcal{R}^m$, then rank $A < m$.
64. If S_1 is contained in a finite set S_2 and S_1 is a generating set for $\mathcal{R}^m$, then S_2 is also a generating set for $\mathcal{R}^m$.

65. Let $\mathbf{u}_1 = \begin{bmatrix} -1 \\ 3 \end{bmatrix}$ and $\mathbf{u}_2 = \begin{bmatrix} 1 \\ -2 \end{bmatrix}$.
    (a) How many vectors are in $\{\mathbf{u}_1, \mathbf{u}_2\}$?
    (b) How many vectors are in the span of $\{\mathbf{u}_1, \mathbf{u}_2\}$?
66. Give three different generating sets for the set of vectors that lie in the xy-plane of $\mathcal{R}^3$.
67. Let A be an $m \times n$ matrix with $m > n$. Explain why $A\mathbf{x} = \mathbf{b}$ is inconsistent for some $\mathbf{b}$ in $\mathcal{R}^m$.
68. What can be said about the number of vectors in a generating set for $\mathcal{R}^m$? Explain your answer.
69. Let $\mathcal{S}_1$ and $\mathcal{S}_2$ be finite subsets of $\mathcal{R}^n$ such that $\mathcal{S}_1$ is contained in $\mathcal{S}_2$. Prove that if $\mathcal{S}_1$ is a generating set for $\mathcal{R}^n$, then so is $\mathcal{S}_2$.
70. Let $\mathbf{u}$ and $\mathbf{v}$ be any vectors in $\mathcal{R}^n$. Prove that the spans of $\{\mathbf{u}, \mathbf{v}\}$ and $\{\mathbf{u}+\mathbf{v}, \mathbf{u}-\mathbf{v}\}$ are equal.
71. Let $\mathbf{u}_1, \mathbf{u}_2, \ldots, \mathbf{u}_k$ be vectors in $\mathcal{R}^n$ and $c_1, c_2, \ldots, c_k$ be nonzero scalars. Prove that Span $\{\mathbf{u}_1, \mathbf{u}_2, \ldots, \mathbf{u}_k\}$ = Span $\{c_1\mathbf{u}_1, c_2\mathbf{u}_2, \ldots, c_k\mathbf{u}_k\}$.
72. Let $\mathbf{u}_1, \mathbf{u}_2, \ldots, \mathbf{u}_k$ be vectors in $\mathcal{R}^n$ and c be a scalar. Prove that the span of $\{\mathbf{u}_1, \mathbf{u}_2, \ldots, \mathbf{u}_k\}$ is equal to the span of $\{\mathbf{u}_1 + c\mathbf{u}_2, \mathbf{u}_2, \ldots, \mathbf{u}_k\}$.
73. Let R be the reduced row echelon form of an $m \times n$ matrix A. Is the span of the columns of R equal to the span of the columns of A? Justify your answer.
74. Let $\mathcal{S}_1$ and $\mathcal{S}_2$ be finite subsets of $\mathcal{R}^n$ such that $\mathcal{S}_1$ is contained in $\mathcal{S}_2$. Use only the definition of *span* to prove that Span $\mathcal{S}_1$ is contained in Span $\mathcal{S}_2$.
75. Let $\mathcal{S}$ be a finite subset of $\mathcal{R}^n$. Prove that if $\mathbf{u}$ and $\mathbf{v}$ are in the span of $\mathcal{S}$, then so is $\mathbf{u} + c\mathbf{v}$ for any scalar c.
76. Let V be the span of a finite subset of $\mathcal{R}^n$. Show that either $V = \{\mathbf{0}\}$ or V contains infinitely many vectors.
77. Let B be a matrix obtained from A by performing a single elementary row operation on A. Prove that the span of the rows of B equals the span of the rows of A. *Hint:* Use Exercises 71 and 72.
78. Prove that every linear combination of the rows of A can be written as a linear combination of the rows of the reduced row echelon form of A. *Hint:* Use Exercise 77.

In Exercises 79–82, use either a calculator with matrix capabilities or computer software such as MATLAB to determine whether each given vector is in the span of

$$\left\{ \begin{bmatrix} 1.2 \\ -0.1 \\ 2.3 \\ 3.1 \\ -1.1 \\ -1.9 \end{bmatrix}, \begin{bmatrix} 3.4 \\ -1.7 \\ 0.0 \\ 2.4 \\ 1.7 \\ 2.6 \end{bmatrix}, \begin{bmatrix} -3.1 \\ 0.0 \\ 2.5 \\ 1.6 \\ -3.2 \\ 1.7 \end{bmatrix}, \begin{bmatrix} 7.7 \\ -1.8 \\ -0.2 \\ 3.9 \\ 3.8 \\ -1.0 \end{bmatrix} \right\}.$$

79. $\begin{bmatrix} 1.0 \\ -1.5 \\ -4.6 \\ -3.8 \\ 3.9 \\ 6.4 \end{bmatrix}$
80. $\begin{bmatrix} -2.6 \\ -1.8 \\ 7.3 \\ 8.7 \\ -5.8 \\ 4.1 \end{bmatrix}$
81. $\begin{bmatrix} 1.5 \\ -1.6 \\ 2.4 \\ 4.0 \\ -1.5 \\ 4.3 \end{bmatrix}$
82. $\begin{bmatrix} -4.1 \\ 1.5 \\ 7.1 \\ 5.4 \\ -7.1 \\ -4.7 \end{bmatrix}$

SOLUTIONS TO THE PRACTICE PROBLEMS

1. Let A be the matrix whose columns are the vectors of $\mathcal{S}$. Then $\mathbf{u}$ is in the span of $\mathcal{S}$ if and only if $A\mathbf{x} = \mathbf{u}$ is consistent. Because the reduced row echelon form of $[A\ \ \mathbf{u}]$ is I_3, this system is inconsistent. Hence $\mathbf{u}$ is not in the span of $\mathcal{S}$. On the other hand, the reduced row echelon form of $[A\ \ \mathbf{v}]$ is

$$\begin{bmatrix} 1 & 0 & 2 \\ 0 & 1 & 3 \\ 0 & 0 & 0 \end{bmatrix}.$$

Thus $A\mathbf{x} = \mathbf{v}$ is consistent, and so $\mathbf{v}$ is in the span of $\mathcal{S}$. In fact, the reduced row echelon form of $[A\ \ \mathbf{v}]$ shows that

$$2\begin{bmatrix} 2 \\ -1 \\ 1 \end{bmatrix} + 3\begin{bmatrix} -1 \\ 1 \\ 0 \end{bmatrix} = \begin{bmatrix} 1 \\ 1 \\ 2 \end{bmatrix}.$$

2. Let A be the matrix whose columns are the vectors in $\mathcal{S}$. The reduced row echelon form of A is

$$\begin{bmatrix} 1 & 0 & 0 & 9 \\ 0 & 1 & 0 & 5 \\ 0 & 0 & 1 & -2 \end{bmatrix}.$$

Thus the rank of A is 3, so $\mathcal{S}$ is a generating set for $\mathcal{R}^3$ by Theorem 1.6.

3. From the reduced row echelon form of A in Practice Problem 2, we see that the last column of A is a linear combination of the first three columns. Thus the vector $\begin{bmatrix} 2 \\ -1 \\ 1 \end{bmatrix}$ can be removed from $\mathcal{S}$ without changing its span. So

$$\mathcal{S}' = \left\{ \begin{bmatrix} 1 \\ 0 \\ 1 \end{bmatrix}, \begin{bmatrix} -1 \\ 1 \\ 2 \end{bmatrix}, \begin{bmatrix} 1 \\ 3 \\ 9 \end{bmatrix} \right\}$$

is a subset of $\mathcal{S}$ that is a generating set for $\mathcal{R}^3$. Moreover, this set is the smallest generating set possible because removing any vector from $\mathcal{S}'$ leaves a set of only 2 vectors. Since the matrix whose columns are the vectors in $\mathcal{S}'$ is a 3×2 matrix, it cannot have rank 3 and so cannot be a generating set for $\mathcal{R}^3$ by Theorem 1.6.

1.7 LINEAR DEPENDENCE AND LINEAR INDEPENDENCE

In Section 1.6, we saw that it is possible to reduce the size of a generating set if some vector in the generating set is a linear combination of the others. In fact, by Theorem 1.7, this vector can be removed without affecting the span. In this section, we consider the problem of recognizing when a generating set cannot be made smaller. Consider, for example, the set $\mathcal{S} = \{\mathbf{u}_1, \mathbf{u}_2, \mathbf{u}_3, \mathbf{u}_4\}$, where

$$\mathbf{u}_1 = \begin{bmatrix} 1 \\ -1 \\ 2 \\ 1 \end{bmatrix}, \mathbf{u}_2 = \begin{bmatrix} 2 \\ 1 \\ -1 \\ -1 \end{bmatrix}, \mathbf{u}_3 = \begin{bmatrix} -1 \\ -8 \\ 13 \\ 8 \end{bmatrix}, \quad \text{and} \quad \mathbf{u}_4 = \begin{bmatrix} 0 \\ 1 \\ -2 \\ 1 \end{bmatrix}.$$

In this case, the reader should check that $\mathbf{u}_4$ is not a linear combination of the vectors $\mathbf{u}_1$, $\mathbf{u}_2$, and $\mathbf{u}_3$. However, this does *not* mean that we cannot find a smaller set having the same span as $\mathcal{S}$ because it is possible that one of $\mathbf{u}_1$, $\mathbf{u}_2$, and $\mathbf{u}_3$ might be a linear combination of the other vectors in $\mathcal{S}$. In fact, this is precisely the situation because

$$\mathbf{u}_3 = 5\mathbf{u}_1 - 3\mathbf{u}_2 + 0\mathbf{u}_4.$$

Thus checking if one of the vectors in a generating set is a linear combination of the others could require us to solve many systems of linear equations. Fortunately, a better method is available.

In the preceding example, in order that we do not have to guess which of $\mathbf{u}_1$, $\mathbf{u}_2$, $\mathbf{u}_3$, and $\mathbf{u}_4$ can be expressed as a linear combination of the others, let us formulate the problem differently. Note that because $\mathbf{u}_3 = 5\mathbf{u}_1 - 3\mathbf{u}_2 + 0\mathbf{u}_4$, we must have

$$-5\mathbf{u}_1 + 3\mathbf{u}_2 + \mathbf{u}_3 - 0\mathbf{u}_4 = \mathbf{0}.$$

Thus, instead of trying to write some $\mathbf{u}_i$ as a linear combination of the others, we can try to write $\mathbf{0}$ as a linear combination of $\mathbf{u}_1$, $\mathbf{u}_2$, $\mathbf{u}_3$, and $\mathbf{u}_4$. Of course, this is always possible if we take each coefficient in the linear combination to be 0. But *if there is a linear combination of* $\mathbf{u}_1$, $\mathbf{u}_2$, $\mathbf{u}_3$, *and* $\mathbf{u}_4$ *that equals* $\mathbf{0}$ *in which not all of the coefficients are* 0, then we can express one of the $\mathbf{u}_i$'s *as a linear combination of the others.* In this case, the equation $-5\mathbf{u}_1 + 3\mathbf{u}_2 + \mathbf{u}_3 - 0\mathbf{u}_4 = \mathbf{0}$ enables us to express any one of $\mathbf{u}_1$, $\mathbf{u}_2$, and $\mathbf{u}_3$ (but *not* $\mathbf{u}_4$) as a linear combination of the others. For example, since $-5\mathbf{u}_1 + 3\mathbf{u}_2 + \mathbf{u}_3 - 0\mathbf{u}_4 = \mathbf{0}$, we have

$$\begin{aligned} -5\mathbf{u}_1 &= -3\mathbf{u}_2 - \mathbf{u}_3 + 0\mathbf{u}_4 \\ \mathbf{u}_1 &= \frac{3}{5}\mathbf{u}_2 + \frac{1}{5}\mathbf{u}_3 + 0\mathbf{u}_4. \end{aligned}$$

We see that at least one of the vectors *depends* on (is a linear combination of) the others. This idea motivates the following definitions.

Definitions A set of k vectors $\{\mathbf{u}_1, \mathbf{u}_2, \ldots, \mathbf{u}_k\}$ in $\mathcal{R}^n$ is called **linearly dependent** if there exist scalars $c_1, c_2, \ldots, c_k$, not all 0, such that

$$c_1\mathbf{u}_1 + c_2\mathbf{u}_2 + \cdots + c_k\mathbf{u}_k = \mathbf{0}.$$

In this case, we also say that **the vectors $\mathbf{u}_1, \mathbf{u}_2, \ldots, \mathbf{u}_k$ are linearly dependent**.

A set of k vectors $\{\mathbf{u}_1, \mathbf{u}_2, \ldots, \mathbf{u}_k\}$ is called **linearly independent** if the only scalars $c_1, c_2, \ldots, c_k$ such that

$$c_1\mathbf{u}_1 + c_2\mathbf{u}_2 + \cdots + c_k\mathbf{u}_k = \mathbf{0}$$

are $c_1 = c_2 = \cdots = c_k = 0$. In this case, we also say that **the vectors $\mathbf{u}_1, \mathbf{u}_2, \ldots, \mathbf{u}_k$ are linearly independent**.

Note that a set is linearly independent if and only if it is not linearly dependent.

Example 1 Show that the sets

$$\mathcal{S}_1 = \left\{ \begin{bmatrix} 2 \\ 3 \end{bmatrix}, \begin{bmatrix} 5 \\ 8 \end{bmatrix}, \begin{bmatrix} 1 \\ 2 \end{bmatrix} \right\} \quad \text{and} \quad \mathcal{S}_2 = \left\{ \begin{bmatrix} 0 \\ 0 \end{bmatrix}, \begin{bmatrix} 1 \\ 0 \end{bmatrix}, \begin{bmatrix} 0 \\ 1 \end{bmatrix} \right\}$$

are linearly dependent.

Solution The equation

$$c_1 \begin{bmatrix} 2 \\ 3 \end{bmatrix} + c_2 \begin{bmatrix} 5 \\ 8 \end{bmatrix} + c_3 \begin{bmatrix} 1 \\ 2 \end{bmatrix} = \begin{bmatrix} 0 \\ 0 \end{bmatrix}$$

is true with $c_1 = 2$, $c_2 = -1$, and $c_3 = 1$. Since not all the coefficients in the preceding linear combination are 0, $\mathcal{S}_1$ is linearly dependent.

Because

$$1 \begin{bmatrix} 0 \\ 0 \end{bmatrix} + 0 \begin{bmatrix} 1 \\ 0 \end{bmatrix} + 0 \begin{bmatrix} 0 \\ 1 \end{bmatrix} = \begin{bmatrix} 0 \\ 0 \end{bmatrix}$$

and at least one of the coefficients in this linear combination is nonzero, $\mathcal{S}_2$ is also linearly dependent.

As Example 1 suggests, *any finite subset $\mathcal{S} = \{\mathbf{0}, \mathbf{u}_1, \mathbf{u}_2, \ldots, \mathbf{u}_k\}$ of $\mathcal{R}^n$ that contains the zero vector is linearly dependent* because

$$1 \cdot \mathbf{0} + 0\mathbf{u}_1 + 0\mathbf{u}_2 + \cdots + 0\mathbf{u}_k = \mathbf{0}$$

is a linear combination of the vectors in $\mathcal{S}$ in which at least one coefficient is nonzero.

! CAUTION While the equation

$$0 \begin{bmatrix} 2 \\ 3 \end{bmatrix} + 0 \begin{bmatrix} 5 \\ 8 \end{bmatrix} + 0 \begin{bmatrix} 1 \\ 2 \end{bmatrix} = \begin{bmatrix} 0 \\ 0 \end{bmatrix}$$

is true, it tells us nothing about the linear independence or dependence of the set $\mathcal{S}_1$ in Example 1. A similar statement is true for *any* set of vectors $\{\mathbf{u}_1, \mathbf{u}_2, \ldots, \mathbf{u}_k\}$:

$$0\mathbf{u}_1 + 0\mathbf{u}_2 + \cdots + 0\mathbf{u}_k = \mathbf{0}.$$

For a set of vectors to be linearly dependent, the equation

$$c_1\mathbf{u}_1 + c_2\mathbf{u}_2 + \cdots + c_k\mathbf{u}_k = \mathbf{0}$$

must be satisfied with at least one *nonzero* coefficient.

Since the equation $c_1\mathbf{u}_1 + c_2\mathbf{u}_2 + \cdots + c_k\mathbf{u}_k = \mathbf{0}$ can be written as a matrix–vector product

$$[\mathbf{u}_1 \ \mathbf{u}_2 \ \ldots \ \mathbf{u}_k]\begin{bmatrix} c_1 \\ c_2 \\ \vdots \\ c_k \end{bmatrix} = \mathbf{0},$$

we have the following useful observation:

> The set $\{\mathbf{u}_1, \mathbf{u}_2, \ldots, \mathbf{u}_k\}$ is linearly dependent if and only if there exists a nonzero solution of $A\mathbf{x} = \mathbf{0}$, where $A = [\mathbf{u}_1 \ \mathbf{u}_2 \ \ldots \ \mathbf{u}_k]$.

Example 2 Determine whether the set

$$\mathcal{S} = \left\{ \begin{bmatrix} 1 \\ 2 \\ 1 \end{bmatrix}, \begin{bmatrix} 1 \\ 0 \\ 1 \end{bmatrix}, \begin{bmatrix} 1 \\ 4 \\ 1 \end{bmatrix}, \begin{bmatrix} 1 \\ 2 \\ 3 \end{bmatrix} \right\}$$

is linearly dependent or linearly independent.

Solution We must determine whether $A\mathbf{x} = \mathbf{0}$ has a nonzero solution, where

$$A = \begin{bmatrix} 1 & 1 & 1 & 1 \\ 2 & 0 & 4 & 2 \\ 1 & 1 & 1 & 3 \end{bmatrix}$$

is the matrix whose columns are the vectors in $\mathcal{S}$. The augmented matrix of $A\mathbf{x} = \mathbf{0}$ is

$$\begin{bmatrix} 1 & 1 & 1 & 1 & 0 \\ 2 & 0 & 4 & 2 & 0 \\ 1 & 1 & 1 & 3 & 0 \end{bmatrix},$$

and its reduced row echelon form is

$$\begin{bmatrix} 1 & 0 & 2 & 0 & 0 \\ 0 & 1 & -1 & 0 & 0 \\ 0 & 0 & 0 & 1 & 0 \end{bmatrix}.$$

Hence the general solution of this system is

$$\begin{aligned} x_1 &= -2x_3 \\ x_2 &= x_3 \\ x_3 & \quad \text{free} \\ x_4 &= 0. \end{aligned}$$

Because the solution of $A\mathbf{x} = \mathbf{0}$ contains a free variable, this system of linear equations has infinitely many solutions, and we can obtain a nonzero solution by choosing any nonzero value of the free variable. Taking $\mathbf{x}_3 = 1$, for instance, we see that

$$\begin{bmatrix} x_1 \\ x_2 \\ x_3 \\ x_4 \end{bmatrix} = \begin{bmatrix} -2 \\ 1 \\ 1 \\ 0 \end{bmatrix}$$

is a nonzero solution of $A\mathbf{x} = \mathbf{0}$. Thus $\mathcal{S}$ is a linearly dependent subset of $\mathcal{R}^3$ since

$$-2\begin{bmatrix}1\\2\\1\end{bmatrix} + 1\begin{bmatrix}1\\0\\1\end{bmatrix} + 1\begin{bmatrix}1\\4\\1\end{bmatrix} + 0\begin{bmatrix}1\\2\\3\end{bmatrix} = \begin{bmatrix}0\\0\\0\end{bmatrix}$$

is a representation of $\mathbf{0}$ as a linear combination of the vectors in $\mathcal{S}$.

Example 3 Determine whether the set

$$\mathcal{S} = \left\{\begin{bmatrix}1\\2\\1\end{bmatrix}, \begin{bmatrix}2\\2\\3\end{bmatrix}, \begin{bmatrix}1\\0\\1\end{bmatrix}\right\}$$

is linearly dependent or linearly independent.

Solution As in Example 2, we must check whether $A\mathbf{x} = \mathbf{0}$ has a nonzero solution, where

$$A = \begin{bmatrix}1 & 2 & 1\\2 & 2 & 0\\1 & 3 & 1\end{bmatrix}.$$

There is a way to do this without actually solving $A\mathbf{x} = \mathbf{0}$ (as we did in Example 2). Note that the system $A\mathbf{x} = \mathbf{0}$ has nonzero solutions if and only if its general solution contains a free variable. Since the reduced row echelon form of A is

$$\begin{bmatrix}1 & 0 & 0\\0 & 1 & 0\\0 & 0 & 1\end{bmatrix},$$

the rank of A is 3, and the nullity of A is $3 - 3 = 0$. Thus the general solution of $A\mathbf{x} = \mathbf{0}$ has no free variables. So $A\mathbf{x} = \mathbf{0}$ has no nonzero solutions, and hence $\mathcal{S}$ is linearly independent.

In Example 3, we showed that a particular set $\mathcal{S}$ is linearly independent without actually solving a system of linear equations. Our next theorem shows that a similar technique can be used for any set whatsoever. Note the relationship between this theorem and Theorem 1.6.

THEOREM 1.8

The following statements about an $m \times n$ matrix A are equivalent:

(a) The columns of A are linearly independent.

(b) The equation $A\mathbf{x} = \mathbf{b}$ has at most one solution for each $\mathbf{b}$ in $\mathcal{R}^m$.

(c) The nullity of A is zero.

(d) The rank of A is n, the number of columns of A.

(e) The columns of the reduced row echelon form of A are distinct standard vectors in $\mathcal{R}^m$.

(f) The only solution of $A\mathbf{x} = \mathbf{0}$ is $\mathbf{0}$.

(g) There is a pivot position in each column of A.

PROOF We have already noted that (a) and (f) are equivalent, and clearly (f) and (g) are equivalent. To complete the proof, we show that (b) implies (c), (c) implies (d), (d) implies (e), (e) implies (f), and (f) implies (b).

(b) *implies* (c) Since $\mathbf{0}$ is a solution of $A\mathbf{x} = \mathbf{0}$, (b) implies that $A\mathbf{x} = \mathbf{0}$ has no nonzero solutions. Thus the general solution of $A\mathbf{x} = \mathbf{0}$ has no free variables. Since the number of free variables is the nullity of A, we see that the nullity of A is zero.

(c) *implies* (d) Because $\text{rank}\, A + \text{nullity}\, A = n$, (d) follows immediately from (c).

(d) *implies* (e) If the rank of A is n, then every column of A is a pivot column, and therefore the reduced row echelon form of A consists entirely of standard vectors. These are necessarily distinct because each column contains the first nonzero entry in some row.

(e) *implies* (f) Let R be the reduced row echelon form of A. If the columns of R are distinct standard vectors in $\mathcal{R}^m$, then $R = [\mathbf{e}_1\ \mathbf{e}_2\ \ldots\ \mathbf{e}_n]$. Clearly, the only solution of $R\mathbf{x} = \mathbf{0}$ is $\mathbf{0}$, and since $A\mathbf{x} = \mathbf{0}$ is equivalent to $R\mathbf{x} = \mathbf{0}$, it follows that the only solution of $A\mathbf{x} = \mathbf{0}$ is $\mathbf{0}$.

(f) *implies* (b) Let $\mathbf{b}$ be any vector in $\mathcal{R}^m$. To show that $A\mathbf{x} = \mathbf{b}$ has at most one solution, we assume that $\mathbf{u}$ and $\mathbf{v}$ are both solutions of $A\mathbf{x} = \mathbf{b}$ and prove that $\mathbf{u} = \mathbf{v}$. Since $\mathbf{u}$ and $\mathbf{v}$ are solutions of $A\mathbf{x} = \mathbf{b}$, we have

$$A(\mathbf{u} - \mathbf{v}) = A\mathbf{u} - A\mathbf{v} = \mathbf{b} - \mathbf{b} = \mathbf{0}.$$

So $\mathbf{u} - \mathbf{v}$ is a solution of $A\mathbf{x} = \mathbf{0}$. Thus (f) implies that $\mathbf{u} - \mathbf{v} = \mathbf{0}$; that is, $\mathbf{u} = \mathbf{v}$. It follows that $A\mathbf{x} = \mathbf{b}$ has at most one solution. ■

Practice Problem 1 ▶ Is some vector in the set

$$\mathcal{S} = \left\{ \begin{bmatrix} -1 \\ 0 \\ 2 \\ 1 \end{bmatrix}, \begin{bmatrix} 1 \\ 1 \\ -1 \\ -1 \end{bmatrix}, \begin{bmatrix} 0 \\ 2 \\ -1 \\ 1 \end{bmatrix}, \begin{bmatrix} -1 \\ 3 \\ 1 \\ 2 \end{bmatrix} \right\}$$

a linear combination of the others? ◀

The equation $A\mathbf{x} = \mathbf{b}$ is called **homogeneous** if $\mathbf{b} = \mathbf{0}$. As Examples 2 and 3 illustrate, in checking if a subset is linearly independent, we are led to a homogeneous equation. Note that, unlike an arbitrary equation, a homogeneous equation must be consistent because $\mathbf{0}$ is a solution of $A\mathbf{x} = \mathbf{0}$. As a result, the important question concerning a homogeneous equation is not *if* it has solutions, but whether $\mathbf{0}$ is the *only* solution. If not, then the system has infinitely many solutions. For example, the general solution of a homogeneous system of linear equations with more variables than equations must have free variables. Hence *a homogeneous system of linear equations with more variables than equations has infinitely many solutions.* According to Theorem 1.8, the number of solutions of $A\mathbf{x} = \mathbf{0}$ determines the linear dependence or independence of the columns of A.

In order to investigate some other properties of the homogeneous equation $A\mathbf{x} = \mathbf{0}$, let us consider this equation for the matrix

$$A = \begin{bmatrix} 1 & -4 & 2 & -1 & 2 \\ 2 & -8 & 3 & 2 & -1 \end{bmatrix}.$$

Since the reduced row echelon form of $[A\ \ \mathbf{0}]$ is

$$\begin{bmatrix} 1 & -4 & 0 & 7 & -8 & 0 \\ 0 & 0 & 1 & -4 & 5 & 0 \end{bmatrix},$$

the general solution of $A\mathbf{x} = \mathbf{0}$ is

$$\begin{aligned} x_1 &= 4x_2 - 7x_4 + 8x_5 \\ x_2 &\quad \text{free} \\ x_3 &= \quad 4x_4 - 5x_5 \\ x_4 &\quad \text{free} \\ x_5 &\quad \text{free.} \end{aligned}$$

Expressing the solutions of $A\mathbf{x} = \mathbf{0}$ in vector form yields

$$\begin{bmatrix} x_1 \\ x_2 \\ x_3 \\ x_4 \\ x_5 \end{bmatrix} = \begin{bmatrix} 4x_2 - 7x_4 + 8x_5 \\ x_2 \\ 4x_4 - 5x_5 \\ x_4 \\ x_5 \end{bmatrix} = x_2 \begin{bmatrix} 4 \\ 1 \\ 0 \\ 0 \\ 0 \end{bmatrix} + x_4 \begin{bmatrix} -7 \\ 0 \\ 4 \\ 1 \\ 0 \end{bmatrix} + x_5 \begin{bmatrix} 8 \\ 0 \\ -5 \\ 0 \\ 1 \end{bmatrix}. \tag{11}$$

Thus the solution of $A\mathbf{x} = \mathbf{0}$ is the span of

$$\mathcal{S} = \left\{ \begin{bmatrix} 4 \\ 1 \\ 0 \\ 0 \\ 0 \end{bmatrix}, \begin{bmatrix} -7 \\ 0 \\ 4 \\ 1 \\ 0 \end{bmatrix}, \begin{bmatrix} 8 \\ 0 \\ -5 \\ 0 \\ 1 \end{bmatrix} \right\}.$$

In a similar manner, for a matrix A, we can express any solution of $A\mathbf{x} = \mathbf{0}$ as a linear combination of vectors in which the coefficients are the free variables in the general solution. We call such a representation a **vector form** of the general solution of $A\mathbf{x} = \mathbf{0}$. The solution set of this equation equals the span of the set of vectors that appear in a vector form of its general solution.

For the preceding set $\mathcal{S}$, we see from equation (11) that the only linear combination of vectors in $\mathcal{S}$ equal to $\mathbf{0}$ is the one in which all of the coefficients are zero. So $\mathcal{S}$ is linearly independent. More generally, the following result is true:

> When a vector form of the general solution of $A\mathbf{x} = \mathbf{0}$ is obtained by the method described in Section 1.3, the vectors that appear in the vector form are linearly independent.

Practice Problem 2 ▶ Determine a vector form for the general solution of

$$\begin{aligned} x_1 - 3x_2 - x_3 + x_4 - x_5 &= 0 \\ 2x_1 - 6x_2 + x_3 - 3x_4 - 9x_5 &= 0 \\ -2x_1 + 6x_2 + 3x_3 + 2x_4 + 11x_5 &= 0. \end{aligned}$$

◀

LINEARLY DEPENDENT AND LINEARLY INDEPENDENT SETS

The following result provides a useful characterization of linearly dependent sets. In Section 2.3, we develop a simple method for implementing Theorem 1.9 to write one of the vectors in a linearly dependent set as a linear combination of the preceding vectors.

THEOREM 1.9

Vectors $\mathbf{u}_1, \mathbf{u}_2, \ldots, \mathbf{u}_k$ in $\mathcal{R}^n$ are linearly dependent if and only if $\mathbf{u}_1 = \mathbf{0}$ or there exists an $i \geq 2$ such that $\mathbf{u}_i$ is a linear combination of the preceding vectors $\mathbf{u}_1, \mathbf{u}_2, \ldots, \mathbf{u}_{i-1}$.

PROOF Suppose first that the vectors $\mathbf{u}_1, \mathbf{u}_2, \ldots, \mathbf{u}_k$ in $\mathcal{R}^n$ are linearly dependent. If $\mathbf{u}_1 = \mathbf{0}$, then we are finished; so suppose $\mathbf{u}_1 \neq \mathbf{0}$. There exist scalars $c_1, c_2, \ldots, c_k$, not all zero, such that

$$c_1\mathbf{u}_1 + c_2\mathbf{u}_2 + \cdots + c_k\mathbf{u}_k = \mathbf{0}.$$

Let i denote the largest index such that $c_i \neq 0$. Note that $i \geq 2$, for otherwise the preceding equation would reduce to $c_1\mathbf{u}_1 = \mathbf{0}$, which is false because $c_1 \neq 0$ and $\mathbf{u}_1 \neq \mathbf{0}$. Hence the preceding equation becomes

$$c_1\mathbf{u}_1 + c_2\mathbf{u}_2 + \cdots + c_i\mathbf{u}_i = \mathbf{0},$$

where $c_i \neq 0$. Solving this equation for $\mathbf{u}_i$, we obtain

$$c_i\mathbf{u}_i = -c_1\mathbf{u}_1 - c_2\mathbf{u}_2 - \cdots - c_{i-1}\mathbf{u}_{i-1}$$

$$\mathbf{u}_i = \frac{-c_1}{c_i}\mathbf{u}_1 - \frac{c_2}{c_i}\mathbf{u}_2 - \cdots - \frac{c_{i-1}}{c_i}\mathbf{u}_{i-1}.$$

Thus $\mathbf{u}_i$ is a linear combination of $\mathbf{u}_1, \mathbf{u}_2, \ldots, \mathbf{u}_{i-1}$.

We leave the proof of the converse as an exercise. ■

The following properties relate to linearly dependent and linearly independent sets.

Properties of Linearly Dependent and Independent Sets

1. A set consisting of a single nonzero vector is linearly independent, but $\{\mathbf{0}\}$ is linearly dependent.
2. A set of two vectors $\{\mathbf{u}_1, \mathbf{u}_2\}$ is linearly dependent if and only if $\mathbf{u}_1 = \mathbf{0}$ or $\mathbf{u}_2$ is in the span of $\{\mathbf{u}_1\}$; that is, if and only if $\mathbf{u}_1 = \mathbf{0}$ or $\mathbf{u}_2$ is a multiple of $\mathbf{u}_1$. Hence *a set of two vectors is linearly dependent if and only if one of the vectors is a multiple of the other.*
3. Let $\mathcal{S} = \{\mathbf{u}_1, \mathbf{u}_2, \ldots, \mathbf{u}_k\}$ be a linearly independent subset of $\mathcal{R}^n$, and $\mathbf{v}$ be in $\mathcal{R}^n$. Then $\mathbf{v}$ does not belong to the span of $\mathcal{S}$ if and only if $\{\mathbf{u}_1, \mathbf{u}_2, \ldots, \mathbf{u}_k, \mathbf{v}\}$ is linearly independent.
4. Every subset of $\mathcal{R}^n$ containing more than n vectors must be linearly dependent.
5. If $\mathcal{S}$ is a subset of $\mathcal{R}^n$ and no vector can be removed from $\mathcal{S}$ without changing its span, then $\mathcal{S}$ is linearly independent.

! CAUTION The result mentioned in item 2 is valid *only* for sets containing two vectors. For example, in $\mathcal{R}^3$, the set $\{\mathbf{e}_1, \mathbf{e}_2, \mathbf{e}_1 + \mathbf{e}_2\}$ is linearly dependent, but no vector in the set is a multiple of another.

Properties 1, 2, and 5 follow from Theorem 1.9.

For a justification of property 3, observe that by Theorem 1.9, $\mathbf{u}_1 \neq \mathbf{0}$, and for $i \geq 2$, no $\mathbf{u}_i$ is in the span of $\{\mathbf{u}_1, \mathbf{u}_2, \ldots, \mathbf{u}_{i-1}\}$. If $\mathbf{v}$ does not belong to the span of $\mathcal{S}$, the vectors $\mathbf{u}_1, \mathbf{u}_2, \ldots, \mathbf{u}_k, \mathbf{v}$ are also linearly independent by Theorem 1.9. Conversely, if the vectors $\mathbf{u}_1, \mathbf{u}_2, \ldots, \mathbf{u}_k, \mathbf{v}$ are linearly independent, then $\mathbf{v}$ is not a linear combination of $\mathbf{u}_1, \mathbf{u}_2, \ldots, \mathbf{u}_k$ by Theorem 1.9. So $\mathbf{v}$ does not belong to the span of $\mathcal{S}$. (See Figure 1.27 for the case that $k = 2$.)

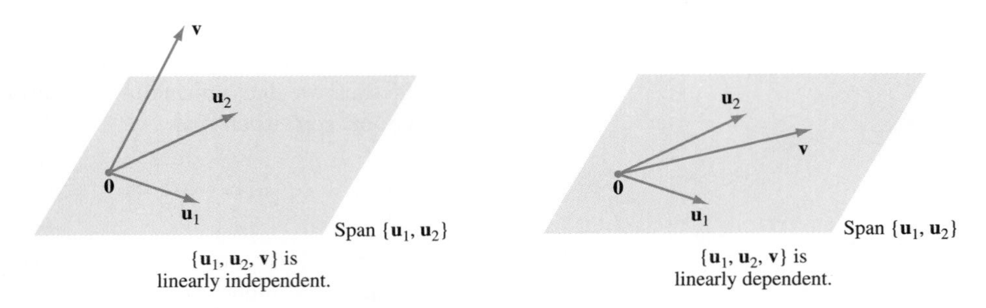

Figure 1.27 Linearly independent and linearly dependent sets of 3 vectors

To justify property 4, consider a set $\{\mathbf{u}_1, \mathbf{u}_2, \ldots, \mathbf{u}_k\}$ of k vectors from $\mathcal{R}^n$, where $k > n$. The $n \times k$ matrix $[\mathbf{u}_1 \ \mathbf{u}_2 \ \ldots \ \mathbf{u}_k]$ cannot have rank k because it has only n rows. Thus the set $\{\mathbf{u}_1, \mathbf{u}_2, \ldots, \mathbf{u}_k\}$ is linearly dependent by Theorem 1.8. However, the next example shows that subsets of $\mathcal{R}^n$ containing n or fewer vectors may be either linearly dependent or linearly independent.

Example 4 Determine by inspection whether the following sets are linearly dependent or linearly independent:

$$\mathcal{S}_1 = \left\{\begin{bmatrix} 3 \\ -1 \\ 7 \end{bmatrix}, \begin{bmatrix} 0 \\ 0 \\ 0 \end{bmatrix}, \begin{bmatrix} -2 \\ 5 \\ 1 \end{bmatrix}\right\}, \quad \mathcal{S}_2 = \left\{\begin{bmatrix} -4 \\ 12 \\ 6 \end{bmatrix}, \begin{bmatrix} -10 \\ 30 \\ 15 \end{bmatrix}\right\},$$

$$\mathcal{S}_3 = \left\{\begin{bmatrix} -3 \\ 7 \\ 0 \end{bmatrix}, \begin{bmatrix} 2 \\ 9 \\ 0 \end{bmatrix}, \begin{bmatrix} -1 \\ 0 \\ 2 \end{bmatrix}\right\}, \quad \text{and} \quad \mathcal{S}_4 = \left\{\begin{bmatrix} 2 \\ 0 \\ 1 \end{bmatrix}, \begin{bmatrix} -1 \\ 3 \\ 2 \end{bmatrix}, \begin{bmatrix} 1 \\ 1 \\ 1 \end{bmatrix}, \begin{bmatrix} 4 \\ -2 \\ 3 \end{bmatrix}\right\}$$

Solution Since $\mathcal{S}_1$ contains the zero vector, it is linearly dependent.

To determine if $\mathcal{S}_2$, a set of two vectors, is linearly dependent or linearly independent, we need only check if either of the vectors in $\mathcal{S}_2$ is a multiple of the other. Because

$$\frac{5}{2}\begin{bmatrix} -4 \\ 12 \\ 6 \end{bmatrix} = \begin{bmatrix} -10 \\ 30 \\ 15 \end{bmatrix},$$

we see that $\mathcal{S}_2$ is linearly dependent.

To see if $\mathcal{S}_3$ is linearly independent, consider the subset $\mathcal{S} = \{\mathbf{u}_1, \mathbf{u}_2\}$, where

$$\mathbf{u}_1 = \begin{bmatrix} -3 \\ 7 \\ 0 \end{bmatrix} \quad \text{and} \quad \mathbf{u}_2 = \begin{bmatrix} 2 \\ 9 \\ 0 \end{bmatrix}.$$

Because $\mathcal{S}$ is a set of two vectors, neither of which is a multiple of the other, $\mathcal{S}$ is linearly independent. Vectors in the span of $\mathcal{S}$ are linear combinations of the vectors in $\mathcal{S}$, and therefore must have 0 as their third component. Since

$$\mathbf{v} = \begin{bmatrix} -1 \\ 0 \\ 2 \end{bmatrix}$$

has a nonzero third component, it does not belong to the span of $\mathcal{S}$. So by property 3 in the preceding list, $\mathcal{S}_3 = \{\mathbf{u}_1, \mathbf{u}_2, \mathbf{v}\}$ is linearly independent.

Finally, the set $\mathcal{S}_4$ is linearly dependent by property 4 because it is a set of 4 vectors from $\mathcal{R}^3$.

Practice Problem 3 ▶ Determine by inspection whether the following sets are linearly dependent or linearly independent:

$$\mathcal{S}_1 = \left\{ \begin{bmatrix} 1 \\ -2 \\ 0 \end{bmatrix} \right\}, \quad \mathcal{S}_2 = \left\{ \begin{bmatrix} 3 \\ -1 \\ 2 \end{bmatrix}, \begin{bmatrix} 6 \\ -2 \\ 4 \end{bmatrix}, \begin{bmatrix} 1 \\ 2 \\ -1 \end{bmatrix} \right\},$$

$$\mathcal{S}_3 = \left\{ \begin{bmatrix} 1 \\ 3 \\ -2 \end{bmatrix}, \begin{bmatrix} 2 \\ 6 \\ -1 \end{bmatrix} \right\}, \quad \text{and} \quad \mathcal{S}_4 = \left\{ \begin{bmatrix} 1 \\ 0 \\ 1 \end{bmatrix}, \begin{bmatrix} -1 \\ 1 \\ 2 \end{bmatrix}, \begin{bmatrix} 2 \\ 1 \\ 3 \end{bmatrix}, \begin{bmatrix} 1 \\ -2 \\ 4 \end{bmatrix} \right\}$$

◀

In this chapter, we introduced matrices and vectors and learned some of their fundamental properties. Since we can write a system of linear equations as an equation involving a matrix and vectors, we can use these arrays to solve any system of linear equations. It is surprising that the number of solutions of the equation $A\mathbf{x} = \mathbf{b}$ is related both to the simple concept of the rank of a matrix and also to the complex concepts of generating sets and linearly independent sets. Yet this is exactly the case, as Theorems 1.6 and 1.8 show. To conclude this chapter, we present the following table, which summarizes the relationships among the ideas that were established in Sections 1.6 and 1.7. We assume that A is an $m \times n$ matrix with reduced row echelon form R. Properties listed in the same row of the table are equivalent.

The rank of A	The number of solutions of $A\mathbf{x} = \mathbf{b}$	The columns of A	The reduced row echelon form R of A
rank $A = m$	$A\mathbf{x} = \mathbf{b}$ has at least one solution for every $\mathbf{b}$ in $\mathcal{R}^m$.	The columns of A are a generating set for $\mathcal{R}^m$.	Every row of R contains a pivot position.
rank $A = n$	$A\mathbf{x} = \mathbf{b}$ has at most one solution for every $\mathbf{b}$ in $\mathcal{R}^m$.	The columns of A are linearly independent.	Every column of R contains a pivot position.

EXERCISES

In Exercises 1–12, determine by inspection whether the given sets are linearly dependent.

1. $\left\{\begin{bmatrix}1\\3\end{bmatrix},\begin{bmatrix}2\\6\end{bmatrix}\right\}$

2. $\left\{\begin{bmatrix}2\\-1\end{bmatrix},\begin{bmatrix}-1\\2\end{bmatrix}\right\}$

3. $\left\{\begin{bmatrix}1\\-3\\0\end{bmatrix},\begin{bmatrix}-2\\6\\0\end{bmatrix}\right\}$

4. $\left\{\begin{bmatrix}3\\-1\\2\end{bmatrix},\begin{bmatrix}0\\0\\0\end{bmatrix},\begin{bmatrix}-2\\5\\1\end{bmatrix}\right\}$

5. $\left\{\begin{bmatrix}0\\0\\-1\end{bmatrix},\begin{bmatrix}0\\2\\1\end{bmatrix},\begin{bmatrix}-3\\7\\2\end{bmatrix}\right\}$

6. $\left\{\begin{bmatrix}1\\-4\end{bmatrix},\begin{bmatrix}2\\3\end{bmatrix},\begin{bmatrix}-5\\6\end{bmatrix}\right\}$

7. $\left\{\begin{bmatrix}-3\\12\end{bmatrix},\begin{bmatrix}1\\-4\end{bmatrix}\right\}$

8. $\left\{\begin{bmatrix}4\\3\end{bmatrix},\begin{bmatrix}-2\\5\end{bmatrix},\begin{bmatrix}2\\1\end{bmatrix}\right\}$

9. $\left\{\begin{bmatrix}5\\3\end{bmatrix}\right\}$

10. $\left\{\begin{bmatrix}3\\7\end{bmatrix},\begin{bmatrix}0\\0\end{bmatrix},\begin{bmatrix}-1\\4\end{bmatrix}\right\}$

11. $\left\{\begin{bmatrix}1\\0\\1\end{bmatrix},\begin{bmatrix}0\\2\\0\end{bmatrix},\begin{bmatrix}1\\6\\1\end{bmatrix}\right\}$

12. $\left\{\begin{bmatrix}1\\-2\\0\end{bmatrix},\begin{bmatrix}1\\0\\1\end{bmatrix},\begin{bmatrix}3\\-1\\4\end{bmatrix},\begin{bmatrix}2\\0\\-1\end{bmatrix}\right\}$

In Exercises 13–22, a set $\mathcal{S}$ is given. Determine by inspection a subset of $\mathcal{S}$ containing the fewest vectors that has the same span as $\mathcal{S}$.

13. $\left\{\begin{bmatrix}1\\-2\\3\end{bmatrix},\begin{bmatrix}-2\\4\\-6\end{bmatrix}\right\}$

14. $\left\{\begin{bmatrix}1\\0\\2\end{bmatrix},\begin{bmatrix}3\\-1\\1\end{bmatrix}\right\}$

15. $\left\{\begin{bmatrix}-3\\2\\0\end{bmatrix},\begin{bmatrix}1\\6\\0\end{bmatrix},\begin{bmatrix}0\\0\\0\end{bmatrix}\right\}$

16. $\left\{\begin{bmatrix}0\\0\\1\end{bmatrix},\begin{bmatrix}0\\1\\2\end{bmatrix},\begin{bmatrix}1\\2\\3\end{bmatrix},\begin{bmatrix}2\\3\\4\end{bmatrix}\right\}$

17. $\left\{\begin{bmatrix}2\\-3\\5\end{bmatrix},\begin{bmatrix}4\\-6\\10\end{bmatrix},\begin{bmatrix}1\\0\\2\end{bmatrix}\right\}$

18. $\left\{\begin{bmatrix}1\\0\\-1\end{bmatrix},\begin{bmatrix}-3\\0\\3\end{bmatrix},\begin{bmatrix}5\\0\\-5\end{bmatrix},\begin{bmatrix}-6\\0\\6\end{bmatrix}\right\}$

19. $\left\{\begin{bmatrix}4\\3\end{bmatrix},\begin{bmatrix}-2\\5\end{bmatrix},\begin{bmatrix}2\\1\end{bmatrix}\right\}$

20. $\left\{\begin{bmatrix}1\\2\\-3\end{bmatrix},\begin{bmatrix}4\\-6\\2\end{bmatrix},\begin{bmatrix}-2\\3\\-1\end{bmatrix},\begin{bmatrix}-3\\-6\\9\end{bmatrix}\right\}$

21. $\left\{\begin{bmatrix}-2\\0\\3\end{bmatrix},\begin{bmatrix}0\\4\\0\end{bmatrix},\begin{bmatrix}-4\\1\\6\end{bmatrix}\right\}$

22. $\left\{\begin{bmatrix}2\\1\\0\end{bmatrix},\begin{bmatrix}3\\2\\1\end{bmatrix},\begin{bmatrix}5\\3\\1\end{bmatrix}\right\}$

In Exercises 23–30, determine whether the given set is linearly independent.

23. $\left\{\begin{bmatrix}1\\-1\\-2\end{bmatrix},\begin{bmatrix}-1\\0\\1\end{bmatrix},\begin{bmatrix}1\\2\\1\end{bmatrix}\right\}$

24. $\left\{\begin{bmatrix}1\\-1\\1\end{bmatrix},\begin{bmatrix}-1\\0\\2\end{bmatrix},\begin{bmatrix}2\\1\\1\end{bmatrix}\right\}$

25. $\left\{\begin{bmatrix}1\\2\\0\\-1\end{bmatrix},\begin{bmatrix}1\\-3\\1\\-2\end{bmatrix},\begin{bmatrix}1\\2\\-2\\3\end{bmatrix}\right\}$

26. $\left\{\begin{bmatrix}-1\\0\\1\\2\end{bmatrix},\begin{bmatrix}-2\\1\\1\\-3\end{bmatrix},\begin{bmatrix}-4\\1\\3\\1\end{bmatrix}\right\}$

27. $\left\{\begin{bmatrix}1\\0\\0\\-2\end{bmatrix},\begin{bmatrix}0\\1\\-1\\0\end{bmatrix},\begin{bmatrix}1\\0\\1\\1\end{bmatrix},\begin{bmatrix}0\\1\\0\\1\end{bmatrix}\right\}$

28. $\left\{\begin{bmatrix}1\\0\\1\\0\end{bmatrix},\begin{bmatrix}-1\\1\\0\\1\end{bmatrix},\begin{bmatrix}1\\-1\\1\\0\end{bmatrix},\begin{bmatrix}3\\-1\\0\\-3\end{bmatrix}\right\}$

29. $\left\{\begin{bmatrix}1\\-1\\-1\\2\end{bmatrix},\begin{bmatrix}-1\\0\\1\\-1\end{bmatrix},\begin{bmatrix}-1\\-4\\1\\3\end{bmatrix},\begin{bmatrix}0\\1\\-2\\1\end{bmatrix}\right\}$

30. $\left\{\begin{bmatrix}-1\\0\\1\\-1\end{bmatrix},\begin{bmatrix}1\\0\\-2\\0\end{bmatrix},\begin{bmatrix}0\\-2\\1\\2\end{bmatrix},\begin{bmatrix}1\\-1\\-1\\2\end{bmatrix}\right\}$

In Exercises 31–38, a linearly dependent set $\mathcal{S}$ is given. Write some vector in $\mathcal{S}$ as a linear combination of the others.

31. $\left\{\begin{bmatrix}-1\\1\\2\end{bmatrix},\begin{bmatrix}3\\-3\\-6\end{bmatrix},\begin{bmatrix}0\\1\\2\end{bmatrix}\right\}$

32. $\left\{\begin{bmatrix}0\\0\\0\end{bmatrix},\begin{bmatrix}-2\\3\\-4\end{bmatrix},\begin{bmatrix}4\\-3\\2\end{bmatrix}\right\}$

33. $\left\{\begin{bmatrix}0\\1\\1\end{bmatrix}, \begin{bmatrix}1\\0\\-1\end{bmatrix}, \begin{bmatrix}4\\5\\1\end{bmatrix}\right\}$

34. $\left\{\begin{bmatrix}1\\2\\-1\end{bmatrix}, \begin{bmatrix}-1\\-3\\2\end{bmatrix}, \begin{bmatrix}4\\6\\-2\end{bmatrix}\right\}$

35. $\left\{\begin{bmatrix}1\\-1\end{bmatrix}, \begin{bmatrix}0\\1\end{bmatrix}, \begin{bmatrix}3\\-2\end{bmatrix}, \begin{bmatrix}1\\4\end{bmatrix}\right\}$

36. $\left\{\begin{bmatrix}1\\0\\3\end{bmatrix}, \begin{bmatrix}2\\-1\\1\end{bmatrix}, \begin{bmatrix}5\\-4\\-5\end{bmatrix}\right\}$

37. $\left\{\begin{bmatrix}1\\2\\-1\end{bmatrix}, \begin{bmatrix}0\\1\\-1\end{bmatrix}, \begin{bmatrix}-1\\-2\\0\end{bmatrix}, \begin{bmatrix}2\\1\\-2\end{bmatrix}\right\}$

38. $\left\{\begin{bmatrix}1\\0\\-1\\-1\\1\end{bmatrix}, \begin{bmatrix}-1\\1\\1\\0\\1\end{bmatrix}, \begin{bmatrix}-1\\-1\\2\\1\\0\end{bmatrix}, \begin{bmatrix}0\\-1\\3\\-2\\7\end{bmatrix}\right\}$

In Exercises 39–50, determine, if possible, a value of r for which the given set is linearly dependent.

39. $\left\{\begin{bmatrix}1\\-1\end{bmatrix}, \begin{bmatrix}-3\\3\end{bmatrix}, \begin{bmatrix}4\\r\end{bmatrix}\right\}$

40. $\left\{\begin{bmatrix}-2\\0\\1\end{bmatrix}, \begin{bmatrix}1\\0\\-3\end{bmatrix}, \begin{bmatrix}-1\\1\\r\end{bmatrix}\right\}$

41. $\left\{\begin{bmatrix}-2\\0\\1\end{bmatrix}, \begin{bmatrix}1\\1\\-3\end{bmatrix}, \begin{bmatrix}-1\\1\\r\end{bmatrix}\right\}$

42. $\left\{\begin{bmatrix}1\\0\\-1\\1\end{bmatrix}, \begin{bmatrix}0\\-1\\2\\1\end{bmatrix}, \begin{bmatrix}-1\\1\\1\\0\end{bmatrix}, \begin{bmatrix}-1\\9\\r\\-2\end{bmatrix}\right\}$

43. $\left\{\begin{bmatrix}2\\1\end{bmatrix}, \begin{bmatrix}5\\3\end{bmatrix}, \begin{bmatrix}r\\0\end{bmatrix}\right\}$

44. $\left\{\begin{bmatrix}2\\1\\0\end{bmatrix}, \begin{bmatrix}5\\3\\0\end{bmatrix}, \begin{bmatrix}r\\0\\r\end{bmatrix}\right\}$

45. $\left\{\begin{bmatrix}2\\-1\end{bmatrix}, \begin{bmatrix}1\\3\end{bmatrix}, \begin{bmatrix}8\\r\end{bmatrix}\right\}$

46. $\left\{\begin{bmatrix}-1\\3\\2\end{bmatrix}, \begin{bmatrix}2\\5\\r\end{bmatrix}\right\}$

47. $\left\{\begin{bmatrix}1\\2\\-1\end{bmatrix}, \begin{bmatrix}2\\1\\-3\end{bmatrix}, \begin{bmatrix}-1\\7\\r\end{bmatrix}\right\}$

48. $\left\{\begin{bmatrix}-1\\2\\1\end{bmatrix}, \begin{bmatrix}0\\1\\2\end{bmatrix}, \begin{bmatrix}1\\2\\r\end{bmatrix}\right\}$

49. $\left\{\begin{bmatrix}1\\2\\3\\-1\end{bmatrix}, \begin{bmatrix}3\\1\\6\\1\end{bmatrix}, \begin{bmatrix}-1\\3\\-2\\r\end{bmatrix}\right\}$

50. $\left\{\begin{bmatrix}0\\-1\\2\\1\end{bmatrix}, \begin{bmatrix}1\\2\\-1\\3\end{bmatrix}, \begin{bmatrix}0\\0\\0\\0\end{bmatrix}, \begin{bmatrix}-1\\0\\r\\-1\end{bmatrix}\right\}$

In Exercises 51–62, write the vector form of the general solution of the given system of linear equations.

51. $x_1 - 4x_2 + 2x_3 = 0$

52. $\begin{aligned} x_1 \qquad + 5x_3 &= 0 \\ x_2 - 3x_3 &= 0 \end{aligned}$

53. $\begin{aligned} x_1 + 3x_2 \qquad + 2x_4 &= 0 \\ x_3 - 6x_4 &= 0 \end{aligned}$

54. $\begin{aligned} x_1 \qquad + 4x_4 &= 0 \\ x_2 - 2x_4 &= 0 \end{aligned}$

55. $\begin{aligned} x_1 \qquad + 4x_3 - 2x_4 &= 0 \\ -x_1 + x_2 - 7x_3 + 7x_4 &= 0 \\ 2x_1 + 3x_2 - x_3 + 11x_4 &= 0 \end{aligned}$

56. $\begin{aligned} x_1 - 2x_2 - x_3 - 4x_4 &= 0 \\ 2x_1 - 4x_2 + 3x_3 + 7x_4 &= 0 \\ -2x_1 + 4x_2 + x_3 + 5x_4 &= 0 \end{aligned}$

57. $\begin{aligned} -x_1 + 2x_3 - 5x_4 + x_5 - x_6 &= 0 \\ x_1 - x_3 + 3x_4 - x_5 + 2x_6 &= 0 \\ x_1 + x_3 - x_4 + x_5 + 4x_6 &= 0 \end{aligned}$

58. $\begin{aligned} -x_1 \qquad - 2x_3 - x_4 - 5x_5 &= 0 \\ -x_2 + 3x_3 + 2x_4 \qquad &= 0 \\ -2x_1 + x_2 + x_3 - x_4 + 8x_5 &= 0 \\ 3x_1 - x_2 - 3x_3 - x_4 - 15x_5 &= 0 \end{aligned}$

59. $\begin{aligned} x_1 + x_2 + x_4 &= 0 \\ x_1 + 2x_2 + 4x_4 &= 0 \\ 2x_1 \qquad - 4x_4 &= 0 \end{aligned}$

60. $\begin{aligned} x_1 - 2x_2 + x_3 + x_4 + 7x_5 &= 0 \\ x_1 - 2x_2 + 2x_3 \qquad + 10x_5 &= 0 \\ 2x_1 - 4x_2 \qquad + 4x_4 + 8x_5 &= 0 \end{aligned}$

61. $\begin{aligned} x_1 + 2x_2 - x_3 \qquad + 2x_5 - x_6 &= 0 \\ 2x_1 + 4x_2 - 2x_3 - x_4 \qquad - 5x_6 &= 0 \\ -x_1 - 2x_2 + x_3 + x_4 + 2x_5 + 4x_6 &= 0 \\ x_4 + 4x_5 + 3x_6 &= 0 \end{aligned}$

62. $\begin{aligned} x_1 - x_2 \qquad - 2x_4 - x_5 + 4x_6 &= 0 \\ 2x_1 - 2x_2 - x_3 - 7x_4 - x_5 + 5x_6 &= 0 \\ -x_1 + x_2 + x_3 + 5x_4 + x_5 - 3x_6 &= 0 \\ x_3 + 3x_4 + x_5 - x_6 &= 0 \end{aligned}$

T&F *In Exercises 63–82, determine whether the statements are true or false.*

63. If $\mathcal{S}$ is linearly independent, then no vector in $\mathcal{S}$ is a linear combination of the others.
64. If the only solution of $A\mathbf{x} = \mathbf{0}$ is $\mathbf{0}$, then the rows of A are linearly independent.
65. If the nullity of A is 0, then the columns of A are linearly dependent.
66. If the columns of the reduced row echelon form of A are distinct standard vectors, then the only solution of $A\mathbf{x} = \mathbf{0}$ is $\mathbf{0}$.
67. If A is an $m \times n$ matrix with rank n, then the columns of A are linearly independent.

68. A homogeneous equation is always consistent.
69. A homogeneous equation always has infinitely many solutions.
70. If a vector form of the general solution of $A\mathbf{x} = \mathbf{0}$ is obtained by the method described in Section 1.3, then the vectors that appear in the vector form are linearly independent.
71. For any vector $\mathbf{v}$, $\{\mathbf{v}\}$ is linearly dependent.
72. A set of vectors in $\mathcal{R}^n$ is linearly dependent if and only if one of the vectors is a multiple of one of the others.
73. If a subset of $\mathcal{R}^n$ is linearly dependent, then it must contain at least n vectors.
74. If the columns of a 3×4 matrix are distinct, then they are linearly dependent.
75. For the system of linear equations $A\mathbf{x} = \mathbf{b}$ to be homogeneous, $\mathbf{b}$ must equal $\mathbf{0}$.
76. If a subset of $\mathcal{R}^n$ contains more than n vectors, then it is linearly dependent.
77. If every column of an $m \times n$ matrix A contains a pivot position, then the matrix equation $A\mathbf{x} = \mathbf{b}$ is consistent for every $\mathbf{b}$ in $\mathcal{R}^n$.
78. If every row of an $m \times n$ matrix A contains a pivot position, then the matrix equation $A\mathbf{x} = \mathbf{b}$ is consistent for every $\mathbf{b}$ in $\mathcal{R}^n$.
79. If $c_1\mathbf{u}_1 + c_2\mathbf{u}_2 + \cdots + c_k\mathbf{u}_k = \mathbf{0}$ for $c_1 = c_2 = \cdots = c_k = 0$, then $\{\mathbf{u}_1, \mathbf{u}_2, \ldots, \mathbf{u}_k\}$ is linearly independent.
80. Any subset of $\mathcal{R}^n$ that contains $\mathbf{0}$ is linearly dependent.
81. The set of standard vectors in $\mathcal{R}^n$ is linearly independent.
82. The largest number of linearly independent vectors in $\mathcal{R}^n$ is n.

---

83. Find a 2×2 matrix A such that $\mathbf{0}$ is the only solution of $A\mathbf{x} = \mathbf{0}$.
84. Find a 2×2 matrix A such that $A\mathbf{x} = \mathbf{0}$ has infinitely many solutions.
85. Find an example of linearly independent subsets $\{\mathbf{u}_1, \mathbf{u}_2\}$ and $\{\mathbf{v}\}$ of $\mathcal{R}^3$ such that $\{\mathbf{u}_1, \mathbf{u}_2, \mathbf{v}\}$ is linearly dependent.
86. Let $\{\mathbf{u}_1, \mathbf{u}_2, \ldots, \mathbf{u}_k\}$ be a linearly independent set of vectors in $\mathcal{R}^n$, and let $\mathbf{v}$ be a vector in $\mathcal{R}^n$ such that $\mathbf{v} = c_1\mathbf{u}_1 + c_2\mathbf{u}_2 + \cdots + c_k\mathbf{u}_k$ for some scalars $c_1, c_2, \ldots, c_k$, with $c_1 \neq 0$. Prove that $\{\mathbf{v}, \mathbf{u}_2, \ldots, \mathbf{u}_k\}$ is linearly independent.
87. Let $\mathbf{u}$ and $\mathbf{v}$ be distinct vectors in $\mathcal{R}^n$. Prove that the set $\{\mathbf{u}, \mathbf{v}\}$ is linearly independent if and only if the set $\{\mathbf{u} + \mathbf{v}, \mathbf{u} - \mathbf{v}\}$ is linearly independent.
88. Let $\mathbf{u}$, $\mathbf{v}$, and $\mathbf{w}$ be distinct vectors in $\mathcal{R}^n$. Prove that $\{\mathbf{u}, \mathbf{v}, \mathbf{w}\}$ is linearly independent if and only if the set $\{\mathbf{u} + \mathbf{v}, \mathbf{u} + \mathbf{w}, \mathbf{v} + \mathbf{w}\}$ is linearly independent.
89. Prove that if $\{\mathbf{u}_1, \mathbf{u}_2, \ldots, \mathbf{u}_k\}$ is a linearly independent subset of $\mathcal{R}^n$ and $c_1, c_2, \ldots, c_k$ are nonzero scalars, then $\{c_1\mathbf{u}_1, c_2\mathbf{u}_2, \ldots, c_k\mathbf{u}_k\}$ is also linearly independent.
90. Complete the proof of Theorem 1.9 by showing that if $\mathbf{u}_1 = \mathbf{0}$ or $\mathbf{u}_i$ is in the span of $\{\mathbf{u}_1, \mathbf{u}_2, \ldots, \mathbf{u}_{i-1}\}$ for some $i \geq 2$, then $\{\mathbf{u}_1, \mathbf{u}_2, \ldots, \mathbf{u}_k\}$ is linearly dependent. *Hint:* Separately consider the case in which $\mathbf{u}_1 = \mathbf{0}$ and the case in which vector $\mathbf{u}_i$ is in the span of $\{\mathbf{u}_1, \mathbf{u}_2, \ldots, \mathbf{u}_{i-1}\}$.
91.[14] Prove that any nonempty subset of a linearly independent subset of $\mathcal{R}^n$ is linearly independent.
92. Prove that if $\mathcal{S}_1$ is a linearly dependent subset of $\mathcal{R}^n$ that is contained in a finite set $\mathcal{S}_2$, then $\mathcal{S}_2$ is linearly dependent.
93. Let $\mathcal{S} = \{\mathbf{u}_1, \mathbf{u}_2, \ldots, \mathbf{u}_k\}$ be a nonempty set of vectors from $\mathcal{R}^n$. Prove that if $\mathcal{S}$ is linearly independent, then every vector in Span $\mathcal{S}$ can be written as $c_1\mathbf{u}_1 + c_2\mathbf{u}_2 + \cdots + c_k\mathbf{u}_k$ for *unique* scalars $c_1, c_2, \ldots, c_k$.
94. State and prove the converse of Exercise 93.
95. Let $\mathcal{S} = \{\mathbf{u}_1, \mathbf{u}_2, \ldots, \mathbf{u}_k\}$ be a nonempty subset of $\mathcal{R}^n$ and A be an $m \times n$ matrix. Prove that if $\mathcal{S}$ is linearly dependent and $\mathcal{S}' = \{A\mathbf{u}_1, A\mathbf{u}_2, \ldots, A\mathbf{u}_k\}$ contains k distinct vectors, then $\mathcal{S}'$ is linearly dependent.
96. Give an example to show that the preceding exercise is false if *linearly dependent* is changed to *linearly independent*.
97. Let $\mathcal{S} = \{\mathbf{u}_1, \mathbf{u}_2, \ldots, \mathbf{u}_k\}$ be a nonempty subset of $\mathcal{R}^n$ and A be an $m \times n$ matrix with rank n. Prove that if $\mathcal{S}$ is a linearly independent set, then the set $\{A\mathbf{u}_1, A\mathbf{u}_2, \ldots, A\mathbf{u}_k\}$ is also linearly independent.
98. Let A and B be $m \times n$ matrices such that B can be obtained by performing a single elementary row operation on A. Prove that if the rows of A are linearly independent, then the rows of B are also linearly independent.
99. Prove that if a matrix is in reduced row echelon form, then its nonzero rows are linearly independent.
100. Prove that the rows of an $m \times n$ matrix A are linearly independent if and only if the rank of A is m. *Hint:* Use Exercises 98 and 99.

In Exercises 101–104, use either a calculator with matrix capabilities or computer software such as MATLAB to determine whether each given set is linearly dependent. In the case that the set is linearly dependent, write some vector in the set as a linear combination of the others.

101. $$\left\{ \begin{bmatrix} 1.1 \\ 2.3 \\ -1.4 \\ 2.7 \\ 3.6 \\ 0.0 \end{bmatrix}, \begin{bmatrix} -1.7 \\ 4.2 \\ 6.2 \\ 0.0 \\ 1.3 \\ -4.0 \end{bmatrix}, \begin{bmatrix} -5.7 \\ 8.1 \\ -4.3 \\ 7.2 \\ 10.5 \\ 2.9 \end{bmatrix}, \begin{bmatrix} -5.0 \\ 2.4 \\ 1.1 \\ 3.4 \\ 3.3 \\ 6.1 \end{bmatrix}, \begin{bmatrix} 2.9 \\ -1.1 \\ 2.6 \\ 1.6 \\ 0.0 \\ 3.2 \end{bmatrix} \right\}$$

102. $$\left\{ \begin{bmatrix} 1.2 \\ -5.4 \\ 3.7 \\ -2.6 \\ 0.3 \\ 1.4 \end{bmatrix}, \begin{bmatrix} -1.7 \\ 4.2 \\ 6.2 \\ 0.0 \\ 1.3 \\ -4.0 \end{bmatrix}, \begin{bmatrix} -5.0 \\ 2.4 \\ 1.1 \\ 3.4 \\ 3.3 \\ 6.1 \end{bmatrix}, \begin{bmatrix} -0.6 \\ 4.2 \\ 2.4 \\ -1.0 \\ 8.3 \\ -2.2 \end{bmatrix}, \begin{bmatrix} 2.4 \\ -1.4 \\ 0.0 \\ 5.6 \\ 2.3 \\ -1.0 \end{bmatrix} \right\}$$

[14] This exercise is used in Section 7.3 (on page 514).

103. $\left\{ \begin{bmatrix} 21 \\ 25 \\ -15 \\ 42 \\ 17 \\ 10 \end{bmatrix}, \begin{bmatrix} 10 \\ -33 \\ 29 \\ 87 \\ -66 \\ 11 \end{bmatrix}, \begin{bmatrix} 32 \\ -21 \\ 15 \\ -11 \\ 25 \\ 16 \end{bmatrix}, \begin{bmatrix} 13 \\ 32 \\ -19 \\ 17 \\ -15 \\ 22 \end{bmatrix}, \begin{bmatrix} 26 \\ 18 \\ -37 \\ 0 \\ -7 \\ 22 \end{bmatrix}, \begin{bmatrix} 16 \\ 18 \\ 21 \\ 19 \\ -15 \\ 24 \end{bmatrix} \right\}$

104. $\left\{ \begin{bmatrix} 21 \\ 25 \\ -15 \\ 42 \\ 17 \\ 10 \end{bmatrix}, \begin{bmatrix} 10 \\ -33 \\ 29 \\ 87 \\ -66 \\ 11 \end{bmatrix}, \begin{bmatrix} -21 \\ 11 \\ 23 \\ -10 \\ 0 \\ 2 \end{bmatrix}, \begin{bmatrix} -14 \\ 3 \\ 15 \\ 0 \\ 45 \\ 15 \end{bmatrix}, \begin{bmatrix} 14 \\ 3 \\ -7 \\ 32 \\ -28 \\ -3 \end{bmatrix}, \begin{bmatrix} -8 \\ 21 \\ 30 \\ -17 \\ 34 \\ 7 \end{bmatrix} \right\}$

SOLUTIONS TO THE PRACTICE PROBLEMS

1. Let A be the matrix whose columns are the vectors in $\mathcal{S}$. Since the reduced row echelon form of A is I_4, the columns of A are linearly independent by Theorem 1.8. Thus $\mathcal{S}$ is linearly independent, and so no vector in $\mathcal{S}$ is a linear combination of the others.

2. The augmented matrix of the given system is

$$\begin{bmatrix} 1 & -3 & -1 & 1 & -1 & 0 \\ 2 & -6 & 1 & -3 & -9 & 0 \\ -2 & 6 & 3 & 2 & 11 & 0 \end{bmatrix}.$$

Since the reduced row echelon form of this matrix is

$$\begin{bmatrix} 1 & -3 & 0 & 0 & -2 & 0 \\ 0 & 0 & 1 & 0 & 1 & 0 \\ 0 & 0 & 0 & 1 & 2 & 0 \end{bmatrix},$$

the general solution of the given system is

$$\begin{aligned} x_1 &= 3x_2 + 2x_5 \\ x_2 & \quad \text{free} \\ x_3 &= -x_5 \\ x_4 &= -2x_5 \\ x_5 & \quad \text{free.} \end{aligned}$$

To obtain its vector form, we express the general solution as a linear combination of vectors in which the coefficients are the free variables.

$$\begin{bmatrix} x_1 \\ x_2 \\ x_3 \\ x_4 \\ x_5 \end{bmatrix} = \begin{bmatrix} 3x_2 + 2x_5 \\ x_2 \\ -x_5 \\ -2x_5 \\ x_5 \end{bmatrix} = x_2 \begin{bmatrix} 3 \\ 1 \\ 0 \\ 0 \\ 0 \end{bmatrix} + x_5 \begin{bmatrix} 2 \\ 0 \\ -1 \\ -2 \\ 1 \end{bmatrix}$$

3. By property 1 on page 81, $\mathcal{S}_1$ is linearly independent.

   By property 2 on page 81, the first two vectors in $\mathcal{S}_2$ are linearly dependent. Therefore $\mathcal{S}_2$ is linearly dependent by Theorem 1.9.

   By property 2 on page 81, $\mathcal{S}_3$ is linearly independent.

   By property 4 on page 81, $\mathcal{S}_4$ is linearly dependent.

CHAPTER 1 REVIEW EXERCISES

In Exercises 1–17, determine whether the statements are true or false.

1. If B is a 3×4 matrix, then its columns are 1×3 vectors.
2. Any scalar multiple of a vector $\mathbf{v}$ in $\mathcal{R}^n$ is a linear combination of $\mathbf{v}$.
3. If a vector $\mathbf{v}$ lies in the span of a finite subset $\mathcal{S}$ of $\mathcal{R}^n$, then $\mathbf{v}$ is a linear combination of the vectors in $\mathcal{S}$.
4. The matrix–vector product of an $m \times n$ matrix A and a vector in $\mathcal{R}^n$ is a linear combination of the columns of A.
5. The rank of the coefficient matrix of a consistent system of linear equations is equal to the number of basic variables in the general solution of the system.
6. The nullity of the coefficient matrix of a consistent system of linear equations is equal to the number of free variables in the general solution of the system.
7. Every matrix can be transformed into one and only one matrix in reduced row echelon form by means of a sequence of elementary row operations.
8. If the last row of the reduced row echelon form of an augmented matrix of a system of linear equations has only one nonzero entry, then the system is inconsistent.
9. If the last row of the reduced row echelon form of an augmented matrix of a system of linear equations has only zero entries, then the system has infinitely many solutions.
10. The zero vector of $\mathcal{R}^n$ lies in the span of any finite subset of $\mathcal{R}^n$.
11. If the rank of an $m \times n$ matrix A is m, then the rows of A are linearly independent.
12. The set of columns of an $m \times n$ matrix A is a generating set for $\mathcal{R}^m$ if and only if the rank of A is m.

13. If the columns of an $m \times n$ matrix are linearly dependent, then the rank of the matrix is less than m.
14. If $\mathcal{S}$ is a linearly independent subset of $\mathcal{R}^n$ and $\mathbf{v}$ is a vector in $\mathcal{R}^n$ such that $\mathcal{S} \cup \{\mathbf{v}\}$ is linearly dependent, then $\mathbf{v}$ is in the span of $\mathcal{S}$.
15. A subset of $\mathcal{R}^n$ containing more than n vectors must be linearly dependent.
16. A subset of $\mathcal{R}^n$ containing fewer than n vectors must be linearly independent.
17. A linearly dependent subset of $\mathcal{R}^n$ must contain more than n vectors.

---

18. Determine whether each of the following phrases is a misuse of terminology. If so, explain what is wrong with each one:
    (a) an inconsistent matrix
    (b) the solution of a matrix
    (c) equivalent matrices
    (d) the nullity of a system of linear equations
    (e) the span of a matrix
    (f) a generating set for a system of linear equations
    (g) a homogeneous matrix
    (h) a linearly independent matrix
19. (a) If A is an $m \times n$ matrix with rank n, what can be said about the number of solutions of $A\mathbf{x} = \mathbf{b}$ for every $\mathbf{b}$ in $\mathcal{R}^m$?
    (b) If A is an $m \times n$ matrix with rank m, what can be said about the number of solutions of $A\mathbf{x} = \mathbf{b}$ for every $\mathbf{b}$ in $\mathcal{R}^m$?

In Exercises 20–27, use the following matrices to compute the given expression, or give a reason why the expression is not defined:

$$A = \begin{bmatrix} 1 & 3 \\ -2 & 4 \\ 0 & 2 \end{bmatrix}, \quad B = \begin{bmatrix} 2 & -1 \\ 0 & 3 \\ 4 & 1 \end{bmatrix}, \quad C = \begin{bmatrix} 1 \\ 5 \end{bmatrix}, \quad \text{and}$$
$$D = [1 \quad -1 \quad 2].$$

20. $A + B^T$
21. $A + B$
22. BC
23. AD^T
24. $2A - 3B$
25. $A^T D^T$
26. $A^T - B$
27. $C^T - 2D$

28. A boat is traveling on a river in a southwesterly direction, parallel to the riverbank, at 10 mph. At the same time, a passenger is walking from the southeast side of the boat to the northwest at 2 mph. Find the velocity and the speed of the passenger with respect to the riverbank.
29. A supermarket chain has 10 stores. For each i such that $1 \le i \le 10$, the 4×1 vector $\mathbf{v}_i$ is defined so that its respective components represent the total value of sales in produce, meats, dairy, and processed foods at store i during January of last year. Provide an interpretation of the vector $(0.1)(\mathbf{v}_1 + \mathbf{v}_2 + \cdots + \mathbf{v}_{10})$.

In Exercises 30–33, compute the matrix–vector products.

30. $\begin{bmatrix} 3 & 1 \\ 0 & -1 \\ 1 & 2 \end{bmatrix} \begin{bmatrix} 4 \\ 1 \end{bmatrix}$
31. $\begin{bmatrix} 1 & 3 & 1 & 2 \\ 1 & -1 & 4 & 0 \end{bmatrix}^T \begin{bmatrix} -1 \\ 1 \end{bmatrix}$
32. $A_{45^\circ} \begin{bmatrix} 2 \\ -1 \end{bmatrix}$
33. $A_{-30^\circ} \begin{bmatrix} 2 \\ -1 \end{bmatrix}$

34. Suppose that
$$\mathbf{v}_1 = \begin{bmatrix} 2 \\ 1 \\ 3 \end{bmatrix} \quad \text{and} \quad \mathbf{v}_2 = \begin{bmatrix} -1 \\ 3 \\ 6 \end{bmatrix}.$$
Represent $3\mathbf{v}_1 - 4\mathbf{v}_2$ as the product of a 3×2 matrix and a vector in $\mathcal{R}^2$.

In Exercises 35–38, determine whether the given vector $\mathbf{v}$ *is in the span of*
$$\mathcal{S} = \left\{ \begin{bmatrix} -1 \\ 5 \\ 2 \end{bmatrix}, \begin{bmatrix} 1 \\ 3 \\ 4 \end{bmatrix}, \begin{bmatrix} 1 \\ -1 \\ 1 \end{bmatrix} \right\}.$$
If so, write $\mathbf{v}$ as a linear combination of the vectors in $\mathcal{S}$.

35. $\mathbf{v} = \begin{bmatrix} 5 \\ 3 \\ 11 \end{bmatrix}$
36. $\mathbf{v} = \begin{bmatrix} 1 \\ 4 \\ 3 \end{bmatrix}$
37. $\mathbf{v} = \begin{bmatrix} 1 \\ 1 \\ 2 \end{bmatrix}$
38. $\mathbf{v} = \begin{bmatrix} 2 \\ 10 \\ 9 \end{bmatrix}$

In Exercises 39–44, determine whether the given system is consistent, and if so, find its general solution.

39. $x_1 + 2x_2 - x_3 = 1$
40. $\begin{aligned} x_1 + x_2 + x_3 &= 3 \\ -2x_1 + 4x_2 + 2x_3 &= 7 \\ 2x_1 - x_2 - 4x_3 &= 2 \end{aligned}$
41. $\begin{aligned} x_1 + 2x_2 + 3x_3 &= 1 \\ 2x_1 + x_2 + x_3 &= 2 \\ x_1 - 4x_2 - 7x_3 &= 4 \end{aligned}$
42. $\begin{aligned} x_1 + 3x_2 + 2x_3 + x_4 &= 2 \\ 2x_1 + x_2 + x_3 - x_4 &= 3 \\ x_1 - 2x_2 - x_3 - 2x_4 &= 4 \end{aligned}$
43. $\begin{aligned} x_1 + x_2 + 2x_3 + x_4 &= 2 \\ 2x_1 + 3x_2 + x_3 - x_4 &= -1 \end{aligned}$
44. $\begin{aligned} 2x_1 + 4x_2 - 2x_3 + 2x_4 &= 4 \\ 2x_1 + x_2 + 4x_3 + 2x_4 &= 1 \\ 4x_1 + 6x_2 + x_3 + 2x_4 &= 1 \end{aligned}$

In Exercises 45–48, find the rank and nullity of the given matrix.

45. $\begin{bmatrix} 1 & 2 & -3 & 0 & 1 \end{bmatrix}$
46. $\begin{bmatrix} 1 & 2 & -3 & 0 & 1 \\ 0 & 0 & 0 & 0 & 0 \end{bmatrix}$

47. $\begin{bmatrix} 1 & 2 & 1 & -1 & 2 \\ 2 & 1 & 0 & 1 & 3 \\ -1 & -3 & 1 & 2 & 4 \end{bmatrix}$

48. $\begin{bmatrix} 2 & 3 & 4 \\ 1 & 2 & 1 \\ -1 & 1 & 2 \\ 3 & 0 & 2 \end{bmatrix}$

49. A company that ships fruit has three kinds of fruit packs. The first pack consists of 10 oranges and 10 grapefruit, the second pack consists of 10 oranges, 15 grapefruit, and 10 apples, and the third pack consists of 5 oranges, 10 grapefruit, and 5 apples. How many of each pack can be made from a stock of 500 oranges, 750 grapefruit, and 300 apples?

In Exercises 50–53, a set of vectors in $\mathcal{R}^n$ is given. Determine whether the set is a generating set for $\mathcal{R}^n$.

50. $\left\{\begin{bmatrix} 1 \\ 1 \\ -1 \\ 1 \end{bmatrix}, \begin{bmatrix} 1 \\ 0 \\ 0 \\ 2 \end{bmatrix}, \begin{bmatrix} 1 \\ 3 \\ -2 \\ 1 \end{bmatrix}\right\}$

51. $\left\{\begin{bmatrix} -1 \\ 1 \\ 1 \end{bmatrix}, \begin{bmatrix} 1 \\ -1 \\ 1 \end{bmatrix}, \begin{bmatrix} 1 \\ 1 \\ -1 \end{bmatrix}\right\}$

52. $\left\{\begin{bmatrix} 1 \\ 0 \\ 1 \end{bmatrix}, \begin{bmatrix} 1 \\ 1 \\ -1 \end{bmatrix}, \begin{bmatrix} 2 \\ 1 \\ 3 \end{bmatrix}\right\}$

53. $\left\{\begin{bmatrix} 1 \\ 2 \\ 1 \end{bmatrix}, \begin{bmatrix} 1 \\ -1 \\ 1 \end{bmatrix}, \begin{bmatrix} 1 \\ 1 \\ 1 \end{bmatrix}, \begin{bmatrix} 0 \\ 1 \\ 0 \end{bmatrix}\right\}$

In Exercises 54–59, an $m \times n$ matrix A is given. Determine whether the equation $A\mathbf{x} = \mathbf{b}$ is consistent for every $\mathbf{b}$ in $\mathcal{R}^n$.

54. $\begin{bmatrix} 1 & 2 \\ 3 & 6 \end{bmatrix}$

55. $\begin{bmatrix} 1 & 1 \\ 3 & 2 \end{bmatrix}$

56. $\begin{bmatrix} 1 & -1 & 1 \\ 2 & 0 & 1 \end{bmatrix}$

57. $\begin{bmatrix} -1 & 1 & 1 \\ 1 & -1 & 1 \\ 1 & 1 & -1 \end{bmatrix}$

58. $\begin{bmatrix} 1 & 2 & 1 \\ 3 & 0 & -3 \\ -1 & 1 & 2 \end{bmatrix}$

59. $\begin{bmatrix} 1 & 2 & 1 \\ 2 & -3 & 1 \\ -1 & 1 & 2 \\ 0 & 1 & 2 \end{bmatrix}$

In Exercises 60–63, determine whether the given set is linearly dependent or linearly independent.

60. $\left\{\begin{bmatrix} 1 \\ 3 \\ 2 \end{bmatrix}, \begin{bmatrix} 1 \\ -1 \\ 2 \end{bmatrix}, \begin{bmatrix} 3 \\ 1 \\ 6 \end{bmatrix}\right\}$

61. $\left\{\begin{bmatrix} 1 \\ -1 \\ 2 \\ 0 \end{bmatrix}, \begin{bmatrix} 0 \\ 1 \\ 2 \\ 3 \end{bmatrix}, \begin{bmatrix} 1 \\ 0 \\ 1 \\ 1 \end{bmatrix}\right\}$

62. $\left\{\begin{bmatrix} 2 \\ 3 \\ 5 \\ 7 \end{bmatrix}, \begin{bmatrix} 4 \\ 6 \\ 10 \\ 14 \end{bmatrix}\right\}$

63. $\left\{\begin{bmatrix} 22.40 \\ 6.02 \\ 6.63 \end{bmatrix}, \begin{bmatrix} 9.11 \\ 1.76 \\ 9.27 \end{bmatrix}, \begin{bmatrix} 3.14 \\ 2.72 \\ 1.41 \end{bmatrix}, \begin{bmatrix} 31 \\ 37 \\ 41 \end{bmatrix}\right\}$

In Exercises 64–67, a linearly dependent set $\mathcal{S}$ is given. Write some vector in $\mathcal{S}$ as a linear combination of the others.

64. $\left\{\begin{bmatrix} 1 \\ -1 \\ 3 \end{bmatrix}, \begin{bmatrix} 1 \\ 2 \\ 1 \end{bmatrix}, \begin{bmatrix} 2 \\ 4 \\ 2 \end{bmatrix}\right\}$

65. $\left\{\begin{bmatrix} 1 \\ 2 \\ 3 \end{bmatrix}, \begin{bmatrix} 1 \\ -1 \\ 2 \end{bmatrix}, \begin{bmatrix} 3 \\ 3 \\ 8 \end{bmatrix}\right\}$

66. $\left\{\begin{bmatrix} 3 \\ 1 \\ 4 \\ 1 \end{bmatrix}, \begin{bmatrix} 3 \\ 0 \\ 5 \\ 1 \end{bmatrix}, \begin{bmatrix} 3 \\ 3 \\ 2 \\ 1 \end{bmatrix}\right\}$

67. $\left\{\begin{bmatrix} 1 \\ -1 \\ 1 \\ 2 \end{bmatrix}, \begin{bmatrix} 1 \\ 0 \\ 1 \\ 0 \end{bmatrix}, \begin{bmatrix} 1 \\ 1 \\ 1 \\ 1 \end{bmatrix}, \begin{bmatrix} 1 \\ -1 \\ 1 \\ -1 \end{bmatrix}\right\}$

In Exercises 68–71, write the vector form of the general solution of the given system of linear equations.

68. $x_1 + 2x_2 - x_3 + x_4 = 0$

69. $$\begin{aligned} x_1 + 2x_2 - x_3 &= 0 \\ x_1 + x_2 + x_3 &= 0 \\ x_1 + 3x_2 - 3x_3 &= 0 \end{aligned}$$

70. $$\begin{aligned} 2x_1 + 5x_2 - x_3 + x_4 &= 0 \\ x_1 + 3x_2 + 2x_3 - x_4 &= 0 \end{aligned}$$

71. $$\begin{aligned} 3x_1 + x_2 - x_3 + x_4 &= 0 \\ 2x_1 + 2x_2 + 4x_3 - 6x_4 &= 0 \\ 2x_1 + x_2 + 3x_3 - x_4 &= 0 \end{aligned}$$

72. Let A be an $m \times n$ matrix, let $\mathbf{b}$ be a vector in $\mathcal{R}^m$, and suppose that $\mathbf{v}$ is a solution of $A\mathbf{x} = \mathbf{b}$.
    (a) Prove that if $\mathbf{w}$ is a solution of $A\mathbf{x} = \mathbf{0}$, then $\mathbf{v} + \mathbf{w}$ is a solution of $A\mathbf{x} = \mathbf{b}$.
    (b) Prove that for any solution $\mathbf{u}$ to $A\mathbf{x} = \mathbf{b}$, there is a solution $\mathbf{w}$ to $A\mathbf{x} = \mathbf{0}$ such that $\mathbf{u} = \mathbf{v} + \mathbf{w}$.

73. Suppose that $\mathbf{w}_1$ and $\mathbf{w}_2$ are linear combinations of vectors $\mathbf{v}_1$ and $\mathbf{v}_2$ in $\mathcal{R}^n$ such that $\mathbf{w}_1$ and $\mathbf{w}_2$ are linearly independent. Prove that $\mathbf{v}_1$ and $\mathbf{v}_2$ are linearly independent.

74. Let A be an $m \times n$ matrix with reduced row echelon form R. Describe the reduced row echelon form of each of the following matrices:
    (a) $[A\ \mathbf{0}]$
    (b) $[\mathbf{a}_1\ \mathbf{a}_2\ \cdots\ \mathbf{a}_k]$ for $k < n$
    (c) cA, where c is a nonzero scalar
    (d) $[I_m\ A]$
    (e) $[A\ cA]$, where c is any scalar

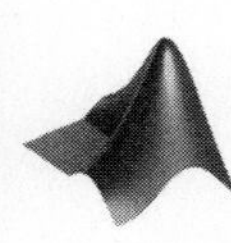

CHAPTER 1 MATLAB EXERCISES

For the following exercises, use MATLAB (or comparable software) or a calculator with matrix capabilities. The MATLAB functions in Tables D.1, D.2, D.3, D.4, and D.5 of Appendix D may be useful.

1. Let
$$A = \begin{bmatrix} 2.1 & 3.2 & 6.1 & -2.3 \\ 1.3 & -2.5 & -1.7 & 1.5 \\ -1.2 & 1.5 & 4.3 & 2.4 \\ 4.1 & 2.0 & 5.1 & 4.2 \\ 6.1 & -1.4 & 3.0 & -1.3 \end{bmatrix}.$$
Use the matrix–vector product of A and a vector to compute each of the following linear combinations of the columns of A:
(a) $1.5\mathbf{a}_1 - 2.2\mathbf{a}_2 + 2.7\mathbf{a}_3 + 4\mathbf{a}_4$
(b) $2\mathbf{a}_1 + 2.1\mathbf{a}_2 - 1.1\mathbf{a}_4$
(c) $3.3\mathbf{a}_2 + 1.2\mathbf{a}_3 - \mathbf{a}_4$

2. Let
$$A = \begin{bmatrix} 1.3 & 2.1 & -3.3 & 4.1 \\ 6.1 & 2.4 & -1.3 & -3.1 \\ -2.2 & 5.1 & 3.2 & 2.1 \\ 2.2 & 6.1 & 7.2 & -5.1 \end{bmatrix},$$
$$B = \begin{bmatrix} 2.1 & -1.1 & 1.2 & 4.2 \\ -4.6 & 8.1 & 9.2 & -3.3 \\ 2.5 & 5.2 & -3.3 & 4.2 \\ -0.7 & 2.8 & -6.3 & 4.7 \end{bmatrix},$$
and
$$\mathbf{v} = \begin{bmatrix} 3.2 \\ -4.6 \\ 1.8 \\ 7.1 \end{bmatrix}.$$
(a) Compute $3A - 2B$.
(b) Compute $A - 4B^T$.
(c) Compute $P = \frac{1}{2}(A + A^T)$.
(d) Compute $Q = \frac{1}{2}(A - A^T)$.
(e) Compute P^T and Q^T to see that P in (c) is symmetric and Q in (d) is skew-symmetric. Then compute $P + Q$. What does it equal?
(f) Compute $A\mathbf{v}$.
(g) Compute $B(A\mathbf{v})$.
(h) Compute $A(B\mathbf{v})$.
(i) Evaluate the linear combination
$$3.5\mathbf{a}_1 - 1.2\mathbf{a}_2 + 4.1\mathbf{a}_3 + 2\mathbf{a}_4,$$
and determine a vector $\mathbf{w}$ such that $A\mathbf{w}$ equals this linear combination.
(j) Let M be the 4×4 matrix whose jth column is $B\mathbf{a}_j$ for $1 \le j \le 4$. Verify that $M\mathbf{e}_j = B(A\mathbf{e}_j)$ for all j, and verify that $M\mathbf{v} = B(A\mathbf{v})$. State and prove the generalization of this result to all vectors in $\mathcal{R}^4$.

3. Let A_θ denote the rotation matrix of θ degrees, as defined in Section 1.2. For the following computations, it is useful to apply the imported MATLAB function `rotdeg`, as described in Table D.5 of Appendix D:
(a) Compute $A_{20°}\begin{bmatrix}1\\3\end{bmatrix}$.
(b) Compute $A_{30°}\left(A_{20°}\begin{bmatrix}1\\3\end{bmatrix}\right)$.
(c) Compute $A_{50°}\begin{bmatrix}1\\3\end{bmatrix}$.
(d) Compute $A_{-20°}\left(A_{20°}\begin{bmatrix}1\\3\end{bmatrix}\right)$.
(e) Make a conjecture about $A_{\theta_1+\theta_2}$ and its relationship to A_{θ_1} and A_{θ_2} for any angles θ_1 and θ_2.
(f) Prove your conjecture.
(g) Make a conjecture about the relationship between A_θ and $A_{-\theta}$ for any angle θ.
(h) Prove your conjecture.

4. Let
$$A = \begin{bmatrix} 1.1 & 2.0 & 4.2 & 2.7 & 1.2 & 0.1 \\ 3.1 & -1.5 & 4.7 & 8.3 & -3.1 & 2.3 \\ 7.1 & -8.5 & 5.7 & 19.5 & -11.7 & 6.7 \\ 2.2 & 4.0 & 8.4 & 6.5 & 2.1 & -3.4 \end{bmatrix}.$$
(a) Use elementary row operations to transform A into reduced row echelon form. (Note that the MATLAB function command `A(i,:)` described in Table D.4 in Appendix D is useful for this purpose. For example, to interchange row i and row j of A, enter the three commands `temp = A(i,:)`, `A(i,:) = A(j,:)`, and `A(j,:) = temp`. To multiply the entries of row i of A by the scalar c, enter the command `A(i,:) = c*A(i,:)`. To add the scalar multiple c of row i to row j of A, enter the command `A(j,:) = A(j,:) + c*A(i,:)`.)
(b) Use technology to compute the reduced row echelon form of A directly. (For example, enter `rref(A)` when using MATLAB.) Compare this to the result obtained in (a).

5. For the matrix A in Exercise 4 and each of the following vectors $\mathbf{b}$, determine whether the system of linear equations $A\mathbf{x} = \mathbf{b}$ is consistent. If so, find its general solution:
(a) $\mathbf{b} = \begin{bmatrix} -1.0 \\ 2.3 \\ 8.9 \\ 1.6 \end{bmatrix}$ (b) $\mathbf{b} = \begin{bmatrix} 1.1 \\ 2.1 \\ 3.2 \\ -1.4 \end{bmatrix}$
(c) $\mathbf{b} = \begin{bmatrix} 3.8 \\ 2.9 \\ 1.1 \\ 12.0 \end{bmatrix}$ (d) $\mathbf{b} = \begin{bmatrix} 1 \\ 1 \\ -1 \\ 2 \end{bmatrix}$

6. Let

$$C = \begin{bmatrix} 0.10 & 0.06 & 0.20 & 0.13 & 0.18 \\ 0.05 & 0.12 & 0.14 & 0.10 & 0.20 \\ 0.12 & 0.21 & 0.06 & 0.14 & 0.15 \\ 0.11 & 0.10 & 0.15 & 0.20 & 0.10 \\ 0.20 & 0.10 & 0.20 & 0.05 & 0.17 \end{bmatrix}$$

and

$$\mathbf{d} = \begin{bmatrix} 80 \\ 100 \\ 150 \\ 50 \\ 60 \end{bmatrix},$$

where C is the input-output matrix for an economy that has been divided into five sectors, and $\mathbf{d}$ is the net production for this economy, where units are in billions of dollars. Find the gross production vector required to produce $\mathbf{d}$, where each component is in units of billions of dollars, rounded to four places after the decimal point.

7. Determine whether each of the sets that follow is linearly dependent or linearly independent. If the set is linearly dependent, write one of the vectors in the set as a linear combination of the other vectors in the set.

(a) $$\mathcal{S}_1 = \left\{ \begin{bmatrix} 1 \\ 2 \\ -1 \\ 3 \\ 2 \\ 1 \end{bmatrix}, \begin{bmatrix} 1 \\ 0 \\ 1 \\ 1 \\ 0 \\ 1 \end{bmatrix}, \begin{bmatrix} 2 \\ 1 \\ -1 \\ 2 \\ 0 \\ 1 \end{bmatrix}, \begin{bmatrix} 3 \\ -1 \\ 1 \\ 2 \\ -1 \\ 1 \end{bmatrix}, \begin{bmatrix} 0 \\ 1 \\ 1 \\ 2 \\ 2 \\ 1 \end{bmatrix} \right\}$$

(b) $$\mathcal{S}_2 = \left\{ \begin{bmatrix} 2 \\ 1 \\ -1 \\ 1 \\ 3 \\ 1 \end{bmatrix}, \begin{bmatrix} 1 \\ 2 \\ 1 \\ 1 \\ 0 \\ -1 \end{bmatrix}, \begin{bmatrix} -1 \\ 1 \\ 4 \\ 1 \\ -1 \\ 2 \end{bmatrix}, \begin{bmatrix} 2 \\ 1 \\ 1 \\ -1 \\ 2 \\ 3 \end{bmatrix}, \begin{bmatrix} -2 \\ 1 \\ 0 \\ 1 \\ -1 \\ -2 \end{bmatrix} \right\}$$

8. Determine whether each of the vectors shown is in the span of $\mathcal{S}_1$ as defined in (a) of Exercise 7. If the vector lies in the span of $\mathcal{S}_1$, then represent it as a linear combination of the vectors in $\mathcal{S}_1$.

(a) $\begin{bmatrix} 14 \\ 2 \\ -1 \\ 12 \\ -1 \\ 5 \end{bmatrix}$ (b) $\begin{bmatrix} 4 \\ 3 \\ -2 \\ 7 \\ 3 \\ 2 \end{bmatrix}$ (c) $\begin{bmatrix} 10 \\ 6 \\ -5 \\ 13 \\ 3 \\ 5 \end{bmatrix}$ (d) $\begin{bmatrix} 1 \\ 6 \\ -5 \\ 4 \\ 3 \\ 1 \end{bmatrix}$

2 INTRODUCTION

Fingerprint recognition was accepted as a valid personal identification method in the early twentieth century and became a standard tool in forensics. The rapid expansion of fingerprint recognition in forensics created enormous collections of fingerprint cards. Large staffs of human fingerprint examiners struggled to provide prompt responses to requests for fingerprint identification. The development of *Automatic Fingerprint Identification Systems (AFIS)* over the past few decades has improved dramatically the productivity of law enforcement agencies and reduced the cost of hiring and training human fingerprint experts. The technology has also made fingerprint recognition a practical tool for unsupervised access control.

Fingerprint identification for law enforcement is based on the location and type of *minutiae*, which are peculiarities in the *friction ridges* (or simply *ridges*) on the fingertips. Analysis usually focuses on a particular pair of minutiae: *ridge endpoints* where a ridge terminates and *bifurcation points* where a ridge splits. An example of each is shown in the center of each of the preceding figures, where the ridges are black against a white background. The identification then hinges on locating the minutiae in the fingerprint, a task originally performed by human examiners.

The AFIS analysis uses a digital image, which is usually acquired by a live-scan fingerprint sensor. The original digital image goes through a preprocessing phase that extracts the ridge pixels, converts them to binary values (pixels are either 0 or 1), and thins the image to obtain a set of one-pixel width curves that are approximately the centerlines of the ridges, as shown in the figure at the right. A few of the minutiae are labeled with blue squares.

Matrices such as $\begin{bmatrix} 0 & 0 & 0 \\ 0 & 1 & 1 \\ 0 & 0 & 0 \end{bmatrix}$ and $\begin{bmatrix} 0 & 0 & 1 \\ 0 & 1 & 0 \\ 0 & 0 & 0 \end{bmatrix}$ are used to detect ridge endpoints. Interpreting the entries as $0 =$ white and $1 =$ black, the matrices match the possible arrangements of pixels at a ridge endpoint.

Similarly, matrices such as $\begin{bmatrix} 1 & 0 & 1 \\ 0 & 1 & 0 \\ 0 & 1 & 0 \end{bmatrix}$ and $\begin{bmatrix} 1 & 0 & 0 \\ 0 & 1 & 1 \\ 1 & 0 & 0 \end{bmatrix}$ are used to find bifurcation endpoints.

The process of detection involves sliding the matrices over the image and looking for 3x3 blocks of image pixels that match one of the matrices. A match indicates an endpoint.

Other uses of matrices whose entries consist of 0s and 1s, called *(0, 1)-matrices*, are examined in Section 2.2.

CHAPTER

2 MATRICES AND LINEAR TRANSFORMATIONS

In Section 1.2, we used matrix–vector products to perform calculations in examples involving rotations of vectors, shifts of population, and mixtures of seed. For these examples, we presented the data—points in the plane, population distributions, and grass seed mixtures—as vectors and then formulated rules for transforming these vectors by matrix–vector products.

In situations of this kind in which a process is repeated, it is often useful to take the product of a matrix and the vector obtained from a previous matrix–vector product. This calculation leads to an extension of the definition of multiplication to include products of matrices of various sizes.

In this chapter, we examine some of the elementary properties and applications of this extended definition of matrix multiplication (Sections 2.1–2.4). Later in the chapter, we study the matrix–vector product from the functional viewpoint of a rule of correspondence. This leads to the definition of *linear transformation* (Sections 2.7 and 2.8). Here we see how the functional properties of linear transformations correspond to the properties of matrix multiplication studied earlier in this chapter.

2.1 MATRIX MULTIPLICATION

In many applications, we need to multiply a matrix–vector product $A\mathbf{v}$ on the left by A again to form a new matrix–vector product $A(A\mathbf{v})$. For instance, in Example 3 in Section 1.2, the product $A\mathbf{p}$ represents the population distribution in a metropolitan area after one year. To find the population distribution after two years, we multiply the product $A\mathbf{p}$ on the left by A to get $A(A\mathbf{p})$. For other problems, we might need to multiply a matrix–vector product $B\mathbf{v}$ by a different matrix C to obtain the product $C(B\mathbf{v})$. The following example illustrates such a product:

Example 1 In the seed example from Section 1.2, recall that

$$\begin{array}{ccccc} & \text{deluxe} & \text{standard} & \text{economy} & \\ B = & \left[\begin{array}{c} .80 \\ .20 \end{array}\right. & \begin{array}{c} .60 \\ .40 \end{array} & \left.\begin{array}{c} .40 \\ .60 \end{array}\right] & \begin{array}{c} \text{bluegrass} \\ \text{rye} \end{array} \end{array}$$

gives the proportions of bluegrass and rye in the deluxe, standard, and economy mixtures of grass seed, and that

$$\mathbf{v} = \begin{bmatrix} 60 \\ 50 \\ 30 \end{bmatrix}$$

gives the number of pounds of each mixture in stock. Then $B\mathbf{v}$ is the vector whose components are the amounts of bluegrass and rye seed, respectively, in a blend obtained by combining the number of pounds of each of the three mixtures in stock. Since

$$B\mathbf{v} = \begin{bmatrix} .80 & .60 & .40 \\ .20 & .40 & .60 \end{bmatrix} \begin{bmatrix} 60 \\ 50 \\ 30 \end{bmatrix} = \begin{bmatrix} 90 \\ 50 \end{bmatrix},$$

we conclude that there are 90 pounds of bluegrass seed and 50 pounds of rye seed in the blend.

Next suppose that we have a seed manual with a table that gives us the germination rates of bluegrass and rye seeds under both wet and dry conditions. The table, given in the form of a matrix A, is as follows:

$$\begin{array}{cc} & \begin{array}{cc} \text{bluegrass} & \text{rye} \end{array} \\ A = & \begin{bmatrix} .80 & .70 \\ .60 & .40 \end{bmatrix} \begin{array}{l} \text{wet} \\ \text{dry} \end{array} \end{array}$$

The $(1, 1)$-entry, which is .80, signifies that 80% of the bluegrass seed germinates under wet conditions, while the $(1, 2)$-entry, which is .70, signifies that 70% of the rye seed germinates under wet conditions. Suppose also that we have a mixture of y_1 pounds of bluegrass seed and y_2 pounds of rye seed. Then $.80y_1 + .70y_2$ is the total weight (in pounds) of the seed that germinates under wet conditions. Similarly, $.60y_1 + .40y_2$ is the total weight of the seed that germinates under dry conditions. Notice that these two expressions are the entries of the matrix–vector product $A\mathbf{v}$, where $\mathbf{v} = \begin{bmatrix} y_1 \\ y_2 \end{bmatrix}$.

Let's combine this with our previous calculation. Since $B\mathbf{v}$ is the vector whose components are the amounts of bluegrass and rye seed in a blend, the components of the matrix–vector product

$$A(B\mathbf{v}) = \begin{bmatrix} .80 & .70 \\ .60 & .40 \end{bmatrix} \begin{bmatrix} 90 \\ 50 \end{bmatrix} = \overset{\substack{\text{lbs. of seed} \\ \text{germinated}}}{\begin{bmatrix} 107 \\ 74 \end{bmatrix}} \begin{array}{l} \text{wet} \\ \text{dry} \end{array}$$

are the amounts of seed that can be expected to germinate under each of the two types of weather conditions. Thus 107 pounds of seed can be expected to germinate under wet conditions, and 74 pounds under dry conditions.

In the preceding example, a matrix–vector product is multiplied on the left by another matrix. An examination of this process leads to an extended definition of matrix multiplication.

Let A be an $m \times n$ matrix and B be an $n \times p$ matrix. Then for any $p \times 1$ vector $\mathbf{v}$, the product $B\mathbf{v}$ is an $n \times 1$ vector, and hence the new product $A(B\mathbf{v})$ is an $m \times 1$ vector. This raises the following question: Is there is an $m \times p$ matrix C such that $A(B\mathbf{v}) = C\mathbf{v}$ for every $p \times 1$ vector $\mathbf{v}$?

By the definition of a matrix–vector product and Theorem 1.3, we have

$$\begin{aligned} A(B\mathbf{v}) &= A(v_1\mathbf{b}_1 + v_2\mathbf{b}_2 + \cdots + v_p\mathbf{b}_p) \\ &= A(v_1\mathbf{b}_1) + A(v_2\mathbf{b}_2) + \cdots + A(v_p\mathbf{b}_p) \end{aligned}$$

$$= v_1 A\mathbf{b}_1 + v_2 A\mathbf{b}_2 + \cdots + v_p A\mathbf{b}_p$$

$$= [A\mathbf{b}_1 \ A\mathbf{b}_2 \ \ldots \ A\mathbf{b}_p]\mathbf{v}.$$

Let C be the $m \times p$ matrix $[A\mathbf{b}_1 \ A\mathbf{b}_2 \ \ldots \ A\mathbf{b}_p]$—that is, the matrix whose jth column is $\mathbf{c}_j = A\mathbf{b}_j$. Then $A(B\mathbf{v}) = C\mathbf{v}$ for all $\mathbf{v}$ in $\mathcal{R}^p$. Furthermore, by Theorem 1.3(e), C is the only matrix with this property. This convenient method of combining the matrices A and B leads to the following definition.

Definition Let A be an $m \times n$ matrix and B be an $n \times p$ matrix. We define the **(matrix) product** AB to be the $m \times p$ matrix whose jth column is $A\mathbf{b}_j$. That is,

$$C = [A\mathbf{b}_1 \ A\mathbf{b}_2 \ \ldots \ A\mathbf{b}_p].$$

In particular, if A is an $m \times n$ matrix and B is an $n \times 1$ column vector, then the matrix product AB is defined and is the same as the matrix–vector product defined in Section 1.2.

In view of this definition and the preceding discussion, we have an *associative law* for the product of two matrices and a vector. (See Figure 2.1.)

> For any $m \times n$ matrix A, any $n \times p$ matrix B, and any $p \times 1$ vector $\mathbf{v}$,
>
> $$(AB)\mathbf{v} = A(B\mathbf{v}).$$

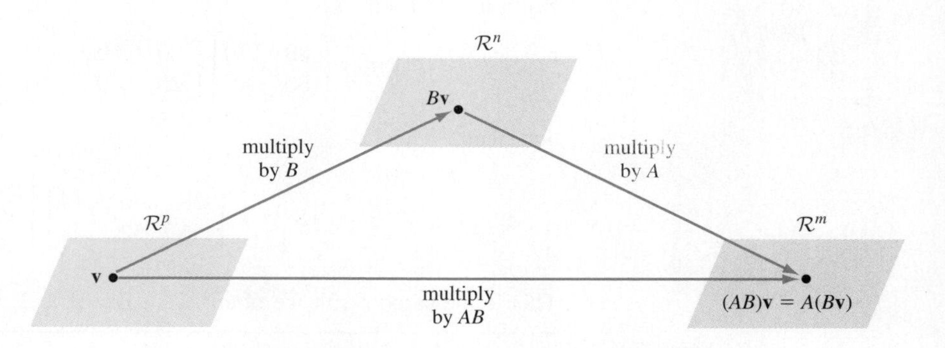

Figure 2.1 The associative law of multiplication

Later in this section, we extend this associative law to the product of any three matrices of compatible sizes. (See Theorem 2.1(b).)

Notice that when the sizes of A and B are written side by side in the same order as the product, that is, $(m \times n)(n \times p)$, the inner dimensions must be equal, and the outer dimensions give the size of the product AB. Symbolically,

$$(m \times n)(n \times p) = (m \times p).$$

Practice Problem 1 ▶ Suppose that A is a 2×4 matrix and B is a 2×3 matrix.

(a) Is the product BA^T defined? If so, what is its size?

(b) Is the product A^TB defined? If so, what is its size? ◀

Example 2 Let

$$A = \begin{bmatrix} 1 & 2 \\ 3 & 4 \\ 5 & 6 \end{bmatrix} \quad \text{and} \quad B = \begin{bmatrix} -1 & 1 \\ 3 & 2 \end{bmatrix}.$$

Notice that A has 2 columns and B has 2 rows. Then AB is the 3×2 matrix with first and second column

$$A\mathbf{b}_1 = \begin{bmatrix} 1 & 2 \\ 3 & 4 \\ 5 & 6 \end{bmatrix} \begin{bmatrix} -1 \\ 3 \end{bmatrix} = \begin{bmatrix} 5 \\ 9 \\ 13 \end{bmatrix} \quad \text{and} \quad A\mathbf{b}_2 = \begin{bmatrix} 1 & 2 \\ 3 & 4 \\ 5 & 6 \end{bmatrix} \begin{bmatrix} 1 \\ 2 \end{bmatrix} = \begin{bmatrix} 5 \\ 11 \\ 17 \end{bmatrix},$$

respectively. Thus

$$AB = [A\mathbf{b}_1 \; A\mathbf{b}_2] = \begin{bmatrix} 5 & 5 \\ 9 & 11 \\ 13 & 17 \end{bmatrix}.$$

Practice Problem 2 ► For $A = \begin{bmatrix} 2 & -1 & 3 \\ 1 & 4 & -2 \end{bmatrix}$ and $B = \begin{bmatrix} -1 & 0 & 2 \\ 0 & -3 & 4 \\ 3 & 1 & -2 \end{bmatrix}$, compute AB. ◄

Example 3 We return to Example 1 in this section and the matrix $A = \begin{bmatrix} .80 & .70 \\ .60 & .40 \end{bmatrix}$. Recall the matrix $B = \begin{bmatrix} .80 & .60 & .40 \\ .20 & .40 & .60 \end{bmatrix}$ and the vector $\mathbf{v} = \begin{bmatrix} 60 \\ 50 \\ 30 \end{bmatrix}$ from the related example in Section 1.2. Then

$$AB = \begin{bmatrix} .80 & .70 \\ .60 & .40 \end{bmatrix} \begin{bmatrix} .80 & .60 & .40 \\ .20 & .40 & .60 \end{bmatrix} = \begin{bmatrix} .78 & .76 & .74 \\ .56 & .52 & .48 \end{bmatrix},$$

and hence

$$(AB)\mathbf{w} = \begin{bmatrix} .78 & .76 & .74 \\ .56 & .52 & .48 \end{bmatrix} \begin{bmatrix} 60 \\ 50 \\ 30 \end{bmatrix} = \begin{bmatrix} 107 \\ 74 \end{bmatrix}.$$

This is the same result we obtained in Example 1, where we computed $A(B\mathbf{w})$.

Example 4 Recall the rotation matrix A_θ described in Section 1.2. Let $\mathbf{v}$ be a nonzero vector, and let α and β be angles. Then the expression $A_\beta(A_\alpha \mathbf{v}) = (A_\beta A_\alpha)\mathbf{v}$ is the result of rotating $\mathbf{v}$ by α followed by β. From Figure 2.2, we see that this is the same as rotating $\mathbf{v}$ by β followed by α, which is also the same as rotating $\mathbf{v}$ by the angle $\alpha + \beta$. Thus

$$(A_\beta A_\alpha)\mathbf{v} = (A_\alpha A_\beta)\mathbf{v} = A_{\alpha+\beta}\mathbf{v}.$$

Since this equation is valid for every vector $\mathbf{v}$ in $\mathcal{R}^2$, we may apply Theorem 1.3(e) to obtain the result

$$A_\beta A_\alpha = A_\alpha A_\beta = A_{\alpha+\beta}.$$

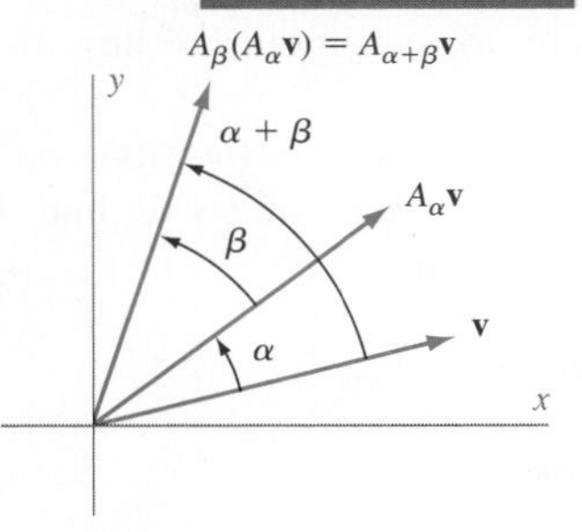

Figure 2.2 Rotating a vector **v** in $\mathcal{R}^2$ twice

Although $A_\beta A_\alpha = A_\alpha A_\beta$ is true for rotation matrices, as shown in Example 4, the matrix products AB and BA are seldom equal.

Matrix Multiplication Is Not Commutative

For arbitrary matrices A and B, AB need not equal BA.

In fact, if A is an $m \times n$ matrix and B is a $n \times p$ matrix, then BA is undefined unless $p = m$. If $p = m$, then AB is an $m \times m$ matrix and BA is an $n \times n$ matrix. Thus, even if both products are defined, AB and BA need not be of the same size. But even if $m = n$, AB might not equal BA, as the next example shows.

Example 5 Let

$$A = \begin{bmatrix} 0 & 0 & 1 \\ 0 & 1 & 0 \\ 1 & 0 & 0 \end{bmatrix} \quad \text{and} \quad B = \begin{bmatrix} 0 & 1 & 0 \\ 1 & 0 & 0 \\ 0 & 0 & 1 \end{bmatrix}.$$

Then

$$(AB)\mathbf{e}_1 = A(B\mathbf{e}_1) = A\mathbf{e}_2 = \mathbf{e}_2 \quad \text{and} \quad (BA)\mathbf{e}_1 = B(A\mathbf{e}_1) = B\mathbf{e}_3 = \mathbf{e}_3,$$

and hence $AB \neq BA$. (See Figure 2.3.)

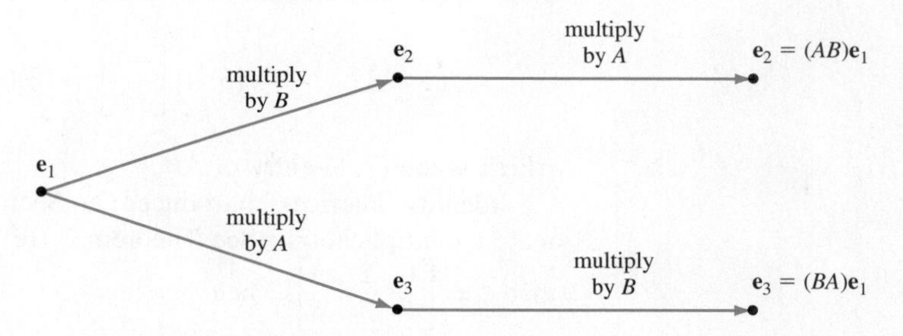

Figure 2.3 Matrix multiplication is not commutative.

Our definition of the matrix product tells how to compute a product AB by finding its columns, as in Example 2. However, there is also a method for computing an individual entry of the product, without calculating the entire column that contains it. This is often called the *row-column rule* and is useful when we need to find only some specific entries of the product. Observe that the (i,j)-entry of AB is the ith component of its jth column, $A\mathbf{b}_j$. This entry equals

$$[a_{i1} \; a_{i2} \; \dots \; a_{in}] \begin{bmatrix} b_{1j} \\ b_{2j} \\ \vdots \\ b_{nj} \end{bmatrix} = a_{i1}b_{1j} + a_{i2}b_{2j} + \cdots + a_{in}b_{nj},$$

which is the product of row i of A and column j of B. We can describe this formula by means of the following diagram:

$$\begin{bmatrix} a_{11} & a_{12} & \cdots & a_{1n} \\ \vdots & \vdots & & \vdots \\ a_{i1} & a_{i2} & \cdots & a_{in} \\ \vdots & \vdots & & \vdots \\ a_{m1} & a_{m2} & \cdots & a_{mn} \end{bmatrix} \begin{bmatrix} b_{11} & \cdots & b_{1j} & \cdots & b_{1p} \\ b_{21} & \cdots & b_{2j} & \cdots & b_{2p} \\ \vdots & & \vdots & & \vdots \\ b_{n1} & \cdots & b_{nj} & \cdots & b_{np} \end{bmatrix}$$

Row-Column Rule for the (i, j)-Entry of a Matrix Product

To compute the (i,j)-entry of the matrix product AB, locate the ith row of A and the jth column of B as in the preceding diagram. Moving across the ith row of A and down the jth column of B, multiply each entry of the row by the corresponding entry of the column. Then sum these products to obtain the (i,j)-entry of AB. In symbols, the (i,j)-entry of AB is

$$a_{i1}b_{1j} + a_{i2}b_{2j} + \cdots + a_{in}b_{nj}.$$

To illustrate this procedure, we compute the (2, 1)-entry of AB in Example 2. In this case, when we multiply each entry of the second row of A by the corresponding entry of the first column of B and sum the results, we get

$$[3 \;\; 4]\begin{bmatrix} -1 \\ 3 \end{bmatrix} = (3)(-1) + (4)(3) = 9,$$

which is the (2, 1)-entry of AB.

Identity matrices, introduced in Section 1.2, leave matrices unchanged under matrix multiplication. (See Theorem 2.1(e) and Exercise 56.) Suppose, for example, that $A = \begin{bmatrix} 1 & 2 & 3 \\ 4 & 5 & 6 \end{bmatrix}$. Then

$$I_2A = \begin{bmatrix} 1 & 0 \\ 0 & 1 \end{bmatrix}\begin{bmatrix} 1 & 2 & 3 \\ 4 & 5 & 6 \end{bmatrix} = \begin{bmatrix} 1 & 2 & 3 \\ 4 & 5 & 6 \end{bmatrix} = A$$

and

$$AI_3 = \begin{bmatrix} 1 & 2 & 3 \\ 4 & 5 & 6 \end{bmatrix}\begin{bmatrix} 1 & 0 & 0 \\ 0 & 1 & 0 \\ 0 & 0 & 1 \end{bmatrix} = \begin{bmatrix} 1 & 2 & 3 \\ 4 & 5 & 6 \end{bmatrix} = A.$$

Also, the product of a matrix and a zero matrix is a zero matrix because each column of the product is a zero vector. (See Theorem 1.3(f).)

The next theorem summarizes various properties of matrix multiplication and illustrates the interplay between matrix multiplication and the other matrix operations.

THEOREM 2.1

Let A and B be $k \times m$ matrices, C be an $m \times n$ matrix, and P and Q be $n \times p$ matrices. Then the following statements are true:

(a) $s(AC) = (sA)C = A(sC)$ for any scalar s.

(b) $A(CP) = (AC)P$. (associative law of matrix multiplication)

(c) $(A + B)C = AC + BC$. (right distributive law)

(d) $C(P + Q) = CP + CQ$. (left distributive law)

(e) $I_k A = A = AI_m$.

(f) The product of any matrix and a zero matrix is a zero matrix.

(g) $(AC)^T = C^T A^T$.

PROOF We prove (b), (c), and (g). The rest are left as exercises.

(b) First observe that both $A(CP)$ and $(AC)P$ are $k \times p$ matrices. Let $\mathbf{u}_j$ denote column j of CP. Since $\mathbf{u}_j = C\mathbf{p}_j$, column j of $A(CP)$ is $A\mathbf{u}_j = A(C\mathbf{p}_j)$. Furthermore, column j of $(AC)P$ is $(AC)\mathbf{p}_j = A(C\mathbf{p}_j)$ by the boxed result on page 97. It follows that the corresponding columns of $A(CP)$ and $(AC)P$ are equal. Therefore $A(CP) = (AC)P$.

(c) Both $(A + B)C$ and $AC + BC$ are $k \times n$ matrices, so we compare the corresponding columns of each matrix. For any j, $A\mathbf{c}_j$ and $B\mathbf{c}_j$ are the jth columns of AC and BC, respectively. But the jth column of $(A + B)C$ is

$$(A + B)\mathbf{c}_j = A\mathbf{c}_j + B\mathbf{c}_j,$$

by Theorem 1.3(c), and this is the jth column of $AC + BC$. This establishes (c).

(g) Both $(AC)^T$ and $C^T A^T$ are $n \times k$ matrices, so we compare the corresponding entries of each matrix. The (i,j)-entry of $(AC)^T$ is the (j,i)-entry of AC, which is

$$[a_{j1} \;\; a_{j2} \;\; \ldots \;\; a_{jm}] \begin{bmatrix} c_{1i} \\ c_{2i} \\ \vdots \\ c_{mi} \end{bmatrix} = a_{j1}c_{1i} + a_{j2}c_{2i} + \cdots + a_{jm}c_{mi}.$$

Also, the (i,j)-entry of $C^T A^T$ is the product of row i of C^T and column j of A^T, which is

$$[c_{1i} \;\; c_{2i} \;\; \ldots \;\; c_{mi}] \begin{bmatrix} a_{j1} \\ a_{j2} \\ \vdots \\ a_{jm} \end{bmatrix} = c_{1i}a_{j1} + c_{2i}a_{j2} + \cdots + c_{mi}a_{jm}.$$

Since the two displayed expressions are equal, the (i,j)-entry of $(AC)^T$ is equal to the (i,j)-entry of $C^T A^T$. This establishes (g). ■

The associative law of matrix multiplication, Theorem 2.1(b), allows us to omit parentheses when writing products of matrices. For this reason, we usually write ABC for a product of the matrices A, B, and C.

If A is an $n \times n$ matrix, we can form products of A with itself any number of times. As with real numbers, we use the exponential notation A^k to denote the product of A with itself k times. By convention, $A^1 = A$ and $A^0 = I_n$.

Example 6 Recall that we can use the stochastic matrix $A = \begin{bmatrix} .85 & .03 \\ .15 & .97 \end{bmatrix}$ in Example 3 of Section 1.2 to study population shifts between the city and suburbs. In that example, if the components of $\mathbf{p}$ are the current populations of the city and suburbs, then the components of $A\mathbf{p}$ are the populations of the city and suburbs for the next year. Extending the argument of that example to subsequent years, we see that $A^2\mathbf{p} = A(A\mathbf{p})$ is the vector whose components are the populations of the city and suburbs after two years. In general, for any positive integer m, $A^m\mathbf{p}$ is the vector whose components are the populations of the city and suburbs after m years. For example, if $\mathbf{p} = \begin{bmatrix} 500 \\ 700 \end{bmatrix}$ (as in Example 3 of Section 1.2), then the vector representing the city and suburb populations in ten years is given by

$$A^{10}\mathbf{p} \approx \begin{bmatrix} 241.2 \\ 958.8 \end{bmatrix}.$$

(Here the entries are rounded.)

A sociologist may need to determine the long-range trend that occurs in populations as a result of these annual shifts. In terms of matrix multiplication, this problem reduces to the study of the vectors $A^m\mathbf{p}$ as m increases. A deeper understanding of matrices, which we will acquire in Chapter 5, provides us with tools to understand the long-term behavior of this population better.

If A and B are matrices with the same number of rows, then we denote by $[A \ \ B]$ the matrix whose columns are the columns of A followed by the columns of B, in order. We call $[A \ \ B]$ an **augmented matrix**. For example, if

$$A = I_2 = \begin{bmatrix} 1 & 0 \\ 0 & 1 \end{bmatrix} \quad \text{and} \quad B = \begin{bmatrix} 2 & 0 & 1 \\ -1 & 3 & 1 \end{bmatrix},$$

then $[A \ \ B]$ is the 2×5 matrix

$$[A \ \ B] = \begin{bmatrix} 1 & 0 & 2 & 0 & 1 \\ 0 & 1 & -1 & 3 & 1 \end{bmatrix}.$$

Occasionally, it is useful to include a vertical line when displaying an augmented matrix $[A \ \ B]$ in order to visually separate the columns of A from those of B. Thus, for these matrices A and B, we may write

$$[A \ \ B] = \left[\begin{array}{cc|ccc} 1 & 0 & 2 & 0 & 1 \\ 0 & 1 & -1 & 3 & 1 \end{array}\right].$$

Because of the way that matrix multiplication is defined, it is easy to see that if P is an $m \times n$ matrix and A and B are matrices with n rows, then $P[A \ \ B] = [PA \ \ PB]$.

Example 7 Let

$$A = I_2 = \begin{bmatrix} 1 & 0 \\ 0 & 1 \end{bmatrix}, \quad B = \begin{bmatrix} 2 & 0 & 1 \\ -1 & 3 & 1 \end{bmatrix}, \quad \text{and} \quad P = \begin{bmatrix} 1 & 2 \\ 2 & -1 \\ 0 & 1 \end{bmatrix}.$$

Use the equation $P[A \ \ B] = [PA \ \ PB]$ to compute the product $P[A \ \ B]$.

Solution Observe that $PA = PI_2 = P$ and

$$PB = \begin{bmatrix} 1 & 2 \\ 2 & -1 \\ 0 & 1 \end{bmatrix} \begin{bmatrix} 2 & 0 & 1 \\ -1 & 3 & 1 \end{bmatrix} = \begin{bmatrix} 0 & 6 & 3 \\ 5 & -3 & 1 \\ -1 & 3 & 1 \end{bmatrix}.$$

Therefore

$$P[A\ \ B] = [PA\ \ PB] = \left[\begin{array}{rr|rrr} 1 & 2 & 0 & 6 & 3 \\ 2 & -1 & 5 & -3 & 1 \\ 0 & 1 & -1 & 3 & 1 \end{array}\right].$$

Practice Problem 3 ▶ Let

$$A = \begin{bmatrix} 1 & 3 & -1 \\ 2 & 5 & 4 \end{bmatrix}, \quad B = \begin{bmatrix} 2 & -2 \\ 0 & 3 \\ -4 & 1 \end{bmatrix}, \quad \text{and} \quad C = \begin{bmatrix} 3 & 0 & -1 \\ 2 & 1 & 5 \\ -6 & 0 & 2 \end{bmatrix}.$$

Find the fourth column of $A[B\ \ C]$ without computing the entire matrix. ◀

SPECIAL MATRICES

We briefly examine matrices with special properties that will be of interest later in this text. We begin with *diagonal matrices*, which are important because of their simplicity. The (i, j)-entry of a matrix A is called a **diagonal entry** if $i = j$. The diagonal entries form the **diagonal** of A. A square matrix A is called a **diagonal matrix** if all its nondiagonal entries are zeros. For example, identity matrices and square zero matrices are diagonal matrices.

If A and B are $n \times n$ diagonal matrices, then AB is also an $n \times n$ diagonal matrix. Moreover, the diagonal entries of AB are the products of the corresponding diagonal entries of A and B. (See Exercise 60.) For example, suppose that

$$A = \begin{bmatrix} 1 & 0 & 0 \\ 0 & 2 & 0 \\ 0 & 0 & 3 \end{bmatrix} \quad \text{and} \quad B = \begin{bmatrix} 3 & 0 & 0 \\ 0 & -1 & 0 \\ 0 & 0 & 2 \end{bmatrix}.$$

Then

$$AB = \begin{bmatrix} 3 & 0 & 0 \\ 0 & -2 & 0 \\ 0 & 0 & 6 \end{bmatrix}.$$

The relationship between diagonal and nondiagonal square matrices will be studied in depth in Chapter 5, where we will see that in many circumstances, an ordinary square matrix can be replaced by a diagonal matrix, thus simplifying theoretical arguments as well as computations.

Observe that any diagonal matrix is equal to its transpose. In general, a (square) matrix A is called **symmetric** if $A^T = A$. For example, diagonal matrices are symmetric. As another example, let

$$A = \begin{bmatrix} 1 & 2 & 4 \\ 2 & 3 & -1 \\ 4 & -1 & 5 \end{bmatrix}.$$

Then

$$A^T = \begin{bmatrix} 1 & 2 & 4 \\ 2 & 3 & -1 \\ 4 & -1 & 5 \end{bmatrix}^T = \begin{bmatrix} 1 & 2 & 4 \\ 2 & 3 & -1 \\ 4 & -1 & 5 \end{bmatrix} = A.$$

So A is symmetric. In general, a square matrix A is symmetric if and only if $a_{ij} = a_{ji}$ for all i and j.

EXERCISES

In Exercises 1–3, decide whether each matrix product AB is defined. If so, find its size.

1. A is a 2×3 matrix, and B^T is a 2×3 matrix.
2. A is a 2×4 matrix, and B is a 4×6 matrix.
3. A^T is a 3×3 matrix, and B is a 2×3 matrix.
4. Give an example of matrices A and B such that BA is defined, but AB is not.

In Exercises 5–20, use the following matrices to compute each expression, or give a reason why the expression is not defined:

$$A = \begin{bmatrix} 1 & -2 \\ 3 & 4 \end{bmatrix} \quad B = \begin{bmatrix} 7 & 4 \\ 1 & 2 \end{bmatrix} \quad C = \begin{bmatrix} 3 & 8 & 1 \\ 2 & 0 & 4 \end{bmatrix}$$

$$\mathbf{x} = \begin{bmatrix} 2 \\ 3 \end{bmatrix} \quad \mathbf{y} = \begin{bmatrix} 1 \\ 3 \\ -5 \end{bmatrix} \quad \mathbf{z} = [7 \ -1]$$

5. $C\mathbf{y}$ 6. $B\mathbf{x}$ 7. $\mathbf{xz}$ 8. $B\mathbf{y}$
9. $AC\mathbf{x}$ 10. $A\mathbf{z}^T$ 11. AB 12. AC
13. BC 14. BA 15. CB^T 16. CB
17. A^3 18. A^2 19. C^2 20. B^2

In Exercises 21–24, use the matrices A, B, C, and **z** *from Exercises 5–20.*

21. Verify that $I_2C = CI_3 = C$.
22. Verify that $(AB)C = A(BC)$.
23. Verify that $(AB)^T = B^TA^T$.
24. Verify that $\mathbf{z}(AC) = (\mathbf{z}A)C$.

In Exercises 25–32, use the following matrices to compute each requested entry or column of the matrix product, without computing the entire matrix:

$$A = \begin{bmatrix} 1 & 2 & 3 \\ 2 & -1 & 4 \\ -3 & -2 & 0 \end{bmatrix}, \ B = \begin{bmatrix} -1 & 0 \\ 4 & 1 \\ 3 & -2 \end{bmatrix}, \ C = \begin{bmatrix} 2 & 1 & -1 \\ 4 & 3 & -2 \end{bmatrix}$$

25. the $(3, 2)$-entry of AB
26. the $(2, 1)$-entry of BC
27. the $(2, 3)$-entry of CA
28. the $(1, 1)$ entry of CB
29. column 2 of AB
30. column 3 of BC
31. column 1 of CA
32. column 2 of CB

T&F *In Exercises 33–50, determine whether the statements are true or false.*

33. The product of two $m \times n$ matrices is defined.
34. For any matrices A and B, if the product AB is defined, then the product BA is also defined.
35. For any matrices A and B, if the products AB and BA are both defined, then $AB = BA$.
36. If A is a square matrix, then A^2 is defined.
37. If A and B are matrices, then both AB and BA are defined if and only if A and B are square matrices.
38. If A is an $m \times n$ matrix and B is an $n \times p$ matrix, then $(AB)^T = A^TB^T$.
39. There exist nonzero matrices A and B for which $AB = BA$.
40. For any matrices A and B for which the product AB is defined, the jth column of AB equals the matrix–vector product of A and the jth column of B.
41. For any matrices A and B for which the product AB is defined, the (i,j)-entry of AB equals $a_{ij}b_{ij}$.
42. For any matrices A and B for which the product AB is defined, the (i,j)-entry of AB equals the sum of the products of corresponding entries from the ith column of A and the jth row of B.
43. If A, B, and C are matrices for which the product $A(BC)$ is defined, then $A(BC) = (AB)C$.
44. If A and B are $m \times n$ matrices and C is an $n \times p$ matrix, then $(A + B)C = AB + BC$.
45. If A and B are $n \times n$ matrices, then the diagonal entries of the product matrix AB are $a_{11}b_{11}, a_{22}b_{22}, \dots, a_{nn}b_{nn}$.
46. If the product AB is defined and either A or B is a zero matrix, then AB is a zero matrix.
47. If the product AB is defined and AB is a zero matrix, then either A or B is a zero matrix.
48. If A_α and A_β are both 2×2 rotation matrices, then $A_\alpha A_\beta$ is a 2×2 rotation matrix.
49. The product of two diagonal matrices is a diagonal matrix.
50. In a symmetric $n \times n$ matrix, the (i,j)- and (j,i)-entries are equal for all $i = 1, 2, \dots n$ and $j = 1, 2, \dots, n$.

51. Let

$$\begin{array}{cc} & \text{From} \\ & \begin{array}{cc}\text{City} & \text{Suburbs}\end{array} \\ \text{To} \begin{array}{c}\text{City} \\ \text{Suburbs}\end{array} & \begin{bmatrix} .85 & .03 \\ .15 & .97 \end{bmatrix} = A \end{array}$$

be the stochastic matrix used in Example 6 to predict population movement between the city and its suburbs. Suppose that 70% of city residents live in single-unit houses (as opposed to multiple-unit or apartment housing) and that 95% of suburb residents live in single-unit houses.

(a) Find a 2×2 matrix B such that if v_1 people live in the city and v_2 people live in the suburbs, then $B\begin{bmatrix} v_1 \\ v_2 \end{bmatrix} = \begin{bmatrix} u_1 \\ u_2 \end{bmatrix}$, where u_1 people live in single-unit houses and u_2 people live in multiple-unit houses.

(b) Explain the significance of $BA\begin{bmatrix} v_1 \\ v_2 \end{bmatrix}$.

52. Of those vehicle owners who live in the city, 60% drive cars, 30% drive vans, and 10% drive recreational vehicles. Of those vehicle owners who live in the suburbs, 30% drive cars, 50% drive vans, and 20% drive recreational vehicles. Of all vehicles (in the city and suburbs), 60% of the cars, 40% of the vans, and 50% of the recreational vehicles are dark in color.

(a) Find a matrix B such that if v_1 vehicle owners live in the city and v_2 vehicle owners live in the suburbs, then $B\begin{bmatrix} v_1 \\ v_2 \end{bmatrix} = \begin{bmatrix} u_1 \\ u_2 \\ u_3 \end{bmatrix}$, where u_1 people drive cars, u_2 people drive vans, and u_3 people drive recreational vehicles.

(b) Find a matrix A such that $A\begin{bmatrix} u_1 \\ u_2 \\ u_3 \end{bmatrix} = \begin{bmatrix} w_1 \\ w_2 \end{bmatrix}$, where u_1, u_2, and u_3 are as in (a), w_1 is the number of people who drive dark vehicles, and w_2 is the number of people who drive light (not dark) vehicles.

(c) Find a matrix C such that $\begin{bmatrix} w_1 \\ w_2 \end{bmatrix} = C\begin{bmatrix} v_1 \\ v_2 \end{bmatrix}$, where v_1 and v_2 are as in (a) and w_1 and w_2 are as in (b).

53. In a certain elementary school, it has been found that of those pupils who buy a hot lunch on a particular school day, 30% buy a hot lunch and 70% bring a bag lunch on the next school day. Furthermore, of those pupils who bring a bag lunch on a particular school day, 40% buy a hot lunch and 60% bring a bag lunch on the next school day.

(a) Find a matrix A such that if u_1 pupils buy a hot lunch and u_2 pupils bring a bag lunch on a particular day, then $A\begin{bmatrix} u_1 \\ u_2 \end{bmatrix} = \begin{bmatrix} v_1 \\ v_2 \end{bmatrix}$, where v_1 pupils buy a hot lunch and v_2 pupils bring a bag lunch on the next school day.

(b) Suppose that $u_1 = 100$ pupils buy a hot lunch and $u_2 = 200$ pupils bring a bag lunch on the first day of school. Compute $A\begin{bmatrix} u_1 \\ u_2 \end{bmatrix}$, $A^2\begin{bmatrix} u_1 \\ u_2 \end{bmatrix}$, and $A^3\begin{bmatrix} u_1 \\ u_2 \end{bmatrix}$. Explain the significance of each result.

(c) To do this problem, you will need a calculator with matrix capabilities or access to computer software such as MATLAB. Using the notation of (b), compute $A^{100}\begin{bmatrix} u_1 \\ u_2 \end{bmatrix} = \begin{bmatrix} w_1 \\ w_2 \end{bmatrix}$. Explain the significance of this result. Now compute $A\begin{bmatrix} w_1 \\ w_2 \end{bmatrix}$, and compare this result with $\begin{bmatrix} w_1 \\ w_2 \end{bmatrix}$. Explain.

54. Prove (a) of Theorem 2.1.

55. Prove (d) of Theorem 2.1.

56. Prove (e) of Theorem 2.1.

57. Prove (f) of Theorem 2.1.

58. Let $A = A_{180°}$, and let B be the matrix that reflects $\mathcal{R}^2$ about the x-axis; that is,

$$B = \begin{bmatrix} 1 & 0 \\ 0 & -1 \end{bmatrix}.$$

Compute BA, and describe geometrically how a vector $\mathbf{v}$ is affected by multiplication by BA.

59. A square matrix A is called **lower triangular** if the (i, j)-entry of A is zero whenever $i < j$. Prove that if A and B are both $n \times n$ lower triangular matrices, then AB is also a lower triangular matrix.

60. Let A be an $n \times n$ matrix.

(a) Prove that A is a diagonal matrix if and only if its jth column equals $a_{jj}\mathbf{e}_j$.

(b) Use (a) to prove that if A and B are $n \times n$ diagonal matrices, then AB is a diagonal matrix whose jth column is $a_{jj}b_{jj}\mathbf{e}_j$.

61. A square matrix A is called **upper triangular** if the (i, j)-entry of A is zero whenever $i > j$. Prove that if A and B are both $n \times n$ upper triangular matrices, then AB is also an upper triangular matrix.

62. Let $A = \begin{bmatrix} 1 & -1 & 2 & -1 \\ -2 & 1 & -1 & 3 \\ -1 & -1 & 4 & 3 \\ -5 & 3 & -4 & 7 \end{bmatrix}$. Find a nonzero 4×2 matrix B with rank 2 such that $AB = O$.

63. Find an example of $n \times n$ matrices A and B such that $AB = O$, but $BA \neq O$.

64. Let A and B be $n \times n$ matrices. Prove or disprove that the ranks of AB and BA are equal.

65. Recall the definition of the *trace* of a matrix, given in Exercise 82 of Section 1.1. Prove that if A is an $m \times n$ matrix and B is an $n \times m$ matrix, then $\text{trace}(AB) = \text{trace}(BA)$.

66. Let $1 \leq r, s \leq n$ be integers, and let E be the $n \times n$ matrix with 1 as the (r, s)-entry and 0s elsewhere. Let B be any $n \times n$ matrix. Describe EB in terms of the entries of B.
67. Prove that if A is a $k \times m$ matrix, B is an $m \times n$ matrix, and C is an $n \times p$ matrix, then $(ABC)^T = C^T B^T A^T$.
68. (a) Let A and B be symmetric matrices of the same size. Prove that AB is symmetric if and only $AB = BA$.
    (b) Find symmetric 2×2 matrices A and B such that $AB \neq BA$.

In Exercises 69–72, use either a calculator with matrix capabilities or computer software such as MATLAB to solve each problem.

69. Let A_θ be the θ-rotation matrix.
    (a) For $\theta = \pi/2$, compute A_θ^2 by hand.
    (b) For $\theta = \pi/3$, compute A_θ^3.
    (c) For $\theta = \pi/8$, compute A_θ^8.
    (d) Use the previous results to make a conjecture about A_θ^k, where $\theta = \pi/k$.
    (e) Draw a sketch to support your conjecture in (d).
70. Let A, B, and C be 4×4 random matrices.
    (a) Illustrate the distributive law $A(B + C) = AB + AC$.
    (b) Check the validity of the equation
$$(A + B)^2 = A^2 + 2AB + B^2.$$
    (c) Make a conjecture about an expression that is equal to $(A + B)^2$ for arbitrary $n \times n$ matrices A and B. Justify your conjecture.
    (d) Make a conjecture about the relationship that must hold between AB and BA for the equation in (b) to hold in general.
    (e) Prove your conjecture in (d).
71. Let A be the stochastic matrix used in the population application in Example 6.
    (a) Verify that $A^{10}\mathbf{p}$ is as given in the example.
    (b) Determine the populations of the city and suburbs after 20 years.
    (c) Determine the populations of the city and suburbs after 50 years.
    (d) Make a conjecture about the eventual populations of the city and suburbs.
72. Let A and B be 4×4 random matrices.
    (a) Compute AB and its rank by finding the reduced row echelon form of AB.
    (b) Compute BA and its rank by finding the reduced row echelon form of BA.
    (c) Compare your answers with your solution to Exercise 64.

SOLUTIONS TO THE PRACTICE PROBLEMS

1. (a) Since B is a 2×3 matrix and A^T is a 4×2 matrix, the product BA^T is not defined.
   (b) Since A^T is a 4×2 matrix and B is a 2×3 matrix, the product $A^T B$ is defined. Its size is 4×3.
2. The matrix AB is a 2×3 matrix. Its first column is
$$\begin{bmatrix} 2 & -1 & 3 \\ 1 & 4 & -2 \end{bmatrix} \begin{bmatrix} -1 \\ 0 \\ 3 \end{bmatrix} = \begin{bmatrix} -2 + 0 + 9 \\ -1 + 0 - 6 \end{bmatrix} = \begin{bmatrix} 7 \\ -7 \end{bmatrix},$$
its second column is
$$\begin{bmatrix} 2 & -1 & 3 \\ 1 & 4 & -2 \end{bmatrix} \begin{bmatrix} 0 \\ -3 \\ 1 \end{bmatrix} = \begin{bmatrix} 0 + 3 + 3 \\ 0 - 12 - 2 \end{bmatrix} = \begin{bmatrix} 6 \\ -14 \end{bmatrix},$$
and its third column is
$$\begin{bmatrix} 2 & -1 & 3 \\ 1 & 4 & -2 \end{bmatrix} \begin{bmatrix} 2 \\ 4 \\ -2 \end{bmatrix} = \begin{bmatrix} 4 - \ \ 4 - 6 \\ 2 + 16 + 4 \end{bmatrix} = \begin{bmatrix} -6 \\ 22 \end{bmatrix}.$$
Hence $AB = \begin{bmatrix} 7 & 6 & -6 \\ -7 & -14 & 22 \end{bmatrix}$.
3. Since the fourth column of $[B \;\; C]$ is $\mathbf{e}_2$, the fourth column of $A[B \;\; C]$ is
$$A\mathbf{e}_2 = \mathbf{a}_2 = \begin{bmatrix} 3 \\ 5 \end{bmatrix}.$$

2.2* APPLICATIONS OF MATRIX MULTIPLICATION

In this section, we present four applications of matrix multiplication.

THE LESLIE MATRIX AND POPULATION CHANGE

The population of a colony of animals depends on the birth and mortality rates for the various age groups of the colony. For example, suppose that the members of

* This section can be omitted without loss of continuity.

a colony of mammals have a life span of less than 3 years. To study the birth rates of the colony, we divide the females into three age groups: those with ages less than 1, those with ages between 1 and 2, and those of age 2. From the mortality rates of the colony, we know that 40% of newborn females survive to age 1 and that 50% of females of age 1 survive to age 2. We need to observe only the rates at which females in each age group give birth to female offspring since there is usually a known relationship between the number of male and female offspring in the colony. Suppose that the females under 1 year of age do not give birth; those with ages between 1 and 2 have, on average, two female offspring; and those of age 2 have, on average, one female offspring. Let x_1, x_2, and x_3 be the numbers of females in the first, second, and third age groups, respectively, at the present time, and let y_1, y_2, and y_3 be the numbers of females in the corresponding groups for the next year. The changes from this year to next year are depicted in Table 2.1.

Table 2.1

Age in years	Current year	Next year
0–1	x_1	y_1
1–2	x_2	y_2
2–3	x_3	y_3

The vector $\mathbf{x} = \begin{bmatrix} x_1 \\ x_2 \\ x_3 \end{bmatrix}$ is the **population distribution** for the female population of the colony in the present year. We can use the preceding information to predict the population distribution for the following year, which is given by the vector $\mathbf{y} = \begin{bmatrix} y_1 \\ y_2 \\ y_3 \end{bmatrix}$.

Note that y_1, the number of females under age 1 in next year's population, is simply equal to the number of female offspring born during the current year. Since there are currently x_2 females of age 1–2, each of which has, on average, 2 female offspring, and x_3 females of age 2–3, each of which has, on average, 1 female offspring, we have the following formula for y_1:

$$y_1 = 2x_2 + x_3$$

The number y_2 is the total number of females in the second age group for next year. Because these females are in the first age group this year, and because only 40% of them will survive to the next year, we have that $y_2 = 0.4x_1$. Similarly, $y_3 = 0.5x_2$. Collecting these three equations, we have

$$\begin{aligned} y_1 &= \phantom{0.4x_1 + {}} 2.0x_2 + 1.0x_3 \\ y_2 &= 0.4x_1 \\ y_3 &= \phantom{0.4x_1 + {}} 0.5x_2 . \end{aligned}$$

These three equations can be represented by the single matrix equation $\mathbf{y} = A\mathbf{x}$, where $\mathbf{x}$ and $\mathbf{y}$ are the population distributions as previously defined and A is the 3×3 matrix

$$A = \begin{bmatrix} 0.0 & 2.0 & 1.0 \\ 0.4 & 0.0 & 0.0 \\ 0.0 & 0.5 & 0.0 \end{bmatrix}.$$

For example, suppose that $\mathbf{x} = \begin{bmatrix} 1000 \\ 1000 \\ 1000 \end{bmatrix}$; that is, there are currently 1000 females in each age group. Then

$$\mathbf{y} = A\mathbf{x} = \begin{bmatrix} 0.0 & 2.0 & 1.0 \\ 0.4 & 0.0 & 0.0 \\ 0.0 & 0.5 & 0.0 \end{bmatrix} \begin{bmatrix} 1000 \\ 1000 \\ 1000 \end{bmatrix} = \begin{bmatrix} 3000 \\ 400 \\ 500 \end{bmatrix}.$$

So one year later there are 3000 females under 1 year of age, 400 females who are between 1 and 2 years old, and 500 females who are 2 years old.

For each positive integer k, let $\mathbf{p}_k$ denote the population distribution k years after a given initial population distribution $\mathbf{p}_0$. In the preceding example,

$$\mathbf{p}_0 = \mathbf{x} = \begin{bmatrix} 1000 \\ 1000 \\ 1000 \end{bmatrix} \quad \text{and} \quad \mathbf{p}_1 = \mathbf{y} = \begin{bmatrix} 3000 \\ 400 \\ 500 \end{bmatrix}.$$

Then, for any positive integer k, we have that $\mathbf{p}_k = A\mathbf{p}_{k-1}$. Thus

$$\mathbf{p}_k = A\mathbf{p}_{k-1} = A^2\mathbf{p}_{k-2} = \cdots = A^k\mathbf{p}_0.$$

In this way, we may predict population trends over the long term. For example, to predict the population distribution after 10 years, we compute $\mathbf{p}_{10} = A^{10}\mathbf{p}_0$. Thus

$$\mathbf{p}_{10} = A^{10}\mathbf{p}_0 = \begin{bmatrix} 1987 \\ 851 \\ 387 \end{bmatrix},$$

where each entry is rounded off to the nearest whole number. If we continue this process in increments of 10 years, we find that (rounding to whole numbers)

$$\mathbf{p}_{20} = \begin{bmatrix} 2043 \\ 819 \\ 408 \end{bmatrix} \quad \text{and} \quad \mathbf{p}_{30} = \mathbf{p}_{40} = \begin{bmatrix} 2045 \\ 818 \\ 409 \end{bmatrix}.$$

It appears that the population stabilizes after 30 years. In fact, for the vector

$$\mathbf{z} = \begin{bmatrix} 2045 \\ 818 \\ 409 \end{bmatrix},$$

we have that $A\mathbf{z} = \mathbf{z}$ precisely. Under this circumstance, the population distribution $\mathbf{z}$ is stable; that is, it does not change from year to year.

In general, whether or not the distribution of an animal population stabilizes for a colony depends on the survival and birth rates of the age groups. (See, for example, Exercises 12–15.) Exercise 10 gives an example of a population for which no nonzero stable population distribution exists.

We can generalize this situation to an arbitrary colony of animals. Suppose that we divide the females of the colony into n age groups, where x_i is the number of members in the ith group. The duration of time in an individual age group need not be a year, but the various durations should be equal. Let $\mathbf{x} = \begin{bmatrix} x_1 \\ x_2 \\ \vdots \\ x_n \end{bmatrix}$ be the population distribution of the females of the colony, p_i be the portion of females in the ith group who survive to the $(i+1)$st group, and b_i be the average number of female offspring of a member of the ith age group. If $\mathbf{y} = \begin{bmatrix} y_1 \\ y_2 \\ \vdots \\ y_n \end{bmatrix}$ is the population for the next time period, then

$$\begin{aligned} y_1 &= b_1x_1 + b_2x_2 + \cdots + b_nx_n \\ y_2 &= p_1x_1 \\ y_3 &= p_2x_2 \\ &\vdots \\ y_n &= p_{n-1}x_{n-1}. \end{aligned}$$

Therefore, for

$$A = \begin{bmatrix} b_1 & b_2 & \cdots & & b_n \\ p_1 & 0 & \cdots & & 0 \\ 0 & p_2 & \cdots & & 0 \\ \vdots & \vdots & & & \vdots \\ 0 & 0 & \cdots & p_{n-1} & 0 \end{bmatrix},$$

we have

$$\mathbf{y} = A\mathbf{x}.$$

The matrix A is called the **Leslie matrix** for the population. The name is due to P. H. Leslie, who introduced this matrix in the 1940s. So if $\mathbf{x}_0$ is the initial population distribution, then the distribution after k time intervals is

$$\mathbf{x}_k = A^k\mathbf{x}_0.$$

Practice Problem 1 ▶ The life span of a certain species of mammal is at most 2 years, but only 25% of the females of this species survive to age 1. Suppose that, on average, the females under 1 year of age give birth to 0.5 females, and the females between 1 and 2 years of age give birth to 2 females.

(a) Write the Leslie matrix for the population.

(b) Suppose that this year there is a population of 200 females under 1 year of age and 200 females between 1 and 2 years of age. Find the population distribution for next year and for 2 years from now.

(c) Suppose this year's population distribution of females is given by the vector $\begin{bmatrix} 400 \\ 100 \end{bmatrix}$. What can you say about all future population distributions?

(d) Suppose that the total population of females this year is 600. What should be the number of females in each age group so that the population distribution remains unchanged from year to year? ◀

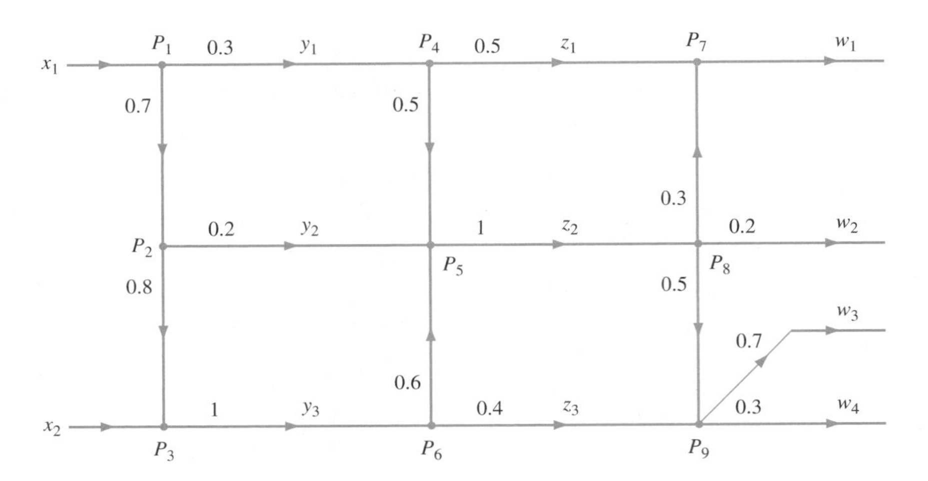

Figure 2.4 Traffic flow along one-way streets

ANALYSIS OF TRAFFIC FLOW

Figure 2.4 represents the flow of traffic through a network of one-way streets, with arrows indicating the direction of traffic flow. The number on any street beyond an intersection is the portion of the traffic entering the street from that intersection. For example, 30% of the traffic leaving intersection P_1 goes to P_4, and the other 70% goes to P_2. Notice that all the traffic leaving P_5 goes to P_8.

Suppose that on a particular day, x_1 cars enter the network from the left of P_1, and x_2 cars enter from the left of P_3. Let w_1, w_2, w_3, and w_4 represent the number of cars leaving the network along the exits to the right. We wish to determine the values of the w_i's. At first glance, this problem seems overwhelming since there are so many routes for the traffic. However, if we decompose the problem into several simpler ones, we can first solve the simpler ones individually and then combine their solutions to obtain the values of the w_i's.

We begin with only the portion of the network involving intersections P_1, P_2, and P_3. Let y_1, y_2, and y_3 each be the number of cars that exit along each of the three eastward routes, respectively. To find y_1, notice that 30% of all cars entering P_1 continue on to P_4. Therefore $y_1 = 0.30x_1$. Also, $0.7x_1$ of the cars turn right at P_1, and of these, 20% turn left at P_2. Because these are the only cars to do so, it follows that $y_2 = (0.2)(0.7)x_1 = 0.14x_1$. Furthermore, since 80% of the cars entering P_2 continue on to P_3, the number of such cars is $(0.8)(0.7)x_1 = 0.56x_1$. Finally, all the cars entering P_3 from the left use the street between P_3 and P_6, so $y_3 = 0.56x_1 + x_2$. Summarizing, we have

$$\begin{aligned} y_1 &= 0.30x_1 \\ y_2 &= 0.14x_1 \\ y_3 &= 0.56x_1 + x_2. \end{aligned}$$

We can express this system of equations by the single matrix equation $\mathbf{y} = A\mathbf{x}$, where

$$\mathbf{y} = \begin{bmatrix} y_1 \\ y_2 \\ y_3 \end{bmatrix} \qquad A = \begin{bmatrix} 0.30 & 0 \\ 0.14 & 0 \\ 0.56 & 1 \end{bmatrix} \qquad \text{and} \qquad \mathbf{x} = \begin{bmatrix} x_1 \\ x_2 \end{bmatrix}.$$

Now consider the next set of intersections P_4, P_5, and P_6. If we let z_1, z_2, and z_3 represent the numbers of cars that exit from the right of P_4, P_5, and P_6, respectively, then by a similar analysis, we have

$$\begin{aligned} z_1 &= 0.5y_1 \\ z_2 &= 0.5y_1 + y_2 + 0.6y_3 \\ z_3 &= 0.4y_3, \end{aligned}$$

or $\mathbf{z} = B\mathbf{y}$, where

$$\mathbf{z} = \begin{bmatrix} z_1 \\ z_2 \\ z_3 \end{bmatrix} \quad \text{and} \quad B = \begin{bmatrix} 0.5 & 0 & 0 \\ 0.5 & 1 & 0.6 \\ 0 & 0 & 0.4 \end{bmatrix}.$$

Finally, if we set

$$\mathbf{w} = \begin{bmatrix} w_1 \\ w_2 \\ w_3 \\ w_4 \end{bmatrix} \quad \text{and} \quad C = \begin{bmatrix} 1 & 0.30 & 0 \\ 0 & 0.20 & 0 \\ 0 & 0.35 & 0.7 \\ 0 & 0.15 & 0.3 \end{bmatrix},$$

then by a similar argument, we have $\mathbf{w} = C\mathbf{z}$. It follows that

$$\mathbf{w} = C\mathbf{z} = C(B\mathbf{y}) = (CB)A\mathbf{x} = (CBA)\mathbf{x}.$$

Let $M = CBA$. Then

$$M = \begin{bmatrix} 1 & 0.30 & 0 \\ 0 & 0.20 & 0 \\ 0 & 0.35 & 0.7 \\ 0 & 0.15 & 0.3 \end{bmatrix} \begin{bmatrix} 0.5 & 0 & 0 \\ 0.5 & 1 & 0.6 \\ 0 & 0 & 0.4 \end{bmatrix} \begin{bmatrix} 0.30 & 0 \\ 0.14 & 0 \\ 0.56 & 1 \end{bmatrix} = \begin{bmatrix} 0.3378 & 0.18 \\ 0.1252 & 0.12 \\ 0.3759 & 0.49 \\ 0.1611 & 0.21 \end{bmatrix}.$$

For example, if 1000 cars enter the traffic pattern at P_1 and 2000 enter at P_3, then, for $\mathbf{x} = \begin{bmatrix} 1000 \\ 2000 \end{bmatrix}$, we have

$$\mathbf{w} = M\mathbf{x} = \begin{bmatrix} 0.3378 & 0.18 \\ 0.1252 & 0.12 \\ 0.3759 & 0.49 \\ 0.1611 & 0.21 \end{bmatrix} \begin{bmatrix} 1000 \\ 2000 \end{bmatrix} = \begin{bmatrix} 697.8 \\ 365.2 \\ 1355.9 \\ 581.1 \end{bmatrix}.$$

Naturally, the actual number of cars traveling on any path is a whole number, unlike the entries of $\mathbf{w}$. Since these calculations are based on percentages, we cannot expect the answers to be exact. For example, approximately 698 cars exit the traffic pattern at P_7, and 365 cars exit the pattern at P_8.

We can apply the same analysis if the quantities studied represent *rates* of traffic flow—for example, the number of cars per hour—rather than the total number of cars.

Finally, we can apply this kind of analysis to other contexts, such as the flow of a fluid through a system of pipes or the movement of money in an economy. For other examples, see the exercises.

Practice Problem 2 ▶ A midwestern supermarket chain imports soy sauce from Japan and South Korea. Of the soy sauce from Japan, 50% is shipped to Seattle and the rest is shipped to San Francisco. Of the soy sauce from South Korea, 60% is shipped to San Francisco and the rest is shipped to Los Angeles. All of the soy sauce shipped to Seattle is sent to

Chicago; 30% of the soy sauce shipped to San Francisco is sent to Chicago and 70% to St. Louis; and all of the soy sauce shipped to Los Angeles is sent to St. Louis. Suppose that soy sauce is shipped from Japan and South Korea at the rates of 10,000 and 5,000 barrels a year, respectively. Find the number of barrels of soy sauce that are sent to Chicago and to St. Louis each year. ◀

(0, 1)-MATRICES

Matrices can be used to study certain relationships between objects. For example, suppose that there are five countries, each of which maintains diplomatic relations with some of the others. To organize these relationships, we use a 5×5 matrix A defined as follows. For $1 \leq i \leq 5$, we let $a_{ii} = 0$, and for $i \neq j$,

$$a_{ij} = \begin{cases} 1 & \text{if country } i \text{ maintains diplomatic relations with country } j \\ 0 & \text{otherwise.} \end{cases}$$

Note that all the entries of A are zeros and ones. Matrices whose only entries are zeros and ones are called **(0, 1)-matrices**, and they are worthy of study in their own right. For purposes of illustration, suppose that

$$A = \begin{bmatrix} 0 & 0 & 1 & 1 & 0 \\ 0 & 0 & 0 & 1 & 1 \\ 1 & 0 & 0 & 0 & 1 \\ 1 & 1 & 0 & 0 & 0 \\ 0 & 1 & 1 & 0 & 0 \end{bmatrix}.$$

In this case, $A = A^T$; that is, A is symmetric. The symmetry occurs here because the underlying relationship is symmetric. (That is, if country i maintains diplomatic relations with country j, then also country j maintains diplomatic relations with country i.) Such symmetry is true of many relationships of interest. Figure 2.5 gives us a visual guide to the relationship, where country i is shown to maintain diplomatic relations with country j if the dots representing the two countries are joined by a line segment. (The diagram in Figure 2.5 is called an *undirected graph*, and the relationships defined in Exercises 21 and 26 lead to *directed graphs.*)

Let us consider the significance of an entry of the matrix $B = A^2$; for example,

$$b_{23} = a_{21}a_{13} + a_{22}a_{23} + a_{23}a_{33} + a_{24}a_{43} + a_{25}a_{53}.$$

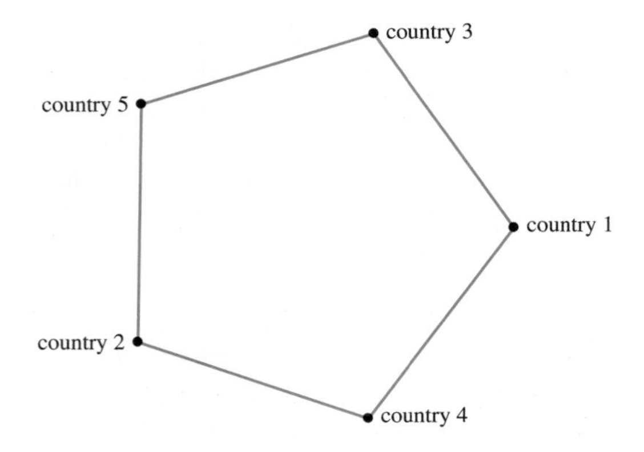

Figure 2.5 Diplomatic relations among countries

A typical term on the right-side of the equation has the form $a_{2k}a_{k3}$. This term is 1 if and only if both factors are 1—that is, if and only if country 2 maintains diplomatic relations with country k and country k maintains diplomatic relations with country 3. Thus b_{23} gives the number of countries that *link* country 2 and country 3. To see all of these entries, we compute

$$B = A^2 = \begin{bmatrix} 2 & 1 & 0 & 0 & 1 \\ 1 & 2 & 1 & 0 & 0 \\ 0 & 1 & 2 & 1 & 0 \\ 0 & 0 & 1 & 2 & 1 \\ 1 & 0 & 0 & 1 & 2 \end{bmatrix}.$$

Since $b_{23} = 1$, there is exactly one country that links countries 2 and 3. A careful examination of the entries of A reveals that $a_{25} = a_{53} = 1$, and hence it is country 5 that serves as the link. (Other deductions are left for the exercises.) We can visualize the (i,j)-entry of A^2 by counting the number of ways to go from country i to country j in Figure 2.5 that use two line segments.

By looking at other powers of A, additional information may be obtained. For example, it can be shown that if A is an $n \times n$ $(0,1)$-matrix and the (i,j)-entry of $A + A^2 + \cdots + A^{n-1}$ is nonzero, then there is a sequence of countries beginning with country i, ending with country j, and such that every pair of consecutive countries in the sequence maintains diplomatic relations. By means of such a sequence, countries i and j can communicate by passing a message only between countries that maintain diplomatic relations. Conversely, if the (i,j)-entry of $A + A^2 + \cdots + A^{n-1}$ is zero, then such communication between countries i and j is impossible.

Example 1 Consider a set of three countries, such that country 3 maintains diplomatic relations with both countries 1 and 2, and countries 1 and 2 do not maintain diplomatic relations with each other. These relationships can be described by the 3×3 $(0,1)$-matrix

$$A = \begin{bmatrix} 0 & 0 & 1 \\ 0 & 0 & 1 \\ 1 & 1 & 0 \end{bmatrix}.$$

In this case, we have

$$A + A^2 = \begin{bmatrix} 1 & 1 & 1 \\ 1 & 1 & 1 \\ 1 & 1 & 2 \end{bmatrix},$$

and so countries 1 and 2 can communicate, even though they do not have diplomatic relations. Here the sequence linking them consists of countries 1, 3, and 2.

A $(0,1)$-matrix can also be used to resolve problems involving scheduling. Suppose, for example, that the administration of a small college with m students wants to plan the times for its n courses. The goal of such planning is to avoid scheduling popular courses at the same time. To minimize the number of time conflicts, the students are surveyed. Each student is asked which courses he or she would like to take during the following semester. The results of this survey may be put in matrix form.

Define the $m \times n$ matrix A as follows:

$$a_{ij} = \begin{cases} 1 & \text{if student } i \text{ wants to take course } j \\ 0 & \text{otherwise.} \end{cases}$$

In this case, the matrix product A^TA provides important information regarding the scheduling of course times. We begin with an interpretation of the entries of this matrix. Let $B = A^T$ and $C = A^TA = BA$. Then, for example,

$$\begin{aligned} c_{12} &= b_{11}a_{12} + b_{12}a_{22} + \cdots + b_{1k}a_{k2} + \cdots + b_{1m}a_{m2} \\ &= a_{11}a_{12} + a_{21}a_{22} + \cdots + a_{k1}a_{k2} + \cdots + a_{m1}a_{m2}. \end{aligned}$$

A typical term on the right side of the equation has the form $a_{k1}a_{k2}$. Now, $a_{k1}a_{k2} = 1$ if and only if $a_{k1} = 1$ and $a_{k2} = 1$; that is, student k wants to take course 1 and course 2. So c_{12} represents the number of students who want to take both courses 1 and 2. In general, for $i \neq j$, c_{ij} is the number of students who want to take both course i and course j. In addition, c_{ii} represents the number of students who desire class i.

Example 2 Suppose that we have a group of 10 students and five courses. The results of the survey concerning course preferences are as follows:

	Course Number				
Student	1	2	3	4	5
1	1	0	1	0	1
2	0	0	0	1	1
3	1	0	0	0	0
4	0	1	1	0	1
5	0	0	0	0	0
6	1	1	0	0	0
7	0	0	1	0	1
8	0	1	0	1	0
9	1	0	1	0	1
10	0	0	0	1	0

Let A be the 10×5 matrix with entries from the previous table. Then

$$A^TA = \begin{bmatrix} 4 & 1 & 2 & 0 & 2 \\ 1 & 3 & 1 & 1 & 1 \\ 2 & 1 & 4 & 0 & 4 \\ 0 & 1 & 0 & 3 & 1 \\ 2 & 1 & 4 & 1 & 5 \end{bmatrix}.$$

From this matrix, we see that there are four students who want both course 3 and course 5. All other pairs of courses are wanted by at most two students. Furthermore, we see that four students want course 1, three students desire course 2, and so on. Thus, the trace (see Exercise 82 of Section 1.1) of A^TA equals the total demand for these five courses (counting students as often as the number of courses they wish to take) if the courses are offered at different times.

Notice that although A is not symmetric, the matrix A^TA is symmetric. Hence we may save computational effort by computing only one of the (i,j)- and (j,i)-entries.

As a final comment, it should be pointed out that many of these facts about (0, 1)-matrices can be adapted to apply to nonsymmetric relationships.

Practice Problem 3 ▶ Suppose that we have four cities with airports. We define the 4×4 matrix A by

$$a_{ij} = \begin{cases} 1 & \text{if there is a nonstop commercial flight from city } i \text{ to city } j \\ 0 & \text{if not.} \end{cases}$$

(a) Prove that A is a (0, 1)-matrix.

(b) Interpret the (2, 3)-entry of A^2.

(c) If

$$A = \begin{bmatrix} 0 & 1 & 1 & 0 \\ 1 & 0 & 0 & 1 \\ 1 & 0 & 0 & 0 \\ 0 & 1 & 0 & 0 \end{bmatrix},$$

compute A^2.

(d) How many flights are there with one "layover" from city 2 to city 3?

(e) How many flights are there with two layovers from city 1 to city 2?

(f) Is it possible to fly between each pair of cities? ◀

AN APPLICATION TO ANTHROPOLOGY

In this application,[1] we see a fascinating use of matrix operations for the study of the marriage laws of the Natchez Indians.

Everyone in this tribe was a member of one of four classes: the Suns, the Nobles, the Honoreds, and the *michy-miche-quipy* (MMQ). There were well-defined rules that determined class membership. The rules depended exclusively on the classes of the parents and required that at least one of the parents be a member of the MMQ. Furthermore, the class of the child depended on the class of the other parent, according to Table 2.2.

We are interested in determining the long-range distributions of these classes—that is, what the relative sizes are of these classes in future generations. It is clear that there will be a problem of survival if the size of the MMQ becomes too small. To simplify matters, we make the following three assumptions:

1. In every generation, each class is divided equally between males and females.
2. Each adult marries exactly once.
3. Each pair of parents has exactly one son and one daughter.

---

[1] This example is taken from Samuel Goldberg, *Introduction to Difference Equations (with Illustrative Examples from Economics, Psychology, and Sociology)*, Dover Publications, Inc., New York, 1986, pp. 238–241.

Table 2.2

Mother is in MMQ		Father is in MMQ	
Father	**Child**	**Mother**	**Child**
Sun	Noble	Sun	Sun
Noble	Honored	Noble	Noble
Honored	MMQ	Honored	Honored
MMQ	MMQ	MMQ	MMQ

Because of assumption 1, we need keep track only of the number of males in each class for every generation. To do this, we introduce the following notation:

$$s_k = \text{number of male Suns in the } k\text{th generation}$$

$$n_k = \text{number of male Nobles in the } k\text{th generation}$$

$$h_k = \text{number of male Honoreds in the } k\text{th generation}$$

$$m_k = \text{number of male MMQ in the } k\text{th generation}$$

Our immediate goal is to find a relationship between the numbers of members in each class of the kth and the $(k-1)$st generations. Since every Sun male must have a Sun mother (and vice versa), we obtain the equation

$$s_k = s_{k-1}. \tag{1}$$

The fact that every Noble male must have a Sun father or a Noble mother yields

$$n_k = s_{k-1} + n_{k-1}. \tag{2}$$

In addition, every Honored male must have a Noble father or an Honored mother. Thus

$$h_k = n_{k-1} + h_{k-1}. \tag{3}$$

Finally, assumption 3 guarantees that the total number of males (and females) remains the same for each generation.

$$s_k + n_k + h_k + m_k = s_{k-1} + n_{k-1} + h_{k-1} + m_{k-1} \tag{4}$$

Substituting the right sides of equations (1), (2), and (3) into (4), we obtain

$$\begin{aligned} s_{k-1} + (s_{k-1} + n_{k-1}) + (n_{k-1} + h_{k-1}) + m_k \\ = s_{k-1} + n_{k-1} + h_{k-1} + m_{k-1}, \end{aligned}$$

which simplifies to

$$m_k = -s_{k-1} - n_{k-1} + m_{k-1}. \tag{5}$$

Equations (1), (2), (3), and (5) relate the numbers of males in different classes of the kth generation to the numbers of males of the previous generation. If we let

$$\mathbf{x}_k = \begin{bmatrix} s_k \\ n_k \\ h_k \\ m_k \end{bmatrix} \quad \text{and} \quad A = \begin{bmatrix} 1 & 0 & 0 & 0 \\ 1 & 1 & 0 & 0 \\ 0 & 1 & 1 & 0 \\ -1 & -1 & 0 & 1 \end{bmatrix},$$

then we may represent all our relationships by the matrix equation

$$\mathbf{x}_k = A\mathbf{x}_{k-1}.$$

Because this equation must hold for all k, we have

$$\mathbf{x}_k = A\mathbf{x}_{k-1} = AA\mathbf{x}_{k-2} = \cdots = A^k\mathbf{x}_0.$$

To evaluate A^k, let $B = A - I_4$. We leave it to the exercises to show that, for any positive integer k,

$$A^k = I + kB + \frac{k(k-1)}{2}B^2, \qquad \text{for } k \geq 2.$$

(See Exercise 24.) Thus, carrying out the matrix multiplication, we obtain

$$\mathbf{x}_k = \begin{bmatrix} s_0 & & & & \\ n_0 & + & ks_0 & & \\ h_0 & + & kn_0 & + & \frac{k(k-1)}{2}s_0 \\ m_0 & - & kn_0 & - & \frac{k(k+1)}{2}s_0 \end{bmatrix}. \tag{6}$$

It is easy to see from equation (6) that if there are initially no Suns or Nobles (i.e., $n_0 = s_0 = 0$), the number of members in each class will remain the same from generation to generation. On the other hand, consider the last entry of $\mathbf{x}_k$. We can conclude that, unless $n_0 = s_0 = 0$, the size of the MMQ will decrease to the point where there are not enough of them to allow the other members to marry. At this point the social order ceases to exist.

EXERCISES

In Exercises 1–9, determine whether the statements are true or false.

1. If A is a Leslie matrix and $\mathbf{v}$ is a population distribution, then each entry of $A\mathbf{v}$ must be greater than the corresponding entry of $\mathbf{v}$.

2. For any population distribution $\mathbf{v}$, if A is the Leslie matrix for the population, then as n grows, $A^n\mathbf{v}$ approaches a specific population distribution.

3. In a Leslie matrix, the (i, j)-entry equals the average number of female offspring of a member of the ith age group.

4. In a Leslie matrix, the $(i+1, i)$-entry equals the portion of females in the ith age group who survive to the next age group.

5. The application in this section on traffic flow relies on the associative law of matrix multiplication.

6. If A and B are matrices and $\mathbf{x}$, $\mathbf{y}$, and $\mathbf{z}$ are vectors such that $\mathbf{y} = A\mathbf{x}$ and $\mathbf{z} = B\mathbf{y}$, then $\mathbf{z} = (AB)\mathbf{x}$.

7. A $(0, 1)$-matrix is a matrix with 0s and 1s as its only entries.

8. A $(0, 1)$-matrix is a square matrix with 0s and 1s as its only entries.

9. Every $(0, 1)$-matrix is a symmetric matrix.

Exercises 10–16 are concerned with Leslie matrices.

10. By observing a certain colony of mice, researchers found that all animals die within 3 years. Of those offspring that are females, 60% live for at least 1 year. Of these, 20% reach their second birthday. The females who are under

1 year of age have, on average, three female offspring. Those females between 1 and 2 years of age have, on average, two female offspring while they are in this age group. None of the females of age 2 give birth.

(a) Construct the Leslie matrix that describes this situation.

(b) Suppose that the current population distribution for females is given by the vector $\begin{bmatrix} 100 \\ 60 \\ 30 \end{bmatrix}$. Find the population distribution for next year. Also, find the population distribution 4 years from now.

(c) Show that there is no nonzero stable population distribution for the colony of mice. *Hint:* Let A be the Leslie matrix, and suppose that $\mathbf{z}$ is a stable population distribution. Then $A\mathbf{z} = \mathbf{z}$. This is equivalent to $(A - I_3)\mathbf{z} = \mathbf{0}$. Solve this homogeneous system of linear equations.

11. Suppose that the females of a certain colony of animals are divided into two age groups, and suppose that the Leslie matrix for this population is

$$\begin{bmatrix} 0 & 1 \\ 1 & 0 \end{bmatrix}.$$

(a) What proportion of the females of the first age group survive to the second age group?

(b) How many female offspring do females of each age group average?

(c) If $\mathbf{x} = \begin{bmatrix} a \\ b \end{bmatrix}$ is the current population distribution for the females of the colony, describe all future population distributions.

In Exercises 12–15, use either a calculator with matrix capabilities or computer software such as MATLAB to solve each problem.

12. A certain colony of lizards has a life span of less than 3 years. Suppose that the females are divided into three age groups: those under age 1, those of age 1, and those of age 2. Suppose further that the proportion of newborn females that survives until age 1 is .5 and that the proportion of one-year-old females that survives until age 2 is q. Assume also that females under age 1 do not give birth, those of age 1 have, on average, 1.2 female offspring, and those of age 2 have, on average, 1 female offspring. Suppose there are initially 450 females of age less than 1, 220 of age 1, and 70 of age 2.

(a) Write a Leslie matrix A for this colony of lizards.

(b) If $q = .3$, what will happen to the population in 50 years?

(c) If $q = .9$, what will happen to the population in 50 years?

(d) Find by trial and error a value of q for which the lizard population reaches a nonzero stable distribution. What is this stable distribution?

(e) For the value of q found in (d), what happens to an initial population of 200 females of age less than 1, 360 of age 1, and 280 of age 2?

(f) For what value of q does $(A - I_3)\mathbf{x} = \mathbf{0}$ have a nonzero solution?

(g) For the value of q found in (f), find the general solution of $(A - I_3)\mathbf{x} = \mathbf{0}$. How does this relate to the stable distributions in (d) and (e)?

13. A certain colony of bats has a life span of less than 3 years. Suppose that the females are divided into three age groups: those under age 1, those of age 1, and those of age 2. Suppose further that the proportion of newborn females that survives until age 1 is q and that the proportion of one-year-old females that survives until age 2 is .5. Assume also that females under age 1 do not give birth, those of age 1 have, on average, 2 female offspring, and those of age 2 have, on average, 1 female offspring. Suppose there are initially 300 females of age less than 1, 180 of age 1, and 130 of age 2.

(a) Write a Leslie matrix A for this colony of bats.

(b) If $q = .8$, what will happen to the population in 50 years?

(c) If $q = .2$, what will happen to the population in 50 years?

(d) Find by trial and error a value of q for which the bat population reaches a nonzero stable distribution. What is this stable distribution?

(e) For the value of q found in (d), what happens to an initial population of 210 females of age less than 1, 240 of age 1, and 180 of age 2?

(f) For what value of q does $(A - I_3)\mathbf{x} = \mathbf{0}$ have a nonzero solution?

(g) For the value of q found in (f), find the general solution of $(A - I_3)\mathbf{x} = \mathbf{0}$. How does this relate to the stable distributions in (d) and (e)?

14. A certain colony of voles has a life span of less than 3 years. Suppose that the females are divided into three age groups: those under age 1, those of age 1, and those of age 2. Suppose further that the proportion of newborn females that survives until age 1 is .1 and that the proportion of one-year-old females that survives until age 2 is .2. Assume also that females under age 1 do not give birth, those of age 1 have, on average, b female offspring, and those of age 2 have, on average, 10 female offspring. Suppose there are initially 150 females of age less than 1, 300 of age 1, and 180 of age 2.

(a) Write a Leslie matrix A for this colony of voles.

(b) If $b = 10$, what will happen to the population in 50 years?

(c) If $b = 4$, what will happen to the population in 50 years?

(d) Find by trial and error a value of b for which the vole population reaches a nonzero stable distribution. What is this stable distribution?

(e) For the value of b found in (d), what happens to an initial population of 80 females of age less than 1, 200 of age 1, and 397 of age 2?

(f) For what value of b does $(A - I_3)\mathbf{x} = \mathbf{0}$ have a nonzero solution?

(g) Let $\mathbf{p} = \begin{bmatrix} p_1 \\ p_2 \\ p_3 \end{bmatrix}$ be an arbitrary population vector, and let b have the value found in (f). Over time, $\mathbf{p}$ approaches a stable population vector $\mathbf{q}$. Express $\mathbf{q}$ in terms of p_1, p_2, and p_3.

15. A certain colony of squirrels has a life span of less than 3 years. Suppose that the females are divided into three age groups: those under age 1, those of age 1, and those of age 2. Suppose further that the proportion of newborn females that survives until age 1 is .2 and that the proportion of one-year-old females that survives until age 2 is .5. Assume also that females under age 1 do not give birth, those of age 1 have, on average, 2 female offspring, and those of age 2 have, on average, b female offspring. Suppose there are initially 240 females of age less than one, 400 of age one, and 320 of age two.

(a) Write a Leslie matrix A for this colony of squirrels.

(b) If $b = 3$, what will happen to the population in 50 years?

(c) If $b = 9$, what will happen to the population in 50 years?

(d) Find by trial and error a value of b for which the squirrel population reaches a nonzero stable distribution. What is this stable distribution?

(e) For the value of b found in (d), what happens to an initial population of 100 females of age less than one, 280 of age one, and 400 of age two?

(f) For what value of b does $(A - I_3)\mathbf{x} = \mathbf{0}$ have a nonzero solution?

(g) Let $\mathbf{p} = \begin{bmatrix} p_1 \\ p_2 \\ p_3 \end{bmatrix}$ be an arbitrary population vector, and let b have the value found in (f). Over time, $\mathbf{p}$ approaches a stable population vector $\mathbf{q}$. Express $\mathbf{q}$ in terms of p_1, p_2, and p_3.

16. The maximum membership term for each member of the Service Club is 3 years. Each first-year and second-year member recruits one new person who begins the membership term in the following year. Of those in their first year of membership, 50% of the members resign, and of those in their second year of membership, 70% resign.

(a) Write a 3×3 matrix A so that if x_i is currently the number of Service Club members in their ith year of membership and y_i is the number in their ith year of membership a year from now, then $\mathbf{y} = A\mathbf{x}$.

(b) Suppose that there are 60 Service Club members in their first year, 20 members in their second year, and 40 members in their third year of membership. Find the distribution of members for next year and for 2 years from now.

Exercises 17 and 18 use the technique developed in the traffic flow application.

17. A certain medical foundation receives money from two sources: donations and interest earned on endowments. Of the donations received, 30% is used to defray the costs of raising funds; only 10% of the interest is used to defray the cost of managing the endowment funds. Of the rest of the money (the net income), 40% is used for research and 60% is used to maintain medical clinics. Of the three expenses (research, clinics, and fundraising), the portions going to materials and personnel are divided according to Table 2.3. Find a matrix M such that if p is the value of donations and q is the value of interest, then $M\begin{bmatrix} p \\ q \end{bmatrix} = \begin{bmatrix} m \\ f \end{bmatrix}$, where m and f are the material and personnel costs of the foundation, respectively.

Table 2.3

	Research	Clinics	Fundraising
material costs	80%	50%	70%
personnel costs	20%	50%	30%

18. Water is pumped into a system of pipes at points P_1 and P_2 shown in Figure 2.6. At each of the junctions P_3, P_4, P_5, P_6, P_7, and P_8, the pipes are split and water flows according to the portions indicated in the diagram. Suppose that water flows into P_1 and P_2 at p and q gallons per minute, respectively, and flows out of P_9, P_{10}, and P_{11} at the rates of a, b, and c gallons per minute, respectively. Find a matrix M such that the vector of outputs is given by

$$\begin{bmatrix} a \\ b \\ c \end{bmatrix} = M\begin{bmatrix} p \\ q \end{bmatrix}.$$

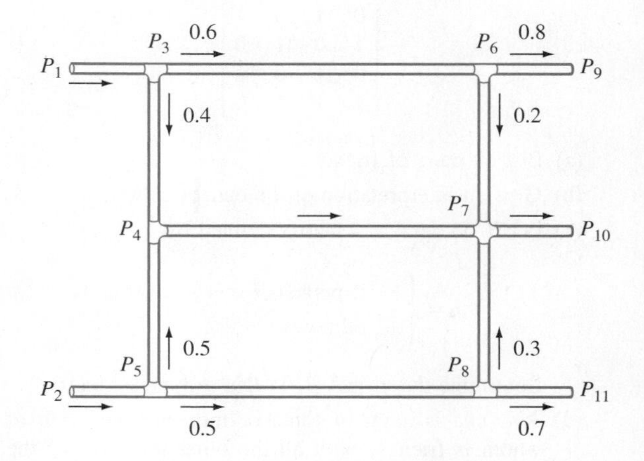

Figure 2.6

Exercises 19–23 are concerned with (0, 1)*-matrices.*

19. With the interpretation of a (0, 1)-matrix found in Practice Problem 3, suppose that we have five cities with the

associated matrix A given in the block form (see Section 2.1)

$$A = \begin{bmatrix} B & O_1 \\ O_2 & C \end{bmatrix},$$

where B is a 3×3 matrix, O_1 is the 3×2 zero matrix, O_2 is the 2×3 zero matrix, and C is a 2×2 matrix.

(a) What does the matrix A tell us about flight connections between the cities?

(b) Use block multiplication to obtain A^2, A^3, and A^k for any positive integer k.

(c) Interpret your result in (b) in terms of flights between the cities.

20. Recall the $(0, 1)$-matrix

$$A = \begin{bmatrix} 0 & 0 & 1 & 1 & 0 \\ 0 & 0 & 0 & 1 & 1 \\ 1 & 0 & 0 & 0 & 1 \\ 1 & 1 & 0 & 0 & 0 \\ 0 & 1 & 1 & 0 & 0 \end{bmatrix}$$

in which the entries describe countries that maintain diplomatic relations with one another.

(a) Which pairs of countries maintain diplomatic relations?

(b) How many countries link country 1 with country 3?

(c) Give an interpretation of the $(1, 4)$-entry of A^3.

21. Suppose that there is a group of four people and an associated 4×4 matrix A defined by

$$a_{ij} = \begin{cases} 1 & \text{if } i \neq j \text{ and person } i \text{ likes person } j \\ 0 & \text{otherwise.} \end{cases}$$

We say that persons i and j are *friends* if they like each other; that is, if $a_{ij} = a_{ji} = 1$. Suppose that A is given by

$$\begin{bmatrix} 0 & 1 & 0 & 1 \\ 1 & 0 & 1 & 0 \\ 0 & 1 & 0 & 1 \\ 1 & 1 & 1 & 0 \end{bmatrix}.$$

(a) List all pairs of friends.

(b) Give an interpretation of the entries of A^2.

(c) Let B be the 4×4 matrix defined by

$$b_{ij} = \begin{cases} 1 & \text{if persons } i \text{ and } j \text{ are friends} \\ 0 & \text{otherwise.} \end{cases}$$

Determine the matrix B. Is B a symmetric matrix?

(d) A *clique* is a set of three or more people, each of whom is friendly with all the other members of the set. Show that person i belongs to a clique if and only if the (i, i)-entry of B^3 is positive.

(e) Use computer software or a calculator that performs matrix arithmetic to count the cliques that exist among the four friends.

Exercises 22 and 23 refer to the scheduling example.

22. Suppose that student preference for a set of courses is given in the following table:

	Course Number				
Student	1	2	3	4	5
1	1	0	0	0	1
2	0	0	1	1	0
3	1	0	1	0	1
4	0	1	0	0	0
5	1	0	0	0	1
6	0	1	0	0	1
7	1	0	1	0	1
8	0	1	1	0	1
9	1	0	0	1	1
10	0	0	1	1	0

(a) Give all pairs of courses that are desired by the most students.

(b) Give all pairs of courses that are desired by the fewest students.

(c) Construct a matrix whose diagonal entries determine the number of students who prefer each course.

23. Let A be the matrix in Example 2.

(a) Justify the following interpretation: For $i \neq j$, the (i, j)-entry of AA^T is the number of classes that are desired by both students i and j.

(b) Show that the $(1, 2)$-entry of AA^T is 1 and the $(9, 1)$-entry of AA^T is 3.

(c) Interpret the answers to (b) in the context of (a) and the data in the scheduling example.

(d) Give an interpretation of the diagonal entries of AA^T.

Exercises 24 and 25 are concerned with the anthropology application.

24. Recall the *binomial formula* for scalars a and b and any positive integer k:

$$(a+b)^k = a^k + ka^{k-1}b + \cdots + \frac{k!}{i!(k-i)!}a^{k-i}b^i + \cdots + b^k,$$

where $i!$ (i factorial) is given by $i! = 1 \cdot 2 \cdots (i-1) \cdot i$.

(a) Suppose that A and B are $m \times m$ matrices that commute; that is, $AB = BA$. Prove that the binomial formula holds for A and B when $k = 2$ and $k = 3$. Specifically, prove that

$$(A+B)^2 = A^2 + 2AB + B^2$$

and

$$(A+B)^3 = A^3 + 3A^2B + 3AB^2 + B^3.$$

(b) Use mathematical induction to extend the results in part (a) to any positive integer k.

(c) For the matrices A and B on page 117, show $B^3 = O$ and that A and B commute. Then use (b) to prove

$$A^k = I_4 + kB + \frac{k(k-1)}{2}B^2 \quad \text{for } k \geq 2.$$

In Exercises 25 and 26, use either a calculator with matrix capabilities or computer software such as MATLAB to solve each problem.

25. In reference to the application in the text involving the Natchez Indians, suppose that initially there are 100 Sun males, 200 Noble males, 300 Honored males, and 8000 MMQ males.

(a) How many males will there be in each class in $k = 1$, 2, and 3 generations?

(b) Use a computer to determine how many males there will be in each class in $k = 9$, 10, and 11 generations.

(c) What do your answers to (b) suggest for the future of the Natchez Indians if they hold to their current marriage laws?

(d) Produce an algebraic proof that for some k there will not be enough MMQ to allow the other members to marry.

26. Suppose that we have a group of six people, each of whom owns a communication device. We define a 6×6 matrix A as follows: For $1 \leq i \leq 6$, let $a_{ii} = 0$; and for $i \neq j$,

$$a_{ij} = \begin{cases} 1 & \text{if person } i \text{ can send a message to person } j \\ 0 & \text{otherwise.} \end{cases}$$

(a) Show that A is a (0, 1)-matrix.

(b) Give an interpretation of what it means for the term $a_{32}a_{21}$ to equal one.

(c) Show that the (3, 1)-entry of A^2 represents the number of ways that person 3 can send a message to person 1 in two *stages*—that is, the number of people to whom person 3 can send a message and who in turn can send a message to person 1. *Hint:* Consider the number of terms that are not equal to zero in the expression

$$a_{31}a_{11} + a_{32}a_{21} + \cdots + a_{36}a_{61}.$$

(d) Generalize your result in (c) to the (i, j)-entry of A^2.

(e) Generalize your result in (d) to the (i, j)-entry of A^m.

Now suppose

$$A = \begin{bmatrix} 0 & 0 & 0 & 1 & 0 & 1 \\ 1 & 0 & 1 & 1 & 0 & 0 \\ 0 & 1 & 0 & 1 & 0 & 0 \\ 1 & 0 & 1 & 0 & 0 & 0 \\ 1 & 1 & 1 & 0 & 0 & 1 \\ 0 & 0 & 1 & 1 & 0 & 0 \end{bmatrix}.$$

(f) Is there any person who cannot receive a message from anyone else in one stage? Justify your answer.

(g) How many ways can person 1 send a message to person 4 in 1, 2, 3, and 4 stages?

(h) The (i, j)-entry of $A + A^2 + \cdots + A^m$ can be shown to equal the number of ways in which person i can send a message to person j in at most m stages. Use this result to determine the number of ways in which person 3 can send a message to person 4 in at most 4 stages.

SOLUTIONS TO THE PRACTICE PROBLEMS

1. (a) $A = \begin{bmatrix} 0.50 & 2 \\ 0.25 & 0 \end{bmatrix}$.

(b) The population distribution of females for this year is given by the vector $\begin{bmatrix} 200 \\ 200 \end{bmatrix}$, and hence the population distribution of females for next year is

$$A\begin{bmatrix} 200 \\ 200 \end{bmatrix} = \begin{bmatrix} 0.50 & 2 \\ 0.25 & 0 \end{bmatrix}\begin{bmatrix} 200 \\ 200 \end{bmatrix} = \begin{bmatrix} 500 \\ 50 \end{bmatrix},$$

and the population distribution of females 2 years from now is

$$A\begin{bmatrix} 500 \\ 50 \end{bmatrix} = \begin{bmatrix} 0.50 & 2 \\ 0.25 & 0 \end{bmatrix}\begin{bmatrix} 500 \\ 50 \end{bmatrix} = \begin{bmatrix} 350 \\ 125 \end{bmatrix}.$$

(c) Since

$$A\begin{bmatrix} 400 \\ 100 \end{bmatrix} = \begin{bmatrix} 0.50 & 2 \\ 0.25 & 0 \end{bmatrix}\begin{bmatrix} 400 \\ 100 \end{bmatrix} = \begin{bmatrix} 400 \\ 100 \end{bmatrix},$$

the population distribution does not change from year to year.

(d) Let x_1 and x_2 be the numbers of females in the first and second age groups, respectively. Since next year's population distribution is the same as this year's, $A\mathbf{x} = \mathbf{x}$, and hence

$$\begin{bmatrix} 0.50 & 2 \\ 0.25 & 0 \end{bmatrix}\begin{bmatrix} x_1 \\ x_2 \end{bmatrix} = \begin{bmatrix} 0.50x_1 + 2x_2 \\ 0.25x_1 \end{bmatrix} = \begin{bmatrix} x_1 \\ x_2 \end{bmatrix}.$$

Thus $x_1 = 4x_2$. But $x_1 + x_2 = 600$, and therefore $4x_2 + x_2 = 5x_2 = 600$, from which it follows that $x_2 = 120$. Finally, $x_1 = 4x_2 = 480$.

2. Let z_1 and z_2 be the rates at which soy sauce is shipped to Chicago and St. Louis, respectively. Then

$$\begin{bmatrix} z_1 \\ z_2 \end{bmatrix} = \begin{bmatrix} 1 & 0.3 & 0 \\ 0 & 0.7 & 1 \end{bmatrix}\begin{bmatrix} 0.5 & 0.0 \\ 0.5 & 0.6 \\ 0 & 0.4 \end{bmatrix}\begin{bmatrix} 10000 \\ 5000 \end{bmatrix} = \begin{bmatrix} 7400 \\ 7600 \end{bmatrix}.$$

3. (a) Clearly, every entry of A is either 0 or 1, so A is a $(0, 1)$-matrix.

(b) The $(2, 3)$-entry of A^2 is

$$a_{21}a_{13} + a_{22}a_{23} + a_{23}a_{33} + a_{24}a_{43}.$$

A typical term has the form $a_{2k}a_{k3}$, which equals 1 or 0. This term equals 1 if and only if $a_{2k} = 1$ and $a_{k3} = 1$. Consequently, this term equals 1 if and only if there is a nonstop flight between city 2 and city k, as well as a nonstop flight between city k and city 3. That is, $a_{2k}a_{k3} = 1$ means that there is a flight with one *layover* (the plane stops at city k) from city 2 to city 3. Therefore we may interpret the $(2, 3)$-entry of A^2 as the number of flights with one layover from city 2 to city 3.

(c) $A^2 = \begin{bmatrix} 2 & 0 & 0 & 1 \\ 0 & 2 & 1 & 0 \\ 0 & 1 & 1 & 0 \\ 1 & 0 & 0 & 1 \end{bmatrix}.$

(d) Because the $(2, 3)$-entry of A^2 is 1, there is one flight with one layover from city 2 to city 3.

(e) We compute A^3 to find the number of flights with two layovers from city 1 to city 2. We have

$$A^3 = \begin{bmatrix} 0 & 3 & 2 & 0 \\ 3 & 0 & 0 & 2 \\ 2 & 0 & 0 & 1 \\ 0 & 2 & 1 & 0 \end{bmatrix}.$$

Because the $(1, 2)$-entry of A^3 is 3, we see that there are three flights with two layovers from city 1 to city 2.

(f) From the entries of A, we see that there are nonstop flights between cities 1 and 2, cities 1 and 3, and cities 2 and 4. From A^2, we see that there are flights between cities 1 and 4, as well as between cities 2 and 3. Finally, from A^3, we discover that there is a flight between cities 3 and 4. We conclude that there are flights between all pairs of cities.

2.3 INVERTIBILITY AND ELEMENTARY MATRICES

In this section, we introduce the concept of *invertible matrix* and examine special invertible matrices that are intimately associated with elementary row operations, the *elementary matrices*.

For any real number $a \neq 0$, there is a unique real number b, called the *multiplicative inverse* of a, with the property that $ab = ba = 1$. For example, if $a = 2$, then $b = 1/2$. In the context of matrices, the identity matrix I_n is a multiplicative identity; so it is natural to ask for what matrices A does there exist a matrix B such that $AB = BA = I_n$. Notice that this last equation is possible only if both A and B are $n \times n$ matrices. This discussion motivates the following definitions:

Definitions An $n \times n$ matrix A is called **invertible** if there exists an $n \times n$ matrix B such that $AB = BA = I_n$. In this case, B is called an **inverse** of A.

If A is an invertible matrix, then its inverse is unique. For if both B and C are inverses of A, then $AB = BA = I_n$ and $AC = CA = I_n$. Hence

$$B = BI_n = B(AC) = (BA)C = I_nC = C.$$

When A is invertible, we denote the unique inverse of A by A^{-1}, so that $AA^{-1} = A^{-1}A = I_n$. Notice the similarity of this statement and $2 \cdot 2^{-1} = 2^{-1} \cdot 2 = 1$, where 2^{-1} is the multiplicative inverse of the real number 2.

Example 1 Let $A = \begin{bmatrix} 1 & 2 \\ 3 & 5 \end{bmatrix}$ and $B = \begin{bmatrix} -5 & 2 \\ 3 & -1 \end{bmatrix}$. Then

$$AB = \begin{bmatrix} 1 & 2 \\ 3 & 5 \end{bmatrix}\begin{bmatrix} -5 & 2 \\ 3 & -1 \end{bmatrix} = \begin{bmatrix} 1 & 0 \\ 0 & 1 \end{bmatrix} = I_2,$$

and

$$BA = \begin{bmatrix} -5 & 2 \\ 3 & -1 \end{bmatrix} \begin{bmatrix} 1 & 2 \\ 3 & 5 \end{bmatrix} = \begin{bmatrix} 1 & 0 \\ 0 & 1 \end{bmatrix} = I_2.$$

So A is invertible, and B is the inverse of A; that is, $A^{-1} = B$.

Practice Problem 1 ▶ If $A = \begin{bmatrix} -1 & 0 & 1 \\ 1 & 2 & -2 \\ 2 & -1 & -1 \end{bmatrix}$ and $B = \begin{bmatrix} 4 & 1 & 2 \\ 3 & 1 & 1 \\ 5 & 1 & 2 \end{bmatrix}$, is $B = A^{-1}$? ◀

Because the roles of the matrices A and B are the same in the preceding definition, it follows that if B is the inverse of A, then A is also the inverse of B. Thus, in Example 1, we also have

$$B^{-1} = A = \begin{bmatrix} 1 & 2 \\ 3 & 5 \end{bmatrix}.$$

Just as the real number 0 has no multiplicative inverse, the $n \times n$ zero matrix O has no inverse because $OB = O \neq I_n$ for any $n \times n$ matrix B. But there are also other square matrices that are not invertible; for example, $A = \begin{bmatrix} 1 & 1 \\ 2 & 2 \end{bmatrix}$. For if $B = \begin{bmatrix} a & b \\ c & d \end{bmatrix}$ is any 2×2 matrix, then

$$AB = \begin{bmatrix} 1 & 1 \\ 2 & 2 \end{bmatrix} \begin{bmatrix} a & b \\ c & d \end{bmatrix} = \begin{bmatrix} a + c & b + d \\ 2a + 2c & 2b + 2d \end{bmatrix}.$$

Since the second row of the matrix on the right equals twice its first row, it cannot be the identity matrix $\begin{bmatrix} 1 & 0 \\ 0 & 1 \end{bmatrix}$. So B cannot be the inverse of A, and hence A is not invertible.

In the next section, we learn which matrices are invertible and how to compute their inverses. In this section, we discuss some elementary properties of invertible matrices.

The inverse of a real number can be used to solve certain equations. For example, the equation $2x = 14$ can be solved by multiplying both sides of the equation by the inverse of 2:

$$\begin{aligned} 2^{-1}(2x) &= 2^{-1}(14) \\ (2^{-1}2)x &= 7 \\ 1x &= 7 \\ x &= 7 \end{aligned}$$

In a similar manner, if A is an invertible $n \times n$ matrix, then we can use A^{-1} to solve matrix equations in which an unknown matrix is multiplied by A. For example, if A is invertible, then we can solve the matrix equation $A\mathbf{x} = \mathbf{b}$ as follows:[2]

$$\begin{aligned} A\mathbf{x} &= \mathbf{b} \\ A^{-1}(A\mathbf{x}) &= A^{-1}\mathbf{b} \end{aligned}$$

[2] Although matrix inverses can be used to solve systems whose coefficient matrices are invertible, the method of solving systems presented in Chapter 1 is far more efficient.

$$(A^{-1}A)\mathbf{x} = A^{-1}\mathbf{b}$$
$$I_n\mathbf{x} = A^{-1}\mathbf{b}$$
$$\mathbf{x} = A^{-1}\mathbf{b}$$

> If A is an invertible $n \times n$ matrix, then for every $\mathbf{b}$ in $\mathcal{R}^n$, $A\mathbf{x} = \mathbf{b}$ has the unique solution $A^{-1}\mathbf{b}$.

In solving a system of linear equations by using the inverse of a matrix A, we observe that A^{-1} "reverses" the action of A; that is, if A is an invertible $n \times n$ matrix and $\mathbf{u}$ is a vector in $\mathcal{R}^n$, then $A^{-1}(A\mathbf{u}) = \mathbf{u}$. (See Figure 2.7.)

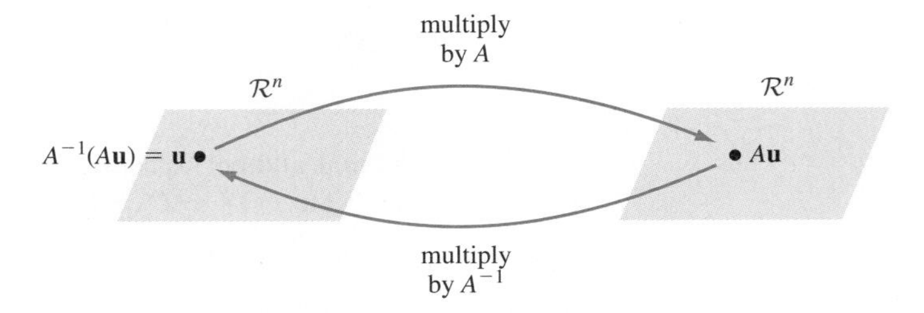

Figure 2.7 Multiplication by a matrix and its inverse

Example 2 Use a matrix inverse to solve the system of linear equations

$$\begin{aligned} x_1 + 2x_2 &= 4 \\ 3x_1 + 5x_2 &= 7. \end{aligned}$$

Solution This system is the same as the matrix equation $A\mathbf{x} = \mathbf{b}$, where

$$A = \begin{bmatrix} 1 & 2 \\ 3 & 5 \end{bmatrix}, \quad \mathbf{x} = \begin{bmatrix} x_1 \\ x_2 \end{bmatrix}, \quad \text{and} \quad \mathbf{b} = \begin{bmatrix} 4 \\ 7 \end{bmatrix}.$$

We saw in Example 1 that A is invertible. Hence we can solve this equation for $\mathbf{x}$ by multiplying both sides of the equation on the left by

$$A^{-1} = \begin{bmatrix} -5 & 2 \\ 3 & -1 \end{bmatrix}$$

as follows:

$$\begin{bmatrix} x_1 \\ x_2 \end{bmatrix} = \mathbf{x} = A^{-1}\mathbf{b} = \begin{bmatrix} -5 & 2 \\ 3 & -1 \end{bmatrix} \begin{bmatrix} 4 \\ 7 \end{bmatrix} = \begin{bmatrix} -6 \\ 5 \end{bmatrix}$$

Therefore $x_1 = -6$ and $x_2 = 5$ is the unique solution of the system.

Practice Problem 2 ▶ Use the answer to Practice Problem 1 to solve the following system of linear equations:

$$\begin{aligned} -x_1 \qquad\quad + x_3 &= 1 \\ x_1 + 2x_2 - 2x_3 &= 2 \\ 2x_1 - x_2 - x_3 &= -1 \end{aligned}$$

◀

Example 3

Recall the rotation matrix

$$A_\theta = \begin{bmatrix} \cos\theta & -\sin\theta \\ \sin\theta & \cos\theta \end{bmatrix}$$

considered in Section 1.2 and Example 4 of Section 2.1. Notice that for $\theta = 0^\circ$, $A_\theta = I_2$. Furthermore, for any angle α,

$$A_\alpha A_{-\alpha} = A_{\alpha+(-\alpha)} = A_{0^\circ} = I_2.$$

Similarly, $A_{-\alpha}A_\alpha = I_2$. Hence A_α satisfies the definition of an invertible matrix with inverse $A_{-\alpha}$. Therefore $(A_\alpha)^{-1} = A_{-\alpha}$.

Another way of viewing $A_{-\alpha}A_\alpha$ is that it represents a rotation by α, followed by a rotation by $-\alpha$, which results in a net rotation of 0°. This is the same as multiplying by the identity matrix.

The following theorem states some useful properties of matrix inverses:

THEOREM 2.2

Let A and B be $n \times n$ matrices.

(a) If A is invertible, then A^{-1} is invertible and $(A^{-1})^{-1} = A$.

(b) If A and B are invertible, then AB is invertible and $(AB)^{-1} = B^{-1}A^{-1}$.

(c) If A is invertible, then A^T is invertible and $(A^T)^{-1} = (A^{-1})^T$.

PROOF The proof of (a) is a simple consequence of the definition of matrix inverse.

(b) Suppose that A and B are invertible. Then

$$(AB)(B^{-1}A^{-1}) = A(BB^{-1})A^{-1} = AI_nA^{-1} = AA^{-1} = I_n.$$

Similarly, $(B^{-1}A^{-1})(AB) = I_n$. Hence AB satisfies the definition of an invertible matrix with inverse $B^{-1}A^{-1}$; that is, $(AB)^{-1} = B^{-1}A^{-1}$.

(c) Suppose that A is invertible. Then $A^{-1}A = I_n$. Using Theorem 2.1(g), we obtain

$$A^T(A^{-1})^T = (A^{-1}A)^T = I_n^T = I_n.$$

Similarly, $(A^{-1})^TA^T = I_n$. Hence A^T satisfies the definition of an invertible matrix with the inverse $(A^{-1})^T$; that is, $(A^T)^{-1} = (A^{-1})^T$. ■

Part (b) of Theorem 2.2 can be easily extended to products of more than two matrices.

Let $A_1, A_2, \ldots, A_k$ be $n \times n$ invertible matrices. Then the product $A_1A_2\cdots A_k$ is invertible, and

$$(A_1A_2\cdots A_k)^{-1} = (A_k)^{-1}(A_{k-1})^{-1}\cdots(A_1)^{-1}.$$

ELEMENTARY MATRICES

It is interesting that every elementary row operation can be performed by matrix multiplication. For example, we can multiply row 2 of the matrix

$$A = \begin{bmatrix} a & b \\ c & d \end{bmatrix}$$

by the scalar k by means of the matrix product

$$\begin{bmatrix} 1 & 0 \\ 0 & k \end{bmatrix} \begin{bmatrix} a & b \\ c & d \end{bmatrix} = \begin{bmatrix} a & b \\ kc & kd \end{bmatrix}.$$

Also, we can interchange rows 1 and 2 by means of the matrix product

$$\begin{bmatrix} 0 & 1 \\ 1 & 0 \end{bmatrix} \begin{bmatrix} a & b \\ c & d \end{bmatrix} = \begin{bmatrix} c & d \\ a & b \end{bmatrix}.$$

We can add k times row 1 of A to row 2 by means of the matrix product

$$\begin{bmatrix} 1 & 0 \\ k & 1 \end{bmatrix} \begin{bmatrix} a & b \\ c & d \end{bmatrix} = \begin{bmatrix} a & b \\ ka + c & kb + d \end{bmatrix}.$$

The matrices

$$\begin{bmatrix} 1 & 0 \\ 0 & k \end{bmatrix}, \quad \begin{bmatrix} 0 & 1 \\ 1 & 0 \end{bmatrix}, \quad \text{and} \quad \begin{bmatrix} 1 & 0 \\ k & 1 \end{bmatrix}$$

are examples of *elementary matrices*. In general, an $n \times n$ matrix E is called an **elementary matrix** if we can obtain E from I_n by a single elementary row operation.

For example, the matrix

$$E = \begin{bmatrix} 1 & 0 & 0 \\ 0 & 1 & 0 \\ 2 & 0 & 1 \end{bmatrix}$$

is an elementary matrix because we can obtain E from I_3 by the elementary row operation of adding two times the first row of I_3 to the third row of I_3. Note that if

$$A = \begin{bmatrix} 1 & 2 \\ 3 & 4 \\ 5 & 6 \end{bmatrix},$$

then

$$EA = \begin{bmatrix} 1 & 0 & 0 \\ 0 & 1 & 0 \\ 2 & 0 & 1 \end{bmatrix} \begin{bmatrix} 1 & 2 \\ 3 & 4 \\ 5 & 6 \end{bmatrix} = \begin{bmatrix} 1 & 2 \\ 3 & 4 \\ 7 & 10 \end{bmatrix}.$$

Hence we can obtain EA from A by adding two times the first row of A to the third row. This is the same elementary row operation that we applied to I_3 to produce E. A similar result holds for each of the three elementary row operations.

> Let A be an $m \times n$ matrix, and let E be an $m \times m$ elementary matrix resulting from an elementary row operation on I_m. Then the product EA can be obtained from A by the identical elementary row operation on A.

Practice Problem 3 ▶ Find an elementary matrix E such that $EA = B$, where

$$A = \begin{bmatrix} 3 & -4 & 1 \\ 2 & 5 & -1 \end{bmatrix} \quad \text{and} \quad B = \begin{bmatrix} 3 & -4 & 1 \\ -4 & 13 & -3 \end{bmatrix}.$$ ◀

As was noted in Section 1.3, any elementary row operation can be reversed. For example, a matrix obtained by adding two times the first row to the third row of A can be changed back into A by adding -2 times the first row to the third row of the new matrix. The concept of a *reverse* operation gives us a way of obtaining the inverse of an elementary matrix. To see how this is done, consider the preceding matrix E. Let

$$F = \begin{bmatrix} 1 & 0 & 0 \\ 0 & 1 & 0 \\ -2 & 0 & 1 \end{bmatrix}$$

be the elementary matrix obtained from I_3 by adding -2 times the first row of I_3 to the third row of I_3. If this elementary row operation is applied to E, then the result is I_3, and therefore $FE = I_3$. Similarly, $EF = I_3$. Hence E is invertible, and $E^{-1} = F$. We can apply the same argument to any elementary matrix to establish the following result:

> Every elementary matrix is invertible. Furthermore, the inverse of an elementary matrix is also an elementary matrix.

The value of elementary matrices is that they allow us to analyze the theoretical properties of elementary row operations using our knowledge about matrix multiplication. Since we can put a matrix into reduced row echelon form by means of elementary row operations, we can carry out this transformation by multiplying the matrix on the left by a sequence of elementary matrices, one for each row operation. Consequently, if A is an $m \times n$ matrix with reduced row echelon form R, there exist elementary matrices $E_1, E_2, \ldots, E_k$ such that

$$E_k E_{k-1} \cdots E_1 A = R.$$

Let $P = E_k E_{k-1} \cdots E_1$. Then P is a product of elementary matrices, so P is an invertible matrix, by the boxed result on page 125. Furthermore, $PA = R$. Thus we have established the following result:

THEOREM 2.3

Let A be an $m \times n$ matrix with reduced row echelon form R. Then there exists an invertible $m \times m$ matrix P such that $PA = R$.

This theorem implies the following important result that justifies the method of row reduction (described in Section 1.4) for solving matrix equations:

Corollary The matrix equation $A\mathbf{x} = \mathbf{b}$ has the same solutions as $R\mathbf{x} = \mathbf{c}$, where $[R\ \ \mathbf{c}]$ is the reduced row echelon form of the augmented matrix $[A\ \ \mathbf{b}]$.

PROOF There is an invertible matrix P such that $P[A\ \ \mathbf{b}] = [R\ \ \mathbf{c}]$ by Theorem 2.3. Therefore

$$[PA\ \ P\mathbf{b}] = P[A\ \ \mathbf{b}] = [R\ \ \mathbf{c}],$$

and hence $PA = R$ and $P\mathbf{b} = \mathbf{c}$. Because P is invertible, it follows that $A = P^{-1}R$ and $\mathbf{b} = P^{-1}\mathbf{c}$.

Suppose that $\mathbf{v}$ is a solution of $A\mathbf{x} = \mathbf{b}$. Then $A\mathbf{v} = \mathbf{b}$, so

$$R\mathbf{v} = (PA)\mathbf{v} = P(A\mathbf{v}) = P\mathbf{b} = \mathbf{c},$$

and therefore $\mathbf{v}$ is a solution of $R\mathbf{x} = \mathbf{c}$. Conversely, suppose that $\mathbf{v}$ is a solution of $R\mathbf{x} = \mathbf{c}$. Then $R\mathbf{v} = \mathbf{c}$, and hence

$$A\mathbf{v} = (P^{-1}R)\mathbf{v} = P^{-1}(R\mathbf{v}) = P^{-1}\mathbf{c} = \mathbf{b}.$$

Therefore $\mathbf{v}$ is a solution of $A\mathbf{x} = \mathbf{b}$. Thus the equations $A\mathbf{x} = \mathbf{b}$ and $R\mathbf{x} = \mathbf{c}$ have the same solutions. ■

As a special case of the preceding corollary, we note that if $\mathbf{b} = \mathbf{0}$, then $\mathbf{c} = \mathbf{0}$. Therefore $A\mathbf{x} = \mathbf{0}$ and $R\mathbf{x} = \mathbf{0}$ are equivalent.

THE COLUMN CORRESPONDENCE PROPERTY

In Section 1.7, we found that if a subset $\mathcal{S}$ of $\mathcal{R}^n$ contains more than n vectors, then the set is linearly dependent, and hence one of the vectors in $\mathcal{S}$ is a linear combination of the others. However, nothing we have learned tells us *which* vector in the set is a linear combination of other vectors. Of course, we could solve systems of linear equations to find out how to do this, but this method could be extremely inefficient.

To illustrate another approach, consider the set

$$\mathcal{S} = \left\{ \begin{bmatrix} 1 \\ -1 \\ 2 \\ -3 \end{bmatrix}, \begin{bmatrix} 2 \\ -2 \\ 4 \\ -6 \end{bmatrix}, \begin{bmatrix} -1 \\ 1 \\ -3 \\ 2 \end{bmatrix}, \begin{bmatrix} 2 \\ 2 \\ 2 \\ 0 \end{bmatrix}, \begin{bmatrix} 1 \\ 3 \\ 0 \\ 3 \end{bmatrix}, \begin{bmatrix} 2 \\ 6 \\ 3 \\ 9 \end{bmatrix} \right\}.$$

Because $\mathcal{S}$ is a subset of $\mathcal{R}^4$ and has five vectors, we know that at least one vector in this set is a linear combination of the others. Thus $A\mathbf{x} = \mathbf{0}$ has nonzero solutions, where

$$A = \begin{bmatrix} 1 & 2 & -1 & 2 & 1 & 2 \\ -1 & -2 & 1 & 2 & 3 & 6 \\ 2 & 4 & -3 & 2 & 0 & 3 \\ -3 & -6 & 2 & 0 & 3 & 9 \end{bmatrix}$$

is the matrix whose columns are the vectors in the set. But as we have just proved, the solutions of $A\mathbf{x} = \mathbf{0}$ are the same as those of $R\mathbf{x} = \mathbf{0}$, where

$$R = \begin{bmatrix} 1 & 2 & 0 & 0 & -1 & -5 \\ 0 & 0 & 1 & 0 & 0 & -3 \\ 0 & 0 & 0 & 1 & 1 & 2 \\ 0 & 0 & 0 & 0 & 0 & 0 \end{bmatrix}$$

is the reduced row echelon form of A (obtained in Section 1.4). Since any solution of $A\mathbf{x} = \mathbf{0}$ gives the coefficients of a linear combination of columns of A that equals $\mathbf{0}$, it follows that the same coefficients yield the equivalent linear combination of the columns

$$\mathcal{S}' = \left\{ \begin{bmatrix} 1 \\ 0 \\ 0 \\ 0 \end{bmatrix}, \begin{bmatrix} 2 \\ 0 \\ 0 \\ 0 \end{bmatrix}, \begin{bmatrix} 0 \\ 1 \\ 0 \\ 0 \end{bmatrix}, \begin{bmatrix} 0 \\ 0 \\ 1 \\ 0 \end{bmatrix}, \begin{bmatrix} -1 \\ 0 \\ 1 \\ 0 \end{bmatrix}, \begin{bmatrix} -5 \\ -3 \\ 2 \\ 0 \end{bmatrix} \right\},$$

of R, and vice versa. As a consequence, if column j of A is a linear combination of the other columns of A, then column j of R is a linear combination of the other columns of R with the same coefficients, and vice versa. For example, in $\mathcal{S}'$ it is obvious that $\mathbf{r}_2 = 2\mathbf{r}_1$ and $\mathbf{r}_5 = -\mathbf{r}_1 + \mathbf{r}_4$. Although it is less obvious, similar relationships hold in $\mathcal{S}$: $\mathbf{a}_2 = 2\mathbf{a}_1$ and $\mathbf{a}_5 = -\mathbf{a}_1 + \mathbf{a}_4$.

The preceding observations may be summarized as the *column correspondence property*.

Column Correspondence Property

Let A be a matrix and R its reduced row echelon form. If column j of R is a linear combination of other columns of R, then column j of A is a linear combination of the corresponding columns of A using the same coefficients, and vice versa.

Practice Problem 4 ▶ For the preceding matrix A, use the column correspondence property to express $\mathbf{a}_6$ as a linear combination of the other columns. ◀

The column correspondence property and the following theorem are proved in Appendix E:

THEOREM 2.4

The following statements are true for any matrix A:

(a) The pivot columns of A are linearly independent.

(b) Each nonpivot column of A is a linear combination of the previous pivot columns of A, where the coefficients of the linear combination are the entries of the corresponding column of the reduced row echelon form of A.

For example, in the previous matrices A and R, the first, third, and fourth columns are the pivot columns, so these columns are linearly independent. Furthermore, the fifth column of R is a linear combination of the preceding pivot columns. In fact, $\mathbf{r}_5 = (-1)\mathbf{r}_1 + 0\mathbf{r}_3 + 1\mathbf{r}_4$. Hence Theorem 2.4(b) guarantees that a similar equation holds for the corresponding columns of A; that is,

$$\mathbf{a}_5 = (-1)\mathbf{a}_1 + 0\mathbf{a}_3 + 1\mathbf{a}_4.$$

Example 4 The reduced row echelon form of a matrix A is

$$R = \begin{bmatrix} 1 & 2 & 0 & -1 \\ 0 & 0 & 1 & 1 \\ 0 & 0 & 0 & 0 \end{bmatrix}.$$

Determine A, given that the first and third columns of A are

$$\mathbf{a}_1 = \begin{bmatrix} 1 \\ 2 \\ 1 \end{bmatrix} \quad \text{and} \quad \mathbf{a}_3 = \begin{bmatrix} 2 \\ 2 \\ 3 \end{bmatrix}.$$

Solution Since the first and third columns of R are the pivot columns, these are also the pivot columns of A. Now observe that the second column of R is $2\mathbf{r}_1$, and hence, by the column correspondence property, the second column of A is

$$\mathbf{a}_2 = 2\mathbf{a}_1 = 2\begin{bmatrix} 1 \\ 2 \\ 1 \end{bmatrix} = \begin{bmatrix} 2 \\ 4 \\ 2 \end{bmatrix}.$$

Furthermore, the fourth column of R is $\mathbf{r}_4 = (-1)\mathbf{r}_1 + \mathbf{r}_3$, and so, again by the column correspondence property,

$$\mathbf{a}_4 = (-1)\mathbf{a}_1 + \mathbf{a}_3 = (-1)\begin{bmatrix} 1 \\ 2 \\ 1 \end{bmatrix} + \begin{bmatrix} 2 \\ 2 \\ 3 \end{bmatrix} = \begin{bmatrix} 1 \\ 0 \\ 2 \end{bmatrix}.$$

Thus

$$A = \begin{bmatrix} 1 & 2 & 2 & 1 \\ 2 & 4 & 2 & 0 \\ 1 & 2 & 3 & 2 \end{bmatrix}.$$

Practice Problem 5 ▶ Suppose that the reduced row echelon form of A is

$$R = \begin{bmatrix} 1 & -3 & 0 & 5 & 3 \\ 0 & 0 & 1 & 2 & -2 \\ 0 & 0 & 0 & 0 & 0 \end{bmatrix}.$$

Determine A if the first and third columns of A are $\mathbf{a}_1 = \begin{bmatrix} 1 \\ -1 \\ 2 \end{bmatrix}$ and $\mathbf{a}_3 = \begin{bmatrix} 2 \\ 0 \\ -1 \end{bmatrix}$, respectively. ◀

EXERCISES

For each of the matrices A and B in Exercises 1–8, determine whether $B = A^{-1}$.

1. $A = \begin{bmatrix} 1 & 2 \\ 1 & -1 \end{bmatrix}$ and $B = \begin{bmatrix} 1 & 0.5 \\ 1 & -1 \end{bmatrix}$

2. $A = \begin{bmatrix} 1 & 2 \\ 3 & 5 \end{bmatrix}$ and $B = \begin{bmatrix} -5 & 2 \\ 3 & -1 \end{bmatrix}$

3. $A = \begin{bmatrix} 1 & 2 & 1 \\ 1 & 1 & 2 \\ 2 & 3 & 4 \end{bmatrix}$ and $B = \begin{bmatrix} 2 & 5 & -3 \\ 0 & -2 & 1 \\ -1 & -1 & 1 \end{bmatrix}$

4. $A = \begin{bmatrix} 1 & 1 & 2 \\ 0 & 1 & 1 \\ 0 & 0 & 1 \end{bmatrix}$ and $B = \begin{bmatrix} 1 & -1 & 1 \\ 1 & 2 & 1 \\ -1 & 0 & -1 \end{bmatrix}$

5. $A = \begin{bmatrix} 1 & -2 & 1 & 0 \\ 2 & -2 & 1 & 0 \\ 1 & -1 & 0 & -1 \\ -1 & 0 & -1 & -1 \end{bmatrix}$ and
$B = \begin{bmatrix} -1 & 1 & 0 & 0 \\ 0 & -1 & 1 & -1 \\ 2 & -3 & 2 & -2 \\ -1 & 2 & -2 & 1 \end{bmatrix}$

6. $A = \begin{bmatrix} 1 & 2 & 2 & 2 \\ 1 & 1 & 1 & 2 \\ 2 & 3 & 2 & 3 \\ 3 & 5 & 7 & 7 \end{bmatrix}$ and

$B = \begin{bmatrix} 2 & 4 & -4 & -9 \\ -1 & -1 & 1 & 4 \\ 1 & 1 & -2 & -3 \\ -1 & -2 & 3 & 4 \end{bmatrix}$

7. $A = \begin{bmatrix} 1 & -1 & 2 & -1 \\ -1 & 2 & -4 & 3 \\ -2 & 2 & -2 & 3 \\ -2 & 1 & -1 & 1 \end{bmatrix}$ and

$B = \begin{bmatrix} 2 & 1 & -1 & 2 \\ 9 & 5 & -4 & 6 \\ 1 & 1 & -1 & 1 \\ -4 & -2 & 1 & -2 \end{bmatrix}$

8. $A = \begin{bmatrix} 1 & 1 & 2 & 2 \\ 2 & 1 & 1 & 2 \\ 2 & 1 & 2 & 3 \\ 2 & 2 & 2 & 4 \end{bmatrix}$ and

$B = \begin{bmatrix} 1 & -1 & 2 & -1 \\ 6 & -4 & 3 & -2 \\ -2 & 2 & -2 & 1 \\ -2 & 1 & -1 & 1 \end{bmatrix}$

For Exercises 9–14, find the value of each matrix expression, where A and B are the invertible 3×3 matrices such that

$$A^{-1} = \begin{bmatrix} 1 & 2 & 3 \\ 2 & 0 & 1 \\ 1 & 1 & -1 \end{bmatrix} \quad \text{and} \quad B^{-1} = \begin{bmatrix} 2 & -1 & 3 \\ 0 & 0 & 4 \\ 3 & -2 & 1 \end{bmatrix}.$$

9. $(A^T)^{-1}$
10. $(B^T)^{-1}$
11. $(AB)^{-1}$
12. $(BA)^{-1}$
13. $(AB^T)^{-1}$
14. $(A^TB^T)^{-1}$

In Exercises 15–22, find the inverse of each elementary matrix.

15. $\begin{bmatrix} 1 & 0 \\ 1 & 1 \end{bmatrix}$

16. $\begin{bmatrix} 1 & -3 \\ 0 & 1 \end{bmatrix}$

17. $\begin{bmatrix} 1 & 0 & 0 \\ -2 & 1 & 0 \\ 0 & 0 & 1 \end{bmatrix}$

18. $\begin{bmatrix} 0 & 0 & 1 \\ 0 & 1 & 0 \\ 1 & 0 & 0 \end{bmatrix}$

19. $\begin{bmatrix} 1 & 0 & 0 & 0 \\ 0 & 4 & 0 & 0 \\ 0 & 0 & 1 & 0 \\ 0 & 0 & 0 & 1 \end{bmatrix}$

20. $\begin{bmatrix} 1 & 0 & 0 \\ 0 & 1 & 0 \\ 0 & 0 & 4 \end{bmatrix}$

21. $\begin{bmatrix} 1 & 0 & 0 & 0 \\ 0 & 0 & 0 & 1 \\ 0 & 0 & 1 & 0 \\ 0 & 1 & 0 & 0 \end{bmatrix}$

22. $\begin{bmatrix} 1 & 0 & 0 & 0 \\ 0 & 1 & 0 & 2 \\ 0 & 0 & 1 & 0 \\ 0 & 0 & 0 & 1 \end{bmatrix}$

In Exercises 23–32, find an elementary matrix E such that $EA = B$.

23. $A = \begin{bmatrix} 1 & 2 \\ 3 & 4 \end{bmatrix}$ and $B = \begin{bmatrix} -1 & -2 \\ 3 & 4 \end{bmatrix}$

24. $A = \begin{bmatrix} -1 & 5 \\ 2 & 3 \end{bmatrix}$ and $B = \begin{bmatrix} -1 & 5 \\ 0 & 13 \end{bmatrix}$

25. $A = \begin{bmatrix} 2 & -3 \\ 7 & 10 \end{bmatrix}$ and $B = \begin{bmatrix} 7 & 10 \\ 2 & -3 \end{bmatrix}$

26. $A = \begin{bmatrix} 1 & 2 & -3 \\ -1 & 4 & 5 \end{bmatrix}$ and $B = \begin{bmatrix} 1 & 2 & -3 \\ 1 & -4 & -5 \end{bmatrix}$

27. $A = \begin{bmatrix} 3 & 2 & -1 \\ -1 & 0 & 6 \end{bmatrix}$ and $B = \begin{bmatrix} -1 & 0 & 6 \\ 3 & 2 & -1 \end{bmatrix}$

28. $A = \begin{bmatrix} -2 & 1 & 4 \\ 1 & -3 & 2 \end{bmatrix}$ and $B = \begin{bmatrix} 0 & -5 & 0 \\ 1 & -3 & 2 \end{bmatrix}$

29. $A = \begin{bmatrix} 1 & -1 & 2 \\ 2 & -3 & 1 \\ 0 & 4 & 5 \end{bmatrix}$ and $B = \begin{bmatrix} 1 & -1 & 2 \\ 2 & -3 & 1 \\ -10 & 19 & 0 \end{bmatrix}$

30. $A = \begin{bmatrix} 1 & 2 & -2 \\ 3 & -1 & 0 \\ -1 & 1 & 6 \end{bmatrix}$ and $B = \begin{bmatrix} 1 & 2 & -2 \\ 3 & -1 & 0 \\ 0 & 3 & 4 \end{bmatrix}$

31. $A = \begin{bmatrix} 1 & 2 & 3 \\ 4 & 5 & 6 \\ 7 & 8 & 9 \end{bmatrix}$ and $B = \begin{bmatrix} 1 & 2 & 3 \\ 7 & 8 & 9 \\ 4 & 5 & 6 \end{bmatrix}$

32. $A = \begin{bmatrix} 1 & 2 & 3 & 4 \\ -1 & 1 & 3 & 2 \\ 2 & -1 & 0 & 4 \end{bmatrix}$ and

$B = \begin{bmatrix} 1 & 2 & 3 & 4 \\ -1 & 1 & 3 & 2 \\ 0 & -5 & -6 & -4 \end{bmatrix}$

In Exercises 33–52, determine whether the statements are true or false.

33. Every square matrix is invertible.
34. Invertible matrices are square.
35. Elementary matrices are invertible.
36. If A and B are matrices such that $AB = I_n$ for some n, then both A and B are invertible.
37. If B and C are inverses of a matrix A, then $B = C$.
38. If A and B are invertible $n \times n$ matrices, then AB^T is invertible.
39. An invertible matrix may have more than one inverse.
40. For any matrices A and B, if A is the inverse of B, then B is the inverse of A.
41. For any matrices A and B, if A is the inverse of B^T, then A is the transpose of B^{-1}.
42. If A and B are invertible $n \times n$ matrices, then AB is also invertible.
43. If A and B are invertible $n \times n$ matrices, then $(AB)^{-1} = A^{-1}B^{-1}$.
44. An elementary matrix is a matrix that can be obtained by a sequence of elementary row operations on an identity matrix.
45. An elementary $n \times n$ matrix has at most $n + 1$ nonzero entries.
46. The product of two elementary $n \times n$ matrices is an elementary $n \times n$ matrix.
47. Every elementary matrix is invertible.

48. If A and B are $m \times n$ matrices and B can be obtained from A by an elementary row operation on A, then there is an elementary $m \times m$ matrix E such that $B = EA$.
49. If R is the reduced row echelon form of a matrix A, then there exists an invertible matrix P such that $PA = R$.
50. Let R be the reduced row echelon form of a matrix A. If column j of R is a linear combination of the previous columns of R, then column j of A is a linear combination of the previous columns of A.
51. The pivot columns of a matrix are linearly dependent.
52. Every column of a matrix is a linear combination of its pivot columns.

53.[3] Let A_α be the α-rotation matrix. Prove that $(A_\alpha)^T = (A_\alpha)^{-1}$.

54. Let $A = \begin{bmatrix} a & b \\ c & d \end{bmatrix}$.

(a) Suppose $ad - bc \neq 0$, and
$$B = \frac{1}{ad - bc}\begin{bmatrix} d & -b \\ -c & a \end{bmatrix}.$$
Show that $AB = BA = I_2$, and hence A is invertible and $B = A^{-1}$.

(b) Prove the converse of (a): If A is invertible, then $ad - bc \neq 0$.

55. Prove that the product of elementary matrices is invertible.

56. (a) Let A be an invertible $n \times n$ matrix, and let $\mathbf{u}$ and $\mathbf{v}$ be vectors in $\mathcal{R}^n$ such that $\mathbf{u} \neq \mathbf{v}$. Prove that $A\mathbf{u} \neq A\mathbf{v}$.

(b) Find a 2×2 matrix A and distinct vectors $\mathbf{u}$ and $\mathbf{v}$ in $\mathcal{R}^2$ such that $A\mathbf{u} = A\mathbf{v}$.

57. Let Q be an invertible $n \times n$ matrix. Prove that the subset $\{\mathbf{u}_1, \mathbf{u}_2, \ldots, \mathbf{u}_k\}$ of $\mathcal{R}^n$ is linearly independent if and only if $\{Q\mathbf{u}_1, Q\mathbf{u}_2, \ldots, Q\mathbf{u}_k\}$ is linearly independent.

58. Prove Theorem 2.2(a).

59. Prove that if A, B, and C are invertible $n \times n$ matrices, then ABC is invertible and $(ABC)^{-1} = C^{-1}B^{-1}A^{-1}$.

60. Let A and B be $n \times n$ matrices such that both A and AB are invertible. Prove that B is invertible, by writing it as the product of two invertible matrices.

61. Let A and B be $n \times n$ matrices such that $AB = I_n$. Prove that the rank of A is n. *Hint:* Theorem 1.6 can be useful.

62. Prove that if A is an $m \times n$ matrix and B is an $n \times p$ matrix, then rank $AB \leq$ rank B. *Hint:* Prove that if the kth column of B is not a pivot column of B, then the kth column of AB is not a pivot column of AB.

63. Prove that if B is an $n \times n$ matrix with rank n, then there exists an $n \times n$ matrix C such that $BC = I_n$. *Hint:* Theorem 1.6 can be useful.

64. Prove that if A and B are $n \times n$ matrices such that $AB = I_n$, then B is invertible and $A = B^{-1}$. *Hint:* Use Exercises 62 and 63.

65. Prove that if an $n \times n$ matrix has rank n, then it is invertible. *Hint:* Use Exercises 63 and 64.

66. Let $M = \begin{bmatrix} A & O_1 \\ O_2 & B \end{bmatrix}$, where A and B are square and O_1 and O_2 are zero matrices. Prove that M is invertible if and only if A and B are both invertible. *Hint:* Think about the reduced row echelon form of M.

In Exercises 67–74, find the matrix A, given the reduced row echelon form R of A and information about certain columns of A.

67. $R = \begin{bmatrix} 1 & 0 & 1 \\ 0 & 1 & 2 \end{bmatrix}$, $\mathbf{a}_1 = \begin{bmatrix} 3 \\ -1 \end{bmatrix}$, and $\mathbf{a}_2 = \begin{bmatrix} 2 \\ 5 \end{bmatrix}$

68. $R = \begin{bmatrix} 1 & 2 & 0 & -3 & 0 & 1 \\ 0 & 0 & 1 & 2 & 0 & 2 \\ 0 & 0 & 0 & 0 & 1 & 3 \\ 0 & 0 & 0 & 0 & 0 & 0 \end{bmatrix}$, $\mathbf{a}_1 = \begin{bmatrix} 2 \\ 0 \\ -1 \\ 1 \end{bmatrix}$,
$\mathbf{a}_3 = \begin{bmatrix} 1 \\ -1 \\ 2 \\ 0 \end{bmatrix}$, and $\mathbf{a}_5 = \begin{bmatrix} 2 \\ 3 \\ 0 \\ 1 \end{bmatrix}$

69. $R = \begin{bmatrix} 1 & -1 & 0 & 0 & 1 \\ 0 & 0 & 1 & 0 & 2 \\ 0 & 0 & 0 & 1 & 3 \end{bmatrix}$, $\mathbf{a}_2 = \begin{bmatrix} 1 \\ -2 \\ 1 \end{bmatrix}$,
$\mathbf{a}_3 = \begin{bmatrix} 1 \\ -1 \\ 0 \end{bmatrix}$, and $\mathbf{a}_4 = \begin{bmatrix} 4 \\ 1 \\ 3 \end{bmatrix}$.

70. $R = \begin{bmatrix} 1 & 2 & 0 & 1 & 0 & 1 \\ 0 & 0 & 1 & -1 & 0 & -1 \\ 0 & 0 & 0 & 0 & 1 & 1 \\ 0 & 0 & 0 & 0 & 0 & 0 \end{bmatrix}$, $\mathbf{a}_2 = \begin{bmatrix} 2 \\ 4 \\ 6 \\ -2 \end{bmatrix}$,
$\mathbf{a}_4 = \begin{bmatrix} 1 \\ 3 \\ -1 \\ 1 \end{bmatrix}$, $\mathbf{a}_6 = \begin{bmatrix} 2 \\ -1 \\ -1 \\ 2 \end{bmatrix}$

71. $R = \begin{bmatrix} 1 & 2 & 0 & 4 \\ 0 & 0 & 1 & 3 \end{bmatrix}$, $\mathbf{a}_2 = \begin{bmatrix} 2 \\ 4 \end{bmatrix}$, and $\mathbf{a}_3 = \begin{bmatrix} 3 \\ 5 \end{bmatrix}$

72. $R = \begin{bmatrix} 1 & -1 & 0 & -2 & -3 & 2 \\ 0 & 0 & 1 & 3 & 4 & -4 \\ 0 & 0 & 0 & 0 & 0 & 0 \end{bmatrix}$, $\mathbf{a}_2 = \begin{bmatrix} -1 \\ -1 \\ -1 \end{bmatrix}$,
and $\mathbf{a}_5 = \begin{bmatrix} -3 \\ 5 \\ 1 \end{bmatrix}$

73. $R = \begin{bmatrix} 1 & -1 & 0 & -2 & 0 & 2 \\ 0 & 0 & 1 & 3 & 0 & -4 \\ 0 & 0 & 0 & 0 & 1 & 1 \end{bmatrix}$, $\mathbf{a}_2 = \begin{bmatrix} -1 \\ 0 \\ -1 \end{bmatrix}$,
$\mathbf{a}_4 = \begin{bmatrix} 1 \\ 6 \\ -2 \end{bmatrix}$, and $\mathbf{a}_6 = \begin{bmatrix} -1 \\ -7 \\ 3 \end{bmatrix}$

---

[3] The result of this exercise is used in Section 4.4 (page 269).

74. $R = \begin{bmatrix} 1 & 0 & 0 & -3 & 1 & 3 \\ 0 & 1 & 0 & 2 & -1 & -2 \\ 0 & 0 & 1 & 0 & 0 & -1 \\ 0 & 0 & 0 & 0 & 0 & 0 \end{bmatrix}$, $\mathbf{a}_1 = \begin{bmatrix} 1 \\ 2 \\ -1 \\ 0 \end{bmatrix}$,
$\mathbf{a}_5 = \begin{bmatrix} 1 \\ 3 \\ -2 \\ -1 \end{bmatrix}$, and $\mathbf{a}_6 = \begin{bmatrix} 4 \\ 9 \\ -6 \\ -3 \end{bmatrix}$

In Exercises 75–78, write the indicated column of

$$A = \begin{bmatrix} 1 & -2 & 1 & -1 & -2 \\ 2 & -4 & 1 & 1 & 1 \\ 3 & -6 & 0 & 6 & 9 \end{bmatrix}$$

as a linear combination of the pivot columns of A.

75. $\mathbf{a}_2$ 76. $\mathbf{a}_3$ 77. $\mathbf{a}_4$ 78. $\mathbf{a}_5$

In Exercises 79–82, write the indicated column of

$$B = \begin{bmatrix} 1 & 0 & 1 & -3 & -1 & 4 \\ 2 & -1 & 3 & -8 & -1 & 9 \\ -1 & 1 & -2 & 5 & 1 & -6 \\ 0 & 1 & -1 & 2 & 1 & -3 \end{bmatrix}$$

as a linear combination of the pivot columns of B.

79. $\mathbf{b}_3$ 80. $\mathbf{b}_4$ 81. $\mathbf{b}_5$ 82. $\mathbf{b}_6$

83. Suppose that $\mathbf{u}$ and $\mathbf{v}$ are linearly independent vectors in $\mathcal{R}^3$. Find the reduced row echelon form of $A = [\mathbf{a}_1\ \mathbf{a}_2\ \mathbf{a}_3\ \mathbf{a}_4]$, given that
$$\mathbf{a}_1 = \mathbf{u}, \quad \mathbf{a}_2 = 2\mathbf{u}, \quad \mathbf{a}_3 = \mathbf{u} + \mathbf{v}, \quad \text{and} \quad \mathbf{a}_4 = \mathbf{v}.$$

84. Let A be an $n \times n$ invertible matrix, and let $\mathbf{e}_j$ be the jth standard vector of $\mathcal{R}^n$.
 (a) Prove that the jth column of A^{-1} is a solution of $A\mathbf{x} = \mathbf{e}_j$.
 (b) Why does the result in (a) imply that A^{-1} is unique?
 (c) Why does the result in (a) imply that rank $A = n$?

85. Let A be a matrix with reduced row echelon form R. Use the column correspondence property to prove the following:
 (a) A column of A is $\mathbf{0}$ if and only if the corresponding column of R is $\mathbf{0}$.
 (b) A set of columns of A is linearly independent if and only if the corresponding set of columns of R is linearly independent.

86. Let R be an $m \times n$ matrix in reduced row echelon form. Find a relationship between the columns of R^T and the columns of $R^T R$. Justify your answer.

87. Let R be an $m \times n$ matrix in reduced row echelon form with rank $R = r$. Prove the following:
 (a) The reduced row echelon form of R^T is the $n \times m$ matrix $[\mathbf{e}_1\ \mathbf{e}_2\ \ldots\ \mathbf{e}_r\ \mathbf{0}\ \ldots\ \mathbf{0}]$, where $\mathbf{e}_j$ is the jth standard vector of $\mathcal{R}^n$ for $1 \le j \le r$.
 (b) rank R^T = rank R.

88. Let A be an $m \times n$ matrix with reduced row echelon form R. Then there exists an invertible matrix P such that $PA = R$ and an invertible matrix Q such that QR^T is the reduced row echelon form of R^T. Describe the matrix PAQ^T in terms of A. Justify your answer.

89. Let A and B be $m \times n$ matrices. Prove that the following conditions are equivalent.
 (a) A and B have the same reduced row echelon form.
 (b) There is an invertible $m \times m$ matrix P such that $B = PA$.

90. Let A be an $n \times n$ matrix. Find a property of A that is equivalent to the statement $AB = AC$ if and only if $B = C$. Justify your answer.

91. Let A be a 2×3 matrix, and let E be an elementary matrix obtained by an elementary row operation on I_2. Prove that the product EA can be obtained from A by the identical elementary row operation on A. *Hint:* Prove this for each of the three kinds of elementary row operations.

92. Let A and B be $m \times n$ matrices. Prove that the following conditions are equivalent:
 (a) A and B have the same reduced row echelon form.
 (b) The system of equations $A\mathbf{x} = \mathbf{0}$ is equivalent to $B\mathbf{x} = \mathbf{0}$.

We can define an elementary column operation in a manner analogous to the definition in Section 1.3 of an elementary row operation. Each of the following operations on a matrix is called an **elementary column operation**: *interchanging any two columns of the matrix, multiplying any column by a nonzero scalar, and adding a multiple of one column of the matrix to another column of the matrix.*

93. Let E be an $n \times n$ matrix. Prove that E is an elementary matrix if and only if E can be obtained from I_n by a single elementary column operation.

94. Prove that if a matrix E is obtained from I_n by a single elementary column operation, then for any $m \times n$ matrix A, the product AE can be obtained from A by the identical elementary column operation on A.

In Exercises 95–99, use either a calculator with matrix capabilities or computer software such as MATLAB to solve each problem.

95. Let
$$A = \begin{bmatrix} 1 & 1 & 0 & -1 \\ 0 & 1 & 1 & 2 \\ 2 & 1 & 0 & -3 \\ -1 & -1 & 1 & 1 \end{bmatrix}.$$
Let B be the matrix obtained by interchanging rows 1 and 3 of A, and let C be the matrix obtained by interchanging rows 2 and 4 of A.
 (a) Show that A is invertible.
 (b) Show that B and C are invertible.
 (c) Compare B^{-1} and C^{-1} with A^{-1}.

(d) Now let A be any invertible $n \times n$ matrix, and let B be the matrix obtained by interchanging rows i and j of A, where $1 \le i < j \le n$. Make a conjecture about the relationship between B^{-1} and A^{-1}.

(e) Prove your conjecture in (d).

96. Let

$$A = \begin{bmatrix} 15 & 30 & 17 & 31 \\ 30 & 66 & 36 & 61 \\ 17 & 36 & 20 & 35 \\ 31 & 61 & 35 & 65 \end{bmatrix}.$$

(a) Show that A is symmetric and invertible.

(b) Show that A^{-1} is symmetric and invertible.

(c) Prove that the inverse of any symmetric and invertible matrix is also symmetric and invertible.

97. Let

$$A = \begin{bmatrix} 1 & 2 & 0 & 3 \\ 2 & 5 & -1 & 8 \\ 2 & 4 & 1 & 6 \\ 3 & 6 & 1 & 8 \end{bmatrix}.$$

(a) Show that A and A^2 are invertible by using their reduced row echelon forms.

(b) Compute the inverse of A^2, and show that it equals $(A^{-1})^2$.

(c) State and verify a similar result to (b) for A^3.

(d) Generalize and prove a result analogous to (c) for the nth power of any invertible matrix.

98. Consider the system $A\mathbf{x} = \mathbf{b}$.

$$\begin{aligned} x_1 + 3x_2 + 2x_3 + x_4 &= 4 \\ x_1 + 2x_2 + 4x_3 &= -3 \\ 2x_1 + 6x_2 + 5x_3 + 2x_4 &= -1 \\ x_1 + 3x_2 + 2x_3 + 2x_4 &= 2 \end{aligned}$$

(a) Compute the inverse of A and use it to solve the system.

(b) Show that your solution is correct by verifying that it satisfies the matrix equation $A\mathbf{x} = \mathbf{b}$.

99. The purpose of this exercise is to illustrate the method of finding an inverse described in Exercise 84. In Exercise 97, you found the inverse of

$$A = \begin{bmatrix} 1 & 2 & 0 & 3 \\ 2 & 5 & -1 & 8 \\ 2 & 4 & 1 & 6 \\ 3 & 6 & 1 & 8 \end{bmatrix}.$$

(a) Solve the system $A\mathbf{x} = \mathbf{e}_1$, and compare your solution with the first column of A^{-1}.

(b) Repeat (a) for the vectors $\mathbf{e}_2$, $\mathbf{e}_3$, and $\mathbf{e}_4$.

SOLUTIONS TO THE PRACTICE PROBLEMS

1. Since

$$AB = \begin{bmatrix} -1 & 0 & 1 \\ 1 & 2 & -2 \\ 2 & -1 & -1 \end{bmatrix} \begin{bmatrix} 4 & 1 & 2 \\ 3 & 1 & 1 \\ 5 & 1 & 2 \end{bmatrix} = \begin{bmatrix} 1 & 0 & 0 \\ 0 & 1 & 0 \\ 0 & 0 & 1 \end{bmatrix}$$

and

$$BA = \begin{bmatrix} 4 & 1 & 2 \\ 3 & 1 & 1 \\ 5 & 1 & 2 \end{bmatrix} \begin{bmatrix} -1 & 0 & 1 \\ 1 & 2 & -2 \\ 2 & -1 & -1 \end{bmatrix} = \begin{bmatrix} 1 & 0 & 0 \\ 0 & 1 & 0 \\ 0 & 0 & 1 \end{bmatrix},$$

$B = A^{-1}$.

2. This system of linear equations can be written as the matrix equation $A\mathbf{x} = \mathbf{b}$, where A is the matrix given in Practice Problem 1, and

$$\mathbf{b} = \begin{bmatrix} 1 \\ 2 \\ -1 \end{bmatrix}.$$

Since the matrix B given in Practice Problem 1 is equal to A^{-1}, we have

$$\begin{bmatrix} x_1 \\ x_2 \\ x_3 \end{bmatrix} = \mathbf{x} = A^{-1}\mathbf{b} = \begin{bmatrix} 4 & 1 & 2 \\ 3 & 1 & 1 \\ 5 & 1 & 2 \end{bmatrix} \begin{bmatrix} 1 \\ 2 \\ -1 \end{bmatrix} = \begin{bmatrix} 4 \\ 4 \\ 5 \end{bmatrix}.$$

Therefore $x_1 = 4$, $x_2 = 4$, and $x_3 = 5$.

3. The matrix B is obtained from A by adding -2 times row 1 of A to row 2. Adding -2 times row 1 of I_2 to row 2 produces the elementary matrix $E = \begin{bmatrix} 1 & 0 \\ -2 & 1 \end{bmatrix}$. This matrix has the property that $EA = B$.

4. From the matrix R, it is clear that $\mathbf{r}_6 = -5\mathbf{r}_1 - 3\mathbf{r}_3 + 2\mathbf{r}_4$. So by the column correspondence property, it follows that $\mathbf{a}_6 = -5\mathbf{a}_1 - 3\mathbf{a}_3 + 2\mathbf{a}_4$.

5. Let $\mathbf{r}_1, \mathbf{r}_2, \mathbf{r}_3, \mathbf{r}_4$, and $\mathbf{r}_5$ denote the columns of R. The pivot columns of R are columns 1 and 3, so the pivot columns of A are columns 1 and 3. Every other column of R is a linear combination of its pivot columns. In fact, $\mathbf{r}_2 = -3\mathbf{r}_1$, $\mathbf{r}_4 = 5\mathbf{r}_1 + 2\mathbf{r}_3$, and $\mathbf{r}_5 = 3\mathbf{r}_1 - 2\mathbf{r}_3$. Thus the column correspondence property implies that column 2 of A is

$$\mathbf{a}_2 = -3\mathbf{a}_1 = -3 \begin{bmatrix} 1 \\ -1 \\ 2 \end{bmatrix} = \begin{bmatrix} -3 \\ 3 \\ -6 \end{bmatrix}.$$

Similarly, the fourth and fifth columns of A are

$$\mathbf{a}_4 = 5\mathbf{a}_1 + 2\mathbf{a}_3 = 5 \begin{bmatrix} 1 \\ -1 \\ 2 \end{bmatrix} + 2 \begin{bmatrix} 2 \\ 0 \\ -1 \end{bmatrix} = \begin{bmatrix} 9 \\ -5 \\ 8 \end{bmatrix}$$

and

$$\mathbf{a}_5 = 3\mathbf{a}_1 - 2\mathbf{a}_3 = 3\begin{bmatrix} 1 \\ -1 \\ 2 \end{bmatrix} - 2\begin{bmatrix} 2 \\ 0 \\ -1 \end{bmatrix} = \begin{bmatrix} -1 \\ -3 \\ 8 \end{bmatrix},$$

respectively. Hence

$$A = \begin{bmatrix} 1 & -3 & 2 & 9 & -1 \\ -1 & 3 & 0 & -5 & -3 \\ 2 & -6 & -1 & 8 & 8 \end{bmatrix}.$$

2.4 THE INVERSE OF A MATRIX

In this section, we learn which matrices are invertible and how to find their inverses. To do this, we apply what we have learned about invertible and elementary matrices. The next theorem tells us when a matrix is invertible.

THEOREM 2.5

Let A be an $n \times n$ matrix. Then A is invertible if and only if the reduced row echelon form of A is I_n.

PROOF First, suppose that A is invertible. Consider any vector $\mathbf{v}$ in $\mathcal{R}^n$ such that $A\mathbf{v} = \mathbf{0}$. Then, by the boxed result on page 124, $\mathbf{v} = A^{-1}\mathbf{0} = \mathbf{0}$. Thus the only solution of $A\mathbf{x} = \mathbf{0}$ is $\mathbf{0}$, and so rank $A = n$ by Theorem 1.8. However, by the second boxed statement on page 48, the reduced row echelon form of A must equal I_n.

Conversely, suppose that the reduced row echelon form of A equals I_n. Then, by Theorem 2.3, there exists an invertible $n \times n$ matrix P such that $PA = I_n$. So

$$A = I_nA = (P^{-1}P)A = P^{-1}(PA) = P^{-1}I_n = P^{-1}.$$

But, by Theorem 2.2, P^{-1} is an invertible matrix, and therefore A is invertible. ■

Theorem 2.5 can be used as a test for matrix invertibility, as follows: To determine if an $n \times n$ matrix is invertible, compute its reduced row echelon form R. If $R = I_n$, then the matrix is invertible; otherwise, if $R \neq I_n$, then the matrix is not invertible.

Example 1 Use Theorem 2.5 to test the following matrices for invertibility:

$$A = \begin{bmatrix} 1 & 2 & 3 \\ 2 & 5 & 6 \\ 3 & 4 & 8 \end{bmatrix} \quad \text{and} \quad B = \begin{bmatrix} 1 & 1 & 2 \\ 2 & 1 & 1 \\ 1 & 0 & -1 \end{bmatrix}$$

You should check that the reduced row echelon form of A is I_3. Therefore A is invertible by Theorem 2.5.

On the other hand, the reduced row echelon form of B is

$$\begin{bmatrix} 1 & 0 & -1 \\ 0 & 1 & 3 \\ 0 & 0 & 0 \end{bmatrix}.$$

So, by Theorem 2.5, B is not invertible.

AN ALGORITHM FOR MATRIX INVERSION

Theorem 2.5 provides not only a method for determining whether a matrix is invertible, but also a method for actually calculating the inverse when it exists. We know that we can transform any $n \times n$ matrix A into a matrix R in reduced row echelon form by means of elementary row operations. Applying the same operations to the $n \times 2n$ matrix $[A \;\; I_n]$ transforms this matrix into an $n \times 2n$ matrix $[R \;\; B]$ for some $n \times n$ matrix B. Hence there is an invertible matrix P such that $P[A \;\; I_n] = [R \;\; B]$. Thus

$$[R \;\; B] = P[A \;\; I_n] = [PA \;\; PI_n] = [PA \;\; P].$$

It follows that $PA = R$, and $P = B$. If $R \neq I_n$, then we know that A is not invertible by Theorem 2.5. On the other hand, if $R = I_n$, then A is invertible, again by Theorem 2.5. Furthermore, since $PA = I_n$ and $P = B$, it follows that $B = A^{-1}$. Thus we have the following algorithm for computing the inverse of a matrix:

Algorithm for Matrix Inversion

Let A be an $n \times n$ matrix. Use elementary row operations to transform $[A \;\; I_n]$ into the form $[R \;\; B]$, where R is a matrix in reduced row echelon form. Then either

(a) $R = I_n$, in which case A is invertible and $B = A^{-1}$; or

(b) $R \neq I_n$, in which case A is not invertible.

Example 2

We use the algorithm for matrix inversion to compute A^{-1} for the invertible matrix A of Example 1. This algorithm requires us to transform $[A \;\; I_3]$ into a matrix of the form $[I_3 \;\; B]$ by means of elementary row operations. For this purpose, we use the Gaussian elimination algorithm in Section 1.4 to transform A into its reduced row echelon form I_3, while applying each row operation to the entire row of the 3×6 matrix.

$$[A \;\; I_3] = \left[\begin{array}{ccc|ccc} 1 & 2 & 3 & 1 & 0 & 0 \\ 2 & 5 & 6 & 0 & 1 & 0 \\ 3 & 4 & 8 & 0 & 0 & 1 \end{array}\right] \xrightarrow[-3\mathbf{r}_1 + \mathbf{r}_3 \to \mathbf{r}_3]{-2\mathbf{r}_1 + \mathbf{r}_2 \to \mathbf{r}_2} \left[\begin{array}{ccc|ccc} 1 & 2 & 3 & 1 & 0 & 0 \\ 0 & 1 & 0 & -2 & 1 & 0 \\ 0 & -2 & -1 & -3 & 0 & 1 \end{array}\right]$$

$$\xrightarrow{2\mathbf{r}_2 + \mathbf{r}_3 \to \mathbf{r}_3} \left[\begin{array}{ccc|ccc} 1 & 2 & 3 & 1 & 0 & 0 \\ 0 & 1 & 0 & -2 & 1 & 0 \\ 0 & 0 & -1 & -7 & 2 & 1 \end{array}\right]$$

$$\xrightarrow{-\mathbf{r}_3 \to \mathbf{r}_3} \left[\begin{array}{ccc|ccc} 1 & 2 & 3 & 1 & 0 & 0 \\ 0 & 1 & 0 & -2 & 1 & 0 \\ 0 & 0 & 1 & 7 & -2 & -1 \end{array}\right]$$

$$\xrightarrow{-3\mathbf{r}_3 + \mathbf{r}_1 \to \mathbf{r}_1} \left[\begin{array}{ccc|ccc} 1 & 2 & 0 & -20 & 6 & 3 \\ 0 & 1 & 0 & -2 & 1 & 0 \\ 0 & 0 & 1 & 7 & -2 & -1 \end{array}\right]$$

$$\xrightarrow{-2\mathbf{r}_2 + \mathbf{r}_1 \to \mathbf{r}_1} \left[\begin{array}{ccc|ccc} 1 & 0 & 0 & -16 & 4 & 3 \\ 0 & 1 & 0 & -2 & 1 & 0 \\ 0 & 0 & 1 & 7 & -2 & -1 \end{array}\right] = [I_3 \;\; B]$$

Thus

$$A^{-1} = B = \begin{bmatrix} -16 & 4 & 3 \\ -2 & 1 & 0 \\ 7 & -2 & -1 \end{bmatrix}.$$

In the context of the preceding discussion, if $[R \;\; B]$ is in reduced row echelon form, then so is R. (See Exercise 73.) This observation is useful if there is a calculator or computer available that can produce the reduced row echelon form of a matrix. In such a case, we simply find the reduced row echelon form $[R \;\; B]$ of $[A \;\; I_n]$. Then, as before, either

(a) $R = I_n$, in which case A is invertible and $B = A^{-1}$; or
(b) $R \neq I_n$, in which case A is not invertible.

Example 3 To illustrate the previous paragraph, we test

$$A = \begin{bmatrix} 1 & 1 \\ 2 & 2 \end{bmatrix}$$

for invertibility. If we are performing calculations by hand, we transform $[A \;\; I_2]$ into a matrix $[R \;\; B]$ such that R is in reduced row echelon form:

$$[A \;\; I_2] = \left[\begin{array}{cc|cc} 1 & 1 & 1 & 0 \\ 2 & 2 & 0 & 1 \end{array}\right] \xrightarrow{-2\mathbf{r}_1+\mathbf{r}_2\to\mathbf{r}_2} \left[\begin{array}{cc|cc} 1 & 1 & 1 & 0 \\ 0 & 0 & -2 & 1 \end{array}\right] = [R \;\; B]$$

Since $R \neq I_2$, A is not invertible.

Note that in this case $[R \;\; B]$ is not in reduced row echelon form because of the -2 in the $(2,3)$-entry. If you use a computer or calculator to put $[A \;\; I_2]$ into reduced row echelon form, some additional steps are performed, resulting in the matrix

$$[R \;\; C] = \left[\begin{array}{cc|cc} 1 & 1 & 0 & 0.5 \\ 0 & 0 & 1 & -0.5 \end{array}\right].$$

Again, we see that A is not invertible because $R \neq I_2$. It is not necessary to perform these additional steps when solving the problem by hand—you can stop as soon as it becomes clear that $R \neq I_2$.

Practice Problem 1 ▶ For each of the matrices

$$A = \begin{bmatrix} 1 & -2 & 1 \\ 2 & -1 & -1 \\ -2 & -5 & 7 \end{bmatrix} \quad \text{and} \quad B = \begin{bmatrix} 1 & 1 & 0 \\ 3 & 4 & 1 \\ -1 & 4 & 4 \end{bmatrix},$$

determine whether the matrix is invertible. If so, find its inverse. ◀

Practice Problem 2 ▶ Consider the system of linear equations

$$\begin{aligned} x_1 - x_2 + 2x_3 &= 2 \\ x_1 + 2x_2 &= 3 \\ -x_2 + x_3 &= -1. \end{aligned}$$

(a) Write the system in the form of a matrix equation $A\mathbf{x} = \mathbf{b}$.
(b) Show that A is invertible, and find A^{-1}.
(c) Solve the system by using the answer to (b).

Although the next theorem includes a very long list of statements, most of it follows easily from previous results. The key idea is that when A is an $n \times n$ matrix, the reduced row echelon form of A equals I_n if and only if A has a pivot position in each row and if and only if A has a pivot position in each column. Thus, in the special case of an $n \times n$ matrix, each statement in Theorems 1.6 and 1.8 is true if and only if the reduced row echelon form of A equals I_n. So all these statements are equivalent to each other.

THEOREM 2.6

(Invertible Matrix Theorem) Let A be an $n \times n$ matrix. The following statements are equivalent:

(a) A is invertible.
(b) The reduced row echelon form of A is I_n.
(c) The rank of A equals n.
(d) The span of the columns of A is $\mathcal{R}^n$.
(e) The equation $A\mathbf{x} = \mathbf{b}$ is consistent for every $\mathbf{b}$ in $\mathcal{R}^n$.
(f) The nullity of A equals zero.
(g) The columns of A are linearly independent.
(h) The only solution of $A\mathbf{x} = \mathbf{0}$ is $\mathbf{0}$.
(i) There exists an $n \times n$ matrix B such that $BA = I_n$.
(j) There exists an $n \times n$ matrix C such that $AC = I_n$.
(k) A is a product of elementary matrices.

PROOF Statements (a) and (b) are equivalent by Theorem 2.5. Because A is an $n \times n$ matrix, statement (b) is equivalent to each of (c), (d), and (e) by Theorem 1.6 on page 70. Similarly, statement (b) is equivalent to each of (f), (g), and (h) by Theorem 1.8 on page 78. Therefore statement (a) is equivalent to each of statements (b) through (h).

Proof that (a) implies (k): The assumption that A is invertible implies that the reduced row echelon form of A is I_n. Then, as in the proof of Theorem 2.5, there exists an invertible $n \times n$ matrix P such that $PA = I_n$ (Theorem 2.3). Therefore $A = P^{-1}$, and the discussion preceding Theorem 2.3 shows that P is a product of invertible matrices. Thus P^{-1} is the product of the inverses of these elementary matrices (in reverse order), each of which is an elementary matrix. It follows that A is a product of elementary matrices, and so (a) implies (k).

Proof that (k) implies (a): Suppose that A is a product of elementary matrices. Since elementary matrices are invertible, A is the product of invertible matrices and hence A is invertible, establishing (a). Thus statements (a) and (k) are equivalent.

It remains to show that (a) is equivalent to (i) and (j).

Clearly, (a) implies (i), with $B = A^{-1}$. Conversely, assume (i). Let $\mathbf{v}$ be any vector in $\mathcal{R}^n$ such that $A\mathbf{v} = \mathbf{0}$. Then

$$\mathbf{v} = I_n\mathbf{v} = (BA)\mathbf{v} = B(A\mathbf{v}) = B\mathbf{0} = \mathbf{0}.$$

It follows that (i) implies (h). But (h) implies (a); so (i) implies (a). Thus statements (a) and (i) are equivalent.

Clearly, (a) implies (j), with $C = A^{-1}$. Conversely, assume (j). Let $\mathbf{b}$ be any vector in $\mathcal{R}^n$ and $\mathbf{v} = C\mathbf{b}$. Then

$$A\mathbf{v} = A(C\mathbf{b}) = (AC)\mathbf{b} = I_n\mathbf{b} = \mathbf{b}.$$

It follows that (j) implies (e). Since (e) implies (a), it follows that (j) implies (a). Hence (a) and (j) are equivalent.

Therefore all of the statements in the Invertible Matrix Theorem are equivalent. ■

Statements (i) and (j) of the Invertible Matrix Theorem are the two conditions in the definition of invertibility. It follows from the Invertible Matrix Theorem that we need verify only one of these conditions to show that a *square* matrix is invertible. For example, suppose that for a given $n \times n$ matrix A, there is an $n \times n$ matrix C such that $AC = I_n$. Then A is invertible by the Invertible Matrix Theorem, and so we may multiply both sides of this equation on the left by A^{-1} to obtain the equation $A^{-1}(AC) = A^{-1}I_n$, which reduces to $C = A^{-1}$. Similarly, if $BA = I_n$ for some matrix B, then A is invertible by the Invertible Matrix Theorem, and

$$B = BI_n = B(AA^{-1}) = (BA)A^{-1} = I_nA^{-1} = A^{-1}.$$

Note that the matrices B and C in statements (i) and (j) of the Invertible Matrix Theorem must be square. There are nonsquare matrices A and C for which the product AC is an identity matrix. For instance, let

$$A = \begin{bmatrix} 1 & 1 & 0 \\ 1 & 2 & 1 \end{bmatrix} \quad \text{and} \quad C = \begin{bmatrix} 2 & 1 \\ -1 & -1 \\ 0 & 2 \end{bmatrix}.$$

Then

$$AC = \begin{bmatrix} 1 & 1 & 0 \\ 1 & 2 & 1 \end{bmatrix} \begin{bmatrix} 2 & 1 \\ -1 & -1 \\ 0 & 2 \end{bmatrix} = \begin{bmatrix} 1 & 0 \\ 0 & 1 \end{bmatrix} = I_2.$$

Of course, A and C are not invertible.

COMPUTING* $A^{-1}B$

When A is an invertible $n \times n$ matrix and B is any $n \times p$ matrix, we can extend the algorithm for matrix inversion to compute $A^{-1}B$. Consider the $n \times (n+p)$ matrix $[A\ \ B]$. Suppose we transform this matrix by means of elementary row operations into the matrix $[I_n\ \ C]$. As in the discussion of the algorithm for matrix inversion, there is an $n \times n$ invertible matrix P such that

$$[I_n\ \ C] = P[A\ \ B] = [PA\ \ PB],$$

from which it follows that $PA = I_n$ and $C = PB$. Therefore $P = A^{-1}$, and hence $C = A^{-1}B$. Thus we have the following algorithm:

> **Algorithm for Computing $\mathbf{A^{-1}B}$**
>
> Let A be an invertible $n \times n$ matrix and B be an $n \times p$ matrix. Suppose that the $n \times (n+p)$ matrix $[A\ \ B]$ is transformed by means of elementary row operations into the matrix $[I_n\ \ C]$ in reduced row echelon form. Then $C = A^{-1}B$.

---

* The remainder of this section may be omitted without loss of continuity.

Example 4 Use the preceding algorithm to compute $A^{-1}B$ for

$$A = \begin{bmatrix} 1 & 2 & 1 \\ 2 & 5 & 1 \\ 2 & 4 & 1 \end{bmatrix} \quad \text{and} \quad B = \begin{bmatrix} 2 & -1 \\ 1 & 3 \\ 0 & 2 \end{bmatrix}.$$

Solution We apply elementary row operations to transform $[A\ B]$ into its reduced row echelon form, $[I_3\ A^{-1}B]$.

$$[A\ B] = \left[\begin{array}{ccc|cc} 1 & 2 & 1 & 2 & -1 \\ 2 & 5 & 1 & 1 & 3 \\ 2 & 4 & 1 & 0 & 2 \end{array}\right] \xrightarrow[-2\mathbf{r}_1 + \mathbf{r}_3 \to \mathbf{r}_3]{-2\mathbf{r}_1 + \mathbf{r}_2 \to \mathbf{r}_2} \left[\begin{array}{ccc|cc} 1 & 2 & 1 & 2 & -1 \\ 0 & 1 & -1 & -3 & 5 \\ 0 & 0 & -1 & -4 & 4 \end{array}\right]$$

$$\xrightarrow{-\mathbf{r}_3 \to \mathbf{r}_3} \left[\begin{array}{ccc|cc} 1 & 2 & 1 & 2 & -1 \\ 0 & 1 & -1 & -3 & 5 \\ 0 & 0 & 1 & 4 & -4 \end{array}\right] \xrightarrow[-\mathbf{r}_3 + \mathbf{r}_1 \to \mathbf{r}_1]{\mathbf{r}_3 + \mathbf{r}_2 \to \mathbf{r}_2} \left[\begin{array}{ccc|cc} 1 & 2 & 0 & -2 & 3 \\ 0 & 1 & 0 & 1 & 1 \\ 0 & 0 & 1 & 4 & -4 \end{array}\right]$$

$$\xrightarrow{-2\mathbf{r}_2 + \mathbf{r}_1 \to \mathbf{r}_1} \left[\begin{array}{ccc|cc} 1 & 0 & 0 & -4 & 1 \\ 0 & 1 & 0 & 1 & 1 \\ 0 & 0 & 1 & 4 & -4 \end{array}\right]$$

Therefore

$$A^{-1}B = \begin{bmatrix} -4 & 1 \\ 1 & 1 \\ 4 & -4 \end{bmatrix}.$$

One of the applications of the algorithm for computing $A^{-1}B$ is to solve several systems of linear equations sharing a common invertible coefficient matrix. Suppose that $A\mathbf{x} = \mathbf{b}_i$, $1 \le i \le k$, is such a collection of systems, and suppose that $\mathbf{x}_i$ is the solution of the ith system. Let

$$X = [\mathbf{x}_1\ \ \mathbf{x}_2\ \ \cdots\ \ \mathbf{x}_k] \quad \text{and} \quad B = [\mathbf{b}_1\ \ \mathbf{b}_2\ \ \cdots\ \ \mathbf{b}_k].$$

Then $AX = B$, and therefore $X = A^{-1}B$.

So if A and B are as in Example 4 and $\mathbf{b}_1$ and $\mathbf{b}_2$ are the columns of B, then the solutions of $A\mathbf{x} = \mathbf{b}_1$ and $A\mathbf{x} = \mathbf{b}_2$ are

$$\mathbf{x}_1 = \begin{bmatrix} -4 \\ 1 \\ 4 \end{bmatrix} \quad \text{and} \quad \mathbf{x}_2 = \begin{bmatrix} 1 \\ 1 \\ -4 \end{bmatrix}.$$

AN INTERPRETATION OF THE INVERSE MATRIX

Consider the system $A\mathbf{x} = \mathbf{b}$, where A is an invertible $n \times n$ matrix. Often, we need to know how a change to the constant $\mathbf{b}$ affects a solution of the system. To examine what happens, let $P = A^{-1}$. Then

$$[\mathbf{e}_1\ \mathbf{e}_2\ \ldots\ \mathbf{e}_n] = I_n = AP = A[\mathbf{p}_1\ \mathbf{p}_2\ \ldots\ \mathbf{p}_n] = [A\mathbf{p}_1\ A\mathbf{p}_2\ \ldots\ A\mathbf{p}_n].$$

It follows that for each j, $A\mathbf{p}_j = \mathbf{e}_j$. Suppose that $\mathbf{u}$ is a solution of $A\mathbf{x} = \mathbf{b}$, and assume that we change the kth component of $\mathbf{b}$ from b_k to $b_k + d$, where d is some scalar. Thus we replace $\mathbf{b}$ by $\mathbf{b} + d\mathbf{e}_k$ to obtain the new system $A\mathbf{x} = \mathbf{b} + d\mathbf{e}_k$. Notice that

$$A(\mathbf{u} + d\mathbf{p}_k) = A\mathbf{u} + dA\mathbf{p}_k = \mathbf{b} + d\mathbf{e}_k;$$

so $\mathbf{u} + d\mathbf{p}_k$ is a solution of $A\mathbf{x} = \mathbf{b} + d\mathbf{e}_k$. This solution differs from the original solution by $d\mathbf{p}_k$; that is, $d\mathbf{p}_k$ measures the change in a solution of $A\mathbf{x} = \mathbf{b}$ when the kth component of $\mathbf{b}$ is increased by d.

An example in which the vector $\mathbf{b}$ and the resulting solution are changed in this way is provided by the Leontief input–output model (discussed in Section 1.5). When an economy is being planned, the gross production vectors corresponding to several different demand vectors may need to be calculated. For example, we may want to compare the effect of increasing the demand vector from

$$\mathbf{d}_1 = \begin{bmatrix} 90 \\ 80 \\ 60 \end{bmatrix} \quad \text{to} \quad \mathbf{d}_2 = \begin{bmatrix} 100 \\ 80 \\ 60 \end{bmatrix}.$$

If C is the input–output matrix for the economy and $I_3 - C$ is invertible,[4] then such comparisons are easily made. For in this case, the solution of $(I_3 - C)\mathbf{x} = \mathbf{d}_1$ is $(I_3 - C)^{-1}\mathbf{d}_1$, as we saw in Section 2.3. Hence the gross production vector needed to meet the demand $\mathbf{d}_2$ is

$$(I_3 - C)^{-1}\mathbf{d}_1 + 10\mathbf{p}_1,$$

where $\mathbf{p}_1$ is the first column of $(I_3 - C)^{-1}$. For the economy in Example 2 of Section 1.5, we have

$$(I_3 - C)^{-1} = \begin{bmatrix} 1.3 & 0.475 & 0.25 \\ 0.6 & 1.950 & 0.50 \\ 0.5 & 0.375 & 1.25 \end{bmatrix} \quad \text{and} \quad (I_3 - C)^{-1}\mathbf{d}_1 = \begin{bmatrix} 170 \\ 240 \\ 150 \end{bmatrix}.$$

So the gross production vector needed to meet the demand $\mathbf{d}_2$ is

$$(I_3 - C)^{-1}\mathbf{d}_1 + 10\mathbf{p}_1 = \begin{bmatrix} 170 \\ 240 \\ 150 \end{bmatrix} + 10 \begin{bmatrix} 1.3 \\ 0.6 \\ 0.5 \end{bmatrix} = \begin{bmatrix} 183 \\ 246 \\ 155 \end{bmatrix}.$$

Example 5 For the input–output matrix and demand in Example 2 of Section 1.5, determine the additional inputs needed to increase the demand for services from \$60 million to \$70 million.

Solution The additional inputs needed to increase the demand for services by one unit are given by the third column of the preceding matrix $(I_3 - C)^{-1}$. Hence an increase of \$10 million in the demand for services requires additional inputs of

$$10 \begin{bmatrix} 0.25 \\ 0.50 \\ 1.25 \end{bmatrix} = \begin{bmatrix} 2.5 \\ 5.0 \\ 12.5 \end{bmatrix},$$

[4] In practice, the sums of the entries in each column of C are less than 1, because each dollar's worth of output normally requires inputs whose total value is less than \$1. In this situation, it can be shown that $I_n - C$ is invertible and has nonnegative entries.

that is, additional inputs of \$2.5 million of agriculture, \$5 million of manufacturing, and \$12.5 million of services.

Practice Problem 3 ▶ Let A be an invertible $n \times n$ matrix and $\mathbf{b}$ be a vector in $\mathcal{R}^n$. Suppose that $\mathbf{x}_1$ is the solution of the equation $A\mathbf{x} = \mathbf{b}$.

(a) Prove that for any vector $\mathbf{c}$ in $\mathcal{R}^n$, the vector $\mathbf{x}_1 + A^{-1}\mathbf{c}$ is the solution of $A\mathbf{x} = \mathbf{b} + \mathbf{c}$.

(b) For the input–output matrix and demand vector in Example 2 of Section 1.5, apply the result in (a) to determine the increase in gross production necessary if the three demands are increased by \$5 million for agriculture, \$4 million for manufacturing, and \$2 million for services. ◀

EXERCISES

In Exercises 1–18, determine whether each matrix is invertible. If so, find its inverse.

1. $\begin{bmatrix} 1 & 3 \\ 1 & 2 \end{bmatrix}$

2. $\begin{bmatrix} 1 & 2 \\ 2 & 4 \end{bmatrix}$

3. $\begin{bmatrix} 1 & -3 \\ -2 & 6 \end{bmatrix}$

4. $\begin{bmatrix} 2 & -4 \\ -3 & 6 \end{bmatrix}$

5. $\begin{bmatrix} 2 & 3 \\ 3 & 5 \end{bmatrix}$

6. $\begin{bmatrix} 6 & -4 \\ -3 & 2 \end{bmatrix}$

7. $\begin{bmatrix} 1 & -2 & 1 \\ 1 & 0 & 1 \\ 1 & -1 & 1 \end{bmatrix}$

8. $\begin{bmatrix} 1 & 3 & 2 \\ 2 & 5 & 5 \\ 1 & 3 & 1 \end{bmatrix}$

9. $\begin{bmatrix} 1 & 1 & 2 \\ 2 & -1 & 1 \\ 2 & 3 & 4 \end{bmatrix}$

10. $\begin{bmatrix} 2 & -1 & 2 \\ 1 & 0 & 3 \\ 0 & 1 & 4 \end{bmatrix}$

11. $\begin{bmatrix} 2 & -1 & 1 \\ 1 & -3 & 2 \\ 1 & 7 & -4 \end{bmatrix}$

12. $\begin{bmatrix} 1 & -1 & 1 \\ 1 & -2 & 0 \\ 2 & -3 & 2 \end{bmatrix}$

13. $\begin{bmatrix} 0 & 2 & -1 \\ 1 & -1 & 2 \\ 2 & -1 & 3 \end{bmatrix}$

14. $\begin{bmatrix} 1 & -2 & 1 \\ 1 & 2 & -1 \\ 1 & 4 & -2 \end{bmatrix}$

15. $\begin{bmatrix} 1 & 0 & 0 & 1 \\ 0 & 1 & 1 & 0 \\ 1 & 0 & 1 & 0 \\ 0 & 1 & 0 & 1 \end{bmatrix}$

16. $\begin{bmatrix} 1 & 2 & 1 & -1 \\ 2 & 5 & 1 & -1 \\ 1 & 3 & 1 & 2 \\ 2 & 4 & 2 & -1 \end{bmatrix}$

17. $\begin{bmatrix} 1 & 1 & 1 & 0 \\ 1 & 1 & 0 & 1 \\ 1 & 0 & 1 & 1 \\ 0 & 1 & 1 & 1 \end{bmatrix}$

18. $\begin{bmatrix} 1 & -1 & 1 & -2 \\ -1 & 3 & -1 & 0 \\ 2 & -2 & -2 & 3 \\ 9 & -5 & -3 & -1 \end{bmatrix}$

In Exercises 19–26, use the algorithm for computing $A^{-1}B$.

19. $A = \begin{bmatrix} 1 & 2 \\ 2 & 3 \end{bmatrix}$ and $B = \begin{bmatrix} 1 & -1 & 2 \\ 1 & 0 & 1 \end{bmatrix}$

20. $A = \begin{bmatrix} -1 & 2 \\ 2 & -3 \end{bmatrix}$ and $B = \begin{bmatrix} 4 & -1 \\ 1 & 2 \end{bmatrix}$

21. $A = \begin{bmatrix} 2 & 2 \\ 2 & 1 \end{bmatrix}$ and $B = \begin{bmatrix} 2 & 4 & 2 & 6 \\ 0 & -2 & 8 & -4 \end{bmatrix}$

22. $A = \begin{bmatrix} 1 & -1 & 1 \\ 2 & -1 & 4 \\ 2 & -2 & 3 \end{bmatrix}$ and $B = \begin{bmatrix} 3 & -2 \\ 1 & -1 \\ 4 & 2 \end{bmatrix}$

23. $A = \begin{bmatrix} -2 & 3 & 7 \\ -1 & 1 & 2 \\ 1 & 1 & 2 \end{bmatrix}$ and $B = \begin{bmatrix} 2 & 0 & 1 & -1 \\ 1 & 2 & -2 & 1 \\ 3 & 1 & 1 & 3 \end{bmatrix}$

24. $A = \begin{bmatrix} 3 & 2 & 4 \\ 4 & 1 & 4 \\ 4 & 2 & 5 \end{bmatrix}$ and $B = \begin{bmatrix} 1 & -1 & 0 & -2 & -3 \\ 1 & -1 & 2 & 4 & 5 \\ 1 & -1 & 1 & 1 & 1 \end{bmatrix}$

25. $A = \begin{bmatrix} 1 & 0 & 1 & 1 \\ 0 & 1 & 1 & -1 \\ 0 & 0 & 1 & -1 \\ 0 & 0 & 0 & 1 \end{bmatrix}$ and $B = \begin{bmatrix} 2 & 1 & -1 \\ 0 & 1 & 1 \\ 1 & 0 & 1 \\ 3 & 1 & 2 \end{bmatrix}$

26. $A = \begin{bmatrix} 5 & 2 & 6 & 2 \\ 0 & 1 & 0 & 0 \\ 4 & 2 & 5 & 2 \\ 0 & 0 & 0 & 1 \end{bmatrix}$ and

$B = \begin{bmatrix} 1 & 0 & -1 & -3 & 1 & 4 \\ 2 & -1 & -1 & -8 & 3 & 9 \\ -1 & 1 & 1 & 5 & -2 & -6 \\ 0 & 1 & 1 & 2 & -1 & -3 \end{bmatrix}$

In Exercises 27–34, a matrix A is given. Determine (a) the reduced row echelon form R of A and (b) an invertible matrix P such that $PA = R$.

27. $\begin{bmatrix} 1 & -1 & 2 \\ -2 & 1 & -1 \end{bmatrix}$

28. $\begin{bmatrix} 1 & 1 & -1 \\ 1 & -1 & 2 \\ 1 & 0 & 1 \end{bmatrix}$

29. $\begin{bmatrix} -1 & 0 & 2 & 1 \\ 0 & 1 & 1 & -1 \\ 2 & 3 & -1 & -5 \end{bmatrix}$

30. $\begin{bmatrix} 1 & -2 & 1 & -1 & -2 \\ 2 & -4 & 1 & 1 & 1 \end{bmatrix}$

31. $\begin{bmatrix} 2 & 1 & 0 & -2 \\ 0 & 1 & -1 & 0 \\ -1 & -2 & 2 & 1 \\ 1 & 3 & 1 & 0 \end{bmatrix}$

32. $\begin{bmatrix} 1 & -1 & 0 & -1 & 2 \\ -1 & 1 & 1 & -2 & 1 \\ 5 & -5 & -3 & 4 & 1 \end{bmatrix}$

33. $\begin{bmatrix} 1 & 0 & 1 & 2 & 1 \\ 0 & 1 & -1 & -1 & 0 \\ 1 & 1 & -2 & 7 & 4 \\ 2 & 1 & 3 & -3 & -1 \end{bmatrix}$

34. $\begin{bmatrix} 1 & 0 & -1 & -3 & 1 & 4 \\ 2 & -1 & -1 & -8 & 3 & 9 \\ -1 & 1 & 1 & 5 & -2 & -6 \\ 0 & 1 & 1 & 2 & -1 & -3 \end{bmatrix}$

T&F *In Exercises 35–54, determine whether the statements are true or false.*

35. A matrix is invertible if and only if its reduced row echelon form is an identity matrix.
36. For any two matrices A and B, if $AB = I_n$ for some positive integer n, then A is invertible.
37. For any two $n \times n$ matrices A and B, if $AB = I_n$, then $BA = I_n$.
38. For any two $n \times n$ matrices A and B, if $AB = I_n$, then A is invertible and $A^{-1} = B$.
39. If an $n \times n$ matrix has rank n, then it is invertible.
40. If an $n \times n$ matrix is invertible, then it has rank n.
41. A square matrix is invertible if and only if its reduced row echelon form has no zero row.
42. If A is an $n \times n$ matrix such that the only solution of $A\mathbf{x} = \mathbf{0}$, then A is invertible.
43. An $n \times n$ matrix is invertible if and only if its columns are linearly independent.
44. An $n \times n$ matrix is invertible if and only if its rows are linearly independent.
45. If a square matrix has a column consisting of all zeros, then it is not invertible.
46. If a square matrix has a row consisting of all zeros, then it is not invertible.
47. Any invertible matrix can be written as a product of elementary matrices.
48. If A and B are invertible $n \times n$ matrices, then $A + B$ is invertible.
49. If A is an $n \times n$ matrix such that $A\mathbf{x} = \mathbf{b}$ is consistent for every $\mathbf{b}$ in $\mathcal{R}^n$, then $A\mathbf{x} = \mathbf{b}$ has a unique solution for every $\mathbf{b}$ in $\mathcal{R}^n$.
50. If A is an invertible $n \times n$ matrix and the reduced row echelon form of $[A\ B]$ is $[I_n\ C]$, then $C = B^{-1}A$.
51. If the reduced row echelon form of $[A\ I_n]$ is $[R\ B]$, then $B = A^{-1}$.
52. If the reduced row echelon form of $[A\ I_n]$ is $[R\ B]$, then B is an invertible matrix.
53. If the reduced row echelon form of $[A\ I_n]$ is $[R\ B]$, then BA equals the reduced row echelon form of A.
54. Suppose that A is an invertible matrix and $\mathbf{u}$ is a solution of $A\mathbf{x} = \begin{bmatrix} 5 \\ 6 \\ 7 \\ 8 \end{bmatrix}$. The solution of $A\mathbf{x} = \begin{bmatrix} 5 \\ 6 \\ 9 \\ 8 \end{bmatrix}$ differs from $\mathbf{u}$ by $2\mathbf{p}_3$, where $\mathbf{p}_3$ is the third column of A^{-1}.

---

55. Prove directly that statement (a) in the Invertible Matrix Theorem implies statements (e) and (h).

In Exercises 56–63, a system of linear equations is given.

(a) Write each system as a matrix equation $A\mathbf{x} = \mathbf{b}$.

(b) Show that A is invertible, and find A^{-1}.

(c) Use A^{-1} to solve each system.

56. $\begin{aligned} x_1 + 2x_2 &= 9 \\ 2x_1 + 3x_2 &= 3 \end{aligned}$

57. $\begin{aligned} -x_1 - 3x_2 &= -6 \\ 2x_1 + 5x_2 &= 4 \end{aligned}$

58. $\begin{aligned} x_1 + x_2 + x_3 &= 4 \\ 2x_1 + x_2 + 4x_3 &= 7 \\ 3x_1 + 2x_2 + 6x_3 &= -1 \end{aligned}$

59. $\begin{aligned} -x_1 + x_3 &= -4 \\ x_1 + 2x_2 - 2x_3 &= 3 \\ 2x_1 - x_2 + x_3 &= 1 \end{aligned}$

60. $\begin{aligned} x_1 + x_2 + x_3 &= -5 \\ 2x_1 + x_2 + x_3 &= -3 \\ 3x_1 + x_3 &= 2 \end{aligned}$

61. $\begin{aligned} 2x_1 + 3x_2 - 4x_3 &= -6 \\ -x_1 - x_2 + 2x_3 &= 5 \\ -x_2 + x_3 &= 3 \end{aligned}$

62. $\begin{aligned} x_1 - x_3 + x_4 &= 3 \\ 2x_1 - x_2 - x_3 &= -2 \\ -x_1 + x_2 + x_3 + x_4 &= 4 \\ x_2 + x_3 + x_4 &= -1 \end{aligned}$

63. $\begin{aligned} x_1 - 2x_2 - x_3 + x_4 &= 4 \\ x_1 + x_2 - x_4 &= -2 \\ -x_1 - x_2 + x_3 + x_4 &= 1 \\ -3x_1 + x_2 + 2x_3 &= -1 \end{aligned}$

64. Let $A = \begin{bmatrix} 1 & 1 \\ 1 & 2 \end{bmatrix}$.
    (a) Verify that $A^2 - 3A + I_2 = O$.
    (b) Let $B = 3I_2 - A$. Use B to prove that A is invertible and $B = A^{-1}$.
65. Let $A = \begin{bmatrix} 1 & -1 & 0 \\ 2 & 3 & -1 \\ -1 & 0 & 1 \end{bmatrix}$.
    (a) Verify that $A^3 - 5A^2 + 9A - 4I_3 = O$.
    (b) Let $B = \frac{1}{4}(A^2 - 5A + 9I_3)$. Use B to prove that A is invertible and $B = A^{-1}$.
    (c) Explain how B in (b) can be obtained from the equation in (a).
66. Let A be an $n \times n$ matrix such that $A^2 = I_n$. Prove that A is invertible and $A^{-1} = A$.

67. Let A be an $n \times n$ matrix such that $A^k = I_n$ for some positive integer k.

    (a) Prove that A is invertible.

    (b) Express A^{-1} as a power of A.

68. Prove that if A is an $m \times n$ matrix and P is an invertible $m \times m$ matrix, then rank PA = rank A. *Hint:* Apply Exercise 62 of Section 2.3 to PA and to $P^{-1}(PA)$.

69. Let B be an $n \times p$ matrix. Prove the following:

    (a) For any $m \times n$ matrix R in reduced row echelon form, rank $RB \leq$ rank R. *Hint:* Use the definition of the reduced row echelon form.

    (b) If A is any $m \times n$ matrix, then rank $AB \leq$ rank A. *Hint:* Use Exercises 68 and 69(a).

70. Prove that if A is an $m \times n$ matrix and Q is an invertible $n \times n$ matrix, then rank AQ = rank A. *Hint:* Apply the result of Exercise 69(b) to AQ and $(AQ)Q^{-1}$.

71. Prove that for any matrix A, rank A^T = rank A. *Hint:* Use Exercise 70, Exercise 87 of Section 2.3, and Theorems 2.2 and 2.3.

72. Use the Invertible Matrix Theorem to prove that for any subset $\mathcal{S}$ of n vectors in $\mathcal{R}^n$, the set $\mathcal{S}$ is linearly independent if and only if $\mathcal{S}$ is a generating set for $\mathcal{R}^n$.

73. Let R and S be matrices with the same number of rows, and suppose that the matrix $[R \;\; S]$ is in reduced row echelon form. Prove that R is in reduced row echelon form.

74. Consider the system of linear equations $A\mathbf{x} = \mathbf{b}$, where

$$A = \begin{bmatrix} 1 & 2 & 3 \\ 2 & 3 & 4 \\ 3 & 4 & 5 \end{bmatrix} \quad \text{and} \quad \mathbf{b} = \begin{bmatrix} 20 \\ 30 \\ 40 \end{bmatrix}.$$

    (a) Solve this matrix equation by using Gaussian elimination.

    (b) On a TI-85 calculator, the value of $A^{-1}\mathbf{b}$ is given as

$$A^{-1}\mathbf{b} = \begin{bmatrix} 8 \\ 10 \\ 4 \end{bmatrix}.$$

    But this is *not* a solution of $A\mathbf{x} = \mathbf{b}$. Why?

75. Repeat Exercise 74 with

$$A = \begin{bmatrix} 1 & 2 & 3 \\ 2 & 3 & 4 \\ 6 & 7 & 8 \end{bmatrix}, \quad \mathbf{b} = \begin{bmatrix} 5 \\ 6 \\ 10 \end{bmatrix}, \quad \text{and} \quad A^{-1}\mathbf{b} = \begin{bmatrix} 0 \\ -8 \\ 3 \end{bmatrix}.$$

76. Repeat Exercise 74 with

$$A = \begin{bmatrix} 1 & 2 & 3 \\ 4 & 5 & 6 \\ 7 & 8 & 9 \end{bmatrix}, \quad \mathbf{b} = \begin{bmatrix} 15 \\ 18 \\ 21 \end{bmatrix}, \quad A^{-1}\mathbf{b} = \begin{bmatrix} -9 \\ 8 \\ 4 \end{bmatrix}.$$

77. In Exercise 19(c) of Section 1.5, how much is required in additional inputs from each sector to increase the net production of oil by \$3 million?

78. In Exercise 20(c) of Section 1.5, how much is required in additional inputs from each sector to increase net production in the nongovernment sector by \$1 million?

79. In Exercise 21(b) of Section 1.5, how much is required in additional inputs from each sector to increase the net production of services by \$40 million?

80. In Exercise 22(b) of Section 1.5, how much is required in additional inputs from each sector to increase the net production of manufacturing by \$24 million?

81. Suppose that the input–output matrix C for an economy is such that $I_n - C$ is invertible and every entry of $(I_n - C)^{-1}$ is positive. If the net production of one particular sector of the economy must be increased, how does this affect the gross production of the economy?

82. Use matrix transposes to modify the algorithm for computing $A^{-1}B$ to devise an algorithm for computing AB^{-1}, and justify your method.

83. Let A be an $m \times n$ matrix with reduced row echelon form R.

    (a) Prove that if rank $A = m$, then there is a unique $m \times m$ matrix P such that $PA = R$. Furthermore, P is invertible. *Hint:* For each j, let $\mathbf{u}_j$ denote the jth pivot column of A. Prove that the $m \times m$ matrix $U = [\mathbf{u}_1 \; \mathbf{u}_1 \; \dots \; \mathbf{u}_m]$ is invertible. Now let $PA = R$, and show that $P = U^{-1}$.

    (b) Prove that if rank $A < m$, then there is more than one invertible $m \times m$ matrix P such that $PA = R$. *Hint:* There is an elementary $m \times m$ matrix E, distinct from I_m, such that $ER = R$.

Let A and B be $n \times n$ matrices. We say that A is **similar** *to B if $B = P^{-1}AP$ for some invertible matrix P. Exercises 84–88 are concerned with this relation.*

84. Let A, B, and C be $n \times n$ matrices. Prove the following statements:

    (a) A is similar to A.

    (b) If A is similar to B, then B is similar to A.

    (c) If A is similar to B and B is similar to C, then A is similar to C.

85. Let A be an $n \times n$ matrix.

    (a) Prove that if A is similar to I_n, then $A = I_n$.

    (b) Prove that if A is similar to O, the $n \times n$ zero matrix, then $A = O$.

    (c) Suppose that $B = cI_n$ for some scalar c. (The matrix B is called a *scalar matrix.*) What can you say about A if A is similar to B?

86. Suppose that A and B are $n \times n$ matrices such that A is similar to B. Prove that if A is invertible, then B is invertible, and A^{-1} is similar to B^{-1}.

87. Suppose that A and B are $n \times n$ matrices such that A is similar to B. Prove that A^T is similar to B^T.

88. Suppose that A and B are $n \times n$ matrices such that A is similar to B. Prove that rank A = rank B. *Hint:* Use Exercises 68 and 70.

In Exercises 89–92, use either a calculator with matrix capabilities or computer software such as MATLAB to solve each problem.

Exercises 89–91 refer to the matrix

$$A = \begin{bmatrix} 2 & 5 & 6 & 1 \\ 3 & 8 & 9 & 2 \\ 2 & 6 & 5 & 2 \\ 3 & 9 & 7 & 4 \end{bmatrix}.$$

89. Show that A is invertible by computing its reduced row echelon form and using Theorem 2.5.

90. Show that A is invertible by solving the system $A\mathbf{x} = \mathbf{0}$ and using the Invertible Matrix Theorem.

91. Show that A is invertible by computing its rank and using the Invertible Matrix Theorem.

92. Show that the matrix

$$P = \begin{bmatrix} 1 & 2 & -1 & 3 \\ 2 & 3 & 2 & 8 \\ 2 & 4 & -1 & 4 \\ 3 & 6 & -2 & 8 \end{bmatrix}$$

is invertible. Illustrate Exercise 68 by creating several random 4×4 matrices A and showing that rank $PA =$ rank A.

SOLUTIONS TO THE PRACTICE PROBLEMS

1. The reduced row echelon form of A is $\begin{bmatrix} 1 & 0 & -1 \\ 0 & 1 & -1 \\ 0 & 0 & 0 \end{bmatrix}$.

Since this matrix is not I_3, Theorem 2.5 implies that A is not invertible.

The reduced row echelon form of B is I_3, so as Theorem 2.5 implies, B is invertible. To compute B^{-1}, we find the reduced row echelon form of $[B \;\; I_3]$.

$$\begin{bmatrix} 1 & 1 & 0 & 1 & 0 & 0 \\ 3 & 4 & 1 & 0 & 1 & 0 \\ -1 & 4 & 4 & 0 & 0 & 1 \end{bmatrix}$$

$$\xrightarrow[\mathbf{r}_1 + \mathbf{r}_3 \to \mathbf{r}_3]{-3\mathbf{r}_1 + \mathbf{r}_2 \to \mathbf{r}_2} \begin{bmatrix} 1 & 1 & 0 & 1 & 0 & 0 \\ 0 & 1 & 1 & -3 & 1 & 0 \\ 0 & 5 & 4 & 1 & 0 & 1 \end{bmatrix}$$

$$\xrightarrow{-5\mathbf{r}_2 + \mathbf{r}_3 \to \mathbf{r}_3} \begin{bmatrix} 1 & 1 & 0 & 1 & 0 & 0 \\ 0 & 1 & 1 & -3 & 1 & 0 \\ 0 & 0 & -1 & 16 & -5 & 1 \end{bmatrix}$$

$$\xrightarrow{-\mathbf{r}_3 \to \mathbf{r}_3} \begin{bmatrix} 1 & 1 & 0 & 1 & 0 & 0 \\ 0 & 1 & 1 & -3 & 1 & 0 \\ 0 & 0 & 1 & -16 & 5 & -1 \end{bmatrix}$$

$$\xrightarrow{-\mathbf{r}_3 + \mathbf{r}_2 \to \mathbf{r}_2} \begin{bmatrix} 1 & 1 & 0 & 1 & 0 & 0 \\ 0 & 1 & 0 & 13 & -4 & 1 \\ 0 & 0 & 1 & -16 & 5 & -1 \end{bmatrix}$$

$$\xrightarrow{-\mathbf{r}_2 + \mathbf{r}_1 \to \mathbf{r}_1} \begin{bmatrix} 1 & 0 & 0 & -12 & 4 & -1 \\ 0 & 1 & 0 & 13 & -4 & 1 \\ 0 & 0 & 1 & -16 & 5 & -1 \end{bmatrix}$$

Thus $B^{-1} = \begin{bmatrix} -12 & 4 & -1 \\ 13 & -4 & 1 \\ -16 & 5 & -1 \end{bmatrix}$.

2. (a) The matrix form of the given system of linear equations is

$$\begin{bmatrix} 1 & -1 & 2 \\ 1 & 2 & 0 \\ 0 & -1 & 1 \end{bmatrix} \begin{bmatrix} x_1 \\ x_2 \\ x_3 \end{bmatrix} = \begin{bmatrix} 2 \\ 3 \\ -1 \end{bmatrix}.$$

(b) Because the reduced row echelon of the 3×3 matrix in (a) is I_3, the matrix is invertible.

(c) The solution of $A\mathbf{x} = \mathbf{b}$ is

$$\mathbf{x} = A^{-1}\mathbf{b} = \begin{bmatrix} 2 & -1 & -4 \\ -1 & 1 & 2 \\ -1 & 1 & 3 \end{bmatrix} \begin{bmatrix} 2 \\ 3 \\ -1 \end{bmatrix} = \begin{bmatrix} 5 \\ -1 \\ -2 \end{bmatrix}.$$

Thus the unique solution of the given system of linear equations is $x_1 = 5$, $x_2 = -1$, and $x_3 = -2$.

3. (a) First observe that $\mathbf{x}_1 = A^{-1}\mathbf{b}$. If $\mathbf{x}_2$ is the solution of $A\mathbf{x} = \mathbf{b} + \mathbf{c}$, then

$$\mathbf{x}_2 = A^{-1}(\mathbf{b} + \mathbf{c}) = A^{-1}\mathbf{b} + A^{-1}\mathbf{c} = \mathbf{x}_1 + A^{-1}\mathbf{c}.$$

(b) In the context of (a), let $A = I_3 - C$, where C is the input–output matrix in Example 2 of Section 1.5 and $\mathbf{b} = \mathbf{d}$, the demand vector used in that example. The increase in the demands is given by the vector $\mathbf{c} = \begin{bmatrix} 5 \\ 4 \\ 2 \end{bmatrix}$, and therefore the increase in the gross production is given by

$$(\mathbf{x}_1 + A^{-1}\mathbf{c}) - \mathbf{x}_1 = \begin{bmatrix} 1.3 & 0.475 & 0.25 \\ 0.6 & 1.950 & 0.50 \\ 0.5 & 0.375 & 1.25 \end{bmatrix} \begin{bmatrix} 5 \\ 4 \\ 2 \end{bmatrix} = \begin{bmatrix} 8.9 \\ 11.8 \\ 6.5 \end{bmatrix}.$$

2.5* PARTITIONED MATRICES AND BLOCK MULTIPLICATION

Suppose that we wish to compute A^3, where

$$A = \begin{bmatrix} 1 & 0 & 0 & 0 \\ 0 & 1 & 0 & 0 \\ 6 & 8 & 5 & 0 \\ -7 & 9 & 0 & 5 \end{bmatrix}.$$

This is a laborious process because A is a 4×4 matrix. However, there is another approach to matrix multiplication that simplifies this calculation. We start by writing A as an array of 2×2 submatrices.

$$A = \left[\begin{array}{cc|cc} 1 & 0 & 0 & 0 \\ 0 & 1 & 0 & 0 \\ \hline 6 & 8 & 5 & 0 \\ -7 & 9 & 0 & 5 \end{array}\right]$$

We can then write A more compactly as

$$A = \begin{bmatrix} I_2 & O \\ B & 5I_2 \end{bmatrix},$$

where

$$I_2 = \begin{bmatrix} 1 & 0 \\ 0 & 1 \end{bmatrix} \quad \text{and} \quad B = \begin{bmatrix} 6 & 8 \\ -7 & 9 \end{bmatrix}.$$

Next, we compute A^2 by the row-column rule, treating I_2, O, B, and $5I_2$ as if they were scalar entries of A.

$$A^2 = \begin{bmatrix} I_2 & O \\ B & 5I_2 \end{bmatrix}\begin{bmatrix} I_2 & O \\ B & 5I_2 \end{bmatrix} = \begin{bmatrix} I_2I_2 + OB & I_2O + O(5I_2) \\ BI_2 + (5I_2)B & BO + (5I_2)(5I_2) \end{bmatrix} = \begin{bmatrix} I_2 & O \\ 6B & 5^2I_2 \end{bmatrix}.$$

Finally,

$$\begin{aligned} A^3 = A^2A &= \begin{bmatrix} I_2 & O \\ 6B & 5^2I_2 \end{bmatrix}\begin{bmatrix} I_2 & O \\ B & 5I_2 \end{bmatrix} \\ &= \begin{bmatrix} I_2I_2 + OB & I_2O + O(5I_2) \\ (6B)I_2 + (5^2I_2)B & (6B)O + (5^2I_2)(5I_2) \end{bmatrix} \\ &= \begin{bmatrix} I_2 & O \\ 31B & 5^3I_2 \end{bmatrix}. \end{aligned}$$

This method for computing A^3 requires only that we multiply the 2×2 matrix B by the scalar 31 and I_2 by 5^3.

We can break up any matrix by drawing horizontal and vertical lines within the matrix, which divides the matrix into an array of submatrices called **blocks**. The resulting array is called a **partition** of the matrix, and the process of forming these blocks is called **partitioning**.

---

* This section can be omitted without loss of continuity.

While there are many ways to partition any given matrix, there is often a natural partition that simplifies matrix multiplication. For example, the matrix

$$A = \begin{bmatrix} 2 & 0 & 1 & -1 \\ 0 & 2 & 2 & 3 \\ 1 & 3 & 0 & 0 \end{bmatrix}$$

can be written as

$$A = \left[\begin{array}{cc|cc} 2 & 0 & 1 & -1 \\ 0 & 2 & 2 & 3 \\ \hline 1 & 3 & 0 & 0 \end{array}\right].$$

The horizontal and vertical lines partition A into an array of four blocks. The first row of the partition consists of the 2×2 matrices

$$2I_2 = \begin{bmatrix} 2 & 0 \\ 0 & 2 \end{bmatrix} \quad \text{and} \quad \begin{bmatrix} 1 & -1 \\ 2 & 3 \end{bmatrix},$$

and the second row consists of the 1×2 matrices $[1 \;\; 3]$ and $O = [0 \;\; 0]$. We can also partition A as

$$\left[\begin{array}{cc|cc} 2 & 0 & 1 & -1 \\ 0 & 2 & 2 & 3 \\ 1 & 3 & 0 & 0 \end{array}\right].$$

In this case, there is only one row and two columns. The blocks of this row are the 3×2 matrices

$$\begin{bmatrix} 2 & 0 \\ 0 & 2 \\ 1 & 3 \end{bmatrix} \quad \text{and} \quad \begin{bmatrix} 1 & -1 \\ 2 & 3 \\ 0 & 0 \end{bmatrix}.$$

The first partition of A contains the submatrices $2I_2$ and O, which are easily multiplied by other matrices, so this partition is usually more desirable.

As shown in the preceding example, partitioning matrices appropriately can simplify matrix multiplication. Two partitioned matrices can be multiplied by treating the blocks as if they were scalars, provided that the products of the individual blocks are defined.

Example 1 Let

$$A = \left[\begin{array}{cc|cc} 1 & 3 & 4 & 2 \\ 0 & 5 & -1 & 6 \\ \hline 1 & 0 & 3 & -1 \end{array}\right] \quad \text{and} \quad B = \left[\begin{array}{cc|c} 1 & 0 & 3 \\ 1 & 2 & 0 \\ \hline 2 & -1 & 2 \\ 0 & 3 & 1 \end{array}\right].$$

We can use the given partition to find the entries in the upper left block of AB by computing

$$\begin{bmatrix} 1 & 3 \\ 0 & 5 \end{bmatrix}\begin{bmatrix} 1 & 0 \\ 1 & 2 \end{bmatrix} + \begin{bmatrix} 4 & 2 \\ -1 & 6 \end{bmatrix}\begin{bmatrix} 2 & -1 \\ 0 & 3 \end{bmatrix} = \begin{bmatrix} 4 & 6 \\ 5 & 10 \end{bmatrix} + \begin{bmatrix} 8 & 2 \\ -2 & 19 \end{bmatrix} = \begin{bmatrix} 12 & 8 \\ 3 & 29 \end{bmatrix}.$$

Similarly, we can find the upper right block of AB by computing

$$\begin{bmatrix} 1 & 3 \\ 0 & 5 \end{bmatrix}\begin{bmatrix} 3 \\ 0 \end{bmatrix} + \begin{bmatrix} 4 & 2 \\ -1 & 6 \end{bmatrix}\begin{bmatrix} 2 \\ 1 \end{bmatrix} = \begin{bmatrix} 3 \\ 0 \end{bmatrix} + \begin{bmatrix} 10 \\ 4 \end{bmatrix} = \begin{bmatrix} 13 \\ 4 \end{bmatrix}.$$

We obtain the lower left block of AB by computing

$$[1 \quad 0]\begin{bmatrix} 1 & 0 \\ 1 & 2 \end{bmatrix} + [3 \quad -1]\begin{bmatrix} 2 & -1 \\ 0 & 3 \end{bmatrix} = [1 \quad 0] + [6 \quad -6] = [7 \quad -6].$$

Finally, we obtain the lower right block of AB by computing

$$[1 \quad 0]\begin{bmatrix} 3 \\ 0 \end{bmatrix} + [3 \quad -1]\begin{bmatrix} 2 \\ 1 \end{bmatrix} = [3] + [5] = [8].$$

Putting these blocks together, we have

$$AB = \left[\begin{array}{cc|c} 12 & 8 & 13 \\ 3 & 29 & 4 \\ \hline 7 & -6 & 8 \end{array}\right].$$

In general, we have the following rule:

Block Multiplication

Suppose two matrices A and B are partitioned into blocks so that the number of blocks in each row of A is the same as the number of blocks in each column of B. Then the matrices can be multiplied according to the usual rules for matrix multiplication, treating the blocks as if they were scalars, provided that the individual products are defined.

TWO ADDITIONAL METHODS FOR COMPUTING A MATRIX PRODUCT

Given two matrices A and B such that the product AB is defined, we have seen how to compute the product by using blocks obtained from partitions of A and B. In this subsection, we look at two specific ways of partitioning A and B that lead to two new methods of computing their product.

By rows Given an $m \times n$ matrix A and an $n \times p$ matrix B, we partition A into an $m \times 1$ array of row vectors $\mathbf{a}'_1, \mathbf{a}'_2, \ldots, \mathbf{a}'_m$ and regard B as a single block in a 1×1 array. In this case,

$$AB = \begin{bmatrix} \mathbf{a}'_1 \\ \mathbf{a}'_2 \\ \vdots \\ \mathbf{a}'_m \end{bmatrix} B = \begin{bmatrix} \mathbf{a}'_1 B \\ \mathbf{a}'_2 B \\ \vdots \\ \mathbf{a}'_m B \end{bmatrix}. \tag{7}$$

Thus the rows of AB are the products of the rows of A with B. More specifically, the ith row of AB is the matrix product of the ith row of A with B.

Example 2 Let

$$A = \begin{bmatrix} 1 & 2 & -1 \\ -1 & 1 & 3 \end{bmatrix} \quad \text{and} \quad B = \begin{bmatrix} -2 & 1 & 0 \\ 1 & -3 & 4 \\ 1 & -1 & -1 \end{bmatrix}.$$

Since

$$\mathbf{a}_1'B = \begin{bmatrix}1 & 2 & -1\end{bmatrix}\begin{bmatrix}-2 & 1 & 0\\ 1 & -3 & 4\\ 1 & -1 & -1\end{bmatrix} = \begin{bmatrix}-1 & -4 & 9\end{bmatrix}$$

and

$$\mathbf{a}_2'B = \begin{bmatrix}-1 & 1 & 3\end{bmatrix}\begin{bmatrix}-2 & 1 & 0\\ 1 & -3 & 4\\ 1 & -1 & -1\end{bmatrix} = \begin{bmatrix}6 & -7 & 1\end{bmatrix},$$

we have $AB = \begin{bmatrix}-1 & -4 & 9\\ 6 & -7 & 1\end{bmatrix}$.

It is interesting to compare the method of computing a product by rows with the definition of a matrix product, which can be thought of as the method of computing a product *by columns*.

By outer products Another method to compute a matrix product is to partition A into columns and B into rows. Suppose that $\mathbf{a}_1, \mathbf{a}_2, \ldots, \mathbf{a}_n$ are the columns of an $m \times n$ matrix A, and $\mathbf{b}_1', \mathbf{b}_2', \ldots, \mathbf{b}_n'$ are the rows of an $n \times p$ matrix B. Then block multiplication gives

$$AB = [\mathbf{a}_1\ \mathbf{a}_2\ \ldots\ \mathbf{a}_n]\begin{bmatrix}\mathbf{b}_1'\\ \mathbf{b}_2'\\ \vdots\\ \mathbf{b}_n'\end{bmatrix} = \mathbf{a}_1\mathbf{b}_1' + \mathbf{a}_2\mathbf{b}_2' + \cdots + \mathbf{a}_n\mathbf{b}_n'. \qquad (8)$$

Thus AB is the sum of matrix products of each column of A with the corresponding row of B.

The terms $\mathbf{a}_i\mathbf{b}_i'$ in equation (8) are matrix products of two vectors, namely, column i of A and row i of B. Such products have an especially simple form. In order to present this result in a more standard notation, we consider the matrix product of $\mathbf{v}$ and $\mathbf{w}^T$, where

$$\mathbf{v} = \begin{bmatrix}v_1\\ v_2\\ \vdots\\ v_m\end{bmatrix} \quad \text{and} \quad \mathbf{w} = \begin{bmatrix}w_1\\ w_2\\ \vdots\\ w_n\end{bmatrix}.$$

It follows from equation (7) that

$$\mathbf{v}\mathbf{w}^T = \begin{bmatrix}v_1\mathbf{w}^T\\ v_2\mathbf{w}^T\\ \vdots\\ v_m\mathbf{w}^T\end{bmatrix}.$$

Thus the rows of the $m \times n$ matrix $\mathbf{v}\mathbf{w}^T$ are all multiples of $\mathbf{w}^T$. If follows (see Exercise 52) that the rank of the matrix $\mathbf{v}\mathbf{w}^T$ is 1 if both $\mathbf{v}$ and $\mathbf{w}$ are nonzero vectors.

Products of the form $\mathbf{v}\mathbf{w}^T$, where $\mathbf{v}$ is in $\mathcal{R}^m$ (regarded as an $m \times 1$ matrix) and $\mathbf{w}$ is in $\mathcal{R}^n$ (regarded as an $n \times 1$ matrix), are called **outer products**. In this terminology, equation (8) states that the product of an $m \times n$ matrix A and an $n \times p$ matrix B is the sum of n matrices of rank at most 1, namely, the outer products of the columns of A with the corresponding rows of B.

In the special case that A is a $1 \times n$ matrix, so that $A = [a_1 \ a_2 \ \ldots \ a_n]$ is a row vector, the product in equation (8) is the linear combination

$$AB = a_1\mathbf{b}'_1 + a_2\mathbf{b}'_2 + \cdots + a_n\mathbf{b}'_n$$

of the rows of B with the corresponding entries of A as the coefficients. For example,

$$[2 \ \ 3]\begin{bmatrix} -1 & 4 \\ 5 & 0 \end{bmatrix} = 2[-1 \ \ 4] + 3[5 \ \ 0] = [13 \ \ 8].$$

Example 3 Use outer products to express the product AB in Example 2 as a sum of matrices of rank 1.

Solution First form the outer products in equation (8) to obtain three matrices of rank 1,

$$\begin{bmatrix} 1 \\ -1 \end{bmatrix}[-2 \ \ 1 \ \ 0] = \begin{bmatrix} -2 & 1 & 0 \\ 2 & -1 & 0 \end{bmatrix},$$

$$\begin{bmatrix} 2 \\ 1 \end{bmatrix}[1 \ \ -3 \ \ 4] = \begin{bmatrix} 2 & -6 & 8 \\ 1 & -3 & 4 \end{bmatrix},$$

and

$$\begin{bmatrix} -1 \\ 3 \end{bmatrix}[1 \ \ -1 \ \ -1] = \begin{bmatrix} -1 & 1 & 1 \\ 3 & -3 & -3 \end{bmatrix}.$$

Then

$$\begin{bmatrix} -2 & 1 & 0 \\ 2 & -1 & 0 \end{bmatrix} + \begin{bmatrix} 2 & -6 & 8 \\ 1 & -3 & 4 \end{bmatrix} + \begin{bmatrix} -1 & 1 & 1 \\ 3 & -3 & -3 \end{bmatrix} = \begin{bmatrix} -1 & -4 & 9 \\ 6 & -7 & 1 \end{bmatrix} = AB,$$

as guaranteed by equation (8).

We summarize the two new methods for computing the matrix product AB.

Two Methods of Computing a Matrix Product *AB*

We assume here that A is an $m \times n$ matrix with rows $\mathbf{a}'_1, \mathbf{a}'_2, \ldots, \mathbf{a}'_m$ and B is an $n \times p$ matrix with rows $\mathbf{b}'_1, \mathbf{b}'_2, \ldots, \mathbf{b}'_n$.

1. **By rows** The ith row of AB is obtained by multiplying the ith row of A by B—that is, $\mathbf{a}'_iB$.
2. **By outer products** The matrix AB is a sum of matrix products of each column of A with the corresponding row of B. Symbolically,

$$AB = \mathbf{a}_1\mathbf{b}'_1 + \mathbf{a}_2\mathbf{b}'_2 + \cdots + \mathbf{a}_n\mathbf{b}'_n.$$

EXERCISES

In Exercises 1–12, compute the product of each partitioned matrix using block multiplication.

1. $[-1\ \ 3\ \ 1]\left[\begin{array}{r|r} 1 & 2 \\ -1 & 1 \\ 0 & 1 \end{array}\right]$

2. $\left[\begin{array}{rrr} 1 & -1 & 0 \\ \hline 0 & 1 & 2 \end{array}\right]\left[\begin{array}{r} 1 \\ 3 \\ 2 \end{array}\right]$

3. $\left[\begin{array}{r|r|r} 1 & -1 & 0 \\ 0 & 1 & 2 \end{array}\right]\left[\begin{array}{r} 1 \\ \hline 3 \\ \hline 2 \end{array}\right]$

4. $\left[\begin{array}{rr|r} 1 & -1 & 0 \\ 0 & 1 & 2 \end{array}\right]\left[\begin{array}{r} 1 \\ 3 \\ \hline 2 \end{array}\right]$

5. $\left[\begin{array}{rr} 2 & 0 \\ 3 & 1 \\ \hline -1 & 5 \\ 1 & 2 \end{array}\right]\left[\begin{array}{rr|rr} -1 & 2 & 3 & 0 \\ 2 & 2 & -1 & 2 \end{array}\right]$

6. $\left[\begin{array}{rr} 2 & 0 \\ 3 & 1 \\ -1 & 5 \\ \hline 1 & 2 \end{array}\right]\left[\begin{array}{rrr|r} -1 & 2 & 3 & 0 \\ 2 & 2 & -1 & 2 \end{array}\right]$

7. $\left[\begin{array}{rr} 2 & 0 \\ \hline 3 & 1 \\ -1 & 5 \\ 1 & 2 \end{array}\right]\left[\begin{array}{r|rrr} -1 & 2 & 3 & 0 \\ 2 & 2 & -1 & 2 \end{array}\right]$

8. $\left[\begin{array}{rr} 0 & 0 \\ 0 & 0 \\ 0 & 0 \\ \hline 2 & 3 \end{array}\right]\left[\begin{array}{rrr|r} 0 & 0 & 0 & 6 \\ 0 & 0 & 0 & -1 \end{array}\right]$

9. $\left[\begin{array}{rr} 3 & 0 \\ 0 & 3 \\ \hline 2 & 0 \\ 0 & 2 \end{array}\right]\left[\begin{array}{rr} 1 & 2 \\ 3 & 4 \end{array}\right]$

10. $\left[\begin{array}{rr|rr} 1 & 2 & 2 & -2 \\ 1 & 1 & -2 & 2 \end{array}\right]\left[\begin{array}{rr} 1 & 0 \\ 0 & 1 \\ \hline 1 & 1 \\ 1 & 1 \end{array}\right]$

11. $\left[\begin{array}{rr|rr} 1 & 1 & 2 & 1 \\ \hline 1 & 0 & 0 & 0 \\ 0 & 1 & 0 & 0 \end{array}\right]\left[\begin{array}{rr|rr} 1 & 0 & 1 & -1 \\ 0 & 1 & -1 & 1 \\ \hline 0 & 0 & 1 & 0 \\ 0 & 0 & 0 & 1 \end{array}\right]$

12. $\left[\begin{array}{c} A_{20^\circ} \\ \hline A_{30^\circ} \end{array}\right]\left[\begin{array}{c|c} A_{40^\circ} & A_{50^\circ} \end{array}\right]$

In Exercises 13–20, compute the indicated row of the given product without computing the entire matrix.

$$A = \begin{bmatrix} 1 & 2 & 3 \\ 2 & -1 & 4 \\ -3 & -2 & 0 \end{bmatrix}, \quad B = \begin{bmatrix} -1 & 0 \\ 4 & 1 \\ 3 & -2 \end{bmatrix},$$

$$C = \begin{bmatrix} 2 & 1 & -1 \\ 4 & 3 & -2 \end{bmatrix}$$

13. row 1 of AB
14. row 1 of CA
15. row 2 of CA
16. row 2 of BC
17. row 3 of BC
18. row 2 of B^TA
19. row 2 of A^2
20. row 3 of A^2

In Exercises 21–28, use the matrices A, B, and C from Exercises 13–20.

21. Use outer products to represent AB as the sum of 3 matrices of rank 1.
22. Use outer products to represent BC as the sum of 2 matrices of rank 1.
23. Use outer products to represent CB as the sum of 3 matrices of rank 1.
24. Use outer products to represent CA as the sum of 3 matrices of rank 1.
25. Use outer products to represent B^TA as the sum of 3 matrices of rank 1.
26. Use outer products to represent AC^T as the sum of 3 matrices of rank 1.
27. Use outer products to represent A^TB as the sum of 3 matrices of rank 1.
28. Use outer products to represent CA^T as the sum of 3 matrices of rank 1.

In Exercises 29–34, determine whether the statements are true or false.

29. The definition of the matrix product AB on page 97 can be regarded as a special case of block multiplication.
30. Let A and B be matrices such that AB is defined, and let A and B be partitioned into blocks so that the number of blocks in each row of A is the same as the number of blocks in each column of B. Then the matrices can be multiplied according to the usual rule for matrix multiplication by treating the blocks as if they were scalars.
31. The outer product $\mathbf{v}\mathbf{w}^T$ is defined only if $\mathbf{v}$ and $\mathbf{w}$ are both in $\mathcal{R}^n$.
32. For any vectors $\mathbf{v}$ and $\mathbf{w}$ in $\mathcal{R}^m$ and $\mathcal{R}^n$, respectively, the outer product $\mathbf{v}\mathbf{w}^T$ is an $m \times n$ matrix.
33. For any vectors $\mathbf{v}$ and $\mathbf{w}$ in $\mathcal{R}^m$ and $\mathcal{R}^n$, respectively, the outer product $\mathbf{v}\mathbf{w}^T$ is an $m \times n$ matrix with rank 1.
34. The product of an $m \times n$ nonzero matrix and an $n \times p$ nonzero matrix can be written as the sum of at most n matrices of rank 1.

In Exercises 35–40, assume that A, B, C, and D are $n \times n$ matrices, O is the $n \times n$ zero matrix, and, in Exercises 35–36, A is invertible. Use block multiplication to find each product.

35. $[A^{-1} \;\; I_n]\begin{bmatrix} A \\ I_n \end{bmatrix}$

36. $\begin{bmatrix} A^{-1} \\ I_n \end{bmatrix}[A \;\; I_n]$

37. $\begin{bmatrix} A & O \\ O & B \end{bmatrix}\begin{bmatrix} O & C \\ D & O \end{bmatrix}$

38. $\begin{bmatrix} I_n & O \\ O & C \end{bmatrix}\begin{bmatrix} A & B \\ O & I_n \end{bmatrix}$

39. $\begin{bmatrix} A & B \\ C & D \end{bmatrix}^T\begin{bmatrix} A & B \\ C & D \end{bmatrix}$

40. $\begin{bmatrix} I_n & A \\ I_n & B \end{bmatrix}\begin{bmatrix} A & B \\ I_n & I_n \end{bmatrix}$

41. Show that if A, B, C, and D are $n \times n$ matrices such that A is invertible,

$$\begin{bmatrix} I_n & O \\ CA^{-1} & I_n \end{bmatrix}\begin{bmatrix} A & O \\ O & D - CA^{-1}B \end{bmatrix}\begin{bmatrix} I_n & A^{-1}B \\ O & I_n \end{bmatrix} = \begin{bmatrix} A & B \\ C & D \end{bmatrix}.$$

In this context, the matrix $D - CA^{-1}B$ is called the **Schur complement** of A.

In Exercises 42–47, assume that A, B, C, and D are $n \times n$ matrices, O is the $n \times n$ zero matrix, and A and D are invertible. Use block multiplication to verify each equation.

42. $\begin{bmatrix} A & O \\ O & D \end{bmatrix}^{-1} = \begin{bmatrix} A^{-1} & O \\ O & D^{-1} \end{bmatrix}$

43. $\begin{bmatrix} O & A \\ D & O \end{bmatrix}^{-1} = \begin{bmatrix} O & D^{-1} \\ A^{-1} & O \end{bmatrix}$

44. $\begin{bmatrix} A & B \\ O & D^{-1} \end{bmatrix}^{-1} = \begin{bmatrix} A^{-1} & -A^{-1}BD \\ O & D \end{bmatrix}$

45. $\begin{bmatrix} C & A \\ D & O \end{bmatrix}^{-1} = \begin{bmatrix} O & D^{-1} \\ A^{-1} & -A^{-1}CD^{-1} \end{bmatrix}$

46. $\begin{bmatrix} O & A \\ D & C \end{bmatrix}^{-1} = \begin{bmatrix} -D^{-1}CA^{-1} & D^{-1} \\ A^{-1} & O \end{bmatrix}$

47. $\begin{bmatrix} I_n & B \\ C & I_n \end{bmatrix}^{-1} = \begin{bmatrix} P & -PB \\ -CP & I_n + CPB \end{bmatrix}$, where $I_n - BC$ is invertible and $P = (I_n - BC)^{-1}$.

48. Let A and B be $n \times n$ matrices and O be the $n \times n$ zero matrix. Use block multiplication to compute $\begin{bmatrix} A & O \\ O & B \end{bmatrix}^k$ for any positive integer k.

49. Let A and B be $n \times n$ matrices and O be the $n \times n$ zero matrix. Use block multiplication to compute $\begin{bmatrix} A & B \\ O & O \end{bmatrix}^k$ for any positive integer k.

50. Let A and B be invertible $n \times n$ matrices. Prove that $\begin{bmatrix} A & B \\ B & A \end{bmatrix}$ is invertible if and only if $A - BA^{-1}B$ is invertible.

51. Prove that if A and B are invertible $n \times n$ matrices, then $\begin{bmatrix} A & O \\ I_n & B \end{bmatrix}$ is invertible. Find the inverse in terms of A^{-1}, B^{-1}, and O.

52. Suppose $\mathbf{a}$ and $\mathbf{b}$ are nonzero vectors in $\mathcal{R}^m$ and $\mathcal{R}^n$, respectively. Prove that the outer product $\mathbf{ab}^T$ has rank 1.

In Exercise 53, use either a calculator with matrix capabilities or computer software such as MATLAB to solve each problem.

53. Suppose that A is a 4×4 matrix in the block form,

$$A = \begin{bmatrix} B & C \\ O & D \end{bmatrix},$$

where the blocks are all 2×2 matrices.

(a) Use a random matrix for A to illustrate that $A^2 = \begin{bmatrix} B^2 & * \\ O & D^2 \end{bmatrix}$, where $*$ represents some 2×2 matrix.

(b) Use a random matrix for A to illustrate that $A^3 = \begin{bmatrix} B^3 & * \\ O & D^3 \end{bmatrix}$, where $*$ represents some 2×2 matrix.

(c) Make a conjecture about the block form of A^k, where k is a positive integer.

(d) Prove your conjecture for $k = 3$.

2.6* THE *LU* DECOMPOSITION OF A MATRIX

In many applications, it is necessary to solve multiple systems of linear equations with the same coefficient matrix. In these situations, using Gaussian elimination on each system involves a great deal of duplication of effort since the augmented matrices for these systems are almost identical. In this section, we examine a method that avoids this duplication.

For now, suppose that an $m \times n$ matrix A can be transformed into a matrix U in row echelon form *without the use of row interchanges.* Then U can be written as

$$U = E_k E_{k-1} \cdots E_1 A,$$

* This section can be omitted without loss of continuity.

where $E_1, \ldots, E_{k-1}, E_k$ are the elementary matrices corresponding to the elementary row operations that transform A into U. Solving this equation for A, we obtain

$$A = (E_k E_{k-1} \cdots E_1)^{-1} U = E_1^{-1} E_2^{-1} \cdots E_k^{-1} U = LU,$$

where

$$L = E_1^{-1} E_2^{-1} \cdots E_k^{-1}. \tag{9}$$

Observe that U is an $m \times n$ matrix and L, which is a product of invertible $m \times m$ matrices, is an invertible $m \times m$ matrix. The matrices L and U have special forms, which we now describe.

Since each elementary row operation used in the process of transforming A into U is the result of adding a multiple of a row to a lower row of a matrix, the corresponding elementary matrix E_p and its inverse E_p^{-1} are of the forms

$$E_p = \begin{array}{r} \\ \\ \text{row } j \rightarrow \\ \\ \text{row } i \rightarrow \\ \\ \end{array} \begin{bmatrix} 1 & & \cdots & & 0 \\ & \ddots & & & \vdots \\ \vdots & & 1 & & \\ & & \vdots & \ddots & \\ & & c & & \ddots \\ 0 & & & & 1 \end{bmatrix}$$

$$\uparrow$$
$$\text{column } j$$

and

$$E_p^{-1} = \begin{array}{r} \\ \\ \text{row } j \rightarrow \\ \\ \text{row } i \rightarrow \\ \\ \end{array} \begin{bmatrix} 1 & & \cdots & & 0 \\ & \ddots & & & \vdots \\ \vdots & & 1 & & \\ & & \vdots & \ddots & \\ & & -c & & \ddots \\ 0 & & & & 1 \end{bmatrix},$$

$$\uparrow$$
$$\text{column } j$$

where c is the multiple and $j < i$ are the rows. Notice that E_p and E_p^{-1} can be obtained from I_m by changing the (i,j)-entry from 0 to c, and from 0 to $-c$, respectively.

Since U is in row echelon form, the entries of U below and to the left of the diagonal entries are all zeros. Any matrix with this description is called an **upper triangular matrix**. Notice that the entries above and to the right of the diagonal entries of each E_p^{-1} are zeros. Any matrix with this description is called a **lower triangular matrix**. Furthermore, the diagonal entries of each E_p^{-1} are ones. A lower triangular matrix whose diagonal entries are all ones is called a **unit lower triangular matrix**. Since the product of unit lower triangular matrices is a unit lower triangular matrix (see Exercise 44), L is also a unit lower triangular matrix. Thus we can factor $A = LU$ into the product of a unit lower triangular matrix L and an upper triangular matrix U.

Example 1 Let

$$A = \begin{bmatrix} 1 & 0 & 0 \\ 0 & 2 & 0 \\ 3 & 4 & 3 \end{bmatrix}, \qquad B = \begin{bmatrix} 1 & 0 \\ 4 & 1 \end{bmatrix}, \qquad \text{and}$$

$$C = \begin{bmatrix} 2 & 0 & 1 & -1 \\ 0 & 0 & 3 & 4 \\ 0 & 0 & 3 & 0 \end{bmatrix}.$$

Both A and B are lower triangular matrices because the entries above and to the right of the diagonal entries are zeros. Both diagonal entries of B are ones, and hence B is a unit lower triangular matrix, whereas A is not. The entries below and to the left of the diagonal entries of C are zeros, and hence C is an upper triangular matrix.

Definition For any matrix A, a factorization $A = LU$, where L is a unit lower triangular matrix and U is an upper triangular matrix, is called an ***LU* decomposition of** A.

If a matrix has an LU decomposition and is also invertible, then the LU decomposition is unique. (See Exercise 46.)

Not every matrix can be transformed into a matrix in row echelon form without the use of row interchanges. For example, to put the matrix $\begin{bmatrix} 0 & 1 \\ 1 & 0 \end{bmatrix}$ into row echelon form, you must interchange its rows. If a matrix cannot be transformed into a matrix in row echelon form without the use of row interchanges, then the matrix has no LU decomposition.

COMPUTING THE *LU* DECOMPOSITION

For the present, we consider a matrix A that has an LU decomposition. We describe a method for finding the matrices L and U, and show how to use them to solve a system of linear equations $A\mathbf{x} = \mathbf{b}$. Given the LU decomposition of A, the number of steps required to solve several systems of linear equations with coefficient matrix A is significantly less than the total number of steps required to use Gaussian elimination on each system separately.

We begin the process of finding an LU decomposition of A by using elementary row operations to transform A into an upper triangular matrix U. Then we use equation (9) to compute L from I_m by applying the elementary row operations corresponding to the E_p^{-1}'s, starting with the last operation and working our way back to the first. We illustrate this process in the following example:

Example 2 Find the LU decomposition of the matrix

$$A = \begin{bmatrix} 1 & -1 & 2 \\ 3 & -1 & 7 \\ 2 & -4 & 5 \end{bmatrix}.$$

Solution First, we use Gaussian elimination to transform A into an upper triangular matrix U in row echelon form without the use of row interchanges. This process consists of three elementary row operations performed in succession: adding -3 times row 1 to row 2 of A, adding -2 times row 1 to row 3 of the resulting matrix, and adding 1 times row 2 to row 3 of the previous matrix to obtain the final result, U. The details are as follows:

$$A = \begin{bmatrix} 1 & -1 & 2 \\ 3 & -1 & 7 \\ 2 & -4 & 5 \end{bmatrix} \xrightarrow{-3\mathbf{r}_1+\mathbf{r}_2\rightarrow\mathbf{r}_2} \begin{bmatrix} 1 & -1 & 2 \\ 0 & 2 & 1 \\ 2 & -4 & 5 \end{bmatrix} \xrightarrow{-2\mathbf{r}_1+\mathbf{r}_3\rightarrow\mathbf{r}_3} \begin{bmatrix} 1 & -1 & 2 \\ 0 & 2 & 1 \\ 0 & -2 & 1 \end{bmatrix}$$

$$\xrightarrow{\mathbf{r}_2+\mathbf{r}_3\rightarrow\mathbf{r}_3} \begin{bmatrix} 1 & -1 & 2 \\ 0 & 2 & 1 \\ 0 & 0 & 2 \end{bmatrix} = U$$

The reverse of the last operation, adding -1 times row 2 to row 3 of a matrix, is the first operation used in the transformation of I_3 into L. We continue to apply the reverse row operations in the opposite order to complete the transformation of I_3 into L.

$$I_3 = \begin{bmatrix} 1 & 0 & 0 \\ 0 & 1 & 0 \\ 0 & 0 & 1 \end{bmatrix} \xrightarrow{-\mathbf{r}_2+\mathbf{r}_3\rightarrow\mathbf{r}_3} \begin{bmatrix} 1 & 0 & 0 \\ 0 & 1 & 0 \\ 0 & -1 & 1 \end{bmatrix} \xrightarrow{2\mathbf{r}_1+\mathbf{r}_3\rightarrow\mathbf{r}_3} \begin{bmatrix} 1 & 0 & 0 \\ 0 & 1 & 0 \\ 2 & -1 & 1 \end{bmatrix}$$

$$\xrightarrow{3\mathbf{r}_1+\mathbf{r}_2\rightarrow\mathbf{r}_2} \begin{bmatrix} 1 & 0 & 0 \\ 3 & 1 & 0 \\ 2 & -1 & 1 \end{bmatrix} = L$$

We can obtain the entries of L below the diagonal directly from the row operations used to transform A into U. In particular, the (i,j)-entry of L is $-c$, where c times row j is added to row i in one of the elementary row operations used to transform A into U. For example, in the first of the three elementary row operations used to transform A into U, -3 times row 1 is added to row 2. Thus the $(2,1)$-entry of L is 3. Since -2 times row 1 is added to row 3 in the second operation, 2 is the $(3,1)$-entry of L. Finally, 1 times row 2 is added to row 3 to complete the transformation of A to U, and hence -1 is the $(3,2)$-entry of L. These entries below the diagonal of L are called **multipliers**.

We summarize the process of obtaining the LU decomposition of a matrix.

The *LU* Decomposition of an $m \times n$ Matrix *A*

(a) Use steps 1, 3, and 4 of Gaussian elimination (as described in Section 1.4) to transform A into a matrix U in row echelon form by means of elementary row operations. If this is impossible, then A has no LU decomposition.

(b) While performing (a), create an $m \times m$ matrix L as follows:

(i) Each diagonal entry of L is 1.

(ii) If some elementary row operation in (a) adds c times row j of a matrix to row i, then $l_{ij} = -c$; otherwise, $l_{ij} = 0$.

Example 3 Find an LU decomposition of the matrix

$$A = \begin{bmatrix} 2 & -2 & 2 & 4 \\ -2 & 4 & 2 & -1 \\ 6 & -2 & 4 & 14 \end{bmatrix}.$$

Solution First, we transform A to U.

$$A = \begin{bmatrix} 2 & -2 & 2 & 4 \\ -2 & 4 & 2 & -1 \\ 6 & -2 & 4 & 14 \end{bmatrix} \xrightarrow{\mathbf{r}_1+\mathbf{r}_2\to\mathbf{r}_2} \begin{bmatrix} 2 & -2 & 2 & 4 \\ 0 & 2 & 4 & 3 \\ 6 & -2 & 4 & 14 \end{bmatrix}$$

$$\xrightarrow{(-3)\mathbf{r}_1+\mathbf{r}_3\to\mathbf{r}_3} \begin{bmatrix} 2 & -2 & 2 & 4 \\ 0 & 2 & 4 & 3 \\ 0 & 4 & -2 & 2 \end{bmatrix} \xrightarrow{(-2)\mathbf{r}_2+\mathbf{r}_3\to\mathbf{r}_3} \begin{bmatrix} 2 & -2 & 2 & 4 \\ 0 & 2 & 4 & 3 \\ 0 & 0 & -10 & -4 \end{bmatrix} = U$$

We are now prepared to obtain L. Of course, the diagonal entries of L are 1s, and the entries above the diagonal are 0s. The entries below the diagonal, the multipliers, can be obtained directly from the labels above the arrows in the transformation of A into U. A label of the form $c\mathbf{r}_j + \mathbf{r}_i \to \mathbf{r}_i$ indicates that the (i,j)-entry of L is $-c$. It follows that

$$L = \begin{bmatrix} 1 & 0 & 0 \\ -1 & 1 & 0 \\ 3 & 2 & 1 \end{bmatrix}.$$

Practice Problem 1 ▶ Find an LU decomposition of

$$A = \begin{bmatrix} 1 & -1 & -2 & -8 \\ -2 & 1 & 2 & 9 \\ 3 & 0 & 2 & 1 \end{bmatrix}.$$ ◀

USING AN *LU* DECOMPOSITION TO SOLVE A SYSTEM OF LINEAR EQUATIONS

Given a system of linear equations of the form $A\mathbf{x} = \mathbf{b}$, where A has an LU decomposition $A = LU$, we can take advantage of this decomposition to reduce the number of steps required to solve the system. Since

$$A\mathbf{x} = LU\mathbf{x} = L(U\mathbf{x}) = \mathbf{b},$$

we can set

$$U\mathbf{x} = \mathbf{y}, \qquad \text{and hence} \qquad L\mathbf{y} = \mathbf{b}.$$

The second system of equations is easily solved for $\mathbf{y}$ because L is a unit lower triangular matrix. Once $\mathbf{y}$ is obtained, the first system can then be easily solved for $\mathbf{x}$ because U is upper triangular. (See Figure 2.8.)

To illustrate this procedure, consider the system

$$\begin{aligned} x_1 - x_2 + 2x_3 &= 2 \\ 3x_1 - x_2 + 7x_3 &= 10 \\ 2x_1 - 4x_2 + 5x_3 &= 4 \end{aligned}$$

with coefficient matrix

$$A = \begin{bmatrix} 1 & -1 & 2 \\ 3 & -1 & 7 \\ 2 & -4 & 5 \end{bmatrix}.$$

The LU decomposition of A, which we obtained in Example 2, is given by

$$L = \begin{bmatrix} 1 & 0 & 0 \\ 3 & 1 & 0 \\ 2 & -1 & 1 \end{bmatrix} \qquad \text{and} \qquad U = \begin{bmatrix} 1 & -1 & 2 \\ 0 & 2 & 1 \\ 0 & 0 & 2 \end{bmatrix}.$$

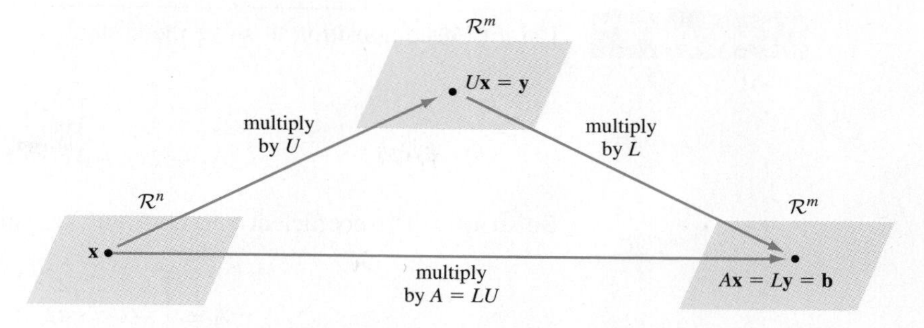

Figure 2.8 Solving a system of linear equations using an *LU* decomposition

The system $A\mathbf{x} = \mathbf{b}$ can be rewritten as $LU\mathbf{x} = \mathbf{b}$. Set $\mathbf{y} = U\mathbf{x}$ so that the system becomes $L\mathbf{y} = \mathbf{b}$, where

$$\mathbf{y} = \begin{bmatrix} y_1 \\ y_2 \\ y_3 \end{bmatrix} \quad \text{and} \quad \mathbf{b} = \begin{bmatrix} 2 \\ 10 \\ 4 \end{bmatrix}.$$

Thus we have the system

$$\begin{aligned} y_1 \qquad\qquad &= 2 \\ 3y_1 + y_2 \qquad &= 10 \\ 2y_1 - y_2 + y_3 &= 4. \end{aligned}$$

The first equation gives us the value of y_1. Substituting this into the second equation, we solve for y_2 to obtain $y_2 = 10 - 3(2) = 4$. Substituting the values for y_1 and y_2 into the third equation, we obtain $y_3 = 4 - 2(2) + 4 = 4$. Thus

$$\mathbf{y} = \begin{bmatrix} 2 \\ 4 \\ 4 \end{bmatrix}.$$

We now solve the system $U\mathbf{x} = \mathbf{y}$, which we can write as

$$\begin{aligned} x_1 - x_2 + 2x_3 &= 2 \\ 2x_2 + x_3 &= 4 \\ 2x_3 &= 4. \end{aligned}$$

Solving the third equation, we obtain $x_3 = 2$. Substituting this into the second equation and solving for x_2, we obtain $x_2 = (4 - 2)/2 = 1$. Finally, substituting the values for x_3 and x_2 into the first equation and solving for x_1 gives $x_1 = 2 + 1 - 2(2) = -1$. Thus

$$\mathbf{x} = \begin{bmatrix} x_1 \\ x_2 \\ x_3 \end{bmatrix} = \begin{bmatrix} -1 \\ 1 \\ 2 \end{bmatrix}.$$

This method of solving $U\mathbf{x} = \mathbf{y}$ is called **back substitution.**

If the coefficient matrix of a system of linear equations is not invertible—for example, if the matrix is not square—we can still solve the system by using an *LU* decomposition. In this case, the process of back substitution is complicated by the presence of free variables. The following example illustrates this situation:

Example 4 Use LU decomposition to solve the system

$$\begin{aligned} 2x_1 - 2x_2 + 2x_3 + 4x_4 &= 6 \\ -2x_1 + 4x_2 + 2x_3 - x_4 &= 4 \\ 6x_1 - 2x_2 + 4x_3 + 14x_4 &= 20. \end{aligned}$$

Solution The coefficient matrix of this system is

$$A = \begin{bmatrix} 2 & -2 & 2 & 4 \\ -2 & 4 & 2 & -1 \\ 6 & -2 & 4 & 14 \end{bmatrix}.$$

An LU decomposition of A was obtained in Example 3; it is given by

$$L = \begin{bmatrix} 1 & 0 & 0 \\ -1 & 1 & 0 \\ 3 & 2 & 1 \end{bmatrix} \quad \text{and} \quad U = \begin{bmatrix} 2 & -2 & 2 & 4 \\ 0 & 2 & 4 & 3 \\ 0 & 0 & -10 & -4 \end{bmatrix}.$$

Solve the system $L\mathbf{y} = \mathbf{b}$, where

$$\mathbf{y} = \begin{bmatrix} y_1 \\ y_2 \\ y_3 \end{bmatrix} \quad \text{and} \quad \mathbf{b} = \begin{bmatrix} 6 \\ 4 \\ 20 \end{bmatrix}.$$

As before, the unique solution of this system can be shown to be

$$\mathbf{y} = \begin{bmatrix} 6 \\ 10 \\ -18 \end{bmatrix}.$$

Next, we use back substitution to solve the system $U\mathbf{x} = \mathbf{y}$, which we can write as

$$\begin{aligned} 2x_1 - 2x_2 + 2x_3 + 4x_4 &= 6 \\ 2x_2 + 4x_3 + 3x_4 &= 10 \\ -10x_3 - 4x_4 &= -18. \end{aligned}$$

We begin with the last equation. In this equation, we solve for the first variable x_3, treating x_4 as a free variable. This yields

$$x_3 = \frac{9}{5} - \frac{2}{5}x_4.$$

Working our way upwards, we substitute this solution into the second equation and solve for the first variable in this equation, x_2.

$$2x_2 = 10 - 4x_3 - 3x_4 = 10 - 4\left(\frac{9}{5} - \frac{2}{5}x_4\right) - 3x_4 = \frac{14}{5} - \frac{7}{5}x_4$$

Hence

$$x_2 = \frac{7}{5} - \frac{7}{10}x_4.$$

Finally, we solve for x_1 in the first equation, substituting the expressions we have already obtained in the previous equations. In this case, there are no new variables other than x_1, and hence no additional free variables. Thus we have

$$\begin{aligned} 2x_1 &= 6 + 2x_2 - 2x_3 - 4x_4 \\ &= 6 + 2\left(\frac{7}{5} - \frac{7}{10}x_4\right) - 2\left(\frac{9}{5} - \frac{2}{5}x_4\right) - 4x_4 \\ &= \frac{26}{5} - \frac{23}{5}x_4, \end{aligned}$$

and hence

$$x_1 = \frac{13}{5} - \frac{23}{10}x_4.$$

Practice Problem 2 ▶ Use your answer to Practice Problem 1 to solve $A\mathbf{x} = \mathbf{b}$, where A is the matrix in Practice Problem 1 and

$$\mathbf{b} = \begin{bmatrix} -3 \\ 5 \\ -8 \end{bmatrix}.$$

◀

WHAT IF A MATRIX HAS NO LU DECOMPOSITION?

We have seen that not every matrix has an LU decomposition. Suppose that A is such a matrix. Then, by means of Gaussian elimination, A can be transformed into an upper triangular matrix U by elementary row operations that include row interchanges. It can be shown that if these row interchanges are applied to A initially, then the resulting matrix C can be transformed into U by means of elementary row operations that do not include row interchanges. Consequently, there is a unit lower triangular matrix L such that $C = LU$. The matrix C has the same rows as A, but in a different sequence. Thus there is a sequence of row interchanges that transforms A into C. Performing this same sequence of row interchanges on the appropriate identity matrix produces a matrix P such that $C = PA$. Any matrix P obtained by permuting the rows of an identity matrix is called a **permutation matrix**. So if A does not have an LU decomposition, there is a permutation matrix P such that PA has an LU decomposition.

In the next example, we illustrate how to find such a permutation matrix for a matrix having no LU decomposition.

Example 5 Let

$$A = \begin{bmatrix} 0 & 2 & 2 & 4 \\ 0 & 2 & 2 & 2 \\ 1 & 2 & 2 & 1 \\ 2 & 6 & 7 & 5 \end{bmatrix}.$$

Find a permutation matrix P and an upper triangular matrix U such that $PA = LU$ is an LU decomposition of PA for some unit lower triangular matrix L.

Solution We begin by transforming A into a matrix U in row echelon form, keeping track of the elementary row operations, as in Example 2.

$$A = \begin{bmatrix} 0 & 2 & 2 & 4 \\ 0 & 2 & 2 & 2 \\ 1 & 2 & 2 & 1 \\ 2 & 6 & 7 & 5 \end{bmatrix} \xrightarrow{\mathbf{r}_1 \leftrightarrow \mathbf{r}_3} \begin{bmatrix} 1 & 2 & 2 & 1 \\ 0 & 2 & 2 & 2 \\ 0 & 2 & 2 & 4 \\ 2 & 6 & 7 & 5 \end{bmatrix} \xrightarrow{-2\mathbf{r}_1 + \mathbf{r}_4 \to \mathbf{r}_4} \begin{bmatrix} 1 & 2 & 2 & 1 \\ 0 & 2 & 2 & 2 \\ 0 & 2 & 2 & 4 \\ 0 & 2 & 3 & 3 \end{bmatrix}$$

$$\xrightarrow{-\mathbf{r}_2 + \mathbf{r}_3 \to \mathbf{r}_3} \begin{bmatrix} 1 & 2 & 2 & 1 \\ 0 & 2 & 2 & 2 \\ 0 & 0 & 0 & 2 \\ 0 & 2 & 3 & 3 \end{bmatrix} \xrightarrow{-\mathbf{r}_2 + \mathbf{r}_4 \to \mathbf{r}_4} \begin{bmatrix} 1 & 2 & 2 & 1 \\ 0 & 2 & 2 & 2 \\ 0 & 0 & 0 & 2 \\ 0 & 0 & 1 & 1 \end{bmatrix}$$

$$\xrightarrow{\mathbf{r}_3 \leftrightarrow \mathbf{r}_4} \begin{bmatrix} 1 & 2 & 2 & 1 \\ 0 & 2 & 2 & 2 \\ 0 & 0 & 1 & 1 \\ 0 & 0 & 0 & 2 \end{bmatrix} = U$$

In this computation, two row interchanges were performed. If we apply these directly to A, we obtain

$$A = \begin{bmatrix} 0 & 2 & 2 & 4 \\ 0 & 2 & 2 & 2 \\ 1 & 2 & 2 & 1 \\ 2 & 6 & 7 & 5 \end{bmatrix} \xrightarrow{\mathbf{r}_1 \leftrightarrow \mathbf{r}_3} \begin{bmatrix} 1 & 2 & 2 & 1 \\ 0 & 2 & 2 & 2 \\ 0 & 2 & 2 & 4 \\ 2 & 6 & 7 & 5 \end{bmatrix} \xrightarrow{\mathbf{r}_3 \leftrightarrow \mathbf{r}_4} \begin{bmatrix} 1 & 2 & 2 & 1 \\ 0 & 2 & 2 & 2 \\ 2 & 6 & 7 & 5 \\ 0 & 2 & 2 & 4 \end{bmatrix} = C.$$

To find P, simply perform the preceding row interchanges on I_4 in the same order:

$$I_4 \xrightarrow{\mathbf{r}_1 \leftrightarrow \mathbf{r}_3} \begin{bmatrix} 0 & 0 & 1 & 0 \\ 0 & 1 & 0 & 0 \\ 1 & 0 & 0 & 0 \\ 0 & 0 & 0 & 1 \end{bmatrix} \xrightarrow{\mathbf{r}_3 \leftrightarrow \mathbf{r}_4} \begin{bmatrix} 0 & 0 & 1 & 0 \\ 0 & 1 & 0 & 0 \\ 0 & 0 & 0 & 1 \\ 1 & 0 & 0 & 0 \end{bmatrix} = P$$

Then $C = PA$ has an LU decomposition.

To complete the process of obtaining an LU decomposition for the matrix PA given in Example 5, we could apply the methods described earlier in this section to PA. However, it is more efficient to take advantage of the work we have already done. The next example illustrates how.

Example 6 For the matrices A, P, and U in Example 5, find a unit lower triangular matrix L such that $PA = LU$.

Solution The method used here is similar to the method in Example 3, but with one complication. If, in the process of transforming A to U, two rows are interchanged, then the multipliers that have already been computed are switched in the same way. So, for example, in the process of transforming A into U in Example 5, we added -2 times row 1 to row 4. This gave us a multiplier 2 in the $(4, 1)$-position. Then several steps later, we interchanged rows 3 and 4, which moved this multiplier to the $(3, 1)$-position. Since there are no more row interchanges, it follows that $l_{31} = 2$.

A simple way of keeping track of these row interchanges is to temporarily place multipliers, as they are created, in the appropriate positions of the intermediate matrices. These replace the actual entries, which, of course, are zeros. Then, when two rows are interchanged, the multipliers in these rows are also interchanged.

So as not to confuse these multipliers with the actual zero entries, we place each multiplier in parentheses. Thus, enhancing the process used in Example 5, we obtain the following sequence of matrices:

$$A = \begin{bmatrix} 0 & 2 & 2 & 4 \\ 0 & 2 & 2 & 2 \\ 1 & 2 & 2 & 1 \\ 2 & 6 & 7 & 5 \end{bmatrix} \xrightarrow{\mathbf{r}_1 \leftrightarrow \mathbf{r}_3} \begin{bmatrix} 1 & 2 & 2 & 1 \\ 0 & 2 & 2 & 2 \\ 0 & 2 & 2 & 4 \\ 2 & 6 & 7 & 5 \end{bmatrix} \xrightarrow{-2\mathbf{r}_1 + \mathbf{r}_4 \to \mathbf{r}_4} \begin{bmatrix} 1 & 2 & 2 & 1 \\ 0 & 2 & 2 & 2 \\ 0 & 2 & 2 & 4 \\ (2) & 2 & 3 & 3 \end{bmatrix}$$

$$\xrightarrow{-\mathbf{r}_2 + \mathbf{r}_3 \to \mathbf{r}_3} \begin{bmatrix} 1 & 2 & 2 & 1 \\ 0 & 2 & 2 & 2 \\ 0 & (1) & 0 & 2 \\ (2) & 2 & 3 & 3 \end{bmatrix} \xrightarrow{-\mathbf{r}_2 + \mathbf{r}_4 \to \mathbf{r}_4} \begin{bmatrix} 1 & 2 & 2 & 1 \\ 0 & 2 & 2 & 2 \\ 0 & (1) & 0 & 2 \\ (2) & (1) & 1 & 1 \end{bmatrix}$$

$$\xrightarrow{\mathbf{r}_3 \leftrightarrow \mathbf{r}_4} \begin{bmatrix} 1 & 2 & 2 & 1 \\ 0 & 2 & 2 & 2 \\ (2) & (1) & 1 & 1 \\ 0 & (1) & 0 & 2 \end{bmatrix}$$

Notice that if the entries in parentheses are replaced by zeros in the last matrix of the previous sequence, we obtain U. Finally, we obtain L by using the entries in parentheses in the last matrix of the sequence. The other nondiagonal entries of L are zeros. Thus

$$L = \begin{bmatrix} 1 & 0 & 0 & 0 \\ 0 & 1 & 0 & 0 \\ 2 & 1 & 1 & 0 \\ 0 & 1 & 0 & 1 \end{bmatrix} \quad \text{and} \quad U = \begin{bmatrix} 1 & 2 & 2 & 1 \\ 0 & 2 & 2 & 2 \\ 0 & 0 & 1 & 1 \\ 0 & 0 & 0 & 2 \end{bmatrix}.$$

It is a simple matter to verify that $LU = PA$.

Practice Problem 3 ▶ Find an LU decomposition of PA, where P is a permutation matrix and

$$A = \begin{bmatrix} 0 & 3 & -6 & 1 \\ -2 & -2 & 2 & 6 \\ 1 & 1 & -1 & -1 \\ 2 & -1 & 2 & -2 \end{bmatrix}.$$

◀

We now use these results to solve a system of linear equations.

Example 7 Use the results of Example 6 to solve the following system of linear equations:

$$\begin{aligned} 2x_2 + 2x_3 + 4x_4 &= -6 \\ 2x_2 + 2x_3 + 2x_4 &= -2 \\ x_1 + 2x_2 + 2x_3 + x_4 &= 3 \\ 2x_1 + 6x_2 + 7x_3 + 5x_4 &= 2 \end{aligned}$$

Solution This system can be written as the matrix equation $A\mathbf{x} = \mathbf{b}$, where

$$A = \begin{bmatrix} 0 & 2 & 2 & 4 \\ 0 & 2 & 2 & 2 \\ 1 & 2 & 2 & 1 \\ 2 & 6 & 7 & 5 \end{bmatrix} \quad \text{and} \quad \mathbf{b} = \begin{bmatrix} -6 \\ -2 \\ 3 \\ 2 \end{bmatrix}.$$

In Example 6, we found a permutation matrix P such that PA has an LU decomposition. Multiplying both sides of the equation $A\mathbf{x} = \mathbf{b}$ on the left by P, we obtain the equivalent equation $PA\mathbf{x} = P\mathbf{b}$. Then $PA = LU$ for the matrices L and U obtained in Example 6. Setting $\mathbf{b}' = P\mathbf{b}$, we have reduced the problem to solving the system $LU = \mathbf{b}'$, where

$$\mathbf{b}' = P\mathbf{b} = \begin{bmatrix} 0 & 0 & 1 & 0 \\ 0 & 1 & 0 & 0 \\ 0 & 0 & 0 & 1 \\ 1 & 0 & 0 & 0 \end{bmatrix} \begin{bmatrix} -6 \\ -2 \\ 3 \\ 2 \end{bmatrix} = \begin{bmatrix} 3 \\ -2 \\ 2 \\ -6 \end{bmatrix}.$$

As in Example 4, we set $\mathbf{y} = U\mathbf{x}$ and solve the system $L\mathbf{y} = \mathbf{b}'$, which has the form

$$\begin{aligned} y_1 \qquad\qquad\qquad &= 3 \\ y_2 \qquad\qquad &= -2 \\ 2y_1 + y_2 + y_3 \qquad &= 2 \\ y_2 \qquad + y_4 &= -6. \end{aligned}$$

We solve this system to obtain

$$\mathbf{y} = \begin{bmatrix} y_1 \\ y_2 \\ y_3 \\ y_4 \end{bmatrix} = \begin{bmatrix} 3 \\ -2 \\ -2 \\ -4 \end{bmatrix}.$$

Finally, to obtain the solution of the original system, we solve $U\mathbf{x} = \mathbf{y}$, which has the form

$$\begin{aligned} x_1 + 2x_2 + 2x_3 + x_4 &= 3 \\ 2x_2 + 2x_3 + 2x_4 &= -2 \\ x_3 + x_4 &= -2 \\ 2x_4 &= -4. \end{aligned}$$

Using back substitution, we solve this system to obtain the desired solution,

$$\mathbf{x} = \begin{bmatrix} x_1 \\ x_2 \\ x_3 \\ x_4 \end{bmatrix} = \begin{bmatrix} 3 \\ 1 \\ 0 \\ -2 \end{bmatrix}.$$

Practice Problem 4 ▶ Use your answer to Practice Problem 3 to solve $A\mathbf{x} = \mathbf{b}$, where A is the matrix in Practice Problem 3 and

$$\mathbf{b} = \begin{bmatrix} -13 \\ -6 \\ 1 \\ 8 \end{bmatrix}.$$

◀

THE RELATIVE EFFICIENCIES OF METHODS FOR SOLVING SYSTEMS OF LINEAR EQUATIONS

Let $A\mathbf{x} = \mathbf{b}$ denote a system of n linear equations in n variables. Suppose that A is invertible and has an LU decomposition. Then we have seen three different methods for solving this system.

1. Use Gaussian elimination to transform the augmented matrix $[A\ \mathbf{b}]$ to reduced row echelon form.
2. Apply elementary row operations to the augmented matrix $[A\ I_n]$ to compute A^{-1}, and then calculate $A^{-1}\mathbf{b}$.
3. Compute the LU decomposition of A, and then use the methods described in this section to solve the system $A\mathbf{x} = \mathbf{b}$.

We can compare the relative efficiencies of these methods by estimating the number of arithmetic operations (additions, subtractions, multiplications, and divisions) used for each method. In calculations performed by computers, which are required for matrices of substantial size, any arithmetic operation is called a **flop** (floating point operation). The total number of flops used to perform a matrix computation by a particular method is called the **flop count** for that method.

Typically, a flop count for a computation involving an $n \times n$ matrix is a polynomial in n. Since these counts are usually rough estimates and are significant only for large values of n, the terms in the polynomial of lower degree are usually ignored, and hence a flop count for a method is usually approximated as a multiple of a power of n.

The table that follows lists approximate flop counts for various matrix computations that can be used in solving a system of n equations in n unknowns. Note that computing A^{-1} to solve $A\mathbf{x} = \mathbf{b}$ is considerably less efficient than using Gaussian elimination or the LU decomposition.

Flop Counts for Various Procedures

In each case, we have a system $A\mathbf{x} = \mathbf{b}$, where A is an $n \times n$ invertible matrix.

Procedure	**Approximate Flop Count**
computing the LU decomposition of A	$\frac{2}{3}n^3$
solving the system using Gaussian elimination	$\frac{2}{3}n^3$
solving the system given the LU decomposition of A	$2n^2$
calculating the inverse of A	$2n^3$

If several systems with the same coefficient matrix are to be solved, then the cost of using Gaussian elimination is approximately $\frac{2}{3}n^3$ flops to solve each system, whereas the cost of using an LU decomposition involves an initial investment of approximately $\frac{2}{3}n^3$ flops for the LU decomposition followed by a much lower cost of approximately $2n^2$ flops to solve each system. For example, the cost of solving n different systems by Gaussian elimination is approximately

$$n\left(\frac{2n^3}{3}\right) = \frac{2n^4}{3}$$

flops, whereas the cost for solving the same n systems by an LU decomposition is approximately

$$\frac{2n^3}{3} + n \cdot 2n^2 = \frac{8n^3}{3}$$

flops.

EXERCISES

In Exercises 1–8, find an LU decomposition of each matrix.

1. $\begin{bmatrix} 2 & 3 & 4 \\ 6 & 8 & 10 \\ -2 & -4 & -3 \end{bmatrix}$

2. $\begin{bmatrix} 2 & -1 & 1 \\ 4 & -1 & 4 \\ -2 & 1 & 2 \end{bmatrix}$

3. $\begin{bmatrix} 1 & -1 & 2 & 1 \\ 2 & -3 & 5 & 4 \\ -3 & 2 & -4 & 0 \end{bmatrix}$

4. $\begin{bmatrix} 1 & -1 & 2 & 4 \\ 3 & -3 & 5 & 9 \end{bmatrix}$

5. $\begin{bmatrix} 1 & -1 & 2 & 1 & 3 \\ -1 & 2 & 0 & -2 & -2 \\ 2 & -1 & 7 & -1 & 1 \end{bmatrix}$

6. $\begin{bmatrix} 3 & 1 & -1 & 1 \\ 6 & 4 & -1 & 4 \\ -3 & -1 & 2 & -1 \\ 3 & 5 & 0 & 3 \end{bmatrix}$

7. $\begin{bmatrix} 1 & 0 & -3 & -1 & -2 & 1 \\ 2 & -1 & -8 & -1 & -5 & 0 \\ -1 & 1 & 5 & 1 & 4 & 2 \\ 0 & 1 & 2 & 1 & 3 & 4 \end{bmatrix}$

8. $\begin{bmatrix} -1 & 2 & 1 & -1 & 3 \\ 1 & -4 & 0 & 5 & -5 \\ -2 & 6 & -1 & -5 & 7 \\ -1 & -4 & 4 & 11 & -2 \end{bmatrix}$

In Exercises 9–16, use the results of Exercises 1–8 to solve each system of linear equations.

9. $\begin{aligned} 2x_1 + 3x_2 + 4x_3 &= 1 \\ 6x_1 + 8x_2 + 10x_3 &= 4 \\ -2x_1 - 4x_2 - 3x_3 &= 0 \end{aligned}$

10. $\begin{aligned} 2x_1 - x_2 + x_3 &= -1 \\ 4x_1 - x_2 + 4x_3 &= -2 \\ -2x_1 + x_2 + 2x_3 &= -2 \end{aligned}$

11. $\begin{aligned} x_1 - x_2 + 2x_3 + x_4 &= 1 \\ 2x_1 - 3x_2 + 5x_3 + 4x_4 &= 8 \\ -3x_1 + 2x_2 - 4x_3 &= 5 \end{aligned}$

12. $\begin{aligned} x_1 - x_2 + 2x_3 + 4x_4 &= 1 \\ 3x_1 - 3x_2 + 5x_3 + 9x_4 &= 5 \end{aligned}$

13. $\begin{aligned} x_1 - x_2 + 2x_3 + x_4 + 3x_5 &= -4 \\ -x_1 + 2x_2 - 2x_4 - 2x_5 &= 9 \\ 2x_1 - x_2 + 7x_3 - x_4 + x_5 &= -2 \end{aligned}$

14. $\begin{aligned} 3x_1 + x_2 - x_3 + x_4 &= 0 \\ 6x_1 + 4x_2 - x_3 + 4x_4 &= 15 \\ -3x_1 - x_2 + 2x_3 - x_4 &= 1 \\ 3x_1 + 5x_2 + 3x_4 &= 21 \end{aligned}$

15. $\begin{aligned} x_1 - 3x_3 - x_4 - 2x_5 + x_6 &= 1 \\ 2x_1 - x_2 - 8x_3 - x_4 - 5x_5 &= 8 \\ -x_1 + x_2 + 5x_3 + x_4 + 4x_5 + 2x_6 &= -5 \\ x_2 + 2x_3 + x_4 + 3x_5 + 4x_6 &= -2 \end{aligned}$

16. $\begin{aligned} -x_1 + 2x_2 + x_3 - x_4 + 3x_5 &= 7 \\ x_1 - 4x_2 + 5x_4 - 5x_5 &= -7 \\ -2x_1 + 6x_2 - x_3 - 5x_4 + 7x_5 &= 6 \\ -x_1 - 4x_2 + 4x_3 + 11x_4 - 2x_5 &= 11 \end{aligned}$

In Exercises 17–24, for each matrix A, find (a) a permutation matrix P such that PA has an LU decomposition and (b) an LU decomposition of PA.

17. $\begin{bmatrix} 1 & -1 & 3 \\ 2 & -2 & 5 \\ -1 & 2 & -1 \end{bmatrix}$

18. $\begin{bmatrix} 0 & 2 & -1 \\ 2 & 6 & 0 \\ 1 & 3 & -1 \end{bmatrix}$

19. $\begin{bmatrix} 1 & 1 & -2 & -1 \\ 2 & 2 & -3 & -1 \\ -1 & -2 & -1 & 1 \end{bmatrix}$

20. $\begin{bmatrix} 0 & -1 & 4 & 3 \\ -2 & -3 & 2 & 2 \\ 1 & 1 & -1 & 1 \end{bmatrix}$

21. $\begin{bmatrix} 0 & 1 & -2 \\ -1 & 2 & -1 \\ 2 & -4 & 3 \\ 1 & -3 & 2 \end{bmatrix}$

22. $\begin{bmatrix} 2 & 4 & -6 & 0 \\ -2 & 1 & 3 & 2 \\ 2 & 9 & -9 & 1 \\ 4 & 3 & -3 & 0 \end{bmatrix}$

23. $\begin{bmatrix} 1 & 2 & 1 & -1 \\ 2 & 4 & 1 & 1 \\ 3 & 2 & -1 & -2 \\ 2 & 5 & 3 & 0 \end{bmatrix}$

24. $\begin{bmatrix} 1 & 2 & 2 & 2 & 1 \\ 2 & 4 & 2 & 1 & 0 \\ 1 & 1 & 1 & 2 & 2 \\ -3 & -2 & 0 & -3 & -5 \end{bmatrix}$

In Exercises 25–32, use the results of Exercises 17–24 to solve each system of linear equations.

25. $\begin{aligned} x_1 - x_2 + 3x_3 &= 6 \\ 2x_1 - 2x_2 + 5x_3 &= 9 \\ -x_1 + 2x_2 - x_3 &= 1 \end{aligned}$

26. $\begin{aligned} 2x_2 - x_3 &= 2 \\ 2x_1 + 6x_2 &= -2 \\ x_1 + 3x_2 - x_3 &= -1 \end{aligned}$

27. $\begin{aligned} x_1 + x_2 - 2x_3 - x_4 &= 1 \\ 2x_1 + 2x_2 - 3x_3 - x_4 &= 5 \\ -x_1 - 2x_2 - x_3 + x_4 &= -1 \end{aligned}$

28. $\begin{aligned} -x_2 + 4x_3 + 3x_4 &= -1 \\ -2x_1 - 3x_2 + 2x_3 + 2x_4 &= 2 \\ x_1 + x_2 - x_3 + x_4 &= 0 \end{aligned}$

29. $\begin{aligned} x_2 - 2x_3 &= 0 \\ -x_1 + 2x_2 - x_3 &= -2 \\ 2x_1 - 4x_2 + 3x_3 &= 5 \\ x_1 - 3x_2 + 2x_3 &= 1 \end{aligned}$

30. $\begin{aligned} 2x_1 + 4x_2 - 6x_3 &= 2 \\ -2x_1 + x_2 + 3x_3 + 2x_4 &= 7 \\ 2x_1 + 9x_2 - 9x_3 + x_4 &= 11 \\ 4x_1 + 3x_2 - 3x_3 &= 7 \end{aligned}$

31. $\begin{aligned} x_1 + 2x_2 + x_3 - x_4 &= 3 \\ 2x_1 + 4x_2 + x_3 + x_4 &= 2 \\ 3x_1 + 2x_2 - x_3 - 2x_4 &= -4 \\ 2x_1 + 5x_2 + 3x_3 &= 7 \end{aligned}$

32. $\begin{aligned} x_1 + 2x_2 + 2x_3 + 2x_4 + x_5 &= 8 \\ 2x_1 + 4x_2 + 2x_3 + x_4 &= 12 \\ x_1 + x_2 + x_3 + 2x_4 + 2x_5 &= 5 \\ -3x_1 - 2x_2 - 3x_4 - 5x_5 &= -8 \end{aligned}$

T&F *In Exercises 33–41, determine whether the statements are true or false.*

33. Every matrix has an LU decomposition.
34. If a matrix A has an LU decomposition, then A can be transformed into a matrix in row echelon form without using any row interchanges.
35. An upper triangular matrix is one in which the entries above and to the right of the diagonal entries are all zeros.
36. In an LU decomposition of A, all the diagonal entries of U are 1s.
37. An LU decomposition of every matrix is unique.
38. The process for solving $U\mathbf{x} = \mathbf{y}$ is called back substitution.
39. Suppose that, in transforming A into a matrix in row echelon form, c times row i of a matrix is added to row j. In an LU decomposition of A, the (i,j)-entry of L is c.
40. Suppose that, in transforming A into a matrix in row echelon form, c times row i of a matrix is added to row j. In an LU decomposition of A, the (i,j)-entry of L is $-c$.
41. For every matrix A, there is a permutation matrix P such that PA has an LU decomposition.

---

42. Let A and B be $n \times n$ upper triangular matrices. Prove that AB is an upper triangular matrix and that its ith diagonal entry is $a_{ii}b_{ii}$.
43. Let U be an invertible upper triangular matrix. Prove that U^{-1} is an upper triangular matrix and that its ith diagonal entry is $1/u_{ii}$.
44. Let A and B be $n \times n$ lower triangular matrices.
    (a) Prove that AB is also a lower triangular matrix.
    (b) Prove that if both A and B are unit lower triangular matrices, then AB is also a unit lower triangular matrix.
45. Prove that a square unit lower triangular matrix L is invertible and that L^{-1} is also a unit lower triangular matrix.
46. Suppose that LU and $L'U'$ are two LU decompositions for an invertible matrix. Prove that $L = L'$ and $U = U'$. Thus an LU decomposition for an invertible matrix is unique. *Hint:* Use the results of Exercises 42–45.
47. Let C be an $n \times n$ matrix and $\mathbf{b}$ be a vector in $\mathcal{R}^n$.
    (a) Show that it requires n multiplications and $n - 1$ additions to compute each component of $C\mathbf{b}$.
    (b) Show that the approximate flop count for computing $C\mathbf{b}$ is $2n^2$.
48. Suppose we are given n systems of n linear equations in n variables, all of which have as their coefficient matrix the same invertible matrix A. Estimate the total flop count for solving all of these systems by first computing A^{-1}, and then computing the product of A^{-1} with the constant vector of each system.
49. Suppose that A is an $m \times n$ matrix and B is an $n \times p$ matrix. Find the exact flop count for computing the product AB.
50. Suppose that A is an $m \times n$ matrix, B is an $n \times p$ matrix, and C is a $p \times q$ matrix. The product ABC can be computed in two ways: (a) First compute AB, and then multiply this (on the right) by C. (b) First compute BC, and then multiply this (on the left) by A. Use Exercise 49 to devise a strategy that compares the two ways so that the more efficient one can be chosen.

In Exercises 51–54, use either a calculator with matrix capabilities or computer software such as MATLAB to solve each problem.[5]

In Exercises 51 and 52, find an LU decomposition of the given matrix.

51. $\begin{bmatrix} 2 & -1 & 3 & 2 & 1 \\ -2 & 2 & -1 & 1 & 4 \\ 4 & 1 & 15 & 12 & 19 \\ 6 & -6 & 9 & -4 & 0 \\ 4 & -2 & 9 & 2 & 9 \end{bmatrix}$

52. $\begin{bmatrix} -3 & 1 & 0 & 2 & 1 \\ -6 & 0 & 1 & 3 & 5 \\ -15 & 7 & 4 & 1 & 12 \\ 0 & -4 & 2 & -6 & 8 \end{bmatrix}$

In Exercises 53 and 54, for each matrix A, find (a) a permutation matrix P such that PA has an LU decomposition and (b) an LU decomposition of PA.

53. $\begin{bmatrix} 0 & 1 & 2 & -1 & 1 \\ 2 & -2 & -1 & 3 & 4 \\ 1 & 1 & 2 & -1 & 2 \\ -1 & 0 & 3 & 0 & 1 \\ 3 & 4 & -1 & 2 & 4 \end{bmatrix}$

54. $\begin{bmatrix} 1 & 2 & -3 & 1 & 4 \\ 3 & 6 & -5 & 4 & 8 \\ 2 & 3 & -3 & 2 & 1 \\ -1 & 2 & 1 & 4 & 2 \\ 3 & 2 & 4 & -4 & 0 \end{bmatrix}$

SOLUTIONS TO THE PRACTICE PROBLEMS

1. First we apply Gaussian elimination to transform A into an upper triangular matrix U in row echelon form without using row interchanges:

$$A = \begin{bmatrix} 1 & -1 & -2 & -8 \\ -2 & 1 & 2 & 9 \\ 3 & 0 & 2 & 1 \end{bmatrix}$$

$$\xrightarrow{2\mathbf{r}_1+\mathbf{r}_2\to\mathbf{r}_2} \begin{bmatrix} 1 & -1 & -2 & -8 \\ 0 & -1 & -2 & -7 \\ 3 & 0 & 2 & 1 \end{bmatrix}$$

$$\xrightarrow{-3\mathbf{r}_1+\mathbf{r}_3\to\mathbf{r}_3} \begin{bmatrix} 1 & -1 & -2 & -8 \\ 0 & -1 & -2 & -7 \\ 0 & 3 & 8 & 25 \end{bmatrix}$$

$$\xrightarrow{3\mathbf{r}_2+\mathbf{r}_3\to\mathbf{r}_3} \begin{bmatrix} 1 & -1 & -2 & -8 \\ 0 & -1 & -2 & -7 \\ 0 & 0 & 2 & 4 \end{bmatrix} = U$$

Then L is the unit lower triangular matrix whose (i,j)-entry, $i > j$, is $-c$, where $c\mathbf{r}_j + \mathbf{r}_i \to \mathbf{r}_i$ is a label over an arrow in the preceding reduction process. Thus

$$L = \begin{bmatrix} 1 & 0 & 0 \\ -2 & 1 & 0 \\ 3 & -3 & 1 \end{bmatrix}.$$

2. Substituting the LU decomposition for A obtained in Practice Problem 1, we write the system as the matrix equation $LU\mathbf{x} = \mathbf{b}$. Setting $\mathbf{y} = U\mathbf{x}$, the system becomes $L\mathbf{y} = \mathbf{b}$, which can be written

$$\begin{aligned} y_1 \qquad\qquad &= -3 \\ -2y_1 + y_2 \qquad &= \ \ 5 \\ 3y_1 - 3y_2 + y_3 &= -8. \end{aligned}$$

We solve this system to obtain

$$\mathbf{y} = \begin{bmatrix} y_1 \\ y_2 \\ y_3 \end{bmatrix} = \begin{bmatrix} -3 \\ -1 \\ -2 \end{bmatrix}.$$

Next we use back substitution to solve the system $U\mathbf{x} = \mathbf{y}$, which can be written

$$\begin{aligned} x_1 - x_2 - 2x_3 - 8x_4 &= -3 \\ -x_2 - 2x_3 - 7x_4 &= -1 \\ 2x_3 + 4x_4 &= -2. \end{aligned}$$

Treating x_4 as a free variable, we obtain the following general solution:

$$\begin{aligned} x_1 &= -2 + x_4 \\ x_2 &= 3 - 3x_4 \\ x_3 &= -1 - 2x_4 \\ x_4 &\quad \text{free} \end{aligned}$$

3. Using the method of Example 6, we have

$$A = \begin{bmatrix} 0 & 3 & -6 & 1 \\ -2 & -2 & 2 & 6 \\ 1 & 1 & -1 & -1 \\ 2 & -1 & 2 & -2 \end{bmatrix}$$

$$\xrightarrow{\mathbf{r}_1\leftrightarrow\mathbf{r}_3} \begin{bmatrix} 1 & 1 & -1 & -1 \\ -2 & -2 & 2 & 6 \\ 0 & 3 & -6 & 1 \\ 2 & -1 & 2 & -2 \end{bmatrix}$$

$$\xrightarrow{2\mathbf{r}_1+\mathbf{r}_2\to\mathbf{r}_2} \begin{bmatrix} 1 & 1 & -1 & -1 \\ (-2) & 0 & 0 & 4 \\ 0 & 3 & -6 & 1 \\ 2 & -1 & 2 & -2 \end{bmatrix}$$

[5] Caution! The MATLAB function `lu` does not compute an *LU* decomposition of a matrix as defined on page 154. (See page 568.)

$$\xrightarrow{-2\mathbf{r}_1+\mathbf{r}_4\to\mathbf{r}_4} \begin{bmatrix} 1 & 1 & -1 & -1 \\ (-2) & 0 & 0 & 4 \\ 0 & 3 & -6 & 1 \\ (2) & -3 & 4 & 0 \end{bmatrix}$$

$$\xrightarrow{\mathbf{r}_2\leftrightarrow\mathbf{r}_4} \begin{bmatrix} 1 & 1 & -1 & -1 \\ (2) & -3 & 4 & 0 \\ 0 & 3 & -6 & 1 \\ (-2) & 0 & 0 & 4 \end{bmatrix}$$

$$\xrightarrow{\mathbf{r}_2+\mathbf{r}_3\to\mathbf{r}_3} \begin{bmatrix} 1 & 1 & -1 & -1 \\ (2) & -3 & 4 & 0 \\ 0 & (-1) & -2 & 1 \\ (-2) & 0 & 0 & 4 \end{bmatrix}.$$

Thus

$$L = \begin{bmatrix} 1 & 0 & 0 & 0 \\ 2 & 1 & 0 & 0 \\ 0 & -1 & 1 & 0 \\ -2 & 0 & 0 & 1 \end{bmatrix}$$

and

$$U = \begin{bmatrix} 1 & 1 & -1 & -1 \\ 0 & -3 & 4 & 0 \\ 0 & 0 & -2 & 1 \\ 0 & 0 & 0 & 4 \end{bmatrix}.$$

Finally, we interchange the first and third rows and the second and fourth rows of I_4 to obtain P:

$$I_4 \xrightarrow{\mathbf{r}_1\leftrightarrow\mathbf{r}_3} \begin{bmatrix} 0 & 0 & 1 & 0 \\ 0 & 1 & 0 & 0 \\ 1 & 0 & 0 & 0 \\ 0 & 0 & 0 & 1 \end{bmatrix}$$

$$\xrightarrow{\mathbf{r}_2\leftrightarrow\mathbf{r}_4} \begin{bmatrix} 0 & 0 & 1 & 0 \\ 0 & 0 & 0 & 1 \\ 1 & 0 & 0 & 0 \\ 0 & 1 & 0 & 0 \end{bmatrix} = P$$

4. Using the permutation matrix P and the LU decomposition of PA obtained in Practice Problem 3, we transform the system of equations $A\mathbf{x} = \mathbf{b}$ into

$$LU\mathbf{x} = PA\mathbf{x} = P\mathbf{b} = \begin{bmatrix} 0 & 0 & 1 & 0 \\ 0 & 0 & 0 & 1 \\ 1 & 0 & 0 & 0 \\ 0 & 1 & 0 & 0 \end{bmatrix} \begin{bmatrix} -13 \\ -6 \\ 1 \\ 8 \end{bmatrix} = \begin{bmatrix} 1 \\ 8 \\ -13 \\ -6 \end{bmatrix}.$$

This system can now be solved by the method used in Practice Problem 2 to obtain the unique solution

$$\mathbf{x} = \begin{bmatrix} x_1 \\ x_2 \\ x_3 \\ x_4 \end{bmatrix} = \begin{bmatrix} 1 \\ 2 \\ 3 \\ -1 \end{bmatrix}.$$

2.7 LINEAR TRANSFORMATIONS AND MATRICES

In Section 1.2, we defined the matrix–vector product $A\mathbf{v}$, where A is an $m \times n$ matrix and $\mathbf{v}$ is in $\mathcal{R}^n$. The correspondence that associates to each vector $\mathbf{v}$ in $\mathcal{R}^n$ the vector $A\mathbf{v}$ in $\mathcal{R}^m$ is an example of a function from $\mathcal{R}^n$ to $\mathcal{R}^m$. We define a function as follows:

Definitions Let $\mathcal{S}_1$ and $\mathcal{S}_2$ be subsets of $\mathcal{R}^n$ and $\mathcal{R}^m$, respectively. A **function** f from $\mathcal{S}_1$ to $\mathcal{S}_2$, written $f\colon \mathcal{S}_1 \to \mathcal{S}_2$, is a rule that assigns to each vector $\mathbf{v}$ in $\mathcal{S}_1$ a unique vector $f(\mathbf{v})$ in $\mathcal{S}_2$. The vector $f(\mathbf{v})$ is called the **image** of $\mathbf{v}$ (under f). The set $\mathcal{S}_1$ is called the **domain** of a function f, and the set $\mathcal{S}_2$ is called the **codomain** of f. The **range** of f is defined to be the set of images $f(\mathbf{v})$ for all $\mathbf{v}$ in $\mathcal{S}_1$.

In Figure 2.9, we see that $\mathbf{u}$ and $\mathbf{v}$ both have $\mathbf{w}$ as their image. So $\mathbf{w} = f(\mathbf{u})$ and $\mathbf{w} = f(\mathbf{v})$.

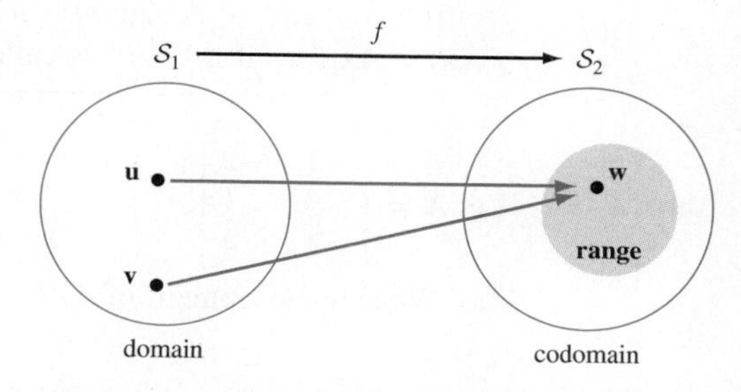

Figure 2.9 The domain, codomain, and range of a function

Example 1 Define $f: \mathcal{R}^3 \to \mathcal{R}^2$ by the rule

$$f\left(\begin{bmatrix} x_1 \\ x_2 \\ x_3 \end{bmatrix}\right) = \begin{bmatrix} x_1 + x_2 + x_3 \\ x_1^2 \end{bmatrix}.$$

Then f is a function whose domain is $\mathcal{R}^3$ and codomain is $\mathcal{R}^2$. Notice that

$$f\left(\begin{bmatrix} 0 \\ 1 \\ 1 \end{bmatrix}\right) = \begin{bmatrix} 2 \\ 0 \end{bmatrix} \quad \text{and} \quad f\left(\begin{bmatrix} 0 \\ 3 \\ -1 \end{bmatrix}\right) = \begin{bmatrix} 2 \\ 0 \end{bmatrix}.$$

So $\begin{bmatrix} 2 \\ 0 \end{bmatrix}$ is the image of both $\begin{bmatrix} 0 \\ 1 \\ 1 \end{bmatrix}$ and $\begin{bmatrix} 0 \\ 3 \\ -1 \end{bmatrix}$. However, not every vector in $\mathcal{R}^2$ is an image of a vector in $\mathcal{R}^3$ because every image must have a nonnegative second component.

Example 2 Let A be the 3×2 matrix

$$A = \begin{bmatrix} 1 & 0 \\ 2 & 1 \\ 1 & -1 \end{bmatrix}.$$

Define the function $T_A: \mathcal{R}^2 \to \mathcal{R}^3$ by

$$T_A(\mathbf{x}) = A\mathbf{x}.$$

Notice that, because A is a 3×2 matrix and $\mathbf{x}$ is a 2×1 vector, the vector $A\mathbf{x}$ has size 3×1. Also, observe that there is a reversal in the order of the size 3×2 of A and the "sizes" of the domain $\mathcal{R}^2$ and the codomain $\mathcal{R}^3$ of T_A.

We can easily obtain a formula for T_A by computing

$$T_A\left(\begin{bmatrix} x_1 \\ x_2 \end{bmatrix}\right) = \begin{bmatrix} 1 & 0 \\ 2 & 1 \\ 1 & -1 \end{bmatrix}\begin{bmatrix} x_1 \\ x_2 \end{bmatrix} = \begin{bmatrix} x_1 \\ 2x_1 + x_2 \\ x_1 - x_2 \end{bmatrix}.$$

A definition similar to that in Example 2 can be given for any $m \times n$ matrix A, in which case we obtain a function T_A with domain $\mathcal{R}^n$ and codomain $\mathcal{R}^m$.

Definition Let A be an $m \times n$ matrix. The function $T_A: \mathcal{R}^n \to \mathcal{R}^m$ defined by $T_A(\mathbf{x}) = A\mathbf{x}$ for all $\mathbf{x}$ in $\mathcal{R}^n$ is called the **matrix transformation induced by A.**

Practice Problem 1 ▶ Let $A = \begin{bmatrix} 1 & -2 \\ 3 & 1 \\ -1 & 4 \end{bmatrix}$.

(a) What is the domain of T_A?

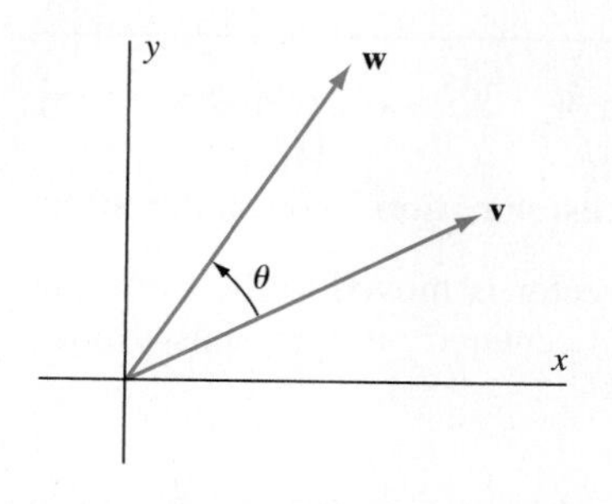

Figure 2.10 Every vector in $\mathcal{R}^2$ is an image.

(b) What is the codomain of T_A?

(c) Compute $T_A\left(\begin{bmatrix}4\\3\end{bmatrix}\right)$.

◀

We have already seen an important example of a matrix transformation in Section 1.2 using the rotation matrix $A = A_\theta$. Here, $T_A\colon \mathcal{R}^2 \to \mathcal{R}^2$ represents the function that rotates a vector counterclockwise by θ. To show that the range of T_A is all of $\mathcal{R}^2$, suppose that $\mathbf{w}$ is any vector in $\mathcal{R}^2$. If we let $\mathbf{v} = A_{-\theta}\mathbf{w}$ (see Figure 2.10), then, as in Example 4 of Section 2.1, we have

$$T_A(\mathbf{v}) = A_\theta(\mathbf{v}) = A_\theta A_{-\theta}\mathbf{w} = A_{\theta-\theta}\mathbf{w} = A_{0^\circ}\mathbf{w} = \mathbf{w}.$$

So every vector $\mathbf{w}$ in $\mathcal{R}^2$ is in the range of T_A.

Example 3

Let A be the matrix

$$\begin{bmatrix}1 & 0 & 0\\0 & 1 & 0\\0 & 0 & 0\end{bmatrix}.$$

So $T_A\colon \mathcal{R}^3 \to \mathcal{R}^3$ is defined by

$$T_A\left(\begin{bmatrix}x_1\\x_2\\x_3\end{bmatrix}\right) = \begin{bmatrix}1 & 0 & 0\\0 & 1 & 0\\0 & 0 & 0\end{bmatrix}\begin{bmatrix}x_1\\x_2\\x_3\end{bmatrix} = \begin{bmatrix}x_1\\x_2\\0\end{bmatrix}.$$

We can see from Figure 2.11 that $T_A(\mathbf{u})$ is the *orthogonal projection of* $\mathbf{u}$ *on the xy-plane*. The range of T_A is the xy-plane in $\mathcal{R}^3$.

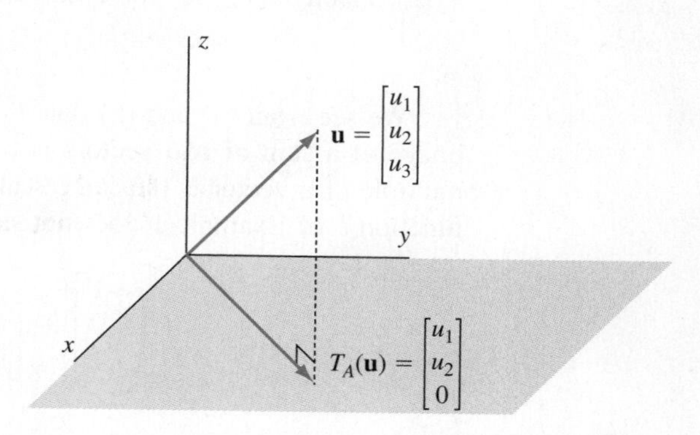

Figure 2.11 The orthogonal projection of **u** on the xy-plane

So far, we have seen that rotations and projections are matrix transformations. In the exercises, we discover that other geometric transformations, namely, reflections, contractions, and dilations, are also matrix transformations. The next example introduces yet another geometric transformation.

Example 4

Let k be a scalar and $A = \begin{bmatrix} 1 & k \\ 0 & 1 \end{bmatrix}$. The function $T_A \colon \mathcal{R}^2 \to \mathcal{R}^2$ is defined by $T_A\left(\begin{bmatrix} x_1 \\ x_2 \end{bmatrix}\right) = \begin{bmatrix} x_1 + kx_2 \\ x_2 \end{bmatrix}$ and is called a **shear transformation**. Notice the effect on the vector $\mathbf{u}$ in Figure 2.12(a). The head of the vector is moved to the right, but at the same height. In Figure 2.12(b), the letter "I" is centered on the y-axis. Notice the effect of the transformation T_A, where $k = 2$.

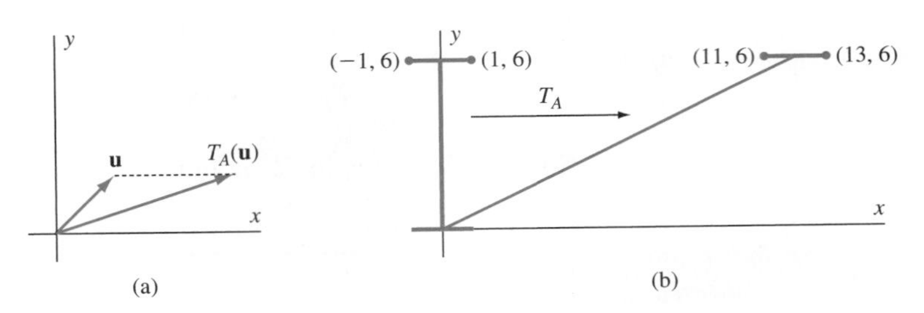

Figure 2.12 A shear transformation

The next result follows immediately from Theorem 1.3.

THEOREM 2.7

For any $m \times n$ matrix A and any vectors $\mathbf{u}$ and $\mathbf{v}$ in $\mathcal{R}^n$, the following statements are true:

(a) $T_A(\mathbf{u} + \mathbf{v}) = T_A(\mathbf{u}) + T_A(\mathbf{v})$.

(b) $T_A(c\mathbf{u}) = cT_A(\mathbf{u})$ for every scalar c.

We see from (a) and (b) that T_A preserves the two vector operations; that is, the image of a sum of two vectors is the sum of the images, and the image of a scalar multiple of a vector is the same scalar multiple of the image. On the other hand, the function f of Example 1 does not satisfy either of these properties. For example,

$$f\left(\begin{bmatrix} 1 \\ 0 \\ 1 \end{bmatrix} + \begin{bmatrix} 2 \\ 0 \\ 0 \end{bmatrix}\right) = f\left(\begin{bmatrix} 3 \\ 0 \\ 1 \end{bmatrix}\right) = \begin{bmatrix} 4 \\ 9 \end{bmatrix},$$

but

$$f\left(\begin{bmatrix} 1 \\ 0 \\ 1 \end{bmatrix}\right) + f\left(\begin{bmatrix} 2 \\ 0 \\ 0 \end{bmatrix}\right) = \begin{bmatrix} 2 \\ 1 \end{bmatrix} + \begin{bmatrix} 2 \\ 4 \end{bmatrix} = \begin{bmatrix} 4 \\ 5 \end{bmatrix}.$$

So (a) is not satisfied. Also,

$$f\left(2\begin{bmatrix} 1 \\ 0 \\ 1 \end{bmatrix}\right) = f\left(\begin{bmatrix} 2 \\ 0 \\ 2 \end{bmatrix}\right) = \begin{bmatrix} 4 \\ 4 \end{bmatrix}, \quad \text{but} \quad 2f\left(\begin{bmatrix} 1 \\ 0 \\ 1 \end{bmatrix}\right) = 2\begin{bmatrix} 2 \\ 1 \end{bmatrix} = \begin{bmatrix} 4 \\ 2 \end{bmatrix}.$$

So (b) is not satisfied.

Functions that do satisfy (a) and (b) of Theorem 2.7 merit their own definition.

Definition A function T from $\mathcal{R}^n$ to $\mathcal{R}^m$, written $T: \mathcal{R}^n \to \mathcal{R}^m$, is called a **linear transformation** (or simply **linear**) if, for all vectors $\mathbf{u}$ and $\mathbf{v}$ in $\mathcal{R}^n$ and all scalars c, both of the following conditions hold:

(i) $T(\mathbf{u}+\mathbf{v}) = T(\mathbf{u}) + T(\mathbf{v})$. (In this case, we say that T **preserves vector addition.**)

(ii) $T(c\mathbf{u}) = cT(\mathbf{u})$. (In this case, we say that T **preserves scalar multiplication.**)

By Theorem 2.7, every matrix transformation is linear.

There are two linear transformations that deserve special attention. The first is the **identity transformation** $I: \mathcal{R}^n \to \mathcal{R}^n$, which is defined by $I(\mathbf{x}) = \mathbf{x}$ for all $\mathbf{x}$ in $\mathcal{R}^n$. It is easy to show that I is linear and its range is all of $\mathcal{R}^n$. The second transformation is the **zero transformation** $T_0: \mathcal{R}^n \to \mathcal{R}^m$, which is defined by $T_0(\mathbf{x}) = \mathbf{0}$ for all $\mathbf{x}$ in $\mathcal{R}^n$. Like the identity transformation, it is easy to show that T_0 is linear. The range of T_0 consists precisely of the zero vector.

The next theorem presents some basic properties of linear transformations.

THEOREM 2.8

For any linear transformation $T: \mathcal{R}^n \to \mathcal{R}^m$, the following statements are true:

(a) $T(\mathbf{0}) = \mathbf{0}$.

(b) $T(-\mathbf{u}) = -T(\mathbf{u})$ for all vectors $\mathbf{u}$ in $\mathcal{R}^n$.

(c) $T(\mathbf{u}-\mathbf{v}) = T(\mathbf{u}) - T(\mathbf{v})$ for all vectors $\mathbf{u}$ and $\mathbf{v}$ in $\mathcal{R}^n$.

(d) $T(a\mathbf{u}+b\mathbf{v}) = aT(\mathbf{u}) + bT(\mathbf{v})$ for all vectors $\mathbf{u}$ and $\mathbf{v}$ in $\mathcal{R}^n$ and all scalars a and b.

PROOF (a) Because T preserves vector addition, we have

$$T(\mathbf{0}) = T(\mathbf{0}+\mathbf{0}) = T(\mathbf{0}) + T(\mathbf{0}).$$

Subtracting $T(\mathbf{0})$ from both sides yields $\mathbf{0} = T(\mathbf{0})$.

(b) Let $\mathbf{u}$ be a vector in $\mathcal{R}^n$. Because T preserves scalar multiplication, we have

$$T(-\mathbf{u}) = T((-1)\mathbf{u}) = (-1)T(\mathbf{u}) = -T(\mathbf{u}).$$

(c) Combining the fact that T preserves vector addition with part (b), we have, for any vectors $\mathbf{u}$ and $\mathbf{v}$ in $\mathcal{R}^n$,

$$T(\mathbf{u}-\mathbf{v}) = T(\mathbf{u}+(-\mathbf{v})) = T(\mathbf{u}) + T(-\mathbf{v}) = T(\mathbf{u}) + (-T(\mathbf{v})) = T(\mathbf{u}) - T(\mathbf{v}).$$

(d) Because T preserves vector addition and scalar multiplication, we have, for any vectors $\mathbf{u}$ and $\mathbf{v}$ in $\mathcal{R}^n$ and scalars a and b, that

$$T(a\mathbf{u}+b\mathbf{v}) = T(a\mathbf{u}) + T(b\mathbf{v}) = aT(\mathbf{u}) + bT(\mathbf{v}).$$

■

We can generalize Theorem 2.8(d) to show that T preserves arbitrary linear combinations.

Let $T: \mathcal{R}^n \to \mathcal{R}^m$ be a linear transformation. If $\mathbf{u}_1, \mathbf{u}_2, \ldots, \mathbf{u}_k$ are vectors in $\mathcal{R}^n$ and $a_1, a_2, \ldots, a_k$ are scalars, then

$$T(a_1\mathbf{u}_1 + a_2\mathbf{u}_2 + \cdots + a_k\mathbf{u}_k) = a_1T(\mathbf{u}_1) + a_2T(\mathbf{u}_2) + \cdots + a_kT(\mathbf{u}_k).$$

Example 5 Suppose that $T\colon \mathcal{R}^2 \to \mathcal{R}^2$ is a linear transformation such that

$$T\left(\begin{bmatrix}1\\1\end{bmatrix}\right) = \begin{bmatrix}2\\3\end{bmatrix} \quad \text{and} \quad T\left(\begin{bmatrix}1\\-1\end{bmatrix}\right) = \begin{bmatrix}4\\-1\end{bmatrix}.$$

(a) Find $T\left(\begin{bmatrix}3\\3\end{bmatrix}\right)$.

(b) Find $T\left(\begin{bmatrix}1\\0\end{bmatrix}\right)$ and $T\left(\begin{bmatrix}0\\-1\end{bmatrix}\right)$, and use the results to determine $T\left(\begin{bmatrix}x_1\\x_2\end{bmatrix}\right)$.

Solution

(a) Since $\begin{bmatrix}3\\3\end{bmatrix} = 3\begin{bmatrix}1\\1\end{bmatrix}$, it follows that

$$T\left(\begin{bmatrix}3\\3\end{bmatrix}\right) = T\left(3\begin{bmatrix}1\\1\end{bmatrix}\right) = 3T\left(\begin{bmatrix}1\\1\end{bmatrix}\right) = 3\begin{bmatrix}2\\3\end{bmatrix} = \begin{bmatrix}6\\9\end{bmatrix}.$$

(b) Observe that $\begin{bmatrix}1\\0\end{bmatrix} = \frac{1}{2}\begin{bmatrix}1\\1\end{bmatrix} + \frac{1}{2}\begin{bmatrix}1\\-1\end{bmatrix}$. Hence

$$\begin{aligned}
T\left(\begin{bmatrix}1\\0\end{bmatrix}\right) &= T\left(\frac{1}{2}\begin{bmatrix}1\\1\end{bmatrix} + \frac{1}{2}\begin{bmatrix}1\\-1\end{bmatrix}\right)\\
&= \frac{1}{2}T\left(\begin{bmatrix}1\\1\end{bmatrix}\right) + \frac{1}{2}T\left(\begin{bmatrix}1\\-1\end{bmatrix}\right)\\
&= \frac{1}{2}\begin{bmatrix}2\\3\end{bmatrix} + \frac{1}{2}\begin{bmatrix}4\\-1\end{bmatrix} = \begin{bmatrix}3\\1\end{bmatrix}.
\end{aligned}$$

Similarly,

$$\begin{aligned}
T\left(\begin{bmatrix}0\\1\end{bmatrix}\right) &= T\left(\frac{1}{2}\begin{bmatrix}1\\1\end{bmatrix} - \frac{1}{2}\begin{bmatrix}1\\-1\end{bmatrix}\right)\\
&= \frac{1}{2}T\left(\begin{bmatrix}1\\1\end{bmatrix}\right) - \frac{1}{2}T\left(\begin{bmatrix}1\\-1\end{bmatrix}\right)\\
&= \frac{1}{2}\begin{bmatrix}2\\3\end{bmatrix} - \frac{1}{2}\begin{bmatrix}4\\-1\end{bmatrix} = \begin{bmatrix}-1\\2\end{bmatrix}.
\end{aligned}$$

Finally,

$$\begin{aligned}
T\left(\begin{bmatrix}x_1\\x_2\end{bmatrix}\right) &= T\left(x_1\begin{bmatrix}1\\0\end{bmatrix} + x_2\begin{bmatrix}0\\1\end{bmatrix}\right)\\
&= x_1T\left(\begin{bmatrix}1\\0\end{bmatrix}\right) + x_2T\left(\begin{bmatrix}0\\1\end{bmatrix}\right)\\
&= x_1\begin{bmatrix}3\\1\end{bmatrix} + x_2\begin{bmatrix}-1\\2\end{bmatrix}\\
&= \begin{bmatrix}3x_1 - x_2\\x_1 + 2x_2\end{bmatrix}.
\end{aligned}$$

Practice Problem 2 ▶ Suppose that $T\colon \mathcal{R}^2 \to \mathcal{R}^3$ is a linear transformation such that

$$T\left(\begin{bmatrix} -1 \\ 0 \end{bmatrix}\right) = \begin{bmatrix} -2 \\ 1 \\ 3 \end{bmatrix} \quad \text{and} \quad T\left(\begin{bmatrix} 0 \\ 2 \end{bmatrix}\right) = \begin{bmatrix} 2 \\ 4 \\ -2 \end{bmatrix}.$$

Determine $T\left(\begin{bmatrix} x_1 \\ x_2 \end{bmatrix}\right)$. ◀

Theorem 2.8(a) can sometimes be used to show that a function is not linear. For example, the function $T\colon \mathcal{R} \to \mathcal{R}$ defined by $T(x) = 2x + 3$ is not linear because $T(0) = 3 \neq 0$. *Note, however, that a function f may satisfy the condition that $f(\mathbf{0}) = \mathbf{0}$, yet not be linear.* For example, the function f in Example 1 is not linear, even though $f(\mathbf{0}) = \mathbf{0}$.

The next example illustrates how to verify that a function is linear.

Example 6

Define $T\colon \mathcal{R}^2 \to \mathcal{R}^2$ by $T\left(\begin{bmatrix} x_1 \\ x_2 \end{bmatrix}\right) = \begin{bmatrix} 2x_1 - x_2 \\ x_1 \end{bmatrix}$. To verify that T is linear, let $\mathbf{u}$ and $\mathbf{v}$ be vectors in $\mathcal{R}^2$. Then $\mathbf{u} = \begin{bmatrix} u_1 \\ u_2 \end{bmatrix}$, $\mathbf{v} = \begin{bmatrix} v_1 \\ v_2 \end{bmatrix}$, and $\mathbf{u} + \mathbf{v} = \begin{bmatrix} u_1 + v_1 \\ u_2 + v_2 \end{bmatrix}$. So

$$T(\mathbf{u} + \mathbf{v}) = T\left(\begin{bmatrix} u_1 + v_1 \\ u_2 + v_2 \end{bmatrix}\right) = \begin{bmatrix} 2(u_1 + v_1) - (u_2 + v_2) \\ u_1 + v_1 \end{bmatrix}.$$

On the other hand,

$$\begin{aligned} T(\mathbf{u}) + T(\mathbf{v}) &= T\left(\begin{bmatrix} u_1 \\ u_2 \end{bmatrix}\right) + T\left(\begin{bmatrix} v_1 \\ v_2 \end{bmatrix}\right) = \begin{bmatrix} 2u_1 - u_2 \\ u_1 \end{bmatrix} + \begin{bmatrix} 2v_1 - v_2 \\ v_1 \end{bmatrix} \\ &= \begin{bmatrix} (2u_1 - u_2) + (2v_1 - v_2) \\ u_1 + v_1 \end{bmatrix} = \begin{bmatrix} 2(u_1 + v_1) - (u_2 + v_2) \\ u_1 + v_1 \end{bmatrix}. \end{aligned}$$

So, $T(\mathbf{u} + \mathbf{v}) = T(\mathbf{u}) + T(\mathbf{v})$.

Now suppose that c is any scalar. Then

$$T(c\mathbf{u}) = T\left(\begin{bmatrix} cu_1 \\ cu_2 \end{bmatrix}\right) = \begin{bmatrix} 2cu_1 - cu_2 \\ cu_1 \end{bmatrix}.$$

Also,

$$cT(\mathbf{u}) = c\begin{bmatrix} 2u_1 - u_2 \\ u_1 \end{bmatrix} = \begin{bmatrix} 2cu_1 - cu_2 \\ cu_1 \end{bmatrix}.$$

Hence $T(c\mathbf{u}) = cT(\mathbf{u})$. Therefore T is linear.

Another way to verify that the transformation T in Example 6 is linear is to find a matrix A such that $T = T_A$, and then appeal to Theorem 2.7. Suppose we let

$$A = \begin{bmatrix} 2 & -1 \\ 1 & 0 \end{bmatrix}.$$

Then

$$T_A\left(\begin{bmatrix}x_1\\x_2\end{bmatrix}\right)=\begin{bmatrix}2 & -1\\1 & 0\end{bmatrix}\begin{bmatrix}x_1\\x_2\end{bmatrix}=\begin{bmatrix}2x_1-x_2\\x_1\end{bmatrix}=T\left(\begin{bmatrix}x_1\\x_2\end{bmatrix}\right).$$

So $T = T_A$.

Now we show that *every* linear transformation with domain $\mathcal{R}^n$ and codomain $\mathcal{R}^m$ is a matrix transformation. This means that if a transformation T is linear, we can produce a corresponding matrix A such that $T = T_A$.

THEOREM 2.9

Let $T: \mathcal{R}^n \to \mathcal{R}^m$ be linear. Then there is a unique $m \times n$ matrix

$$A = [T(\mathbf{e}_1)\ T(\mathbf{e}_2)\ \dots\ T(\mathbf{e}_n)],$$

whose columns are the images under T of the standard vectors for $\mathcal{R}^n$, such that $T(\mathbf{v}) = A\mathbf{v}$ for all $\mathbf{v}$ in $\mathcal{R}^n$.

PROOF Let $A = [T(\mathbf{e}_1)\ T(\mathbf{e}_2)\ \dots\ T(\mathbf{e}_n)]$. We show that $T = T_A$. Notice that

$$\mathbf{v}=\begin{bmatrix}v_1\\v_2\\\vdots\\v_n\end{bmatrix}=v_1\mathbf{e}_1+v_2\mathbf{e}_2+\cdots+v_n\mathbf{e}_n$$

for any $\mathbf{v}$ in $\mathcal{R}^n$; so we have

$$\begin{aligned}T(\mathbf{v}) &= T(v_1\mathbf{e}_1+v_2\mathbf{e}_2+\cdots+v_n\mathbf{e}_n)\\&= v_1T(\mathbf{e}_1)+v_2T(\mathbf{e}_2)+\cdots+v_nT(\mathbf{e}_n)\\&= v_1\mathbf{a}_1+v_2\mathbf{a}_2+\cdots+v_n\mathbf{a}_n\\&= A\mathbf{v}\\&= T_A(\mathbf{v}).\end{aligned}$$

Therefore $T = T_A$.

To prove uniqueness, suppose that $T_A = T_B$ for some $m \times n$ matrix B. Then $A\mathbf{v} = B\mathbf{v}$ for every vector $\mathbf{v}$ in $\mathcal{R}^n$, and therefore $A = B$ by Theorem 1.3(e). ■

Let $T: \mathcal{R}^n \to \mathcal{R}^m$ be a linear transformation. We call the $m \times n$ matrix

$$A = [T(\mathbf{e}_1)\ T(\mathbf{e}_2)\ \dots\ T(\mathbf{e}_n)]$$

the **standard matrix** of T. Note that, by Theorem 2.9, the standard matrix A of T has the property that $T(\mathbf{v}) = A\mathbf{v}$ for every $\mathbf{v}$ in $\mathcal{R}^n$.

Example 7

Let $T: \mathcal{R}^3 \to \mathcal{R}^2$ be defined by $T\left(\begin{bmatrix}x_1\\x_2\\x_3\end{bmatrix}\right)=\begin{bmatrix}3x_1-4x_2\\2x_1+x_3\end{bmatrix}$. It is straightforward to show that T is linear. To find the standard matrix of T, we compute its columns

$T(\mathbf{e}_1)$, $T(\mathbf{e}_2)$, and $T(\mathbf{e}_3)$. We have $T(\mathbf{e}_1) = \begin{bmatrix} 3 \\ 2 \end{bmatrix}$, $T(\mathbf{e}_2) = \begin{bmatrix} -4 \\ 0 \end{bmatrix}$, and $T(\mathbf{e}_3) = \begin{bmatrix} 0 \\ 1 \end{bmatrix}$. So the standard matrix of T is

$$\begin{bmatrix} 3 & -4 & 0 \\ 2 & 0 & 1 \end{bmatrix}.$$

Example 8 Let $U\colon \mathcal{R}^2 \to \mathcal{R}^2$ be defined by $U\left(\begin{bmatrix} x_1 \\ x_2 \end{bmatrix}\right) = \begin{bmatrix} x_1 \\ -x_2 \end{bmatrix}$. Then U is the *reflection of $\mathcal{R}^2$ about the x-axis.* (See Figure 2.13.) It is straightforward to show that U is a linear transformation. Observe that $U(\mathbf{e}_1) = \mathbf{e}_1$ and $U(\mathbf{e}_2) = -\mathbf{e}_2$. Hence the standard matrix of U is

$$\begin{bmatrix} 1 & 0 \\ 0 & -1 \end{bmatrix}.$$

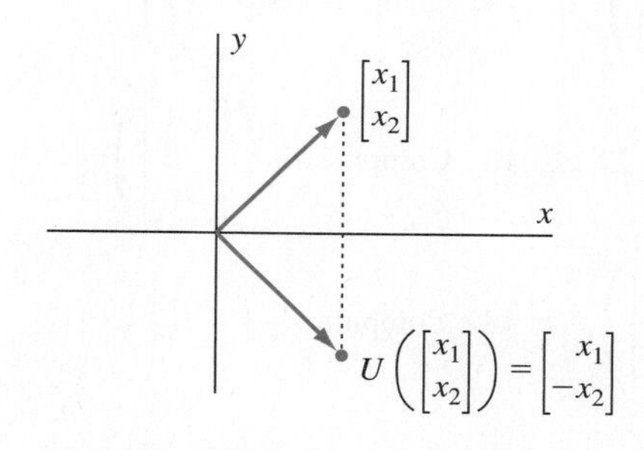

Figure 2.13 Reflection of $\mathcal{R}^2$ about the x-axis

Practice Problem 3 ▶ Determine the standard matrix of the linear transformation $T\colon \mathcal{R}^3 \to \mathcal{R}^2$ defined by

$$T\left(\begin{bmatrix} x_1 \\ x_2 \\ x_3 \end{bmatrix}\right) = \begin{bmatrix} 2x_1 - 5x_3 \\ -3x_2 + 4x_3 \end{bmatrix}.$$

◀

In the next section, we illustrate the close relationship between a linear transformation and its standard matrix.

EXERCISES

Exercises 1–20 refer to the following matrices:

$$A = \begin{bmatrix} 2 & -3 & 1 \\ 4 & 0 & -2 \end{bmatrix}, \quad B = \begin{bmatrix} 1 & 5 & 0 \\ 2 & -1 & 3 \\ 0 & 4 & -2 \end{bmatrix}, \quad \text{and}$$

$$C = \begin{bmatrix} 1 & 2 \\ 0 & -2 \\ 4 & 1 \end{bmatrix}$$

1. Give the domain and codomain of the matrix transformation induced by A.
2. Give the domain and codomain of the matrix transformation induced by B.
3. Give the domain and codomain of the matrix transformation induced by C.

4. Give the domain and codomain of the matrix transformation induced by A^T.
5. Give the domain and codomain of the matrix transformation induced by B^T.
6. Give the domain and codomain of the matrix transformation induced by C^T.

7. Compute $T_A\left(\begin{bmatrix} 3 \\ -1 \\ 2 \end{bmatrix}\right)$.

8. Compute $T_B\left(\begin{bmatrix} 1 \\ 0 \\ 1 \end{bmatrix}\right)$.

9. Compute $T_C\left(\begin{bmatrix} 2 \\ 3 \end{bmatrix}\right)$.

10. Compute $T_A\left(\begin{bmatrix} 2 \\ -1 \\ 2 \end{bmatrix}\right)$.

11. Compute $T_B\left(\begin{bmatrix} -4 \\ 2 \\ 1 \end{bmatrix}\right)$.

12. Compute $T_C\left(\begin{bmatrix} -1 \\ 4 \end{bmatrix}\right)$.

13. Compute $T_A\left(\begin{bmatrix} 4 \\ 0 \\ -3 \end{bmatrix}\right)$.

14. Compute $T_B\left(\begin{bmatrix} 3 \\ 0 \\ 2 \end{bmatrix}\right)$.

15. Compute $T_C\left(\begin{bmatrix} 5 \\ -3 \end{bmatrix}\right)$.

16. Compute $T_A\left(\begin{bmatrix} -1 \\ -2 \\ -3 \end{bmatrix}\right)$.

17. Compute $T_B\left(\begin{bmatrix} -3 \\ 0 \\ -1 \end{bmatrix}\right)$.

18. Compute $T_C\left(\begin{bmatrix} -1 \\ -2 \end{bmatrix}\right)$.

19. Compute $T_{(A+C^T)}\left(\begin{bmatrix} 2 \\ 1 \\ 1 \end{bmatrix}\right)$ and

$$T_A\left(\begin{bmatrix} 2 \\ 1 \\ 1 \end{bmatrix}\right) + T_{C^T}\left(\begin{bmatrix} 2 \\ 1 \\ 1 \end{bmatrix}\right).$$

20. Compute $T_A(\mathbf{e}_1)$ and $T_A(\mathbf{e}_3)$.

In Exercises 21–24, identify the values of n and m for each linear transformation $T: \mathcal{R}^n \to \mathcal{R}^m$.

21. T is defined by $T\left(\begin{bmatrix} x_1 \\ x_2 \\ x_3 \end{bmatrix}\right) = \begin{bmatrix} 2x_1 \\ x_1 - x_2 \end{bmatrix}$.

22. T is defined by $T\left(\begin{bmatrix} x_1 \\ x_2 \end{bmatrix}\right) = \begin{bmatrix} x_1 + x_2 \\ x_1 - x_2 \\ x_2 \end{bmatrix}$.

23. T is defined by $T\left(\begin{bmatrix} x_1 \\ x_2 \end{bmatrix}\right) = \begin{bmatrix} x_1 - 4x_2 \\ 2x_1 - 3x_2 \\ 0 \\ x_2 \end{bmatrix}$.

24. T is defined by $T\left(\begin{bmatrix} x_1 \\ x_2 \\ x_3 \\ x_4 \end{bmatrix}\right) = \begin{bmatrix} 5x_1 - 4x_2 + x_3 - 2x_4 \\ -2x_2 + 4x_4 \\ 3x_1 - 5x_3 \end{bmatrix}$.

In Exercises 25–34, linear transformations are given. Compute their standard matrices.

25. $T: \mathcal{R}^2 \to \mathcal{R}^2$ defined by $T\left(\begin{bmatrix} x_1 \\ x_2 \end{bmatrix}\right) = \begin{bmatrix} x_2 \\ x_1 + x_2 \end{bmatrix}$

26. $T: \mathcal{R}^2 \to \mathcal{R}^2$ defined by $T\left(\begin{bmatrix} x_1 \\ x_2 \end{bmatrix}\right) = \begin{bmatrix} 2x_1 + 3x_2 \\ 4x_1 + 5x_2 \end{bmatrix}$

27. $T: \mathcal{R}^3 \to \mathcal{R}^2$ defined by $T\left(\begin{bmatrix} x_1 \\ x_2 \\ x_3 \end{bmatrix}\right) = \begin{bmatrix} x_1 + x_2 + x_3 \\ 2x_1 \end{bmatrix}$

28. $T: \mathcal{R}^2 \to \mathcal{R}^3$ defined by $T\left(\begin{bmatrix} x_1 \\ x_2 \end{bmatrix}\right) = \begin{bmatrix} 3x_2 \\ 2x_1 - x_2 \\ x_1 + x_2 \end{bmatrix}$

29. $T: \mathcal{R}^2 \to \mathcal{R}^4$ defined by $T\left(\begin{bmatrix} x_1 \\ x_2 \end{bmatrix}\right) = \begin{bmatrix} x_1 - x_2 \\ 2x_1 - 3x_2 \\ 0 \\ x_2 \end{bmatrix}$

30. $T: \mathcal{R}^3 \to \mathcal{R}^3$ defined by $T\left(\begin{bmatrix} x_1 \\ x_2 \\ x_3 \end{bmatrix}\right) = \begin{bmatrix} x_1 - 2x_3 \\ -3x_1 + 4x_2 \\ 0 \end{bmatrix}$

31. $T: \mathcal{R}^2 \to \mathcal{R}^4$ defined by $T\left(\begin{bmatrix} x_1 \\ x_2 \end{bmatrix}\right) = \begin{bmatrix} x_1 - x_2 \\ 0 \\ 3x_1 \\ x_2 \end{bmatrix}$

32. $T: \mathcal{R}^4 \to \mathcal{R}^3$ defined by $T\left(\begin{bmatrix} x_1 \\ x_2 \\ x_3 \\ x_4 \end{bmatrix}\right) = \begin{bmatrix} 2x_1 - x_2 + 3x_4 \\ -x_1 + 2x_4 \\ 3x_2 - x_3 \end{bmatrix}$

33. $T: \mathcal{R}^3 \to \mathcal{R}^3$ defined by $T(\mathbf{v}) = \mathbf{v}$ for all $\mathbf{v}$ in $\mathcal{R}^3$
34. $T: \mathcal{R}^3 \to \mathcal{R}^2$ defined by $T(\mathbf{v}) = \mathbf{0}$ for all $\mathbf{v}$ in $\mathcal{R}^3$

T&F *In Exercises 35–54, determine whether the statements are true or false.*

35. Every function from $\mathcal{R}^n$ to $\mathcal{R}^m$ has a standard matrix.
36. Every matrix transformation is linear.
37. A function from $\mathcal{R}^n$ to $\mathcal{R}^m$ that preserves scalar multiplication is linear.
38. The image of the zero vector under any linear transformation is the zero vector.
39. If $T: \mathcal{R}^3 \to \mathcal{R}^2$ is linear, then its standard matrix has size 3×2.
40. The zero transformation is linear.
41. A function is uniquely determined by the images of the standard vectors in its domain.
42. The first column of the standard matrix of a linear transformation is the image of the first standard vector under the transformation.
43. The domain of a function f is the set of all images $f(\mathbf{x})$.
44. The codomain of any function is contained in its range.
45. If f is a function and $f(\mathbf{u}) = f(\mathbf{v})$, then $\mathbf{u} = \mathbf{v}$.
46. The matrix transformation induced by a matrix A is a linear transformation.

47. Every linear transformation $T: \mathcal{R}^n \to \mathcal{R}^m$ is a matrix transformation.
48. Every linear transformation $T: \mathcal{R}^n \to \mathcal{R}^m$ is the matrix transformation induced by its standard matrix.
49. The projection of a vector in $\mathcal{R}^3$ on the xy-plane in $\mathcal{R}^3$ is the matrix transformation induced by $\begin{bmatrix} 1 & 0 & 0 \\ 0 & 1 & 0 \\ 0 & 0 & 0 \end{bmatrix}$.
50. Every linear transformation preserves linear combinations.
51. If $T(\mathbf{u}+\mathbf{v}) = T(\mathbf{u}) + T(\mathbf{v})$ for all vectors $\mathbf{u}$ and $\mathbf{v}$ in the domain of T, then T is said to preserve vector addition.
52. Every function $f: \mathcal{R}^n \to \mathcal{R}^m$ preserves scalar multiplication.
53. If $f: \mathcal{R}^n \to \mathcal{R}^m$ and $g: \mathcal{R}^n \to \mathcal{R}^m$ are functions such that $f(\mathbf{e}_i) = g(\mathbf{e}_i)$ for every standard vector $\mathbf{e}_i$, then $f(\mathbf{v}) = g(\mathbf{v})$ for every $\mathbf{v}$ in $\mathcal{R}^n$.
54. If T and U are linear transformations whose standard matrices are equal, then T and U are equal.

---

55. If T is the identity transformation, what is true about the domain and the codomain of T?
56. Suppose that T is linear and $T\left(\begin{bmatrix} 4 \\ -2 \end{bmatrix}\right) = \begin{bmatrix} -6 \\ 16 \end{bmatrix}$. Determine $T\left(\begin{bmatrix} -2 \\ 1 \end{bmatrix}\right)$ and $T\left(\begin{bmatrix} 8 \\ -4 \end{bmatrix}\right)$.
57. Suppose that T is linear and $T\left(\begin{bmatrix} 8 \\ 2 \end{bmatrix}\right) = \begin{bmatrix} 2 \\ -4 \\ 6 \end{bmatrix}$. Determine $T\left(\begin{bmatrix} 16 \\ 4 \end{bmatrix}\right)$ and $T\left(\begin{bmatrix} -4 \\ -1 \end{bmatrix}\right)$. Justify your answers.
58. Suppose that T is linear and $T\left(\begin{bmatrix} -2 \\ 6 \\ 4 \end{bmatrix}\right) = \begin{bmatrix} -4 \\ 2 \end{bmatrix}$. Determine $T\left(\begin{bmatrix} 1 \\ -3 \\ -2 \end{bmatrix}\right)$ and $T\left(\begin{bmatrix} -4 \\ 12 \\ 8 \end{bmatrix}\right)$.
59. Suppose that T is linear and $T\left(\begin{bmatrix} 3 \\ 6 \\ 9 \end{bmatrix}\right) = \begin{bmatrix} 12 \\ -9 \\ -3 \end{bmatrix}$. Determine $T\left(\begin{bmatrix} -4 \\ -8 \\ -12 \end{bmatrix}\right)$ and $T\left(\begin{bmatrix} 5 \\ 10 \\ 15 \end{bmatrix}\right)$.
60. Suppose that T is linear, such that $T\left(\begin{bmatrix} 2 \\ 0 \end{bmatrix}\right) = \begin{bmatrix} -4 \\ 6 \end{bmatrix}$ and $T\left(\begin{bmatrix} 0 \\ 3 \end{bmatrix}\right) = \begin{bmatrix} 9 \\ -6 \end{bmatrix}$. Determine $T\left(\begin{bmatrix} 1 \\ 2 \end{bmatrix}\right)$. Justify your answer.
61. Suppose that T is linear, such that $T\left(\begin{bmatrix} -3 \\ 0 \end{bmatrix}\right) = \begin{bmatrix} 6 \\ 3 \\ 9 \end{bmatrix}$ and $T\left(\begin{bmatrix} 0 \\ 4 \end{bmatrix}\right) = \begin{bmatrix} 8 \\ 0 \\ -4 \end{bmatrix}$. Determine $T\left(\begin{bmatrix} -2 \\ 6 \end{bmatrix}\right)$. Justify your answer.
62. Suppose that T is linear, such that $T\left(\begin{bmatrix} 1 \\ 2 \end{bmatrix}\right) = \begin{bmatrix} -2 \\ 0 \\ 1 \end{bmatrix}$ and $T\left(\begin{bmatrix} 0 \\ 3 \end{bmatrix}\right) = \begin{bmatrix} 6 \\ -3 \\ 3 \end{bmatrix}$. Determine $T\left(\begin{bmatrix} -3 \\ 3 \end{bmatrix}\right)$. Justify your answer.
63. Suppose that T is linear, such that $T\left(\begin{bmatrix} 2 \\ 3 \end{bmatrix}\right) = \begin{bmatrix} 1 \\ 2 \end{bmatrix}$ and $T\left(\begin{bmatrix} -4 \\ 0 \end{bmatrix}\right) = \begin{bmatrix} -5 \\ 1 \end{bmatrix}$. Determine $T\left(\begin{bmatrix} -2 \\ 3 \end{bmatrix}\right)$. Justify your answer.
64. Suppose that $T: \mathcal{R}^2 \to \mathcal{R}^2$ is a linear transformation such that $T(\mathbf{e}_1) = \begin{bmatrix} 2 \\ 3 \end{bmatrix}$ and $T(\mathbf{e}_2) = \begin{bmatrix} 4 \\ 1 \end{bmatrix}$. Determine $T\left(\begin{bmatrix} 5 \\ 6 \end{bmatrix}\right)$. Justify your answer.
65. Suppose that $T: \mathcal{R}^2 \to \mathcal{R}^2$ is a linear transformation such that $T(\mathbf{e}_1) = \begin{bmatrix} 2 \\ 3 \end{bmatrix}$ and $T(\mathbf{e}_2) = \begin{bmatrix} 4 \\ 1 \end{bmatrix}$. Determine $T\left(\begin{bmatrix} x_1 \\ x_2 \end{bmatrix}\right)$ for any $\begin{bmatrix} x_1 \\ x_2 \end{bmatrix}$ in $\mathcal{R}^2$. Justify your answer.
66. Suppose that $T: \mathcal{R}^2 \to \mathcal{R}^2$ is a linear transformation such that $T(\mathbf{e}_1) = \begin{bmatrix} 3 \\ -1 \end{bmatrix}$ and $T(\mathbf{e}_2) = \begin{bmatrix} -1 \\ 2 \end{bmatrix}$. Determine $T\left(\begin{bmatrix} x_1 \\ x_2 \end{bmatrix}\right)$ for any $\begin{bmatrix} x_1 \\ x_2 \end{bmatrix}$ in $\mathcal{R}^2$. Justify your answer.
67. Suppose that $T: \mathcal{R}^3 \to \mathcal{R}^3$ is a linear transformation such that

$$T(\mathbf{e}_1) = \begin{bmatrix} -1 \\ 0 \\ 2 \end{bmatrix}, \quad T(\mathbf{e}_2) = \begin{bmatrix} 3 \\ -1 \\ 0 \end{bmatrix}, \quad \text{and } T(\mathbf{e}_3) = \begin{bmatrix} 0 \\ -3 \\ 2 \end{bmatrix}.$$

Determine $T\left(\begin{bmatrix} x_1 \\ x_2 \\ x_3 \end{bmatrix}\right)$ for any $\begin{bmatrix} x_1 \\ x_2 \\ x_3 \end{bmatrix}$ in $\mathcal{R}^3$. Justify your answer.
68. Suppose that $T: \mathcal{R}^3 \to \mathcal{R}^2$ is a linear transformation such that

$$T(\mathbf{e}_1) = \begin{bmatrix} -2 \\ 1 \end{bmatrix}, \quad T(\mathbf{e}_2) = \begin{bmatrix} 0 \\ -3 \end{bmatrix}, \quad \text{and} \quad T(\mathbf{e}_3) = \begin{bmatrix} 2 \\ 4 \end{bmatrix}.$$

Determine $T\left(\begin{bmatrix} x_1 \\ x_2 \\ x_3 \end{bmatrix}\right)$ for any $\begin{bmatrix} x_1 \\ x_2 \\ x_3 \end{bmatrix}$ in $\mathcal{R}^3$. Justify your answer.
69. Suppose that $T: \mathcal{R}^2 \to \mathcal{R}^2$ is a linear transformation such that

$$T\left(\begin{bmatrix} 1 \\ -2 \end{bmatrix}\right) = \begin{bmatrix} 2 \\ 1 \end{bmatrix} \quad \text{and} \quad T\left(\begin{bmatrix} -1 \\ 3 \end{bmatrix}\right) = \begin{bmatrix} 3 \\ 0 \end{bmatrix}.$$

Determine $T\left(\begin{bmatrix} x_1 \\ x_2 \end{bmatrix}\right)$ for any $\begin{bmatrix} x_1 \\ x_2 \end{bmatrix}$ in $\mathcal{R}^2$. Justify your answer.

70. Suppose that $T: \mathcal{R}^2 \to \mathcal{R}^3$ is a linear transformation such that

$$T\left(\begin{bmatrix} 3 \\ -5 \end{bmatrix}\right) = \begin{bmatrix} 1 \\ -1 \\ 2 \end{bmatrix} \quad \text{and} \quad T\left(\begin{bmatrix} -1 \\ 2 \end{bmatrix}\right) = \begin{bmatrix} 3 \\ 0 \\ -2 \end{bmatrix}.$$

Determine $T\left(\begin{bmatrix} x_1 \\ x_2 \end{bmatrix}\right)$ for any $\begin{bmatrix} x_1 \\ x_2 \end{bmatrix}$ in $\mathcal{R}^2$. Justify your answer.

71. Suppose that $T: \mathcal{R}^3 \to \mathcal{R}^3$ is a linear transformation such that

$$T\left(\begin{bmatrix} -1 \\ 1 \\ 1 \end{bmatrix}\right) = \begin{bmatrix} 1 \\ 2 \\ 3 \end{bmatrix}, \quad T\left(\begin{bmatrix} 1 \\ -1 \\ 1 \end{bmatrix}\right) = \begin{bmatrix} -3 \\ 0 \\ 1 \end{bmatrix}, \quad \text{and}$$

$$T\left(\begin{bmatrix} 1 \\ 1 \\ -1 \end{bmatrix}\right) = \begin{bmatrix} 5 \\ 4 \\ 3 \end{bmatrix}.$$

Determine $T\left(\begin{bmatrix} x_1 \\ x_2 \\ x_3 \end{bmatrix}\right)$ for any $\begin{bmatrix} x_1 \\ x_2 \\ x_3 \end{bmatrix}$ in $\mathcal{R}^3$. Justify your answer.

In Exercises 72–80, a function $T: \mathcal{R}^n \to \mathcal{R}^m$ is given. Either prove that T is linear, or explain why T is not linear.

72. $T: \mathcal{R}^2 \to \mathcal{R}^2$ defined by $T\left(\begin{bmatrix} x_1 \\ x_2 \end{bmatrix}\right) = \begin{bmatrix} 2x_1 \\ x_2^2 \end{bmatrix}$

73. $T: \mathcal{R}^2 \to \mathcal{R}^2$ defined by $T\left(\begin{bmatrix} x_1 \\ x_2 \end{bmatrix}\right) = \begin{bmatrix} 0 \\ 2x_1 \end{bmatrix}$

74. $T: \mathcal{R}^2 \to \mathcal{R}^2$ defined by $T\left(\begin{bmatrix} x_1 \\ x_2 \end{bmatrix}\right) = \begin{bmatrix} 1 \\ 2x_1 \end{bmatrix}$

75. $T: \mathcal{R}^3 \to \mathcal{R}$ defined by $T\left(\begin{bmatrix} x_1 \\ x_2 \\ x_3 \end{bmatrix}\right) = x_1 + x_2 + x_3 - 1$

76. $T: \mathcal{R}^3 \to \mathcal{R}$ defined by $T\left(\begin{bmatrix} x_1 \\ x_2 \\ x_3 \end{bmatrix}\right) = x_1 + x_2 + x_3$

77. $T: \mathcal{R}^2 \to \mathcal{R}^2$ defined by $T\left(\begin{bmatrix} x_1 \\ x_2 \end{bmatrix}\right) = \begin{bmatrix} x_1 + x_2 \\ 2x_1 - x_2 \end{bmatrix}$

78. $T: \mathcal{R}^2 \to \mathcal{R}^2$ defined by $T\left(\begin{bmatrix} x_1 \\ x_2 \end{bmatrix}\right) = \begin{bmatrix} x_2 \\ |x_1| \end{bmatrix}$

79. $T: \mathcal{R} \to \mathcal{R}^2$ defined by $T(x) = \begin{bmatrix} \sin x \\ x \end{bmatrix}$

80. $T: \mathcal{R}^2 \to \mathcal{R}^2$ defined by $T\left(\begin{bmatrix} x_1 \\ x_2 \end{bmatrix}\right) = \begin{bmatrix} ax_1 \\ bx_2 \end{bmatrix}$, where a and b are scalars

81. Prove that the identity transformation $I: \mathcal{R}^n \to \mathcal{R}^n$ equals T_{I_n} and hence is linear.

82. Prove that the zero transformation $T_0: \mathcal{R}^n \to \mathcal{R}^m$ equals T_O and hence is linear.

Definitions Let $T,\ U: \mathcal{R}^n \to \mathcal{R}^m$ be functions and c be a scalar. Define $(T + U): \mathcal{R}^n \to \mathcal{R}^m$ and $cT: \mathcal{R}^n \to \mathcal{R}^m$ by

$$(T + U)(\mathbf{x}) = T(\mathbf{x}) + U(\mathbf{x}) \quad \text{and} \quad (cT)(\mathbf{x}) = cT(\mathbf{x})$$

for all $\mathbf{x}$ in $\mathcal{R}^n$.

The preceding definitions are used in Exercises 83–86:

83. Prove that if T is linear and c is a scalar, then cT is linear.

84. Prove that if T and U are linear, then $T + U$ is linear.

85. Suppose that c is a scalar. Use Exercise 83 to prove that if T is linear and has standard matrix A, then the standard matrix of cT is cA.

86. Use Exercise 84 to prove that if T and U are linear with standard matrices A and B, respectively, then the standard matrix of $T + U$ is $A + B$.

87. Let $T: \mathcal{R}^2 \to \mathcal{R}^2$ be a linear transformation. Prove that there exist unique scalars a, b, c, and d such that $T\left(\begin{bmatrix} x_1 \\ x_2 \end{bmatrix}\right) = \begin{bmatrix} ax_1 + bx_2 \\ cx_1 + dx_2 \end{bmatrix}$ for every vector $\begin{bmatrix} x_1 \\ x_2 \end{bmatrix}$ in $\mathcal{R}^2$.
Hint: Use Theorem 2.9.

88. State and prove a generalization of Exercise 87.

89. Define $T: \mathcal{R}^2 \to \mathcal{R}^2$ by $T\left(\begin{bmatrix} x_1 \\ x_2 \end{bmatrix}\right) = \begin{bmatrix} x_1 \\ 0 \end{bmatrix}$. T represents the *orthogonal projection of $\mathcal{R}^2$ on the x-axis.*
    (a) Prove that T is linear.
    (b) Find the standard matrix of T.
    (c) Prove that $T(T(\mathbf{v})) = T(\mathbf{v})$ for every $\mathbf{v}$ in $\mathcal{R}^2$.

90. Define $T: \mathcal{R}^3 \to \mathcal{R}^3$ by $T\left(\begin{bmatrix} x_1 \\ x_2 \\ x_3 \end{bmatrix}\right) = \begin{bmatrix} 0 \\ x_2 \\ x_3 \end{bmatrix}$. T represents the *orthogonal projection of $\mathcal{R}^3$ on the yz-plane.*
    (a) Prove that T is linear.
    (b) Find the standard matrix of T.
    (c) Prove that $T(T(\mathbf{v})) = T(\mathbf{v})$ for every $\mathbf{v}$ in $\mathcal{R}^3$.

91. Define the linear transformation $T: \mathcal{R}^2 \to \mathcal{R}^2$ by $T\left(\begin{bmatrix} x_1 \\ x_2 \end{bmatrix}\right) = \begin{bmatrix} -x_1 \\ x_2 \end{bmatrix}$. T represents the *reflection of $\mathcal{R}^2$ about the y-axis* the *reflection of $\mathcal{R}^2$ about the y-axis.*
    (a) Show that T is a matrix transformation.
    (b) Determine the range of T.

92. Define the linear transformation $T: \mathcal{R}^3 \to \mathcal{R}^3$ by $T\left(\begin{bmatrix} x_1 \\ x_2 \\ x_3 \end{bmatrix}\right) = \begin{bmatrix} x_1 \\ x_2 \\ -x_3 \end{bmatrix}$. T represents the *reflection of $\mathcal{R}^3$ about the xy-plane.*
    (a) Show that T is a matrix transformation.
    (b) Determine the range of T.

93. A linear transformation $T: \mathcal{R}^n \to \mathcal{R}^n$ defined by $T(\mathbf{x}) = k\mathbf{x}$, where $0 < k < 1$, is called a *contraction.*
    (a) Show that T is a matrix transformation.
    (b) Determine the range of T.

94. A linear transformation $T: \mathcal{R}^n \to \mathcal{R}^n$ defined by $T(\mathbf{x}) = k\mathbf{x}$, where $k > 1$, is called a *dilation*.

    (a) Show that T is a matrix transformation.

    (b) Determine the range of T.

95. Let $T: \mathcal{R}^n \to \mathcal{R}^m$ be a linear transformation. Prove that $T(\mathbf{u}) = T(\mathbf{v})$ if and only if $T(\mathbf{u} - \mathbf{v}) = \mathbf{0}$.

96. Find functions $f: \mathcal{R}^2 \to \mathcal{R}^2$ and $g: \mathcal{R}^2 \to \mathcal{R}^2$ such that $f(\mathbf{e}_1) = g(\mathbf{e}_1)$ and $f(\mathbf{e}_2) = g(\mathbf{e}_2)$, but $f(\mathbf{v}) \neq g(\mathbf{v})$ for some $\mathbf{v}$ in $\mathcal{R}^2$.

97. Let A be an invertible $n \times n$ matrix. Determine $T_{A^{-1}}(T_A(\mathbf{v}))$ and $T_A(T_{A^{-1}}(\mathbf{v}))$ for all $\mathbf{v}$ in $\mathcal{R}^n$.

98. Let A be an $m \times n$ matrix and B an $n \times p$ matrix. Prove that $T_{AB}(\mathbf{v}) = T_A(T_B(\mathbf{v}))$ for all $\mathbf{v}$ in $\mathcal{R}^p$.

99. Let $T: \mathcal{R}^n \to \mathcal{R}^m$ be a linear transformation with standard matrix A. Prove that the columns of A form a generating set for the range of T.

100. For a linear transformation $T: \mathcal{R}^n \to \mathcal{R}^m$, prove that its range is $\mathcal{R}^m$ if and only if the rank of its standard matrix is m.

101. Let $T: \mathcal{R}^n \to \mathcal{R}^m$ be a linear transformation and $\mathcal{S} = \{\mathbf{v}_1, \mathbf{v}_2, \ldots, \mathbf{v}_k\}$ be a subset of $\mathcal{R}^n$. Prove that if the set $\{T(\mathbf{v}_1), T(\mathbf{v}_2), \ldots, T(\mathbf{v}_k)\}$ is a linearly independent subset of $\mathcal{R}^m$, then $\mathcal{S}$ is a linearly independent subset of $\mathcal{R}^n$.

In Exercises 102 and 103, use either a calculator with matrix capabilities or computer software such as MATLAB to solve each problem.

102. Suppose that $T: \mathcal{R}^4 \to \mathcal{R}^4$ is a linear transformation such that

$$T\left(\begin{bmatrix}1\\2\\0\\-1\end{bmatrix}\right) = \begin{bmatrix}0\\1\\1\\0\end{bmatrix}, T\left(\begin{bmatrix}1\\1\\1\\-1\end{bmatrix}\right) = \begin{bmatrix}-2\\1\\3\\2\end{bmatrix},$$

$$T\left(\begin{bmatrix}0\\1\\0\\1\end{bmatrix}\right) = \begin{bmatrix}4\\6\\0\\-3\end{bmatrix},$$

and

$$T\left(\begin{bmatrix}-1\\2\\-3\\1\end{bmatrix}\right) = \begin{bmatrix}0\\0\\0\\0\end{bmatrix}.$$

    (a) Find a rule for T.

    (b) Is T uniquely determined by these four images? Why or why not?

103. Suppose that $T: \mathcal{R}^4 \to \mathcal{R}^4$ is the linear transformation defined by the rule

$$T\left(\begin{bmatrix}x_1\\x_2\\x_3\\x_4\end{bmatrix}\right) = \begin{bmatrix}x_1 + x_2 + x_3 + 2x_4\\x_1 + 2x_2 - 3x_3 + 4x_4\\x_2 + 2x_4\\x_1 + 5x_2 - x_3\end{bmatrix}.$$

Determine if the vector $\begin{bmatrix}2\\-1\\0\\3\end{bmatrix}$ is in the range of T.

SOLUTIONS TO THE PRACTICE PROBLEMS

1. (a) Because A is a 3×2 matrix, the domain of T_A is $\mathcal{R}^2$.

   (b) The codomain of T_A is $\mathcal{R}^3$.

   (c) We have

$$T_A\left(\begin{bmatrix}4\\3\end{bmatrix}\right) = A\begin{bmatrix}4\\3\end{bmatrix} = \begin{bmatrix}1&-2\\3&1\\-1&4\end{bmatrix}\begin{bmatrix}4\\3\end{bmatrix} = \begin{bmatrix}-2\\15\\8\end{bmatrix}.$$

2. Since $\mathbf{e}_1 = (-1)\begin{bmatrix}-1\\0\end{bmatrix}$ and $\mathbf{e}_2 = \frac{1}{2}\begin{bmatrix}0\\2\end{bmatrix}$, it follows that

$$\begin{aligned}T\left(\begin{bmatrix}x_1\\x_2\end{bmatrix}\right) &= T(x_1\mathbf{e}_1 + x_2\mathbf{e}_2)\\ &= T\left(x_1(-1)\begin{bmatrix}-1\\0\end{bmatrix} + x_2\left(\frac{1}{2}\right)\begin{bmatrix}0\\2\end{bmatrix}\right)\\ &= -x_1T\left(\begin{bmatrix}-1\\0\end{bmatrix}\right) + \frac{1}{2}x_2T\left(\begin{bmatrix}0\\2\end{bmatrix}\right)\\ &= -x_1\begin{bmatrix}-2\\1\\3\end{bmatrix} + \frac{1}{2}x_2\begin{bmatrix}2\\4\\-2\end{bmatrix}\\ &= \begin{bmatrix}2x_1 + x_2\\-x_1 + 2x_2\\-3x_1 - x_2\end{bmatrix}.\end{aligned}$$

3. Since

$$T(\mathbf{e}_1) = \begin{bmatrix}2\\0\end{bmatrix}, \quad T(\mathbf{e}_2) = \begin{bmatrix}0\\-3\end{bmatrix}, \quad \text{and}$$

$$T(\mathbf{e}_3) = \begin{bmatrix}-5\\4\end{bmatrix},$$

the standard matrix of T is

$$\begin{bmatrix}2&0&-5\\0&-3&4\end{bmatrix}.$$

2.8 COMPOSITION AND INVERTIBILITY OF LINEAR TRANSFORMATIONS

In this section, we use the standard matrix to study some basic properties of a linear transformation. We begin by determining whether a transformation is onto and one-to-one, which is closely related to the existence and uniqueness of solutions of systems of linear equations.

ONTO AND ONE-TO-ONE FUNCTIONS

Once we have found the standard matrix of a linear transformation T, we can use it to find a generating set for the range of T. For example, suppose that $T\colon \mathcal{R}^3 \to \mathcal{R}^2$ is defined by

$$T\left(\begin{bmatrix} x_1 \\ x_2 \\ x_3 \end{bmatrix}\right) = \begin{bmatrix} 3x_1 - 4x_2 \\ 2x_1 + x_3 \end{bmatrix}.$$

In Example 7 of Section 2.7, we saw that the standard matrix of T is

$$A = [T(\mathbf{e}_1)\ T(\mathbf{e}_2)\ T(\mathbf{e}_3)] = \begin{bmatrix} 3 & -4 & 0 \\ 2 & 0 & 1 \end{bmatrix}.$$

Now $\mathbf{w}$ is in the range of T if and only if $\mathbf{w} = T(\mathbf{v})$ for some $\mathbf{v}$ in $\mathcal{R}^2$. Writing $\mathbf{v} = v_1\mathbf{e}_1 + v_2\mathbf{e}_2 + v_3\mathbf{e}_3$, we see that

$$\mathbf{w} = T(\mathbf{v}) = T(v_1\mathbf{e}_1 + v_2\mathbf{e}_2 + v_3\mathbf{e}_3) = v_1T(\mathbf{e}_1) + v_2T(\mathbf{e}_2) + v_3T(\mathbf{e}_3),$$

which is a linear combination of the columns of A. Likewise, it is clear from the same computation that every linear combination of the columns of A is in the range of T. We conclude that the range of T equals the span of

$$\left\{\begin{bmatrix} 3 \\ 2 \end{bmatrix}, \begin{bmatrix} -4 \\ 0 \end{bmatrix}, \begin{bmatrix} 0 \\ 1 \end{bmatrix}\right\}.$$

This argument can be generalized to prove the following result:

> The range of a linear transformation equals the span of the columns of its standard matrix.

In what follows, we obtain some additional properties about a linear transformation from its standard matrix. First, however, we recall some properties of functions.

Definition A function $f\colon \mathcal{R}^n \to \mathcal{R}^m$ is said to be **onto** if its range is all of $\mathcal{R}^m$; that is, if every vector in $\mathcal{R}^m$ is an image.

From the preceding boxed statement, it follows that a linear transformation is onto if and only if the columns of its standard matrix form a generating set for its codomain. Thus we can illustrate that the reflection of $\mathcal{R}^2$ about the x-axis in Example 8 of Section 2.7 is onto by showing that the columns $\mathbf{e}_1, -\mathbf{e}_2$ of its standard matrix form a generating set for $\mathcal{R}^2$. Because $\mathbf{e}_1$ and $-\mathbf{e}_2$ are nonparallel vectors in

$\mathcal{R}^2$, every vector in $\mathcal{R}^2$ is a linear combination of $\mathbf{e}_1$ and $-\mathbf{e}_2$. So the columns of the standard matrix of U form a generating set for $\mathcal{R}^2$.

We can also use the rank of the standard matrix A to determine if T is onto. If A is an $m \times n$ matrix, then, by Theorem 1.6, the columns of A form a generating set for $\mathcal{R}^m$ if and only if rank $A = m$.

Example 1 Determine if the linear transformation $T: \mathcal{R}^3 \to \mathcal{R}^3$ defined by

$$T\left(\begin{bmatrix} x_1 \\ x_2 \\ x_3 \end{bmatrix}\right) = \begin{bmatrix} x_1 + 2x_2 + 4x_3 \\ x_1 + 3x_2 + 6x_3 \\ 2x_1 + 5x_2 + 10x_3 \end{bmatrix}$$

is onto.

Solution For T to be onto, the rank of its standard matrix must equal 3. But the standard matrix A of T and its reduced row echelon form R are

$$A = \begin{bmatrix} 1 & 2 & 4 \\ 1 & 3 & 6 \\ 2 & 5 & 10 \end{bmatrix} \quad \text{and} \quad R = \begin{bmatrix} 1 & 0 & 0 \\ 0 & 1 & 2 \\ 0 & 0 & 0 \end{bmatrix}.$$

So rank $A = 2 \neq 3$, and therefore T is not onto.

We now state the first of several theorems that relate a linear transformation to its standard matrix. Its proof follows from our preceding observations and from Theorem 1.6.

THEOREM 2.10

Let $T: \mathcal{R}^n \to \mathcal{R}^m$ be a linear transformation with standard matrix A. The following conditions are equivalent:

(a) T is onto; that is, the range of T is $\mathcal{R}^m$.

(b) The columns of A form a generating set for $\mathcal{R}^m$.

(c) rank $A = m$.

There is a close relationship between the range of a matrix transformation and the consistency of a system of linear equations. For example, consider the system

$$\begin{aligned} x_1 \quad\quad &= 1 \\ 2x_1 + x_2 &= 3 \\ x_1 - x_2 &= 1. \end{aligned}$$

The system is equivalent to the matrix equation $A\mathbf{x} = \mathbf{b}$, where

$$A = \begin{bmatrix} 1 & 0 \\ 2 & 1 \\ 1 & -1 \end{bmatrix}, \quad \mathbf{x} = \begin{bmatrix} x_1 \\ x_2 \end{bmatrix}, \quad \text{and} \quad \mathbf{b} = \begin{bmatrix} 1 \\ 3 \\ 1 \end{bmatrix}.$$

Because we can write the matrix equation as $T_A(\mathbf{x}) = \mathbf{b}$, the system has a solution if and only if $\mathbf{b}$ is in the range of T_A.

Now suppose that we define $T: \mathcal{R}^3 \to \mathcal{R}^2$ by

$$T\left(\begin{bmatrix} x_1 \\ x_2 \\ x_3 \end{bmatrix}\right) = \begin{bmatrix} x_1 + \ \ x_2 + x_3 \\ x_1 + 3x_2 - x_3 \end{bmatrix} \quad \text{and} \quad \mathbf{w} = \begin{bmatrix} 2 \\ 8 \end{bmatrix}.$$

It is easy to see that $T = T_B$,where $B = \begin{bmatrix} 1 & 1 & 1 \\ 1 & 3 & -1 \end{bmatrix}$; so T is linear. Suppose that we want to determine if $\mathbf{w}$ is in the range of T. This question is equivalent to asking if there exists a vector $\mathbf{x}$ such that $T(\mathbf{x}) = \mathbf{w}$, or, in other words, if the following system is consistent:

$$\begin{aligned} x_1 + \ \ x_2 + x_3 &= 2 \\ x_1 + 3x_2 - x_3 &= 8 \end{aligned}$$

Using Gaussian elimination, we obtain the general solution

$$\begin{aligned} x_1 &= -2x_3 - 1 \\ x_2 &= \quad\ \ x_3 + 3 \\ x_3 & \quad \text{free.} \end{aligned}$$

We conclude that the system is consistent and there are infinitely many vectors whose image is $\mathbf{w}$. For example, for the cases $x_3 = 0$ and $x_3 = 1$, we obtain the vectors

$$\begin{bmatrix} -1 \\ 3 \\ 0 \end{bmatrix} \quad \text{and} \quad \begin{bmatrix} -3 \\ 4 \\ 1 \end{bmatrix},$$

respectively. Alternatively, we could have observed that A has rank 2 and then appealed to Theorem 2.10 to conclude that every vector in $\mathcal{R}^2$, including $\mathbf{w}$, is in the range of T.

Another important property of a function is that of being *one-to-one.*

Definition A function $f: \mathcal{R}^n \to \mathcal{R}^m$ is said to be **one-to-one** if every pair of distinct vectors in $\mathcal{R}^n$ has distinct images. That is, if $\mathbf{u}$ and $\mathbf{v}$ are distinct vectors in $\mathcal{R}^n$, then $f(\mathbf{u})$ and $f(\mathbf{v})$ are distinct vectors in $\mathcal{R}^m$.

In Figure 2.14(a), we see that distinct vectors $\mathbf{u}$ and $\mathbf{v}$ have distinct images, which is necessary for f to be one-to-one. In Figure 2.14(b), f is not one-to-one because there exist distinct vectors $\mathbf{u}$ and $\mathbf{v}$ that have the same image $\mathbf{w}$.

Suppose that $T: \mathcal{R}^n \to \mathcal{R}^m$ is a one-to-one linear transformation. If $\mathbf{w}$ is a nonzero vector in $\mathcal{R}^n$, then $T(\mathbf{w}) \neq T(\mathbf{0}) = \mathbf{0}$, and hence $\mathbf{0}$ is the only vector in $\mathcal{R}^n$ whose image under T is the zero vector of $\mathcal{R}^m$.

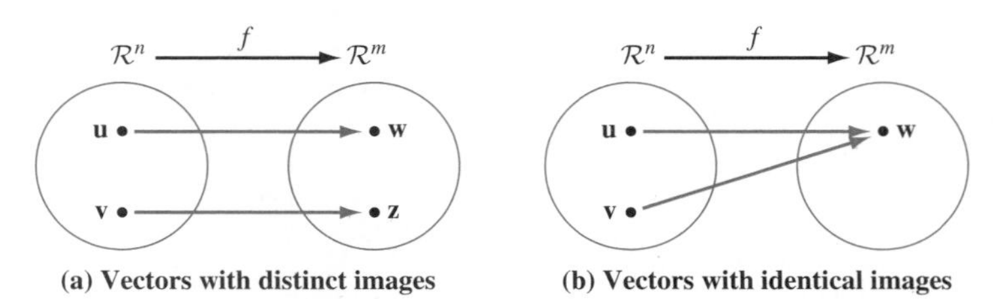

(a) Vectors with distinct images **(b) Vectors with identical images**

Figure 2.14

Conversely, if $\mathbf{0}$ is the only vector whose image under T is the zero vector, then T must be one-to-one. For suppose that $\mathbf{u}$ and $\mathbf{v}$ are vectors with $T(\mathbf{u}) = T(\mathbf{v})$. Then $T(\mathbf{u}-\mathbf{v}) = T(\mathbf{u}) - T(\mathbf{v}) = \mathbf{0}$. So $\mathbf{u}-\mathbf{v}=\mathbf{0}$, or $\mathbf{u}=\mathbf{v}$, and hence T is one-to-one.

Definition Let $T\colon \mathcal{R}^n \to \mathcal{R}^m$ be linear. The **null space** of T is the set of all $\mathbf{v}$ in $\mathcal{R}^n$ such that $T(\mathbf{v}) = \mathbf{0}$.

The discussion preceding the definition proves the following result:

> A linear transformation is one-to-one if and only if its null space contains only $\mathbf{0}$.

Note also that if A is the standard matrix of a linear transformation T, then the null space of T is the set of solutions of $A\mathbf{x} = \mathbf{0}$.

Example 2 Suppose that $T\colon \mathcal{R}^3 \to \mathcal{R}^2$ is defined by

$$T\left(\begin{bmatrix} x_1 \\ x_2 \\ x_3 \end{bmatrix}\right) = \begin{bmatrix} x_1 - x_2 + 2x_3 \\ -x_1 + x_2 - 3x_3 \end{bmatrix}.$$

Find a generating set for the null space of T.

Solution The standard matrix of T is

$$A = \begin{bmatrix} 1 & -1 & 2 \\ -1 & 1 & -3 \end{bmatrix}.$$

Because the null space of T is the set of solutions of $A\mathbf{x} = \mathbf{0}$, we must compute the reduced row echelon form of A, which is

$$\begin{bmatrix} 1 & -1 & 0 \\ 0 & 0 & 1 \end{bmatrix}.$$

This matrix corresponds to the system

$$\begin{aligned} x_1 - x_2 &= 0 \\ x_3 &= 0. \end{aligned}$$

So every solution has the form

$$\begin{bmatrix} x_1 \\ x_2 \\ x_3 \end{bmatrix} = \begin{bmatrix} x_2 \\ x_2 \\ 0 \end{bmatrix} = x_2 \begin{bmatrix} 1 \\ 1 \\ 0 \end{bmatrix}.$$

Thus a generating set for the null space of T is $\left\{\begin{bmatrix} 1 \\ 1 \\ 0 \end{bmatrix}\right\}$.

Using Theorem 1.8, we can give additional statements that are equivalent to the statement that a linear transformation is one-to-one.

THEOREM 2.11

Let $T\colon \mathcal{R}^n \to \mathcal{R}^m$ be a linear transformation with standard matrix A. Then the following statements are equivalent:

(a) T is one-to-one.

(b) The null space of T consists only of the zero vector.

(c) The columns of A are linearly independent.

(d) rank $A = n$.

Example 3 Determine whether the linear transformation $T\colon \mathcal{R}^3 \to \mathcal{R}^3$ in Example 1,

$$T\left(\begin{bmatrix} x_1 \\ x_2 \\ x_3 \end{bmatrix}\right) = \begin{bmatrix} x_1 + 2x_2 + 4x_3 \\ x_1 + 3x_2 + 6x_3 \\ 2x_1 + 5x_2 + 10x_3 \end{bmatrix},$$

is one-to-one.

Solution For T to be one-to-one, the rank of its standard matrix A must be 3. However, we saw in Example 1 that rank $A = 2$. Thus Theorem 2.11(d) does not hold, and so T is not one-to-one.

Finally, we relate all three topics: linear transformations, matrices, and systems of linear equations. Suppose that we begin with the system $A\mathbf{x} = \mathbf{b}$, where A is an $m \times n$ matrix, $\mathbf{x}$ is a vector in $\mathcal{R}^n$, and $\mathbf{b}$ is a vector in $\mathcal{R}^m$. We may write the system in the equivalent form $T_A(\mathbf{x}) = \mathbf{b}$. The following list compares the nature of solutions of $A\mathbf{x} = \mathbf{b}$ and properties of T_A:

(a) $A\mathbf{x} = \mathbf{b}$ has a solution if and only if $\mathbf{b}$ is in the range of T_A.

(b) $A\mathbf{x} = \mathbf{b}$ has a solution for every $\mathbf{b}$ if and only if T_A is onto.

(c) $A\mathbf{x} = \mathbf{b}$ has at most one solution for every $\mathbf{b}$ if and only if T_A is one-to-one.

Practice Problem 1 ▶ Let $T\colon \mathcal{R}^2 \to \mathcal{R}^3$ be the linear transformation defined by

$$T\left(\begin{bmatrix} x_1 \\ x_2 \end{bmatrix}\right) = \begin{bmatrix} 3x_1 - x_2 \\ -x_1 + 2x_2 \\ 2x_1 \end{bmatrix}.$$

(a) Determine a generating set for the range of T.

(b) Determine a generating set for the null space of T.

(c) Is T onto?

(d) Is T one-to-one? ◀

Example 4 Let

$$A = \begin{bmatrix} 0 & 0 & 1 & 3 & 3 \\ 2 & 3 & 1 & 5 & 2 \\ 4 & 6 & 1 & 7 & 2 \\ 4 & 6 & 1 & 7 & 1 \end{bmatrix}.$$

Is the system $A\mathbf{x} = \mathbf{b}$ consistent for every $\mathbf{b}$? If $A\mathbf{x} = \mathbf{b}_1$ is consistent for some $\mathbf{b}_1$, is the solution unique?

Solution Although these questions can be answered by computing the rank of A and using Theorems 1.6 and 1.8, we give here an alternative solution using the matrix transformation $T_A\colon \mathcal{R}^5 \to \mathcal{R}^4$, which we denote by T. First, note that the reduced row echelon form of A is

$$R = \begin{bmatrix} 1 & 1.5 & 0 & 1 & 0 \\ 0 & 0 & 1 & 3 & 0 \\ 0 & 0 & 0 & 0 & 1 \\ 0 & 0 & 0 & 0 & 0 \end{bmatrix}.$$

Because rank $A = 3 \neq 4$, we see that T is not onto by Theorem 2.10. So there exists a vector $\mathbf{b}_0$ in $\mathcal{R}^4$ that is not in the range of T. It follows that $A\mathbf{x} = \mathbf{b}_0$ is not consistent. Also, Theorem 2.11 shows T is not one-to-one, so there exists a nonzero solution $\mathbf{u}$ of $A\mathbf{x} = \mathbf{0}$. Therefore if for some $\mathbf{b}_1$ we have $A\mathbf{v} = \mathbf{b}_1$, then we also have

$$A(\mathbf{v} + \mathbf{u}) = A\mathbf{v} + A\mathbf{u} = \mathbf{b}_1 + \mathbf{0} = \mathbf{b}_1.$$

So the solution of $A\mathbf{x} = \mathbf{b}_1$ is *never* unique.

COMPOSITION OF LINEAR TRANSFORMATIONS

Recall that if $f\colon \mathcal{S}_1 \to \mathcal{S}_2$ and $g\colon \mathcal{S}_2 \to \mathcal{S}_3$, then the *composition* $g \circ f\colon \mathcal{S}_1 \to \mathcal{S}_3$ is defined by $(g \circ f)(\mathbf{u}) = g(f(\mathbf{u}))$ for all $\mathbf{u}$ in $\mathcal{S}_1$. (See Figure 2.15.)

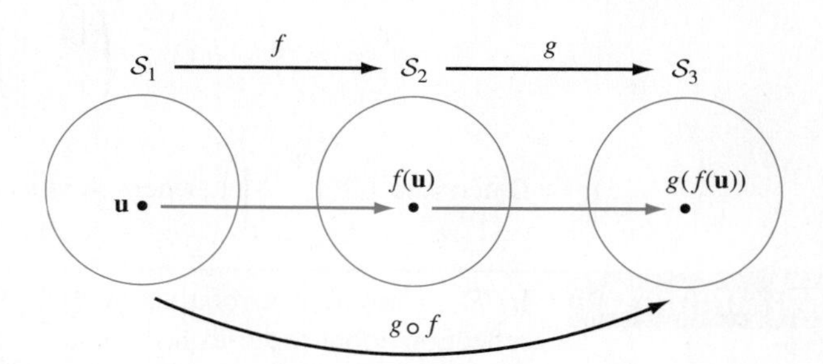

Figure 2.15 The composition of functions

Example 5 Suppose that $f\colon \mathcal{R}^2 \to \mathcal{R}^3$ and $g\colon \mathcal{R}^3 \to \mathcal{R}^2$ are the functions defined by

$$f\left(\begin{bmatrix} x_1 \\ x_2 \end{bmatrix}\right) = \begin{bmatrix} x_1^2 \\ x_1x_2 \\ x_1 + x_2 \end{bmatrix} \quad \text{and} \quad g\left(\begin{bmatrix} x_1 \\ x_2 \\ x_3 \end{bmatrix}\right) = \begin{bmatrix} x_1 - x_3 \\ 3x_2 \end{bmatrix}.$$

Then $g \circ f\colon \mathcal{R}^2 \to \mathcal{R}^2$ is defined by

$$(g \circ f)\left(\begin{bmatrix} x_1 \\ x_2 \end{bmatrix}\right) = g\left(f\left(\begin{bmatrix} x_1 \\ x_2 \end{bmatrix}\right)\right) = g\left(\begin{bmatrix} x_1^2 \\ x_1x_2 \\ x_1 + x_2 \end{bmatrix}\right) = \begin{bmatrix} x_1^2 - (x_1 + x_2) \\ 3x_1x_2 \end{bmatrix}.$$

In linear algebra, it is customary to drop the "circle" notation and write the composition of linear transformations $T: \mathcal{R}^n \to \mathcal{R}^m$ and $U: \mathcal{R}^m \to \mathcal{R}^p$ as UT rather than $U \circ T$. In this case, UT has domain $\mathcal{R}^n$ and codomain $\mathcal{R}^p$.

Suppose that we have an $m \times n$ matrix A and a $p \times m$ matrix B, so that BA is a $p \times n$ matrix. The corresponding matrix transformations are $T_A: \mathcal{R}^n \to \mathcal{R}^m$, $T_B: \mathcal{R}^m \to \mathcal{R}^p$, and $T_{BA}: \mathcal{R}^n \to \mathcal{R}^p$. For any $\mathbf{v}$ in $\mathcal{R}^n$, we have

$$T_{BA}(\mathbf{v}) = (BA)\mathbf{v} = B(A\mathbf{v}) = B(T_A(\mathbf{v})) = T_B(T_A(\mathbf{v})) = T_B T_A(\mathbf{v}).$$

It follows that the matrix transformation T_{BA} is the composition of T_B and T_A; that is,

$$T_B T_A = T_{BA}.$$

This result can be restated as the following theorem:

THEOREM 2.12

If $T: \mathcal{R}^n \to \mathcal{R}^m$ and $U: \mathcal{R}^m \to \mathcal{R}^p$ are linear transformations with standard matrices A and B, respectively, then the composition $UT: \mathcal{R}^n \to \mathcal{R}^p$ is also linear, and its standard matrix is BA.

PROOF Theorem 2.9 implies that $T = T_A$ and $U = T_B$. So, by our previous observation, we have $UT = T_B T_A = T_{BA}$, which is a matrix transformation and hence is linear. Furthermore, since $UT = T_{BA}$, the matrix BA is the standard matrix of UT. ■

Practice Problem 2 ► Let $U: \mathcal{R}^3 \to \mathcal{R}^2$ be the linear transformation defined by

$$U\left(\begin{bmatrix} x_1 \\ x_2 \\ x_3 \end{bmatrix}\right) = \begin{bmatrix} x_2 - 4x_3 \\ 2x_1 + 3x_3 \end{bmatrix}.$$

Determine $UT\left(\begin{bmatrix} x_1 \\ x_2 \end{bmatrix}\right)$, where T is as in Practice Problem 1. ◄

Example 6

In $\mathcal{R}^2$, show that a rotation by $180°$ followed by a reflection about the x-axis is a reflection about the y-axis.

Solution Let T and U denote the given rotation and reflection, respectively. We want to show that UT is a reflection about the y-axis. Let A and B be the standard matrices of T and U, respectively. Then $A = A_{180°}$, and B is computed in Example 8 of Section 2.7. So

$$A = \begin{bmatrix} -1 & 0 \\ 0 & -1 \end{bmatrix} \quad \text{and} \quad B = \begin{bmatrix} 1 & 0 \\ 0 & -1 \end{bmatrix}.$$

Then $T = T_A$, $U = T_B$, and $BA = \begin{bmatrix} -1 & 0 \\ 0 & 1 \end{bmatrix}$. Thus, for any $\mathbf{u} = \begin{bmatrix} u_1 \\ u_2 \end{bmatrix}$, we have

$$\begin{aligned} UT\left(\begin{bmatrix} u_1 \\ u_2 \end{bmatrix}\right) &= T_B T_A\left(\begin{bmatrix} u_1 \\ u_2 \end{bmatrix}\right) = T_{BA}\left(\begin{bmatrix} u_1 \\ u_2 \end{bmatrix}\right) \\ &= BA\begin{bmatrix} u_1 \\ u_2 \end{bmatrix} = \begin{bmatrix} -1 & 0 \\ 0 & 1 \end{bmatrix}\begin{bmatrix} u_1 \\ u_2 \end{bmatrix} = \begin{bmatrix} -u_1 \\ u_2 \end{bmatrix}, \end{aligned}$$

which represents a reflection about the y-axis. (See Figure 2.16.)

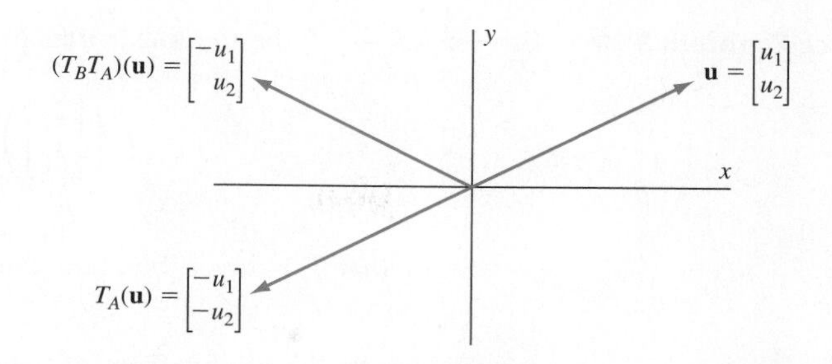

Figure 2.16 The rotation of **u** by 180° and its reflection about the *y*-axis

---

We can use the preceding result to obtain an interesting relationship between an invertible matrix and the corresponding matrix transformation. First, recall that a function $f\colon \mathcal{S}_1 \to \mathcal{S}_2$ is *invertible* if there exists a function $g\colon \mathcal{S}_2 \to \mathcal{S}_1$ such that $g(f(\mathbf{v})) = \mathbf{v}$ for all $\mathbf{v}$ in $\mathcal{S}_1$ and $f(g(\mathbf{v})) = \mathbf{v}$ for all $\mathbf{v}$ in $\mathcal{S}_2$. If f is invertible, then the function g is unique and is called the *inverse* of f; it is denoted by f^{-1}. (Note that the symbol f^{-1} should not be confused with the reciprocal $1/f$, which is usually undefined.) It can be shown that a function is invertible if and only if it is one-to-one and onto.

Now suppose that A is an $n \times n$ invertible matrix. Then for all $\mathbf{v}$ in $\mathcal{R}^n$, we have

$$T_A T_{A^{-1}}(\mathbf{v}) = T_A(T_{A^{-1}}(\mathbf{v})) = T_A(A^{-1}\mathbf{v}) = A(A^{-1}\mathbf{v}) = (AA^{-1})\mathbf{v} = I_n\mathbf{v} = \mathbf{v}.$$

Likewise, $T_{A^{-1}}T_A(\mathbf{v}) = \mathbf{v}$. We conclude that T_A is invertible and

$$T_A^{-1} = T_{A^{-1}}.$$

It is easy to see, for example, that if T_A represents rotation in $\mathcal{R}^2$ by θ, then T_A^{-1} represents rotation by $-\theta$. Using the previous result, we can show this also by computing the rotation matrix $A_{-\theta}$, as we did in Example 3 of Section 2.3.

If $T\colon \mathcal{R}^n \to \mathcal{R}^n$ is linear and invertible, then it is also one-to-one. By Theorem 2.11, its standard matrix A has rank n and hence is also invertible. Thus we have $T^{-1} = T_{A^{-1}}$. In particular, T^{-1} is a matrix transformation, and hence linear.

The following theorem summarizes this discussion:

THEOREM 2.13

Let $T\colon \mathcal{R}^n \to \mathcal{R}^n$ be a linear transformation with standard matrix A. Then T is invertible if and only if A is invertible, in which case $T^{-1} = T_{A^{-1}}$. Thus T^{-1} is linear, and its standard matrix is A^{-1}.

Example 7

Suppose that $A = \begin{bmatrix} 1 & 2 \\ 3 & 5 \end{bmatrix}$ so that $T_A\left(\begin{bmatrix} v_1 \\ v_2 \end{bmatrix}\right) = \begin{bmatrix} v_1 + 2v_2 \\ 3v_1 + 5v_2 \end{bmatrix}$. We saw in Example 1 of Section 2.3 that $A^{-1} = \begin{bmatrix} -5 & 2 \\ 3 & -1 \end{bmatrix}$. By Theorem 2.13, we have

$$T_A^{-1}\left(\begin{bmatrix} v_1 \\ v_2 \end{bmatrix}\right) = T_{A^{-1}}\left(\begin{bmatrix} v_1 \\ v_2 \end{bmatrix}\right) = \begin{bmatrix} -5 & 2 \\ 3 & -1 \end{bmatrix}\begin{bmatrix} v_1 \\ v_2 \end{bmatrix} = \begin{bmatrix} -5v_1 + 2v_2 \\ 3v_1 - v_2 \end{bmatrix}.$$

Practice Problem 3 ▶ Let $T\colon \mathcal{R}^2 \to \mathcal{R}^2$ be the linear transformation defined by

$$T\left(\begin{bmatrix} x_1 \\ x_2 \end{bmatrix}\right) = \begin{bmatrix} x_1 + 4x_2 \\ 2x_1 + 7x_2 \end{bmatrix}.$$

Show that T is invertible, and determine $T^{-1}\left(\begin{bmatrix} x_1 \\ x_2 \end{bmatrix}\right)$. ◀

We list the highlights from this section in the next table. Let $T\colon \mathcal{R}^n \to \mathcal{R}^m$ be a linear transformation with standard matrix A, which has size $m \times n$. *Properties listed in the same row of the table are equivalent.*

Property of T	The number of solutions of $A\mathbf{x} = \mathbf{b}$	Property of the columns of A	Property of the rank of A
T is onto.	$A\mathbf{x} = \mathbf{b}$ has at least one solution for every $\mathbf{b}$ in $\mathcal{R}^m$.	The columns of A are a generating set for $\mathcal{R}^m$.	rank $A = m$
T is one-to-one.	$A\mathbf{x} = \mathbf{b}$ has at most one solution for every $\mathbf{b}$ in $\mathcal{R}^m$.	The columns of A are linearly independent.	rank $A = n$
T is invertible.	$A\mathbf{x} = \mathbf{b}$ has a unique solution for every $\mathbf{b}$ in $\mathcal{R}^m$.	The columns of A are a linearly independent generating set for $\mathcal{R}^m$.	rank $A = m = n$

EXERCISES

In Exercises 1–12, find a generating set for the range of each linear transformation T.

1. $T\colon \mathcal{R}^2 \to \mathcal{R}^2$ defined by $T\left(\begin{bmatrix} x_1 \\ x_2 \end{bmatrix}\right) = \begin{bmatrix} 2x_1 + 3x_2 \\ 4x_1 + 5x_2 \end{bmatrix}$

2. $T\colon \mathcal{R}^2 \to \mathcal{R}^2$ defined by $T\left(\begin{bmatrix} x_1 \\ x_2 \end{bmatrix}\right) = \begin{bmatrix} x_2 \\ x_1 + x_2 \end{bmatrix}$

3. $T\colon \mathcal{R}^2 \to \mathcal{R}^3$ defined by $T\left(\begin{bmatrix} x_1 \\ x_2 \end{bmatrix}\right) = \begin{bmatrix} 3x_2 \\ 2x_1 - x_2 \\ x_1 + x_2 \end{bmatrix}$

4. $T\colon \mathcal{R}^3 \to \mathcal{R}^2$ defined by $T\left(\begin{bmatrix} x_1 \\ x_2 \\ x_3 \end{bmatrix}\right) = \begin{bmatrix} x_1 + x_2 + x_3 \\ 2x_1 \end{bmatrix}$

5. $T\colon \mathcal{R}^3 \to \mathcal{R}^3$ defined by
$$T\left(\begin{bmatrix} x_1 \\ x_2 \\ x_3 \end{bmatrix}\right) = \begin{bmatrix} 2x_1 + x_2 + x_3 \\ 2x_1 + 2x_2 + 3x_3 \\ 4x_1 + x_2 \end{bmatrix}$$

6. $T\colon \mathcal{R}^3 \to \mathcal{R}^3$ defined by
$$T\left(\begin{bmatrix} x_1 \\ x_2 \\ x_3 \end{bmatrix}\right) = \begin{bmatrix} 5x_1 - 4x_2 + x_3 \\ x_1 - 2x_2 \\ x_1 + x_3 \end{bmatrix}$$

7. $T\colon \mathcal{R}^2 \to \mathcal{R}^2$ defined by $T\left(\begin{bmatrix} x_1 \\ x_2 \end{bmatrix}\right) = \begin{bmatrix} x_1 \\ 0 \end{bmatrix}$

8. $T\colon \mathcal{R}^2 \to \mathcal{R}^4$ defined by $T\left(\begin{bmatrix} x_1 \\ x_2 \end{bmatrix}\right) = \begin{bmatrix} x_1 - 4x_2 \\ 2x_1 - 3x_2 \\ 0 \\ x_2 \end{bmatrix}$

9. $T\colon \mathcal{R}^3 \to \mathcal{R}^3$ defined by $T\left(\begin{bmatrix} x_1 \\ x_2 \\ x_3 \end{bmatrix}\right) = \begin{bmatrix} x_1 \\ x_2 \\ 0 \end{bmatrix}$

10. $T\colon \mathcal{R}^3 \to \mathcal{R}^3$ defined by $T(\mathbf{v}) = \mathbf{v}$ for all $\mathbf{v}$ in $\mathcal{R}^3$

11. $T\colon \mathcal{R}^3 \to \mathcal{R}^2$ defined by $T(\mathbf{v}) = \mathbf{0}$ for all $\mathbf{v}$ in $\mathcal{R}^3$

12. $T\colon \mathcal{R}^3 \to \mathcal{R}^3$ defined by $T(\mathbf{v}) = 4\mathbf{v}$ for all $\mathbf{v}$ in $\mathcal{R}^3$

In Exercises 13–23, find a generating set for the null space of each linear transformation T, and use your answer to determine whether T is one-to-one.

13. $T\colon \mathcal{R}^2 \to \mathcal{R}^2$ defined by $T\left(\begin{bmatrix} x_1 \\ x_2 \end{bmatrix}\right) = \begin{bmatrix} x_2 \\ x_1 + x_2 \end{bmatrix}$

14. $T\colon \mathcal{R}^2 \to \mathcal{R}^2$ defined by $T\left(\begin{bmatrix} x_1 \\ x_2 \end{bmatrix}\right) = \begin{bmatrix} 2x_1 + 3x_2 \\ 4x_1 + 5x_2 \end{bmatrix}$

15. $T\colon \mathcal{R}^3 \to \mathcal{R}^2$ defined by $T\left(\begin{bmatrix} x_1 \\ x_2 \\ x_3 \end{bmatrix}\right) = \begin{bmatrix} x_1 + x_2 + x_3 \\ 2x_1 \end{bmatrix}$

16. $T\colon \mathcal{R}^2 \to \mathcal{R}^3$ defined by $T\left(\begin{bmatrix} x_1 \\ x_2 \end{bmatrix}\right) = \begin{bmatrix} 3x_2 \\ 2x_1 - x_2 \\ x_1 + x_2 \end{bmatrix}$

17. $T\colon \mathcal{R}^3 \to \mathcal{R}^3$ defined by
$$T\left(\begin{bmatrix} x_1 \\ x_2 \\ x_3 \end{bmatrix}\right) = \begin{bmatrix} x_1 + 2x_2 + x_3 \\ x_1 + 3x_2 + 2x_3 \\ 2x_1 + 5x_2 + 3x_3 \end{bmatrix}$$

18. $T\colon \mathcal{R}^3 \to \mathcal{R}^3$ defined by $T\left(\begin{bmatrix} x_1 \\ x_2 \\ x_3 \end{bmatrix}\right) = \begin{bmatrix} 2x_1 + 3x_2 \\ x_1 - x_3 \\ x_1 + x_2 + 4x_3 \end{bmatrix}$

19. $T\colon \mathcal{R}^3 \to \mathcal{R}^3$ defined by $T(\mathbf{v}) = \mathbf{v}$ for all $\mathbf{v}$ in $\mathcal{R}^3$

20. $T\colon \mathcal{R}^3 \to \mathcal{R}^2$ defined by $T(\mathbf{v}) = \mathbf{0}$ for all $\mathbf{v}$ in $\mathcal{R}^3$

21. $T\colon \mathcal{R}^2 \to \mathcal{R}^2$ defined by $T\left(\begin{bmatrix} x_1 \\ x_2 \end{bmatrix}\right) = \begin{bmatrix} x_1 \\ 0 \end{bmatrix}$

22. $T\colon \mathcal{R}^3 \to \mathcal{R}$ defined by $T\left(\begin{bmatrix} x_1 \\ x_2 \\ x_3 \end{bmatrix}\right) = x_1 + 2x_3$

23. $T\colon \mathcal{R}^4 \to \mathcal{R}^3$ defined by
$$T\left(\begin{bmatrix} x_1 \\ x_2 \\ x_3 \\ x_4 \end{bmatrix}\right) = \begin{bmatrix} 2x_1 + x_2 + x_3 - x_4 \\ x_1 + x_2 + 2x_3 + 2x_4 \\ x_1 - x_3 - 3x_4 \end{bmatrix}$$

In Exercises 24–31, find the standard matrix of each linear transformation T, and use it to determine whether T is one-to-one.

24. $T\colon \mathcal{R}^2 \to \mathcal{R}^2$ defined by $T\left(\begin{bmatrix} x_1 \\ x_2 \end{bmatrix}\right) = \begin{bmatrix} x_2 \\ x_1 + x_2 \end{bmatrix}$

25. $T\colon \mathcal{R}^2 \to \mathcal{R}^2$ defined by $T\left(\begin{bmatrix} x_1 \\ x_2 \end{bmatrix}\right) = \begin{bmatrix} 2x_1 + 3x_2 \\ 4x_1 + 5x_2 \end{bmatrix}$

26. $T\colon \mathcal{R}^3 \to \mathcal{R}^2$ defined by $T\left(\begin{bmatrix} x_1 \\ x_2 \\ x_3 \end{bmatrix}\right) = \begin{bmatrix} x_1 + x_2 + x_3 \\ 2x_1 \end{bmatrix}$

27. $T\colon \mathcal{R}^2 \to \mathcal{R}^3$ defined by $T\left(\begin{bmatrix} x_1 \\ x_2 \end{bmatrix}\right) = \begin{bmatrix} 3x_2 \\ 2x_1 - x_2 \\ x_1 + x_2 \end{bmatrix}$

28. $T\colon \mathcal{R}^3 \to \mathcal{R}^3$ defined by $T\left(\begin{bmatrix} x_1 \\ x_2 \\ x_3 \end{bmatrix}\right) = \begin{bmatrix} x_1 - 2x_3 \\ -3x_1 + 4x_2 \\ 0 \end{bmatrix}$

29. $T\colon \mathcal{R}^3 \to \mathcal{R}^3$ defined by $T\left(\begin{bmatrix} x_1 \\ x_2 \\ x_3 \end{bmatrix}\right) = \begin{bmatrix} x_1 - x_2 \\ x_2 - x_3 \\ x_1 - x_3 \end{bmatrix}$

30. $T\colon \mathcal{R}^4 \to \mathcal{R}^4$ defined by
$$T\left(\begin{bmatrix} x_1 \\ x_2 \\ x_3 \\ x_4 \end{bmatrix}\right) = \begin{bmatrix} x_1 - x_2 + x_3 + x_4 \\ -2x_1 + x_2 - x_3 - x_4 \\ 2x_1 + 3x_2 - 6x_3 + 5x_4 \\ -x_1 + 2x_2 - x_3 - 5x_4 \end{bmatrix}$$

31. $T\colon \mathcal{R}^5 \to \mathcal{R}^4$ defined by
$$T\left(\begin{bmatrix} x_1 \\ x_2 \\ x_3 \\ x_4 \\ x_5 \end{bmatrix}\right) = \begin{bmatrix} x_1 + 2x_2 + 2x_3 + x_4 + 8x_5 \\ x_1 + 2x_2 + x_3 + 6x_5 \\ x_1 + x_2 + x_3 + 2x_4 + 5x_5 \\ 3x_1 + 2x_2 + 5x_4 + 8x_5 \end{bmatrix}$$

In Exercises 32–40, find the standard matrix of each linear transformation T, and use it to determine whether T is onto.

32. $T\colon \mathcal{R}^2 \to \mathcal{R}^2$ defined by $T\left(\begin{bmatrix} x_1 \\ x_2 \end{bmatrix}\right) = \begin{bmatrix} x_2 \\ x_1 + x_2 \end{bmatrix}$

33. $T\colon \mathcal{R}^2 \to \mathcal{R}^2$ defined by $T\left(\begin{bmatrix} x_1 \\ x_2 \end{bmatrix}\right) = \begin{bmatrix} 2x_1 + 3x_2 \\ 4x_1 + 5x_2 \end{bmatrix}$

34. $T\colon \mathcal{R}^3 \to \mathcal{R}^2$ defined by $T\left(\begin{bmatrix} x_1 \\ x_2 \\ x_3 \end{bmatrix}\right) = \begin{bmatrix} x_1 + x_2 + x_3 \\ 2x_1 \end{bmatrix}$

35. $T\colon \mathcal{R}^2 \to \mathcal{R}^3$ defined by $T\left(\begin{bmatrix} x_1 \\ x_2 \end{bmatrix}\right) = \begin{bmatrix} 3x_2 \\ 2x_1 - x_2 \\ x_1 + x_2 \end{bmatrix}$

36. $T\colon \mathcal{R}^3 \to \mathcal{R}$ defined by $T\left(\begin{bmatrix} x_1 \\ x_2 \\ x_3 \end{bmatrix}\right) = 2x_1 - 5x_2 + 4x_3$

37. $T\colon \mathcal{R}^3 \to \mathcal{R}^3$ defined by
$$T\left(\begin{bmatrix} x_1 \\ x_2 \\ x_3 \end{bmatrix}\right) = \begin{bmatrix} x_2 - 2x_3 \\ x_1 - x_3 \\ -x_1 + 2x_2 - 3x_3 \end{bmatrix}$$

38. $T\colon \mathcal{R}^4 \to \mathcal{R}^4$ defined by
$$T\left(\begin{bmatrix} x_1 \\ x_2 \\ x_3 \\ x_4 \end{bmatrix}\right) = \begin{bmatrix} x_1 - x_2 + 2x_3 \\ -2x_1 + x_2 - 7x_3 \\ x_1 - x_2 + 2x_3 \\ -x_1 + 2x_2 + x_3 \end{bmatrix}$$

39. $T\colon \mathcal{R}^4 \to \mathcal{R}^4$ defined by
$$T\left(\begin{bmatrix} x_1 \\ x_2 \\ x_3 \\ x_4 \end{bmatrix}\right) = \begin{bmatrix} x_1 - 2x_2 + 2x_3 - x_4 \\ -x_1 + x_2 + 3x_3 + 2x_4 \\ x_1 - x_2 - 6x_3 - x_4 \\ x_1 - 2x_2 + 5x_3 - 5x_4 \end{bmatrix}$$

40. $T\colon \mathcal{R}^4 \to \mathcal{R}^4$ defined by
$$T\left(\begin{bmatrix} x_1 \\ x_2 \\ x_3 \\ x_4 \end{bmatrix}\right) = \begin{bmatrix} x_1 + 2x_2 + 2x_3 + x_4 \\ x_1 + 2x_2 + x_3 \\ x_1 + x_2 + x_3 + 2x_4 \\ 3x_1 + 2x_2 + 5x_4 \end{bmatrix}$$

T&F *In Exercises 41–60, determine whether the statements are true or false.*

41. A linear transformation with codomain $\mathcal{R}^m$ is onto if and only if the rank of its standard matrix equals m.

42. A linear transformation is onto if and only if the columns of its standard matrix form a generating set for its range.
43. A linear transformation is onto if and only if the columns of its standard matrix are linearly independent.
44. A linear transformation is one-to-one if and only if every vector in its range is the image of a unique vector in its domain.
45. A linear transformation is one-to-one if and only if its null space consists only of the zero vector.
46. A linear transformation is invertible if and only if its standard matrix is invertible.
47. The system $A\mathbf{x} = \mathbf{b}$ is consistent for all $\mathbf{b}$ if and only if the transformation T_A is one-to-one.
48. Let A be an $m \times n$ matrix. The system $A\mathbf{x} = \mathbf{b}$ is consistent for all $\mathbf{b}$ in $\mathcal{R}^m$ if and only if the columns of A form a generating set for $\mathcal{R}^m$.
49. A function is onto if its range equals its domain.
50. A function is onto if its range equals its codomain.
51. The set $\{T(\mathbf{e}_1), T(\mathbf{e}_2), \ldots, T(\mathbf{e}_n)\}$ is a generating set for the range of any function $T: \mathcal{R}^n \to \mathcal{R}^m$.
52. A linear transformation $T: \mathcal{R}^n \to \mathcal{R}^m$ is onto if and only if the rank of its standard matrix is n.
53. The null space of a linear transformation $T: \mathcal{R}^n \to \mathcal{R}^m$ is the set of vectors in $\mathcal{R}^n$ whose image is $\mathbf{0}$.
54. A function $T: \mathcal{R}^n \to \mathcal{R}^m$ is one-to-one if the only vector $\mathbf{v}$ in $\mathcal{R}^n$ whose image is $\mathbf{0}$ is $\mathbf{v} = \mathbf{0}$.
55. A linear transformation $T: \mathcal{R}^n \to \mathcal{R}^m$ is one-to-one if and only if the rank of its standard matrix is m.
56. If the composition UT of two linear transformations $T: \mathcal{R}^n \to \mathcal{R}^m$ and $U: \mathcal{R}^p \to \mathcal{R}^q$ is defined, then $m = p$.
57. The composition of linear transformations is a linear transformation.
58. If $T: \mathcal{R}^n \to \mathcal{R}^m$ and $U: \mathcal{R}^p \to \mathcal{R}^n$ are linear transformations with standard matrices A and B, respectively, the standard matrix of TU equals BA.
59. For every invertible linear transformation T, the function T^{-1} is a linear transformation.
60. If A is the standard matrix of an invertible linear transformation T, then the standard matrix of T^{-1} is A^{-1}.

---

61. Suppose that $T: \mathcal{R}^2 \to \mathcal{R}^2$ is the linear transformation that rotates a vector by $90°$.
    (a) What is the null space of T?
    (b) Is T one-to-one?
    (c) What is the range of T?
    (d) Is T onto?
62. Suppose that $T: \mathcal{R}^2 \to \mathcal{R}^2$ is the reflection of $\mathcal{R}^2$ about the x-axis. (See Exercise 73 of Section 1.2.)
    (a) What is the null space of T?
    (b) Is T one-to-one?
    (c) What is the range of T?
    (d) Is T onto?
63. Define $T: \mathcal{R}^2 \to \mathcal{R}^2$ by $T\left(\begin{bmatrix} x_1 \\ x_2 \end{bmatrix}\right) = \begin{bmatrix} 0 \\ x_2 \end{bmatrix}$, which is the projection of $\begin{bmatrix} x_1 \\ x_2 \end{bmatrix}$ on the y-axis.
    (a) What is the null space of T?
    (b) Is T one-to-one?
    (c) What is the range of T?
    (d) Is T onto?
64. Define $T: \mathcal{R}^3 \to \mathcal{R}^3$ by $T\left(\begin{bmatrix} x_1 \\ x_2 \\ x_3 \end{bmatrix}\right) = \begin{bmatrix} 0 \\ 0 \\ x_3 \end{bmatrix}$, which is the projection of $\begin{bmatrix} x_1 \\ x_2 \\ x_3 \end{bmatrix}$ on the z-axis.
    (a) What is the null space of T?
    (b) Is T one-to-one?
    (c) What is the range of T?
    (d) Is T onto?
65. Define $T: \mathcal{R}^3 \to \mathcal{R}^3$ by $T\left(\begin{bmatrix} x_1 \\ x_2 \\ x_3 \end{bmatrix}\right) = \begin{bmatrix} x_1 \\ x_2 \\ 0 \end{bmatrix}$, which is the projection of $\begin{bmatrix} x_1 \\ x_2 \\ x_3 \end{bmatrix}$ on the xy-plane.
    (a) What is the null space of T?
    (b) Is T one-to-one?
    (c) What is the range of T?
    (d) Is T onto?
66. Define $T: \mathcal{R}^3 \to \mathcal{R}^3$ by $T\left(\begin{bmatrix} x_1 \\ x_2 \\ x_3 \end{bmatrix}\right) = \begin{bmatrix} x_1 \\ x_2 \\ -x_3 \end{bmatrix}$. (See Exercise 92 in Section 2.7.)
    (a) What is the null space of T?
    (b) Is T one-to-one?
    (c) What is the range of T?
    (d) Is T onto?
67. Suppose that $T: \mathcal{R}^2 \to \mathcal{R}^2$ is linear and has the property that $T(\mathbf{e}_1) = \begin{bmatrix} 3 \\ 1 \end{bmatrix}$ and $T(\mathbf{e}_2) = \begin{bmatrix} 4 \\ 2 \end{bmatrix}$.
    (a) Determine whether T is one-to-one.
    (b) Determine whether T is onto.
68. Suppose that $T: \mathcal{R}^2 \to \mathcal{R}^2$ is linear and has the property that $T(\mathbf{e}_1) = \begin{bmatrix} 3 \\ 1 \end{bmatrix}$ and $T(\mathbf{e}_2) = \begin{bmatrix} 6 \\ 2 \end{bmatrix}$.
    (a) Determine whether T is one-to-one.
    (b) Determine whether T is onto.

Exercises 69–75 are concerned with the linear transformations $T: \mathcal{R}^2 \to \mathcal{R}^3$ and $U: \mathcal{R}^3 \to \mathcal{R}^2$ defined as

$$T\left(\begin{bmatrix} x_1 \\ x_2 \end{bmatrix}\right) = \begin{bmatrix} x_1 + x_2 \\ x_1 - 3x_2 \\ 4x_1 \end{bmatrix}$$

and

$$U\left(\begin{bmatrix}x_1\\x_2\\x_3\end{bmatrix}\right)=\begin{bmatrix}x_1-x_2+4x_3\\x_1+3x_2\end{bmatrix}.$$

69. Determine the domain, the codomain, and the rule for UT.
70. Use the rule for UT obtained in Exercise 69 to find the standard matrix of UT.
71. Determine the standard matrices A and B of T and U, respectively.
72. Compute the product BA of the matrices found in Exercise 71, and illustrate Theorem 2.12 by comparing your answer with the result obtained in Exercise 70.
73. Determine the domain, the codomain, and the rule for TU.
74. Use the rule for TU obtained in Exercise 73 to find the standard matrix of TU.
75. Compute the product AB of the matrices found in Exercise 71, and illustrate Theorem 2.12 by comparing your answer with the result obtained in Exercise 74.

Exercises 76–82 are concerned with the linear transformations $T\colon \mathcal{R}^2 \to \mathcal{R}^2$ *and* $U\colon \mathcal{R}^2 \to \mathcal{R}^2$ *defined as*

$$T\left(\begin{bmatrix}x_1\\x_2\end{bmatrix}\right)=\begin{bmatrix}x_1+2x_2\\3x_1-x_2\end{bmatrix}\quad\text{and}\quad U\left(\begin{bmatrix}x_1\\x_2\end{bmatrix}\right)=\begin{bmatrix}2x_1-x_2\\5x_2\end{bmatrix}.$$

76. Determine the domain, the codomain, and the rule for UT.
77. Use the rule for UT obtained in Exercise 76 to find the standard matrix of UT.
78. Determine the standard matrices A and B of T and U, respectively.
79. Compute the product BA of the matrices found in Exercise 78, and illustrate Theorem 2.12 by comparing your answer with the result obtained in Exercise 77.
80. Determine the domain, the codomain, and the rule for TU.
81. Use the rule for TU obtained in Exercise 80 to find the standard matrix of TU.
82. Compute the product AB of the matrices found in Exercise 78, and illustrate Theorem 2.12 by comparing your answer with the result obtained in Exercise 81.

In Exercises 83–90, an invertible linear transformation T is defined. Determine a similar definition for the inverse T^{-1} of each linear transformation.

83. $T\colon \mathcal{R}^2 \to \mathcal{R}^2$ defined by $T\left(\begin{bmatrix}x_1\\x_2\end{bmatrix}\right)=\begin{bmatrix}2x_1-x_2\\x_1+x_2\end{bmatrix}$

84. $T\colon \mathcal{R}^2 \to \mathcal{R}^2$ defined by $T\left(\begin{bmatrix}x_1\\x_2\end{bmatrix}\right)=\begin{bmatrix}x_1+3x_2\\2x_1+x_2\end{bmatrix}$

85. $T\colon \mathcal{R}^3 \to \mathcal{R}^3$ defined by
$$T\left(\begin{bmatrix}x_1\\x_2\\x_3\end{bmatrix}\right)=\begin{bmatrix}-x_1+x_2+3x_3\\2x_1-x_3\\-x_1+2x_2+5x_3\end{bmatrix}$$

86. $T\colon \mathcal{R}^3 \to \mathcal{R}^3$ defined by
$$T\left(\begin{bmatrix}x_1\\x_2\\x_3\end{bmatrix}\right)=\begin{bmatrix}x_2-2x_3\\x_1-x_3\\-x_1+2x_2-2x_3\end{bmatrix}$$

87. $T\colon \mathcal{R}^3 \to \mathcal{R}^3$ defined by
$$T\left(\begin{bmatrix}x_1\\x_2\\x_3\end{bmatrix}\right)=\begin{bmatrix}4x_1+x_2-x_3\\-x_1-x_2\\-5x_1-3x_2+x_3\end{bmatrix}$$

88. $T\colon \mathcal{R}^3 \to \mathcal{R}^3$ defined by
$$T\left(\begin{bmatrix}x_1\\x_2\\x_3\end{bmatrix}\right)=\begin{bmatrix}x_1-x_2+2x_3\\-x_1+2x_2-3x_3\\2x_1+x_3\end{bmatrix}$$

89. $T\colon \mathcal{R}^4 \to \mathcal{R}^4$ defined by
$$T\left(\begin{bmatrix}x_1\\x_2\\x_3\\x_4\end{bmatrix}\right)=\begin{bmatrix}2x_1-3x_2-6x_3+3x_4\\3x_1-x_2-3x_3+3x_4\\-3x_1+3x_2+5x_3-3x_4\\-3x_1+6x_2+9x_3-4x_4\end{bmatrix}$$

90. $T\colon \mathcal{R}^4 \to \mathcal{R}^4$ defined by
$$T\left(\begin{bmatrix}x_1\\x_2\\x_3\\x_4\end{bmatrix}\right)=\begin{bmatrix}8x_1-2x_2+2x_3-4x_4\\-9x_1+7x_2-9x_3\\-16x_1+8x_2-10x_3+4x_4\\13x_1-5x_2+5x_3-6x_4\end{bmatrix}$$

91. Prove that the composition of two one-to-one linear transformations is one-to-one. Is the result true if the transformations are not linear? Justify your answer.
92. Prove that if two linear transformations are onto, then their composition is onto. Is the result true if the transformations are not linear? Justify your answer.
93. In $\mathcal{R}^2$, show that the composition of two reflections about the x-axis is the identity transformation.
94. In $\mathcal{R}^2$, show that a reflection about the y-axis followed by a rotation by 180° is equal to a reflection about the x-axis.
95. In $\mathcal{R}^2$, show that the composition of the projection on the x-axis followed by the reflection about the y-axis is equal to the composition of the reflection about the y-axis followed by the projection on the x-axis.
96. Prove that the composition of two shear transformations is a shear transformation. (See Example 4 of Section 2.7.)
97. Suppose that $T\colon \mathcal{R}^n \to \mathcal{R}^m$ is linear and one-to-one. Let $\{\mathbf{v}_1, \mathbf{v}_2, \ldots, \mathbf{v}_k\}$ be a linearly independent subset of $\mathcal{R}^n$.
    (a) Prove that the set $\{T(\mathbf{v}_1), T(\mathbf{v}_2), \ldots, T(\mathbf{v}_k)\}$ is a linearly independent subset of $\mathcal{R}^m$.
    (b) Show by example that (a) is false if T is not one-to-one.
98. Use Theorem 2.12 to prove that matrix multiplication is associative.

In Exercises 99 and 100, use either a calculator with matrix capabilities or computer software such as MATLAB to solve each problem.

99. The linear transformations $T, U\colon \mathcal{R}^4 \to \mathcal{R}^4$ are defined as follows:

$$T\left(\begin{bmatrix}x_1\\x_2\\x_3\\x_4\end{bmatrix}\right)=\begin{bmatrix}x_1+3x_2-2x_3+x_4\\3x_1+4x_3+x_4\\2x_1-x_2+2x_4\\x_3+x_4\end{bmatrix}$$

and

$$U\left(\begin{bmatrix} x_1 \\ x_2 \\ x_3 \\ x_4 \end{bmatrix}\right) = \begin{bmatrix} x_2 - 3x_4 \\ 2x_1 + x_3 - x_4 \\ x_1 - 2x_2 + 4x_4 \\ 5x_2 + x_3 \end{bmatrix}$$

(a) Compute the standard matrices A and B of T and U, respectively.

(b) Compute the product AB.

(c) Use your answer to (b) to write a rule for TU.

100. Define the linear transformation $T\colon \mathcal{R}^4 \to \mathcal{R}^4$ by the rule

$$T\left(\begin{bmatrix} x_1 \\ x_2 \\ x_3 \\ x_4 \end{bmatrix}\right) = \begin{bmatrix} 2x_1 + 4x_2 + x_3 + 6x_4 \\ 3x_1 + 7x_2 - x_3 + 11x_4 \\ x_1 + 2x_2 + 2x_4 \\ 2x_1 + 5x_2 - x_3 + 8x_4 \end{bmatrix}.$$

(a) Find the standard matrix A of T.

(b) Show that A is invertible and find its inverse.

(c) Use your answer to (b) to find the rule for T^{-1}.

SOLUTIONS TO THE PRACTICE PROBLEMS

1. The standard matrix of T is $A = \begin{bmatrix} 3 & -1 \\ -1 & 2 \\ 2 & 0 \end{bmatrix}$.

(a) Since the columns of A form a generating set for the range of T, the desired generating set is

$$\left\{\begin{bmatrix} 3 \\ -1 \\ 2 \end{bmatrix}, \begin{bmatrix} -1 \\ 2 \\ 0 \end{bmatrix}\right\}.$$

(b) The null space of A is the solution set of $A\mathbf{x} = \mathbf{0}$. Since the reduced row echelon form of A is $\begin{bmatrix} 1 & 0 \\ 0 & 1 \\ 0 & 0 \end{bmatrix}$, we see that the general solution of $A\mathbf{x} = \mathbf{0}$ is

$$x_1 = 0$$
$$x_2 = 0.$$

Hence the null space of T is $\{\mathbf{0}\}$, so a generating set for the null space of T is $\{\mathbf{0}\}$.

(c) From (b), we see that the rank of A is 2. Since Theorem 2.10 implies that T is onto if and only if the rank of A is 3, we see that T is not onto.

(d) From (b), we see that the null space of T is $\{\mathbf{0}\}$. Hence T is one-to-one by Theorem 2.11.

2. The standard matrices of T and U are

$$A = \begin{bmatrix} 3 & -1 \\ -1 & 2 \\ 2 & 0 \end{bmatrix} \quad \text{and} \quad B = \begin{bmatrix} 0 & 1 & -4 \\ 2 & 0 & 3 \end{bmatrix},$$

respectively. Hence, by Theorem 2.12(a), the standard matrix of UT is

$$BA = \begin{bmatrix} -9 & 2 \\ 12 & -2 \end{bmatrix}.$$

Therefore

$$UT\left(\begin{bmatrix} x_1 \\ x_2 \end{bmatrix}\right) = \begin{bmatrix} -9 & 2 \\ 12 & -2 \end{bmatrix}\begin{bmatrix} x_1 \\ x_2 \end{bmatrix} = \begin{bmatrix} -9x_1 + 2x_2 \\ 12x_1 - 2x_2 \end{bmatrix}.$$

3. The standard matrix of T is $A = \begin{bmatrix} 1 & 4 \\ 2 & 7 \end{bmatrix}$. Because the rank of A is 2, A is invertible. In fact, $A^{-1} = \begin{bmatrix} -7 & 4 \\ 2 & -1 \end{bmatrix}$. Hence, by Theorem 2.13, T is invertible, and

$$T^{-1}\left(\begin{bmatrix} x_1 \\ x_2 \end{bmatrix}\right) = \begin{bmatrix} -7 & 4 \\ 2 & -1 \end{bmatrix}\begin{bmatrix} x_1 \\ x_2 \end{bmatrix} = \begin{bmatrix} -7x_1 + 4x_2 \\ 2x_1 - x_2 \end{bmatrix}.$$

CHAPTER 2 REVIEW EXERCISES

In Exercises 1–21, determine whether the statements are true or false.

1. A symmetric matrix equals its transpose.
2. If a symmetric matrix is written in block form, then the blocks are also symmetric matrices.
3. The product of square matrices is always defined.
4. The transpose of an invertible matrix is invertible.
5. It is possible for an invertible matrix to have two distinct inverses.
6. The sum of an invertible matrix and its inverse is the zero matrix.
7. The columns of an invertible matrix are linearly independent.
8. If a matrix is invertible, then its rank equals the number of its rows.

9. A matrix is invertible if and only if its reduced row echelon form is an identity matrix.
10. If A is an $n \times n$ matrix and the system $A\mathbf{x} = \mathbf{b}$ is consistent for some $\mathbf{b}$, then A is invertible.
11. The range of a linear transformation is contained in the codomain of the linear transformation.
12. The null space of a linear transformation is contained in the codomain of the linear transformation.
13. Linear transformations preserve linear combinations.
14. Linear transformations preserve linearly independent sets.
15. Every linear transformation has a standard matrix.
16. The zero transformation is the only linear transformation whose standard matrix is the zero matrix.
17. If a linear transformation is one-to-one, then it is invertible.
18. If a linear transformation is onto, then its range equals its codomain.
19. If a linear transformation is one-to-one, then its range consists exactly of the zero vector.
20. If a linear transformation is onto, then the rows of its standard matrix form a generating set for its codomain.
21. If a linear transformation is one-to-one, then the columns of its standard matrix form a linearly independent set.

---

22. Determine whether each phrase is a misuse of terminology. If so, explain what is wrong.
    (a) the range of a matrix
    (b) the standard matrix of a function $f: \mathcal{R}^n \to \mathcal{R}^n$
    (c) a generating set for the range of a linear transformation
    (d) the null space of a system of linear equations
    (e) a one-to-one matrix
23. Let A be an $m \times n$ matrix and B be a $p \times q$ matrix.
    (a) Under what conditions is the matrix product BA defined?
    (b) If BA is defined, what size is it?

In Exercises 24–35, use the given matrices to compute each expression, or give a reason why the expression is not defined.

$$A = \begin{bmatrix} 2 & 1 \\ 4 & -1 \end{bmatrix}, \quad B = \begin{bmatrix} 2 & 3 \\ 4 & 6 \end{bmatrix}, \quad C = \begin{bmatrix} 2 & -1 \\ 3 & 5 \\ 0 & 1 \end{bmatrix},$$

$$\mathbf{u} = \begin{bmatrix} 3 \\ 2 \\ -1 \end{bmatrix}, \quad \mathbf{v} = \begin{bmatrix} 1 & -2 & 2 \end{bmatrix}, \quad \text{and} \quad \mathbf{w} = \begin{bmatrix} 3 \\ 4 \end{bmatrix}$$

24. $A\mathbf{w}$ 25. ABA 26. $A\mathbf{u}$
27. $C\mathbf{w}$ 28. $\mathbf{v}C$ 29. $\mathbf{v}A$
30. A^TB 31. $A^{-1}B^T$ 32. $B^{-1}\mathbf{w}$
33. $AC^T\mathbf{u}$ 34. B^3 35. $\mathbf{u}^2$

In Exercises 36 and 37, compute the product of the matrices in block form.

36. $$\left[\begin{array}{cc|cc} 1 & 0 & 3 & 1 \\ 0 & 1 & 2 & 4 \\ \hline 0 & 0 & 2 & 1 \\ 0 & 0 & -1 & 3 \end{array}\right] \left[\begin{array}{cc|cc} 1 & -1 & 0 & 0 \\ 2 & 1 & 0 & 0 \\ \hline 0 & 0 & 2 & 0 \\ 0 & 0 & 0 & 2 \end{array}\right]$$

37. $$\left[\begin{array}{c|c} I_2 & -I_2 \end{array}\right] \left[\begin{array}{c} 1 \\ 3 \\ \hline -7 \\ -4 \end{array}\right]$$

In Exercises 38 and 39, determine whether each matrix is invertible. If so, find its inverse; if not, explain why not.

38. $\begin{bmatrix} 1 & 0 & 2 \\ 2 & -1 & 3 \\ 4 & 1 & 8 \end{bmatrix}$ 39. $\begin{bmatrix} 2 & -1 & 3 \\ 1 & 2 & -4 \\ 4 & 3 & 5 \end{bmatrix}$

40. Let A and B be square matrices of the same size. Prove that if the first row of A is zero, then the first row of AB is zero.
41. Let A and B be square matrices of the same size. Prove that if the first column of B is zero, then the first column of AB is zero.
42. Give examples of 2×2 matrices A and B such that A and B are invertible, and $(A+B)^{-1} \neq A^{-1} + B^{-1}$.

In Exercises 43 and 44, systems of equations are given. First use the appropriate matrix inverse to solve each system, and then use Gaussian elimination to check your answer.

43. $$\begin{aligned} 2x_1 + x_2 &= 3 \\ x_1 + x_2 &= 5 \end{aligned}$$

44. $$\begin{aligned} x_1 + x_2 + x_3 &= 3 \\ x_1 + 3x_2 + 4x_3 &= -1 \\ 2x_1 + 4x_2 + x_3 &= 2 \end{aligned}$$

45. Suppose that the reduced row echelon form R and three columns of A are given by
$$R = \begin{bmatrix} 1 & 2 & 0 & 0 & -2 \\ 0 & 0 & 1 & 0 & 3 \\ 0 & 0 & 0 & 1 & 1 \end{bmatrix}, \quad \mathbf{a}_1 = \begin{bmatrix} 3 \\ 5 \\ 2 \end{bmatrix},$$
$$\mathbf{a}_3 = \begin{bmatrix} 2 \\ 0 \\ -1 \end{bmatrix}, \quad \text{and} \quad \mathbf{a}_4 = \begin{bmatrix} 2 \\ -1 \\ 3 \end{bmatrix}.$$
Determine the matrix A.

Exercises 46–49 refer to the following matrices:

$$A = \begin{bmatrix} 2 & -1 & 3 \\ 4 & 0 & -2 \end{bmatrix} \quad \text{and} \quad B = \begin{bmatrix} 4 & 2 \\ 1 & -3 \\ 0 & 1 \end{bmatrix}$$

46. Find the range and codomain of the matrix transformation T_A.
47. Find the range and codomain of the matrix transformation T_B.

48. Compute $T_A\left(\begin{bmatrix}2\\0\\3\end{bmatrix}\right)$.

49. Compute $T_B\left(\begin{bmatrix}4\\2\end{bmatrix}\right)$.

In Exercises 50–53, a linear transformation is given. Compute its standard matrix.

50. $T\colon \mathcal{R}^2 \to \mathcal{R}^2$ defined by $T\left(\begin{bmatrix}x_1\\x_2\end{bmatrix}\right) = \begin{bmatrix}3x_1 - x_2\\4x_1\end{bmatrix}$

51. $T\colon \mathcal{R}^3 \to \mathcal{R}^2$ defined by $T\left(\begin{bmatrix}x_1\\x_2\\x_3\end{bmatrix}\right) = \begin{bmatrix}2x_1 - x_3\\4x_1\end{bmatrix}$

52. $T\colon \mathcal{R}^2 \to \mathcal{R}^2$ defined by $T(\mathbf{v}) = 6\mathbf{v}$ for all $\mathbf{v}$ in $\mathcal{R}^2$

53. $T\colon \mathcal{R}^2 \to \mathcal{R}^2$ defined by $T(\mathbf{v}) = 2\mathbf{v} + U(\mathbf{v})$ for all $\mathbf{v}$ in $\mathcal{R}^2$, where $U\colon \mathcal{R}^2 \to \mathcal{R}^2$is the linear transformation defined by $U\left(\begin{bmatrix}x_1\\x_2\end{bmatrix}\right) = \begin{bmatrix}2x_1 + x_2\\3x_1\end{bmatrix}$

In Exercises 54–57, a function $T\colon \mathcal{R}^n \to \mathcal{R}^m$ is given. Either prove that T is linear or explain why T is not linear.

54. $T\colon \mathcal{R}^2 \to \mathcal{R}^2$ defined by $T\left(\begin{bmatrix}x_1\\x_2\end{bmatrix}\right) = \begin{bmatrix}x_1 + 1\\x_2\end{bmatrix}$

55. $T\colon \mathcal{R}^2 \to \mathcal{R}^2$ defined by $T\left(\begin{bmatrix}x_1\\x_2\end{bmatrix}\right) = \begin{bmatrix}2x_2\\x_1\end{bmatrix}$

56. $T\colon \mathcal{R}^2 \to \mathcal{R}^2$ defined by $T\left(\begin{bmatrix}x_1\\x_2\end{bmatrix}\right) = \begin{bmatrix}x_1x_2\\x_1\end{bmatrix}$

57. $T\colon \mathcal{R}^3 \to \mathcal{R}^2$ defined by $T\left(\begin{bmatrix}x_1\\x_2\\x_3\end{bmatrix}\right) = \begin{bmatrix}x_1 + x_2\\x_3\end{bmatrix}$

In Exercises 58 and 59, find a generating set for the range of each linear transformation T.

58. $T\colon \mathcal{R}^2 \to \mathcal{R}^3$ defined by $T\left(\begin{bmatrix}x_1\\x_2\end{bmatrix}\right) = \begin{bmatrix}x_1 + x_2\\0\\2x_1 - x_2\end{bmatrix}$

59. $T\colon \mathcal{R}^3 \to \mathcal{R}^2$ defined by $T\left(\begin{bmatrix}x_1\\x_2\\x_3\end{bmatrix}\right) = \begin{bmatrix}x_1 + 2x_2\\x_2 - x_3\end{bmatrix}$

In Exercises 60 and 61, find a generating set for the null space of each linear transformation T. Use your answer to determine whether T is one-to-one.

60. $T\colon \mathcal{R}^2 \to \mathcal{R}^3$ defined by $T\left(\begin{bmatrix}x_1\\x_2\end{bmatrix}\right) = \begin{bmatrix}x_1 + x_2\\0\\2x_1 - x_2\end{bmatrix}$

61. $T\colon \mathcal{R}^3 \to \mathcal{R}^2$ defined by $T\left(\begin{bmatrix}x_1\\x_2\\x_3\end{bmatrix}\right) = \begin{bmatrix}x_1 + 2x_2\\x_2 - x_3\end{bmatrix}$

In Exercises 62 and 63, find the standard matrix of each linear transformation T, and use it to determine whether T is one-to-one.

62. $T\colon \mathcal{R}^3 \to \mathcal{R}^2$ defined by $T\left(\begin{bmatrix}x_1\\x_2\\x_3\end{bmatrix}\right) = \begin{bmatrix}x_1 + 2x_2\\x_2 - x_3\end{bmatrix}$

63. $T\colon \mathcal{R}^2 \to \mathcal{R}^3$ defined by $T\left(\begin{bmatrix}x_1\\x_2\end{bmatrix}\right) = \begin{bmatrix}x_1 + x_2\\0\\2x_1 - x_2\end{bmatrix}$

In Exercises 64 and 65, find the standard matrix of each linear transformation T, and use it to determine whether T is onto.

64. $T\colon \mathcal{R}^3 \to \mathcal{R}^2$ defined by $T\left(\begin{bmatrix}x_1\\x_2\\x_3\end{bmatrix}\right) = \begin{bmatrix}2x_1 + x_3\\x_1 + x_2 - x_3\end{bmatrix}$

65. $T\colon \mathcal{R}^2 \to \mathcal{R}^3$ defined by $T\left(\begin{bmatrix}x_1\\x_2\end{bmatrix}\right) = \begin{bmatrix}3x_1 - x_2\\x_2\\x_1 + x_2\end{bmatrix}$

Exercises 66–72 are concerned with the linear transformations $T\colon \mathcal{R}^3 \to \mathcal{R}^2$ and $U\colon \mathcal{R}^2 \to \mathcal{R}^3$ defined by

$$T\left(\begin{bmatrix}x_1\\x_2\\x_3\end{bmatrix}\right) = \begin{bmatrix}2x_1 + x_3\\x_1 + x_2 - x_3\end{bmatrix}$$

and

$$U\left(\begin{bmatrix}x_1\\x_2\end{bmatrix}\right) = \begin{bmatrix}3x_1 - x_2\\x_2\\x_1 + x_2\end{bmatrix}.$$

66. Determine the domain, codomain, and the rule for UT.

67. Use the rule for UT obtained in Exercise 66 to find the standard matrix of UT.

68. Determine the standard matrices A and B of T and U, respectively.

69. Compute the product BA of the matrices found in Exercise 68, and illustrate Theorem 2.12 by comparing this answer with the result of Exercise 67.

70. Determine the domain, codomain, and the rule for TU.

71. Use the rule for TU obtained in Exercise 70 to find the standard matrix of TU.

72. Compute the product AB of the matrices found in Exercise 68, and illustrate Theorem 2.12 by comparing this answer with the result of Exercise 71.

In Exercises 73 and 74, an invertible linear transformation T is defined. Determine a similar definition for T^{-1}.

73. $T\colon \mathcal{R}^2 \to \mathcal{R}^2$ defined by $T\left(\begin{bmatrix}x_1\\x_2\end{bmatrix}\right) = \begin{bmatrix}x_1 + 2x_2\\-x_1 + 3x_2\end{bmatrix}$

74. $T\colon \mathcal{R}^3 \to \mathcal{R}^3$ defined by $T\left(\begin{bmatrix}x_1\\x_2\\x_3\end{bmatrix}\right) = \begin{bmatrix}x_1 + x_2 + x_3\\x_1 + 3x_2 + 4x_3\\2x_1 + 4x_2 + x_3\end{bmatrix}$

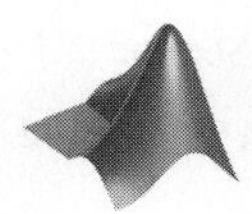

CHAPTER 2 MATLAB EXERCISES

For the following exercises, use MATLAB (or comparable software) or a calculator with matrix capabilities. The MATLAB functions in Tables D.1, D.2, D.3, D.4, and D.5 of Appendix D may be useful.

1. Let

$$A = \begin{bmatrix} 1 & -1 & 2 & 1 & 3 \\ 0 & 1 & 1 & 0 & 1 \\ 1 & 2 & 0 & -1 & 3 \\ 4 & 0 & 1 & 0 & -2 \\ -1 & 1 & 2 & 1 & -3 \end{bmatrix},$$

$$B = \begin{bmatrix} 1 & -1 & 2 & 3 & 1 \\ 1 & 0 & 1 & 2 & 2 \\ -1 & 0 & 2 & 1 & -1 \end{bmatrix},$$

$$C = \begin{bmatrix} 2 & -1 & 1 & 2 & 3 \\ 1 & -1 & 0 & 1 & 2 \\ 3 & 1 & 2 & 3 & -1 \end{bmatrix},$$

$$D = \begin{bmatrix} 2 & 3 & -1 \\ 1 & 0 & 4 \\ -1 & 0 & 2 \\ 2 & 1 & 1 \\ 1 & 2 & 3 \end{bmatrix}, \quad \text{and} \quad \mathbf{v} = \begin{bmatrix} 1 \\ 3 \\ 1 \\ -1 \\ 4 \end{bmatrix}.$$

Compute each of the following matrices or vectors.

(a) AD (b) DB (c) $(AB^T)C$

(d) $A(B^TC)$ (e) $D(B - 2C)$ (f) $A\mathbf{v}$

(g) $C(A\mathbf{v})$ (h) $(CA)\mathbf{v}$ (i) A^3

2. Suppose that

$$A = \begin{bmatrix} 0 & .3 & .5 & .6 & .3 & 0 \\ .7 & 0 & 0 & 0 & 0 & 0 \\ 0 & .9 & 0 & 0 & 0 & 0 \\ 0 & 0 & .8 & 0 & 0 & 0 \\ 0 & 0 & 0 & .3 & 0 & 0 \\ 0 & 0 & 0 & 0 & .1 & 0 \end{bmatrix}$$

is the Leslie matrix for a colony of an animal species living in an isolated region, where each time period is equal to one year.

(a) Compute A^{10}, A^{100}, and A^{500}. On the basis of your results, what do you predict will be the ultimate fate of the colony?

(b) Now suppose that each year, a particular population of the species immigrates into the region from the outside, and that the distribution of females in this immigration is described by the vector $\mathbf{b}$. Assume that the original population of females in the region is given by the population distribution $\mathbf{x}_0$, and after n years, the population distribution is $\mathbf{x}_n$.

(i) Prove that, for any positive integer n,

$$\mathbf{x}_n = A\mathbf{x}_{n-1} + \mathbf{b}.$$

(ii) Assume that

$$\mathbf{x}_0 = \begin{bmatrix} 3.1 \\ 2.2 \\ 4.3 \\ 2.4 \\ 1.8 \\ 0.0 \end{bmatrix} \quad \text{and} \quad \mathbf{b} = \begin{bmatrix} 0 \\ 1.1 \\ 2.1 \\ 0 \\ 0 \\ 0 \end{bmatrix},$$

where each unit represents 1000 females. Compute the population distributions $\mathbf{x}_1$, $\mathbf{x}_2$, $\mathbf{x}_3$, $\mathbf{x}_4$, and $\mathbf{x}_5$.

(iii) Prove that, for any positive integer n,

$$\mathbf{x}_n = A^n\mathbf{x}_0 + (A - I_6)^{-1}(A^n - I_6)\mathbf{b},$$

and use this formula to recompute $\mathbf{x}_5$, requested in (ii). *Hint for the proof:* Apply either the formula for the sum of terms of a geometric series or the principle of mathematical induction.

(iv) Suppose that after n years, the population is stable, that is, $\mathbf{x}_{n+1} = \mathbf{x}_n$. Prove that

$$\mathbf{x}_n = (I_6 - A)^{-1}\mathbf{b}.$$

Use this result to predict the stable distribution of females for the vector $\mathbf{b}$ given in (ii).

3. Let

$$A = \begin{bmatrix} 0 & 1 & 0 & 0 & 0 & 1 & 0 & 0 \\ 1 & 0 & 0 & 0 & 0 & 0 & 0 & 0 \\ 0 & 0 & 0 & 1 & 1 & 0 & 0 & 0 \\ 0 & 0 & 1 & 0 & 0 & 0 & 1 & 0 \\ 0 & 0 & 1 & 0 & 0 & 0 & 0 & 0 \\ 1 & 0 & 0 & 0 & 0 & 0 & 0 & 1 \\ 0 & 0 & 0 & 1 & 0 & 0 & 0 & 0 \\ 0 & 0 & 0 & 0 & 0 & 1 & 0 & 0 \end{bmatrix}$$

be the matrix that describes flights between a set of 8 airports. For each i and j, $a_{ij} = 1$ if there is a regularly scheduled flight from airport i to airport j, and $a_{ij} = 0$ otherwise. Notice that if there is a regularly scheduled flight from airport i to airport j, then there is a regularly scheduled flight in the opposite direction, and hence A is symmetric. Use the method described on page 113 to divide the airports into subsets so that one can fly, with possible connecting flights, between any two airports within a subset, but not to an airport outside of the subset.

4. Let

$$A = \begin{bmatrix} 1 & 2 & 3 & 2 & 1 & 1 & 1 \\ 2 & 4 & 1 & -1 & 2 & 1 & -2 \\ 1 & 2 & 2 & 1 & 1 & -1 & 2 \\ 1 & 2 & 1 & 0 & 1 & 1 & -1 \\ -1 & -2 & 1 & 2 & 1 & 2 & -2 \end{bmatrix}.$$

(a) Use the MATLAB function `rref(A)` to obtain the reduced row echelon form R of A.

(b) Use the function `null(A, 'r')`, described in Table D.2 of Appendix D, to obtain a matrix S whose columns form a basis for Null A.

(c) Compare S and R, and outline a procedure that can be applied in general to obtain S from R without solving the equation $R\mathbf{x} = \mathbf{0}$. Apply your procedure to the matrix R in (a) to obtain the matrix S in (b).

5. Read Example 3 of Appendix D, which describes a method for obtaining an invertible matrix P for a given matrix A such that PA is the reduced row echelon form of A.

   (a) Justify the method.

   (b) Apply the method to the matrix A in Exercise 6.

6. Let

$$A = \begin{bmatrix} 1 & 2 & 3 & 1 & 3 & 2 \\ 2 & 0 & 2 & -1 & 0 & 3 \\ -1 & 1 & 0 & 0 & -1 & 0 \\ 2 & 1 & 3 & 1 & 4 & 2 \\ 4 & 4 & 8 & 1 & 6 & 6 \end{bmatrix}.$$

   (a) Compute M, the matrix of pivot columns of A listed in the same order. For example, you can use the imported MATLAB function `pvtcol(A)`, described in Table D.5 of Appendix D, to obtain M.

   (b) Compute R, the reduced row echelon form of A, and obtain the matrix S from R by deleting the zero rows of R.

   (c) Prove that the matrix product MS is defined.

   (d) Verify that $MS = A$.

   (e) Show that for any matrix A, if M and S are obtained from A by the methods described in (a) and (b), then $MS = A$.

7. Let

$$A = \begin{bmatrix} 1 & 1 & 1 & 1 & 2 \\ 1 & 2 & 1 & 1 & 2 \\ -1 & -1 & 0 & -1 & -2 \\ 2 & 3 & 2 & 2 & 5 \\ 2 & 2 & 2 & 3 & 4 \end{bmatrix}$$

and

$$B = \begin{bmatrix} 1 & -1 & 2 & 4 & 0 & 2 & 1 \\ 0 & 1 & -2 & 3 & 4 & -1 & -1 \\ -3 & 1 & 0 & 2 & -1 & 4 & 2 \\ 0 & 0 & 2 & 1 & -1 & 3 & 2 \\ 2 & -1 & 1 & 0 & 2 & 1 & 3 \end{bmatrix}.$$

   (a) Compute $A^{-1}B$ by using the algorithm in Section 2.4.

   (b) Compute $A^{-1}B$ directly by computing A^{-1} and the product of A^{-1} and B.

   (c) Compare your results in (a) and (b).

8. Let

$$A = \begin{bmatrix} 1 & -1 & 2 & 0 & -2 & 4 \\ 2 & 1 & 1 & -2 & 1 & 3 \\ 1 & 0 & -1 & 3 & -3 & 2 \\ 2 & -1 & 1 & 1 & 2 & 3 \end{bmatrix} \quad \text{and} \quad \mathbf{b} = \begin{bmatrix} 5 \\ 4 \\ 9 \\ 11 \end{bmatrix}.$$

   (a) Obtain an LU decomposition of A.

   (b) Use the LU decomposition obtained in (a) to solve the system $A\mathbf{x} = \mathbf{b}$.

9. Let $T\colon \mathcal{R}^6 \to \mathcal{R}^6$ be the linear transformation defined by the rule

$$T\left(\begin{bmatrix} x_1 \\ x_2 \\ x_3 \\ x_4 \\ x_5 \\ x_6 \end{bmatrix}\right) = \begin{bmatrix} x_1 + 2x_2 + x_4 - 3x_5 - 2x_6 \\ x_2 - x_4 \\ x_1 + x_3 + 3x_6 \\ 2x_1 + 4x_2 + 3x_4 - 6x_5 - 4x_6 \\ 3x_1 + 2x_2 + 2x_3 + x_4 - 2x_5 - 4x_6 \\ 4x_1 + 4x_2 + 2x_3 + 2x_4 - 5x_5 + 3x_6 \end{bmatrix}.$$

   (a) Find the standard matrix A of T.

   (b) Show that A is invertible and find its inverse.

   (c) Use your answer to (b) to find the rule for T^{-1}.

10. Let $U\colon \mathcal{R}^6 \to \mathcal{R}^4$ be the linear transformation defined by the rule

$$U\left(\begin{bmatrix} x_1 \\ x_2 \\ x_3 \\ x_4 \\ x_5 \\ x_6 \end{bmatrix}\right) = \begin{bmatrix} x_1 + 2x_3 + x_6 \\ 2x_1 - x_2 + x_4 \\ 3x_2 - x_5 \\ 2x_1 + x_2 - x_3 + x_6 \end{bmatrix},$$

let T be the linear transformation in Exercise 9, and let A be the standard matrix of T.

   (a) Find the standard matrix B of U.

   (b) Use A and B to find the standard matrix of UT.

   (c) Find the rule for UT.

   (d) Find the rule for UT^{-1}.

3 INTRODUCTION

Geographic information systems (GIS) have fundamentally changed the way cartographers and geographical analysts do their jobs. A GIS is a computer system capable of capturing, storing, analyzing, and displaying geographic data and geographically referenced information (data identified according to location). Calculations and analyses that would take hours or days if done by hand can be done in minutes with a GIS.

The figure at the left, which shows the boundary of a county (heavy line) and the boundary of a state forest (light line), illustrates a relatively simple GIS analysis. In this analysis, the goal is to answer the question: How much of the forest is in the county? In particular, we need to compute the area of the shaded region, whose boundary is a polygon (a figure composed of connected line segments). This problem can be solved using determinants.

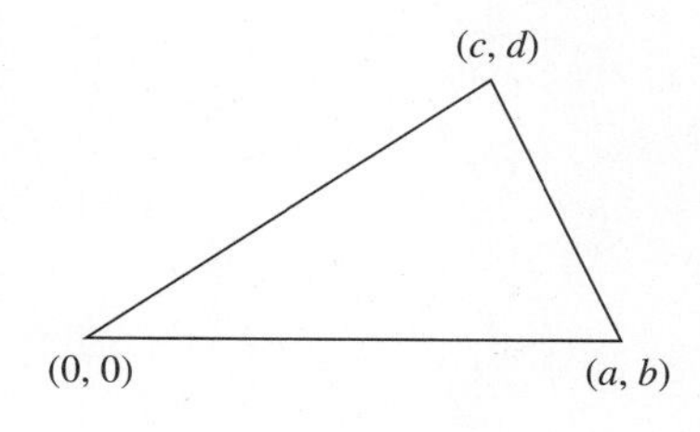

As explained in Section 3.1, the area of the parallelogram having the vectors $\mathbf{u}$ and $\mathbf{v}$ as adjacent sides is $|\det[\mathbf{u}\ \mathbf{v}]|$. Because the area of the triangle having $\mathbf{u}$ and $\mathbf{v}$ as adjacent sides is half that of the parallelogram with $\mathbf{u}$ and $\mathbf{v}$ as adjacent sides, the area of this triangle is $\frac{1}{2}|\det[\mathbf{u}\ \mathbf{v}]|$. Thus if a triangle has vertices at the points (0, 0), (a, b), and (c, d) as shown in the figure at the left, then its area is $\pm\frac{1}{2}\det A$, where $A = \begin{bmatrix} a & c \\ b & d \end{bmatrix}$. The value of $\det A$ is positive if the vertices (0, 0), (a, b), and (c, d) are in a counterclockwise order, and negative if the positions of (a, b) and (c, d) are reversed.

This geometric interpretation of the determinant can be used to find the areas of complex polygonal regions, such as the intersection of the forest and the county. A simple example is shown in the figure at the right. Translate the polygon so that one of its vertices P_0 is at the origin, and label its successive vertices in a counter-clockwise order around the boundary. Then half the sum of the determinants

$$\det \begin{bmatrix} x_i & x_{i+1} \\ y_i & y_{i+1} \end{bmatrix}$$

for $i = 1, 2, \ldots, 6$ gives the area of the polygon.

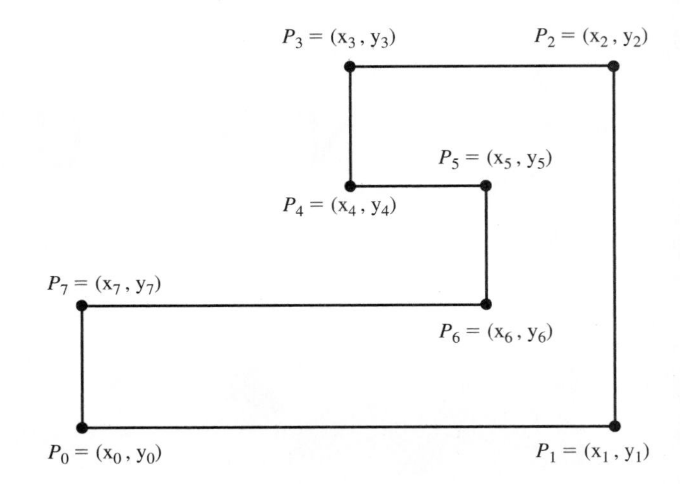

CHAPTER

3 DETERMINANTS

The *determinant*[1] of a square matrix is a scalar that provides information about the matrix, such as whether or not the matrix is invertible. Determinants were first considered in the late seventeenth century. For over one hundred years thereafter, they were studied principally because of their connection to systems of linear equations. The best-known result involving determinants and systems of linear equations is *Cramer's rule*, presented in Section 3.2.

In recent years, the use of determinants as a computational tool has greatly diminished. This is primarily because the size of the systems of linear equations that arise in applications has increased greatly, requiring the use of high-speed computers and efficient computational algorithms to obtain solutions. Since calculations with determinants are usually inefficient, they are normally avoided. Although determinants can be used to compute the areas and volumes of geometric objects, our principal use of determinants in this book is to determine the *eigenvalues* of a square matrix (to be discussed in Chapter 5).

3.1 COFACTOR EXPANSION

We begin this section by showing that we can assign a scalar to a 2×2 matrix that tells us whether or not the matrix is invertible. We then generalize this to $n \times n$ matrices.

Consider the 2×2 matrices

$$A = \begin{bmatrix} a & b \\ c & d \end{bmatrix} \quad \text{and} \quad C = \begin{bmatrix} d & -b \\ -c & a \end{bmatrix}.$$

We see that

$$AC = \begin{bmatrix} a & b \\ c & d \end{bmatrix}\begin{bmatrix} d & -b \\ -c & a \end{bmatrix} = \begin{bmatrix} ad-bc & 0 \\ 0 & ad-bc \end{bmatrix} = (ad-bc)\begin{bmatrix} 1 & 0 \\ 0 & 1 \end{bmatrix},$$

and similarly,

$$CA = \begin{bmatrix} d & -b \\ -c & a \end{bmatrix}\begin{bmatrix} a & b \\ c & d \end{bmatrix} = \begin{bmatrix} ad-bc & 0 \\ 0 & ad-bc \end{bmatrix} = (ad-bc)\begin{bmatrix} 1 & 0 \\ 0 & 1 \end{bmatrix}.$$

[1] Although work with determinants can be found in ancient China and in the writings of Gabriel Cramer in 1750, the study of determinants dates mainly from the early nineteenth century. In an 84-page paper presented to the French Institut in 1812, Augustin-Louis Cauchy (1789–1857) introduced the term *determinant* and proved many of the well-known results about determinants. He also used modern double-subscript notation for matrices and showed how to evaluate a determinant by cofactor expansion.

Thus if $ad - bc \neq 0$, the matrix $\dfrac{1}{ad - bc}C$ is the inverse of A, and so A is invertible.

Conversely, suppose that $ad - bc = 0$. The previous calculations show that AC and CA equal O, the 2×2 zero matrix. We show by contradiction that A is not invertible. For if A were invertible, then

$$C = CI_2 = C(AA^{-1}) = (CA)A^{-1} = OA^{-1} = O,$$

and so all the entries of C are equal to 0. It follows that all the entries of A equal 0; that is, $A = O$, contradicting that A is invertible.

We summarize these results.

> The matrix $A = \begin{bmatrix} a & b \\ c & d \end{bmatrix}$ is invertible if and only if $ad - bc \neq 0$, in which case
>
> $$A^{-1} = \frac{1}{ad - bc}\begin{bmatrix} d & -b \\ -c & a \end{bmatrix}.$$

Thus the scalar $ad - bc$ *determines* whether or not the preceding matrix A is invertible. We call the scalar $ad - bc$ the **determinant** of A and denote it by $\det A$ or $|A|$.

Example 1 For

$$A = \begin{bmatrix} 1 & 2 \\ 3 & 4 \end{bmatrix} \quad \text{and} \quad B = \begin{bmatrix} 1 & 2 \\ 3 & 6 \end{bmatrix},$$

the determinants are

$$\det A = 1 \cdot 4 - 2 \cdot 3 = -2 \quad \text{and} \quad \det B = 1 \cdot 6 - 2 \cdot 3 = 0.$$

Thus A is invertible, but B is not.

Since A is invertible, we can compute its inverse

$$A^{-1} = \frac{1}{-2}\begin{bmatrix} 4 & -2 \\ -3 & 1 \end{bmatrix} = \begin{bmatrix} -2 & 1 \\ \frac{3}{2} & -\frac{1}{2} \end{bmatrix}.$$

Practice Problem 1 ▶ Evaluate the determinant of $\begin{bmatrix} 8 & 3 \\ -6 & 5 \end{bmatrix}$. Is this matrix invertible? If so, compute A^{-1}. ◀

Example 2 Determine the scalars c for which $A - cI_2$ is not invertible, where

$$A = \begin{bmatrix} 11 & 12 \\ -8 & -9 \end{bmatrix}.$$

Solution The matrix $A - cI_2$ has the form

$$A - cI_2 = \begin{bmatrix} 11 & 12 \\ -8 & -9 \end{bmatrix} - c\begin{bmatrix} 1 & 0 \\ 0 & 1 \end{bmatrix} = \begin{bmatrix} 11 - c & 12 \\ -8 & -9 - c \end{bmatrix}.$$

Although we can use elementary row operations to determine the values of c for which this matrix is not invertible, the presence of the unknown scalar c makes the calculations difficult. By using the determinant instead, we obtain an easier computation. Since

$$\begin{aligned}\det(A - cI_2) &= \det\begin{bmatrix} 11-c & 12 \\ -8 & -9-c \end{bmatrix} \\ &= (11-c)(-9-c) - 12(-8) \\ &= (c^2 - 2c - 99) + 96 \\ &= c^2 - 2c - 3 \\ &= (c+1)(c-3),\end{aligned}$$

we see that $\det(A - cI_2) = 0$ if and only if $c = -1$ or $c = 3$. Thus $A - cI_2$ is not invertible when $c = -1$ or $c = 3$. Calculate $A - (-1)I_2$ and $A + 3I_2$ to verify that these matrices are not invertible.

Practice Problem 2 ▶ Determine the scalars c for which $A - cI_2$ is not invertible, where

$$A = \begin{bmatrix} 4 & 6 \\ -1 & -3 \end{bmatrix}.$$

◀

Our principal use of determinants in this book is to calculate the scalars c for which a matrix $A - cI_n$ is not invertible, as in Example 2. In order to have such a test for $n \times n$ matrices, we must extend the definition of determinants to square matrices of any size so that a nonzero determinant is equivalent to invertibility. For 1×1 matrices, the appropriate definition is not hard to discover. Since the product of 1×1 matrices satisfies $[a][b] = [ab]$, we see that $[a]$ is invertible if and only if $a \neq 0$. Hence for a 1×1 matrix $[a]$, we define the **determinant** of $[a]$ by $\det[a] = a$.

Unfortunately, for $n \geq 3$, the determinant of an $n \times n$ matrix A is more complicated to define. To begin, we need additional notation. First, we define the $(n-1) \times (n-1)$ matrix A_{ij} to be the matrix obtained from A by deleting row i and column j.

$$A_{ij} = \begin{bmatrix} a_{11} & \cdots & a_{1j} & \cdots & a_{1n} \\ \vdots & & \vdots & & \vdots \\ a_{i1} & \cdots & a_{ij} & \cdots & a_{in} \\ \vdots & & \vdots & & \vdots \\ a_{n1} & \cdots & a_{nj} & \cdots & a_{nn} \end{bmatrix} \longleftarrow \text{row } i$$

$$\uparrow \text{ column } j$$

Thus, for example, if

$$A = \begin{bmatrix} 1 & 2 & 3 \\ 4 & 5 & 6 \\ 7 & 9 & 8 \end{bmatrix},$$

then

$$A_{12} = \begin{bmatrix} 4 & 6 \\ 7 & 8 \end{bmatrix}, \quad A_{21} = \begin{bmatrix} 2 & 3 \\ 9 & 8 \end{bmatrix}, \quad A_{23} = \begin{bmatrix} 1 & 2 \\ 7 & 9 \end{bmatrix}, \text{ and } A_{33} = \begin{bmatrix} 1 & 2 \\ 4 & 5 \end{bmatrix}.$$

We can express the determinant of a 2×2 matrix by using these matrices. For if

$$A = \begin{bmatrix} a & b \\ c & d \end{bmatrix},$$

then $A_{11} = [d]$ and $A_{12} = [c]$. Thus

$$\det A = ad - bc = a \cdot \det A_{11} - b \cdot \det A_{12}. \tag{1}$$

Notice that this representation expresses the determinant of the 2×2 matrix A in terms of the determinants of the 1×1 matrices A_{ij}.

Using equation (1) as a motivation, we define the **determinant** of an $n \times n$ matrix A for $n \geq 3$ by

$$\det A = a_{11} \cdot \det A_{11} - a_{12} \cdot \det A_{12} + \cdots + (-1)^{1+n} a_{1n} \cdot \det A_{1n}. \tag{2}$$

We denote the determinant of A by det A or $|A|$. Note that the expression on the right side of equation (2) is an alternating sum of products of entries from the first row of A multiplied by the determinant of the corresponding matrix A_{1j}. If we let $c_{ij} = (-1)^{i+j} \cdot \det A_{ij}$, then our definition of the determinant of A can be written as

$$\det A = a_{11}c_{11} + a_{12}c_{12} + \cdots + a_{1n}c_{1n}. \tag{3}$$

The number c_{ij} is called the ***(i, j)*-cofactor** of A, and equation (3) is called the **cofactor expansion** of A along the first row.

Equations (1) and (2) define the determinant of an $n \times n$ matrix recursively. That is, the determinant of a matrix is defined in terms of determinants of smaller matrices. For example, if we want to compute the determinant of a 4×4 matrix A, equation (2) enables us to express the determinant of A in terms of determinants of 3×3 matrices. The determinants of these 3×3 matrices can then be expressed by equation (2) in terms of determinants of 2×2 matrices, and the determinants of these 2×2 matrices can finally be evaluated by equation (1).

Example 3 Evaluate the determinant of A by using the cofactor expansion along the first row, if

$$A = \begin{bmatrix} 1 & 2 & 3 \\ 4 & 5 & 6 \\ 7 & 9 & 8 \end{bmatrix}.$$

Solution The cofactor expansion of A along the first row yields

$$\begin{aligned}
\det A &= a_{11}c_{11} + a_{12}c_{12} + a_{13}c_{13} \\
&= 1(-1)^{1+1} \det A_{11} + 2(-1)^{1+2} \det A_{12} + 3(-1)^{1+3} \det A_{13} \\
&= 1(-1)^{1+1} \det \begin{bmatrix} 5 & 6 \\ 9 & 8 \end{bmatrix} + 2(-1)^{1+2} \det \begin{bmatrix} 4 & 6 \\ 7 & 8 \end{bmatrix} + 3(-1)^{1+3} \det \begin{bmatrix} 4 & 5 \\ 7 & 9 \end{bmatrix} \\
&= 1(1)(5 \cdot 8 - 6 \cdot 9) + 2(-1)(4 \cdot 8 - 6 \cdot 7) + 3(1)(4 \cdot 9 - 5 \cdot 7) \\
&= 1(1)(-14) + (2)(-1)(-10) + 3(1)(1) \\
&= -14 + 20 + 3 \\
&= 9.
\end{aligned}$$

As we illustrate in Example 5, it is often more efficient to evaluate a determinant by cofactor expansion along a row other than the first row. The important result presented next enables us to do this. (For a proof of this theorem, see [4].)

THEOREM 3.1

For any $i = 1, 2, \ldots, n$, we have

$$\det A = a_{i1}c_{i1} + a_{i2}c_{i2} + \cdots + a_{in}c_{in},$$

where c_{ij} denotes the (i,j)-cofactor of A.

The expression $a_{i1}c_{i1} + a_{i2}c_{i2} + \cdots + a_{in}c_{in}$ in Theorem 3.1 is called the **cofactor expansion** of A along row i. Thus the determinant of an $n \times n$ matrix can be evaluated using a cofactor expansion along *any* row.

Example 4 To illustrate Theorem 3.1, we compute the determinant of the matrix

$$A = \begin{bmatrix} 1 & 2 & 3 \\ 4 & 5 & 6 \\ 7 & 9 & 8 \end{bmatrix}$$

in Example 3 by using the cofactor expansion along the second row.

Solution Using the cofactor expansion of A along the second row, we have

$$\begin{aligned} \det A &= a_{21}c_{21} + a_{22}c_{22} + a_{23}c_{23} \\ &= 4(-1)^{2+1} \det A_{21} + 5(-1)^{2+2} \det A_{22} + 6(-1)^{2+3} \det A_{23} \\ &= 4(-1)^{2+1} \det \begin{bmatrix} 2 & 3 \\ 9 & 8 \end{bmatrix} + 5(-1)^{2+2} \det \begin{bmatrix} 1 & 3 \\ 7 & 8 \end{bmatrix} + 6(-1)^{2+3} \det \begin{bmatrix} 1 & 2 \\ 7 & 9 \end{bmatrix} \\ &= 4(-1)(2 \cdot 8 - 3 \cdot 9) + 5(1)(1 \cdot 8 - 3 \cdot 7) + 6(-1)(1 \cdot 9 - 2 \cdot 7) \\ &= 4(-1)(-11) + 5(1)(-13) + (6)(-1)(-5) \\ &= 44 - 65 + 30 \\ &= 9. \end{aligned}$$

Note that we obtained the same value for $\det A$ as in Example 3.

Practice Problem 3 ► Evaluate the determinant of

$$\begin{bmatrix} 1 & 3 & -3 \\ -3 & -5 & 2 \\ -4 & 4 & -6 \end{bmatrix}$$

by using the cofactor expansion along the second row. ◄

Example 5 Let

$$M = \begin{bmatrix} 1 & 2 & 3 & 8 & 5 \\ 4 & 5 & 6 & 9 & 1 \\ 7 & 9 & 8 & 4 & 7 \\ 0 & 0 & 0 & 1 & 0 \\ 0 & 0 & 0 & 0 & 1 \end{bmatrix}.$$

Since the last row of M has only one nonzero entry, the cofactor expansion of M along the last row has only one nonzero term. Thus the cofactor expansion of M along the last row involves only one-fifth the work of the cofactor expansion along the first row. Using the last row, we obtain

$$\det M = 0 + 0 + 0 + 0 + 1(-1)^{5+5} \det M_{55} = \det M_{55} = \det \begin{bmatrix} 1 & 2 & 3 & 8 \\ 4 & 5 & 6 & 9 \\ 7 & 9 & 8 & 4 \\ 0 & 0 & 0 & 1 \end{bmatrix}.$$

Once again, we use the cofactor expansion along the last row to compute $\det M_{55}$.

$$\det M = \det \begin{bmatrix} 1 & 2 & 3 & 8 \\ 4 & 5 & 6 & 9 \\ 7 & 9 & 8 & 4 \\ 0 & 0 & 0 & 1 \end{bmatrix} = 0 + 0 + 0 + 1(-1)^{4+4} \det \begin{bmatrix} 1 & 2 & 3 \\ 4 & 5 & 6 \\ 7 & 9 & 8 \end{bmatrix} = \det A$$

Here, A is the matrix in Example 3. Thus $\det M = \det A = 9$.

Note that M has the form

$$M = \begin{bmatrix} A & B \\ O & I_2 \end{bmatrix},$$

where

$$A = \begin{bmatrix} 1 & 2 & 3 \\ 4 & 5 & 6 \\ 7 & 9 & 8 \end{bmatrix} \quad \text{and} \quad B = \begin{bmatrix} 8 & 5 \\ 9 & 1 \\ 4 & 7 \end{bmatrix}.$$

More generally, the approach used in the preceding paragraph can be used to show that for any $m \times m$ matrix A and $m \times n$ matrix B,

$$\det \begin{bmatrix} A & B \\ O & I_n \end{bmatrix} = \det A.$$

Evaluating the determinant of an arbitrary matrix by cofactor expansion is extremely inefficient. In fact, it can be shown that the cofactor expansion of an arbitrary $n \times n$ matrix requires approximately $e \cdot n!$ arithmetic operations, where e is the base of the natural logarithm. Suppose that we have a computer capable of performing 1 billion arithmetic operations per second and we use it to evaluate the cofactor expansion of a 20×20 matrix (which in applied problems is a relatively small matrix).

For such a computer, the amount of time required to perform this calculation is approximately

$$\begin{aligned}\frac{e \cdot 20!}{10^9} \text{ seconds} &> 6.613 \cdot 10^9 \text{ seconds}\\ &> 1.837 \cdot 10^6 \text{ hours}\\ &> 76{,}542 \text{ days}\\ &> 209 \text{ years.}\end{aligned}$$

If determinants are to be of practical value, we must have an efficient method for evaluating them. The key to developing such a method is to observe that we can easily evaluate the determinant of a matrix such as

$$B = \begin{bmatrix} 3 & -4 & -7 & -5 \\ 0 & 8 & -2 & 6 \\ 0 & 0 & 9 & -1 \\ 0 & 0 & 0 & 4 \end{bmatrix}.$$

If we repeatedly evaluate the determinant by a cofactor expansion along the last row, we obtain

$$\begin{aligned}\det B = \det \begin{bmatrix} 3 & -4 & -7 & -5 \\ 0 & 8 & -2 & 6 \\ 0 & 0 & 9 & -1 \\ 0 & 0 & 0 & 4 \end{bmatrix} &= 4(-1)^{4+4} \cdot \det \begin{bmatrix} 3 & -4 & -7 \\ 0 & 8 & -2 \\ 0 & 0 & 9 \end{bmatrix}\\ &= 4 \cdot 9(-1)^{3+3} \cdot \det \begin{bmatrix} 3 & -4 \\ 0 & 8 \end{bmatrix}\\ &= 4 \cdot 9 \cdot 8(-1)^{2+2} \cdot \det [3]\\ &= 4 \cdot 9 \cdot 8 \cdot 3.\end{aligned}$$

In Section 2.6, we defined a matrix to be **upper triangular** if all its entries to the left and below the diagonal entries are zero, and to be **lower triangular** if all its entries to the right and above the diagonal entries are zero. The previous matrix B is an upper triangular 4×4 matrix, and its determinant equals the product of its diagonal entries. The determinant of any such matrix can be computed in a similar fashion.

THEOREM 3.2

The determinant of an upper triangular $n \times n$ matrix or a lower triangular $n \times n$ matrix equals the product of its diagonal entries.

An important consequence of Theorem 3.2 is that $\det I_n = 1$.

Example 6 Compute the determinants of

$$A = \begin{bmatrix} -2 & 0 & 0 & 0 \\ 8 & 7 & 0 & 0 \\ -6 & -1 & -3 & 0 \\ 4 & 3 & 9 & 5 \end{bmatrix} \quad \text{and} \quad B = \begin{bmatrix} 2 & 3 & 4 \\ 0 & 5 & 6 \\ 0 & 0 & 7 \end{bmatrix}.$$

Solution Since A is a lower triangular 4×4 matrix and B is an upper triangular 3×3 matrix, we have

$$\det A = (-2)(7)(-3)(5) = 210 \qquad \text{and} \qquad \det B = 2 \cdot 5 \cdot 7 = 70.$$

Practice Problem 4 ▶ Evaluate the determinant of the matrix

$$\begin{bmatrix} 4 & 0 & 0 & 0 \\ -2 & -1 & 0 & 0 \\ 8 & 7 & -2 & 0 \\ 9 & -5 & 6 & 3 \end{bmatrix}.$$ ◀

GEOMETRIC APPLICATIONS OF THE DETERMINANT*

Two vectors $\mathbf{u}$ and $\mathbf{v}$ in $\mathcal{R}^2$ determine a parallelogram having $\mathbf{u}$ and $\mathbf{v}$ as adjacent sides. (See Figure 3.1.) We call this the **parallelogram determined by u and v**. Note that if this parallelogram is rotated through the angle θ in Figure 3.1, we obtain the parallelogram determined by $\mathbf{u}'$ and $\mathbf{v}'$ in Figure 3.2.

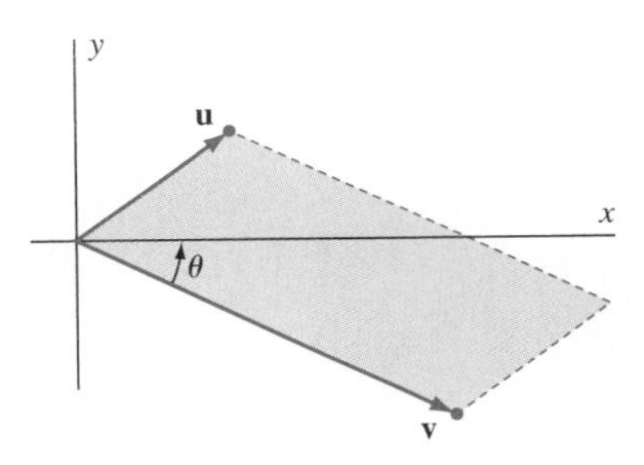

Figure 3.1 The parallelogram determined by **u** and **v**

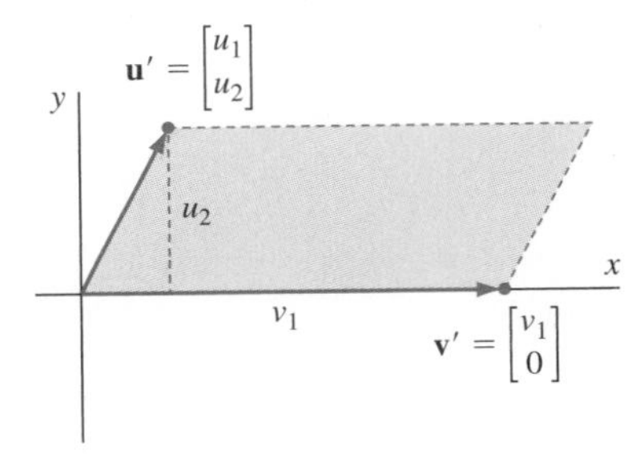

Figure 3.2 A rotation of the parallelogram determined by **u** and **v**

Suppose that

$$\mathbf{u}' = \begin{bmatrix} u_1 \\ u_2 \end{bmatrix} \qquad \text{and} \qquad \mathbf{v}' = \begin{bmatrix} v_1 \\ 0 \end{bmatrix}.$$

Then the parallelogram determined by $\mathbf{u}'$ and $\mathbf{v}'$ has base v_1 and height u_2, so its area is

$$v_1 u_2 = \left| \det \begin{bmatrix} u_1 & v_1 \\ u_2 & 0 \end{bmatrix} \right| = |\det [\mathbf{u}' \ \ \mathbf{v}']|.$$

Because a rotation maps a parallelogram into a congruent parallelogram, the parallelogram determined by $\mathbf{u}$ and $\mathbf{v}$ has the same area as that determined by $\mathbf{u}'$ and $\mathbf{v}'$. Recall that multiplying by the rotation matrix A_θ rotates a vector by the angle θ. Using the facts that $\det AB = (\det A)(\det B)$ for any 2×2 matrices A and B (Exercise 71) and $\det A_\theta = 1$ (Exercise 65), we see that the area of the parallelogram

* The remainder of this section may be omitted without loss of continuity.

determined by **u** and **v** is

$$\begin{aligned}|\det[\mathbf{u}' \ \ \mathbf{v}']| &= |\det[A_\theta\mathbf{u} \ \ A_\theta\mathbf{v}]|\\ &= |\det(A_\theta[\mathbf{u} \ \ \mathbf{v}])|\\ &= |(\det A_\theta)(\det[\mathbf{u} \ \ \mathbf{v}])|\\ &= |(1)(\det[\mathbf{u} \ \ \mathbf{v}])|\\ &= |\det[\mathbf{u} \ \ \mathbf{v}]|.\end{aligned}$$

> The area of the parallelogram determined by **u** and **v** is $|\det[\mathbf{u} \ \ \mathbf{v}]|$.

Moreover, a corresponding result can be proved for $\mathcal{R}^n$ by using the appropriate n-dimensional analog of area.

Example 7 The area of the parallelogram in $\mathcal{R}^2$ determined by the vectors

$$\begin{bmatrix}-2\\3\end{bmatrix} \quad \text{and} \quad \begin{bmatrix}1\\5\end{bmatrix}$$

is

$$\left|\det\begin{bmatrix}-2 & 1\\3 & 5\end{bmatrix}\right| = |(-2)(5) - (1)(3)| = |-13| = 13.$$

Example 8 The volume of the parallelepiped in $\mathcal{R}^3$ determined by the vectors

$$\begin{bmatrix}1\\1\\1\end{bmatrix}, \quad \begin{bmatrix}1\\-2\\1\end{bmatrix}, \quad \text{and} \quad \begin{bmatrix}1\\0\\-1\end{bmatrix}$$

is

$$\left|\det\begin{bmatrix}1 & 1 & 1\\1 & -2 & 0\\1 & 1 & -1\end{bmatrix}\right| = 6.$$

Note that the object in question is a rectangular parallelepiped with sides of length $\sqrt{3}$, $\sqrt{6}$, and $\sqrt{2}$. (See Figure 3.3.) Hence, by the familiar formula for volume, its volume should be $\sqrt{3}\cdot\sqrt{6}\cdot\sqrt{2} = 6$, as the determinant calculation shows.

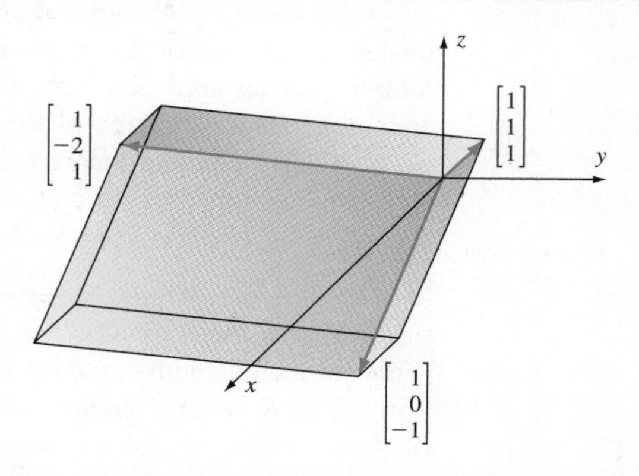

Figure 3.3 The parallelepiped determined by 3 vectors in $\mathcal{R}^3$

Practice Problem 5 ▶ What is the area of the parallelogram in $\mathcal{R}^2$ determined by the vectors

$$\begin{bmatrix} 4 \\ 3 \end{bmatrix} \quad \text{and} \quad \begin{bmatrix} 2 \\ 5 \end{bmatrix} \; ?$$

◀

The points on or within the parallelogram in $\mathcal{R}^2$ determined by $\mathbf{u}$ and $\mathbf{v}$ can be written in the form $a\mathbf{u} + b\mathbf{v}$, where a and b are scalars such that $0 \le a \le 1$ and $0 \le b \le 1$. (See Figure 3.4.) If $T\colon \mathcal{R}^2 \to \mathcal{R}^2$ is a linear transformation, then

$$T(a\mathbf{u} + b\mathbf{v}) = aT(\mathbf{u}) + bT(\mathbf{v}).$$

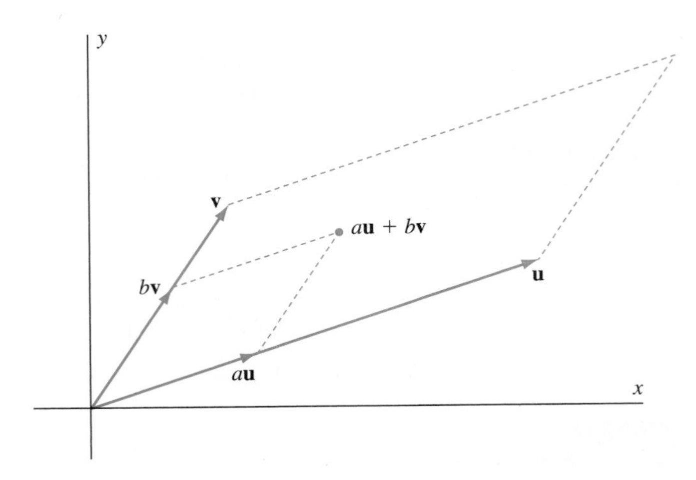

Figure 3.4 Points within the parallelogram determined by **u** and **v**

Hence the image under T of the parallelogram determined by $\mathbf{u}$ and $\mathbf{v}$ is the parallelogram determined by $T(\mathbf{u})$ and $T(\mathbf{v})$. The area of this parallelogram is

$$|\det[T(\mathbf{u}) \;\; T(\mathbf{v})]| = |\det[A\mathbf{u} \;\; A\mathbf{v}]| = |\det A[\mathbf{u} \;\; \mathbf{v}]| = |\det A| \cdot |\det[\mathbf{u} \;\; \mathbf{v}]|,$$

where A is the standard matrix of T. Thus the area of the image parallelogram is $|\det A|$ times larger than that of the parallelogram determined by $\mathbf{u}$ and $\mathbf{v}$. (If T is not invertible, then $\det A = 0$, and the parallelogram determined by $T(\mathbf{u})$ and $T(\mathbf{v})$ is degenerate.)

More generally, the area of any "sufficiently nice" region R in $\mathcal{R}^2$ can be approximated as the sum of the areas of rectangles. In fact, as the lengths of the sides of these rectangles approach zero, the sum of the areas of the rectangles approaches the area of R. Hence the area of the image of R under T equals $|\det A|$ times the area of R. A corresponding theorem can be proved about the volume of a region in $\mathcal{R}^3$ with similar properties. In fact, by the appropriate n-dimensional analog of volume, the following result is true:

> If R is a "sufficiently nice" region in $\mathcal{R}^n$ and $T\colon \mathcal{R}^n \to \mathcal{R}^n$ is an invertible linear transformation with standard matrix A, then the n-dimensional volume of the image of R under T equals $|\det A|$ times the n-dimensional volume of R.

This result plays a crucial role when we make a change of variables in calculus.

EXERCISES

In Exercises 1–8, compute the determinant of each matrix.

1. $\begin{bmatrix} 6 & 2 \\ -3 & -1 \end{bmatrix}$
2. $\begin{bmatrix} 4 & 5 \\ 3 & -7 \end{bmatrix}$
3. $\begin{bmatrix} -2 & 9 \\ 1 & 8 \end{bmatrix}$
4. $\begin{bmatrix} 9 & -2 \\ 3 & 4 \end{bmatrix}$
5. $\begin{bmatrix} -5 & -6 \\ 10 & 12 \end{bmatrix}$
6. $\begin{bmatrix} -7 & 8 \\ 4 & -5 \end{bmatrix}$
7. $\begin{bmatrix} 4 & 3 \\ -2 & -1 \end{bmatrix}$
8. $\begin{bmatrix} 4 & -2 \\ 3 & -1 \end{bmatrix}$

In Exercises 9–12, compute each indicated cofactor of the matrix

$$A = \begin{bmatrix} 9 & -2 & 4 \\ -1 & 6 & 3 \\ 7 & 8 & -5 \end{bmatrix}.$$

9. the $(1, 2)$-cofactor
10. the $(2, 3)$-cofactor
11. the $(3, 1)$-cofactor
12. the $(3, 3)$-cofactor

In Exercises 13–20, compute the determinant of each matrix A by a cofactor expansion along the indicated row.

13. $\begin{bmatrix} 2 & -1 & 3 \\ 1 & 4 & -2 \\ -1 & 0 & 1 \end{bmatrix}$ along the first row
14. $\begin{bmatrix} 1 & -2 & 2 \\ 2 & -1 & 3 \\ 0 & 1 & -1 \end{bmatrix}$ along the second row
15. $\begin{bmatrix} 1 & -2 & 2 \\ 2 & -1 & 3 \\ 0 & 1 & -1 \end{bmatrix}$ along the third row
16. $\begin{bmatrix} 2 & -1 & 3 \\ 1 & 4 & -2 \\ -1 & 0 & 1 \end{bmatrix}$ along the third row
17. $\begin{bmatrix} 1 & 4 & -3 \\ 5 & 0 & 0 \\ 2 & 0 & -1 \end{bmatrix}$ along the second row
18. $\begin{bmatrix} 4 & 1 & 0 \\ 0 & 3 & -2 \\ 2 & 0 & 5 \end{bmatrix}$ along the first row
19. $\begin{bmatrix} 1 & 2 & 1 & -1 \\ 0 & -1 & 0 & 1 \\ 4 & -3 & 2 & -1 \\ 0 & 3 & 0 & -2 \end{bmatrix}$ along the second row
20. $\begin{bmatrix} 0 & -1 & 0 & 1 \\ -2 & 3 & 1 & 4 \\ 1 & -2 & 2 & 3 \\ 0 & 1 & 0 & -2 \end{bmatrix}$ along the fourth row

In Exercises 21–28, compute each determinant by any legitimate method.

21. $\begin{bmatrix} 4 & -1 & 2 \\ 0 & 3 & 7 \\ 0 & 0 & 5 \end{bmatrix}$
22. $\begin{bmatrix} 8 & 0 & 0 \\ -1 & -2 & 0 \\ 4 & 5 & 3 \end{bmatrix}$
23. $\begin{bmatrix} -6 & 0 & 0 \\ 7 & -3 & 2 \\ 2 & 9 & 4 \end{bmatrix}$
24. $\begin{bmatrix} 7 & 1 & 8 \\ 0 & -3 & 4 \\ 0 & 0 & -2 \end{bmatrix}$
25. $\begin{bmatrix} 2 & 3 & 4 \\ 5 & 6 & 1 \\ 7 & 0 & 0 \end{bmatrix}$
26. $\begin{bmatrix} 5 & 1 & 1 \\ 0 & 2 & 0 \\ 6 & -4 & 3 \end{bmatrix}$
27. $\begin{bmatrix} -2 & -1 & -5 & 1 \\ 0 & 0 & 0 & 4 \\ 0 & -2 & 0 & 5 \\ 3 & 1 & 6 & -2 \end{bmatrix}$
28. $\begin{bmatrix} 4 & 2 & 2 & -3 \\ 6 & -1 & 1 & 5 \\ 0 & -3 & 0 & 0 \\ 2 & -5 & 0 & 0 \end{bmatrix}$

In Exercises 29–36, compute the area of each parallelogram determined by $\mathbf{u}$ and $\mathbf{v}$.

29. $\mathbf{u} = \begin{bmatrix} 3 \\ 5 \end{bmatrix}, \mathbf{v} = \begin{bmatrix} -2 \\ 7 \end{bmatrix}$
30. $\mathbf{u} = \begin{bmatrix} -3 \\ 6 \end{bmatrix}, \mathbf{v} = \begin{bmatrix} 8 \\ -5 \end{bmatrix}$
31. $\mathbf{u} = \begin{bmatrix} 6 \\ 4 \end{bmatrix}, \mathbf{v} = \begin{bmatrix} 3 \\ 2 \end{bmatrix}$
32. $\mathbf{u} = \begin{bmatrix} -1 \\ 2 \end{bmatrix}, \mathbf{v} = \begin{bmatrix} 4 \\ 5 \end{bmatrix}$
33. $\mathbf{u} = \begin{bmatrix} 4 \\ 3 \end{bmatrix}, \mathbf{v} = \begin{bmatrix} 6 \\ -1 \end{bmatrix}$
34. $\mathbf{u} = \begin{bmatrix} 4 \\ -2 \end{bmatrix}, \mathbf{v} = \begin{bmatrix} -2 \\ 5 \end{bmatrix}$
35. $\mathbf{u} = \begin{bmatrix} 6 \\ -1 \end{bmatrix}, \mathbf{v} = \begin{bmatrix} 4 \\ 3 \end{bmatrix}$
36. $\mathbf{u} = \begin{bmatrix} -2 \\ 4 \end{bmatrix}, \mathbf{v} = \begin{bmatrix} 5 \\ -2 \end{bmatrix}$

In Exercises 37–44, find each value of c for which the matrix is not invertible.

37. $\begin{bmatrix} 3 & 6 \\ c & 4 \end{bmatrix}$
38. $\begin{bmatrix} 9 & -18 \\ 4 & c \end{bmatrix}$
39. $\begin{bmatrix} c & 3 \\ 6 & -2 \end{bmatrix}$
40. $\begin{bmatrix} c & -1 \\ 2 & 5 \end{bmatrix}$
41. $\begin{bmatrix} c & -2 \\ -8 & c \end{bmatrix}$
42. $\begin{bmatrix} c & -3 \\ 4 & c \end{bmatrix}$
43. $\begin{bmatrix} c & 5 \\ -2 & c \end{bmatrix}$
44. $\begin{bmatrix} c & 9 \\ 4 & c \end{bmatrix}$

In Exercises 45–64, determine whether the statements are true or false.

45. The determinant of a matrix is a matrix of the same size.
46. $\det \begin{bmatrix} a & b \\ c & d \end{bmatrix} = ad + bc$.
47. If the determinant of a 2×2 matrix equals zero, then the matrix is invertible.
48. If a 2×2 matrix is invertible, then its determinant equals zero.
49. If B is a matrix obtained by multiplying each entry of some row of a 2×2 matrix A by the scalar k, then $\det B = k \det A$.
50. For $n \geq 2$, the (i, j)-cofactor of an $n \times n$ matrix A is the determinant of the $(n-1) \times (n-1)$ matrix obtained by deleting row i and column j from A.
51. For $n \geq 2$, the (i, j)-cofactor of an $n \times n$ matrix A equals $(-1)^{i+j}$ times the determinant of the $(n-1) \times (n-1)$ matrix obtained by deleting row i and column j from A.
52. The determinant of an $n \times n$ matrix can be evaluated by a cofactor expansion along any row.
53. Cofactor expansion is an efficient method for evaluating the determinant of a matrix.
54. The determinant of a matrix with integer entries must be an integer.

55. The determinant of a matrix with positive entries must be positive.
56. If some row of a square matrix consists only of zero entries, then the determinant of the matrix equals zero.
57. An upper triangular matrix must be square.
58. A matrix in which all the entries to the left and below the diagonal entries equal zero is called a lower triangular matrix.
59. A 4×4 upper triangular matrix has at most 10 nonzero entries.
60. The transpose of a lower triangular matrix is an upper triangular matrix.
61. The determinant of an upper triangular $n \times n$ matrix or a lower triangular $n \times n$ matrix equals the sum of its diagonal entries.
62. The determinant of I_n equals 1.
63. The area of the parallelogram determined by $\mathbf{u}$ and $\mathbf{v}$ is $\det [\mathbf{u}\ \ \mathbf{v}]$.
64. If $T: \mathcal{R}^2 \to \mathcal{R}^2$ is a linear transformation, then $\det [T(\mathbf{u})\ \ T(\mathbf{v})] = \det [\mathbf{u}\ \ \mathbf{v}]$ for any vectors $\mathbf{u}$ and $\mathbf{v}$ in $\mathcal{R}^2$.

---

65. Show that the determinant of the rotation matrix A_θ is 1.
66. Show that if A is a 2×2 matrix in which every entry is 0 or 1, then the determinant of A equals 0, 1, or -1.
67. Show that the conclusion of Exercise 66 is false for 3×3 matrices by calculating

$$\det \begin{bmatrix} 1 & 0 & 1 \\ 1 & 1 & 0 \\ 0 & 1 & 1 \end{bmatrix}.$$

68. Prove that if a 2×2 matrix has identical rows, then its determinant is zero.
69. Prove that, for any 2×2 matrix A, $\det A^T = \det A$.
70. Let A be a 2×2 matrix and k be a scalar. How does $\det kA$ compare with $\det A$? Justify your answer.
71. Prove that, for any 2×2 matrices A and B, $\det AB = (\det A)(\det B)$.
72. What is the determinant of an $n \times n$ matrix with a zero row? Justify your answer.

For each elementary matrix E in Exercises 73–76, and for the matrix

$$A = \begin{bmatrix} a & b \\ c & d \end{bmatrix},$$

verify that $\det EA = (\det E)(\det A)$.

73. $\begin{bmatrix} 1 & 0 \\ 0 & k \end{bmatrix}$ 74. $\begin{bmatrix} 0 & 1 \\ 1 & 0 \end{bmatrix}$ 75. $\begin{bmatrix} 1 & 0 \\ k & 1 \end{bmatrix}$ 76. $\begin{bmatrix} 1 & k \\ 0 & 1 \end{bmatrix}$

77. Prove that

$$\det \begin{bmatrix} a & b \\ c + kp & d + kq \end{bmatrix} = \det \begin{bmatrix} a & b \\ c & d \end{bmatrix} + k \cdot \det \begin{bmatrix} a & b \\ p & q \end{bmatrix}.$$

78. The TI-85 calculator gives

$$\det \begin{bmatrix} 1 & 2 & 3 \\ 2 & 3 & 4 \\ 3 & 4 & 5 \end{bmatrix} = -4 \times 10^{-13}.$$

Why must this answer be wrong? *Hint:* State a general fact about the determinant of a square matrix in which all the entries are integers.

79. Use a determinant to express the area of the triangle in $\mathcal{R}^2$ having vertices $\mathbf{0}$, $\mathbf{u}$, and $\mathbf{v}$.
80. Calculate the determinant of $\begin{bmatrix} O & I_m \\ I_n & O' \end{bmatrix}$ if O and O' are zero matrices.

In Exercises 81–84, use either a calculator with matrix capabilities or computer software such as MATLAB to solve each problem.

81. (a) Generate random 4×4 matrices A and B. Evaluate $\det A$, $\det B$, and $\det (A + B)$.
    (b) Repeat (a) with random 5×5 matrices.
    (c) Does $\det (A + B) = \det A + \det B$ appear to be true for all $n \times n$ matrices A and B?
82. (a) Generate random 4×4 matrices A and B. Evaluate $\det A$, $\det B$, and $\det (AB)$.
    (b) Repeat (a) with random 5×5 matrices.
    (c) Does $\det (AB) = (\det A)(\det B)$ appear to be true for all $n \times n$ matrices A and B?
83. (a) Generate a random 4×4 matrix A. Evaluate $\det A$ and $\det A^T$.
    (b) Repeat (a) with a random 5×5 matrix.
    (c) Do you suspect that $\det A = \det A^T$ might be true for all $n \times n$ matrices?
84. (a) Let

$$A = \begin{bmatrix} 0 & -1 & 2 & 2 \\ 1 & -1 & 0 & -2 \\ 2 & 1 & 0 & 1 \\ -1 & 1 & 2 & -1 \end{bmatrix}.$$

Show that A is invertible, and evaluate $\det A$ and $\det A^{-1}$.

    (b) Repeat (a) with a random invertible 5×5 matrix.
    (c) Make a conjecture about $\det A$ and $\det A^{-1}$ for any invertible matrix A.

SOLUTIONS TO THE PRACTICE PROBLEMS

1. We have $\det\begin{bmatrix} 8 & 3 \\ -6 & 5 \end{bmatrix} = 8(5) - 3(-6) = 40 + 18 = 58.$ Because its determinant is nonzero, the matrix is invertible. Furthermore,

$$A^{-1} = \frac{1}{8 \cdot 5 - 3 \cdot (-6)} \begin{bmatrix} 5 & -3 \\ 6 & 8 \end{bmatrix} = \frac{1}{58} \begin{bmatrix} 5 & -3 \\ 6 & 8 \end{bmatrix}.$$

2. The matrix $A - cI_2$ has the form

$$A - cI_2 = \begin{bmatrix} 4 & 6 \\ -1 & -3 \end{bmatrix} - c\begin{bmatrix} 1 & 0 \\ 0 & 1 \end{bmatrix} = \begin{bmatrix} 4-c & 6 \\ -1 & -3-c \end{bmatrix}.$$

Thus

$$\begin{aligned} \det(A - cI_2) &= (4-c)(-3-c) - 6(-1) \\ &= c^2 - c - 6 = (c-3)(c+2). \end{aligned}$$

Since $A - cI_2$ is not invertible if and only if we have $\det(A - cI_2) = 0$, we see that $A - cI_2$ is not invertible when $c - 3 = 0$ or $c + 2 = 0$, that is, when $c = 3$ or $c = -2$.

3. Let c_{ij} denote the (i,j)-cofactor of A. The cofactor expansion along the second row gives the following value for $\det A$:

$$\begin{aligned} \det A &= (-3)c_{21} + (-5)c_{22} + 2c_{23} \\ &= -3(-1)^{2+1} \cdot \det\begin{bmatrix} 3 & -3 \\ 4 & -6 \end{bmatrix} \\ &\quad + (-5)(-1)^{2+2} \cdot \det\begin{bmatrix} 1 & -3 \\ -4 & -6 \end{bmatrix} \\ &\quad + 2(-1)^{2+3} \cdot \det\begin{bmatrix} 1 & 3 \\ -4 & 4 \end{bmatrix} \\ &= -3(-1)[3(-6) - (-3)(4)] \\ &\quad + (-5)(1)[1(-6) - (-3)(-4)] \\ &\quad + 2(-1)[1(4) - 3(-4)] \\ &= 3(-6) + (-5)(-18) + (-2)(16) \\ &= 40 \end{aligned}$$

4. Because this matrix is lower triangular, its determinant is $4(-1)(-2)(3) = 24$, the product of its diagonal entries.

5. The area of the parallelogram in $\mathcal{R}^2$ determined by the vectors

$$\begin{bmatrix} 4 \\ 3 \end{bmatrix} \quad \text{and} \quad \begin{bmatrix} 2 \\ 5 \end{bmatrix}$$

is

$$\left|\det\begin{bmatrix} 4 & 2 \\ 3 & 5 \end{bmatrix}\right| = |4(5) - 2(3)| = |14| = 14.$$

3.2 PROPERTIES OF DETERMINANTS

We have seen that, for arbitrary matrices, the computation of determinants by cofactor expansion is quite inefficient. Theorem 3.2, however, provides a simple and efficient method for evaluating the determinant of an upper triangular matrix. Fortunately, the forward pass of the Gaussian elimination algorithm in Section 1.4 transforms any matrix into an upper triangular matrix by a sequence of elementary row operations. If we knew the effect of these elementary row operations on the determinant of a matrix, we could then evaluate the determinant efficiently by using Theorem 3.2.

For example, the following sequence of elementary row operations transforms

$$A = \begin{bmatrix} 1 & 2 & 3 \\ 4 & 5 & 6 \\ 7 & 9 & 8 \end{bmatrix}$$

into an upper triangular matrix U:

$$A = \begin{bmatrix} 1 & 2 & 3 \\ 4 & 5 & 6 \\ 7 & 9 & 8 \end{bmatrix} \xrightarrow{-4\mathbf{r}_1+\mathbf{r}_2\to\mathbf{r}_2} \begin{bmatrix} 1 & 2 & 3 \\ 0 & -3 & -6 \\ 7 & 9 & 8 \end{bmatrix} \xrightarrow{-7\mathbf{r}_1+\mathbf{r}_3\to\mathbf{r}_3} \begin{bmatrix} 1 & 2 & 3 \\ 0 & -3 & -6 \\ 0 & -5 & -13 \end{bmatrix}$$

$$\xrightarrow{-\frac{5}{3}\mathbf{r}_2+\mathbf{r}_3\to\mathbf{r}_3} \begin{bmatrix} 1 & 2 & 3 \\ 0 & -3 & -6 \\ 0 & 0 & -3 \end{bmatrix} = U$$

The three elementary row operations used in this transformation are row addition operations (operations that add a multiple of some row to another). Theorem 3.3 shows that this type of elementary row operation leaves the determinant unchanged. Hence the determinant of each matrix in the preceding sequence is the same, so

$$\det A = \det U = 1(-3)(-3) = 9.$$

This calculation is much more efficient than the one in Example 3 of Section 3.1.

The following theorem enables us to use elementary row operations to evaluate determinants:

THEOREM 3.3

Let A be an $n \times n$ matrix.

(a) If B is a matrix obtained by interchanging two rows of A, then $\det B = -\det A$.

(b) If B is a matrix obtained by multiplying each entry of some row of A by a scalar k, then $\det B = k \cdot \det A$.

(c) If B is a matrix obtained by adding a multiple of some row of A to a different row, then $\det B = \det A$.

(d) For any $n \times n$ elementary matrix E, we have $\det EA = (\det E)(\det A)$.

Parts (a), (b), and (c) of Theorem 3.3 describe how the determinant of a matrix changes when an elementary row operation is performed on the matrix. Its proof is found at the end of this section. Note that if $A = I_n$ in Theorem 3.3, then (a), (b), and (c) give the value of the determinant of each type of elementary matrix. In particular, $\det E = 1$ if E performs a row addition operation, and $\det E = -1$ if E performs a row interchange operation.

Suppose that an $n \times n$ matrix is transformed into an upper triangular matrix U by a sequence of elementary row operations other than scaling operations. (This can always be done by using steps 1–4 of the Gaussian elimination algorithm. The elementary row operations that occur are interchange operations in step 2 and row addition operations in step 3.) We saw in Section 2.3 that each of these elementary row operations can be implemented by multiplying by an elementary matrix. Thus there is a sequence of elementary matrices $E_1, E_2, \ldots, E_k$ such that

$$E_k \cdots E_2 E_1 A = U.$$

By Theorem 3.3(d), we have

$$(\det E_k) \cdots (\det E_2)(\det E_1)(\det A) = \det U.$$

Thus

$$(-1)^r \det A = \det U,$$

where r is the number of row interchange operations that occur in the transformation of A into U. Since U is an upper triangular matrix, its determinant is the product $u_{11}u_{22} \cdots u_{nn}$ of its diagonal entries, by Theorem 3.2. Hence we have the following important result, which provides an efficient method for evaluating a determinant:

If an $n \times n$ matrix A is transformed into an upper triangular matrix U by elementary row operations other than scaling operations, then

$$\det A = (-1)^r u_{11} u_{22} \cdots u_{nn},$$

where r is the number of row interchanges performed.

Example 1 Use elementary row operations to compute the determinant of

$$A = \begin{bmatrix} 0 & 1 & 3 & -3 \\ 0 & 0 & 4 & -2 \\ -2 & 0 & 4 & -7 \\ 4 & -4 & 4 & 15 \end{bmatrix}.$$

Solution We apply steps 1–4 of the Gaussian elimination algorithm to transform A into an upper triangular matrix U.

$$A = \begin{bmatrix} 0 & 1 & 3 & -3 \\ 0 & 0 & 4 & -2 \\ -2 & 0 & 4 & -7 \\ 4 & -4 & 4 & 15 \end{bmatrix} \xrightarrow{\mathbf{r}_1 \leftrightarrow \mathbf{r}_3} \begin{bmatrix} -2 & 0 & 4 & -7 \\ 0 & 0 & 4 & -2 \\ 0 & 1 & 3 & -3 \\ 4 & -4 & 4 & 15 \end{bmatrix}$$

$$\xrightarrow{2\mathbf{r}_1 + \mathbf{r}_4 \to \mathbf{r}_4} \begin{bmatrix} -2 & 0 & 4 & -7 \\ 0 & 0 & 4 & -2 \\ 0 & 1 & 3 & -3 \\ 0 & -4 & 12 & 1 \end{bmatrix} \xrightarrow{\mathbf{r}_2 \leftrightarrow \mathbf{r}_3} \begin{bmatrix} -2 & 0 & 4 & -7 \\ 0 & 1 & 3 & -3 \\ 0 & 0 & 4 & -2 \\ 0 & -4 & 12 & 1 \end{bmatrix}$$

$$\xrightarrow{4\mathbf{r}_2 + \mathbf{r}_4 \to \mathbf{r}_4} \begin{bmatrix} -2 & 0 & 4 & -7 \\ 0 & 1 & 3 & -3 \\ 0 & 0 & 4 & -2 \\ 0 & 0 & 24 & -11 \end{bmatrix} \xrightarrow{-6\mathbf{r}_3 + \mathbf{r}_4 \to \mathbf{r}_4} \begin{bmatrix} -2 & 0 & 4 & -7 \\ 0 & 1 & 3 & -3 \\ 0 & 0 & 4 & -2 \\ 0 & 0 & 0 & 1 \end{bmatrix} = U$$

Since U is an upper triangular matrix, we have $\det U = (-2) \cdot 1 \cdot 4 \cdot 1 = -8$. During the transformation of A into U, two row interchanges were performed. Thus

$$\det A = (-1)^2 \cdot \det U = -8.$$

Practice Problem 1 ▶ Use elementary row operations to evaluate the determinant of

$$A = \begin{bmatrix} 1 & 3 & -3 \\ -3 & -9 & 2 \\ -4 & 4 & -6 \end{bmatrix}.$$

◀

In Section 3.1, we mentioned that the cofactor expansion of an arbitrary $n \times n$ matrix requires approximately $e \cdot n!$ arithmetic operations. By contrast, evaluating the determinant of an $n \times n$ matrix with the use of elementary row operations requires only about $\frac{2}{3}n^3$ arithmetic operations. Thus a computer capable of performing 1 billion arithmetic operations per second could calculate the determinant of a 20×20 matrix in about 5 millionths of a second by using elementary row operations, compared with the more than 209 years it would need to evaluate the determinant by a cofactor expansion.

FOUR PROPERTIES OF DETERMINANTS

Several useful results about determinants can be proved from Theorem 3.3. Among these is the desired test for invertibility of a matrix.

THEOREM 3.4

Let A and B be square matrices of the same size. The following statements are true:

(a) A is invertible if and only if $\det A \neq 0$.

(b) $\det AB = (\det A)(\det B)$.

(c) $\det A^T = \det A$.

(d) If A is invertible, then $\det A^{-1} = \dfrac{1}{\det A}$.

PROOF We first prove (a), (b), and (c) for an invertible $n \times n$ matrix A. If A is invertible, there are elementary matrices $E_1, E_2, \ldots, E_k$ such that $A = E_k \cdots E_2E_1$ by the Invertible Matrix Theorem on page 138. Hence, by repeated applications of Theorem 3.3(d), we obtain

$$\det A = (\det E_k) \cdots (\det E_2)(\det E_1).$$

Since the determinant of an elementary matrix is nonzero, we have $\det A \neq 0$. This proves (a) for an invertible matrix. Moreover, for any $n \times n$ matrix B, repeated applications of Theorem 3.3(d) give

$$\begin{aligned}
(\det A)(\det B) &= (\det E_k) \cdots (\det E_2)(\det E_1)(\det B) \\
&= (\det E_k) \cdots (\det E_2)(\det E_1 B) \\
&\;\;\vdots \\
&= \det (E_k \cdots E_2 E_1 B) \\
&= \det AB.
\end{aligned}$$

This proves (b) when A is invertible. Furthermore, we also have

$$A^T = (E_k \cdots E_2E_1)^T = E_1^T E_2^T \cdots E_k^T.$$

We leave as an exercise the proof that $\det E^T = \det E$ for every elementary matrix E. It follows that

$$\begin{aligned}
\det A^T &= \det (E_1^T E_2^T \cdots E_k^T) \\
&= (\det E_1^T)(\det E_2^T) \cdots (\det E_k^T) \\
&= (\det E_1)(\det E_2) \cdots (\det E_k) \\
&= (\det E_k) \cdots (\det E_2)(\det E_1) \\
&= \det (E_k \cdots E_2E_1) \\
&= \det A,
\end{aligned}$$

proving (c) for an invertible matrix.

Now we prove (a), (b), and (c) in the case that A is an $n \times n$ matrix that is not invertible. By Theorem 2.3, there exists an invertible matrix P such that $PA = R$, where R is the reduced row echelon form of A. Since the rank of A is not n by the Invertible Matrix Theorem, the $n \times n$ matrix R must contain a row of zeros. Performing the cofactor expansion of R along this row yields $\det R = 0$. Because P^{-1} is invertible, (b) implies that

$$\det A = \det(P^{-1}R) = (\det P^{-1})(\det R) = (\det P^{-1}) \cdot 0 = 0,$$

completing the proof of (a).

To prove (b), first observe that since A is not invertible, AB is not invertible, because otherwise, for $C = B(AB)^{-1}$, we obtain $AC = I_n$, which is not possible by the Invertible Matrix Theorem. Therefore

$$\det AB = 0 = 0 \cdot \det B = (\det A)(\det B),$$

completing the proof of (b).

For the proof of (c), observe, by Theorem 2.2, that A^T is not invertible. (Otherwise, $(A^T)^T = A$ would be invertible.) Hence $\det A^T = 0 = \det A$ by (a). This completes the proof of (c).

The proof of (d) follows from (b) and the fact that $\det I_n = 1$. We leave the details as an exercise. ■

As we have said, our principal reason for studying determinants is that they provide a means for testing whether a matrix is invertible, namely, the result of Theorem 3.4(a). This fact is essential to Chapter 5, where it is used to determine the eigenvalues of a matrix. The next example illustrates how this test can be used.

Example 2 For what scalar c is the matrix

$$A = \begin{bmatrix} 1 & -1 & 2 \\ -1 & 0 & c \\ 2 & 1 & 7 \end{bmatrix}$$

not invertible?

Solution To answer this question, we compute the determinant of A. The following sequence of row addition operations transforms A into an upper triangular matrix:

$$\begin{bmatrix} 1 & -1 & 2 \\ -1 & 0 & c \\ 2 & 1 & 7 \end{bmatrix} \xrightarrow{\mathbf{r}_1+\mathbf{r}_2 \to \mathbf{r}_2} \begin{bmatrix} 1 & -1 & 2 \\ 0 & -1 & c+2 \\ 2 & 1 & 7 \end{bmatrix} \xrightarrow{-2\mathbf{r}_1+\mathbf{r}_3 \to \mathbf{r}_3} \begin{bmatrix} 1 & -1 & 2 \\ 0 & -1 & c+2 \\ 0 & 3 & 3 \end{bmatrix}$$

$$\xrightarrow{3\mathbf{r}_2+\mathbf{r}_3 \to \mathbf{r}_3} \begin{bmatrix} 1 & -1 & 2 \\ 0 & -1 & c+2 \\ 0 & 0 & 3c+9 \end{bmatrix}$$

Hence $\det A = 1(-1)(3c+9) = -3c-9$. Theorem 3.4(a) states that A is not invertible if and only if its determinant is 0. Thus A is not invertible if and only if $0 = -3c - 9$, that is, if and only if $c = -3$.

Practice Problem 2 ▶ For what value of c is

$$B = \begin{bmatrix} 1 & -1 & 2 \\ -1 & 0 & c \\ 2 & 1 & 4 \end{bmatrix}$$

not invertible? ◀

For those familiar with partitioned matrices, the following example illustrates how Theorem 3.4(b) can be used:

Example 3 Suppose that a matrix M can be partitioned as

$$\begin{bmatrix} A & B \\ O & C \end{bmatrix},$$

where A is an $m \times m$ matrix, C is an $n \times n$ matrix, and O is the $n \times m$ zero matrix. Show that $\det M = (\det A)(\det C)$.

Solution Using block multiplication, we see that

$$\begin{bmatrix} I_m & O' \\ O & C \end{bmatrix} \begin{bmatrix} A & B \\ O & I_n \end{bmatrix} = \begin{bmatrix} A & B \\ O & C \end{bmatrix} = M,$$

where O' is the $m \times n$ zero matrix. Therefore, by Theorem 3.4,

$$\det \begin{bmatrix} I_m & O' \\ O & C \end{bmatrix} \cdot \det \begin{bmatrix} A & B \\ O & I_n \end{bmatrix} = \det M.$$

As in Example 5 of Section 3.1, it can be shown that

$$\det \begin{bmatrix} A & B \\ O & I_n \end{bmatrix} = \det A.$$

A similar argument (using cofactor expansion along the first row) yields

$$\det \begin{bmatrix} I_m & O' \\ O & C \end{bmatrix} = \det C.$$

Hence

$$\det M = \det \begin{bmatrix} I_m & O' \\ O & C \end{bmatrix} \cdot \det \begin{bmatrix} A & B \\ O & I_n \end{bmatrix} = (\det C)(\det A).$$

Several important theoretical results follow from Theorem 3.4(c). For example, we can evaluate the determinant of a matrix A by computing the determinant of its transpose instead. Thus we can evaluate the determinant of A by a cofactor expansion along the rows of A^T. But the rows of A^T are the columns of A, so the determinant of A can be evaluated by cofactor expansion along *any column* of A, as well as any row.

CAUTION Let A and B be arbitrary $n \times n$ matrices. Although $\det AB = (\det A)(\det B)$ by Theorem 3.4(b), the corresponding property for matrix addition is *not* true. Consider, for instance, the matrices

$$A = \begin{bmatrix} 1 & 0 \\ 0 & 0 \end{bmatrix} \quad \text{and} \quad B = \begin{bmatrix} 0 & 0 \\ 0 & 1 \end{bmatrix}.$$

Clearly, $\det A = \det B = 0$, whereas $\det(A+B) = \det I_2 = 1$. Therefore

$$\det(A+B) \neq \det A + \det B.$$

So the determinant of a sum of matrices need not be equal to the sum of their determinants.

CRAMER'S RULE*

One of the original motivations for studying determinants was that they provide a method for solving systems of linear equations having an invertible coefficient matrix. The following result was published in 1750 by the Swiss mathematician Gabriel Cramer (1704–1752):

THEOREM 3.5

(Cramer's Rule[2]) Let A be an invertible $n \times n$ matrix, $\mathbf{b}$ be in $\mathcal{R}^n$, and M_i be the matrix obtained from A by replacing column i of A by $\mathbf{b}$. Then $A\mathbf{x} = \mathbf{b}$ has a unique solution $\mathbf{u}$ in which the components are given by

$$u_i = \frac{\det M_i}{\det A} \quad \text{for } i = 1, 2, \ldots, n.$$

PROOF Since A is invertible, $A\mathbf{x} = \mathbf{b}$ has the unique solution $\mathbf{u} = A^{-1}\mathbf{b}$, as we saw in Section 2.3. For each i, let U_i denote the matrix obtained by replacing column i of I_n by

$$\mathbf{u} = \begin{bmatrix} u_1 \\ u_2 \\ \vdots \\ u_n \end{bmatrix}.$$

Then

$$\begin{aligned} AU_i &= A[\mathbf{e}_1 \ \cdots \ \mathbf{e}_{i-1} \ \mathbf{u} \ \mathbf{e}_{i+1} \ \cdots \ \mathbf{e}_n] \\ &= [A\mathbf{e}_1 \ \cdots \ A\mathbf{e}_{i-1} \ A\mathbf{u} \ A\mathbf{e}_{i+1} \ \cdots \ A\mathbf{e}_n] \\ &= [\mathbf{a}_1 \ \cdots \ \mathbf{a}_{i-1} \ \mathbf{b} \ \mathbf{a}_{i+1} \ \cdots \ \mathbf{a}_n] \\ &= M_i. \end{aligned}$$

* The remainder of this section may be omitted without loss of continuity.

[2] Cramer's rule was first stated in its most general form in a 1750 paper by the Swiss mathematician Gabriel Cramer (1704–1752), where it was used to find the equations of curves in the plane passing through given points.

Evaluating U_i by cofactor expansion along row i produces

$$\det U_i = u_i \cdot \det I_{n-1} = u_i.$$

Hence, by Theorem 3.4(b), we have

$$\det M_i = \det AU_i = (\det A) \cdot (\det U_i) = (\det A) \cdot u_i.$$

Because $\det A \neq 0$ by Theorem 3.4(a), it follows that

$$u_i = \frac{\det M_i}{\det A}.$$

■

Example 4 Use Cramer's rule to solve the system of equations

$$\begin{aligned} x_1 + 2x_2 + 3x_3 &= 2 \\ x_1 + \phantom{2x_2 + {}} x_3 &= 3 \\ x_1 + x_2 - x_3 &= 1. \end{aligned}$$

Solution The coefficient matrix of this system is

$$A = \begin{bmatrix} 1 & 2 & 3 \\ 1 & 0 & 1 \\ 1 & 1 & -1 \end{bmatrix}.$$

Since $\det A = 6$, A is invertible by Theorem 3.4(a), and hence Cramer's rule can be used. In the notation of Theorem 3.5, we have

$$M_1 = \begin{bmatrix} 2 & 2 & 3 \\ 3 & 0 & 1 \\ 1 & 1 & -1 \end{bmatrix}, \quad M_2 = \begin{bmatrix} 1 & 2 & 3 \\ 1 & 3 & 1 \\ 1 & 1 & -1 \end{bmatrix}, \quad \text{and} \quad M_3 = \begin{bmatrix} 1 & 2 & 2 \\ 1 & 0 & 3 \\ 1 & 1 & 1 \end{bmatrix}.$$

Therefore the unique solution of the given system is the vector with components

$$u_1 = \frac{\det M_1}{\det A} = \frac{15}{6} = \frac{5}{2}, \quad u_2 = \frac{\det M_2}{\det A} = \frac{-6}{6} = -1, \quad \text{and} \quad u_3 = \frac{\det M_3}{\det A} = \frac{3}{6} = \frac{1}{2}.$$

It is readily checked that these values satisfy each of the equations in the given system.

Practice Problem 3 ► Solve the following system of linear equations using Cramer's rule:

$$\begin{aligned} 3x_1 + 8x_2 &= 4 \\ 2x_1 + 6x_2 &= 2 \end{aligned}$$

◄

We noted earlier that evaluating the determinant of an $n \times n$ matrix with the use of elementary row operations requires about $\frac{2}{3}n^3$ arithmetic operations. On the other hand, we saw in Section 2.6 that an entire system of n linear equations in n unknowns can be solved by Gaussian elimination with about the same number of operations. Thus Cramer's rule is not an efficient method for solving systems of linear equations; moreover, it can be used only in the special case where the coefficient matrix is invertible. Nevertheless, Cramer's rule is useful for certain theoretical purposes. It can be used, for example, to analyze how the solution of $A\mathbf{x} = \mathbf{b}$ is influenced by changes in $\mathbf{b}$.

We conclude this section with a proof of Theorem 3.3.

Proof of Theorem 3.3. Let A be an $n \times n$ matrix with rows $\mathbf{a}_1', \mathbf{a}_2', \ldots, \mathbf{a}_n'$, respectively.

(a) Suppose that B is the matrix obtained from A by interchanging rows r and s, where $r < s$. We begin by establishing the result in the case that $s = r + 1$. In this case, $a_{rj} = b_{sj}$ and $A_{rj} = B_{sj}$ for each j. Thus each cofactor in the cofactor expansion of B along row s is the negative of the corresponding cofactor in the cofactor expansion of A along row r. It follows that $\det B = -\det A$.

Now suppose that $s > r + 1$. Beginning with rows r and $r + 1$, successively interchange $\mathbf{a}_r'$ with the following row until the rows of A are in the order

$$\mathbf{a}_1', \ldots, \mathbf{a}_{r-1}', \mathbf{a}_{r+1}', \ldots, \mathbf{a}_s', \mathbf{a}_r', \mathbf{a}_{s+1}', \ldots, \mathbf{a}_n'.$$

A total of $s - r$ interchanges are necessary to produce this ordering. Now successively interchange $\mathbf{a}_s'$ with the preceding row until the rows are in the order

$$\mathbf{a}_1', \ldots, \mathbf{a}_{r-1}', \mathbf{a}_s', \mathbf{a}_{r+1}', \ldots, \mathbf{a}_{s-1}', \mathbf{a}_r', \mathbf{a}_{s+1}', \ldots, \mathbf{a}_n'.$$

This process requires another $s - r - 1$ interchanges of adjacent rows and produces the matrix B. Thus the preceding paragraph shows that

$$\det B = (-1)^{s-r}(-1)^{s-r-1} \cdot \det A = (-1)^{2(s-r)-1} \cdot \det A = -\det A.$$

(b) Suppose that B is a matrix obtained by multiplying each entry of row r of A by a scalar k. Comparing the cofactor expansion of B along row r with that of A, it is easy to see that $\det B = k \cdot \det A$. We leave the details to the reader.

(c) We first show that if C is an $n \times n$ matrix having two identical rows, then $\det C = 0$. Suppose that rows r and s of C are equal, and let M be obtained from C by interchanging rows r and s. Then $\det M = -\det C$ by (a). But since rows r and s of C are equal, we also have $C = M$. Thus $\det C = \det M$. Combining the two equations involving $\det M$, we obtain $\det C = -\det C$. Therefore $\det C = 0$.

Now suppose that B is obtained from A by adding k times row s of A to row r, where $r \neq s$. Let C be the $n \times n$ matrix obtained from A by replacing $\mathbf{a}_r' = [u_1, u_2, \ldots, u_n]$ by $\mathbf{a}_s' = [v_1, v_2, \ldots, v_n]$. Since A, B, and C differ only in row r, we have $A_{rj} = B_{rj} = C_{rj}$ for every j. Using the cofactor expansion of B along row r, we obtain

$$\begin{aligned}
\det B &= (u_1 + kv_1)(-1)^{r+1} \det B_{r1} + \cdots + (u_n + kv_n)(-1)^{r+n} \det B_{rn} \\
&= \left(u_1(-1)^{r+1} \det B_{r1} + \cdots + u_n(-1)^{r+n} \det B_{rn}\right) \\
&\qquad + k\left(v_1(-1)^{r+1} \det B_{r1} + \cdots + v_n(-1)^{r+n} \det B_{rn}\right) \\
&= \left[u_1(-1)^{r+1} \det A_{r1} + \cdots + u_n(-1)^{r+n} \det A_{rn}\right] \\
&\qquad + k\left[v_1(-1)^{r+1} \det C_{r1} + \cdots + v_n(-1)^{r+n} \det C_{rn}\right].
\end{aligned}$$

In this equation, the first expression in brackets is the cofactor expansion of A along row r, and the second is the cofactor expansion of C along row r. Thus we have

$$\det B = \det A + k \cdot \det C.$$

However, C is a matrix with two identical rows (namely, rows r and s, which are both equal to $\mathbf{a}_s'$). Since $\det C = 0$ by the preceding paragraph, it follows that $\det B = \det A$.

(d) Let E be an elementary matrix obtained by interchanging two rows of I_n. Then $\det EA = -\det A$ by (a). Since $\det E = -1$, we have $\det EA = (\det E)(\det A)$. Similar arguments establish (d) for the other two types of elementary matrices. ■

EXERCISES

In Exercises 1–10, evaluate the determinant of each matrix using a cofactor expansion along the indicated column.

1. $\begin{bmatrix} 1 & 0 & -1 \\ -1 & 0 & 4 \\ 2 & 3 & -2 \end{bmatrix}$ second column

2. $\begin{bmatrix} 1 & -2 & 2 \\ 2 & -1 & 3 \\ 0 & 1 & -1 \end{bmatrix}$ first column

3. $\begin{bmatrix} 2 & -1 & 3 \\ 1 & 4 & -2 \\ -1 & 0 & 1 \end{bmatrix}$ second column

4. $\begin{bmatrix} -1 & 2 & 1 \\ 5 & -9 & -2 \\ 3 & -1 & 2 \end{bmatrix}$ third column

5. $\begin{bmatrix} 1 & 3 & 2 \\ 2 & 2 & 3 \\ 3 & 1 & 1 \end{bmatrix}$ third column

6. $\begin{bmatrix} 1 & 3 & 2 \\ 2 & 2 & 3 \\ 3 & 1 & 1 \end{bmatrix}$ first column

7. $\begin{bmatrix} 0 & 2 & 0 \\ 1 & 1 & 2 \\ 0 & -1 & 1 \end{bmatrix}$ first column

8. $\begin{bmatrix} 0 & 2 & 0 \\ 1 & 1 & 2 \\ 0 & -1 & 1 \end{bmatrix}$ third column

9. $\begin{bmatrix} 3 & 2 & 1 \\ 1 & 0 & -1 \\ -2 & -1 & 1 \end{bmatrix}$ second column

10. $\begin{bmatrix} 3 & 2 & 1 \\ 1 & 0 & -1 \\ -2 & -1 & 1 \end{bmatrix}$ third column

In Exercises 11–24, evaluate the determinant of each matrix using elementary row operations.

11. $\begin{bmatrix} 0 & 0 & 5 \\ 0 & 3 & 7 \\ 4 & -1 & -2 \end{bmatrix}$

12. $\begin{bmatrix} -6 & 0 & 0 \\ 2 & 9 & 4 \\ 7 & -3 & 0 \end{bmatrix}$

13. $\begin{bmatrix} 1 & -2 & 2 \\ 0 & 5 & -1 \\ 2 & -4 & 1 \end{bmatrix}$

14. $\begin{bmatrix} -2 & 1 & -2 \\ 4 & -2 & -1 \\ 0 & 3 & 6 \end{bmatrix}$

15. $\begin{bmatrix} 3 & -2 & 1 \\ 0 & 0 & 5 \\ -9 & 4 & 2 \end{bmatrix}$

16. $\begin{bmatrix} -2 & 6 & 1 \\ 0 & 0 & 3 \\ 4 & -1 & 2 \end{bmatrix}$

17. $\begin{bmatrix} 1 & 4 & 2 \\ 2 & -1 & 3 \\ -1 & 3 & 1 \end{bmatrix}$

18. $\begin{bmatrix} -1 & 2 & 1 \\ 5 & -9 & -2 \\ 3 & -1 & 2 \end{bmatrix}$

19. $\begin{bmatrix} 1 & 2 & 1 \\ 1 & 1 & 2 \\ 3 & 4 & 8 \end{bmatrix}$

20. $\begin{bmatrix} 3 & 4 & 2 \\ 2 & -1 & 3 \\ -1 & 3 & 1 \end{bmatrix}$

21. $\begin{bmatrix} 1 & -1 & 2 & 1 \\ 2 & -1 & -1 & 4 \\ -4 & 5 & -10 & -6 \\ 3 & -2 & 10 & -1 \end{bmatrix}$

22. $\begin{bmatrix} 2 & 1 & 5 & 2 \\ 2 & 1 & 8 & 1 \\ 2 & -1 & 5 & 3 \\ 4 & -2 & 10 & 3 \end{bmatrix}$

23. $\begin{bmatrix} 0 & 4 & -1 & 1 \\ -3 & 1 & 1 & 2 \\ 1 & 0 & -2 & 3 \\ 2 & 3 & 0 & 1 \end{bmatrix}$

24. $\begin{bmatrix} 1 & -1 & 2 & -1 \\ 2 & -2 & -3 & 8 \\ -3 & 4 & 1 & -1 \\ -2 & 6 & -4 & 18 \end{bmatrix}$

For each of the matrices in Exercises 25–38, determine the value(s) of c for which the given matrix is not invertible.

25. $\begin{bmatrix} 4 & c \\ 3 & -6 \end{bmatrix}$

26. $\begin{bmatrix} 3 & 9 \\ 5 & c \end{bmatrix}$

27. $\begin{bmatrix} c & 6 \\ 2 & c+4 \end{bmatrix}$

28. $\begin{bmatrix} c & c-1 \\ -8 & c-6 \end{bmatrix}$

29. $\begin{bmatrix} 1 & 2 & -1 \\ 0 & -1 & c \\ 3 & 4 & 7 \end{bmatrix}$

30. $\begin{bmatrix} 1 & 2 & -6 \\ 2 & 4 & c \\ -3 & -5 & 7 \end{bmatrix}$

31. $\begin{bmatrix} 1 & -1 & 2 \\ -1 & 0 & 4 \\ 2 & 1 & c \end{bmatrix}$

32. $\begin{bmatrix} 1 & 2 & c \\ -2 & -2 & 4 \\ 1 & 6 & -12 \end{bmatrix}$

33. $\begin{bmatrix} 1 & 2 & -1 \\ 2 & 3 & c \\ 0 & c & -15 \end{bmatrix}$

34. $\begin{bmatrix} -1 & 1 & 1 \\ 3 & -2 & -c \\ 0 & c & -10 \end{bmatrix}$

35. $\begin{bmatrix} 1 & 0 & -1 & 1 \\ 0 & -1 & 2 & -1 \\ 1 & -1 & 1 & -1 \\ -1 & 1 & c & 0 \end{bmatrix}$

36. $\begin{bmatrix} 1 & 0 & -1 & 1 \\ 0 & -1 & 2 & -1 \\ 1 & -1 & 1 & -1 \\ -1 & 1 & 0 & c \end{bmatrix}$

37. $\begin{bmatrix} 1 & 0 & -1 & 1 \\ 0 & -1 & 2 & -1 \\ 1 & -1 & c & -1 \\ -1 & 1 & 0 & 2 \end{bmatrix}$

38. $\begin{bmatrix} 1 & 0 & -1 & 1 \\ 0 & -1 & 2 & -1 \\ 1 & -1 & 1 & c \\ -1 & 1 & 0 & 2 \end{bmatrix}$

T&F *In Exercises 39–58, determine whether the statements are true or false.*

39. The determinant of a square matrix equals the product of its diagonal entries.

40. Performing a row addition operation on a square matrix does not change its determinant.
41. Performing a scaling operation on a square matrix does not change its determinant.
42. Performing an interchange operation on a square matrix changes its determinant by a factor of -1.
43. For any $n \times n$ matrices A and B, we have $\det(A+B) = \det A + \det B$.
44. For any $n \times n$ matrices A and B, $\det AB = (\det A)(\det B)$.
45. If A is any invertible matrix, then $\det A = 0$.
46. For any square matrix A, $\det A^T = -\det A$.
47. The determinant of any square matrix can be evaluated by a cofactor expansion along any column.
48. The determinant of any square matrix equals the product of the diagonal entries of its reduced row echelon form.
49. If $\det A \neq 0$, then A is an invertible matrix.
50. The determinant of the $n \times n$ identity matrix is 1.
51. If A is any square matrix and c is any scalar, then $\det cA = c \det A$.
52. Cramer's rule can be used to solve any system of n linear equations in n variables.
53. To solve a system of 5 linear equations in 5 variables with Cramer's rule, the determinants of six 5×5 matrices must be evaluated.
54. If A is an invertible matrix, then $\det A^{-1} = \dfrac{1}{\det A}$.
55. If A is a 4×4 matrix, then $\det(-A) = \det A$.
56. If A is a 5×5 matrix, then $\det(-A) = \det A$.
57. For any square matrix A and any positive integer k, $\det(A^k) = (\det A)^k$.
58. If an $n \times n$ matrix A is transformed into an upper triangular matrix U by only row interchanges and row addition operations, then $\det A = u_{11}u_{22}\cdots u_{nn}$.

In Exercises 59–66, solve each system using Cramer's rule.

59. $\begin{aligned} x_1 + 2x_2 &= 6 \\ 3x_1 + 4x_2 &= -3 \end{aligned}$

60. $\begin{aligned} 2x_1 + 3x_2 &= 7 \\ 3x_1 + 4x_2 &= 6 \end{aligned}$

61. $\begin{aligned} 2x_1 + 4x_2 &= -2 \\ 7x_1 + 12x_2 &= 5 \end{aligned}$

62. $\begin{aligned} 3x_1 + 2x_2 &= -6 \\ 6x_1 + 5x_2 &= 9 \end{aligned}$

63. $\begin{aligned} x_1 \qquad - 2x_3 &= 6 \\ -x_1 + x_2 + 3x_3 &= -5 \\ 2x_2 + x_3 &= 4 \end{aligned}$

64. $\begin{aligned} -x_1 + 2x_2 + x_3 &= -3 \\ x_2 + 2x_3 &= -1 \\ x_1 - x_2 + 3x_3 &= 4 \end{aligned}$

65. $\begin{aligned} 2x_1 - x_2 + x_3 &= -5 \\ x_1 \qquad - x_3 &= 2 \\ -x_1 + 3x_2 + 2x_3 &= 1 \end{aligned}$

66. $\begin{aligned} -2x_1 + 3x_2 + x_3 &= -2 \\ 3x_1 + x_2 - x_3 &= 1 \\ -x_1 + 2x_2 + x_3 &= -1 \end{aligned}$

67. Give an example to show that $\det kA \neq k \det A$ for some matrix A and scalar k.
68. Evaluate $\det kA$ if A is an $n \times n$ matrix and k is a scalar. Justify your answer.
69. Prove that if A is an invertible matrix, then $\det A^{-1} = \dfrac{1}{\det A}$.
70. Under what conditions is $\det(-A) = -\det A$? Justify your answer.
71. Let A and B be $n \times n$ matrices such that B is invertible. Prove that $\det(B^{-1}AB) = \det A$.
72. An $n \times n$ matrix A is called *nilpotent* if, for some positive integer k, $A^k = O$, where O is the $n \times n$ zero matrix. Prove that if A is nilpotent, then $\det A = 0$.
73. An $n \times n$ matrix Q is called *orthogonal* if $Q^TQ = I_n$. Prove that if Q is orthogonal, then $\det Q = \pm 1$.
74. A square matrix A is called *skew-symmetric* if $A^T = -A$. Prove that if A is a skew-symmetric $n \times n$ matrix and n is odd, then A is not invertible. What if n is even?
75. The matrix

$$A = \begin{bmatrix} 1 & a & a^2 \\ 1 & b & b^2 \\ 1 & c & c^2 \end{bmatrix}$$

is called a *Vandermonde matrix.* Show that

$$\det A = (b-a)(c-a)(c-b).$$

76. Use properties of determinants to show that the equation of the line in $\mathcal{R}^2$ passing through the points (x_1, y_1) and (x_2, y_2) can be written

$$\det \begin{bmatrix} 1 & x_1 & y_1 \\ 1 & x_2 & y_2 \\ 1 & x & y \end{bmatrix} = 0.$$

77. Let $\mathcal{B} = \{\mathbf{b}_1, \mathbf{b}_2, \ldots, \mathbf{b}_n\}$ be a subset of $\mathcal{R}^n$ containing n distinct vectors, and let $B = [\mathbf{b}_1\ \mathbf{b}_2\ \ldots\ \mathbf{b}_n]$. Prove that $\mathcal{B}$ is linearly independent if and only if $\det B \neq 0$.
78. Let A be an $n \times n$ matrix with rows $\mathbf{a}'_1, \mathbf{a}'_2, \ldots, \mathbf{a}'_n$ and B be the $n \times n$ matrix with rows $\mathbf{a}'_n, \ldots, \mathbf{a}'_2, \mathbf{a}'_1$. How are the determinants of A and B related? Justify your answer.
79. Complete the proof of Theorem 3.3(b).
80. Complete the proof of Theorem 3.3(d).
81. Prove that $\det E^T = \det E$ for every elementary matrix E.
82. Let A be an $n \times n$ matrix and b_{jk} denote the (k,j)-cofactor of A.

    (a) Prove that if P is the matrix obtained from A by replacing column k by $\mathbf{e}_j$, then $\det P = b_{kj}$.

    (b) Show that for each j, we have

$$A \begin{bmatrix} b_{1j} \\ b_{2j} \\ \vdots \\ b_{nj} \end{bmatrix} = (\det A) \cdot \mathbf{e}_j.$$

Hint: Apply Cramer's rule to $A\mathbf{x} = \mathbf{e}_j$.

    (c) Deduce that if B is the $n \times n$ matrix whose (i,j)-entry is b_{ij}, then $AB = (\det A)I_n$. This matrix B is called the *classical adjoint* of A.

    (d) Show that if $\det A \neq 0$, then $A^{-1} = \dfrac{1}{\det A}B$.

In Exercises 83–85, use either a calculator with matrix capabilities or computer software such as MATLAB to solve the problem.

83. (a) Use elementary row operations other than scaling operations to transform

$$A = \begin{bmatrix} 0.0 & -3.0 & -2 & -5 \\ 2.4 & 3.0 & -6 & 9 \\ -4.8 & 6.3 & 4 & -2 \\ 9.6 & 1.5 & 5 & 9 \end{bmatrix}$$

into an upper triangular matrix U.

(b) Use the boxed result on page 213 to compute $\det A$.

84. (a) Solve $A\mathbf{x} = \mathbf{b}$ by using Cramer's rule, where

$$A = \begin{bmatrix} 0 & 1 & 2 & -1 \\ 1 & 2 & 1 & -2 \\ 2 & -1 & 0 & 3 \\ 3 & 0 & -3 & 1 \end{bmatrix} \quad \text{and} \quad \mathbf{b} = \begin{bmatrix} 24 \\ -16 \\ 8 \\ 10 \end{bmatrix}.$$

(b) How many determinants of 4×4 matrices are evaluated in (a)?

85. Compute the classical adjoint (as defined in Exercise 82) of the matrix A in Exercise 84.

SOLUTIONS TO THE PRACTICE PROBLEMS

1. The following sequence of elementary row operations transforms A into an upper triangular matrix U:

$$\begin{bmatrix} 1 & 3 & -3 \\ -3 & -9 & 2 \\ -4 & 4 & -6 \end{bmatrix} \xrightarrow[4\mathbf{r}_1+\mathbf{r}_3 \to \mathbf{r}_3]{3\mathbf{r}_1+\mathbf{r}_2 \to \mathbf{r}_2} \begin{bmatrix} 1 & 3 & -3 \\ 0 & 0 & -7 \\ 0 & 16 & -18 \end{bmatrix}$$

$$\xrightarrow{\mathbf{r}_2 \leftrightarrow \mathbf{r}_3} \begin{bmatrix} 1 & 3 & -3 \\ 0 & 16 & -18 \\ 0 & 0 & -7 \end{bmatrix} = U$$

Since one row interchange operation was performed, we have

$$\det A = (-1)^1 \cdot \det U = (-1)(1)(16)(-7) = 112.$$

2. The following sequence of elementary row operations transforms B into an upper triangular matrix:

$$\begin{bmatrix} 1 & -1 & 2 \\ -1 & 0 & c \\ 2 & 1 & 4 \end{bmatrix} \xrightarrow[-2\mathbf{r}_1+\mathbf{r}_3 \to \mathbf{r}_3]{\mathbf{r}_1+\mathbf{r}_2 \to \mathbf{r}_2} \begin{bmatrix} 1 & -1 & 2 \\ 0 & -1 & c+2 \\ 0 & 3 & 0 \end{bmatrix}$$

$$\xrightarrow{3\mathbf{r}_2+\mathbf{r}_3 \to \mathbf{r}_3} \begin{bmatrix} 1 & -1 & 2 \\ 0 & -1 & c+2 \\ 0 & 0 & 3c+6 \end{bmatrix}$$

Because no row interchanges were performed, the determinant of B is the product of the diagonal entries of the previous matrix, which is $-(3c+6) = -3(c+2)$. Since a matrix is invertible if and only if its determinant is nonzero, the only value for which B is not invertible is $c = -2$.

3. The coefficient matrix of this system is

$$A = \begin{bmatrix} 3 & 8 \\ 2 & 6 \end{bmatrix}.$$

Since $\det A = 3(6) - 8(2) = 2$, matrix A is invertible by Theorem 3.4(a), and hence Cramer's rule can be used. In the notation of Theorem 3.5, we have

$$M_1 = \begin{bmatrix} 4 & 8 \\ 2 & 6 \end{bmatrix} \quad \text{and} \quad M_2 = \begin{bmatrix} 3 & 4 \\ 2 & 2 \end{bmatrix}.$$

Therefore the unique solution of the given system is the vector with components

$$u_1 = \frac{\det M_1}{\det A} = \frac{8}{2} = 4 \quad \text{and} \quad u_2 = \frac{\det M_2}{\det A} = \frac{-2}{2} = -1.$$

CHAPTER 3 REVIEW EXERCISES

In Exercises 1–11, determine whether the statements are true or false.

1. $\det \begin{bmatrix} a & b \\ c & d \end{bmatrix} = bc - ad.$
2. For $n \geq 2$, the (i,j)-cofactor of an $n \times n$ matrix A is the determinant of the $(n-1) \times (n-1)$ matrix obtained by deleting row j and column i from A.
3. If A is an $n \times n$ matrix and c_{ij} denotes the (i,j)-cofactor of A, then $\det A = a_{i1}c_{i1} + a_{i2}c_{i2} + \cdots + a_{in}c_{in}$ for any $i = 1, 2, \ldots, n$.
4. For all $n \times n$ matrices A and B, we have $\det(A+B) = \det A + \det B$.
5. For all $n \times n$ matrices A and B, $\det AB = (\det A)(\det B)$.
6. If B is obtained by interchanging two rows of an $n \times n$ matrix A, then $\det B = \det A$.
7. An $n \times n$ matrix is invertible if and only if its determinant is 0.
8. For any square matrix A, $\det A^T = \det A$.
9. For any invertible matrix A, $\det A^{-1} = -\det A$.

10. For any square matrix A and scalar c, $\det cA = c \det A$.
11. If A is an upper or lower triangular $n \times n$ matrix, then $\det A = a_{11} + a_{22} + \cdots + a_{nn}$.

In Exercises 12–15, compute each indicated cofactor of the matrix

$$\begin{bmatrix} 1 & -1 & 2 \\ -1 & 2 & -1 \\ 2 & 1 & 3 \end{bmatrix}.$$

12. the (1, 2)-cofactor
13. the (2, 1)-cofactor
14. the (2, 3)-cofactor
15. the (3, 1)-cofactor

In Exercises 16–19, compute the determinant of the matrix in Exercises 12–15, using a cofactor expansion along each indicated row or column.

16. row 1
17. row 3
18. column 2
19. column 1

In Exercises 20–23, evaluate the determinant of each matrix by any legitimate method.

20. $\begin{bmatrix} 5 & 6 \\ 3 & 2 \end{bmatrix}$

21. $\begin{bmatrix} -5.0 & 3.0 \\ 3.5 & -2.1 \end{bmatrix}$

22. $\begin{bmatrix} 1 & -1 & 2 \\ 2 & -1 & 3 \\ 3 & -1 & 4 \end{bmatrix}$

23. $\begin{bmatrix} 1 & -3 & 1 \\ 4 & -2 & 1 \\ 2 & 5 & -1 \end{bmatrix}$

24. (a) Perform a sequence of elementary row operations on

$$A = \begin{bmatrix} 0 & 3 & -6 & 1 \\ -2 & -2 & 2 & 6 \\ 1 & 1 & -1 & -1 \\ 2 & -1 & 2 & -2 \end{bmatrix}$$

to transform it into an upper triangular matrix.

(b) Use your answer to (a) to compute $\det A$.

In Exercises 25–28, use a determinant to find all values of the scalar c for which each matrix is not invertible.

25. $\begin{bmatrix} c-17 & -13 \\ 20 & c+16 \end{bmatrix}$

26. $\begin{bmatrix} 1 & c+1 \\ 2 & 3c+4 \end{bmatrix}$

27. $\begin{bmatrix} c+4 & -1 & c+5 \\ -3 & 3 & -4 \\ c+6 & -3 & c+7 \end{bmatrix}$

28. $\begin{bmatrix} -1 & c-1 & 1-c \\ -c-2 & 2c-3 & 4-c \\ -c-2 & c-1 & 2 \end{bmatrix}$

29. Compute the area of the parallelogram in $\mathcal{R}^2$ determined by the vectors

$$\begin{bmatrix} 3 \\ 7 \end{bmatrix} \quad \text{and} \quad \begin{bmatrix} 4 \\ 1 \end{bmatrix}.$$

30. Compute the volume of the parallelepiped in $\mathcal{R}^3$ determined by the vectors

$$\begin{bmatrix} 1 \\ 0 \\ 2 \end{bmatrix}, \quad \begin{bmatrix} -1 \\ 2 \\ 1 \end{bmatrix}, \quad \text{and} \quad \begin{bmatrix} 3 \\ 1 \\ -1 \end{bmatrix}.$$

In Exercises 31–32, solve each system of linear equations by Cramer's rule.

31. $\begin{aligned} 2x_1 + x_2 &= 5 \\ -4x_1 + 3x_2 &= -6 \end{aligned}$

32. $\begin{aligned} x_1 - x_2 + 2x_3 &= 7 \\ -x_1 + 2x_2 - x_3 &= -3 \\ 2x_1 + x_2 + 2x_3 &= 4 \end{aligned}$

Let A be a 3×3 matrix such that $\det A = 5$. Evaluate the determinant of each matrix given in Exercises 33–40.

33. A^T
34. A^{-1}
35. $2A$
36. A^3

37. $\begin{bmatrix} a_{11} - 3a_{21} & a_{12} - 3a_{22} & a_{13} - 3a_{23} \\ a_{21} & a_{22} & a_{23} \\ a_{31} & a_{32} & a_{33} \end{bmatrix}$

38. $\begin{bmatrix} a_{11} & a_{12} & a_{13} \\ -2a_{21} & -2a_{22} & -2a_{23} \\ a_{31} & a_{32} & a_{33} \end{bmatrix}$

39. $\begin{bmatrix} a_{11} + 5a_{31} & a_{12} + 5a_{32} & a_{13} + 5a_{33} \\ 4a_{21} & 4a_{22} & 4a_{23} \\ a_{31} - 2a_{21} & a_{32} - 2a_{22} & a_{33} - 2a_{23} \end{bmatrix}$

40. $\begin{bmatrix} a_{31} & a_{32} & a_{33} \\ a_{21} & a_{22} & a_{23} \\ a_{11} & a_{12} & a_{13} \end{bmatrix}$

41. A square matrix B is called *idempotent* if $B^2 = B$. What can be said about the determinant of an idempotent matrix?
42. Suppose that an $n \times n$ matrix can be expressed in the form $A = PDP^{-1}$, where P is an invertible matrix and D is a diagonal matrix. Prove that the determinant of A equals the product of the diagonal entries of D.
43. Show that the equation

$$\det \begin{bmatrix} 1 & x & y \\ 1 & x_1 & y_1 \\ 0 & 1 & m \end{bmatrix} = 0$$

yields the equation of the line through the point (x_1, y_1) with slope m.

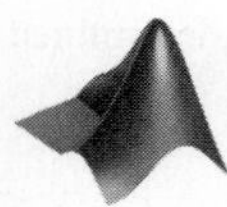

CHAPTER 3 MATLAB EXERCISES

For the following exercises, use MATLAB (or comparable software) or a calculator with matrix capabilities. The MATLAB functions in Tables D.1, D.2, D.3, D.4, and D.5 of Appendix D may be useful.

In Exercises 1 and 2, a matrix A is given. Use elementary row operations other than scaling to transform A into an upper triangular matrix, and then use the boxed result on page 213 to compute $\det A$.

1. $A = \begin{bmatrix} -0.8 & 3.5 & -1.4 & 2.5 & 6.7 & -2.0 \\ -6.5 & -2.0 & -1.4 & 3.2 & 1.7 & -6.5 \\ 5.7 & 7.9 & 1.0 & 2.2 & -1.3 & 5.7 \\ -2.1 & -3.1 & 0.0 & -1.0 & 3.5 & -2.1 \\ 0.2 & 8.8 & -2.8 & 5.0 & 11.4 & -2.2 \\ 4.8 & 10.3 & -0.4 & 3.7 & 8.9 & 3.6 \end{bmatrix}$

2. $A = \begin{bmatrix} 0 & 1 & 2 & -2 & 3 & 1 \\ 1 & 1 & 2 & -2 & 1 & 2 \\ 2 & 2 & 4 & -5 & 6 & 3 \\ 2 & 3 & 6 & -6 & -4 & 5 \\ -2 & -6 & 5 & -5 & 4 & 4 \\ 1 & 3 & 5 & -5 & -3 & 6 \end{bmatrix}$

3. Let

$$A = \begin{bmatrix} 8 & 3 & 3 & 14 & 6 \\ 3 & 3 & 2 & -6 & 0 \\ 2 & 0 & 1 & 5 & 2 \\ 1 & 1 & 0 & -1 & 1 \end{bmatrix}, \quad \mathbf{v} = \begin{bmatrix} 4 & 1 & 1 & 10 & 4 \end{bmatrix},$$

$$\text{and} \quad \mathbf{w} = \begin{bmatrix} 2 & 1 & 2 & -4 & 1 \end{bmatrix}.$$

For any 1×5 row vector $\mathbf{x}$, let $\begin{bmatrix} \mathbf{x} \\ A \end{bmatrix}$ denote the 5×5 matrix whose rows are $\mathbf{x}$ followed by the rows of A in the same order.

(a) Compute $\det \begin{bmatrix} \mathbf{v} \\ A \end{bmatrix}$ and $\det \begin{bmatrix} \mathbf{w} \\ A \end{bmatrix}$.

(b) Compute $\det \begin{bmatrix} \mathbf{v}+\mathbf{w} \\ A \end{bmatrix}$ and $\det \begin{bmatrix} \mathbf{v} \\ A \end{bmatrix} + \det \begin{bmatrix} \mathbf{w} \\ A \end{bmatrix}$.

(c) Compute $\det \begin{bmatrix} 3\mathbf{v}-2\mathbf{w} \\ A \end{bmatrix}$ and $3\det \begin{bmatrix} \mathbf{v} \\ A \end{bmatrix} - 2\det \begin{bmatrix} \mathbf{w} \\ A \end{bmatrix}$.

(d) Use the results of (b) and (c) to make a conjecture about any function $T\colon \mathcal{R}^n \to \mathcal{R}$ defined by $T(\mathbf{x}) = \det \begin{bmatrix} \mathbf{x} \\ C \end{bmatrix}$, where C is an $(n-1) \times n$ matrix.

(e) Prove your conjecture in (d).

(f) In (a)–(e), we considered the case where row 1 of a matrix varies and the other rows remain fixed. State and prove a result for the general case, in which row i varies and all of the other rows remain fixed.

(g) What happens if *rows* are replaced by *columns* in (f)? Investigate this situation, first by experimentation, and then by formulating and proving a conjecture.

4 INTRODUCTION

Solid modeling (three-dimensional geometric modeling) systems have become an indispensable tool for mechanical designers. They create a virtual three-dimensional representation of mechanical components for machine design and analysis and also provide tools for visualizing components and computing volumes and surface areas. Engineering drawings can be created semi-automatically from such a model, and tool paths for machining parts can be generated from it.

The models are constructed by a variety of techniques. For instance, a polyhedron is constructed by specifying the coordinates of its vertices and how the vertices are connected to form the faces of the polyhedron. If the polyhedron is positioned simply relative to the coordinate system, the vertex coordinates can be easy to compute. For example, a cube centered at the origin with an edge length of 2 that has faces parallel to the coordinate planes has vertex coordinates of $(\pm 1, \pm 1, \pm 1)$. (See the figure at the left.)

If the polyhedron is not in such a simple position, as in the figure on the next page, the construction is more complicated. Here again the goal is to construct a cube with an edge length of 2 centered at the origin having its top and bottom faces parallel to the xy-plane. In this polyhedron, however, one pair of faces is perpendicular to the line $y = x$ and the third pair is perpendicular to the line $y = -x$. In this case, the new cube can be constructed by rotating the original cube by 45° about the z-axis. For less trivial orientations, multiple rotations might be required, which can be difficult to visualize and specify.

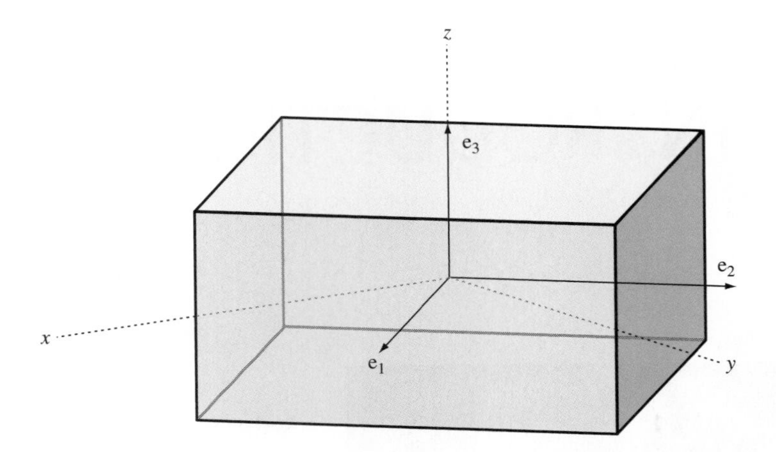

A simple alternative is to specify the vertices in terms of basis vectors (see Section 4.2). Specifically, we can use the standard vectors $\mathbf{e}_1, \mathbf{e}_2, \mathbf{e}_3$, which are perpendicular to the indicated faces in the second figure. As described in Section 4.4, the coordinates of the vertices relative to these basis vectors are $(\pm 1, \pm 1, \pm 1)$. Thus the vertices of the cube are $\pm \mathbf{e}_1$, $\pm \mathbf{e}_2$, $\pm \mathbf{e}_3$. More generally, if the cube is centered at a point $\mathbf{p}$, its vertices are $\mathbf{p} \pm \mathbf{e}_1$, $\mathbf{p} \pm \mathbf{e}_2$, $\mathbf{p} \pm \mathbf{e}_3$.

CHAPTER

4 SUBSPACES AND THEIR PROPERTIES

In many applications, it is necessary to study vectors only in a subset of $\mathcal{R}^n$, which is simpler than working with all vectors in $\mathcal{R}^n$. For instance, suppose that A is an $m \times n$ matrix and $\mathbf{u}$ and $\mathbf{v}$ are solutions of $A\mathbf{x} = \mathbf{0}$. Then

$$A(\mathbf{u} + \mathbf{v}) = A\mathbf{u} + A\mathbf{v} = \mathbf{0} + \mathbf{0} = \mathbf{0},$$

so that $\mathbf{u} + \mathbf{v}$ is a solution of $A\mathbf{x} = \mathbf{0}$. Likewise, for any scalar s, $s\mathbf{u}$ is a solution of $A\mathbf{x} = \mathbf{0}$. Therefore the solution set of $A\mathbf{x} = \mathbf{0}$ has the following *closure* properties: (a) the sum of any pair of vectors in the set lies in the set, and (b) every scalar multiple of a vector in the set lies in the set. Subsets that have these properties are called *subspaces* of $\mathcal{R}^n$. As another example, a plane passing through the origin of $\mathcal{R}^3$ is a subspace of $\mathcal{R}^3$. In Section 4.1, we define subspaces and give several examples.

The techniques learned in Chapter 1 enable us to find generating sets for particular subspaces. A generating set of the smallest size, called a *basis* for the subspace, is particularly useful in representing vectors in the subspace. In Section 4.2, we show that any two bases of a subspace have the same number of vectors. This number is called the *dimension* of the subspace.

In Sections 4.4 and 4.5, we investigate different coordinate systems on $\mathcal{R}^n$ and describe examples in which it is preferable to use a coordinate system other than the usual one.

4.1 SUBSPACES

When we add two vectors in $\mathcal{R}^n$ or multiply a vector in $\mathcal{R}^n$ by a scalar, the resulting vectors are also in $\mathcal{R}^n$. In other words, $\mathcal{R}^n$ is *closed* under the operations of vector addition and scalar multiplication. In this section, we study the subsets of $\mathcal{R}^n$ that possess this type of closure.

Definition A set W of vectors in $\mathcal{R}^n$ is called a **subspace** of $\mathcal{R}^n$ if it has the following three properties:

1. The zero vector belongs to W.
2. Whenever $\mathbf{u}$ and $\mathbf{v}$ belong to W, then $\mathbf{u} + \mathbf{v}$ belongs to W. (In this case, we say that W is **closed under (vector) addition.**)
3. Whenever $\mathbf{u}$ belongs to W and c is a scalar, then $c\mathbf{u}$ belongs to W. (In this case, we say that W is **closed under scalar multiplication.**)

The next two examples describe two special subspaces of $\mathcal{R}^n$.

Example 1 The set $\mathcal{R}^n$ is a subspace of itself because the zero vector belongs to $\mathcal{R}^n$, the sum of any two vectors in $\mathcal{R}^n$ is also in $\mathcal{R}^n$, and every scalar multiple of a vector in $\mathcal{R}^n$ belongs to $\mathcal{R}^n$.

Example 2 The set W consisting of only the zero vector in $\mathcal{R}^n$ is a subspace of $\mathcal{R}^n$ called the **zero subspace**. Clearly, $\mathbf{0}$ is in W. Moreover, if $\mathbf{u}$ and $\mathbf{v}$ are vectors in W, then $\mathbf{u} = \mathbf{0}$ and $\mathbf{v} = \mathbf{0}$, so $\mathbf{u} + \mathbf{v} = \mathbf{0} + \mathbf{0} = \mathbf{0}$. Hence $\mathbf{u} + \mathbf{v}$ is in W, so W is closed under vector addition. Finally, if $\mathbf{u}$ is in W and c is a scalar, then $c\mathbf{u} = c\mathbf{0} = \mathbf{0}$, so $c\mathbf{u}$ is in W. Hence W is also closed under scalar multiplication.

A subspace of $\mathcal{R}^n$ other than $\{\mathbf{0}\}$ is called a **nonzero subspace**. Examples 3 and 4 show how to verify that two nonzero subspaces of $\mathcal{R}^3$ satisfy the three properties in the definition of a subspace.

Example 3 We show that the set $W = \left\{ \begin{bmatrix} w_1 \\ w_2 \\ w_3 \end{bmatrix} \in \mathcal{R}^3 \colon 6w_1 - 5w_2 + 4w_3 = 0 \right\}$ is a subspace of $\mathcal{R}^3$.

1. Since $6(0) - 5(0) + 4(0) = 0$, the components of $\mathbf{0}$ satisfy the equation that defines W. Hence $\mathbf{0} = \begin{bmatrix} 0 \\ 0 \\ 0 \end{bmatrix}$ is in W.

2. Let $\mathbf{u} = \begin{bmatrix} u_1 \\ u_2 \\ u_3 \end{bmatrix}$ and $\mathbf{v} = \begin{bmatrix} v_1 \\ v_2 \\ v_3 \end{bmatrix}$ be in W. Then $6u_1 - 5u_2 + 4u_3 = 0$, and also $6v_1 - 5v_2 + 4v_3 = 0$. Now $\mathbf{u} + \mathbf{v} = \begin{bmatrix} u_1 + v_1 \\ u_2 + v_2 \\ u_3 + v_3 \end{bmatrix}$. Since

$$\begin{aligned} 6(u_1 + v_1) - 5(u_2 + v_2) + 4(u_3 + v_3) &= (6u_1 - 5u_2 + 4u_3) + (6v_1 - 5v_2 + 4v_3) \\ &= 0 + 0 \\ &= 0, \end{aligned}$$

we see that the components of $\mathbf{u} + \mathbf{v}$ satisfy the equation defining W. Therefore $\mathbf{u} + \mathbf{v}$ is in W, so W is closed under vector addition.

3. Let $\mathbf{u} = \begin{bmatrix} u_1 \\ u_2 \\ u_3 \end{bmatrix}$ be in W. For any scalar c, we have $c\mathbf{u} = c \begin{bmatrix} u_1 \\ u_2 \\ u_3 \end{bmatrix} = \begin{bmatrix} cu_1 \\ cu_2 \\ cu_3 \end{bmatrix}$. Because

$$6(cu_1) - 5(cu_2) + 4(cu_3) = c(6u_1 - 5u_2 + 4u_3) = c(0) = 0,$$

the components of $c\mathbf{u}$ satisfy the equation defining W. Therefore $c\mathbf{u}$ is in W, so W is also closed under scalar multiplication.

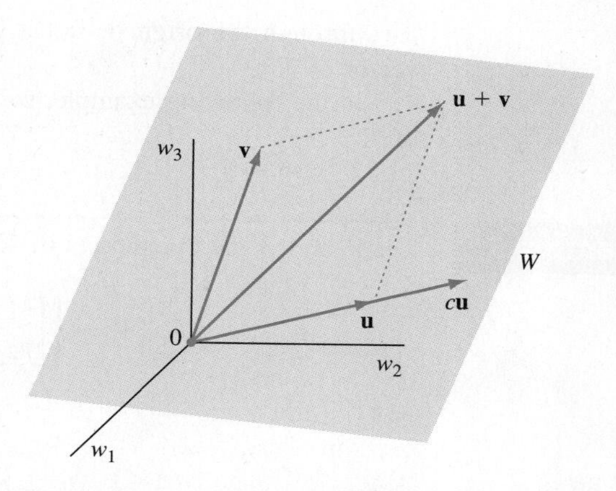

Figure 4.1 The subspace W is a plane through the origin.

Since W is a subset of $\mathcal{R}^3$ that contains the zero vector and is closed under both vector addition and scalar multiplication, W is a subspace of $\mathcal{R}^3$. (See Figure 4.1.) Geometrically, W is a plane through the origin of $\mathcal{R}^3$.

Example 4 Let $\mathbf{w}$ be a nonzero vector in $\mathcal{R}^3$. Show that the set W of all multiples of $\mathbf{w}$ is a subspace of $\mathcal{R}^3$.

Solution First, $\mathbf{0} = 0\mathbf{w}$ is in W. Next, let $\mathbf{u}$ and $\mathbf{v}$ be vectors in W. Then $\mathbf{u} = a\mathbf{w}$ and $\mathbf{v} = b\mathbf{w}$ for some scalars a and b. Since

$$\mathbf{u} + \mathbf{v} = a\mathbf{w} + b\mathbf{w} = (a + b)\mathbf{w},$$

we see that $\mathbf{u} + \mathbf{v}$ is a multiple of $\mathbf{w}$. Hence $\mathbf{u} + \mathbf{v}$ is in W, so W is closed under vector addition. Finally, for any scalar c, $c\mathbf{u} = c(a\mathbf{w}) = (ca)\mathbf{w}$ is a multiple of $\mathbf{w}$. Thus $c\mathbf{u}$ is in W, so W is also closed under scalar multiplication. Therefore W is a subspace of $\mathcal{R}^3$. Note that W can be depicted as a line in $\mathcal{R}^3$ through the origin. (See Figure 4.2.)

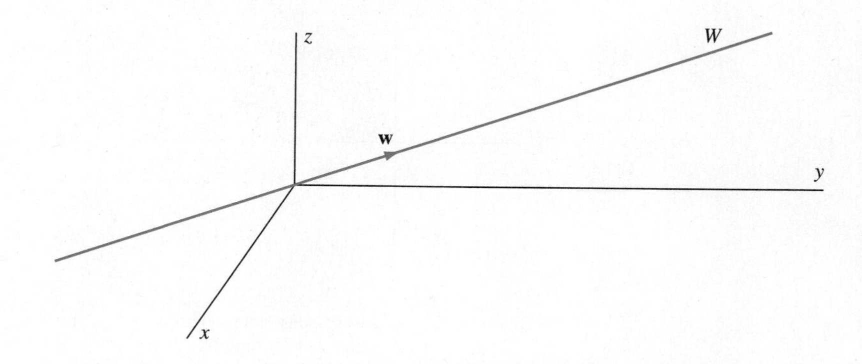

Figure 4.2 W is the set of all multiples of $\mathbf{w}$.

Example 4 shows that the set of vectors in $\mathcal{R}^3$ that lie along a line through the origin is a subspace of $\mathcal{R}^3$. However, the set of vectors on a line in $\mathcal{R}^3$ that does not pass through the origin is not a subspace, for such a set does not contain the zero vector of $\mathcal{R}^3$.

In the following example, we consider two subsets of $\mathcal{R}^2$ that are not subspaces of $\mathcal{R}^2$:

Example 5 Let V and W be the subsets of $\mathcal{R}^2$ defined by

$$V = \left\{ \begin{bmatrix} v_1 \\ v_2 \end{bmatrix} \in \mathcal{R}^2 : v_1 \geq 0 \text{ and } v_2 \geq 0 \right\}$$

and

$$W = \left\{ \begin{bmatrix} w_1 \\ w_2 \end{bmatrix} \in \mathcal{R}^2 : w_1^2 = w_2^2 \right\}.$$

The vectors in V are those that lie in the first quadrant of $\mathcal{R}^2$, and the nonnegative parts of the x- and y-axes. (See Figure 4.3(a).) Clearly, $\mathbf{0} = \begin{bmatrix} 0 \\ 0 \end{bmatrix}$ is in V. Suppose that $\mathbf{u} = \begin{bmatrix} u_1 \\ u_2 \end{bmatrix}$ and $\mathbf{v} = \begin{bmatrix} v_1 \\ v_2 \end{bmatrix}$ are in V. Then $u_1 \geq 0$, $u_2 \geq 0$, $v_1 \geq 0$, and $v_2 \geq 0$. Hence $u_1 + v_1 \geq 0$ and $u_2 + v_2 \geq 0$, so that

$$\mathbf{u} + \mathbf{v} = \begin{bmatrix} u_1 + v_1 \\ u_2 + v_2 \end{bmatrix}$$

is in V. Thus V is closed under vector addition. This conclusion can also be seen by the parallelogram law. However, V is *not* closed under scalar multiplication, because $\mathbf{u} = \begin{bmatrix} 1 \\ 2 \end{bmatrix}$ belongs to V, but $(-1)\mathbf{u} = \begin{bmatrix} -1 \\ -2 \end{bmatrix}$ does not. Thus V is not a subspace of $\mathcal{R}^2$.

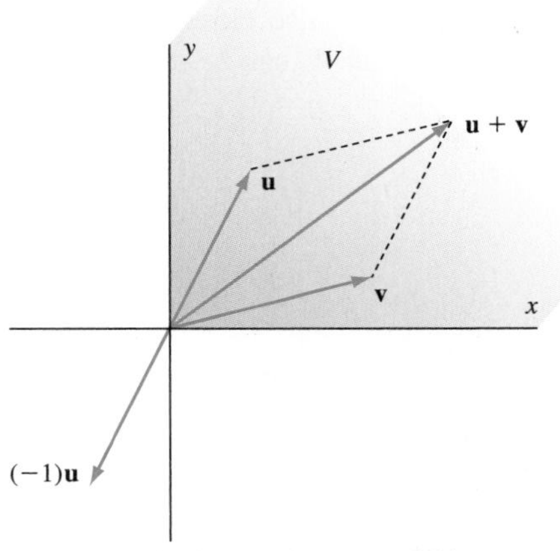

V is closed under vector addition, but not under scalar multiplication.

(a)

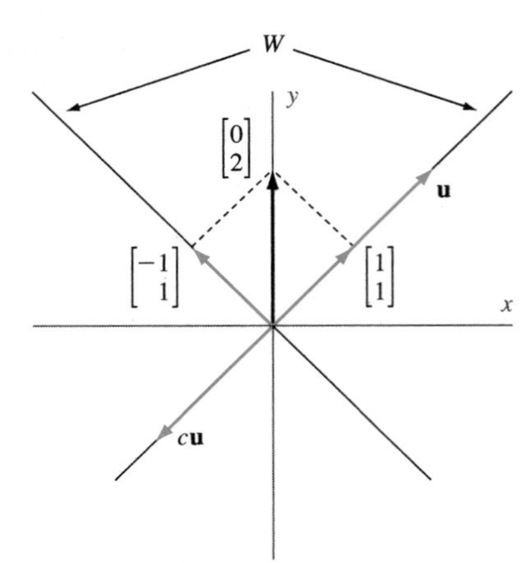

W is closed under scalar multiplication, but not under vector addition.

(b)

Figure 4.3

Consider a vector $\mathbf{u} = \begin{bmatrix} u_1 \\ u_2 \end{bmatrix}$ in W. Since $u_1^2 = u_2^2$, it follows that $u_2 = \pm u_1$. Hence the vector $\mathbf{u}$ lies along one of the lines $y = x$ or $y = -x$. (See Figure 4.3(b).) Clearly, $\mathbf{0} = \begin{bmatrix} 0 \\ 0 \end{bmatrix}$ belongs to W. Moreover, if $\mathbf{u} = \begin{bmatrix} u_1 \\ u_2 \end{bmatrix}$ is in W, then $u_1^2 = u_2^2$. So, for any scalar c, $c\mathbf{u} = \begin{bmatrix} cu_1 \\ cu_2 \end{bmatrix}$ is in W because $(cu_1)^2 = c^2u_1^2 = c^2u_2^2 = (cu_2)^2$. Thus W is closed under scalar multiplication. However, W is *not* closed under vector addition. For example, the vectors $\begin{bmatrix} 1 \\ 1 \end{bmatrix}$ and $\begin{bmatrix} -1 \\ 1 \end{bmatrix}$ belong to W, but $\begin{bmatrix} 1 \\ 1 \end{bmatrix} + \begin{bmatrix} -1 \\ 1 \end{bmatrix} = \begin{bmatrix} 0 \\ 2 \end{bmatrix}$ does not. Thus W is not a subspace of $\mathcal{R}^2$.

Our first theorem generalizes Example 4.

THEOREM 4.1

The span of a finite nonempty subset of $\mathcal{R}^n$ is a subspace of $\mathcal{R}^n$.

PROOF Let $\mathcal{S} = \{\mathbf{w}_1, \mathbf{w}_2, \ldots, \mathbf{w}_k\}$. Since

$$0\mathbf{w}_1 + 0\mathbf{w}_2 + \cdots + 0\mathbf{w}_k = \mathbf{0},$$

we see that $\mathbf{0}$ belongs to the span of $\mathcal{S}$. Let $\mathbf{u}$ and $\mathbf{v}$ belong to the span of $\mathcal{S}$. Then

$$\mathbf{u} = a_1\mathbf{w}_1 + a_2\mathbf{w}_2 + \cdots + a_k\mathbf{w}_k \quad \text{and} \quad \mathbf{v} = b_1\mathbf{w}_1 + b_2\mathbf{w}_2 + \cdots + b_k\mathbf{w}_k$$

for some scalars $a_1, a_2, \ldots, a_k$ and $b_1, b_2, \ldots, b_k$. Since

$$\begin{aligned} \mathbf{u} + \mathbf{v} &= (a_1\mathbf{w}_1 + a_2\mathbf{w}_2 + \cdots + a_k\mathbf{w}_k) + (b_1\mathbf{w}_1 + b_2\mathbf{w}_2 + \cdots + b_k\mathbf{w}_k) \\ &= (a_1 + b_1)\mathbf{w}_1 + (a_2 + b_2)\mathbf{w}_2 + \cdots + (a_k + b_k)\mathbf{w}_k, \end{aligned}$$

it follows that $\mathbf{u} + \mathbf{v}$ belongs to the span of $\mathcal{S}$. Hence the span of $\mathcal{S}$ is closed under vector addition. Furthermore, for any scalar c,

$$\begin{aligned} c\mathbf{u} &= c(a_1\mathbf{w}_1 + a_2\mathbf{w}_2 + \cdots + a_k\mathbf{w}_k) \\ &= (c_1a_1)\mathbf{w}_1 + (c_2a_2)\mathbf{w}_2 + \cdots + (c_ka_k)\mathbf{w}_k, \end{aligned}$$

so $c\mathbf{u}$ belongs to the span of $\mathcal{S}$. Thus the span of $\mathcal{S}$ is also closed under scalar multiplication, and therefore the span of $\mathcal{S}$ is a subspace of $\mathcal{R}^n$. ■

Example 6 We can use Theorem 4.1 to show that the set of vectors of the form

$$W = \left\{ \begin{bmatrix} 2a - 3b \\ b \\ -a + 4b \end{bmatrix} \in \mathcal{R}^3 \colon a \text{ and } b \text{ are scalars} \right\}$$

is a subspace of $\mathcal{R}^3$. Simply observe that

$$\begin{bmatrix} 2a - 3b \\ b \\ -a + 4b \end{bmatrix} = a \begin{bmatrix} 2 \\ 0 \\ -1 \end{bmatrix} + b \begin{bmatrix} -3 \\ 1 \\ 4 \end{bmatrix},$$

so $W = \text{Span}\, \mathcal{S}$, where

$$\mathcal{S} = \left\{ \begin{bmatrix} 2 \\ 0 \\ -1 \end{bmatrix}, \begin{bmatrix} -3 \\ 1 \\ 4 \end{bmatrix} \right\}.$$

Therefore W is a subspace of $\mathcal{R}^3$, by Theorem 4.1.

In Example 6, we found a generating set $\mathcal{S}$ for W, thereby showing that W is a subspace. It follows from Theorem 4.1 that *the only sets of vectors in* $\mathcal{R}^n$ *that have generating sets are the subspaces of* $\mathcal{R}^n$. Note also that a generating set for a subspace V must consist of vectors *from* V. So in Example 6, even though every vector in the subspace W is a linear combination of the standard vectors in $\mathcal{R}^3$, the set of standard vectors is *not* a generating set for W because $\mathbf{e}_1$ is not in W.

Practice Problem 1 ▶ Show that

$$V = \left\{ \begin{bmatrix} -s \\ 2t \\ 3s - t \end{bmatrix} \in \mathcal{R}^3 : s \text{ and } t \text{ are scalars} \right\}$$

is a subspace of $\mathcal{R}^3$ by finding a generating set for V. ◀

SUBSPACES ASSOCIATED WITH A MATRIX

There are several important subspaces associated with a matrix. The first one that we consider is the *null space*, a term we introduced earlier for linear transformations.

Definition The **null space** of a matrix A is the solution set of $A\mathbf{x} = \mathbf{0}$. It is denoted by Null A.

For an $m \times n$ matrix A, the null space of A is the set

$$\text{Null}\, A = \{\mathbf{v} \in \mathcal{R}^n : A\mathbf{v} = \mathbf{0}\}.$$

For example, the null space of the matrix

$$\begin{bmatrix} 1 & -5 & 3 \\ 2 & -9 & -6 \end{bmatrix}$$

equals the solution set of the homogeneous system of linear equations

$$\begin{aligned} x_1 - 5x_2 + 3x_3 &= 0 \\ 2x_1 - 9x_2 - 6x_3 &= 0. \end{aligned}$$

More generally, the solution set of any homogeneous system of linear equations equals the null space of the coefficient matrix of that system.

The set W in Example 3 is such a solution set. (In this case, it is the solution set of a single equation in 3 variables, $6x_1 - 5x_2 + 4x_3 = 0$.) We saw in Example 3 that W is a subspace of $\mathcal{R}^3$. Such sets are always subspaces, as the next theorem shows.

THEOREM 4.2

If A is an $m \times n$ matrix, then Null A is a subspace of $\mathcal{R}^n$.

PROOF Since A is an $m \times n$ matrix, the vectors in Null A, which are the solutions of $A\mathbf{x} = \mathbf{0}$, belong to $\mathcal{R}^n$. Clearly, $\mathbf{0}$ is in Null A because $A\mathbf{0} = \mathbf{0}$. Suppose that $\mathbf{u}$ and $\mathbf{v}$ belong to Null A. Then $A\mathbf{u} = \mathbf{0}$ and $A\mathbf{v} = \mathbf{0}$. Hence, by Theorem 1.3(b), we have

$$A(\mathbf{u} + \mathbf{v}) = A\mathbf{u} + A\mathbf{v} = \mathbf{0} + \mathbf{0} = \mathbf{0}.$$

This argument proves that $\mathbf{u} + \mathbf{v}$ is in Null A, so Null A is closed under vector addition. Moreover, for any scalar c, we have by Theorem 1.3(c) that

$$A(c\mathbf{u}) = c(A\mathbf{u}) = c\mathbf{0} = \mathbf{0}.$$

Thus $c\mathbf{u}$ is in Null A, so Null A is also closed under scalar multiplication. Therefore Null A is a subspace of $\mathcal{R}^n$. ■

Another important subspace associated with a matrix is its *column space*.

Definition The **column space** of a matrix A is the span of its columns. It is denoted by Col A.

For example, if

$$A = \begin{bmatrix} 1 & -5 & 3 \\ 2 & -9 & -6 \end{bmatrix},$$

then

$$\text{Col } A = \text{Span}\left\{\begin{bmatrix} 1 \\ 2 \end{bmatrix}, \begin{bmatrix} -5 \\ -9 \end{bmatrix}, \begin{bmatrix} 3 \\ -6 \end{bmatrix}\right\}.$$

Recall from the boxed result on page 68 that a vector $\mathbf{b}$ is a linear combination of the columns of an $m \times n$ matrix A if and only if the matrix equation $A\mathbf{x} = \mathbf{b}$ is consistent. Hence

$$\text{Col } A = \{A\mathbf{v}\colon \mathbf{v} \text{ is in } \mathcal{R}^n\}.$$

It follows from Theorem 4.1 that the column space of an $m \times n$ matrix is a subspace of $\mathcal{R}^m$. Since the null space of A is a subspace of $\mathcal{R}^n$, the column space and null space of an $m \times n$ matrix are contained in different spaces if $m \neq n$. Even if $m = n$, however, these two subspaces are rarely equal.

Example 7 Find a generating set for the column space of the matrix

$$A = \begin{bmatrix} 1 & 2 & 1 & -1 \\ 2 & 4 & 0 & -8 \\ 0 & 0 & 2 & 6 \end{bmatrix}.$$

Is $\mathbf{u} = \begin{bmatrix} 2 \\ 1 \\ 1 \end{bmatrix}$ in Col A? Is $\mathbf{v} = \begin{bmatrix} 2 \\ 1 \\ 3 \end{bmatrix}$ in Col A?

Solution The column space of A is the span of the columns of A. Hence one generating set for Col A is

$$\left\{ \begin{bmatrix} 1 \\ 2 \\ 0 \end{bmatrix}, \begin{bmatrix} 2 \\ 4 \\ 0 \end{bmatrix}, \begin{bmatrix} 1 \\ 0 \\ 2 \end{bmatrix}, \begin{bmatrix} -1 \\ -8 \\ 6 \end{bmatrix} \right\}.$$

To see whether the vector $\mathbf{u}$ lies in the column space of A, we must determine whether $A\mathbf{x} = \mathbf{u}$ is consistent. Since the reduced row echelon form of $[A \;\; \mathbf{u}]$ is

$$\begin{bmatrix} 1 & 2 & 0 & -4 & 0 \\ 0 & 0 & 1 & 3 & 0 \\ 0 & 0 & 0 & 0 & 1 \end{bmatrix},$$

we see that the system is inconsistent, and hence $\mathbf{u}$ is not in Col A. (See Figure 4.4.) On the other hand, the reduced row echelon form of $[A \;\; \mathbf{v}]$ is

$$\begin{bmatrix} 1 & 2 & 0 & -4 & 0.5 \\ 0 & 0 & 1 & 3 & 1.5 \\ 0 & 0 & 0 & 0 & 0 \end{bmatrix}.$$

Thus the system $A\mathbf{x} = \mathbf{v}$ is consistent, so $\mathbf{v}$ is in Col A. (See Figure 4.4.)

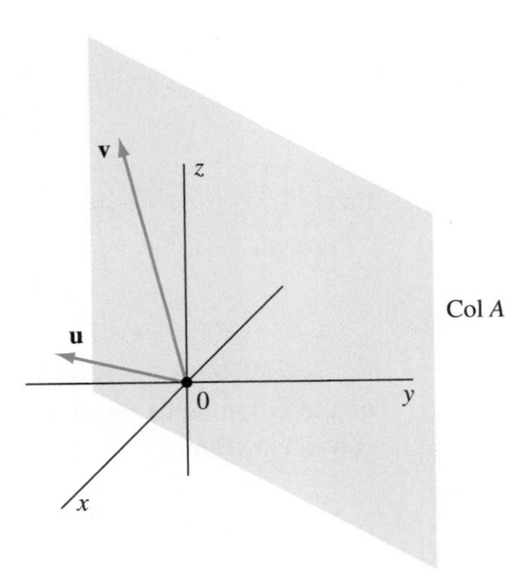

Figure 4.4 The vector **v** is in the column space of A, but **u** is not.

Example 8

Find a generating set for the null space of the matrix A in Example 7. Is $\mathbf{u} = \begin{bmatrix} 2 \\ -3 \\ 3 \\ -1 \end{bmatrix}$ in Null A? Is $\mathbf{v} = \begin{bmatrix} 5 \\ -3 \\ 2 \\ 1 \end{bmatrix}$ in Null A?

Solution Unlike the calculation of a generating set for the column space of A in Example 7, there is no easy way to obtain a generating set for the null space of A directly from the entries of A. Instead, we must solve $A\mathbf{x} = \mathbf{0}$. Because the reduced row echelon form of A is

$$\begin{bmatrix} 1 & 2 & 0 & -4 \\ 0 & 0 & 1 & 3 \\ 0 & 0 & 0 & 0 \end{bmatrix},$$

the vector form of the general solution of $A\mathbf{x} = \mathbf{0}$ is

$$\begin{bmatrix} x_1 \\ x_2 \\ x_3 \\ x_4 \end{bmatrix} = \begin{bmatrix} -2x_2 + 4x_4 \\ x_2 \\ -3x_4 \\ x_4 \end{bmatrix} = x_2 \begin{bmatrix} -2 \\ 1 \\ 0 \\ 0 \end{bmatrix} + x_4 \begin{bmatrix} 4 \\ 0 \\ -3 \\ 1 \end{bmatrix}.$$

It follows that

$$\text{Null } A = \text{Span} \left\{ \begin{bmatrix} -2 \\ 1 \\ 0 \\ 0 \end{bmatrix}, \begin{bmatrix} 4 \\ 0 \\ -3 \\ 1 \end{bmatrix} \right\}.$$

So the span of the set of vectors in the vector form of the general solution of $A\mathbf{x} = \mathbf{0}$ equals Null A.

To see if the vector $\mathbf{u}$ lies in the null space of A, we must determine whether $A\mathbf{u} = \mathbf{0}$. An easy calculation confirms this; so $\mathbf{u}$ belongs to Null A. On the other hand,

$$A\mathbf{v} = \begin{bmatrix} 1 & 2 & 1 & -1 \\ 2 & 4 & 0 & -8 \\ 0 & 0 & 2 & 6 \end{bmatrix} \begin{bmatrix} 5 \\ -3 \\ 2 \\ 1 \end{bmatrix} = \begin{bmatrix} 0 \\ -10 \\ 10 \end{bmatrix}.$$

Since $A\mathbf{v} \neq \mathbf{0}$, we see that $\mathbf{v}$ is not in Null A.

Practice Problem 2 ▶ Find a generating set for the column space and null space of

$$A = \begin{bmatrix} 1 & 2 & -1 \\ -1 & -3 & 4 \end{bmatrix}.$$

◀

In this book, the subspaces that we consider usually arise as the span of a given set of vectors or as the solution set of a homogeneous system of linear equations. As Examples 7 and 8 illustrate, when we define a subspace by giving a generating set, there is no work involved in obtaining a generating set, but we must solve a system of linear equations to check whether a particular vector belongs to the subspace. On the other hand, if V is a subspace that is the solution set of a homogeneous system of linear equations, then we must solve a system of linear equations in order to find a generating set for V. But we can easily check whether a particular vector lies in V by verifying that its components satisfy the linear system defining the subspace.

Like the column space of a matrix, the **row space** of a matrix is defined to be the span of its rows. The row space of a matrix A is denoted by Row A. So for the matrix

$$\begin{bmatrix} 1 & 2 & 1 & -1 \\ 2 & 4 & 0 & -8 \\ 0 & 0 & 2 & 6 \end{bmatrix}$$

in Example 7, we have

$$\text{Row } A = \text{Span}\left\{ \begin{bmatrix} 1 \\ 2 \\ 1 \\ -1 \end{bmatrix}, \begin{bmatrix} 2 \\ 4 \\ 0 \\ -8 \end{bmatrix}, \begin{bmatrix} 0 \\ 0 \\ 2 \\ 6 \end{bmatrix} \right\}.$$

Recall that, for any matrix A, the rows of A are the columns of A^T. Hence Row $A =$ Col A^T, and so the row space of an $m \times n$ matrix is a subspace of $\mathcal{R}^n$. Usually, the three subspaces Null A, Col A, and Row A are distinct.

SUBSPACES ASSOCIATED WITH A LINEAR TRANSFORMATION

In Section 2.8, we saw that the range of a linear transformation is the span of the columns of its standard matrix. We have just defined the span of the columns of a matrix to be its column space. Thus we can reformulate this statement from Section 2.8 as follows:

> The range of a linear transformation is the same as the column space of its standard matrix.

As a consequence of this result, the range of a linear transformation $T\colon \mathcal{R}^n \to \mathcal{R}^m$ is a subspace of $\mathcal{R}^m$.

Example 9 Determine a generating set for the range of the linear transformation $T\colon \mathcal{R}^4 \to \mathcal{R}^3$ defined by

$$T\left(\begin{bmatrix} x_1 \\ x_2 \\ x_3 \\ x_4 \end{bmatrix}\right) = \begin{bmatrix} x_1 + 2x_2 + x_3 - x_4 \\ 2x_1 + 4x_2 - 8x_4 \\ 2x_3 + 6x_4 \end{bmatrix}.$$

Solution The standard matrix of T is

$$A = \begin{bmatrix} 1 & 2 & 1 & -1 \\ 2 & 4 & 0 & -8 \\ 0 & 0 & 2 & 6 \end{bmatrix}.$$

Since the range of T is the same as the column space of A, a generating set for the range of T is

$$\left\{ \begin{bmatrix} 1 \\ 2 \\ 0 \end{bmatrix}, \begin{bmatrix} 2 \\ 4 \\ 0 \end{bmatrix}, \begin{bmatrix} 1 \\ 0 \\ 2 \end{bmatrix}, \begin{bmatrix} -1 \\ -8 \\ -6 \end{bmatrix} \right\},$$

the set of columns of A.

We also learned in Section 2.8 that the null space of a linear transformation is the solution set of $A\mathbf{x} = \mathbf{0}$, where A is the standard matrix of T. We can now restate this result as follows:

> The null space of a linear transformation is the same as the null space of its standard matrix.

This result implies that the null space of a linear transformation $T: \mathcal{R}^n \to \mathcal{R}^m$ is a subspace of $\mathcal{R}^n$.

Example 10 Determine a generating set for the null space of the linear transformation in Example 9.

Solution The standard matrix of T is given in Example 9. Its reduced row echelon form is the matrix

$$\begin{bmatrix} 1 & 2 & 0 & -4 \\ 0 & 0 & 1 & 3 \\ 0 & 0 & 0 & 0 \end{bmatrix}$$

in Example 8. From the latter example, we see that

$$\text{Null } A = \text{Span}\left\{ \begin{bmatrix} -2 \\ 1 \\ 0 \\ 0 \end{bmatrix}, \begin{bmatrix} 4 \\ 0 \\ -3 \\ 1 \end{bmatrix} \right\}.$$

Practice Problem 3 ▶ For the linear transformation $T: \mathcal{R}^4 \to \mathcal{R}^3$ defined by

$$T\left(\begin{bmatrix} x_1 \\ x_2 \\ x_3 \\ x_4 \end{bmatrix}\right) = \begin{bmatrix} x_1 + x_3 + 2x_4 \\ -x_2 + x_3 + x_4 \\ 2x_1 + 3x_2 - x_3 + x_4 \end{bmatrix},$$

find generating sets for the null space and range. ◀

EXERCISES

In Exercises 1–10, find a generating set for each subspace.

1. $\left\{ \begin{bmatrix} 0 \\ s \end{bmatrix} \in \mathcal{R}^2 : s \text{ is a scalar} \right\}$

2. $\left\{ \begin{bmatrix} 2s \\ -3s \end{bmatrix} \in \mathcal{R}^2 : s \text{ is a scalar} \right\}$

3. $\left\{ \begin{bmatrix} 4s \\ -s \end{bmatrix} \in \mathcal{R}^2 : s \text{ is a scalar} \right\}$

4. $\left\{\begin{bmatrix} 4t \\ s+t \\ -3s+t \end{bmatrix} \in \mathcal{R}^3 \colon s \text{ and } t \text{ are scalars}\right\}$

5. $\left\{\begin{bmatrix} -s+t \\ 2s-t \\ s+3t \end{bmatrix} \in \mathcal{R}^3 \colon s \text{ and } t \text{ are scalars}\right\}$

6. $\left\{\begin{bmatrix} -r+3s \\ 0 \\ s-t \\ r-2t \end{bmatrix} \in \mathcal{R}^4 \colon r, s, \text{ and } t \text{ are scalars}\right\}$

7. $\left\{\begin{bmatrix} -r+s \\ 4s-3t \\ 0 \\ 3r-t \end{bmatrix} \in \mathcal{R}^4 \colon r, s, \text{ and } t \text{ are scalars}\right\}$

8. $\left\{\begin{bmatrix} r-s+3t \\ 2r-t \\ -r+3s+2t \\ -2r+s+t \end{bmatrix} \in \mathcal{R}^4 \colon r, s, \text{ and } t \text{ are scalars}\right\}$

9. $\left\{\begin{bmatrix} 2s-5t \\ 3r+s-2t \\ r-4s+3t \\ -r+2s \end{bmatrix} \in \mathcal{R}^4 \colon r, s, \text{ and } t \text{ are scalars}\right\}$

10. $\left\{\begin{bmatrix} -r+4t \\ r-s+2t \\ 3t \\ r-t \end{bmatrix} \in \mathcal{R}^4 \colon r, s, \text{ and } t \text{ are scalars}\right\}$

In Exercises 11–18, determine whether each vector belongs to Null A, where

$$A = \begin{bmatrix} 1 & -2 & -1 & 0 \\ 0 & 1 & 3 & -2 \\ -2 & 3 & -1 & 2 \end{bmatrix}.$$

11. $\begin{bmatrix} 1 \\ 1 \\ -1 \\ -1 \end{bmatrix}$ 12. $\begin{bmatrix} 1 \\ 0 \\ 1 \\ 2 \end{bmatrix}$ 13. $\begin{bmatrix} -1 \\ 2 \\ -2 \\ -2 \end{bmatrix}$

14. $\begin{bmatrix} 2 \\ 0 \\ 2 \\ 3 \end{bmatrix}$ 15. $\begin{bmatrix} 1 \\ -1 \\ 3 \\ 4 \end{bmatrix}$ 16. $\begin{bmatrix} 1 \\ -3 \\ 5 \\ 6 \end{bmatrix}$

17. $\begin{bmatrix} 3 \\ 1 \\ 1 \\ 2 \end{bmatrix}$ 18. $\begin{bmatrix} 3 \\ 2 \\ -1 \\ 1 \end{bmatrix}$

In Exercises 19–26, determine whether each vector belongs to Col A, where A is the matrix used in Exercises 11–18.

19. $\begin{bmatrix} 2 \\ -1 \\ 3 \end{bmatrix}$ 20. $\begin{bmatrix} -1 \\ 3 \\ -1 \end{bmatrix}$ 21. $\begin{bmatrix} 1 \\ -4 \\ 2 \end{bmatrix}$

22. $\begin{bmatrix} -1 \\ 2 \\ 1 \end{bmatrix}$ 23. $\begin{bmatrix} 1 \\ 2 \\ -4 \end{bmatrix}$ 24. $\begin{bmatrix} 1 \\ -3 \\ 3 \end{bmatrix}$

25. $\begin{bmatrix} 5 \\ -4 \\ -6 \end{bmatrix}$ 26. $\begin{bmatrix} 2 \\ -1 \\ 1 \end{bmatrix}$

In Exercises 27–34, find a generating set for the null space of each matrix.

27. $\begin{bmatrix} -1 & 1 & 2 \\ 1 & -2 & 3 \end{bmatrix}$

28. $\begin{bmatrix} 1 & 2 & 0 \\ 0 & -1 & 1 \\ 1 & 0 & 2 \end{bmatrix}$

29. $\begin{bmatrix} 1 & 1 & -1 & 4 \\ 2 & 1 & -3 & 5 \\ -2 & 0 & 4 & -2 \end{bmatrix}$

30. $\begin{bmatrix} 1 & 1 & 1 \\ 0 & -1 & -3 \\ 1 & 1 & 1 \\ 0 & -2 & -6 \end{bmatrix}$

31. $\begin{bmatrix} 1 & 1 & 2 & 1 \\ -1 & 0 & -5 & 3 \\ 1 & 1 & 2 & 1 \\ -1 & 0 & -5 & 3 \end{bmatrix}$

32. $\begin{bmatrix} 1 & 1 & 0 & 2 & 1 \\ 3 & 2 & 1 & 6 & 3 \\ 0 & -1 & 1 & -1 & -1 \end{bmatrix}$

33. $\begin{bmatrix} 1 & -3 & 0 & 1 & -2 & -2 \\ 2 & -6 & -1 & 0 & 2 & 5 \\ -1 & 3 & 2 & 3 & -1 & 2 \end{bmatrix}$

34. $\begin{bmatrix} 1 & 0 & -1 & -3 & 1 & 4 \\ 2 & -1 & -1 & -8 & 3 & 9 \\ -1 & 1 & 1 & 5 & -2 & -6 \\ 0 & 1 & 1 & 2 & -1 & -3 \end{bmatrix}$

In Exercises 35–42, find generating sets for the range and null space of each linear transformation.

35. $T\left(\begin{bmatrix} x_1 \\ x_2 \\ x_3 \end{bmatrix}\right) = [x_1 + 2x_2 - x_3]$

36. $T\left(\begin{bmatrix} x_1 \\ x_2 \end{bmatrix}\right) = \begin{bmatrix} x_1 + 2x_2 \\ 2x_1 + 4x_2 \end{bmatrix}$

37. $T\left(\begin{bmatrix} x_1 \\ x_2 \end{bmatrix}\right) = \begin{bmatrix} x_1 + x_2 \\ x_1 - x_2 \\ x_1 \\ x_2 \end{bmatrix}$

38. $T\left(\begin{bmatrix} x_1 \\ x_2 \\ x_3 \end{bmatrix}\right) = \begin{bmatrix} x_1 - 2x_2 + 3x_3 \\ -2x_1 + 4x_2 - 6x_3 \end{bmatrix}$

39. $T\left(\begin{bmatrix} x_1 \\ x_2 \\ x_3 \end{bmatrix}\right) = \begin{bmatrix} x_1 + x_2 - x_3 \\ 0 \\ 2x_1 - x_3 \end{bmatrix}$

40. $T\left(\begin{bmatrix} x_1 \\ x_2 \\ x_3 \end{bmatrix}\right) = \begin{bmatrix} x_1 + x_2 \\ x_2 + x_3 \\ x_1 - x_3 \\ x_1 + 2x_2 + x_3 \end{bmatrix}$

41. $T\left(\begin{bmatrix} x_1 \\ x_2 \\ x_3 \end{bmatrix}\right) = \begin{bmatrix} x_1 - x_2 - 5x_3 \\ -x_1 + 2x_2 + 7x_3 \\ 2x_1 - x_2 - 8x_3 \\ 2x_2 + 4x_3 \end{bmatrix}$

42. $T\left(\begin{bmatrix} x_1 \\ x_2 \\ x_3 \\ x_4 \end{bmatrix}\right) = \begin{bmatrix} x_1 - x_2 - 3x_3 - 2x_4 \\ -x_1 + 2x_2 + 4x_3 + 5x_4 \\ x_1 - 2x_3 + x_4 \\ x_1 + x_2 - x_3 + 4x_4 \end{bmatrix}$

T&F *In Exercises 43–62, determine whether the statements are true or false.*

43. If V is a subspace of $\mathcal{R}^n$ and $\mathbf{v}$ is in V, then $c\mathbf{v}$ is in V for every scalar c.
44. Every subspace of $\mathcal{R}^n$ contains $\mathbf{0}$.
45. The subspace $\{\mathbf{0}\}$ is called the null space.
46. The span of a finite nonempty subset of $\mathcal{R}^n$ is a subspace of $\mathcal{R}^n$.
47. The null space of an $m \times n$ matrix is contained in $\mathcal{R}^n$.
48. The column space of an $m \times n$ matrix is contained in $\mathcal{R}^n$.
49. The row space of an $m \times n$ matrix is contained in $\mathcal{R}^m$.
50. The row space of an $m \times n$ matrix A is the set $\{A\mathbf{v}\colon \mathbf{v} \text{ is in } R^n\}$.
51. For any matrix A, the row space of A^T equals the column space of A.
52. The null space of every linear transformation is a subspace.
53. The range of a function need not be a subspace.
54. The range of a linear transformation is a subspace.
55. The range of a linear transformation equals the row space of its standard matrix.
56. The null space of a linear transformation equals the null space of its standard matrix.
57. Every nonzero subspace of $\mathcal{R}^n$ contains infinitely many vectors.
58. A subspace of $\mathcal{R}^n$ must be closed under vector addition.
59. $\mathcal{R}^n$ contains at least two subspaces.
60. A vector $\mathbf{v}$ is in Null A if and only if $A\mathbf{v} = \mathbf{0}$.
61. A vector $\mathbf{v}$ is in Col A if and only if $A\mathbf{x} = \mathbf{v}$ is consistent.
62. A vector $\mathbf{v}$ is in Row A if and only if $A^T\mathbf{x} = \mathbf{v}$ is consistent.

---

63. Find a generating set containing exactly two vectors for the column space of the matrix in Exercise 27.
64. Find a generating set containing exactly two vectors for the column space of the matrix in Exercise 28.
65. Find a generating set containing exactly four vectors for the column space of the matrix in Exercise 32.
66. Find a generating set containing exactly four vectors for the column space of the matrix in Exercise 33.

In Exercises 67–70, for the column space of each matrix, find a generating set containing exactly the number of vectors specified.

67. $\begin{bmatrix} 1 & -3 & 5 \\ -2 & 4 & -1 \end{bmatrix}$, 2 vectors

68. $\begin{bmatrix} -1 & 6 & -7 \\ 5 & -3 & 8 \\ 4 & -2 & 3 \end{bmatrix}$, 3 vectors

69. $\begin{bmatrix} -2 & -1 & -1 & 3 \\ 4 & 1 & 5 & -4 \\ 5 & 2 & 4 & -5 \\ -1 & 0 & -2 & 1 \end{bmatrix}$, 3 vectors

70. $\begin{bmatrix} 1 & 0 & 4 \\ 1 & -1 & 7 \\ 0 & 1 & -3 \\ 1 & 1 & 1 \end{bmatrix}$, 2 vectors

71. Determine the null space, column space, and row space of the $m \times n$ zero matrix.
72. Let R be the reduced row echelon form of A. Is Null $A =$ Null R? Justify your answer.
73. Let R be the reduced row echelon form of A. Is Col $A =$ Col R? Justify your answer.
74. Let R be the reduced row echelon form of A. Prove that Row $A =$ Row R.
75. Give an example of a nonzero matrix for which the row space equals the column space.
76. Give an example of a matrix for which the null space equals the column space.
77. Prove that the intersection of two subspaces of $\mathcal{R}^n$ is a subspace of $\mathcal{R}^n$.
78. Let
$$V = \left\{\begin{bmatrix} v_1 \\ v_2 \end{bmatrix} \in \mathcal{R}^2\colon v_1 = 0\right\}$$
and
$$W = \left\{\begin{bmatrix} v_1 \\ v_2 \end{bmatrix} \in \mathcal{R}^2\colon v_2 = 0\right\}.$$
(a) Prove that both V and W are subspaces of $\mathcal{R}^2$.
(b) Show that $V \cup W$ is *not* a subspace of $\mathcal{R}^2$.
79. Let $\mathcal{S}$ be a nonempty subset of $\mathcal{R}^n$. Prove that $\mathcal{S}$ is a subspace of $\mathcal{R}^n$ if and only if, for all vectors $\mathbf{u}$ and $\mathbf{v}$ in $\mathcal{S}$ and all scalars c, the vector $\mathbf{u} + c\mathbf{v}$ is in $\mathcal{S}$.
80. Prove that if V is a subspace of $\mathcal{R}^n$ containing vectors $\mathbf{u}_1, \mathbf{u}_2, \ldots, \mathbf{u}_k$, then V contains the span of $\{\mathbf{u}_1, \mathbf{u}_2, \ldots, \mathbf{u}_k\}$. For this reason, the span of $\{\mathbf{u}_1, \mathbf{u}_2, \ldots, \mathbf{u}_k\}$ is called the *smallest* subspace of $\mathcal{R}^n$ containing the vectors $\mathbf{u}_1, \mathbf{u}_2, \ldots, \mathbf{u}_k$.

In Exercises 81–88, show that each set is not a subspace of the appropriate $\mathcal{R}^n$.

81. $\left\{\begin{bmatrix} u_1 \\ u_2 \end{bmatrix} \in \mathcal{R}^2\colon u_1u_2 = 0\right\}$

82. $\left\{\begin{bmatrix} u_1 \\ u_2 \end{bmatrix} \in \mathcal{R}^2\colon 2u_1^2 + 3u_2^2 = 12\right\}$

83. $\left\{\begin{bmatrix} 3s - 2 \\ 2s + 4t \\ -t \end{bmatrix} \in \mathcal{R}^3\colon s \text{ and } t \text{ are scalars}\right\}$

84. $\left\{\begin{bmatrix} u_1 \\ u_2 \end{bmatrix} \in \mathcal{R}^2\colon u_1^2 + u_2^2 \leq 1\right\}$

85. $\left\{ \begin{bmatrix} u_1 \\ u_2 \\ u_3 \end{bmatrix} \in \mathcal{R}^3 \colon u_1 > u_2 \text{ and } u_3 < 0 \right\}$

86. $\left\{ \begin{bmatrix} u_1 \\ u_2 \\ u_3 \end{bmatrix} \in \mathcal{R}^3 \colon u_1 \geq u_2 \geq u_3 \right\}$

87. $\left\{ \begin{bmatrix} u_1 \\ u_2 \\ u_3 \end{bmatrix} \in \mathcal{R}^3 \colon u_1 = u_2 u_3 \right\}$

88. $\left\{ \begin{bmatrix} u_1 \\ u_2 \\ u_3 \end{bmatrix} \in \mathcal{R}^3 \colon u_1 u_2 = u_3^2 \right\}$

In Exercises 89–94, use the definition of a subspace, as in Example 3, to prove that each set is a subspace of the appropriate $\mathcal{R}^n$.

89. $\left\{ \begin{bmatrix} u_1 \\ u_2 \end{bmatrix} \in \mathcal{R}^2 \colon u_1 - 3u_2 = 0 \right\}$

90. $\left\{ \begin{bmatrix} u_1 \\ u_2 \end{bmatrix} \in \mathcal{R}^2 \colon 5u_1 + 4u_2 = 0 \right\}$

91. $\left\{ \begin{bmatrix} u_1 \\ u_2 \\ u_3 \end{bmatrix} \in \mathcal{R}^3 \colon 2u_1 + 5u_2 - 4u_3 = 0 \right\}$

92. $\left\{ \begin{bmatrix} u_1 \\ u_2 \\ u_3 \end{bmatrix} \in \mathcal{R}^3 \colon -u_1 + 7u_2 + 2u_3 = 0 \right\}$

93. $\left\{ \begin{bmatrix} u_1 \\ u_2 \\ u_3 \\ u_4 \end{bmatrix} \in \mathcal{R}^4 \colon 3u_1 - u_2 + 6u_4 = 0 \text{ and } u_3 = 0 \right\}$

94. $\left\{ \begin{bmatrix} u_1 \\ u_2 \\ u_3 \\ u_4 \end{bmatrix} \in \mathcal{R}^4 \colon u_1 + 5u_3 = 0 \text{ and } 4u_2 - 3u_4 = 0 \right\}$

95. Let $T \colon \mathcal{R}^n \to \mathcal{R}^m$ be a linear transformation. Use the definition of a subspace to prove that the null space of T is a subspace of $\mathcal{R}^n$.

96. Let $T \colon \mathcal{R}^n \to \mathcal{R}^m$ be a linear transformation. Use the definition of a subspace to prove that the range of T is a subspace of $\mathcal{R}^m$.

97. Let $T \colon \mathcal{R}^n \to \mathcal{R}^m$ be a linear transformation. Prove that if V is a subspace of $\mathcal{R}^n$, then $\{T(\mathbf{u}) \in \mathcal{R}^m \colon \mathbf{u} \text{ is in } V\}$ is a subspace of $\mathcal{R}^m$.

98. Let $T \colon \mathcal{R}^n \to \mathcal{R}^m$ be a linear transformation. Prove that if W is a subspace of $\mathcal{R}^m$, then $\{\mathbf{u} \colon T(\mathbf{u}) \text{ is in } W\}$ is a subspace of $\mathcal{R}^n$.

99. Let A and B be two $m \times n$ matrices. Use the definition of a subspace to prove that $V = \{\mathbf{v} \in \mathcal{R}^n \colon A\mathbf{v} = B\mathbf{v}\}$ is a subspace of $\mathcal{R}^n$.

100. Let V and W be two subspaces of $\mathcal{R}^n$. Use the definition of a subspace to prove that

$$S = \{\mathbf{s} \in \mathcal{R}^n \colon \mathbf{s} = \mathbf{v} + \mathbf{w} \text{ for some } \mathbf{v} \text{ in } V \text{ and } \mathbf{w} \text{ in } W\}$$

is a subspace of $\mathcal{R}^n$.

In Exercises 101–103, use either a calculator with matrix capabilities or computer software such as MATLAB to solve each problem.

101. Let

$$A = \begin{bmatrix} -1 & 0 & 2 & 1 & 1 \\ 1 & 1 & 1 & 0 & 0 \\ 1 & -1 & -5 & 3 & -2 \\ 1 & 1 & 1 & -1 & 0 \\ 0 & 1 & 3 & -2 & 1 \end{bmatrix},$$

$$\mathbf{u} = \begin{bmatrix} 3.0 \\ 1.8 \\ -10.3 \\ 2.3 \\ 6.3 \end{bmatrix}, \quad \text{and} \quad \mathbf{v} = \begin{bmatrix} -.6 \\ 1.4 \\ -1.6 \\ 1.2 \\ 1.8 \end{bmatrix}.$$

(a) Is $\mathbf{u}$ in the column space of A?

(b) Is $\mathbf{v}$ in the column space of A?

102. Let A be the matrix in Exercise 101, and let

$$\mathbf{u} = \begin{bmatrix} 0.5 \\ -1.6 \\ -2.1 \\ 0.0 \\ 4.7 \end{bmatrix} \quad \text{and} \quad \mathbf{v} = \begin{bmatrix} 2.4 \\ -6.3 \\ 3.9 \\ 0.0 \\ -5.4 \end{bmatrix}.$$

(a) Is $\mathbf{u}$ in the null space of A?

(b) Is $\mathbf{v}$ in the null space of A?

103. Let A be the matrix in Exercise 101, and let

$$\mathbf{u} = \begin{bmatrix} -5.1 \\ -2.2 \\ 3.6 \\ 8.2 \\ 2.9 \end{bmatrix} \quad \text{and} \quad \mathbf{v} = \begin{bmatrix} -5.6 \\ -1.4 \\ 3.5 \\ 2.9 \\ 4.2 \end{bmatrix}.$$

(a) Is $\mathbf{u}$ in the row space of A?

(b) Is $\mathbf{v}$ in the row space of A?

SOLUTIONS TO THE PRACTICE PROBLEMS

1. The vectors in V can be written in the form

$$s\begin{bmatrix} -1 \\ 0 \\ 3 \end{bmatrix} + t\begin{bmatrix} 0 \\ 2 \\ -1 \end{bmatrix}.$$

Hence

$$\left\{ \begin{bmatrix} -1 \\ 0 \\ 3 \end{bmatrix}, \begin{bmatrix} 0 \\ 2 \\ -1 \end{bmatrix} \right\}$$

is a generating set for V.

2. The set

$$\left\{\begin{bmatrix}1\\-1\end{bmatrix}, \begin{bmatrix}2\\-3\end{bmatrix}, \begin{bmatrix}-1\\4\end{bmatrix}\right\}$$

consisting of the columns of A is a generating set for the column space of A. To find a generating set for the null space of A, we must solve the equation $A\mathbf{x} = \mathbf{0}$. Since the reduced row echelon form of A is

$$\begin{bmatrix}1 & 0 & 5\\0 & 1 & -3\end{bmatrix},$$

the vector form of the general solution is

$$\begin{bmatrix}x_1\\x_2\\x_3\end{bmatrix} = \begin{bmatrix}-5x_3\\3x_3\\x_3\end{bmatrix} = x_3\begin{bmatrix}-5\\3\\1\end{bmatrix}.$$

Hence

$$\left\{\begin{bmatrix}-5\\3\\1\end{bmatrix}\right\}$$

is a generating set for the null space of A.

3. The standard matrix of T is

$$A = \begin{bmatrix}1 & 0 & 1 & 2\\0 & -1 & 1 & 1\\2 & 3 & -1 & 1\end{bmatrix}.$$

The null space of T is the same as the null space of A, so it is the solution set of $A\mathbf{x} = \mathbf{0}$. Since the reduced row echelon form of A is

$$\begin{bmatrix}1 & 0 & 1 & 2\\0 & 1 & -1 & -1\\0 & 0 & 0 & 0\end{bmatrix},$$

the solutions of $A\mathbf{x} = \mathbf{0}$ can be written as

$$\begin{bmatrix}x_1\\x_2\\x_3\\x_4\end{bmatrix} = \begin{bmatrix}-x_3 - 2x_4\\x_3 + x_4\\x_3\\x_4\end{bmatrix} = x_3\begin{bmatrix}-1\\1\\1\\0\end{bmatrix} + x_4\begin{bmatrix}-2\\1\\0\\1\end{bmatrix}.$$

Thus

$$\left\{\begin{bmatrix}-1\\1\\1\\0\end{bmatrix}, \begin{bmatrix}-2\\1\\0\\1\end{bmatrix}\right\}$$

is a generating set for the null space of T.

The range of T is the same as the column space of A. Hence the set

$$\left\{\begin{bmatrix}1\\0\\2\end{bmatrix}, \begin{bmatrix}0\\-1\\3\end{bmatrix}, \begin{bmatrix}1\\1\\-1\end{bmatrix}, \begin{bmatrix}2\\1\\1\end{bmatrix}\right\}$$

of columns of A is a generating set for the range of T.

4.2 BASIS AND DIMENSION

In the previous section, we saw how to describe subspaces in terms of generating sets. To do so, we write each vector in the subspace as a linear combination of the vectors in the generating set. While there are many generating sets for a given nonzero subspace, it is best to use a generating set that contains as few vectors as possible. Such a generating set, which must be linearly independent, is called a *basis* for the subspace.

Definition Let V be a nonzero subspace of $\mathcal{R}^n$. A **basis** (*plural, bases*) for V is a linearly independent generating set for V.

For example, the set of standard vectors $\{\mathbf{e}_1, \mathbf{e}_2, \ldots, \mathbf{e}_n\}$ in $\mathcal{R}^n$ is both a linearly independent set and a generating set for $\mathcal{R}^n$. Hence $\{\mathbf{e}_1, \mathbf{e}_2, \ldots, \mathbf{e}_n\}$ is a basis for $\mathcal{R}^n$. We call this basis the **standard basis** for $\mathcal{R}^n$ and denote it by $\mathcal{E}$. (See Figure 4.5.) However, there are many other possible bases for $\mathcal{R}^n$. For any angle θ such that $0^\circ < \theta < 360^\circ$, the vectors $A_\theta\mathbf{e}_1$ and $A_\theta\mathbf{e}_2$ obtained by rotating the vectors $\mathbf{e}_1$ and $\mathbf{e}_2$ by θ form a basis for $\mathcal{R}^2$. (See Figure 4.6.) There are also other bases for $\mathcal{R}^2$ in which the vectors are not perpendicular. Thus there are infinitely many bases for $\mathcal{R}^2$. In many applications, it is natural and convenient to describe vectors in terms of a basis other than the standard basis for $\mathcal{R}^n$.

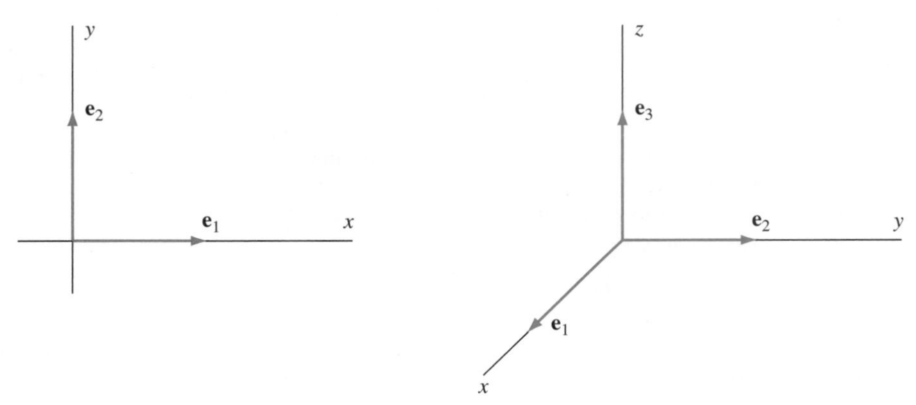

The standard basis for $\mathcal{R}^2$

The standard basis for $\mathcal{R}^3$

Figure 4.5

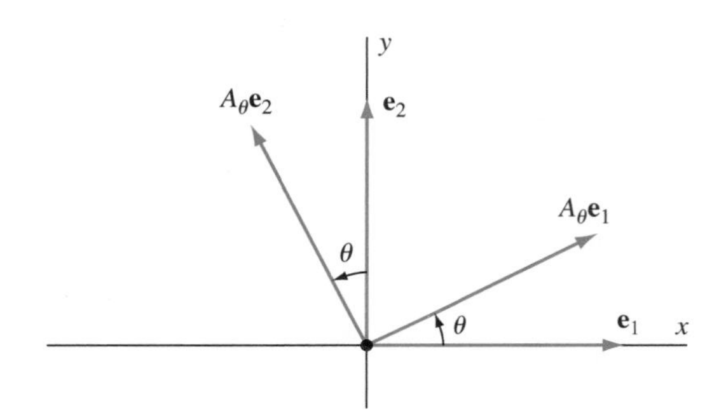

Figure 4.6 Rotated standard vectors

We can restate some of our previous results about generating sets and linearly independent sets, using the concept of a basis. For example, recall that the pivot columns of a matrix are those corresponding to the columns containing leading ones in the reduced row echelon form of the matrix. We can restate Theorem 2.4 (which states that the pivot columns of a matrix are linearly independent and form a generating set for its column space), as follows:

> The pivot columns of a matrix form a basis for its column space.

The following example illustrates this fact:

Example 1 Find a basis for Col A if

$$A = \begin{bmatrix} 1 & 2 & -1 & 2 & 1 & 2 \\ -1 & -2 & 1 & 2 & 3 & 6 \\ 2 & 4 & -3 & 2 & 0 & 3 \\ -3 & -6 & 2 & 0 & 3 & 9 \end{bmatrix}.$$

Solution In Example 1 of Section 1.4, we showed that the reduced row echelon form of A is

$$\begin{bmatrix} 1 & 2 & 0 & 0 & -1 & -5 \\ 0 & 0 & 1 & 0 & 0 & -3 \\ 0 & 0 & 0 & 1 & 1 & 2 \\ 0 & 0 & 0 & 0 & 0 & 0 \end{bmatrix}.$$

Since the leading ones of this matrix are in columns one, three, and four, the pivot columns of A are

$$\begin{bmatrix} 1 \\ -1 \\ 2 \\ -3 \end{bmatrix}, \quad \begin{bmatrix} -1 \\ 1 \\ -3 \\ 2 \end{bmatrix}, \quad \text{and} \quad \begin{bmatrix} 2 \\ 2 \\ 2 \\ 0 \end{bmatrix}.$$

As previously mentioned, these vectors form a basis for Col A. Note that it is the pivot columns of A, and not those of the reduced row echelon form of A, that form a basis for Col A.

! *CAUTION* As seen in Example 1, the column space of a matrix is usually different from that of its reduced row echelon form. In fact, the column space of a matrix in reduced row echelon form always has a basis consisting of standard vectors, which is not usually the case for other matrices.

We can apply the method in Example 1 to find a basis for any subspace if we know a finite generating set for the subspace. For if $\mathcal{S}$ is a finite generating set for a subspace of $\mathcal{R}^n$, and A is a matrix whose columns are the vectors in $\mathcal{S}$, then the pivot columns of A constitute a basis for Col A, which is the span of $\mathcal{S}$. Note that this basis is also contained in $\mathcal{S}$. For example, if W is the span of

$$\mathcal{S} = \left\{ \begin{bmatrix} 1 \\ -1 \\ 2 \\ -3 \end{bmatrix}, \begin{bmatrix} 2 \\ -2 \\ 4 \\ -6 \end{bmatrix}, \begin{bmatrix} -1 \\ 1 \\ -3 \\ 2 \end{bmatrix}, \begin{bmatrix} 2 \\ 2 \\ 2 \\ 0 \end{bmatrix}, \begin{bmatrix} 1 \\ 3 \\ 0 \\ 3 \end{bmatrix}, \begin{bmatrix} 2 \\ 6 \\ 3 \\ 9 \end{bmatrix} \right\},$$

then we can find a basis for W by forming the matrix whose columns are the vectors in $\mathcal{S}$ and choosing its pivot columns.

THEOREM 4.3

(Reduction Theorem) Let $\mathcal{S}$ be a finite generating set for a nonzero subspace V of $\mathcal{R}^n$. Then $\mathcal{S}$ can be reduced to a basis for V by removing vectors from $\mathcal{S}$.

PROOF Let V be a subspace of $\mathcal{R}^n$ and $\mathcal{S} = \{\mathbf{u}_1, \mathbf{u}_2, \ldots, \mathbf{u}_k\}$ be a generating set for V. If $A = [\mathbf{u}_1 \ \mathbf{u}_2 \ \ldots \ \mathbf{u}_k]$, then Col $A = \text{Span}\{\mathbf{u}_1, \mathbf{u}_2, \ldots, \mathbf{u}_k\} = V$. Since the pivot columns of A form a basis for Col A, the set consisting of the pivot columns of A is a basis for V. This basis is clearly contained in $\mathcal{S}$. ■

Example 2 Find a basis for Span $\mathcal{S}$ consisting of vectors in $\mathcal{S}$ if

$$\mathcal{S} = \left\{ \begin{bmatrix} 1 \\ 2 \\ 1 \\ 1 \end{bmatrix}, \begin{bmatrix} 2 \\ 4 \\ 1 \\ 1 \end{bmatrix}, \begin{bmatrix} 1 \\ -1 \\ 0 \\ 1 \end{bmatrix}, \begin{bmatrix} 2 \\ 1 \\ 1 \\ 2 \end{bmatrix}, \begin{bmatrix} 1 \\ -1 \\ 2 \\ 1 \end{bmatrix} \right\}.$$

Solution Let

$$A = \begin{bmatrix} 1 & 2 & 1 & 2 & 1 \\ 2 & 4 & -1 & 1 & -1 \\ 1 & 2 & 0 & 1 & 2 \\ 1 & 2 & 1 & 2 & 1 \end{bmatrix},$$

the matrix whose columns are the vectors in $\mathcal{S}$. It can be shown that the reduced row echelon form of A is

$$\begin{bmatrix} 1 & 2 & 0 & 1 & 0 \\ 0 & 0 & 1 & 1 & 0 \\ 0 & 0 & 0 & 0 & 1 \\ 0 & 0 & 0 & 0 & 0 \end{bmatrix}.$$

Since the leading ones of this matrix are in columns 1, 3, and 5, the corresponding columns of A form a basis for $\mathcal{S}$. Thus the set

$$\left\{ \begin{bmatrix} 1 \\ 2 \\ 1 \\ 1 \end{bmatrix}, \begin{bmatrix} 1 \\ -1 \\ 0 \\ 1 \end{bmatrix}, \begin{bmatrix} 1 \\ -1 \\ 2 \\ 1 \end{bmatrix} \right\}$$

is a basis for the span of $\mathcal{S}$, consisting of vectors of $\mathcal{S}$.

If $\{\mathbf{u}_1, \mathbf{u}_2, \ldots, \mathbf{u}_k\}$ is a generating set for $\mathcal{R}^n$, then $[\mathbf{u}_1 \ \mathbf{u}_2 \ \ldots \ \mathbf{u}_k]$ must have a pivot position in each row, by Theorem 1.6. Since no two pivot positions can lie in the same column, it follows that this matrix must have at least n columns. Thus $k \geq n$; that is, *a generating set for $\mathcal{R}^n$ must contain at least n vectors.*

Now suppose that $\{\mathbf{v}_1, \mathbf{v}_2, \ldots, \mathbf{v}_j\}$ is a linearly independent subset of $\mathcal{R}^n$. As noted in Section 1.7, every subset of $\mathcal{R}^n$ containing more than n vectors must be linearly dependent. Hence, in order that $\{\mathbf{v}_1, \mathbf{v}_2, \ldots, \mathbf{v}_j\}$ be linearly independent, we must have $j \leq n$. That is, *a linearly independent subset of $\mathcal{R}^n$ must contain at most n vectors.*

Combining the observations in the two preceding paragraphs, we see that *every basis for $\mathcal{R}^n$ must contain exactly n vectors.* In summary, we have the following result:

> Let $\mathcal{S}$ be a finite subset of $\mathcal{R}^n$. Then the following are true:
>
> 1. If $\mathcal{S}$ is a generating set for $\mathcal{R}^n$, then $\mathcal{S}$ contains at least n vectors.
> 2. If $\mathcal{S}$ is linearly independent, then $\mathcal{S}$ contains at most n vectors.
> 3. If $\mathcal{S}$ is a basis for $\mathcal{R}^n$, then $\mathcal{S}$ contains exactly n vectors.

Our next theorem shows that every nonzero subspace of $\mathcal{R}^n$ has a basis.

THEOREM 4.4

(Extension Theorem) Let $\mathcal{S}$ be a linearly independent subset of a nonzero subspace V of $\mathcal{R}^n$. Then $\mathcal{S}$ can be extended to a basis for V by inclusion of additional vectors. In particular, every nonzero subspace has a basis.

PROOF Let $\mathcal{S} = \{\mathbf{u}_1, \mathbf{u}_2, \ldots, \mathbf{u}_k\}$ be a linearly independent subset of V. If the span of $\mathcal{S}$ is V, then $\mathcal{S}$ is a basis for V that contains $\mathcal{S}$, and we are done. Otherwise, there exists a vector $\mathbf{v}_1$ in V that is not in the span of $\mathcal{S}$. Property 3 on page 81 implies that $\mathcal{S}' = \{\mathbf{u}_1, \mathbf{u}_2, \ldots, \mathbf{u}_k, \mathbf{v}_1\}$ is linearly independent. If the span of $\mathcal{S}'$ is V, then $\mathcal{S}'$ is a basis for V that contains $\mathcal{S}$, and again we are done. Otherwise, there exists a vector $\mathbf{v}_2$ in V that is not in the span of $\mathcal{S}'$. As before, $\mathcal{S}'' = \{\mathbf{u}_1, \mathbf{u}_2, \ldots, \mathbf{u}_k, \mathbf{v}_1, \mathbf{v}_2\}$ is linearly independent. We continue this process of selecting larger linearly independent subsets of V containing $\mathcal{S}$ until one of them is a generating set for V (and hence a basis for V that contains $\mathcal{S}$). Note that this process must stop in at most n steps, because every subset of $\mathcal{R}^n$ containing more than n vectors is linearly dependent by property 4 on page 81.

To prove that V actually has a basis, let $\mathbf{u}$ be a nonzero vector in V. Applying the Extension Theorem to $\mathcal{S} = \{\mathbf{u}\}$, which is a linearly independent subset of V (property 1 on page 81), we see that V has a basis. ■

We have seen that a nonzero subspace of $\mathcal{R}^n$ has infinitely many bases. Although the vectors in two bases for a nonzero subspace may be different, the next theorem shows that the *number* of vectors in each basis for a particular subspace must be the same.

THEOREM 4.5

Let V be a nonzero subspace of $\mathcal{R}^n$. Then any two bases for V contain the same number of vectors.

PROOF Suppose that $\{\mathbf{u}_1, \mathbf{u}_2, \ldots, \mathbf{u}_k\}$ and $\{\mathbf{v}_1, \mathbf{v}_2, \ldots, \mathbf{v}_p\}$ are bases for V, and let $A = [\mathbf{u}_1\ \mathbf{u}_2\ \ldots\ \mathbf{u}_k]$ and $B = [\mathbf{v}_1\ \mathbf{v}_2\ \ldots\ \mathbf{v}_p]$. Because $\{\mathbf{u}_1, \mathbf{u}_2, \ldots, \mathbf{u}_k\}$ is a generating set for V, there are vectors $\mathbf{c}_i$ in $\mathcal{R}^k$ for $i = 1, 2, \ldots, p$ such that $A\mathbf{c}_i = \mathbf{v}_i$. Let $C = [\mathbf{c}_1\ \mathbf{c}_2\ \ldots\ \mathbf{c}_p]$. Then C is a $k \times p$ matrix such that $AC = B$. Now suppose that $C\mathbf{x} = \mathbf{0}$ for some vector $\mathbf{x}$ in $\mathcal{R}^p$. Then $B\mathbf{x} = AC\mathbf{x} = \mathbf{0}$. But the columns of B are linearly independent, and hence $\mathbf{x} = \mathbf{0}$ by Theorem 1.8. Applying Theorem 1.8 to the matrix C, we conclude that the columns of C are linearly independent vectors in $\mathcal{R}^k$. Because a set of more than k vectors from $\mathcal{R}^k$ is linearly dependent (property 4 on page 81), we have that $p \le k$. Reversing the roles of the two bases in the preceding argument shows that $k \le p$ also. Therefore $k = p$; that is, the two bases contain the same number of vectors. ■

The Reduction and Extension Theorems give two different characteristics of a basis for a subspace. The Reduction Theorem tells us that vectors can be deleted from a generating set to form a basis. In fact, we learned from Theorem 1.7 that a vector can be deleted from a generating set without changing the set's span if that vector is a linear combination of the other vectors in the generating set. Furthermore, we saw in item 5 on page 81 that if no vector can be removed from a generating set without changing the set's span, then the set must be linearly independent.

> A basis is a generating set for a subspace containing the fewest possible vectors.

On the other hand, if we adjoin an additional vector to a basis, then the larger set cannot be contained in a basis, because any two bases for a subspace must contain the same number of vectors. Thus the Extension Theorem implies that the larger set cannot be linearly independent.

> A basis is a linearly independent subset of a subspace that is as large as possible.

According to Theorem 4.5, the size of every basis for a subspace is the same. This permits the following definition:

Definition The number of vectors in a basis for a nonzero subspace V of $\mathcal{R}^n$ is called the **dimension** of V and is denoted by $\dim V$. It is convenient to define the dimension of the zero subspace of $\mathcal{R}^n$ to be 0.

Since the standard basis for $\mathcal{R}^n$ contains n vectors, $\dim \mathcal{R}^n = n$. In Section 4.3, we discuss the dimensions of several of the types of subspaces that we have previously encountered.

Practice Problem 1 ▶ Is $\left\{\begin{bmatrix} 0 \\ -1 \\ 1 \end{bmatrix}, \begin{bmatrix} -1 \\ 1 \\ 2 \end{bmatrix}\right\}$ a basis for $\mathcal{R}^3$? Justify your answer. ◀

The next result follows from the preceding two theorems. It contains important information about the size of linearly independent subsets of a subspace.

THEOREM 4.6

Let V be a subspace of $\mathcal{R}^n$ with dimension k. Then every linearly independent subset of V contains at most k vectors; or equivalently, any finite subset of V containing more than k vectors is linearly dependent.

PROOF Let $\{\mathbf{v}_1, \mathbf{v}_2, \ldots, \mathbf{v}_p\}$ be a linearly independent subset of V. By the Extension Theorem, this set can be extended to a basis $\{\mathbf{v}_1, \mathbf{v}_2, \ldots, \mathbf{v}_p, \ldots, \mathbf{v}_k\}$ for V. It follows that $p \le k$. ■

Example 3 Find a basis for the subspace

$$V = \left\{\begin{bmatrix} x_1 \\ x_2 \\ x_3 \\ x_4 \end{bmatrix} \in \mathcal{R}^4 : x_1 - 3x_2 + 5x_3 - 6x_4 = 0\right\}$$

of $\mathcal{R}^4$, and determine the dimension of V. (Since V is defined as the set of solutions of a homogeneous system of linear equations, V is, in fact, a subspace.)

Solution The vectors in V are solutions of $x_1 - 3x_2 + 5x_3 - 6x_4 = 0$, a system of 1 linear equation in 4 variables. To solve this system, we apply the technique described in Section 1.3. Since

$$x_1 = 3x_2 - 5x_3 + 6x_4,$$

the vector form of the general solution of this system is

$$\begin{bmatrix} x_1 \\ x_2 \\ x_3 \\ x_4 \end{bmatrix} = \begin{bmatrix} 3x_2 - 5x_3 + 6x_4 \\ x_2 \\ x_3 \\ x_4 \end{bmatrix} = x_2 \begin{bmatrix} 3 \\ 1 \\ 0 \\ 0 \end{bmatrix} + x_3 \begin{bmatrix} -5 \\ 0 \\ 1 \\ 0 \end{bmatrix} + x_4 \begin{bmatrix} 6 \\ 0 \\ 0 \\ 1 \end{bmatrix}.$$

As noted in Section 1.7, the set

$$\mathcal{S} = \left\{ \begin{bmatrix} 3 \\ 1 \\ 0 \\ 0 \end{bmatrix}, \begin{bmatrix} -5 \\ 0 \\ 1 \\ 0 \end{bmatrix}, \begin{bmatrix} 6 \\ 0 \\ 0 \\ 1 \end{bmatrix} \right\}$$

containing the vectors in the vector form is both a generating set for V and a linearly independent set. Therefore $\mathcal{S}$ is a basis for V. Since $\mathcal{S}$ is a basis for V containing 3 vectors, $\dim V = 3$.

Practice Problem 2 ▶ Find a basis for the column space and null space of

$$\begin{bmatrix} -1 & 2 & 1 & -1 \\ 2 & -4 & -3 & 0 \\ 1 & -2 & 0 & 3 \end{bmatrix}.$$ ◀

To find the dimension of a subspace, we must usually determine a basis for that subspace. In this book, a subspace is almost always defined as either

(a) the span of a given set of vectors, or

(b) the solution set of a homogeneous system of linear equations.

Recall that a basis for a subspace as defined in (a) can be found by the technique of Example 1. When a subspace is defined as in (b), we can obtain a basis by solving the system of linear equations. The method was demonstrated in Example 3.

CONFIRMING THAT A SET IS A BASIS FOR A SUBSPACE

As we mentioned earlier in this section, a nonzero subspace has many bases. Depending on the application, some bases are more useful than others. For example, we might want to find a basis whose vectors are perpendicular to each other. While the techniques of this section enable us to find a basis for any subspace, we might not produce a basis with the most desirable properties. In Chapters 5 and 6, we describe methods for finding bases with special properties. In the remainder of this section, we show how to determine whether or not a set of vectors forms a basis for a subspace.

Consider, for example, the subspace

$$V = \left\{ \begin{bmatrix} v_1 \\ v_2 \\ v_3 \end{bmatrix} \in \mathcal{R}^3 \colon v_1 - v_2 + 2v_3 = 0 \right\}$$

of $\mathcal{R}^3$ and the set

$$\mathcal{S} = \left\{ \begin{bmatrix} 1 \\ 1 \\ 0 \end{bmatrix}, \begin{bmatrix} -1 \\ 1 \\ 1 \end{bmatrix} \right\}.$$

By using the method of Example 3, we can find a basis for V. Since this basis contains two vectors, the dimension of V is 2. We would like to show that $\mathcal{S}$, which contains two perpendicular vectors, is also a basis for V. (In Section 6.2, we discuss a method for transforming any basis into one whose vectors are perpendicular to each other.) By definition, $\mathcal{S}$ is a basis for V if $\mathcal{S}$ is both linearly independent and a generating set for V. In this case, it is clear that $\mathcal{S}$ is linearly independent because $\mathcal{S}$ contains two vectors, neither of which is a multiple of the other. It remains to show only that $\mathcal{S}$ is a generating set for V, that is, to show that $\mathcal{S}$ is a subset of V such that every vector in V is a linear combination of the vectors in $\mathcal{S}$. Checking that $\mathcal{S}$ is a subset of V is straightforward. Because $1-1+2(0)=0$ and $-1-1+2(1)=0$, the vectors in $\mathcal{S}$ satisfy the equation defining V. Hence both

$$\begin{bmatrix}1\\1\\0\end{bmatrix} \quad \text{and} \quad \begin{bmatrix}-1\\1\\1\end{bmatrix}$$

are in V. Unfortunately, checking that *every* vector in V is a linear combination of the vectors in $\mathcal{S}$ is more tedious. This requires us to show that, for every

$$\begin{bmatrix}v_1\\v_2\\v_3\end{bmatrix}$$

such that $v_1 - v_2 + 2v_3 = 0$, there are scalars c_1 and c_2 such that

$$c_1\begin{bmatrix}1\\1\\0\end{bmatrix} + c_2\begin{bmatrix}-1\\1\\1\end{bmatrix} = \begin{bmatrix}v_1\\v_2\\v_3\end{bmatrix}.$$

The following result removes the need for this type of calculation:

THEOREM 4.7

Let V be a k-dimensional subspace of $\mathcal{R}^n$. Suppose that $\mathcal{S}$ is a subset of V with exactly k vectors. Then $\mathcal{S}$ is a basis for V if either $\mathcal{S}$ is linearly independent or $\mathcal{S}$ is a generating set for V.

PROOF Suppose that $\mathcal{S}$ is linearly independent. By the Extension Theorem, there is a basis $\mathcal{B}$ for V that contains $\mathcal{S}$. Because both $\mathcal{B}$ and $\mathcal{S}$ contain k vectors, $\mathcal{B} = \mathcal{S}$. Thus $\mathcal{S}$ is a basis for V.

Now suppose that $\mathcal{S}$ is a generating set for V. By the Reduction Theorem, some subset $\mathcal{C}$ of $\mathcal{S}$ is a basis for V. But because V has dimension k, every basis for V must contain exactly k vectors by Theorem 4.5. Hence we must have $\mathcal{C} = \mathcal{S}$, so $\mathcal{S}$ is a basis for V. ■

This theorem gives us three straightforward steps to show that a given set $\mathcal{B}$ is a basis for a subspace V:

Steps to Show that a Set $\mathcal{B}$ is a Basis for a Subspace V of $\mathcal{R}^n$

1. Show that $\mathcal{B}$ is contained in V.
2. Show that $\mathcal{B}$ is linearly independent (or that $\mathcal{B}$ is a generating set for V).

3. Compute the dimension of V, and confirm that the number of vectors in $\mathcal{B}$ equals the dimension of V.

In the preceding example, we showed that $\mathcal{S}$ is a linearly independent subset of V containing two vectors. Because $\dim V = 2$, all three of these steps are satisfied, and hence $\mathcal{S}$ is a basis for V. Therefore we need not verify that $\mathcal{S}$ is a generating set for V.

Two more examples of this technique follow.

Example 4 Show that

$$\mathcal{B} = \left\{ \begin{bmatrix} 1 \\ -1 \\ 1 \\ 0 \end{bmatrix}, \begin{bmatrix} 1 \\ 0 \\ 1 \\ -1 \end{bmatrix}, \begin{bmatrix} 0 \\ 1 \\ 1 \\ -1 \end{bmatrix} \right\}$$

is a basis for

$$V = \left\{ \begin{bmatrix} v_1 \\ v_2 \\ v_3 \\ v_4 \end{bmatrix} \in \mathcal{R}^4 \colon v_1 + v_2 + v_4 = 0 \right\}.$$

Solution Clearly, the components of the three vectors in $\mathcal{B}$ all satisfy the equation $v_1 + v_2 + v_4 = 0$. Thus $\mathcal{B}$ is a subset of V, so step 1 is satisfied.

Because the reduced row echelon form of

$$\begin{bmatrix} 1 & 1 & 0 \\ -1 & 0 & 1 \\ 1 & 1 & 1 \\ 0 & -1 & -1 \end{bmatrix} \quad \text{is} \quad \begin{bmatrix} 1 & 0 & 0 \\ 0 & 1 & 0 \\ 0 & 0 & 1 \\ 0 & 0 & 0 \end{bmatrix},$$

it follows that $\mathcal{B}$ is linearly independent, so step 2 is satisfied.

As in Example 3, we find that

$$\left\{ \begin{bmatrix} -1 \\ 1 \\ 0 \\ 0 \end{bmatrix}, \begin{bmatrix} 0 \\ 0 \\ 1 \\ 0 \end{bmatrix}, \begin{bmatrix} -1 \\ 0 \\ 0 \\ 1 \end{bmatrix} \right\}$$

is a basis for V. Hence the dimension of V is 3. But $\mathcal{B}$ contains three vectors, so step 3 is satisfied, and hence $\mathcal{B}$ is a basis for V.

Practice Problem 3 ▶ Show that

$$\left\{ \begin{bmatrix} -1 \\ 1 \\ -2 \\ 1 \end{bmatrix}, \begin{bmatrix} 0 \\ 3 \\ -4 \\ 2 \end{bmatrix} \right\}$$

is a basis for the null space of the matrix in Practice Problem 1. ◀

Example 5 Let W be the span of $\mathcal{S}$, where

$$\mathcal{S} = \left\{ \begin{bmatrix} 1 \\ 1 \\ 1 \\ 2 \end{bmatrix}, \begin{bmatrix} -1 \\ 3 \\ 1 \\ -1 \end{bmatrix}, \begin{bmatrix} 3 \\ 1 \\ -1 \\ 1 \end{bmatrix}, \begin{bmatrix} 1 \\ 1 \\ -1 \\ -1 \end{bmatrix} \right\}.$$

Show that a basis for W is

$$\mathcal{B} = \left\{ \begin{bmatrix} 1 \\ 2 \\ 0 \\ 0 \end{bmatrix}, \begin{bmatrix} 1 \\ 0 \\ 0 \\ 1 \end{bmatrix}, \begin{bmatrix} 0 \\ 1 \\ 1 \\ 1 \end{bmatrix} \right\}.$$

Solution Let

$$B = \begin{bmatrix} 1 & 1 & 0 \\ 2 & 0 & 1 \\ 0 & 0 & 1 \\ 0 & 1 & 1 \end{bmatrix} \quad \text{and} \quad A = \begin{bmatrix} 1 & -1 & 3 & 1 \\ 1 & 3 & 1 & 1 \\ 1 & 1 & -1 & -1 \\ 2 & -1 & 1 & -1 \end{bmatrix}.$$

You should check that the equation $A\mathbf{x} = \mathbf{b}$ is consistent for each $\mathbf{b}$ in $\mathcal{B}$. Therefore $\mathcal{B}$ is a subset of W, so step 1 is satisfied.

We can easily verify that the reduced row echelon form of B is

$$\begin{bmatrix} 1 & 0 & 0 \\ 0 & 1 & 0 \\ 0 & 0 & 1 \\ 0 & 0 & 0 \end{bmatrix},$$

so $\mathcal{B}$ is linearly independent. Thus step 2 is satisfied.

Since the reduced row echelon form of A is

$$\begin{bmatrix} 1 & 0 & 0 & -\frac{2}{3} \\ 0 & 1 & 0 & \frac{1}{3} \\ 0 & 0 & 1 & \frac{2}{3} \\ 0 & 0 & 0 & 0 \end{bmatrix},$$

we see that the first 3 columns of A are its pivot columns and hence are a basis for Col $A = W$. Thus $\dim W = 3$, which equals the number of vectors contained in $\mathcal{B}$, and step 3 is satisfied. Hence $\mathcal{B}$ is a basis for W.

EXERCISES

In Exercises 1–8, find a basis for (a) the column space and (b) the null space of each matrix.

1. $\begin{bmatrix} 1 & -3 & 4 & -2 \\ -1 & 3 & -4 & 2 \end{bmatrix}$

2. $\begin{bmatrix} 1 & 0 & -2 & 1 \\ 2 & -1 & -3 & 4 \end{bmatrix}$

3. $\begin{bmatrix} 1 & 2 & 4 \\ -1 & -1 & -1 \\ -1 & 0 & 2 \end{bmatrix}$

4. $\begin{bmatrix} 1 & 3 & -2 \\ -1 & -3 & 2 \\ 2 & 6 & -4 \end{bmatrix}$

5. $\begin{bmatrix} 1 & -2 & 0 & 2 \\ -1 & 2 & 1 & -3 \\ 2 & -4 & 3 & 1 \end{bmatrix}$

6. $\begin{bmatrix} 1 & 1 & -1 & -2 \\ -1 & -2 & 1 & 3 \\ 2 & 3 & 1 & 4 \end{bmatrix}$

7. $\begin{bmatrix} -1 & 1 & 2 & 2 \\ 2 & 0 & -5 & 3 \\ 1 & -1 & -1 & -1 \\ 0 & 1 & -2 & 2 \end{bmatrix}$

8. $\begin{bmatrix} 1 & -1 & 2 & 1 \\ 3 & -3 & 5 & 4 \\ 0 & 0 & 3 & -3 \\ 2 & -2 & 1 & 5 \end{bmatrix}$

In Exercises 9–16, a linear transformation T is given. (a) Find a basis for the range of T. (b) If the null space of T is nonzero, find a basis for the null space of T.

9. $T\left(\begin{bmatrix} x_1 \\ x_2 \\ x_3 \end{bmatrix}\right) = \begin{bmatrix} x_1 + 2x_2 + x_3 \\ 2x_1 + 3x_2 + 3x_3 \\ x_1 + 2x_2 + 4x_3 \end{bmatrix}$

10. $T\left(\begin{bmatrix} x_1 \\ x_2 \\ x_3 \end{bmatrix}\right) = \begin{bmatrix} x_1 + 2x_2 - x_3 \\ x_1 + x_2 \\ x_2 - x_3 \end{bmatrix}$

11. $T\left(\begin{bmatrix} x_1 \\ x_2 \\ x_3 \\ x_4 \end{bmatrix}\right) = \begin{bmatrix} x_1 - 2x_2 + x_3 + x_4 \\ 2x_1 - 5x_2 + x_3 + 3x_4 \\ x_1 - 3x_2 + 2x_4 \end{bmatrix}$

12. $T\left(\begin{bmatrix} x_1 \\ x_2 \\ x_3 \\ x_4 \end{bmatrix}\right) = \begin{bmatrix} x_1 + 2x_3 + x_4 \\ x_1 + 3x_3 + 2x_4 \\ -x_1 + x_3 \end{bmatrix}$

13. $T\left(\begin{bmatrix} x_1 \\ x_2 \\ x_3 \\ x_4 \end{bmatrix}\right) = \begin{bmatrix} x_1 + x_2 + 2x_3 - x_4 \\ 2x_1 + x_2 + x_3 \\ 0 \\ 3x_1 + x_2 + x_4 \end{bmatrix}$

14. $T\left(\begin{bmatrix} x_1 \\ x_2 \\ x_3 \\ x_4 \end{bmatrix}\right) = \begin{bmatrix} -2x_1 - x_2 + x_4 \\ 0 \\ x_1 + 2x_2 + 3x_3 + 4x_4 \\ 2x_1 + 3x_2 + 4x_3 + 5x_4 \end{bmatrix}$

15. $T\left(\begin{bmatrix} x_1 \\ x_2 \\ x_3 \\ x_4 \\ x_5 \end{bmatrix}\right) = \begin{bmatrix} x_1 + 2x_2 + 3x_3 + 4x_5 \\ 3x_1 + x_2 - x_3 - 3x_5 \\ 7x_1 + 4x_2 + x_3 - 2x_5 \end{bmatrix}$

16. $T\left(\begin{bmatrix} x_1 \\ x_2 \\ x_3 \\ x_4 \\ x_5 \end{bmatrix}\right) = \begin{bmatrix} -x_1 + x_2 + 4x_3 + 6x_4 + 9x_5 \\ x_1 + x_2 + 2x_3 + 4x_4 + 3x_5 \\ 3x_1 + x_2 + 2x_4 - 3x_5 \\ x_1 + 2x_2 + 5x_3 + 9x_4 + 9x_5 \end{bmatrix}$

In Exercises 17–32, find a basis for each subspace.

17. $\left\{\begin{bmatrix} s \\ -2s \end{bmatrix} \in \mathcal{R}^2 \colon s \text{ is a scalar}\right\}$

18. $\left\{\begin{bmatrix} 2s \\ -s + 4t \\ s - 3t \end{bmatrix} \in \mathcal{R}^3 \colon s \text{ and } t \text{ are scalars}\right\}$

19. $\left\{\begin{bmatrix} 5r - 3s \\ 2r \\ 0 \\ -4s \end{bmatrix} \in \mathcal{R}^4 \colon r \text{ and } s \text{ are scalars}\right\}$

20. $\left\{\begin{bmatrix} 5r - 3s \\ 2r + 6s \\ 4s - 7t \\ 3r - s + 9t \end{bmatrix} \in \mathcal{R}^4 \colon r, s, \text{ and } t \text{ are scalars}\right\}$

21. $\left\{\begin{bmatrix} x_1 \\ x_2 \\ x_3 \end{bmatrix} \in \mathcal{R}^3 \colon x_1 - 3x_2 + 5x_3 = 0\right\}$

22. $\left\{\begin{bmatrix} x_1 \\ x_2 \\ x_3 \end{bmatrix} \in \mathcal{R}^3 \colon -x_1 + x_2 + 2x_3 = 0 \text{ and } 2x_1 - 3x_2 + 4x_3 = 0\right\}$

23. $\left\{\begin{bmatrix} x_1 \\ x_2 \\ x_3 \\ x_4 \end{bmatrix} \in \mathcal{R}^4 \colon x_1 - 2x_2 + 3x_3 - 4x_4 = 0\right\}$

24. $\left\{\begin{bmatrix} x_1 \\ x_2 \\ x_3 \\ x_4 \end{bmatrix} \in \mathcal{R}^4 \colon x_1 - x_2 + 2x_3 + x_4 = 0 \text{ and } 2x_1 - 3x_2 - 5x_3 - x_4 = 0\right\}$

25. Span $\left\{\begin{bmatrix} 1 \\ 2 \\ 1 \end{bmatrix}, \begin{bmatrix} 2 \\ 1 \\ 3 \end{bmatrix}, \begin{bmatrix} 1 \\ -4 \\ 3 \end{bmatrix}\right\}$

26. Span $\left\{\begin{bmatrix} 1 \\ 1 \\ -1 \end{bmatrix}, \begin{bmatrix} 2 \\ 2 \\ -2 \end{bmatrix}, \begin{bmatrix} 1 \\ 2 \\ 0 \end{bmatrix}, \begin{bmatrix} -1 \\ 1 \\ 3 \end{bmatrix}\right\}$

27. Span $\left\{\begin{bmatrix} 1 \\ -1 \\ 3 \end{bmatrix}, \begin{bmatrix} 0 \\ -1 \\ 1 \end{bmatrix}, \begin{bmatrix} 2 \\ 3 \\ 1 \end{bmatrix}, \begin{bmatrix} 1 \\ -2 \\ 0 \end{bmatrix}, \begin{bmatrix} 4 \\ -7 \\ -9 \end{bmatrix}\right\}$

28. Span $\left\{\begin{bmatrix} 2 \\ 3 \\ -5 \end{bmatrix}, \begin{bmatrix} 8 \\ -12 \\ 20 \end{bmatrix}, \begin{bmatrix} 1 \\ 0 \\ -2 \end{bmatrix}, \begin{bmatrix} 0 \\ 2 \\ -1 \end{bmatrix}, \begin{bmatrix} 7 \\ 2 \\ 0 \end{bmatrix}\right\}$

29. Span $\left\{\begin{bmatrix} 1 \\ 0 \\ -1 \\ 2 \end{bmatrix}, \begin{bmatrix} 1 \\ 1 \\ -2 \\ 1 \end{bmatrix}, \begin{bmatrix} -2 \\ 3 \\ -1 \\ -7 \end{bmatrix}, \begin{bmatrix} 1 \\ -1 \\ 0 \\ 3 \end{bmatrix}, \begin{bmatrix} 0 \\ 1 \\ -1 \\ 2 \end{bmatrix}\right\}$

30. Span $\left\{\begin{bmatrix} 0 \\ 2 \\ 3 \\ 1 \end{bmatrix}, \begin{bmatrix} 1 \\ 1 \\ 1 \\ 3 \end{bmatrix}, \begin{bmatrix} 3 \\ 1 \\ 0 \\ 8 \end{bmatrix}, \begin{bmatrix} 1 \\ 0 \\ 1 \\ -1 \end{bmatrix}, \begin{bmatrix} -6 \\ 2 \\ 3 \\ -7 \end{bmatrix}\right\}$

31. Span $\left\{\begin{bmatrix} -2 \\ 4 \\ 5 \\ -1 \end{bmatrix}, \begin{bmatrix} 3 \\ -4 \\ -5 \\ 1 \end{bmatrix}, \begin{bmatrix} 1 \\ 5 \\ 4 \\ -2 \end{bmatrix}, \begin{bmatrix} -1 \\ 1 \\ 2 \\ 0 \end{bmatrix}\right\}$

32. Span $\left\{\begin{bmatrix} 1 \\ 3 \\ 3 \\ 1 \end{bmatrix}, \begin{bmatrix} 1 \\ -1 \\ -1 \\ 1 \end{bmatrix}, \begin{bmatrix} 0 \\ 0 \\ 0 \\ 0 \end{bmatrix}, \begin{bmatrix} 1 \\ 0 \\ 0 \\ 1 \end{bmatrix}, \begin{bmatrix} 2 \\ -5 \\ -5 \\ 2 \end{bmatrix}\right\}$

T&F *In Exercises 33–52, determine whether the statements are true or false.*

33. Every nonzero subspace of $\mathcal{R}^n$ has a unique basis.
34. Every nonzero subspace of $\mathcal{R}^n$ has a basis.
35. A basis for a subspace is a generating set that is as large as possible.
36. If $\mathcal{S}$ is a linearly independent set and Span $\mathcal{S} = V$, then $\mathcal{S}$ is a basis for V.

37. Every finite generating set for a subspace contains a basis for the subspace.
38. A basis for a subspace is a linearly independent subset of the subspace that is as large as possible.
39. Every basis for a particular subspace contains the same number of vectors.
40. The columns of any matrix form a basis for its column space.
41. The pivot columns of the reduced row echelon form of A form a basis for the column space of A.
42. The vectors in the vector form of the general solution of $A\mathbf{x} = \mathbf{0}$ form a basis for the null space of A.
43. If V is a subspace of dimension k, then every generating set for V contains exactly k vectors.
44. If V is a subspace of dimension k, then every generating set for V contains at least k vectors.
45. If S is a linearly independent set of k vectors from a subspace V of dimension k, then S is a basis for V.
46. If V is a subspace of dimension k, then every set containing more than k vectors from V is linearly dependent.
47. The dimension of $\mathcal{R}^n$ is n.
48. The vectors in the standard basis for $\mathcal{R}^n$ are the standard vectors of $\mathcal{R}^n$.
49. Every linearly independent subset of a subspace is contained in a basis for the subspace.
50. Every subspace of $\mathcal{R}^n$ has a basis composed of standard vectors.
51. A basis for the null space of a linear transformation is also a basis for the null space of its standard matrix.
52. A basis for the range of a linear transformation is also a basis for the column space of its standard matrix.

---

53. Explain why $\left\{\begin{bmatrix}1\\-1\\2\\1\end{bmatrix}, \begin{bmatrix}1\\3\\-1\\4\end{bmatrix}, \begin{bmatrix}2\\1\\5\\-3\end{bmatrix}\right\}$ is not a generating set for $\mathcal{R}^4$.
54. Explain why $\left\{\begin{bmatrix}1\\-3\\4\end{bmatrix}, \begin{bmatrix}-2\\5\\3\end{bmatrix}, \begin{bmatrix}-1\\6\\-4\end{bmatrix}, \begin{bmatrix}5\\3\\-1\end{bmatrix}\right\}$ is not linearly independent.
55. Explain why $\left\{\begin{bmatrix}-4\\6\\2\end{bmatrix}, \begin{bmatrix}2\\-3\\7\end{bmatrix}\right\}$ is not a basis for $\mathcal{R}^3$.
56. Explain why $\left\{\begin{bmatrix}1\\-3\\3\end{bmatrix}, \begin{bmatrix}-1\\2\\1\end{bmatrix}\right\}$ is not a generating set for $\mathcal{R}^3$.
57. Explain why $\left\{\begin{bmatrix}1\\-1\end{bmatrix}, \begin{bmatrix}-2\\5\end{bmatrix}, \begin{bmatrix}-1\\3\end{bmatrix}, \begin{bmatrix}4\\-3\end{bmatrix}\right\}$ is not linearly independent.
58. Explain why $\left\{\begin{bmatrix}1\\-3\end{bmatrix}, \begin{bmatrix}-2\\1\end{bmatrix}, \begin{bmatrix}1\\-1\end{bmatrix}\right\}$ is not a basis for $\mathcal{R}^2$.
59. Show that $\left\{\begin{bmatrix}1\\2\\1\end{bmatrix}, \begin{bmatrix}-1\\3\\2\end{bmatrix}\right\}$ is a basis for the subspace in Exercise 21.
60. Show that $\left\{\begin{bmatrix}1\\0\\1\\-3\end{bmatrix}, \begin{bmatrix}2\\3\\-2\\5\end{bmatrix}\right\}$ is a basis for the subspace in Exercise 24.
61. Show that $\left\{\begin{bmatrix}1\\-3\\2\\2\end{bmatrix}, \begin{bmatrix}2\\-2\\0\\9\end{bmatrix}, \begin{bmatrix}1\\-6\\5\\2\end{bmatrix}\right\}$ is a basis for the subspace in Exercise 29.
62. Show that $\left\{\begin{bmatrix}-2\\1\\4\\-8\end{bmatrix}, \begin{bmatrix}-2\\5\\7\\1\end{bmatrix}, \begin{bmatrix}-1\\1\\5\\-9\end{bmatrix}\right\}$ is a basis for the subspace in Exercise 30.
63. Show that $\left\{\begin{bmatrix}0\\1\\1\\1\end{bmatrix}, \begin{bmatrix}2\\2\\1\\1\end{bmatrix}\right\}$ is a basis for the null space of the matrix in Exercise 5.
64. Show that $\left\{\begin{bmatrix}0\\3\\1\\1\end{bmatrix}, \begin{bmatrix}-1\\2\\1\\1\end{bmatrix}\right\}$ is a basis for the null space of the matrix in Exercise 8.
65. Show that $\left\{\begin{bmatrix}1\\3\\-2\\4\end{bmatrix}, \begin{bmatrix}-2\\1\\3\\-3\end{bmatrix}, \begin{bmatrix}-3\\9\\2\\3\end{bmatrix}\right\}$ is a basis for the column space of the matrix in Exercise 7.
66. Show that $\left\{\begin{bmatrix}1\\1\\6\\-4\end{bmatrix}, \begin{bmatrix}0\\1\\-3\\3\end{bmatrix}\right\}$ is a basis for the column space of the matrix in Exercise 8.
67. What is the dimension of Span $\{\mathbf{v}\}$, where $\mathbf{v} \neq \mathbf{0}$? Justify your answer.
68. What is the dimension of the subspace $\left\{\begin{bmatrix}v_1\\v_2\\\vdots\\v_n\end{bmatrix} \in \mathcal{R}^n : v_1 = 0\right\}$? Justify your answer.
69. What is the dimension of the subspace $\left\{\begin{bmatrix}v_1\\v_2\\\vdots\\v_n\end{bmatrix} \in \mathcal{R}^n : v_1 = 0 \text{ and } v_2 = 0\right\}$? Justify your answer.
70. Find the dimension of the subspace

$$\left\{\begin{bmatrix}v_1\\v_2\\\vdots\\v_n\end{bmatrix} \in \mathcal{R}^n : v_1 + v_2 + \cdots + v_n = 0\right\}.$$

Justify your answer.

71. Let $\mathcal{A} = \{\mathbf{u}_1, \mathbf{u}_2, \ldots, \mathbf{u}_k\}$ be a basis for a k-dimensional subspace V of $\mathcal{R}^n$. For any nonzero scalars $c_1, c_2, \ldots, c_k$, prove that $\mathcal{B} = \{c_1\mathbf{u}_1, c_2\mathbf{u}_2, \ldots, c_k\mathbf{u}_k\}$ is also a basis for V.

72. Let $\mathcal{A} = \{\mathbf{u}_1, \mathbf{u}_2, \ldots, \mathbf{u}_k\}$ be a basis for a k-dimensional subspace V of $\mathcal{R}^n$. Prove that

$$\mathcal{B} = \{\mathbf{u}_1, \mathbf{u}_1 + \mathbf{u}_2, \mathbf{u}_1 + \mathbf{u}_3, \ldots, \mathbf{u}_1 + \mathbf{u}_k\}$$

is also a basis for V.

73. Let $\mathcal{A} = \{\mathbf{u}_1, \mathbf{u}_2, \ldots, \mathbf{u}_k\}$ be a basis for a k-dimensional subspace V of $\mathcal{R}^n$. Prove that $\{\mathbf{v}, \mathbf{u}_2, \mathbf{u}_3, \ldots, \mathbf{u}_k\}$ is also a basis for V, where $\mathbf{v} = \mathbf{u}_1 + \mathbf{u}_2 + \cdots + \mathbf{u}_k$.

74. Let $\mathcal{A} = \{\mathbf{u}_1, \mathbf{u}_2, \ldots, \mathbf{u}_k\}$ be a basis for a k-dimensional subspace V of $\mathcal{R}^n$, and let $\mathcal{B} = \{\mathbf{v}_1, \mathbf{v}_2, \ldots, \mathbf{v}_k\}$, where

$$\mathbf{v}_i = \mathbf{u}_i + \mathbf{u}_{i+1} + \cdots + \mathbf{u}_k \quad \text{for } i = 1, 2, \ldots, k.$$

Prove that $\mathcal{B}$ is also a basis for V.

75. Let $T\colon \mathcal{R}^n \to \mathcal{R}^m$ be a linear transformation and $\{\mathbf{u}_1, \mathbf{u}_2, \ldots, \mathbf{u}_n\}$ be a basis for $\mathcal{R}^n$.

   (a) Prove that $\mathcal{S} = \{T(\mathbf{u}_1), T(\mathbf{u}_2), \ldots, T(\mathbf{u}_n)\}$ is a generating set for the range of T.

   (b) Give an example to show that $\mathcal{S}$ need not be a basis for the range of T.

76. Let $T\colon \mathcal{R}^n \to \mathcal{R}^m$ be a one-to-one linear transformation and V be a subspace of $\mathcal{R}^n$. Recall from Exercise 97 of Section 4.1 that $W = \{T(\mathbf{u})\colon \mathbf{u} \text{ is in } V\}$ is a subspace of $\mathcal{R}^m$.

   (a) Prove that if $\{\mathbf{u}_1, \mathbf{u}_2, \ldots, \mathbf{u}_k\}$ is a basis for V, $\{T(\mathbf{u}_1), T(\mathbf{u}_2), \ldots, T(\mathbf{u}_k)\}$ is a basis for W.

   (b) Prove that $\dim V = \dim W$.

77. Let V and W be nonzero subspaces of $\mathcal{R}^n$ such that each vector $\mathbf{u}$ in $\mathcal{R}^n$ can be uniquely expressed in the form $\mathbf{u} = \mathbf{v} + \mathbf{w}$ for some $\mathbf{v}$ in V and some $\mathbf{w}$ in W.

   (a) Prove that $\mathbf{0}$ is the only vector in both V and W.

   (b) Prove that $\dim V + \dim W = n$.

78. Let V be a subspace of $\mathcal{R}^n$. According to Theorem 4.4, a linearly independent subset $\mathcal{L} = \{\mathbf{u}_1, \mathbf{u}_2, \ldots, \mathbf{u}_m\}$ of V is contained in a basis for V. Show that if $\mathcal{S} = \{\mathbf{b}_1, \mathbf{b}_2, \ldots, \mathbf{b}_k\}$ is any generating set for V, then the pivot columns of the matrix $[\mathbf{u}_1\ \mathbf{u}_2\ \ldots\ \mathbf{u}_m\ \mathbf{b}_1\ \mathbf{b}_2\ \ldots\ \mathbf{b}_k]$ form a basis for V that contains $\mathcal{L}$.

In Exercises 79–82, use the procedure described in Exercise 78 to find a basis for the subspace V that contains the given linearly independent subset $\mathcal{L}$ of V.

79. $\mathcal{L} = \left\{\begin{bmatrix} 2 \\ 3 \\ 0 \end{bmatrix}\right\}, \quad V = \mathcal{R}^3$

80. $\mathcal{L} = \left\{\begin{bmatrix} -1 \\ -1 \\ 6 \\ -7 \end{bmatrix}, \begin{bmatrix} 5 \\ -9 \\ -2 \\ -1 \end{bmatrix}\right\},$

$$V = \text{Span}\left\{\begin{bmatrix} 1 \\ -2 \\ 0 \\ 1 \end{bmatrix}, \begin{bmatrix} 1 \\ -1 \\ -2 \\ 3 \end{bmatrix}, \begin{bmatrix} 0 \\ 1 \\ -2 \\ 10 \end{bmatrix}\right\}$$

81. $\mathcal{L} = \left\{\begin{bmatrix} 0 \\ 2 \\ 1 \\ 0 \end{bmatrix}\right\}, \quad V = \text{Null}\begin{bmatrix} 1 & -1 & 2 & 1 \\ 2 & -2 & 4 & 2 \\ -3 & 3 & -6 & -3 \end{bmatrix}$

82. $\mathcal{L} = \left\{\begin{bmatrix} 0 \\ 0 \\ 1 \\ 0 \end{bmatrix}\right\}, \quad V = \text{Col}\begin{bmatrix} 1 & -1 & -3 & 1 \\ -1 & 1 & 3 & 2 \\ -3 & 1 & -1 & -1 \\ 2 & -2 & -6 & 1 \end{bmatrix}$

83. Let $V = \left\{\begin{bmatrix} v_1 \\ v_2 \\ v_3 \end{bmatrix} \in \mathcal{R}^3\colon v_1 - v_2 + v_3 = 0\right\}$ and

$$\mathcal{S} = \left\{\begin{bmatrix} 1 \\ -1 \\ 2 \end{bmatrix}, \begin{bmatrix} 2 \\ -1 \\ 3 \end{bmatrix}, \begin{bmatrix} 2 \\ 1 \\ 2 \end{bmatrix}\right\}.$$

   (a) Show that $\mathcal{S}$ is linearly independent.

   (b) Show that

$$(-9v_1 + 6v_2)\begin{bmatrix} 1 \\ -1 \\ 2 \end{bmatrix} + (7v_1 - 5v_2)\begin{bmatrix} 2 \\ -1 \\ 3 \end{bmatrix} + (-2v_1 + 2v_2)\begin{bmatrix} 2 \\ 1 \\ 2 \end{bmatrix} = \begin{bmatrix} v_1 \\ v_2 \\ v_3 \end{bmatrix}$$

   for every vector $\begin{bmatrix} v_1 \\ v_2 \\ v_3 \end{bmatrix}$ in V.

   (c) Determine whether $\mathcal{S}$ is a basis for V. Justify your answer.

84. Let $V = \left\{\begin{bmatrix} v_1 \\ v_2 \\ v_3 \\ v_4 \end{bmatrix} \in \mathcal{R}^4\colon 3v_1 - v_3 = 0 \text{ and } v_4 = 0\right\}$ and

$$\mathcal{S} = \left\{\begin{bmatrix} 1 \\ 3 \\ 1 \\ 2 \end{bmatrix}, \begin{bmatrix} 2 \\ 5 \\ 3 \\ 3 \end{bmatrix}, \begin{bmatrix} 1 \\ -1 \\ 3 \\ 0 \end{bmatrix}\right\}.$$

   (a) Show that $\mathcal{S}$ is linearly independent.

   (b) Show that

$$(9v_1 - 1.5v_2 - 3.5v_3)\begin{bmatrix} 1 \\ 3 \\ 1 \\ 2 \end{bmatrix} + (-5v_1 + v_2 + 2v_3)\begin{bmatrix} 2 \\ 5 \\ 3 \\ 3 \end{bmatrix} + (2v_1 - 0.5v_2 - 0.5v_3)\begin{bmatrix} 1 \\ -1 \\ 3 \\ 0 \end{bmatrix} = \begin{bmatrix} v_1 \\ v_2 \\ v_3 \\ v_4 \end{bmatrix}$$

   for every vector $\begin{bmatrix} v_1 \\ v_2 \\ v_3 \\ v_4 \end{bmatrix}$ in V.

(c) Determine whether $\mathcal{S}$ is a basis for V. Justify your answer.

In Exercises 85–88, use either a calculator with matrix capabilities or computer software such as MATLAB to solve each problem.

85. Let

$$A = \begin{bmatrix} 0.1 & 0.2 & 0.34 & 0.5 & -0.09 \\ 0.7 & 0.9 & 1.23 & -0.5 & -1.98 \\ -0.5 & 0.5 & 1.75 & -0.5 & -2.50 \end{bmatrix}.$$

(a) Find a basis for the column space of A.

(b) Find a basis for the null space of A.

86. Show that

$$\left\{ \begin{bmatrix} 29.0 \\ -57.1 \\ 16.0 \\ 4.9 \\ -7.0 \end{bmatrix}, \begin{bmatrix} -26.6 \\ 53.8 \\ -7.0 \\ -9.1 \\ 13.0 \end{bmatrix} \right\}$$

is a basis for the null space of the matrix A in Exercise 85.

87. Show that

$$\left\{ \begin{bmatrix} 1.1 \\ -7.8 \\ -9.0 \end{bmatrix}, \begin{bmatrix} -2.7 \\ 7.6 \\ 4.0 \end{bmatrix}, \begin{bmatrix} 2.5 \\ -4.5 \\ -6.5 \end{bmatrix} \right\}$$

is a basis for the column space of the matrix A in Exercise 85.

88. Let

$$A = \begin{bmatrix} -0.1 & -0.21 & 0.2 & 0.58 & 0.4 & 0.61 \\ 0.3 & 0.63 & -0.1 & -0.59 & -0.5 & -0.81 \\ 1.2 & 2.52 & 0.6 & -0.06 & 0.6 & 0.12 \\ -0.6 & -1.26 & 0.2 & 1.18 & -0.2 & 0.30 \end{bmatrix}.$$

(a) Compute the rank of A, the dimension of Col A, and the dimension of Row A.

(b) Use the result of (a) to make a conjecture about the relationships among the rank of A, the dimension of Col A, and the dimension of Row A, for an arbitrary matrix A.

(c) Test your conjecture, using random 4×7 and 6×3 matrices.

SOLUTIONS TO THE PRACTICE PROBLEMS

1. Since the given set contains 2 vectors, and $\mathcal{R}^3$ has dimension 3, this set cannot be a basis for $\mathcal{R}^3$.

2. The reduced row echelon form of the given matrix A is

$$\begin{bmatrix} 1 & -2 & 0 & 3 \\ 0 & 0 & 1 & 2 \\ 0 & 0 & 0 & 0 \end{bmatrix}.$$

Hence a basis for the column space of A is

$$\left\{ \begin{bmatrix} -1 \\ 2 \\ 1 \end{bmatrix}, \begin{bmatrix} 1 \\ -3 \\ 0 \end{bmatrix} \right\},$$

the set consisting of the pivot columns of A. Thus the dimension of Col A is 2.

Solving the homogeneous system of linear equations having the reduced row echelon form of A as its coefficient matrix, we obtain the vector form of the general solution of $A\mathbf{x} = \mathbf{0}$, which is

$$\begin{bmatrix} x_1 \\ x_2 \\ x_3 \\ x_4 \end{bmatrix} = x_2 \begin{bmatrix} 2 \\ 1 \\ 0 \\ 0 \end{bmatrix} + x_4 \begin{bmatrix} -3 \\ 0 \\ -2 \\ 1 \end{bmatrix}.$$

The set of vectors in this representation,

$$\left\{ \begin{bmatrix} 2 \\ 1 \\ 0 \\ 0 \end{bmatrix}, \begin{bmatrix} -3 \\ 0 \\ -2 \\ 1 \end{bmatrix} \right\},$$

is a basis for the null space of A. Thus the dimension of Null A is also 2.

3. Let A be the matrix in Practice Problem 1. Since each of the vectors in the given set $\mathcal{B}$ is a solution of $A\mathbf{x} = \mathbf{0}$, $\mathcal{B}$ is a subset of the null space of A. Moreover, $\mathcal{B}$ is linearly independent since neither vector in $\mathcal{B}$ is a multiple of the other. Because Practice Problem 1 shows that Null A has dimension 2, it follows from Theorem 4.7 that $\mathcal{B}$ is a basis for Null A.

4.3 THE DIMENSION OF SUBSPACES ASSOCIATED WITH A MATRIX

In this section, we investigate the dimension of several important subspaces, including those defined in Section 4.1.

The first example illustrates an important general result.

Example 1 For the matrix

$$A = \begin{bmatrix} 1 & 2 & -1 & 2 & 1 & 2 \\ -1 & -2 & 1 & 2 & 3 & 6 \\ 2 & 4 & -3 & 2 & 0 & 3 \\ -3 & -6 & 2 & 0 & 3 & 9 \end{bmatrix}$$

in Example 1 of Section 4.2, we saw that the set of pivot columns of A,

$$\mathcal{B} = \left\{ \begin{bmatrix} 1 \\ -1 \\ 2 \\ -3 \end{bmatrix}, \begin{bmatrix} -1 \\ 1 \\ -3 \\ 2 \end{bmatrix}, \begin{bmatrix} 2 \\ 2 \\ 2 \\ 0 \end{bmatrix} \right\},$$

is a basis for Col A. Hence the dimension of Col A is 3.

As mentioned in Example 1, the pivot columns of any matrix form a basis for its column space. Hence the dimension of the column space of a matrix equals the number of pivot columns in the matrix. However, the number of pivot columns in a matrix is its rank.

> The dimension of the column space of a matrix equals the rank of the matrix.

The dimensions of the other subspaces associated with a matrix are also determined by the rank of the matrix. For instance, when finding the general solution of a homogeneous system $A\mathbf{x} = \mathbf{0}$, we have seen that the number of free variables that appear is the nullity of A. As in Example 3 of Section 4.2, each free variable in the vector form of the general solution is multiplied by a vector in a basis for the solution set. Hence the dimension of the solution set of $A\mathbf{x} = \mathbf{0}$ is the nullity of A.

> The dimension of the null space of a matrix equals the nullity of the matrix.

The preceding boxed result can sometimes help to find the dimension of a subspace without actually determining a basis for the subspace. In Example 4 of Section 4.2, for instance, we showed that

$$\mathcal{B} = \left\{ \begin{bmatrix} 1 \\ -1 \\ 1 \\ 0 \end{bmatrix}, \begin{bmatrix} 1 \\ 0 \\ 1 \\ -1 \end{bmatrix}, \begin{bmatrix} 0 \\ 1 \\ 1 \\ -1 \end{bmatrix} \right\}$$

is a basis for

$$V = \left\{ \begin{bmatrix} v_1 \\ v_2 \\ v_3 \\ v_4 \end{bmatrix} \in \mathcal{R}^4 \colon v_1 + v_2 + v_4 = 0 \right\}.$$

Our approach required that we know the dimension of V, so we solved the equation $v_1 + v_2 + v_4 = 0$ to obtain a basis for V. There is an easier method, however, because

$V = \text{Null}\ [1\ \ 1\ \ 0\ \ 1]$. Since $[1\ \ 1\ \ 0\ \ 1]$ is in reduced row echelon form, its rank is 1 and hence its nullity is $4 - 1 = 3$. Thus $\dim V = 3$.

Example 2 Show that

$$\mathcal{B} = \left\{ \begin{bmatrix} -2 \\ 1 \\ 1 \\ 2 \\ 1 \end{bmatrix}, \begin{bmatrix} 3 \\ -6 \\ -2 \\ -2 \\ -1 \end{bmatrix} \right\}$$

is a basis for the null space of

$$A = \begin{bmatrix} 3 & 1 & -2 & 1 & 5 \\ 1 & 0 & 1 & 0 & 1 \\ -5 & -2 & 5 & -5 & -3 \\ -2 & -1 & 3 & 2 & -10 \end{bmatrix}.$$

Solution Because each vector in $\mathcal{B}$ is a solution of $A\mathbf{x} = \mathbf{0}$, the set $\mathcal{B}$ is contained in Null A. Moreover, neither vector in $\mathcal{B}$ is a multiple of the other, so $\mathcal{B}$ is linearly independent. Since the reduced row echelon form of A is

$$\begin{bmatrix} 1 & 0 & 1 & 0 & 1 \\ 0 & 1 & -5 & 0 & 4 \\ 0 & 0 & 0 & 1 & -2 \\ 0 & 0 & 0 & 0 & 0 \end{bmatrix},$$

the rank of A is 3 and its nullity is $5 - 3 = 2$. Hence Null A has dimension 2, so $\mathcal{B}$ is a basis for Null A by Theorem 4.7.

Practice Problem 1 ► Show that $\left\{ \begin{bmatrix} 0 \\ -2 \\ 1 \\ -1 \\ 1 \end{bmatrix}, \begin{bmatrix} 2 \\ 1 \\ -1 \\ 1 \\ -1 \end{bmatrix} \right\}$ is a basis for the null space of the matrix

$$\begin{bmatrix} 1 & 2 & 3 & -2 & -1 \\ 0 & 0 & 1 & 3 & 2 \\ 2 & 4 & 7 & 0 & 1 \\ 3 & 6 & 11 & 1 & 2 \end{bmatrix}.$$ ◄

We now know how the dimensions of the column space and the null space of a matrix are related to the rank of the matrix. So it is natural to turn our attention to the other subspace associated with a matrix, its row space. Because the row space of a matrix A equals the column space of A^T, we can obtain a basis for Row A as follows:

(a) Form the transpose of A, whose columns are the rows of A.

(b) Find the pivot columns of A^T, which form a basis for the column space of A^T.

This approach shows us that the dimension of Row A is the rank of A^T. In order to express the dimension of Row A in terms of the rank of A, we need another method for finding a basis for Row A.

It is important to note that, unlike the column space of a matrix, the row space is unaffected by elementary row operations. Consider the matrix A and its reduced row echelon form R, given by

$$A = \begin{bmatrix} 1 & 1 \\ 2 & 2 \end{bmatrix} \quad \text{and} \quad R = \begin{bmatrix} 1 & 1 \\ 0 & 0 \end{bmatrix}.$$

Notice that

$$\text{Row } A = \text{Row } R = \text{Span}\left\{\begin{bmatrix} 1 \\ 1 \end{bmatrix}\right\},$$

but

$$\text{Col } A = \text{Span}\left\{\begin{bmatrix} 1 \\ 2 \end{bmatrix}\right\} \neq \text{Col } R = \text{Span}\left\{\begin{bmatrix} 1 \\ 0 \end{bmatrix}\right\}.$$

It follows from the next theorem that the row space of a matrix and the row space of its reduced row echelon form are always equal.

THEOREM 4.8

The nonzero rows of the reduced row echelon form of a matrix constitute a basis for the row space of the matrix.

PROOF Let R be the reduced row echelon form of a matrix A. Since R is in reduced row echelon form, the leading entry of each nonzero row of R is the only nonzero entry in its column. Thus no nonzero row of R is a linear combination of other rows. Hence the nonzero rows of R are linearly independent, and clearly they are also a generating set for Row R. Therefore the nonzero rows of R are a basis for Row R.

To complete the proof, we need show only that Row A = Row R. Because R is obtained from A by elementary row operations, each row of R is a linear combination of the rows of A. Thus Row R is contained in Row A. Since elementary row operations are reversible, each row of A must also be a linear combination of the rows of R. Therefore Row A is also contained in Row R. It follows that Row A = Row R, completing the proof. ■

Example 3 Recall that for the matrix

$$A = \begin{bmatrix} 3 & 1 & -2 & 1 & 5 \\ 1 & 0 & 1 & 0 & 1 \\ -5 & -2 & 5 & -5 & -3 \\ -2 & -1 & 3 & 2 & -10 \end{bmatrix}$$

in Example 2, the reduced row echelon form is

$$\begin{bmatrix} 1 & 0 & 1 & 0 & 1 \\ 0 & 1 & -5 & 0 & 4 \\ 0 & 0 & 0 & 1 & -2 \\ 0 & 0 & 0 & 0 & 0 \end{bmatrix}.$$

Hence a basis for Row A is

$$\left\{\begin{bmatrix} 1 \\ 0 \\ 1 \\ 0 \\ 1 \end{bmatrix}, \begin{bmatrix} 0 \\ 1 \\ -5 \\ 0 \\ 4 \end{bmatrix}, \begin{bmatrix} 0 \\ 0 \\ 0 \\ 1 \\ -2 \end{bmatrix}\right\}.$$

Thus the dimension of Row A is 3, which is the rank of A.

As Example 3 illustrates, Theorem 4.8 yields the following important fact.

> The dimension of the row space of a matrix equals its rank.

We have noted that the row and column spaces of a matrix are rarely equal. (For instance, the row space of the matrix A in Example 3 is a subspace of $\mathcal{R}^5$, whereas its column space is a subspace of $\mathcal{R}^4$.) Nevertheless, the results of this section show that their dimensions are always the same. It follows that

$$\dim(\text{Row } A) = \dim(\text{Col } A) = \dim(\text{Row } A^T),$$

and thus we have the following result:

> The rank of any matrix equals the rank of its transpose.

We can easily extend the results in this section from matrices to linear transformations. Recall from Section 4.1 that the null space of a linear transformation $T: \mathcal{R}^n \to \mathcal{R}^m$ is equal to that of its standard matrix A, and the range of T is equal to the column space of A. Hence the dimension of the null space of T is the nullity of A, and the dimension of the range of T is the rank of A. It follows that *the sum of the dimensions of the null space and range of T equals the dimension of the domain of T*. (See Exercise 71.)

Practice Problem 2 ▶ Find the dimensions of the column space, null space, and row space of the matrix in Practice Problem 1. ◀

SUBSPACES CONTAINED WITHIN SUBSPACES

Suppose that both V and W are subspaces of $\mathcal{R}^n$ such that V is contained in W. Because V is contained in W, it is natural to expect that the dimension of V would be less than or equal to the dimension of W. This expectation is indeed correct, as our next result shows.

THEOREM 4.9

If V and W are subspaces of $\mathcal{R}^n$ with V contained in W, then $\dim V \leq \dim W$. Moreover, if V and W also have the same dimension, then $V = W$.

PROOF It is easy to verify the theorem if V is the zero subspace. Assume, therefore, that V is a nonzero subspace, and let $\mathcal{B}$ be a basis for V. By the Extension Theorem, $\mathcal{B}$ is contained in a basis for W, so $\dim V \leq \dim W$.

Suppose also that both V and W have dimension k. Then $\mathcal{B}$ is a linearly independent subset of W that contains k vectors, and $\mathcal{B}$ is a basis for W by Theorem 4.7. Therefore $V = \text{Span } \mathcal{B} = W$. ■

Theorem 4.9 enables us to characterize the subspaces of $\mathcal{R}^n$. For example, this theorem shows that a subspace of $\mathcal{R}^3$ must have dimension 0, 1, 2, or 3. First, a subspace of dimension 0 must be the zero subspace. Second, a subspace of dimension 1 must have a basis $\{\mathbf{u}\}$ consisting of a single nonzero vector, and thus the subspace must consist of all vectors that are multiples of $\mathbf{u}$. As noted in Example 4 in Section 4.1, we can depict such a set as a line through the origin of $\mathcal{R}^3$. Third, a subspace of dimension 2 must have a basis $\{\mathbf{u}, \mathbf{v}\}$ consisting of two vectors, neither of which is a multiple of the other. In this case, the subspace consists of all vectors in $\mathcal{R}^3$ having the form $a\mathbf{u} + b\mathbf{v}$ for some scalars a and b. As in Example 2 of Section 1.6, we can visualize such a set as a plane through the origin of $\mathcal{R}^3$. Finally, a subspace of $\mathcal{R}^3$ having dimension 3 must be $\mathcal{R}^3$ itself by Theorem 4.9. (See Figure 4.7.)

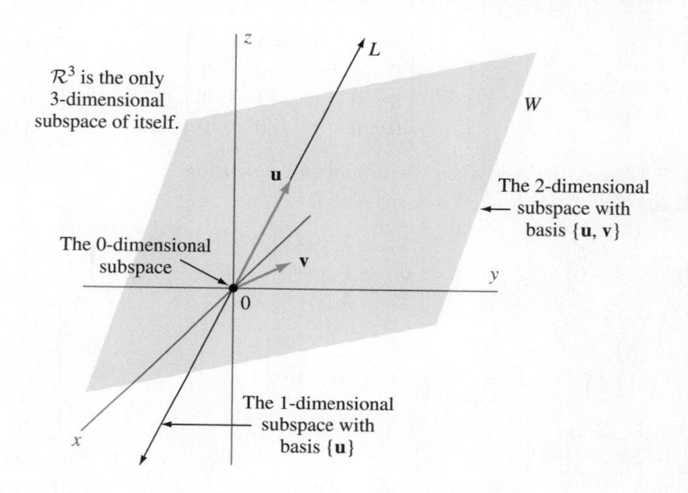

Figure 4.7 Subspaces of $\mathcal{R}^3$

We conclude this section with a table summarizing some of the facts from Sections 4.1 and 4.3 concerning important subspaces associated with an $m \times n$ matrix A. (This table also applies to a linear transformation $T: \mathcal{R}^n \to \mathcal{R}^m$ by taking A to be the standard matrix of T.)

The Dimensions of the Subspaces Associated with an $m \times n$ Matrix A

Subspace	Containing Space	Dimension
Col A	$\mathcal{R}^m$	rank A
Null A	$\mathcal{R}^n$	nullity $A = n - \text{rank } A$
Row A	$\mathcal{R}^m$	rank A

You can use the following procedures to obtain bases for these subspaces:

Bases for the Subspaces Associated with a Matrix A

Col A: The pivot columns of A form a basis for Col A.

Null A: The vectors in the vector form of the solution of $A\mathbf{x} = \mathbf{0}$ constitute a basis for Null A. (See page 80.)

Row A: The nonzero rows of the reduced row echelon form of A constitute a basis for Row A. (See page 257.)

EXERCISES

In Exercises 1–4, the reduced row echelon form of a matrix A is given. Determine the dimension of (a) Col A, (b) Null A, (c) Row A, and (d) Null A^T.

1. $\begin{bmatrix} 1 & -3 & 0 & 2 \\ 0 & 0 & 1 & -4 \\ 0 & 0 & 0 & 0 \end{bmatrix}$

2. $\begin{bmatrix} 1 & 0 & -2 & 0 \\ 0 & 1 & 5 & 0 \\ 0 & 0 & 0 & 1 \end{bmatrix}$

3. $\begin{bmatrix} 1 & -1 & 0 & 2 & 0 \\ 0 & 0 & 1 & 6 & 0 \\ 0 & 0 & 0 & 0 & 1 \end{bmatrix}$

4. $\begin{bmatrix} 1 & 0 & 0 & -4 & 2 \\ 0 & 1 & 0 & 2 & -1 \\ 0 & 0 & 1 & -3 & 1 \\ 0 & 0 & 0 & 0 & 0 \end{bmatrix}$

In Exercises 5–12, a matrix A is given. Determine the dimension of (a) Col A, (b) Null A, (c) Row A, and (d) Null A^T.

5. $\begin{bmatrix} 2 & -8 & -4 & 6 \end{bmatrix}$

6. $\begin{bmatrix} 1 & 2 & -3 \\ 0 & -1 & -1 \\ 1 & 4 & -1 \end{bmatrix}$

7. $\begin{bmatrix} 1 & -1 & 2 \\ 2 & -3 & 1 \end{bmatrix}$

8. $\begin{bmatrix} 1 & -2 & 3 \\ -3 & 6 & -9 \end{bmatrix}$

9. $\begin{bmatrix} 1 & 1 & 2 & 1 \\ -1 & -2 & 2 & -2 \\ 2 & 3 & 0 & 3 \end{bmatrix}$

10. $\begin{bmatrix} -1 & 2 & 1 & -1 & -2 \\ 2 & -4 & 1 & 5 & 7 \\ 2 & -4 & -3 & 1 & 3 \end{bmatrix}$

11. $\begin{bmatrix} 1 & 1 & 1 \\ 1 & -1 & 5 \\ 2 & 1 & 4 \\ 0 & 2 & -4 \end{bmatrix}$

12. $\begin{bmatrix} 0 & -1 & 1 \\ 1 & 2 & -3 \\ 3 & 1 & -2 \\ -1 & 0 & 4 \end{bmatrix}$

In Exercises 13–16, a subspace is given. Determine its dimension.

13. $\left\{ \begin{bmatrix} -2s \\ s \end{bmatrix} \in \mathcal{R}^2 : s \text{ is a scalar} \right\}$

14. $\left\{ \begin{bmatrix} s \\ 0 \\ 2s \end{bmatrix} \in \mathcal{R}^3 : s \text{ is a scalar} \right\}$

15. $\left\{ \begin{bmatrix} -3s + 4t \\ s - 2t \\ 2s \end{bmatrix} \in \mathcal{R}^3 : s \text{ and } t \text{ are scalars} \right\}$

16. $\left\{ \begin{bmatrix} s + 2t \\ 0 \\ 3t \end{bmatrix} \in \mathcal{R}^3 : s \text{ and } t \text{ are scalars} \right\}$

In Exercises 17–24, a matrix A is given. Find a basis for Row A.

17. $\begin{bmatrix} 1 & -1 & 1 \\ 0 & 1 & 2 \end{bmatrix}$

18. $\begin{bmatrix} 1 & -1 & 0 & -2 \\ 1 & -1 & 2 & 4 \end{bmatrix}$

19. $\begin{bmatrix} -1 & 1 & 1 & -2 \\ 2 & -2 & -2 & 4 \\ 2 & -1 & -1 & 3 \end{bmatrix}$

20. $\begin{bmatrix} 1 & -2 & 1 & -1 & -2 \\ 3 & -6 & 3 & -3 & -6 \\ 2 & -4 & 1 & 1 & 1 \end{bmatrix}$

21. $\begin{bmatrix} 1 & 0 & -1 & -3 & 1 & 4 \\ 2 & -1 & -1 & -8 & 3 & 9 \\ -1 & 1 & 0 & 5 & -2 & -5 \\ 0 & 1 & 1 & 2 & -1 & -3 \end{bmatrix}$

22. $\begin{bmatrix} -1 & 1 & 1 & 5 & -2 & -6 \\ 2 & -1 & -1 & -8 & 3 & 9 \\ 0 & 1 & -1 & 2 & -1 & -1 \\ 1 & 0 & -1 & -3 & 1 & 4 \end{bmatrix}$

23. $\begin{bmatrix} 1 & 0 & -1 & 1 & 3 \\ 2 & -1 & -1 & 3 & -8 \\ 0 & 1 & -1 & -1 & 2 \\ -1 & 1 & 1 & -2 & 5 \\ 1 & -1 & 1 & 2 & -5 \end{bmatrix}$

24. $\begin{bmatrix} 1 & 0 & -1 & -3 & 1 & 4 \\ 2 & -1 & -1 & -8 & 3 & 9 \\ 1 & 1 & -2 & -1 & 0 & 3 \\ -1 & 3 & -2 & 9 & -4 & -7 \\ 0 & 1 & 1 & 2 & -1 & -3 \end{bmatrix}$

In Exercises 25–32, use the method described on page 256 to find a basis for Row A.

25. Exercise 17
26. Exercise 18
27. Exercise 19
28. Exercise 20
29. Exercise 21
30. Exercise 22
31. Exercise 23
32. Exercise 24

In Exercises 33–40, determine the dimension of the (a) range and (b) null space of each linear transformation T. Use this information to determine whether T is one-to-one or onto.

33. $T\left(\begin{bmatrix} x_1 \\ x_2 \end{bmatrix}\right) = \begin{bmatrix} x_1 + x_2 \\ 2x_1 + x_2 \end{bmatrix}$

34. $T\left(\begin{bmatrix} x_1 \\ x_2 \end{bmatrix}\right) = \begin{bmatrix} x_1 - 3x_2 \\ -3x_1 + 9x_2 \end{bmatrix}$

35. $T\left(\begin{bmatrix} x_1 \\ x_2 \\ x_3 \end{bmatrix}\right) = \begin{bmatrix} -x_1 + 2x_2 + x_3 \\ x_1 - 2x_2 - x_3 \end{bmatrix}$

36. $T\left(\begin{bmatrix} x_1 \\ x_2 \\ x_3 \end{bmatrix}\right) = \begin{bmatrix} -x_1 + x_2 + 2x_3 \\ x_1 - 3x_3 \end{bmatrix}$

37. $T\left(\begin{bmatrix} x_1 \\ x_2 \end{bmatrix}\right) = \begin{bmatrix} x_1 \\ 2x_1 + x_2 \\ -x_2 \end{bmatrix}$

38. $T\left(\begin{bmatrix} x_1 \\ x_2 \end{bmatrix}\right) = \begin{bmatrix} x_1 - x_2 \\ -2x_1 + 3x_2 \\ x_1 \end{bmatrix}$

39. $T\left(\begin{bmatrix} x_1 \\ x_2 \\ x_3 \end{bmatrix}\right) = \begin{bmatrix} -x_1 - x_2 + x_3 \\ x_1 + 2x_2 + x_3 \end{bmatrix}$

40. $T\left(\begin{bmatrix} x_1 \\ x_2 \\ x_3 \end{bmatrix}\right) = \begin{bmatrix} -x_1 - x_2 + x_3 \\ x_1 + 2x_2 + x_3 \\ x_1 + x_2 \end{bmatrix}$

In Exercises 41–60, determine whether the statements are true or false.

41. If V and W are subspaces of $\mathcal{R}^n$ having the same dimension, then $V = W$.
42. If V is a subspace of $\mathcal{R}^n$ having dimension n, then $V = \mathcal{R}^n$.
43. If V is a subspace of $\mathcal{R}^n$ having dimension 0, then $V = \{\mathbf{0}\}$.
44. The dimension of the null space of a matrix equals the rank of the matrix.
45. The dimension of the column space of a matrix equals the nullity of the matrix.
46. The dimension of the row space of a matrix equals the rank of the matrix.
47. The row space of any matrix equals the row space of its reduced row echelon form.
48. The column space of any matrix equals the column space of its reduced row echelon form.
49. The null space of any matrix equals the null space of its reduced row echelon form.
50. The nonzero rows of a matrix form a basis for its row space.
51. If the row space of a matrix A has dimension k, then the first k rows of A form a basis for its row space.
52. The row space of any matrix equals its column space.
53. The dimension of the row space of any matrix equals the dimension of its column space.
54. The rank of any matrix equals the rank of its transpose.
55. The nullity of any matrix equals the nullity of its transpose.
56. For any $m \times n$ matrix A, the dimension of the null space of A plus the dimension of the column space of A equals m.
57. For any $m \times n$ matrix A, the dimension of the null space of A plus the dimension of the row space of A equals n.
58. If T is a linear transformation, then the dimension of the range of T plus the dimension of the null space of T equals the dimension of the domain of T.
59. If V is a subspace of W, then the dimension of V is less than or equal to the dimension of W.
60. If W is a subspace of $\mathcal{R}^n$ and V is a subspace of W having the same dimension as W, then $V = W$.

In Exercises 61–68, use the results of this section to show that $\mathcal{B}$ is a basis for each subspace V.

61. $\mathcal{B}=\{\mathbf{e}_1, \mathbf{e}_2\}$, $V = \left\{\begin{bmatrix} 2s - t \\ s + 3t \end{bmatrix} \in \mathcal{R}^2 \colon s \text{ and } t \text{ are scalars}\right\}$

62. $\mathcal{B}=\{\mathbf{e}_1, \mathbf{e}_2\}$, $V = \left\{\begin{bmatrix} -2t \\ 5s + 3t \end{bmatrix} \in \mathcal{R}^2 \colon s \text{ and } t \text{ are scalars}\right\}$

63. $\mathcal{B} = \left\{\begin{bmatrix} 3 \\ 1 \\ 0 \end{bmatrix}, \begin{bmatrix} 2 \\ 1 \\ -1 \end{bmatrix}\right\}$,

$$V = \left\{\begin{bmatrix} 4t \\ s + t \\ -3s + t \end{bmatrix} \in \mathcal{R}^3 \colon s \text{ and } t \text{ are scalars}\right\}$$

64. $\mathcal{B} = \left\{\begin{bmatrix} 0 \\ 1 \\ 4 \end{bmatrix}, \begin{bmatrix} 4 \\ -7 \\ 0 \end{bmatrix}\right\}$,

$$V = \left\{\begin{bmatrix} -s + t \\ 2s - t \\ s + 3t \end{bmatrix} \in \mathcal{R}^3 \colon s \text{ and } t \text{ are scalars}\right\}$$

65. $\mathcal{B} = \left\{\begin{bmatrix} 1 \\ 0 \\ 0 \\ 0 \end{bmatrix}, \begin{bmatrix} 1 \\ 0 \\ -1 \\ -1 \end{bmatrix}, \begin{bmatrix} 1 \\ 0 \\ 1 \\ 2 \end{bmatrix}\right\}$,

$$V = \left\{\begin{bmatrix} -r + 3s \\ 0 \\ s - t \\ r - 2t \end{bmatrix} \in \mathcal{R}^4 \colon r, s, \text{ and } t \text{ are scalars}\right\}$$

66. $\mathcal{B} = \left\{\begin{bmatrix} 0 \\ 1 \\ 0 \\ 2 \end{bmatrix}, \begin{bmatrix} 1 \\ -1 \\ 0 \\ 0 \end{bmatrix}, \begin{bmatrix} 1 \\ -1 \\ 0 \\ 1 \end{bmatrix}\right\}$,

$$V = \left\{\begin{bmatrix} -r + s \\ 4s - 3t \\ 0 \\ 3r - t \end{bmatrix} \in \mathcal{R}^4 \colon r, s, \text{ and } t \text{ are scalars}\right\}$$

67. $\mathcal{B} = \left\{\begin{bmatrix} 4 \\ 1 \\ 1 \\ -1 \end{bmatrix}, \begin{bmatrix} 0 \\ 9 \\ -3 \\ -8 \end{bmatrix}, \begin{bmatrix} 3 \\ 0 \\ 15 \\ 4 \end{bmatrix}\right\}$,

$$V = \left\{\begin{bmatrix} r - s + 3t \\ 2r - t \\ -r + 3s + 2t \\ -2r + s + t \end{bmatrix} \in \mathcal{R}^4 \colon r, s, \text{ and } t \text{ are scalars}\right\}$$

68. $\mathcal{B} = \left\{\begin{bmatrix} 1 \\ 0 \\ 5 \\ -4 \end{bmatrix}, \begin{bmatrix} -4 \\ 2 \\ -5 \\ 5 \end{bmatrix}, \begin{bmatrix} 1 \\ 9 \\ 8 \\ -7 \end{bmatrix}\right\}$,

$$V = \left\{\begin{bmatrix} 2s - 5t \\ 3r + s - 2t \\ r - 4s + 3t \\ -r + 2s \end{bmatrix} \in \mathcal{R}^4 \colon r, s, \text{ and } t \text{ are scalars}\right\}$$

69. (a) Find bases for the row space and null space of the matrix in Exercise 9.
(b) Show that the union of the two bases in (a) is a basis for $\mathcal{R}^4$.

70. (a) Find bases for the row space and null space of the matrix in Exercise 11.
(b) Show that the union of the two bases in (a) is a basis for $\mathcal{R}^3$.

71. Let $T: \mathcal{R}^n \to \mathcal{R}^m$ be a linear transformation. Prove the *Dimension Theorem:* The sum of the dimensions of the null space and range of T is n.

72. Is there a 3×3 matrix whose null space and column space are equal? Justify your answer.

73. Prove that, for any $m \times n$ matrix A and any $n \times p$ matrix B, the column space of AB is contained in the column space of A.

74. Prove that, for any $m \times n$ matrix A and any $n \times p$ matrix B, the null space of B is contained in the null space of AB.

75. Use Exercise 73 to prove that, for any $m \times n$ matrix A and any $n \times p$ matrix B, $\operatorname{rank} AB \leq \operatorname{rank} A$.

76. Use Exercise 75 to prove that, for any $m \times n$ matrix A and any $n \times n$ matrix B, $\operatorname{nullity} A \leq \operatorname{nullity} AB$.

77. Use Exercise 75 to prove that, for any $m \times n$ matrix A and any $n \times p$ matrix B, $\operatorname{rank} AB \leq \operatorname{rank} B$. *Hint:* The ranks of AB and $(AB)^T$ are equal.

78. Use Exercise 77 to prove that, for any $m \times n$ matrix A and any $n \times p$ matrix B, $\operatorname{nullity} B \leq \operatorname{nullity} AB$.

79. Find 2-dimensional subspaces V and W of $\mathcal{R}^5$ such that the only vector belonging to both V and W is $\mathbf{0}$.

80. Prove that if V and W are 3-dimensional subspaces of $\mathcal{R}^5$, then there is some nonzero vector that belongs to both V and W.

81. Let V be a k-dimensional subspace of $\mathcal{R}^n$ having a basis $\{\mathbf{u}_1, \mathbf{u}_2, \ldots, \mathbf{u}_k\}$. Define a function $T: \mathcal{R}^k \to \mathcal{R}^n$ by

$$T\left(\begin{bmatrix} x_1 \\ x_2 \\ \vdots \\ x_k \end{bmatrix}\right) = x_1\mathbf{u}_1 + x_2\mathbf{u}_2 + \cdots + x_k\mathbf{u}_k.$$

(a) Prove that T is a linear transformation.

(b) Prove that T is one-to-one.

(c) Prove that the range of T is V.

82. Let V be a subspace of $\mathcal{R}^n$ and $\mathbf{u}$ be a vector in $\mathcal{R}^n$ that is not in V. Define

$$W = \{\mathbf{v} + c\mathbf{u} \colon \mathbf{v} \text{ is in } V \text{ and } c \text{ is a scalar}\}.$$

(a) Prove that W is a subspace of $\mathcal{R}^n$.

(b) Determine the dimension of W. Justify your answer.

83. (a) Show that for any vector $\mathbf{u}$ in $\mathcal{R}^n$, $\mathbf{u}^T\mathbf{u} = 0$ if and only if $\mathbf{u} = \mathbf{0}$.

(b) Prove that for any matrix A, if $\mathbf{u}$ is in Row A and $\mathbf{v}$ is in Null A, then $\mathbf{u}^T\mathbf{v} = 0$. *Hint:* The row space of A equals the column space of A^T.

(c) Show that, for any matrix A, if $\mathbf{u}$ belongs to both Row A and Null A, then $\mathbf{u} = \mathbf{0}$.

84. Show that, for any $m \times n$ matrix A, the union of a basis for Row A and a basis for Null A is a basis for $\mathcal{R}^n$. *Hint:* Use Exercise 83(c).

85. Let

$$A = \begin{bmatrix} 1 & 0 & -1 & -2 \\ -1 & 1 & 2 & 1 \\ 1 & 3 & 2 & -5 \\ -1 & 6 & 7 & -4 \end{bmatrix}.$$

(a) Find a 4×4 matrix B with rank 2 such that $AB = O$.

(b) Prove that if C is a 4×4 matrix such that $AC = O$, then $\operatorname{rank} C \leq 2$.

In Exercises 86–88, use either a calculator with matrix capabilities or computer software such as MATLAB to solve each problem.

86. Let

$$\mathcal{B} = \left\{ \begin{bmatrix} -1 \\ 1 \\ 1 \\ 1 \\ 0 \end{bmatrix}, \begin{bmatrix} 0 \\ 1 \\ -1 \\ 1 \\ 1 \end{bmatrix}, \begin{bmatrix} 1 \\ 0 \\ 3 \\ -1 \\ -2 \end{bmatrix} \right\}.$$

Show that $\mathcal{B}$ is linearly independent and hence is a basis for a subspace W of $\mathcal{R}^5$.

87. Let

$$\mathcal{A}_1 = \left\{ \begin{bmatrix} 1.0 \\ 2.0 \\ 1.0 \\ 1.0 \\ 0.1 \end{bmatrix}, \begin{bmatrix} -1.0 \\ 3.0 \\ 4.0 \\ 2.0 \\ -0.6 \end{bmatrix}, \begin{bmatrix} 2.0 \\ -1.0 \\ -1.0 \\ -1.4 \\ -0.3 \end{bmatrix} \right\}$$

and

$$\mathcal{A}_2 = \left\{ \begin{bmatrix} 2 \\ 1 \\ 0 \\ 0 \\ 0 \end{bmatrix}, \begin{bmatrix} -3 \\ 0 \\ 1 \\ 1 \\ 0 \end{bmatrix}, \begin{bmatrix} 1 \\ 0 \\ -2 \\ 0 \\ 1 \end{bmatrix} \right\}.$$

(a) Is $\mathcal{A}_1$ a basis for the subspace W in Exercise 86? Justify your answer.

(b) Is $\mathcal{A}_2$ a basis for the subspace W in Exercise 86? Justify your answer.

(c) Let B, A_1, and A_2 be the matrices whose columns are the vectors in $\mathcal{B}$, $\mathcal{A}_1$, and $\mathcal{A}_2$, respectively. Compute the reduced row echelon forms of B, A_1, A_2, B^T, A_1^T, and A_2^T.

88. Let $\mathcal{B}$ be a basis for a subspace W of $\mathcal{R}^n$, and B be the matrix whose columns are the vectors in $\mathcal{B}$. Suppose that $\mathcal{A}$ is a set of vectors in $\mathcal{R}^n$, and A is the matrix whose columns are the vectors in $\mathcal{A}$. Use the results of Exercise 87 to devise a test for $\mathcal{A}$ to be a basis for W. (The test should involve the matrices A and B.) Then prove that your test is valid.

SOLUTIONS TO THE PRACTICE PROBLEMS

1. Let A denote the given matrix. Clearly,

$$\begin{bmatrix} 1 & 2 & 3 & -2 & -1 \\ 0 & 0 & 1 & 3 & 2 \\ 2 & 4 & 7 & 0 & 1 \\ 3 & 6 & 11 & 1 & 2 \end{bmatrix} \begin{bmatrix} 0 \\ -2 \\ 1 \\ -1 \\ 1 \end{bmatrix} = \begin{bmatrix} 0 \\ 0 \\ 0 \\ 0 \end{bmatrix}$$

and

$$\begin{bmatrix} 1 & 2 & 3 & -2 & -1 \\ 0 & 0 & 1 & 3 & 2 \\ 2 & 4 & 7 & 0 & 1 \\ 3 & 6 & 11 & 1 & 2 \end{bmatrix} \begin{bmatrix} 2 \\ 1 \\ -1 \\ 1 \\ -1 \end{bmatrix} = \begin{bmatrix} 0 \\ 0 \\ 0 \\ 0 \end{bmatrix}.$$

Thus both

$$\begin{bmatrix} 0 \\ -2 \\ 1 \\ -1 \\ 1 \end{bmatrix} \quad \text{and} \quad \begin{bmatrix} 2 \\ 1 \\ -1 \\ 1 \\ -1 \end{bmatrix}$$

belong to Null A. Moreover, neither of these vectors is a multiple of the other, and so they are linearly independent. Since Null A has dimension 2, it follows from Theorem 4.7 that

$$\left\{ \begin{bmatrix} 0 \\ -2 \\ 1 \\ -1 \\ 1 \end{bmatrix}, \begin{bmatrix} 2 \\ 1 \\ -1 \\ 1 \\ -1 \end{bmatrix} \right\}$$

is a basis for Null A.

2. The reduced row echelon form of the given matrix A is

$$\begin{bmatrix} 1 & 2 & 0 & 0 & 4 \\ 0 & 0 & 1 & 0 & -1 \\ 0 & 0 & 0 & 1 & 1 \\ 0 & 0 & 0 & 0 & 0 \end{bmatrix},$$

and so its rank is 3. The dimensions of both the column space and the row space of A equal the rank of A, so these dimensions are 3. The dimension of the null space of A is the nullity of A, which is

$$5 - \operatorname{rank} A = 5 - 3 = 2.$$

4.4 COORDINATE SYSTEMS

In previous sections, we have usually expressed vectors in terms of the standard basis for $\mathcal{R}^n$. For example,

$$\begin{bmatrix} 5 \\ 2 \\ 8 \end{bmatrix} = 5 \begin{bmatrix} 1 \\ 0 \\ 0 \end{bmatrix} + 2 \begin{bmatrix} 0 \\ 1 \\ 0 \end{bmatrix} + 8 \begin{bmatrix} 0 \\ 0 \\ 1 \end{bmatrix}.$$

However, in many applications, it is more natural to express vectors in terms of some basis other than the standard one. For example, consider the ellipse shown in Figure 4.8. With respect to the usual xy-coordinate system, it can be shown that the ellipse has the equation

$$13x^2 - 10xy + 13y^2 = 72.$$

Observe that the axes of symmetry of the ellipse are the lines $y = x$ and $y = -x$, and the lengths of the semimajor and semiminor axes are 3 and 2, respectively. Consider a new $x'y'$-coordinate system in which the x'-axis is the line $y = x$ and the y'-axis is the line $y = -x$. Because the ellipse is in standard position with respect to this coordinate system, the equation of this ellipse in the $x'y'$-coordinate system is

$$\frac{(x')^2}{3^2} + \frac{(y')^2}{2^2} = 1.$$

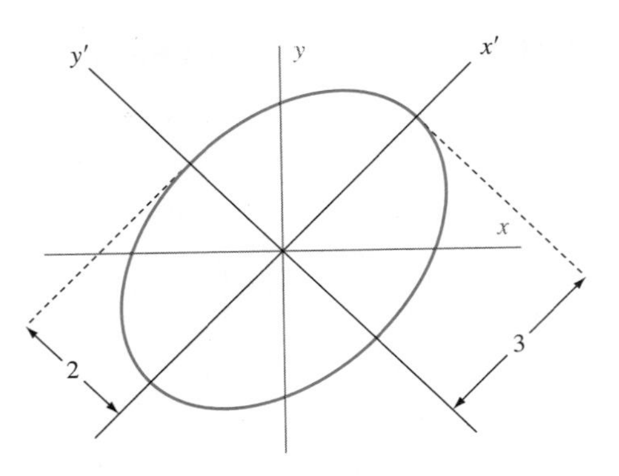

Figure 4.8 The ellipse with equation $13x^2 - 10xy + 13y^2 = 72$

But how can we use this simple equation of the ellipse in the $x'y'$-coordinate system to derive the equation of the same ellipse in the usual xy-coordinate system?

To answer this question requires a careful examination of what is meant by a coordinate system. When we say that $\mathbf{v} = \begin{bmatrix} 8 \\ 4 \end{bmatrix}$, we mean that $\mathbf{v} = 8\mathbf{e}_1 + 4\mathbf{e}_2$. That is, the components of $\mathbf{v}$ are the coefficients used to represent $\mathbf{v}$ as a linear combination of the standard vectors. (See Figure 4.9(a).)

Any basis for a subspace provides a means for establishing a coordinate system on the subspace. For the ellipse in Figure 4.8, the standard basis does not lead to a simple equation. A better choice is $\{\mathbf{u}_1, \mathbf{u}_2\}$, where $\mathbf{u}_1 = \begin{bmatrix} 1 \\ 1 \end{bmatrix}$ is a vector that lies along the x'-axis and $\mathbf{u}_2 = \begin{bmatrix} -1 \\ 1 \end{bmatrix}$ is a vector that lies along the y'-axis. (Note that since $\mathbf{u}_1$ and $\mathbf{u}_2$ are not multiples of one another, $\mathcal{B} = \{\mathbf{u}_1, \mathbf{u}_2\}$ is a linearly independent subset of two vectors from $\mathcal{R}^2$. Hence, by Theorem 4.7, $\{\mathbf{u}_1, \mathbf{u}_2\}$ is a basis for $\mathcal{R}^2$.) For this basis, we can write the vector $\mathbf{v} = \begin{bmatrix} 8 \\ 4 \end{bmatrix}$ in Figure 4.9(a) as $\mathbf{v} = 6\mathbf{u}_1 + (-2)\mathbf{u}_2$. Thus the coordinates of $\mathbf{v}$ are 6 and -2 with respect to $\mathcal{B}$, as shown in Figure 4.9(b). In this case, the coordinates of $\mathbf{v}$ are different from those in the usual coordinate system for $\mathcal{R}^2$.

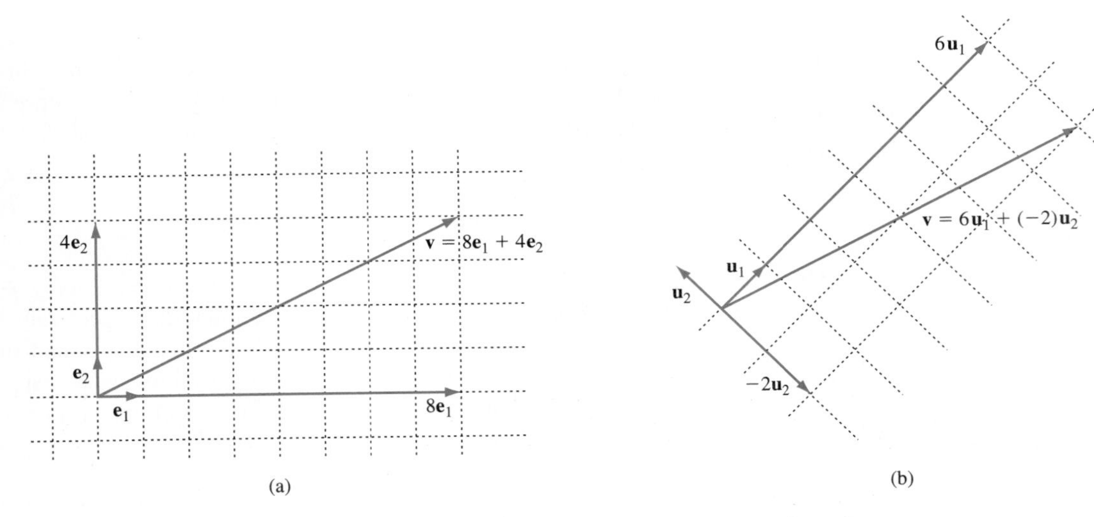

Figure 4.9 Two coordinate systems for $\mathcal{R}^2$

In order to identify the vector $\mathbf{v}$ with the coefficients 6 and -2 in the linear combination $\mathbf{v} = 6\mathbf{u}_1 + (-2)\mathbf{u}_2$, it is important that this is the only possible way to represent $\mathbf{v}$ in terms of $\mathbf{u}_1$ and $\mathbf{u}_2$.

The following theorem assures us that whenever we have a basis for $\mathcal{R}^n$, we can write any vector in $\mathcal{R}^n$ as a unique linear combination of the basis vectors:

THEOREM 4.10

Let $\mathcal{B} = \{\mathbf{b}_1, \mathbf{b}_2, \ldots, \mathbf{b}_k\}$ be a basis for a subspace V of $\mathcal{R}^n$. Any vector $\mathbf{v}$ in V can be uniquely represented as a linear combination of the vectors in $\mathcal{B}$; that is, there are unique scalars $a_1, a_2, \ldots, a_k$ such that $\mathbf{v} = a_1\mathbf{b}_1 + a_2\mathbf{b}_2 + \cdots + a_k\mathbf{b}_k$.

PROOF Let $\mathbf{v}$ be in V. Since $\mathcal{B}$ is a generating set for V, every vector in V is a linear combination of $\mathbf{b}_1, \mathbf{b}_2, \ldots, \mathbf{b}_k$. Hence there exist scalars $a_1, a_2, \ldots, a_k$ such that $\mathbf{v} = a_1\mathbf{b}_1 + a_2\mathbf{b}_2 + \cdots + a_k\mathbf{b}_k$.

Now let $c_1, c_2, \ldots, c_k$ be scalars such that $\mathbf{v} = c_1\mathbf{b}_1 + c_2\mathbf{b}_2 + \cdots + c_k\mathbf{b}_k$. To show that the coefficients in the linear combination are unique, we prove that each c_i equals the corresponding a_i. We have

$$\begin{aligned}
\mathbf{0} &= \mathbf{v} - \mathbf{v} \\
&= (a_1\mathbf{b}_1 + a_2\mathbf{b}_2 + \cdots + a_k\mathbf{b}_k) - (c_1\mathbf{b}_1 + c_2\mathbf{b}_2 + \cdots + c_k\mathbf{b}_k) \\
&= (a_1 - c_1)\mathbf{b}_1 + (a_2 - c_2)\mathbf{b}_2 + \cdots + (a_k - c_k)\mathbf{b}_k.
\end{aligned}$$

Because $\mathcal{B}$ is linearly independent, this equation implies that $a_1 - c_1 = 0$, $a_2 - c_2 = 0, \cdots, a_k - c_k = 0$. Thus $a_1 = c_1, a_2 = c_2, \cdots, a_k = c_k$, which proves that the representation of $\mathbf{v}$ as a linear combination of the vectors in $\mathcal{B}$ is unique. ■

The conclusion of Theorem 4.10 is of great practical value. A nonzero subspace V of $\mathcal{R}^n$ contains infinitely many vectors. But V has a finite basis $\mathcal{B}$, and we can uniquely express every vector in V as a linear combination of the vectors in $\mathcal{B}$. Thus we are able to study the infinitely many vectors in V as linear combinations of the finite number of vectors in $\mathcal{B}$.

COORDINATE VECTORS

We have seen that we can uniquely represent each vector in $\mathcal{R}^n$ as a linear combination of the vectors in any basis for $\mathcal{R}^n$. We can use this representation to introduce a coordinate system into $\mathcal{R}^n$ by making the following definition:

Definition Let $\mathcal{B} = \{\mathbf{b}_1, \mathbf{b}_2, \ldots, \mathbf{b}_n\}$ be a basis[1] for $\mathcal{R}^n$. For each $\mathbf{v}$ in $\mathcal{R}^n$, there are unique scalars $c_1, c_2, \ldots, c_n$ such that $\mathbf{v} = c_1\mathbf{b}_1 + c_2\mathbf{b}_2 + \cdots + c_n\mathbf{b}_n$. The vector

$$\begin{bmatrix} c_1 \\ c_2 \\ \vdots \\ c_n \end{bmatrix}$$

---

[1] In order for the definition of a coordinate vector to be unambiguous, we must assume that the vectors in $\mathcal{B}$ are listed in the specific sequence $\mathbf{b}_1, \mathbf{b}_2, \ldots, \mathbf{b}_n$. When working with coordinate vectors, we always make this assumption. If we wish to emphasize this ordering, we refer to $\mathcal{B}$ as an *ordered basis*.

in $\mathcal{R}^n$ is called the **coordinate vector** of $\mathbf{v}$ relative to $\mathcal{B}$, or the $\mathcal{B}$**-coordinate vector** of $\mathbf{v}$. We denote the $\mathcal{B}$-coordinate vector of $\mathbf{v}$ by $[\mathbf{v}]_{\mathcal{B}}$.

Example 1

Let

$$\mathcal{B} = \left\{ \begin{bmatrix} 1 \\ 1 \\ 1 \end{bmatrix}, \begin{bmatrix} 1 \\ -1 \\ 1 \end{bmatrix}, \begin{bmatrix} 1 \\ 2 \\ 2 \end{bmatrix} \right\}.$$

Since $\mathcal{B}$ is a linearly independent set of 3 vectors in $\mathcal{R}^3$, $\mathcal{B}$ is a basis for $\mathcal{R}^3$. Calculate $\mathbf{u}$ if

$$[\mathbf{u}]_{\mathcal{B}} = \begin{bmatrix} 3 \\ 6 \\ -2 \end{bmatrix}.$$

Solution Since the components of $[\mathbf{u}]_{\mathcal{B}}$ are the coefficients that express $\mathbf{u}$ as a linear combination of the vectors in $\mathcal{B}$, we have

$$\mathbf{u} = 3\begin{bmatrix} 1 \\ 1 \\ 1 \end{bmatrix} + 6\begin{bmatrix} 1 \\ -1 \\ 1 \end{bmatrix} + (-2)\begin{bmatrix} 1 \\ 2 \\ 2 \end{bmatrix} = \begin{bmatrix} 7 \\ -7 \\ 5 \end{bmatrix}.$$

Practice Problem 1 ▶ Let $\mathcal{B} = \{\mathbf{b}_1, \mathbf{b}_2\}$ be the basis for $\mathcal{R}^2$ pictured in Figure 4.10. Referring to that figure, find the coordinate vectors $[\mathbf{u}]_{\mathcal{B}}$ and $[\mathbf{v}]_{\mathcal{B}}$.

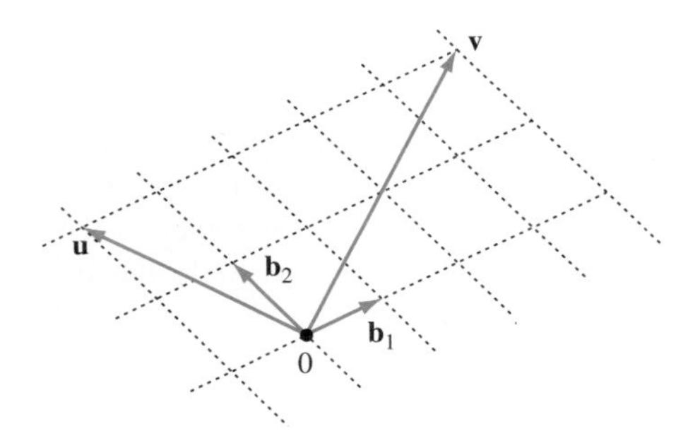

Figure 4.10 ◀

Example 2

For

$$\mathbf{v} = \begin{bmatrix} 1 \\ -4 \\ 4 \end{bmatrix} \quad \text{and} \quad \mathcal{B} = \left\{ \begin{bmatrix} 1 \\ 1 \\ 1 \end{bmatrix}, \begin{bmatrix} 1 \\ -1 \\ 1 \end{bmatrix}, \begin{bmatrix} 1 \\ 2 \\ 2 \end{bmatrix} \right\},$$

determine $[\mathbf{v}]_{\mathcal{B}}$.

Solution To find the $\mathcal{B}$-coordinate vector of $\mathbf{v}$, we must write $\mathbf{v}$ as a linear combination of the vectors in $\mathcal{B}$. As we learned in Chapter 1, this requires us to find scalars c_1, c_2, and c_3 such that

$$c_1\begin{bmatrix} 1 \\ 1 \\ 1 \end{bmatrix} + c_2\begin{bmatrix} 1 \\ -1 \\ 1 \end{bmatrix} + c_3\begin{bmatrix} 1 \\ 2 \\ 2 \end{bmatrix} = \begin{bmatrix} 1 \\ -4 \\ 4 \end{bmatrix}.$$

From this equation, we obtain a system of linear equations with augmented matrix

$$\begin{bmatrix} 1 & 1 & 1 & 1 \\ 1 & -1 & 2 & -4 \\ 1 & 1 & 2 & 4 \end{bmatrix}.$$

Since the reduced row echelon form of this matrix is

$$\begin{bmatrix} 1 & 0 & 0 & -6 \\ 0 & 1 & 0 & 4 \\ 0 & 0 & 1 & 3 \end{bmatrix},$$

we see that the desired scalars are $c_1 = -6$, $c_2 = 4$, $c_3 = 3$. Thus

$$[\mathbf{v}]_{\mathcal{B}} = \begin{bmatrix} -6 \\ 4 \\ 3 \end{bmatrix}$$

is the $\mathcal{B}$-coordinate vector of $\mathbf{v}$.

We can easily compute coordinate vectors relative to the standard basis $\mathcal{E}$ for $\mathcal{R}^n$. Because we can write every vector

$$\mathbf{v} = \begin{bmatrix} v_1 \\ v_2 \\ \vdots \\ v_n \end{bmatrix}$$

in $\mathcal{R}^n$ as $\mathbf{v} = v_1\mathbf{e}_1 + v_2\mathbf{e}_2 + \cdots + v_n\mathbf{e}_n$, we see that $[\mathbf{v}]_{\mathcal{E}} = \mathbf{v}$.

It is also easy to compute the $\mathcal{B}$-coordinate vector of any vector in a basis $\mathcal{B}$. For if $\mathcal{B} = \{\mathbf{b}_1, \mathbf{b}_2, \ldots, \mathbf{b}_n\}$ is a basis for $\mathcal{R}^n$, we have

$$\mathbf{b}_i = 0\mathbf{b}_1 + 0\mathbf{b}_2 + \cdots + 0\mathbf{b}_{i-1} + 1\mathbf{b}_i + 0\mathbf{b}_{i+1} + \cdots + 0\mathbf{b}_n,$$

and so $[\mathbf{b}_i]_{\mathcal{B}} = \mathbf{e}_i$.

In general, the following result provides a simple method for calculating coordinate vectors relative to an arbitrary basis for $\mathcal{R}^n$:

THEOREM 4.11

Let $\mathcal{B}$ be a basis for $\mathcal{R}^n$ and B be the matrix whose columns are the vectors in $\mathcal{B}$. Then B is invertible, and for every vector $\mathbf{v}$ in $\mathcal{R}^n$, $B[\mathbf{v}]_{\mathcal{B}} = \mathbf{v}$, or equivalently, $[\mathbf{v}]_{\mathcal{B}} = B^{-1}\mathbf{v}$.

PROOF Let $\mathcal{B} = \{\mathbf{b}_1, \mathbf{b}_2, \ldots, \mathbf{b}_n\}$ be a basis for $\mathcal{R}^n$ and $\mathbf{v}$ be a vector in $\mathcal{R}^n$. If

$$[\mathbf{v}]_{\mathcal{B}} = \begin{bmatrix} c_1 \\ c_2 \\ \vdots \\ c_n \end{bmatrix},$$

then

$$\begin{aligned}\mathbf{v} &= c_1\mathbf{b}_1 + c_2\mathbf{b}_2 + \cdots + c_n\mathbf{b}_n \\ &= [\mathbf{b}_1\ \mathbf{b}_2\ \ldots\ \mathbf{b}_n]\begin{bmatrix} c_1 \\ c_2 \\ \vdots \\ c_n \end{bmatrix} \\ &= B[\mathbf{v}]_{\mathcal{B}},\end{aligned}$$

where $B = [\mathbf{b}_1\ \mathbf{b}_2\ \ldots\ \mathbf{b}_n]$. Since $\mathcal{B}$ is a basis, the columns of B are linearly independent. Hence B is invertible by the Invertible Matrix Theorem, and thus $B[\mathbf{v}]_{\mathcal{B}} = \mathbf{v}$ is equivalent to $[\mathbf{v}]_{\mathcal{B}} = B^{-1}\mathbf{v}$. ■

As an alternative to the calculation in Example 1, we can compute the vector **u** by using Theorem 4.11. The result is

$$\mathbf{u} = B[\mathbf{u}]_{\mathcal{B}} = \begin{bmatrix} 1 & 1 & 1 \\ 1 & -1 & 2 \\ 1 & 1 & 2 \end{bmatrix}\begin{bmatrix} 3 \\ 6 \\ -2 \end{bmatrix} = \begin{bmatrix} 7 \\ -7 \\ 5 \end{bmatrix},$$

in agreement with Example 1.

Practice Problem 2 ▶ Verify that $\mathcal{B} = \left\{\begin{bmatrix} 1 \\ 1 \\ 0 \end{bmatrix}, \begin{bmatrix} 1 \\ 1 \\ 1 \end{bmatrix}, \begin{bmatrix} 3 \\ 2 \\ 1 \end{bmatrix}\right\}$ is a basis for $\mathcal{R}^3$. Then find **u** if $[\mathbf{u}]_{\mathcal{B}} = \begin{bmatrix} 5 \\ -2 \\ -1 \end{bmatrix}$. ◀

We can also use Theorem 4.11 to compute coordinate vectors, as the next example demonstrates.

Example 3 Let

$$\mathcal{B} = \left\{\begin{bmatrix} 1 \\ 1 \\ 1 \end{bmatrix}, \begin{bmatrix} 1 \\ -1 \\ 1 \end{bmatrix}, \begin{bmatrix} 1 \\ 2 \\ 2 \end{bmatrix}\right\} \quad \text{and} \quad \mathbf{v} = \begin{bmatrix} 1 \\ -4 \\ 4 \end{bmatrix},$$

as in Example 2, and let B be the matrix whose columns are the vectors in $\mathcal{B}$. Then

$$[\mathbf{v}]_{\mathcal{B}} = B^{-1}\mathbf{v} = \begin{bmatrix} -6 \\ 4 \\ 3 \end{bmatrix}.$$

Of course, this result agrees with that in Example 2.

Practice Problem 3 ▶ Find $[\mathbf{v}]_{\mathcal{B}}$ if $\mathbf{v} = \begin{bmatrix} -2 \\ 1 \\ 3 \end{bmatrix}$ and $\mathcal{B}$ is defined as in Practice Problem 1. ◀

CHANGING COORDINATES

To obtain an equation of the ellipse in Figure 4.8, we must be able to switch between the $x'y'$-coordinate system and the xy-coordinate system. This requires that we be able to convert coordinate vectors relative to an arbitrary basis $\mathcal{B}$ into coordinate vectors relative to the standard basis, or vice versa. In other words, we must know the relationship between $[\mathbf{v}]_{\mathcal{B}}$ and $\mathbf{v}$ for an arbitrary vector $\mathbf{v}$ in $\mathcal{R}^n$. According to Theorem 4.11, this relationship is $[\mathbf{v}]_{\mathcal{B}} = B^{-1}\mathbf{v}$, where B is the matrix whose columns are the vectors in $\mathcal{B}$.

Although a change of basis could be useful between any two bases for $\mathcal{R}^n$, the examples we give next arise as rotations of the usual coordinate axes. Such changes of bases are especially important in Chapter 6.

Consider the basis $\mathcal{B} = \{\mathbf{b}_1, \mathbf{b}_2\}$ obtained by rotating the vectors in the standard basis through 45°. We can compute the components of these vectors by using the 45°-rotation matrix, as in Section 1.2:

$$\mathbf{b}_1 = A_{45^\circ}\mathbf{e}_1 \qquad \text{and} \qquad \mathbf{b}_2 = A_{45^\circ}\mathbf{e}_2$$

In order to write the x', y'-equation

$$\frac{(x')^2}{3^2} + \frac{(y')^2}{2^2} = 1$$

as an equation in the usual xy-coordinate system, we must use the relationship between a vector and its $\mathcal{B}$-coordinates. Let

$$\mathbf{v} = \begin{bmatrix} x \\ y \end{bmatrix}, \qquad [\mathbf{v}]_{\mathcal{B}} = \begin{bmatrix} x' \\ y' \end{bmatrix},$$

and

$$B = [\mathbf{b}_1\ \mathbf{b}_2] = [A_{45^\circ}\mathbf{e}_1\ A_{45^\circ}\mathbf{e}_2] = A_{45^\circ}[\mathbf{e}_1\ \mathbf{e}_2] = A_{45^\circ}I_2 = A_{45^\circ}.$$

Since B is a rotation matrix, we have $B^{-1} = B^T$ by Exercise 53 in Section 2.3. Hence $[\mathbf{v}]_{\mathcal{B}} = B^{-1}\mathbf{v} = B^T\mathbf{v}$, and so

$$\begin{bmatrix} x' \\ y' \end{bmatrix} = B^T \begin{bmatrix} x \\ y \end{bmatrix} = \begin{bmatrix} \frac{\sqrt{2}}{2} & \frac{\sqrt{2}}{2} \\ -\frac{\sqrt{2}}{2} & \frac{\sqrt{2}}{2} \end{bmatrix} \begin{bmatrix} x \\ y \end{bmatrix} = \begin{bmatrix} \frac{\sqrt{2}}{2}x + \frac{\sqrt{2}}{2}y \\ -\frac{\sqrt{2}}{2}x + \frac{\sqrt{2}}{2}y \end{bmatrix}.$$

Therefore substituting

$$x' = \frac{\sqrt{2}}{2}x + \frac{\sqrt{2}}{2}y$$

$$y' = -\frac{\sqrt{2}}{2}x + \frac{\sqrt{2}}{2}y$$

into an equation in the $x'y'$-coordinate system converts it into one in the xy-coordinate system. So the equation of the given ellipse in the standard coordinate system is

$$\frac{\left(\frac{\sqrt{2}}{2}x + \frac{\sqrt{2}}{2}y\right)^2}{3^2} + \frac{\left(-\frac{\sqrt{2}}{2}x + \frac{\sqrt{2}}{2}y\right)^2}{2^2} = 1,$$

which simplifies to

$$13x^2 - 10xy + 13y^2 = 72.$$

Example 4 Write the equation $-\sqrt{3}x^2 + 2xy + \sqrt{3}y^2 = 12$ in terms of the $x'y'$-coordinate system, where the x'-axis and the y'-axis are obtained by rotating the x-axis and y-axis through the angle 30°.

Solution Again, a change of coordinates is required. This time, however, we must change from the xy-coordinate system to the $x'y'$-coordinate system. Consider the basis $\mathcal{B} = \{\mathbf{b}_1, \mathbf{b}_2\}$ obtained by rotating the vectors in the standard basis through 30°. As we did earlier, let

$$\mathbf{v} = \begin{bmatrix} x \\ y \end{bmatrix}, \qquad [\mathbf{v}]_{\mathcal{B}} = \begin{bmatrix} x' \\ y' \end{bmatrix}, \qquad \text{and} \qquad B = [\mathbf{b}_1\ \mathbf{b}_2] = A_{30^\circ}.$$

Since $\mathbf{v} = B[\mathbf{v}]_{\mathcal{B}}$, we have

$$\begin{bmatrix} x \\ y \end{bmatrix} = B\begin{bmatrix} x' \\ y' \end{bmatrix} = \begin{bmatrix} \frac{\sqrt{3}}{2} & -\frac{1}{2} \\ \frac{1}{2} & \frac{\sqrt{3}}{2} \end{bmatrix}\begin{bmatrix} x' \\ y' \end{bmatrix} = \begin{bmatrix} \frac{\sqrt{3}}{2}x' - \frac{1}{2}y' \\ \frac{1}{2}x' + \frac{\sqrt{3}}{2}y' \end{bmatrix}.$$

Hence the equations relating the two coordinate systems are

$$x = \frac{\sqrt{3}}{2}x' - \frac{1}{2}y'$$

$$y = \frac{1}{2}x' + \frac{\sqrt{3}}{2}y'.$$

These substitutions change the equation $-\sqrt{3}x^2 + 2xy + \sqrt{3}y^2 = 12$ into $4x'y' = 12$; that is, $x'y' = 3$. From this equation, we see that the graph of the given equation is a hyperbola (shown in Figure 4.11).

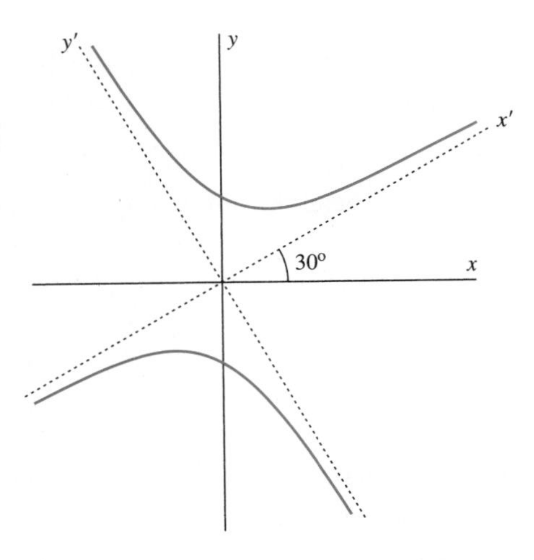

Figure 4.11 The hyperbola with equation $-\sqrt{3}x^2 + 2xy + \sqrt{3}y^2 = 12$

EXERCISES

In Exercises 1–6, the $\mathcal{B}$-coordinate vector of $\mathbf{v}$ *is given. Find* $\mathbf{v}$ *if*

$$\mathcal{B} = \left\{ \begin{bmatrix} 1 \\ -1 \end{bmatrix}, \begin{bmatrix} -1 \\ 2 \end{bmatrix} \right\}.$$

1. $[\mathbf{v}]_{\mathcal{B}} = \begin{bmatrix} 4 \\ 3 \end{bmatrix}$
2. $[\mathbf{v}]_{\mathcal{B}} = \begin{bmatrix} -3 \\ 2 \end{bmatrix}$
3. $[\mathbf{v}]_{\mathcal{B}} = \begin{bmatrix} 1 \\ 6 \end{bmatrix}$
4. $[\mathbf{v}]_{\mathcal{B}} = \begin{bmatrix} -1 \\ 4 \end{bmatrix}$
5. $[\mathbf{v}]_{\mathcal{B}} = \begin{bmatrix} 2 \\ 5 \end{bmatrix}$
6. $[\mathbf{v}]_{\mathcal{B}} = \begin{bmatrix} 5 \\ 2 \end{bmatrix}$

In Exercises 7–10, the $\mathcal{B}$-coordinate vector of $\mathbf{v}$ *is given. Find* $\mathbf{v}$ *if*

$$\mathcal{B} = \left\{ \begin{bmatrix} 0 \\ 1 \\ 1 \end{bmatrix}, \begin{bmatrix} -1 \\ 0 \\ 1 \end{bmatrix}, \begin{bmatrix} 1 \\ 1 \\ 1 \end{bmatrix} \right\}.$$

7. $[\mathbf{v}]_{\mathcal{B}} = \begin{bmatrix} 2 \\ -1 \\ 3 \end{bmatrix}$
8. $[\mathbf{v}]_{\mathcal{B}} = \begin{bmatrix} 3 \\ 1 \\ -4 \end{bmatrix}$
9. $[\mathbf{v}]_{\mathcal{B}} = \begin{bmatrix} -1 \\ 5 \\ -2 \end{bmatrix}$
10. $[\mathbf{v}]_{\mathcal{B}} = \begin{bmatrix} 3 \\ -4 \\ 2 \end{bmatrix}$

11. (a) Prove that $\mathcal{B} = \left\{ \begin{bmatrix} 1 \\ 3 \end{bmatrix}, \begin{bmatrix} -2 \\ 1 \end{bmatrix} \right\}$ is a basis for $\mathcal{R}^2$.

    (b) Find $[\mathbf{v}]_{\mathcal{B}}$ if $\mathbf{v} = 5\begin{bmatrix} 1 \\ 3 \end{bmatrix} - 3\begin{bmatrix} -2 \\ 1 \end{bmatrix}$.

12. (a) Prove that $\mathcal{B} = \left\{ \begin{bmatrix} 3 \\ -2 \end{bmatrix}, \begin{bmatrix} -1 \\ 2 \end{bmatrix} \right\}$ is a basis for $\mathcal{R}^2$.

    (b) Find $[\mathbf{v}]_{\mathcal{B}}$ if $\mathbf{v} = -2\begin{bmatrix} 3 \\ -2 \end{bmatrix} + 4\begin{bmatrix} -1 \\ 2 \end{bmatrix}$.

13. (a) Prove that $\mathcal{B} = \left\{ \begin{bmatrix} -1 \\ 0 \\ 1 \end{bmatrix}, \begin{bmatrix} 2 \\ 1 \\ -1 \end{bmatrix}, \begin{bmatrix} 1 \\ -3 \\ 2 \end{bmatrix} \right\}$ is a basis for $\mathcal{R}^3$.

    (b) Find $[\mathbf{v}]_{\mathcal{B}}$ if $\mathbf{v} = 3\begin{bmatrix} -1 \\ 0 \\ 1 \end{bmatrix} - \begin{bmatrix} 1 \\ -3 \\ 2 \end{bmatrix}$.

14. (a) Prove that $\mathcal{B} = \left\{ \begin{bmatrix} 1 \\ -1 \\ 1 \end{bmatrix}, \begin{bmatrix} -1 \\ 1 \\ 1 \end{bmatrix}, \begin{bmatrix} 1 \\ 1 \\ 1 \end{bmatrix} \right\}$ is a basis for $\mathcal{R}^3$.

    (b) Find $[\mathbf{v}]_{\mathcal{B}}$ if $\mathbf{v} = \begin{bmatrix} -1 \\ 1 \\ 1 \end{bmatrix} - 4\begin{bmatrix} 1 \\ 1 \\ 1 \end{bmatrix}$.

In Exercises 15–18, a vector is given. Find its $\mathcal{B}$-coordinate vector relative to the basis $\mathcal{B}$ in Exercises 1–6.

15. $\begin{bmatrix} -4 \\ 3 \end{bmatrix}$
16. $\begin{bmatrix} -1 \\ 2 \end{bmatrix}$
17. $\begin{bmatrix} 5 \\ -3 \end{bmatrix}$
18. $\begin{bmatrix} 3 \\ 2 \end{bmatrix}$

In Exercises 19–22, a vector is given. Find its $\mathcal{B}$-coordinate vector relative to the basis $\mathcal{B}$ in Exercises 7–10.

19. $\begin{bmatrix} 4 \\ 3 \\ 2 \end{bmatrix}$
20. $\begin{bmatrix} -2 \\ 6 \\ 3 \end{bmatrix}$
21. $\begin{bmatrix} 1 \\ -3 \\ -2 \end{bmatrix}$
22. $\begin{bmatrix} -1 \\ 5 \\ 2 \end{bmatrix}$

23. Find the unique representation of $\mathbf{u} = \begin{bmatrix} a \\ b \end{bmatrix}$ as a linear combination of
$$\mathbf{b}_1 = \begin{bmatrix} 3 \\ -1 \end{bmatrix} \text{ and } \mathbf{b}_2 = \begin{bmatrix} -2 \\ 1 \end{bmatrix}.$$

24. Find the unique representation of $\mathbf{u} = \begin{bmatrix} a \\ b \end{bmatrix}$ as a linear combination of
$$\mathbf{b}_1 = \begin{bmatrix} 2 \\ -1 \end{bmatrix} \text{ and } \mathbf{b}_2 = \begin{bmatrix} -1 \\ 1 \end{bmatrix}.$$

25. Find the unique representation of $\mathbf{u} = \begin{bmatrix} a \\ b \end{bmatrix}$ as a linear combination of
$$\mathbf{b}_1 = \begin{bmatrix} -2 \\ 3 \end{bmatrix} \text{ and } \mathbf{b}_2 = \begin{bmatrix} 3 \\ -5 \end{bmatrix}.$$

26. Find the unique representation of $\mathbf{u} = \begin{bmatrix} a \\ b \end{bmatrix}$ as a linear combination of
$$\mathbf{b}_1 = \begin{bmatrix} 3 \\ 1 \end{bmatrix} \text{ and } \mathbf{b}_2 = \begin{bmatrix} 2 \\ 1 \end{bmatrix}.$$

27. Find the unique representation of $\mathbf{u} = \begin{bmatrix} a \\ b \\ c \end{bmatrix}$ as a linear combination of
$$\mathbf{b}_1 = \begin{bmatrix} 1 \\ -1 \\ 1 \end{bmatrix}, \quad \mathbf{b}_2 = \begin{bmatrix} -1 \\ 2 \\ 1 \end{bmatrix}, \text{ and } \mathbf{b}_3 = \begin{bmatrix} 1 \\ 0 \\ 2 \end{bmatrix}.$$

28. Find the unique representation of $\mathbf{u} = \begin{bmatrix} a \\ b \\ c \end{bmatrix}$ as a linear combination of
$$\mathbf{b}_1 = \begin{bmatrix} 1 \\ -1 \\ 2 \end{bmatrix}, \quad \mathbf{b}_2 = \begin{bmatrix} 1 \\ 0 \\ 2 \end{bmatrix}, \text{ and } \mathbf{b}_3 = \begin{bmatrix} 0 \\ 1 \\ 1 \end{bmatrix}.$$

29. Find the unique representation of $\mathbf{u} = \begin{bmatrix} a \\ b \\ c \end{bmatrix}$ as a linear combination of
$$\mathbf{b}_1 = \begin{bmatrix} 1 \\ 0 \\ 1 \end{bmatrix}, \quad \mathbf{b}_2 = \begin{bmatrix} -1 \\ 1 \\ 0 \end{bmatrix}, \text{ and } \mathbf{b}_3 = \begin{bmatrix} -2 \\ 0 \\ -1 \end{bmatrix}.$$

30. Find the unique representation of $\mathbf{u} = \begin{bmatrix} a \\ b \\ c \end{bmatrix}$ as a linear combination of

$$\mathbf{b}_1 = \begin{bmatrix} -1 \\ 1 \\ 2 \end{bmatrix}, \quad \mathbf{b}_2 = \begin{bmatrix} 2 \\ -1 \\ -1 \end{bmatrix}, \quad \text{and} \quad \mathbf{b}_3 = \begin{bmatrix} 0 \\ 1 \\ 2 \end{bmatrix}.$$

T&F *In Exercises 31–50, determine whether the statements are true or false.*

31. If $\mathcal{S}$ is a generating set for a subspace V of $\mathcal{R}^n$, then every vector in V can be uniquely represented as a linear combination of the vectors in $\mathcal{S}$.
32. The components of the $\mathcal{B}$-coordinate vector of $\mathbf{u}$ are the coefficients that express $\mathbf{u}$ as a linear combination of the vectors in $\mathcal{B}$.
33. If $\mathcal{E}$ is the standard basis for $\mathcal{R}^n$, then $[\mathbf{v}]_{\mathcal{E}} = \mathbf{v}$ for all $\mathbf{v}$ in $\mathcal{R}^n$.
34. The $\mathcal{B}$-coordinate vector of a vector in $\mathcal{B}$ is a standard vector.
35. If $\mathcal{B}$ is a basis for $\mathcal{R}^n$ and B is the matrix whose columns are the vectors in $\mathcal{B}$, then $B^{-1}\mathbf{v} = [\mathbf{v}]_{\mathcal{B}}$ for all $\mathbf{v}$ in $\mathcal{R}^n$.
36. If $\mathcal{B} = \{\mathbf{b}_1, \mathbf{b}_2, \ldots, \mathbf{b}_n\}$ is any basis for $\mathcal{R}^n$, then, for every vector $\mathbf{v}$ in $\mathcal{R}^n$, there are unique scalars such that $\mathbf{v} = c_1\mathbf{b}_1 + c_2\mathbf{b}_2 + \cdots + c_n\mathbf{b}_n$.
37. If the columns of an $n \times n$ matrix form a basis for $\mathcal{R}^n$, then the matrix is invertible.
38. If B is a matrix whose columns are the vectors in a basis for $\mathcal{R}^n$, then, for any vector $\mathbf{v}$ in $\mathcal{R}^n$, the system $B\mathbf{x} = \mathbf{v}$ has a unique solution.
39. If B is a matrix whose columns are the vectors in a basis $\mathcal{B}$ for $\mathcal{R}^n$, then, for any vector $\mathbf{v}$ in $\mathcal{R}^n$, $[\mathbf{v}]_{\mathcal{B}}$ is a solution of $B\mathbf{x} = \mathbf{v}$.
40. If B is a matrix whose columns are the vectors in a basis $\mathcal{B}$ for $\mathcal{R}^n$, then, for any vector $\mathbf{v}$ in $\mathcal{R}^n$, the reduced row echelon form of $[B \;\; \mathbf{v}]$ is $[I_n \;\; [\mathbf{v}]_{\mathcal{B}}]$.
41. If $\mathcal{B}$ is any basis for $\mathcal{R}^n$, then $[\mathbf{0}]_{\mathcal{B}} = \mathbf{0}$.
42. If $\mathbf{u}$ and $\mathbf{v}$ are any vectors in $\mathcal{R}^n$ and $\mathcal{B}$ is any basis for $\mathcal{R}^n$, then $[\mathbf{u} + \mathbf{v}]_{\mathcal{B}} = [\mathbf{u}]_{\mathcal{B}} + [\mathbf{v}]_{\mathcal{B}}$.
43. If $\mathbf{v}$ is any vector in $\mathcal{R}^n$, $\mathcal{B}$ is any basis for $\mathcal{R}^n$, and c is a scalar, then $[c\mathbf{v}]_{\mathcal{B}} = c[\mathbf{v}]_{\mathcal{B}}$.
44. Suppose that the x'-, y'-axes are obtained by rotating the usual x-, y-axes through the angle θ. Then $\begin{bmatrix} x' \\ y' \end{bmatrix} = A_\theta \begin{bmatrix} x \\ y \end{bmatrix}$.
45. Suppose that the x'-, y'-axes are obtained by rotating the usual x-, y-axes through the angle θ. Then $\begin{bmatrix} x \\ y \end{bmatrix} = A_\theta \begin{bmatrix} x' \\ y' \end{bmatrix}$.
46. If A_θ is a rotation matrix, then $A_\theta^T = A_\theta^{-1}$.
47. The graph of an equation of the form $\dfrac{(x')^2}{a^2} + \dfrac{(y')^2}{b^2} = 1$ is an ellipse.
48. The graph of an equation of the form $\dfrac{(x')^2}{a^2} - \dfrac{(y')^2}{b^2} = 1$ is a parabola.
49. The graph of an ellipse with center at the origin can be written in the form $\dfrac{(x')^2}{a^2} + \dfrac{(y')^2}{b^2} = 1$ by a suitable rotation of the coordinate axes.
50. The graph of a hyperbola with center at the origin can be written in the form $\dfrac{(x')^2}{a^2} - \dfrac{(y')^2}{b^2} = 1$ by a suitable rotation of the coordinate axes.

---

51. Let $\mathcal{B} = \{\mathbf{b}_1, \mathbf{b}_2\}$, where $\mathbf{b}_1 = \begin{bmatrix} 1 \\ 2 \end{bmatrix}$ and $\mathbf{b}_2 = \begin{bmatrix} 2 \\ 3 \end{bmatrix}$.
    (a) Show that $\mathcal{B}$ is a basis for $\mathcal{R}^2$.
    (b) Determine the matrix $A = [\,[\mathbf{e}_1]_{\mathcal{B}} \;\; [\mathbf{e}_2]_{\mathcal{B}}\,]$.
    (c) What is the relationship between A and $B = [\mathbf{b}_1 \;\; \mathbf{b}_2]$?
52. Let $\mathcal{B} = \{\mathbf{b}_1, \mathbf{b}_2\}$, where $\mathbf{b}_1 = \begin{bmatrix} 1 \\ 3 \end{bmatrix}$ and $\mathbf{b}_2 = \begin{bmatrix} -1 \\ -2 \end{bmatrix}$.
    (a) Show that $\mathcal{B}$ is a basis for $\mathcal{R}^2$.
    (b) Determine the matrix $A = [\,[\mathbf{e}_1]_{\mathcal{B}} \;\; [\mathbf{e}_2]_{\mathcal{B}}\,]$.
    (c) What is the relationship between A and $B = [\mathbf{b}_1 \;\; \mathbf{b}_2]$?
53. Let $\mathcal{B} = \{\mathbf{b}_1, \mathbf{b}_2, \mathbf{b}_3\}$, where $\mathbf{b}_1 = \begin{bmatrix} 2 \\ 1 \\ -1 \end{bmatrix}$, $\mathbf{b}_2 = \begin{bmatrix} -1 \\ -1 \\ 1 \end{bmatrix}$, and $\mathbf{b}_3 = \begin{bmatrix} -1 \\ -2 \\ 1 \end{bmatrix}$.
    (a) Show that $\mathcal{B}$ is a basis for $\mathcal{R}^3$.
    (b) Determine the matrix $A = [\,[\mathbf{e}_1]_{\mathcal{B}} \;\; [\mathbf{e}_2]_{\mathcal{B}} \;\; [\mathbf{e}_3]_{\mathcal{B}}\,]$.
    (c) What is the relationship between A and $B = [\mathbf{b}_1 \;\; \mathbf{b}_2 \;\; \mathbf{b}_3]$?
54. Let $\mathcal{B} = \{\mathbf{b}_1, \mathbf{b}_2, \mathbf{b}_3\}$, where

$$\mathbf{b}_1 = \begin{bmatrix} 1 \\ -2 \\ 1 \end{bmatrix}, \mathbf{b}_2 = \begin{bmatrix} -1 \\ 3 \\ 0 \end{bmatrix}, \text{ and } \mathbf{b}_3 = \begin{bmatrix} 0 \\ 2 \\ 1 \end{bmatrix}.$$

    (a) Show that $\mathcal{B}$ is a basis for $\mathcal{R}^3$.
    (b) Determine the matrix $A = [\,[\mathbf{e}_1]_{\mathcal{B}} \;\; [\mathbf{e}_2]_{\mathcal{B}} \;\; [\mathbf{e}_3]_{\mathcal{B}}\,]$.
    (c) What is the relationship between A and $B = [\mathbf{b}_1 \;\; \mathbf{b}_2 \;\; \mathbf{b}_3]$?

In Exercises 55–58, an angle θ is given. Let $\mathbf{v} = \begin{bmatrix} x \\ y \end{bmatrix}$ and $[\mathbf{v}]_{\mathcal{B}} = \begin{bmatrix} x' \\ y' \end{bmatrix}$, where $\mathcal{B}$ is the basis for $\mathcal{R}^2$ obtained by rotating $\mathbf{e}_1$ and $\mathbf{e}_2$ through the angle θ. Write equations expressing x' and y' in terms of x and y.

55. $\theta = 30°$ 56. $\theta = 60°$ 57. $\theta = 135°$ 58. $\theta = 330°$

In Exercises 59–62, a basis $\mathcal{B}$ for $\mathcal{R}^2$ is given. If $\mathbf{v} = \begin{bmatrix} x \\ y \end{bmatrix}$ and $[\mathbf{v}]_{\mathcal{B}} = \begin{bmatrix} x' \\ y' \end{bmatrix}$, write equations expressing x' and y' in terms of x and y.

59. $\left\{\begin{bmatrix}1\\-2\end{bmatrix}, \begin{bmatrix}-3\\5\end{bmatrix}\right\}$

60. $\left\{\begin{bmatrix}3\\4\end{bmatrix}, \begin{bmatrix}2\\3\end{bmatrix}\right\}$

61. $\left\{\begin{bmatrix}1\\-2\end{bmatrix}, \begin{bmatrix}-1\\1\end{bmatrix}\right\}$

62. $\left\{\begin{bmatrix}3\\-5\end{bmatrix}, \begin{bmatrix}2\\-3\end{bmatrix}\right\}$

In Exercises 63–66, a basis $\mathcal{B}$ for $\mathcal{R}^3$ is given. If $\mathbf{v} = \begin{bmatrix}x\\y\\z\end{bmatrix}$ *and* $[\mathbf{v}]_{\mathcal{B}} = \begin{bmatrix}x'\\y'\\z'\end{bmatrix}$, *write equations expressing x', y', and z' in terms of x, y, and z.*

63. $\left\{\begin{bmatrix}1\\0\\1\end{bmatrix}, \begin{bmatrix}1\\1\\0\end{bmatrix}, \begin{bmatrix}0\\-2\\1\end{bmatrix}\right\}$

64. $\left\{\begin{bmatrix}-1\\-1\\1\end{bmatrix}, \begin{bmatrix}-2\\-2\\1\end{bmatrix}, \begin{bmatrix}1\\0\\1\end{bmatrix}\right\}$

65. $\left\{\begin{bmatrix}0\\1\\2\end{bmatrix}, \begin{bmatrix}-1\\0\\1\end{bmatrix}, \begin{bmatrix}-2\\-1\\1\end{bmatrix}\right\}$

66. $\left\{\begin{bmatrix}-1\\2\\0\end{bmatrix}, \begin{bmatrix}1\\2\\1\end{bmatrix}, \begin{bmatrix}1\\1\\1\end{bmatrix}\right\}$

In Exercises 67–70, an angle θ is given. Let $\mathbf{v} = \begin{bmatrix}x\\y\end{bmatrix}$ *and* $[\mathbf{v}]_{\mathcal{B}} = \begin{bmatrix}x'\\y'\end{bmatrix}$, *where $\mathcal{B}$ is the basis for $\mathcal{R}^2$ obtained by rotating $\mathbf{e}_1$ and $\mathbf{e}_2$ through the angle θ. Write equations expressing x and y in terms of x' and y'.*

67. $\theta = 60^\circ$ 68. $\theta = 45^\circ$ 69. $\theta = 135^\circ$ 70. $\theta = 330^\circ$

In Exercises 71–74, a basis $\mathcal{B}$ for $\mathcal{R}^2$ is given. If $\mathbf{v} = \begin{bmatrix}x\\y\end{bmatrix}$ *and* $[\mathbf{v}]_{\mathcal{B}} = \begin{bmatrix}x'\\y'\end{bmatrix}$, *write equations expressing x and y in terms of x' and y'.*

71. $\left\{\begin{bmatrix}1\\2\end{bmatrix}, \begin{bmatrix}3\\4\end{bmatrix}\right\}$

72. $\left\{\begin{bmatrix}2\\-1\end{bmatrix}, \begin{bmatrix}1\\3\end{bmatrix}\right\}$

73. $\left\{\begin{bmatrix}-1\\3\end{bmatrix}, \begin{bmatrix}3\\5\end{bmatrix}\right\}$

74. $\left\{\begin{bmatrix}3\\2\end{bmatrix}, \begin{bmatrix}2\\4\end{bmatrix}\right\}$

In Exercises 75–78, a basis $\mathcal{B}$ for $\mathcal{R}^3$ is given. If $\mathbf{v} = \begin{bmatrix}x\\y\\z\end{bmatrix}$ *and* $[\mathbf{v}]_{\mathcal{B}} = \begin{bmatrix}x'\\y'\\z'\end{bmatrix}$, *write equations expressing x, y, and z in terms of x', y', and z'.*

75. $\left\{\begin{bmatrix}1\\3\\0\end{bmatrix}, \begin{bmatrix}-1\\1\\1\end{bmatrix}, \begin{bmatrix}0\\-1\\1\end{bmatrix}\right\}$

76. $\left\{\begin{bmatrix}2\\-1\\1\end{bmatrix}, \begin{bmatrix}0\\-1\\1\end{bmatrix}, \begin{bmatrix}1\\-1\\2\end{bmatrix}\right\}$

77. $\left\{\begin{bmatrix}1\\-1\\1\end{bmatrix}, \begin{bmatrix}-1\\3\\2\end{bmatrix}, \begin{bmatrix}-1\\1\\1\end{bmatrix}\right\}$

78. $\left\{\begin{bmatrix}-1\\1\\1\end{bmatrix}, \begin{bmatrix}-1\\2\\2\end{bmatrix}, \begin{bmatrix}1\\-1\\1\end{bmatrix}\right\}$

In Exercises 79–86, an equation of a conic section is given in the $x'y'$-coordinate system. Determine the equation of the conic section in the usual xy-coordinate system if the x'-axis and the y'-axis are obtained by rotating the usual x-axis and y-axis through the given angle θ.

79. $\dfrac{(x')^2}{4^2} + \dfrac{(y')^2}{5^2} = 1,\ \theta = 60^\circ$

80. $\dfrac{(x')^2}{2^2} - \dfrac{(y')^2}{3^2} = 1,\ \theta = 45^\circ$

81. $\dfrac{(x')^2}{3^2} - \dfrac{(y')^2}{5^2} = 1,\ \theta = 135^\circ$

82. $\dfrac{(x')^2}{6} + \dfrac{(y')^2}{4} = 1,\ \theta = 150^\circ$

83. $\dfrac{(x')^2}{3^2} - \dfrac{(y')^2}{2^2} = 1,\ \theta = 120^\circ$

84. $\dfrac{(x')^2}{2^2} + \dfrac{(y')^2}{5^2} = 1,\ \theta = 330^\circ$

85. $\dfrac{(x')^2}{3^2} - \dfrac{(y')^2}{4^2} = 1,\ \theta = 240^\circ$

86. $\dfrac{(x')^2}{3^2} + \dfrac{(y')^2}{2^2} = 1,\ \theta = 300^\circ$

In Exercises 87–94, an equation of a conic section is given in the xy-coordinate system. Determine the equation of the conic section in the $x'y'$-coordinate system if the x'-axis and the y'-axis are obtained by rotating the usual x-axis and y-axis through the given angle θ.

87. $-3x^2 + 14xy - 3y^2 = 20,\ \theta = 45^\circ$

88. $6x^2 - 2\sqrt{3}xy + 4y^2 = 21,\ \theta = 60^\circ$

89. $15x^2 - 2\sqrt{3}xy + 13y^2 = 48,\ \theta = 150^\circ$

90. $x^2 - 6xy + y^2 = 12,\ \theta = 135^\circ$

91. $9x^2 + 14\sqrt{3}xy - 5y^2 = 240,\ \theta = 30^\circ$

92. $35x^2 - 2\sqrt{3}xy + 33y^2 = 720,\ \theta = 60^\circ$

93. $17x^2 - 6\sqrt{3}xy + 11y^2 = 40,\ \theta = 150^\circ$

94. $2x^2 - xy + 2y^2 = 12,\ \theta = 135^\circ$

95. Let $\mathcal{A} = \{\mathbf{u}_1, \mathbf{u}_2, \ldots, \mathbf{u}_n\}$ be a basis for $\mathcal{R}^n$ and $c_1, c_2, \ldots, c_n$ be nonzero scalars. Recall from Exercise 71 of Section 4.2 that $\mathcal{B} = \{c_1\mathbf{u}_1, c_2\mathbf{u}_2, \ldots, c_n\mathbf{u}_n\}$ is also a basis for $\mathcal{R}^n$. If $\mathbf{v}$ is a vector in $\mathcal{R}^n$ and

$$[\mathbf{v}]_{\mathcal{A}} = \begin{bmatrix}a_1\\a_2\\\vdots\\a_n\end{bmatrix},$$

compute $[\mathbf{v}]_{\mathcal{B}}$.

96. Let $\mathcal{A} = \{\mathbf{u}_1, \mathbf{u}_2, \ldots, \mathbf{u}_n\}$ be a basis for $\mathcal{R}^n$. Recall that Exercise 72 of Section 4.2 shows that

$$\mathcal{B} = \{\mathbf{u}_1, \mathbf{u}_1 + \mathbf{u}_2, \ldots, \mathbf{u}_1 + \mathbf{u}_n\}$$

is also a basis for $\mathcal{R}^n$. If $\mathbf{v}$ is a vector in $\mathcal{R}^n$ and

$$[\mathbf{v}]_{\mathcal{A}} = \begin{bmatrix} a_1 \\ a_2 \\ \vdots \\ a_n \end{bmatrix},$$

compute $[\mathbf{v}]_{\mathcal{B}}$.

97. Let $\mathcal{A} = \{\mathbf{u}_1, \mathbf{u}_2, \ldots, \mathbf{u}_n\}$ be a basis for $\mathcal{R}^n$. Recall that Exercise 73 of Section 4.2 shows that

$$\mathcal{B} = \{\mathbf{u}_1 + \mathbf{u}_2 + \cdots + \mathbf{u}_n, \mathbf{u}_2, \mathbf{u}_3, \ldots, \mathbf{u}_n\}$$

is also a basis for $\mathcal{R}^n$. If $\mathbf{v}$ is a vector in $\mathcal{R}^n$ and

$$[\mathbf{v}]_{\mathcal{A}} = \begin{bmatrix} a_1 \\ a_2 \\ \vdots \\ a_n \end{bmatrix},$$

compute $[\mathbf{v}]_{\mathcal{B}}$.

98. Let $\mathcal{A} = \{\mathbf{u}_1, \mathbf{u}_2, \ldots, \mathbf{u}_n\}$ be a basis for $\mathcal{R}^n$. Recall that Exercise 74 of Section 4.2 shows that

$$\mathcal{B} = \{\mathbf{v}_1, \mathbf{v}_2, \ldots, \mathbf{v}_k\},$$

where

$$\mathbf{v}_i = \mathbf{u}_i + \mathbf{u}_{i+1} + \cdots + \mathbf{u}_k \quad \text{for } i = 1, 2, \ldots, k$$

is also a basis for $\mathcal{R}^n$. If $\mathbf{v}$ is a vector in $\mathcal{R}^n$ and

$$[\mathbf{v}]_{\mathcal{A}} = \begin{bmatrix} a_1 \\ a_2 \\ \vdots \\ a_n \end{bmatrix},$$

compute $[\mathbf{v}]_{\mathcal{B}}$.

99. Let $\mathcal{A}$ and $\mathcal{B}$ be two bases for $\mathcal{R}^n$. If $[\mathbf{v}]_{\mathcal{A}} = [\mathbf{v}]_{\mathcal{B}}$ for some nonzero vector $\mathbf{v}$ in $\mathcal{R}^n$, must $\mathcal{A} = \mathcal{B}$? Justify your answer.

100. Let $\mathcal{A}$ and $\mathcal{B}$ be two bases for $\mathcal{R}^n$. If $[\mathbf{v}]_{\mathcal{A}} = [\mathbf{v}]_{\mathcal{B}}$ for every vector $\mathbf{v}$ in $\mathcal{R}^n$, must $\mathcal{A} = \mathcal{B}$? Justify your answer.

101. Prove that if $\mathcal{S}$ is linearly dependent, then every vector in the span of $\mathcal{S}$ can be written as a linear combination of the vectors in $\mathcal{S}$ in more than one way.

102. Let $\mathcal{A}$ and $\mathcal{B}$ be two bases for $\mathcal{R}^n$. Express $[\mathbf{v}]_{\mathcal{A}}$ in terms of $[\mathbf{v}]_{\mathcal{B}}$.

103. (a) Let $\mathcal{B}$ be a basis for $\mathcal{R}^n$. Prove that the function $T\colon \mathcal{R}^n \to \mathcal{R}^n$ defined for all $\mathbf{v}$ in $\mathcal{R}^n$ by $T(\mathbf{v}) = [\mathbf{v}]_{\mathcal{B}}$ is a linear transformation.

(b) Prove that T is one-to-one and onto.

104. What is the standard matrix of the linear transformation T in Exercise 103?

105. Let V be a subspace of $\mathcal{R}^n$ and $\mathcal{B} = \{\mathbf{u}_1, \mathbf{u}_2, \ldots, \mathbf{u}_k\}$ be a subset of V. Prove that if every vector $\mathbf{v}$ in V can be uniquely represented as a linear combination of the vectors in $\mathcal{B}$ (that is, if $\mathbf{v} = a_1\mathbf{u}_1 + a_2\mathbf{u}_2 + \cdots + a_k\mathbf{u}_k$ for unique scalars $a_1, a_2, \ldots, a_k$), then $\mathcal{B}$ is a basis for V. (This is the converse of Theorem 4.10.)

106. Let $V = \left\{ \begin{bmatrix} v_1 \\ v_2 \\ v_3 \end{bmatrix} \in \mathcal{R}^3 \colon -2v_1 + v_2 + v_3 = 0 \right\}$ and

$$\mathcal{S} = \left\{ \begin{bmatrix} -1 \\ 1 \\ 1 \end{bmatrix}, \begin{bmatrix} 1 \\ 0 \\ 1 \end{bmatrix}, \begin{bmatrix} 1 \\ -2 \\ -2 \end{bmatrix} \right\}.$$

(a) Show that $\mathcal{S}$ is linearly independent.

(b) Show that

$$(2v_1 - 5v_2) \begin{bmatrix} -1 \\ 1 \\ 1 \end{bmatrix} + (2v_1 - 2v_2) \begin{bmatrix} 1 \\ 0 \\ 1 \end{bmatrix} + (v_1 - 3v_2) \begin{bmatrix} 1 \\ -2 \\ -2 \end{bmatrix} = \begin{bmatrix} v_1 \\ v_2 \\ v_3 \end{bmatrix}$$

for every vector $\begin{bmatrix} v_1 \\ v_2 \\ v_3 \end{bmatrix}$ in $\mathcal{S}$.

(c) Is $\mathcal{S}$ a basis for V? Justify your answer.

107. Let $\mathcal{B}$ be a basis for $\mathcal{R}^n$ and $\{\mathbf{u}_1, \mathbf{u}_2, \ldots, \mathbf{u}_k\}$ be a subset of $\mathcal{R}^n$. Prove that $\{\mathbf{u}_1, \mathbf{u}_2, \ldots, \mathbf{u}_k\}$ is linearly independent if and only if $\{[\mathbf{u}_1]_{\mathcal{B}}, [\mathbf{u}_2]_{\mathcal{B}}, \ldots, [\mathbf{u}_k]_{\mathcal{B}}\}$ is linearly independent.

108. Let $\mathcal{B}$ be a basis for $\mathcal{R}^n$, $\{\mathbf{u}_1, \mathbf{u}_2, \ldots, \mathbf{u}_k\}$ be a subset of $\mathcal{R}^n$, and $\mathbf{v}$ be a vector in $\mathcal{R}^n$. Prove that $\mathbf{v}$ is a linear combination of $\{\mathbf{u}_1, \mathbf{u}_2, \ldots, \mathbf{u}_k\}$ if and only if $[\mathbf{v}]_{\mathcal{B}}$ is a linear combination of $\{[\mathbf{u}_1]_{\mathcal{B}}, [\mathbf{u}_2]_{\mathcal{B}}, \ldots, [\mathbf{u}_k]_{\mathcal{B}}\}$.

In Exercises 109–112, use either a calculator with matrix capabilities or computer software such as MATLAB to solve each problem.

109. Let

$$\mathcal{B} = \left\{ \begin{bmatrix} 0 \\ 25 \\ -21 \\ 23 \\ 12 \end{bmatrix}, \begin{bmatrix} 14 \\ 73 \\ -66 \\ 64 \\ 42 \end{bmatrix}, \begin{bmatrix} -6 \\ -56 \\ 47 \\ -50 \\ -29 \end{bmatrix}, \begin{bmatrix} -14 \\ -68 \\ 60 \\ -59 \\ -39 \end{bmatrix}, \begin{bmatrix} -12 \\ -118 \\ 102 \\ -106 \\ -62 \end{bmatrix} \right\}$$

and $\mathbf{v} = \begin{bmatrix} -2 \\ 3 \\ 1 \\ 2 \\ -1 \end{bmatrix}$.

(a) Show that $\mathcal{B}$ is a basis for $\mathcal{R}^5$.

(b) Find $[\mathbf{v}]_{\mathcal{B}}$.

110. For the basis $\mathcal{B}$ in Exercise 109, find a nonzero vector $\mathbf{u}$ in $\mathcal{R}^5$ such that $\mathbf{u} = [\mathbf{u}]_{\mathcal{B}}$.

111. For the basis $\mathcal{B}$ in Exercise 109, find a nonzero vector $\mathbf{v}$ in $\mathcal{R}^5$ such that $[\mathbf{v}]_{\mathcal{B}} = .5\mathbf{v}$.

112. Let $\mathcal{B}$ and $\mathbf{v}$ be as in Exercise 109, and let

$$\mathbf{u}_1 = \begin{bmatrix} 1 \\ 0 \\ -1 \\ 1 \\ 0 \end{bmatrix}, \quad \mathbf{u}_2 = \begin{bmatrix} 2 \\ -1 \\ 0 \\ 1 \\ 1 \end{bmatrix}, \quad \text{and} \quad \mathbf{u}_3 = \begin{bmatrix} 1 \\ 0 \\ -6 \\ 0 \\ -2 \end{bmatrix}.$$

(a) Show that $\mathbf{v}$ is a linear combination of $\mathbf{u}_1$, $\mathbf{u}_2$, and $\mathbf{u}_3$.

(b) Show that $[\mathbf{v}]_{\mathcal{B}}$ is a linear combination of $[\mathbf{u}_1]_{\mathcal{B}}$, $[\mathbf{u}_2]_{\mathcal{B}}$, and $[\mathbf{u}_3]_{\mathcal{B}}$.

(c) Make a conjecture that generalizes the results of (a) and (b).

SOLUTIONS TO THE PRACTICE PROBLEMS

1. Based on Figure 4.10, we see that

$$\mathbf{u} = (-1)\mathbf{b}_1 + 2\mathbf{b}_2 \quad \text{and} \quad \mathbf{v} = 4\mathbf{b}_1 + 2\mathbf{b}_2.$$

It follows that

$$[\mathbf{u}]_{\mathcal{B}} = \begin{bmatrix} -1 \\ 2 \end{bmatrix} \quad \text{and} \quad [\mathbf{v}]_{\mathcal{B}} = \begin{bmatrix} 4 \\ 2 \end{bmatrix}.$$

2. Let B be the matrix whose columns are the vectors in $\mathcal{B}$. Since the reduced row echelon form of B is I_3, the columns of B are linearly independent. Thus $\mathcal{B}$ is a linearly independent set of 3 vectors from $\mathcal{R}^3$, and hence it is a basis for $\mathcal{R}^3$.

By Theorem 4.11, we have

$$\mathbf{u} = B[\mathbf{u}]_{\mathcal{B}} = \begin{bmatrix} 1 & 1 & 3 \\ 1 & 1 & 2 \\ 0 & 1 & 1 \end{bmatrix} \begin{bmatrix} 5 \\ -2 \\ -1 \end{bmatrix} = \begin{bmatrix} 0 \\ 1 \\ -3 \end{bmatrix}.$$

3. By Theorem 4.11, we also have

$$[\mathbf{v}]_{\mathcal{B}} = B^{-1}\mathbf{v} = \begin{bmatrix} 1 & 1 & 3 \\ 1 & 1 & 2 \\ 0 & 1 & 1 \end{bmatrix}^{-1} \begin{bmatrix} -2 \\ 1 \\ 3 \end{bmatrix} = \begin{bmatrix} 1 \\ 6 \\ -3 \end{bmatrix}.$$

4.5 MATRIX REPRESENTATIONS OF LINEAR OPERATORS

Our knowledge of coordinate systems is useful in the study of linear transformations from $\mathcal{R}^n$ to $\mathcal{R}^n$. A linear transformation where the domain and codomain equal $\mathcal{R}^n$ is called a **linear operator on** $\mathcal{R}^n$. Most of the linear transformations that we encounter from now on are linear operators.

Recall the reflection U of $\mathcal{R}^2$ about the x-axis defined by

$$U\left(\begin{bmatrix} x_1 \\ x_2 \end{bmatrix}\right) = \begin{bmatrix} x_1 \\ -x_2 \end{bmatrix},$$

given in Example 8 of Section 2.7. (See Figure 2.13 on page 175.) In that example, we computed the standard matrix of U from the standard basis for $\mathcal{R}^2$ by using that $U(\mathbf{e}_1) = \mathbf{e}_1$ and $U(\mathbf{e}_2) = -\mathbf{e}_2$. The resulting matrix is

$$\begin{bmatrix} 1 & 0 \\ 0 & -1 \end{bmatrix}.$$

The reflection of $\mathcal{R}^2$ about the x-axis is a special case of a *reflection* of $\mathcal{R}^2$ about a line.

In general, a **reflection** about a line $\mathcal{L}$ through the origin of $\mathcal{R}^2$ is a function $T\colon \mathcal{R}^2 \to \mathcal{R}^2$ defined as follows: Let $\mathbf{v}$ be a vector in $\mathcal{R}^2$ with endpoint P. (See Figure 4.12.) Construct a line from P perpendicular to $\mathcal{L}$, and let F denote the point of intersection of this perpendicular line with $\mathcal{L}$. Extend segment $\overline{PF}$ through F to a point P' such that $\overline{PF} = \overline{FP'}$. The vector with endpoint P' is $T(\mathbf{v})$. It can be shown that reflections are linear operators.

For such a reflection, we can select nonzero vectors $\mathbf{b}_1$ and $\mathbf{b}_2$ so that $\mathbf{b}_1$ is in the direction of $\mathcal{L}$ and $\mathbf{b}_2$ is in a direction perpendicular to $\mathcal{L}$. It then follows

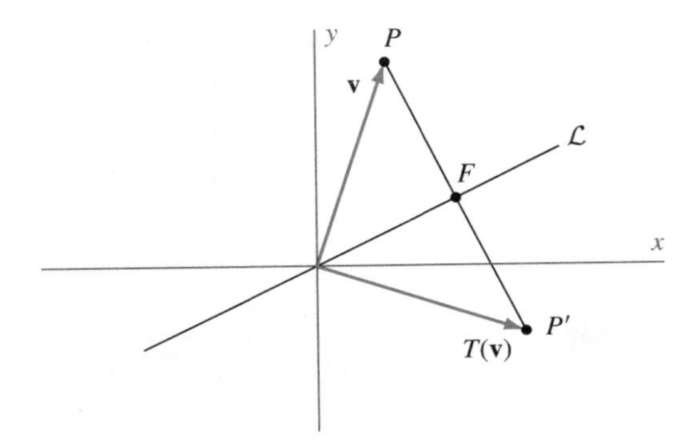

Figure 4.12 The reflection of the vector **v** about the line $\mathcal{L}$ through the origin of $\mathcal{R}^2$

that $T(\mathbf{b}_1) = \mathbf{b}_1$ and $T(\mathbf{b}_2) = -\mathbf{b}_2$. (See Figure 4.13.) Observe that $\mathcal{B} = \{\mathbf{b}_1, \mathbf{b}_2\}$ is a basis for $\mathcal{R}^2$ because $\mathcal{B}$ is a linearly independent subset of $\mathcal{R}^2$ consisting of 2 vectors. Moreover, we can describe the action of the reflection T on $\mathcal{B}$. In particular, since

$$T(\mathbf{b}_1) = 1\mathbf{b}_1 + 0\mathbf{b}_2 \qquad \text{and} \qquad T(\mathbf{b}_2) = 0\mathbf{b}_1 + (-1)\mathbf{b}_2,$$

the coordinate vectors of $T(\mathbf{b}_1)$ and $T(\mathbf{b}_2)$ relative to $\mathcal{B}$ are given by

$$[T(\mathbf{b}_1)]_{\mathcal{B}} = \begin{bmatrix} 1 \\ 0 \end{bmatrix} \qquad \text{and} \qquad [T(\mathbf{b}_2)]_{\mathcal{B}} = \begin{bmatrix} 0 \\ -1 \end{bmatrix}.$$

We can use these columns to form a matrix

$$[\,[T(\mathbf{b}_1)]_{\mathcal{B}}\ \ [T(\mathbf{b}_2)]_{\mathcal{B}}\,] = \begin{bmatrix} 1 & 0 \\ 0 & -1 \end{bmatrix}$$

that captures the behavior of T in relation to the basis $\mathcal{B}$. In the case of the previously described reflection U about the x-axis, $\mathcal{B}$ is the standard basis for $\mathcal{R}^2$, and the matrix $[\,[U(\mathbf{b}_1)]_{\mathcal{B}}\ \ [U(\mathbf{b}_2)]_{\mathcal{B}}\,]$ is the standard matrix of U.

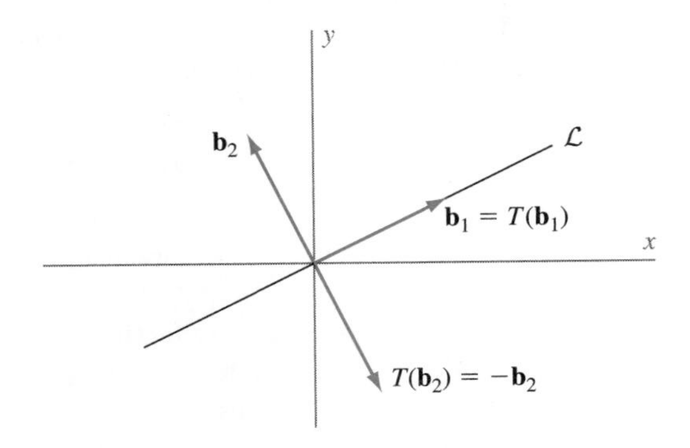

Figure 4.13 The images of the basis vectors $\mathbf{b}_1$ and $\mathbf{b}_2$ under T

A similar approach works for any linear operator on $\mathcal{R}^n$ in which the images of vectors in a particular basis are given. This motivates the following definition:

Definition Let T be a linear operator[2] on $\mathcal{R}^n$ and $\mathcal{B} = \{\mathbf{b}_1, \mathbf{b}_2, \ldots, \mathbf{b}_n\}$ be a basis for $\mathcal{R}^n$. The matrix

$$[\,[T(\mathbf{b}_1)]_{\mathcal{B}}\ [T(\mathbf{b}_2)]_{\mathcal{B}}\ \ldots\ [T(\mathbf{b}_n)]_{\mathcal{B}}\,]$$

is called the **matrix representation of T with respect to $\mathcal{B}$**, or the **$\mathcal{B}$-matrix of T**. It is denoted by $[T]_{\mathcal{B}}$.

Notice that the jth column of the $\mathcal{B}$-matrix of T is the $\mathcal{B}$-coordinate vector of $T(\mathbf{b}_j)$, the image of the jth vector in $\mathcal{B}$. Also, when $\mathcal{B} = \mathcal{E}$, the $\mathcal{B}$-matrix of T is

$$[T]_{\mathcal{E}} = [\,[T(\mathbf{b}_1)]_{\mathcal{E}}\ [T(\mathbf{b}_2)]_{\mathcal{E}}\ \ldots\ [T(\mathbf{b}_n)]_{\mathcal{E}}\,] = [T(\mathbf{e}_1)\ T(\mathbf{e}_2)\ \ldots\ T(\mathbf{e}_n)],$$

which is the standard matrix of T. So this definition extends the notion of a standard matrix to the context of an arbitrary basis for $\mathcal{R}^n$.

For the reflection T about line $\mathcal{L}$ and the basis $\mathcal{B} = \{\mathbf{b}_1, \mathbf{b}_2\}$ described earlier in this section, we have seen that the $\mathcal{B}$-matrix of T is

$$[T]_{\mathcal{B}} = [\,[T(\mathbf{b}_1)]_{\mathcal{B}}\ [T(\mathbf{b}_2)]_{\mathcal{B}}\,] = \begin{bmatrix} 1 & 0 \\ 0 & -1 \end{bmatrix}.$$

Calling $[T]_{\mathcal{B}}$ a *matrix representation* of T suggests that this matrix describes the action of T in some way. Recall that if A is the standard matrix of a linear operator T on $\mathcal{R}^n$, then $T(\mathbf{v}) = A\mathbf{v}$ for all vectors $\mathbf{v}$ in $\mathcal{R}^n$. Since $[\mathbf{v}]_{\mathcal{E}} = \mathbf{v}$ for every $\mathbf{v}$ in $\mathcal{R}^n$, we see that $[T]_{\mathcal{E}}[\mathbf{v}]_{\mathcal{E}} = A\mathbf{v} = T(\mathbf{v}) = [T(\mathbf{v})]_{\mathcal{E}}$. An analogous result is true for $[T]_{\mathcal{B}}$: If T is a linear operator on $\mathcal{R}^n$ and $\mathcal{B}$ is a basis for $\mathcal{R}^n$, then the $\mathcal{B}$-matrix of T is the unique $n \times n$ matrix such that

$$[T(\mathbf{v})]_{\mathcal{B}} = [T]_{\mathcal{B}}[\mathbf{v}]_{\mathcal{B}}$$

for all vectors $\mathbf{v}$ in $\mathcal{R}^n$. (See Exercise 100.)

Example 1 Let T be the linear operator on $\mathcal{R}^3$ defined by

$$T\left(\begin{bmatrix} x_1 \\ x_2 \\ x_3 \end{bmatrix}\right) = \begin{bmatrix} 3x_1 + x_3 \\ x_1 + x_2 \\ -x_1 - x_2 + 3x_3 \end{bmatrix}.$$

Calculate the matrix representation of T with respect to the basis $\mathcal{B} = \{\mathbf{b}_1, \mathbf{b}_2, \mathbf{b}_3\}$, where

$$\mathbf{b}_1 = \begin{bmatrix} 1 \\ 1 \\ 1 \end{bmatrix}, \quad \mathbf{b}_2 = \begin{bmatrix} 1 \\ 2 \\ 3 \end{bmatrix}, \quad \text{and} \quad \mathbf{b}_3 = \begin{bmatrix} 2 \\ 1 \\ 1 \end{bmatrix}.$$

---

[2] The definition of *matrix representation of T* generalizes to the case of any linear transformation $T\colon \mathcal{R}^n \to \mathcal{R}^m$. (See Exercises 101 and 102.)

Solution Applying T to each of the vectors in $\mathcal{B}$, we obtain

$$T(\mathbf{b}_1) = \begin{bmatrix} 4 \\ 2 \\ 1 \end{bmatrix}, \quad T(\mathbf{b}_2) = \begin{bmatrix} 6 \\ 3 \\ 6 \end{bmatrix}, \quad \text{and} \quad T(\mathbf{b}_3) = \begin{bmatrix} 7 \\ 3 \\ 0 \end{bmatrix}.$$

We must now compute the *coordinate vectors* of these images with respect to $\mathcal{B}$. Let $B = [\mathbf{b}_1 \ \mathbf{b}_2 \ \mathbf{b}_3]$. Then

$$[T(\mathbf{b}_1)]_{\mathcal{B}} = B^{-1}T(\mathbf{b}_1) = \begin{bmatrix} 3 \\ -1 \\ 1 \end{bmatrix}, \qquad [T(\mathbf{b}_2)]_{\mathcal{B}} = B^{-1}T(\mathbf{b}_2) = \begin{bmatrix} -9 \\ 3 \\ 6 \end{bmatrix},$$

and

$$[T(\mathbf{b}_3)]_{\mathcal{B}} = B^{-1}T(\mathbf{b}_3) = \begin{bmatrix} 8 \\ -3 \\ 1 \end{bmatrix},$$

from which it follows that the $\mathcal{B}$-matrix of T is

$$[T]_{\mathcal{B}} = \begin{bmatrix} 3 & -9 & 8 \\ -1 & 3 & -3 \\ 1 & 6 & 1 \end{bmatrix}.$$

As in Section 4.4, it is natural to ask how the matrix representation of T with respect to $\mathcal{B}$ is related to the standard matrix of T (which is the matrix representation of T with respect to the standard basis for $\mathcal{R}^n$). The answer is provided by the next theorem.

THEOREM 4.12

Let T be a linear operator on $\mathcal{R}^n$, $\mathcal{B}$ a basis for $\mathcal{R}^n$, B the matrix whose columns are the vectors in $\mathcal{B}$, and A the standard matrix of T. Then $[T]_{\mathcal{B}} = B^{-1}AB$, or equivalently, $A = B[T]_{\mathcal{B}}B^{-1}$.

PROOF Let $\mathcal{B} = \{\mathbf{b}_1, \mathbf{b}_2, \ldots, \mathbf{b}_n\}$ and $B = [\mathbf{b}_1 \ \mathbf{b}_2 \ \ldots \ \mathbf{b}_n]$. Recall that $T(\mathbf{u}) = A\mathbf{u}$ and $[\mathbf{v}]_{\mathcal{B}} = B^{-1}\mathbf{v}$ for all $\mathbf{u}$ and $\mathbf{v}$ in $\mathcal{R}^n$. Hence

$$\begin{aligned}
[T]_{\mathcal{B}} &= [\,[T(\mathbf{b}_1)]_{\mathcal{B}} \ [T(\mathbf{b}_2)]_{\mathcal{B}} \ \ldots \ [T(\mathbf{b}_n)]_{\mathcal{B}}\,] \\
&= [\,[A\mathbf{b}_1]_{\mathcal{B}} \ [A\mathbf{b}_2]_{\mathcal{B}} \ \ldots \ [A\mathbf{b}_n]_{\mathcal{B}}\,] \\
&= [B^{-1}(A\mathbf{b}_1) \ B^{-1}(A\mathbf{b}_2) \ \ldots \ B^{-1}(A\mathbf{b}_n)] \\
&= [(B^{-1}A)\mathbf{b}_1 \ (B^{-1}A)\mathbf{b}_2 \ \ldots \ (B^{-1}A)\mathbf{b}_n] \\
&= B^{-1}A[\mathbf{b}_1 \ \mathbf{b}_2 \ \ldots \ \mathbf{b}_n] \\
&= B^{-1}AB.
\end{aligned}$$

Thus $[T]_{\mathcal{B}} = B^{-1}AB$, which is equivalent to

$$B[T]_{\mathcal{B}}B^{-1} = B(B^{-1}AB)B^{-1} = A.$$

■

If two square matrices A and B are such that $B = P^{-1}AP$ for some invertible matrix P, then A is said to be **similar** to B. It is easily seen that A is similar to B if and only if B is similar to A. (See Exercise 84 of Section 2.4.) Hence we usually describe this situation by saying that *A and B are similar.*

Theorem 4.12 shows that the $\mathcal{B}$-matrix representation of a linear operator on $\mathcal{R}^n$ is similar to its standard matrix. Theorem 4.12 not only gives us the relationship between $[T]_{\mathcal{B}}$ and the standard matrix of T, it also provides a practical method for computing one of these matrices from the other. The following examples illustrate these computations:

Example 2 Calculate $[T]_{\mathcal{B}}$ by Theorem 4.12 if T and $\mathcal{B}$ are the linear operator and the basis given in Example 1:

$$T\left(\begin{bmatrix} x_1 \\ x_2 \\ x_3 \end{bmatrix}\right) = \begin{bmatrix} 3x_1 + x_3 \\ x_1 + x_2 \\ -x_1 - x_2 + 3x_3 \end{bmatrix} \quad \text{and} \quad \mathcal{B} = \left\{\begin{bmatrix} 1 \\ 1 \\ 1 \end{bmatrix}, \begin{bmatrix} 1 \\ 2 \\ 3 \end{bmatrix}, \begin{bmatrix} 2 \\ 1 \\ 1 \end{bmatrix}\right\}$$

Solution The standard matrix of T is

$$A = [T(\mathbf{e}_1)\ \ T(\mathbf{e}_2)\ \ T(\mathbf{e}_3)] = \begin{bmatrix} 3 & 0 & 1 \\ 1 & 1 & 0 \\ -1 & -1 & 3 \end{bmatrix}.$$

Taking B to be the matrix whose columns are the vectors in $\mathcal{B}$, we have

$$[T]_{\mathcal{B}} = B^{-1}AB = \begin{bmatrix} 3 & -9 & 8 \\ -1 & 3 & -3 \\ 1 & 6 & 1 \end{bmatrix}.$$

Note that our answer agrees with that in Example 1.

Practice Problem 1 ▶ Find the $\mathcal{B}$-matrix representation of the linear operator T on $\mathcal{R}^3$, where

$$T\left(\begin{bmatrix} x_1 \\ x_2 \\ x_3 \end{bmatrix}\right) = \begin{bmatrix} -x_1 + 2x_3 \\ x_1 + x_2 \\ -x_2 + x_3 \end{bmatrix} \quad \text{and} \quad \mathcal{B} = \left\{\begin{bmatrix} 1 \\ 1 \\ 0 \end{bmatrix}, \begin{bmatrix} 1 \\ 1 \\ 1 \end{bmatrix}, \begin{bmatrix} 3 \\ 2 \\ 1 \end{bmatrix}\right\}.$$ ◀

Example 3 Let T be a linear operator on $\mathcal{R}^3$ such that

$$T\left(\begin{bmatrix} 1 \\ 1 \\ 0 \end{bmatrix}\right) = \begin{bmatrix} 1 \\ 2 \\ -1 \end{bmatrix}, \quad T\left(\begin{bmatrix} 1 \\ 0 \\ 1 \end{bmatrix}\right) = \begin{bmatrix} 3 \\ -1 \\ 1 \end{bmatrix}, \quad \text{and} \quad T\left(\begin{bmatrix} 0 \\ 1 \\ 1 \end{bmatrix}\right) = \begin{bmatrix} 2 \\ 0 \\ 1 \end{bmatrix}.$$

Find the standard matrix of T.

Solution Let A denote the standard matrix of T,

$$\mathbf{b}_1 = \begin{bmatrix} 1 \\ 1 \\ 0 \end{bmatrix}, \quad \mathbf{b}_2 = \begin{bmatrix} 1 \\ 0 \\ 1 \end{bmatrix}, \quad \text{and} \quad \mathbf{b}_3 = \begin{bmatrix} 0 \\ 1 \\ 1 \end{bmatrix},$$

and

$$\mathbf{c}_1 = \begin{bmatrix} 1 \\ 2 \\ -1 \end{bmatrix}, \quad \mathbf{c}_2 = \begin{bmatrix} 3 \\ -1 \\ 1 \end{bmatrix}, \quad \text{and} \quad \mathbf{c}_3 = \begin{bmatrix} 2 \\ 0 \\ 1 \end{bmatrix}.$$

Observe that $\mathcal{B} = \{\mathbf{b}_1, \mathbf{b}_2, \mathbf{b}_3\}$ is linearly independent and thus is a basis for $\mathcal{R}^3$. Hence we are given the images of the vectors in a basis for $\mathcal{R}^3$. Let $B = [\mathbf{b}_1 \ \mathbf{b}_2 \ \mathbf{b}_3]$ and $C = [\mathbf{c}_1 \ \mathbf{c}_2 \ \mathbf{c}_3]$. Then $A\mathbf{b}_1 = T(\mathbf{b}_1) = \mathbf{c}_1$, $A\mathbf{b}_2 = T(\mathbf{b}_2) = \mathbf{c}_2$, and $A\mathbf{b}_3 = T(\mathbf{b}_3) = \mathbf{c}_3$. Thus

$$AB = A[\mathbf{b}_1 \ \mathbf{b}_2 \ \mathbf{b}_3] = [A\mathbf{b}_1 \ A\mathbf{b}_2 \ A\mathbf{b}_3] = [\mathbf{c}_1 \ \mathbf{c}_2 \ \mathbf{c}_3] = C.$$

Because the columns of B are linearly independent, B is invertible by the Invertible Matrix Theorem. Thus

$$A = A(BB^{-1}) = (AB)B^{-1} = CB^{-1} = \begin{bmatrix} 1 & 0 & 2 \\ .5 & 1.5 & -1.5 \\ -.5 & -.5 & 1.5 \end{bmatrix}.$$

Therefore the standard matrix of T, and hence T itself, is uniquely determined by the images of the vectors in a basis for $\mathcal{R}^3$.

Example 3 suggests that a linear operator is uniquely determined by its action on a basis because we are able to determine the standard matrix of the operator in this example solely from this information. This is indeed the case, as Exercise 98 shows.

To conclude this section, we apply Theorem 4.12 to find an explicit formula for the reflection T of $\mathcal{R}^2$ about the line $\mathcal{L}$, with equation $y = \frac{1}{2}x$. In the earlier discussion of this problem, we have seen that nonzero vectors $\mathbf{b}_1$ and $\mathbf{b}_2$ in $\mathcal{R}^2$ must be selected so that $\mathbf{b}_1$ lies on $\mathcal{L}$ and $\mathbf{b}_2$ is perpendicular to $\mathcal{L}$. (See Figure 4.13.) One selection that works is

$$\mathbf{b}_1 = \begin{bmatrix} 2 \\ 1 \end{bmatrix} \quad \text{and} \quad \mathbf{b}_2 = \begin{bmatrix} -1 \\ 2 \end{bmatrix},$$

because $\mathbf{b}_1$ lies on $\mathcal{L}$, which has slope $\frac{1}{2}$, and $\mathbf{b}_2$ lies on the line perpendicular to $\mathcal{L}$, which has slope -2. Let $\mathcal{B} = \{\mathbf{b}_1, \mathbf{b}_2\}$, $B = [\mathbf{b}_1 \ \mathbf{b}_2]$, and A be the standard matrix of T. Recall that

$$[T]_{\mathcal{B}} = [\,[T(\mathbf{b}_1)]_{\mathcal{B}} \ T(\mathbf{b}_2)]_{\mathcal{B}}\,] = \begin{bmatrix} 1 & 0 \\ 0 & -1 \end{bmatrix}.$$

Then, by Theorem 4.12,

$$A = B[T]_{\mathcal{B}}B^{-1} = \begin{bmatrix} .6 & .8 \\ .8 & -.6 \end{bmatrix}.$$

It follows that the reflection of $\mathcal{R}^2$ about the line with equation $y = \frac{1}{2}x$ is given by

$$T\left(\begin{bmatrix} x_1 \\ x_2 \end{bmatrix}\right) = A\begin{bmatrix} x_1 \\ x_2 \end{bmatrix} = \begin{bmatrix} .6 & .8 \\ .8 & -.6 \end{bmatrix}\begin{bmatrix} x_1 \\ x_2 \end{bmatrix} = \begin{bmatrix} .6x_1 + .8x_2 \\ .8x_1 - .6x_2 \end{bmatrix}.$$

Practice Problem 2 ▶ Let $\mathcal{B}$ be the basis in Practice Problem 1, and let U be the linear operator on $\mathcal{R}^3$ such that

$$[U]_{\mathcal{B}} = \begin{bmatrix} 3 & 0 & 0 \\ 0 & 2 & 0 \\ 0 & 0 & 1 \end{bmatrix}.$$

Determine an explicit formula for $U(\mathbf{x})$. ◀

EXERCISES

In Exercises 1–10, determine $[T]_{\mathcal{B}}$ for each linear operator T and basis $\mathcal{B}$.

1. $T\left(\begin{bmatrix} x_1 \\ x_2 \end{bmatrix}\right) = \begin{bmatrix} 2x_1 + x_2 \\ x_1 - x_2 \end{bmatrix}$ and $\mathcal{B} = \left\{\begin{bmatrix} 2 \\ 1 \end{bmatrix}, \begin{bmatrix} 1 \\ 0 \end{bmatrix}\right\}$

2. $T\left(\begin{bmatrix} x_1 \\ x_2 \end{bmatrix}\right) = \begin{bmatrix} x_1 - x_2 \\ x_2 \end{bmatrix}$ and $\mathcal{B} = \left\{\begin{bmatrix} 2 \\ 3 \end{bmatrix}, \begin{bmatrix} 1 \\ 1 \end{bmatrix}\right\}$

3. $T\left(\begin{bmatrix} x_1 \\ x_2 \end{bmatrix}\right) = \begin{bmatrix} x_1 + 2x_2 \\ x_1 + x_2 \end{bmatrix}$ and $\mathcal{B} = \left\{\begin{bmatrix} 1 \\ 1 \end{bmatrix}, \begin{bmatrix} 2 \\ 1 \end{bmatrix}\right\}$

4. $T\left(\begin{bmatrix} x_1 \\ x_2 \end{bmatrix}\right) = \begin{bmatrix} -2x_1 + x_2 \\ x_1 + 3x_2 \end{bmatrix}$ and $\mathcal{B} = \left\{\begin{bmatrix} 1 \\ 3 \end{bmatrix}, \begin{bmatrix} 2 \\ 5 \end{bmatrix}\right\}$

5. $T\left(\begin{bmatrix} x_1 \\ x_2 \\ x_3 \end{bmatrix}\right) = \begin{bmatrix} x_1 + x_2 \\ x_2 - 2x_3 \\ 2x_1 - x_2 + 3x_3 \end{bmatrix}$ and
$\mathcal{B} = \left\{\begin{bmatrix} 1 \\ 1 \\ 1 \end{bmatrix}, \begin{bmatrix} 2 \\ 3 \\ 2 \end{bmatrix}, \begin{bmatrix} 1 \\ 2 \\ 2 \end{bmatrix}\right\}$

6. $T\left(\begin{bmatrix} x_1 \\ x_2 \\ x_3 \end{bmatrix}\right) = \begin{bmatrix} x_1 + x_3 \\ x_2 - x_3 \\ 2x_1 - x_2 \end{bmatrix}$ and
$\mathcal{B} = \left\{\begin{bmatrix} 0 \\ -1 \\ 1 \end{bmatrix}, \begin{bmatrix} 1 \\ 0 \\ -1 \end{bmatrix}, \begin{bmatrix} 1 \\ 1 \\ -1 \end{bmatrix}\right\}$

7. $T\left(\begin{bmatrix} x_1 \\ x_2 \\ x_3 \end{bmatrix}\right) = \begin{bmatrix} 4x_2 \\ x_1 + 2x_3 \\ -2x_2 + 3x_3 \end{bmatrix}$ and
$\mathcal{B} = \left\{\begin{bmatrix} 1 \\ 0 \\ 1 \end{bmatrix}, \begin{bmatrix} 1 \\ -2 \\ 0 \end{bmatrix}, \begin{bmatrix} -1 \\ 3 \\ 1 \end{bmatrix}\right\}$

8. $T\left(\begin{bmatrix} x_1 \\ x_2 \\ x_3 \end{bmatrix}\right) = \begin{bmatrix} x_1 - 2x_2 + 4x_3 \\ 3x_1 \\ -3x_2 + 2x_3 \end{bmatrix}$ and
$\mathcal{B} = \left\{\begin{bmatrix} 1 \\ -2 \\ 1 \end{bmatrix}, \begin{bmatrix} 0 \\ -1 \\ 1 \end{bmatrix}, \begin{bmatrix} 1 \\ -5 \\ 3 \end{bmatrix}\right\}$

9. $T\left(\begin{bmatrix} x_1 \\ x_2 \\ x_3 \\ x_4 \end{bmatrix}\right) = \begin{bmatrix} x_1 + x_2 \\ x_2 - x_3 \\ x_1 + 2x_4 \\ x_2 - x_3 + 3x_4 \end{bmatrix}$ and
$\mathcal{B} = \left\{\begin{bmatrix} 1 \\ -1 \\ 2 \\ 3 \end{bmatrix}, \begin{bmatrix} 1 \\ -2 \\ 1 \\ 4 \end{bmatrix}, \begin{bmatrix} 1 \\ -2 \\ 0 \\ 3 \end{bmatrix}, \begin{bmatrix} 0 \\ 1 \\ 1 \\ -2 \end{bmatrix}\right\}$

10. $T\left(\begin{bmatrix} x_1 \\ x_2 \\ x_3 \\ x_4 \end{bmatrix}\right) = \begin{bmatrix} x_1 - x_2 + x_3 + 2x_4 \\ 2x_1 - 3x_4 \\ x_1 + x_2 + x_3 \\ -3x_3 + x_4 \end{bmatrix}$ and
$\mathcal{B} = \left\{\begin{bmatrix} 1 \\ 1 \\ 1 \\ 2 \end{bmatrix}, \begin{bmatrix} 2 \\ 3 \\ 3 \\ 3 \end{bmatrix}, \begin{bmatrix} 1 \\ 3 \\ 4 \\ 1 \end{bmatrix}, \begin{bmatrix} 4 \\ 5 \\ 8 \\ 8 \end{bmatrix}\right\}$

In Exercises 11–18, determine the standard matrix of the linear operator T using the given basis $\mathcal{B}$ and the matrix representation of T with respect to $\mathcal{B}$.

11. $[T]_{\mathcal{B}} = \begin{bmatrix} 1 & 4 \\ -3 & 5 \end{bmatrix}$ and $\mathcal{B} = \left\{\begin{bmatrix} -1 \\ 0 \end{bmatrix}, \begin{bmatrix} 3 \\ 1 \end{bmatrix}\right\}$

12. $[T]_{\mathcal{B}} = \begin{bmatrix} 2 & 0 \\ 1 & -1 \end{bmatrix}$ and $\mathcal{B} = \left\{\begin{bmatrix} 1 \\ -2 \end{bmatrix}, \begin{bmatrix} -2 \\ 3 \end{bmatrix}\right\}$

13. $[T]_{\mathcal{B}} = \begin{bmatrix} -2 & -1 \\ 1 & 3 \end{bmatrix}$ and $\mathcal{B} = \left\{\begin{bmatrix} 1 \\ -2 \end{bmatrix}, \begin{bmatrix} -3 \\ 5 \end{bmatrix}\right\}$

14. $[T]_{\mathcal{B}} = \begin{bmatrix} 3 & 1 \\ -2 & 4 \end{bmatrix}$ and $\mathcal{B} = \left\{\begin{bmatrix} 1 \\ 2 \end{bmatrix}, \begin{bmatrix} 1 \\ 1 \end{bmatrix}\right\}$

15. $[T]_{\mathcal{B}} = \begin{bmatrix} 1 & 0 & -3 \\ -2 & 1 & 2 \\ -1 & 1 & 1 \end{bmatrix}$ and
$\mathcal{B} = \left\{\begin{bmatrix} -2 \\ -1 \\ 1 \end{bmatrix}, \begin{bmatrix} -1 \\ -2 \\ 1 \end{bmatrix}, \begin{bmatrix} -1 \\ -1 \\ 1 \end{bmatrix}\right\}$

16. $[T]_{\mathcal{B}} = \begin{bmatrix} -1 & 1 & -2 \\ 0 & 2 & 1 \\ 1 & 2 & 0 \end{bmatrix}$ and
$\mathcal{B} = \left\{\begin{bmatrix} 1 \\ 0 \\ 1 \end{bmatrix}, \begin{bmatrix} 1 \\ -2 \\ 0 \end{bmatrix}, \begin{bmatrix} -1 \\ 3 \\ 1 \end{bmatrix}\right\}$

17. $[T]_{\mathcal{B}} = \begin{bmatrix} 1 & 0 & -1 \\ 0 & 2 & 1 \\ -1 & 1 & 0 \end{bmatrix}$ and

$$\mathcal{B}=\left\{\begin{bmatrix}1\\0\\1\end{bmatrix},\begin{bmatrix}-1\\1\\0\end{bmatrix},\begin{bmatrix}2\\0\\1\end{bmatrix}\right\}$$

18. $[T]_{\mathcal{B}}=\begin{bmatrix}-1&2&1\\1&0&-2\\1&1&-1\end{bmatrix}$ and

$$\mathcal{B}=\left\{\begin{bmatrix}-1\\1\\2\end{bmatrix},\begin{bmatrix}-2\\1\\1\end{bmatrix},\begin{bmatrix}0\\1\\2\end{bmatrix}\right\}$$

T&F *In Exercises 19–38, determine whether the statements are true or false.*

19. A linear transformation $T\colon \mathcal{R}^n \to \mathcal{R}^m$ that is one-to-one is called a linear operator on $\mathcal{R}^n$.
20. The matrix representation of a linear operator on $\mathcal{R}^n$ with respect to a basis for $\mathcal{R}^n$ is an $n\times n$ matrix.
21. If T is a linear operator on $\mathcal{R}^n$ and $\mathcal{B}=\{\mathbf{b}_1,\mathbf{b}_2,\dots,\mathbf{b}_n\}$ is a basis for $\mathcal{R}^n$, then column j of $[T]_{\mathcal{B}}$ is the $\mathcal{B}$-coordinate vector of $T(\mathbf{b}_j)$.
22. If T is a linear operator on $\mathcal{R}^n$ and $\mathcal{B}=\{\mathbf{b}_1,\mathbf{b}_2,\dots,\mathbf{b}_n\}$ is a basis for $\mathcal{R}^n$, then the matrix representation of T with respect to $\mathcal{B}$ is
$$[T(\mathbf{b}_1)\ T(\mathbf{b}_2)\ \cdots\ T(\mathbf{b}_n)].$$
23. If $\mathcal{E}$ is the standard basis for $\mathcal{R}^n$, then $[T]_{\mathcal{E}}$ is the standard matrix of T.
24. If T is a linear operator on $\mathcal{R}^n$, $\mathcal{B}$ is a basis for $\mathcal{R}^n$, B is the matrix whose columns are the vectors in $\mathcal{B}$, and A is the standard matrix of T, then $[T]_{\mathcal{B}}=B^{-1}A$.
25. If T is a linear operator on $\mathcal{R}^n$, $\mathcal{B}$ is a basis for $\mathcal{R}^n$, B is the matrix whose columns are the vectors in $\mathcal{B}$, and A is the standard matrix of T, then $[T]_{\mathcal{B}}=BAB^{-1}$.
26. If $\mathcal{B}$ is a basis for $\mathcal{R}^n$ and T is the identity operator on $\mathcal{R}^n$, then $[T]_{\mathcal{B}}=I_n$.
27. If T is a reflection of $\mathcal{R}^2$ about a line L, then $T(\mathbf{v})=-\mathbf{v}$ for every vector $\mathbf{v}$ on L.
28. If T is a reflection of $\mathcal{R}^2$ about a line L, then $T(\mathbf{v})=\mathbf{0}$ for every vector $\mathbf{v}$ on L.
29. If T is a reflection of $\mathcal{R}^2$ about a line, then there exists a basis $\mathcal{B}$ for $\mathcal{R}^n$ such that $[T]_{\mathcal{B}}=\begin{bmatrix}1&0\\0&0\end{bmatrix}$.
30. If T is a reflection of $\mathcal{R}^2$ about a line L, then there exists a basis $\mathcal{B}$ for $\mathcal{R}^n$ such that $[T]_{\mathcal{B}}=\begin{bmatrix}1&0\\0&-1\end{bmatrix}$.
31. If T, $\mathcal{B}$, and L are as in Exercise 30, then $\mathcal{B}$ consists of two vectors on the line L.
32. An $n\times n$ matrix A is said to be similar to an $n\times n$ matrix B if $B=P^TAP$.
33. An $n\times n$ matrix A is similar to an $n\times n$ matrix B if B is similar to A.
34. If T is a linear operator on $\mathcal{R}^n$ and $\mathcal{B}$ is a basis for $\mathcal{R}^n$, then the $\mathcal{B}$-matrix of T is similar to the standard matrix of T.
35. The only matrix that is similar to the $n\times n$ zero matrix is the $n\times n$ zero matrix.
36. The only matrix that is similar to I_n is I_n.
37. If T is a linear operator on $\mathcal{R}^n$ and $\mathcal{B}$ is a basis for $\mathcal{R}^n$, then $[T]_{\mathcal{B}}[\mathbf{v}]_{\mathcal{B}}=T(\mathbf{v})$.
38. If T is a linear operator on $\mathcal{R}^n$ and $\mathcal{B}$ is a basis for $\mathcal{R}^n$, then $[T]_{\mathcal{B}}$ is the unique $n\times n$ matrix such that $[T]_{\mathcal{B}}[\mathbf{v}]_{\mathcal{B}}=[T(\mathbf{v})]_{\mathcal{B}}$ for all $\mathbf{v}$ in $\mathcal{R}^n$.

---

39. Let $\mathcal{B}=\{\mathbf{b}_1,\mathbf{b}_2\}$ be a basis for $\mathcal{R}^2$ and T be a linear operator on $\mathcal{R}^2$ such that $T(\mathbf{b}_1)=\mathbf{b}_1+4\mathbf{b}_2$ and $T(\mathbf{b}_2)=-3\mathbf{b}_1$. Determine $[T]_{\mathcal{B}}$.
40. Let $\mathcal{B}=\{\mathbf{b}_1,\mathbf{b}_2\}$ be a basis for $\mathcal{R}^2$ and T be a linear operator on $\mathcal{R}^2$ such that $T(\mathbf{b}_1)=2\mathbf{b}_1-5\mathbf{b}_2$ and $T(\mathbf{b}_2)=-\mathbf{b}_1+3\mathbf{b}_2$. Determine $[T]_{\mathcal{B}}$.
41. Let $\mathcal{B}=\{\mathbf{b}_1,\mathbf{b}_2\}$ be a basis for $\mathcal{R}^2$ and T be a linear operator on $\mathcal{R}^2$ such that $T(\mathbf{b}_1)=3\mathbf{b}_1-5\mathbf{b}_2$ and $T(\mathbf{b}_2)=2\mathbf{b}_1+4\mathbf{b}_2$. Determine $[T]_{\mathcal{B}}$.
42. Let $\mathcal{B}=\{\mathbf{b}_1,\mathbf{b}_2,\mathbf{b}_3\}$ be a basis for $\mathcal{R}^3$ and T be a linear operator on $\mathcal{R}^3$ such that $T(\mathbf{b}_1)=\mathbf{b}_1-2\mathbf{b}_2+3\mathbf{b}_3$, $T(\mathbf{b}_2)=6\mathbf{b}_2-\mathbf{b}_3$, and $T(\mathbf{b}_3)=5\mathbf{b}_1+2\mathbf{b}_2-4\mathbf{b}_3$. Determine $[T]_{\mathcal{B}}$.
43. Let $\mathcal{B}=\{\mathbf{b}_1,\mathbf{b}_2,\mathbf{b}_3\}$ be a basis for $\mathcal{R}^3$ and T be a linear operator on $\mathcal{R}^3$ such that $T(\mathbf{b}_1)=-5\mathbf{b}_2+4\mathbf{b}_3$, $T(\mathbf{b}_2)=2\mathbf{b}_1-7\mathbf{b}_3$, and $T(\mathbf{b}_3)=3\mathbf{b}_1+\mathbf{b}_3$. Determine $[T]_{\mathcal{B}}$.
44. Let $\mathcal{B}=\{\mathbf{b}_1,\mathbf{b}_2,\mathbf{b}_3\}$ be a basis for $\mathcal{R}^3$ and T be a linear operator on $\mathcal{R}^3$ such that $T(\mathbf{b}_1)=2\mathbf{b}_1+5\mathbf{b}_2$, $T(\mathbf{b}_2)=-\mathbf{b}_1+3\mathbf{b}_2$, and $T(\mathbf{b}_3)=\mathbf{b}_2-2\mathbf{b}_3$. Determine $[T]_{\mathcal{B}}$.
45. Let $\mathcal{B}=\{\mathbf{b}_1,\mathbf{b}_2,\mathbf{b}_3,\mathbf{b}_4\}$ a basis for $\mathcal{R}^4$ and T a linear operator on $\mathcal{R}^4$ such that $T(\mathbf{b}_1)=\mathbf{b}_1-\mathbf{b}_2+\mathbf{b}_3-\mathbf{b}_4$, $T(\mathbf{b}_2)=2\mathbf{b}_2-\mathbf{b}_4$, $T(\mathbf{b}_3)=-3\mathbf{b}_1+5\mathbf{b}_3$, and $T(\mathbf{b}_4)=4\mathbf{b}_2-\mathbf{b}_3+3\mathbf{b}_4$. Determine $[T]_{\mathcal{B}}$.
46. Let $\mathcal{B}=\{\mathbf{b}_1,\mathbf{b}_2,\mathbf{b}_3,\mathbf{b}_4\}$ be a basis for $\mathcal{R}^4$ and T be a linear operator on $\mathcal{R}^4$ such that $T(\mathbf{b}_1)=-\mathbf{b}_2+\mathbf{b}_4$, $T(\mathbf{b}_2)=\mathbf{b}_1-2\mathbf{b}_3$, $T(\mathbf{b}_3)=2\mathbf{b}_1-3\mathbf{b}_4$, and $T(\mathbf{b}_4)=-\mathbf{b}_2+2\mathbf{b}_3+\mathbf{b}_4$. Determine $[T]_{\mathcal{B}}$.

In Exercises 47–54, determine (a) $[T]_{\mathcal{B}}$, (b) the standard matrix of T, and (c) an explicit formula for $T(\mathbf{x})$ from the given information.

47. $\mathcal{B}=\left\{\begin{bmatrix}1\\1\end{bmatrix},\begin{bmatrix}1\\2\end{bmatrix}\right\}$, $T\left(\begin{bmatrix}1\\1\end{bmatrix}\right)=\begin{bmatrix}1\\2\end{bmatrix}$, $T\left(\begin{bmatrix}1\\2\end{bmatrix}\right)=3\begin{bmatrix}1\\1\end{bmatrix}$
48. $\mathcal{B}=\left\{\begin{bmatrix}1\\3\end{bmatrix},\begin{bmatrix}1\\0\end{bmatrix}\right\}$, $T\left(\begin{bmatrix}1\\3\end{bmatrix}\right)=\begin{bmatrix}1\\3\end{bmatrix}-2\begin{bmatrix}1\\0\end{bmatrix}$, $T\left(\begin{bmatrix}1\\0\end{bmatrix}\right)=2\begin{bmatrix}1\\3\end{bmatrix}-\begin{bmatrix}1\\0\end{bmatrix}$
49. $\mathcal{B}=\left\{\begin{bmatrix}-1\\2\end{bmatrix},\begin{bmatrix}1\\-1\end{bmatrix}\right\}$, $T\left(\begin{bmatrix}-1\\2\end{bmatrix}\right)=3\begin{bmatrix}-1\\2\end{bmatrix}-\begin{bmatrix}1\\-1\end{bmatrix}$, $T\left(\begin{bmatrix}1\\-1\end{bmatrix}\right)=2\begin{bmatrix}-1\\2\end{bmatrix}$
50. $\mathcal{B}=\left\{\begin{bmatrix}1\\2\end{bmatrix},\begin{bmatrix}1\\3\end{bmatrix}\right\}$, $T\left(\begin{bmatrix}1\\2\end{bmatrix}\right)=-\begin{bmatrix}1\\2\end{bmatrix}+4\begin{bmatrix}1\\3\end{bmatrix}$, $T\left(\begin{bmatrix}1\\3\end{bmatrix}\right)=3\begin{bmatrix}1\\2\end{bmatrix}-2\begin{bmatrix}1\\3\end{bmatrix}$

51. $\mathcal{B} = \left\{ \begin{bmatrix} 1 \\ 0 \\ 1 \end{bmatrix}, \begin{bmatrix} 0 \\ 1 \\ 0 \end{bmatrix}, \begin{bmatrix} 1 \\ 1 \\ 0 \end{bmatrix} \right\}$, $T\left(\begin{bmatrix} 1 \\ 0 \\ 1 \end{bmatrix}\right) = -\begin{bmatrix} 0 \\ 1 \\ 0 \end{bmatrix}$,

$T\left(\begin{bmatrix} 0 \\ 1 \\ 0 \end{bmatrix}\right) = 2\begin{bmatrix} 1 \\ 1 \\ 0 \end{bmatrix}$, $T\left(\begin{bmatrix} 1 \\ 1 \\ 0 \end{bmatrix}\right) = \begin{bmatrix} 1 \\ 0 \\ 1 \end{bmatrix} + 2\begin{bmatrix} 0 \\ 1 \\ 0 \end{bmatrix}$

52. $\mathcal{B} = \left\{ \begin{bmatrix} 1 \\ 1 \\ -1 \end{bmatrix}, \begin{bmatrix} 0 \\ 1 \\ 1 \end{bmatrix}, \begin{bmatrix} 1 \\ 2 \\ 3 \end{bmatrix} \right\}$,

$T\left(\begin{bmatrix} 1 \\ 1 \\ -1 \end{bmatrix}\right) = \begin{bmatrix} 0 \\ 1 \\ 1 \end{bmatrix} + 2\begin{bmatrix} 1 \\ 2 \\ 3 \end{bmatrix}$,

$T\left(\begin{bmatrix} 0 \\ 1 \\ 1 \end{bmatrix}\right) = 4\begin{bmatrix} 1 \\ 1 \\ -1 \end{bmatrix} - \begin{bmatrix} 1 \\ 2 \\ 3 \end{bmatrix}$,

$T\left(\begin{bmatrix} 1 \\ 2 \\ 3 \end{bmatrix}\right) = -\begin{bmatrix} 1 \\ 1 \\ -1 \end{bmatrix} + 3\begin{bmatrix} 0 \\ 1 \\ 1 \end{bmatrix} + 2\begin{bmatrix} 1 \\ 2 \\ 3 \end{bmatrix}$

53. $\mathcal{B} = \left\{ \begin{bmatrix} 1 \\ 0 \\ 1 \end{bmatrix}, \begin{bmatrix} -1 \\ 1 \\ 0 \end{bmatrix}, \begin{bmatrix} -2 \\ 0 \\ -1 \end{bmatrix} \right\}$,

$T\left(\begin{bmatrix} 1 \\ 0 \\ 1 \end{bmatrix}\right) = 3\begin{bmatrix} -1 \\ 1 \\ 0 \end{bmatrix} - 2\begin{bmatrix} -2 \\ 0 \\ -1 \end{bmatrix}$,

$T\left(\begin{bmatrix} -1 \\ 1 \\ 0 \end{bmatrix}\right) = -1\begin{bmatrix} 1 \\ 0 \\ 1 \end{bmatrix} + 4\begin{bmatrix} -2 \\ 0 \\ -1 \end{bmatrix}$,

$T\left(\begin{bmatrix} -2 \\ 0 \\ -1 \end{bmatrix}\right) = 2\begin{bmatrix} 1 \\ 0 \\ 1 \end{bmatrix} + 5\begin{bmatrix} -1 \\ 1 \\ 0 \end{bmatrix}$

54. $\mathcal{B} = \left\{ \begin{bmatrix} 0 \\ 1 \\ 1 \end{bmatrix}, \begin{bmatrix} 1 \\ 0 \\ 2 \end{bmatrix}, \begin{bmatrix} 1 \\ -1 \\ 2 \end{bmatrix} \right\}$,

$T\left(\begin{bmatrix} 0 \\ 1 \\ 1 \end{bmatrix}\right) = 3\begin{bmatrix} 0 \\ 1 \\ 1 \end{bmatrix} - 2\begin{bmatrix} 1 \\ 0 \\ 2 \end{bmatrix} + \begin{bmatrix} 1 \\ -1 \\ 2 \end{bmatrix}$,

$T\left(\begin{bmatrix} 1 \\ 0 \\ 2 \end{bmatrix}\right) = -1\begin{bmatrix} 0 \\ 1 \\ 1 \end{bmatrix} + 3\begin{bmatrix} 1 \\ 0 \\ 2 \end{bmatrix}$,

$T\left(\begin{bmatrix} 1 \\ -1 \\ 2 \end{bmatrix}\right) = 5\begin{bmatrix} 0 \\ 1 \\ 1 \end{bmatrix} - 2\begin{bmatrix} 1 \\ 0 \\ 2 \end{bmatrix} - \begin{bmatrix} 1 \\ -1 \\ 2 \end{bmatrix}$

55. Find $T(3\mathbf{b}_1 - 2\mathbf{b}_2)$ for the operator T and the basis $\mathcal{B}$ in Exercise 39.

56. Find $T(-\mathbf{b}_1 + 4\mathbf{b}_2)$ for the operator T and the basis $\mathcal{B}$ in Exercise 40.

57. Find $T(\mathbf{b}_1 - 3\mathbf{b}_2)$ for the operator T and the basis $\mathcal{B}$ in Exercise 41.

58. Find $T(\mathbf{b}_2 - 2\mathbf{b}_3)$ for the operator T and the basis $\mathcal{B}$ in Exercise 42.

59. Find $T(2\mathbf{b}_1 - \mathbf{b}_2)$ for the operator T and the basis $\mathcal{B}$ in Exercise 43.

60. Find $T(\mathbf{b}_1 + 3\mathbf{b}_2 - 2\mathbf{b}_3)$ for the operator T and the basis $\mathcal{B}$ in Exercise 44.

61. Find $T(-\mathbf{b}_1 + 2\mathbf{b}_2 - 3\mathbf{b}_3)$ for the operator T and the basis $\mathcal{B}$ in Exercise 45.

62. Find $T(\mathbf{b}_1 - \mathbf{b}_3 + 2\mathbf{b}_4)$ for the operator T and the basis $\mathcal{B}$ in Exercise 46.

63. Let I be the identity operator on $\mathcal{R}^n$, and let $\mathcal{B}$ be any basis for $\mathcal{R}^n$. Determine the matrix representation of I with respect to $\mathcal{B}$.

64. Let T be the zero operator on $\mathcal{R}^n$, and let $\mathcal{B}$ be any basis for $\mathcal{R}^n$. Determine the matrix representation of T with respect to $\mathcal{B}$.

In Exercises 65–68, find an explicit description of the reflection T of $\mathcal{R}^2$ about the line with each equation.

65. $y = \frac{1}{3}x$ 66. $y = 2x$ 67. $y = -2x$ 68. $y = mx$

The **orthogonal projection** *of $\mathcal{R}^2$ on line $\mathcal{L}$ through the origin is a function $U: \mathcal{R}^2 \to \mathcal{R}^2$ defined in the following manner: Let $\mathbf{v}$ be a vector in $\mathcal{R}^2$ with endpoint P. Construct a line from P perpendicular to $\mathcal{L}$, and let F denote the point of intersection of this perpendicular line with $\mathcal{L}$. The vector with endpoint F is $U(\mathbf{v})$. (See Figure 4.14).*

It can be shown that orthogonal projections of $\mathcal{R}^2$ on a line containing $\mathbf{0}$ are linear.

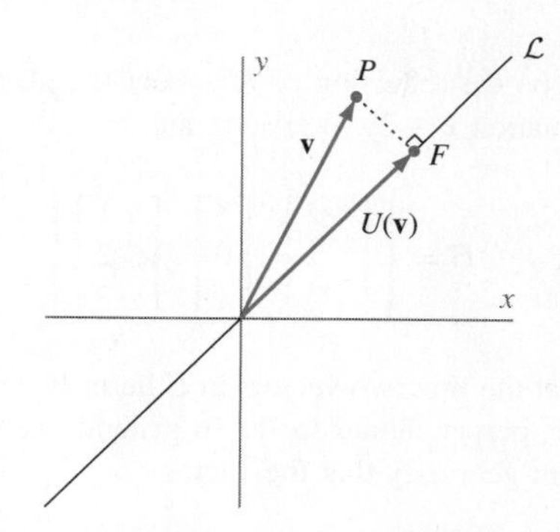

Figure 4.14 The orthogonal projection of a vector **v** on a line through the origin of $\mathcal{R}^2$

69. Find $U\left(\begin{bmatrix} x_1 \\ x_2 \end{bmatrix}\right)$, where U is the orthogonal projection of $\mathcal{R}^2$ on the line with equation $y = x$. *Hint:* First, find $[U]_{\mathcal{B}}$, where $\mathcal{B} = \left\{ \begin{bmatrix} 1 \\ 1 \end{bmatrix}, \begin{bmatrix} -1 \\ 1 \end{bmatrix} \right\}$.

70. Find $U\left(\begin{bmatrix} x_1 \\ x_2 \end{bmatrix}\right)$, where U is the orthogonal projection of $\mathcal{R}^2$ on the line with equation $y = -\frac{1}{2}x$.

71. Find $U\left(\begin{bmatrix} x_1 \\ x_2 \end{bmatrix}\right)$, where U is the orthogonal projection of $\mathcal{R}^2$ on the line with equation $y = -3x$.

72. Find $U\left(\begin{bmatrix} x_1 \\ x_2 \end{bmatrix}\right)$, where U is the orthogonal projection of $\mathcal{R}^2$ on the line with equation $y = mx$.

Let W be a plane through the origin of $\mathcal{R}^3$, and let $\mathbf{v}$ be a vector in $\mathcal{R}^3$ with endpoint P. Construct a line from P perpendicular to W, and let F denote the point of intersection of this perpendicular line with W. Denote the vector with endpoint F as $U_W(\mathbf{v})$. Now extend the perpendicular from P to F an equal distance to a point P' on the other side of W, and denote the vector with endpoint P' as $T_W(\mathbf{v})$. In Chapter 6, it is shown that the functions U_W and T_W are linear operators on $\mathcal{R}^3$. We call U_W the **orthogonal projection** *of $\mathcal{R}^3$ on W and T_W the* **reflection** *of $\mathcal{R}^3$ about W. (See Figure 4.15).*

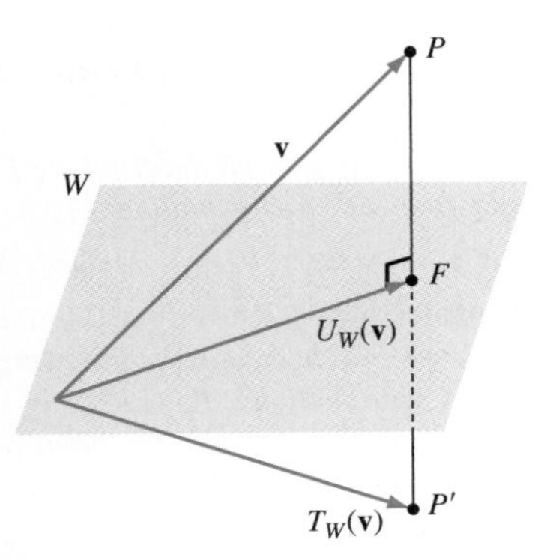

Figure 4.15 The orthogonal projection of a vector $\mathbf{v}$ on a subspace W of $\mathcal{R}^3$ and the reflection of a vector about W

73. Let T_W be the reflection of $\mathcal{R}^3$ about the plane W in $\mathcal{R}^3$ with equation $x + 2y - 3z = 0$, and let

$$\mathcal{B} = \left\{ \begin{bmatrix} -2 \\ 1 \\ 0 \end{bmatrix}, \begin{bmatrix} 3 \\ 0 \\ 1 \end{bmatrix}, \begin{bmatrix} 1 \\ 2 \\ -3 \end{bmatrix} \right\}.$$

Note that the first two vectors in $\mathcal{B}$ lie in W, and the third vector is perpendicular to W. In general, we can apply a fact from geometry that the vector

$$\begin{bmatrix} a \\ b \\ c \end{bmatrix},$$

whose components are the coefficients of the equation of the plane $ax + by + cz = d$, is normal (perpendicular) to the plane.

(a) Find $T_W(\mathbf{v})$ for each vector $\mathbf{v}$ in $\mathcal{B}$.

(b) Show that $\mathcal{B}$ is a basis for $\mathcal{R}^3$.

(c) Find $[T_W]_{\mathcal{B}}$.

(d) Find the standard matrix of T_W.

(e) Determine an explicit formula for $T_W\left(\begin{bmatrix} x_1 \\ x_2 \\ x_3 \end{bmatrix}\right)$.

In each of Exercises 74–80, find an explicit formula for $T_W\left(\begin{bmatrix} x_1 \\ x_2 \\ x_3 \end{bmatrix}\right)$, the reflection of $\mathcal{R}^3$ about the plane W defined by the given equation.

74. $2x - y + z = 0$
75. $x - 4y + 3z = 0$
76. $x + 2y - 5z = 0$
77. $x + 6y - 2z = 0$
78. $x - 3y + 5z = 0$
79. $x - 2y - 4z = 0$
80. $x + 5y + 7z = 0$

81. Let W and $\mathcal{B}$ be as in Exercise 73, and let U_W be the orthogonal projection of $\mathcal{R}^3$ on W.
    (a) Find $U_W(\mathbf{v})$ for each vector $\mathbf{v}$ in $\mathcal{B}$.
    (b) Find $[U_W]_{\mathcal{B}}$.
    (c) Find the standard matrix of U_W.
    (d) Determine an explicit formula for $U_W\left(\begin{bmatrix} x_1 \\ x_2 \\ x_3 \end{bmatrix}\right)$.

In each of Exercises 82–88, find an explicit formula for $U_W\left(\begin{bmatrix} x_1 \\ x_2 \\ x_3 \end{bmatrix}\right)$, the orthogonal projection of $\mathcal{R}^3$ on the plane W defined by the given equation.

82. $x + y - 2z = 0$
83. $x - 2y + 5z = 0$
84. $x + 4y - 3z = 0$
85. $x - 3y - 5z = 0$
86. $x + 6y + 2z = 0$
87. $x - 5y + 7z = 0$
88. $x + 2y - 4z = 0$

89. Let $\mathcal{B}$ be a basis for $\mathcal{R}^n$ and T be a linear operator on $\mathcal{R}^n$. Prove that T is invertible if and only if $[T]_{\mathcal{B}}$ is invertible.

90. Let $\mathcal{B}$ be a basis for $\mathcal{R}^n$ and T and U be linear operators on $\mathcal{R}^n$. Prove that $[UT]_{\mathcal{B}} = [U]_{\mathcal{B}}[T]_{\mathcal{B}}$.

91. Let $\mathcal{B}$ be a basis for $\mathcal{R}^n$ and T be a linear operator on $\mathcal{R}^n$. Prove that the dimension of the range of T equals the rank of $[T]_{\mathcal{B}}$.

92. Let $\mathcal{B}$ be a basis for $\mathcal{R}^n$ and T be a linear operator on $\mathcal{R}^n$. Prove that the dimension of the null space of T equals the nullity of $[T]_{\mathcal{B}}$.

93. Let $\mathcal{B}$ be a basis for $\mathcal{R}^n$ and T and U be linear operators on $\mathcal{R}^n$. Prove that $[T + U]_{\mathcal{B}} = [T]_{\mathcal{B}} + [U]_{\mathcal{B}}$. (See page 178 for the definition of $T + U$.)

94. Let $\mathcal{B}$ be a basis for $\mathcal{R}^n$ and T be a linear operator on $\mathcal{R}^n$. Prove that $[cT]_{\mathcal{B}} = c[T]_{\mathcal{B}}$ for any scalar c. (See page 178 for the definition of cT.)

95. Let T be a linear operator on $\mathcal{R}^n$, and let $\mathcal{A}$ and $\mathcal{B}$ be two bases for $\mathcal{R}^n$. Prove that $[T]_{\mathcal{A}}$ and $[T]_{\mathcal{B}}$ are similar.

96. Let A and B be similar matrices. Find bases $\mathcal{A}$ and $\mathcal{B}$ for $\mathcal{R}^n$ such that $[T_A]_{\mathcal{A}} = A$ and $[T_A]_{\mathcal{B}} = B$. (This proves that similar matrices are matrix representations of the same linear operator.)

97. Show that if A is the standard matrix of a reflection of $\mathcal{R}^2$ about a line, then $\det A = -1$.

98. Let $\mathcal{B} = \{\mathbf{b}_1, \mathbf{b}_2, \ldots, \mathbf{b}_n\}$ be a basis for $\mathcal{R}^n$, and let $\mathbf{c}_1, \mathbf{c}_2, \ldots, \mathbf{c}_n$ be (not necessarily distinct) vectors in $\mathcal{R}^n$.
    (a) Show that the matrix transformation T induced by CB^{-1} satisfies $T(\mathbf{b}_j) = \mathbf{c}_j$ for $j = 1, 2, \ldots, n$.
    (b) Prove that the linear transformation in (a) is the unique linear transformation such that $T(\mathbf{b}_j) = \mathbf{c}_j$ for $j = 1, 2, \ldots, n$.
    (c) Extend these results to an arbitrary linear transformation $T: \mathcal{R}^n \to \mathcal{R}^m$.
99. Let T be a linear operator on $\mathcal{R}^n$ and $\mathcal{B} = \{\mathbf{b}_1, \mathbf{b}_2, \ldots, \mathbf{b}_n\}$ be a basis for $\mathcal{R}^n$. Prove that $[T]_{\mathcal{B}}$ is an upper triangular matrix (see the definition given in Exercise 61 of Section 2.1) if and only if $T(\mathbf{b}_j)$ is a linear combination of $\mathbf{b}_1, \ldots, \mathbf{b}_j$ for every j, $1 \le j \le n$.
100. Let T be a linear operator on $\mathcal{R}^n$ and $\mathcal{B}$ be an ordered basis for $\mathcal{R}^n$. Prove the following results:
    (a) For every vector $\mathbf{v}$ in $\mathcal{R}^n$, $[T(\mathbf{v})]_{\mathcal{B}} = [T]_{\mathcal{B}}[\mathbf{v}]_{\mathcal{B}}$.
    (b) If C is an $n \times n$ matrix such that $[T(\mathbf{v})]_{\mathcal{B}} = C[\mathbf{v}]_{\mathcal{B}}$ for every vector $\mathbf{v}$ in $\mathcal{R}^n$, then $C = [T]_{\mathcal{B}}$.

The following definition of matrix representation of a linear transformation is used in Exercises 101 and 102:

Definition Let $T: \mathcal{R}^n \to \mathcal{R}^m$ be a linear transformation, and let $\mathcal{B} = \{\mathbf{b}_1, \mathbf{b}_2, \ldots, \mathbf{b}_n\}$ and $\mathcal{C} = \{\mathbf{c}_1, \mathbf{c}_2, \ldots, \mathbf{c}_m\}$ be bases for $\mathcal{R}^n$ and $\mathcal{R}^m$, respectively. The matrix

$$[\,[T(\mathbf{b}_1)]_{\mathcal{C}}\ \ [T(\mathbf{b}_2)]_{\mathcal{C}}\ \ \ldots\ \ [T(\mathbf{b}_n)]_{\mathcal{C}}\,]$$

is called the **matrix representation of** T **with respect to** $\mathcal{B}$ **and** $\mathcal{C}$. It is denoted by $[T]_{\mathcal{B}}^{\mathcal{C}}$.

101. Let

$$\mathcal{B} = \left\{ \begin{bmatrix} 1 \\ 1 \\ 1 \end{bmatrix}, \begin{bmatrix} 1 \\ -1 \\ 1 \end{bmatrix}, \begin{bmatrix} 1 \\ 1 \\ -1 \end{bmatrix} \right\} \quad \text{and} \quad \mathcal{C} = \left\{ \begin{bmatrix} 1 \\ 2 \end{bmatrix}, \begin{bmatrix} 2 \\ 3 \end{bmatrix} \right\}.$$

    (a) Prove that $\mathcal{B}$ and $\mathcal{C}$ are bases for $\mathcal{R}^3$ and $\mathcal{R}^2$, respectively.
    (b) Let $T: \mathcal{R}^3 \to \mathcal{R}^2$ be the linear transformation defined by

$$T\left(\begin{bmatrix} x_1 \\ x_2 \\ x_3 \end{bmatrix}\right) = \begin{bmatrix} x_1 + 2x_2 - x_3 \\ x_1 - x_2 + 2x_3 \end{bmatrix}.$$

    Find $[T]_{\mathcal{B}}^{\mathcal{C}}$.
102. Let $T: \mathcal{R}^n \to \mathcal{R}^m$ be a linear transformation, and $\mathcal{B} = \{\mathbf{b}_1, \mathbf{b}_2, \ldots, \mathbf{b}_n\}$ and $\mathcal{C} = \{\mathbf{c}_1, \mathbf{c}_2, \ldots, \mathbf{c}_m\}$ be bases for $\mathcal{R}^n$ and $\mathcal{R}^m$, respectively. Let B and C be the matrices whose columns are the vectors in $\mathcal{B}$ and $\mathcal{C}$, respectively. Prove the following results:
    (a) If A is the standard matrix of T, then $[T]_{\mathcal{B}}^{\mathcal{C}} = C^{-1}AB$.
    (b) If $U: \mathcal{R}^n \to \mathcal{R}^m$ is linear and s is any scalar, then
        (i) $[T + U]_{\mathcal{B}}^{\mathcal{C}} = [T]_{\mathcal{B}}^{\mathcal{C}} + [U]_{\mathcal{B}}^{\mathcal{C}}$;
        (ii) $[sT]_{\mathcal{B}}^{\mathcal{C}} = s[T]_{\mathcal{B}}^{\mathcal{C}}$ (see page 178 for the definitions of $T + U$ and sT);
        (iii) $[T(\mathbf{v})]_{\mathcal{C}} = [T]_{\mathcal{B}}^{\mathcal{C}}[\mathbf{v}]_{\mathcal{B}}$, for any vector $\mathbf{v}$ in $\mathcal{R}^n$.
    (c) Let $U: \mathcal{R}^m \to \mathcal{R}^p$ be linear, and let $\mathcal{D}$ be a basis for $\mathcal{R}^p$. Then

$$[UT]_{\mathcal{B}}^{\mathcal{D}} = [U]_{\mathcal{C}}^{\mathcal{D}}[T]_{\mathcal{B}}^{\mathcal{C}}.$$

    (d) Let $\mathcal{B}$ and $\mathcal{C}$ be the bases in Exercise 101, and let $U: \mathcal{R}^3 \to \mathcal{R}^2$ be a linear transformation such that

$$[U]_{\mathcal{B}}^{\mathcal{C}} = \begin{bmatrix} 1 & -2 & 4 \\ 3 & -3 & 1 \end{bmatrix}.$$

    Use (a) to find an explicit formula for $U(\mathbf{x})$.

In Exercises 103–107, use either a calculator with matrix capabilities or computer software such as MATLAB to solve each problem.

103. Let T and U be the linear operators on $\mathcal{R}^4$ defined by

$$T\left(\begin{bmatrix} x_1 \\ x_2 \\ x_3 \\ x_4 \end{bmatrix}\right) = \begin{bmatrix} x_1 - 2x_2 \\ x_3 \\ -x_1 + 3x_3 \\ 2x_2 - x_4 \end{bmatrix}$$

and

$$U\left(\begin{bmatrix} x_1 \\ x_2 \\ x_3 \\ x_4 \end{bmatrix}\right) = \begin{bmatrix} x_2 - x_3 + 2x_4 \\ -2x_1 + 3x_4 \\ 2x_2 - x_3 \\ 3x_1 + x_4 \end{bmatrix},$$

and let $\mathcal{B} = \{\mathbf{b}_1, \mathbf{b}_2, \mathbf{b}_3, \mathbf{b}_4\}$, where

$$\mathbf{b}_1 = \begin{bmatrix} 0 \\ 1 \\ 1 \\ 1 \end{bmatrix}, \ \mathbf{b}_2 = \begin{bmatrix} 0 \\ 1 \\ 2 \\ -1 \end{bmatrix}, \ \mathbf{b}_3 = \begin{bmatrix} 1 \\ 1 \\ -1 \\ 0 \end{bmatrix}, \text{ and } \mathbf{b}_4 = \begin{bmatrix} 1 \\ 0 \\ -2 \\ -2 \end{bmatrix}.$$

    (a) Compute $[T]_{\mathcal{B}}$, $[U]_{\mathcal{B}}$, and $[UT]_{\mathcal{B}}$.
    (b) Determine a relationship among $[T]_{\mathcal{B}}$, $[U]_{\mathcal{B}}$, and $[UT]_{\mathcal{B}}$.
104. Let T and U be linear operators on $\mathcal{R}^n$ and $\mathcal{B}$ be a basis for $\mathcal{R}^n$. Use the result of Exercise 103(b) to conjecture a relationship among $[T]_{\mathcal{B}}$, $[U]_{\mathcal{B}}$, and $[UT]_{\mathcal{B}}$, and then prove that your conjecture is true.
105. Let $\mathcal{B}$ and $\mathbf{b}_1$, $\mathbf{b}_2$, $\mathbf{b}_3$, $\mathbf{b}_4$ be as defined in Exercise 103.
    (a) Compute $[T]_{\mathcal{B}}$, where T is the linear operator on $\mathcal{R}^4$ such that $T(\mathbf{b}_1) = \mathbf{b}_2$, $T(\mathbf{b}_2) = \mathbf{b}_3$, $T(\mathbf{b}_3) = \mathbf{b}_4$, and $T(\mathbf{b}_4) = \mathbf{b}_1$.
    (b) Determine an explicit formula for $T(\mathbf{x})$, where $\mathbf{x}$ is an arbitrary vector in $\mathcal{R}^4$.
106. Let $\mathcal{B} = \{\mathbf{b}_1, \mathbf{b}_2, \mathbf{b}_3, \mathbf{b}_4\}$ be as in Exercise 103, and let T be the linear operator on $\mathcal{R}^4$ defined by

$$T\left(\begin{bmatrix} x_1 \\ x_2 \\ x_3 \\ x_4 \end{bmatrix}\right) = \begin{bmatrix} x_1 + 2x_2 - 3x_3 - 2x_4 \\ -x_1 - 2x_2 + 4x_3 + 6x_4 \\ 2x_1 + 3x_2 - 5x_3 - 4x_4 \\ -x_1 + x_2 - x_3 - x_4 \end{bmatrix}.$$

(a) Determine an explicit formula for $T^{-1}(\mathbf{x})$, where $\mathbf{x}$ is an arbitrary vector in $\mathcal{R}^4$.

(b) Compute $[T]_\mathcal{B}$ and $[T^{-1}]_\mathcal{B}$.

(c) Determine a relationship between $[T]_\mathcal{B}$ and $[T^{-1}]_\mathcal{B}$.

107. Let T be an invertible linear operator on $\mathcal{R}^n$ and $\mathcal{B}$ be a basis for $\mathcal{R}^n$. Use the result of Exercise 106(c) to conjecture a relationship between $[T]_\mathcal{B}$ and $[T^{-1}]_\mathcal{B}$, and then prove that your conjecture is true.

SOLUTIONS TO THE PRACTICE PROBLEMS

1. The standard matrix of T is

$$A = \begin{bmatrix} -1 & 0 & 2 \\ 1 & 1 & 0 \\ 0 & -1 & 1 \end{bmatrix}.$$

Let

$$B = \begin{bmatrix} 1 & 1 & 3 \\ 1 & 1 & 2 \\ 0 & 1 & 1 \end{bmatrix}$$

be the matrix whose columns are the vectors in $\mathcal{B}$. Then the $\mathcal{B}$-matrix representation of T is

$$[T]_\mathcal{B} = B^{-1}AB = \begin{bmatrix} 6 & 3 & 12 \\ 2 & 1 & 5 \\ -3 & -1 & -6 \end{bmatrix}.$$

2. The standard matrix of U is

$$A = B[U]_\mathcal{B}B^{-1} = \begin{bmatrix} -2 & 5 & -1 \\ -3 & 6 & -1 \\ -1 & 1 & 2 \end{bmatrix}.$$

Hence

$$U\left(\begin{bmatrix} x_1 \\ x_2 \\ x_3 \end{bmatrix}\right) = A\begin{bmatrix} x_1 \\ x_2 \\ x_3 \end{bmatrix} = \begin{bmatrix} -2x_1 + 5x_2 - x_3 \\ -3x_1 + 6x_2 - x_3 \\ -x_1 + x_2 + 2x_3 \end{bmatrix}.$$

CHAPTER 4 REVIEW EXERCISES

In Exercises 1–25, determine whether the statements are true or false.

1. If $\mathbf{u}_1, \mathbf{u}_2, \ldots, \mathbf{u}_k$ are vectors in a subspace V of $\mathcal{R}^n$, then every linear combination of $\mathbf{u}_1, \mathbf{u}_2, \ldots, \mathbf{u}_k$ belongs to V.
2. The span of a finite nonempty subset of $\mathcal{R}^n$ is a subspace of $\mathcal{R}^n$.
3. The null space of an $m \times n$ matrix is contained in $\mathcal{R}^m$.
4. The column space of an $m \times n$ matrix is contained in $\mathcal{R}^n$.
5. The row space of an $m \times n$ matrix is contained in $\mathcal{R}^m$.
6. The range of every linear transformation is a subspace.
7. The null space of every linear transformation equals the null space of its standard matrix.
8. The range of every linear transformation equals the row space of its standard matrix.
9. Every nonzero subspace of $\mathcal{R}^n$ has a unique basis.
10. It is possible for different bases for a particular subspace to contain different numbers of vectors.
11. Every finite generating set for a nonzero subspace contains a basis for the subspace.
12. The pivot columns of every matrix form a basis for its column space.
13. The vectors in the vector form of the general solution of $A\mathbf{x} = \mathbf{0}$ constitute a basis for the null space of A.
14. No subspace of $\mathcal{R}^n$ has dimension greater than n.
15. There is only one subspace of $\mathcal{R}^n$ having dimension n.
16. There is only one subspace of $\mathcal{R}^n$ having dimension 0.
17. The dimension of the null space of a matrix equals the rank of the matrix.
18. The dimension of the column space of a matrix equals the rank of the matrix.
19. The dimension of the row space of a matrix equals the nullity of the matrix.
20. The column space of any matrix equals the column space of its reduced row echelon form.
21. The null space of any matrix equals the null space of its reduced row echelon form.
22. If $\mathcal{B}$ is a basis for $\mathcal{R}^n$ and B is the matrix whose columns are the vectors in $\mathcal{B}$, then $B^{-1}\mathbf{v} = [\mathbf{v}]_\mathcal{B}$ for all $\mathbf{v}$ in $\mathcal{R}^n$.
23. If T is a linear operator on $\mathcal{R}^n$, $\mathcal{B}$ is a basis for $\mathcal{R}^n$, B is the matrix whose columns are the vectors in $\mathcal{B}$, and A is the standard matrix of T, then $[T]_\mathcal{B} = BAB^{-1}$.
24. If T is a linear operator on $\mathcal{R}^n$ and $\mathcal{B}$ is a basis for $\mathcal{R}^n$, then $[T]_\mathcal{B}$ is the unique $n \times n$ matrix such that $[T]_\mathcal{B}[\mathbf{v}]_\mathcal{B} = [T(\mathbf{v})]_\mathcal{B}$ for all $\mathbf{v}$ in $\mathcal{R}^n$.
25. If T is a reflection of $\mathcal{R}^2$ about a line, then there exists a basis $\mathcal{B}$ for $\mathcal{R}^2$ such that $[T]_\mathcal{B} = \begin{bmatrix} 1 & 0 \\ 0 & -1 \end{bmatrix}$.

26. Determine whether each phrase is a misuse of terminology. If so, explain what is wrong.

(a) a basis for a matrix

(b) the rank of a subspace

(c) the dimension of a square matrix

(d) the dimension of the zero subspace

(e) the dimension of a basis for a subspace

(f) the column space of a linear transformation

(g) the dimension of a linear transformation

(h) the coordinate vector of a linear operator

27. Let V be a subspace of $\mathcal{R}^n$ with dimension k, and let S be a subset of V. What can be said about m, the number of vectors in S, under the following conditions?

(a) S is linearly independent.

(b) S is linearly dependent.

(c) S is a generating set for V.

28. Let A be the standard matrix of a linear transformation $T\colon \mathcal{R}^5 \to \mathcal{R}^7$. If the range of T has dimension 2, determine the dimension of each of the following subspaces:

(a) Col A (b) Null A (c) Row A
(d) Null A^T (e) the null space of T

In Exercises 29 and 30, determine whether the given set is a subspace of $\mathcal{R}^4$. Justify your answer.

29. $\left\{\begin{bmatrix} u_1 \\ u_2 \\ u_3 \\ u_4 \end{bmatrix} \text{ in } \mathcal{R}^4\colon u_1^2 = u_3^3,\ u_2 = 0, \text{ and } u_4 = 0\right\}$

30. $\left\{\begin{bmatrix} u_1 \\ u_2 \\ u_3 \\ u_4 \end{bmatrix} \text{ in } \mathcal{R}^4\colon u_2 = 5u_3,\ u_1 = 0, \text{ and } u_4 = 0\right\}$

In Exercises 31 and 32, find bases for (a) the null space if it is nonzero, (b) the column space, and (c) the row space of each matrix.

31. $\begin{bmatrix} 1 & 2 & -1 \\ -1 & -1 & -1 \\ 2 & 1 & 4 \\ 1 & 4 & -5 \end{bmatrix}$ 32. $\begin{bmatrix} -1 & 1 & 2 & 2 & 1 \\ 2 & -2 & -1 & -3 & 2 \\ 1 & -1 & 1 & 1 & 2 \\ 1 & -1 & 4 & 8 & 3 \end{bmatrix}$

In Exercises 33 and 34, a linear transformation T is given. (a) Find a basis for the range of T. (b) If the null space of T is nonzero, find a basis for the null space of T.

33. $T\colon \mathcal{R}^3 \to \mathcal{R}^4$ defined by

$$T\left(\begin{bmatrix} x_1 \\ x_2 \\ x_3 \end{bmatrix}\right) = \begin{bmatrix} x_2 - 2x_3 \\ -x_1 + 3x_2 + x_3 \\ x_1 - 4x_2 + x_3 \\ 2x_1 - x_2 + 3x_3 \end{bmatrix}$$

34. $T\colon \mathcal{R}^4 \to \mathcal{R}^2$ defined by

$$T\left(\begin{bmatrix} x_1 \\ x_2 \\ x_3 \\ x_4 \end{bmatrix}\right) = \begin{bmatrix} x_1 - 2x_2 + x_3 - 3x_4 \\ -2x_1 + 3x_2 - 3x_3 + 2x_4 \end{bmatrix}$$

35. Prove that $\left\{\begin{bmatrix} -1 \\ 2 \\ 2 \\ -1 \end{bmatrix}, \begin{bmatrix} 1 \\ 5 \\ 3 \\ -2 \end{bmatrix}\right\}$ is a basis for the null space of the linear transformation in Exercise 34.

36. Prove that $\left\{\begin{bmatrix} 1 \\ 0 \\ -1 \\ -5 \end{bmatrix}, \begin{bmatrix} 1 \\ -7 \\ -4 \\ -3 \end{bmatrix}, \begin{bmatrix} 1 \\ -5 \\ -1 \\ 5 \end{bmatrix}\right\}$ is a basis for the column space of the matrix in Exercise 32.

37. Let $\mathcal{B} = \left\{\begin{bmatrix} 0 \\ -1 \\ 1 \end{bmatrix}, \begin{bmatrix} 1 \\ 0 \\ -1 \end{bmatrix}, \begin{bmatrix} -1 \\ -1 \\ 1 \end{bmatrix}\right\}$.

(a) Prove that $\mathcal{B}$ is a basis for $\mathcal{R}^3$.

(b) Find $\mathbf{v}$ if $[\mathbf{v}]_{\mathcal{B}} = \begin{bmatrix} 4 \\ -3 \\ -2 \end{bmatrix}$.

(c) Find $[\mathbf{w}]_{\mathcal{B}}$ if $\mathbf{w} = \begin{bmatrix} -2 \\ 5 \\ 3 \end{bmatrix}$.

38. Let $\mathcal{B} = \{\mathbf{b}_1, \mathbf{b}_2, \mathbf{b}_3\}$ be a basis for $\mathcal{R}^3$ and T be the linear operator on $\mathcal{R}^3$ such that

$$T(\mathbf{b}_1) = -2\mathbf{b}_2 + \mathbf{b}_3, \quad T(\mathbf{b}_2) = 4\mathbf{b}_1 - 3\mathbf{b}_3, \quad \text{and}$$
$$T(\mathbf{b}_3) = 5\mathbf{b}_1 - 4\mathbf{b}_2 + 2\mathbf{b}_3.$$

(a) Determine $[T]_{\mathcal{B}}$.

(b) Express $T(\mathbf{v})$ as a linear combination of the vectors in $\mathcal{B}$ if $\mathbf{v} = 3\mathbf{b}_1 - \mathbf{b}_2 - 2\mathbf{b}_3$.

39. Determine (a) $[T]_{\mathcal{B}}$, (b) the standard matrix of T, and (c) an explicit formula for $T(\mathbf{x})$ from the given information about the linear operator T on $\mathcal{R}^2$.

$$\mathcal{B} = \left\{\begin{bmatrix} 1 \\ -2 \end{bmatrix}, \begin{bmatrix} -2 \\ 3 \end{bmatrix}\right\},$$

$$T\left(\begin{bmatrix} 1 \\ -2 \end{bmatrix}\right) = \begin{bmatrix} 3 \\ 4 \end{bmatrix}, \quad \text{and} \quad T\left(\begin{bmatrix} -2 \\ 3 \end{bmatrix}\right) = \begin{bmatrix} -1 \\ 1 \end{bmatrix}$$

40. Let T be the linear operator on $\mathcal{R}^2$ and $\mathcal{B}$ be the basis for $\mathcal{R}^2$ defined by

$$T\left(\begin{bmatrix} x_1 \\ x_2 \end{bmatrix}\right) = \begin{bmatrix} 2x_1 - x_2 \\ x_1 - 2x_2 \end{bmatrix} \quad \text{and} \quad \mathcal{B} = \left\{\begin{bmatrix} 1 \\ 2 \end{bmatrix}, \begin{bmatrix} 3 \\ 7 \end{bmatrix}\right\}.$$

Determine $[T]_{\mathcal{B}}$.

41. Determine an explicit description of $T(\mathbf{x})$, using the given basis $\mathcal{B}$ and the matrix representation of T with respect to $\mathcal{B}$.

$$[T]_{\mathcal{B}} = \begin{bmatrix} 1 & 2 & -1 \\ -1 & 3 & 2 \\ 2 & 1 & 2 \end{bmatrix} \quad \text{and} \quad \mathcal{B} = \left\{\begin{bmatrix} 2 \\ 1 \\ 1 \end{bmatrix}, \begin{bmatrix} 1 \\ 2 \\ 1 \end{bmatrix}, \begin{bmatrix} 1 \\ 1 \\ 1 \end{bmatrix}\right\}$$

42. Let T be the linear operator on $\mathcal{R}^3$ such that

$$T\left(\begin{bmatrix} 1 \\ 0 \\ 1 \end{bmatrix}\right) = \begin{bmatrix} 2 \\ 1 \\ -2 \end{bmatrix}, \quad T\left(\begin{bmatrix} 0 \\ -1 \\ 1 \end{bmatrix}\right) = \begin{bmatrix} 1 \\ 3 \\ -1 \end{bmatrix},$$

and

$$T\left(\begin{bmatrix} -1 \\ 1 \\ -1 \end{bmatrix}\right) = \begin{bmatrix} -2 \\ 1 \\ 3 \end{bmatrix}.$$

Find an explicit description of $T(\mathbf{x})$.

In Exercises 43 and 44, an equation of a conic section is given in the $x'y'$-coordinate system. Determine the equation of the conic section in the usual xy-coordinate system if the x'-axis and the

y'-axis are obtained by rotating the usual x-axis and y-axis through the given angle θ.

43. $\dfrac{(x')^2}{2^2} + \dfrac{(y')^2}{3^2} = 1, \qquad \theta = 120°$

44. $-\sqrt{3}(x')^2 + 2x'y' + \sqrt{3}(y')^2 = 12, \qquad \theta = 330°$

In Exercises 45 and 46, an equation of a conic section is given in the xy-coordinate system. Determine the equation of the conic section in the $x'y'$-coordinate system if the x'-axis and the y'-axis are obtained by rotating the usual x-axis and y-axis through the given angle θ.

45. $29x^2 - 42xy + 29y^2 = 200, \qquad \theta = 315°$

46. $-39x^2 - 50\sqrt{3}xy + 11y^2 = 576, \qquad \theta = 210°$

47. Find an explicit description of the reflection T of $\mathcal{R}^2$ about the line with equation $y = -\frac{3}{2}x$.

48. Find an explicit description of the orthogonal projection U of $\mathcal{R}^2$ on the line with equation $y = -\frac{3}{2}x$.

49. Prove that if $\{\mathbf{v}_1, \mathbf{v}_2, \ldots, \mathbf{v}_n\}$ is a basis for $\mathcal{R}^n$ and A is an invertible $n \times n$ matrix, then $\{A\mathbf{v}_1, A\mathbf{v}_2, \ldots, A\mathbf{v}_n\}$ is also a basis for $\mathcal{R}^n$.

50. Let V and W be subspaces of $\mathcal{R}^n$. Prove that $V \cup W$ is a subspace of $\mathcal{R}^n$ if and only if V is contained in W or W is contained in V.

51. Let $\mathcal{B}$ be a basis for $\mathcal{R}^n$ and T be an invertible linear operator on $\mathcal{R}^n$. Prove that $[T^{-1}]_{\mathcal{B}} = ([T]_{\mathcal{B}})^{-1}$.

*Let V and W be subsets of $\mathcal{R}^n$. We define the **sum** of V and W, denoted $V + W$, as*

$$V + W = \{\mathbf{u} \text{ in } \mathcal{R}^n : \mathbf{u} = \mathbf{v} + \mathbf{w} \text{ for some } \mathbf{v} \text{ in } V \text{ and some } \mathbf{w} \text{ in } W\}.$$

In Exercises 52–54, use the preceding definition.

52. Prove that if V and W are subspaces of $\mathcal{R}^n$, then $V + W$ is also a subspace of $\mathcal{R}^n$.

53. Let

$$V = \left\{\begin{bmatrix} v_1 \\ v_2 \\ v_3 \end{bmatrix} \text{ in } \mathcal{R}^3 : v_1 + v_2 = 0 \text{ and } 2v_1 - v_3 = 0\right\}$$

and

$$W = \left\{\begin{bmatrix} w_1 \\ w_2 \\ w_3 \end{bmatrix} \text{ in } \mathcal{R}^3 : w_1 - 2w_3 = 0 \text{ and } w_2 + w_3 = 0\right\}.$$

Find a basis for $V + W$.

54. Let S_1 and S_2 be subsets of $\mathcal{R}^n$, and let $S = S_1 \cup S_2$. Prove that if $V = \operatorname{Span} S_1$ and $W = \operatorname{Span} S_2$, then $\operatorname{Span} S = V + W$.

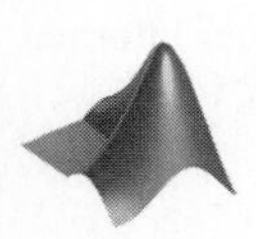

CHAPTER 4 MATLAB EXERCISES

For the following exercises, use MATLAB (or comparable software) or a calculator with matrix capabilities. The MATLAB functions in Tables D.1, D.2, D.3, D.4, and D.5 of Appendix D may be useful.

1. Let

$$A = \begin{bmatrix} 1.1 & 0.0 & 2.2 & -1.3 & -0.2 \\ 2.1 & -1.5 & 2.7 & 2.2 & 4.3 \\ -1.2 & 4.1 & 1.7 & 1.4 & 0.2 \\ 2.2 & 2.1 & 6.5 & 2.1 & 4.3 \\ 1.3 & 1.2 & 3.8 & -1.7 & -0.4 \\ 3.1 & -4.0 & 2.2 & -1.1 & 2.0 \end{bmatrix}.$$

For each of the following parts, use Theorem 1.5 to determine whether each vector belongs to Col A.

(a) $\begin{bmatrix} -1.5 \\ 11.0 \\ -10.7 \\ 0.1 \\ -5.7 \\ 12.9 \end{bmatrix}$ (b) $\begin{bmatrix} 3.5 \\ 2.0 \\ -3.8 \\ 2.3 \\ 4.3 \\ 2.2 \end{bmatrix}$

(c) $\begin{bmatrix} 1.1 \\ -2.8 \\ 4.1 \\ 2.0 \\ 4.0 \\ -3.7 \end{bmatrix}$ (d) $\begin{bmatrix} 4.8 \\ -3.2 \\ 3.0 \\ 4.4 \\ 8.4 \\ 0.4 \end{bmatrix}$

2. Let

$$A = \begin{bmatrix} 1.2 & 2.3 & 1.2 & 4.7 & -5.8 \\ -1.1 & 3.2 & -3.1 & -1.0 & -3.3 \\ 2.3 & 1.1 & 2.1 & 5.5 & -4.3 \\ -1.2 & 1.4 & -1.4 & -1.2 & -1.4 \\ 1.1 & -4.1 & 5.1 & 2.1 & 3.1 \\ 0.1 & -2.1 & 1.2 & -0.8 & 3.0 \end{bmatrix}.$$

For each of the following parts, determine whether each vector belongs to Null A.

(a) $\begin{bmatrix} 2.6 \\ 0.8 \\ 1.7 \\ -2.6 \\ -0.9 \end{bmatrix}$ (b) $\begin{bmatrix} -3.4 \\ 5.6 \\ 1.1 \\ 3.4 \\ 4.5 \end{bmatrix}$

(c) $\begin{bmatrix} 1.5 \\ -1.2 \\ 2.4 \\ -0.3 \\ 3.7 \end{bmatrix}$ (d) $\begin{bmatrix} 1.3 \\ -0.7 \\ 0.3 \\ -1.3 \\ -1.0 \end{bmatrix}$

3. Let A be the matrix of Exercise 2.

   (a) Find a basis for the column space of A consisting of columns of A.

   (b) Use Exercise 78 of Section 4.2 to extend this basis to a basis for $\mathcal{R}^6$.

(c) Find a basis for Null A.

(d) Find a basis for Row A.

4. Let

$$A = \begin{bmatrix} 1.3 & 2.1 & 0.5 & 2.9 \\ 2.2 & -1.4 & 5.8 & -3.0 \\ -1.2 & 1.3 & -3.7 & 3.8 \\ 4.0 & 2.7 & 5.3 & 1.4 \\ 1.7 & 4.1 & -0.7 & 6.5 \\ -3.1 & 1.0 & -7.2 & 5.1 \end{bmatrix}.$$

(a) Find a basis for the column space of A consisting of columns of A.

(b) Use Exercise 78 of Section 4.2 to extend this basis to a basis for $\mathcal{R}^6$.

(c) Find a basis for Null A.

(d) Find a basis for Row A.

5. Let

$$\mathcal{B} = \left\{ \begin{bmatrix} 1.1 \\ 3.3 \\ -1.7 \\ 2.2 \\ 0.7 \\ 6.1 \end{bmatrix}, \begin{bmatrix} 2.1 \\ -1.3 \\ 2.4 \\ 1.5 \\ 4.2 \\ 2.2 \end{bmatrix}, \begin{bmatrix} -1.2 \\ 4.1 \\ 4.6 \\ -4.2 \\ 1.6 \\ -3.1 \end{bmatrix}, \begin{bmatrix} 3.1 \\ 4.3 \\ -3.2 \\ 3.1 \\ 3.8 \\ 0.4 \end{bmatrix}, \begin{bmatrix} 4.5 \\ 2.5 \\ 5.3 \\ 1.3 \\ -1.4 \\ 2.5 \end{bmatrix}, \begin{bmatrix} 5.3 \\ -4.5 \\ 1.8 \\ 4.1 \\ -2.4 \\ -2.3 \end{bmatrix} \right\}.$$

(a) Show that $\mathcal{B}$ is a basis for $\mathcal{R}^6$.

(b) For each of the following parts, represent the given vector as a linear combination of the vectors in $\mathcal{B}$.

$$\text{(i)} \begin{bmatrix} 7.4 \\ 5.1 \\ -10.8 \\ 14.0 \\ -8.0 \\ 26.6 \end{bmatrix} \qquad \text{(ii)} \begin{bmatrix} -4.2 \\ 5.3 \\ -20.0 \\ 2.9 \\ 7.5 \\ -8.2 \end{bmatrix} \qquad \text{(iii)} \begin{bmatrix} -19.3 \\ 6.6 \\ -30.2 \\ -7.7 \\ 2.2 \\ -18.9 \end{bmatrix}$$

(c) For each vector in (b), find the corresponding coordinate vector relative to $\mathcal{B}$.

Exercises 6 and 7 apply the results of Exercise 98 of Section 4.5.

6. Let $\mathcal{B} = \{\mathbf{b}_1, \mathbf{b}_2, \mathbf{b}_3, \mathbf{b}_4, \mathbf{b}_5\}$ be the basis for $\mathcal{R}^5$ given by

$$\mathcal{B} = \left\{ \begin{bmatrix} -1.4 \\ 10.0 \\ 9.0 \\ 4.4 \\ 4.0 \end{bmatrix}, \begin{bmatrix} -1.9 \\ 3.0 \\ 4.0 \\ 2.9 \\ 1.0 \end{bmatrix}, \begin{bmatrix} 2.3 \\ 2.5 \\ 1.0 \\ -2.3 \\ 0.0 \end{bmatrix}, \begin{bmatrix} -3.1 \\ 2.0 \\ 4.0 \\ 4.1 \\ 1.0 \end{bmatrix}, \begin{bmatrix} 0.7 \\ 8.0 \\ 5.0 \\ 1.3 \\ 3.0 \end{bmatrix} \right\}.$$

Let T be a linear operator T on $\mathcal{R}^5$ such that

$$T(\mathbf{b}_1) = \begin{bmatrix} 1 \\ -1 \\ 2 \\ 1 \\ 1 \end{bmatrix}, \quad T(\mathbf{b}_2) = \begin{bmatrix} 0 \\ 0 \\ 1 \\ 1 \\ -2 \end{bmatrix}, \quad T(\mathbf{b}_3) = \begin{bmatrix} -2 \\ 1 \\ 0 \\ 1 \\ 2 \end{bmatrix},$$

$$T(\mathbf{b}_4) = \begin{bmatrix} 3 \\ 1 \\ 0 \\ 1 \\ -1 \end{bmatrix}, \quad T(\mathbf{b}_5) = \begin{bmatrix} 1 \\ 0 \\ 1 \\ -1 \\ 2 \end{bmatrix}.$$

Find the standard matrix of the linear operator T from the given information.

7. Find the standard matrix of the linear transformation $U : \mathcal{R}^6 \to \mathcal{R}^4$ such that

$$U\left(\begin{bmatrix} 2 \\ 1 \\ -1 \\ 0 \\ 0 \\ -1 \end{bmatrix}\right) = \begin{bmatrix} 1 \\ -1 \\ 0 \\ 2 \end{bmatrix}, \quad U\left(\begin{bmatrix} 0 \\ -1 \\ 0 \\ 1 \\ -2 \\ 0 \end{bmatrix}\right) = \begin{bmatrix} 0 \\ -1 \\ 1 \\ 2 \end{bmatrix},$$

$$U\left(\begin{bmatrix} -4 \\ -2 \\ 1 \\ 2 \\ -4 \\ 2 \end{bmatrix}\right) = \begin{bmatrix} 1 \\ 1 \\ -2 \\ 3 \end{bmatrix}, \quad U\left(\begin{bmatrix} 0 \\ -2 \\ 0 \\ 1 \\ -2 \\ 0 \end{bmatrix}\right) = \begin{bmatrix} -2 \\ 3 \\ 0 \\ 1 \end{bmatrix},$$

$$U\left(\begin{bmatrix} 0 \\ 1 \\ 0 \\ 0 \\ 1 \\ 0 \end{bmatrix}\right) = \begin{bmatrix} 1 \\ 0 \\ 0 \\ -1 \end{bmatrix}, \quad U\left(\begin{bmatrix} -1 \\ 1 \\ 0 \\ 2 \\ 0 \\ 1 \end{bmatrix}\right) = \begin{bmatrix} 1 \\ 0 \\ 2 \\ 0 \end{bmatrix}.$$

8. It is clear that any subspace W of $\mathcal{R}^n$ can be described as the column space of a matrix A. Simply choose a finite generating set for W and let A be the matrix whose columns are the vectors in this set (in any order). What is less clear is that W can also be described as the null space of a matrix. The method is described next.

Let W be a subspace of $\mathcal{R}^n$. Choose a basis for W and extend it to a basis $\mathcal{B} = \{\mathbf{b}_1, \mathbf{b}_2, \ldots, \mathbf{b}_n\}$ for $\mathcal{R}^n$, where the first k vectors in $\mathcal{B}$ constitute the original basis for W. Let T be the linear operator on $\mathcal{R}^n$ (whose existence is guaranteed by Exercise 98 of Section 4.5) defined by

$$T(\mathbf{b}_j) = \begin{cases} \mathbf{0} & \text{if } j \le k \\ \mathbf{b}_j & \text{if } j > k. \end{cases}$$

Prove that $W = \text{Null } A$, where A is the standard matrix of T.

9. Let W be the subspace of $\mathcal{R}^5$ with basis

$$\left\{\begin{bmatrix}1\\3\\-1\\0\\2\end{bmatrix}, \begin{bmatrix}-1\\0\\1\\2\\1\end{bmatrix}, \begin{bmatrix}0\\2\\0\\2\\3\end{bmatrix}\right\}.$$

Use the method described in Exercise 8 to find a matrix A such that $W = \text{Null } A$.

10. An advantage of representing subspaces of $\mathcal{R}^n$ as null spaces of matrices (see Exercise 8) is that these matrices can be used to describe the intersection of two subspaces, which is a subspace of $\mathcal{R}^n$ (see Exercise 77 of Section 4.1).

(a) Let V and W be subspaces of $\mathcal{R}^n$, and suppose that A and B are matrices such that $V = \text{Null } A$ and $W = \text{Null } B$. Notice that A and B each consist of n columns. (Why?) Let $C = \begin{bmatrix}A\\B\end{bmatrix}$; that is, C is the matrix whose rows consist of the rows of A followed by the rows of B. Prove that

$$\text{Null } C = \text{Null } A \cap \text{Null } B = V \cap W.$$

(b) Let

$$V = \text{Span}\left\{\begin{bmatrix}1\\2\\1\\-1\end{bmatrix}, \begin{bmatrix}2\\1\\0\\1\end{bmatrix}, \begin{bmatrix}1\\3\\1\\0\end{bmatrix}\right\}$$

and

$$W = \text{Span}\left\{\begin{bmatrix}1\\-1\\1\\1\end{bmatrix}, \begin{bmatrix}0\\1\\1\\1\end{bmatrix}, \begin{bmatrix}1\\0\\1\\2\end{bmatrix}\right\}.$$

Use (a) and the MATLAB function `null(A, 'r')` described in Table D.2 of Appendix D to find a basis for $V \cap W$.

5 INTRODUCTION

The control of vibrations in mechanical systems, such as cars, power plants, or bridges, is an important design consideration. The consequences of uncontrolled vibrations range from discomfort to the actual failure of the system with damage to one or more components. The failures can be spectacular, as in the collapse of the Angers Bridge (Maine River, France) in 1850 and the Tacoma Narrows Bridge (Washington State, United States) in 1940. The vibration problem and its solutions can be explained and understood by means of *eigenvalues* and *eigenvectors* of the differential equations that model the system (Section 5.5).

It is convenient to model a mechanical system as a mass attached to a spring. The mass-spring system has a natural frequency determined by the size of the mass and the stiffness of the spring. This frequency is natural in the sense that if the mass is struck, it will vibrate up and down at this frequency. A final element of the model is a periodic external force, $F_0 \sin \omega t$, applied to the main mass. In a car, this could come from the engine or from regular defects on the highway. In the Angers Bridge collapse, the applied force came from soldiers marching in step. In the Tacoma Narrows Bridge collapse[3], the applied force came from wind-induced vibrations. If the frequency, ω, of the external force equals or is close to the natural frequency of the system, a phenomenon called *resonance* occurs when the variations of the applied force are in step with, and reinforce, the motions of the main

3 Video clips from this collapse are viewable at http://www.pbs.org/wgbh/nova/bridge/tacoma3.html and http://www.enm.bris.ac.uk/anm/tacoma/tacoma.html#mpeg.

mass. As a consequence, those motions can become quite large. In the case of the bridges mentioned above, the motions led to the collapse of the bridges.

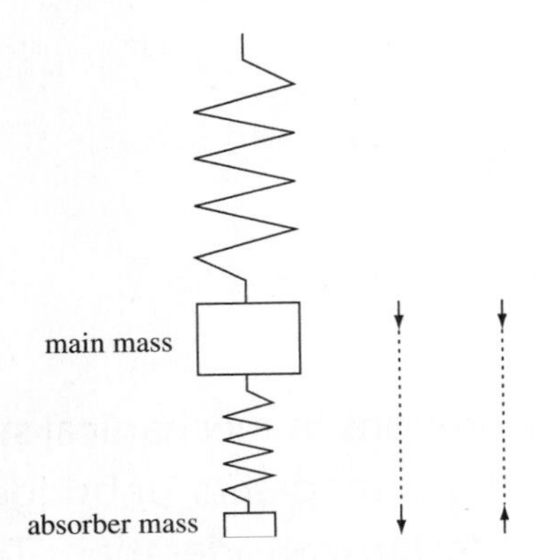

The solution to the vibration problem is either to redesign the system to shift its natural frequencies away from that of the applied forces or to try to minimize the response of the system to those forces. A classic approach to minimizing response (the *Den Hartog vibration absorber*) is to attach an additional mass and spring to the main mass. This new system has two natural frequencies, neither of which is the same as that of the original system. One natural frequency corresponds to a mode of vibration where the masses move in the same direction (the left pair of arrows in the figure). The other natural frequency corresponds to a mode of vibration where the masses move in opposite directions (the right pair of arrows in the figure). If the size of the absorber mass and the stiffness of the new spring are adjusted properly, these two modes of vibration can be combined so that the net displacement of the main mass is zero when the frequency of the applied force, ω, equals the natural frequency of the original system. In this case, the absorber mass absorbs all of the vibration from the energy applied to the main mass by the external force.

CHAPTER 5

EIGENVALUES, EIGENVECTORS, AND DIAGONALIZATION

In many applications, it is important to understand how vectors in $\mathcal{R}^n$ are transformed when they are multiplied by a square matrix. In this chapter, we see that in many circumstances a problem can be reformulated so that the original matrix can be replaced by a diagonal matrix, which simplifies the problem. We have already seen an example of this in Section 4.5, where we learned how to describe a reflection in the plane in terms of a diagonal matrix. One important class of problems in which this approach is also useful involves sequences of matrix–vector products of the form $A\mathbf{p}, A^2\mathbf{p}, A^3\mathbf{p}, \ldots$. For example, such a sequence arises in the study of long-term population trends considered in Example 6 of Section 2.1.

The central theme of this chapter is the investigation of matrices that can be replaced by diagonal matrices, the *diagonalizable* matrices. We begin this investigation with the introduction of special scalars and vectors, called *eigenvalues* and *eigenvectors*, respectively, that provide us with the tools necessary to describe diagonalizable matrices.

5.1 EIGENVALUES AND EIGENVECTORS

In Section 4.5, we discussed the reflection T of $\mathcal{R}^2$ about the line with equation $y = \frac{1}{2}x$. Recall that the vectors

$$\mathbf{b}_1 = \begin{bmatrix} 2 \\ 1 \end{bmatrix} \quad \text{and} \quad \mathbf{b}_2 = \begin{bmatrix} -1 \\ 2 \end{bmatrix}$$

played an essential role in determining the rule for T. The key to this computation is that $T(\mathbf{b}_1)$ is a multiple of $\mathbf{b}_1$ and $T(\mathbf{b}_2)$ is a multiple of $\mathbf{b}_2$. (See Figure 5.1.)

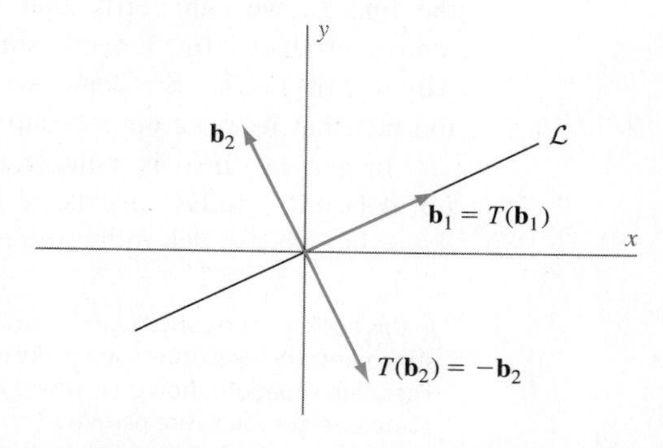

Figure 5.1 The images of the basis vectors $\mathbf{b}_1$ and $\mathbf{b}_2$ are multiples of the vectors.

Nonzero vectors that are mapped to a multiple of themselves play an important role in understanding the behavior of linear operators and square matrices.

Definitions Let T be a linear operator on $\mathcal{R}^n$. A *nonzero* vector $\mathbf{v}$ in $\mathcal{R}^n$ is called an **eigenvector** of T if $T(\mathbf{v})$ is a multiple of $\mathbf{v}$; that is, $T(\mathbf{v}) = \lambda\mathbf{v}$ for some scalar λ. The scalar[1] λ is called the **eigenvalue** of T that corresponds to $\mathbf{v}$.

For the reflection T of $\mathcal{R}^2$ about line $\mathcal{L}$ and the vectors $\mathbf{b}_1$ and $\mathbf{b}_2$, where $\mathbf{b}_1$ is in the direction of $\mathcal{L}$ and $\mathbf{b}_2$ is in a direction perpendicular to $\mathcal{L}$, we have

$$T(\mathbf{b}_1) = \mathbf{b}_1 = 1\mathbf{b}_1 \quad \text{and} \quad T(\mathbf{b}_2) = -\mathbf{b}_2 = (-1)\mathbf{b}_2.$$

Therefore $\mathbf{b}_1$ is an eigenvector of T with corresponding eigenvalue 1, and $\mathbf{b}_2$ is an eigenvector of T with corresponding eigenvalue -1. No nonzero vectors other than multiples of $\mathbf{b}_1$ and $\mathbf{b}_2$ have this property.

Since the action of a linear operator is the same as multiplication by its standard matrix, the concepts of eigenvector and eigenvalue can be defined similarly for square matrices.

Definitions Let A be an $n \times n$ matrix. A *nonzero* vector $\mathbf{v}$ in $\mathcal{R}^n$ is called an **eigenvector** of A if $A\mathbf{v} = \lambda\mathbf{v}$ for some scalar[2] λ. The scalar λ is called the **eigenvalue** of A that corresponds to $\mathbf{v}$.

For example, let $A = \begin{bmatrix} .6 & .8 \\ .8 & -.6 \end{bmatrix}$. In Section 4.5, we showed that A is the standard matrix of the reflection of $\mathcal{R}^2$ about the line $\mathcal{L}$ with the equation $y = \frac{1}{2}x$. Consider the vectors

$$\mathbf{u}_1 = \begin{bmatrix} -5 \\ 5 \end{bmatrix}, \quad \mathbf{u}_2 = \begin{bmatrix} 7 \\ 6 \end{bmatrix}, \quad \mathbf{b}_1 = \begin{bmatrix} 2 \\ 1 \end{bmatrix}, \quad \text{and} \quad \mathbf{b}_2 = \begin{bmatrix} 1 \\ -2 \end{bmatrix}.$$

It is a simple matter to verify directly that

$$A\mathbf{u}_1 = \begin{bmatrix} 1 \\ -7 \end{bmatrix}, \quad A\mathbf{u}_2 = \begin{bmatrix} 9 \\ 2 \end{bmatrix}, \quad A\mathbf{b}_1 = \begin{bmatrix} 2 \\ 1 \end{bmatrix} = 1\mathbf{b}_1, \quad \text{and} \quad A\mathbf{b}_2 = \begin{bmatrix} -1 \\ 2 \end{bmatrix} = (-1)\mathbf{b}_2.$$

Therefore neither $\mathbf{u}_1$ nor $\mathbf{u}_2$ is an eigenvector of A. However, $\mathbf{b}_1$ is an eigenvector of A with corresponding eigenvalue 1, and $\mathbf{b}_2$ is an eigenvector of A with corresponding eigenvalue -1. Because A is the standard matrix of the reflection T of $\mathcal{R}^2$ about the line $\mathcal{L}$, we can verify that $\mathbf{b}_1$ is an eigenvector of A without calculating the matrix product $A\mathbf{b}_1$. Indeed, since $\mathbf{b}_1$ is a vector in the direction of $\mathcal{L}$, we have $A\mathbf{b}_1 = T(\mathbf{b}_1) = \mathbf{b}_1$. Similarly, we can also verify that $\mathbf{b}_2$ is an eigenvector of A from the fact that $\mathbf{b}_2$ is an eigenvector of T.

In general, if T is a linear operator on $\mathcal{R}^n$, then T is a matrix transformation. Let A be the standard matrix of T. Since the equation $T(\mathbf{v}) = \lambda\mathbf{v}$ can be rewritten as $A\mathbf{v} = \lambda\mathbf{v}$, we can determine the eigenvalues and eigenvectors of T from A.

---

[1] In this book, we are concerned primarily with scalars that are real numbers. Therefore, unless an explicit statement is made to the contrary, the term *eigenvalue* should be interpreted as meaning *real eigenvalue.* There are situations, however, where it is useful to allow eigenvalues to be complex numbers. When complex eigenvalues are permitted, the definition of an eigenvector must be changed to allow vectors in $\mathcal{C}^n$, the set of n-tuples whose components are complex numbers.

[2] See footnote 1.

> The eigenvectors and corresponding eigenvalues of a linear operator are the same as those of its standard matrix.

In view of the relationship between a linear operator and its standard matrix, eigenvalues and eigenvectors of linear operators can be studied simultaneously with those of matrices. Example 1 shows how to verify that a given vector $\mathbf{v}$ is an eigenvector of a matrix A.

Example 1 For

$$\mathbf{v} = \begin{bmatrix} 1 \\ -1 \\ 1 \end{bmatrix} \quad \text{and} \quad A = \begin{bmatrix} 5 & 2 & 1 \\ -2 & 1 & -1 \\ 2 & 2 & 4 \end{bmatrix},$$

show that $\mathbf{v}$ is an eigenvector of A.

Solution Because $\mathbf{v}$ is nonzero, to verify that $\mathbf{v}$ is an eigenvector of A, we need only show that $A\mathbf{v}$ is a multiple of $\mathbf{v}$. Since

$$A\mathbf{v} = \begin{bmatrix} 5 & 2 & 1 \\ -2 & 1 & -1 \\ 2 & 2 & 4 \end{bmatrix} \begin{bmatrix} 1 \\ -1 \\ 1 \end{bmatrix} = \begin{bmatrix} 4 \\ -4 \\ 4 \end{bmatrix} = 4 \begin{bmatrix} 1 \\ -1 \\ 1 \end{bmatrix} = 4\mathbf{v},$$

we see that $\mathbf{v}$ is an eigenvector of A with corresponding eigenvalue 4.

Practice Problem 1 ▶ Show that

$$\mathbf{u} = \begin{bmatrix} -2 \\ 1 \\ 2 \end{bmatrix} \quad \text{and} \quad \mathbf{v} = \begin{bmatrix} 1 \\ -3 \\ 4 \end{bmatrix}$$

are eigenvectors of

$$A = \begin{bmatrix} 5 & 2 & 1 \\ -2 & 1 & -1 \\ 2 & 2 & 4 \end{bmatrix}.$$

To what eigenvalues do $\mathbf{u}$ and $\mathbf{v}$ correspond? ◀

An eigenvector $\mathbf{v}$ of a matrix A is associated with exactly one eigenvalue. For if $\lambda_1\mathbf{v} = A\mathbf{v} = \lambda_2\mathbf{v}$, then $\lambda_1 = \lambda_2$ because $\mathbf{v} \neq \mathbf{0}$. In contrast, if $\mathbf{v}$ is an eigenvector of A corresponding to eigenvalue λ, then every nonzero multiple of $\mathbf{v}$ is also an eigenvector of A corresponding to λ. For if $c \neq 0$, then

$$A(c\mathbf{v}) = cA\mathbf{v} = c\lambda\mathbf{v} = \lambda(c\mathbf{v}).$$

The process of finding the eigenvectors of an $n \times n$ matrix that correspond to a particular eigenvalue is also straightforward. Note that $\mathbf{v}$ is an eigenvector of A corresponding to the eigenvalue λ if and only if $\mathbf{v}$ is a nonzero vector such that

$$\begin{aligned} A\mathbf{v} &= \lambda\mathbf{v} \\ A\mathbf{v} - \lambda\mathbf{v} &= \mathbf{0} \\ A\mathbf{v} - \lambda I_n\mathbf{v} &= \mathbf{0} \\ (A - \lambda I_n)\mathbf{v} &= \mathbf{0}. \end{aligned}$$

Thus $\mathbf{v}$ is a nonzero solution of the system of linear equations $(A - \lambda I_n)\mathbf{x} = \mathbf{0}$.

Let A be an $n \times n$ matrix with eigenvalue λ. The eigenvectors of A corresponding to λ are the nonzero solutions of $(A - \lambda I_n)\mathbf{x} = \mathbf{0}$.

In this context, the set of solutions of $(A - \lambda I_n)\mathbf{x} = \mathbf{0}$ is called the **eigenspace of A corresponding to the eigenvalue λ**. This is just the null space of $A - \lambda I_n$, and hence it is a subspace of $\mathcal{R}^n$. Note that the eigenspace of A corresponding to λ consists of the zero vector and all the eigenvectors corresponding to λ.

Similarly, if λ is an eigenvalue of a linear operator T on $\mathcal{R}^n$, the set of vectors $\mathbf{v}$ in $\mathcal{R}^n$ such that $T(\mathbf{v}) = \lambda\mathbf{v}$ is called the **eigenspace** of T corresponding to λ. (See Figure 5.2.)

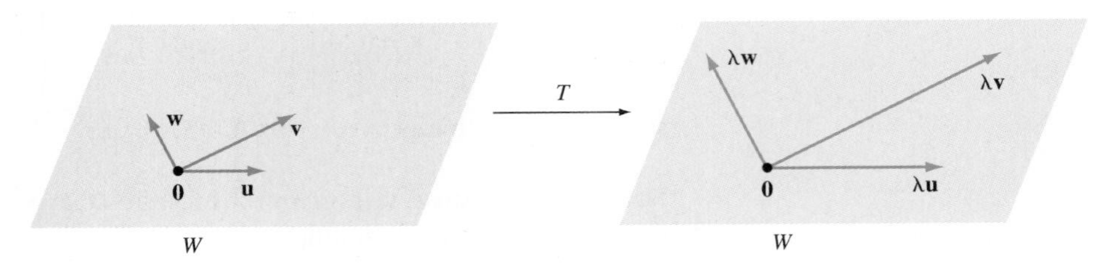

Figure 5.2 W is the eigenspace of T corresponding to eigenvalue λ.

In Section 5.4, we see that under certain conditions the bases for the various eigenspaces of a linear operator on $\mathcal{R}^n$ can be combined to form a basis for $\mathcal{R}^n$. This basis enables us to find a very simple matrix representation of the operator.

Example 2 Show that 3 and -2 are eigenvalues of the linear operator T on $\mathcal{R}^2$ defined by

$$T\left(\begin{bmatrix} x_1 \\ x_2 \end{bmatrix}\right) = \begin{bmatrix} -2x_2 \\ -3x_1 + x_2 \end{bmatrix},$$

and find bases for the corresponding eigenspaces.

Solution The standard matrix of T is

$$A = \begin{bmatrix} 0 & -2 \\ -3 & 1 \end{bmatrix}.$$

To show that 3 is an eigenvalue of T, we show that 3 is an eigenvalue of A. Thus we must find a nonzero vector $\mathbf{u}$ such that $A\mathbf{u} = 3\mathbf{u}$. In other words, we must show that the solution set of the system of equations $(A - 3I_2)\mathbf{x} = \mathbf{0}$, which is the null space of $A - 3I_2$, contains nonzero vectors. The reduced row echelon form of $A - 3I_2$ is

$$\begin{bmatrix} 1 & \frac{2}{3} \\ 0 & 0 \end{bmatrix}.$$

Because this is a matrix of nullity

$$2 - \text{rank}\,(A - 3I_2) = 2 - 1 = 1,$$

nonzero solutions exist, and the eigenspace corresponding to the eigenvalue 3 has dimension equal to 1. Furthermore, we see that the eigenvectors of A corresponding to the eigenvalue 3 have the form

$$\begin{bmatrix} x_1 \\ x_2 \end{bmatrix} = \begin{bmatrix} -\frac{2}{3}x_2 \\ x_2 \end{bmatrix} = x_2 \begin{bmatrix} -\frac{2}{3} \\ 1 \end{bmatrix}$$

for $x_2 \neq 0$. It follows that

$$\left\{\begin{bmatrix} -\frac{2}{3} \\ 1 \end{bmatrix}\right\}$$

is a basis for the eigenspace of A corresponding to the eigenvalue 3. Notice that by taking $x_2 = 3$ in the preceding calculations, we obtain another basis for this eigenspace (one consisting of a vector with integer components), namely,

$$\left\{\begin{bmatrix} -2 \\ 3 \end{bmatrix}\right\}.$$

In a similar fashion, we must show that there is a nonzero vector $\mathbf{v}$ such that $A\mathbf{v} = (-2)\mathbf{v}$. In this case, we must show that the system of equations $(A + 2I_2)\mathbf{x} = \mathbf{0}$ has nonzero solutions. Since the reduced row echelon form of $A + 2I_2$ is

$$\begin{bmatrix} 1 & -1 \\ 0 & 0 \end{bmatrix},$$

another matrix of nullity 1, the eigenspace corresponding to the eigenvalue -2 also has dimension equal to 1. From the vector form of the general solution of this system, we see that the eigenvectors corresponding to the eigenvalue -2 have the form

$$\begin{bmatrix} x_1 \\ x_2 \end{bmatrix} = \begin{bmatrix} x_2 \\ x_2 \end{bmatrix} = x_2 \begin{bmatrix} 1 \\ 1 \end{bmatrix}$$

for $x_2 \neq 0$. Thus a basis for the eigenspace of A corresponding to the eigenvalue -2 is

$$\left\{\begin{bmatrix} 1 \\ 1 \end{bmatrix}\right\}.$$

(See Figures 5.3 and 5.4.)

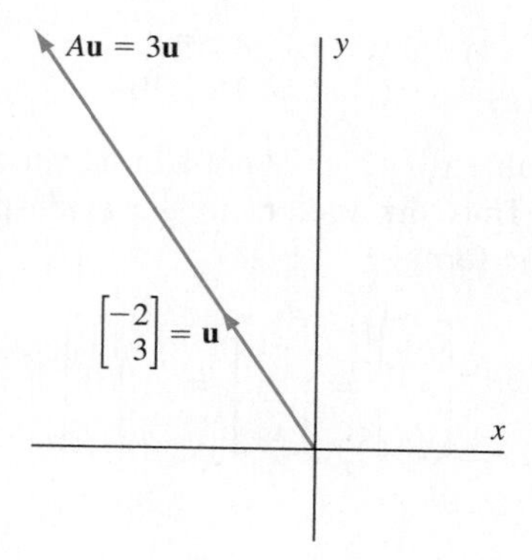

Figure 5.3 A basis vector for the eigenspace of A corresponding to eigenvalue 3

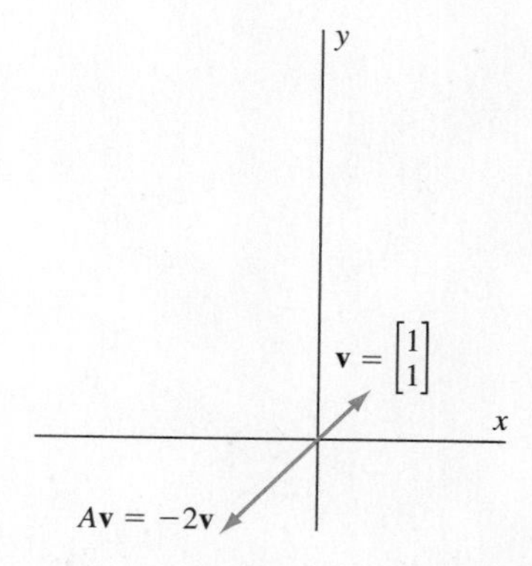

Figure 5.4 A basis vector for the eigenspace of A corresponding to eigenvalue -2

Since 3 and -2 are eigenvalues of A, they are also eigenvalues of T. Moreover, the corresponding eigenspaces of T have the same bases as those of A, namely,

$$\left\{\begin{bmatrix} -2 \\ 3 \end{bmatrix}\right\} \quad \text{and} \quad \left\{\begin{bmatrix} 1 \\ 1 \end{bmatrix}\right\}.$$

Practice Problem 2 ▶ Show that 1 is an eigenvalue of the linear operator on $\mathcal{R}^3$ defined by

$$T\left(\begin{bmatrix} x_1 \\ x_2 \\ x_3 \end{bmatrix}\right) = \begin{bmatrix} x_1 + 2x_2 \\ -x_1 - x_2 + x_3 \\ x_2 + x_3 \end{bmatrix},$$

and find a basis for the corresponding eigenspace. ◀

The two eigenspaces in Example 2 each have dimension 1. This need not always be the case, as our next example shows.

Example 3 Show that 3 is an eigenvalue of

$$B = \begin{bmatrix} 3 & 0 & 0 \\ 0 & 1 & 2 \\ 0 & 2 & 1 \end{bmatrix},$$

and find a basis for the corresponding eigenspace.

Solution As in Example 2, we must show that the null space of $B - 3I_3$ contains nonzero vectors. The reduced row echelon form of $B - 3I_3$ is

$$\begin{bmatrix} 0 & 1 & -1 \\ 0 & 0 & 0 \\ 0 & 0 & 0 \end{bmatrix}.$$

Hence the vectors in the eigenspace of B corresponding to the eigenvalue 3 satisfy $x_2 - x_3 = 0$, and so the general solution of $(B - 3I_3)\mathbf{x} = \mathbf{0}$ is

$$\begin{aligned} &x_1 \quad \text{free} \\ &x_2 = x_3 \\ &x_3 \quad \text{free.} \end{aligned}$$

(Notice that the variable x_1, which is not a basic variable in the equation $x_2 - x_3 = 0$, is a free variable.) Thus the vectors in the eigenspace of B corresponding to the eigenvalue 3 have the form

$$\begin{bmatrix} x_1 \\ x_2 \\ x_3 \end{bmatrix} = \begin{bmatrix} x_1 \\ x_3 \\ x_3 \end{bmatrix} = x_1 \begin{bmatrix} 1 \\ 0 \\ 0 \end{bmatrix} + x_3 \begin{bmatrix} 0 \\ 1 \\ 1 \end{bmatrix}.$$

Therefore

$$\left\{ \begin{bmatrix} 1 \\ 0 \\ 0 \end{bmatrix}, \begin{bmatrix} 0 \\ 1 \\ 1 \end{bmatrix} \right\}$$

is a basis for the eigenspace of B corresponding to the eigenvalue 3.

Not all square matrices and linear operators on $\mathcal{R}^n$ have real eigenvalues. (Such matrices and operators have no eigenvectors with components that are real numbers either.) Consider, for example, the linear operator T on $\mathcal{R}^2$ that rotates a vector by $90°$. If this operator had a real eigenvalue λ, then there would be a nonzero vector $\mathbf{v}$ in $\mathcal{R}^2$ such that $T(\mathbf{v}) = \lambda\mathbf{v}$. But for any nonzero vector $\mathbf{v}$, the vector $T(\mathbf{v})$ obtained by rotating $\mathbf{v}$ through 90° is not a multiple of $\mathbf{v}$. (See Figure 5.5.) Hence $\mathbf{v}$ cannot be an eigenvector of T, so T has no real eigenvalues. Note that this argument also shows that the standard matrix of T, which is the 90°-rotation matrix

$$\begin{bmatrix} 0 & -1 \\ 1 & 0 \end{bmatrix},$$

has no real eigenvalues.

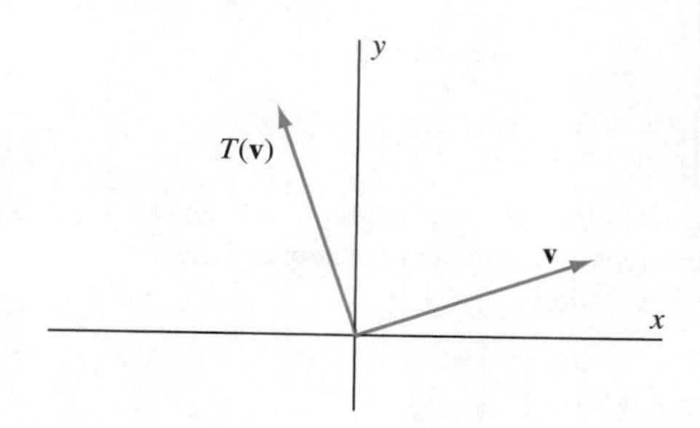

Figure 5.5 The image of **v** is not a multiple of **v**.

EXERCISES

In Exercises 1–12, a matrix and a vector are given. Show that the vector is an eigenvector of the matrix and determine the corresponding eigenvalue.

1. $\begin{bmatrix} -10 & -8 \\ 24 & 18 \end{bmatrix}, \begin{bmatrix} 1 \\ -2 \end{bmatrix}$

2. $\begin{bmatrix} 12 & -14 \\ 7 & -9 \end{bmatrix}, \begin{bmatrix} 1 \\ 1 \end{bmatrix}$

3. $\begin{bmatrix} -5 & -4 \\ 8 & 7 \end{bmatrix}, \begin{bmatrix} 1 \\ -2 \end{bmatrix}$

4. $\begin{bmatrix} 15 & 24 \\ -4 & -5 \end{bmatrix}, \begin{bmatrix} -2 \\ 1 \end{bmatrix}$

5. $\begin{bmatrix} 19 & -7 \\ 42 & -16 \end{bmatrix}, \begin{bmatrix} 1 \\ 3 \end{bmatrix}$

6. $\begin{bmatrix} -9 & -8 & 5 \\ 7 & 6 & -5 \\ -6 & -6 & 4 \end{bmatrix}, \begin{bmatrix} 3 \\ -2 \\ 1 \end{bmatrix}$

7. $\begin{bmatrix} 4 & 6 & -5 \\ 9 & 7 & -11 \\ 8 & 8 & -11 \end{bmatrix}, \begin{bmatrix} -1 \\ 2 \\ 1 \end{bmatrix}$

8. $\begin{bmatrix} -3 & 14 & 10 \\ -2 & 5 & 2 \\ 2 & -10 & -7 \end{bmatrix}, \begin{bmatrix} -3 \\ -1 \\ 2 \end{bmatrix}$

9. $\begin{bmatrix} 2 & -6 & 6 \\ 1 & 9 & -6 \\ -2 & 16 & -13 \end{bmatrix}, \begin{bmatrix} -1 \\ 1 \\ 2 \end{bmatrix}$

10. $\begin{bmatrix} -5 & -1 & 2 \\ 2 & -1 & -2 \\ -7 & -2 & 2 \end{bmatrix}, \begin{bmatrix} 1 \\ -2 \\ 1 \end{bmatrix}$

11. $\begin{bmatrix} 5 & 6 & 12 \\ 3 & 2 & 6 \\ -3 & -3 & -7 \end{bmatrix}, \begin{bmatrix} -2 \\ -1 \\ 1 \end{bmatrix}$

12. $\begin{bmatrix} 6 & 5 & 15 \\ 5 & 6 & 15 \\ -5 & -5 & -14 \end{bmatrix}, \begin{bmatrix} -1 \\ -1 \\ 1 \end{bmatrix}$

In Exercises 13–24, a matrix and a scalar λ are given. Show that λ is an eigenvalue of the matrix and determine a basis for its eigenspace.

13. $\begin{bmatrix} 10 & 7 \\ -14 & -11 \end{bmatrix}, \lambda = 3$

14. $\begin{bmatrix} -11 & 14 \\ -7 & 10 \end{bmatrix}, \lambda = -4$

15. $\begin{bmatrix} 11 & 18 \\ -3 & -4 \end{bmatrix}, \lambda = 5$

16. $\begin{bmatrix} -11 & 5 \\ -30 & 14 \end{bmatrix}, \lambda = -1$

17. $\begin{bmatrix} -2 & -5 & 2 \\ 4 & 7 & -2 \\ -3 & -3 & 5 \end{bmatrix}, \lambda = 3$

18. $\begin{bmatrix} 6 & 9 & -10 \\ 6 & 3 & -4 \\ 7 & 7 & -9 \end{bmatrix}, \lambda = 5$

19. $\begin{bmatrix} -3 & 12 & 6 \\ -3 & 6 & 0 \\ 3 & -9 & -3 \end{bmatrix}, \lambda = 0$

20. $\begin{bmatrix} 3 & -2 & 2 \\ -4 & 1 & -2 \\ -5 & 1 & -2 \end{bmatrix}, \lambda = 2$

21. $\begin{bmatrix} -13 & -4 & 8 \\ 24 & 7 & -16 \\ -12 & -4 & 7 \end{bmatrix}$, $\lambda = -1$

22. $\begin{bmatrix} -2 & -2 & -4 \\ -1 & -1 & -2 \\ 1 & 1 & 2 \end{bmatrix}$, $\lambda = 0$

23. $\begin{bmatrix} 4 & -3 & -3 \\ -3 & 4 & 3 \\ 3 & -3 & -2 \end{bmatrix}$, $\lambda = 1$

24. $\begin{bmatrix} 5 & 3 & 9 \\ 3 & 5 & 9 \\ -3 & -3 & -7 \end{bmatrix}$, $\lambda = 2$

In Exercises 25–32, a linear operator and a vector are given. Show that the vector is an eigenvector of the operator and determine the corresponding eigenvalue.

25. $T\left(\begin{bmatrix} x_1 \\ x_2 \end{bmatrix}\right) = \begin{bmatrix} -3x_1 - 6x_2 \\ 12x_1 + 14x_2 \end{bmatrix}$, $\begin{bmatrix} -2 \\ 3 \end{bmatrix}$

26. $T\left(\begin{bmatrix} x_1 \\ x_2 \end{bmatrix}\right) = \begin{bmatrix} 8x_1 - 2x_2 \\ 6x_1 + x_2 \end{bmatrix}$, $\begin{bmatrix} 1 \\ 2 \end{bmatrix}$

27. $T\left(\begin{bmatrix} x_1 \\ x_2 \end{bmatrix}\right) = \begin{bmatrix} -12x_1 - 12x_2 \\ 20x_1 + 19x_2 \end{bmatrix}$, $\begin{bmatrix} -3 \\ 4 \end{bmatrix}$

28. $T\left(\begin{bmatrix} x_1 \\ x_2 \end{bmatrix}\right) = \begin{bmatrix} 14x_1 - 6x_2 \\ 18x_1 - 7x_2 \end{bmatrix}$, $\begin{bmatrix} 2 \\ 3 \end{bmatrix}$

29. $T\left(\begin{bmatrix} x_1 \\ x_2 \\ x_3 \end{bmatrix}\right) = \begin{bmatrix} -8x_1 + 9x_2 - 3x_3 \\ -5x_1 + 6x_2 - 3x_3 \\ -x_1 + x_2 - 2x_3 \end{bmatrix}$, $\begin{bmatrix} 3 \\ 2 \\ 1 \end{bmatrix}$

30. $T\left(\begin{bmatrix} x_1 \\ x_2 \\ x_3 \end{bmatrix}\right) = \begin{bmatrix} -2x_1 - x_2 - 3x_3 \\ -3x_1 - 4x_2 - 9x_3 \\ x_1 + x_2 + 2x_3 \end{bmatrix}$, $\begin{bmatrix} -1 \\ -3 \\ 1 \end{bmatrix}$

31. $T\left(\begin{bmatrix} x_1 \\ x_2 \\ x_3 \end{bmatrix}\right) = \begin{bmatrix} 6x_1 + x_2 - 2x_3 \\ -6x_1 + x_2 + 6x_3 \\ -2x_1 - x_2 + 6x_3 \end{bmatrix}$, $\begin{bmatrix} -1 \\ 3 \\ 1 \end{bmatrix}$

32. $T\left(\begin{bmatrix} x_1 \\ x_2 \\ x_3 \end{bmatrix}\right) = \begin{bmatrix} 4x_1 + 9x_2 + 8x_3 \\ -2x_1 - x_2 - 2x_3 \\ 2x_1 - 3x_2 - 2x_3 \end{bmatrix}$, $\begin{bmatrix} -1 \\ 0 \\ 1 \end{bmatrix}$

In Exercises 33–40, a linear operator and a scalar λ are given. Show that λ is an eigenvalue of the operator and determine a basis for its eigenspace.

33. $T\left(\begin{bmatrix} x_1 \\ x_2 \end{bmatrix}\right) = \begin{bmatrix} x_1 - 2x_2 \\ 6x_1 - 6x_2 \end{bmatrix}$, $\lambda = -2$

34. $T\left(\begin{bmatrix} x_1 \\ x_2 \end{bmatrix}\right) = \begin{bmatrix} 4x_1 + 6x_2 \\ -12x_1 - 13x_2 \end{bmatrix}$, $\lambda = -5$

35. $T\left(\begin{bmatrix} x_1 \\ x_2 \end{bmatrix}\right) = \begin{bmatrix} 20x_1 + 8x_2 \\ -24x_1 - 8x_2 \end{bmatrix}$, $\lambda = 8$

36. $T\left(\begin{bmatrix} x_1 \\ x_2 \end{bmatrix}\right) = \begin{bmatrix} -x_1 + 2x_2 \\ -6x_1 + 6x_2 \end{bmatrix}$, $\lambda = 3$

37. $T\left(\begin{bmatrix} x_1 \\ x_2 \\ x_3 \end{bmatrix}\right) = \begin{bmatrix} x_1 - x_2 - 3x_3 \\ -3x_1 - x_2 - 9x_3 \\ x_1 + x_2 + 5x_3 \end{bmatrix}$, $\lambda = 2$

38. $T\left(\begin{bmatrix} x_1 \\ x_2 \\ x_3 \end{bmatrix}\right) = \begin{bmatrix} 4x_1 - 2x_2 - 5x_3 \\ 3x_1 - x_2 - 5x_3 \\ 4x_1 - 4x_2 - 3x_3 \end{bmatrix}$, $\lambda = -3$

39. $T\left(\begin{bmatrix} x_1 \\ x_2 \\ x_3 \end{bmatrix}\right) = \begin{bmatrix} x_1 + 4x_2 + 5x_3 \\ 2x_1 + 6x_2 + 2x_3 \\ -2x_1 - 10x_2 - 6x_3 \end{bmatrix}$, $\lambda = 3$

40. $T\left(\begin{bmatrix} x_1 \\ x_2 \\ x_3 \end{bmatrix}\right) = \begin{bmatrix} 5x_1 + 2x_2 - 4x_3 \\ -12x_1 - 5x_2 + 12x_3 \\ -4x_1 - 2x_2 + 5x_3 \end{bmatrix}$, $\lambda = 1$

T&F *In Exercises 41–60, determine whether the statements are true or false.*

41. If $A\mathbf{v} = \lambda\mathbf{v}$ for some vector $\mathbf{v}$, then λ is an eigenvalue of the matrix A.
42. If $A\mathbf{v} = \lambda\mathbf{v}$ for some vector $\mathbf{v}$, then $\mathbf{v}$ is an eigenvector of the matrix A.
43. A scalar λ is an eigenvalue of an $n \times n$ matrix A if and only if the equation $(A - \lambda I_n)\mathbf{x} = \mathbf{0}$ has a nonzero solution.
44. If $\mathbf{v}$ is an eigenvector of a matrix, then there is a unique eigenvalue of the matrix that corresponds to $\mathbf{v}$.
45. If λ is an eigenvalue of a linear operator, then there are infinitely many eigenvectors of the operator that correspond to λ.
46. The eigenspace of an $n \times n$ matrix A corresponding to an eigenvalue λ is the column space of $A - \lambda I_n$.
47. The eigenvalues of a linear operator on $\mathcal{R}^n$ are the same as those of its standard matrix.
48. The eigenspaces of a linear operator on $\mathcal{R}^n$ are the same as those of its standard matrix.
49. Every linear operator on $\mathcal{R}^n$ has real eigenvalues.
50. Only square matrices have eigenvalues.
51. Every vector in the eigenspace of a matrix A corresponding to an eigenvalue λ is an eigenvector corresponding to λ.
52. The linear operator on $\mathcal{R}^2$ that rotates a vector through the angle θ, where $0° < \theta < 180°$, has no eigenvectors.
53. The standard matrix of the linear operator on $\mathcal{R}^2$ that rotates a vector through the angle θ, where $0° < \theta < 180°$, has no eigenvalues.
54. If a nonzero vector $\mathbf{v}$ is in the null space of a linear operator T, then $\mathbf{v}$ is an eigenvector of T.
55. If $\mathbf{v}$ is an eigenvector of a matrix A, then $c\mathbf{v}$ is also an eigenvector for any scalar c.
56. If $\mathbf{v}$ is an eigenvector of a matrix A, then $c\mathbf{v}$ is also an eigenvector for any nonzero scalar c.
57. If A and B are $n \times n$ matrices and λ is an eigenvalue of both A and B, then λ is an eigenvalue of $A + B$.
58. If A and B are $n \times n$ matrices and $\mathbf{v}$ is an eigenvector of both A and B, then $\mathbf{v}$ is an eigenvector of $A + B$.
59. If A and B are $n \times n$ matrices and λ is an eigenvalue of both A and B, then λ is an eigenvalue of AB.

60. If A and B are $n \times n$ matrices and $\mathbf{v}$ is an eigenvector of both A and B, then $\mathbf{v}$ is an eigenvector of AB.

---

61. What are the eigenvalues of the identity operator on $\mathcal{R}^n$? Justify your answer. Describe each eigenspace.
62. What are the eigenvalues of the zero operator on $\mathcal{R}^n$? Justify your answer. Describe each eigenspace.
63. Prove that if $\mathbf{v}$ is an eigenvector of a matrix A, then for any nonzero scalar c, $c\mathbf{v}$ is also an eigenvector of A.
64. Prove that if $\mathbf{v}$ is an eigenvector of a matrix A, then there is a unique scalar λ such that $A\mathbf{v} = \lambda\mathbf{v}$.
65. Suppose that 0 is an eigenvalue of a matrix A. Give another name for the eigenspace of A corresponding to 0.
66. Prove that a square matrix is invertible if and only if 0 is not an eigenvalue.
67. Prove that if λ is an eigenvalue of an invertible matrix A, then $\lambda \neq 0$ and $1/\lambda$ is an eigenvalue of A^{-1}.
68. Suppose that A is a square matrix in which the sum of the entries of each row equals the same scalar r. Show that r is an eigenvalue of A by finding an eigenvector of A corresponding to r.
69. Prove that if λ is an eigenvalue of a matrix A, then λ^2 is an eigenvalue of A^2.
70. State and prove a generalization of Exercise 69.
71. Determine necessary and sufficient conditions on a vector $\mathbf{v}$ such that the span of $\{A\mathbf{v}\}$ equals the span of $\{\mathbf{v}\}$.
72. An $n \times n$ matrix A is called *nilpotent* if, for some positive integer k, $A^k = O$, where O is the $n \times n$ zero matrix. Prove that 0 is the only eigenvalue of a nilpotent matrix.
73. Let $\mathbf{v}_1$ and $\mathbf{v}_2$ be eigenvectors of a linear operator T on $\mathcal{R}^n$, and let λ_1 and λ_2, respectively, be the corresponding eigenvalues. Prove that if $\lambda_1 \neq \lambda_2$, then $\{\mathbf{v}_1, \mathbf{v}_2\}$ is linearly independent.
74. Let $\mathbf{v}_1$, $\mathbf{v}_2$, and $\mathbf{v}_3$ be eigenvectors of a linear operator T on $\mathcal{R}^n$, and let λ_1, λ_2, and λ_3, respectively, be the corresponding eigenvalues. Prove that if these eigenvalues are distinct, then $\{\mathbf{v}_1, \mathbf{v}_2, \mathbf{v}_3\}$ is linearly independent. *Hint:* Letting

$$c_1\mathbf{v}_1 + c_2\mathbf{v}_2 + c_3\mathbf{v}_3 = \mathbf{0}$$

for scalars c_1, c_2, and c_3, apply T to both sides of this equation. Then multiply both sides of this equation by λ_3, and subtract it from the first equation. Now use the result of Exercise 73.

75. Let T be a linear operator on $\mathcal{R}^2$ with an eigenspace of dimension 2. Prove that $T = \lambda I$ for some scalar λ.

In Exercises 76–82, use either a calculator with matrix capabilities or computer software such as MATLAB to solve each problem.

76. Let

$$A = \begin{bmatrix} -1.9 & 14.4 & -8.4 & 34.8 \\ 1.6 & -2.7 & 3.2 & -1.6 \\ 1.2 & -8.0 & 4.7 & -18.2 \\ 1.6 & -1.6 & 3.2 & -2.7 \end{bmatrix}.$$

Show that

$$\mathbf{v}_1 = \begin{bmatrix} -9 \\ 1 \\ 5 \\ 1 \end{bmatrix}, \quad \mathbf{v}_2 = \begin{bmatrix} -2 \\ 0 \\ 1 \\ 0 \end{bmatrix}, \quad \mathbf{v}_3 = \begin{bmatrix} -3 \\ 1 \\ 2 \\ 0 \end{bmatrix}, \quad \text{and}$$

$$\mathbf{v}_4 = \begin{bmatrix} -3 \\ -5 \\ 0 \\ 2 \end{bmatrix}$$

are eigenvectors of A. What are the eigenvalues corresponding to each of these eigenvectors?

77. Are the eigenvalues of A (determined in Exercise 76) also eigenvalues of $3A$? If so, find an eigenvector corresponding to each eigenvalue.
78. Are $\mathbf{v}_1$, $\mathbf{v}_2$, $\mathbf{v}_3$, and $\mathbf{v}_4$ in Exercise 76 also eigenvectors of $3A$? If so, what eigenvalue corresponds to each of these eigenvectors?
79. (a) Based on the results of Exercises 76–78, make a conjecture about the relationship between the eigenvalues of an $n \times n$ matrix B and those of cB, where c is a nonzero scalar.

    (b) Based on the results of Exercises 76–78, make a conjecture about the relationship between the eigenvectors of an $n \times n$ matrix B and those of cB, where c is a nonzero scalar.

    (c) Justify the conjectures made in (a) and (b).
80. Are $\mathbf{v}_1$, $\mathbf{v}_2$, $\mathbf{v}_3$, and $\mathbf{v}_4$ in Exercise 76 also eigenvectors of A^T? If so, what eigenvalue corresponds to each of these eigenvectors?
81. Are the eigenvalues of A (determined in Exercise 76) also eigenvalues of A^T? If so, find an eigenvector corresponding to each eigenvalue.
82. Based on the results of Exercises 80 and 81, make a conjecture about any possible relationship between the eigenvalues or eigenvectors of an $n \times n$ matrix B and those of B^T.

SOLUTIONS TO THE PRACTICE PROBLEMS

1. Since

$$A\mathbf{u} = \begin{bmatrix} 5 & 2 & 1 \\ -2 & 1 & -1 \\ 2 & 2 & 4 \end{bmatrix} \begin{bmatrix} -2 \\ 1 \\ 2 \end{bmatrix} = \begin{bmatrix} -6 \\ 3 \\ 6 \end{bmatrix} = 3 \begin{bmatrix} -2 \\ 1 \\ 2 \end{bmatrix} = 3\mathbf{u},$$

$\mathbf{u}$ is an eigenvector of A corresponding to the eigenvalue 3. Likewise,

$$A\mathbf{v} = \begin{bmatrix} 5 & 2 & 1 \\ -2 & 1 & -1 \\ 2 & 2 & 4 \end{bmatrix} \begin{bmatrix} 1 \\ -3 \\ 4 \end{bmatrix} = \begin{bmatrix} 3 \\ -9 \\ 12 \end{bmatrix} = 3 \begin{bmatrix} 1 \\ -3 \\ 4 \end{bmatrix} = 3\mathbf{v},$$

so $\mathbf{v}$ is also an eigenvector of A corresponding to the eigenvalue 3.

2. The standard matrix of T is

$$A = \begin{bmatrix} 1 & 2 & 0 \\ -1 & -1 & 1 \\ 0 & 1 & 1 \end{bmatrix},$$

and the row echelon form of $A - I_3$ is

$$\begin{bmatrix} 1 & 0 & -1 \\ 0 & 1 & 0 \\ 0 & 0 & 0 \end{bmatrix}.$$

Therefore the vector form of the general solution of $(A - I_3)\mathbf{x} = \mathbf{0}$ is

$$\begin{bmatrix} x_1 \\ x_2 \\ x_3 \end{bmatrix} = \begin{bmatrix} x_3 \\ 0 \\ x_3 \end{bmatrix} = x_3 \begin{bmatrix} 1 \\ 0 \\ 1 \end{bmatrix},$$

so

$$\left\{ \begin{bmatrix} 1 \\ 0 \\ 1 \end{bmatrix} \right\}$$

is a basis for the eigenspace of T corresponding to eigenvalue 1.

5.2 THE CHARACTERISTIC POLYNOMIAL

In Section 5.1, we learned how to find the eigenvalue corresponding to a given eigenvector and the eigenvectors for a given eigenvalue. But ordinarily, we know neither the eigenvalues nor the eigenvectors of a matrix. Suppose, for instance, that we want to find the eigenvalues and the eigenspaces of an $n \times n$ matrix A. If λ is an eigenvalue of A, there must be a nonzero vector $\mathbf{v}$ in $\mathcal{R}^n$ such that $A\mathbf{v} = \lambda\mathbf{v}$. Thus, as on page 295, $\mathbf{v}$ is a nonzero solution of $(A - \lambda I_n)\mathbf{x} = \mathbf{0}$. But in order for the homogeneous system of linear equations $(A - \lambda I_n)\mathbf{x} = \mathbf{0}$ to have nonzero solutions, the rank of $A - \lambda I_n$ must be less than n. Hence, by the Invertible Matrix Theorem, the $n \times n$ matrix $A - \lambda I_n$ is not invertible, so its determinant must be 0. Because these steps are all reversible, we have the following result:

> The eigenvalues of a square matrix A are the values of t that satisfy
>
> $$\det(A - tI_n) = 0.$$

The equation $\det(A - tI_n) = 0$ is called the **characteristic equation** of A, and $\det(A - tI_n)$ is called the **characteristic polynomial** of A. Thus the eigenvalues of the matrix A are the (real) roots of the characteristic polynomial of A.

Example 1 Determine the eigenvalues of

$$A = \begin{bmatrix} -4 & -3 \\ 3 & 6 \end{bmatrix},$$

and then find a basis for each eigenspace.

Solution We begin by forming the matrix

$$A - tI_2 = \begin{bmatrix} -4-t & -3 \\ 3 & 6-t \end{bmatrix}.$$

The characteristic polynomial of A is the determinant of this matrix, which is

$$\begin{aligned} \det(A - tI_2) &= (-4-t)(6-t) - (-3)\cdot 3 \\ &= (-24 - 2t + t^2) + 9 \\ &= t^2 - 2t - 15 \\ &= (t+3)(t-5). \end{aligned}$$

Therefore the roots of the characteristic polynomial are -3 and 5; so these are the eigenvalues of A.

As in Section 5.1, we solve $(A + 3I_2)\mathbf{x} = \mathbf{0}$ and $(A - 5I_2)\mathbf{x} = \mathbf{0}$ to find bases for the eigenspaces. Since the reduced row echelon form of $A + 3I_2$ is

$$\begin{bmatrix} 1 & 3 \\ 0 & 0 \end{bmatrix},$$

the vector form of the general solution of $(A + 3I_2)\mathbf{x} = \mathbf{0}$ is

$$\begin{bmatrix} x_1 \\ x_2 \end{bmatrix} = \begin{bmatrix} -3x_2 \\ x_2 \end{bmatrix} = x_2 \begin{bmatrix} -3 \\ 1 \end{bmatrix}.$$

Hence

$$\left\{ \begin{bmatrix} -3 \\ 1 \end{bmatrix} \right\}$$

is a basis for the eigenspace of A corresponding to the eigenvalue -3.

In a similar manner, the reduced row echelon form of $A - 5I_2$, which is

$$\begin{bmatrix} 1 & \frac{1}{3} \\ 0 & 0 \end{bmatrix},$$

produces the basis

$$\left\{ \begin{bmatrix} -1 \\ 3 \end{bmatrix} \right\}$$

for the eigenspace of A corresponding to the eigenvalue 5.

The characteristic polynomial of the 2×2 matrix in Example 1 is $t^2 - 2t - 15 = (t + 3)(t - 5)$, a polynomial of degree 2. In general, *the characteristic polynomial of an $n \times n$ matrix is a polynomial of degree n.*

! CAUTION Note that the reduced row echelon form of the matrix A in Example 1 is I_2, which has the characteristic polynomial $(t - 1)^2$. Thus the characteristic polynomial of a matrix is not usually equal to the characteristic polynomial of its reduced row echelon form. In general, therefore, the eigenvalues of a matrix and its reduced row echelon form are not the same. Likewise, the eigenvectors of a matrix and its reduced row echelon form are not usually the same. Consequently, there is no way to apply elementary row operations to a matrix in hopes of finding its eigenvalues or eigenvectors.

Computing the characteristic polynomial of a 2×2 matrix is straightforward, as Example 1 shows. On the other hand, calculating the characteristic polynomial of a larger matrix by hand can be quite tedious. Although a programmable calculator or computer software can be used to determine the characteristic polynomial of matrices that are not too large, there is no way to find the precise roots of the characteristic polynomial of an arbitrary matrix that is larger than 4×4. Hence, for large matrices, a numerical method is usually used to approximate the eigenvalues. Because of the difficulty in computing the characteristic polynomial, in this book we usually use only 2×2 and 3×3 matrices in our examples and exercises.

It follows from Theorem 3.2, however, that there are some matrices for which the eigenvalues can be easily determined:

> The eigenvalues of an upper triangular or lower triangular matrix are its diagonal entries.

Example 2 Determine the eigenvalues of the matrix

$$A = \begin{bmatrix} -3 & -1 & -7 & 1 \\ 0 & 6 & 9 & -2 \\ 0 & 0 & -5 & 3 \\ 0 & 0 & 0 & 8 \end{bmatrix}.$$

Solution Because A is an upper triangular matrix, its eigenvalues are its diagonal entries, which are -3, 6, -5, and 8.

Practice Problem 1 ▶ Determine the eigenvalues of the matrix

$$\begin{bmatrix} 4 & 0 & 0 & 0 \\ -2 & -1 & 0 & 0 \\ 8 & 7 & -2 & 0 \\ 9 & -5 & 6 & 3 \end{bmatrix}.$$ ◀

Recall that the problem of finding eigenvalues and eigenvectors of a linear operator can be replaced by the corresponding problem for its standard matrix. In the context of a linear operator T, the characteristic equation of the standard matrix of T is called the **characteristic equation** of T, and the characteristic polynomial of the standard matrix of T is called the **characteristic polynomial** of T. Thus *the characteristic polynomial of a linear operator T on $\mathcal{R}^n$ is a polynomial of degree n whose roots are the eigenvalues of T.*

Example 3 We have noted on page 298 that the linear operator T on $\mathcal{R}^2$ that rotates a vector by 90° has no real eigenvalues. Equivalently, the 90°-rotation matrix has no real eigenvalues. In fact, the characteristic polynomial of T, which is also the characteristic polynomial of the 90°-rotation matrix, is given by

$$\det\left(A_{90^\circ} - tI_2\right) = \det\left(\begin{bmatrix} 0 & -1 \\ 1 & 0 \end{bmatrix} - tI_2\right) = \det\begin{bmatrix} -t & -1 \\ 1 & -t \end{bmatrix} = t^2 + 1,$$

which has no real roots. This confirms the observation made in Section 5.1 that T, and hence the 90°-rotation matrix, has no real eigenvalues.

THE MULTIPLICITY OF AN EIGENVALUE

Consider the matrix

$$A = \begin{bmatrix} -1 & 0 & 0 \\ 0 & 1 & 2 \\ 0 & 2 & 1 \end{bmatrix}.$$

Using the cofactor expansion along the first row, we see that

$$\det(A - tI_3) = \det\begin{bmatrix} -1-t & 0 & 0 \\ 0 & 1-t & 2 \\ 0 & 2 & 1-t \end{bmatrix}$$

$$= (-1-t)\cdot\det\begin{bmatrix} 1-t & 2 \\ 2 & 1-t \end{bmatrix}$$

$$\begin{aligned} &= (-1-t)[(1-t)^2 - 4] \\ &= (-1-t)(t^2 - 2t - 3) \\ &= -(t+1)(t+1)(t-3) \\ &= -(t+1)^2(t-3). \end{aligned}$$

Hence the eigenvalues of A are -1 and 3. A similar calculation shows that the characteristic polynomial of

$$B = \begin{bmatrix} 3 & 0 & 0 \\ 0 & 1 & 2 \\ 0 & 2 & 1 \end{bmatrix}$$

is $-(t+1)(t-3)^2$. Therefore the eigenvalues of B are also -1 and 3. But, as we explain next, the status of the eigenvalues -1 and 3 is different in A and B.

If λ is an eigenvalue of an $n \times n$ matrix M, then the largest positive integer k such that $(t - \lambda)^k$ is a factor of the characteristic polynomial of M is called the **multiplicity**[3] of λ. Thus, if

$$\det(M - tI_n) = (t-5)^2(t+6)(t-7)^3(t-8)^4,$$

then the eigenvalues of M are 5, which has multiplicity 2; -6, which has multiplicity 1; 7, which has multiplicity 3; and 8, which has multiplicity 4.

Practice Problem 2 ▶ If the characteristic polynomial of a matrix is

$$-(t-3)(t+5)^2(t-8)^4,$$

determine the eigenvalues of the matrix and their multiplicities. ◀

For the preceding matrices A and B, the eigenvalues -1 and 3 have different multiplicities. For A, the multiplicity of -1 is 2 and the multiplicity of 3 is 1, whereas for B the multiplicity of -1 is 1 and the multiplicity of 3 is 2. It is instructive to investigate the eigenspaces of A and B corresponding to the same eigenvalue, say, 3. Since the reduced row echelon form of $A - 3I_3$ is

$$\begin{bmatrix} 1 & 0 & 0 \\ 0 & 1 & -1 \\ 0 & 0 & 0 \end{bmatrix},$$

the vector form of the general solution of $(A - 3I_3)\mathbf{x} = \mathbf{0}$ is

$$\begin{bmatrix} x_1 \\ x_2 \\ x_3 \end{bmatrix} = \begin{bmatrix} 0 \\ x_3 \\ x_3 \end{bmatrix} = x_3 \begin{bmatrix} 0 \\ 1 \\ 1 \end{bmatrix}.$$

Hence

$$\left\{ \begin{bmatrix} 0 \\ 1 \\ 1 \end{bmatrix} \right\}$$

[3] Some authors use the term *algebraic multiplicity*. In this case, the dimension of the eigenspace corresponding to λ is usually called the *geometric multiplicity* of λ.

is a basis for the eigenspace of A corresponding to the eigenvalue 3. Therefore this eigenspace has dimension 1. On the other hand, in Example 3 of Section 5.1, we saw that

$$\left\{ \begin{bmatrix} 1 \\ 0 \\ 0 \end{bmatrix}, \begin{bmatrix} 0 \\ 1 \\ 1 \end{bmatrix} \right\}$$

is a basis for the eigenspace of B corresponding to the eigenvalue 3. Therefore this eigenspace has dimension 2.

For these matrices A and B, the dimension of the eigenspace corresponding to the eigenvalue 3 equals the multiplicity of the eigenvalue. This need not always happen, but there is a connection between the dimension of an eigenspace and the multiplicity of the corresponding eigenvalue. It is described in our next theorem, whose proof we omit.[4]

THEOREM 5.1

Let λ be an eigenvalue of a matrix A. The dimension of the eigenspace of A corresponding to λ is less than or equal to the multiplicity of λ.

Example 4

Determine the eigenvalues, their multiplicities, and a basis for each eigenspace of the linear operator T on $\mathcal{R}^3$ defined by

$$T\left(\begin{bmatrix} x_1 \\ x_2 \\ x_3 \end{bmatrix}\right) = \begin{bmatrix} -x_1 \\ 2x_1 - x_2 - x_3 \\ -x_3 \end{bmatrix}.$$

Solution The standard matrix of T is

$$A = \begin{bmatrix} -1 & 0 & 0 \\ 2 & -1 & -1 \\ 0 & 0 & -1 \end{bmatrix}.$$

Hence the characteristic polynomial of T is

$$\begin{aligned} \det(A - tI_3) &= \det \begin{bmatrix} -1-t & 0 & 0 \\ 2 & -1-t & -1 \\ 0 & 0 & -1-t \end{bmatrix} \\ &= (-1-t) \cdot \det \begin{bmatrix} -1-t & -1 \\ 0 & -1-t \end{bmatrix} \\ &= (-1-t)^3 \\ &= -(t+1)^3. \end{aligned}$$

Therefore the only eigenvalue of T is -1, and its multiplicity is 3. The eigenspace of T corresponding to -1 is the solution set of $(A + I_3)\mathbf{x} = \mathbf{0}$. Since the reduced row echelon form of $A + I_3$ is

$$\begin{bmatrix} 1 & 0 & -.5 \\ 0 & 0 & 0 \\ 0 & 0 & 0 \end{bmatrix},$$

[4] For a proof of Theorem 5.1, see [4, page 264].

we see that

$$\left\{ \begin{bmatrix} 0 \\ 1 \\ 0 \end{bmatrix}, \begin{bmatrix} 1 \\ 0 \\ 2 \end{bmatrix} \right\}$$

is a basis for the eigenspace of T corresponding to -1. Note that this eigenspace is 2-dimensional and the multiplicity of -1 is 3, in agreement with Theorem 5.1.

Practice Problem 3 ▶ Determine the eigenvalues of

$$A = \begin{bmatrix} 1 & -1 & -1 \\ 4 & -3 & -5 \\ 0 & 0 & 2 \end{bmatrix},$$

their multiplicities, and a basis for each eigenspace. ◀

THE EIGENVALUES OF SIMILAR MATRICES

Recall that two matrices A and B are called *similar* if there exists an invertible matrix P such that $B = P^{-1}AP$. By Theorem 3.4, we have

$$\begin{aligned} \det(B - tI_n) &= \det(P^{-1}AP - tP^{-1}I_nP) \\ &= \det(P^{-1}AP - P^{-1}(tI_n)P) \\ &= \det(P^{-1}(A - tI_n)P) \\ &= (\det P^{-1})[\det(A - tI_n)](\det P) \\ &= \left(\frac{1}{\det P}\right)[\det(A - tI_n)](\det P) \\ &= \det(A - tI_n). \end{aligned}$$

Thus the characteristic polynomial of A is the same as that of B. Therefore the following statements are true (see Exercise 84):

> Similar matrices have the same characteristic polynomial and hence have the same eigenvalues and multiplicities. In addition, their eigenspaces corresponding to the same eigenvalue have the same dimension.

In Section 5.3, we investigate matrices that are similar to a diagonal matrix.

COMPLEX EIGENVALUES*

We have seen in Example 3 that not all $n \times n$ matrices or linear operators on $\mathcal{R}^n$ have real eigenvalues and eigenvectors. The characteristic polynomial of such a matrix

* The remainder of this section is used only in the description of harmonic motion (an optional topic in Section 5.5).

must have no real roots. However, it is a consequence of the fundamental theorem of algebra that every $n \times n$ matrix has complex eigenvalues. (See Appendix C.) In fact, the fundamental theorem of algebra implies that the characteristic polynomial of every $n \times n$ matrix can be written in the form

$$c(t - \lambda_1)(t - \lambda_2)\cdots(t - \lambda_n)$$

for some complex numbers $c, \lambda_1, \lambda_2, \ldots, \lambda_n$. Thus, if we count each eigenvalue as often as its multiplicity, every $n \times n$ matrix has exactly n complex eigenvalues. However, some or all of these may not be real numbers.

There are applications (in such disciplines as physics and electrical engineering) where complex eigenvalues provide useful information about real-world problems. For the most part, the mathematical theory is no different for complex numbers than for real numbers. In the complex case, however, we must allow complex entries in matrices and vectors. Thus the set of all $n \times 1$ matrices with complex entries, denoted by $\mathcal{C}^n$, replaces the usual set of vectors $\mathcal{R}^n$, and the set $\mathcal{C}$ of complex numbers replaces $\mathcal{R}$ as the set of scalars.

Example 5 illustrates the calculations required to find eigenvalues and eigenvectors involving complex numbers. However, with the exception of an application discussed in Section 5.5 and designated exercises in Sections 5.2 and 5.3, in this book we restrict our attention to real eigenvalues and eigenvectors having real components.

Example 5 Determine the complex eigenvalues and a basis for each eigenspace of

$$A = \begin{bmatrix} 1 & -10 \\ 2 & 5 \end{bmatrix}.$$

Solution The characteristic polynomial of A is

$$\det(A - tI_2) = \det \begin{bmatrix} 1-t & -10 \\ 2 & 5-t \end{bmatrix} = (1-t)(5-t) + 20 = t^2 - 6t + 25.$$

Applying the quadratic formula, we find that the roots of the characteristic polynomial of A are

$$t = \frac{6 \pm \sqrt{(-6)^2 - 4(1)(25)}}{2} = \frac{6 \pm \sqrt{-64}}{2} = \frac{6 \pm 8i}{2} = 3 \pm 4i.$$

Hence the eigenvalues of A are $3 + 4i$ and $3 - 4i$. As with real eigenvalues, we find the eigenvectors in $\mathcal{C}^2$ corresponding to $3 + 4i$ by solving $(A - (3 + 4i)I_2)\mathbf{x} = \mathbf{0}$. The reduced row echelon form of $A - (3 + 4i)I_2$ is

$$\begin{bmatrix} 1 & 1-2i \\ 0 & 0 \end{bmatrix}.$$

Thus the vectors in the eigenspace corresponding to $3 + 4i$ have the form

$$\begin{bmatrix} x_1 \\ x_2 \end{bmatrix} = \begin{bmatrix} (-1+2i)x_2 \\ x_2 \end{bmatrix} = x_2 \begin{bmatrix} -1+2i \\ 1 \end{bmatrix},$$

so a basis for the eigenspace corresponding to $3 + 4i$ is

$$\left\{ \begin{bmatrix} -1+2i \\ 1 \end{bmatrix} \right\}.$$

Similarly, the reduced row echelon form of $A - (3 - 4i)I_2$ is

$$\begin{bmatrix} 1 & 1+2i \\ 0 & 0 \end{bmatrix}.$$

Hence the vectors in the eigenspace corresponding to $3 - 4i$ have the form

$$\begin{bmatrix} x_1 \\ x_2 \end{bmatrix} = \begin{bmatrix} (-1-2i)x_2 \\ x_2 \end{bmatrix} = x_2 \begin{bmatrix} -1-2i \\ 1 \end{bmatrix},$$

and thus a basis for the eigenspace corresponding to $3 - 4i$ is

$$\left\{ \begin{bmatrix} -1-2i \\ 1 \end{bmatrix} \right\}.$$

Practice Problem 4 ▶ Determine the complex eigenvalues and a basis for each eigenspace of the 90°-rotation matrix

$$A = \begin{bmatrix} 0 & -1 \\ 1 & 0 \end{bmatrix}.$$

◀

When the entries of A are real numbers, the characteristic polynomial of A has real coefficients. Under these conditions, if some nonreal number is a root of the characteristic polynomial of A, then its complex conjugate can also be shown to be a root. Thus *the nonreal eigenvalues of a real matrix occur in complex conjugate pairs.* Moreover, if $\mathbf{v}$ is an eigenvector of A corresponding to a nonreal eigenvalue λ, then the complex conjugate of $\mathbf{v}$ (the vector whose components are the complex conjugates of the components of $\mathbf{v}$) can be shown to be an eigenvector of A corresponding to the complex conjugate of λ. Note this relationship in Example 5.

EXERCISES

In Exercises 1–12, a matrix and its characteristic polynomial are given. Find the eigenvalues of each matrix and determine a basis for each eigenspace.

1. $\begin{bmatrix} 3 & -3 \\ 2 & 8 \end{bmatrix}$, $(t-5)(t-6)$

2. $\begin{bmatrix} -7 & 1 \\ -6 & -2 \end{bmatrix}$, $(t+4)(t+5)$

3. $\begin{bmatrix} -10 & 6 \\ -15 & 9 \end{bmatrix}$, $t(t+1)$

4. $\begin{bmatrix} -9 & -7 \\ 14 & 12 \end{bmatrix}$, $(t+2)(t-5)$

5. $\begin{bmatrix} 6 & -5 & -4 \\ 5 & -3 & -5 \\ 4 & -5 & -2 \end{bmatrix}$, $-(t+3)(t-2)^2$

6. $\begin{bmatrix} -2 & -6 & -6 \\ -3 & 2 & -2 \\ 3 & 2 & 6 \end{bmatrix}$, $-(t+2)(t-4)^2$

7. $\begin{bmatrix} 6 & -4 & -4 \\ -8 & 2 & 4 \\ 8 & -4 & -6 \end{bmatrix}$, $-(t-6)(t+2)^2$

8. $\begin{bmatrix} -5 & 6 & 1 \\ -1 & 2 & 1 \\ -8 & 6 & 4 \end{bmatrix}$, $-(t+4)(t-2)(t-3)$

9. $\begin{bmatrix} 0 & 2 & 1 \\ 1 & -1 & -1 \\ 4 & 4 & -3 \end{bmatrix}$, $-(t+3)(t+2)(t-1)$

10. $\begin{bmatrix} 3 & 2 & 2 \\ -2 & -1 & -2 \\ 2 & 2 & 3 \end{bmatrix}$, $-(t-3)(t-1)^2$

11. $\begin{bmatrix} -1 & 4 & -4 & -4 \\ 5 & -2 & 1 & 6 \\ 0 & 0 & -1 & 0 \\ 5 & -5 & 5 & 9 \end{bmatrix}$, $(t-3)(t-4)(t+1)^2$

12. $\begin{bmatrix} 1 & 6 & -6 & -6 \\ 6 & 7 & -6 & -12 \\ 3 & 3 & -2 & -6 \\ 3 & 9 & -9 & -11 \end{bmatrix}$, $(t+5)(t+2)(t-1)^2$

In Exercises 13–24, find the eigenvalues of each matrix and determine a basis for each eigenspace.

13. $\begin{bmatrix} 1 & 3 \\ 0 & -4 \end{bmatrix}$

14. $\begin{bmatrix} 8 & 2 \\ -12 & -2 \end{bmatrix}$

15. $\begin{bmatrix} -3 & -4 \\ 12 & 11 \end{bmatrix}$

16. $\begin{bmatrix} -2 & 0 \\ 3 & -1 \end{bmatrix}$

17. $\begin{bmatrix} -7 & 5 & 4 \\ 0 & -3 & 0 \\ -8 & 9 & 5 \end{bmatrix}$

18. $\begin{bmatrix} -3 & -12 & 0 \\ 0 & 3 & 0 \\ -4 & -8 & 1 \end{bmatrix}$

19. $\begin{bmatrix} -1 & 0 & 0 \\ 2 & 5 & 0 \\ 1 & -2 & -1 \end{bmatrix}$

20. $\begin{bmatrix} 3 & 0 & 0 \\ 9 & 3 & 10 \\ -5 & 0 & -2 \end{bmatrix}$

21. $\begin{bmatrix} -4 & 0 & 2 \\ 2 & 4 & -8 \\ 2 & 0 & -4 \end{bmatrix}$

22. $\begin{bmatrix} -4 & 7 & 7 \\ 0 & 3 & 7 \\ 0 & 0 & -4 \end{bmatrix}$

23. $\begin{bmatrix} -1 & -2 & -1 & 4 \\ 0 & 1 & 2 & 0 \\ 0 & 0 & -2 & -1 \\ 0 & 0 & 0 & 2 \end{bmatrix}$

24. $\begin{bmatrix} 1 & 0 & 0 & 0 \\ 9 & -2 & -3 & 3 \\ -6 & 0 & 1 & -3 \\ -6 & 0 & 0 & -2 \end{bmatrix}$

In Exercises 25–32, a linear operator and its characteristic polynomial are given. Find the eigenvalues of each operator and determine a basis for each eigenspace.

25. $T\left(\begin{bmatrix} x_1 \\ x_2 \end{bmatrix}\right) = \begin{bmatrix} -x_1 + 6x_2 \\ -8x_1 + 13x_2 \end{bmatrix}$, $(t-5)(t-7)$

26. $T\left(\begin{bmatrix} x_1 \\ x_2 \end{bmatrix}\right) = \begin{bmatrix} -x_1 + 2x_2 \\ -4x_1 - 7x_2 \end{bmatrix}$, $(t+5)(t+3)$

27. $T\left(\begin{bmatrix} x_1 \\ x_2 \end{bmatrix}\right) = \begin{bmatrix} -10x_1 - 24x_2 \\ 8x_1 + 18x_2 \end{bmatrix}$, $(t-6)(t-2)$

28. $T\left(\begin{bmatrix} x_1 \\ x_2 \end{bmatrix}\right) = \begin{bmatrix} -x_1 + 2x_2 \\ -10x_1 + 8x_2 \end{bmatrix}$, $(t-4)(t-3)$

29. $T\left(\begin{bmatrix} x_1 \\ x_2 \\ x_3 \end{bmatrix}\right) = \begin{bmatrix} -2x_2 + 4x_3 \\ -3x_1 + x_2 + 3x_3 \\ -x_1 + x_2 + 5x_3 \end{bmatrix}$,
$-(t+2)(t-4)^2$

30. $T\left(\begin{bmatrix} x_1 \\ x_2 \\ x_3 \end{bmatrix}\right) = \begin{bmatrix} -8x_1 - 5x_2 - 7x_3 \\ 6x_2 + 3x_2 + 7x_3 \\ 8x_1 + 8x_2 - 9x_3 \end{bmatrix}$,
$-(t+3)(t+2)(t+9)$

31. $T\left(\begin{bmatrix} x_1 \\ x_2 \\ x_3 \end{bmatrix}\right) = \begin{bmatrix} 3x_1 + 2x_2 - 2x_3 \\ 2x_1 + 6x_2 - 4x_3 \\ 3x_1 + 6x_2 - 4x_3 \end{bmatrix}$,
$-(t-1)(t-2)^2$

32. $T\left(\begin{bmatrix} x_1 \\ x_2 \\ x_3 \end{bmatrix}\right) = \begin{bmatrix} 3x_1 + 4x_2 - 4x_3 \\ 8x_1 + 7x_2 - 8x_3 \\ 8x_1 + 8x_2 - 9x_3 \end{bmatrix}$,
$-(t+1)^2(t-3)$

In Exercises 33–40, find the eigenvalues of each linear operator and determine a basis for each eigenspace.

33. $T\left(\begin{bmatrix} x_1 \\ x_2 \end{bmatrix}\right) = \begin{bmatrix} -4x_1 + x_2 \\ -2x_1 - x_2 \end{bmatrix}$

34. $T\left(\begin{bmatrix} x_1 \\ x_2 \end{bmatrix}\right) = \begin{bmatrix} 6x_1 - x_2 \\ 6x_1 + x_2 \end{bmatrix}$

35. $T\left(\begin{bmatrix} x_1 \\ x_2 \end{bmatrix}\right) = \begin{bmatrix} 2x_2 \\ -10x_1 + 9x_2 \end{bmatrix}$

36. $T\left(\begin{bmatrix} x_1 \\ x_2 \end{bmatrix}\right) = \begin{bmatrix} -5x_1 - 8x_2 \\ 12x_1 + 15x_2 \end{bmatrix}$

37. $T\left(\begin{bmatrix} x_1 \\ x_2 \\ x_3 \end{bmatrix}\right) = \begin{bmatrix} 7x_1 - 10x_2 \\ 5x_1 - 8x_2 \\ -x_1 + x_2 + 2x_3 \end{bmatrix}$

38. $T\left(\begin{bmatrix} x_1 \\ x_2 \\ x_3 \end{bmatrix}\right) = \begin{bmatrix} -6x_1 - 5x_2 + 5x_3 \\ -x_2 \\ -10x_1 - 10x_2 + 9x_3 \end{bmatrix}$

39. $T\left(\begin{bmatrix} x_1 \\ x_2 \\ x_3 \end{bmatrix}\right) = \begin{bmatrix} -3x_1 \\ -8x_1 + x_2 \\ -12x_1 + x_3 \end{bmatrix}$

40. $T\left(\begin{bmatrix} x_1 \\ x_2 \\ x_3 \end{bmatrix}\right) = \begin{bmatrix} -4x_1 + 6x_2 \\ 2x_2 \\ -5x_1 + 5x_2 + x_3 \end{bmatrix}$

41. Show that $\begin{bmatrix} 6 & -7 \\ 4 & -3 \end{bmatrix}$ has no real eigenvalues.

42. Show that $\begin{bmatrix} 4 & -5 \\ 3 & -2 \end{bmatrix}$ has no real eigenvalues.

43. Show that the linear operator $T\left(\begin{bmatrix} x_1 \\ x_2 \end{bmatrix}\right) = \begin{bmatrix} x_1 + 3x_2 \\ -2x_1 + 5x_2 \end{bmatrix}$ has no real eigenvalues.

44. Show that the linear operator $T\left(\begin{bmatrix} x_1 \\ x_2 \end{bmatrix}\right) = \begin{bmatrix} 2x_1 - 3x_2 \\ 2x_1 + 4x_2 \end{bmatrix}$ has no real eigenvalues.

In Exercises 45–52, use complex numbers to determine the eigenvalues of each linear operator and determine a basis for each eigenspace.

45. $\begin{bmatrix} 1-10i & -4i \\ 24i & 1+10i \end{bmatrix}$

46. $\begin{bmatrix} 1-i & -4 \\ 1 & 1-i \end{bmatrix}$

47. $\begin{bmatrix} 5 & 51 \\ -3 & 11 \end{bmatrix}$

48. $\begin{bmatrix} 2 & -1 \\ 1 & 2 \end{bmatrix}$

49. $\begin{bmatrix} 2i & 1+2i & -6-i \\ 0 & 4 & 3i \\ 0 & 0 & 1 \end{bmatrix}$

50. $\begin{bmatrix} 2 & -i & -i+1 \\ 0 & 0 & i \\ 0 & 0 & i \end{bmatrix}$

51. $\begin{bmatrix} i & 0 & 1-5i \\ 0 & 1 & \frac{1}{2} \\ 0 & 0 & 2 \end{bmatrix}$

52. $\begin{bmatrix} 2i & 0 & 0 \\ 0 & 0 & 0 \\ 0 & -i & 1 \end{bmatrix}$

In Exercises 53–72, determine whether the statements are true or false.

53. If two matrices have the same characteristic polynomial, then they have the same eigenvectors.

54. If two matrices have the same characteristic polynomial, then they have the same eigenvalues.

55. The characteristic polynomial of an $n \times n$ matrix is a polynomial of degree n.

56. The eigenvalues of a matrix are equal to those of its reduced row echelon form.

57. The eigenvectors of a matrix are equal to those of its reduced row echelon form.

58. An $n \times n$ matrix has n distinct eigenvalues.

59. Every $n \times n$ matrix has an eigenvector in $\mathcal{R}^n$.

60. Every square matrix has a complex eigenvalue.

61. The characteristic polynomial of an $n \times n$ matrix can be written $c(t - \lambda_1)(t - \lambda_2)\cdots(t - \lambda_n)$ for some real numbers $c, \lambda_1, \lambda_2, \ldots, \lambda_n$.

62. The characteristic polynomial of an $n \times n$ matrix can be written $c(t - \lambda_1)(t - \lambda_2)\cdots(t - \lambda_n)$ for some complex numbers $c, \lambda_1, \lambda_2, \ldots, \lambda_n$.

63. If $(t - 4)^2$ divides the characteristic polynomial of A, then 4 is an eigenvalue of A with multiplicity 2.

64. The multiplicity of an eigenvalue equals the dimension of the corresponding eigenspace.

65. If λ is an eigenvalue of multiplicity 1 for a matrix A, then the dimension of the eigenspace of A corresponding to λ is 1.

66. The nonreal eigenvalues of a matrix occur in complex conjugate pairs.

67. The nonreal eigenvalues of a real matrix occur in complex conjugate pairs.

68. If A is an $n \times n$ matrix, then the sum of the multiplicities of the eigenvalues of A equals n.

69. The scalar 1 is an eigenvalue of I_n.

70. The only eigenvalue of I_n is 1.

71. A square zero matrix has no eigenvalues.

72. The eigenvalues of an $n \times n$ matrix A are the solutions of $\det(A - tI_n) = 0$.

---

73. Let A be an $n \times n$ matrix, and suppose that, for a particular scalar c, the reduced row echelon form of $A - cI_n$ is I_n. What can be said about c?

74. If $f(t)$ is the characteristic polynomial of a square matrix A, what is $f(0)$?

75. Suppose that the characteristic polynomial of an $n \times n$ matrix A is

$$a_n t^n + a_{n-1} t^{n-1} + \cdots + a_1 t + a_0.$$

Determine the characteristic polynomial of $-A$.

76. What is the coefficient of t^n in the characteristic polynomial of an $n \times n$ matrix?

77. Suppose that A is a 4×4 matrix with no nonreal eigenvalues and exactly two real eigenvalues, 5 and -9. Let W_1 and W_2 be the eigenspaces of A corresponding to 5 and -9, respectively. Write all the possible characteristic polynomials of A that are consistent with the following information:
   (a) $\dim W_1 = 3$
   (b) $\dim W_2 = 1$
   (c) $\dim W_1 = 2$

78. Suppose that A is a 5×5 matrix with no nonreal eigenvalues and exactly three real eigenvalues, 4, 6, and 7. Let W_1, W_2, and W_3 be the eigenspaces corresponding to 4, 6, and 7, respectively. Write all the possible characteristic polynomials of A that are consistent with the following information:
   (a) $\dim W_2 = 3$
   (b) $\dim W_1 = 2$
   (c) $\dim W_1 = 1$ and $\dim W_2 = 2$
   (d) $\dim W_2 = 2$ and $\dim W_3 = 2$

79. Show that if A is an upper triangular or a lower triangular matrix, then λ is an eigenvalue of A with multiplicity k if and only if λ appears exactly k times on the diagonal of A.

80. Show that the rotation matrix A_θ has no real eigenvalues if $0^\circ < \theta < 180^\circ$.

81. (a) Determine a basis for each eigenspace of
$$A = \begin{bmatrix} 3 & 2 \\ -1 & 0 \end{bmatrix}.$$
   (b) Determine a basis for each eigenspace of $-3A$.
   (c) Determine a basis for each eigenspace of $5A$.
   (d) Establish a relationship between the eigenvectors of any square matrix B and those of cB for any scalar $c \neq 0$.
   (e) Establish a relationship between the eigenvalues of a square matrix B and those of cB for any scalar $c \neq 0$.

82. (a) Determine a basis for each eigenspace of
$$A = \begin{bmatrix} 5 & -2 \\ 1 & 8 \end{bmatrix}.$$
   (b) Determine a basis for each eigenspace of $A + 4I_2$.
   (c) Determine a basis for each eigenspace of $A - 6I_2$.
   (d) Establish a relationship between the eigenvectors of any $n \times n$ matrix B and those of $B + cI_n$ for any scalar c.
   (e) Establish a relationship between the eigenvalues of any $n \times n$ matrix B and those of $B + cI_n$ for any scalar c.

83. (a) Determine the characteristic polynomial of A^T, where A is the matrix in Exercise 82.
   (b) Establish a relationship between the characteristic polynomial of any square matrix B and that of B^T.
   (c) What does (b) imply about the relationship between the eigenvalues of a square matrix B and those of B^T?
   (d) Is there a relationship between the eigenvectors of a square matrix B and those of B^T?

84. Let A and B be $n \times n$ matrices such that $B = P^{-1}AP$, and let λ be an eigenvalue of A (and hence of B). Prove the following results:

    (a) A vector $\mathbf{v}$ in $\mathcal{R}^n$ is in the eigenspace of A corresponding to λ if and only if $P^{-1}\mathbf{v}$ is in the eigenspace of B corresponding to λ.

    (b) If $\{\mathbf{v}_1, \mathbf{v}_2, \ldots, \mathbf{v}_k\}$ is a basis for the eigenspace of A corresponding to λ, then $\{P^{-1}\mathbf{v}_1, P^{-1}\mathbf{v}_2, \ldots, P^{-1}\mathbf{v}_k\}$ is a basis for the eigenspace of B corresponding to λ.

    (c) The eigenspaces of A and B that correspond to the same eigenvalue have the same dimension.

85. Let A be a symmetric 2×2 matrix. Prove that A has real eigenvalues.

86. (a) The characteristic polynomial of $A = \begin{bmatrix} a & b \\ c & d \end{bmatrix}$ has the form $t^2 + rt + s$ for some scalars r and s. Determine r and s in terms of a, b, c, and d.

    (b) Show that $A^2 + rA + sI_2 = O$, the 2×2 zero matrix. (A similar result is true for any square matrix. It is called the *Cayley-Hamilton theorem.*)

In Exercises 87–91, use either a calculator with matrix capabilities or computer software such as MATLAB to solve each problem.

87. Compute the characteristic polynomial of

$$\begin{bmatrix} 1 & \frac{1}{2} & \frac{1}{3} \\ \frac{1}{2} & \frac{1}{3} & \frac{1}{4} \\ \frac{1}{3} & \frac{1}{4} & \frac{1}{5} \end{bmatrix}.$$

(This matrix is called the 3×3 *Hilbert matrix.* Computations with Hilbert matrices are subject to significant roundoff errors.)

88. Compute the characteristic polynomial of

$$\begin{bmatrix} 0 & 0 & 0 & -17 \\ 1 & 0 & 0 & -18 \\ 0 & 1 & 0 & -19 \\ 0 & 0 & 1 & -20 \end{bmatrix}.$$

89. Use the result of Exercise 88 to find a 4×4 matrix whose characteristic polynomial is $t^4 - 11t^3 + 23t^2 + 7t - 5$.

90. Let A be a random 4×4 matrix.

    (a) Compute the characteristic polynomials of A and A^T.

    (b) Formulate a conjecture about the characteristic polynomials of B and B^T, where B is an arbitrary $n \times n$ matrix. Test your conjecture using an arbitrary 5×5 matrix.

    (c) Prove that your conjecture in (b) is valid.

91. Let

$$A = \begin{bmatrix} 6.5 & -3.5 \\ 7.0 & -4.0 \end{bmatrix}.$$

    (a) Find the eigenvalues of A and an eigenvector corresponding to each eigenvalue.

    (b) Show that A is invertible, and then find the eigenvalues of A^{-1} and an eigenvector corresponding to each eigenvalue.

    (c) Use the results of (a) and (b) to formulate a conjecture about the relationship between the eigenvectors and eigenvalues of an invertible $n \times n$ matrix and those of its inverse.

    (d) Test your conjecture in (c) on the invertible matrix

$$\begin{bmatrix} 3 & -2 & 2 \\ -4 & 8 & -10 \\ -5 & 2 & -4 \end{bmatrix}.$$

    (e) Prove that your conjecture in (c) is valid.

SOLUTIONS TO THE PRACTICE PROBLEMS

1. Because the given matrix is a diagonal matrix, its eigenvalues are its diagonal entries, which are 4, -1, -2, and 3.

2. The eigenvalues of the matrix are the roots of its characteristic polynomial, which is $-(t-3)(t+5)^2(t-8)^4$. Thus the eigenvalues of the matrix are 3, -5, and 8. The multiplicity of an eigenvalue λ is the number of factors of $t - \lambda$ that appear in the characteristic polynomial. Hence 3 is an eigenvalue of multiplicity 1, -5 is an eigenvalue of multiplicity 2, and 8 is an eigenvalue of multiplicity 4.

3. Form the matrix

$$B = A - tI_3 = \begin{bmatrix} 1-t & -1 & -1 \\ 4 & -3-t & -5 \\ 0 & 0 & 2-t \end{bmatrix}.$$

To evaluate the determinant of B, we use cofactor expansion along the third row. Then

$$\begin{aligned} \det B &= (-1)^{3+1}b_{31} \cdot \det B_{31} + (-1)^{3+2}b_{32} \cdot \det B_{32} \\ &\quad + (-1)^{3+3}b_{33} \cdot \det B_{33} \\ &= 0 + 0 + (-1)^6(2-t) \cdot \det \begin{bmatrix} 1-t & -1 \\ 4 & -3-t \end{bmatrix} \\ &= (2-t)[(1-t)(-3-t)+4] \\ &= (2-t)[(t^2+2t-3)+4] \\ &= (2-t)(t^2+2t+1) \\ &= -(t-2)(t+1)^2. \end{aligned}$$

Hence the eigenvalues of A are -1, which has multiplicity 2, and 2, which has multiplicity 1. Because the reduced row echelon form of $A + I_3$ is

$$\begin{bmatrix} 1 & -.5 & 0 \\ 0 & 0 & 1 \\ 0 & 0 & 0 \end{bmatrix},$$

we see that the vector form of the general solution of $(A + I_3)\mathbf{x} = \mathbf{0}$ is

$$\begin{bmatrix} x_1 \\ x_2 \\ x_3 \end{bmatrix} = \begin{bmatrix} .5x_2 \\ x_2 \\ 0 \end{bmatrix} = x_2 \begin{bmatrix} .5 \\ 1 \\ 0 \end{bmatrix}.$$

Taking $x_2 = 2$, we obtain the basis

$$\left\{ \begin{bmatrix} 1 \\ 2 \\ 0 \end{bmatrix} \right\}$$

for the eigenspace of A corresponding to the eigenvalue -1. Also,

$$\begin{bmatrix} 1 & 0 & 0 \\ 0 & 1 & 1 \\ 0 & 0 & 0 \end{bmatrix}$$

is the reduced row echelon form of $A - 2I_3$. Therefore a basis for the eigenspace of A corresponding to the eigenvalue 2 is

$$\left\{ \begin{bmatrix} 0 \\ -1 \\ 1 \end{bmatrix} \right\}.$$

4. The characteristic polynomial of A is

$$\det (A - tI_2) = \det \begin{bmatrix} -t & -1 \\ 1 & -t \end{bmatrix} = t^2 + 1 = (t + i)(t - i).$$

Hence A has eigenvalues $-i$ and i. Since the reduced row echelon form of $A + iI_2$ is

$$\begin{bmatrix} 1 & i \\ 0 & 0 \end{bmatrix},$$

a basis for the eigenspace corresponding to the eigenvalue $-i$ is

$$\left\{ \begin{bmatrix} -i \\ 1 \end{bmatrix} \right\}.$$

Furthermore, the reduced row echelon form of $A - iI_2$ is

$$\begin{bmatrix} 1 & -i \\ 0 & 0 \end{bmatrix},$$

so a basis for the eigenspace corresponding to the eigenvalue i is

$$\left\{ \begin{bmatrix} i \\ 1 \end{bmatrix} \right\}.$$

5.3 DIAGONALIZATION OF MATRICES

In Example 6 of Section 2.1, we considered a metropolitan area in which the current populations (in thousands) of the city and suburbs are given by

$$\begin{matrix} \text{City} \\ \text{Suburbs} \end{matrix} \begin{bmatrix} 500 \\ 700 \end{bmatrix} = \mathbf{p},$$

and the population shifts between the city and suburbs are described by the following matrix:

$$\text{To} \begin{matrix} \\ \\ \text{City} \\ \text{Suburbs} \end{matrix} \begin{matrix} \text{From} \\ \begin{matrix} \text{City} & \text{Suburbs} \end{matrix} \\ \begin{bmatrix} .85 & .03 \\ .15 & .97 \end{bmatrix} \end{matrix} = A$$

We saw in this example that the populations of the city and suburbs after m years are given by the matrix–vector product $A^m\mathbf{p}$.

In this section, we discuss a technique for computing $A^m\mathbf{p}$. Note that when m is a large positive integer, a direct computation of $A^m\mathbf{p}$ involves considerable work. This calculation would be quite easy, however, if A were a diagonal matrix such as

$$D = \begin{bmatrix} .82 & 0 \\ 0 & 1 \end{bmatrix}.$$

For in this case, the powers of D are diagonal matrices and hence can be easily determined by the method described in Section 2.1. In fact,

$$D^m = \begin{bmatrix} (.82)^m & 0 \\ 0 & 1^m \end{bmatrix} = \begin{bmatrix} (.82)^m & 0 \\ 0 & 1 \end{bmatrix}.$$

Although $A \neq D$, it can be checked that $A = PDP^{-1}$, where

$$P = \begin{bmatrix} -1 & 1 \\ 1 & 5 \end{bmatrix}.$$

This relationship enables us to compute the powers of A in terms of those of D. For example,

$$A^2 = (PDP^{-1})(PDP^{-1}) = PD(P^{-1}P)DP^{-1} = PDDP^{-1} = PD^2P^{-1},$$

and

$$A^3 = (PDP^{-1})(PDP^{-1})(PDP^{-1}) = PD^3P^{-1}.$$

In a similar manner, it can be shown that

$$\begin{aligned} A^m &= PD^mP^{-1} \\ &= \begin{bmatrix} -1 & 1 \\ 1 & 5 \end{bmatrix} \begin{bmatrix} (.82)^m & 0 \\ 0 & 1 \end{bmatrix} \begin{bmatrix} -1 & 1 \\ 1 & 5 \end{bmatrix}^{-1} \\ &= \begin{bmatrix} -1 & 1 \\ 1 & 5 \end{bmatrix} \begin{bmatrix} (.82)^m & 0 \\ 0 & 1 \end{bmatrix} \begin{bmatrix} -\frac{5}{6} & \frac{1}{6} \\ \frac{1}{6} & \frac{1}{6} \end{bmatrix} \\ &= \frac{1}{6} \begin{bmatrix} 1 + 5(.82)^m & 1 - (.82)^m \\ 5 - 5(.82)^m & 5 + (.82)^m \end{bmatrix}. \end{aligned}$$

Hence

$$\begin{aligned} A^m\mathbf{p} &= \frac{1}{6} \begin{bmatrix} 1 + 5(.82)^m & 1 - (.82)^m \\ 5 - 5(.82)^m & 5 + (.82)^m \end{bmatrix} \begin{bmatrix} 500 \\ 700 \end{bmatrix} \\ &= \frac{1}{6} \begin{bmatrix} 1200 + 1800(.82)^m \\ 6000 - 1800(.82)^m \end{bmatrix} \\ &= \begin{bmatrix} 200 + 300(.82)^m \\ 1000 - 300(.82)^m \end{bmatrix}. \end{aligned}$$

Because $\lim_{m \to \infty} (.82)^m = 0$, we see that the limit of $A^m\mathbf{p}$ is

$$\begin{bmatrix} 200 \\ 1000 \end{bmatrix}.$$

Hence after many years, the population of the metropolitan area will consist of about 200 thousand city dwellers and 1 million suburbanites.

In the previous computation, note that calculating PD^mP^{-1} requires only 2 matrix multiplications instead of the $m-1$ multiplications needed to compute A^m directly. This simplification of the calculation is possible because A can be written in the form PDP^{-1} for some diagonal matrix D and some invertible matrix P.

Definition An $n \times n$ matrix A is called **diagonalizable** if $A = PDP^{-1}$ for some diagonal $n \times n$ matrix D and some invertible $n \times n$ matrix P.

Because the equation $A = PDP^{-1}$ can be written as $P^{-1}AP = D$, we see that a diagonalizable matrix is one that is similar to a diagonal matrix. It follows that if $A = PDP^{-1}$ for some diagonal matrix D, then the eigenvalues of A are the diagonal entries of D.

The matrix

$$A = \begin{bmatrix} .85 & .03 \\ .15 & .97 \end{bmatrix}$$

in Example 6 of Section 2.1 is diagonalizable because $A = PDP^{-1}$ for the matrices

$$P = \begin{bmatrix} -1 & 1 \\ 1 & 5 \end{bmatrix} \quad \text{and} \quad D = \begin{bmatrix} .82 & 0 \\ 0 & 1 \end{bmatrix}.$$

Notice that the eigenvalues of A are .82 and 1.

Every diagonal matrix is diagonalizable. (See Exercise 79.) However, not every matrix is diagonalizable, as the following example shows:

Example 1 Show that the matrix

$$A = \begin{bmatrix} 0 & 1 \\ 0 & 0 \end{bmatrix}$$

is not diagonalizable.

Solution Suppose, to the contrary, that $A = PDP^{-1}$, where P is an invertible 2×2 matrix and D is a diagonal 2×2 matrix. Because A is upper triangular, the only eigenvalue of A is 0, which has multiplicity 2. Thus the diagonal matrix D must be $D = O$. Hence $A = PDP^{-1} = POP^{-1} = O$, a contradiction.

The following theorem tells us when a matrix A is diagonalizable and how to find an invertible matrix P and a diagonal matrix D such that $A = PDP^{-1}$:

THEOREM 5.2

An $n \times n$ matrix A is diagonalizable if and only if there is a basis for $\mathcal{R}^n$ consisting of eigenvectors of A.

Furthermore, $A = PDP^{-1}$, where D is a diagonal matrix and P is an invertible matrix, if and only if the columns of P are a basis for $\mathcal{R}^n$ consisting of eigenvectors of A and the diagonal entries of D are the eigenvalues corresponding to the respective columns of P.

PROOF Suppose first that A is diagonalizable. Then $A = PDP^{-1}$ for some diagonal matrix D and invertible matrix P. Let $\lambda_1, \lambda_2, \ldots, \lambda_n$ denote the diagonal entries of D. Since P is invertible, its columns must be linearly independent and hence form a basis for $\mathcal{R}^n$. Rewriting $A = PDP^{-1}$ as $AP = PD$, we have that the jth columns of these matrices are equal; that is,

$$A\mathbf{p}_j = P\mathbf{d}_j = P(\lambda_j \mathbf{e}_j) = \lambda_j(P\mathbf{e}_j) = \lambda_j \mathbf{p}_j.$$

Therefore each column of P is an eigenvector of A, and the diagonal entries of D are the corresponding eigenvalues of A.

Conversely, suppose that $\{\mathbf{p}_1, \mathbf{p}_2, \ldots, \mathbf{p}_n\}$ is a basis for $\mathcal{R}^n$ consisting of eigenvectors of A and that λ_j is the eigenvalue of A corresponding to $\mathbf{p}_j$. Let P denote the matrix whose columns are $\mathbf{p}_1, \mathbf{p}_2, \ldots, \mathbf{p}_n$ and D denote the diagonal matrix whose diagonal entries are $\lambda_1, \lambda_2, \ldots, \lambda_n$. Then the jth column of AP equals $A\mathbf{p}_j$, and the jth column of PD is $P(\lambda_j \mathbf{e}_j) = \lambda_j(P\mathbf{e}_j) = \lambda_j \mathbf{p}_j$. But $A\mathbf{p}_j = \lambda_j \mathbf{p}_j$ for every j because $\mathbf{p}_j$ is an eigenvector of A corresponding to the eigenvalue λ_j. Thus $AP = PD$. Because the columns of P are linearly independent, P is invertible by the Invertible Matrix Theorem. So multiplying the preceding equation on the right by P^{-1} gives $A = PDP^{-1}$, proving that A is diagonalizable. ■

Theorem 5.2 shows us how to *diagonalize* a matrix such as

$$A = \begin{bmatrix} .85 & .03 \\ .15 & .97 \end{bmatrix}$$

in the population example discussed earlier. The characteristic polynomial of A is

$$\begin{aligned} \det(A - tI_2) &= \det \begin{bmatrix} .85 - t & .03 \\ .15 & .97 - t \end{bmatrix} \\ &= (.85 - t)(.97 - t) - .03(.15) \\ &= t^2 - 1.82t + .82 \\ &= (t - .82)(t - 1), \end{aligned}$$

so the eigenvalues of A are .82 and 1. Since the reduced row echelon form of $A - .82I_2$ is

$$\begin{bmatrix} 1 & 1 \\ 0 & 0 \end{bmatrix},$$

we see that

$$\mathcal{B}_1 = \left\{ \begin{bmatrix} -1 \\ 1 \end{bmatrix} \right\}$$

is a basis for the eigenspace of A corresponding to .82. Likewise,

$$\begin{bmatrix} 1 & -.2 \\ 0 & 0 \end{bmatrix}$$

is the reduced row echelon form of $A - I_2$, and thus

$$\mathcal{B}_2 = \left\{ \begin{bmatrix} 1 \\ 5 \end{bmatrix} \right\}$$

is a basis for the eigenspace of A corresponding to 1. The set

$$\mathcal{B} = \left\{ \begin{bmatrix} -1 \\ 1 \end{bmatrix}, \begin{bmatrix} 1 \\ 5 \end{bmatrix} \right\}$$

obtained by combining $\mathcal{B}_1$ and $\mathcal{B}_2$ is linearly independent, since neither of its vectors is a multiple of the other. Therefore $\mathcal{B}$ is a basis for $\mathcal{R}^2$, and it consists of eigenvectors of A. So Theorem 5.2 guarantees that A is diagonalizable. Notice that the columns of the matrix

$$P = \begin{bmatrix} -1 & 1 \\ 1 & 5 \end{bmatrix}$$

in the diagonalization of A are the vectors in this basis, and the diagonal matrix

$$D = \begin{bmatrix} .82 & 0 \\ 0 & 1 \end{bmatrix}$$

has as its diagonal entries the eigenvalues of A corresponding to the respective columns of P.

The matrices P and D such that $PDP^{-1} = A$ are not unique. For example, taking

$$P = \begin{bmatrix} 2 & -3 \\ 10 & 3 \end{bmatrix} \quad \text{and} \quad D = \begin{bmatrix} 1 & 0 \\ 0 & .82 \end{bmatrix}$$

also gives $PDP^{-1} = A$, because these matrices satisfy the hypotheses of Theorem 5.2. Note, however, that although the matrix D in Theorem 5.2 is not unique, any two such matrices differ only in the order in which the eigenvalues of A are listed along the diagonal of D.

When applying Theorem 5.2 to show that a matrix is diagonalizable, we normally use the following result: *If the bases for distinct eigenspaces are combined, then the resulting set is linearly independent.* This result is a consequence of Theorem 5.3, and is proved in [4, page 267].

THEOREM 5.3

A set of eigenvectors of a square matrix that correspond to distinct eigenvalues is linearly independent.

PROOF Let A be an $n \times n$ matrix with eigenvectors $\mathbf{v}_1, \mathbf{v}_2, \ldots, \mathbf{v}_m$ having corresponding distinct eigenvalues $\lambda_1, \lambda_2, \ldots, \lambda_m$. We give a proof by contradiction. Assume that this set of eigenvectors is linearly dependent. Since eigenvectors are nonzero, Theorem 1.9 shows that there is a smallest index k $(2 \le k \le m)$ such that $\mathbf{v}_k$ is a linear combination of $\mathbf{v}_1, \mathbf{v}_2, \ldots, \mathbf{v}_{k-1}$, say,

$$\mathbf{v}_k = c_1\mathbf{v}_1 + c_2\mathbf{v}_2 + \cdots + c_{k-1}\mathbf{v}_{k-1} \tag{1}$$

for some scalars $c_1, c_2, \ldots, c_{k-1}$. Because $A\mathbf{v}_i = \lambda_i\mathbf{v}_i$ for each i, when we multiply both sides of equation (1) by A, we obtain

$$\begin{aligned} A\mathbf{v}_k &= A(c_1\mathbf{v}_1 + c_2\mathbf{v}_2 + \cdots + c_{k-1}\mathbf{v}_{k-1}) \\ &= c_1A\mathbf{v}_1 + c_2A\mathbf{v}_2 + \cdots + c_{k-1}A\mathbf{v}_{k-1}; \end{aligned}$$

that is,

$$\lambda_k \mathbf{v}_k = c_1\lambda_1\mathbf{v}_1 + c_2\lambda_2\mathbf{v}_2 + \cdots + c_{k-1}\lambda_{k-1}\mathbf{v}_{k-1}. \tag{2}$$

Now multiply both sides of equation (1) by λ_k and subtract the result from equation (2) to obtain

$$\mathbf{0} = c_1(\lambda_1 - \lambda_k)\mathbf{v}_1 + c_2(\lambda_2 - \lambda_k)\mathbf{v}_2 + \cdots + c_{k-1}(\lambda_{k-1} - \lambda_k)\mathbf{v}_{k-1}. \tag{3}$$

By our choice of k, the set $\{\mathbf{v}_1, \mathbf{v}_2, \ldots, \mathbf{v}_{k-1}\}$ is linearly independent, and thus

$$c_1(\lambda_1 - \lambda_k) = c_2(\lambda_2 - \lambda_k) = \cdots = c_{k-1}(\lambda_{k-1} - \lambda_k) = 0.$$

But the scalars $\lambda_i - \lambda_k$ are nonzero because $\lambda_1, \lambda_2, \ldots, \lambda_m$ are distinct; so

$$c_1 = c_2 = \cdots = c_{k-1} = 0.$$

Thus equation (1) implies that $\mathbf{v}_k = \mathbf{0}$, which contradicts the definition of an eigenvector. Therefore the set of eigenvectors $\{\mathbf{v}_1, \mathbf{v}_2, \ldots, \mathbf{v}_m\}$ is linearly independent. ■

It follows from Theorem 5.3 that an $n \times n$ matrix having n distinct eigenvalues must have n linearly independent eigenvectors.

> Every $n \times n$ matrix having n distinct eigenvalues is diagonalizable.

The technique used to produce the invertible matrix P and the diagonal matrix D on page 317 works for any diagonalizable matrix.

> **Algorithm for Matrix Diagonalization**
>
> Let A be a diagonalizable $n \times n$ matrix. Combining bases for each eigenspace of A forms a basis $\mathcal{B}$ for $\mathcal{R}^n$ consisting of eigenvectors of A. Therefore, if P is the matrix whose columns are the vectors in $\mathcal{B}$ and D is a diagonal matrix whose diagonal entries are eigenvalues of A corresponding to the respective columns of P, then $A = PDP^{-1}$.

Example 2 Show that the matrix

$$A = \begin{bmatrix} -1 & 0 & 0 \\ 0 & 1 & 2 \\ 0 & 2 & 1 \end{bmatrix}$$

is diagonalizable, and find an invertible matrix P and a diagonal matrix D such that $A = PDP^{-1}$.

Solution On pages 304–306 in Section 5.2, we computed the characteristic polynomial $-(t+1)^2(t-3)$ of A, and we also found that

$$\mathcal{B}_1 = \left\{ \begin{bmatrix} 0 \\ 1 \\ 1 \end{bmatrix} \right\}$$

is a basis for the eigenspace of A corresponding to eigenvalue 3. Similarly,

$$\mathcal{B}_2 = \left\{ \begin{bmatrix} 1 \\ 0 \\ 0 \end{bmatrix}, \begin{bmatrix} 0 \\ 1 \\ -1 \end{bmatrix} \right\}$$

is a basis for the eigenspace corresponding to -1. The combined set

$$\mathcal{B} = \left\{ \begin{bmatrix} 0 \\ 1 \\ 1 \end{bmatrix}, \begin{bmatrix} 1 \\ 0 \\ 0 \end{bmatrix}, \begin{bmatrix} 0 \\ 1 \\ -1 \end{bmatrix} \right\}$$

is therefore linearly independent by the comment on page 317. It follows from the Algorithm for Matrix Diagonalization that A is diagonalizable and $A = PDP^{-1}$, where

$$P = \begin{bmatrix} 0 & 1 & 0 \\ 1 & 0 & 1 \\ 1 & 0 & -1 \end{bmatrix} \quad \text{and} \quad D = \begin{bmatrix} 3 & 0 & 0 \\ 0 & -1 & 0 \\ 0 & 0 & -1 \end{bmatrix}.$$

Practice Problem 1 ▶ The characteristic polynomial of

$$A = \begin{bmatrix} -4 & -6 & 0 \\ 3 & 5 & 0 \\ 3 & 3 & 2 \end{bmatrix}$$

is $-(t+1)(t-2)^2$. Show that A is diagonalizable by finding an invertible matrix P and a diagonal matrix D such that $A = PDP^{-1}$. ◀

WHEN IS A MATRIX DIAGONALIZABLE?

As we saw in Example 1, not all square matrices are diagonalizable. Theorem 5.2 tells us that an $n \times n$ matrix is diagonalizable when there are n linearly independent eigenvectors of A. For this to occur, two different conditions must be satisfied; they are given next.[5]

Test for a Diagonalizable Matrix Whose Characteristic Polynomial Is Known

An $n \times n$ matrix A is diagonalizable if and only if both of the following conditions are true[6]:

1. The total number of eigenvalues of A, when each eigenvalue is counted as often as its multiplicity, is equal to n.
2. For each eigenvalue λ of A, the dimension of the corresponding eigenspace, which is $n - \text{rank}\,(A - \lambda I_n)$, is equal to the multiplicity of λ.

Note that, by Theorem 5.1, the eigenspace corresponding to an eigenvalue of multiplicity 1 must have dimension 1. Hence *condition (2) need be checked only*

---

[5] For a proof of this result, see [4, page 268].

[6] It follows from the fundamental theorem of algebra that the first condition is always satisfied if complex eigenvalues are allowed.

for eigenvalues of multiplicity greater than 1. Therefore, to check if the matrix A in Practice Problem 3 of Section 5.2 is diagonalizable, it is necessary to check condition (2) only for the eigenvalue -1 (the one with multiplicity greater than 1).

Example 3 Determine whether each of the following matrices is diagonalizable (using real eigenvalues):

$$A = \begin{bmatrix} 0 & 2 & 1 \\ -2 & 0 & -2 \\ 0 & 0 & -1 \end{bmatrix} \qquad B = \begin{bmatrix} -7 & -3 & -6 \\ 0 & -4 & 0 \\ 3 & 3 & 2 \end{bmatrix}$$

$$C = \begin{bmatrix} -6 & -3 & 1 \\ 5 & 2 & -1 \\ 2 & 3 & -5 \end{bmatrix} \qquad M = \begin{bmatrix} -3 & 2 & 1 \\ 3 & -4 & -3 \\ -8 & 8 & 6 \end{bmatrix}$$

The respective characteristic polynomials of these matrices are

$$-(t+1)(t^2+4), \quad -(t+1)(t+4)^2, \quad -(t+1)(t+4)^2, \quad \text{and} \quad -(t+1)(t^2-4).$$

Solution The only eigenvalue of A is -1, which has multiplicity 1. Because A is a 3×3 matrix, the sum of the multiplicities of the eigenvalues of A must be 3 in order to be diagonalizable. Thus A is not diagonalizable because it has too few eigenvalues.

The eigenvalues of B are -1, which has multiplicity 1, and -4, which has multiplicity 2. Thus B has 3 eigenvalues if we count each eigenvalue as often as its multiplicity. Therefore B is diagonalizable if and only if the dimension of each eigenspace equals the multiplicity of the corresponding eigenvalue. As noted, it is only the eigenvalues with multiplicities greater than 1 that must be checked. So B is diagonalizable if and only if the dimension of the eigenspace corresponding to the eigenvalue -4 is 2. Because the reduced row echelon form of $B - (-4)I_3$ is

$$\begin{bmatrix} 1 & 1 & 2 \\ 0 & 0 & 0 \\ 0 & 0 & 0 \end{bmatrix},$$

a matrix of rank 1, this eigenspace has dimension

$$3 - \text{rank}\,(B - (-4)I_3) = 3 - 1 = 2.$$

Since this equals the multiplicity of the eigenvalue -4, B is diagonalizable.

The eigenvalues of C are also -1, which has multiplicity 1, and -4, which has multiplicity 2. So we see that C is diagonalizable if and only if the dimension of its eigenspace corresponding to eigenvalue -4 is 2. But the reduced row echelon form of $C - (-4)I_3$ is

$$\begin{bmatrix} 1 & 0 & 1 \\ 0 & 1 & -1 \\ 0 & 0 & 0 \end{bmatrix},$$

a matrix of rank 2. So the dimension of the eigenspace corresponding to eigenvalue -4 is

$$3 - \text{rank}\,(C - (-4)I_3) = 3 - 2 = 1,$$

which is less than the multiplicity of the eigenvalue -4. Therefore C is not diagonalizable.

Finally, the characteristic polynomial of M is

$$-(t+1)(t^2-4) = -(t+1)(t+2)(t-2).$$

We see that M has 3 distinct eigenvalues (-1, -2, and 2), and so it is diagonalizable by the boxed result on page 318.

Practice Problem 2 ▶ Determine whether each of the given matrices is diagonalizable. If so, express it in the form PDP^{-1}, where P is an invertible matrix and D is a diagonal matrix.

$$A = \begin{bmatrix} 2 & 2 & 1 \\ 0 & 0 & 3 \\ 0 & -1 & 0 \end{bmatrix} \qquad B = \begin{bmatrix} 5 & 5 & -6 \\ 0 & -1 & 0 \\ 3 & 2 & -4 \end{bmatrix}$$

The characteristic polynomials of A and B are $-(t-2)(t^2+3)$ and $-(t-2)(t+1)^2$, respectively. ◀

In Section 5.4, we consider what it means for a linear operator to be diagonalizable.

EXERCISES

In Exercises 1–12, a matrix A and its characteristic polynomial are given. Find, if possible, an invertible matrix P and a diagonal matrix D such that $A = PDP^{-1}$. Otherwise, explain why A is not diagonalizable.

1. $\begin{bmatrix} 7 & 6 \\ -1 & 2 \end{bmatrix}$ $(t-4)(t-5)$

2. $\begin{bmatrix} -2 & 7 \\ -1 & 2 \end{bmatrix}$ t^2+3

3. $\begin{bmatrix} 8 & 9 \\ -4 & -4 \end{bmatrix}$ $(t-2)^2$

4. $\begin{bmatrix} 9 & 15 \\ -6 & -10 \end{bmatrix}$ $t(t+1)$

5. $\begin{bmatrix} 3 & 2 & -2 \\ -8 & 0 & -5 \\ -8 & -2 & -3 \end{bmatrix}$ $-(t+5)(t-2)(t-3)$

6. $\begin{bmatrix} -9 & 8 & -8 \\ -4 & 3 & -4 \\ 2 & -2 & 1 \end{bmatrix}$ $-(t+3)(t+1)^2$

7. $\begin{bmatrix} 3 & -5 & 6 \\ 1 & 3 & -6 \\ 0 & 3 & -5 \end{bmatrix}$ $-(t-1)(t^2+2)$

8. $\begin{bmatrix} -2 & 6 & 3 \\ -2 & -8 & -2 \\ 4 & 6 & -1 \end{bmatrix}$ $-(t+5)(t+4)(t+2)$

9. $\begin{bmatrix} 1 & -2 & 2 \\ 8 & 11 & -8 \\ 4 & 4 & -1 \end{bmatrix}$ $-(t-5)(t-3)^2$

10. $\begin{bmatrix} 5 & 1 & 2 \\ 1 & 4 & 1 \\ -3 & -2 & 0 \end{bmatrix}$ $-(t-3)^3$

11. $\begin{bmatrix} -1 & 0 & 0 & 0 \\ 0 & -1 & 0 & 0 \\ 5 & 5 & 4 & -5 \\ 0 & 0 & 0 & -1 \end{bmatrix}$ $(t+1)^3(t-4)$

12. $\begin{bmatrix} -8 & 0 & -10 & 0 \\ -5 & 2 & -5 & 0 \\ 5 & 0 & 7 & 0 \\ -5 & 0 & -5 & 2 \end{bmatrix}$ $(t+3)(t-2)^3$

In Exercises 13–20, a matrix A is given. Find, if possible, an invertible matrix P and a diagonal matrix D such that $A = PDP^{-1}$. Otherwise, explain why A is not diagonalizable.

13. $\begin{bmatrix} 16 & -9 \\ 25 & -14 \end{bmatrix}$

14. $\begin{bmatrix} -1 & 2 \\ 3 & 4 \end{bmatrix}$

15. $\begin{bmatrix} 6 & 6 \\ -2 & -1 \end{bmatrix}$

16. $\begin{bmatrix} 1 & 5 \\ -1 & -1 \end{bmatrix}$

17. $\begin{bmatrix} -1 & 2 & -1 \\ 0 & -3 & 1 \\ 0 & 0 & 2 \end{bmatrix}$

18. $\begin{bmatrix} -3 & 0 & -5 \\ 0 & 2 & 0 \\ 2 & 0 & 3 \end{bmatrix}$

19. $\begin{bmatrix} 0 & 0 & 0 \\ 1 & 1 & 0 \\ 0 & -1 & 0 \end{bmatrix}$

20. $\begin{bmatrix} 2 & 0 & -1 \\ 1 & 3 & -1 \\ 2 & 0 & 5 \end{bmatrix}$

In Exercises 21–28, use complex numbers to find an invertible matrix P and a diagonal matrix D such that $A = PDP^{-1}$.

21. $\begin{bmatrix} 2 & -1 \\ 1 & 2 \end{bmatrix}$

22. $\begin{bmatrix} 5 & 51 \\ -3 & 11 \end{bmatrix}$

23. $\begin{bmatrix} 1-i & -4 \\ 1 & 1-i \end{bmatrix}$

24. $\begin{bmatrix} 1-10i & -4i \\ 24i & 1+10i \end{bmatrix}$

25. $\begin{bmatrix} 0 & -1 & 1 \\ 3 & 3 & -2 \\ 2 & 1 & 1 \end{bmatrix}$

26. $\begin{bmatrix} 1 & -1+i & 1-2i \\ 0 & i & -i \\ 0 & 0 & 0 \end{bmatrix}$

27. $\begin{bmatrix} 2i & 0 & 0 \\ 0 & 1 & -i \\ 0 & 0 & 0 \end{bmatrix}$

28. $\begin{bmatrix} 2i & -6-i & 1+2i \\ 0 & 1 & 0 \\ 0 & 3i & 4 \end{bmatrix}$

In Exercises 29–48, determine whether the statements are true or false.

29. Every $n \times n$ matrix is diagonalizable.
30. An $n \times n$ matrix A is diagonalizable if and only if there is a basis for $\mathcal{R}^n$ consisting of eigenvectors of A.
31. If P is an invertible $n \times n$ matrix and D is a diagonal $n \times n$ matrix such that $A = PDP^{-1}$, then the columns of P form a basis for $\mathcal{R}^n$ consisting of eigenvectors of A.
32. If P is an invertible matrix and D is a diagonal matrix such that $A = PDP^{-1}$, then the eigenvalues of A are the diagonal entries of D.
33. If A is a diagonalizable matrix, then there exists a unique diagonal matrix D such that $A = PDP^{-1}$.
34. If an $n \times n$ matrix has n distinct eigenvectors, then it is diagonalizable.
35. Every diagonalizable $n \times n$ matrix has n distinct eigenvalues.
36. If $\mathcal{B}_1, \mathcal{B}_2, \ldots, \mathcal{B}_k$ are bases for distinct eigenspaces of a matrix A, then $\mathcal{B}_1 \cup \mathcal{B}_2 \cup \cdots \cup \mathcal{B}_k$ is linearly independent.
37. If the sum of the multiplicities of the eigenvalues of an $n \times n$ matrix A equals n, then A is diagonalizable.
38. If, for each eigenvalue λ of A, the multiplicity of λ equals the dimension of the corresponding eigenspace, then A is diagonalizable.
39. If A is a diagonalizable 6×6 matrix having two distinct eigenvalues with multiplicities 2 and 4, then the corresponding eigenspaces of A must be 2-dimensional and 4-dimensional.
40. If λ is an eigenvalue of A, then the dimension of the eigenspace corresponding to λ equals the rank of $A - \lambda I_n$.
41. A diagonal $n \times n$ matrix has n distinct eigenvalues.
42. A diagonal matrix is diagonalizable.
43. The standard vectors are eigenvectors of a diagonal matrix.
44. Let A and P be $n \times n$ matrices. If the columns of P form a set of n linearly independent eigenvectors of A, then PAP^{-1} is a diagonal matrix.
45. If S is a set of distinct eigenvectors of a matrix, then S is linearly independent.
46. If S is a set of eigenvectors of a matrix A that correspond to distinct eigenvalues of A, then S is linearly independent.
47. If the characteristic polynomial of a matrix A factors into a product of linear factors, then A is diagonalizable.
48. If, for each eigenvalue λ of a matrix A, the dimension of the eigenspace of A corresponding to λ equals the multiplicity of λ, then A is diagonalizable.

---

49. A 3×3 matrix has eigenvalues -4, 2, and 5. Is the matrix diagonalizable? Justify your answer.
50. A 4×4 matrix has eigenvalues -3, -1, 2, and 5. Is the matrix diagonalizable? Justify your answer.
51. A 4×4 matrix has eigenvalues -3, -1, and 2. The eigenvalue -1 has multiplicity 2.
    (a) Under what conditions is the matrix diagonalizable? Justify your answer.
    (b) Under what conditions is it not diagonalizable? Justify your answer.
52. A 5×5 matrix has eigenvalues -4, which has multiplicity 3, and 6, which has multiplicity 2. The eigenspace corresponding to the eigenvalue 6 has dimension 2.
    (a) Under what conditions is the matrix diagonalizable? Justify your answer.
    (b) Under what conditions is it not diagonalizable? Justify your answer.
53. A 5×5 matrix has eigenvalues -3, which has multiplicity 4, and 7, which has multiplicity 1.
    (a) Under what conditions is the matrix diagonalizable? Justify your answer.
    (b) Under what conditions is it not diagonalizable? Justify your answer.
54. Let A be a 4×4 matrix with exactly the eigenvalues 2 and 7, and corresponding eigenspaces W_1 and W_2. For each of the parts shown, either write the characteristic polynomial of A, or state why there is insufficient information to determine the characteristic polynomial.
    (a) $\dim W_1 = 3$.
    (b) $\dim W_2 = 2$.
    (c) A is diagonalizable and $\dim W_2 = 2$.
55. Let A be a 5×5 matrix with exactly the eigenvalues 4, 5, and 8, and corresponding eigenspaces W_1, W_2, and W_3. For each of the given parts, either write the characteristic polynomial of A, or state why there is insufficient information to determine the characteristic polynomial.
    (a) $\dim W_1 = 2$ and $\dim W_3 = 2$.
    (b) A is diagonalizable and $\dim W_2 = 2$.
    (c) A is diagonalizable, $\dim W_1 = 1$, and $\dim W_2 = 2$.
56. Let $A = \begin{bmatrix} 1 & -2 \\ 1 & -2 \end{bmatrix}$ and $B = \begin{bmatrix} 2 & 0 \\ 1 & 0 \end{bmatrix}$.
    (a) Show that AB and BA have the same eigenvalues.
    (b) Is AB diagonalizable? Justify your answer.
    (c) Is BA diagonalizable? Justify your answer.

In Exercises 57–62, an $n \times n$ matrix A, a basis for $\mathcal{R}^n$ consisting of eigenvectors of A, and the corresponding eigenvalues are given. Calculate A^k for an arbitrary positive integer k.

57. $\begin{bmatrix} 2 & 2 \\ -1 & 5 \end{bmatrix}$; $\left\{\begin{bmatrix} 1 \\ 1 \end{bmatrix}, \begin{bmatrix} 2 \\ 1 \end{bmatrix}\right\}$; 4, 3

58. $\begin{bmatrix} -4 & 1 \\ -2 & -1 \end{bmatrix}$; $\left\{\begin{bmatrix} 1 \\ 2 \end{bmatrix}, \begin{bmatrix} 1 \\ 1 \end{bmatrix}\right\}$; -2, -3

59. $\begin{bmatrix} 5 & 6 \\ -1 & 0 \end{bmatrix}$; $\left\{\begin{bmatrix} 2 \\ -1 \end{bmatrix}, \begin{bmatrix} -3 \\ 1 \end{bmatrix}\right\}$; 2, 3

60. $\begin{bmatrix} 7 & 5 \\ -10 & -8 \end{bmatrix}$; $\left\{\begin{bmatrix} -1 \\ 2 \end{bmatrix}, \begin{bmatrix} -1 \\ 1 \end{bmatrix}\right\}$; -3, 2

61. $\begin{bmatrix} -3 & -8 & 0 \\ 4 & 9 & 0 \\ 0 & 0 & 5 \end{bmatrix}$; $\left\{\begin{bmatrix} -1 \\ 1 \\ 0 \end{bmatrix}, \begin{bmatrix} 0 \\ 0 \\ 1 \end{bmatrix}, \begin{bmatrix} -2 \\ 1 \\ 0 \end{bmatrix}\right\}$; 5, 5, 1

62. $\begin{bmatrix} -1 & 0 & 2 \\ 0 & 2 & 0 \\ -4 & 0 & 5 \end{bmatrix}$; $\left\{\begin{bmatrix} 1 \\ 0 \\ 1 \end{bmatrix}, \begin{bmatrix} 0 \\ 1 \\ 0 \end{bmatrix}, \begin{bmatrix} 1 \\ 0 \\ 2 \end{bmatrix}\right\}$; 1, 2, 3

In Exercises 63–72, a matrix and its characteristic polynomial are given. Determine all values of the scalar c for which each matrix is not diagonalizable.

63. $\begin{bmatrix} 1 & 0 & -1 \\ -2 & c & -2 \\ 2 & 0 & 4 \end{bmatrix}$
$-(t-c)(t-2)(t-3)$

64. $\begin{bmatrix} -7 & -1 & 2 \\ 0 & c & 0 \\ -10 & 3 & 3 \end{bmatrix}$
$-(t-c)(t+3)(t+2)$

65. $\begin{bmatrix} c & 0 & 0 \\ -1 & 1 & 4 \\ 3 & -2 & -1 \end{bmatrix}$
$-(t-c)(t^2+7)$

66. $\begin{bmatrix} 0 & 0 & -2 \\ -4 & c & -4 \\ 4 & 0 & 6 \end{bmatrix}$
$-(t-c)(t-2)(t-4)$

67. $\begin{bmatrix} 1 & -1 & 0 \\ 6 & 6 & 0 \\ 0 & 0 & c \end{bmatrix}$
$-(t-c)(t-3)(t-4)$

68. $\begin{bmatrix} 2 & -4 & -1 \\ 3 & -2 & 1 \\ 0 & 0 & c \end{bmatrix}$
$-(t-c)(t^2+8)$

69. $\begin{bmatrix} -3 & 0 & -2 \\ -6 & c & -2 \\ 1 & 0 & 0 \end{bmatrix}$
$-(t-c)(t+2)(t+1)$

70. $\begin{bmatrix} 3 & 0 & 0 \\ 0 & c & 0 \\ 1 & 0 & -2 \end{bmatrix}$
$-(t-c)(t+2)(t-3)$

71. $\begin{bmatrix} c & -9 & -3 & -15 \\ 0 & -7 & 0 & -6 \\ 0 & 7 & 2 & 13 \\ 0 & 4 & 0 & 3 \end{bmatrix}$
$(t-c)(t+3)(t+1)(t-2)$

72. $\begin{bmatrix} c & 6 & 2 & 10 \\ 0 & -12 & 0 & -15 \\ 0 & -11 & 1 & -15 \\ 0 & 10 & 0 & 13 \end{bmatrix}$
$(t-c)(t+2)(t-1)(t-3)$

73. Find a 2×2 matrix having eigenvalues -3 and 5, with corresponding eigenvectors $\begin{bmatrix} 1 \\ 1 \end{bmatrix}$ and $\begin{bmatrix} 1 \\ 3 \end{bmatrix}$.

74. Find a 2×2 matrix having eigenvalues 7 and -4, with corresponding eigenvectors $\begin{bmatrix} -1 \\ 3 \end{bmatrix}$ and $\begin{bmatrix} 1 \\ -2 \end{bmatrix}$.

75. Find a 3×3 matrix having eigenvalues 3, -2, and 1, with corresponding eigenvectors $\begin{bmatrix} -1 \\ 0 \\ 1 \end{bmatrix}$, $\begin{bmatrix} -1 \\ 1 \\ 1 \end{bmatrix}$, and $\begin{bmatrix} -2 \\ 0 \\ 1 \end{bmatrix}$.

76. Find a 3×3 matrix having eigenvalues 3, 2, and 2, with corresponding eigenvectors $\begin{bmatrix} 2 \\ 1 \\ 1 \end{bmatrix}$, $\begin{bmatrix} 1 \\ 0 \\ 1 \end{bmatrix}$, and $\begin{bmatrix} 1 \\ 1 \\ 1 \end{bmatrix}$.

77. Give an example of diagonalizable $n \times n$ matrices A and B such that $A + B$ is not diagonalizable.

78. Give an example of diagonalizable $n \times n$ matrices A and B such that AB is not diagonalizable.

79. Show that every diagonal $n \times n$ matrix is diagonalizable.

80. (a) Let A be an $n \times n$ matrix having a single eigenvalue c. Show that if A is diagonalizable, then $A = cI_n$.
(b) Use (a) to explain why $\begin{bmatrix} 2 & 1 \\ 0 & 2 \end{bmatrix}$ is not diagonalizable.

81. If A is a diagonalizable matrix, prove that A^T is diagonalizable.

82. If A is an invertible matrix that is diagonalizable, prove that A^{-1} is diagonalizable.

83. If A is a diagonalizable matrix, prove that A^2 is diagonalizable.

84. If A is a diagonalizable matrix, prove that A^k is diagonalizable for any positive integer k.

85. Suppose that A and B are similar matrices such that $B = P^{-1}AP$ for some invertible matrix P.
(a) Show that A is diagonalizable if and only if B is diagonalizable.
(b) How are the eigenvalues of A related to the eigenvalues of B? Justify your answer.
(c) How are the eigenvectors of A related to the eigenvectors of B? Justify your answer.

86. A matrix B is called a *cube root* of a matrix A if $B^3 = A$. Prove that every diagonalizable matrix has a cube root.

87. Prove that if a nilpotent matrix is diagonalizable, then it must be the zero matrix. *Hint:* Use Exercise 72 of Section 5.1.

88. Let A be a diagonalizable $n \times n$ matrix. Prove that if the characteristic polynomial of A is $f(t) = a_n t^n + a_{n-1}t^{n-1} + \cdots + a_1 t + a_0$, then $f(A) = O$, where $f(A) = a_n A^n + a_{n-1}A^{n-1} + \cdots + a_1 A + a_0 I_n$. (This result is called the *Cayley-Hamilton theorem.*[7]) *Hint:* If $A = PDP^{-1}$, show that $f(A) = Pf(D)P^{-1}$.

[7] The Cayley-Hamilton theorem first appeared in 1858. Arthur Cayley (1821–1895) was an English mathematician who contributed greatly to the development of both algebra and geometry. He was one of the first to study matrices, and this work contributed to the development of quantum mechanics. The Irish mathematician William Rowan Hamilton (1805–1865) is perhaps best known for his use of algebra in optics. His 1833 paper first gave a formal structure to ordered pairs of real numbers that yielded the system of complex numbers and led to his later development of *quaternions*.

89. The **trace** of a square matrix is the sum of its diagonal entries.

(a) Prove that if A is a diagonalizable matrix, then the trace of A equals the sum of the eigenvalues of A. *Hint:* For all $n \times n$ matrices A and B, show that the trace of AB equals the trace of BA.

(b) Let A be a diagonalizable $n \times n$ matrix with characteristic polynomial $(-1)^n(t-\lambda_1)(t-\lambda_2)\cdots(t-\lambda_n)$. Prove that the coefficient of t^{n-1} in this polynomial is $(-1)^{n-1}$ times the trace of A.

(c) For A as in (b), what is the constant term of the characteristic polynomial of A?

In Exercises 90–94, use either a calculator with matrix capabilities or computer software such as MATLAB to solve each problem.

For each of the matrices in Exercises 90–93, find, if possible, an invertible matrix P and a diagonal matrix D such that $A = PDP^{-1}$. If no such matrices exist, explain why not.

90. $\begin{bmatrix} 2 & 1 & 1 & 1 \\ 1 & 2 & 1 & 1 \\ -2 & 2 & 2 & 3 \\ 0 & 2 & 1 & 2 \end{bmatrix}$

91. $\begin{bmatrix} -4 & -5 & -7 & -4 \\ -1 & -6 & -4 & -3 \\ 1 & 1 & 1 & 1 \\ 1 & 7 & 5 & 3 \end{bmatrix}$

92. $\begin{bmatrix} 7 & 6 & 24 & -2 & 14 \\ 6 & 5 & 18 & 0 & 12 \\ -8 & -6 & -25 & 2 & -14 \\ -12 & -8 & -36 & 3 & -20 \\ 6 & 4 & 18 & -2 & 9 \end{bmatrix}$

93. $\begin{bmatrix} 4 & 13 & -5 & -29 & -17 \\ -3 & -11 & 0 & 32 & 24 \\ 0 & -3 & 7 & 3 & -3 \\ -2 & -5 & -5 & 18 & 17 \\ 1 & 2 & 5 & -10 & -11 \end{bmatrix}$

94. Let

$$A = \begin{bmatrix} 1.00 & 4.0 & c \\ 0.16 & 0.0 & 0 \\ 0.00 & -0.5 & 0 \end{bmatrix} \quad \text{and} \quad \mathbf{u} = \begin{bmatrix} 1 \\ 3 \\ -5 \end{bmatrix}.$$

(a) If $c = 8.1$, what appears to happen to the vector $A^m\mathbf{u}$ as m increases?

(b) What are the eigenvalues of A when $c = 8.1$?

(c) If $c = 8.0$, what appears to happen to the vector $A^m\mathbf{u}$ as m increases?

(d) What are the eigenvalues of A when $c = 8.0$?

(e) If $c = 7.9$, what appears to happen to the vector $A^m\mathbf{u}$ as m increases?

(f) What are the eigenvalues of A when $c = 7.9$?

(g) Let B be an $n \times n$ matrix having n distinct eigenvalues, all of which have absolute value less than 1. Let $\mathbf{u}$ be any vector in $\mathcal{R}^n$. Based on your answers to (a) through (f), make a conjecture about the behavior of $B^m\mathbf{u}$ as m increases. Then prove that your conjecture is valid.

SOLUTIONS TO THE PRACTICE PROBLEMS

1. The given matrix A has two eigenvalues: -1, which has multiplicity 1, and 2, which has multiplicity 2. For the eigenvalue 2, we see that the reduced row echelon form of $A - 2I_3$ is

$$\begin{bmatrix} 1 & 1 & 0 \\ 0 & 0 & 0 \\ 0 & 0 & 0 \end{bmatrix}.$$

Therefore the vector form of the general solution of $(A - 2I_3)\mathbf{x} = \mathbf{0}$ is

$$\begin{bmatrix} x_1 \\ x_2 \\ x_3 \end{bmatrix} = \begin{bmatrix} -x_2 \\ x_2 \\ x_3 \end{bmatrix} = x_2 \begin{bmatrix} -1 \\ 1 \\ 0 \end{bmatrix} + x_3 \begin{bmatrix} 0 \\ 0 \\ 1 \end{bmatrix},$$

so

$$\left\{ \begin{bmatrix} -1 \\ 1 \\ 0 \end{bmatrix}, \begin{bmatrix} 0 \\ 0 \\ 1 \end{bmatrix} \right\}$$

is a basis for the eigenspace of A corresponding to the eigenvalue 2. Likewise, from the reduced row echelon form of $A + I_3$, which is

$$\begin{bmatrix} 1 & 0 & 2 \\ 0 & 1 & -1 \\ 0 & 0 & 0 \end{bmatrix},$$

we see that

$$\left\{ \begin{bmatrix} -2 \\ 1 \\ 1 \end{bmatrix} \right\}$$

is a basis for the eigenspace of A corresponding to the eigenvalue -1.

Take

$$P = \begin{bmatrix} -1 & 0 & -2 \\ 1 & 0 & 1 \\ 0 & 1 & 1 \end{bmatrix},$$

the matrix whose columns are the vectors in the eigenspace bases. The corresponding diagonal matrix D is

$$D = \begin{bmatrix} 2 & 0 & 0 \\ 0 & 2 & 0 \\ 0 & 0 & -1 \end{bmatrix},$$

the matrix whose diagonal entries are the eigenvalues that correspond to the respective columns of P. Then $A = PDP^{-1}$.

2. The characteristic polynomial of A shows that the only eigenvalue of A is 2, and that its multiplicity is 1. Because A is a 3×3 matrix, the sum of the multiplicities of the eigenvalues of A must be 3 for A to be diagonalizable. Hence A is not diagonalizable.

   The matrix B has two eigenvalues: 2, which has multiplicity 1: and -1, which has multiplicity 2. So the sum of the multiplicities of the eigenvalues of B is 3. Thus there are enough eigenvalues for B to be diagonalizable. Checking the eigenvalue -1 (the eigenvalue with multiplicity greater than 1), we see that the reduced row echelon form of $B - (-1)I_3 = B + I_3$ is

$$\begin{bmatrix} 1 & 0 & -1 \\ 0 & 1 & 0 \\ 0 & 0 & 0 \end{bmatrix}.$$

   Since the rank of this matrix is 2, the dimension of the eigenspace corresponding to eigenvalue -1 is $3 - 2 = 1$, which is less than the multiplicity of the eigenvalue. Therefore B is not diagonalizable.

5.4* DIAGONALIZATION OF LINEAR OPERATORS

In Section 5.3, we defined a diagonalizable matrix and saw that an $n \times n$ matrix is diagonalizable if and only if there is a basis for $\mathcal{R}^n$ consisting of eigenvectors of the matrix (Theorem 5.2). We now define a linear operator on $\mathcal{R}^n$ to be **diagonalizable** if there is a basis for $\mathcal{R}^n$ consisting of eigenvectors of the operator.

Since the eigenvalues and eigenvectors of a linear operator are the same as those of its standard matrix, the procedure for finding a basis of eigenvectors for a linear operator is the same as that for a matrix. Moreover, a basis of eigenvectors of the operator or its standard matrix is also a basis of eigenvectors of the other. Therefore *a linear operator is diagonalizable if and only if its standard matrix is diagonalizable.* So the algorithm for matrix diagonalization on page 318 can be used to obtain a basis of eigenvectors for a diagonalizable linear operator, and the test for a diagonalizable matrix on page 319 can be used to identify a linear operator that is diagonalizable.

Example 1 Find, if possible, a basis for $\mathcal{R}^3$ consisting of eigenvectors of the linear operator T on $\mathcal{R}^3$ defined by

$$T\left(\begin{bmatrix} x_1 \\ x_2 \\ x_3 \end{bmatrix}\right) = \begin{bmatrix} 8x_1 + 9x_2 \\ -6x_1 - 7x_2 \\ 3x_1 + 3x_2 - x_3 \end{bmatrix}.$$

Solution The standard matrix of T is

$$A = \begin{bmatrix} 8 & 9 & 0 \\ -6 & -7 & 0 \\ 3 & 3 & -1 \end{bmatrix}.$$

Since T is diagonalizable if and only if A is, we must determine the eigenvalues and eigenspaces of A. The characteristic polynomial of A is $-(t+1)^2(t-2)$. Thus the eigenvalues of A are -1, which has multiplicity 2, and 2, which has multiplicity 1.

The reduced row echelon form of $A + I_3$ is

$$\begin{bmatrix} 1 & 1 & 0 \\ 0 & 0 & 0 \\ 0 & 0 & 0 \end{bmatrix},$$

* This section can be omitted without loss of continuity.

so the eigenspace corresponding to the eigenvalue -1 has

$$\mathcal{B}_1 = \left\{ \begin{bmatrix} -1 \\ 1 \\ 0 \end{bmatrix}, \begin{bmatrix} 0 \\ 0 \\ 1 \end{bmatrix} \right\}$$

as a basis. Since -1 is the only eigenvalue of multiplicity greater than 1 and the dimension of its eigenspace equals its multiplicity, A (and hence T) is diagonalizable. To obtain a basis of eigenvectors, we need to examine the reduced row echelon form of $A - 2I_3$, which is

$$\begin{bmatrix} 1 & 0 & -3 \\ 0 & 1 & 2 \\ 0 & 0 & 0 \end{bmatrix}.$$

It follows that

$$\mathcal{B}_2 = \left\{ \begin{bmatrix} 3 \\ -2 \\ 1 \end{bmatrix} \right\}$$

is a basis for the eigenspace of A corresponding to the eigenvalue 2. Then

$$\left\{ \begin{bmatrix} -1 \\ 1 \\ 0 \end{bmatrix}, \begin{bmatrix} 0 \\ 0 \\ 1 \end{bmatrix}, \begin{bmatrix} 3 \\ -2 \\ 1 \end{bmatrix} \right\}$$

is a basis for $\mathcal{R}^3$ consisting of eigenvectors of A. This set is also a basis for $\mathcal{R}^3$ consisting of eigenvectors of T.

Example 2 Find, if possible, a basis for $\mathcal{R}^3$ consisting of eigenvectors of the linear operator T on $\mathcal{R}^3$ defined by

$$T\left(\begin{bmatrix} x_1 \\ x_2 \\ x_3 \end{bmatrix}\right) = \begin{bmatrix} -x_1 + x_2 + 2x_3 \\ x_1 - x_2 \\ 0 \end{bmatrix}.$$

Solution The standard matrix of T is

$$A = \begin{bmatrix} -1 & 1 & 2 \\ 1 & -1 & 0 \\ 0 & 0 & 0 \end{bmatrix}.$$

To determine if A is diagonalizable, we must find the characteristic polynomial of A, which is $-t^2(t+2)$. Thus A has the eigenvalues 0, which has multiplicity 2, and -2, which has multiplicity 1. According to the test for a diagonalizable matrix, A is diagonalizable if and only if the eigenspace corresponding to the eigenvalue of multiplicity 2 has dimension 2. Therefore we must examine the eigenspace corresponding to the eigenvalue 0. Because the reduced row echelon form of $A - 0I_3 = A$ is

$$\begin{bmatrix} 1 & -1 & 0 \\ 0 & 0 & 1 \\ 0 & 0 & 0 \end{bmatrix},$$

we see that the rank of $A - 0I_3$ is 2. Thus the eigenspace corresponding to the eigenvalue 0 has dimension 1, and so A, and hence T, is not diagonalizable.

Practice Problem 1 ▶ The characteristic polynomial of the linear operator

$$T\left(\begin{bmatrix} x_1 \\ x_2 \\ x_3 \end{bmatrix}\right) = \begin{bmatrix} x_1 + 2x_2 + x_3 \\ 2x_2 \\ -x_1 + 2x_2 + 3x_3 \end{bmatrix}$$

is $-(t-2)^3$. Determine if this linear operator is diagonalizable. If so, find a basis for $\mathcal{R}^3$ consisting of eigenvectors of T. ◀

Practice Problem 2 ▶ Determine if the linear operator

$$T\left(\begin{bmatrix} x_1 \\ x_2 \end{bmatrix}\right) = \begin{bmatrix} -7x_1 - 10x_2 \\ 3x_1 + 4x_2 \end{bmatrix}$$

is diagonalizable. If so, find a basis for $\mathcal{R}^2$ consisting of eigenvectors of T. ◀

Let T be a linear operator on $\mathcal{R}^n$ for which there is a basis $\mathcal{B} = \{\mathbf{v}_1, \mathbf{v}_2, \ldots, \mathbf{v}_n\}$ consisting of eigenvectors of T. Then for each i, we have $T(\mathbf{v}_i) = \lambda_i \mathbf{v}_i$, where λ_i is the eigenvalue corresponding to $\mathbf{v}_i$. Therefore $[T(\mathbf{v}_i)]_{\mathcal{B}} = \lambda_i \mathbf{e}_i$ for each i, and so

$$[T]_{\mathcal{B}} = [\,[T(\mathbf{v}_1)]_{\mathcal{B}}\ [T(\mathbf{v}_2)]_{\mathcal{B}}\ \ldots\ [T(\mathbf{v}_n)]_{\mathcal{B}}\,] = [\lambda_1 \mathbf{e}_1\ \lambda_2 \mathbf{e}_2\ \ldots\ \lambda_n \mathbf{e}_n]$$

is a diagonal matrix. The converse of this result is also true, which explains the use of the term *diagonalizable* for such a linear operator.

> A linear operator T on $\mathcal{R}^n$ is diagonalizable if and only if there is a basis $\mathcal{B}$ for $\mathcal{R}^n$ such that $[T]_{\mathcal{B}}$, the $\mathcal{B}$-matrix of T, is a diagonal matrix. Such a basis $\mathcal{B}$ must consist of eigenvectors of T.

Recall from Theorem 4.12 that the $\mathcal{B}$-matrix of T is given by $[T]_{\mathcal{B}} = B^{-1}AB$, where B is the matrix whose columns are the vectors in $\mathcal{B}$ and A is the standard matrix of T. Thus, if we take

$$\mathcal{B} = \left\{ \begin{bmatrix} -1 \\ 1 \\ 0 \end{bmatrix}, \begin{bmatrix} 0 \\ 0 \\ 1 \end{bmatrix}, \begin{bmatrix} 3 \\ -2 \\ 1 \end{bmatrix} \right\}$$

in Example 1, we have

$$[T]_{\mathcal{B}} = B^{-1}AB = \begin{bmatrix} -1 & 0 & 0 \\ 0 & -1 & 0 \\ 0 & 0 & 2 \end{bmatrix},$$

which is the diagonal matrix whose diagonal entries are the eigenvalues corresponding to the respective columns of B.

A REFLECTION OPERATOR

We conclude this section with an example of a diagonalizable linear operator of geometric interest, the *reflection* of $\mathcal{R}^3$ about a 2-dimensional subspace. Although previously defined in the Exercises for Section 4.5, we repeat the definition here.

Let W be a 2-dimensional subspace of $\mathcal{R}^3$, that is, a plane containing the origin, and consider the mapping $T_W : \mathcal{R}^3 \to \mathcal{R}^3$ defined as follows: For a vector $\mathbf{u}$ in $\mathcal{R}^3$ with endpoint P (see Figure 5.6), drop a perpendicular from P to W, and extend this perpendicular an equal distance to the point P' on the other side of W. Then $T_W(\mathbf{u})$ is the vector with endpoint P'.

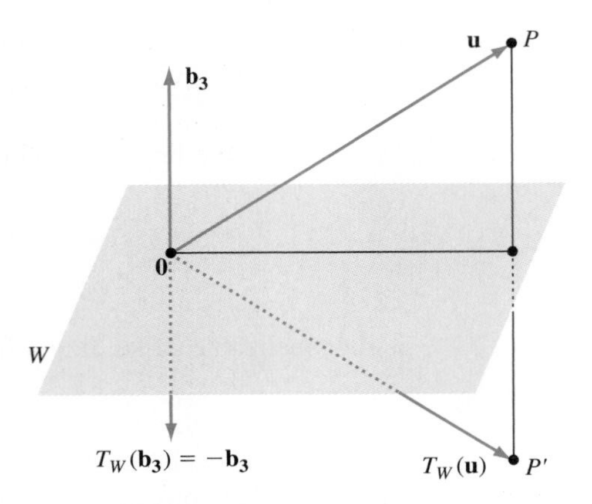

Figure 5.6 The reflection of $\mathcal{R}^3$ about a subspace W

It will be shown in Chapter 6 that T_W is linear. (See Exercise 84 in Section 6.3.) Assuming that T_W is linear, we show that reflections are diagonalizable. Choose any two linearly independent vectors $\mathbf{b}_1$ and $\mathbf{b}_2$ in W, and choose a third nonzero vector $\mathbf{b}_3$ perpendicular to W, as shown in Figure 5.6. Since $\mathbf{b}_3$ is not a linear combination of $\mathbf{b}_1$ and $\mathbf{b}_2$, the set $\mathcal{B} = \{\mathbf{b}_1, \mathbf{b}_2, \mathbf{b}_3\}$ is linearly independent and hence is a basis for $\mathcal{R}^3$. Furthermore, $T_W(\mathbf{b}_1) = \mathbf{b}_1$ and $T_W(\mathbf{b}_2) = \mathbf{b}_2$ because a vector in W coincides with its reflection. In addition, $T_W(\mathbf{b}_3) = -\mathbf{b}_3$ because T_W reflects $\mathbf{b}_3$ through W an equal distance to the other side. (See Figure 5.6.) It follows that $\mathbf{b}_1$ and $\mathbf{b}_2$ are eigenvectors of T_W with corresponding eigenvalue 1, and $\mathbf{b}_3$ is an eigenvector of T_W with corresponding eigenvalue -1. Therefore T_W is diagonalizable. In fact, its columns are

$$[T_W(\mathbf{b}_1)]_{\mathcal{B}} = \begin{bmatrix} 1 \\ 0 \\ 0 \end{bmatrix}, \qquad [T_W(\mathbf{b}_2)]_{\mathcal{B}} = \begin{bmatrix} 0 \\ 1 \\ 0 \end{bmatrix}, \qquad \text{and} \qquad [T_W(\mathbf{b}_3)]_{\mathcal{B}} = \begin{bmatrix} 0 \\ 0 \\ -1 \end{bmatrix}.$$

So

$$[T_W]_{\mathcal{B}} = \begin{bmatrix} 1 & 0 & 0 \\ 0 & 1 & 0 \\ 0 & 0 & -1 \end{bmatrix}. \tag{4}$$

Example 3 Find an explicit formula for the reflection operator T_W of $\mathcal{R}^3$ about the plane W, where

$$W = \left\{ \begin{bmatrix} x_1 \\ x_2 \\ x_3 \end{bmatrix} \in \mathcal{R}^3 : x_1 - x_2 + x_3 = 0 \right\}.$$

Solution As in Chapter 4, we can obtain a basis for W by solving the equation that defines it, namely, $x_1 - x_2 + x_3 = 0$:

$$\begin{bmatrix} x_1 \\ x_2 \\ x_3 \end{bmatrix} = \begin{bmatrix} x_2 - x_3 \\ x_2 \\ x_3 \end{bmatrix} = x_2 \begin{bmatrix} 1 \\ 1 \\ 0 \end{bmatrix} + x_3 \begin{bmatrix} -1 \\ 0 \\ 1 \end{bmatrix}$$

Therefore

$$\mathcal{B}_1 = \left\{ \begin{bmatrix} 1 \\ 1 \\ 0 \end{bmatrix}, \begin{bmatrix} -1 \\ 0 \\ 1 \end{bmatrix} \right\}$$

is a basis for W.

Recall from analytic geometry that the vector $\begin{bmatrix} a \\ b \\ c \end{bmatrix}$ is perpendicular (normal) to the plane with the equation $ax + by + cz = d$. So setting $\mathbf{b}_3 = \begin{bmatrix} 1 \\ -1 \\ 1 \end{bmatrix}$, we obtain a nonzero vector perpendicular to W. Adjoin $\mathbf{b}_3$ to $\mathcal{B}_1$ to obtain a basis

$$\mathcal{B} = \left\{ \begin{bmatrix} 1 \\ 1 \\ 0 \end{bmatrix}, \begin{bmatrix} -1 \\ 0 \\ 1 \end{bmatrix}, \begin{bmatrix} 1 \\ -1 \\ 1 \end{bmatrix} \right\}$$

for $\mathcal{R}^3$ consisting of eigenvectors of T_W. Then $[T_W]_{\mathcal{B}}$ is as in Equation (4).

By Theorem 4.12, the standard matrix of A is given by

$$A = B[T_W]_{\mathcal{B}}B^{-1} = \begin{bmatrix} \frac{1}{3} & \frac{2}{3} & -\frac{2}{3} \\ \frac{2}{3} & \frac{1}{3} & \frac{2}{3} \\ -\frac{2}{3} & \frac{2}{3} & \frac{1}{3} \end{bmatrix},$$

where $B = [\mathbf{b}_1\ \mathbf{b}_2\ \mathbf{b}_3]$. So

$$T_W\left(\begin{bmatrix} x_1 \\ x_2 \\ x_3 \end{bmatrix}\right) = A \begin{bmatrix} x_1 \\ x_2 \\ x_3 \end{bmatrix} = \begin{bmatrix} \frac{1}{3}x_2 + \frac{2}{3}x_2 - \frac{2}{3}x_3 \\ \frac{2}{3}x_1 + \frac{1}{3}x_2 + \frac{2}{3}x_3 \\ -\frac{2}{3}x_1 + \frac{2}{3}x_2 + \frac{1}{3}x_3 \end{bmatrix}$$

is an explicit formula for T_W.

Practice Problem 3 ▶ Find an explicit formula for the reflection operator of $\mathcal{R}^3$ about the plane with equation $x - 2y + 3z = 0$. ◀

EXERCISES

In Exercises 1–8, a linear operator T on $\mathcal{R}^3$ and a basis $\mathcal{B}$ for $\mathcal{R}^3$ are given. Compute $[T]_{\mathcal{B}}$, and determine whether $\mathcal{B}$ is a basis for $\mathcal{R}^3$ consisting of eigenvectors of T.

1. $T\left(\begin{bmatrix} x_1 \\ x_2 \\ x_3 \end{bmatrix}\right) = \begin{bmatrix} 2x_3 \\ -3x_1 + 3x_2 + 2x_3 \\ 4x_1 \end{bmatrix},$

$$\mathcal{B} = \left\{ \begin{bmatrix} 1 \\ 1 \\ 2 \end{bmatrix}, \begin{bmatrix} 0 \\ 1 \\ 0 \end{bmatrix}, \begin{bmatrix} 1 \\ 1 \\ 1 \end{bmatrix} \right\}$$

2. $T\left(\begin{bmatrix} x_1 \\ x_2 \\ x_3 \end{bmatrix}\right) = \begin{bmatrix} -x_1 + x_2 - x_3 \\ x_1 - x_2 + 3x_3 \\ 2x_1 - 2x_2 + 6x_3 \end{bmatrix}$,

$$\mathcal{B} = \left\{ \begin{bmatrix} -1 \\ 1 \\ 2 \end{bmatrix}, \begin{bmatrix} 0 \\ 1 \\ 2 \end{bmatrix}, \begin{bmatrix} 2 \\ 1 \\ 0 \end{bmatrix} \right\}$$

3. $T\left(\begin{bmatrix} x_1 \\ x_2 \\ x_3 \end{bmatrix}\right) = \begin{bmatrix} -x_2 - 2x_3 \\ 2x_2 \\ x_1 + x_2 + 3x_3 \end{bmatrix}$,

$$\mathcal{B} = \left\{ \begin{bmatrix} -1 \\ 0 \\ 1 \end{bmatrix}, \begin{bmatrix} -1 \\ 1 \\ 1 \end{bmatrix}, \begin{bmatrix} -2 \\ 0 \\ 1 \end{bmatrix} \right\}$$

4. $T\left(\begin{bmatrix} x_1 \\ x_2 \\ x_3 \end{bmatrix}\right) = \begin{bmatrix} 7x_1 + 5x_2 + 4x_3 \\ -4x_1 - 2x_2 - 2x_3 \\ -8x_1 - 7x_2 - 5x_3 \end{bmatrix}$,

$$\mathcal{B} = \left\{ \begin{bmatrix} -1 \\ 0 \\ 2 \end{bmatrix}, \begin{bmatrix} -1 \\ -1 \\ 3 \end{bmatrix}, \begin{bmatrix} 1 \\ -2 \\ 1 \end{bmatrix} \right\}$$

5. $T\left(\begin{bmatrix} x_1 \\ x_2 \\ x_3 \end{bmatrix}\right) = \begin{bmatrix} -4x_1 + 2x_2 - 2x_3 \\ -7x_1 - 3x_2 - 7x_3 \\ 7x_1 + x_2 + 5x_3 \end{bmatrix}$,

$$\mathcal{B} = \left\{ \begin{bmatrix} 0 \\ 1 \\ -1 \end{bmatrix}, \begin{bmatrix} -1 \\ 0 \\ 1 \end{bmatrix}, \begin{bmatrix} -1 \\ -1 \\ 1 \end{bmatrix} \right\}$$

6. $T\left(\begin{bmatrix} x_1 \\ x_2 \\ x_3 \end{bmatrix}\right) = \begin{bmatrix} -5x_1 - 2x_2 \\ 5x_1 - 6x_3 \\ 4x_1 + 4x_2 + 7x_3 \end{bmatrix}$,

$$\mathcal{B} = \left\{ \begin{bmatrix} 1 \\ -4 \\ 2 \end{bmatrix}, \begin{bmatrix} 0 \\ -1 \\ 1 \end{bmatrix}, \begin{bmatrix} 2 \\ -4 \\ 1 \end{bmatrix} \right\}$$

7. $T\left(\begin{bmatrix} x_1 \\ x_2 \\ x_3 \end{bmatrix}\right) = \begin{bmatrix} -3x_1 + 5x_2 - 5x_3 \\ 2x_1 - 3x_2 + 2x_3 \\ 2x_1 - 5x_2 + 4x_3 \end{bmatrix}$,

$$\mathcal{B} = \left\{ \begin{bmatrix} -1 \\ 0 \\ 1 \end{bmatrix}, \begin{bmatrix} 0 \\ 1 \\ 1 \end{bmatrix}, \begin{bmatrix} -1 \\ 1 \\ 1 \end{bmatrix} \right\}$$

8. $T\left(\begin{bmatrix} x_1 \\ x_2 \\ x_3 \end{bmatrix}\right) = \begin{bmatrix} -x_1 + x_2 + 3x_3 \\ 2x_1 + 6x_3 \\ -x_1 - x_2 - 5x_3 \end{bmatrix}$,

$$\mathcal{B} = \left\{ \begin{bmatrix} -2 \\ -1 \\ 1 \end{bmatrix}, \begin{bmatrix} -1 \\ -2 \\ 1 \end{bmatrix}, \begin{bmatrix} -1 \\ -3 \\ 1 \end{bmatrix} \right\}$$

In Exercises 9–20, a linear operator T on $\mathcal{R}^n$ and its characteristic polynomial are given. Find, if possible, a basis for $\mathcal{R}^n$ consisting of eigenvectors of T. If no such basis exists, explain why.

9. $T\left(\begin{bmatrix} x_1 \\ x_2 \end{bmatrix}\right) = \begin{bmatrix} 7x_1 - 6x_2 \\ 9x_1 - 7x_2 \end{bmatrix}$, $t^2 + 5$

10. $T\left(\begin{bmatrix} x_1 \\ x_2 \end{bmatrix}\right) = \begin{bmatrix} x_1 + x_2 \\ -9x_1 - 5x_2 \end{bmatrix}$, $(t + 2)^2$

11. $T\left(\begin{bmatrix} x_1 \\ x_2 \end{bmatrix}\right) = \begin{bmatrix} 7x_1 - 5x_2 \\ 10x_1 - 8x_2 \end{bmatrix}$, $(t + 3)(t - 2)$

12. $T\left(\begin{bmatrix} x_1 \\ x_2 \end{bmatrix}\right) = \begin{bmatrix} -7x_1 - 4x_2 \\ 8x_1 + 5x_2 \end{bmatrix}$, $(t + 3)(t - 1)$

13. $T\left(\begin{bmatrix} x_1 \\ x_2 \\ x_3 \end{bmatrix}\right) = \begin{bmatrix} -5x_1 \\ 7x_1 + 2x_2 \\ -7x_1 + x_2 + 3x_3 \end{bmatrix}$,
$-(t + 5)(t - 2)(t - 3)$

14. $T\left(\begin{bmatrix} x_1 \\ x_2 \\ x_3 \end{bmatrix}\right) = \begin{bmatrix} -3x_1 \\ 4x_1 + x_2 \\ x_3 \end{bmatrix}$, $-(t + 3)(t - 1)^2$

15. $T\left(\begin{bmatrix} x_1 \\ x_2 \\ x_3 \end{bmatrix}\right) = \begin{bmatrix} -x_1 - x_2 \\ -x_2 \\ x_1 + x_2 \end{bmatrix}$, $-t(t + 1)^2$

16. $T\left(\begin{bmatrix} x_1 \\ x_2 \\ x_3 \end{bmatrix}\right) = \begin{bmatrix} 3x_1 + 2x_2 \\ x_2 \\ 4x_1 - 3x_2 \end{bmatrix}$, $-t(t - 1)(t - 3)$

17. $T\left(\begin{bmatrix} x_1 \\ x_2 \\ x_3 \end{bmatrix}\right) = \begin{bmatrix} 6x_1 - 9x_2 + 9x_3 \\ -3x_2 + 7x_3 \\ 4x_3 \end{bmatrix}$,
$-(t + 3)(t - 4)(t - 6)$

18. $T\left(\begin{bmatrix} x_1 \\ x_2 \\ x_3 \end{bmatrix}\right) = \begin{bmatrix} -x_1 \\ -x_2 \\ x_1 - 2x_2 - x_3 \end{bmatrix}$, $-(t + 1)^3$

19. $T\left(\begin{bmatrix} x_1 \\ x_2 \\ x_3 \\ x_4 \end{bmatrix}\right) = \begin{bmatrix} -7x_1 - 4x_2 + 4x_3 - 4x_4 \\ x_2 \\ -8x_1 - 4x_2 + 5x_3 - 4x_4 \\ x_4 \end{bmatrix}$
$(t + 3)(t - 1)^3$

20. $T\left(\begin{bmatrix} x_1 \\ x_2 \\ x_3 \\ x_4 \end{bmatrix}\right) = \begin{bmatrix} 3x_1 - 5x_3 \\ 3x_2 - 5x_3 \\ -2x_3 \\ 5x_3 + 3x_4 \end{bmatrix}$, $(t + 2)(t - 3)^3$

In Exercises 21–28, a linear operator T on $\mathcal{R}^n$ is given. Find, if possible, a basis $\mathcal{B}$ for $\mathcal{R}^n$ such that $[T]_{\mathcal{B}}$ is a diagonal matrix. If no such basis exists, explain why.

21. $T\left(\begin{bmatrix} x_1 \\ x_2 \end{bmatrix}\right) = \begin{bmatrix} x_1 - x_2 \\ 3x_1 - x_2 \end{bmatrix}$

22. $T\left(\begin{bmatrix} x_1 \\ x_2 \end{bmatrix}\right) = \begin{bmatrix} -x_1 + 3x_2 \\ -4x_1 + 6x_2 \end{bmatrix}$

23. $T\left(\begin{bmatrix} x_1 \\ x_2 \end{bmatrix}\right) = \begin{bmatrix} -2x_1 + 3x_2 \\ 4x_1 - 3x_2 \end{bmatrix}$

24. $T\left(\begin{bmatrix} x_1 \\ x_2 \end{bmatrix}\right) = \begin{bmatrix} 11x_1 - 9x_2 \\ 16x_1 - 13x_2 \end{bmatrix}$

25. $T\left(\begin{bmatrix} x_1 \\ x_2 \\ x_3 \end{bmatrix}\right) = \begin{bmatrix} -x_1 \\ 3x_1 - x_2 + 3x_3 \\ 3x_1 + 2x_3 \end{bmatrix}$

26. $T\left(\begin{bmatrix} x_1 \\ x_2 \\ x_3 \end{bmatrix}\right) = \begin{bmatrix} 4x_1 - 5x_2 \\ -x_2 \\ -x_3 \end{bmatrix}$

27. $T\left(\begin{bmatrix} x_1 \\ x_2 \\ x_3 \end{bmatrix}\right) = \begin{bmatrix} x_1 \\ -x_1 + x_2 - x_3 \\ x_3 \end{bmatrix}$

28. $T\left(\begin{bmatrix} x_1 \\ x_2 \\ x_3 \end{bmatrix}\right) = \begin{bmatrix} 3x_1 - x_2 - 3x_3 \\ 3x_2 - 4x_3 \\ -x_3 \end{bmatrix}$

In Exercises 29–48, determine whether the statements are true or false.

29. If a linear operator on $\mathcal{R}^n$ is diagonalizable, then its standard matrix is a diagonal matrix.
30. For every linear operator on $\mathcal{R}^n$, there is a basis $\mathcal{B}$ for $\mathcal{R}^n$ such that $[T]_{\mathcal{B}}$ is a diagonal matrix.
31. A linear operator on $\mathcal{R}^n$ is diagonalizable if and only if its standard matrix is diagonalizable.
32. If T is a diagonalizable linear operator on $\mathcal{R}^n$, there is a unique basis $\mathcal{B}$ such that $[T]_{\mathcal{B}}$ is a diagonal matrix.
33. If T is a diagonalizable linear operator on $\mathcal{R}^n$, there is a unique diagonal matrix D such that $[T]_{\mathcal{B}} = D$.
34. Let W be a 2-dimensional subspace of $\mathcal{R}^3$. The reflection of $\mathcal{R}^3$ about W is one-to-one.
35. Let W be a 2-dimensional subspace of $\mathcal{R}^3$. The reflection of $\mathcal{R}^3$ about W is onto.
36. If T is a linear operator on $\mathcal{R}^n$ and $\mathcal{B}$ is a basis for $\mathcal{R}^n$ such that $[T]_{\mathcal{B}}$ is a diagonal matrix, then $\mathcal{B}$ consists of eigenvectors of T.
37. The characteristic polynomial of a linear operator T on $\mathcal{R}^n$ is a polynomial of degree n.
38. If the characteristic polynomial of a linear operator T on $\mathcal{R}^n$ factors into a product of linear factors, then T is diagonalizable.
39. If the characteristic polynomial of a linear operator T on $\mathcal{R}^n$ does not factor into a product of linear factors, then T is not diagonalizable.
40. If, for each eigenvalue λ of a linear operator T on $\mathcal{R}^n$, the dimension of the eigenspace of T corresponding to λ equals the multiplicity of λ, then T is diagonalizable.
41. Let W be a two-dimensional subspace of $\mathcal{R}^3$. If $T_W : \mathcal{R}^3 \to \mathcal{R}^3$ is the reflection of $\mathcal{R}^3$ about W, then each nonzero vector in W is an eigenvector of T_W corresponding to the eigenvalue -1.
42. Let W be a two-dimensional subspace of $\mathcal{R}^3$. If $T_W : \mathcal{R}^3 \to \mathcal{R}^3$ is the reflection of $\mathcal{R}^3$ about W, then each nonzero vector that is perpendicular to W is an eigenvector of T_W corresponding to the eigenvalue 0.
43. If T is a diagonalizable linear operator having 0 as an eigenvalue of multiplicity m, then the dimension of the null space of T equals m.
44. If T is a linear operator on $\mathcal{R}^n$, then the sum of the multiplicities of the eigenvalues of T equals n.
45. If T is a diagonalizable linear operator on $\mathcal{R}^n$, then the sum of the multiplicities of the eigenvalues of T equals n.
46. If T is a linear operator on $\mathcal{R}^n$ having n distinct eigenvalues, then T is diagonalizable.
47. If $\mathcal{B}_1, \mathcal{B}_2, \ldots, \mathcal{B}_k$ are bases for distinct eigenspaces of a linear operator T, then $\mathcal{B}_1 \cup \mathcal{B}_2 \cup \cdots \cup \mathcal{B}_k$ is a linearly independent set.
48. If $\mathcal{B}_1, \mathcal{B}_2, \ldots, \mathcal{B}_k$ are bases for all the distinct eigenspaces of a linear operator T, then $\mathcal{B}_1 \cup \mathcal{B}_2 \cup \cdots \cup \mathcal{B}_k$ is a basis for $\mathcal{R}^n$ consisting of eigenvectors of T.

In Exercises 49–58, a linear operator and its characteristic polynomial are given. Determine all the values of the scalar c for which the given linear operator on $\mathcal{R}^3$ is not diagonalizable.

49. $T\left(\begin{bmatrix} x_1 \\ x_2 \\ x_3 \end{bmatrix}\right) = \begin{bmatrix} 12x_1 + 10x_3 \\ -5x_1 + cx_2 - 5x_3 \\ -5x_1 - 3x_3 \end{bmatrix}$
$-(t-c)(t-2)(t-7)$

50. $T\left(\begin{bmatrix} x_1 \\ x_2 \\ x_3 \end{bmatrix}\right) = \begin{bmatrix} x_1 + 2x_2 - x_3 \\ cx_2 \\ 6x_1 - x_2 + 6x_3 \end{bmatrix}$
$-(t-c)(t-3)(t-4)$

51. $T\left(\begin{bmatrix} x_1 \\ x_2 \\ x_3 \end{bmatrix}\right) = \begin{bmatrix} cx_1 \\ -x_1 - 3x_2 - x_3 \\ -8x_1 + x_2 - 5x_3 \end{bmatrix}$
$-(t-c)(t+4)^2$

52. $T\left(\begin{bmatrix} x_1 \\ x_2 \\ x_3 \end{bmatrix}\right) = \begin{bmatrix} -4x_1 + x_2 \\ -4x_2 \\ cx_3 \end{bmatrix}$
$-(t-c)(t+4)^2$

53. $T\left(\begin{bmatrix} x_1 \\ x_2 \\ x_3 \end{bmatrix}\right) = \begin{bmatrix} cx_1 \\ 2x_1 - 3x_2 + 2x_3 \\ -3x_1 - x_3 \end{bmatrix}$
$-(t-c)(t+3)(t+1)$

54. $T\left(\begin{bmatrix} x_1 \\ x_2 \\ x_3 \end{bmatrix}\right) = \begin{bmatrix} -4x_1 - 2x_2 \\ cx_2 \\ 4x_1 + 4x_2 - 2x_3 \end{bmatrix}$
$-(t-c)(t+4)(t+2)$

55. $T\left(\begin{bmatrix} x_1 \\ x_2 \\ x_3 \end{bmatrix}\right) = \begin{bmatrix} -5x_1 + 9x_2 + 3x_3 \\ cx_2 \\ -9x_1 + 13x_2 + 5x_3 \end{bmatrix}$
$-(t-c)(t^2+2)$

56. $T\left(\begin{bmatrix} x_1 \\ x_2 \\ x_3 \end{bmatrix}\right) = \begin{bmatrix} cx_1 \\ 10x_2 - 2x_3 \\ 6x_2 + 3x_3 \end{bmatrix}$
$-(t-c)(t-6)(t-7)$

57. $T\left(\begin{bmatrix} x_1 \\ x_2 \\ x_3 \end{bmatrix}\right) = \begin{bmatrix} -7x_1 + 2x_2 \\ -10x_1 + 2x_2 \\ cx_3 \end{bmatrix}$
$-(t-c)(t+3)(t+2)$

58. $T\left(\begin{bmatrix} x_1 \\ x_2 \\ x_3 \end{bmatrix}\right) = \begin{bmatrix} 3x_1 + 7x_3 \\ x_1 + cx_2 + 2x_3 \\ -2x_1 - 3x_3 \end{bmatrix}$
$-(t-c)(t^2+5)$

In Exercises 59–64, the equation of a plane W through the origin of $\mathcal{R}^3$ is given. Determine an explicit formula for the reflection T_W of $\mathcal{R}^3$ about W.

59. $x + y + z = 0$
60. $2x + y + z = 0$
61. $x + 2y - z = 0$
62. $x + z = 0$
63. $x + 8y - 5z = 0$
64. $3x - 4y + 5z = 0$

In Exercises 65 and 66, the standard matrix of a reflection of $\mathcal{R}^3$ about a 2-dimensional subspace W is given. Find an equation for W.

65. $\frac{1}{9}\begin{bmatrix} 1 & -8 & -4 \\ -8 & 1 & -4 \\ -4 & -4 & 7 \end{bmatrix}$
66. $\frac{1}{3}\begin{bmatrix} 2 & -2 & -1 \\ -2 & -1 & -2 \\ -1 & -2 & 2 \end{bmatrix}$

Exercises 67–74 use the definition of the orthogonal projection U_W of $\mathcal{R}^3$ on a 2-dimensional subspace W, which is given in the exercises of Section 4.5.

67. Let W be a 2-dimensional subspace of $\mathcal{R}^3$.
   (a) Prove that there exists a basis $\mathcal{B}$ for $\mathcal{R}^3$ such that
   $$[U_W]_{\mathcal{B}} = \begin{bmatrix} 1 & 0 & 0 \\ 0 & 1 & 0 \\ 0 & 0 & 0 \end{bmatrix}.$$
   (b) Prove that $[T_W]_{\mathcal{B}} = \begin{bmatrix} 1 & 0 & 0 \\ 0 & 1 & 0 \\ 0 & 0 & -1 \end{bmatrix}$, where $\mathcal{B}$ is the basis in (a).
   (c) Prove that $[U_W]_{\mathcal{B}} = \frac{1}{2}([T_W]_{\mathcal{B}} + I_3)$, where $\mathcal{B}$ is the basis in (a).
   (d) Use (c) and Exercise 59 to find an explicit formula for the orthogonal projection U_W, where W is the plane with equation $x + y + z = 0$.

In Exercises 68–74, find an explicit formula for the orthogonal projection U_W of $\mathcal{R}^3$ on the plane W in each specified exercise. Use Exercise 67(c) and the basis obtained in the specified exercise to obtain your answer.

68. Exercise 60
69. Exercise 61
70. Exercise 62
71. Exercise 63
72. Exercise 64
73. Exercise 65
74. Exercise 66

75. Let $\{\mathbf{u}, \mathbf{v}, \mathbf{w}\}$ be a basis for $\mathcal{R}^3$, and let T be the linear operator on $\mathcal{R}^3$ defined by
$$T(a\mathbf{u} + b\mathbf{v} + c\mathbf{w}) = a\mathbf{u} + b\mathbf{v}$$
for all scalars a, b, and c.
   (a) Find the eigenvalues of T and determine a basis for each eigenspace.
   (b) Is T diagonalizable? Justify your answer.

76. Let $\{\mathbf{u}, \mathbf{v}, \mathbf{w}\}$ be a basis for $\mathcal{R}^3$, and let T be the linear operator on $\mathcal{R}^3$ defined by
$$T(a\mathbf{u} + b\mathbf{v} + c\mathbf{w}) = a\mathbf{u} + b\mathbf{v} - c\mathbf{w}$$
for all scalars a, b, and c.
   (a) Find the eigenvalues of T and determine a basis for each eigenspace.
   (b) Is T diagonalizable? Justify your answer.

77. Let T be a linear operator on $\mathcal{R}^n$ and $\mathcal{B}$ be a basis for $\mathcal{R}^n$ such that $[T]_{\mathcal{B}}$ is a diagonal matrix. Prove that $\mathcal{B}$ must consist of eigenvectors of T.
78. If T and U are diagonalizable linear operators on $\mathcal{R}^n$, must $T + U$ be a diagonalizable linear operator on $\mathcal{R}^n$? Justify your answer.
79. If T is a diagonalizable linear operator on $\mathcal{R}^n$, must cT be a diagonalizable linear operator on $\mathcal{R}^n$ for any scalar c? Justify your answer.
80. If T and U are diagonalizable linear operators on $\mathcal{R}^n$, must TU be a diagonalizable linear operator on $\mathcal{R}^n$? Justify your answer.
81. Let T be a linear operator on $\mathcal{R}^n$, and suppose that $\mathbf{v}_1, \mathbf{v}_2, \ldots, \mathbf{v}_k$ are eigenvectors of T corresponding to distinct nonzero eigenvalues. Prove that the set $\{T(\mathbf{v}_1), T(\mathbf{v}_2), \ldots, T(\mathbf{v}_k)\}$ is linearly independent.
82. Let T and U be diagonalizable linear operators on $\mathcal{R}^n$. Prove that if there is a basis for $\mathcal{R}^n$ consisting of eigenvectors of both T and U, then $TU = UT$.
83. Let T and U be linear operators on $\mathcal{R}^n$. If $T^2 = U$ (where $T^2 = TT$), then T is called a *square root* of U. Show that if U is diagonalizable and has only nonnegative eigenvalues, then U has a square root.
84. Let T be a linear operator on $\mathcal{R}^n$ and $\mathcal{B}_1, \mathcal{B}_2, \ldots, \mathcal{B}_k$ be bases for all the distinct eigenspaces of T. Prove that T is diagonalizable if and only if $\mathcal{B}_1 \cup \mathcal{B}_2 \cup \cdots \cup \mathcal{B}_k$ is a generating set for $\mathcal{R}^n$.

In Exercises 85 and 86, use either a calculator with matrix capabilities or computer software such as MATLAB to find a basis for $\mathcal{R}^5$ consisting of eigenvectors of each linear operator T; or explain why no such basis exists.

85. T is the linear operator on $\mathcal{R}^5$ defined by

$$T\begin{bmatrix}x_1\\x_2\\x_3\\x_4\\x_5\end{bmatrix}=\begin{bmatrix}-11x_1-9x_2+13x_3+18x_4-9x_5\\6x_1+5x_2-6x_3-8x_4+4x_5\\6x_1+3x_2-4x_3-6x_4+3x_5\\-2x_1+2x_3+3x_4-2x_5\\14x_1+12x_2-14x_3-20x_4+9x_5\end{bmatrix}.$$

86. T is the linear operator on $\mathcal{R}^5$ defined by

$$T\begin{bmatrix}x_1\\x_2\\x_3\\x_4\\x_5\end{bmatrix}=\begin{bmatrix}-2x_1-4x_2-9x_3-5x_4-16x_5\\x_1+4x_2+6x_3+5x_4+12x_5\\4x_1+10x_2+20x_3+14x_4+37x_5\\3x_1+2x_2+3x_3+2x_4+4x_5\\-4x_1-6x_2-12x_3-8x_4-21x_5\end{bmatrix}.$$

SOLUTIONS TO THE PRACTICE PROBLEMS

1. Since the characteristic polynomial of T is $-(t-2)^3$, the only eigenvalue of T is 2, which has multiplicity 3. The standard matrix of T is

$$A=\begin{bmatrix}1&2&1\\0&2&0\\-1&2&3\end{bmatrix},$$

and the reduced row echelon form of $A-2I_3$ is

$$\begin{bmatrix}1&-2&-1\\0&0&0\\0&0&0\end{bmatrix},$$

a matrix of rank 1. Since $3-1<3$, the dimension of the eigenspace corresponding to eigenvalue 2 is less than its multiplicity. Hence the second condition in the test for a diagonalizable matrix is not true for A, and A (and thus T) is not diagonalizable.

2. The standard matrix of T is

$$A=\begin{bmatrix}-7&-10\\3&4\end{bmatrix},$$

and its characteristic polynomial is $(t+1)(t+2)$. So A has two eigenvalues (-1 and -2), each of multiplicity 1. Thus A, and hence T, is diagonalizable.

To find a basis for $\mathcal{R}^2$ consisting of eigenvectors of T, we find bases for each of the eigenspaces of A. Since the reduced row echelon form of $A+2I_2$ is

$$\begin{bmatrix}1&2\\0&0\end{bmatrix},$$

we see that

$$\left\{\begin{bmatrix}-2\\1\end{bmatrix}\right\}$$

is a basis for the eigenspace of A corresponding to -2. Also, the reduced row echelon form of $A+I_2$ is

$$\begin{bmatrix}1&\frac{5}{3}\\0&0\end{bmatrix}.$$

Hence

$$\left\{\begin{bmatrix}-5\\3\end{bmatrix}\right\}$$

is a basis for the eigenspace of A corresponding to -1. Combining these eigenspace bases, we obtain the set

$$\left\{\begin{bmatrix}-2\\1\end{bmatrix},\begin{bmatrix}-5\\3\end{bmatrix}\right\},$$

which is a basis for $\mathcal{R}^2$ consisting of eigenvectors of A and T.

3. We must construct a basis $\mathcal{B}$ for $\mathcal{R}^3$ consisting of two vectors from W and one vector perpendicular to W. Solving the equation $x-2y+3z=0$, we obtain

$$\begin{bmatrix}x\\y\\z\end{bmatrix}=\begin{bmatrix}2y-3z\\y\\z\end{bmatrix}=y\begin{bmatrix}2\\1\\0\end{bmatrix}+z\begin{bmatrix}-3\\0\\1\end{bmatrix}.$$

Thus

$$\left\{\begin{bmatrix}2\\1\\0\end{bmatrix},\begin{bmatrix}-3\\0\\1\end{bmatrix}\right\}$$

is a basis for W. In addition, the vector

$$\begin{bmatrix}1\\-2\\3\end{bmatrix}$$

is normal to the plane W. Combining these three vectors, we obtain the desired basis

$$\mathcal{B}=\left\{\begin{bmatrix}2\\1\\0\end{bmatrix},\begin{bmatrix}-3\\0\\1\end{bmatrix},\begin{bmatrix}1\\-2\\3\end{bmatrix}\right\}$$

for $\mathcal{R}^3$.

For the orthogonal projection operator U_W on W, the vectors in $\mathcal{B}$ are eigenvectors corresponding to the eigenvalues 1, 1, and 0, respectively. Therefore

$$[U_W]_{\mathcal{B}}=\begin{bmatrix}1&0&0\\0&1&0\\0&0&0\end{bmatrix}.$$

Letting B denote the matrix whose columns are the vectors in $\mathcal{B}$, we see by Theorem 4.12 that the standard matrix

A of U_W is

$$A = B[U_W]_{\mathcal{B}}B^{-1} = \frac{1}{14}\begin{bmatrix} 13 & 2 & -3 \\ 2 & 10 & 6 \\ -3 & 6 & 5 \end{bmatrix}.$$

So the formula for U_W is

$$U_W\left(\begin{bmatrix} x_1 \\ x_2 \\ x_3 \end{bmatrix}\right) = A\begin{bmatrix} x_1 \\ x_2 \\ x_3 \end{bmatrix} = \frac{1}{14}\begin{bmatrix} 13x_1 + 2x_2 - 3x_3 \\ 2x_1 + 10x_2 + 6x_3 \\ -3x_1 + 6x_2 + 5x_3 \end{bmatrix}.$$

5.5* APPLICATIONS OF EIGENVALUES

In this section, we discuss four applications involving eigenvalues.

MARKOV CHAINS

Markov chains have been used to analyze situations as diverse as land use in Toronto, Canada [3], economic development in New Zealand [6], and the game of Monopoly [1] and [2]. This concept is named after the Russian mathematician Andrei Markov (1856–1922), who developed the fundamentals of the theory at the beginning of the twentieth century.

A **Markov chain** is a process that consists of a finite number of **states** and known probabilities p_{ij}, where p_{ij} represents the probability of moving from state j to state i. Note that this probability depends only on the present state j and the future state i. The movement of population between the city and suburbs described in Example 6 of Section 2.1 is an example of a Markov chain with two states (living in the city and living in the suburbs), where p_{ij} represents the probability of moving from one location to another during the coming year. Other possible examples include political affiliation (Democrat, Republican, or Independent), where p_{ij} represents the probability of a son having affiliation i if his father has affiliation j; cholesterol level (high, normal, and low), where p_{ij} represents the probability of moving from one level to another in a specific amount of time; or market share of competing products, where p_{ij} represents the probability that a consumer switches brands in a certain amount of time.

Consider a Markov chain with n states, where the probability of moving from state j to state i during a certain period of time is p_{ij} for $1 \le i, j \le n$. The $n \times n$ matrix A with (i,j)-entry equal to p_{ij} is called the **transition matrix** of this Markov chain. It is a stochastic matrix, that is, a matrix with nonnegative entries whose column sums are all 1. A Markov chain often has the property that it is possible to move from any state to any other over several periods. In such a case, the transition matrix of the Markov chain is called **regular**. It can be shown that the transition matrix of a Markov chain is regular if and only if some power of it contains no zero entries. Thus, if

$$A = \begin{bmatrix} .5 & 0 & .3 \\ 0 & .4 & .7 \\ .5 & .6 & 0 \end{bmatrix},$$

then A is regular because

$$A^2 = \begin{bmatrix} .40 & .18 & .15 \\ .35 & .58 & .28 \\ .25 & .24 & .57 \end{bmatrix}$$

has no zero entries. On the other hand,

$$B = \begin{bmatrix} .5 & 0 & .3 \\ 0 & 1 & .7 \\ .5 & 0 & 0 \end{bmatrix}$$

* This section can be omitted without loss of continuity.

is not a regular transition matrix because, for every positive integer k, B^k contains at least one zero entry, for example, the $(1, 2)$-entry.

Suppose that A is the transition matrix of a Markov chain with n states. If $\mathbf{p}$ is a vector in $\mathcal{R}^n$ whose components represent the probabilities of being in each state of the Markov chain at a particular time, then the components of $\mathbf{p}$ are nonnegative numbers whose sum equals 1. Such a vector is called a **probability vector**. In this case, the vector $A^m\mathbf{p}$ is a probability vector for every positive integer m, and the components of $A^m\mathbf{p}$ give the probabilities of being in each state after m periods.

As we saw in Section 5.3, the behavior of the vectors $A^m\mathbf{p}$ is often of interest in the study of a Markov chain. When A is a regular transition matrix, the behavior of these vectors can be easily described. A proof of the following theorem can be found in [4, page 300]:

THEOREM 5.4

If A is a regular $n \times n$ transition matrix and $\mathbf{p}$ is a probability vector in $\mathcal{R}^n$, then

(a) 1 is an eigenvalue of A;

(b) there is a unique probability vector $\mathbf{v}$ of A that is also an eigenvector corresponding to eigenvalue 1;

(c) the vectors $A^m\mathbf{p}$ approach $\mathbf{v}$ for $m = 1, 2, 3, \ldots$.

A probability vector $\mathbf{v}$ such that $A\mathbf{v} = \mathbf{v}$ is called a **steady-state vector**. Such a vector is a probability vector that is also an eigenvector of A corresponding to eigenvalue 1. Theorem 5.4 asserts that a regular Markov chain has a unique steady-state vector, and moreover, the vectors $A^m\mathbf{p}$ approach $\mathbf{v}$ for $m = 1, 2, 3, \ldots$, no matter what the original probability vector $\mathbf{p}$ is.

To illustrate these ideas, consider the following example:

Example 1

Suppose that Amy jogs or rides her bicycle every day for exercise. If she jogs today, then tomorrow she will flip a fair coin and jog if it lands heads and ride her bicycle if it lands tails. If she rides her bicycle one day, then she will always jog the next day. This situation can be modeled by a Markov chain with two states (jog and ride) and the transition matrix

$$\begin{array}{cc} & \text{today} \\ & \begin{array}{cc}\text{jog} & \text{ride}\end{array} \\ \text{tomorrow} \begin{array}{c}\text{jog} \\ \text{ride}\end{array} & \begin{bmatrix} .5 & 1 \\ .5 & 0 \end{bmatrix} \end{array} = A.$$

For example, the $(1, 1)$-entry of A is .5 because if Amy jogs today, there is a .5 probability that she will jog tomorrow.

Suppose that Amy decides to jog on Monday. By taking $\mathbf{p} = \begin{bmatrix} 1 \\ 0 \end{bmatrix}$ as the original probability vector, we obtain

$$A\mathbf{p} = \begin{bmatrix} .5 & 1 \\ .5 & 0 \end{bmatrix} \begin{bmatrix} 1 \\ 0 \end{bmatrix} = \begin{bmatrix} .5 \\ .5 \end{bmatrix}$$

and

$$A^2\mathbf{p} = A(A\mathbf{p}) = \begin{bmatrix} .5 & 1 \\ .5 & 0 \end{bmatrix} \begin{bmatrix} .5 \\ .5 \end{bmatrix} = \begin{bmatrix} .75 \\ .25 \end{bmatrix}.$$

Therefore on Tuesday the probabilities that Amy will jog or ride her bicycle are both .5, and on Wednesday the probability that she will jog is .75 and the probability that she will ride her bicycle is .25. Since $A^2 = \begin{bmatrix} .75 & .5 \\ .25 & .5 \end{bmatrix}$ has no zero entries, A is a regular transition matrix. Thus A has a unique steady-state vector $\mathbf{v}$, and the vectors $A^m\mathbf{p}$ $(m = 1, 2, 3, \ldots)$ approach $\mathbf{v}$ by Theorem 5.4. The steady-state vector is a solution of $A\mathbf{v} = \mathbf{v}$, that is, $(A - I_2)\mathbf{v} = \mathbf{0}$. Since the reduced row echelon form of $A - I_2$ is

$$\begin{bmatrix} 1 & -2 \\ 0 & 0 \end{bmatrix},$$

we see that the solutions of $(A - I_2)\mathbf{v} = \mathbf{0}$ have the form

$$\begin{aligned} v_1 &= 2v_2 \\ v_2 &\text{ free.} \end{aligned}$$

In order that $\mathbf{v} = \begin{bmatrix} v_1 \\ v_2 \end{bmatrix}$ be a probability vector, we must have $v_1 + v_2 = 1$. Hence

$$\begin{aligned} 2v_2 + v_2 &= 1 \\ 3v_2 &= 1 \\ v_2 &= \tfrac{1}{3}. \end{aligned}$$

Thus the unique steady-state vector for A is $\mathbf{v} = \begin{bmatrix} \frac{2}{3} \\ \frac{1}{3} \end{bmatrix}$. Hence, over the long run, Amy jogs $\frac{2}{3}$ of the time and rides her bicycle $\frac{1}{3}$ of the time.

Practice Problem 1 ▶ A car survey has found that 80% of those who were driving a car five years ago are now driving a car, 10% are now driving a minivan, and 10% are now driving a sport utility vehicle. Of those who were driving a minivan five years ago, 20% are now driving a car, 70% are now driving a minivan, and 10% are now driving a sport utility vehicle. Finally, of those who were driving a sport utility vehicle five years ago, 10% are now driving a car, 30% are now driving a minivan, and 60% are now driving a sport utility vehicle.

(a) Determine the transition matrix for this Markov chain.

(b) Suppose that 70% of those questioned were driving cars five years ago, 20% were driving minivans, and 10% were driving sport utility vehicles. Estimate the percentage of these persons driving each type of vehicle now.

(c) Under the conditions in (b), estimate the percentage of these persons who will be driving each type of vehicle five years from now.

(d) Determine the percentage of these persons driving each type of vehicle in the long run, assuming that the present trend continues indefinitely. ◀

GOOGLE SEARCHES

In December 2003, the most popular search engine for Web searches was Google, which performed about 35% of the month's Web searches. Although Google does not reveal the precise details of how its search engine prioritizes the websites it lists for a user's search, one of the reasons for Google's success is its PageRank™ algorithm, which was created by its developers, Larry Page and Sergey Brin, while they were graduate students at Stanford University.

Intuitively, the PageRank™ algorithm ranks Web pages in the following way: Consider a dedicated Web surfer who is viewing a page on the Web. The surfer moves to a new page by either randomly choosing a link from the current page (this happens 85% of the time) or moves to a new page by choosing a page at random from all of the other pages on the Web (this happens 15% of the time). We will see that the process of moving from one page to another produces a Markov chain, where the pages are the states, and the steady-state vector gives the proportions of times pages are visited. These proportions give the ranks of the pages. It follows that a page with many links to and from pages of high rank will itself be ranked high. In response to a user's search, Google determines which websites are relevant and then lists them in order of rank.

To investigate this process formally, we introduce some notation. Let n denote the number of pages that are considered by a Google search. (In November 2004, n was about 8.058 billion.) Let A be the $n \times n$ transition matrix associated with the Markov chain, where each of the n pages is a state and a_{ij} is the probability of moving from page j to page i for $1 \le i, j \le n$. To determine the entries a_{ij} of A, we make two assumptions:

1. If the current page has links to other pages, then a certain portion p (usually about 0.85) of the time, the surfer moves from the current page to the next page by choosing one of these links at random. The complementary portion of time, $1-p$, the surfer randomly selects one of the pages on the Web.
2. If the current page does not link to other pages, then the surfer randomly selects one of the pages on the Web.

To aid in the calculation of A, we begin with a matrix that describes the page links. Let C be the $n \times n$ matrix defined by

$$c_{ij} = \begin{cases} 1 & \text{if there is a link from page } j \text{ to page } i \\ 0 & \text{otherwise.} \end{cases}$$

For any j, let s_j denote the number of pages to which j links; that is, s_j equals the sum of the entries of column j of C. We determine a_{ij}, the probability that the reader moves from page j to page i, as follows. If page j has no links at all (that is, $s_j = 0$) then the probability that the surfer picks page i randomly from the Web is simply $1/n$. Now suppose that page j has links to other pages; that is, $s_j \neq 0$. Then to move to page i from page j, there are two cases:

Case 1. The surfer chooses a link on page *j*, and this link takes the surfer to page *i*.

The probability that the surfer chooses a link on page j is p, and the probability that this link connects to page i is $1/s_j = c_{ij}/s_j$. So the probability of this case is $p(c_{ij}/s_j)$.

Case 2. The surfer decides to choose a page on the Web randomly, and the page chosen is page *i*.

The probability that the surfer randomly chooses a page on the Web is $1-p$, and the probability that page i is the chosen page is $1/n$. So the probability of this case is $(1-p)/n$.

Thus, if $s_j \neq 0$, the probability that the surfer moves to page i from page j is

$$\frac{pc_{ij}}{s_j} + \frac{(1-p)}{n}.$$

So we have

$$a_{ij} = \begin{cases} \dfrac{pc_{ij}}{s_j} + \dfrac{1-p}{n} & \text{if } s_j \neq 0 \\[2ex] \dfrac{1}{n} & \text{if } s_j = 0. \end{cases}$$

For a simple method of obtaining A from C, we introduce the $n \times n$ matrix $M = \begin{bmatrix} \mathbf{m}_1 & \mathbf{m}_2 & \dots & \mathbf{m}_n \end{bmatrix}$, where

$$\mathbf{m}_j = \begin{cases} \dfrac{1}{s_j}\mathbf{c}_j & \text{if } s_j \neq 0 \\[2ex] \dfrac{1}{n}\begin{bmatrix} 1 \\ 1 \\ \vdots \\ 1 \end{bmatrix} & \text{if } s_j = 0. \end{cases}$$

It follows that $A = pM + \dfrac{1-p}{n}W$, where W is the $n \times n$ matrix whose entries are all equal to 1. (See Exercise 34.)

Observe that $a_{ij} > 0$ for all i and j. (For $n = 8.058$ billion and $p = 0.85$, the smallest possible value for a_{ij} is equal to $(1-p)/n \approx 0.186 \times 10^{-10}$.) Also observe that the sum of the entries in each column of A is 1, and hence A is a regular transition matrix. So, by Theorem 5.4, there exists a unique steady-state vector $\mathbf{v}$ of A. Over time, the components of $\mathbf{v}$ describe the distribution of surfers visiting the various pages. The components of $\mathbf{v}$ are used to rank the Web pages listed by Google.

Example 2

Suppose that a search engine considers only 10 Web pages, which are linked in the following manner:

Page	Links to pages
1	5 and 10
2	1 and 8
3	1, 4, 5, 6, and 7
4	1, 3, 5, and 10
5	2, 7, 8, and 10
6	no links
7	2 and 4
8	1, 3, 4, and 7
9	1 and 3
10	9

The transition matrix for the Markov chain associated with this search engine, with $p = 0.85$ as the probability that the algorithm follows an outgoing link from one page to the next, is:

$$\begin{bmatrix} 0.0150 & 0.4400 & 0.1850 & 0.2275 & 0.0150 & 0.1000 & 0.0150 & 0.2275 & 0.4400 & 0.0150 \\ 0.0150 & 0.0150 & 0.0150 & 0.0150 & 0.2275 & 0.1000 & 0.4400 & 0.0150 & 0.0150 & 0.0150 \\ 0.0150 & 0.0150 & 0.0150 & 0.2275 & 0.0150 & 0.1000 & 0.0150 & 0.2275 & 0.4400 & 0.0150 \\ 0.0150 & 0.0150 & 0.1850 & 0.0150 & 0.0150 & 0.1000 & 0.4400 & 0.2275 & 0.0150 & 0.0150 \\ 0.4400 & 0.0150 & 0.1850 & 0.2275 & 0.0150 & 0.1000 & 0.0150 & 0.0150 & 0.0150 & 0.0150 \\ 0.0150 & 0.0150 & 0.1850 & 0.0150 & 0.0150 & 0.1000 & 0.0150 & 0.0150 & 0.0150 & 0.0150 \\ 0.0150 & 0.0150 & 0.1850 & 0.0150 & 0.2275 & 0.1000 & 0.0150 & 0.2275 & 0.0150 & 0.0150 \\ 0.0150 & 0.4400 & 0.0150 & 0.0150 & 0.2275 & 0.1000 & 0.0150 & 0.0150 & 0.0150 & 0.0150 \\ 0.0150 & 0.0150 & 0.0150 & 0.0150 & 0.0150 & 0.1000 & 0.0150 & 0.0150 & 0.0150 & 0.8650 \\ 0.4400 & 0.0150 & 0.0150 & 0.2275 & 0.2275 & 0.1000 & 0.0150 & 0.0150 & 0.0150 & 0.0150 \end{bmatrix}$$

The steady-state vector for this Markov chain is approximately

$$\begin{bmatrix} 0.1583 \\ 0.0774 \\ 0.1072 \\ 0.0860 \\ 0.1218 \\ 0.0363 \\ 0.0785 \\ 0.0769 \\ 0.1282 \\ 0.1295 \end{bmatrix}.$$

The components of this vector give the rankings for the pages. Here, page 1 has the highest rank of .1583, page 10 has the second-highest rank of .1295, page 9 has the third-highest rank of .1282, etc. Notice that, even though page 3 has the greatest number of links (5 links to other pages and 3 links from other pages), its ranking is lower than that of page 9, which has only 3 links (2 links to other pages, and 1 link from another page). This fact illustrates that Google's method of ranking the pages takes into account not just the number of links to and from a page, but also the rankings of the pages that are linked to each page.

SYSTEMS OF DIFFERENTIAL EQUATIONS

The decay of radioactive material and the unrestricted growth of bacteria and other organisms are examples of processes in which the quantity of a substance changes at every instant in proportion to the amount present. If $y = f(t)$ represents the amount of such a substance present at time t, and k represents the constant of proportionality, then this type of growth is described by the differential equation $f'(t) = kf(t)$, or

$$y' = ky. \tag{5}$$

In calculus, it is shown that the *general solution* of equation (5) is

$$y = ae^{kt},$$

where a is an arbitrary constant. That is, if we substitute ae^{kt} for y (and its derivative ake^{kt} for y') in equation (5), we obtain an identity. To find the value of a in the general solution, we need an *initial condition*. For instance, we need to know how much of the substance is present at a particular time, say, $t = 0$. If 3 units of the substance are present initially, then $y(0) = 3$. Therefore

$$3 = y(0) = ae^{k(0)} = a \cdot 1 = a,$$

and the *particular solution* of equation (5) is $y = 3e^{kt}$.

Now suppose that we have a system of three differential equations:

$$\begin{aligned} y_1' &= 3y_1 \\ y_2' &= 4y_2 \\ y_3' &= 5y_3 \end{aligned}$$

This system is just as easy to solve as equation (5), because each of the three equations can be solved independently. Its general solution is

$$\begin{aligned} y_1 &= ae^{3t} \\ y_2 &= be^{4t} \\ y_3 &= ce^{5t}. \end{aligned}$$

Moreover, if there are initial conditions $y_1(0) = 10$, $y_2(0) = 12$, and $y_3(0) = 15$, then the particular solution is given by

$$\begin{aligned} y_1 &= 10e^{3t} \\ y_2 &= 12e^{4t} \\ y_3 &= 15e^{5t}. \end{aligned}$$

As for systems of linear equations, the preceding system of differential equations can be represented by the matrix equation

$$\begin{bmatrix} y_1' \\ y_2' \\ y_3' \end{bmatrix} = \begin{bmatrix} 3 & 0 & 0 \\ 0 & 4 & 0 \\ 0 & 0 & 5 \end{bmatrix} \begin{bmatrix} y_1 \\ y_2 \\ y_3 \end{bmatrix}.$$

Letting

$$\mathbf{y} = \begin{bmatrix} y_1 \\ y_2 \\ y_3 \end{bmatrix}, \qquad \mathbf{y}' = \begin{bmatrix} y_1' \\ y_2' \\ y_3' \end{bmatrix}, \qquad \text{and} \qquad D = \begin{bmatrix} 3 & 0 & 0 \\ 0 & 4 & 0 \\ 0 & 0 & 5 \end{bmatrix},$$

we can represent the system as the matrix equation

$$\mathbf{y}' = D\mathbf{y}$$

with initial condition

$$\mathbf{y}(0) = \begin{bmatrix} 10 \\ 12 \\ 15 \end{bmatrix}.$$

More generally, the system of linear differential equations

$$\begin{aligned} y_1' &= a_{11}y_1 + a_{12}y_2 + \cdots + a_{1n}y_n \\ y_2' &= a_{21}y_1 + a_{22}y_2 + \cdots + a_{2n}y_n \\ &\vdots \\ y_n' &= a_{n1}y_1 + a_{n2}y_2 + \cdots + a_{nn}y_n \end{aligned}$$

can be written as

$$\mathbf{y}' = A\mathbf{y}, \tag{6}$$

where A is an $n \times n$ matrix. For example, such a system could describe the numbers of animals of three species that are dependent upon one another, so that the growth rate of each species depends on the present number of animals of all three species.

This type of system arises in the context of *predator-prey models,* where y_1 and y_2 might represent the numbers of rabbits and foxes in an ecosystem (see Exercise 67) or the numbers of food fish and sharks.

To obtain a solution of equation (6), we make an appropriate *change of variable.* Define $\mathbf{z} = P^{-1}\mathbf{y}$ (or equivalently, $\mathbf{y} = P\mathbf{z}$), where P is an invertible matrix. It is not hard to prove that $\mathbf{y}' = P\mathbf{z}'$. (See Exercise 66.) Therefore substituting $P\mathbf{z}$ for $\mathbf{y}$ and $P\mathbf{z}'$ for $\mathbf{y}'$ in equation (6) yields

$$P\mathbf{z}' = AP\mathbf{z},$$

or

$$\mathbf{z}' = P^{-1}AP\mathbf{z}.$$

Thus, if there is an invertible matrix P such that $P^{-1}AP$ is a diagonal matrix D, then we obtain the system $\mathbf{z}' = D\mathbf{z}$, which is of the same simple form as the one just solved. Moreover, the solution of equation (6) can be obtained easily from $\mathbf{z}$ because $\mathbf{y} = P\mathbf{z}$.

If A is diagonalizable, we can choose P to be a matrix whose columns form a basis for $\mathcal{R}^n$ consisting of eigenvectors of A. Of course, the diagonal entries of D are the eigenvalues of A. This method for solving a system of differential equations with a diagonalizable coefficient matrix is summarized as follows:

Solution of $\mathbf{y}' = A\mathbf{y}$ When A Is Diagonalizable

1. Find the eigenvalues of A and a basis for each eigenspace.
2. Let P be a matrix whose columns consist of basis vectors for each eigenspace of A, and let D be the diagonal matrix whose diagonal entries are the eigenvalues of A corresponding to the respective columns of P.
3. Solve the system $\mathbf{z}' = D\mathbf{z}$.
4. The solution of the original system is $\mathbf{y} = P\mathbf{z}$.

Example 3 Consider the system

$$\begin{aligned} y_1' &= 4y_1 + y_2 \\ y_2' &= 3y_1 + 2y_2. \end{aligned}$$

The matrix form of this system is $\mathbf{y}' = A\mathbf{y}$, where

$$\mathbf{y} = \begin{bmatrix} y_1 \\ y_2 \end{bmatrix} \quad \text{and} \quad A = \begin{bmatrix} 4 & 1 \\ 3 & 2 \end{bmatrix}.$$

Using the techniques of Section 5.3, we see that A is diagonalizable because it has distinct eigenvalues 1 and 5. Moreover,

$$\left\{\begin{bmatrix} -1 \\ 3 \end{bmatrix}\right\} \quad \text{and} \quad \left\{\begin{bmatrix} 1 \\ 1 \end{bmatrix}\right\}$$

are bases for the corresponding eigenspaces of A. Hence we take

$$P = \begin{bmatrix} -1 & 1 \\ 3 & 1 \end{bmatrix} \quad \text{and} \quad D = \begin{bmatrix} 1 & 0 \\ 0 & 5 \end{bmatrix}.$$

Now solve the system $\mathbf{z}' = D\mathbf{z}$, which is

$$\begin{aligned} z_1' &= z_1 \\ z_2' &= 5z_2. \end{aligned}$$

The solution of this system is

$$\mathbf{z} = \begin{bmatrix} ae^t \\ be^{5t} \end{bmatrix}.$$

Thus the general solution of the original system is

$$\mathbf{y} = P\mathbf{z} = \begin{bmatrix} -1 & 1 \\ 3 & 1 \end{bmatrix} \begin{bmatrix} ae^t \\ be^{5t} \end{bmatrix} = \begin{bmatrix} -ae^t + be^{5t} \\ 3ae^t + be^{5t} \end{bmatrix},$$

or

$$\begin{aligned} y_1 &= -ae^t + be^{5t} \\ y_2 &= 3ae^t + be^{5t}. \end{aligned}$$

Note that it is not necessary to compute P^{-1}.

If, additionally, we are given the initial conditions $y_1(0) = 120$ and $y_2(0) = 40$, then we can find the particular solution of the system. To do so, we must solve the system of linear equations

$$\begin{aligned} 120 = y_1(0) &= -ae^0 + be^{5(0)} = -a + b \\ 40 = y_2(0) &= 3ae^0 + be^{5(0)} = 3a + b \end{aligned}$$

for a and b. Since the solution of this system is $a = -20$ and $b = 100$, the particular solution of the original system of differential equations is

$$\begin{aligned} y_1 &= 20e^t + 100e^{5t} \\ y_2 &= -60e^t + 100e^{5t}. \end{aligned}$$

Practice Problem 2 ▶ Consider the following system of differential equations:

$$\begin{aligned} y_1' &= -5y_1 - 4y_2 \\ y_2' &= 8y_1 - 7y_2 \end{aligned}$$

(a) Find the general solution of the system.

(b) Find the particular solution of the system that satisfies the initial conditions $y_1(0) = 1$ and $y_2(0) = 4$. ◀

It should be noted that, with only a slight modification, the procedure we have presented for solving $\mathbf{y}' = A\mathbf{y}$ can also be used to solve a nonhomogeneous system $\mathbf{y}' = A\mathbf{y} + \mathbf{b}$, where $\mathbf{b} \neq \mathbf{0}$.

However, this procedure for solving $\mathbf{y}' = A\mathbf{y}$ cannot be used if A is not diagonalizable. In such a case, a similar technique can be developed from the *Jordan canonical form* of A. (See [4, pages 515–516].)

Under some circumstances, the system of differential equations (6) can be used to solve a higher-order differential equation. We illustrate this technique by solving the third-order differential equation

$$y''' - 6y'' + 5y' + 12y = 0.$$

By making the substitutions $y_1 = y$, $y_2 = y'$, and $y_3 = y''$, we obtain the system

$$\begin{aligned} y_1' &= \qquad\qquad y_2 \\ y_2' &= \qquad\qquad\qquad\quad y_3 \\ y_3' &= -12y_1 - 5y_2 + 6y_3. \end{aligned}$$

The matrix form of this system is

$$\begin{bmatrix} y_1' \\ y_2' \\ y_3' \end{bmatrix} = \begin{bmatrix} 0 & 1 & 0 \\ 0 & 0 & 1 \\ -12 & -5 & 6 \end{bmatrix} \begin{bmatrix} y_1 \\ y_2 \\ y_3 \end{bmatrix}.$$

The characteristic equation of the 3×3 matrix

$$A = \begin{bmatrix} 0 & 1 & 0 \\ 0 & 0 & 1 \\ -12 & -5 & 6 \end{bmatrix}$$

is $-(t^3 - 6t^2 + 5t + 12) = 0$, which resembles the original differential equation $y''' - 6y'' + 5y' + 12y = 0$. This similarity is no accident. (See Exercise 68.) Since A has distinct eigenvalues -1, 3, and 4, it must be diagonalizable. Using the method previously described, we can solve for y_1, y_2, and y_3. Of course, in this case we are interested only in $y_1 = y$. The general solution of the given third-order equation is

$$y = ae^{-t} + be^{3t} + ce^{4t}.$$

HARMONIC MOTION*

Many real-world problems involve a differential equation that can be solved by the preceding method. Consider, for instance, a body of weight w that is suspended from a spring. (See Figure 5.7.) Suppose that the body is moved from its resting position and set in motion. Let $y(t)$ denote the distance of the body from its resting point at time t, where positive distances are measured downward. If k is the spring constant, g is the acceleration due to gravity (32 feet per second per second), and $-by'(t)$ is the *damping force*,[8] then the motion of the body satisfies the differential equation

$$\frac{w}{g}y''(t) + by'(t) + ky(t) = 0.$$

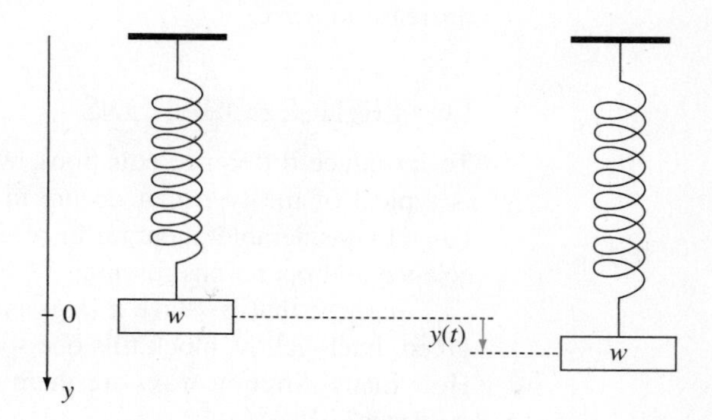

Figure 5.7 A weight suspended from a spring

* This section requires knowledge of complex numbers. (See Appendix C.)

[8] The damping force is a frictional force that reflects the viscosity of the medium in which the motion occurs. It is proportional to the velocity, but acts in the opposite direction.

For example, if the body weighs 8 pounds, the spring constant is $k = 2.125$ pounds per foot, and the damping force constant is $b = 0.75$, then the previous differential equation reduces to the form

$$y'' + 3y' + 8.5y = 0.$$

By making the substitutions $y_1 = y$ and $y_2 = y'$, we obtain the system

$$\begin{aligned} y_1' &= \phantom{-8.5y_1 - {}} y_2 \\ y_2' &= -8.5y_1 - 3y_2, \end{aligned}$$

or, in matrix form,

$$\begin{bmatrix} y_1' \\ y_2' \end{bmatrix} = \begin{bmatrix} 0 & 1 \\ -8.5 & -3 \end{bmatrix} \begin{bmatrix} y_1 \\ y_2 \end{bmatrix}.$$

The characteristic polynomial of the preceding matrix is

$$t^2 + 3t + 8.5,$$

which has the nonreal roots $-1.5 + 2.5i$ and $-1.5 - 2.5i$.

The general solution of the differential equation can be shown to be

$$y = ae^{(-1.5+2.5i)t} + be^{(-1.5-2.5i)t}.$$

Using Euler's formula (see Appendix C), we obtain

$$y = ae^{-1.5t}(\cos 2.5t + i \sin 2.5t) + be^{-1.5t}(\cos 2.5t - i \sin 2.5t),$$

which can be written as

$$y = ce^{-1.5t} \cos 2.5t + de^{-1.5t} \sin 2.5t.$$

The constants c and d can be determined from initial conditions, such as the initial displacement of the body from its resting position, $y(0)$, and its initial velocity, $y'(0)$. From this solution, it can be shown that the body oscillates with amplitudes that decrease to zero.

DIFFERENCE EQUATIONS

To introduce difference equations, we begin with a counting problem. This problem is typical of the type that occurs in the study of *combinatorial analysis*, which has gained considerable attention in recent years because of its applications to computer science and operations research.

Suppose that we have a large number of blocks of three colors: yellow, red, and green. Each yellow block fills one space, and each red or green block fills two spaces. How many different ways are there of arranging the blocks in a line so that they fill n spaces?

Denote the answer to this question by r_n, and let Y, R, and G represent a yellow, a red, and a green block, respectively. For convenience, we let $r_0 = 1$. The following table lists the possible arrangements in the cases $n = 0, 1, 2,$ and 3:

n	Arrangements	r_n
0		1
1	Y	1
2	YY, R, G	3
3	YYY, YR, YG, RY, GY	5

Now suppose that we have to fill n spaces. Consider the three possible cases:

Case 1. The last block is yellow.
In this case, the yellow block fills the last space, and we can fill the first $n-1$ spaces in r_{n-1} ways.

Case 2. The last block is red.
In this case, the red block fills the last two spaces, and we can fill the first $n-2$ spaces in r_{n-2} ways.

Case 3. The last block is green.
This is similar to case 2, so the total number of ways to fill the first $n-2$ spaces is r_{n-2}.

Putting these cases together, we have

$$r_n = r_{n-1} + 2r_{n-2}. \tag{7}$$

Notice that this equation agrees with the table for $n = 2$ and $n = 3$:

$$r_2 = r_1 + 2r_0 = 1 + 2 \cdot 1 = 3$$

and

$$r_3 = r_2 + 2r_1 = 3 + 2 \cdot 1 = 5.$$

With this formula, we can easily determine that the number of arrangements for $n = 4$ is

$$r_4 = r_3 + 2r_2 = 5 + 2 \cdot 3 = 11.$$

Equation (7) is an example of a **difference equation**, or **recurrence relation**. Difference equations are analogous to differential equations, except that the independent variable is treated as *discrete* in a difference equation and as *continuous* in a differential equation.

But how do we find a formula expressing r_n as a function of n? One way is to rewrite equation (7) as a matrix equation. First, write

$$r_n = r_n$$
$$r_{n+1} = r_n + 2r_{n-1}.$$

(The second equation is formed by replacing n by $n+1$ in equation (7).) The system can now be written in the matrix form

$$\begin{bmatrix} r_n \\ r_{n+1} \end{bmatrix} = \begin{bmatrix} 0 & 1 \\ 2 & 1 \end{bmatrix} \begin{bmatrix} r_{n-1} \\ r_n \end{bmatrix},$$

or $\mathbf{s}_n = A\mathbf{s}_{n-1}$, where

$$\mathbf{s}_n = \begin{bmatrix} r_n \\ r_{n+1} \end{bmatrix} \quad \text{and} \quad A = \begin{bmatrix} 0 & 1 \\ 2 & 1 \end{bmatrix}.$$

Furthermore, from the solutions for $n = 0$ and $n = 1$, we have

$$\mathbf{s}_0 = \begin{bmatrix} r_0 \\ r_1 \end{bmatrix} = \begin{bmatrix} 1 \\ 1 \end{bmatrix}.$$

Thus

$$\mathbf{s}_n = A\mathbf{s}_{n-1} = A^2\mathbf{s}_{n-2} = \cdots = A^n\mathbf{s}_0 = A^n\begin{bmatrix} 1 \\ 1 \end{bmatrix}.$$

To compute $\mathbf{s}_n$, we must compute powers of a matrix, a problem we considered in Section 5.3 for the case in which the matrix is diagonalizable. Using the methods developed earlier in this chapter, we find matrices

$$P = \begin{bmatrix} 1 & 1 \\ -1 & 2 \end{bmatrix} \quad \text{and} \quad D = \begin{bmatrix} -1 & 0 \\ 0 & 2 \end{bmatrix}$$

such that $A = PDP^{-1}$. Then, as in Section 5.3, we have $A^n = PD^nP^{-1}$. Thus

$$\mathbf{s}_n = PD^nP^{-1}\mathbf{s}_0,$$

or

$$\begin{bmatrix} r_n \\ r_{n+1} \end{bmatrix} = \begin{bmatrix} 1 & 1 \\ -1 & 2 \end{bmatrix} \begin{bmatrix} (-1)^n & 0 \\ 0 & 2^n \end{bmatrix} \begin{bmatrix} \frac{2}{3} & -\frac{1}{3} \\ \frac{1}{3} & \frac{1}{3} \end{bmatrix} \begin{bmatrix} 1 \\ 1 \end{bmatrix}$$

$$= \frac{1}{3}\begin{bmatrix} (-1)^n + 2^{n+1} \\ (-1)^{n+1} + 2^{n+2} \end{bmatrix}.$$

Therefore

$$r_n = \frac{(-1)^n + 2^{n+1}}{3}.$$

As a check, observe that this formula gives $r_0 = 1$, $r_1 = 1$, $r_2 = 3$, and $r_3 = 5$, which agree with the values obtained previously. It is now easy with a calculator to compute r_n for larger values of n. For example, r_{10} is 683, r_{20} is 669,051, and r_{32} is almost 3 billion!

In general, a **kth-order homogeneous linear difference equation** (or **recurrence relation**) is an equation of the form

$$r_n = a_1 r_{n-1} + a_2 r_{n-2} + \cdots + a_k r_{n-k}, \tag{8}$$

where the a_i's are scalars, n and k are positive integers such that $n > k$, and $a_k \neq 0$.

Equation (7) enables us to compute successive values of r_n if we know two consecutive values. In the block problem, we found that $r_0 = 1$ and $r_1 = 1$ by enumerating the possibilities for filling 0 spaces and 1 space. Such a set of consecutive values of r_n is called a set of **initial conditions**. More generally, in equation (8), we need to know k consecutive values of r_n to have a set of initial conditions. Thus the number of consecutive values required equals the order of the difference equation.

As in the preceding example, we can represent the kth-order equation (8) by a matrix equation of the form $\mathbf{s}_n = A\mathbf{s}_{n-1}$, where A is a $k \times k$ matrix and $\mathbf{s}_n$ is a vector in $\mathcal{R}^k$. (See Exercise 82.) It can be shown (see Exercise 83) that if A has k distinct eigenvalues, $\lambda_1, \lambda_2, \ldots, \lambda_k$, then the general solution of equation (8) has the form

$$r_n = b_1\lambda_1^n + b_2\lambda_2^n + \cdots + b_k\lambda_k^n. \tag{9}$$

The b_i's are determined by the initial conditions, which are given by the components of the vector $\mathbf{s}_0$. Furthermore, the λ_i's, which are the distinct eigenvalues of A, can also be obtained as solutions of the equation

$$\lambda^k = a_1\lambda^{k-1} + a_2\lambda^{k-2} + \cdots + a_{k-1}\lambda + a_k. \tag{10}$$

(See Exercise 84.)

Equations (9) and (10) offer us an alternative method for finding r_n *without* computing either the characteristic polynomial or the eigenvectors of A. We illustrate this method with another example.

It is known that rabbits reproduce at a very rapid rate. For the sake of simplicity, we assume that a pair of rabbits does not produce any offspring during the first month of their lives, but that they produce exactly one pair (male and female) each month thereafter. Suppose that, initially, we have one male–female pair of newborn rabbits and that no rabbits die. How many pairs of rabbits will there be after n months?

Let r_n denote the number of pairs of rabbits after n months. Let's try to answer this question for $n = 0, 1, 2,$ and 3. After zero months, we have only the original pair. Similarly, after 1 month, we still have only the original pair. So $r_1 = r_0 = 1$. After 2 months, we have the original pair and their offspring; that is, $r_2 = 2$. After 3 months, we have all the previous pairs and, in addition, the new pair of offspring of the original pair; that is, $r_3 = r_2 + r_1 = 2 + 1 = 3$. In general, after n months, we have the pairs we had last month and the offspring of those pairs that are over 1 month old. Thus

$$r_n = r_{n-1} + r_{n-2}, \tag{11}$$

a second-order difference equation. The numbers generated by equation (11) are 1, 1, 2, 3, 5, 8, 13, 21, 34, $\ldots$. Each number is the sum of the preceding two numbers. A sequence with this property is called a **Fibonacci sequence**. It occurs in a variety of settings, including the number of spirals of various plants.

Now we use equations (9) and (10) to obtain a formula for r_n. By equations (11) and (10), we have

$$\lambda^2 = \lambda + 1,$$

which has solutions $(1 \pm \sqrt{5})/2$. Thus, by equation (9), there are scalars b_1 and b_2 such that

$$r_n = b_1\left(\frac{1}{2} + \frac{\sqrt{5}}{2}\right)^n + b_2\left(\frac{1}{2} - \frac{\sqrt{5}}{2}\right)^n.$$

To find b_1 and b_2, we use the initial conditions

$$\begin{aligned} 1 = r_0 &= \qquad (1)\,b_1 + \qquad (1)\,b_2 \\ 1 = r_1 &= \left(\tfrac{1}{2} + \tfrac{\sqrt{5}}{2}\right)b_1 + \left(\tfrac{1}{2} - \tfrac{\sqrt{5}}{2}\right)b_2. \end{aligned}$$

This system has the solution

$$b_1 = \frac{1}{\sqrt{5}}\left(\frac{1}{2} + \frac{\sqrt{5}}{2}\right) \quad \text{and} \quad b_2 = -\frac{1}{\sqrt{5}}\left(\frac{1}{2} - \frac{\sqrt{5}}{2}\right).$$

Thus, in general,

$$r_n = \frac{1}{\sqrt{5}}\left(\frac{1}{2}+\frac{\sqrt{5}}{2}\right)^{n+1} - \frac{1}{\sqrt{5}}\left(\frac{1}{2}-\frac{\sqrt{5}}{2}\right)^{n+1}.$$

This complicated formula is a surprise, because r_n is a positive integer for every value of n. To find the fiftieth Fibonacci number, we compute r_{50}, which is over 20 billion!

Our final example, which can also be solved by differential equations, involves an application to heat loss.

Example 4 The water in a hot tub loses heat to the surrounding air so that the difference between the temperature of the water and the temperature of the surrounding air is reduced by 5% each minute. The temperature of the water is 120°F now, and the temperature of the surrounding air is a constant 70°F. Let r_n denote the temperature difference at the end of n minutes. Then

$$r_n = .95r_{n-1} \text{ for each } n \qquad \text{and} \qquad r_0 = 120 - 70 = 50°\text{F}.$$

By equations (9) and (10), $r_n = b\lambda^n$ and $\lambda = 0.95$, and hence $r_n = b(.95)^n$. Furthermore, $50 = r_0 = b(.95)^0 = b$, and thus $r_n = 50(.95)^n$. For example, at the end of 10 minutes, $r_n = 50(.95)^{10} \approx 30°$F, and so the water temperature is approximately $70 + 30 = 100°$F.

In Exercises 85–87, we show how to find a formula for the solution of the first-order nonhomogeneous equation $r_n = ar_{n-1} + c$, where a and c are scalars. This equation occurs frequently in financial applications, such as annuities. (See Exercise 88.)

Practice Problem 3 ▶ In the back room of a bookstore there are a large number of copies of three books: a novel by Nabokov, a novel by Updike, and a calculus book. The novels are each 1 inch thick, and the calculus book is 2 inches thick. Find r_n, the number of ways of arranging these copies in a stack n inches high. ◀

EXERCISES

In Exercises 1–12, determine whether the statements are true or false.

1. The row sums of the transition matrix of a Markov chain are all 1.
2. If the transition matrix of a Markov chain contains zero entries, then it is not regular.
3. If A is the transition matrix of a Markov chain and $\mathbf{p}$ is any probability vector, then $A\mathbf{p}$ is a probability vector.
4. If A is the transition matrix of a Markov chain and $\mathbf{p}$ is any probability vector, then as m approaches infinity, the vectors $A^m\mathbf{p}$ approach a probability vector.
5. If A is the transition matrix of a regular Markov chain, then as m approaches infinity, the vectors $A^m\mathbf{p}$ approach the same probability vector for every probability vector $\mathbf{p}$.
6. Every regular transition matrix has 1 as an eigenvalue.
7. Every regular transition matrix has a unique probability vector that is an eigenvector corresponding to eigenvalue 1.
8. The general solution of $y' = ky$ is $y = ke^t$.
9. If $P^{-1}AP$ is a diagonal matrix D, then the change of variable $\mathbf{z} = P\mathbf{y}$ transforms the matrix equation $\mathbf{y}' = A\mathbf{y}$ into $\mathbf{z}' = D\mathbf{z}$.
10. A differential equation $a_3y''' + a_2y'' + a_1y' + a_0y = 0$, where a_3, a_2, a_1, a_0 are scalars, can be written as a system of linear differential equations.
11. If $A = PDP^{-1}$, where P is an invertible matrix and D is a diagonal matrix, then the solution of $\mathbf{y}' = A\mathbf{y}$ is $P^{-1}\mathbf{z}$, where $\mathbf{z}$ is a solution of $\mathbf{z}' = D\mathbf{z}$.
12. In a Fibonacci sequence, each term after the first two is the sum of the two preceding terms.

In Exercises 13–20, determine whether each transition matrix is regular.

13. $\begin{bmatrix} 0.25 & 0 \\ 0.75 & 1 \end{bmatrix}$

14. $\begin{bmatrix} 0 & .5 \\ 1 & .5 \end{bmatrix}$

15. $\begin{bmatrix} .5 & 0 & .7 \\ .5 & 0 & .3 \\ 0 & 1 & 0 \end{bmatrix}$

16. $\begin{bmatrix} .9 & .5 & .4 \\ 0 & .5 & 0 \\ .1 & 0 & .6 \end{bmatrix}$

17. $\begin{bmatrix} .8 & 0 & 0 \\ .2 & .7 & .1 \\ 0 & .3 & .9 \end{bmatrix}$

18. $\begin{bmatrix} .2 & .7 & .1 \\ .8 & 0 & 0 \\ 0 & .3 & .9 \end{bmatrix}$

19. $\begin{bmatrix} .6 & 0 & 0 & .1 \\ 0 & .5 & .2 & 0 \\ .4 & 0 & 0 & .9 \\ 0 & .5 & .8 & 0 \end{bmatrix}$

20. $\begin{bmatrix} .6 & 0 & .1 & 0 \\ 0 & .5 & 0 & .2 \\ .4 & 0 & .9 & 0 \\ 0 & .5 & 0 & .8 \end{bmatrix}$

In Exercises 21–28, a regular transition matrix is given. Determine its steady-state vector.

21. $\begin{bmatrix} .9 & .3 \\ .1 & .7 \end{bmatrix}$

22. $\begin{bmatrix} .6 & .1 \\ .4 & .9 \end{bmatrix}$

23. $\begin{bmatrix} .5 & .1 & .2 \\ .2 & .6 & .1 \\ .3 & .3 & .7 \end{bmatrix}$

24. $\begin{bmatrix} .7 & .1 & .6 \\ 0 & .9 & 0 \\ .3 & 0 & .4 \end{bmatrix}$

25. $\begin{bmatrix} .8 & 0 & .1 \\ 0 & .4 & .9 \\ .2 & .6 & 0 \end{bmatrix}$

26. $\begin{bmatrix} .7 & 0 & .2 \\ 0 & .4 & .8 \\ .3 & .6 & 0 \end{bmatrix}$

27. $\begin{bmatrix} .6 & 0 & 0 & .1 \\ 0 & .5 & .2 & 0 \\ .4 & .2 & .8 & 0 \\ 0 & .3 & 0 & .9 \end{bmatrix}$

28. $\begin{bmatrix} .6 & 0 & 0 & .1 \\ 0 & .5 & .2 & 0 \\ .4 & 0 & 0 & .9 \\ 0 & .5 & .8 & 0 \end{bmatrix}$

29. When Alison goes to her favorite ice cream store, she buys either a root beer float or a chocolate sundae. If she bought a float on her last visit, there is a .25 probability that she will buy a float on her next visit. If she bought a sundae on her last visit, there is a .5 probability that she will buy a float on her next visit.
    (a) Assuming that this information describes a Markov chain, write a transition matrix for this situation.
    (b) If Alison bought a sundae on her next-to-last visit, what is the probability that she will buy a float on her next visit?
    (c) Over the long run, on what proportion of Alison's trips does she buy a sundae?

30. Suppose that the probability that the child of a college-educated parent also becomes college-educated is .75, and that the probability that the child of a non-college-educated parent becomes college-educated is .35.
    (a) Assuming that this information describes a Markov chain, write a transition matrix for this situation.
    (b) If 30% of the parents are college-educated, what (approximate) proportion of the population will be college-educated in one, two, and three generations?
    (c) Without any knowledge of the present proportion of college-educated parents, determine the eventual proportion of college-educated people.

31. A supermarket sells three brands of baking powder. Of those who bought brand A last, 70% will buy brand A the next time they buy baking powder, 10% will buy brand B, and 20% will buy brand C. Of those who bought brand B last, 10% will buy brand A the next time they buy baking powder, 60% will buy brand B, and 30% will buy brand C. Of those who bought brand C last, 10% will buy each of brands A and B the next time they buy baking powder, and 80% will buy brand C.
    (a) Assuming that the information describes a Markov chain, write a transition matrix for this situation.
    (b) If a customer last bought brand B, what is the probability that his or her next purchase of baking powder is brand B?
    (c) If a customer last bought brand A, what is the probability that his or her second future purchase of baking powder is brand C?
    (d) Over the long run, what proportion of the supermarket's baking powder sales are accounted for by each brand?

32. Suppose that a particular region with a constant population is divided into three areas: the city, suburbs, and country. The probability that a person living in the city moves to the suburbs (in one year) is .10, and the probability that someone moves to the country is .50. The probability is .20 that a person living in the suburbs moves to the city and is .10 for a move to the country. The probability is .20 that a person living in the country moves to the city and is .20 for a move to the suburbs. Suppose initially that 50% of the people live in the city, 30% live in the suburbs, and 20% live in the country.
    (a) Determine the transition matrix for the three states.
    (b) Determine the percentage of people living in each area after 1, 2, and 3 years.
    (c) Use either a calculator with matrix capabilities or computer software such as MATLAB to find the percentage of people living in each area after 5 and 8 years.
    (d) Determine the eventual percentages of people in each area.

33. Prove that the sum of the entries of each column of the matrix A in the subsection on Google searches is equal to 1.

34. Verify that the matrix A in the subsection on Google searches satisfies the equation

$$A = pM + \frac{1-p}{n}W,$$

where M and W are as defined in the subsection.

In Exercise 35, use either a calculator with matrix capabilities or computer software such as MATLAB.

35. In [5], Gabriel and Neumann found that a Markov chain could be used to describe the occurrence of rainfall in Tel Aviv during the rainy seasons from 1923–24 to 1949–50. A day was classified as *wet* if at least 0.1 mm of precipitation fell at a certain location in Tel Aviv during the period from 8 AM one day to 8 AM the next day; otherwise, the day was classified as *dry*. The data for November follow:

	Current day wet	Current day dry
Next day wet	117	80
Next day dry	78	535

(a) Assuming that the preceding information describes a Markov chain, write a transition matrix for this situation.

(b) If a November day was dry, what is the probability that the following day will be dry?

(c) If a Tuesday in November was dry, what is the probability that the following Thursday will be dry?

(d) If a Wednesday in November was wet, what is the probability that the following Saturday will be dry?

(e) Over the long run, what is the probability that a November day in Tel Aviv is wet?

36. A company leases rental cars at three Chicago offices (located at Midway Airport, O'Hare Field, and the Loop). Its records show that 60% of the cars rented at Midway are returned there and 20% are returned to each of the other locations. Also, 80% of the cars rented at O'Hare are returned there and 10% are returned to each of the other locations. Finally, 70% of the cars rented at the Loop are returned there, 10% are returned to Midway, and 20% are returned to O'Hare.

(a) Assuming that the preceding information describes a Markov chain, write a transition matrix for this situation.

(b) If a car is rented at O'Hare, what is the probability that it will be returned to the Loop?

(c) If a car is rented at Midway, what is the probability that it will be returned to the Loop after its second rental?

(d) Over the long run, if all of the cars are returned, what proportion of the company's fleet will be located at each office?

37. Suppose that the transition matrix of a Markov chain is

$$A = \begin{bmatrix} .90 & .1 & .3 \\ .05 & .8 & .3 \\ .05 & .1 & .4 \end{bmatrix}.$$

(a) What are the probabilities that an object in the first state will next move to each of the other states?

(b) What are the probabilities that an object in the second state will next move to each of the other states?

(c) What are the probabilities that an object in the third state will next move to each of the other states?

(d) Use your answers to (a), (b), and (c) to predict the steady-state vector for A.

(e) Verify your prediction in (d).

38. Give an example of a 3×3 regular transition matrix A such that A, A^2, and A^3 all contain zero entries.

39. Let A be an $n \times n$ stochastic matrix, and let $\mathbf{u}$ be the vector in $\mathcal{R}^n$ with all components equal to 1.

(a) Compute $A^T\mathbf{u}$.

(b) What does (a) imply about the eigenvalues of A^T?

(c) Prove that $\det(A - I_n) = 0$.

(d) What does (c) imply about the eigenvalues of A?

40. Use ideas from Exercise 37 to construct two regular 3×3 stochastic matrices having

$$\begin{bmatrix} .4 \\ .2 \\ .4 \end{bmatrix}$$

as their steady-state vector.

41. Prove that if A is a stochastic matrix and $\mathbf{p}$ is a probability vector, then $A\mathbf{p}$ is a probability vector.

42. Let A be the 2×2 stochastic matrix $\begin{bmatrix} a & 1-b \\ 1-a & b \end{bmatrix}$.

(a) Determine the eigenvalues of A.

(b) Determine a basis for each eigenspace of A.

(c) Under what conditions is A diagonalizable?

43. Let A be an $n \times n$ stochastic matrix.

(a) Let $\mathbf{v}$ be any vector in $\mathcal{R}^n$ and k be an index such that $|v_j| \le |v_k|$ for each j. Prove that the absolute value of every component of $A^T\mathbf{v}$ is less than or equal to $|v_k|$.

(b) Use (a) to show that if $\mathbf{v}$ is an eigenvector of A^T that corresponds to an eigenvalue λ, then $|\lambda| \cdot |v_k| \le |v_k|$.

(c) Deduce that if λ is an eigenvalue of A, then $|\lambda| \le 1$.

44. Prove that if A and B are stochastic matrices, then AB is a stochastic matrix.

In Exercises 45–52, find the general solution of each system of differential equations.

45. $\begin{aligned} y_1' &= 3y_1 + 2y_2 \\ y_2' &= 3y_1 - 2y_2 \end{aligned}$

46. $\begin{aligned} y_1' &= y_1 + 2y_2 \\ y_2' &= -y_1 + 4y_2 \end{aligned}$

47. $\begin{aligned} y_1' &= 2y_1 + 4y_2 \\ y_2' &= -6y_1 - 8y_2 \end{aligned}$

48. $\begin{aligned} y_1' &= -5y_1 + 6y_2 \\ y_2' &= -15y_1 + 14y_2 \end{aligned}$

49. $\begin{aligned} y_1' &= 2y_1 \\ y_2' &= 3y_1 + 2y_2 + 3y_3 \\ y_3' &= -3y_1 - y_3 \end{aligned}$

50. $\begin{aligned} y_1' &= y_1 + 2y_2 - y_3 \\ y_2' &= y_1 + y_3 \\ y_3' &= 4y_1 - 4y_2 + 5y_3 \end{aligned}$

51. $\begin{aligned} y_1' &= -3y_1 + y_2 + y_3 \\ y_2' &= 8y_1 - 2y_2 - 4y_3 \\ y_3' &= -10y_1 + 2y_2 + 4y_3 \end{aligned}$

52. $\begin{aligned} y_1' &= 12y_1 - 10y_2 - 10y_3 \\ y_2' &= 10y_1 - 8y_2 - 10y_3 \\ y_3' &= 5y_1 - 5y_2 - 3y_3 \end{aligned}$

In Exercises 53–60, find the particular solution of each system of differential equations that satisfies the given initial conditions.

53. $\begin{aligned} y_1' &= y_1 + y_2 \\ y_2' &= 4y_1 + y_2 \end{aligned}$
with $y_1(0) = 15,\ y_2(0) = -10$

54. $\begin{aligned} y_1' &= 2y_1 + 2y_2 \\ y_2' &= -y_1 + 5y_2 \end{aligned}$
with $y_1(0) = 7,\ y_2(0) = 5$

55. $\begin{aligned} y_1' &= 8y_1 + 2y_2 \\ y_2' &= -4y_1 + 2y_2 \end{aligned}$
with $y_1(0) = 2,\ y_2(0) = 1$

56. $\begin{aligned} y_1' &= -5y_1 - 8y_2 \\ y_2' &= 4y_1 + 7y_2 \end{aligned}$
with $y_1(0) = 1,\ y_2(0) = -3$

57. $\begin{aligned} y_1' &= 6y_1 - 5y_2 - 7y_3 \\ y_2' &= y_1 \qquad\quad - y_3 \\ y_3' &= 3y_1 - 3y_2 - 4y_3 \end{aligned}$
with $y_1(0) = 0,\ y_2(0) = 2,\ y_3(0) = 1$

58. $\begin{aligned} y_1' &= y_1 \qquad\quad + 2y_3 \\ y_2' &= 2y_1 + 3y_2 - 2y_3 \\ y_3' &= \qquad\qquad 3y_3 \end{aligned}$
with $y_1(0) = -1,\ y_2(0) = 1,\ y_3(0) = 2$

59. $\begin{aligned} y_1' &= -3y_1 + 2y_2 \\ y_2' &= -7y_1 + 9y_2 + 3y_3 \\ y_3' &= 13y_1 - 20y_2 - 8y_3 \end{aligned}$
with $y_1(0) = -4,\ y_2(0) = -5,\ y_3(0) = 3$

60. $\begin{aligned} y_1' &= 5y_1 - 2y_2 - 2y_3 \\ y_2' &= 18y_1 - 7y_2 - 6y_3 \\ y_3' &= -6y_1 + 2y_2 + y_3 \end{aligned}$
with $y_1(0) = 4,\ y_2(0) = 5,\ y_3(0) = 8$

61. Convert the second-order differential equation

$$y'' - 2y' - 3y = 0$$

into a system of differential equations, and then find its general solution.

62. Convert the third-order differential equation

$$y''' - 2y'' - 8y' = 0$$

into a system of differential equations, and then find its general solution.

63. Convert the third-order differential equation

$$y''' - 2y'' - y' + 2y = 0$$

into a system of differential equations, and then find the particular solution such that $y(0) = 2$, $y'(0) = -3$, and $y''(0) = 5$.

64. Find the general solution of the differential equation that describes the harmonic motion of a 4-pound weight attached to a spring, where the spring constant is 1.5 pounds per foot and the damping force constant is 0.5.

65. Find the general solution of the differential equation that describes the harmonic motion of a 10-pound weight attached to a spring, where the spring constant is 1.25 pounds per foot and the damping force constant is 0.625.

66. Let $\mathbf{z}$ be an $n \times 1$ column vector of differentiable functions, and let P be any $n \times n$ matrix. Prove that if $\mathbf{y} = P\mathbf{z}$, then $\mathbf{y}' = P\mathbf{z}'$.

67. Let y_1 denote the number of rabbits in a certain area at time t and y_2 denote the number of foxes in this area at time t. Suppose that at time 0 there are 900 rabbits and 300 foxes in this area, and assume that the system of differential equations

$$\begin{aligned} y_1' &= 2y_1 - 4y_2 \\ y_2' &= y_1 - 3y_2 \end{aligned}$$

expresses the rate at which the number of animals of each species changes.

(a) Find the particular solution of this system.

(b) Approximately how many of each species will be present at times $t = 1$, 2, and 3? For each of these times, compute the ratio of foxes to rabbits.

(c) Approximately what is the eventual ratio of foxes to rabbits in this area? Does this number depend on the initial numbers of rabbits and foxes in the area?

68. Show that the characteristic polynomial of

$$\begin{bmatrix} 0 & 1 & 0 \\ 0 & 0 & 1 \\ -c & -b & -a \end{bmatrix}$$

is $-t^3 - at^2 - bt - c$.

69. Show that if λ_1, λ_2, and λ_3 are distinct roots of the polynomial $t^3 + at^2 + bt + c$, then $y = ae^{\lambda_1 t} + be^{\lambda_2 t} + ce^{\lambda_3 t}$ is the general solution of $y''' + ay'' + by' + cy = 0$. *Hint:* Express the differential equation as a system of differential equations $\mathbf{y}' = A\mathbf{y}$, and show that

$$\left\{ \begin{bmatrix} 1 \\ \lambda_1 \\ \lambda_1^2 \end{bmatrix}, \begin{bmatrix} 1 \\ \lambda_2 \\ \lambda_2^2 \end{bmatrix}, \begin{bmatrix} 1 \\ \lambda_3 \\ \lambda_3^2 \end{bmatrix} \right\}$$

is a basis for $\mathcal{R}^3$ consisting of eigenvectors of A.

In Exercises 70–78, use either of the two methods developed in this section to find a formula for r_n. Then use your result to find r_6.

70. $r_n = 2r_{n-1}$; $r_0 = 5$

71. $r_n = -3r_{n-1}$; $r_0 = 8$

72. $r_n = r_{n-1} + 2r_{n-2}$; $r_0 = 7$ and $r_1 = 2$

73. $r_n = 3r_{n-1} + 4r_{n-2}$; $r_0 = 1$ and $r_1 = 1$

74. $r_n = 3r_{n-1} - 2r_{n-2}$; $r_0 = 1$ and $r_1 = 3$
75. $r_n = -r_{n-1} + 6r_{n-2}$; $r_0 = 8$ and $r_1 = 1$
76. $r_n = -5r_{n-1} - 4r_{n-2}$; $r_0 = 3$ and $r_1 = 15$
77. $r_n = r_{n-1} + 2r_{n-2}$; $r_0 = 9$ and $r_1 = 0$
78. $r_n = 2r_{n-1} + r_{n-2} - 2r_{n-3}$; $r_0 = 3$, $r_1 = 1$, and $r_2 = 3$
79. Suppose that we have a large number of blocks. The blocks are of five colors: red, yellow, green, orange, and blue. Each of the red and yellow blocks weighs one ounce; and each of the green, orange, and blue blocks weighs two ounces. Let r_n be the number of ways the blocks can be arranged in a stack that weighs n ounces.
   (a) Determine r_0, r_1, r_2, and r_3 by listing the possibilities.
   (b) Write the difference equation involving r_n.
   (c) Use (b) to find a formula for r_n.
   (d) Use your answer to (c) to check your answers in (a).
80. Suppose that a bank pays interest on savings of 8% compounded annually. Use the appropriate difference equation to determine how much money would be in a savings account after n years if initially there was \$1000. What is the value of the account after 5 years, 10 years, and 15 years?
81. Write the third-order difference equation

$$r_n = 4r_{n-1} - 2r_{n-2} + 5r_{n-3}$$

in matrix notation, $\mathbf{s}_n = A\mathbf{s}_{n-1}$, as we did in this section.
82. Justify the matrix form of equation (8) given in this section: $\mathbf{s}_n = A\mathbf{s}_{n-1}$, where

$$\mathbf{s}_n = \begin{bmatrix} r_n \\ r_{n+1} \\ \vdots \\ r_{n+1} \\ r_{n+k-1} \end{bmatrix}$$

and

$$A = \begin{bmatrix} 0 & 1 & 0 & \cdots & 0 & 0 \\ 0 & 0 & 1 & \cdots & 0 & 0 \\ \vdots & \vdots & & & \vdots & \vdots \\ 0 & 0 & 0 & \cdots & 0 & 1 \\ a_k & a_{k-1} & a_{k-2} & \cdots & a_2 & a_1 \end{bmatrix}.$$

83. Consider a kth-order difference equation in the form of equation (8) with a set of k initial conditions, and let the matrix form of the equation be $\mathbf{s}_n = A\mathbf{s}_{n-1}$. Suppose, furthermore, that A has k distinct eigenvalues $\lambda_1, \lambda_2, \ldots, \lambda_k$, and that $\mathbf{v}_1, \mathbf{v}_2, \ldots, \mathbf{v}_k$ are corresponding eigenvectors.
   (a) Prove that there exist scalars $t_1, t_2, \ldots, t_k$ such that

$$\mathbf{s}_0 = t_1\mathbf{v}_1 + t_2\mathbf{v}_2 + \cdots + t_k\mathbf{v}_k.$$

   (b) Prove that for any positive integer n,

$$\mathbf{s}_n = \lambda_1^n t_1\mathbf{v}_1 + \lambda_2^n t_2\mathbf{v}_2 + \cdots + \lambda_k^n t_k\mathbf{v}_k.$$

   (c) Derive equation (9) by comparing the last components of the vector equation in (b).
84. Prove that a scalar λ is an eigenvalue of the matrix A in Exercise 82 if and only if λ is a solution of equation (10). *Hint:* Let

$$\mathbf{w}_\lambda = \begin{bmatrix} 1 \\ \lambda \\ \lambda^2 \\ \vdots \\ \lambda^{k-1} \end{bmatrix},$$

and prove each of the following results:
   (i) If λ is an eigenvalue of A, then $\mathbf{w}_\lambda$ is a corresponding eigenvector, and hence λ is a solution of equation (10).
   (ii) If λ is a solution of equation (10), then $\mathbf{w}_\lambda$ is an eigenvector of A, and λ is the corresponding eigenvalue.

In Exercises 85–87, we examine the **nonhomogeneous** *first-order difference equation, which is of the form*

$$r_n = ar_{n-1} + c,$$

where a and c are constants. For the purpose of these exercises, we let

$$\mathbf{s}_n = \begin{bmatrix} 1 \\ r_n \end{bmatrix} \quad \text{and} \quad A = \begin{bmatrix} 1 & 0 \\ c & a \end{bmatrix}.$$

85. Prove that $\mathbf{s}_n = A^n\mathbf{s}_0$ for any positive integer n.
86. For this exercise, we assume that $a = 1$.
   (a) Prove that $A^n = \begin{bmatrix} 1 & 0 \\ nc & 1 \end{bmatrix}$ for any positive integer n.
   (b) Use (a) to derive the solution $r_n = r_0 + nc$.
87. For this exercise, we assume that $a \neq 1$.
   (a) Prove that 1 and a are eigenvalues of A.
   (b) Prove that there exist eigenvectors $\mathbf{v}_1$ and $\mathbf{v}_2$ of A corresponding to 1 and a, respectively, and scalars t_1 and t_2 such that

$$\mathbf{s}_0 = t_1\mathbf{v}_1 + t_2\mathbf{v}_2.$$

   (c) Use (b) to prove that there exist scalars b_1 and b_2 such that $r_n = b_1 + a^n b_2$, where

$$b_1 = \frac{-c}{a-1} \quad \text{and} \quad b_2 = r_0 + \frac{c}{a-1}.$$

88. An investor opened a savings account on March 1 with an initial deposit of \$5000. Each year thereafter, he added \$2000 to the account on March 1. If the account earns interest at the rate of 6% per year, find a formula for the value of this account after n years. *Hint:* Use Exercise 87(c).

In Exercises 89–91, use either a calculator with matrix capabilities or computer software such as MATLAB to solve each problem.

89. Solve the system of differential equations

$$\begin{aligned} y_1' &= 3.2y_1 + 4.1y_2 + 7.7y_3 + 3.7y_4 \\ y_2' &= -0.3y_1 + 1.2y_2 + 0.2y_3 + 0.5y_4 \\ y_3' &= -1.8y_1 - 1.8y_2 - 4.4y_3 - 1.8y_4 \\ y_4' &= 1.7y_1 - 0.7y_2 + 2.9y_3 + 0.4y_4 \end{aligned}$$

subject to the initial conditions $y_1(0) = 1$, $y_2(0) = -4$, $y_3(0) = 2$, $y_4(0) = 3$.

90. In [3], Bourne examined the changes in land use in Toronto, Canada, during the years 1952–1962. Land was classified in ten ways:

1. low-density residential
2. high-density residential
3. office
4. general commercial
5. automobile commercial
6. parking
7. warehousing
8. industry
9. transportation
10. vacant

The following transition matrix shows the changes in land use from 1952 to 1962:

Use in 1952 (columns); Use in 1962 (rows)

$$\begin{array}{cc} & \begin{array}{cccccccccc} 1 & 2 & 3 & 4 & 5 & 6 & 7 & 8 & 9 & 10 \end{array} \\ \begin{array}{c} 1 \\ 2 \\ 3 \\ 4 \\ 5 \\ 6 \\ 7 \\ 8 \\ 9 \\ 10 \end{array} & \begin{bmatrix} .13 & .02 & .00 & .02 & .00 & .08 & .01 & .01 & .01 & .25 \\ .34 & .41 & .07 & .01 & .00 & .05 & .03 & .02 & .18 & .08 \\ .10 & .05 & .43 & .09 & .11 & .14 & .02 & .02 & .14 & .03 \\ .04 & .04 & .05 & .30 & .07 & .08 & .12 & .03 & .04 & .03 \\ .04 & .00 & .01 & .09 & .70 & .12 & .03 & .03 & .10 & .05 \\ .22 & .04 & .28 & .27 & .06 & .39 & .11 & .08 & .39 & .15 \\ .03 & .00 & .14 & .05 & .00 & .04 & .38 & .18 & .03 & .22 \\ .02 & .00 & .00 & .08 & .01 & .00 & .21 & .61 & .03 & .13 \\ .00 & .00 & .00 & .01 & .00 & .01 & .01 & .00 & .08 & .00 \\ .08 & .44 & .02 & .08 & .05 & .09 & .08 & .02 & .00 & .06 \end{bmatrix} \end{array}$$

Assume that the trend in land use changes from 1952 to 1962 continues indefinitely.

(a) Suppose that at some time the percentage of land use for each purpose is as follows: 10%, 20%, 25%, 0%, 0%, 5%, 15%, 10%, 10%, and 5%, respectively. What percentage of land will be used for each purpose two decades later?

(b) Show that the transition matrix is regular.

(c) After many decades, what percentage of land will be used for each purpose?

91. A search engine considers only 10 Web pages, which are linked in the following manner:

Page	Links
1	2, 6, 8, and 9
2	4 and 7
3	1, 4, 5, 8, and 10
4	no links
5	2 and 10
6	5 and 9
7	1, 5, 6, and 9
8	4 and 8
9	5 and 10
10	no links

(a) Apply the PageRank algorithm to find the transition matrix for the Markov chain associated with this search engine, based on $p = 0.85$, the portion of the time a surfer moves from the current page to the next page by randomly choosing a link, provided that one exists.

(b) Find the steady-state vector for the Markov chain obtained in (a), and use it to rank the pages.

SOLUTIONS TO THE PRACTICE PROBLEMS

1. (a) The Markov chain has three states, which correspond to the type of vehicle driven—car, van, or sport utility vehicle (suv). The transition matrix for this Markov chain is

Five Years Ago

$$\text{Now}\quad \begin{array}{c} \text{car} \\ \text{van} \\ \text{suv} \end{array} \overset{\begin{array}{ccc} \text{car} & \text{van} & \text{suv} \end{array}}{\begin{bmatrix} .8 & .2 & .1 \\ .1 & .7 & .3 \\ .1 & .1 & .6 \end{bmatrix}} = A.$$

(b) The probability vector that gives the probability of driving each type of vehicle five years ago is

$$\mathbf{p} = \begin{bmatrix} .70 \\ .20 \\ .10 \end{bmatrix}.$$

Hence the probability that someone in the survey is now driving each type of vehicle is given by

$$A\mathbf{p} = \begin{bmatrix} .8 & .2 & .1 \\ .1 & .7 & .3 \\ .1 & .1 & .6 \end{bmatrix} \begin{bmatrix} .70 \\ .20 \\ .10 \end{bmatrix} = \begin{bmatrix} .61 \\ .24 \\ .15 \end{bmatrix}.$$

Therefore 61% of those surveyed are now driving cars, 24% are now driving minivans, and 15% are now driving sport utility vehicles.

(c) The probability that five years from now someone in the survey will be driving each type of vehicle is given by

$$A(A\mathbf{p}) = \begin{bmatrix} .8 & .2 & .1 \\ .1 & .7 & .3 \\ .1 & .1 & .6 \end{bmatrix} \begin{bmatrix} .61 \\ .24 \\ .15 \end{bmatrix} = \begin{bmatrix} .551 \\ .274 \\ .175 \end{bmatrix}.$$

So we estimate that in five years 55.1% of those surveyed will drive cars, 27.4% minivans, and 17.5% sport utility vehicles.

(d) Note that A is a regular transition matrix, so, by Theorem 5.4, A has a steady-state vector $\mathbf{v}$. This vector is a solution of $A\mathbf{v} = \mathbf{v}$, that is, of the equation $(A - I_3)\mathbf{v} = \mathbf{0}$. Since the reduced row echelon form of A is

$$\begin{bmatrix} 1 & 0 & -2.25 \\ 0 & 1 & -1.75 \\ 0 & 0 & 0 \end{bmatrix},$$

we see that the solutions of $(A - I_3)\mathbf{v} = \mathbf{0}$ have the form

$$\begin{aligned} v_1 &= 2.25v_3 \\ v_2 &= 1.75v_3 \\ v_3 &\quad \text{free.} \end{aligned}$$

In order that $\mathbf{v} = \begin{bmatrix} v_1 \\ v_2 \\ v_3 \end{bmatrix}$ be a probability vector, we must have $v_1 + v_2 + v_3 = 1$. Hence

$$\begin{aligned} 2.25v_3 + 1.75v_3 + v_3 &= 1 \\ 5v_3 &= 1 \\ v_3 &= .2. \end{aligned}$$

So $v_1 = .45$, $v_2 = .35$, and $v_3 = .2$. Thus we expect that, in the long run, 45% of those surveyed will drive cars, 35% minivans, and 20% sport utility vehicles.

2. (a) The matrix form of the given system of differential equations is $\mathbf{y}' = A\mathbf{y}$, where

$$\mathbf{y} = \begin{bmatrix} y_1 \\ y_2 \end{bmatrix} \quad \text{and} \quad A = \begin{bmatrix} -5 & -4 \\ 8 & 7 \end{bmatrix}.$$

The characteristic polynomial of A is $(t + 1)(t - 3)$, so A has eigenvalues -1 and 3. Since each eigenvalue of A has multiplicity 1, A is diagonalizable. In the usual manner, we find that

$$\left\{\begin{bmatrix} -1 \\ 1 \end{bmatrix}\right\} \quad \text{and} \quad \left\{\begin{bmatrix} -1 \\ 2 \end{bmatrix}\right\}$$

are bases for the eigenspaces of A. Take

$$P = \begin{bmatrix} -1 & -1 \\ 1 & 2 \end{bmatrix} \quad \text{and} \quad D = \begin{bmatrix} -1 & 0 \\ 0 & 3 \end{bmatrix}.$$

Then the change of variable $\mathbf{y} = P\mathbf{z}$ transforms $\mathbf{y}' = A\mathbf{y}$ into $\mathbf{z}' = D\mathbf{z}$, which is

$$\begin{aligned} z_1' &= -z_1 \\ z_2' &= 3z_2. \end{aligned}$$

Hence

$$\mathbf{z} = \begin{bmatrix} z_1 \\ z_2 \end{bmatrix} = \begin{bmatrix} ae^{-t} \\ be^{3t} \end{bmatrix}.$$

Therefore the general solution of the given system of differential equations is

$$\mathbf{y} = P\mathbf{z} = \begin{bmatrix} -1 & -1 \\ 1 & 2 \end{bmatrix}\begin{bmatrix} ae^{-t} \\ be^{3t} \end{bmatrix} = \begin{bmatrix} -ae^{-t} - be^{3t} \\ ae^{-t} + 2be^{3t} \end{bmatrix};$$

that is,

$$\begin{aligned} y_1 &= -ae^{-t} - be^{3t} \\ y_2 &= ae^{-t} + 2be^{3t}. \end{aligned}$$

(b) In order to satisfy the initial conditions $y_1(0) = 1$ and $y_2(0) = 4$, the constants a and b must satisfy the system of linear equations

$$\begin{aligned} 1 = y_1(0) &= -ae^0 - be^{3(0)} = -a - b \\ 4 = y_2(0) &= ae^0 + 2be^{3(0)} = a + 2b. \end{aligned}$$

It is easily checked that $a = -6$ and $b = 5$, so the desired particular solution is

$$\begin{aligned} y_1 &= 6e^{-t} - 5e^{3t} \\ y_2 &= -6e^{-t} + 10e^{3t}. \end{aligned}$$

3. As in the block example discussed earlier, $r_0 = 1$ because there is one "empty" stack. Furthermore, $r_1 = 2$ because each of the two novels is 1 inch high. Now suppose that we have to pile a stack n inches high. There are three cases to consider:

Case 1. The bottom book is the novel by Nabokov. Since this novel is 1 inch thick, there are r_{n-1} ways of stacking the rest of the books.

Case 2. The bottom book is the novel by Updike. This is similar to case 1, so there are r_{n-1} ways of stacking the rest of the books.

Case 3. The bottom book is a calculus book. Since this book is 2 inches thick, there are r_{n-2} ways of stacking the rest of the books.

Combining these three cases, we have the second-order difference equation

$$r_n = 2r_{n-1} + r_{n-2}.$$

We use equation (10) to obtain

$$\lambda^2 = 2\lambda + 1,$$

which has solutions $1 \pm \sqrt{2}$. Thus, by equation (9), there are scalars b_1 and b_2 such that

$$r_n = b_1(1 + \sqrt{2})^n + b_2(1 - \sqrt{2})^n.$$

To find b_1 and b_2, we use the initial conditions

$$\begin{aligned} 1 = r(0) &= (1)b_1 + (1)b_2 \\ 2 = r(1) &= (1 + \sqrt{2})b_1 + (1 - \sqrt{2})b_2. \end{aligned}$$

This system has the solution

$$b_1 = \frac{\sqrt{2} + 1}{2\sqrt{2}} \quad \text{and} \quad b_2 = \frac{\sqrt{2} - 1}{2\sqrt{2}}.$$

Thus, in general,

$$r_n = \frac{1}{2\sqrt{2}}\left[(1 + \sqrt{2})^{n+1} - (1 - \sqrt{2})^{n+1}\right].$$

CHAPTER 5 REVIEW EXERCISES

T&F *In Exercises 1–17, determine whether the statements are true or false.*

1. A scalar λ is an eigenvalue of an $n \times n$ matrix A if and only if $\det(A - \lambda I_n) = 0$.
2. If λ is an eigenvalue of a matrix, then there is a unique eigenvector of the matrix that corresponds to λ.
3. If $\mathbf{v}$ is an eigenvector of a matrix, then there is a unique eigenvalue of the matrix that corresponds to $\mathbf{v}$.
4. The eigenspace of an $n \times n$ matrix A corresponding to eigenvalue λ is the null space of $A - \lambda I_n$.
5. The eigenvalues of a linear operator on $\mathcal{R}^n$ are the same as those of its standard matrix.
6. The eigenspaces of a linear operator on $\mathcal{R}^n$ are the same as those of its standard matrix.
7. Every linear operator on $\mathcal{R}^n$ has real eigenvalues.
8. Every $n \times n$ matrix has n distinct eigenvalues.
9. Every diagonalizable $n \times n$ matrix has n distinct eigenvalues.
10. If two $n \times n$ matrices have the same characteristic polynomial, then they have the same eigenvectors.
11. The multiplicity of an eigenvalue need not equal the dimension of the corresponding eigenspace.
12. An $n \times n$ matrix A is diagonalizable if and only if there is a basis for $\mathcal{R}^n$ consisting of eigenvectors of A.
13. If P is an invertible $n \times n$ matrix and D is a diagonal $n \times n$ matrix such that $A = P^{-1}DP$, then the columns of P form a basis for $\mathcal{R}^n$ consisting of eigenvectors of A.
14. If P is an invertible $n \times n$ matrix and D is a diagonal $n \times n$ matrix such that $A = PDP^{-1}$, then the eigenvalues of A are the diagonal entries of D.
15. If λ is an eigenvalue of an $n \times n$ matrix A, then the dimension of the eigenspace corresponding to λ is the nullity of $A - \lambda I_n$.
16. A linear operator on $\mathcal{R}^n$ is diagonalizable if and only if its standard matrix is diagonalizable.
17. If 0 is an eigenvalue of a matrix A, then A is not invertible.

---

18. Show that $\begin{bmatrix} 1 & 2 \\ -3 & -2 \end{bmatrix}$ has no real eigenvalues.

In Exercises 19–22, determine the eigenvalues of each matrix and a basis for each eigenspace.

19. $\begin{bmatrix} 5 & 6 \\ -2 & -2 \end{bmatrix}$

20. $\begin{bmatrix} 1 & -9 \\ 1 & -5 \end{bmatrix}$

21. $\begin{bmatrix} -2 & 0 & 2 \\ 1 & -1 & 0 \\ 0 & 0 & -2 \end{bmatrix}$

22. $\begin{bmatrix} -1 & 0 & 0 \\ 1 & 0 & 1 \\ -1 & -1 & -2 \end{bmatrix}$

In Exercises 23–26, a matrix A is given. Find, if possible, an invertible matrix P and a diagonal matrix D such that $A = PDP^{-1}$. If no such matrices exist, explain why.

23. $\begin{bmatrix} 1 & 2 \\ -3 & 8 \end{bmatrix}$

24. $\begin{bmatrix} -1 & 1 \\ -1 & -3 \end{bmatrix}$

25. $\begin{bmatrix} 1 & 0 & 0 \\ -2 & 0 & 1 \\ 2 & -1 & -2 \end{bmatrix}$

26. $\begin{bmatrix} -2 & 0 & 0 \\ -4 & 2 & 0 \\ 4 & -3 & -1 \end{bmatrix}$

In Exercises 27–30, a linear operator T on $\mathcal{R}^n$ is given. Find, if possible, a basis for $\mathcal{R}^n$ consisting of eigenvectors of T. If no such basis exists, explain why.

27. $T\left(\begin{bmatrix} x_1 \\ x_2 \end{bmatrix}\right) = \begin{bmatrix} 4x_1 + 2x_2 \\ -4x_1 - 5x_2 \end{bmatrix}$

28. $T\left(\begin{bmatrix} x_1 \\ x_2 \end{bmatrix}\right) = \begin{bmatrix} x_1 - 2x_2 \\ 4x_1 - x_2 \end{bmatrix}$

29. $T\left(\begin{bmatrix} x_1 \\ x_2 \\ x_3 \end{bmatrix}\right) = \begin{bmatrix} 2x_1 \\ 2x_2 \\ -3x_1 + 3x_2 - x_3 \end{bmatrix}$

30. $T\left(\begin{bmatrix} x_1 \\ x_2 \\ x_3 \end{bmatrix}\right) = \begin{bmatrix} x_1 \\ 3x_1 + x_2 - 3x_3 \\ 3x_1 - 2x_3 \end{bmatrix}$

In Exercises 31–34, a matrix and its characteristic polynomial are given. Determine all values of the scalar c for which each matrix is not diagonalizable.

31. $\begin{bmatrix} 1 & 0 & 1 \\ 0 & c & 0 \\ -2 & 0 & 4 \end{bmatrix}$ $\quad -(t-c)(t-2)(t-3)$

32. $\begin{bmatrix} 5 & 1 & -3 \\ 0 & c & 0 \\ 6 & 2 & -4 \end{bmatrix}$ $\quad -(t-c)(t+1)(t-2)$

33. $\begin{bmatrix} c & -1 & 2 \\ 0 & -10 & -8 \\ 0 & 12 & 10 \end{bmatrix}$ $\quad -(t-c)(t-2)(t+2)$

34. $\begin{bmatrix} 3 & 1 & 0 \\ -1 & 1 & 0 \\ 0 & 0 & c \end{bmatrix}$ $\quad -(t-c)(t-2)^2$

In Exercises 35 and 36, a matrix A is given. Find A^k for an arbitrary positive integer k.

35. $\begin{bmatrix} 5 & -6 \\ 3 & -4 \end{bmatrix}$

36. $\begin{bmatrix} 11 & 8 \\ -12 & -9 \end{bmatrix}$

37. Let T be the linear operator on $\mathcal{R}^3$ defined by
$$T\left(\begin{bmatrix} x_1 \\ x_2 \\ x_3 \end{bmatrix}\right) = \begin{bmatrix} -4x_1 - 3x_2 - 3x_3 \\ -x_2 \\ 6x_1 + 6x_2 + 5x_3 \end{bmatrix}.$$
Find a basis $\mathcal{B}$ such that $[T]_{\mathcal{B}}$ is a diagonal matrix.

38. Find a 3×3 matrix having eigenvalues -1, 2, and 3 with corresponding eigenvectors $\begin{bmatrix} -1 \\ 1 \\ 1 \end{bmatrix}$, $\begin{bmatrix} -2 \\ 1 \\ 2 \end{bmatrix}$, and $\begin{bmatrix} -1 \\ 1 \\ 2 \end{bmatrix}$.

39. Prove that $\begin{bmatrix} a & 1 & 0 \\ 0 & a & 0 \\ 0 & 0 & b \end{bmatrix}$ is not diagonalizable for any scalars a and b.

40. Suppose that A is an $n \times n$ matrix having two distinct eigenvalues, λ_1 and λ_2, where λ_1 has multiplicity 1. State and prove a necessary and sufficient condition for A to be diagonalizable.

41. Prove that $I_n - A$ is invertible if and only if 1 is *not* an eigenvalue of A.

42. Two $n \times n$ matrices A and B are called *simultaneously diagonalizable* if there exists an invertible matrix P such that both $P^{-1}AP$ and $P^{-1}BP$ are diagonal matrices. Prove that if A and B are simultaneously diagonalizable, then $AB = BA$.

43. Let T be a linear operator on $\mathcal{R}^n$, $\mathcal{B}$ be a basis for $\mathcal{R}^n$, and A be the standard matrix of T. Prove that $[T]_{\mathcal{B}}$ and A have the same characteristic polynomial.

44. Let T be a linear operator on $\mathcal{R}^n$. A subspace W of $\mathcal{R}^n$ is called *T-invariant* if $T(\mathbf{w})$ is in W for each $\mathbf{w}$ in W. Prove that if V is an eigenspace of T, then V is T-invariant.

CHAPTER 5 MATLAB EXERCISES

For the following exercises, use MATLAB (or comparable software) or a calculator with matrix capabilities. The MATLAB functions in Tables D.1, D.2, D.3, D.4, and D.5 of Appendix D may be useful.

1. For each of the following matrices A, find, if possible, an invertible matrix P and a diagonal matrix D such that $A = PDP^{-1}$. If no such matrices exist, explain why.

(a) $\begin{bmatrix} -2 & -5 & 14 & -7 & 6 \\ -4 & 1 & -2 & 2 & 2 \\ 2 & 1 & -2 & 3 & -3 \\ -2 & -2 & 8 & -3 & 4 \\ -6 & -6 & 18 & -10 & 11 \end{bmatrix}$

(b) $\begin{bmatrix} -3 & 4 & -9 & 15 \\ 2 & 3 & -4 & 4 \\ -11 & 10 & -22 & 39 \\ -7 & 6 & -14 & 25 \end{bmatrix}$

(c) $\begin{bmatrix} -39 & -12 & -14 & -40 \\ -11 & -2 & -4 & -11 \\ 23 & 7 & 9 & 23 \\ 34 & 10 & 12 & 35 \end{bmatrix}$

(d) $\begin{bmatrix} 0 & 1 & 1 & -2 & 2 \\ -1 & 2 & 1 & -2 & 2 \\ -1 & 1 & 2 & -2 & 2 \\ -1 & 1 & 0 & 0 & 2 \\ 1 & -1 & 0 & 0 & 0 \end{bmatrix}$

2. Let A be an $n \times n$ matrix, and let $1 \le i < j \le n$. Let B be the matrix obtained from A by interchanging columns i and j of A, and let C be the matrix obtained from A by interchanging rows i and j of A. Use MATLAB to investigate the relationship between the eigenvalues and eigenvectors of B and those of C.
Hint: Experiment with

$$A = \begin{bmatrix} -117 & -80 & -46 & -30 & -2 \\ -12 & -11 & -5 & -2 & 0 \\ 258 & 182 & 102 & 64 & 4 \\ 107 & 72 & 42 & 28 & 2 \\ -215 & -146 & -85 & -57 & -5 \end{bmatrix}.$$

3. Find the matrix with eigenvectors

$$\begin{bmatrix} 1 \\ 1 \\ 0 \\ 2 \\ 3 \end{bmatrix}, \begin{bmatrix} 2 \\ 1 \\ 0 \\ 2 \\ 6 \end{bmatrix}, \begin{bmatrix} -1 \\ 0 \\ 1 \\ 0 \\ -3 \end{bmatrix}, \begin{bmatrix} 2 \\ 1 \\ 0 \\ 1 \\ 6 \end{bmatrix}, \begin{bmatrix} 2 \\ 1 \\ 0 \\ 2 \\ 7 \end{bmatrix}$$

and corresponding eigenvalues $1, -1, 2, 3$, and 0.

4. A search engine considers only 10 Web pages, which are linked in the following manner:

Page	Links to pages
1	2, 5, 7, and 10
2	1, 8, and 9
3	7 and 10
4	5 and 6
5	1, 4, 7, and 9
6	4, 7, and 8
7	1, 3, 5, and 6
8	no links
9	2 and 5
10	1 and 3

(a) Apply the PageRank algorithm to find the transition matrix for the Markov chain associated with this search engine, based on $p = 0.85$, the portion of the time a surfer moves from the current page to the next page by randomly choosing a link when one exists.

(b) Find the steady-state vector for the Markov chain obtained in (a), and use it to rank the pages.

5. For each of the following linear operators T, find a basis for the domain of T consisting of eigenvectors of T, or explain why no such basis exists.

(a) $T\left(\begin{bmatrix} x_1 \\ x_2 \\ x_3 \\ x_4 \end{bmatrix}\right) = \begin{bmatrix} x_1 + x_4 \\ 2x_1 + x_2 + x_3 + 3x_4 \\ 3x_1 + x_3 + 4x_4 \\ x_1 + x_2 + 2x_3 + 4x_4 \end{bmatrix}$

(b) $T\left(\begin{bmatrix} x_1 \\ x_2 \\ x_3 \\ x_4 \\ x_5 \end{bmatrix}\right) = \begin{bmatrix} 12x_1 + x_2 + 13x_3 + 12x_4 + 14x_5 \\ 10x_1 + 2x_2 + 6x_3 + 11x_4 + 7x_5 \\ -3x_1 - x_2 - x_3 - 4x_4 - 3x_5 \\ -13x_1 - 2x_2 - 11x_3 - 14x_4 - 13x_5 \\ 3x_1 + 2x_2 - 3x_3 + 5x_4 \end{bmatrix}$

6. Let

$$\mathbf{v}_1 = \begin{bmatrix} 1 \\ 1 \\ 3 \\ 2 \end{bmatrix}, \quad \mathbf{v}_2 = \begin{bmatrix} -1 \\ 0 \\ 0 \\ -3 \end{bmatrix}, \quad \mathbf{v}_3 = \begin{bmatrix} 2 \\ 1 \\ -2 \\ 1 \end{bmatrix}, \quad \mathbf{v}_4 = \begin{bmatrix} 3 \\ 2 \\ 1 \\ 1 \end{bmatrix},$$

and let

$$\mathcal{B} = \{\mathbf{v}_1, \mathbf{v}_2, \mathbf{v}_3, \mathbf{v}_4\}.$$

(a) Show that $\mathcal{B}$ is a basis for $\mathcal{R}^4$.

(b) Find the rule for the unique linear operator T on $\mathcal{R}^4$ such that

$$T(\mathbf{v}_1) = 2\mathbf{v}_1, \quad T(\mathbf{v}_2) = 3\mathbf{v}_2, \quad T(\mathbf{v}_3) = -\mathbf{v}_3, \quad \text{and}$$
$$T(\mathbf{v}_4) = \mathbf{v}_3 - \mathbf{v}_4.$$

(c) Determine whether T is diagonalizable. If T is diagonalizable, find a basis for $\mathcal{R}^4$ of eigenvectors of T.

7. Let $\mathcal{B}$ be the basis given in Exercise 6, and let T be the linear operator on $\mathcal{R}^4$ such that

$$T(\mathbf{v}_1) = \mathbf{v}_2, \quad T(\mathbf{v}_2) = 2\mathbf{v}_1, \quad T(\mathbf{v}_3) = -\mathbf{v}_3, \quad \text{and}$$
$$T(\mathbf{v}_4) = 2\mathbf{v}_4.$$

(a) Find the rule for T.

(b) Determine whether T is diagonalizable. If T is diagonalizable, find a basis for $\mathcal{R}^4$ of eigenvectors of T.

8. Let

$$A = \begin{bmatrix} 0.1 & 0.2 & 0.4 & 0.3 & 0.2 \\ 0.2 & 0.2 & 0.2 & 0.1 & 0.3 \\ 0.1 & 0.1 & 0.1 & 0.2 & 0.4 \\ 0.5 & 0.3 & 0.1 & 0.1 & 0.1 \\ 0.1 & 0.2 & 0.2 & 0.3 & 0 \end{bmatrix} \quad \text{and} \quad \mathbf{p} = \begin{bmatrix} 8 \\ 2 \\ 3 \\ 5 \\ 7 \end{bmatrix}.$$

(a) Show that A is a regular transition matrix.

(b) Find the steady-state vector $\mathbf{v}$ for A.

(c) Compute $A\mathbf{p}$, $A^{10}\mathbf{p}$, and $A^{100}\mathbf{p}$.

(d) Compare the last vector, $A^{100}\mathbf{p}$, with $25\mathbf{v}$. Explain.

9. Find the solution of the finite difference equation

$$r_n = 3r_{n-1} + 5r_{n-2} - 15r_{n-3} - 4r_{n-4} + 12r_{n-5}$$

with initial conditions $r_0 = 5$, $r_1 = 1$, $r_2 = 3$, $r_3 = 1$, and $r_4 = 3$.

6 INTRODUCTION

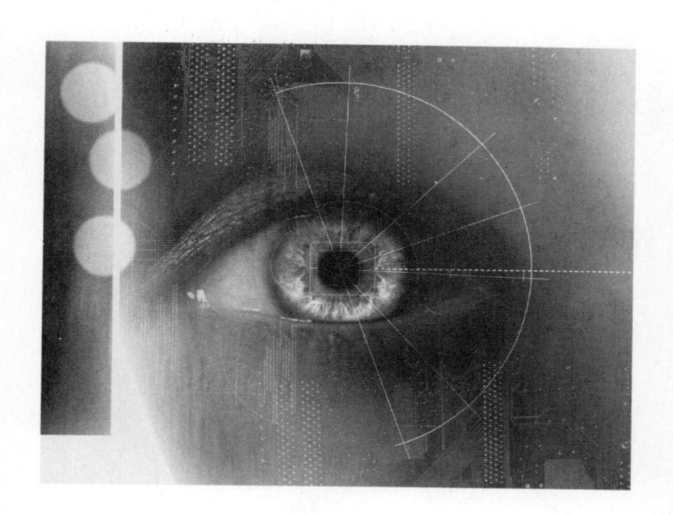

Identity verification is increasingly important in our modern and highly mobile society. Applications of identity verification range from national security to banking. Most of us no longer conduct transactions in person at a bank where the tellers know our names. Today it is not unusual for a person to obtain money at an Automated Teller Machine (ATM) far from where the money was deposited.

Increasingly, *biometric authentication* is used to verify the identity of a person. Biometric authentication is the automatic identification or identity verification of living humans based on behavioral and physiological characteristics. In contrast to common identification methods such as ID cards and Personal Identification Numbers (PINs), biometric identifiers cannot be lost or stolen and are more difficult to forge. The most commonly used biometric methods are:

- Fingerprints
- Hand Geometry
- Iris Recognition
- Voice Recognition
- Facial Recognition

The earliest facial recognition technique is the *eigenfaces method*, which is based on *principal component analysis* (see Section 6.8).

Each facial image is a matrix of pixels in which every pixel is represented by a numerical value corresponding to its intensity. In the eigenfaces method, each image is converted initially to a single long vector (see the figure on the left). Principal component analysis, based on the eigenvalues and eigenvectors of the *covariance matrix*, yields a new and relatively small set of vectors (the *eigenfaces*) that captures most of the variation in the original set of vectors (images).

The figure in the upper left corner corresponds to the first principal component of an original image; the ghost-like images in the rest of the figure correspond to its other principal components. The original image can be expressed as a weighted sum of vectors (eigenfaces) from the principal component analysis. These weights provide the identity of the person: Different images of the same person have approximately the same weights, and the image of another person will have significantly different weights.

CHAPTER

6 ORTHOGONALITY

Until now, we have focused our attention on two operations with vectors, namely, addition and scalar multiplication. In this chapter, we consider such geometric concepts as *length* and *perpendicularity* of vectors. By combining the geometry of vectors with matrices and linear transformations, we obtain powerful techniques for solving a wide variety of problems. For example, we apply these new tools to such areas as least-squares approximation, the graphing of conic sections, computer graphics, and statistical analyses. The key to most of these solutions is the construction of a basis of perpendicular eigenvectors for a given matrix or linear transformation.

To do this, we show how to convert any basis for a subspace of $\mathcal{R}^n$ into one in which all of the vectors are perpendicular to each other. Once this is done, we determine conditions that guarantee that there is a basis for $\mathcal{R}^n$ consisting of perpendicular eigenvectors of a matrix or a linear transformation. Surprisingly, for a matrix, a necessary and sufficient condition that such a basis exists is that the matrix be symmetric.

6.1 THE GEOMETRY OF VECTORS

In this section, we introduce the concepts of length and perpendicularity of vectors in $\mathcal{R}^n$. Many familiar geometric properties seen in earlier courses extend to this more general space. In particular, the Pythagorean theorem, which relates the squared lengths of sides of a right triangle, also holds in $\mathcal{R}^n$. To show that many of these results hold in $\mathcal{R}^n$, we define and develop the notion of *dot product.* The dot product is fundamental in the sense that, from it, we can define length and perpendicularity.

Perhaps the most basic concept of geometry is length. In Figure 6.1(a), an application of the Pythagorean theorem suggests that we define the *length* of the vector $\mathbf{u}$ to be $\sqrt{u_1^2 + u_2^2}$.

This definition easily extends to any vector $\mathbf{v}$ in $\mathcal{R}^n$ by defining its **norm (length),** denoted by $\|\mathbf{v}\|$, by

$$\|\mathbf{v}\| = \sqrt{v_1^2 + v_2^2 + \cdots + v_n^2}.$$

A vector whose norm is 1 is called a **unit vector**. Using the definition of vector norm, we can now define the **distance** between two vectors $\mathbf{u}$ and $\mathbf{v}$ in $\mathcal{R}^n$ as $\|\mathbf{u} - \mathbf{v}\|$. (See Figure 6.1(b).)

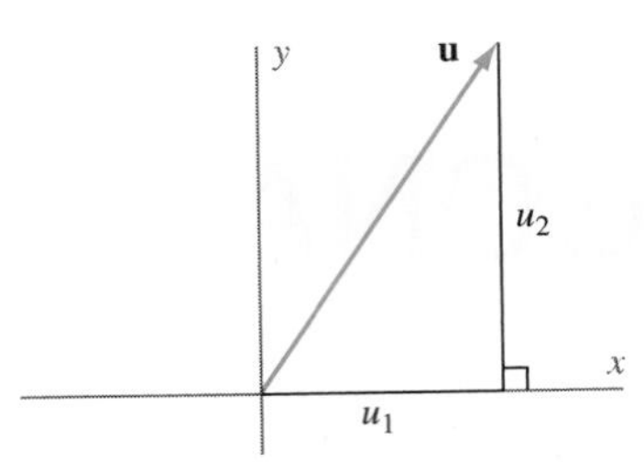

(a) The length of a vector **u** in $\mathcal{R}^2$

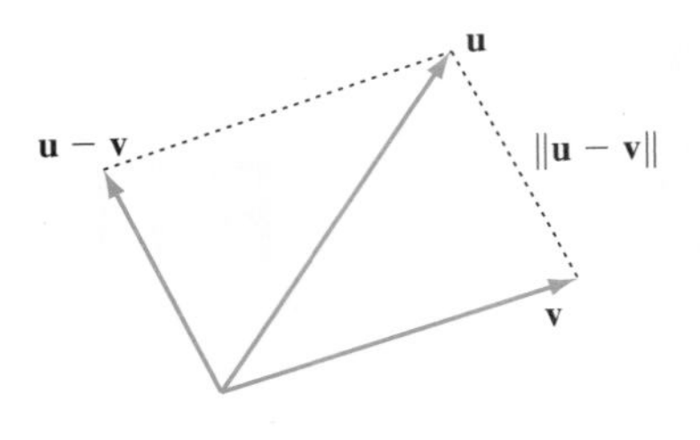

(b) The distance between vectors **u** and **v** in $\mathcal{R}^n$

Figure 6.1

Example 1 Find $\|\mathbf{u}\|$, $\|\mathbf{v}\|$, and the distance between **u** and **v** if

$$\mathbf{u} = \begin{bmatrix} 1 \\ 2 \\ 3 \end{bmatrix} \quad \text{and} \quad \mathbf{v} = \begin{bmatrix} 2 \\ -3 \\ 0 \end{bmatrix}.$$

Solution By definition,

$$\|\mathbf{u}\| = \sqrt{1^2 + 2^2 + 3^2} = \sqrt{14}, \quad \|\mathbf{v}\| = \sqrt{2^2 + (-3)^2 + 0^2} = \sqrt{13},$$

and the distance between **u** and **v** is

$$\|\mathbf{u} - \mathbf{v}\| = \sqrt{(1-2)^2 + (2-(-3))^2 + (3-0)^2} = \sqrt{35}.$$

Practice Problem 1 ▶ Let

$$\mathbf{u} = \begin{bmatrix} 1 \\ -2 \\ 2 \end{bmatrix} \quad \text{and} \quad \mathbf{v} = \begin{bmatrix} 6 \\ 2 \\ 3 \end{bmatrix}.$$

(a) Compute $\|\mathbf{u}\|$ and $\|\mathbf{v}\|$.

(b) Determine the distance between **u** and **v**.

(c) Show that both $\dfrac{1}{\|\mathbf{u}\|}\mathbf{u}$ and $\dfrac{1}{\|\mathbf{v}\|}\mathbf{v}$ are unit vectors. ◀

Just as we used the Pythagorean theorem in $\mathcal{R}^2$ to motivate the definition of the norm of a vector, we use this theorem again to examine what it means for two vectors **u** and **v** in $\mathcal{R}^2$ to be perpendicular. According to the Pythagorean theorem (see Figure 6.2), we see that **u** and **v** are perpendicular if and only if

$$\begin{aligned} \|\mathbf{v} - \mathbf{u}\|^2 &= \|\mathbf{u}\|^2 + \|\mathbf{v}\|^2 \\ (v_1 - u_1)^2 + (v_2 - u_2)^2 &= u_1^2 + u_2^2 + v_1^2 + v_2^2 \\ v_1^2 - 2u_1v_1 + u_1^2 + v_2^2 - 2u_2v_2 + u_2^2 &= u_1^2 + u_2^2 + v_1^2 + v_2^2 \\ -2u_1v_1 - 2u_2v_2 &= 0 \\ u_1v_1 + u_2v_2 &= 0. \end{aligned}$$

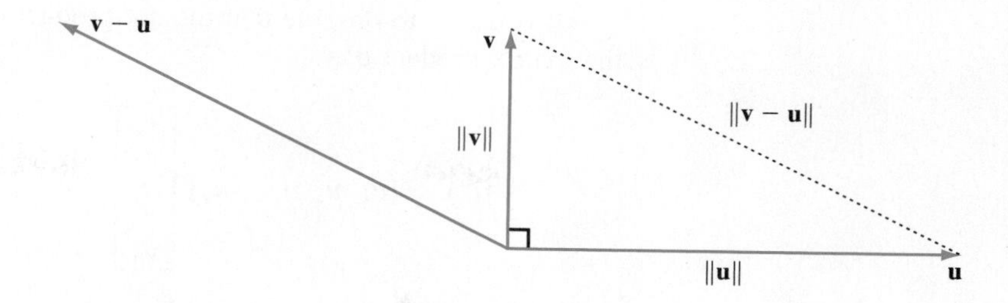

Figure 6.2 The Pythagorean theorem

The expression $u_1v_1 + u_2v_2$ in the last equation is called the *dot product* of **u** and **v**, and is denoted by $\mathbf{u} \cdot \mathbf{v}$. So **u** and **v** are perpendicular if and only if their dot product equals zero.

Using this observation, we define the **dot product** of vectors **u** and **v** in $\mathcal{R}^n$ by

$$\mathbf{u} \cdot \mathbf{v} = u_1v_1 + u_2v_2 + \cdots + u_nv_n.$$

We say that **u** and **v** are **orthogonal (perpendicular)** if $\mathbf{u} \cdot \mathbf{v} = 0$.

Notice that, in $\mathcal{R}^n$, the dot product of two vectors is a scalar, and the dot product of **0** with every vector is zero. Hence **0** is orthogonal to every vector in $\mathcal{R}^n$. Also, as noted, the property of being orthogonal in $\mathcal{R}^2$ and $\mathcal{R}^3$ is equivalent to the usual geometric definition of perpendicularity.

Example 2 Let

$$\mathbf{u} = \begin{bmatrix} 2 \\ -1 \\ 3 \end{bmatrix}, \qquad \mathbf{v} = \begin{bmatrix} 1 \\ 4 \\ -2 \end{bmatrix}, \qquad \text{and} \qquad \mathbf{w} = \begin{bmatrix} -8 \\ 3 \\ 2 \end{bmatrix}.$$

Determine which pairs of these vectors are orthogonal.

Solution We need only check which pairs have dot products equal to zero.

$$\mathbf{u} \cdot \mathbf{v} = (2)(1) + (-1)(4) + (3)(-2) = -8$$

$$\mathbf{u} \cdot \mathbf{w} = (2)(-8) + (-1)(3) + (3)(2) = -13$$

$$\mathbf{v} \cdot \mathbf{w} = (1)(-8) + (4)(3) + (-2)(2) = 0$$

We see that **v** and **w** are the only orthogonal vectors.

Practice Problem 2 ▶ Determine which pairs of the vectors

$$\mathbf{u} = \begin{bmatrix} -2 \\ -5 \\ 3 \end{bmatrix}, \qquad \mathbf{v} = \begin{bmatrix} 1 \\ -1 \\ 2 \end{bmatrix}, \qquad \text{and} \qquad \mathbf{w} = \begin{bmatrix} -3 \\ 1 \\ 2 \end{bmatrix}$$

are orthogonal. ◀

It is useful to observe that the dot product of **u** and **v** can also be represented as the matrix product $\mathbf{u}^T\mathbf{v}$.

$$\mathbf{u}^T\mathbf{v} = [u_1 \ \ u_2 \ \ \cdots \ \ u_n]\begin{bmatrix} v_1 \\ v_2 \\ \vdots \\ v_n \end{bmatrix} = u_1v_1 + u_2v_2 + \cdots + u_nv_n = \mathbf{u}\cdot\mathbf{v}$$

Notice that we have treated the 1×1 matrix $\mathbf{u}^T\mathbf{v}$ as a scalar by writing it as $u_1v_1 + u_2v_2 + \cdots + u_nv_n$ instead of $[u_1v_1 + u_2v_2 + \cdots + u_nv_n]$.

One useful consequence of identifying a dot product as a matrix product is that it enables us to "move" a matrix from one side of a dot product to the other. More precisely, if A is an $m \times n$ matrix, **u** is in $\mathcal{R}^n$, and **v** is in $\mathcal{R}^m$, then

$$A\mathbf{u}\cdot\mathbf{v} = \mathbf{u}\cdot A^T\mathbf{v}.$$

This follows because

$$A\mathbf{u}\cdot\mathbf{v} = (A\mathbf{u})^T\mathbf{v} = (\mathbf{u}^TA^T)\mathbf{v} = \mathbf{u}^T(A^T\mathbf{v}) = \mathbf{u}\cdot A^T\mathbf{v}.$$

Just as there are arithmetic properties of vector addition and scalar multiplication, there are arithmetic properties for the dot product and norm.

THEOREM 6.1

For all vectors **u**, **v**, and **w** in $\mathcal{R}^n$ and every scalar c,

(a) $\mathbf{u}\cdot\mathbf{u} = \|\mathbf{u}\|^2$.

(b) $\mathbf{u}\cdot\mathbf{u} = 0$ if and only if $\mathbf{u} = \mathbf{0}$.

(c) $\mathbf{u}\cdot\mathbf{v} = \mathbf{v}\cdot\mathbf{u}$.

(d) $\mathbf{u}\cdot(\mathbf{v}+\mathbf{w}) = \mathbf{u}\cdot\mathbf{v} + \mathbf{u}\cdot\mathbf{w}$.

(e) $(\mathbf{v}+\mathbf{w})\cdot\mathbf{u} = \mathbf{v}\cdot\mathbf{u} + \mathbf{w}\cdot\mathbf{u}$.

(f) $(c\mathbf{u})\cdot\mathbf{v} = c(\mathbf{u}\cdot\mathbf{v}) = \mathbf{u}\cdot(c\mathbf{v})$.

(g) $\|c\mathbf{u}\| = |c|\,\|\mathbf{u}\|$.

PROOF We prove parts (d) and (g) and leave the rest as exercises.

(d) Using matrix properties, we have

$$\begin{aligned} \mathbf{u}\cdot(\mathbf{v}+\mathbf{w}) &= \mathbf{u}^T(\mathbf{v}+\mathbf{w}) \\ &= \mathbf{u}^T\mathbf{v} + \mathbf{u}^T\mathbf{w} \\ &= \mathbf{u}\cdot\mathbf{v} + \mathbf{u}\cdot\mathbf{w}. \end{aligned}$$

(g) By (a) and (f), we have

$$\begin{aligned} \|c\mathbf{u}\|^2 &= (c\mathbf{u})\cdot(c\mathbf{u}) \\ &= c^2\mathbf{u}\cdot\mathbf{u} \\ &= c^2\|\mathbf{u}\|^2. \end{aligned}$$

By taking the square root of both sides and using $\sqrt{c^2} = |c|$, we obtain $\|c\mathbf{u}\| = |c|\|\mathbf{u}\|$. ■

Because of Theorem 6.1(f), there is no ambiguity in writing $c\mathbf{u}\cdot\mathbf{v}$ for any of the three expressions in (f).

Note that, by Theorem 6.1(g), any nonzero vector $\mathbf{v}$ can be **normalized**, that is, transformed into a unit vector by multiplying it by the scalar $\dfrac{1}{\|\mathbf{v}\|}$. For if $\mathbf{u} = \dfrac{1}{\|\mathbf{v}\|}\mathbf{v}$, then

$$\|\mathbf{u}\| = \left\|\frac{1}{\|\mathbf{v}\|}\mathbf{v}\right\| = \left|\frac{1}{\|\mathbf{v}\|}\right| \|\mathbf{v}\| = \frac{1}{\|\mathbf{v}\|}\|\mathbf{v}\| = 1.$$

This theorem allows us to treat expressions with dot products and norms just as we would algebraic expressions. For example, compare the similarity of the algebraic result

$$(2x + 3y)^2 = 4x^2 + 12xy + 9y^2$$

with

$$\|2\mathbf{u} + 3\mathbf{v}\|^2 = 4\|\mathbf{u}\|^2 + 12\mathbf{u}\cdot\mathbf{v} + 9\|\mathbf{v}\|^2.$$

The proof of the preceding equality relies heavily on Theorem 6.1:

$$\begin{aligned}
\|2\mathbf{u} + 3\mathbf{v}\|^2 &= (2\mathbf{u} + 3\mathbf{v})\cdot(2\mathbf{u} + 3\mathbf{v}) && \text{by (a)}\\
&= (2\mathbf{u})\cdot(2\mathbf{u} + 3\mathbf{v}) + (3\mathbf{v})\cdot(2\mathbf{u} + 3\mathbf{v}) && \text{by (e)}\\
&= (2\mathbf{u})\cdot(2\mathbf{u}) + (2\mathbf{u})\cdot(3\mathbf{v}) + (3\mathbf{v})\cdot(2\mathbf{u}) + (3\mathbf{v})\cdot(3\mathbf{v}) && \text{by (d)}\\
&= 4(\mathbf{u}\cdot\mathbf{u}) + 6(\mathbf{u}\cdot\mathbf{v}) + 6(\mathbf{v}\cdot\mathbf{u}) + 9(\mathbf{v}\cdot\mathbf{v}) && \text{by (f)}\\
&= 4\|\mathbf{u}\|^2 + 6(\mathbf{u}\cdot\mathbf{v}) + 6(\mathbf{u}\cdot\mathbf{v}) + 9\|\mathbf{v}\|^2 && \text{by (a) and (c)}\\
&= 4\|\mathbf{u}\|^2 + 12(\mathbf{u}\cdot\mathbf{v}) + 9\|\mathbf{v}\|^2
\end{aligned}$$

As noted earlier, we can write the last expression as $4\|\mathbf{u}\|^2 + 12\mathbf{u}\cdot\mathbf{v} + 9\|\mathbf{v}\|^2$. From now on, we will omit these steps when computing with dot products and norms.

! CAUTION Expressions such as $\mathbf{u}^2$ and $\mathbf{uv}$ are *not* defined.

It is easy to extend (d) and (e) of Theorem 6.1 to linear combinations, namely,

$$\mathbf{u}\cdot(c_1\mathbf{v}_1 + c_2\mathbf{v}_2 + \cdots + c_p\mathbf{v}_p) = c_1\mathbf{u}\cdot\mathbf{v}_1 + c_2\mathbf{u}\cdot\mathbf{v}_2 + \cdots + c_p\mathbf{u}\cdot\mathbf{v}_p$$

and

$$(c_1\mathbf{v}_1 + c_2\mathbf{v}_2 + \cdots + c_p\mathbf{v}_p)\cdot\mathbf{u} = c_1\mathbf{v}_1\cdot\mathbf{u} + c_2\mathbf{v}_2\cdot\mathbf{u} + \cdots + c_p\mathbf{v}_p\cdot\mathbf{u}.$$

As an application of these arithmetic properties, we show that the Pythagorean theorem holds in $\mathcal{R}^n$.

THEOREM 6.2

(Pythagorean Theorem in $\mathcal{R}^n$) Let $\mathbf{u}$ and $\mathbf{v}$ be vectors in $\mathcal{R}^n$. Then $\mathbf{u}$ and $\mathbf{v}$ are orthogonal if and only if

$$\|\mathbf{u} + \mathbf{v}\|^2 = \|\mathbf{u}\|^2 + \|\mathbf{v}\|^2.$$

PROOF Applying the arithmetic of dot products and norms to the vectors **u** and **v**, we have

$$\|\mathbf{u}+\mathbf{v}\|^2 = \|\mathbf{u}\|^2 + 2\mathbf{u}\cdot\mathbf{v} + \|\mathbf{v}\|^2.$$

Because **u** and **v** are orthogonal if and only if $\mathbf{u}\cdot\mathbf{v} = 0$, the result follows immediately. ■

ORTHOGONAL PROJECTION OF A VECTOR ON A LINE

Suppose we want to find the distance from a point P to the line $\mathcal{L}$ given in Figure 6.3. It is clear that if we can determine the vector **w**, then the desired distance is given by $\|\mathbf{u}-\mathbf{w}\|$. The vector **w** is called the **orthogonal projection of u on** $\mathcal{L}$. To find **w** in terms of **u** and $\mathcal{L}$, let **v** be *any* nonzero vector along $\mathcal{L}$, and let $\mathbf{z} = \mathbf{u} - \mathbf{w}$. Then $\mathbf{w} = c\mathbf{v}$ for some scalar c. Notice that **z** and **v** are orthogonal; that is,

$$0 = \mathbf{z}\cdot\mathbf{v} = (\mathbf{u}-\mathbf{w})\cdot\mathbf{v} = (\mathbf{u}-c\mathbf{v})\cdot\mathbf{v} = \mathbf{u}\cdot\mathbf{v} - c\mathbf{v}\cdot\mathbf{v} = \mathbf{u}\cdot\mathbf{v} - c\|\mathbf{v}\|^2.$$

So $c = \dfrac{\mathbf{u}\cdot\mathbf{v}}{\|\mathbf{v}\|^2}$, and thus $\mathbf{w} = \dfrac{\mathbf{u}\cdot\mathbf{v}}{\|\mathbf{v}\|^2}\mathbf{v}$. Therefore the distance from P to $\mathcal{L}$ is given by

$$\|\mathbf{u}-\mathbf{w}\| = \left\|\mathbf{u} - \frac{\mathbf{u}\cdot\mathbf{v}}{\|\mathbf{v}\|^2}\mathbf{v}\right\|.$$

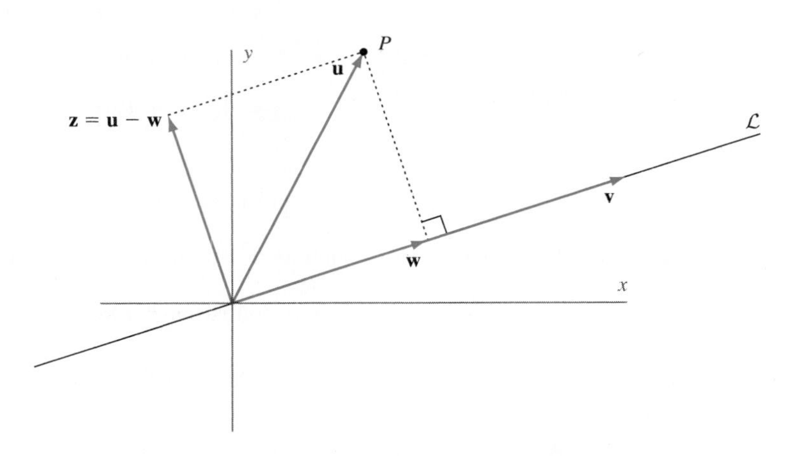

Figure 6.3 The vector **w** is the orthogonal projection of **u** on $\mathcal{L}$.

Example 3 Find the distance from the point (4, 1) to the line whose equation is $y = \frac{1}{2}x$.

Solution Following our preceding derivation, we let

$$\mathbf{u} = \begin{bmatrix}4\\1\end{bmatrix}, \qquad \mathbf{v} = \begin{bmatrix}2\\1\end{bmatrix}, \qquad \text{and} \qquad \frac{\mathbf{u}\cdot\mathbf{v}}{\|\mathbf{v}\|^2}\mathbf{v} = \frac{9}{5}\begin{bmatrix}2\\1\end{bmatrix}.$$

Then the desired distance is $\left\|\begin{bmatrix}4\\1\end{bmatrix} - \dfrac{9}{5}\begin{bmatrix}2\\1\end{bmatrix}\right\| = \dfrac{1}{5}\left\|\begin{bmatrix}2\\-4\end{bmatrix}\right\| = \dfrac{2}{5}\sqrt{5}.$

Practice Problem 3 ▶ Find the orthogonal projection of **u** on the line through the origin with direction **v**, where **u** and **v** are as in Practice Problem 2. ◀

We return to a more general case in Section 6.4, where we apply our results to solve an interesting statistical problem.

AN APPLICATION OF THE DOT PRODUCT TO GEOMETRY*

Recall that a *rhombus* is a parallelogram with all sides of equal length. We use Theorem 6.1 to prove the following result from geometry:

The diagonals of a parallelogram are orthogonal if and only if the parallelogram is a rhombus.

The diagonals of the rhombus are $\mathbf{u}+\mathbf{v}$ and $\mathbf{u}-\mathbf{v}$. (See Figure 6.4.) Applying the arithmetic of dot products and norms, we obtain

$$(\mathbf{u}+\mathbf{v})\cdot(\mathbf{u}-\mathbf{v}) = \|\mathbf{u}\|^2 - \|\mathbf{v}\|^2.$$

From this result, we see that the diagonals are orthogonal if and only if the preceding dot product is zero. This occurs if and only if $\|\mathbf{u}\|^2 = \|\mathbf{v}\|^2$, that is, the sides have equal lengths.

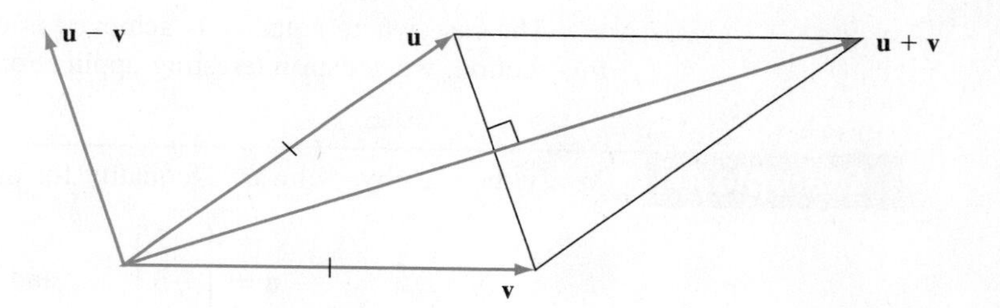

Figure 6.4 The diagonals of a rhombus

THE CAUCHY–SCHWARZ AND TRIANGLE INEQUALITIES

Recall that, in every triangle, the length of any side is less than the sum of the lengths of the other two sides. This simple result may be stated in the language of norms of vectors. Referring to Figure 6.5, we see that this statement is a consequence of the *triangle inequality* in $\mathcal{R}^2$:

$$\|\mathbf{u}+\mathbf{v}\| \le \|\mathbf{u}\| + \|\mathbf{v}\|$$

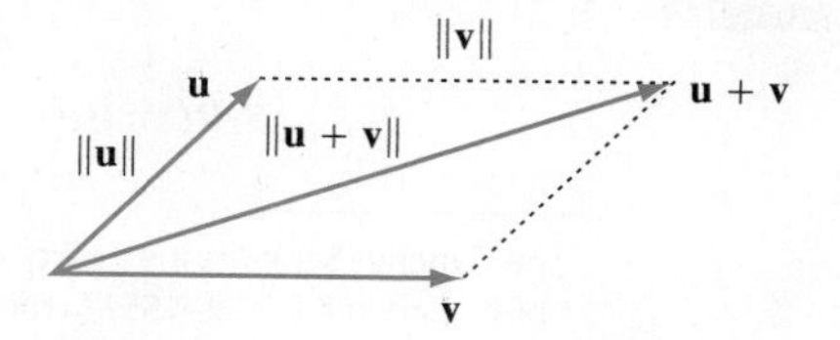

Figure 6.5 The triangle inequality

---

* The remainder of this section may be omitted without loss of continuity. However, the Cauchy–Schwarz and the triangle inequalities are frequently used in later courses.

What we show is that this inequality holds not just for vectors in $\mathcal{R}^2$, but also for vectors in $\mathcal{R}^n$. We begin with the *Cauchy–Schwarz inequality.*

THEOREM 6.3

(Cauchy–Schwarz Inequality[1]) For any vectors $\mathbf{u}$ and $\mathbf{v}$ in $\mathcal{R}^n$, we have

$$|\mathbf{u} \cdot \mathbf{v}| \leq \|\mathbf{u}\| \cdot \|\mathbf{v}\|.$$

PROOF If $\mathbf{u} = \mathbf{0}$ or $\mathbf{v} = \mathbf{0}$, the result is immediate. So assume that neither $\mathbf{u}$ nor $\mathbf{v}$ is zero, and let

$$\mathbf{w} = \frac{1}{\|\mathbf{u}\|}\mathbf{u} \quad \text{and} \quad \mathbf{z} = \frac{1}{\|\mathbf{v}\|}\mathbf{v}.$$

Then $\mathbf{w} \cdot \mathbf{w} = \mathbf{z} \cdot \mathbf{z} = 1$, and hence

$$0 \leq \|\mathbf{w} \pm \mathbf{z}\|^2 = (\mathbf{w} \pm \mathbf{z}) \cdot (\mathbf{w} \pm \mathbf{z}) = \mathbf{w} \cdot \mathbf{w} \pm 2(\mathbf{w} \cdot \mathbf{z}) + \mathbf{z} \cdot \mathbf{z} = 2 \pm 2(\mathbf{w} \cdot \mathbf{z}).$$

It follows that $\pm \mathbf{w} \cdot \mathbf{z} \leq 1$, and so $|\mathbf{w} \cdot \mathbf{z}| \leq 1$. Therefore

$$|\mathbf{u} \cdot \mathbf{v}| = |(\|\mathbf{u}\|\mathbf{w}) \cdot (\|\mathbf{v}\|\mathbf{z})| = \|\mathbf{u}\|\ \|\mathbf{v}\|\ |\mathbf{w} \cdot \mathbf{z}| \leq \|\mathbf{u}\|\ \|\mathbf{v}\|.$$ ■

The case where equality is achieved is examined in the exercises. At the end of this section, we see an interesting application of the Cauchy–Schwarz inequality.

Example 4 Verify the Cauchy–Schwarz inequality for the vectors

$$\mathbf{u} = \begin{bmatrix} 2 \\ -3 \\ 4 \end{bmatrix} \quad \text{and} \quad \mathbf{v} = \begin{bmatrix} 1 \\ -2 \\ -5 \end{bmatrix}.$$

Solution We have $\mathbf{u} \cdot \mathbf{v} = -12$, $\|\mathbf{u}\| = \sqrt{29}$, and $\|\mathbf{v}\| = \sqrt{30}$. So

$$|\mathbf{u} \cdot \mathbf{v}|^2 = 144 \leq 870 = (29)(30) = \|\mathbf{u}\|^2 \cdot \|\mathbf{v}\|^2.$$

Taking square roots confirms the Cauchy–Schwarz inequality for these vectors.

Example 5 contains another consequence of the Cauchy–Schwarz inequality.

Example 5 For any real numbers a_1, a_2, a_3, b_1, b_2, and b_3, show that

$$|a_1b_1 + a_2b_2 + a_3b_3| \leq \sqrt{a_1^2 + a_2^2 + a_3^2}\ \sqrt{b_1^2 + b_2^2 + b_3^2}.$$

[1] The Cauchy–Schwarz inequality was developed independently by the French mathematician Augustin-Louis Cauchy (1789–1857), the German Amandus Schwarz (1843–1921), and the Russian Viktor Yakovlevich Bunyakovsky (1804–1899). The result first appeared in Cauchy's 1821 text for an analysis course at the École Polytechnique in Paris. It was later proved for functions by Bunyakovsky in 1859 and by Schwarz in 1884.

Solution By applying the Cauchy–Schwarz inequality to $\mathbf{u} = \begin{bmatrix} a_1 \\ a_2 \\ a_3 \end{bmatrix}$ and $\mathbf{v} = \begin{bmatrix} b_1 \\ b_2 \\ b_3 \end{bmatrix}$, we obtain the desired inequality.

Our next result is the promised generalization to $\mathcal{R}^n$ of the triangle inequality.

THEOREM 6.4

(Triangle Inequality) For any vectors $\mathbf{u}$ and $\mathbf{v}$ in $\mathcal{R}^n$, we have

$$\|\mathbf{u} + \mathbf{v}\| \leq \|\mathbf{u}\| + \|\mathbf{v}\|.$$

PROOF Applying the Cauchy–Schwarz inequality, we obtain

$$\|\mathbf{u} + \mathbf{v}\|^2 = \|\mathbf{u}\|^2 + 2\mathbf{u} \cdot \mathbf{v} + \|\mathbf{v}\|^2 \leq \|\mathbf{u}\|^2 + 2\|\mathbf{u}\| \cdot \|\mathbf{v}\| + \|\mathbf{v}\|^2 = (\|\mathbf{u}\| + \|\mathbf{v}\|)^2.$$

Taking square roots of both sides yields the triangle inequality. ■

The case where equality is achieved is examined in the exercises.

Example 6 Verify the triangle inequality for the vectors $\mathbf{u}$ and $\mathbf{v}$ in Example 4.

Solution Since $\mathbf{u} + \mathbf{v} = \begin{bmatrix} 3 \\ -5 \\ -1 \end{bmatrix}$, it follows that $\|\mathbf{u} + \mathbf{v}\| = \sqrt{35}$. Recalling that $\|\mathbf{u}\| = \sqrt{29}$ and $\|\mathbf{v}\| = \sqrt{30}$, the triangle inequality follows from the observation that

$$\|\mathbf{u} + \mathbf{v}\| = \sqrt{35} < \sqrt{36} = 6 < 5 + 5 = \sqrt{25} + \sqrt{25} < \sqrt{29} + \sqrt{30} = \|\mathbf{u}\| + \|\mathbf{v}\|.$$

COMPUTING AVERAGE CLASS SIZE

A private school for the training of computer programmers advertised that its average class size is 20. After receiving complaints of false advertising, an investigator for the Office of Consumer Affairs obtained a list of the 60 students enrolled in the school. He polled each student and learned the student's class size. He added these numbers and divided the total by 60. The result was 27.6, a figure significantly higher than the advertised number 20. As a result, he initiated a complaint against the school. However, the complaint was withdrawn by his supervisor after she did some work of her own.

Using the same enrollment list, the supervisor polled all 60 students. She found that the students were divided among three classes. The first class had 25 students, the second class had 3 students, and the third class had 32 students. Notice that the sum of these three enrollments is 60. She then divided 60 by 3 to obtain a class average of 20, confirming the advertised class average.

To see why there is a difference between the results of these two computations, we apply linear algebra to a more general situation. Suppose that we have a total of m students who are divided into n classes consisting of $v_1, v_2, \dots, v_n$ students, respectively. Using this notation, we see that the average of the class sizes is given by

$$\overline{v} = \frac{1}{n}(v_1 + v_2 + \cdots + v_n) = \frac{m}{n}.$$

This is the method used by the supervisor.

Now consider the method used by the investigator. A student in the ith class responds that his or her class size is v_i. Because there are v_i students who give this response, the poll yields a sum of $v_i v_i = v_i^2$ that is contributed by the ith class. Because this is done for each class, the total sum of the responses is

$$v_1^2 + v_2^2 + \cdots + v_n^2.$$

Since m students are polled, this sum is divided by m to obtain the investigator's "average," say v^*, given by

$$v^* = \frac{1}{m}(v_1^2 + v_2^2 + \cdots + v_n^2).$$

To see the relationship between $\overline{v}$ and v^*, we define the vectors $\mathbf{u}$ and $\mathbf{v}$ in $\mathcal{R}^n$ by

$$\mathbf{u} = \begin{bmatrix} 1 \\ 1 \\ \vdots \\ 1 \end{bmatrix} \quad \text{and} \quad \mathbf{v} = \begin{bmatrix} v_1 \\ v_2 \\ \vdots \\ v_n \end{bmatrix}.$$

Then

$$\mathbf{u} \cdot \mathbf{u} = n, \qquad \mathbf{u} \cdot \mathbf{v} = m, \qquad \text{and} \qquad \mathbf{v} \cdot \mathbf{v} = v_1^2 + v_2^2 + \cdots + v_n^2.$$

Hence

$$\overline{v} = \frac{m}{n} = \frac{\mathbf{u} \cdot \mathbf{v}}{\mathbf{u} \cdot \mathbf{u}}, \qquad \text{and} \qquad v^* = \frac{1}{m}(v_1^2 + v_2^2 + \cdots + v_n^2) = \frac{\mathbf{v} \cdot \mathbf{v}}{\mathbf{u} \cdot \mathbf{v}}.$$

By the Cauchy–Schwarz inequality,

$$(\mathbf{u} \cdot \mathbf{v})^2 \le \|\mathbf{u}\|^2 \|\mathbf{v}\|^2 = (\mathbf{u} \cdot \mathbf{u})(\mathbf{v} \cdot \mathbf{v}),$$

and so, dividing both sides of this inequality by $(\mathbf{u} \cdot \mathbf{v})(\mathbf{u} \cdot \mathbf{u})$, we have

$$\overline{v} = \frac{\mathbf{u} \cdot \mathbf{v}}{\mathbf{u} \cdot \mathbf{u}} \le \frac{\mathbf{v} \cdot \mathbf{v}}{\mathbf{u} \cdot \mathbf{v}} = v^*.$$

Consequently, we always have that $\overline{v} \le v^*$. It can be shown that $\overline{v} = v^*$ if and only if all of the class sizes are equal. (See Exercise 124.)

EXERCISES

In Exercises 1–8, two vectors **u** *and* **v** *are given. Compute the norms of the vectors and the distance d between them.*

1. $\mathbf{u} = \begin{bmatrix} 5 \\ -3 \end{bmatrix}$ and $\mathbf{v} = \begin{bmatrix} 2 \\ 4 \end{bmatrix}$

2. $\mathbf{u} = \begin{bmatrix} 1 \\ 2 \end{bmatrix}$ and $\mathbf{v} = \begin{bmatrix} 3 \\ 7 \end{bmatrix}$

3. $\mathbf{u} = \begin{bmatrix} 1 \\ -1 \end{bmatrix}$ and $\mathbf{v} = \begin{bmatrix} 2 \\ 1 \end{bmatrix}$

4. $\mathbf{u} = \begin{bmatrix} 1 \\ 3 \\ 1 \end{bmatrix}$ and $\mathbf{v} = \begin{bmatrix} -1 \\ 4 \\ 2 \end{bmatrix}$

5. $\mathbf{u} = \begin{bmatrix} 1 \\ -1 \\ 3 \end{bmatrix}$ and $\mathbf{v} = \begin{bmatrix} 2 \\ 1 \\ 0 \end{bmatrix}$

6. $\mathbf{u} = \begin{bmatrix} 1 \\ 2 \\ 1 \\ -1 \end{bmatrix}$ and $\mathbf{v} = \begin{bmatrix} 2 \\ 3 \\ 2 \\ 0 \end{bmatrix}$

7. $\mathbf{u} = \begin{bmatrix} 1 \\ -1 \\ -2 \\ 1 \end{bmatrix}$ and $\mathbf{v} = \begin{bmatrix} 2 \\ 3 \\ 1 \\ 1 \end{bmatrix}$

8. $\mathbf{u} = \begin{bmatrix} 1 \\ 0 \\ -2 \\ 1 \end{bmatrix}$ and $\mathbf{v} = \begin{bmatrix} -1 \\ 2 \\ 1 \\ 3 \end{bmatrix}$

In Exercises 9–16, two vectors are given. Compute the dot product of the vectors, and determine whether the vectors are orthogonal.

9. $\mathbf{u} = \begin{bmatrix} 3 \\ -2 \end{bmatrix}$ and $\mathbf{v} = \begin{bmatrix} 4 \\ 6 \end{bmatrix}$

10. $\mathbf{u} = \begin{bmatrix} 1 \\ 2 \end{bmatrix}$ and $\mathbf{v} = \begin{bmatrix} 3 \\ 7 \end{bmatrix}$

11. $\mathbf{u} = \begin{bmatrix} 1 \\ -1 \end{bmatrix}$ and $\mathbf{v} = \begin{bmatrix} 2 \\ 1 \end{bmatrix}$

12. $\mathbf{u} = \begin{bmatrix} 1 \\ 3 \\ 1 \end{bmatrix}$ and $\mathbf{v} = \begin{bmatrix} -1 \\ 4 \\ 2 \end{bmatrix}$

13. $\mathbf{u} = \begin{bmatrix} 1 \\ -2 \\ 3 \end{bmatrix}$ and $\mathbf{v} = \begin{bmatrix} 2 \\ 1 \\ 0 \end{bmatrix}$

14. $\mathbf{u} = \begin{bmatrix} 1 \\ 2 \\ -3 \\ -1 \end{bmatrix}$ and $\mathbf{v} = \begin{bmatrix} 2 \\ 3 \\ 2 \\ 0 \end{bmatrix}$

15. $\mathbf{u} = \begin{bmatrix} 1 \\ -1 \\ -2 \\ 1 \end{bmatrix}$ and $\mathbf{v} = \begin{bmatrix} 2 \\ 3 \\ 1 \\ 1 \end{bmatrix}$

16. $\mathbf{u} = \begin{bmatrix} -1 \\ 3 \\ -2 \\ 4 \end{bmatrix}$ and $\mathbf{v} = \begin{bmatrix} -1 \\ -1 \\ 3 \\ 2 \end{bmatrix}$

In Exercises 17–24, two orthogonal vectors **u** *and* **v** *are given. Compute the quantities* $\|\mathbf{u}\|^2$, $\|\mathbf{v}\|^2$, *and* $\|\mathbf{u}+\mathbf{v}\|^2$. *Use your results to illustrate the Pythagorean theorem.*

17. $\mathbf{u} = \begin{bmatrix} -2 \\ 4 \end{bmatrix}$ and $\mathbf{v} = \begin{bmatrix} 6 \\ 3 \end{bmatrix}$

18. $\mathbf{u} = \begin{bmatrix} 3 \\ 1 \end{bmatrix}$ and $\mathbf{v} = \begin{bmatrix} -1 \\ 3 \end{bmatrix}$

19. $\mathbf{u} = \begin{bmatrix} 2 \\ 3 \end{bmatrix}$ and $\mathbf{v} = \begin{bmatrix} 0 \\ 0 \end{bmatrix}$

20. $\mathbf{u} = \begin{bmatrix} 2 \\ 6 \end{bmatrix}$ and $\mathbf{v} = \begin{bmatrix} 9 \\ -3 \end{bmatrix}$

21. $\mathbf{u} = \begin{bmatrix} 1 \\ 3 \\ 2 \end{bmatrix}$ and $\mathbf{v} = \begin{bmatrix} -1 \\ 1 \\ -1 \end{bmatrix}$

22. $\mathbf{u} = \begin{bmatrix} 1 \\ -1 \\ 2 \end{bmatrix}$ and $\mathbf{v} = \begin{bmatrix} -2 \\ 0 \\ 1 \end{bmatrix}$

23. $\mathbf{u} = \begin{bmatrix} 1 \\ 2 \\ 3 \end{bmatrix}$ and $\mathbf{v} = \begin{bmatrix} -11 \\ 4 \\ 1 \end{bmatrix}$

24. $\mathbf{u} = \begin{bmatrix} 2 \\ -1 \\ 4 \end{bmatrix}$ and $\mathbf{v} = \begin{bmatrix} -3 \\ 2 \\ 2 \end{bmatrix}$

In Exercises 25–32, two vectors **u** *and* **v** *are given. Compute the quantities* $\|\mathbf{u}\|$, $\|\mathbf{v}\|$, *and* $\|\mathbf{u}+\mathbf{v}\|$. *Use your results to illustrate the triangle inequality.*

25. $\mathbf{u} = \begin{bmatrix} 3 \\ 2 \end{bmatrix}$ and $\mathbf{v} = \begin{bmatrix} -6 \\ -4 \end{bmatrix}$

26. $\mathbf{u} = \begin{bmatrix} 2 \\ 1 \end{bmatrix}$ and $\mathbf{v} = \begin{bmatrix} 3 \\ -2 \end{bmatrix}$

27. $\mathbf{u} = \begin{bmatrix} 4 \\ 2 \end{bmatrix}$ and $\mathbf{v} = \begin{bmatrix} 3 \\ -1 \end{bmatrix}$

28. $\mathbf{u} = \begin{bmatrix} -2 \\ 5 \end{bmatrix}$ and $\mathbf{v} = \begin{bmatrix} 3 \\ 1 \end{bmatrix}$

29. $\mathbf{u} = \begin{bmatrix} 1 \\ -4 \\ 2 \end{bmatrix}$ and $\mathbf{v} = \begin{bmatrix} 3 \\ 1 \\ 1 \end{bmatrix}$

30. $\mathbf{u} = \begin{bmatrix} 2 \\ -3 \\ 1 \end{bmatrix}$ and $\mathbf{v} = \begin{bmatrix} 1 \\ 1 \\ 2 \end{bmatrix}$

31. $\mathbf{u} = \begin{bmatrix} 2 \\ -1 \\ 3 \end{bmatrix}$ and $\mathbf{v} = \begin{bmatrix} 4 \\ 0 \\ 1 \end{bmatrix}$

32. $\mathbf{u} = \begin{bmatrix} 2 \\ -3 \\ 1 \end{bmatrix}$ and $\mathbf{v} = \begin{bmatrix} -4 \\ 6 \\ -2 \end{bmatrix}$

In Exercises 33–40, two vectors $\mathbf{u}$ *and* $\mathbf{v}$ *are given. Compute the quantities* $\|\mathbf{u}\|$, $\|\mathbf{v}\|$, *and* $\mathbf{u}\cdot\mathbf{v}$. *Use your results to illustrate the Cauchy–Schwarz inequality.*

33. $\mathbf{u} = \begin{bmatrix} -2 \\ 3 \end{bmatrix}$ and $\mathbf{v} = \begin{bmatrix} 5 \\ 3 \end{bmatrix}$

34. $\mathbf{u} = \begin{bmatrix} 2 \\ 5 \end{bmatrix}$ and $\mathbf{v} = \begin{bmatrix} 3 \\ 4 \end{bmatrix}$

35. $\mathbf{u} = \begin{bmatrix} 4 \\ 1 \end{bmatrix}$ and $\mathbf{v} = \begin{bmatrix} 0 \\ -2 \end{bmatrix}$

36. $\mathbf{u} = \begin{bmatrix} -3 \\ 4 \end{bmatrix}$ and $\mathbf{v} = \begin{bmatrix} 1 \\ 2 \end{bmatrix}$

37. $\mathbf{u} = \begin{bmatrix} 6 \\ -1 \\ 2 \end{bmatrix}$ and $\mathbf{v} = \begin{bmatrix} 1 \\ 4 \\ -1 \end{bmatrix}$

38. $\mathbf{u} = \begin{bmatrix} 0 \\ 1 \\ 1 \end{bmatrix}$ and $\mathbf{v} = \begin{bmatrix} -2 \\ 1 \\ 3 \end{bmatrix}$

39. $\mathbf{u} = \begin{bmatrix} 4 \\ 2 \\ 1 \end{bmatrix}$ and $\mathbf{v} = \begin{bmatrix} 2 \\ -1 \\ -1 \end{bmatrix}$

40. $\mathbf{u} = \begin{bmatrix} 3 \\ -1 \\ 2 \end{bmatrix}$ and $\mathbf{v} = \begin{bmatrix} 1 \\ 3 \\ -1 \end{bmatrix}$

In Exercises 41–48, a vector $\mathbf{u}$ *and a line* $\mathcal{L}$ *in* $\mathcal{R}^2$ *are given. Compute the orthogonal projection* $\mathbf{w}$ *of* $\mathbf{u}$ *on* $\mathcal{L}$, *and use it to compute the distance* d *from the endpoint of* $\mathbf{u}$ *to* $\mathcal{L}$.

41. $\mathbf{u} = \begin{bmatrix} 5 \\ 0 \end{bmatrix}$ and $y = 0$

42. $\mathbf{u} = \begin{bmatrix} 2 \\ 3 \end{bmatrix}$ and $y = 2x$

43. $\mathbf{u} = \begin{bmatrix} 3 \\ 4 \end{bmatrix}$ and $y = -x$

44. $\mathbf{u} = \begin{bmatrix} 3 \\ 4 \end{bmatrix}$ and $y = -2x$

45. $\mathbf{u} = \begin{bmatrix} 4 \\ 1 \end{bmatrix}$ and $y = 3x$

46. $\mathbf{u} = \begin{bmatrix} -3 \\ 2 \end{bmatrix}$ and $y = x$

47. $\mathbf{u} = \begin{bmatrix} 2 \\ 5 \end{bmatrix}$ and $y = -3x$

48. $\mathbf{u} = \begin{bmatrix} 6 \\ 5 \end{bmatrix}$ and $y = -4x$

For Exercises 49–54, suppose that $\mathbf{u}$, $\mathbf{v}$, *and* $\mathbf{w}$ *are vectors in* $\mathcal{R}^n$ *such that* $\|\mathbf{u}\| = 2$, $\|\mathbf{v}\| = 3$, $\|\mathbf{w}\| = 5$, $\mathbf{u}\cdot\mathbf{v} = -1$, $\mathbf{u}\cdot\mathbf{w} = 1$, *and* $\mathbf{v}\cdot\mathbf{w} = -4$.

49. Compute $(\mathbf{u}+\mathbf{v})\cdot\mathbf{w}$.

50. Compute $\|4\mathbf{w}\|$.

51. Compute $\|\mathbf{u}+\mathbf{v}\|^2$.

52. Compute $(\mathbf{u}+\mathbf{w})\cdot\mathbf{v}$.

53. Compute $\|\mathbf{v}-4\mathbf{w}\|^2$.

54. Compute $\|2\mathbf{u}+3\mathbf{v}\|^2$.

For Exercises 55–60, suppose that $\mathbf{u}$, $\mathbf{v}$, *and* $\mathbf{w}$ *are vectors in* $\mathcal{R}^n$ *such that* $\mathbf{u}\cdot\mathbf{u} = 14$, $\mathbf{u}\cdot\mathbf{v} = 7$, $\mathbf{u}\cdot\mathbf{w} = -20$, $\mathbf{v}\cdot\mathbf{v} = 21$, $\mathbf{v}\cdot\mathbf{w} = -5$, *and* $\mathbf{w}\cdot\mathbf{w} = 30$.

55. Compute $\|\mathbf{v}\|^2$.

56. Compute $\|3\mathbf{u}\|$.

57. Compute $\mathbf{v}\cdot\mathbf{u}$.

58. Compute $\mathbf{w}\cdot(\mathbf{u}+\mathbf{v})$.

59. Compute $\|2\mathbf{u}-\mathbf{v}\|^2$.

60. Compute $\|\mathbf{v}+3\mathbf{w}\|$.

T&F *In Exercises 61–80, determine whether the statements are true or false.*

61. Vectors must be of the same size for their dot product to be defined.

62. The dot product of two vectors in $\mathcal{R}^n$ is a vector in $\mathcal{R}^n$.

63. The norm of a vector equals the dot product of the vector with itself.

64. The norm of a multiple of a vector is the same multiple of the norm of the vector.

65. The norm of a sum of vectors is the sum of the norms of the vectors.

66. The squared norm of a sum of orthogonal vectors is the sum of the squared norms of the vectors.

67. The orthogonal projection of a vector on a line is a vector that lies along the line.

68. The norm of a vector is always a nonnegative real number.

69. If the norm of $\mathbf{v}$ equals 0, then $\mathbf{v} = \mathbf{0}$.

70. If $\mathbf{u}\cdot\mathbf{v} = 0$, then $\mathbf{u} = \mathbf{0}$ or $\mathbf{v} = \mathbf{0}$.

71. For all vectors $\mathbf{u}$ and $\mathbf{v}$ in $\mathcal{R}^n$, $|\mathbf{u}\cdot\mathbf{v}| = \|\mathbf{u}\|\cdot\|\mathbf{v}\|$.

72. For all vectors $\mathbf{u}$ and $\mathbf{v}$ in $\mathcal{R}^n$, $\mathbf{u}\cdot\mathbf{v} = \mathbf{v}\cdot\mathbf{u}$.

73. The distance between vectors $\mathbf{u}$ and $\mathbf{v}$ in $\mathcal{R}^n$ is $\|\mathbf{u}-\mathbf{v}\|$.

74. For all vectors $\mathbf{u}$ and $\mathbf{v}$ in $\mathcal{R}^n$ and every scalar c,
$$(c\mathbf{u})\cdot\mathbf{v} = \mathbf{u}\cdot(c\mathbf{v}).$$

75. For all vectors $\mathbf{u}$, $\mathbf{v}$, and $\mathbf{w}$ in $\mathcal{R}^n$,
$$\mathbf{u}\cdot(\mathbf{v}+\mathbf{w}) = \mathbf{u}\cdot\mathbf{v}+\mathbf{u}\cdot\mathbf{w}.$$

76. If A is an $n\times n$ matrix and $\mathbf{u}$ and $\mathbf{v}$ are vectors in $\mathcal{R}^n$, then $A\mathbf{u}\cdot\mathbf{v} = \mathbf{u}\cdot A\mathbf{v}$.

77. For every vector $\mathbf{v}$ in $\mathcal{R}^n$, $\|\mathbf{v}\| = \|-\mathbf{v}\|$.

78. If $\mathbf{u}$ and $\mathbf{v}$ are orthogonal vectors in $\mathcal{R}^n$, then
$$\|\mathbf{u}+\mathbf{v}\| = \|\mathbf{u}\| + \|\mathbf{v}\|.$$

79. If $\mathbf{w}$ is the orthogonal projection of $\mathbf{u}$ on a line through the origin of $\mathcal{R}^2$, then $\mathbf{u}-\mathbf{w}$ is orthogonal to every vector on the line.

80. If $\mathbf{w}$ is the orthogonal projection of $\mathbf{u}$ on a line through the origin of $\mathcal{R}^2$, then $\mathbf{w}$ is the vector on the line closest to $\mathbf{u}$.

---

81. Prove (a) of Theorem 6.1.

82. Prove (b) of Theorem 6.1.

83. Prove (c) of Theorem 6.1.

84. Prove (e) of Theorem 6.1.

85. Prove (f) of Theorem 6.1.

86. Prove that if $\mathbf{u}$ is orthogonal to both $\mathbf{v}$ and $\mathbf{w}$, then $\mathbf{u}$ is orthogonal to every linear combination of $\mathbf{v}$ and $\mathbf{w}$.

87. Let $\{\mathbf{v},\mathbf{w}\}$ be a basis for a subspace W of $\mathcal{R}^n$, and define
$$\mathbf{z} = \mathbf{w} - \frac{\mathbf{v}\cdot\mathbf{w}}{\mathbf{v}\cdot\mathbf{v}}\mathbf{v}.$$
Prove that $\{\mathbf{v},\mathbf{z}\}$ is a basis for W consisting of orthogonal vectors.

88. Prove that the Cauchy–Schwarz inequality is an equality if and only if $\mathbf{u}$ is a multiple of $\mathbf{v}$ or $\mathbf{v}$ is a multiple of $\mathbf{u}$.

89. Prove that the triangle inequality is an equality if and only if $\mathbf{u}$ is a nonnegative multiple of $\mathbf{v}$ or $\mathbf{v}$ is a nonnegative multiple of $\mathbf{u}$.

90. Use the triangle inequality to prove that $|\,\|\mathbf{v}\| - \|\mathbf{w}\|\,| \leq \|\mathbf{v} - \mathbf{w}\|$ for all vectors $\mathbf{v}$ and $\mathbf{w}$ in $\mathcal{R}^n$.

91. Prove $(\mathbf{u}+\mathbf{v})\cdot\mathbf{w} = \mathbf{u}\cdot\mathbf{w} + \mathbf{v}\cdot\mathbf{w}$ for all vectors $\mathbf{u}$, $\mathbf{v}$, and $\mathbf{w}$ in $\mathcal{R}^n$.

92. Let $\mathbf{z}$ be a vector in $\mathcal{R}^n$. Let $W = \{\mathbf{u} \in \mathcal{R}^n : \mathbf{u}\cdot\mathbf{z} = 0\}$. Prove that W is a subspace of $\mathcal{R}^n$.

93. Let $\mathcal{S}$ be a subset of $\mathcal{R}^n$ and

$$W = \{\mathbf{u} \in \mathcal{R}^n : \mathbf{u}\cdot\mathbf{z} = 0 \text{ for all } \mathbf{z} \text{ in } \mathcal{S}\}.$$

Prove that W is a subspace of $\mathcal{R}^n$.

94. Let W denote the set of all vectors that lie along the line with equation $y = 2x$. Find a vector $\mathbf{z}$ in $\mathcal{R}^2$ such that $W = \{\mathbf{u} \in \mathcal{R}^2 : \mathbf{u}\cdot\mathbf{z} = 0\}$. Justify your answer.

95. Prove the *parallelogram law* for vectors in $\mathcal{R}^n$:

$$\|\mathbf{u}+\mathbf{v}\|^2 + \|\mathbf{u}-\mathbf{v}\|^2 = 2\|\mathbf{u}\|^2 + 2\|\mathbf{v}\|^2.$$

96. Prove that if $\mathbf{u}$ and $\mathbf{v}$ are orthogonal nonzero vectors in $\mathcal{R}^n$, then they are linearly independent.

97.[2] Let A be any $m \times n$ matrix.

(a) Prove that A^TA and A have the same null space. *Hint:* Let $\mathbf{v}$ be a vector in $\mathcal{R}^n$ such that $A^TA\mathbf{v} = \mathbf{0}$. Observe that $A^TA\mathbf{v}\cdot\mathbf{v} = A\mathbf{v}\cdot A\mathbf{v} = 0$.

(b) Use (a) to prove that $\operatorname{rank} A^TA = \operatorname{rank} A$.

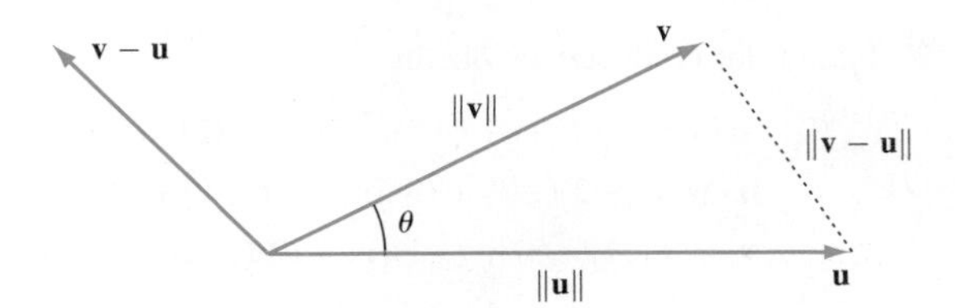

Figure 6.6

98.[3] Let $\mathbf{u}$ and $\mathbf{v}$ be nonzero vectors in $\mathcal{R}^2$ or $\mathcal{R}^3$, and let θ be the angle between $\mathbf{u}$ and $\mathbf{v}$. Then $\mathbf{u}$, $\mathbf{v}$, and $\mathbf{v}-\mathbf{u}$ determine a triangle. (See Figure 6.6.) The relationship between the lengths of the sides of this triangle and θ is called the *law of cosines*. It states that

$$\|\mathbf{v}-\mathbf{u}\|^2 = \|\mathbf{u}\|^2 + \|\mathbf{v}\|^2 - 2\|\mathbf{u}\|\,\|\mathbf{v}\|\cos\theta.$$

Use the law of cosines and Theorem 6.1 to derive the formula

$$\mathbf{u}\cdot\mathbf{v} = \|\mathbf{u}\|\,\|\mathbf{v}\|\cos\theta.$$

In Exercises 99–106, use the formula in Exercise 98 to determine the angle between the vectors $\mathbf{u}$ *and* $\mathbf{v}$.

99. $\mathbf{u} = \begin{bmatrix}-3\\1\end{bmatrix}$ and $\mathbf{v} = \begin{bmatrix}4\\2\end{bmatrix}$

100. $\mathbf{u} = \begin{bmatrix}1\\2\end{bmatrix}$ and $\mathbf{v} = \begin{bmatrix}-1\\3\end{bmatrix}$

101. $\mathbf{u} = \begin{bmatrix}-2\\4\end{bmatrix}$ and $\mathbf{v} = \begin{bmatrix}1\\-2\end{bmatrix}$

102. $\mathbf{u} = \begin{bmatrix}-1\\1\end{bmatrix}$ and $\mathbf{v} = \begin{bmatrix}3\\1\end{bmatrix}$

103. $\mathbf{u} = \begin{bmatrix}-1\\2\\1\end{bmatrix}$ and $\mathbf{v} = \begin{bmatrix}1\\1\\2\end{bmatrix}$

104. $\mathbf{u} = \begin{bmatrix}2\\1\\-3\end{bmatrix}$ and $\mathbf{v} = \begin{bmatrix}1\\-3\\2\end{bmatrix}$

105. $\mathbf{u} = \begin{bmatrix}1\\-2\\1\end{bmatrix}$ and $\mathbf{v} = \begin{bmatrix}-1\\1\\0\end{bmatrix}$

106. $\mathbf{u} = \begin{bmatrix}1\\2\\1\end{bmatrix}$ and $\mathbf{v} = \begin{bmatrix}1\\1\\0\end{bmatrix}$

Let $\mathbf{u}$ *and* $\mathbf{v}$ *be vectors in* $\mathcal{R}^3$. *Define* $\mathbf{u}\times\mathbf{v}$ *to be the vector* $\begin{bmatrix}u_2v_3 - u_3v_2\\ u_3v_1 - u_1v_3\\ u_1v_2 - u_2v_1\end{bmatrix}$, *which is called the* **cross product** *of* $\mathbf{u}$ *and* $\mathbf{v}$.

For Exercises 107–120, use the preceding definition of the cross product.

107. For every vector $\mathbf{u}$ in $\mathcal{R}^3$, prove that $\mathbf{u}\times\mathbf{u} = \mathbf{0}$.

108. Prove that $\mathbf{u}\times\mathbf{v} = -(\mathbf{v}\times\mathbf{u})$ for all vectors $\mathbf{u}$ and $\mathbf{v}$ in $\mathcal{R}^3$.

109. For every vector $\mathbf{u}$ in $\mathcal{R}^3$, prove that $\mathbf{u}\times\mathbf{0} = \mathbf{0}\times\mathbf{u} = \mathbf{0}$.

110. For all vectors $\mathbf{u}$ and $\mathbf{v}$ in $\mathcal{R}^3$, prove that $\mathbf{u}$ and $\mathbf{v}$ are parallel if and only if $\mathbf{u}\times\mathbf{v} = \mathbf{0}$.

111. For all vectors $\mathbf{u}$ and $\mathbf{v}$ in $\mathcal{R}^3$ and all scalars c, prove that

$$c(\mathbf{u}\times\mathbf{v}) = c\mathbf{u}\times\mathbf{v} = \mathbf{u}\times c\mathbf{v}.$$

112. For all vectors $\mathbf{u}$, $\mathbf{v}$, and $\mathbf{w}$ in $\mathcal{R}^3$, prove that

$$\mathbf{u}\times(\mathbf{v}+\mathbf{w}) = \mathbf{u}\times\mathbf{v} + \mathbf{u}\times\mathbf{w}.$$

113. For all vectors $\mathbf{u}$, $\mathbf{v}$, and $\mathbf{w}$ in $\mathcal{R}^3$, prove that

$$(\mathbf{u}+\mathbf{v})\times\mathbf{w} = \mathbf{u}\times\mathbf{w} + \mathbf{v}\times\mathbf{w}.$$

114. For all vectors $\mathbf{u}$ and $\mathbf{v}$ in $\mathcal{R}^3$, prove that $\mathbf{u}\times\mathbf{v}$ is orthogonal to both $\mathbf{u}$ and $\mathbf{v}$.

115. For all vectors $\mathbf{u}$, $\mathbf{v}$, and $\mathbf{w}$ in $\mathcal{R}^3$, prove that

$$(\mathbf{u}\times\mathbf{v})\cdot\mathbf{w} = \mathbf{u}\cdot(\mathbf{v}\times\mathbf{w}).$$

116. For all vectors $\mathbf{u}$, $\mathbf{v}$, and $\mathbf{w}$ in $\mathcal{R}^3$, prove that

$$\mathbf{u}\times(\mathbf{v}\times\mathbf{w}) = (\mathbf{u}\cdot\mathbf{w})\mathbf{v} - (\mathbf{u}\cdot\mathbf{v})\mathbf{w}.$$

---

[2] This exercise is used in Section 6.7 (on page 439).

[3] This exercise is used in Section 6.9 (on page 471).

117. For all vectors $\mathbf{u}$, $\mathbf{v}$, and $\mathbf{w}$ in $\mathcal{R}^3$, prove that

$$(\mathbf{u} \times \mathbf{v}) \times \mathbf{w} = (\mathbf{w} \cdot \mathbf{u})\mathbf{v} - (\mathbf{w} \cdot \mathbf{v})\mathbf{u}.$$

118. For all vectors $\mathbf{u}$ and $\mathbf{v}$ in $\mathcal{R}^3$, prove that

$$\|\mathbf{u} \times \mathbf{v}\|^2 = \|\mathbf{u}\|^2\|\mathbf{v}\|^2 - (\mathbf{u} \cdot \mathbf{v})^2.$$

119. For all vectors $\mathbf{u}$ and $\mathbf{v}$ in $\mathcal{R}^3$, prove that $\|\mathbf{u} \times \mathbf{v}\| = \|\mathbf{u}\|\|\mathbf{v}\| \sin\theta$, where θ is the angle between $\mathbf{u}$ and $\mathbf{v}$. *Hint:* Use Exercises 98 and 118.

120. For all vectors $\mathbf{u}$, $\mathbf{v}$, and $\mathbf{w}$ in $\mathcal{R}^3$, prove the *Jacobi identity*:

$$(\mathbf{u} \times \mathbf{v}) \times \mathbf{w} + (\mathbf{v} \times \mathbf{w}) \times \mathbf{u} + (\mathbf{w} \times \mathbf{u}) \times \mathbf{v} = \mathbf{0}$$

Exercises 121–124 refer to the application regarding the two methods of computing average class size given in this section. In Exercises 121–123, data are given for students enrolled in a three-section seminar course. Compute the average $\bar{v}$ determined by the supervisor and the average v^* determined by the investigator.

121. Section 1 contains 8 students, section 2 contains 12 students, and section 3 contains 6 students.

122. Section 1 contains 15 students, and each of sections 2 and 3 contains 30 students.

123. Each of the three sections contains 22 students.

124. Use Exercise 88 to prove that the two averaging methods for determining class size are equal if and only if all of the class sizes are equal.

In Exercise 125, use either a calculator with matrix capabilities or computer software such as MATLAB to solve the problem.

125. In every triangle, the length of any side is less than the sum of the lengths of the other two sides. When this observation is generalized to $\mathcal{R}^n$, we obtain the *triangle inequality* (Theorem 6.4), which states

$$\|\mathbf{u} + \mathbf{v}\| \leq \|\mathbf{u}\| + \|\mathbf{v}\|$$

for any vectors $\mathbf{u}$ and $\mathbf{v}$ in $\mathcal{R}^n$. Let

$$\mathbf{u} = \begin{bmatrix} 1 \\ 2 \\ 3 \\ 4 \end{bmatrix}, \quad \mathbf{v} = \begin{bmatrix} -8 \\ -6 \\ 4 \\ 5 \end{bmatrix}, \quad \mathbf{v}_1 = \begin{bmatrix} 2.01 \\ 4.01 \\ 6.01 \\ 8.01 \end{bmatrix}, \quad \text{and}$$

$$\mathbf{v}_2 = \begin{bmatrix} 3.01 \\ 6.01 \\ 9.01 \\ 12.01 \end{bmatrix}.$$

(a) Verify the triangle inequality for $\mathbf{u}$ and $\mathbf{v}$.

(b) Verify the triangle inequality for $\mathbf{u}$ and $\mathbf{v}_1$.

(c) Verify the triangle inequality for $\mathbf{u}$ and $\mathbf{v}_2$.

(d) From what you have observed in (b) and (c), make a conjecture about when equality occurs in the triangle inequality.

(e) Interpret your conjecture in (d) geometrically in $\mathcal{R}^2$.

SOLUTIONS TO THE PRACTICE PROBLEMS

1. (a) We have $\|\mathbf{u}\| = \sqrt{1^2 + (-2)^2 + 2^2} = 3$ and $\|\mathbf{v}\| = \sqrt{6^2 + 2^2 + 3^2} = 7$.

(b) We have $\|\mathbf{u} - \mathbf{v}\| = \left\| \begin{bmatrix} -5 \\ -4 \\ -1 \end{bmatrix} \right\|$

$$= \sqrt{(-5)^2 + (-4)^2 + (-1)^2} = \sqrt{42}.$$

(c) We have

$$\left\| \frac{1}{\|\mathbf{u}\|}\mathbf{u} \right\| = \left\| \frac{1}{3} \begin{bmatrix} 1 \\ -2 \\ 2 \end{bmatrix} \right\| = \left\| \begin{bmatrix} \frac{1}{3} \\ -\frac{2}{3} \\ \frac{2}{3} \end{bmatrix} \right\| = \sqrt{\frac{1}{9} + \frac{4}{9} + \frac{4}{9}} = 1$$

and

$$\left\| \frac{1}{\|\mathbf{v}\|}\mathbf{v} \right\| = \left\| \frac{1}{7} \begin{bmatrix} 6 \\ 2 \\ 3 \end{bmatrix} \right\| = \left\| \begin{bmatrix} \frac{6}{7} \\ \frac{2}{7} \\ \frac{3}{7} \end{bmatrix} \right\| = \sqrt{\frac{36}{49} + \frac{4}{49} + \frac{9}{49}} = 1.$$

2. Taking dot products, we obtain

$$\mathbf{u} \cdot \mathbf{v} = (-2)(1) + (-5)(-1) + (3)(2) = 9$$
$$\mathbf{u} \cdot \mathbf{w} = (-2)(-3) + (-5)(1) + (3)(2) = 7$$
$$\mathbf{v} \cdot \mathbf{w} = (1)(-3) + (-1)(1) + (2)(2) = 0.$$

So $\mathbf{u}$ and $\mathbf{w}$ are orthogonal, but $\mathbf{u}$ and $\mathbf{v}$ are not orthogonal, and $\mathbf{v}$ and $\mathbf{w}$ are not orthogonal.

3. Let $\mathbf{w}$ be the required orthogonal projection. Then

$$\mathbf{w} = \frac{\mathbf{u} \cdot \mathbf{v}}{\|\mathbf{v}\|^2}\mathbf{v} = \frac{(-2)(1) + (-5)(-1) + (3)(2)}{1^2 + (-1)^2 + 2^2} \begin{bmatrix} 1 \\ -1 \\ 2 \end{bmatrix}$$

$$= \frac{3}{2} \begin{bmatrix} 1 \\ -1 \\ 2 \end{bmatrix}.$$

6.2 ORTHOGONAL VECTORS

It is easy to extend the property of orthogonality to any set of vectors. We say that a subset of $\mathcal{R}^n$ is an **orthogonal set** if every pair of distinct vectors in the set is orthogonal. The subset is called an **orthonormal set** if it is an orthogonal set consisting entirely of unit vectors.

For example, the set

$$\mathcal{S} = \left\{ \begin{bmatrix} 1 \\ 2 \\ 3 \end{bmatrix}, \begin{bmatrix} 1 \\ 1 \\ -1 \end{bmatrix}, \begin{bmatrix} 5 \\ -4 \\ 1 \end{bmatrix} \right\}$$

is an orthogonal set because the dot product of every pair of distinct vectors in $\mathcal{S}$ is equal to zero. Also, the standard basis for $\mathcal{R}^n$ is an orthogonal set that is also an orthonormal set. Note that any set consisting of just one vector is an orthogonal set.

Practice Problem 1 ▶ Determine whether each of the sets

$$\mathcal{S}_1 = \left\{ \begin{bmatrix} 1 \\ -2 \end{bmatrix}, \begin{bmatrix} 2 \\ 1 \end{bmatrix} \right\} \quad \text{and} \quad \mathcal{S}_2 = \left\{ \begin{bmatrix} 1 \\ 1 \\ 2 \end{bmatrix}, \begin{bmatrix} 1 \\ 1 \\ -1 \end{bmatrix}, \begin{bmatrix} 2 \\ 0 \\ -1 \end{bmatrix} \right\}$$

is an orthogonal set. ◀

Our first result asserts that in most circumstances orthogonal sets are linearly independent.

THEOREM 6.5

Any orthogonal set of nonzero vectors is linearly independent.

PROOF Let $\{\mathbf{v}_1, \mathbf{v}_2, \ldots, \mathbf{v}_k\}$ be an orthogonal subset of $\mathcal{R}^n$ consisting of k nonzero vectors, and let $c_1, c_2, \ldots, c_k$ be scalars such that

$$c_1\mathbf{v}_1 + c_2\mathbf{v}_2 + \cdots + c_k\mathbf{v}_k = \mathbf{0}.$$

Then, for any $\mathbf{v}_i$, we have

$$\begin{aligned} 0 &= \mathbf{0} \cdot \mathbf{v}_i \\ &= (c_1\mathbf{v}_1 + c_2\mathbf{v}_2 + \cdots + c_i\mathbf{v}_i + \cdots + c_k\mathbf{v}_k) \cdot \mathbf{v}_i \\ &= c_1\mathbf{v}_1 \cdot \mathbf{v}_i + c_2\mathbf{v}_2 \cdot \mathbf{v}_i + \cdots + c_i\mathbf{v}_i \cdot \mathbf{v}_i + \cdots + c_k\mathbf{v}_k \cdot \mathbf{v}_i \\ &= c_i(\mathbf{v}_i \cdot \mathbf{v}_i) \\ &= c_i\|\mathbf{v}_i\|^2. \end{aligned}$$

But $\|\mathbf{v}_i\|^2 \neq 0$ because $\mathbf{v}_i \neq \mathbf{0}$, and hence $c_i = 0$. We conclude that $\mathbf{v}_1, \mathbf{v}_2, \ldots, \mathbf{v}_k$ are linearly independent. ■

An orthogonal set that is also a basis for a subspace of $\mathcal{R}^n$ is called an **orthogonal basis** for the subspace. Likewise, a basis that is also an orthonormal set is called an **orthonormal basis**. For example, the standard basis for $\mathcal{R}^n$ is an orthonormal basis for $\mathcal{R}^n$.

Replacing a vector in an orthogonal set by a scalar multiple of the vector results in a new set that is also an orthogonal set. If the scalar is nonzero and the orthogonal set consists of nonzero vectors, then the new set is linearly independent and is a generating set for the same subspace as the original set. So multiplying vectors in an orthogonal basis by nonzero scalars produces a new orthogonal basis for the same subspace. (See Exercise 53.) For example, consider the orthogonal set $\{\mathbf{v}_1, \mathbf{v}_2, \mathbf{v}_3\}$, where

$$\mathbf{v}_1 = \begin{bmatrix} 1 \\ 1 \\ 1 \\ 1 \end{bmatrix}, \quad \mathbf{v}_2 = \begin{bmatrix} 1 \\ 0 \\ -1 \\ 0 \end{bmatrix}, \quad \text{and} \quad \mathbf{v}_3 = \frac{1}{4}\begin{bmatrix} 1 \\ -1 \\ 1 \\ -1 \end{bmatrix}.$$

This set is an orthogonal basis for the subspace $W = \text{Span}\,\{\mathbf{v}_1, \mathbf{v}_2, \mathbf{v}_3\}$. In order to eliminate the fractional components in $\mathbf{v}_3$, we may replace it with the vector $4\mathbf{v}_3$ to obtain another orthogonal set $\{\mathbf{v}_1, \mathbf{v}_2, 4\mathbf{v}_3\}$, which is also an orthogonal basis for W. In particular, if we normalize each vector in an orthogonal basis, we obtain an orthonormal basis for the same subspace. So

$$\left\{ \frac{1}{2}\begin{bmatrix} 1 \\ 1 \\ 1 \\ 1 \end{bmatrix}, \frac{1}{\sqrt{2}}\begin{bmatrix} 1 \\ 0 \\ -1 \\ 0 \end{bmatrix}, \frac{1}{2}\begin{bmatrix} 1 \\ -1 \\ 1 \\ -1 \end{bmatrix} \right\}$$

is an orthonormal basis for W.

If $\mathcal{S} = \{\mathbf{v}_1, \mathbf{v}_2, \ldots, \mathbf{v}_k\}$ is an orthogonal basis for a subspace V of $\mathcal{R}^n$, we can adapt the proof of Theorem 6.5 to obtain a simple method of representing any vector in V as a linear combination of the vectors in $\mathcal{S}$. This method uses dot products, in contrast to the tedious task of solving systems of linear equations.

Consider any vector $\mathbf{u}$ in V, and suppose that

$$\mathbf{u} = c_1\mathbf{v}_1 + c_2\mathbf{v}_2 + \cdots + c_k\mathbf{v}_k.$$

To obtain c_i, we observe that

$$\begin{aligned} \mathbf{u} \cdot \mathbf{v}_i &= (c_1\mathbf{v}_1 + c_2\mathbf{v}_2 + \cdots + c_i\mathbf{v}_i + \cdots + c_k\mathbf{v}_k) \cdot \mathbf{v}_i \\ &= c_1\mathbf{v}_1 \cdot \mathbf{v}_i + c_2\mathbf{v}_2 \cdot \mathbf{v}_i + \cdots + c_i\mathbf{v}_i \cdot \mathbf{v}_i + \cdots + c_k\mathbf{v}_k \cdot \mathbf{v}_i \\ &= c_i(\mathbf{v}_i \cdot \mathbf{v}_i) \\ &= c_i\|\mathbf{v}_i\|^2, \end{aligned}$$

and hence

$$c_i = \frac{\mathbf{u} \cdot \mathbf{v}_i}{\|\mathbf{v}_i\|^2}.$$

Summarizing, we have the following result:

Representation of a Vector in Terms of an Orthogonal or Orthonormal Basis

Let $\{\mathbf{v}_1, \mathbf{v}_2, \ldots, \mathbf{v}_k\}$ be an orthogonal basis for a subspace V of $\mathcal{R}^n$, and let $\mathbf{u}$ be a vector in V. Then

$$\mathbf{u} = \frac{\mathbf{u} \cdot \mathbf{v}_1}{\|\mathbf{v}_1\|^2}\mathbf{v}_1 + \frac{\mathbf{u} \cdot \mathbf{v}_2}{\|\mathbf{v}_2\|^2}\mathbf{v}_2 + \cdots + \frac{\mathbf{u} \cdot \mathbf{v}_k}{\|\mathbf{v}_k\|^2}\mathbf{v}_k.$$

If, in addition, the orthogonal basis is an orthonormal basis for V, then

$$\mathbf{u} = (\mathbf{u} \cdot \mathbf{v}_1)\mathbf{v}_1 + (\mathbf{u} \cdot \mathbf{v}_2)\mathbf{v}_2 + \cdots + (\mathbf{u} \cdot \mathbf{v}_k)\mathbf{v}_k.$$

Example 1 Recall that $\mathcal{S} = \{\mathbf{v}_1, \mathbf{v}_2, \mathbf{v}_3\}$, where

$$\mathbf{v}_1 = \begin{bmatrix} 1 \\ 2 \\ 3 \end{bmatrix}, \quad \mathbf{v}_2 = \begin{bmatrix} 1 \\ 1 \\ -1 \end{bmatrix}, \quad \text{and} \quad \mathbf{v}_3 = \begin{bmatrix} 5 \\ -4 \\ 1 \end{bmatrix}$$

is an orthogonal subset of $\mathcal{R}^3$. Since the vectors in $\mathcal{S}$ are nonzero, Theorem 6.5 tells us that $\mathcal{S}$ is linearly independent. Therefore, by Theorem 4.7, $\mathcal{S}$ is a basis for $\mathcal{R}^3$. So $\mathcal{S}$ is an orthogonal basis for $\mathcal{R}^3$.

Let $\mathbf{u} = \begin{bmatrix} 3 \\ 2 \\ 1 \end{bmatrix}$. We now use the method previously described to obtain the coefficients that represent $\mathbf{u}$ as a linear combination of the vectors of $\mathcal{S}$. Suppose that

$$\mathbf{u} = c_1\mathbf{v}_1 + c_2\mathbf{v}_2 + c_3\mathbf{v}_3.$$

Then

$$c_1 = \frac{\mathbf{u} \cdot \mathbf{v}_1}{\|\mathbf{v}_1\|^2} = \frac{10}{14}, \quad c_2 = \frac{\mathbf{u} \cdot \mathbf{v}_2}{\|\mathbf{v}_2\|^2} = \frac{4}{3}, \quad \text{and} \quad c_3 = \frac{\mathbf{u} \cdot \mathbf{v}_3}{\|\mathbf{v}_3\|^2} = \frac{8}{42}.$$

The reader should verify that

$$\mathbf{u} = \frac{10}{14}\mathbf{v}_1 + \frac{4}{3}\mathbf{v}_2 + \frac{8}{42}\mathbf{v}_3.$$

Practice Problem 2 ► Let

$$\mathcal{S} = \left\{ \begin{bmatrix} 1 \\ 1 \\ 2 \end{bmatrix}, \begin{bmatrix} 1 \\ 1 \\ -1 \end{bmatrix}, \begin{bmatrix} 1 \\ -1 \\ 0 \end{bmatrix} \right\}.$$

(a) Verify that $\mathcal{S}$ is an orthogonal basis for $\mathcal{R}^3$.

(b) Let $\mathbf{u} = \begin{bmatrix} 2 \\ -4 \\ 7 \end{bmatrix}$. Use the preceding boxed formula to obtain the coefficients that represent $\mathbf{u}$ as a linear combination of the vectors of $\mathcal{S}$. ◄

Since we have seen the advantages of using orthogonal bases for subspaces, two natural questions arise:

1. Does every subspace of $\mathcal{R}^n$ have an orthogonal basis?
2. If a subspace of $\mathcal{R}^n$ has an orthogonal basis, how can it be found?

The next theorem not only provides a positive answer to the first question, but also gives a method for converting any linearly independent set into an orthogonal set with the same span. This method is called the *Gram–Schmidt (orthogonalization) process*.[4] We have the following important consequence:

> Every subspace of $\mathcal{R}^n$ has an orthogonal and hence an orthonormal basis.

[4] A modification of this procedure, usually called the *modified Gram–Schmidt process*, is more computationally efficient.

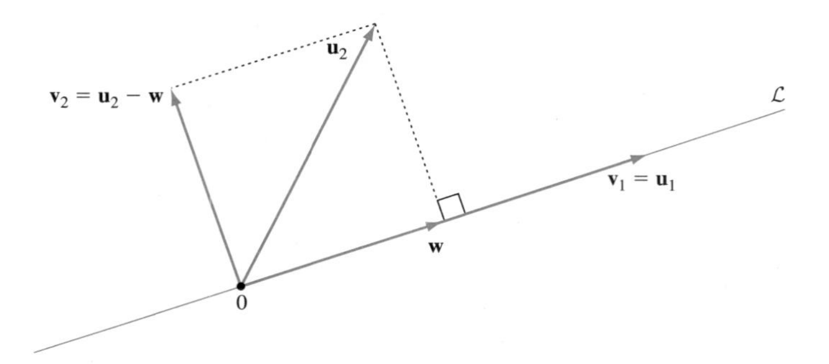

Figure 6.7 The vector **w** is the orthogonal projection of $\mathbf{u}_2$ on the line $\mathcal{L}$ through $\mathbf{v}_1$.

The Gram–Schmidt process is an extension of the procedure in Section 6.1 for finding the orthogonal projection of a vector on a line. Its purpose is to replace a basis $\{\mathbf{u}_1, \mathbf{u}_2, \ldots, \mathbf{u}_k\}$ for a subspace W of $\mathcal{R}^n$ with an orthogonal basis $\{\mathbf{v}_1, \mathbf{v}_2, \ldots, \mathbf{v}_k\}$ for W. For this purpose, we can choose $\mathbf{v}_1 = \mathbf{u}_1$ and compute $\mathbf{v}_2$ using the method outlined in Section 6.1. Let $\mathbf{w} = \dfrac{\mathbf{u}_2 \cdot \mathbf{v}_1}{\|\mathbf{v}_1\|^2}\mathbf{v}_1$, the orthogonal projection of $\mathbf{u}_2$ on the line $\mathcal{L}$ through $\mathbf{v}_1$, and let $\mathbf{v}_2 = \mathbf{u}_2 - \mathbf{w}$. (See Figure 6.7.) Then $\mathbf{v}_1$ and $\mathbf{v}_2$ are nonzero orthogonal vectors such that Span $\{\mathbf{v}_1, \mathbf{v}_2\}$ = Span $\{\mathbf{u}_1, \mathbf{u}_2\}$.

Figure 6.8 visually suggests the next step in the process, in which the vector $\mathbf{u}_3$ is replaced by $\mathbf{v}_3$. See if you can relate this figure to the proof of the next theorem.

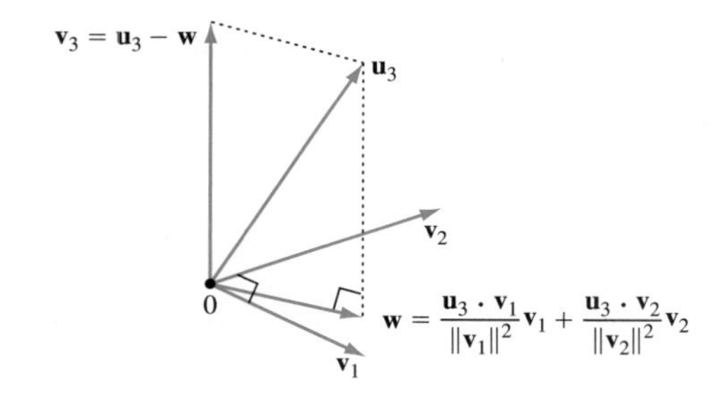

Figure 6.8 The construction of $\mathbf{v}_3$ by the Gram–Schmidt process

THEOREM 6.6

(The Gram–Schmidt Process[5]) Let $\{\mathbf{u}_1, \mathbf{u}_2, \ldots, \mathbf{u}_k\}$ be a basis for a subspace W of $\mathcal{R}^n$. Define

$$\begin{aligned}
\mathbf{v}_1 &= \mathbf{u}_1,\\
\mathbf{v}_2 &= \mathbf{u}_2 - \frac{\mathbf{u}_2 \cdot \mathbf{v}_1}{\|\mathbf{v}_1\|^2}\mathbf{v}_1,\\
\mathbf{v}_3 &= \mathbf{u}_3 - \frac{\mathbf{u}_3 \cdot \mathbf{v}_1}{\|\mathbf{v}_1\|^2}\mathbf{v}_1 - \frac{\mathbf{u}_3 \cdot \mathbf{v}_2}{\|\mathbf{v}_2\|^2}\mathbf{v}_2,\\
&\vdots\\
\mathbf{v}_k &= \mathbf{u}_k - \frac{\mathbf{u}_k \cdot \mathbf{v}_1}{\|\mathbf{v}_1\|^2}\mathbf{v}_1 - \frac{\mathbf{u}_k \cdot \mathbf{v}_2}{\|\mathbf{v}_2\|^2}\mathbf{v}_2 - \cdots - \frac{\mathbf{u}_k \cdot \mathbf{v}_{k-1}}{\|\mathbf{v}_{k-1}\|^2}\mathbf{v}_{k-1}.
\end{aligned}$$

[5] The Gram–Schmidt process first appeared in an 1833 paper by the Danish mathematician Jorgen P. Gram (1850–1916). A later paper by the German mathematician Erhard Schmidt (1876–1959) contained a detailed proof of the result.

Then $\{\mathbf{v}_1, \mathbf{v}_2, \ldots, \mathbf{v}_i\}$ is an orthogonal set of nonzero vectors such that

$$\text{Span}\,\{\mathbf{v}_1, \mathbf{v}_2, \ldots, \mathbf{v}_i\} = \text{Span}\,\{\mathbf{u}_1, \mathbf{u}_2, \ldots, \mathbf{u}_i\}$$

for each i. So $\{\mathbf{v}_1, \mathbf{v}_2, \ldots, \mathbf{v}_k\}$ is an orthogonal basis for W.

PROOF For $i = 1, 2, \ldots, k$, let $\mathcal{S}_i = \{\mathbf{u}_1, \mathbf{u}_2, \ldots, \mathbf{u}_i\}$ and $\mathcal{S}'_i = \{\mathbf{v}_1, \mathbf{v}_2, \ldots, \mathbf{v}_i\}$. Each $\mathcal{S}_i$ consists of vectors in a basis for W, so each $\mathcal{S}_i$ is a linearly independent set. We first prove that $\mathcal{S}'_i$ is an orthogonal set of nonzero vectors such that $\text{Span}\,\mathcal{S}'_i = \text{Span}\,\mathcal{S}_i$ for each i.

Note that $\mathbf{v}_1 = \mathbf{u}_1 \neq \mathbf{0}$ because $\mathbf{u}_1$ is in $\mathcal{S}_1$, which is linearly independent. Therefore $\mathcal{S}'_1 = \{\mathbf{v}_1\}$ is an orthogonal set of nonzero vectors such that $\text{Span}\,\mathcal{S}'_1 = \text{Span}\,\mathcal{S}_1$. For some $i = 2, 3, \ldots, k$, suppose that $\mathcal{S}'_{i-1}$ is an orthogonal set of nonzero vectors such that $\text{Span}\,\mathcal{S}'_{i-1} = \text{Span}\,\mathcal{S}_{i-1}$. Since

$$\mathbf{v}_i = \mathbf{u}_i - \frac{\mathbf{u}_i \cdot \mathbf{v}_1}{\|\mathbf{v}_1\|^2}\mathbf{v}_1 - \frac{\mathbf{u}_i \cdot \mathbf{v}_2}{\|\mathbf{v}_2\|^2}\mathbf{v}_2 - \cdots - \frac{\mathbf{u}_i \cdot \mathbf{v}_{i-1}}{\|\mathbf{v}_{i-1}\|^2}\mathbf{v}_{i-1},$$

$\mathbf{v}_i$ is a linear combination of vectors contained in the span of $\mathcal{S}_i$. Furthermore, $\mathbf{v}_i \neq \mathbf{0}$ because, otherwise, $\mathbf{u}_i$ would be contained in the span of $\mathcal{S}_{i-1}$, which is not the case because $\mathcal{S}_i$ is linearly independent. Next, observe that, for any $j < i$,

$$\begin{aligned}
\mathbf{v}_i \cdot \mathbf{v}_j &= \left(\mathbf{u}_i - \frac{\mathbf{u}_i \cdot \mathbf{v}_1}{\|\mathbf{v}_1\|^2}\mathbf{v}_1 - \frac{\mathbf{u}_i \cdot \mathbf{v}_2}{\|\mathbf{v}_2\|^2}\mathbf{v}_2 - \cdots - \frac{\mathbf{u}_i \cdot \mathbf{v}_j}{\|\mathbf{v}_j\|^2}\mathbf{v}_j - \cdots - \frac{\mathbf{u}_i \cdot \mathbf{v}_{i-1}}{\|\mathbf{v}_{i-1}\|^2}\mathbf{v}_{i-1}\right) \cdot \mathbf{v}_j \\
&= \mathbf{u}_i \cdot \mathbf{v}_j - \frac{\mathbf{u}_i \cdot \mathbf{v}_1}{\|\mathbf{v}_1\|^2}\mathbf{v}_1 \cdot \mathbf{v}_j - \frac{\mathbf{u}_i \cdot \mathbf{v}_2}{\|\mathbf{v}_2\|^2}\mathbf{v}_2 \cdot \mathbf{v}_j - \cdots \\
&\qquad - \frac{\mathbf{u}_i \cdot \mathbf{v}_j}{\|\mathbf{v}_j\|^2}\mathbf{v}_j \cdot \mathbf{v}_j - \cdots - \frac{\mathbf{u}_i \cdot \mathbf{v}_{i-1}}{\|\mathbf{v}_{i-1}\|^2}\mathbf{v}_{i-1} \cdot \mathbf{v}_j \\
&= \mathbf{u}_i \cdot \mathbf{v}_j - \frac{\mathbf{u}_i \cdot \mathbf{v}_j}{\|\mathbf{v}_j\|^2}\mathbf{v}_j \cdot \mathbf{v}_j \\
&= \mathbf{u}_i \cdot \mathbf{v}_j - \mathbf{u}_i \cdot \mathbf{v}_j \\
&= 0.
\end{aligned}$$

It follows that $\mathcal{S}'_i$ is an orthogonal set of nonzero vectors contained in the span of $\mathcal{S}_i$. But $\mathcal{S}'_i$ is linearly independent by Theorem 6.5, and so $\mathcal{S}'_i$ is a basis for the span of $\mathcal{S}_i$ by Theorem 4.7. Thus $\text{Span}\,\mathcal{S}'_i = \text{Span}\,\mathcal{S}_i$. In particular, when $i = k$, we see that $\text{Span}\,\mathcal{S}'_k$ is a linearly independent set of orthogonal vectors such that $\text{Span}\,\mathcal{S}'_k = \text{Span}\,\mathcal{S}_k = W$. That is, $\mathcal{S}'_k$ is an orthogonal basis for W. ■

Example 2 Let W be the span of $\mathcal{S} = \{\mathbf{u}_1, \mathbf{u}_2, \mathbf{u}_3\}$, where

$$\mathbf{u}_1 = \begin{bmatrix} 1 \\ 1 \\ 1 \\ 1 \end{bmatrix}, \quad \mathbf{u}_2 = \begin{bmatrix} 2 \\ 1 \\ 0 \\ 1 \end{bmatrix}, \quad \text{and} \quad \mathbf{u}_3 = \begin{bmatrix} 1 \\ 1 \\ 2 \\ 1 \end{bmatrix}$$

are linearly independent vectors in $\mathcal{R}^4$. Apply the Gram–Schmidt process to $\mathcal{S}$ to obtain an orthogonal basis $\mathcal{S}'$ for W.

Solution Let

$$\mathbf{v}_1 = \mathbf{u}_1 = \begin{bmatrix} 1 \\ 1 \\ 1 \\ 1 \end{bmatrix},$$

$$\mathbf{v}_2 = \mathbf{u}_2 - \frac{\mathbf{u}_2 \cdot \mathbf{v}_1}{\|\mathbf{v}_1\|^2}\mathbf{v}_1 = \begin{bmatrix} 2 \\ 1 \\ 0 \\ 1 \end{bmatrix} - \frac{4}{4}\begin{bmatrix} 1 \\ 1 \\ 1 \\ 1 \end{bmatrix} = \begin{bmatrix} 1 \\ 0 \\ -1 \\ 0 \end{bmatrix},$$

and

$$\mathbf{v}_3 = \mathbf{u}_3 - \frac{\mathbf{u}_3 \cdot \mathbf{v}_1}{\|\mathbf{v}_1\|^2}\mathbf{v}_1 - \frac{\mathbf{u}_3 \cdot \mathbf{v}_2}{\|\mathbf{v}_2\|^2}\mathbf{v}_2 = \begin{bmatrix} 1 \\ 1 \\ 2 \\ 1 \end{bmatrix} - \frac{5}{4}\begin{bmatrix} 1 \\ 1 \\ 1 \\ 1 \end{bmatrix} - \frac{(-1)}{2}\begin{bmatrix} 1 \\ 0 \\ -1 \\ 0 \end{bmatrix} = \frac{1}{4}\begin{bmatrix} 1 \\ -1 \\ 1 \\ -1 \end{bmatrix}.$$

Then $\mathcal{S}' = \{\mathbf{v}_1, \mathbf{v}_2, \mathbf{v}_3\}$ is an orthogonal basis for W.

Practice Problem 3 ▶ Let

$$W = \text{Span}\left\{ \begin{bmatrix} 1 \\ -1 \\ -1 \\ 1 \end{bmatrix}, \begin{bmatrix} -1 \\ 3 \\ -3 \\ 5 \end{bmatrix}, \begin{bmatrix} 1 \\ 6 \\ 3 \\ -4 \end{bmatrix} \right\}.$$

Apply the Gram–Schmidt process to obtain an orthogonal basis for W. ◀

Example 3 Find an orthonormal basis for the subspace W in Example 2, and write

$$\mathbf{u} = \begin{bmatrix} 2 \\ 3 \\ 5 \\ 3 \end{bmatrix}$$

as a linear combination of the vectors in this basis.

Solution As noted previously, normalizing $\mathbf{v}_1$, $\mathbf{v}_2$, and $\mathbf{v}_3$ produces the orthonormal basis

$$\{\mathbf{w}_1, \mathbf{w}_2, \mathbf{w}_3\} = \left\{ \frac{1}{2}\begin{bmatrix} 1 \\ 1 \\ 1 \\ 1 \end{bmatrix}, \frac{1}{\sqrt{2}}\begin{bmatrix} 1 \\ 0 \\ -1 \\ 0 \end{bmatrix}, \frac{1}{2}\begin{bmatrix} 1 \\ -1 \\ 1 \\ -1 \end{bmatrix} \right\}$$

for W. To represent $\mathbf{u}$ as a linear combination of $\mathbf{w}_1$, $\mathbf{w}_2$, and $\mathbf{w}_3$, we use the boxed formula on page 376. Since

$$\mathbf{u} \cdot \mathbf{w}_1 = \frac{13}{2}, \qquad \mathbf{u} \cdot \mathbf{w}_2 = \frac{-3}{\sqrt{2}}, \qquad \text{and} \qquad \mathbf{u} \cdot \mathbf{w}_3 = \frac{1}{2},$$

we see that

$$\mathbf{u} = (\mathbf{u} \cdot \mathbf{w}_1)\mathbf{w}_1 + (\mathbf{u} \cdot \mathbf{w}_2)\mathbf{w}_2 + (\mathbf{u} \cdot \mathbf{w}_3)\mathbf{w}_3$$

$$= \frac{13}{2}\mathbf{w}_1 + \left(\frac{-3}{\sqrt{2}}\right)\mathbf{w}_2 + \frac{1}{2}\mathbf{w}_3.$$

Practice Problem 4 ▶ Let

$$A = \begin{bmatrix} 1 & -1 & 1 \\ -1 & 3 & 6 \\ -1 & -3 & 3 \\ 1 & 5 & -4 \end{bmatrix}.$$

(a) Use the result of Practice Problem 3 to find an orthonormal basis $\mathcal{B}$ for the column space of A.

(b) Let

$$\mathbf{u} = \begin{bmatrix} 1 \\ 4 \\ 7 \\ -10 \end{bmatrix},$$

which is in the column space of A. Write $\mathbf{u}$ as a linear combination of the vectors in the basis $\mathcal{B}$. ◀

THE *QR* FACTORIZATION OF A MATRIX*

Although we have developed methods for solving systems of linear equations and finding eigenvalues, many of these approaches suffer from roundoff errors when applied to larger matrices. It can be shown that if the matrix under consideration can be factored as a product of matrices with desirable properties, then these methods can be modified to produce more reliable results. In what follows, we discuss one of these factorizations.

We illustrate the process with a 4×3 matrix A having linearly independent columns $\mathbf{a}_1$, $\mathbf{a}_2$, $\mathbf{a}_3$. First, apply the Gram–Schmidt process to these vectors to obtain orthogonal vectors $\mathbf{v}_1$, $\mathbf{v}_2$, and $\mathbf{v}_3$. Then normalize these vectors to obtain the orthonormal set $\{\mathbf{w}_1, \mathbf{w}_2, \mathbf{w}_3\}$. Note that

$$\mathbf{a}_1 \text{ is in } \operatorname{Span}\,\{\mathbf{a}_1\} = \operatorname{Span}\,\{\mathbf{w}_1\},$$

$$\mathbf{a}_2 \text{ is in } \operatorname{Span}\,\{\mathbf{a}_1, \mathbf{a}_2\} = \operatorname{Span}\,\{\mathbf{w}_1, \mathbf{w}_2\}, \text{ and}$$

$$\mathbf{a}_3 \text{ is in } \operatorname{Span}\,\{\mathbf{a}_1, \mathbf{a}_2, \mathbf{a}_3\} = \operatorname{Span}\,\{\mathbf{w}_1, \mathbf{w}_2, \mathbf{w}_3\}.$$

So we may write

$$\mathbf{a}_1 = r_{11}\mathbf{w}_1$$

$$\mathbf{a}_2 = r_{12}\mathbf{w}_1 + r_{22}\mathbf{w}_2$$

$$\mathbf{a}_3 = r_{13}\mathbf{w}_1 + r_{23}\mathbf{w}_2 + r_{33}\mathbf{w}_3$$

---

* The remainder of this section may be omitted without loss of continuity.

for some scalars $r_{11}, r_{12}, r_{22}, r_{13}, r_{23}, r_{33}$. Define the 4×3 matrix $Q = [\mathbf{w}_1\ \mathbf{w}_2\ \mathbf{w}_3]$. Note that we have

$$\mathbf{a}_1 = Q\begin{bmatrix} r_{11} \\ 0 \\ 0 \end{bmatrix}, \quad \mathbf{a}_2 = Q\begin{bmatrix} r_{12} \\ r_{22} \\ 0 \end{bmatrix}, \quad \text{and} \quad \mathbf{a}_3 = Q\begin{bmatrix} r_{13} \\ r_{23} \\ r_{33} \end{bmatrix}.$$

If we let

$$\mathbf{r}_1 = \begin{bmatrix} r_{11} \\ 0 \\ 0 \end{bmatrix}, \quad \mathbf{r}_2 = \begin{bmatrix} r_{12} \\ r_{22} \\ 0 \end{bmatrix}, \quad \text{and} \quad \mathbf{r}_3 = \begin{bmatrix} r_{13} \\ r_{23} \\ r_{33} \end{bmatrix},$$

then we have

$$A = [\mathbf{a}_1\ \mathbf{a}_2\ \mathbf{a}_3] = [Q\mathbf{r}_1\ Q\mathbf{r}_2\ Q\mathbf{r}_3] = QR,$$

where R is the upper triangular matrix

$$R = \begin{bmatrix} r_{11} & r_{12} & r_{13} \\ 0 & r_{22} & r_{23} \\ 0 & 0 & r_{33} \end{bmatrix}.$$

Notice that, by the boxed result on page 376, we have $r_{ij} = \mathbf{a}_j \cdot \mathbf{w}_i$ for all i and j.

This result can be extended to any $m \times n$ matrix with linearly independent columns, that is, with rank n.

The *QR* Factorization of a Matrix

Let A be an $m \times n$ matrix with linearly independent columns. There exists an $m \times n$ matrix Q whose columns form an orthonormal set in $\mathcal{R}^m$, and an $n \times n$ upper triangular matrix R such that $A = QR$. Furthermore, R can be chosen to have positive diagonal entries.[6]

In general, suppose that A is an $m \times n$ matrix with linearly independent columns. Any factorization $A = QR$, where Q is an $m \times n$ matrix whose columns form an orthonormal set in $\mathcal{R}^m$ and R is an $n \times n$ upper triangular matrix, is called a ***QR* factorization of** A.

It can be shown (see Exercise 66) that for any QR factorization of A, the columns of Q form an orthonormal basis for $\text{Col}\,A$.

Example 4 Find a QR factorization of the matrix

$$A = \begin{bmatrix} 1 & 2 & 1 \\ 1 & 1 & 1 \\ 1 & 0 & 2 \\ 1 & 1 & 1 \end{bmatrix}.$$

Solution We begin by letting

$$\mathbf{a}_1 = \begin{bmatrix} 1 \\ 1 \\ 1 \\ 1 \end{bmatrix}, \quad \mathbf{a}_2 = \begin{bmatrix} 2 \\ 1 \\ 0 \\ 1 \end{bmatrix}, \quad \text{and} \quad \mathbf{a}_3 = \begin{bmatrix} 1 \\ 1 \\ 2 \\ 1 \end{bmatrix}.$$

[6] See Exercises 59–61.

To find vectors $\mathbf{w}_1$, $\mathbf{w}_2$, and $\mathbf{w}_3$ with the desired properties, we use the results of Examples 2 and 3. Let

$$\mathbf{w}_1 = \frac{1}{2}\begin{bmatrix}1\\1\\1\\1\end{bmatrix}, \quad \mathbf{w}_2 = \frac{1}{\sqrt{2}}\begin{bmatrix}1\\0\\-1\\0\end{bmatrix}, \quad \text{and} \quad \mathbf{w}_3 = \frac{1}{2}\begin{bmatrix}1\\-1\\1\\-1\end{bmatrix}.$$

So

$$Q = \begin{bmatrix}\frac{1}{2} & \frac{1}{\sqrt{2}} & \frac{1}{2}\\ \frac{1}{2} & 0 & -\frac{1}{2}\\ \frac{1}{2} & -\frac{1}{\sqrt{2}} & \frac{1}{2}\\ \frac{1}{2} & 0 & -\frac{1}{2}\end{bmatrix}.$$

As noted, we can quickly compute the entries of R:

$$r_{11} = \mathbf{a}_1 \cdot \mathbf{w}_1 = 2 \qquad r_{12} = \mathbf{a}_2 \cdot \mathbf{w}_1 = 2 \qquad r_{13} = \mathbf{a}_3 \cdot \mathbf{w}_1 = \frac{5}{2}$$

$$r_{22} = \mathbf{a}_2 \cdot \mathbf{w}_2 = \sqrt{2} \qquad r_{23} = \mathbf{a}_3 \cdot \mathbf{w}_2 = -\frac{1}{\sqrt{2}} \qquad r_{33} = \mathbf{a}_3 \cdot \mathbf{w}_3 = \frac{1}{2}$$

So

$$R = \begin{bmatrix}2 & 2 & \frac{5}{2}\\ 0 & \sqrt{2} & -\frac{1}{\sqrt{2}}\\ 0 & 0 & \frac{1}{2}\end{bmatrix}.$$

The reader can verify that $A = QR$.

Practice Problem 5 ▶ Find the QR factorization of the matrix A in Practice Problem 4. ◀

AN APPLICATION OF QR FACTORIZATION TO SYSTEMS OF LINEAR EQUATIONS

Suppose we are given the consistent system of linear equations $A\mathbf{x} = \mathbf{b}$, where A is an $m \times n$ matrix with linearly independent columns. Let $A = QR$ be a QR factorization of A. Using the result that $Q^TQ = I_m$ (shown in Section 6.5), we obtain the equivalent systems

$$\begin{aligned} A\mathbf{x} &= \mathbf{b}\\ QR\mathbf{x} &= \mathbf{b}\\ Q^TQR\mathbf{x} &= Q^T\mathbf{b}\\ I_mR\mathbf{x} &= Q^T\mathbf{b}\\ R\mathbf{x} &= Q^T\mathbf{b}. \end{aligned}$$

Notice that, because the coefficient matrix R in the last system is upper triangular, this system is easy to solve.

Example 5 Solve the system

$$\begin{aligned} x_1 + 2x_2 + x_3 &= 1 \\ x_1 + x_2 + x_3 &= 3 \\ x_1 \quad\quad + 2x_3 &= 6 \\ x_1 + x_2 + x_3 &= 3. \end{aligned}$$

Solution The coefficient matrix is

$$A = \begin{bmatrix} 1 & 2 & 1 \\ 1 & 1 & 1 \\ 1 & 0 & 2 \\ 1 & 1 & 1 \end{bmatrix}.$$

Using the results of Example 4, we have $A = QR$, where

$$Q = \begin{bmatrix} \frac{1}{2} & \frac{1}{\sqrt{2}} & \frac{1}{2} \\ \frac{1}{2} & 0 & -\frac{1}{2} \\ \frac{1}{2} & -\frac{1}{\sqrt{2}} & \frac{1}{2} \\ \frac{1}{2} & 0 & -\frac{1}{2} \end{bmatrix} \quad \text{and} \quad R = \begin{bmatrix} 2 & 2 & \frac{5}{2} \\ 0 & \sqrt{2} & -\frac{1}{\sqrt{2}} \\ 0 & 0 & \frac{1}{2} \end{bmatrix}.$$

So an equivalent system is $R\mathbf{x} = Q^T\mathbf{b}$, or

$$\begin{aligned} 2x_1 + 2x_2 + \tfrac{5}{2}x_3 &= \tfrac{13}{2} \\ \sqrt{2}x_2 - \tfrac{\sqrt{2}}{2}x_3 &= -\tfrac{5\sqrt{2}}{2} \\ \tfrac{1}{2}x_3 &= \tfrac{1}{2}. \end{aligned}$$

Solving the third equation, we obtain $x_3 = 1$. Substituting this into the second equation and solving for x_2, we obtain

$$\sqrt{2}x_2 - \frac{\sqrt{2}}{2}x_3 = -\frac{5\sqrt{2}}{2},$$

or $x_2 = -2$. Finally, substituting the values for x_3 and x_2 into the first equation and solving for x_1 gives

$$2x_1 + 2(-2) + \frac{5}{2} = \frac{13}{2},$$

or $x_1 = 4$. So the solution of the given system is

$$\begin{bmatrix} 4 \\ -2 \\ 1 \end{bmatrix}.$$

Practice Problem 6 ▶ Use the preceding method and the solution of Practice Problem 5 to solve the system

$$\begin{aligned} x_1 - x_2 + x_3 &= 6 \\ -x_1 + 3x_2 + 6x_3 &= 13 \\ -x_1 - 3x_2 + 3x_3 &= 10 \\ x_1 + 5x_2 - 4x_3 &= -15. \end{aligned}$$ ◀

EXERCISES

In Exercises 1–8, determine whether each set is orthogonal.

1. $\left\{\begin{bmatrix}-2\\3\end{bmatrix}, \begin{bmatrix}2\\3\end{bmatrix}\right\}$

2. $\left\{\begin{bmatrix}1\\1\end{bmatrix}, \begin{bmatrix}1\\-1\end{bmatrix}\right\}$

3. $\left\{\begin{bmatrix}1\\2\\1\end{bmatrix}, \begin{bmatrix}1\\-1\\1\end{bmatrix}, \begin{bmatrix}2\\-1\\0\end{bmatrix}\right\}$

4. $\left\{\begin{bmatrix}1\\0\\1\end{bmatrix}, \begin{bmatrix}-1\\0\\1\end{bmatrix}, \begin{bmatrix}0\\-1\\0\end{bmatrix}\right\}$

5. $\left\{\begin{bmatrix}2\\1\\-5\end{bmatrix}, \begin{bmatrix}2\\1\\1\end{bmatrix}, \begin{bmatrix}3\\-1\\1\end{bmatrix}\right\}$

6. $\left\{\begin{bmatrix}1\\-2\\3\end{bmatrix}, \begin{bmatrix}1\\2\\1\end{bmatrix}, \begin{bmatrix}-1\\1\\1\end{bmatrix}\right\}$

7. $\left\{\begin{bmatrix}1\\2\\3\\-3\end{bmatrix}, \begin{bmatrix}1\\1\\-1\\0\end{bmatrix}, \begin{bmatrix}3\\-3\\0\\-1\end{bmatrix}\right\}$

8. $\left\{\begin{bmatrix}2\\1\\-1\\1\end{bmatrix}, \begin{bmatrix}1\\1\\3\\0\end{bmatrix}, \begin{bmatrix}1\\-1\\0\\1\end{bmatrix}\right\}$

In Exercises 9–16, (a) apply the Gram–Schmidt process to replace the given linearly independent set $\mathcal{S}$ by an orthogonal set of nonzero vectors with the same span, and (b) obtain an orthonormal set with the same span as $\mathcal{S}$.

9. $\left\{\begin{bmatrix}1\\1\\1\end{bmatrix}, \begin{bmatrix}5\\-1\\2\end{bmatrix}\right\}$

10. $\left\{\begin{bmatrix}1\\-2\\1\end{bmatrix}, \begin{bmatrix}1\\-1\\0\end{bmatrix}\right\}$

11. $\left\{\begin{bmatrix}1\\-2\\-1\end{bmatrix}, \begin{bmatrix}7\\7\\5\end{bmatrix}\right\}$

12. $\left\{\begin{bmatrix}-1\\3\\4\end{bmatrix}, \begin{bmatrix}-7\\11\\3\end{bmatrix}\right\}$

13. $\left\{\begin{bmatrix}0\\1\\1\\1\end{bmatrix}, \begin{bmatrix}1\\0\\1\\1\end{bmatrix}, \begin{bmatrix}1\\1\\0\\1\end{bmatrix}\right\}$

14. $\left\{\begin{bmatrix}1\\-1\\0\\2\end{bmatrix}, \begin{bmatrix}1\\1\\1\\3\end{bmatrix}, \begin{bmatrix}3\\1\\1\\5\end{bmatrix}\right\}$

15. $\left\{\begin{bmatrix}1\\0\\-1\\1\end{bmatrix}, \begin{bmatrix}2\\1\\-1\\0\end{bmatrix}, \begin{bmatrix}2\\-1\\-1\\3\end{bmatrix}\right\}$

16. $\left\{\begin{bmatrix}1\\-1\\0\\1\\1\end{bmatrix}, \begin{bmatrix}2\\-1\\0\\3\\2\end{bmatrix}, \begin{bmatrix}1\\-1\\1\\1\\1\end{bmatrix}, \begin{bmatrix}3\\1\\1\\1\\1\end{bmatrix}\right\}$

In Exercises 17–24, an orthogonal set $\mathcal{S}$ and a vector $\mathbf{v}$ in Span $\mathcal{S}$ are given. Use dot products (not systems of linear equations) to represent $\mathbf{v}$ as a linear combination of the vectors in $\mathcal{S}$.

17. $\mathcal{S} = \left\{\begin{bmatrix}2\\1\end{bmatrix}, \begin{bmatrix}-1\\2\end{bmatrix}\right\}$ and $\mathbf{v} = \begin{bmatrix}1\\8\end{bmatrix}$

18. $\mathcal{S} = \left\{\begin{bmatrix}-1\\1\end{bmatrix}, \begin{bmatrix}1\\1\end{bmatrix}\right\}$ and $\mathbf{v} = \begin{bmatrix}5\\-1\end{bmatrix}$

19. $\mathcal{S} = \left\{\begin{bmatrix}-1\\3\\-2\end{bmatrix}, \begin{bmatrix}-1\\1\\2\end{bmatrix}, \begin{bmatrix}1\\1\\1\end{bmatrix}\right\}$ and $\mathbf{v} = \begin{bmatrix}7\\-1\\2\end{bmatrix}$

20. $\mathcal{S} = \left\{\begin{bmatrix}1\\1\\1\end{bmatrix}, \begin{bmatrix}1\\2\\-3\end{bmatrix}\right\}$ and $\mathbf{v} = \begin{bmatrix}2\\1\\6\end{bmatrix}$

21. $\mathcal{S} = \left\{\begin{bmatrix}1\\0\\1\end{bmatrix}, \begin{bmatrix}1\\2\\-1\end{bmatrix}, \begin{bmatrix}1\\-1\\-1\end{bmatrix}\right\}$ and $\mathbf{v} = \begin{bmatrix}3\\1\\2\end{bmatrix}$

22. $\mathcal{S} = \left\{\begin{bmatrix}1\\-2\\0\\-1\end{bmatrix}, \begin{bmatrix}2\\1\\1\\0\end{bmatrix}, \begin{bmatrix}1\\0\\-2\\1\end{bmatrix}\right\}$ and $\mathbf{v} = \begin{bmatrix}6\\9\\9\\0\end{bmatrix}$

23. $\mathcal{S} = \left\{\begin{bmatrix}1\\-1\\-1\\1\end{bmatrix}, \begin{bmatrix}2\\1\\1\\0\end{bmatrix}, \begin{bmatrix}-1\\1\\1\\3\end{bmatrix}\right\}$ and $\mathbf{v} = \begin{bmatrix}1\\5\\5\\-7\end{bmatrix}$

24. $\mathcal{S} = \left\{\begin{bmatrix}1\\1\\1\\1\end{bmatrix}, \begin{bmatrix}1\\-1\\1\\-1\end{bmatrix}, \begin{bmatrix}1\\-1\\-1\\1\end{bmatrix}, \begin{bmatrix}1\\1\\-1\\-1\end{bmatrix}\right\}$ and $\mathbf{v} = \begin{bmatrix}2\\1\\-1\\2\end{bmatrix}$

In Exercises 25–32, let A be the matrix whose columns are the vectors in each indicated exercise.

(a) Determine the matrices Q and R in a QR factorization of A.
(b) Verify that $A = QR$.

25. Exercise 9
26. Exercise 10
27. Exercise 11
28. Exercise 12
29. Exercise 13
30. Exercise 14
31. Exercise 15
32. Exercise 16

In Exercises 33–40, solve the system $A\mathbf{x} = \mathbf{b}$ using the QR factorization of A obtained in each indicated exercise.

33. Exercise 25, $\mathbf{b} = \begin{bmatrix}-3\\3\\0\end{bmatrix}$

34. Exercise 26, $\mathbf{b} = \begin{bmatrix}6\\-8\\2\end{bmatrix}$

35. Exercise 27, $\mathbf{b} = \begin{bmatrix}-11\\-20\\-13\end{bmatrix}$

36. Exercise 28, $\mathbf{b} = \begin{bmatrix}13\\-19\\-2\end{bmatrix}$

37. Exercise 29, $\mathbf{b} = \begin{bmatrix}4\\1\\-1\\2\end{bmatrix}$

38. Exercise 30, $\mathbf{b} = \begin{bmatrix}8\\0\\1\\11\end{bmatrix}$

39. Exercise 31, $\mathbf{b} = \begin{bmatrix}0\\-7\\-1\\11\end{bmatrix}$

40. Exercise 32, $\mathbf{b} = \begin{bmatrix}8\\-4\\4\\6\\6\end{bmatrix}$

In Exercises 41–52, determine whether the statements are true or false.

41. Any orthogonal subset of $\mathcal{R}^n$ is linearly independent.
42. Every nonzero subspace of $\mathcal{R}^n$ has an orthogonal basis.
43. Any subset of $\mathcal{R}^n$ consisting of a single vector is an orthogonal set.
44. If $\mathcal{S}$ is an orthogonal set of n nonzero vectors in $\mathcal{R}^n$, then $\mathcal{S}$ is a basis for $\mathcal{R}^n$.
45. If $\{\mathbf{v}_1, \mathbf{v}_2, \ldots, \mathbf{v}_k\}$ is an orthonormal basis for a subspace W and $\mathbf{w}$ is a vector in W, then
$$\mathbf{w} = (\mathbf{w} \cdot \mathbf{v}_1)\mathbf{v}_1 + (\mathbf{w} \cdot \mathbf{v}_2)\mathbf{v}_2 + \cdots + (\mathbf{w} \cdot \mathbf{v}_k)\mathbf{v}_k.$$
46. For any nonzero vector $\mathbf{v}$, $\frac{1}{\|\mathbf{v}\|}\mathbf{v}$ is a unit vector.
47. The set of standard vectors $\mathbf{e}_1, \mathbf{e}_2, \ldots, \mathbf{e}_n$ is an orthonormal basis for $\mathcal{R}^n$.
48. Every orthonormal subset is linearly independent.
49. Combining the vectors in two orthonormal subsets of $\mathcal{R}^n$ produces another orthonormal subset of $\mathcal{R}^n$.
50. If $\mathbf{x}$ is orthogonal to $\mathbf{y}$ and $\mathbf{y}$ is orthogonal to $\mathbf{z}$, then $\mathbf{x}$ is orthogonal to $\mathbf{z}$.
51. The Gram–Schmidt process transforms a linearly independent set into an orthogonal set.
52. In the QR factorization of a matrix, both factors are upper triangular matrices.

---

53. Let $\{\mathbf{v}_1, \mathbf{v}_2, \ldots, \mathbf{v}_k\}$ be an orthogonal subset of $\mathcal{R}^n$. Prove that, for any scalars $c_1, c_2, \ldots, c_k$, the set $\{c_1\mathbf{v}_1, c_2\mathbf{v}_2, \ldots, c_k\mathbf{v}_k\}$ is also orthogonal.
54. Suppose that $\mathcal{S}$ is a nonempty *orthogonal* subset of $\mathcal{R}^n$ consisting of nonzero vectors, and suppose that $\mathcal{S}'$ is obtained by applying the Gram–Schmidt process to $\mathcal{S}$. Prove that $\mathcal{S}' = \mathcal{S}$.
55. Let $\{\mathbf{w}_1, \mathbf{w}_2, \ldots, \mathbf{w}_n\}$ be an orthonormal basis for $\mathcal{R}^n$. Prove that, for any vectors $\mathbf{u}$ and $\mathbf{v}$ in $\mathcal{R}^n$,
   (a) $\mathbf{u} + \mathbf{v} = (\mathbf{u} \cdot \mathbf{w}_1 + \mathbf{v} \cdot \mathbf{w}_1)\mathbf{w}_1 + \cdots + (\mathbf{u} \cdot \mathbf{w}_n + \mathbf{v} \cdot \mathbf{w}_n)\mathbf{w}_n$.
   (b) $\mathbf{u} \cdot \mathbf{v} = (\mathbf{u} \cdot \mathbf{w}_1)(\mathbf{v} \cdot \mathbf{w}_1) + \cdots + (\mathbf{u} \cdot \mathbf{w}_n)(\mathbf{v} \cdot \mathbf{w}_n)$. (This result is known as *Parseval's identity*.)
   (c) $\|\mathbf{u}\|^2 = (\mathbf{u} \cdot \mathbf{w}_1)^2 + (\mathbf{u} \cdot \mathbf{w}_2)^2 + \cdots + (\mathbf{u} \cdot \mathbf{w}_n)^2$.
56.[7] Suppose that $\{\mathbf{v}_1, \mathbf{v}_2, \ldots, \mathbf{v}_k\}$ is an orthonormal subset of $\mathcal{R}^n$. Combine Theorem 4.4 on page 245 with the Gram–Schmidt process to prove that this set can be extended to an orthonormal basis $\{\mathbf{v}_1, \mathbf{v}_2, \ldots, \mathbf{v}_k, \ldots, \mathbf{v}_n\}$ for $\mathcal{R}^n$.
57. Let $\mathcal{S} = \{\mathbf{v}_1, \mathbf{v}_2, \ldots, \mathbf{v}_k\}$ be an orthonormal subset of $\mathcal{R}^n$, and let $\mathbf{u}$ be a vector in $\mathcal{R}^n$. Use Exercise 56 to prove that
   (a) $(\mathbf{u} \cdot \mathbf{v}_1)^2 + (\mathbf{u} \cdot \mathbf{v}_2)^2 + \cdots + (\mathbf{u} \cdot \mathbf{v}_k)^2 \leq \|\mathbf{u}\|^2$.
   (b) the inequality in (a) is an equality if and only if $\mathbf{u}$ lies in Span $\mathcal{S}$.
58. If Q is an $n \times n$ upper triangular matrix whose columns form an orthonormal basis for $\mathcal{R}^n$, then it can be shown that Q is a diagonal matrix. Prove this in the special case that $n = 3$.

For Exercises 59–65, we assume a knowledge of the QR factorization of a matrix.

59. Show that in a QR factorization of A, the matrix R must be invertible. *Hint:* Use Exercise 77 of Section 4.3.
60. Use the preceding exercise to prove that the matrix R in a QR factorization has nonzero diagonal entries.
61. Use the preceding exercise to prove that the matrix R in a QR factorization can be chosen to have positive diagonal entries.
62. Suppose that A is an $m \times n$ matrix whose columns form an orthonormal set. Identify the matrices Q and R of the QR factorization of A.
63. Let Q be an $n \times n$ matrix. Prove that the columns of Q form an orthonormal basis for $\mathcal{R}^n$ if and only if $Q^TQ = I_n$.
64. Let P and Q be $n \times n$ matrices whose columns form an orthonormal basis for $\mathcal{R}^n$. Prove that the columns of PQ also form an orthonormal basis for $\mathcal{R}^n$. *Hint:* Use Exercise 63.
65. Suppose that A is an invertible $n \times n$ matrix. Let $A = QR$ and $A = Q'R'$ be two QR factorizations in which the diagonal entries of R are positive. It can be shown that $Q = Q'$ and $R = R'$. Prove this in the special case that $n = 3$.
66. Prove that, for any QR factorization of a matrix A, the columns of Q form an orthonormal basis for Col A.

In Exercises 67 and 68, a matrix A is given. Use either a calculator with matrix capabilities or computer software such as MATLAB to solve each problem.

(a) Verify that A has linearly independent columns by computing its rank.

(b) Find matrices Q and R in a QR factorization of A.

(c) Verify that A is approximately equal to the product QR. (Note that roundoff error is likely.)

(d) Verify that Q^TQ is approximately equal to the identity matrix.

67. $A = \begin{bmatrix} 5.3000 & 7.1000 & 8.4000 \\ -4.4000 & 11.0000 & 8.0000 \\ -12.0000 & 13.0000 & 7.0000 \\ 9.0000 & 8.7000 & -6.1000 \\ 2.6000 & -7.4000 & 8.9000 \end{bmatrix}$

68. $A = \begin{bmatrix} 2.0000 & -3.4000 & 5.6000 & 2.6000 \\ 0.0000 & 7.3000 & 5.4000 & 8.2000 \\ 9.0000 & 11.0000 & -5.0000 & 8.0000 \\ -5.3000 & 4.0000 & 5.0000 & 9.0000 \\ -13.0000 & 7.0000 & 8.0000 & 1.0000 \end{bmatrix}$

---

[7] This exercise is used in Section 6.7 (on page 440).

SOLUTIONS TO THE PRACTICE PROBLEMS

1. (a) Since the dot product of the two vectors in $\mathcal{S}_1$ is

$$(1)(2) + (-2)(1) = 0,$$

and these are the only distinct vectors in $\mathcal{S}_1$, we conclude that $\mathcal{S}_1$ is an orthogonal set.

(b) Taking the dot product of the second and third vectors in $\mathcal{S}_2$, we have

$$(1)(2) + (1)(0) + (-1)(-1) = 3 \neq 0,$$

and hence these vectors are not orthogonal. It follows that $\mathcal{S}_2$ is not an orthogonal set.

2. (a) We compute the dot products

$$\begin{bmatrix}1\\1\\2\end{bmatrix} \cdot \begin{bmatrix}1\\1\\-1\end{bmatrix} = (1)(1) + (1)(1) + (2)(-1) = 0$$

$$\begin{bmatrix}1\\1\\2\end{bmatrix} \cdot \begin{bmatrix}1\\-1\\0\end{bmatrix} = (1)(1) + (1)(-1) + (2)(0) = 0$$

$$\begin{bmatrix}1\\1\\-1\end{bmatrix} \cdot \begin{bmatrix}1\\-1\\0\end{bmatrix} = (1)(1) + (1)(-1)(-1)(0) = 0,$$

which proves that $\mathcal{S}$ is an orthogonal set. It follows that $\mathcal{S}$ is linearly independent by Theorem 6.5. Since this set consists of three vectors, it is a basis for $\mathcal{R}^3$.

(b) For $\mathbf{u} = \begin{bmatrix}2\\-4\\7\end{bmatrix}$ and $\mathcal{S} = \{\mathbf{v}_1, \mathbf{v}_2, \mathbf{v}_3\}$, we have

$$\mathbf{u} = c_1\mathbf{v}_1 + c_2\mathbf{v}_2 + c_3\mathbf{v}_3,$$

where

$$c_1 = \frac{\mathbf{u} \cdot \mathbf{v}_1}{\|\mathbf{v}_1\|^2} = 2, \quad c_2 = \frac{\mathbf{u} \cdot \mathbf{v}_2}{\|\mathbf{v}_2\|^2} = -3, \quad \text{and}$$

$$c_3 = \frac{\mathbf{u} \cdot \mathbf{v}_3}{\|\mathbf{v}_3\|^2} = 3.$$

3. We apply the Gram–Schmidt process to the vectors

$$\mathbf{u}_1 = \begin{bmatrix}1\\-1\\-1\\1\end{bmatrix}, \quad \mathbf{u}_2 = \begin{bmatrix}-1\\3\\-3\\5\end{bmatrix}, \quad \text{and}$$

$$\mathbf{u}_3 = \begin{bmatrix}1\\6\\3\\-4\end{bmatrix}$$

to obtain the orthogonal set $\{\mathbf{v}_1, \mathbf{v}_2, \mathbf{v}_3\}$, where

$$\mathbf{v}_1 = \mathbf{u}_1 = \begin{bmatrix}1\\-1\\-1\\1\end{bmatrix},$$

$$\mathbf{v}_2 = \mathbf{u}_2 - \frac{\mathbf{u}_2 \cdot \mathbf{v}_1}{\|\mathbf{v}_1\|^2}\mathbf{v}_1 = \begin{bmatrix}-1\\3\\-3\\5\end{bmatrix} - \frac{4}{4}\begin{bmatrix}1\\-1\\-1\\1\end{bmatrix} = \begin{bmatrix}-2\\4\\-2\\4\end{bmatrix},$$

and

$$\mathbf{v}_3 = \mathbf{u}_3 - \frac{\mathbf{u}_3 \cdot \mathbf{v}_1}{\|\mathbf{v}_1\|^2}\mathbf{v}_1 - \frac{\mathbf{u}_3 \cdot \mathbf{v}_2}{\|\mathbf{v}_2\|^2}\mathbf{v}_2$$

$$= \begin{bmatrix}1\\6\\3\\-4\end{bmatrix} - \frac{(-12)}{4}\begin{bmatrix}1\\-1\\-1\\1\end{bmatrix} - 0\begin{bmatrix}-2\\4\\-2\\4\end{bmatrix} = \begin{bmatrix}4\\3\\0\\-1\end{bmatrix}.$$

4. (a) Using the basis obtained in Practice Problem 3, we have

$$\mathcal{B} = \left\{\frac{1}{\|\mathbf{v}_1\|}\mathbf{v}_1, \frac{1}{\|\mathbf{v}_2\|}\mathbf{v}_2, \frac{1}{\|\mathbf{v}_3\|}\mathbf{v}_3\right\}$$

$$= \left\{\frac{1}{2}\begin{bmatrix}1\\-1\\-1\\1\end{bmatrix}, \frac{1}{\sqrt{10}}\begin{bmatrix}-1\\2\\-1\\2\end{bmatrix}, \frac{1}{\sqrt{26}}\begin{bmatrix}4\\3\\0\\-1\end{bmatrix}\right\}.$$

(b) For each i, let $\mathbf{w}_i = \frac{1}{\|\mathbf{v}_i\|}\mathbf{v}_i$. Then

$$\begin{aligned}\mathbf{u} &= (\mathbf{u} \cdot \mathbf{w}_1)\mathbf{w}_1 + (\mathbf{u} \cdot \mathbf{w}_2)\mathbf{w}_2 + (\mathbf{u} \cdot \mathbf{w}_3)\mathbf{w}_3 \\ &= (-10)\mathbf{w}_1 + (-2\sqrt{10})\mathbf{w}_2 + \sqrt{26}\mathbf{w}_3.\end{aligned}$$

5. Using the result of Practice Problem 4(a), we have

$$\mathbf{w}_1 = \frac{1}{2}\begin{bmatrix}1\\-1\\-1\\1\end{bmatrix}, \quad \mathbf{w}_2 = \frac{1}{\sqrt{10}}\begin{bmatrix}-1\\2\\-1\\2\end{bmatrix}, \quad \text{and}$$

$$\mathbf{w}_3 = \frac{1}{\sqrt{26}}\begin{bmatrix}4\\3\\0\\-1\end{bmatrix}.$$

Thus

$$Q = [\mathbf{w}_1\ \mathbf{w}_2\ \mathbf{w}_3] = \begin{bmatrix} \frac{1}{2} & -\frac{1}{\sqrt{10}} & \frac{4}{\sqrt{26}} \\ -\frac{1}{2} & \frac{2}{\sqrt{10}} & \frac{3}{\sqrt{26}} \\ -\frac{1}{2} & -\frac{1}{\sqrt{10}} & 0 \\ \frac{1}{2} & \frac{2}{\sqrt{10}} & -\frac{1}{\sqrt{26}} \end{bmatrix}.$$

To compute R, observe that

$$r_{11} = \mathbf{a}_1 \cdot \mathbf{w}_1 = 2 \qquad r_{12} = \mathbf{a}_2 \cdot \mathbf{w}_1 = 2$$

$$r_{13} = \mathbf{a}_3 \cdot \mathbf{w}_1 = -6 \qquad r_{22} = \mathbf{a}_2 \cdot \mathbf{w}_2 = 2\sqrt{10}$$

$$r_{23} = \mathbf{a}_3 \cdot \mathbf{w}_2 = 0 \qquad r_{33} = \mathbf{a}_3 \cdot \mathbf{w}_3 = \sqrt{26},$$

and hence

$$R = \begin{bmatrix} 2 & 2 & -6 \\ 0 & 2\sqrt{10} & 0 \\ 0 & 0 & \sqrt{26} \end{bmatrix}.$$

6. Using the Q and R in the solution of Practice Problem 5, we form the equivalent system $R\mathbf{x} = Q^T\mathbf{b}$, or

$$\begin{aligned} 2x_1 + \quad 2x_2 - \quad 6x_3 &= \quad -16 \\ 2\sqrt{10}x_2 \qquad\qquad &= -2\sqrt{10} \\ \sqrt{26}x_3 &= \quad 3\sqrt{26}. \end{aligned}$$

Solving the third and second equations, we obtain $x_3 = 3$ and $x_2 = -1$. Finally, substituting the values of x_3 and x_2 into the first equation and solving for x_1 gives

$$2x_1 + 2(-1) - 6(3) = -16,$$

or $x_1 = 2$. So the solution of the system is

$$\begin{bmatrix} 2 \\ -1 \\ 3 \end{bmatrix}.$$

6.3 ORTHOGONAL PROJECTIONS

Many practical applications require us to approximate a given vector $\mathbf{u}$ in $\mathcal{R}^n$ by a vector in a particular subspace W. In such a situation, we obtain the best possible approximation by choosing the vector $\mathbf{w}$ in W that is closest to $\mathbf{u}$. When W is a plane in $\mathcal{R}^3$, as in Figure 6.9, we find $\mathbf{w}$ by dropping a perpendicular from the endpoint of $\mathbf{u}$ to W. Notice that we can find $\mathbf{w}$ if we can find the vector $\mathbf{z} = \mathbf{u} - \mathbf{w}$, as depicted in Figure 6.9. Because $\mathbf{z}$ is orthogonal to every vector in W, it makes sense to study the set of all vectors that are orthogonal to every vector in a given set. Consider, for example, the set

$$\mathcal{S} = \left\{ \begin{bmatrix} -1 \\ -2 \end{bmatrix}, \begin{bmatrix} 2 \\ 4 \end{bmatrix} \right\}.$$

What vectors are orthogonal to every vector in $\mathcal{S}$?

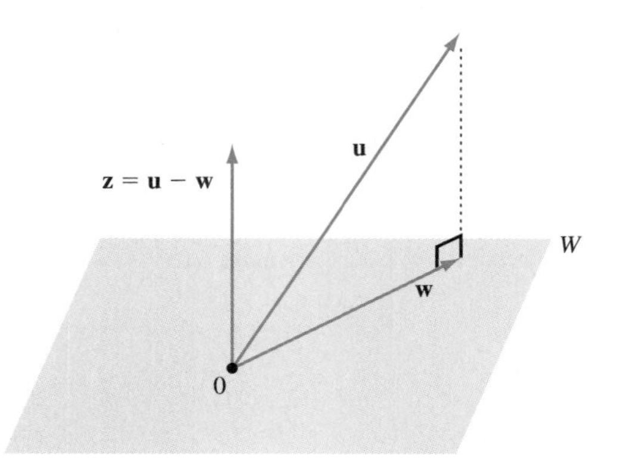

Figure 6.9 The best approximation of **u** by a vector **w** in W

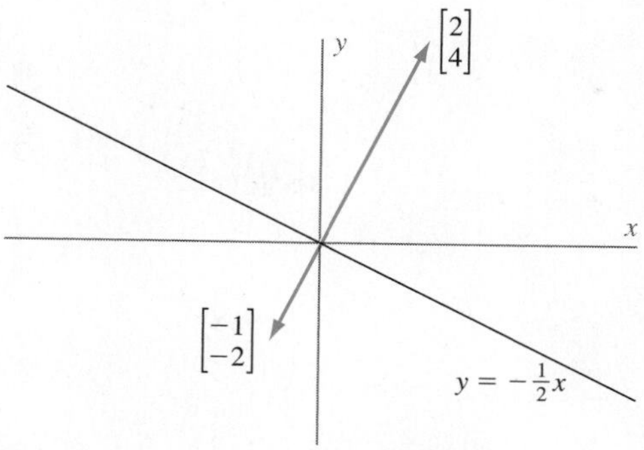

Figure 6.10 The vectors orthogonal to $\begin{bmatrix} 2 \\ 4 \end{bmatrix}$ and $\begin{bmatrix} 1 \\ -2 \end{bmatrix}$ lie along the line $y = -\frac{1}{2}x$.

Note that the two vectors in $\mathcal{S}$ lie along the line with equation $y = 2x$. (See Figure 6.10.) Hence the vectors orthogonal to the vectors in $\mathcal{S}$ lie along the line with equation $y = -\frac{1}{2}x$. In this context, the set of vectors that lie along the line $y = -\frac{1}{2}x$ is called the *orthogonal complement* of $\mathcal{S}$. More generally, we have the following definition:

Definition The **orthogonal complement** of a nonempty subset $\mathcal{S}$ of $\mathcal{R}^n$, denoted by $\mathcal{S}^\perp$ (read "$\mathcal{S}$ perp"), is the set of all vectors in $\mathcal{R}^n$ that are orthogonal to every vector in $\mathcal{S}$. That is,

$$\mathcal{S}^\perp = \{\mathbf{v} \in \mathcal{R}^n : \mathbf{v} \cdot \mathbf{u} = 0 \text{ for every } \mathbf{u} \text{ in } \mathcal{S}\}.$$

For example, if $\mathcal{S} = \mathcal{R}^n$, then $\mathcal{S}^\perp = \{\mathbf{0}\}$; and if $\mathcal{S} = \{\mathbf{0}\}$, then $\mathcal{S}^\perp = \mathcal{R}^n$.

Example 1 Let W denote the xy-plane viewed as a subspace of $\mathcal{R}^3$; that is,

$$W = \left\{ \begin{bmatrix} u_1 \\ u_2 \\ 0 \end{bmatrix} : u_1 \text{ and } u_2 \text{ are real numbers} \right\}.$$

A vector $\mathbf{v} = \begin{bmatrix} v_1 \\ v_2 \\ v_3 \end{bmatrix}$ lies in $W^\perp$ if and only if $v_1 = v_2 = 0$. For if $v_1 = v_2 = 0$, then $\mathbf{v} \cdot \mathbf{u} = 0$ for all $\mathbf{u}$ in W, and hence $\mathbf{v}$ is in $W^\perp$. And if $\mathbf{v}$ is in $W^\perp$, then

$$v_1 = \mathbf{e}_1 \cdot \mathbf{v} = 0$$

because $\mathbf{e}_1$ is in W. Similarly, $v_2 = 0$ because $\mathbf{e}_2$ is in W. Thus

$$W^\perp = \left\{ \begin{bmatrix} 0 \\ 0 \\ v_3 \end{bmatrix} : v_3 \text{ is a real number} \right\},$$

and so $W^\perp$ can be identified with the z-axis.

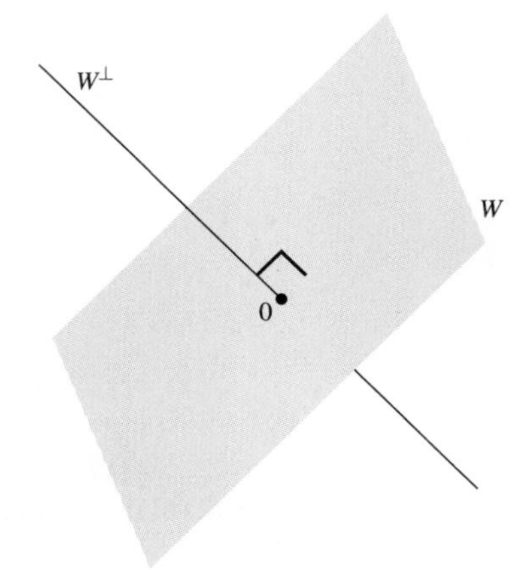

Figure 6.11 The orthogonal complement of a 2-dimensional subspace W of $\mathcal{R}^3$

More generally, in Figure 6.11, we see a 2-dimensional subspace W of $\mathcal{R}^3$, which is a plane containing $\mathbf{0}$. Its orthogonal complement $W^{\perp}$ is a line through $\mathbf{0}$, perpendicular to W.

If $\mathcal{S}$ is any nonempty subset of $\mathcal{R}^n$, then $\mathbf{0}$ is in $\mathcal{S}^{\perp}$ because $\mathbf{0}$ is orthogonal to every vector in $\mathcal{S}$. Moreover, if $\mathbf{v}$ and $\mathbf{w}$ are in $\mathcal{S}^{\perp}$, then, for every vector $\mathbf{u}$ in $\mathcal{S}$,

$$(\mathbf{v}+\mathbf{w})\cdot\mathbf{u} = \mathbf{v}\cdot\mathbf{u} + \mathbf{w}\cdot\mathbf{u} = 0 + 0 = 0,$$

and therefore $\mathbf{v}+\mathbf{w}$ is in $\mathcal{S}^{\perp}$. So $\mathcal{S}^{\perp}$ is closed under vector addition. A similar argument shows that $\mathcal{S}^{\perp}$ is closed under scalar multiplication. Therefore $\mathcal{S}^{\perp}$ is a subspace of $\mathcal{R}^n$. Hence we have the following result:

> The orthogonal complement of any nonempty subset of $\mathcal{R}^n$ is a subspace of $\mathcal{R}^n$.

In Figure 6.10, the span of

$$\mathcal{S} = \left\{ \begin{bmatrix} -1 \\ -2 \end{bmatrix}, \begin{bmatrix} 2 \\ 4 \end{bmatrix} \right\}$$

can be visualized as the line with equation $y = 2x$. We have already seen that $\mathcal{S}^{\perp}$ consists of the vectors that lie along the line $y = -\frac{1}{2}x$. Note that $(\text{Span}\,\mathcal{S})^{\perp}$ also consists of the vectors that lie along the line $y = -\frac{1}{2}x$, and hence $\mathcal{S}^{\perp} = (\text{Span}\,\mathcal{S})^{\perp}$. A similar result is true in general. (See Exercise 57.)

> For any nonempty subset $\mathcal{S}$ of $\mathcal{R}^n$, we have $\mathcal{S}^{\perp} = (\text{Span}\,\mathcal{S})^{\perp}$. In particular, the orthogonal complement of a basis for a subspace is the same as the orthogonal complement of the subspace.

The next example shows how orthogonal complements arise in the study of systems of linear equations.

Example 2 Find a basis for the orthogonal complement of $W = \text{Span}\,\{\mathbf{u}_1, \mathbf{u}_2\}$, where

$$\mathbf{u}_1 = \begin{bmatrix} 1 \\ 1 \\ -1 \\ 4 \end{bmatrix} \quad \text{and} \quad \mathbf{u}_2 = \begin{bmatrix} 1 \\ -1 \\ 1 \\ 2 \end{bmatrix}.$$

Solution A vector $\mathbf{v} = \begin{bmatrix} v_1 \\ v_2 \\ v_3 \\ v_4 \end{bmatrix}$ lies in $W^\perp$ if and only if $\mathbf{u}_1 \cdot \mathbf{v} = 0$ and $\mathbf{u}_2 \cdot \mathbf{v} = 0$. Notice that these two equations can be written as the homogeneous system of linear equations

$$\begin{aligned} v_1 + v_2 - v_3 + 4v_4 &= 0 \\ v_1 - v_2 + v_3 + 2v_4 &= 0. \end{aligned} \qquad (1)$$

From the reduced row echelon form of the augmented matrix, we see that the vector form of the general solution of system (1) is

$$\begin{bmatrix} v_1 \\ v_2 \\ v_3 \\ v_4 \end{bmatrix} = \begin{bmatrix} -3v_4 \\ v_3 - v_4 \\ v_3 \\ v_4 \end{bmatrix} = v_3 \begin{bmatrix} 0 \\ 1 \\ 1 \\ 0 \end{bmatrix} + v_4 \begin{bmatrix} -3 \\ -1 \\ 0 \\ 1 \end{bmatrix}.$$

Thus

$$\mathcal{B} = \left\{ \begin{bmatrix} 0 \\ 1 \\ 1 \\ 0 \end{bmatrix}, \begin{bmatrix} -3 \\ -1 \\ 0 \\ 1 \end{bmatrix} \right\}$$

is a basis for $W^\perp$.

Let A denote the coefficient matrix of system (1). Notice that the vectors $\mathbf{u}_1$ and $\mathbf{u}_2$ in Example 2 are the rows of A, and hence W is the row space of A. Furthermore, the set of solutions of system (1) is the null space of A. It follows that

$$W^\perp = (\text{Row}\, A)^\perp = \text{Null}\, A.$$

This observation is valid for any matrix.

> For any matrix A, the orthogonal complement of the row space of A is the null space of A; that is,
>
> $$(\text{Row}\, A)^\perp = \text{Null}\, A.$$

(See Figure 6.12.)

Applying this result to the matrix A^T, we see that

$$(\text{Col}\, A)^\perp = (\text{Row}\, A^T)^\perp = \text{Null}\, A^T.$$

Practice Problem 1 ▶ Let $W = \text{Span}\,\{\mathbf{u}_1, \mathbf{u}_2\}$, where

$$\mathbf{u}_1 = \begin{bmatrix} 1 \\ 0 \\ -1 \\ 1 \end{bmatrix} \quad \text{and} \quad \mathbf{u}_2 = \begin{bmatrix} -1 \\ 1 \\ 3 \\ -4 \end{bmatrix}.$$

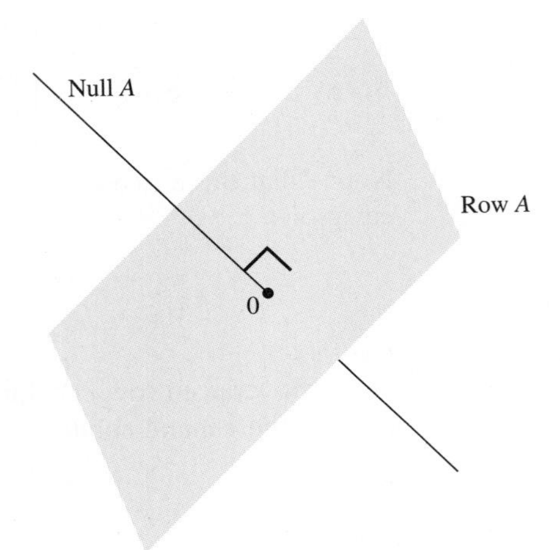

Figure 6.12 The null space of A is the orthogonal complement of the row space of A.

Find a basis for $W^\perp$ by first determining the matrix A such that $W = \text{Row}\, A$, and then using the result that $W^\perp = \text{Null}\, A$. ◀

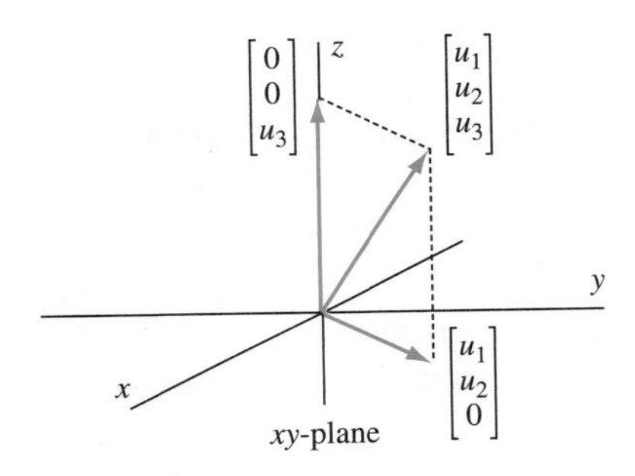

Figure 6.13 Every vector in $\mathcal{R}^3$ is the sum of vectors in the xy-plane and its orthogonal complement.

In the notation of Example 1, any vector $\begin{bmatrix} u_1 \\ u_2 \\ u_3 \end{bmatrix}$ in $\mathcal{R}^3$ can be written as the sum of the vector $\begin{bmatrix} u_1 \\ u_2 \\ 0 \end{bmatrix}$ in W and the vector $\begin{bmatrix} 0 \\ 0 \\ u_3 \end{bmatrix}$ in $W^\perp$. So, in some sense, a vector in $\mathcal{R}^3$ is subdivided into two pieces, one in W and the other in $W^\perp$. (See Figure 6.13.)

The next theorem tells us that this is true in general.

THEOREM 6.7

(Orthogonal Decomposition Theorem) Let W be a subspace of $\mathcal{R}^n$. Then, for any vector $\mathbf{u}$ in $\mathcal{R}^n$, there exist unique vectors $\mathbf{w}$ in W and $\mathbf{z}$ in $W^\perp$ such that $\mathbf{u} = \mathbf{w} + \mathbf{z}$. In addition, if $\{\mathbf{v}_1, \mathbf{v}_2, \ldots, \mathbf{v}_k\}$ is an orthonormal basis for W, then

$$\mathbf{w} = (\mathbf{u} \cdot \mathbf{v}_1)\mathbf{v}_1 + (\mathbf{u} \cdot \mathbf{v}_2)\mathbf{v}_2 + \cdots + (\mathbf{u} \cdot \mathbf{v}_k)\mathbf{v}_k.$$

PROOF Consider any vector $\mathbf{u}$ in $\mathcal{R}^n$. Choose $\mathcal{B} = \{\mathbf{v}_1, \mathbf{v}_2, \ldots, \mathbf{v}_k\}$ to be an orthonormal basis for W, and let

$$\mathbf{w} = (\mathbf{u} \cdot \mathbf{v}_1)\mathbf{v}_1 + (\mathbf{u} \cdot \mathbf{v}_2)\mathbf{v}_2 + \cdots + (\mathbf{u} \cdot \mathbf{v}_k)\mathbf{v}_k. \tag{2}$$

Then $\mathbf{w}$ is in W because it is a linear combination of the basis vectors for W. (Notice that equation (2) resembles the equation given in the representation of a vector in terms of an orthonormal basis. Indeed, $\mathbf{u} = \mathbf{w}$ if and only if $\mathbf{u}$ is in W.)

Let $\mathbf{z} = \mathbf{u} - \mathbf{w}$. Then clearly, $\mathbf{u} = \mathbf{w} + \mathbf{z}$. We show that $\mathbf{z}$ is in $W^\perp$. It suffices to show that $\mathbf{z}$ is orthogonal to every vector in $\mathcal{B}$. From equation (2) and the computation on page 376, we see that $\mathbf{w} \cdot \mathbf{v}_i = \mathbf{u} \cdot \mathbf{v}_i$ for any $\mathbf{v}_i$ in $\mathcal{B}$. Therefore

$$\mathbf{z} \cdot \mathbf{v}_i = (\mathbf{u} - \mathbf{w}) \cdot \mathbf{v}_i = \mathbf{u} \cdot \mathbf{v}_i - \mathbf{w} \cdot \mathbf{v}_i = \mathbf{u} \cdot \mathbf{v}_i - \mathbf{u} \cdot \mathbf{v}_i = 0.$$

Hence $\mathbf{z}$ is in $W^\perp$.

Next, we show that this representation is unique. Suppose that $\mathbf{u} = \mathbf{w}' + \mathbf{z}'$, where $\mathbf{w}'$ is in W and $\mathbf{z}'$ is in $W^\perp$. Then $\mathbf{w} + \mathbf{z} = \mathbf{w}' + \mathbf{z}'$, and hence $\mathbf{w} - \mathbf{w}' = \mathbf{z}' - \mathbf{z}$. But $\mathbf{w} - \mathbf{w}'$ is in W, and $\mathbf{z}' - \mathbf{z}$ is in $W^\perp$. Thus $\mathbf{w} - \mathbf{w}'$ lies in both W and $W^\perp$. This means that $\mathbf{w} - \mathbf{w}'$ is orthogonal to itself. But, by Theorem 6.1(b), $\mathbf{0}$ is the only vector with this property. Hence $\mathbf{w} - \mathbf{w}' = \mathbf{0}$, and therefore $\mathbf{w} = \mathbf{w}'$. It follows that $\mathbf{z} = \mathbf{z}'$, and thus we conclude that the representation is unique. ■

Suppose that we combine a basis for W with a basis for $W^\perp$. Using the orthogonal decomposition theorem, we can show that the resulting set is a basis for $\mathcal{R}^n$. One simple consequence of this observation is the following useful result:

> For any subspace W of $\mathcal{R}^n$,
>
> $$\dim W + \dim W^\perp = n.$$

ORTHOGONAL PROJECTIONS ON SUBSPACES

The orthogonal decomposition theorem gives us a computational method for representing a given vector as the sum of a vector in a subspace and a vector in the orthogonal complement of the subspace.

Definitions Let W be a subspace of $\mathcal{R}^n$ and $\mathbf{u}$ be a vector in $\mathcal{R}^n$. The **orthogonal projection of u on W** is the unique vector $\mathbf{w}$ in W such that $\mathbf{u} - \mathbf{w}$ is in $W^\perp$.

Furthermore, the function $U_W\colon \mathcal{R}^n \to \mathcal{R}^n$ such that $U_W(\mathbf{u})$ is the orthogonal projection of $\mathbf{u}$ on W for every $\mathbf{u}$ in $\mathcal{R}^n$ is called the **orthogonal projection operator** on W.

In the case that $n = 3$ and W is a 2-dimensional subspace of $\mathcal{R}^3$, the orthogonal projection operator U_W defined here coincides with the orthogonal projection as discussed in the exercises of Sections 4.5 and 5.4. In fact, for any vector $\mathbf{u}$ in $\mathcal{R}^3$ that is not in W, the vector $\mathbf{u} - U_W(\mathbf{u})$ is orthogonal to W, and hence the line segment connecting the endpoint of $\mathbf{u}$ to the endpoint of $U_W(\mathbf{u})$ is perpendicular to W and has length equal to $\|\mathbf{u} - U_W(\mathbf{u})\|$. (See Figure 6.14.)

Similarly, in the case that $n = 2$ and W is a line through the origin (a 1-dimensional subspace), U_W coincides with the orthogonal projection of $\mathcal{R}^2$ on W, as defined on page 283.

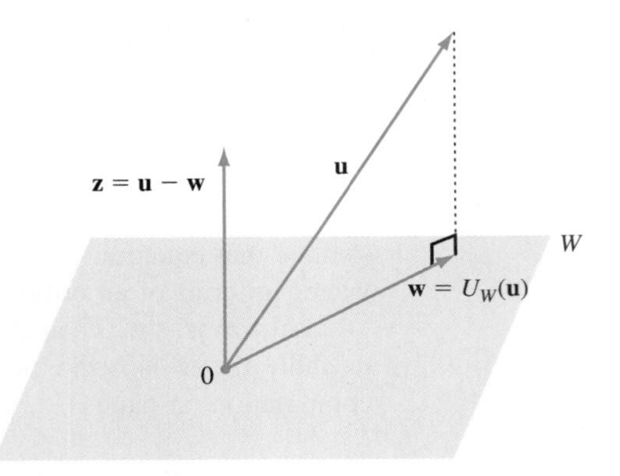

Figure 6.14 The vector **w** is the orthogonal projection of **u** on W.

We now show that any orthogonal projection U_W of $\mathcal{R}^n$ is linear. Let $\mathbf{u}_1$ and $\mathbf{u}_2$ be vectors in $\mathcal{R}^n$, and suppose that $U_W(\mathbf{u}_1) = \mathbf{w}_1$ and $U_W(\mathbf{u}_2) = \mathbf{w}_2$. Then there are unique vectors $\mathbf{z}_1$ and $\mathbf{z}_2$ in $W^\perp$ such that $\mathbf{u}_1 = \mathbf{w}_1 + \mathbf{z}_1$ and $\mathbf{u}_2 = \mathbf{w}_2 + \mathbf{z}_2$. Thus

$$\mathbf{u}_1 + \mathbf{u}_2 = (\mathbf{w}_1 + \mathbf{w}_2) + (\mathbf{z}_1 + \mathbf{z}_2).$$

Since $\mathbf{w}_1 + \mathbf{w}_2$ is in W and $\mathbf{z}_1 + \mathbf{z}_2$ is in $W^\perp$, it follows that

$$U_W(\mathbf{u}_1 + \mathbf{u}_2) = \mathbf{w}_1 + \mathbf{w}_2 = U_W(\mathbf{u}_1) + U_W(\mathbf{u}_2).$$

Hence U_W preserves vector addition. Similarly, U_W preserves scalar multiplication, and therefore U_W is linear.

Example 3

Find the orthogonal projection $\mathbf{w} = U_W(\mathbf{u})$ of $\mathbf{u} = \begin{bmatrix} 1 \\ 3 \\ 4 \end{bmatrix}$ on the 2-dimensional subspace W of $\mathcal{R}^3$ defined by

$$x_1 - x_2 + 2x_3 = 0.$$

Then find the vector $\mathbf{z}$ in $W^\perp$ such that $\mathbf{u} = \mathbf{w} + \mathbf{z}$.

Solution First, observe that

$$\mathcal{B} = \{\mathbf{v}_1, \mathbf{v}_2\} = \left\{ \frac{1}{\sqrt{2}} \begin{bmatrix} 1 \\ 1 \\ 0 \end{bmatrix}, \frac{1}{\sqrt{3}} \begin{bmatrix} -1 \\ 1 \\ 1 \end{bmatrix} \right\}$$

is an orthonormal basis for W. (An orthonormal basis such as $\mathcal{B}$ can be obtained by applying the Gram–Schmidt process to an ordinary basis for W.) We use $\mathcal{B}$ to find $\mathbf{w}$, as in the proof of the orthogonal decomposition theorem. By equation (2), we have

$$\begin{aligned} \mathbf{w} = U_W(\mathbf{u}) &= (\mathbf{u} \cdot \mathbf{v}_1)\mathbf{v}_1 + (\mathbf{u} \cdot \mathbf{v}_2)\mathbf{v}_2 \\ &= \frac{4}{\sqrt{2}}\mathbf{v}_1 + \frac{6}{\sqrt{3}}\mathbf{v}_2 \end{aligned}$$

$$= 2\begin{bmatrix}1\\1\\0\end{bmatrix} + 2\begin{bmatrix}-1\\1\\1\end{bmatrix}$$

$$= \begin{bmatrix}0\\4\\2\end{bmatrix}.$$

Therefore

$$\mathbf{z} = \mathbf{u} - \mathbf{w} = \begin{bmatrix}1\\3\\4\end{bmatrix} - \begin{bmatrix}0\\4\\2\end{bmatrix} = \begin{bmatrix}1\\-1\\2\end{bmatrix}.$$

Note that $\mathbf{z}$ is orthogonal to $\mathbf{v}_1$ and $\mathbf{v}_2$, confirming that $\mathbf{z}$ is in $W^\perp$.

The standard matrix of an orthogonal projection operator U_W on a subspace W of $\mathcal{R}^n$ is called the **orthogonal projection matrix** for W and is denoted P_W. The columns of an orthogonal projection matrix P_W are the images of the standard vectors under U_W—that is, the orthogonal projections of the standard vectors—which can be computed by the method of Example 3. However, Theorem 6.8 gives an alternative method for computing P_W that does not require an orthonormal basis for W. Its proof uses the following result:

Lemma Let C be a matrix whose columns are linearly independent. Then C^TC is invertible.

PROOF Suppose that $C^TC\mathbf{b} = \mathbf{0}$. Recall from Section 6.1 that the dot product of two vectors $\mathbf{u}$ and $\mathbf{v}$ in $\mathcal{R}^n$ can be represented as the matrix product $\mathbf{u} \cdot \mathbf{v} = \mathbf{u}^T\mathbf{v}$. Therefore

$$\|C\mathbf{b}\|^2 = (C\mathbf{b}) \cdot (C\mathbf{b}) = (C\mathbf{b})^T C\mathbf{b} = \mathbf{b}^T C^T C\mathbf{b} = \mathbf{b}^T(C^TC\mathbf{b}) = \mathbf{b}^T\mathbf{0} = 0,$$

and so $C\mathbf{b} = \mathbf{0}$. Since the columns of C are linearly independent, it follows that $\mathbf{b} = \mathbf{0}$. Thus $\mathbf{0}$ is the only solution of $C^TC\mathbf{x} = \mathbf{0}$, and hence C^TC is invertible by the Invertible Matrix Theorem on page 138. ■

THEOREM 6.8

Let C be an $n \times k$ matrix whose columns form a basis for a subspace W of $\mathcal{R}^n$. Then

$$P_W = C(C^TC)^{-1}C^T.$$

PROOF Let $\mathbf{u}$ be any vector in $\mathcal{R}^n$, and let $\mathbf{w} = U_W(\mathbf{u})$, the orthogonal projection of $\mathbf{u}$ on W. Since $W = \operatorname{Col} C$, we have $\mathbf{w} = C\mathbf{v}$ for some $\mathbf{v}$ in $\mathcal{R}^k$. Consequently, $\mathbf{u} - \mathbf{w}$ is in

$$W^\perp = (\operatorname{Col} C)^\perp = (\operatorname{Row} C^T)^\perp = \operatorname{Null} C^T.$$

Hence

$$\mathbf{0} = C^T(\mathbf{u} - \mathbf{w}) = C^T\mathbf{u} - C^T\mathbf{w} = C^T\mathbf{u} - C^T C\mathbf{v}.$$

Thus

$$C^T C\mathbf{v} = C^T\mathbf{u}.$$

By the lemma, $C^T C$ is invertible, and hence $\mathbf{v} = (C^T C)^{-1} C^T\mathbf{u}$. Therefore the orthogonal projection of $\mathbf{u}$ on W is

$$U_W(\mathbf{u}) = \mathbf{w} = C\mathbf{v} = C(C^T C)^{-1} C^T\mathbf{u}.$$

Since this is true for every vector $\mathbf{u}$ in $\mathcal{R}^n$, it follows that $C(C^T C)^{-1} C^T$ is the standard matrix of U_W. That is, $P_W = C(C^T C)^{-1} C^T$. ■

Example 4 Find P_W, where W is the 2-dimensional subspace of $\mathcal{R}^3$ with equation

$$x_1 - x_2 + 2x_3 = 0.$$

Solution Observe that a vector $\mathbf{w}$ is in W if and only

$$\mathbf{w} = \begin{bmatrix} x_1 \\ x_2 \\ x_3 \end{bmatrix} = \begin{bmatrix} x_2 - 2x_3 \\ x_2 \\ x_3 \end{bmatrix} = x_2 \begin{bmatrix} 1 \\ 1 \\ 0 \end{bmatrix} + x_3 \begin{bmatrix} -2 \\ 0 \\ 1 \end{bmatrix},$$

and hence

$$\left\{ \begin{bmatrix} 1 \\ 1 \\ 0 \end{bmatrix}, \begin{bmatrix} -2 \\ 0 \\ 1 \end{bmatrix} \right\}$$

is a basis for W. Let

$$C = \begin{bmatrix} 1 & -2 \\ 1 & 0 \\ 0 & 1 \end{bmatrix},$$

the matrix whose columns are the basis vectors just computed. Then

$$P_W = C(C^T C)^{-1} C^T = \frac{1}{6} \begin{bmatrix} 5 & 1 & -2 \\ 1 & 5 & 2 \\ -2 & 2 & 2 \end{bmatrix}.$$

We can use the orthogonal projection matrix P_W computed in Example 4 to find the orthogonal projection of the vector $\mathbf{u} = \begin{bmatrix} 1 \\ 3 \\ 4 \end{bmatrix}$ of Example 3:

$$\mathbf{w} = P_W\mathbf{u} = \frac{1}{6} \begin{bmatrix} 5 & 1 & -2 \\ 1 & 5 & 2 \\ -2 & 2 & 2 \end{bmatrix} \begin{bmatrix} 1 \\ 3 \\ 4 \end{bmatrix} = \begin{bmatrix} 0 \\ 4 \\ 2 \end{bmatrix}$$

This calculation agrees with the result computed in that example. Note that, unlike Example 3, this calculation does not require the use of the Gram–Schmidt process to find an orthonormal basis for W.

Figure 6.14 suggests that, for a vector $\mathbf{u}$ in $\mathcal{R}^3$ and a 2-dimensional subspace W of $\mathcal{R}^3$, the orthogonal projection $U_W(\mathbf{u})$ is the vector in W that is closest to $\mathbf{u}$. We now show that this statement is true in general.

Let W be a subspace of $\mathcal{R}^n$, $\mathbf{w} = U_W(\mathbf{u})$, and $\mathbf{w}'$ be any vector in W. Since $\mathbf{u} - \mathbf{w}$ is in $W^\perp$, it is orthogonal to $\mathbf{w} - \mathbf{w}'$, which lies in W. Thus, by the Pythagorean theorem in $\mathcal{R}^n$ (Theorem 6.2),

$$\|\mathbf{u} - \mathbf{w}'\|^2 = \|(\mathbf{u} - \mathbf{w}) + (\mathbf{w} - \mathbf{w}')\|^2 = \|\mathbf{u} - \mathbf{w}\|^2 + \|\mathbf{w} - \mathbf{w}'\|^2 \geq \|\mathbf{u} - \mathbf{w}\|^2.$$

Moreover, the final inequality is a strict inequality if $\mathbf{w} \neq \mathbf{w}'$.

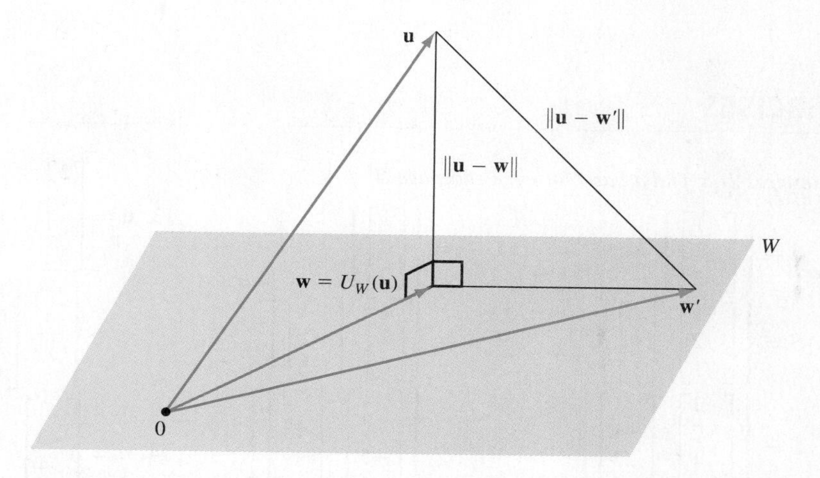

Figure 6.15 The vector $U_W(\mathbf{u})$ is the vector in W that is closest to $\mathbf{u}$.

Figure 6.15 gives us a visual understanding of this inequality. In this figure, notice that the line segment connecting the endpoint of $\mathbf{u}$ to the endpoint of $\mathbf{w}$, which has length $\|\mathbf{u} - \mathbf{w}\|$, is the leg of a right triangle whose hypotenuse is the segment connecting the endpoint of $\mathbf{u}$ to the endpoint of $\mathbf{w}'$. Furthermore, the length of this hypotenuse is $\|\mathbf{u} - \mathbf{w}'\|$. Since the length of the hypotenuse of a right triangle is greater than the length of either of its legs, we see that $\|\mathbf{u} - \mathbf{w}'\| > \|\mathbf{u} - \mathbf{w}\|$.

We now state this important result:

Closest Vector Property

Let W be a subspace of $\mathcal{R}^n$ and $\mathbf{u}$ be a vector in $\mathcal{R}^n$. Among all vectors in W, the vector closest to $\mathbf{u}$ is the orthogonal projection $U_W(\mathbf{u})$ of $\mathbf{u}$ on W.

We define the **distance from a vector u in $\mathcal{R}^n$ to a subspace W of $\mathcal{R}^n$** to be the distance between $\mathbf{u}$ and the orthogonal projection of $\mathbf{u}$ on W. So the distance between $\mathbf{u}$ and W is the minimum distance between $\mathbf{u}$ and every vector in W.

In the context of Example 3, the distance between $\mathbf{u}$ and W is

$$\|\mathbf{u} - \mathbf{w}\| = \|\mathbf{z}\| = \left\| \begin{bmatrix} 1 \\ -1 \\ 2 \end{bmatrix} \right\| = \sqrt{6}.$$

Practice Problem 2 ▶ Let

$$W = \text{Span}\left\{\begin{bmatrix}1\\1\\-1\\1\end{bmatrix}, \begin{bmatrix}3\\2\\-1\\0\end{bmatrix}\right\} \quad \text{and} \quad \mathbf{u} = \begin{bmatrix}0\\7\\4\\7\end{bmatrix}.$$

(a) Use the method in Example 3 to find the vectors $\mathbf{w}$ in W and $\mathbf{z}$ in $W^\perp$ such that $\mathbf{u} = \mathbf{w} + \mathbf{z}$.

(b) Find the orthogonal projection matrix P_W, and then use it to find the orthogonal projection of $\mathbf{u}$ on W.

(c) Find the distance from $\mathbf{u}$ to W. ◀

EXERCISES

In Exercises 1–8, find a basis for each subspace $\mathcal{S}^\perp$.

1. $\mathcal{S} = \left\{\begin{bmatrix}1\\-1\\2\end{bmatrix}\right\}$

2. $\mathcal{S} = \left\{\begin{bmatrix}1\\0\\2\end{bmatrix}\right\}$

3. $\mathcal{S} = \left\{\begin{bmatrix}-1\\2\\1\end{bmatrix}, \begin{bmatrix}2\\1\\3\end{bmatrix}\right\}$

4. $\mathcal{S} = \left\{\begin{bmatrix}1\\1\\1\end{bmatrix}, \begin{bmatrix}1\\-1\\-1\end{bmatrix}\right\}$

5. $\mathcal{S} = \left\{\begin{bmatrix}1\\-2\\1\\1\end{bmatrix}, \begin{bmatrix}1\\-1\\3\\2\end{bmatrix}\right\}$

6. $\mathcal{S} = \left\{\begin{bmatrix}1\\-1\\-5\\-1\end{bmatrix}, \begin{bmatrix}2\\-1\\-7\\0\end{bmatrix}\right\}$

7. $\mathcal{S} = \left\{\begin{bmatrix}1\\-1\\-3\\4\end{bmatrix}, \begin{bmatrix}2\\-1\\-4\\7\end{bmatrix}\right\}$

8. $\mathcal{S} = \left\{\begin{bmatrix}1\\-1\\1\\1\end{bmatrix}, \begin{bmatrix}1\\1\\-1\\1\end{bmatrix}, \begin{bmatrix}1\\1\\1\\-1\end{bmatrix}\right\}$

In Exercises 9–16, a vector $\mathbf{u}$ in $\mathcal{R}^n$ and an orthonormal basis $\mathcal{S}$ for a subspace W of $\mathcal{R}^n$ are given.

(a) Use the method in Example 3 to obtain the unique vectors $\mathbf{w}$ in W and $\mathbf{z}$ in $W^\perp$ such that $\mathbf{u} = \mathbf{w} + \mathbf{z}$.

(b) Find the orthogonal projection of $\mathbf{u}$ on W.

(c) Find the distance from $\mathbf{u}$ to W.

9. $\mathbf{u} = \begin{bmatrix}1\\3\end{bmatrix}$ and $\mathcal{S} = \left\{\frac{1}{\sqrt{2}}\begin{bmatrix}1\\-1\end{bmatrix}\right\}$

10. $\mathbf{u} = \begin{bmatrix}2\\3\\-1\end{bmatrix}$ and $\mathcal{S} = \left\{\frac{1}{\sqrt{2}}\begin{bmatrix}1\\1\\0\end{bmatrix}, \frac{1}{\sqrt{3}}\begin{bmatrix}1\\-1\\1\end{bmatrix}\right\}$

11. $\mathbf{u} = \begin{bmatrix}1\\4\\-1\end{bmatrix}$ and $\mathcal{S} = \left\{\frac{1}{\sqrt{6}}\begin{bmatrix}-1\\2\\1\end{bmatrix}, \frac{1}{\sqrt{3}}\begin{bmatrix}1\\1\\-1\end{bmatrix}\right\}$

12. $\mathbf{u} = \begin{bmatrix}3\\1\\1\end{bmatrix}$ and $\mathcal{S} = \left\{\frac{1}{3}\begin{bmatrix}2\\-1\\-2\end{bmatrix}, \frac{1}{3}\begin{bmatrix}1\\-2\\2\end{bmatrix}\right\}$

13. $\mathbf{u} = \begin{bmatrix}2\\4\\1\\3\end{bmatrix}$ and

$$\mathcal{S} = \left\{\frac{1}{\sqrt{3}}\begin{bmatrix}1\\0\\1\\1\end{bmatrix}, \frac{1}{\sqrt{3}}\begin{bmatrix}1\\1\\0\\-1\end{bmatrix}, \frac{1}{\sqrt{3}}\begin{bmatrix}-1\\1\\1\\0\end{bmatrix}\right\}$$

14. $\mathbf{u} = \begin{bmatrix}3\\-2\\4\\1\end{bmatrix}$ and $\mathcal{S} = \left\{\frac{1}{2}\begin{bmatrix}1\\-1\\1\\-1\end{bmatrix}, \frac{1}{2}\begin{bmatrix}-1\\-1\\1\\1\end{bmatrix}\right\}$

15. $\mathbf{u} = \begin{bmatrix}0\\5\\-3\\4\end{bmatrix}$ and $\mathcal{S} = \left\{\frac{1}{\sqrt{6}}\begin{bmatrix}1\\0\\-2\\1\end{bmatrix}, \frac{1}{\sqrt{12}}\begin{bmatrix}1\\3\\1\\1\end{bmatrix}\right\}$

16. $\mathbf{u} = \begin{bmatrix}3\\-1\\-1\\7\end{bmatrix}$, and

$$\mathcal{S} = \left\{\frac{1}{\sqrt{10}}\begin{bmatrix}1\\-1\\-2\\2\end{bmatrix}, \frac{1}{\sqrt{10}}\begin{bmatrix}2\\-2\\1\\-1\end{bmatrix}, \frac{1}{2}\begin{bmatrix}1\\1\\-1\\-1\end{bmatrix}\right\}$$

In Exercises 17–32, a vector $\mathbf{u}$ in $\mathcal{R}^n$ and a subspace W of $\mathcal{R}^n$ are given.

(a) Find the orthogonal projection matrix P_W.

(b) Use your result to obtain the unique vectors $\mathbf{w}$ in W and $\mathbf{z}$ in $W^\perp$ such that $\mathbf{u} = \mathbf{w} + \mathbf{z}$.

(c) Find the distance from $\mathbf{u}$ to W.

17. $\mathbf{u} = \begin{bmatrix}-10\\5\end{bmatrix}$ and $W = \text{Span}\left\{\begin{bmatrix}-3\\4\end{bmatrix}\right\}$

18. $\mathbf{u} = \begin{bmatrix}1\\3\\7\end{bmatrix}$ and W is the solution set of the equation

$$x_1 - 2x_2 + 3x_3 = 0.$$

19. $\mathbf{u} = \begin{bmatrix} 1 \\ 2 \\ -1 \end{bmatrix}$ and W is the solution set of the system of equations

$$\begin{aligned} x_1 + x_2 - x_3 &= 0 \\ x_1 - x_2 + 3x_3 &= 0. \end{aligned}$$

20. $\mathbf{u} = \begin{bmatrix} -6 \\ 4 \\ 5 \end{bmatrix}$ and $W = \text{Col} \begin{bmatrix} 1 & 3 \\ -1 & 1 \\ 2 & 5 \end{bmatrix}$

21. $\mathbf{u} = \begin{bmatrix} 1 \\ 1 \\ 2 \\ 6 \end{bmatrix}$ and $W = \text{Col} \begin{bmatrix} 1 & 1 & 5 \\ -1 & 2 & 1 \\ -1 & 1 & -1 \\ 2 & -1 & 4 \end{bmatrix}$

22. $\mathbf{u} = \begin{bmatrix} -3 \\ 7 \\ -1 \\ 5 \end{bmatrix}$ and W is the solution set of

$$x_1 - x_2 + 2x_3 + x_4 = 0.$$

23. $\mathbf{u} = \begin{bmatrix} 2 \\ 0 \\ -3 \\ 5 \end{bmatrix}$ and $W = \text{Col} \begin{bmatrix} 1 & 1 & 1 \\ -1 & -3 & -7 \\ 0 & -1 & -2 \\ 2 & 1 & 2 \end{bmatrix}$

24. $\mathbf{u} = \begin{bmatrix} 7 \\ 4 \\ 1 \\ 2 \end{bmatrix}$ and $W = \text{Span} \left\{ \begin{bmatrix} 1 \\ 2 \\ 1 \\ -1 \end{bmatrix}, \begin{bmatrix} 1 \\ 3 \\ 2 \\ 2 \end{bmatrix} \right\}$

25. $\mathbf{u} = \begin{bmatrix} 3 \\ 1 \\ -1 \end{bmatrix}$ and W is the solution set of

$$x_1 + 2x_2 - x_3 = 0.$$

26. $\mathbf{u} = \begin{bmatrix} 1 \\ 3 \\ -2 \end{bmatrix}$ and W is the solution set of

$$\begin{aligned} x_1 + 2x_2 - 3x_3 &= 0 \\ x_1 + x_2 - 3x_3 &= 0. \end{aligned}$$

27. $\mathbf{u} = \begin{bmatrix} 8 \\ 0 \\ 2 \end{bmatrix}$ and W is the solution set of

$$\begin{aligned} x_1 + x_2 - x_3 &= 0 \\ x_1 + 2x_2 + 3x_3 &= 0. \end{aligned}$$

28. $\mathbf{u} = \begin{bmatrix} 1 \\ 3 \\ -3 \\ 1 \end{bmatrix}$ and W is the solution set of

$$x_1 + x_2 - x_3 - x_4 = 0.$$

29. $\mathbf{u} = \begin{bmatrix} 1 \\ 5 \\ 1 \\ -1 \end{bmatrix}$ and W is the solution set of

$$\begin{aligned} x_1 + x_2 - x_3 + x_4 &= 0 \\ x_1 - x_2 + 3x_3 + x_4 &= 0. \end{aligned}$$

30. $\mathbf{u} = \begin{bmatrix} 1 \\ 1 \\ -5 \\ 1 \end{bmatrix}$ and $W = \text{Null} \begin{bmatrix} 2 & -2 & 3 & 4 \\ 1 & -1 & 1 & 1 \end{bmatrix}$

31. $\mathbf{u} = \begin{bmatrix} 2 \\ 3 \end{bmatrix}$ and $W = \text{Col} \begin{bmatrix} 2 & -2 & 3 & 4 \\ 1 & -1 & 1 & 1 \end{bmatrix}$

32. $\mathbf{u} = \begin{bmatrix} 4 \\ 1 \\ 3 \\ -1 \end{bmatrix}$ and $W = \text{Row} \begin{bmatrix} 2 & -2 & 3 & 4 \\ 1 & -1 & 1 & 1 \end{bmatrix}$

In Exercises 33–56, determine whether the statements are true or false.

33. For any nonempty subset $\mathcal{S}$ of $\mathcal{R}^n$, $(\mathcal{S}^\perp)^\perp = \mathcal{S}$.
34. If F and G are subsets of $\mathcal{R}^n$ and $F^\perp = G^\perp$, then $F = G$.
35. The orthogonal complement of any nonempty subset of $\mathcal{R}^n$ is a subspace of $\mathcal{R}^n$.
36. For any matrix A, $(\text{Col}\,A)^\perp = \text{Null}\,A$.
37. For any matrix A, $(\text{Null}\,A)^\perp = \text{Row}\,A$.
38. Let W be a subspace of $\mathcal{R}^n$. If $\{\mathbf{w}_1, \mathbf{w}_2, \ldots, \mathbf{w}_k\}$ is an orthonormal basis for W and $\{\mathbf{z}_1, \mathbf{z}_2, \ldots, \mathbf{z}_m\}$ is an orthonormal basis for $W^\perp$, then
$$\{\mathbf{w}_1, \mathbf{w}_2, \ldots, \mathbf{w}_k, \mathbf{z}_1, \mathbf{z}_2, \ldots, \mathbf{z}_m\}$$
is an orthonormal basis for $\mathcal{R}^n$.
39. For any subspace W of $\mathcal{R}^n$, the only vector in both W and $W^\perp$ is $\mathbf{0}$.
40. For any subspace W of $\mathcal{R}^n$ and any vector $\mathbf{u}$ in $\mathcal{R}^n$, there is a unique vector in W that is closest to $\mathbf{u}$.
41. Let W be a subspace of $\mathcal{R}^n$, $\mathbf{u}$ be any vector in $\mathcal{R}^n$, and $\mathbf{w}$ be the orthogonal projection of $\mathbf{u}$ on W. Then $\mathbf{u} - \mathbf{w}$ is in $W^\perp$.
42. For any subspace W of $\mathcal{R}^n$, $\dim W = \dim W^\perp$.
43. If $\{\mathbf{w}_1, \mathbf{w}_2, \ldots, \mathbf{w}_k\}$ is a basis for W and $\mathbf{u}$ is a vector in $\mathcal{R}^n$, then the orthogonal projection of $\mathbf{u}$ on W is $(\mathbf{u} \cdot \mathbf{w}_1)\mathbf{w}_1 + (\mathbf{u} \cdot \mathbf{w}_2)\mathbf{w}_2 + \cdots + (\mathbf{u} \cdot \mathbf{w}_k)\mathbf{w}_k$.
44. For any subspace W of $\mathcal{R}^n$ and any vector $\mathbf{v}$ in $\mathcal{R}^n$, the distance from $\mathbf{v}$ to W equals $\|\mathbf{v} - \mathbf{w}\|$, where $\mathbf{w}$ is the orthogonal projection of $\mathbf{v}$ on W.
45. If $\mathbf{u}$ is in $\mathcal{R}^n$ and W is a subspace of $\mathcal{R}^n$, then $P_W\mathbf{u}$ is the vector in W that is closest to $\mathbf{u}$.
46. Every orthogonal projection matrix is invertible.
47. If W is a subspace of $\mathcal{R}^n$, then, for any vector $\mathbf{u}$ in $\mathcal{R}^n$, the vector $\mathbf{u} - P_W\mathbf{u}$ is orthogonal to every vector in W.
48. In order for $C(C^TC)^{-1}C^T$ to equal the orthogonal projection matrix for a subspace W, the columns of C must form an orthonormal basis for W.
49. If C is a matrix whose columns form a generating set for a subspace W of $\mathcal{R}^n$ and $\mathbf{u}$ is a vector in $\mathcal{R}^n$, then the orthogonal projection of $\mathbf{u}$ on W is $C(C^TC)^{-1}C^T\mathbf{u}$.
50. If C is a matrix whose columns form a basis for a subspace W of $\mathcal{R}^n$, then the matrix C^TC is invertible.

51. If C is a matrix whose columns form a basis for a subspace W of $\mathcal{R}^n$, then $C(C^TC)^{-1}C^T = I_n$.
52. If B and C are matrices whose columns form bases for a subspace W of $\mathcal{R}^n$, then $B(B^TB)^{-1}B^T = C(C^TC)^{-1}C^T$.
53. If W is a subspace of $\mathcal{R}^n$ and $\mathbf{u}$ is a vector in $\mathcal{R}^n$, then $\mathbf{u} - P_W\mathbf{u}$ is in $W^\perp$.
54. If W is a subspace of $\mathcal{R}^n$ and $\mathbf{u}$ is a vector in $\mathcal{R}^n$, then the distance from $\mathbf{u}$ to W is $\|P_W\mathbf{u}\|$.
55. If W is a subspace of $\mathcal{R}^n$, then $\text{Null}\, P_W = W^\perp$.
56. If W is a subspace of $\mathcal{R}^n$ and $\mathbf{u}$ is a vector in W, then $P_W\mathbf{u} = \mathbf{u}$.

---

57. Let $\mathcal{S}$ be a nonempty finite subset of $\mathcal{R}^n$, and suppose that $W = \text{Span}\, \mathcal{S}$. Prove that $W^\perp = \mathcal{S}^\perp$.
58. Let W be a subspace of $\mathcal{R}^n$, and let $\mathcal{B}_1$ and $\mathcal{B}_2$ be bases for W and $W^\perp$, respectively. Apply the orthogonal decomposition theorem to prove that
    (a) $\mathcal{B}_1 \cup \mathcal{B}_2$ is a basis for $\mathcal{R}^n$.
    (b) $\dim W + \dim W^\perp = n$.
59. Suppose that $\{\mathbf{v}_1, \mathbf{v}_2, \ldots, \mathbf{v}_n\}$ is an orthogonal basis for $\mathcal{R}^n$. For any k, where $1 \le k < n$, define $W = \text{Span}\,\{\mathbf{v}_1, \mathbf{v}_2, \ldots, \mathbf{v}_k\}$. Prove that $\{\mathbf{v}_{k+1}, \mathbf{v}_{k+2}, \ldots, \mathbf{v}_n\}$ is an orthogonal basis for $W^\perp$.
60. Prove that for any subspace W of $\mathcal{R}^n$, $(W^\perp)^\perp = W$.
61. Prove the following statements for any matrix A:
    (a) $(\text{Row}\, A)^\perp = \text{Null}\, A$
    (b) $(\text{Col}\, A)^\perp = \text{Null}\, A^T$
62. Prove that if $\mathcal{S}_1$ and $\mathcal{S}_2$ are subsets of $\mathcal{R}^n$ such that $\mathcal{S}_1$ is contained in $\mathcal{S}_2$, then $\mathcal{S}_2^\perp$ is contained in $\mathcal{S}_1^\perp$.
63. Prove that for any nonempty finite subset $\mathcal{S}$ of $\mathcal{R}^n$, $(\mathcal{S}^\perp)^\perp = \text{Span}\, \mathcal{S}$.
64. Use the fact that $(\text{Row}\, A)^\perp = \text{Null}\, A$ for any matrix A to give another proof that $\dim W + \dim W^\perp = n$ for any subspace W of $\mathcal{R}^n$. (*Hint:* Let A be a $k \times n$ matrix whose rows constitute a basis for W.)
65. Let A be an $n \times n$ matrix. Prove that if $\mathbf{v}$ is a vector in both $\text{Row}\, A$ and $\text{Null}\, A$, then $\mathbf{v} = \mathbf{0}$.
66. Let V and W be subspaces of $\mathcal{R}^n$ such that every vector in V is orthogonal to every vector in W. Prove that $\dim V + \dim W \le n$.
67.[8] Let W be a subspace of $\mathcal{R}^n$.
    (a) Prove that $(P_W)^2 = P_W$.
    (b) Prove that $(P_W)^T = P_W$.
68. Let W be a subspace of $\mathcal{R}^n$. Prove that, for any $\mathbf{u}$ in $\mathcal{R}^n$, $P_W\mathbf{u} = \mathbf{u}$ if and only if $\mathbf{u}$ is in W.
69. Let W be a subspace of $\mathcal{R}^n$. Prove that, for any $\mathbf{u}$ in $\mathcal{R}^n$, $P_W\mathbf{u} = \mathbf{0}$ if and only if $\mathbf{u}$ is in $W^\perp$.
70. Let W be a subspace of $\mathcal{R}^n$. Prove that a vector $\mathbf{u}$ in $\mathcal{R}^n$ is an eigenvector of P_W if and only if $\mathbf{u}$ is an eigenvector of $P_{W^\perp}$.
71. Let W be a subspace of $\mathcal{R}^n$. Prove that $(P_W\mathbf{u}) \cdot \mathbf{v} = \mathbf{u} \cdot (P_W\mathbf{v})$ for every $\mathbf{u}$ and $\mathbf{v}$ in $\mathcal{R}^n$.
72. Let W be a subspace of $\mathcal{R}^n$. Prove that $P_W P_{W^\perp} = P_{W^\perp} P_W = O$, and hence $P_{W^\perp} = I_n - P_W$.
73. Let W be a subspace of $\mathcal{R}^n$. Prove that $P_W + P_{W^\perp} = I_n$.
74. Let V and W be subspaces of $\mathcal{R}^n$ such that for any $\mathbf{v}$ in V and $\mathbf{w}$ in W, the vectors $\mathbf{v}$ and $\mathbf{w}$ are orthogonal. Prove that $P_V + P_W$ is an orthogonal projection matrix. Describe the subspace Z of $\mathcal{R}^n$ such that $P_Z = P_V + P_W$.
75. Suppose that $\mathcal{B} = \{\mathbf{v}_1, \mathbf{v}_2, \ldots, \mathbf{v}_k\}$ is an orthonormal basis for a subspace W of $\mathcal{R}^n$. Let C be the $n \times k$ matrix whose columns are the vectors in $\mathcal{B}$. Prove the following statements:
    (a) $C^TC = I_k$.
    (b) $P_W = CC^T$.
76. Show that for any vector $\mathbf{u}$ in $\mathcal{R}^n$, the orthogonal projection of $\mathbf{u}$ on $\{\mathbf{0}\}$ is $\mathbf{0}$.
77. Let W be a subspace of $\mathcal{R}^n$ having dimension k, where $0 < k < n$.
    (a) Prove that 1 and 0 are the only eigenvalues of P_W.
    (b) Prove that W and $W^\perp$ are eigenspaces of P_W corresponding to the eigenvalues 1 and 0, respectively.
    (c) Let $\mathcal{B}_1$ and $\mathcal{B}_2$ be bases for W and $W^\perp$, respectively. Recall from Exercise 58 that $\mathcal{B} = \mathcal{B}_1 \cup \mathcal{B}_2$ is a basis for $\mathcal{R}^n$. Prove that if B is the matrix whose columns are the vectors in $\mathcal{B}$, and if D is the diagonal $n \times n$ matrix whose first k diagonal entries are 1s and whose other entries are 0s, then $P_W = BDB^{-1}$.
78. Let $V = \text{Row}\, A$, where

$$A = \begin{bmatrix} -1 & 1 & 0 & -1 \\ 0 & 1 & -2 & 1 \\ -3 & 1 & 4 & -5 \\ 1 & 1 & -4 & 3 \end{bmatrix}.$$

    Use the method described in Exercise 77 to find P_V. *Hint:* Obtain a basis for $V^\perp$ as in Example 2 of Section 6.2.
79. (a) Let $W = \text{Null}\, C$, where C is an $m \times n$ matrix of rank m. Prove that

$$P_{W^\perp} = C^T(CC^T)^{-1}C$$

    and

$$P_W = I_n - C^T(CC^T)^{-1}C.$$

    Hint: Observe that $W^\perp = \text{Row}\, C$.
    (b) Let $W = \text{Null}\, A$, where A is the matrix in Exercise 78. Use (a) to compute P_W. **Caution:** Because A is a 4×4 matrix of rank 2, A must be replaced by an appropriate 2×4 matrix of rank 2. This can be done by replacing A by the 2×4 matrix of nonzero rows of the reduced row echelon form of A.

---

[8] This exercise is used in Section 6.6 (on page 432).

(c) Let $W = \text{Null}\,A$, where A is the matrix in Exercise 78. Use the method described in Exercise 77 to find P_W. *Hint:* Obtain a basis for $W^\perp$ as in Example 2 of Section 6.2. Compare your result with the result obtained in (b).

80. Let V and W be as in Exercises 78 and 79. Compute $P_V + P_W$. What accounts for your answer?

81. Let $W = \text{Col}\,A$, where

$$A = \begin{bmatrix} 1 & 0 & 5 & -3 \\ 0 & 1 & 2 & 4 \\ -1 & -2 & -9 & -5 \\ 1 & 1 & 7 & 1 \end{bmatrix}.$$

Use the method described in Exercise 77 to find P_W.

82.[9] Suppose that P is an $n \times n$ matrix such that $P^2 = P^T = P$. Prove that P is the orthogonal projection matrix P_W, where $W = \text{Col}\,P = \{P\mathbf{u}\colon \mathbf{u} \text{ is in } \mathcal{R}^n\}$. *Hint:* Show that for any $\mathbf{u}$ in $\mathcal{R}^n$, $\mathbf{u} = P\mathbf{u} + (I_n - P)\mathbf{u}$, $P\mathbf{u}$ is in W, and $(I_n - P)\mathbf{u}$ is in $W^\perp$.

83. Let W be a 1-dimensional subspace of $\mathcal{R}^n$, $\mathbf{v}$ be a nonzero vector in W, and A the $n \times n$ matrix with $a_{ij} = v_i v_j$ for all i, j. Prove that $P_W = \dfrac{1}{\|\mathbf{v}\|^2} A$.

84. Let W be a 2-dimensional subspace of $\mathcal{R}^3$. Use the definition of the reflection operator T_W in Section 5.4 to prove the following:

(a) $T_W(\mathbf{u}) = 2U_W(\mathbf{u}) - \mathbf{u}$ for every vector $\mathbf{u}$ in $\mathcal{R}^3$.

(b) T_W is linear.

In Exercises 85–88, use either a calculator with matrix capabilities or computer software such as MATLAB to solve each problem.

85. Let $W = \text{Span}\,S$, where

$$S = \left\{ \begin{bmatrix} 0 \\ -3 \\ 9 \\ 0 \\ -4 \end{bmatrix}, \begin{bmatrix} -8 \\ 9 \\ -8 \\ 0 \\ 2 \end{bmatrix}, \begin{bmatrix} -4 \\ 8 \\ 1 \\ -1 \\ 8 \end{bmatrix}, \begin{bmatrix} -9 \\ 5 \\ 5 \\ 6 \\ -7 \end{bmatrix} \right\} \text{ and } \mathbf{u} = \begin{bmatrix} -9 \\ 4 \\ 7 \\ 2 \\ 4 \end{bmatrix}.$$

(a) Find an orthonormal basis for W.

(b) Use your answer to (a) to find the orthogonal projection of $\mathbf{u}$ on W.

(c) Use your answer to (b) to find the distance from $\mathbf{u}$ to W.

86. Find a basis for the subspace $W^\perp$, where W is given in Exercise 85.

87. Find P_W, where W is given in Exercise 85. Use your answer to find the orthogonal projection of the vector $\mathbf{u}$ in Exercise 85, and compare the result with your answer to Exercise 85(b).

88. Let

$$\mathbf{u} = \begin{bmatrix} 6 \\ -4 \\ -2 \\ 1 \\ -1 \end{bmatrix} \quad \text{and}$$

$$W = \text{Span}\left\{ \begin{bmatrix} -9 \\ 5 \\ 5 \\ 6 \\ -7 \end{bmatrix}, \begin{bmatrix} -9 \\ 4 \\ 7 \\ 2 \\ 4 \end{bmatrix}, \begin{bmatrix} 4 \\ 9 \\ 7 \\ -5 \\ -4 \end{bmatrix} \right\}.$$

Compute the orthogonal projection matrix P_W, and use it to find the distance from $\mathbf{u}$ to W.

SOLUTIONS TO THE PRACTICE PROBLEMS

1. $A = \begin{bmatrix} 1 & 0 & -1 & 1 \\ -1 & 1 & 3 & -4 \end{bmatrix}$. Thus $W^\perp = \text{Null}\,A$ is the set of solutions of the homogeneous system of linear equations $A\mathbf{x} = \mathbf{0}$. The vector form of the general solution of this system is

$$\begin{bmatrix} x_1 \\ x_2 \\ x_3 \\ x_4 \end{bmatrix} = x_3 \begin{bmatrix} 1 \\ -2 \\ 1 \\ 0 \end{bmatrix} + x_4 \begin{bmatrix} -1 \\ 3 \\ 0 \\ 1 \end{bmatrix}.$$

Therefore

$$\left\{ \begin{bmatrix} 1 \\ -2 \\ 1 \\ 0 \end{bmatrix}, \begin{bmatrix} -1 \\ 3 \\ 0 \\ 1 \end{bmatrix} \right\}$$

is a basis for $W^\perp$.

2. (a) A vector $\mathbf{x} = \begin{bmatrix} x_1 \\ x_2 \\ x_3 \\ x_4 \end{bmatrix}$ is in $W^\perp$ if and only if it is a solution of the homogeneous system of linear equations

$$\begin{aligned} x_1 + x_2 - x_3 + x_4 &= 0 \\ 3x_1 + 2x_2 - x_3 &= 0. \end{aligned}$$

The vector form of the general solution of this system is

$$\begin{bmatrix} x_1 \\ x_2 \\ x_3 \\ x_4 \end{bmatrix} = x_3 \begin{bmatrix} -1 \\ 2 \\ 1 \\ 0 \end{bmatrix} + x_4 \begin{bmatrix} 2 \\ -3 \\ 0 \\ 1 \end{bmatrix}.$$

[9] This exercise is used in Section 6.7 (on page 447).

Thus

$$\left\{\begin{bmatrix}-1\\2\\1\\0\end{bmatrix}, \begin{bmatrix}2\\-3\\0\\1\end{bmatrix}\right\}$$

is a basis for $W^{\perp}$.

Next, we apply the methods used in Practice Problem 1 to obtain an orthonormal basis $\{\mathbf{w}_1, \mathbf{w}_2\}$ for W, where

$$\mathbf{w}_1 = \frac{1}{2}\begin{bmatrix}1\\1\\-1\\1\end{bmatrix} \quad \text{and} \quad \mathbf{w}_2 = \frac{1}{2\sqrt{5}}\begin{bmatrix}3\\1\\1\\-3\end{bmatrix}.$$

Thus

$$\begin{aligned}\mathbf{w} &= (\mathbf{u}\cdot\mathbf{w}_1)\mathbf{w}_1 + (\mathbf{u}\cdot\mathbf{w}_2)\mathbf{w}_2\\ &= (5)\cdot\frac{1}{2}\begin{bmatrix}1\\1\\-1\\1\end{bmatrix} + (-\sqrt{5})\cdot\frac{1}{2\sqrt{5}}\begin{bmatrix}3\\1\\1\\-3\end{bmatrix}\\ &= \begin{bmatrix}1\\2\\-3\\4\end{bmatrix}.\end{aligned}$$

Finally,

$$\mathbf{z} = \mathbf{u} - \mathbf{w} = \begin{bmatrix}0\\7\\4\\7\end{bmatrix} - \begin{bmatrix}1\\2\\-3\\4\end{bmatrix} = \begin{bmatrix}-1\\5\\7\\3\end{bmatrix}.$$

(b) Let

$$C = \begin{bmatrix}1 & 3\\1 & 2\\-1 & -1\\1 & 0\end{bmatrix}.$$

Then

$$\begin{aligned}P_W &= C(C^TC)^{-1}C^T\\ &= \begin{bmatrix}1 & 3\\1 & 2\\-1 & -1\\1 & 0\end{bmatrix}\begin{bmatrix}0.7 & -0.3\\-0.3 & 0.2\end{bmatrix}\begin{bmatrix}1 & 1 & -1 & 1\\3 & 2 & -1 & 0\end{bmatrix}\\ &= \begin{bmatrix}0.7 & 0.4 & -0.1 & -0.2\\0.4 & 0.3 & -0.2 & 0.1\\-0.1 & -0.2 & 0.3 & -0.4\\-0.2 & 0.1 & -0.4 & 0.7\end{bmatrix}.\end{aligned}$$

Observe that the product $P_W\mathbf{u}$ gives the same result as obtained in (a).

(c) The distance from $\mathbf{u}$ to W is the distance between $\mathbf{u}$ and the orthogonal projection of $\mathbf{u}$ on W, which is

$$\|\mathbf{z}\| = \left\|\begin{bmatrix}-1\\5\\7\\3\end{bmatrix}\right\| = \sqrt{84}.$$

6.4 LEAST-SQUARES APPROXIMATION AND ORTHOGONAL PROJECTION MATRICES

In almost all areas of empirical research, there is an interest in finding simple mathematical relationships between variables. In economics, the variables might be the gross domestic product, the unemployment rate, and the annual deficit. In the life sciences, the variables of interest might be the incidence of smoking and heart disease. In sociology, it might be birth order and frequency of juvenile delinquency.

Many relationships in science are *deterministic*; that is, information about one variable completely determines the value of another variable. For example, the relationship between force f and acceleration a of an object of mass m is given by the equation $f = ma$ (Newton's second law). Another example is the height of a freely falling object and the time that it has been falling. On the other hand, the relationship between the height and the weight of an individual is not deterministic. There are many people with the same height, but different weights. Yet, in hospitals, there exist charts that give the recommended weights for given heights. Relationships that are not deterministic are often called *probabilistic* or *stochastic.*

We can apply what we know about orthogonal projections to identify relationships between variables. We begin with a given set of data $(x_1, y_1), (x_2, y_2), \ldots, (x_n, y_n)$

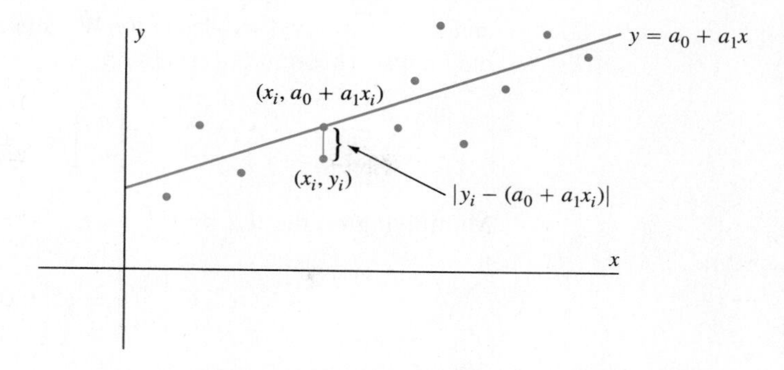

Figure 6.16 A plot of the data

obtained by empirical measurements. For example, we might have a randomly selected sample of n people, where x_i represents the number of years of education and y_i represents the annual income of the ith person. The data are plotted as in Figure 6.16. Notice that there is an approximately linear (straight line) relationship between x and y. To obtain this relationship, we would like to find the line $y = a_0 + a_1x$ that *best fits* the data. The usual criterion that statisticians use for defining the line of best fit is that the sum of the squared vertical distances of the data from it is smaller than from any other line. From Figure 6.16, we see that we must find a_0 and a_1 so that the quantity

$$E = [y_1 - (a_0 + a_1x_1)]^2 + [y_2 - (a_0 + a_1x_2)]^2 + \cdots + [y_n - (a_0 + a_1x_n)]^2 \qquad (3)$$

is minimized. The technique to find this line is called the **method of least squares,**[10] E is called the **error sum of squares**, and the line for which E is minimized is called the **least-squares line.**

To find the least-squares line, we let

$$\mathbf{v}_1 = \begin{bmatrix} 1 \\ 1 \\ \vdots \\ 1 \end{bmatrix}, \quad \mathbf{v}_2 = \begin{bmatrix} x_1 \\ x_2 \\ \vdots \\ x_n \end{bmatrix}, \quad \mathbf{y} = \begin{bmatrix} y_1 \\ y_2 \\ \vdots \\ y_n \end{bmatrix}, \quad \text{and} \quad C = [\mathbf{v}_1 \ \mathbf{v}_2].$$

With this notation, equation (3) can be rewritten in the notation of vectors as

$$E = \|\mathbf{y} - (a_0\mathbf{v}_1 + a_1\mathbf{v}_2)\|^2. \qquad (4)$$

(See Exercise 33.) Notice that $\sqrt{E} = \|\mathbf{y} - (a_0\mathbf{v}_1 + a_1\mathbf{v}_2)\|$ is the distance between $\mathbf{y}$ and the vector $a_0\mathbf{v}_1 + a_1\mathbf{v}_2$, which lies in $W = \text{Span}\,\{\mathbf{v}_1, \mathbf{v}_2\}$. So to minimize E, we need only choose the vector in W that is nearest to $\mathbf{y}$. But from the closest vector property, this vector is the orthogonal projection of $\mathbf{y}$ on W. Thus we want

$$a_0\mathbf{v}_1 + a_1\mathbf{v}_2 = C\begin{bmatrix} a_0 \\ a_1 \end{bmatrix} = P_W\mathbf{y},$$

the orthogonal projection of $\mathbf{y}$ on W.

For any reasonable set of data, the x_i's are not all equal, and hence $\mathbf{v}_1$ and $\mathbf{v}_2$ are not multiples of one another. Thus the vectors $\mathbf{v}_1$ and $\mathbf{v}_2$ are linearly independent,

[10] The method of least squares first appeared in a paper by Adrien Marie Legendre (1752–1833), entitled *Nouvelles Méthodes pour la détermination des orbites des comètes.*

and so $\mathcal{B} = \{\mathbf{v}_1, \mathbf{v}_2\}$ is a basis for W. Since the columns of C form a basis for W, we may apply Theorem 6.8 to obtain

$$C\begin{bmatrix} a_0 \\ a_1 \end{bmatrix} = C(C^TC)^{-1}C^T\mathbf{y}.$$

Multiplying on the left by C^T gives

$$C^TC\begin{bmatrix} a_0 \\ a_1 \end{bmatrix} = C^TC(C^TC)^{-1}C^T\mathbf{y} = C^T\mathbf{y}.$$

The matrix equation $C^TC\mathbf{x} = C^T\mathbf{y}$ corresponds to a system of linear equations called the **normal equations**. Thus the line of best fit occurs when $\begin{bmatrix} a_0 \\ a_1 \end{bmatrix}$ is the solution of the normal equations. Since C^TC is invertible by the lemma preceding Theorem 6.8, we see that the least-squares line has the equation $y = a_0 + a_1x$, where

$$\begin{bmatrix} a_0 \\ a_1 \end{bmatrix} = (C^TC)^{-1}C^T\mathbf{y}.$$

Example 1

In the manufacture of refrigerators, it is necessary to finish connecting rods. If the weight of the finished rod is above a certain amount, the rod must be discarded. As the finishing process is expensive, it would be of considerable value to the manufacturer to be able to estimate the relationship between the finished weight and the initial rough weight. Then, those rods whose rough weights are too high could be discarded before they are finished. From past experience, the manufacturer knows that this relationship is approximately linear.

From a sample of five rods, we let x_i and y_i denote the rough weight and the finished weight, respectively, of the ith rod. The data are given in the following table:

Rough weight x_i (in pounds)	Finished weight y_i (in pounds)
2.60	2.00
2.72	2.10
2.75	2.10
2.67	2.03
2.68	2.04

From this information, we let

$$C = \begin{bmatrix} 1 & 2.60 \\ 1 & 2.72 \\ 1 & 2.75 \\ 1 & 2.67 \\ 1 & 2.68 \end{bmatrix} \quad \text{and} \quad \mathbf{y} = \begin{bmatrix} 2.00 \\ 2.10 \\ 2.10 \\ 2.03 \\ 2.04 \end{bmatrix}.$$

Then

$$C^TC = \begin{bmatrix} 5.0000 & 13.4200 \\ 13.4200 & 36.0322 \end{bmatrix} \quad \text{and} \quad C^T\mathbf{y} = \begin{bmatrix} 10.2700 \\ 27.5743 \end{bmatrix},$$

and the solution of the normal equations is

$$\begin{bmatrix} a_0 \\ a_1 \end{bmatrix} \approx \begin{bmatrix} 0.056 \\ 0.745 \end{bmatrix}.$$

Thus the approximate relationship between the finished weight y and the rough weight x is given by the equation of the least-squares line

$$y = 0.056 + 0.745x.$$

For example, if the rough weight of a rod is 2.65 pounds, then the finished weight is approximately

$$0.056 + 0.745(2.65) \approx 2.030 \text{ pounds}.$$

Practice Problem 1 ▶ Find the equation of the least-squares line for the data $(1, 62)$, $(3, 54)$, $(4, 50)$, $(5, 48)$, and $(7, 40)$. ◀

The method we have developed for finding the best fit to data points (x_1, y_1), (x_2, y_2), $\ldots$, (x_n, y_n) by a linear polynomial $a_0 + a_1x$ can be modified to find the best fit by a quadratic polynomial $y = a_0 + a_1x + a_2x^2$. The only change in the method is that the new error sum of squares is

$$E = [y_1 - (a_0 + a_1x_1 + a_2x_1^2)]^2 + \cdots + [y_n - (a_0 + a_1x_n + a_2x_n^2)]^2.$$

In this case, let

$$\mathbf{v}_1 = \begin{bmatrix} 1 \\ 1 \\ \vdots \\ 1 \end{bmatrix}, \quad \mathbf{v}_2 = \begin{bmatrix} x_1 \\ x_2 \\ \vdots \\ x_n \end{bmatrix}, \quad \mathbf{v}_3 = \begin{bmatrix} x_1^2 \\ x_2^2 \\ \vdots \\ x_n^2 \end{bmatrix}, \quad \text{and} \quad \mathbf{y} = \begin{bmatrix} y_1 \\ y_2 \\ \vdots \\ y_n \end{bmatrix}.$$

Assuming that the x_i's are distinct and $n \geq 3$, which in practice is always the case, the vectors $\mathbf{v}_1$, $\mathbf{v}_2$, and $\mathbf{v}_3$ are linearly independent (see Exercise 34), and hence they form a basis for a 3-dimensional subspace W of $\mathcal{R}^n$. So we let C be the $n \times 3$ matrix $C = [\mathbf{v}_1\ \mathbf{v}_2\ \mathbf{v}_3]$. As in the linear case, we can obtain the normal equations

$$C^TC \begin{bmatrix} a_0 \\ a_1 \\ a_2 \end{bmatrix} = c^T\mathbf{y},$$

whose solution is

$$\begin{bmatrix} a_0 \\ a_1 \\ a_2 \end{bmatrix} = (C^TC)^{-1}C^T\mathbf{y}.$$

Example 2 It is known from physics that if a ball is thrown upward at a velocity of v_0 feet per second from a building of height s_0 feet, then the height of the ball after t seconds is given by $s = s_0 + v_0t + \frac{1}{2}gt^2$, where g represents the acceleration due to gravity. To provide an empirical estimate of g, a ball is thrown upward from a building 100 feet high at a velocity of 30 feet per second. The height of the ball is observed at the times given in the following table:

Time (in seconds)	Height (in feet)
0	100
1	118
2	92
3	48
3.5	7

For these data, we let

$$C = \begin{bmatrix} 1 & 0 & 0 \\ 1 & 1 & 1 \\ 1 & 2 & 4 \\ 1 & 3 & 9 \\ 1 & 3.5 & 12.25 \end{bmatrix} \quad \text{and} \quad \mathbf{y} = \begin{bmatrix} 100 \\ 118 \\ 92 \\ 48 \\ 7 \end{bmatrix}.$$

Thus the quadratic polynomial $y = a_0 + a_1x + a_2x^2$ of best fit satisfies

$$\begin{bmatrix} s_0 \\ v_0 \\ \frac{1}{2}g \end{bmatrix} = \begin{bmatrix} a_0 \\ a_1 \\ a_2 \end{bmatrix} = (C^TC)^{-1}C^T\mathbf{y} \approx \begin{bmatrix} 101.00 \\ 29.77 \\ -16.11 \end{bmatrix}.$$

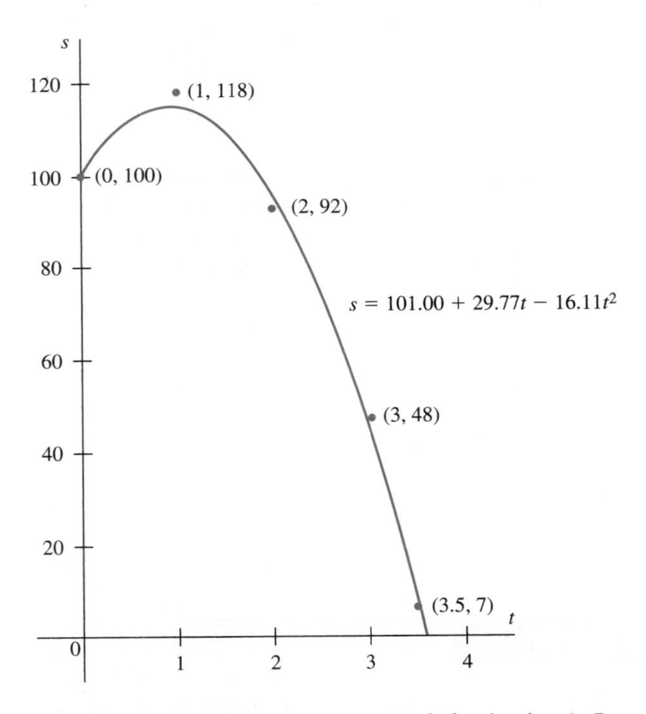

Figure 6.17 The quadratic polynomial of best fit for the data in Example 2

This yields the approximate relationship

$$s = 101.00 + 29.77t - 16.11t^2.$$

(See Figure 6.17.) Setting $\frac{1}{2}g = -16.11$, we obtain -32.22 feet per second per second as the estimate for g.

It should be pointed out that the same method may be extended to find the best-fitting polynomial[11] of any desired maximum degree, provided that the data set is sufficiently large. Furthermore, by using the appropriate change of variable, many more complicated relationships may be estimated by the same type of matrix computations.

The material treated in the rest of this section will be revisited from a different perspective in Section 6.7.

INCONSISTENT SYSTEMS OF LINEAR EQUATIONS*

The preceding examples are special cases of inconsistent systems of linear equations for which it is desirable to obtain approximate solutions. In general, a system of linear equations $A\mathbf{x} = \mathbf{b}$ arising from the application of a theoretical model to real data may be inconsistent because the entries of A and $\mathbf{b}$ that are obtained from empirical measurements are not precise or because the model only approximates reality. In these circumstances, we are interested in obtaining a vector $\mathbf{z}$ for which $\|A\mathbf{z} - \mathbf{b}\|$ is a minimum. Let W denote the set of all vectors of the form $A\mathbf{u}$. Then W is the column space of A. By the closest vector property, the vector in W that is closest to $\mathbf{b}$ is the orthogonal projection of $\mathbf{b}$ on W, which can be computed as $P_W\mathbf{b}$. Thus a vector $\mathbf{z}$ minimizes $\|A\mathbf{z} - \mathbf{b}\|$ if and only if it is a solution of the system of linear equations

$$A\mathbf{x} = P_W\mathbf{b},$$

which is guaranteed to be consistent. (See Figure 6.18.)

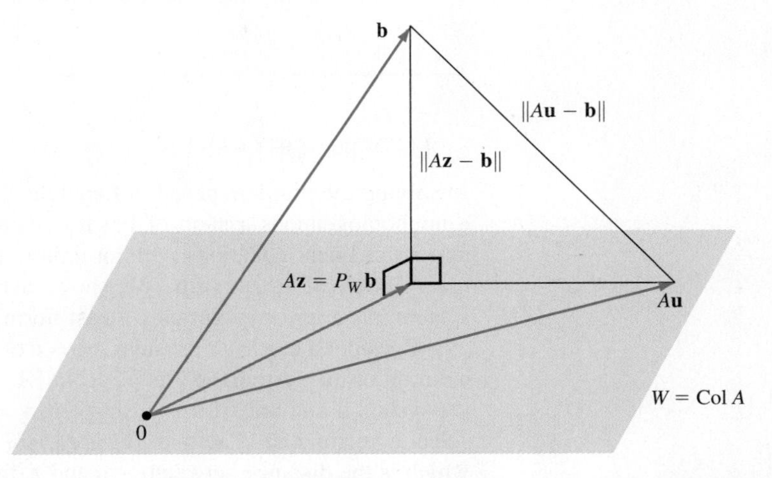

Figure 6.18 The vector $\mathbf{z}$ minimizes $\|A\mathbf{z} - \mathbf{b}\|$ if and only if it is a solution of the system of linear equations $A\mathbf{x} = P_W\mathbf{b}$.

[11] Caution! The MATLAB function `polyfit` returns the coefficients of the polynomial of best fit with terms written in *descending* order (rather than in ascending order, as in this book).

* The remainder of this section may be omitted without loss of continuity.

Example 3 Given the inconsistent system of linear equations $A\mathbf{x} = \mathbf{b}$, with

$$A = \begin{bmatrix} 1 & 1 & 1 \\ 2 & 1 & 4 \\ -1 & 0 & -3 \\ 3 & 2 & 5 \end{bmatrix} \quad \text{and} \quad \mathbf{b} = \begin{bmatrix} 1 \\ 7 \\ -4 \\ 8 \end{bmatrix},$$

use the method of least squares to describe the vectors $\mathbf{z}$ for which $\|A\mathbf{z} - \mathbf{b}\|$ is a minimum.

Solution By computing the reduced row echelon form of A, we see that the rank of A is 2 and that the first two columns of A are linearly independent. Thus the first two columns of A form a basis for $W = \operatorname{Col} A$. Let C be the 4×2 matrix with these two vectors as its columns. Then

$$P_W\mathbf{b} = C(C^TC)^{-1}C^T = \frac{1}{3}\begin{bmatrix} 1 & 0 & 1 & 1 \\ 0 & 1 & -1 & 1 \\ 1 & -1 & 2 & 0 \\ 1 & 1 & 0 & 2 \end{bmatrix}\begin{bmatrix} 1 \\ 7 \\ -4 \\ 8 \end{bmatrix} = \frac{1}{3}\begin{bmatrix} 5 \\ 19 \\ -14 \\ 24 \end{bmatrix}.$$

As noted, the vectors that minimize $\|A\mathbf{z} - \mathbf{b}\|$ are the solutions to $A\mathbf{x} = P_W\mathbf{b}$. The general solution of this system is

$$\frac{1}{3}\begin{bmatrix} 14 \\ -9 \\ 0 \end{bmatrix} + x_3\begin{bmatrix} -3 \\ 2 \\ 1 \end{bmatrix}.$$

So these are the vectors that minimize $\|A\mathbf{z} - \mathbf{b}\|$. Note that, for each of these vectors, we have

$$\|A\mathbf{z} - \mathbf{b}\| = \|P_W\mathbf{b} - \mathbf{b}\| = \left\| \frac{1}{3}\begin{bmatrix} 5 \\ 19 \\ -14 \\ 24 \end{bmatrix} - \begin{bmatrix} 1 \\ 7 \\ -4 \\ 8 \end{bmatrix} \right\| = \frac{2}{\sqrt{3}}.$$

SOLUTIONS OF LEAST NORM

In solving the problem posed in Example 3, we obtained an infinite set of solutions of a nonhomogeneous system of linear equations. In general, given a nonhomogeneous system of linear equations with an infinite set of solutions, it is often useful to select the solution of least norm. We show, using orthogonal projections, that any such system has a unique solution of least norm.

Consider a consistent system $A\mathbf{x} = \mathbf{c}$ of linear equations with $\mathbf{c} \neq \mathbf{0}$. Let $\mathbf{v}_0$ be any solution of the system, and let $Z = \operatorname{Null} A$. By Exercise 35, a vector $\mathbf{v}$ is a solution of the system if and only if it is of the form $\mathbf{v} = \mathbf{v}_0 + \mathbf{z}$, where $\mathbf{z}$ is in Z. Here, we wish to select a vector $\mathbf{z}$ in Z so that $\|\mathbf{v}_0 + \mathbf{z}\|$ is a minimum. Since $\|\mathbf{v}_0 + \mathbf{z}\| = \|-\mathbf{v}_0 - \mathbf{z}\|$, which is the distance between $-\mathbf{v}_0$ and $\mathbf{z}$, the vector in Z that minimizes this distance is, of course, the orthogonal projection of $-\mathbf{v}_0$ on Z; that is, $\mathbf{z} = P_Z(-\mathbf{v}_0) = -P_Z\mathbf{v}_0$. Thus $\mathbf{v}_0 + \mathbf{z} = \mathbf{v}_0 - P_Z\mathbf{v}_0$ is the unique solution of the system of least norm.

Example 4 Find the solution of least norm to the equation $A\mathbf{x} = P_W\mathbf{b}$ in Example 3.

Solution Based on the vector form of the solution given in Example 3, a vector $\mathbf{v}$ is a solution if and only if $\mathbf{v} = \mathbf{v}_0 + \mathbf{z}$, for $\mathbf{z}$ in Z, where

$$\mathbf{v}_0 = \frac{1}{3}\begin{bmatrix} 14 \\ -9 \\ 0 \end{bmatrix} \quad \text{and} \quad Z = \text{Null}\,A = \text{Span}\left\{\begin{bmatrix} -3 \\ 2 \\ 1 \end{bmatrix}\right\}.$$

Setting $C = \begin{bmatrix} -3 \\ 2 \\ 1 \end{bmatrix}$, we compute the orthogonal projection matrix

$$P_Z = C(C^TC)^{-1}C^T = \frac{1}{14}\begin{bmatrix} 9 & -6 & 3 \\ -6 & 4 & 2 \\ -3 & 2 & 1 \end{bmatrix}.$$

Thus

$$\mathbf{v}_0 - P_Z\mathbf{v}_0 = (I_3 - P_Z)\mathbf{v}_0 = \frac{1}{21}\begin{bmatrix} 8 \\ -3 \\ 30 \end{bmatrix}$$

is the solution of least norm.

EXERCISES

In Exercises 1–8, find the equation of the least-squares line for the given data.

1. $(1, 14), (3, 17), (5, 19), (7, 20)$
2. $(1, 30), (2, 27), (4, 21), (7, 14)$
3. $(1, 5), (2, 6), (3, 8), (4, 10), (5, 11)$
4. $(1, 2), (2, 4), (3, 7), (4, 8), (5, 10)$
5. $(1, 40), (3, 36), (7, 23), (8, 21), (10, 13)$
6. $(1, 19), (2, 17), (3, 16), (4, 14), (5, 12)$
7. $(1, 4), (4, 24), (5, 30), (8, 32), (12, 36)$
8. $(1, 21), (3, 32), (9, 38), (12, 41), (15, 51)$
9. Suppose that a spring whose natural length is L inches is attached to a wall. A force y is applied to the free end of the spring, stretching the spring s inches beyond its natural length. *Hooke's law* states (within certain limits) that $y = ks$, where k is a constant called the *spring constant.* Now suppose that after the force y is applied, the new length of the spring is x. Then $s = x - L$, and Hooke's law yields

$$y = ks = k(x - L) = a + kx,$$

where $a = -kL$. Apply the method of least squares to the following data to estimate k and L:

Length x in inches	Force y in pounds
3.5	1.0
4.0	2.2
4.5	2.8
5.0	4.3

In Exercises 10–15, use the method of least squares to find the polynomial of degree at most n that best fits the given data.

10. $n = 2$ with data $(0, 2), (1, 2), (2, 4), (3, 8)$
11. $n = 2$ with data $(0, 3), (1, 3), (2, 5), (3, 9)$
12. $n = 2$ with data $(0, 1), (1, 2), (2, 3), (3, 4)$
13. $n = 2$ with data $(0, 2), (1, 3), (2, 5), (3, 8)$
14. $n = 3$ with data $(-2, -5), (-1, -1), (0, -1), (1, 1), (2, 11)$
15. $n = 3$ with data $(-2, -4), (-1, -5), (0, 5), (1, -3), (2, 12)$

In Exercises 16–19, an inconsistent system of linear equations $A\mathbf{x} = \mathbf{b}$ *is given. Use the method of least squares to obtain the vectors* $\mathbf{z}$ *for which* $\|A\mathbf{z} - \mathbf{b}\|$ *is a minimum.*

16. $A = \begin{bmatrix} 1 & 1 \\ 1 & 2 \\ 3 & 1 \end{bmatrix}$ and $\mathbf{b} = \begin{bmatrix} 3 \\ 5 \\ 4 \end{bmatrix}$

17. $A = \begin{bmatrix} 1 & 2 & -1 \\ 1 & -1 & 2 \\ 2 & 1 & 1 \end{bmatrix}$ and $\mathbf{b} = \begin{bmatrix} 1 \\ 3 \\ 1 \end{bmatrix}$

18. $A = \begin{bmatrix} 1 & 1 & 0 & 3 \\ 0 & 1 & 0 & 1 \\ 1 & -1 & 1 & 2 \\ 0 & -1 & 1 & 0 \end{bmatrix}$ and $\mathbf{b} = \begin{bmatrix} 1 \\ 2 \\ 1 \\ 0 \end{bmatrix}$

19. $A = \begin{bmatrix} -1 & 1 & 1 & 0 \\ 2 & 1 & 4 & 3 \\ 0 & -1 & -1 & 0 \\ 0 & 2 & 4 & 2 \\ 1 & 1 & 3 & 2 \end{bmatrix}$ and $\mathbf{b} = \begin{bmatrix} 1 \\ 0 \\ 1 \\ 1 \\ 1 \end{bmatrix}$

In Exercises 20–23, a system of linear equations $A\mathbf{x} = \mathbf{b}$ *is given. Use the method of least squares to obtain the solution of least norm.*

20. $A = \begin{bmatrix} 1 & 1 & 2 \\ 3 & -1 & -2 \end{bmatrix}$ and $\mathbf{b} = \begin{bmatrix} 3 \\ 1 \end{bmatrix}$

21. $A = \begin{bmatrix} 1 & 2 & -1 \\ -3 & -5 & 2 \\ 2 & 3 & -1 \end{bmatrix}$ and $\mathbf{b} = \begin{bmatrix} -1 \\ 0 \\ 1 \end{bmatrix}$

22. $A = \begin{bmatrix} 1 & -3 & 2 \end{bmatrix}$ and $\mathbf{b} = [5]$

23. $A = \begin{bmatrix} 2 & -1 & 1 & 1 \\ 1 & 1 & -1 & 2 \\ 1 & -2 & 2 & -1 \end{bmatrix}$ and $\mathbf{b} = \begin{bmatrix} 4 \\ -1 \\ 5 \end{bmatrix}$

In Exercises 24–27, find the vector $\mathbf{z}$ *of least norm for which* $\|A\mathbf{x} - \mathbf{b}\|$ *is a minimum, where* $A\mathbf{x} = \mathbf{b}$ *is the inconsistent system of linear equations given in each specified exercise.*

24. Exercise 16
25. Exercise 17
26. Exercise 18
27. Exercise 19

In Exercises 28–32, determine whether the statements are true or false.

28. For a given set of data plotted in the xy-plane, the least-squares line is the unique line in the plane that minimizes the sum of the vertical distances from the data points to the line.

29. If $\begin{bmatrix} a_0 \\ a_1 \end{bmatrix}$ is a solution of the normal equations for the data, then $y = a_0 + a_1 x$ is the equation of the least-squares line.

30. The method of least squares can be used only to approximate data with a straight line.

31. For any inconsistent system of linear equations $A\mathbf{x} = \mathbf{b}$, the vector $\mathbf{z}$ for which $\|A\mathbf{z} - \mathbf{b}\|$ is a minimum is unique.

32. Every consistent system of linear equations $A\mathbf{x} = \mathbf{b}$ has a unique solution of least norm.

33. Let E be the error sum of squares for the data $(x_1, y_1), (x_2, y_2), \ldots, (x_n, y_n)$, as in equation (4). Prove that $E = \|\mathbf{y} - (a_0\mathbf{v}_1 + a_1\mathbf{v}_2)\|^2$, where

$$\mathbf{v}_1 = \begin{bmatrix} 1 \\ 1 \\ \vdots \\ 1 \end{bmatrix}, \quad \mathbf{v}_2 = \begin{bmatrix} x_1 \\ x_2 \\ \vdots \\ x_n \end{bmatrix}, \quad \text{and} \quad \mathbf{y} = \begin{bmatrix} y_1 \\ y_2 \\ \vdots \\ y_n \end{bmatrix}.$$

34. Prove that for any set of data $(x_1, y_1), (x_2, y_2), \ldots, (x_n, y_n)$, where the x_i's are distinct and $n \geq 3$, the vectors

$$\mathbf{v}_1 = \begin{bmatrix} 1 \\ 1 \\ \vdots \\ 1 \end{bmatrix}, \quad \mathbf{v}_2 = \begin{bmatrix} x_1 \\ x_2 \\ \vdots \\ x_n \end{bmatrix}, \quad \text{and} \quad \mathbf{v}_3 = \begin{bmatrix} x_1^2 \\ x_2^2 \\ \vdots \\ x_n^2 \end{bmatrix}$$

form a linearly independent subset of $\mathcal{R}^n$.

35. Suppose that $A\mathbf{x} = \mathbf{c}$ is a consistent system of linear equations with $\mathbf{c} \neq \mathbf{0}$, and $\mathbf{v}_0$ is a solution of the system. Prove that a vector $\mathbf{v}$ is a solution of the system if and only if $\mathbf{v} = \mathbf{v}_0 + \mathbf{z}$ for some vector $\mathbf{z}$ in Null A.

Exercises 36 and 37 require familiarity with the material on QR factorization in Section 6.2.

36. Let $A = QR$ be a QR factorization of a matrix with linearly independent columns. Prove that $P_W = QQ^T$, where $W = \text{Col}\,A$.

37. Consider the consistent system $A\mathbf{x} = \mathbf{b}$, where A has linearly independent columns. In Section 6.2, we learned how to use a QR factorization of A to solve this by finding the solution to the related system $R\mathbf{x} = Q^T\mathbf{b}$, which is always consistent. Show that if $A\mathbf{x} = \mathbf{b}$ is *inconsistent,* the solution of $R\mathbf{x} = Q^T\mathbf{b}$ minimizes $\|A\mathbf{x} - \mathbf{b}\|$.

In Exercises 38–41, use either a calculator with matrix capabilities or computer software such as MATLAB.

38. A space vehicle is launched from a space platform near a space station. The vehicle moves in a direction away from the station at a constant acceleration so that t seconds after launch, its distance y from the station (in meters) is given by the formula $y = a + bt + ct^2$. (Here, a is the distance from the station to the platform at the time of launch, b is the speed of the platform relative to the space station, and $2c$ is the acceleration of the vehicle.) Use the method of least squares to obtain the best quadratic fit to the data that follows:

t	5	10	15	20	25	30
y	140	290	560	910	1400	2000

39. Use the method of least squares to find the best cubic fit for the points $(-2, -4)$, $(-1, 1)$, $(0, 1)$, $(2, 10)$, and $(3, 26)$.

40. The accompanying table gives the approximate values of the function $y = 10 \sin x$ over the interval $[0, 2\pi]$. We use the method of least squares to approximate this function by linear and cubic polynomials.

x	$y = 10\sin x$
0.00000	0.00000
0.62832	5.87786
1.25664	9.51057
1.88496	9.51055
2.51328	5.87781
3.14160	−0.00007
3.76992	−5.87792
4.39824	−9.51060
5.02656	−9.51053
5.65488	−5.87775
6.28320	0.00014

(a) Use the method of least squares to find the equation of the least-squares line for the data in the table.

(b) Compute the error sum of squares associated with (a).

(c) Graph $y = 10\sin x$ and the least-squares line obtained in (a) using the same set of axes.

(d) Use the method of least squares to produce the best cubic fit for the data.

(e) Compute the error sum of squares associated with (d).

(f) Graph $y = 10\sin x$ and the cubic polynomial obtained in (d) using the same set of axes.

41. Suppose that a mathematical model predicts that two quantities, x and y, are related by the equation $y = a\cos x + b\sin x$, where x is in units of degrees. As a result of experiments, the following table of data is obtained:

x	5	10	15	20	25	30
y	2.8	2.6	2.4	2.1	1.9	1.6

Use the method of least squares to estimate the values of a and b rounded to two significant figures. *Hint:* Let $\mathbf{v}_1$ and $\mathbf{v}_2$ be the vectors in $\mathcal{R}^6$ whose entries are the cosines and the sines of the angles $5°, 10°, \ldots, 30°$, respectively, and let $\mathbf{y}$ be the vector in $\mathcal{R}^6$ whose entries are the corresponding values of y in the table. Let $A = [\mathbf{v}_1\ \mathbf{v}_2]$. Use the method of least squares to find the vector $\mathbf{z}$ that minimizes $\|A\mathbf{z} - \mathbf{y}\|$.

SOLUTIONS TO THE PRACTICE PROBLEMS

1. Let

$$C = \begin{bmatrix} 1 & 1 \\ 1 & 3 \\ 1 & 4 \\ 1 & 5 \\ 1 & 7 \end{bmatrix} \quad \text{and} \quad \mathbf{y} = \begin{bmatrix} 62 \\ 54 \\ 50 \\ 48 \\ 40 \end{bmatrix}.$$

Then $y = a_0 + a_1 x$, where

$$\begin{bmatrix} a_0 \\ a_1 \end{bmatrix} = (C^T C)^{-1} C^T \mathbf{y} = \begin{bmatrix} 65.2 \\ -3.6 \end{bmatrix}.$$

Hence the equation of the least-squares line is

$$y = 65.2 - 3.6x.$$

2. (a) Let

$$C = \begin{bmatrix} 1 & 1 \\ -1 & 1 \\ 2 & -3 \\ -1 & 2 \end{bmatrix}.$$

Then

$$P_W = C(C^T C)^{-1} C^T$$
$$= \frac{1}{41}\begin{bmatrix} 38 & -8 & 1 & 7 \\ -8 & 6 & -11 & 5 \\ 1 & -11 & 27 & -16 \\ 7 & 5 & -16 & 11 \end{bmatrix}.$$

(b) The vector in W that is closest to $\mathbf{u}$ is given by

$$P_W\mathbf{u} = \frac{1}{41}\begin{bmatrix} 38 & -8 & 1 & 7 \\ -8 & 6 & -11 & 5 \\ 1 & -11 & 27 & -16 \\ 7 & 5 & -16 & 11 \end{bmatrix}\begin{bmatrix} 4 \\ 0 \\ -3 \\ 8 \end{bmatrix}$$
$$= \begin{bmatrix} 5 \\ 1 \\ -5 \\ 4 \end{bmatrix}.$$

6.5 ORTHOGONAL MATRICES AND OPERATORS

In Chapter 2, we studied the functions from $\mathcal{R}^n$ to $\mathcal{R}^n$ that preserve the operations of vector addition and scalar multiplication. Now that we have introduced the concept of the norm of a vector, it is natural to ask which linear operators on $\mathcal{R}^n$ also preserve norms; that is, which operators T satisfy $\|T(\mathbf{u})\| = \|\mathbf{u}\|$ for every vector $\mathbf{u}$ in $\mathcal{R}^n$.

These linear operators and their standard matrices are extremely useful in numerical calculations because they do not magnify any roundoff or experimental error. Because such operators on $\mathcal{R}^2$ preserve the angle between nonzero vectors (see Exercise 66), it follows that they also preserve many familiar properties from geometry.

It is clear that an arbitrary operator on $\mathcal{R}^n$ does not have this property. For if an operator U on $\mathcal{R}^n$ has an eigenvalue λ other than ± 1 with corresponding eigenvector $\mathbf{v}$, then $\|U(\mathbf{v})\| = \|\lambda\mathbf{v}\| = |\lambda| \cdot \|\mathbf{v}\| \neq \|\mathbf{v}\|$. There are, however, familiar operators that do have this property, as our first example shows.

Example 1 Let T be the linear operator on $\mathcal{R}^2$ that rotates a vector through an angle θ. Clearly, $T(\mathbf{v})$ has the same length as $\mathbf{v}$ for every $\mathbf{v}$ in $\mathcal{R}^2$, and therefore $\|T(\mathbf{v})\| = \|\mathbf{v}\|$ for every $\mathbf{v}$ in $\mathcal{R}^2$.

A linear operator that rotates every vector in $\mathcal{R}^2$ through a particular angle is called a **rotation operator**, or simply, a **rotation**. Clearly, a linear operator on $\mathcal{R}^2$ is a rotation if and only if its standard matrix is a rotation matrix.

Because of the connection between linear operators and their standard matrices, we can study linear operators on $\mathcal{R}^n$ that preserve norms by studying the $n \times n$ matrices Q such that $\|Q\mathbf{u}\| = \|\mathbf{u}\|$ for every $\mathbf{u}$ in $\mathcal{R}^n$. Consider an arbitrary column $\mathbf{q}_j$ of such a matrix. Since

$$\|\mathbf{q}_j\| = \|Q\mathbf{e}_j\| = \|\mathbf{e}_j\| = 1, \tag{5}$$

the norm of every column of Q is 1. Moreover, if $i \neq j$, we have

$$\|\mathbf{q}_i + \mathbf{q}_j\|^2 = \|Q\mathbf{e}_i + Q\mathbf{e}_j\|^2 = \|Q(\mathbf{e}_i + \mathbf{e}_j)\|^2 = \|\mathbf{e}_i + \mathbf{e}_j\|^2 = 2 = \|\mathbf{q}_i\|^2 + \|\mathbf{q}_j\|^2. \tag{6}$$

Hence $\mathbf{q}_i$ and $\mathbf{q}_j$ are orthogonal by Theorem 6.2. It follows that the columns of Q form an orthonormal set of distinct vectors, and so constitute an orthonormal basis for $\mathcal{R}^n$.

Because of this result, we say that an $n \times n$ matrix is an **orthogonal matrix** (or simply, **orthogonal**) if its columns form an orthonormal basis for $\mathcal{R}^n$. A linear operator on $\mathcal{R}^n$ is called an **orthogonal operator** (or simply, **orthogonal**) if its standard matrix is an orthogonal matrix.

To verify that an $n \times n$ matrix Q is orthogonal, it suffices to show that the columns of Q are distinct and form an orthonormal set.

Example 2 Consider the θ-rotation matrix

$$A_\theta = \begin{bmatrix} \cos\theta & -\sin\theta \\ \sin\theta & \cos\theta \end{bmatrix}.$$

Since

$$\begin{bmatrix} \cos\theta \\ \sin\theta \end{bmatrix} \cdot \begin{bmatrix} -\sin\theta \\ \cos\theta \end{bmatrix} = (\cos\theta)(-\sin\theta) + (\sin\theta)(\cos\theta) = 0,$$

$$\begin{bmatrix} \cos\theta \\ \sin\theta \end{bmatrix} \cdot \begin{bmatrix} \cos\theta \\ \sin\theta \end{bmatrix} = \cos^2\theta + \sin^2\theta = 1,$$

and

$$\begin{bmatrix} -\sin\theta \\ \cos\theta \end{bmatrix} \cdot \begin{bmatrix} -\sin\theta \\ \cos\theta \end{bmatrix} = \sin^2\theta + \cos^2\theta = 1,$$

A_θ is an orthogonal matrix because its columns form an orthonormal set of two distinct vectors in $\mathcal{R}^2$.

The following theorem lists several conditions that are equivalent to a matrix being orthogonal:

THEOREM 6.9

The following conditions about an $n \times n$ matrix Q are equivalent:

(a) Q is orthogonal.

(b) $Q^T Q = I_n$.

(c) Q is invertible, and $Q^T = Q^{-1}$.

(d) $Q\mathbf{u} \cdot Q\mathbf{v} = \mathbf{u} \cdot \mathbf{v}$ for any $\mathbf{u}$ and $\mathbf{v}$ in $\mathcal{R}^n$. (Q preserves dot products.)

(e) $\|Q\mathbf{u}\| = \|\mathbf{u}\|$ for any $\mathbf{u}$ in $\mathcal{R}^n$. (Q preserves norms.)

PROOF We show that (a) $\Rightarrow$ (b) $\Rightarrow$ (c) $\Rightarrow$ (d) $\Rightarrow$ (e) $\Rightarrow$ (a) to establish the equivalence of these conditions.

To prove that (a) implies (b), suppose that Q is orthogonal. Then the columns of Q form an orthonormal basis for $\mathcal{R}^n$. Next observe that the (i,j)-entry of $Q^T Q$ is the dot product of the ith row of Q^T and $\mathbf{q}_j$. But the ith row of Q^T equals $\mathbf{q}_i$, and hence the (i,j)-entry of $Q^T Q$ equals $\mathbf{q}_i \cdot \mathbf{q}_j$. Since $\mathbf{q}_i \cdot \mathbf{q}_j = 1$ if $i = j$, and $\mathbf{q}_i \cdot \mathbf{q}_j = 0$ if $i \neq j$, we see that $Q^T Q = I_n$.

To prove that (b) implies (c), suppose that $Q^T Q = I_n$. Then Q is invertible and $Q^T = Q^{-1}$ by the Invertible Matrix Theorem.

To prove that (c) implies (d), assume that (c) is true. Then for any $\mathbf{u}$ and $\mathbf{v}$ in $\mathcal{R}^n$,

$$Q\mathbf{u} \cdot Q\mathbf{v} = \mathbf{u} \cdot Q^T Q\mathbf{v} = \mathbf{u} \cdot Q^{-1} Q\mathbf{v} = \mathbf{u} \cdot \mathbf{v}.$$

To prove that (d) implies (e), assume that (d) is true. Then for any $\mathbf{u}$ in $\mathcal{R}^n$,

$$\|Q\mathbf{u}\| = \sqrt{Q\mathbf{u} \cdot Q\mathbf{u}} = \sqrt{\mathbf{u} \cdot \mathbf{u}} = \|\mathbf{u}\|.$$

The proof that (e) implies (a) follows from equations (5) and (6). ■

Theorem 6.9 shows that an $n \times n$ matrix Q is orthogonal if and only if $Q^T = Q^{-1}$. By the Invertible Matrix Theorem, this condition can be checked by showing that $Q^T Q = I_n$ or $QQ^T = I_n$. Normally we use one of these simple conditions to prove that a matrix is orthogonal. For instance, we have

$$A_\theta^T A_\theta = \begin{bmatrix} \cos\theta & \sin\theta \\ -\sin\theta & \cos\theta \end{bmatrix} \begin{bmatrix} \cos\theta & -\sin\theta \\ \sin\theta & \cos\theta \end{bmatrix} = \begin{bmatrix} 1 & 0 \\ 0 & 1 \end{bmatrix} = I_2.$$

Thus A_θ is an orthogonal 2×2 matrix, confirming the result of Example 2. Notice also that the equation $QQ^T = I_n$ is equivalent to the condition that the *rows* of Q form an orthonormal basis for $\mathcal{R}^n$. (See Exercise 46.)

Practice Problem 1 ▶ Determine whether each of the following matrices is orthogonal:

$$(a)\ \begin{bmatrix} .7 & -.3 \\ .3 & .7 \end{bmatrix} \qquad (b)\ \begin{bmatrix} .3\sqrt{2} & -.8 & .3\sqrt{2} \\ .4\sqrt{2} & .6 & .4\sqrt{2} \\ .5\sqrt{2} & 0 & -.5\sqrt{2} \end{bmatrix}$$ ◀

The following general result lists some important properties of orthogonal matrices:

THEOREM 6.10

Let P and Q be $n \times n$ orthogonal matrices.

(a) $\det Q = \pm 1$.
(b) PQ is an orthogonal matrix.
(c) Q^{-1} is an orthogonal matrix.
(d) Q^T is an orthogonal matrix.

PROOF (a) Since Q is an orthogonal matrix, $Q^TQ = I_n$ by Theorem 6.9(b), and hence

$$1 = \det I_n = \det\,(Q^TQ) = (\det Q^T)(\det Q) = (\det Q)(\det Q) = (\det Q)^2.$$

Therefore $\det Q = \pm 1$.

(b) Because P and Q are orthogonal, they are invertible, and hence PQ is invertible. Therefore, by Theorem 6.9(c),

$$(PQ)^T = Q^TP^T = Q^{-1}P^{-1} = (PQ)^{-1},$$

and hence PQ is an orthogonal matrix, also by Theorem 6.9(c).

(c) By the preceding remarks, $QQ^T = I_n$, and hence

$$(Q^{-1})^TQ^{-1} = (Q^T)^{-1}Q^{-1} = (QQ^T)^{-1} = (I_n)^{-1} = I_n$$

by Theorem 2.2. Thus, by Theorem 6.9(b), Q^{-1} is an orthogonal matrix.

(d) This follows immediately from (c) and Theorem 6.9(c). ■

Since a linear operator is orthogonal if and only if its standard matrix is orthogonal, we can restate parts of Theorem 6.9 for orthogonal operators.

> If T is a linear operator on $\mathcal{R}^n$, then the following statements are equivalent:
>
> (a) T is an orthogonal operator.
> (b) $T(\mathbf{u})\cdot T(\mathbf{v}) = \mathbf{u}\cdot\mathbf{v}$ for all $\mathbf{u}$ and $\mathbf{v}$ in $\mathcal{R}^n$. (T preserves dot products.)
> (c) $\|T(\mathbf{u})\| = \|\mathbf{u}\|$ for all $\mathbf{u}$ in $\mathcal{R}^n$. (T preserves norms.)

Likewise, we can restate parts of Theorem 6.10 for orthogonal operators.

> If T and U are orthogonal operators on $\mathcal{R}^n$, then TU and T^{-1} are orthogonal operators on $\mathcal{R}^n$.

It follows from Example 1 and the first boxed statement just presented that rotations of the plane are orthogonal operators. It is clear geometrically that reflections of the plane about a line through the origin, as defined in Section 4.5, also preserve norms. Thus they are also orthogonal operators.

The following example illustrates how Theorems 6.9 and 6.10 can be used to create orthogonal operators with specific properties:

Example 3 Find an orthogonal operator T on $\mathcal{R}^3$ such that

$$T\left(\begin{bmatrix} \frac{1}{\sqrt{2}} \\ 0 \\ -\frac{1}{\sqrt{2}} \end{bmatrix}\right) = \begin{bmatrix} 0 \\ 1 \\ 0 \end{bmatrix}.$$

Solution Let

$$\mathbf{v} = \begin{bmatrix} \frac{1}{\sqrt{2}} \\ 0 \\ -\frac{1}{\sqrt{2}} \end{bmatrix}.$$

Suppose that T is such an operator with standard matrix A. Then A is an orthogonal matrix, and $A^T A = I_n$ by Theorem 6.9(b). Moreover, $A\mathbf{v} = T(\mathbf{v}) = \mathbf{e}_2$. Thus T satisfies $T(\mathbf{v}) = \mathbf{e}_2$ if and only if

$$\mathbf{v} = I_n\mathbf{v} = A^T A\mathbf{v} = A^T\mathbf{e}_2,$$

which is the second column of A^T. Hence it suffices to choose A so that A^T is an orthogonal matrix whose second column is $\mathbf{v}$. Since the columns of an orthogonal matrix form an orthonormal basis for $\mathcal{R}^3$, we must construct an orthonormal basis for $\mathcal{R}^3$ containing $\mathbf{v}$. One way to do this is to determine an orthonormal basis for $\{\mathbf{v}\}^\perp$. Now the vectors in $\{\mathbf{v}\}^\perp$ satisfy

$$\frac{1}{\sqrt{2}}x_1 - \frac{1}{\sqrt{2}}x_3 = 0,$$

or equivalently,

$$x_1 - x_3 = 0.$$

Thus a basis for the solution space of this equation is

$$\left\{\begin{bmatrix} 1 \\ 0 \\ 1 \end{bmatrix}, \begin{bmatrix} 0 \\ 1 \\ 0 \end{bmatrix}\right\},$$

which is an orthogonal set, and so

$$\left\{\begin{bmatrix} \frac{1}{\sqrt{2}} \\ 0 \\ \frac{1}{\sqrt{2}} \end{bmatrix}, \begin{bmatrix} 0 \\ 1 \\ 0 \end{bmatrix}\right\}$$

is an orthonormal basis for $\{\mathbf{v}\}^\perp$. (Note that if the basis for the solution space is not an orthogonal set, we can use the Gram–Schmidt process to replace it by an orthogonal basis, and then by an orthonormal basis.) Hence one acceptable choice for A^T is

$$A^T = \begin{bmatrix} \frac{1}{\sqrt{2}} & \frac{1}{\sqrt{2}} & 0 \\ 0 & 0 & 1 \\ \frac{1}{\sqrt{2}} & -\frac{1}{\sqrt{2}} & 0 \end{bmatrix},$$

in which case

$$A = \begin{bmatrix} \frac{1}{\sqrt{2}} & 0 & \frac{1}{\sqrt{2}} \\ \frac{1}{\sqrt{2}} & 0 & -\frac{1}{\sqrt{2}} \\ 0 & 1 & 0 \end{bmatrix}.$$

Therefore one possibility for T is the matrix transformation induced by A.

Practice Problem 2 ▶ Find an orthogonal operator T on $\mathcal{R}^3$ such that

$$T\left(\begin{bmatrix} \frac{1}{\sqrt{3}} \\ -\frac{1}{\sqrt{3}} \\ \frac{1}{\sqrt{3}} \end{bmatrix}\right) = \begin{bmatrix} 1 \\ 0 \\ 0 \end{bmatrix}.$$

◀

ORTHOGONAL OPERATORS ON THE EUCLIDEAN PLANE*

We have noted that rotations and reflections are orthogonal operators on $\mathcal{R}^2$. We now show that these are the *only* orthogonal operators on $\mathcal{R}^2$ and can be distinguished by computing the determinants of their standard matrices.

THEOREM 6.11

Let T be an orthogonal linear operator on $\mathcal{R}^2$ with standard matrix Q.

(a) If $\det Q = 1$, then T is a rotation.

(b) If $\det Q = -1$, then T is a reflection.

PROOF Suppose that $Q = \begin{bmatrix} a & c \\ b & d \end{bmatrix}$. Since Q is an orthogonal matrix, its columns are unit vectors, and so $a^2 + b^2 = 1$ and $c^2 + d^2 = 1$. Thus there exist angles θ and μ such that $a = \cos\theta$, $b = \sin\theta$, $c = \cos\mu$, and $d = \sin\mu$. Since the two columns of Q are orthogonal, θ and μ can be chosen so that they differ by 90°; that is, $\mu = \theta \pm 90°$. (See Figure 6.19.) We consider each case separately.

Case 1. $\mu = \theta + 90°$.
In this case,

$$\cos\mu = \cos(\theta + 90°) = -\sin\theta \qquad \text{and} \qquad \sin\mu = \sin(\theta + 90°) = \cos\theta.$$

---

* The remainder of this section may be omitted without loss of continuity.

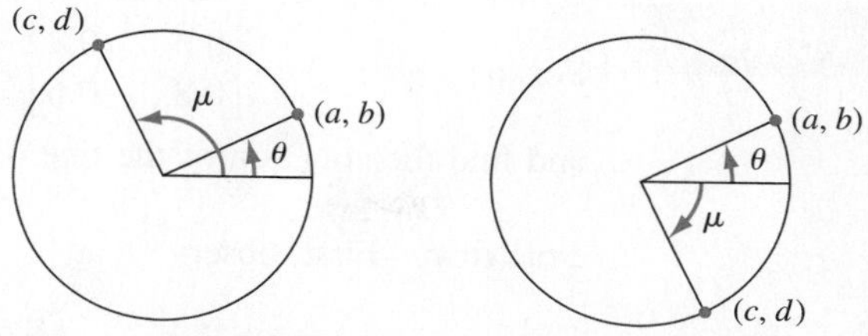

Figure 6.19 The angles θ and μ differ by 90°.

Therefore $Q = \begin{bmatrix} \cos\theta & -\sin\theta \\ \sin\theta & \cos\theta \end{bmatrix}$, which we recognize as the rotation matrix A_θ. Furthermore,

$$\det Q = \det \begin{bmatrix} \cos\theta & -\sin\theta \\ \sin\theta & \cos\theta \end{bmatrix} = \cos^2\theta + \sin^2\theta = 1.$$

Case 2. $\mu = \theta - 90°$.
In this case,

$$\cos\mu = \cos(\theta - 90^\circ) = \sin\theta \qquad \text{and} \qquad \sin\mu = \sin(\theta - 90^\circ) = -\cos\theta.$$

Therefore $Q = \begin{bmatrix} \cos\theta & \sin\theta \\ \sin\theta & -\cos\theta \end{bmatrix}$, and so the characteristic polynomial of Q is

$$\begin{aligned} \det(Q - tI_2) &= \det \begin{bmatrix} \cos\theta - t & \sin\theta \\ \sin\theta & -\cos\theta - t \end{bmatrix} \\ &= (\cos\theta - t)(-\cos\theta - t) - \sin^2\theta \\ &= t^2 - \cos^2\theta - \sin^2\theta \\ &= t^2 - 1 \\ &= (t+1)(t-1). \end{aligned}$$

It follows that Q, and hence T, has the eigenvalues 1 and -1. Let $\mathbf{b}_1$ and $\mathbf{b}_2$ be eigenvectors corresponding to the eigenvalues 1 and -1 , respectively. Then $T(\mathbf{b}_1) = \mathbf{b}_1$ and $T(\mathbf{b}_2) = -\mathbf{b}_2$. Moreover, because T preserves dot products,

$$\mathbf{b}_1 \cdot \mathbf{b}_2 = T(\mathbf{b}_1) \cdot T(\mathbf{b}_2) = \mathbf{b}_1 \cdot (-\mathbf{b}_2) = -\mathbf{b}_1 \cdot \mathbf{b}_2.$$

Therefore $2\mathbf{b}_1 \cdot \mathbf{b}_2 = 0$, and hence $\mathbf{b}_1 \cdot \mathbf{b}_2 = 0$. So $\mathbf{b}_1$ and $\mathbf{b}_2$ are orthogonal. Now let $\mathcal{L}$ be the line containing $\mathbf{0}$ in the direction of $\mathbf{b}_1$. Then $\mathbf{b}_2$ is a nonzero vector in a direction perpendicular to $\mathcal{L}$. It follows that T is a reflection about $\mathcal{L}$. Furthermore,

$$\det Q = \det \begin{bmatrix} \cos\theta & \sin\theta \\ \sin\theta & -\cos\theta \end{bmatrix} = -\cos^2\theta - \sin^2\theta = -1.$$

To summarize, we have shown that, under case 1, T is a rotation and $\det Q = 1$, and that, under case 2, T is a reflection and $\det Q = -1$. Since these are the only two cases, the result is established. ■

Example 4

For the matrix $Q = \begin{bmatrix} 0.6 & 0.8 \\ 0.8 & -0.6 \end{bmatrix}$, verify that Q is the standard matrix of a reflection, and find the equation of the line of reflection $\mathcal{L}$.

Solution First, observe that

$$Q^TQ = \begin{bmatrix} 0.6 & 0.8 \\ 0.8 & -0.6 \end{bmatrix} \begin{bmatrix} 0.6 & 0.8 \\ 0.8 & -0.6 \end{bmatrix} = I_2,$$

and hence Q is an orthogonal matrix by Theorem 6.9(b). Next, observe that

$$\det Q = -0.6^2 - 0.8^2 = -1,$$

and hence Q is the standard matrix of a reflection by Theorem 6.11. To determine the equation of $\mathcal{L}$, we first find an eigenvector of Q corresponding to the eigenvalue 1. Such a vector is a nonzero solution of the homogeneous system of equations

$$(Q - I_2)\mathbf{x} = \mathbf{0};$$

that is,

$$\begin{aligned} -0.4x_1 + 0.8x_2 &= 0 \\ 0.8x_1 - 1.6x_2 &= 0. \end{aligned}$$

The vector $\mathbf{b} = \begin{bmatrix} 2 \\ 1 \end{bmatrix}$ is such a solution. Notice that $\mathbf{b}$ lies on the line with equation $y = 0.5x$, which is, therefore, the equation of $\mathcal{L}$.

Example 5

For the matrix $Q = \begin{bmatrix} -0.6 & 0.8 \\ -0.8 & -0.6 \end{bmatrix}$, verify that Q is the standard matrix of a rotation, and find the angle of rotation.

Solution Observe that $Q^TQ = I_2$ and $\det Q = 1$. Hence Q is an orthogonal matrix and is the standard matrix of a rotation by Theorem 6.11. Thus Q is a rotation matrix, and so

$$Q = \begin{bmatrix} -0.6 & 0.8 \\ -0.8 & -0.6 \end{bmatrix} = \begin{bmatrix} \cos\theta & -\sin\theta \\ \sin\theta & \cos\theta \end{bmatrix} = A_\theta$$

where θ is the angle of rotation. Equating corresponding entries in the first column, we see that

$$\cos\theta = -0.6 \qquad \text{and} \qquad \sin\theta = -0.8.$$

It follows that θ is in the third quadrant, and

$$\theta = 180^\circ + \cos^{-1}(0.6) \approx 233.2^\circ.$$

Practice Problem 3 ▶ Show that each of the given functions $T\colon \mathcal{R}^2 \to \mathcal{R}^2$ is an orthogonal operator on $\mathcal{R}^2$. Then determine whether it is a rotation or a reflection. If it is a rotation, give the angle of rotation; if it is a reflection, give the line of reflection.

(a) $T\left(\begin{bmatrix} x_1 \\ x_2 \end{bmatrix}\right) = \frac{1}{13}\begin{bmatrix} 5x_1 - 12x_2 \\ 12x_1 + 5x_2 \end{bmatrix}$ (b) $T\left(\begin{bmatrix} x_1 \\ x_2 \end{bmatrix}\right) = \frac{1}{61}\begin{bmatrix} -60x_1 + 11x_2 \\ 11x_1 + 60x_2 \end{bmatrix}$ ◀

We have seen that the composition of two rotations on $\mathcal{R}^2$ is also a rotation. But what about the composition of two reflections, or the composition of a reflection and a rotation? The next theorem, which is an easy consequence of Theorem 6.11, answers these questions.

THEOREM 6.12

Let T and U be orthogonal operators on $\mathcal{R}^2$.

(a) If both T and U are reflections, then TU is a rotation.

(b) If one of T or U is a reflection and the other is a rotation, then TU is a reflection.

PROOF Let P and Q be the standard matrices of T and U, respectively. Then PQ is the standard matrix of TU. Furthermore, TU is an orthogonal operator, since both T and U are orthogonal operators.

(a) Since both T and U are reflections, $\det P = \det Q = -1$ by Theorem 6.11. Hence

$$\det(PQ) = (\det P)(\det Q) = (-1)(-1) = 1,$$

and therefore PQ is a rotation by Theorem 6.11.

(b) The proof of (b) is similar and is left as an exercise. ∎

RIGID MOTIONS

A function $F\colon \mathcal{R}^n \to \mathcal{R}^n$ is called a **rigid motion** if

$$\|F(\mathbf{u}) - F(\mathbf{v})\| = \|\mathbf{u} - \mathbf{v}\|$$

for all $\mathbf{u}$ and $\mathbf{v}$ in $\mathcal{R}^n$. In geometric terms, a rigid motion preserves the distance between vectors.

Any orthogonal operator is a rigid motion because if T is an orthogonal operator on $\mathcal{R}^n$, then for any $\mathbf{u}$ and $\mathbf{v}$ in $\mathcal{R}^n$,

$$\|T(\mathbf{u}) - T(\mathbf{v})\| = \|T(\mathbf{u} - \mathbf{v})\| = \|\mathbf{u} - \mathbf{v}\|.$$

Furthermore, *any rigid motion that is also linear is an orthogonal operator* because if F is a linear rigid motion, then $F(\mathbf{0}) = \mathbf{0}$, and hence, for any vector $\mathbf{v}$ in $\mathcal{R}^n$,

$$\|F(\mathbf{v})\| = \|F(\mathbf{v}) - \mathbf{0}\| = \|F(\mathbf{v}) - F(\mathbf{0})\| = \|\mathbf{v} - \mathbf{0}\| = \|\mathbf{v}\|.$$

Therefore F is an orthogonal operator by Theorem 6.9(e).

There is one kind of rigid motion that is not usually linear, namely, a *translation.* For any $\mathbf{b}$ in $\mathcal{R}^n$, the function $F_{\mathbf{b}}\colon \mathcal{R}^n \to \mathcal{R}^n$ defined by $F_{\mathbf{b}}(\mathbf{v}) = \mathbf{v} + \mathbf{b}$ is called the **translation by b**. If $\mathbf{b} \neq \mathbf{0}$, then F is not linear because $F_{\mathbf{b}}(\mathbf{0}) = \mathbf{b} \neq \mathbf{0}$. However, $F_{\mathbf{b}}$ is a rigid motion, because for any $\mathbf{u}$ and $\mathbf{v}$ in $\mathcal{R}^n$,

$$\|F_{\mathbf{b}}(\mathbf{u}) - F_{\mathbf{b}}(\mathbf{v})\| = \|(\mathbf{u} + \mathbf{b}) - (\mathbf{v} + \mathbf{b})\| = \|\mathbf{u} - \mathbf{v}\|.$$

We can use function composition to combine rigid motions to create new ones, because *the composition of two rigid motions on $\mathcal{R}^n$ is a rigid motion on $\mathcal{R}^n$*. (See Exercise 56.) It follows, for example, that if $F_{\mathbf{b}}$ is a translation and T is an orthogonal operator on $\mathcal{R}^n$, then the composition $F_{\mathbf{b}}T$ is a rigid motion. It is remarkable that the converse of this is also true; that is, any rigid motion on $\mathcal{R}^n$ can be represented as the composition of an orthogonal operator followed by a translation. To establish this result, we first prove the following theorem:

THEOREM 6.13

Let $T: \mathcal{R}^n \to \mathcal{R}^n$ be a rigid motion such that $T(\mathbf{0}) = \mathbf{0}$.

(a) $\|T(\mathbf{u})\| = \|\mathbf{u}\|$ for every $\mathbf{u}$ in $\mathcal{R}^n$.

(b) $T(\mathbf{u}) \cdot T(\mathbf{v}) = \mathbf{u} \cdot \mathbf{v}$ for all $\mathbf{u}$ and $\mathbf{v}$ in $\mathcal{R}^n$.

(c) T is linear.

(d) T is an orthogonal operator.

PROOF The proof of (a) is left as an exercise.

(b) Let $\mathbf{u}$ and $\mathbf{v}$ be in $\mathcal{R}^n$. Observe that

$$\|T(\mathbf{u}) - T(\mathbf{v})\|^2 = \|T(\mathbf{u})\|^2 - 2T(\mathbf{u}) \cdot T(\mathbf{v}) + \|T(\mathbf{v})\|^2,$$

and

$$\|\mathbf{u} - \mathbf{v}\|^2 = \|\mathbf{u}\|^2 - 2\mathbf{u} \cdot \mathbf{v} + \|\mathbf{v}\|^2.$$

Since T is a rigid motion, $\|T(\mathbf{u}) - T(\mathbf{v})\|^2 = \|\mathbf{u} - \mathbf{v}\|^2$. Hence (b) follows from the two preceding equations and (a).

(c) Let $\mathbf{u}$ and $\mathbf{v}$ be in $\mathcal{R}^n$. Then, by (a) and (b), we have

$$\begin{aligned}
\|T(\mathbf{u}+\mathbf{v}) - T(\mathbf{u}) - T(\mathbf{v})\|^2 \\
&= [T(\mathbf{u}+\mathbf{v}) - T(\mathbf{u}) - T(\mathbf{v})] \cdot [T(\mathbf{u}+\mathbf{v}) - T(\mathbf{u}) - T(\mathbf{v})] \\
&= \|T(\mathbf{u}+\mathbf{v})\|^2 + \|T(\mathbf{u})\|^2 + \|T(\mathbf{v})\|^2 - 2T(\mathbf{u}+\mathbf{v}) \cdot T(\mathbf{u}) \\
&\qquad - 2T(\mathbf{u}+\mathbf{v}) \cdot T(\mathbf{v}) + 2T(\mathbf{u}) \cdot T(\mathbf{v}) \\
&= \|\mathbf{u}+\mathbf{v}\|^2 + \|\mathbf{u}\|^2 + \|\mathbf{v}\|^2 - 2(\mathbf{u}+\mathbf{v}) \cdot \mathbf{u} \\
&\qquad - 2(\mathbf{u}+\mathbf{v}) \cdot \mathbf{v} + 2\mathbf{u} \cdot \mathbf{v}.
\end{aligned}$$

We leave it to the reader to show that the last expression equals 0. Therefore $T(\mathbf{u}+\mathbf{v}) - T(\mathbf{u}) - T(\mathbf{v}) = \mathbf{0}$, and so $T(\mathbf{u}+\mathbf{v}) = T(\mathbf{u}) + T(\mathbf{v})$. Thus T preserves vector addition. Similarly (see Exercise 58), T preserves scalar multiplication, and hence T is linear.

(d) Part (d) follows from (c) and (a). ■

Consider any rigid motion F on $\mathcal{R}^n$, and let $T: \mathcal{R}^n \to \mathcal{R}^n$ be defined by

$$T(\mathbf{v}) = F(\mathbf{v}) - F(\mathbf{0}).$$

Then T is a rigid motion, and $T(\mathbf{0}) = F(\mathbf{0}) - F(\mathbf{0}) = \mathbf{0}$. Therefore T is an orthogonal operator by Theorem 6.13. Furthermore,

$$F(\mathbf{v}) = T(\mathbf{v}) + F(\mathbf{0})$$

for any $\mathbf{v}$ in $\mathcal{R}^n$. Thus, setting $\mathbf{b} = F(\mathbf{0})$, we obtain

$$F(\mathbf{v}) = F_{\mathbf{b}}T(\mathbf{v})$$

for any $\mathbf{v}$ in $\mathcal{R}^n$, and hence F is the composition $F = F_{\mathbf{b}}T$. Combining this observation with Theorem 6.11 yields the following result:

> Any rigid motion on $\mathcal{R}^n$ is the composition of an orthogonal operator followed by a translation. Hence any rigid motion on $\mathcal{R}^2$ is the composition of a rotation or a reflection, followed by a translation.

EXERCISES

In Exercises 1–8, determine whether the given matrix is orthogonal.

1. $\frac{1}{3}\begin{bmatrix} 2 & -1 & -2 \\ 2 & 2 & 1 \end{bmatrix}$
2. $\begin{bmatrix} 1 & 1 \\ 1 & -1 \end{bmatrix}$
3. $\begin{bmatrix} 0.6 & 0.4 \\ 0.4 & -0.6 \end{bmatrix}$
4. I_5
5. $\begin{bmatrix} 0 & 1 & 0 \\ 0 & 0 & 1 \\ 1 & 0 & 0 \end{bmatrix}$
6. $\frac{1}{\sqrt{3}}\begin{bmatrix} 1 & 1 & 1 \\ 1 & -1 & 1 \\ 1 & 0 & -2 \end{bmatrix}$
7. $\frac{1}{\sqrt{2}}\begin{bmatrix} 1 & 1 \\ 0 & 0 \\ 1 & -1 \end{bmatrix}$
8. $\begin{bmatrix} \frac{2}{3} & \frac{\sqrt{2}}{2} & \frac{\sqrt{2}}{6} \\ \frac{2}{3} & -\frac{\sqrt{2}}{2} & \frac{\sqrt{2}}{6} \\ \frac{1}{3} & 0 & \frac{-2\sqrt{2}}{3} \end{bmatrix}$

In Exercises 9–16, determine whether each orthogonal matrix is the standard matrix of a rotation or of a reflection. If the operator is a rotation, determine the angle of rotation. If the operator is a reflection, determine the equation of the line of reflection.

9. $\frac{1}{\sqrt{2}}\begin{bmatrix} 1 & 1 \\ 1 & -1 \end{bmatrix}$
10. $\frac{1}{\sqrt{2}}\begin{bmatrix} 1 & -1 \\ 1 & 1 \end{bmatrix}$
11. $\frac{1}{2}\begin{bmatrix} \sqrt{3} & -1 \\ 1 & \sqrt{3} \end{bmatrix}$
12. $\frac{1}{2}\begin{bmatrix} -\sqrt{3} & 1 \\ 1 & \sqrt{3} \end{bmatrix}$
13. $\frac{1}{13}\begin{bmatrix} 5 & 12 \\ 12 & -5 \end{bmatrix}$
14. $\begin{bmatrix} 0 & 1 \\ 1 & 0 \end{bmatrix}$
15. $\begin{bmatrix} 0 & 1 \\ -1 & 0 \end{bmatrix}$
16. $\frac{1}{2}\begin{bmatrix} -1 & \sqrt{3} \\ \sqrt{3} & 1 \end{bmatrix}$

In Exercises 17–36, determine whether the statements are true or false.

17. The rows of an $n \times n$ orthogonal matrix form an orthonormal basis for $\mathcal{R}^n$.
18. If $T\colon \mathcal{R}^n \to \mathcal{R}^n$ is a function such that $\|T(\mathbf{u}) - T(\mathbf{v})\| = \|\mathbf{u} - \mathbf{v}\|$ for all vectors $\mathbf{u}$ and $\mathbf{v}$ in $\mathcal{R}^n$, then T is an orthogonal operator.
19. Every linear operator preserves dot products.
20. If a linear operator preserves dot products, then it preserves norms.
21. If P is an orthogonal matrix, then P^T is an orthogonal matrix.
22. If P and Q are $n \times n$ orthogonal matrices, then PQ^T is an orthogonal matrix.
23. If P and Q are $n \times n$ orthogonal matrices, then $P + Q$ is an orthogonal matrix.
24. If P is an $n \times n$ matrix such that $\det P = \pm 1$, then P is an orthogonal matrix.
25. If P and Q are $n \times n$ orthogonal matrices, then PQ is an orthogonal matrix.
26. If the columns of an $n \times n$ matrix form an orthogonal basis for $\mathcal{R}^n$, then the matrix is an orthogonal matrix.
27. For any subspace W of $\mathcal{R}^n$, the matrix P_W is an orthogonal matrix.
28. If P is a matrix such that $P^T = P^{-1}$, then P is an orthogonal matrix.
29. Every orthogonal matrix is invertible.
30. The linear operator on $\mathcal{R}^2$ that rotates a vector by an angle θ is an orthogonal operator.
31. If Q is the standard matrix of an orthogonal linear operator T on $\mathcal{R}^2$ and $\det Q = -1$, then T is a rotation.
32. Every rigid motion is an orthogonal operator.
33. Every rigid motion is a linear operator.
34. Every orthogonal operator is a rigid motion.
35. The composition of two rigid motions on $\mathcal{R}^n$ is a rigid motion on $\mathcal{R}^n$.
36. Every rigid motion on $\mathcal{R}^n$ is the composition of an orthogonal operator followed by a translation.

---

37. Find an orthogonal operator T on $\mathcal{R}^3$ such that

$$T\left(\frac{1}{7}\begin{bmatrix} 3 \\ -2 \\ 6 \end{bmatrix}\right) = \begin{bmatrix} 0 \\ 0 \\ 1 \end{bmatrix}.$$

38. Find an orthogonal operator T on $\mathcal{R}^3$ such that $T(\mathbf{v}) = \mathbf{w}$, where

$$\mathbf{v} = \frac{1}{\sqrt{10}}\begin{bmatrix} 3 \\ 1 \\ 0 \end{bmatrix} \quad \text{and} \quad \mathbf{w} = \frac{1}{\sqrt{5}}\begin{bmatrix} 0 \\ -2 \\ 1 \end{bmatrix}.$$

39. Let $0^\circ < \theta < 180^\circ$ be a particular angle, and suppose that T is the linear operator on $\mathcal{R}^3$ such that

$$T(\mathbf{e}_1) = \cos\theta\, \mathbf{e}_1 + \sin\theta\, \mathbf{e}_2$$
$$T(\mathbf{e}_2) = -\sin\theta\, \mathbf{e}_1 + \cos\theta\, \mathbf{e}_2$$
$$T(\mathbf{e}_3) = \mathbf{e}_3.$$

(a) Prove that T is an orthogonal operator.

(b) Find the eigenvalues of T and a basis for each eigenspace.

(c) Give a geometric description of T.

40. Suppose that $\{\mathbf{v}_1, \mathbf{v}_2, \ldots, \mathbf{v}_k\}$ and $\{\mathbf{w}_1, \mathbf{w}_2, \ldots, \mathbf{w}_k\}$ are orthonormal subsets of $\mathcal{R}^n$, each containing k vectors. The following sequence of steps can be used to obtain an orthogonal operator T on $\mathcal{R}^n$ such that $T(\mathbf{v}_i) = \mathbf{w}_i$ for $i = 1, 2, \ldots, k$:

(i) Extend $\{\mathbf{v}_1, \mathbf{v}_2, \ldots, \mathbf{v}_k\}$ and $\{\mathbf{w}_1, \mathbf{w}_2, \ldots, \mathbf{w}_k\}$ to orthonormal bases $\mathcal{B} = \{\mathbf{v}_1, \mathbf{v}_2, \ldots, \mathbf{v}_n\}$ and $\mathcal{C} = \{\mathbf{w}_1, \mathbf{w}_2, \ldots, \mathbf{w}_n\}$, respectively, for $\mathcal{R}^n$.

(ii) Let B and C be the $n \times n$ matrices whose columns are the vectors in $\mathcal{B}$ and $\mathcal{C}$, listed in the same order.

(iii) Let $A = CB^T$, and $T = T_A$, the matrix transformation induced by A.

Prove that the resulting operator T satisfies the stated requirements; that is, T is an orthogonal operator on $\mathcal{R}^n$ such that $T(\mathbf{v}_i) = \mathbf{w}_i$ for $i = 1, 2, \ldots, k$.

41. Apply the result of Exercise 40 to obtain an orthogonal operator T on $\mathcal{R}^3$ such that $T(\mathbf{v}_1) = \mathbf{w}_1$ and $T(\mathbf{v}_2) = \mathbf{w}_2$, where

$$\mathbf{v}_1 = \frac{1}{3}\begin{bmatrix} 1 \\ 2 \\ 2 \end{bmatrix}, \quad \mathbf{v}_2 = \frac{1}{3}\begin{bmatrix} 2 \\ 1 \\ -2 \end{bmatrix}, \quad \mathbf{w}_1 = \frac{1}{7}\begin{bmatrix} 2 \\ 3 \\ 6 \end{bmatrix}, \quad \text{and}$$
$$\mathbf{w}_2 = \frac{1}{7}\begin{bmatrix} 6 \\ 2 \\ -3 \end{bmatrix}.$$

42. Let T be the linear operator on $\mathcal{R}^3$ defined by

$$T\left(\begin{bmatrix} x_1 \\ x_2 \\ x_3 \end{bmatrix}\right) = \begin{bmatrix} -x_1 \\ x_2 \\ x_3 \end{bmatrix}.$$

Prove that T is an orthogonal operator.

43. Let Q be a 2×2 orthogonal matrix such that $Q \neq I_2$ and $Q \neq -I_2$. Prove that Q is diagonalizable if and only if Q is a reflection.

44. Let W be a subspace of $\mathcal{R}^n$. Let T be the linear operator on $\mathcal{R}^n$ defined by $T(\mathbf{v}) = \mathbf{w} - \mathbf{z}$, where $\mathbf{v} = \mathbf{w} + \mathbf{z}$, $\mathbf{w}$ is in W, and $\mathbf{z}$ is in $W^\perp$. (See Theorem 6.7.)

(a) Prove that T is an orthogonal operator.

(b) Let $U : \mathcal{R}^n \to \mathcal{R}^n$ be the function defined by $U(\mathbf{v}) = \frac{1}{2}(\mathbf{v} + T(\mathbf{v}))$ for $\mathbf{v}$ in $\mathcal{R}^n$. Prove that the standard matrix of U is P_W, the orthogonal projection matrix for W.

45. Let $\{\mathbf{v}, \mathbf{w}\}$ be an orthonormal basis for $\mathcal{R}^2$, and let $T : \mathcal{R}^2 \to \mathcal{R}^2$ be the function defined by

$$T(\mathbf{u}) = (\mathbf{u} \cdot \mathbf{v} \cos\theta + \mathbf{u} \cdot \mathbf{w} \sin\theta)\mathbf{v} + (-\mathbf{u} \cdot \mathbf{v} \sin\theta + \mathbf{u} \cdot \mathbf{w} \cos\theta)\mathbf{w}.$$

Prove that T is an orthogonal operator.

46. Let Q be an $n \times n$ matrix. Prove that Q is an orthogonal matrix if and only if the rows of Q form an orthonormal basis for $\mathcal{R}^n$. *Hint:* Interpret the (i,j)-entry of QQ^T as the dot product of the ith and jth rows of Q.

47. Use Theorem 6.10 to prove that if T and U are orthogonal operators on $\mathcal{R}^n$, then both TU and T^{-1} are orthogonal operators.

48. Prove Theorem 6.12(b).

49.[12] Prove that if Q is an orthogonal matrix and λ is a (real) eigenvalue of Q, then $\lambda = \pm 1$.

50. Let U be a reflection and T be a rotation of $\mathcal{R}^2$. Prove the following equalities:

(a) $U^2 = I$, where I is the identity transformation on $\mathcal{R}^2$, and so $U^{-1} = U$.

(b) $TUT = U$. *Hint:* Consider TU.

(c) $UTU = T^{-1}$.

51. Let T be an orthogonal operator on $\mathcal{R}^2$.

(a) Prove that if T is a rotation, then T^{-1} is also a rotation. How is the angle of rotation of T^{-1} related to the angle of rotation of T?

(b) Prove that if T is a reflection, then T^{-1} is also a reflection. How is the line of reflection of T^{-1} related to the line of reflection of T?

52. Let U be a reflection of $\mathcal{R}^2$, and let T be the linear operator on $\mathcal{R}^2$ that rotates a vector by an angle θ. By Theorem 6.12, TU is a reflection. If U reflects about the line $\mathcal{L}$, we can describe the line about which TU reflects in terms of $\mathcal{L}$ and θ. To do so, let S be the linear operator on $\mathcal{R}^2$ that rotates a vector by the angle $\theta/2$, and let $\mathbf{b}$ be a nonzero vector parallel to $\mathcal{L}$, so that $\mathbf{b}$ is an eigenvector of U corresponding to the eigenvalue 1.

(a) Prove that $S(\mathbf{b})$ is an eigenvector of TU corresponding to eigenvalue 1. *Hint:* Show that $TS^{-1} = S$, and use Exercise 50.

(b) Prove that if $\mathcal{L}'$ is obtained by rotating $\mathcal{L}$ by the angle $\theta/2$, then TU is the reflection about $\mathcal{L}'$.

53. Let W be a 1-dimensional subspace of $\mathcal{R}^2$. Regard W as a line containing the origin. Let P_W be the orthogonal projection matrix on W, and let $Q_W = 2P_W - I_2$. Prove the following results:

(a) $Q_W^T = Q_W$.

(b) $Q_W^2 = I_2$.

(c) Q_W is an orthogonal matrix.

[12] This exercise is used in Section 6.9 (on page 477).

(d) $Q_W \mathbf{w} = \mathbf{w}$ for all $\mathbf{w}$ in W.

(e) $Q_W \mathbf{v} = -\mathbf{v}$ for all $\mathbf{v}$ in $W^{\perp}$.

(f) Q_W is the standard matrix of the reflection of $\mathcal{R}^2$ about W.

54. Let T be a linear operator on $\mathcal{R}^n$, and suppose that $\{\mathbf{v}_1, \mathbf{v}_2, \ldots, \mathbf{v}_n\}$ is an orthonormal basis for $\mathcal{R}^n$. Prove that T is an orthogonal operator if and only if $\{T(\mathbf{v}_1), T(\mathbf{v}_2), \ldots, T(\mathbf{v}_n)\}$ is also an orthonormal basis for $\mathcal{R}^n$.

55. Suppose that $\{\mathbf{v}_1, \mathbf{v}_2, \ldots, \mathbf{v}_n\}$ and $\{\mathbf{w}_1, \mathbf{w}_2, \ldots, \mathbf{w}_n\}$ are orthonormal bases for $\mathcal{R}^n$. Prove that there exists a unique orthogonal operator T on $\mathcal{R}^n$ such that $T(\mathbf{v}_i) = \mathbf{w}_i$ for $1 \le i \le n$. (This is the converse of Exercise 54.)

56. Prove that the composition of two rigid motions on $\mathcal{R}^n$ is a rigid motion.

57. Prove Theorem 6.13(a).

58. Complete the proof of Theorem 6.13(c) by showing that T preserves scalar multiplication.

59. Let $F: \mathcal{R}^n \to \mathcal{R}^n$ be a rigid motion. By the final result of this section, there exists an $n \times n$ orthogonal matrix Q and a vector $\mathbf{b}$ in $\mathcal{R}^n$ such that

$$F(\mathbf{v}) = Q\mathbf{v} + \mathbf{b}$$

for all $\mathbf{v}$ in $\mathcal{R}^n$. Prove that Q and $\mathbf{b}$ are unique.

60. Suppose that F and G are rigid motions on $\mathcal{R}^n$. By Exercise 59, there exist unique orthogonal matrices P and Q and unique vectors $\mathbf{a}$ and $\mathbf{b}$ such that

$$F(\mathbf{v}) = Q\mathbf{v} + \mathbf{b} \quad \text{and} \quad G(\mathbf{v}) = P\mathbf{v} + \mathbf{a}$$

for all $\mathbf{v}$ in $\mathcal{R}^n$. By Exercise 56, the composition of F and G is a rigid motion, and hence by Exercise 59 there exist a unique orthogonal matrix R and a unique vector $\mathbf{c}$ such that $F(G(\mathbf{v})) = R\mathbf{v} + \mathbf{c}$ for all $\mathbf{v}$ in $\mathcal{R}^n$. Find R and $\mathbf{c}$ in terms of P, Q, $\mathbf{a}$, and $\mathbf{b}$.

In Exercises 61–64, a rigid motion $F: \mathcal{R}^2 \to \mathcal{R}^2$ is given. Use the given information to find the orthogonal matrix Q and the vector $\mathbf{b}$ such that $F(\mathbf{v}) = Q\mathbf{v} + \mathbf{b}$ for all $\mathbf{v}$ in $\mathcal{R}^2$.

61. $F\left(\begin{bmatrix}1\\0\end{bmatrix}\right) = \begin{bmatrix}2\\4\end{bmatrix}$, $F\left(\begin{bmatrix}0\\1\end{bmatrix}\right) = \begin{bmatrix}1\\3\end{bmatrix}$, and $F\left(\begin{bmatrix}1\\1\end{bmatrix}\right) = \begin{bmatrix}2\\3\end{bmatrix}$

62. $F\left(\begin{bmatrix}2\\1\end{bmatrix}\right) = \begin{bmatrix}1\\2\end{bmatrix}$, $F\left(\begin{bmatrix}1\\3\end{bmatrix}\right) = \begin{bmatrix}2\\0\end{bmatrix}$, and $F\left(\begin{bmatrix}7\\1\end{bmatrix}\right) = \begin{bmatrix}4\\6\end{bmatrix}$

63. $F\left(\begin{bmatrix}3\\-1\end{bmatrix}\right) = \begin{bmatrix}3\\4\end{bmatrix}$, $F\left(\begin{bmatrix}1\\3\end{bmatrix}\right) = \begin{bmatrix}-1\\6\end{bmatrix}$ and $F\left(\begin{bmatrix}2\\1\end{bmatrix}\right) = \begin{bmatrix}1\\5\end{bmatrix}$

64. $F\left(\begin{bmatrix}1\\2\end{bmatrix}\right) = \begin{bmatrix}5\\3\end{bmatrix}$, $F\left(\begin{bmatrix}3\\1\end{bmatrix}\right) = \begin{bmatrix}3\\4\end{bmatrix}$ and $F\left(\begin{bmatrix}-2\\1\end{bmatrix}\right) = \begin{bmatrix}6\\0\end{bmatrix}$

65. Let $T: \mathcal{R}^n \to \mathcal{R}^n$ be a function such that $T(\mathbf{u}) \cdot T(\mathbf{v}) = \mathbf{u} \cdot \mathbf{v}$ for all $\mathbf{u}$ and $\mathbf{v}$ in $\mathcal{R}^n$. Prove that T is linear, and hence is an orthogonal operator. *Hint:* Apply Theorem 6.13.

66. Use Exercise 98 of Section 6.1 to prove that if T is an orthogonal operator on $\mathcal{R}^2$, then T preserves the angle between any two nonzero vectors. That is, for any nonzero vectors $\mathbf{u}$ and $\mathbf{v}$ in $\mathcal{R}^2$, the angle between $T(\mathbf{u})$ and $T(\mathbf{v})$ equals the angle between $\mathbf{u}$ and $\mathbf{v}$.

67. Let E_n be the $n \times n$ matrix, all of whose entries are ones. Let $A_n = I_n - \frac{2}{n}E_n$.

(a) Determine A_n for $n = 2, 3, 6$.

(b) Compute $A_n^T A_n$ for $n = 2, 3, 6$, and use Theorem 6.9(b) to conclude that A_n is an orthogonal matrix.

(c) Prove that A_n is symmetric for all n.

(d) Prove that A_n is an orthogonal matrix for all n. *Hint:* First prove that $E_n^2 = nE_n$.

68. In $\mathcal{R}^2$, let $\mathcal{L}$ be the line through the origin that makes an angle θ with the positive half of the x-axis, and let U be the reflection of $\mathcal{R}^2$ about $\mathcal{L}$. Prove that the standard matrix of U is

$$\begin{bmatrix}\cos 2\theta & \sin 2\theta\\ \sin 2\theta & -\cos 2\theta\end{bmatrix}.$$

69. In $\mathcal{R}^2$, let $\mathcal{L}$ be the line through the origin with slope m, and let U be the reflection of $\mathcal{R}^2$ about $\mathcal{L}$. Prove that the standard matrix of U is

$$\frac{1}{1+m^2}\begin{bmatrix}1-m^2 & 2m\\ 2m & m^2-1\end{bmatrix}.$$

In Exercises 70–73, use either a calculator with matrix capabilities or computer software such as MATLAB to solve each problem.

70. Find the standard matrix of the reflection about the line in $\mathcal{R}^2$ that contains the origin and the point with coordinates $(2.43, -1.31)$.

71. Find the standard matrix of the reflection about the line in $\mathcal{R}^2$ that contains the origin and the point with coordinates $(3.27, 1.14)$.

72. According to Theorem 6.12, the composition of two reflections is a rotation. Find the angle, to the nearest degree, that a vector is rotated if it is reflected about the line with equation $y = 3.21x$ and the result is then reflected about the line with equation $y = 1.54x$.

73. According to Theorem 6.12, the composition of two reflections is a rotation. Find the angle, to the nearest degree, that a vector is rotated if it is reflected about the line with equation $y = 1.23x$ and the result is then reflected about the line with equation $y = -0.24x$.

SOLUTIONS TO THE PRACTICE PROBLEMS

1. (a) The product of this matrix and its transpose is

$$\begin{bmatrix} .7 & -.3 \\ .3 & .7 \end{bmatrix}\begin{bmatrix} .7 & -.3 \\ .3 & .7 \end{bmatrix}^T = \begin{bmatrix} .7 & -.3 \\ .3 & .7 \end{bmatrix}\begin{bmatrix} .7 & .3 \\ -.3 & .7 \end{bmatrix}$$
$$= \begin{bmatrix} .58 & 0 \\ 0 & .58 \end{bmatrix} \neq I_2,$$

and hence the matrix is not orthogonal.

(b) The product of this matrix and its transpose is

$$\begin{bmatrix} .3\sqrt{2} & -.8 & .3\sqrt{2} \\ .4\sqrt{2} & .6 & .4\sqrt{2} \\ .5\sqrt{2} & 0 & -.5\sqrt{2} \end{bmatrix}\begin{bmatrix} .3\sqrt{2} & .4\sqrt{2} & .5\sqrt{2} \\ -.8 & .6 & 0 \\ .3\sqrt{2} & .4\sqrt{2} & -.5\sqrt{2} \end{bmatrix}$$
$$= I_3,$$

and hence this matrix is orthogonal.

2. Let A be the standard matrix of such an operator, and let

$$\mathbf{v} = \begin{bmatrix} \frac{1}{\sqrt{3}} \\ -\frac{1}{\sqrt{3}} \\ \frac{1}{\sqrt{3}} \end{bmatrix}.$$

Then, as in Example 3, $\mathbf{v} = A^T\mathbf{e}_1$, which is the first column of A^T. We choose the second and third columns of A^T so that the three columns form an orthonormal basis for $\mathcal{R}^3$. These columns are orthogonal to $\mathbf{v}$, and hence satisfy

$$x_1 - x_2 + x_3 = 0.$$

A basis for the solution space of this equation is

$$\left\{\begin{bmatrix} 1 \\ 1 \\ 0 \end{bmatrix}, \begin{bmatrix} -1 \\ 0 \\ 1 \end{bmatrix}\right\},$$

which, unfortunately, is not an orthogonal set. We apply the Gram–Schmidt process to this set and normalize the resulting orthogonal set to obtain the orthonormal basis

$$\left\{\begin{bmatrix} \frac{1}{\sqrt{2}} \\ \frac{1}{\sqrt{2}} \\ 0 \end{bmatrix}, \begin{bmatrix} -\frac{1}{\sqrt{6}} \\ \frac{1}{\sqrt{6}} \\ \frac{2}{\sqrt{6}} \end{bmatrix}\right\}$$

for $\{\mathbf{v}\}^\perp$. Hence one acceptable choice for A^T is

$$A^T = \begin{bmatrix} \frac{1}{\sqrt{3}} & \frac{1}{\sqrt{2}} & -\frac{1}{\sqrt{6}} \\ -\frac{1}{\sqrt{3}} & \frac{1}{\sqrt{2}} & \frac{1}{\sqrt{6}} \\ \frac{1}{\sqrt{3}} & 0 & \frac{2}{\sqrt{6}} \end{bmatrix},$$

in which case

$$A = \begin{bmatrix} \frac{1}{\sqrt{3}} & -\frac{1}{\sqrt{3}} & \frac{1}{\sqrt{3}} \\ \frac{1}{\sqrt{2}} & \frac{1}{\sqrt{2}} & 0 \\ -\frac{1}{\sqrt{6}} & \frac{1}{\sqrt{6}} & \frac{2}{\sqrt{6}} \end{bmatrix}.$$

Therefore one possibility for T is the matrix transformation induced by A.

3. (a) The standard matrix of T is

$$\begin{bmatrix} \frac{5}{13} & -\frac{12}{13} \\ \frac{12}{13} & \frac{5}{13} \end{bmatrix},$$

which has determinant equal to 1. Thus T is a rotation, and its standard matrix is a rotation matrix A_θ, where θ is the angle of rotation. Hence

$$\begin{bmatrix} \frac{5}{13} & -\frac{12}{13} \\ \frac{12}{13} & \frac{5}{13} \end{bmatrix} = \begin{bmatrix} \cos\theta & -\sin\theta \\ \sin\theta & \cos\theta \end{bmatrix}.$$

Comparing the corresponding entries of the first column, we have that

$$\cos\theta = \frac{5}{13} \quad \text{and} \quad \sin\theta = \frac{12}{13}.$$

Thus θ can be chosen as the angle in the first quadrant with

$$\theta = \cos^{-1}\left(\frac{5}{13}\right) \approx 67.4^\circ.$$

(b) The standard matrix of T is

$$Q = \frac{1}{61}\begin{bmatrix} -60 & 11 \\ 11 & 60 \end{bmatrix},$$

which has determinant equal to -1. Hence T is a reflection. To determine the line of reflection, we first find an eigenvector of Q corresponding to the eigenvalue 1. One such eigenvector is $\mathbf{b} = \begin{bmatrix} 1 \\ 11 \end{bmatrix}$, which lies on the line with equation $y = 11x$. This is the line of reflection.

6.6 SYMMETRIC MATRICES

We have seen in Sections 5.3 and 5.4 that diagonalizable matrices and operators have important properties that allow us to solve difficult computational problems. For example, in the case of an $n \times n$ diagonalizable matrix A, the existence of an invertible matrix P and a diagonal matrix D such that $A = PDP^{-1}$ allows us to compute powers of A very easily because $A^m = PD^mP^{-1}$ for any positive integer m. Recall that the columns of P form a basis for $\mathcal{R}^n$ consisting of eigenvectors of A, and the diagonal entries of D are the corresponding eigenvalues.

Now suppose that the columns of P also form an orthonormal basis for $\mathcal{R}^n$; that is, P is an orthogonal matrix. By Theorem 6.9, $P^T = P^{-1}$. Therefore

$$A^T = (PDP^{-1})^T = (PDP^T)^T = (P^T)^T D^T P^T = PDP^T = PDP^{-1} = A.$$

It follows that $A^T = A$. Recall from Section 2.1 that such a matrix is called *symmetric*.

The preceding calculation shows that if there is an orthonormal basis for $\mathcal{R}^n$ consisting of eigenvectors of a matrix, then the matrix must be symmetric. The next result is useful in proving the converse.

THEOREM 6.14

If $\mathbf{u}$ and $\mathbf{v}$ are eigenvectors of a symmetric matrix that correspond to distinct eigenvalues, then $\mathbf{u}$ and $\mathbf{v}$ are orthogonal.

PROOF Let A be a symmetric matrix. Suppose that $\mathbf{u}$ and $\mathbf{v}$ are eigenvectors of A associated with distinct eigenvalues λ and μ, respectively. Then

$$A\mathbf{u} \cdot \mathbf{v} = \lambda\mathbf{u} \cdot \mathbf{v} = \lambda(\mathbf{u} \cdot \mathbf{v}).$$

Also, by a result in Section 6.1,

$$A\mathbf{u} \cdot \mathbf{v} = \mathbf{u} \cdot A^T\mathbf{v} = \mathbf{u} \cdot A\mathbf{v} = \mathbf{u} \cdot \mu\mathbf{v} = \mu(\mathbf{u} \cdot \mathbf{v}).$$

So $\lambda(\mathbf{u} \cdot \mathbf{v}) = \mu(\mathbf{u} \cdot \mathbf{v})$. Because λ and μ are distinct, we have $\mathbf{u} \cdot \mathbf{v} = 0$; that is, $\mathbf{u}$ and $\mathbf{v}$ are orthogonal. ∎

Consider a symmetric 2×2 matrix $A = \begin{bmatrix} a & b \\ b & c \end{bmatrix}$. Its characteristic polynomial is

$$\det(A - tI_2) = \det \begin{bmatrix} a - t & b \\ b & c - t \end{bmatrix} = (a - t)(c - t) - b^2 = t^2 - (a + c)t + ac - b^2.$$

To check if this quadratic polynomial has real roots, we compute its discriminant

$$(a + c)^2 - 4(ac - b^2) = (a - c)^2 + 4b^2.$$

Because this is a sum of two squares, it is nonnegative for all choices of a, b, and c. Therefore the eigenvalues of A are real.

Case 1. The discriminant is positive.
In this case, the eigenvalues of A are distinct and, by Theorem 6.14, any two corresponding eigenvectors are orthogonal.

Case 2. The discriminant is zero.
If $(a-c)^2 + 4b^2 = 0$, then $a = c$ and $b = 0$. Thus

$$A = \begin{bmatrix} c & 0 \\ 0 & c \end{bmatrix} = cI_2.$$

In this case, we may choose the two standard vectors as orthogonal eigenvectors of A.

In either case, multiplying each eigenvector by the reciprocal of its norm produces an orthonormal basis for $\mathcal{R}^2$ consisting of eigenvectors of A.

Example 1

For the symmetric matrix $A = \begin{bmatrix} 2 & -2 \\ -2 & 5 \end{bmatrix}$, find an orthogonal matrix P such that P^TAP is a diagonal matrix.

Solution We need to find an orthonormal basis for $\mathcal{R}^2$ consisting of eigenvectors of A. Using the methods of Chapter 5, we find that the eigenvalues of A are 6 and 1 with corresponding eigenvectors $\begin{bmatrix} -1 \\ 2 \end{bmatrix}$ and $\begin{bmatrix} 2 \\ 1 \end{bmatrix}$. Notice that these two vectors are orthogonal, as predicted by Theorem 6.14. By multiplying each of these vectors by the reciprocal of its norm, we obtain the orthonormal basis $\left\{ \frac{1}{\sqrt{5}} \begin{bmatrix} -1 \\ 2 \end{bmatrix}, \frac{1}{\sqrt{5}} \begin{bmatrix} 2 \\ 1 \end{bmatrix} \right\}$ for $\mathcal{R}^2$. So, for

$$P = \begin{bmatrix} \frac{-1}{\sqrt{5}} & \frac{2}{\sqrt{5}} \\ \frac{2}{\sqrt{5}} & \frac{1}{\sqrt{5}} \end{bmatrix} = \frac{1}{\sqrt{5}} \begin{bmatrix} -1 & 2 \\ 2 & 1 \end{bmatrix} \quad \text{and} \quad D = \begin{bmatrix} 6 & 0 \\ 0 & 1 \end{bmatrix},$$

we have $P^TAP = D$.

More generally, the following theorem is true:

THEOREM 6.15

An $n \times n$ matrix A is symmetric if and only if there is an orthonormal basis for $\mathcal{R}^n$ consisting of eigenvectors of A. In this case, there exists an orthogonal matrix P and a diagonal matrix D such that $P^TAP = D$.

The proof of Theorem 6.15 requires knowledge of complex numbers; it can be found in Appendix C.

We conclude from Theorem 6.14 that the vectors in any eigenspace of a symmetric $n \times n$ matrix A are orthogonal to the vectors in any other eigenspace of A. So if we combine all the vectors from orthonormal bases for the distinct eigenspaces of A, we obtain an orthonormal basis for $\mathcal{R}^n$ consisting of eigenvectors of A.

Example 2

For the matrix

$$A = \begin{bmatrix} 4 & 2 & 2 \\ 2 & 4 & 2 \\ 2 & 2 & 4 \end{bmatrix},$$

find an orthogonal matrix P such that P^TAP is a diagonal matrix D.

Solution As in Example 1, we know that because A is symmetric, such a matrix P exists. We compute the characteristic polynomial of A to be $-(t-2)^2(t-8)$. It can be shown that the vectors

$$\begin{bmatrix} -1 \\ 1 \\ 0 \end{bmatrix} \quad \text{and} \quad \begin{bmatrix} -1 \\ 0 \\ 1 \end{bmatrix}$$

form a basis for the eigenspace corresponding to the eigenvalue 2. Because these vectors are not orthogonal, we apply the Gram–Schmidt process to these two vectors and obtain the orthogonal vectors

$$\begin{bmatrix} -1 \\ 1 \\ 0 \end{bmatrix} \quad \text{and} \quad -\frac{1}{2}\begin{bmatrix} 1 \\ 1 \\ -2 \end{bmatrix}.$$

These two vectors form an orthogonal basis for the eigenspace corresponding to the eigenvalue 2. Furthermore, we can choose any eigenvector corresponding to the eigenvalue 8, for example $\begin{bmatrix} 1 \\ 1 \\ 1 \end{bmatrix}$, because by Theorem 6.14 it must be orthogonal to the two preceding vectors. So the set

$$\left\{ \begin{bmatrix} \frac{-1}{\sqrt{2}} \\ \frac{1}{\sqrt{2}} \\ 0 \end{bmatrix}, \begin{bmatrix} \frac{1}{\sqrt{6}} \\ \frac{1}{\sqrt{6}} \\ \frac{-2}{\sqrt{6}} \end{bmatrix}, \begin{bmatrix} \frac{1}{\sqrt{3}} \\ \frac{1}{\sqrt{3}} \\ \frac{1}{\sqrt{3}} \end{bmatrix} \right\}$$

is an orthonormal basis of eigenvectors of A. Consequently, one possible choice of the orthogonal matrix P and diagonal matrix D is

$$P = \begin{bmatrix} \frac{-1}{\sqrt{2}} & \frac{1}{\sqrt{6}} & \frac{1}{\sqrt{3}} \\ \frac{1}{\sqrt{2}} & \frac{1}{\sqrt{6}} & \frac{1}{\sqrt{3}} \\ 0 & \frac{-2}{\sqrt{6}} & \frac{1}{\sqrt{3}} \end{bmatrix} \quad \text{and} \quad D = \begin{bmatrix} 2 & 0 & 0 \\ 0 & 2 & 0 \\ 0 & 0 & 8 \end{bmatrix}.$$

Practice Problem 1 ▶ For the matrix

$$A = \begin{bmatrix} 2 & 4 & 4 \\ 4 & 17 & -1 \\ 4 & -1 & 17 \end{bmatrix},$$

find an orthogonal matrix P and a diagonal matrix D such that $P^TAP = D$. ◀

QUADRATIC FORMS

Historically, the conic sections have played an important role in physics. For example, ellipses describe the motion of the planets, hyperbolas are used in the manufacture of telescopes, and parabolas describe the paths of projectiles. In the plane, the equations of all the conic sections (the circle, ellipse, parabola, and hyperbola) can be obtained from

$$ax^2 + 2bxy + cy^2 + dx + ey + f = 0 \tag{7}$$

by making various choices for the coefficients.[13] For example, $a = c = 1$, $b = d = e = 0$, and $f = -9$ yields the equation

$$x^2 + y^2 = 9,$$

which represents a circle with radius 3 and center at the origin. If we change d to 8 and complete the square, we obtain

$$(x + 4)^2 + y^2 = 25,$$

which represents a circle with radius 5 and center at the point $(-4, 0)$.

In the case that the coefficient of the xy-term of the equation of a conic section is zero, its major axis is parallel to the x- or the y-axis. Figure 6.20 shows representative graphs and the forms of the corresponding equations for an ellipse and a hyperbola centered at the origin and having the x-axis as the major axis.

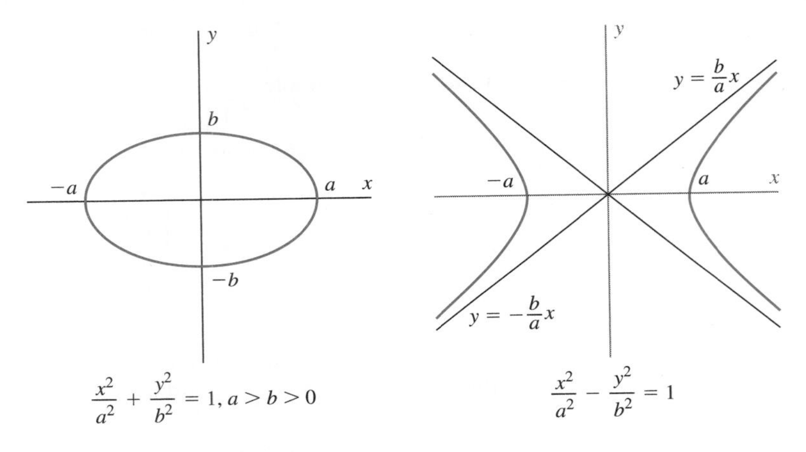

Figure 6.20 The x-axis is the major axis of the conic section.

If the coefficient of the xy-term of the equation of a conic section is not zero, the major axis is not parallel to either of the coordinate axes. (See Figure 6.21.) In this case, it is always possible to rotate the x- and y-axes to new x'- and y'-axes so that the major axis of the conic section is parallel to one of these new axes. When the equation of the conic section is written in the $x'y'$-coordinate system by the techniques of Section 4.4, the coefficient of the $x'y'$-term of the equation is zero. We can use our knowledge of orthogonal and symmetric matrices to discover an appropriate angle θ by which to rotate the original coordinate system.

We begin by considering the **associated quadratic form** of equation (7), namely,

$$ax^2 + 2bxy + cy^2.$$

We assume that $b \neq 0$, so the coefficient of the xy-term of equation (7) is not zero. If we let

$$A = \begin{bmatrix} a & b \\ b & c \end{bmatrix} \quad \text{and} \quad \mathbf{v} = \begin{bmatrix} x \\ y \end{bmatrix},$$

[13] The coefficient $2b$ is used for computational purposes.

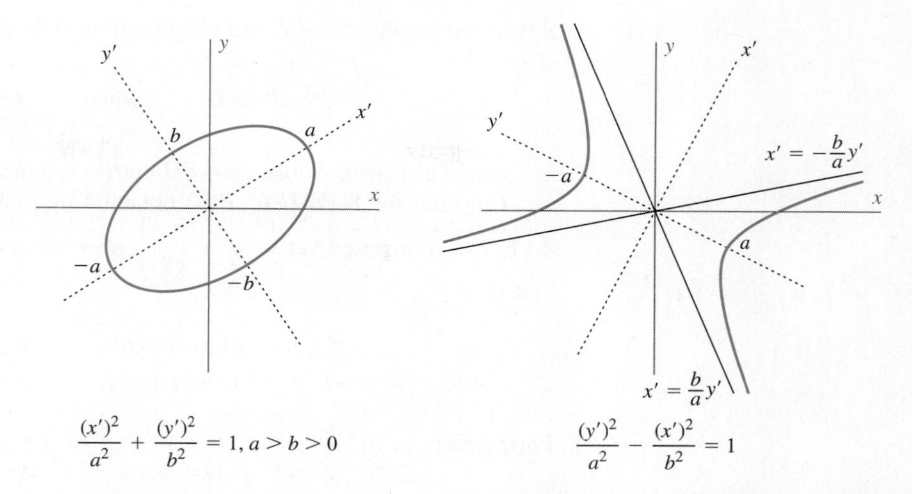

Figure 6.21 The x'-axis is the major axis of the ellipse, and the y'-axis is the major axis of the hyperbola.

then the associated quadratic form can be written as $\mathbf{v}^T A\mathbf{v}$. For example, the form $3x^2 + 4xy + 6y^2$ can be written as

$$[x \ \ y]\begin{bmatrix} 3 & 2 \\ 2 & 6 \end{bmatrix}\begin{bmatrix} x \\ y \end{bmatrix}.$$

We now show how to choose the appropriate angle of rotation θ that satisfies $0^\circ < \theta < 90^\circ$. The method involves finding an orthonormal basis for $\mathcal{R}^2$ so that the x'- and y'-axes corresponding to this basis are parallel to the axes of symmetry of the conic section.

Because A is symmetric, it follows from Theorem 6.15 that there is an orthonormal basis $\{\mathbf{b}_1, \mathbf{b}_2\}$ for $\mathcal{R}^2$ consisting of eigenvectors of A. Necessarily, one of the eigenvectors $\mathbf{b}_1$, $\mathbf{b}_2$, $-\mathbf{b}_1$, and $-\mathbf{b}_2$ must lie in the first quadrant; that is, both of its components must be positive. (See Figure 6.22.) Since this vector is a unit vector, it has the form

$$\begin{bmatrix} \cos\theta \\ \sin\theta \end{bmatrix}$$

for some angle θ ($0^\circ < \theta < 90^\circ$). Let P be the rotation matrix A_θ; that is,

$$P = \begin{bmatrix} \cos\theta & -\sin\theta \\ \sin\theta & \cos\theta \end{bmatrix}.$$

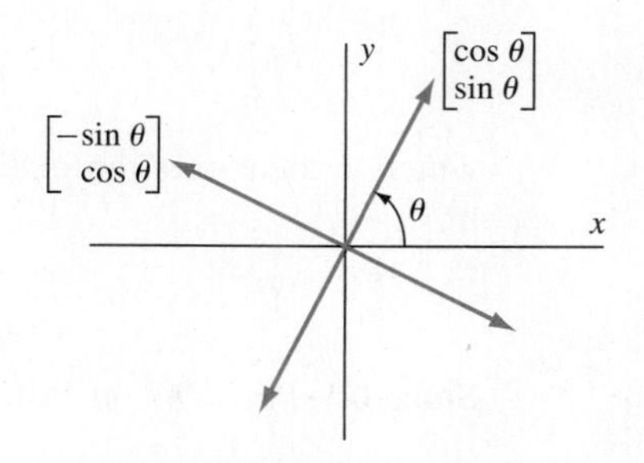

Figure 6.22 The eigenvectors $\mathbf{b}_1$, $\mathbf{b}_2$, $-\mathbf{b}_1$, and $-\mathbf{b}_2$

Because the columns of P are eigenvectors of A, we see that

$$P^T AP = D, \qquad \text{where} \qquad D = \begin{bmatrix} \lambda_1 & 0 \\ 0 & \lambda_2 \end{bmatrix}$$

is a diagonal matrix whose diagonal entries are the eigenvalues of A.

Consider the basis $\{P\mathbf{e}_1, P\mathbf{e}_2\}$ obtained by rotating $\mathbf{e}_1$ and $\mathbf{e}_2$ by θ. By Theorem 4.11, the coordinate vector $\mathbf{v}' = \begin{bmatrix} x' \\ y' \end{bmatrix}$ of $\mathbf{v}$ relative to this basis satisfies $\mathbf{v} = P\mathbf{v}'$; that is,

$$\begin{aligned} x &= (\cos\theta)x' - (\sin\theta)y' \\ y &= (\sin\theta)x' + (\cos\theta)y'. \end{aligned}$$

Furthermore,

$$\begin{aligned} ax^2 + 2bxy + cy^2 &= \mathbf{v}^T A\mathbf{v} \\ &= (P\mathbf{v}')^T A(P\mathbf{v}') \\ &= (\mathbf{v}')^T P^T AP\mathbf{v}' \\ &= (\mathbf{v}')^T D\mathbf{v}' \\ &= \lambda_1(x')^2 + \lambda_2(y')^2, \end{aligned}$$

and hence

$$ax^2 + 2bxy + cy^2 = \lambda_1(x')^2 + \lambda_2(y')^2. \tag{8}$$

So, by converting to the variables x' and y', we may rewrite the associated quadratic form with no $x'y'$-term.

To see how all of this works in practice, consider the equation

$$2x^2 - 4xy + 5y^2 = 36.$$

The associated quadratic form is $2x^2 - 4xy + 5y^2$, and so we let

$$A = \begin{bmatrix} 2 & -2 \\ -2 & 5 \end{bmatrix}.$$

Recall from Example 1 that the eigenvalues of A are 6 and 1 with corresponding unit eigenvectors $\frac{1}{\sqrt{5}}\begin{bmatrix} -1 \\ 2 \end{bmatrix}$ and $\frac{1}{\sqrt{5}}\begin{bmatrix} 2 \\ 1 \end{bmatrix}$. The second of these has both components positive, so we take

$$P = \begin{bmatrix} \frac{2}{\sqrt{5}} & -\frac{1}{\sqrt{5}} \\ \frac{1}{\sqrt{5}} & \frac{2}{\sqrt{5}} \end{bmatrix},$$

which is the θ-rotation matrix in which

$$\cos\theta = \frac{2}{\sqrt{5}} \qquad \text{and} \qquad \sin\theta = \frac{1}{\sqrt{5}}.$$

Since $0^\circ < \theta < 90^\circ$, it follows that

$$\theta = \cos^{-1}\left(\frac{2}{\sqrt{5}}\right) \approx 63.4^\circ.$$

In addition, by making the change of variable $\mathbf{v} = P\mathbf{v}'$, that is,

$$x = \frac{2}{\sqrt{5}}x' - \frac{1}{\sqrt{5}}y'$$
$$y = \frac{1}{\sqrt{5}}x' + \frac{2}{\sqrt{5}}y',$$

it follows from equation (8) that

$$2x^2 - 4xy + 5y^2 = (x')^2 + 6(y')^2.$$

Thus the simplified equation is

$$(x')^2 + 6(y')^2 = 36,$$

or

$$\frac{(x')^2}{36} + \frac{(y')^2}{6} = 1.$$

From this form, we see that the equation represents an ellipse whose axes of symmetry are obtained by rotating the usual x- and y-axes by θ. (See Figure 6.23.)

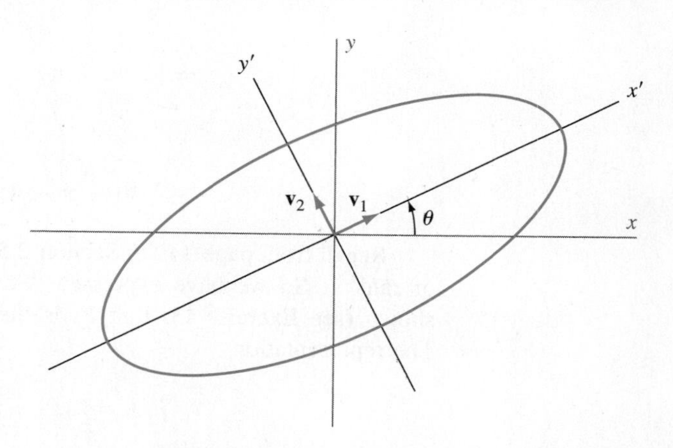

Figure 6.23 The graph of $2x^2 - 4xy + 5y^2 = 36$

Practice Problem 2 ▶ (a) Find a symmetric matrix A so that the associated quadratic form of the equation $-4x^2 + 24xy - 11y^2 = 20$ can be written as $\mathbf{v}^T A\mathbf{v}$.

(b) Find a rotation of the x- and y-axes to x'- and y'-axes that transforms the equation in (a) into one having no $x'y'$-term. Give the angle of rotation, identify the type of conic section, and sketch its graph. ◀

SPECTRAL DECOMPOSITION OF A MATRIX

It is interesting to note that every symmetric matrix can be decomposed into a sum of very simple matrices. With this decomposition in hand, it is an easy task to prove deep results about matrices.

Consider an $n \times n$ symmetric matrix A and an orthonormal basis $\{\mathbf{u}_1, \mathbf{u}_2, \ldots, \mathbf{u}_n\}$ for $\mathcal{R}^n$ consisting of eigenvectors of A. Suppose that $\lambda_1, \lambda_2, \ldots, \lambda_n$ are the corresponding eigenvalues. Let $P = [\mathbf{u}_1 \;\; \mathbf{u}_2 \;\; \ldots \;\; \mathbf{u}_n]$, and let D denote the $n \times n$ diagonal matrix with diagonal entries $\lambda_1, \lambda_2, \ldots, \lambda_n$, respectively. Then

$$A = PDP^T$$

$$= P[\lambda_1\mathbf{e}_1 \ \ \lambda_2\mathbf{e}_2 \ \ \dots \ \ \lambda_n\mathbf{e}_n]\begin{bmatrix}\mathbf{u}_1^T\\ \mathbf{u}_2^T\\ \vdots\\ \mathbf{u}_n^T\end{bmatrix}$$

$$= [P(\lambda_1\mathbf{e}_1) \ \ P(\lambda_2\mathbf{e}_2) \ \ \dots \ \ P(\lambda_n\mathbf{e}_n)]\begin{bmatrix}\mathbf{u}_1^T\\ \mathbf{u}_2^T\\ \vdots\\ \mathbf{u}_n^T\end{bmatrix}$$

$$= [\lambda_1 P\mathbf{e}_1 \ \ \lambda_2 P\mathbf{e}_2 \ \ \dots \ \ \lambda_n P\mathbf{e}_n]\begin{bmatrix}\mathbf{u}_1^T\\ \mathbf{u}_2^T\\ \vdots\\ \mathbf{u}_n^T\end{bmatrix}$$

$$= [\lambda_1\mathbf{u}_1 \ \ \lambda_2\mathbf{u}_2 \ \ \dots \ \ \lambda_n\mathbf{u}_n]\begin{bmatrix}\mathbf{u}_1^T\\ \mathbf{u}_2^T\\ \vdots\\ \mathbf{u}_n^T\end{bmatrix}$$

$$= \lambda_1\mathbf{u}_1\mathbf{u}_1^T + \lambda_2\mathbf{u}_2\mathbf{u}_2^T + \cdots + \lambda_n\mathbf{u}_n\mathbf{u}_n^T.$$

Recall from page 149 in Section 2.5 that the matrix product $P_i = \mathbf{u}_i\mathbf{u}_i^T$ is a matrix of rank 1. So we have expressed A as a sum of n matrices of rank 1. It can be shown (see Exercise 43) that P_i is the orthogonal projection matrix for Span $\{\mathbf{u}_i\}$. The representation

$$A = \lambda_1 P_1 + \lambda_2 P_2 + \cdots + \lambda_n P_n$$

is called a **spectral decomposition** of A.

By Exercise 67 of Section 6.3, we have that each P_i is symmetric and satisfies $P_i^2 = P_i$. Consequently, a number of other properties follow easily. They are given in the next theorem. The proofs of parts (b), (c), and (d) are left as exercises. (See Exercises 44–46.)

THEOREM 6.16

(Spectral Decomposition Theorem) Let A be an $n \times n$ symmetric matrix, and let $\{\mathbf{u}_1, \mathbf{u}_2, \dots, \mathbf{u}_n\}$ be an orthonormal basis for $\mathcal{R}^n$ consisting of eigenvectors of A with corresponding eigenvalues $\lambda_1, \lambda_2, \dots, \lambda_n$. Then there exist symmetric matrices $P_1, P_2, \dots, P_n$ such that the following results hold:

(a) $A = \lambda_1 P_1 + \lambda_2 P_2 + \cdots + \lambda_n P_n$.

(b) $\text{rank}\, P_i = 1$ for all i.

(c) $P_i P_i = P_i$ for all i, and $P_i P_j = O$ if $i \neq j$.

(d) $P_i\mathbf{u}_i = \mathbf{u}_i$ for all i, and $P_i\mathbf{u}_j = \mathbf{0}$ if $i \neq j$.

Example 3

Find a spectral decomposition of the matrix $A = \begin{bmatrix} 3 & -4 \\ -4 & -3 \end{bmatrix}$ in Example 1.

Solution Using the results of Example 1, we let $\mathbf{u}_1 = \frac{1}{\sqrt{5}}\begin{bmatrix} -2 \\ 1 \end{bmatrix}$, $\mathbf{u}_2 = \frac{1}{\sqrt{5}}\begin{bmatrix} 1 \\ 2 \end{bmatrix}$, $\lambda_1 = 5$, and $\lambda_2 = -5$. So

$$P_1 = \mathbf{u}_1\mathbf{u}_1^T = \begin{bmatrix} \frac{4}{5} & -\frac{2}{5} \\ -\frac{2}{5} & \frac{1}{5} \end{bmatrix} \quad \text{and} \quad P_2 = \mathbf{u}_2\mathbf{u}_2^T = \begin{bmatrix} \frac{1}{5} & \frac{2}{5} \\ \frac{2}{5} & \frac{4}{5} \end{bmatrix}.$$

Therefore

$$A = \lambda_1 P_1 + \lambda_2 P_2 = 5\begin{bmatrix} \frac{4}{5} & -\frac{2}{5} \\ -\frac{2}{5} & \frac{1}{5} \end{bmatrix} + (-5)\begin{bmatrix} \frac{1}{5} & \frac{2}{5} \\ \frac{2}{5} & \frac{4}{5} \end{bmatrix}.$$

Practice Problem 3 ▶ Find a spectral decomposition of

$$A = \begin{bmatrix} 4 & 1 & -1 \\ 1 & 4 & -1 \\ -1 & -1 & 4 \end{bmatrix}.$$ ◀

SPECTRAL APPROXIMATION

Suppose that we wish to send a huge data set of information either quickly or by means of a transmission method that does not allow long messages. Assume that the data can be put into the form of a symmetric matrix[14] A. We show that the spectral decomposition of A allows us to reduce the amount of data we need to send with little loss of information. To illustrate the technique, suppose that

$$A = \begin{bmatrix} 153 & -142 & 56 & 256 & 37 \\ -142 & 182 & -86 & -276 & -44 \\ 56 & -86 & 55 & 117 & 22 \\ 256 & -276 & 117 & 475 & 68 \\ 37 & -44 & 22 & 68 & 11 \end{bmatrix}.$$

Using MATLAB, we find an orthogonal matrix P and a diagonal matrix D such that $P^TAP = D$.[15] We have

$$P = \begin{bmatrix} 0.4102 & -0.4886 & -0.7235 & 0.2200 & 0.1454 \\ -0.4536 & -0.5009 & -0.1590 & -0.7096 & -0.1211 \\ 0.2001 & 0.6693 & -0.4671 & -0.5211 & 0.1493 \\ 0.7572 & -0.2275 & 0.4472 & -0.4157 & -0.0467 \\ 0.1125 & 0.1033 & -0.1821 & 0.0614 & -0.9694 \end{bmatrix}$$

[14] If A is not symmetric, we can apply the technique that follows to the singular value decomposition of A. (See Section 6.7.)

[15] MATLAB often produces matrices with the columns in a different order than is listed here. For convenience, in this subsection we list the eigenvalues and eigenvectors so that the eigenvalues are in order of decreasing magnitude. Also, we occasionally replace an eigenvector produced by MATLAB with a nonzero multiple of itself, which, of course, is still an eigenvector corresponding to the same eigenvalue.

and

$$D = \begin{bmatrix} 820.0273 & 0 & 0 & 0 & 0 \\ 0 & 42.1027 & 0 & 0 & 0 \\ 0 & 0 & 9.0352 & 0 & 0 \\ 0 & 0 & 0 & 4.9926 & 0 \\ 0 & 0 & 0 & 0 & -0.1578 \end{bmatrix}.$$

Using the notation of Theorem 6.16, we obtain a spectral decomposition

$$A = 820.0273P_1 + 42.1027P_2 + 9.0352P_3 + 4.9926P_4 - 0.1578P_5,$$

where each P_i is an orthogonal projection matrix.

Notice the wide variation in the magnitudes of the coefficients (eigenvalues), with relatively small coefficients of P_2, P_3, P_4, and P_5. This observation suggests that we approximate A by the matrix $A_1 = 820.0273P_1$, a matrix of rank 1, whereas A has rank 5. We obtain

$$A_1 = \begin{bmatrix} 137.9793 & -152.5673 & 67.2916 & 254.6984 & 37.8456 \\ -152.5673 & 168.6976 & -74.4061 & -281.6266 & -41.8468 \\ 67.2916 & -74.4061 & 32.8177 & 124.2148 & 18.4570 \\ 254.6984 & -281.6266 & 124.2148 & 470.1522 & 69.8598 \\ 37.8456 & -41.8468 & 18.4570 & 69.8598 & 10.3805 \end{bmatrix}.$$

How well does A_1 approximate A? One criterion of closeness to consider is the relative "size" of the *error matrix*

$$E_1 = A - A_1 = \begin{bmatrix} 15.0207 & 10.5673 & -11.2916 & 1.3016 & -0.8456 \\ 10.5673 & 13.3024 & -11.5939 & 5.6266 & -2.1532 \\ -11.2916 & -11.5939 & 22.1823 & -7.2148 & 3.5430 \\ 1.3016 & 5.6266 & -7.2148 & 4.8478 & -1.8598 \\ -0.8456 & -2.1532 & 3.5430 & -1.8598 & 0.6195 \end{bmatrix}.$$

To quantify the relative size of E_1 compared with A, we use the *Frobenius norm* of a matrix, which is defined on page 534. This norm is obtained in MATLAB with the command `norm`(E1,'fro'). The value returned is 43.3500. The norm of A is computed as 821.1723. So if we use A_1 to approximate A, we have "lost" only $43.3500/821.1723 = 5.28\%$ of the original information, and have replaced the matrix A of rank 5 with the matrix A_1 of rank 1. In this case, only the first column of A_1 needs to be transmitted, along with the multiples of that column that yield columns 2 through 5.

For a smaller loss of information, we could use the matrix

$$A_2 = 820.0273P_1 + 42.1027P_2,$$

a matrix of rank 2. We compute

$$A_2 = \begin{bmatrix} 148.0324 & -142.2625 & 53.5226 & 259.3786 & 35.7200 \\ -142.2625 & 179.2604 & -88.5199 & -276.8293 & -44.0256 \\ 53.5226 & -88.5199 & 51.6763 & 117.8046 & 21.3683 \\ 259.3786 & -276.8293 & 117.8046 & 472.3311 & 68.8702 \\ 35.7200 & -44.0256 & 21.3683 & 68.8702 & 10.8299 \end{bmatrix}.$$

Let $E_2 = A - A_2$. Then

$$E_2 = \begin{bmatrix} 4.9676 & 0.2625 & 2.4774 & -3.3786 & 1.2800 \\ 0.2625 & 2.7396 & 2.5199 & 0.8293 & 0.0256 \\ 2.4774 & 2.5199 & 3.3237 & -0.8046 & 0.6317 \\ -3.3786 & 0.8293 & -0.8046 & 2.6689 & -0.8702 \\ 1.2800 & 0.0256 & 0.6317 & -0.8702 & 0.1701 \end{bmatrix}.$$

The norm of E_2 is 10.3240. So we have lost only $10.3240/821.1723 = 1.26\%$ of the original information in this case, and have replaced the matrix A of rank 5 with the matrix A_2 of rank 2.

Note that, because the rank of A_2 is 2, we need only transmit two linearly independent columns and the coefficients in each of the three linear combinations that represent the other three columns.

A number of interesting consequences of the spectral decomposition are given in the exercises.

EXERCISES

In Exercises 1–12, answer the following parts for each equation of a conic section.

(a) Find a symmetric matrix A so that the associated quadratic form may be written as $\mathbf{v}^T A\mathbf{v}$.

Find a rotation of the x- and y-axes to x′- and y′-axes that transforms the given equation into one having no x′y′-term.

(a) Give the angle of rotation.

(b) Give the equations that relate x′ and y′ to x and y.

(c) Give the transformed equation.

(d) Identify the type of conic section.

1. $2x^2 - 14xy + 50y^2 - 255 = 0$
2. $2x^2 + 2xy + 2y^2 - 1 = 0$
3. $x^2 - 12xy - 4y^2 = 40$
4. $3x^2 - 4xy + 3y^2 - 5 = 0$
5. $5x^2 + 4xy + 5y^2 - 9 = 0$
6. $11x^2 + 24xy + 4y^2 - 15 = 0$
7. $x^2 + 4xy + y^2 - 7 = 0$
8. $4x^2 + 6xy - 4y^2 = 180$
9. $2x^2 - 12xy - 7y^2 = 200$
10. $6x^2 + 5xy - 6y^2 = 26$
11. $x^2 + 2xy + y^2 + 8x + y = 0$
12. $52x^2 + 72xy + 73y^2 - 160x - 130y - 25 = 0$

In Exercises 13–20, a symmetric matrix A is given. Find an orthonormal basis of eigenvectors and their corresponding eigenvalues. Use this information to obtain a spectral decomposition of each matrix.

13. $\begin{bmatrix} 3 & 1 \\ 1 & 3 \end{bmatrix}$
14. $\begin{bmatrix} 7 & 6 \\ 6 & -2 \end{bmatrix}$
15. $\begin{bmatrix} 1 & 2 \\ 2 & 1 \end{bmatrix}$
16. $\begin{bmatrix} 1 & -1 \\ -1 & 1 \end{bmatrix}$
17. $\begin{bmatrix} 3 & 2 & 2 \\ 2 & 2 & 0 \\ 2 & 0 & 4 \end{bmatrix}$
18. $\begin{bmatrix} 0 & 2 & 2 \\ 2 & 0 & 2 \\ 2 & 2 & 0 \end{bmatrix}$
19. $\begin{bmatrix} -1 & 0 & 0 \\ 0 & 0 & 2 \\ 0 & 2 & 3 \end{bmatrix}$
20. $\begin{bmatrix} -2 & 0 & -36 \\ 0 & -3 & 0 \\ -36 & 0 & -23 \end{bmatrix}$

In Exercises 21–40, determine whether the statements are true or false.

21. Every symmetric matrix is diagonalizable.
22. If P is a matrix whose columns are eigenvectors of a symmetric matrix, then P is orthogonal.
23. If A is an $n \times n$ matrix and there exists an orthonormal basis for $\mathcal{R}^n$ consisting of eigenvectors of A, then A is symmetric.
24. Eigenvectors of a matrix that correspond to distinct eigenvalues are orthogonal.
25. Distinct eigenvectors of a symmetric matrix are orthogonal.
26. By a suitable rotation of the xy-axes to $x'y'$-axes, the equation of any conic section with center at the origin can be written without an $x'y'$-term.
27. The associated quadratic form of an equation of any conic section can be written as $\mathbf{v}^T A\mathbf{v}$, where A is a 2×2 matrix and $\mathbf{v}$ is in $\mathcal{R}^2$.
28. Every symmetric matrix can be written as a sum of orthogonal projection matrices.
29. Every symmetric matrix can be written as a sum of multiples of orthogonal projection matrices.
30. Every symmetric matrix can be written as a sum of multiples of orthogonal projection matrices of rank 1.
31. Every symmetric matrix can be written as a sum of multiples of orthogonal projection matrices, in which the multiples are the eigenvalues of the matrix.
32. If the equation of a conic section is written without an xy-term by a rotation of the coordinate axes through an angle θ, where $0° \le \theta < 360°$, then θ is unique.
33. Eigenvectors of a symmetric matrix that correspond to distinct eigenvalues are orthogonal.
34. The spectral decomposition of a symmetric matrix is unique, except for the order of the terms.
35. Every matrix has a spectral decomposition.
36. To rotate the coordinate axes in order to remove the xy-term of the equation $ax^2 + 2bxy + cy^2 = d$, we must determine the eigenvectors of $\begin{bmatrix} a & b \\ c & d \end{bmatrix}$.
37. If a rotation of the x- and y-axes is used to write the equation $ax^2 + 2bxy + cy^2 = d$ as $a'(x')^2 + b'(y')^2 = d$, then the scalars a' and b' are eigenvalues of $\begin{bmatrix} a & b \\ b & c \end{bmatrix}$.

38. If $\mathcal{B}_1, \mathcal{B}_2, \ldots, \mathcal{B}_k$ are orthonormal bases for the distinct eigenspaces of a symmetric $n \times n$ matrix, then their union $\mathcal{B}_1 \cup \mathcal{B}_2 \cup \cdots \cup \mathcal{B}_k$ is an orthonormal basis for $\mathcal{R}^n$.

39. If P is a 2×2 orthogonal matrix whose columns form an orthonormal basis for $\mathcal{R}^2$, then P is a rotation matrix.

40. If P is a 2×2 orthogonal matrix whose columns form an orthonormal basis for $\mathcal{R}^2$ consisting of eigenvectors of $A = \begin{bmatrix} a & b \\ b & c \end{bmatrix}$, then the change of variable $\begin{bmatrix} x \\ y \end{bmatrix} = P\begin{bmatrix} x' \\ y' \end{bmatrix}$ changes $ax^2 + bxy + cy^2 = d$ into $\lambda_1(x')^2 + \lambda_2(y')^2 = d$, where λ_1 and λ_2 are the eigenvalues of A.

---

41. Show that a spectral decomposition is not unique by finding two different spectral decompositions with different orthogonal projection matrices for the matrix $2I_2$.

42. Show that a spectral decomposition is not unique by finding two different spectral decompositions with different orthogonal projection matrices for the matrix in Exercise 19.

43. Let $\mathbf{u}$ be a unit vector in $\mathcal{R}^n$, and let P be the matrix $\mathbf{u}\mathbf{u}^T$. Prove that P is the orthogonal projection matrix for the subspace Span $\{\mathbf{u}\}$.

44. Prove Theorem 6.16(b).

45. Prove Theorem 6.16(c).

46. Prove Theorem 6.16(d).

In Exercises 47–54, let A be an $n \times n$ symmetric matrix with spectral decomposition $A = \lambda_1 P_1 + \lambda_2 P_2 + \cdots + \lambda_n P_n$. Assume that $\mu_1, \mu_2, \ldots, \mu_k$ are all the distinct eigenvalues of A and that Q_j denotes the sum of all the P_i's that are associated with μ_j.

47. Prove $A = \mu_1 Q_1 + \mu_1 Q_2 + \cdots + \mu_k Q_k$.

48. Prove $Q_j Q_j = Q_j$ for all j, and $Q_i Q_j = O$ if $j \neq i$.

49. Prove Q_j is symmetric for all j.

50. Prove that, for all j, Q_j is the orthogonal projection matrix for the eigenspace associated with μ_j.

51. Suppose that $\{\mathbf{w}_1, \mathbf{w}_2, \ldots, \mathbf{w}_s\}$ is an orthonormal basis for the eigenspace corresponding to μ_j. Represent Q_j as a sum of matrices, each of rank 1.

52. Prove that the rank of Q_j equals the dimension of the eigenspace associated with μ_j.

53. Use the given spectral decomposition of A to compute a spectral decomposition of A^s, where s is any positive integer greater than 1.

54. Use the given spectral decomposition of A to find a spectral decomposition of a matrix C such that $C^3 = A$.

Exercises 55–57 use the definition: For an $n \times n$ matrix B and a polynomial $g(t) = a_n t^n + a_{n-1} t^{n-1} + \cdots + a_1 t + a_0$, define the matrix $g(B)$ by

$$g(B) = a_n B^n + a_{n-1} B^{n-1} + \cdots + a_1 B + a_0 I_n.$$

55. Use Exercises 47 and 48 to show that for any polynomial g,

$$g(A) = g(\mu_1)Q_1 + g(\mu_2)Q_2 + \cdots + g(\mu_k)Q_k,$$

where A, μ_i, and Q_i are as in Exercises 47–54.

56. Use Exercise 55 to prove a special case of the *Cayley–Hamilton theorem:* If f is the characteristic polynomial of a symmetric matrix A, then $f(A) = O$.

57. Let A, μ_i, and Q_i be as in Exercises 47–54. It can be shown that for any j, $1 \leq j \leq k$, there is a polynomial f_j such that $f_j(\mu_j) = 1$ and $f_j(\mu_i) = 0$ if $i \neq j$. (See [4, pages 51–52].) Use this result along with Exercise 55 to show that $Q_j = f_j(A)$. So the Q_j's are uniquely determined by the properties given in Exercises 47 and 48.

58. Use Exercise 57 to prove that an $n \times n$ matrix B commutes with A (that is, $AB = BA$) if and only if B commutes with each Q_j.

An $n \times n$ matrix C is said to be **positive definite** *if C is symmetric and $\mathbf{v}^T C \mathbf{v} > 0$ for every nonzero vector $\mathbf{v}$ in $\mathcal{R}^n$. We say C is* **positive semidefinite** *if C is symmetric and $\mathbf{v}^T C \mathbf{v} \geq 0$ for every vector $\mathbf{v}$ in $\mathcal{R}^n$.*

In Exercises 59–73, we assume the preceding definitions. The equation $\mathbf{v}^T A \mathbf{v} = \mathbf{v} \cdot A\mathbf{v}$ is often useful in solving these exercises.

59. Suppose that A is a symmetric matrix. Prove that A is positive definite if and only if all of its eigenvalues are positive.

60.[16] State and prove a characterization of positive semidefinite matrices analogous to that in Exercise 59.

61. Suppose that A is invertible and positive definite. Prove that A^{-1} is positive definite.

62. Suppose that A is positive definite and $c > 0$. Prove that cA is positive definite.

63. State and prove a result analogous to Exercise 62 if A is positive semidefinite.

64. Suppose that A and B are positive definite $n \times n$ matrices. Prove that $A + B$ is positive definite.

65. State and prove a result analogous to Exercise 64 if A and B are positive semidefinite.

66. Suppose that $A = QBQ^T$, where Q is an orthogonal matrix and B is positive definite. Prove that A is positive definite.

67. State and prove a result analogous to Exercise 66 if B is positive semidefinite.

68. Prove that if A is positive definite, then there exists a positive definite matrix B such that $B^2 = A$.

69. State and prove a result analogous to Exercise 68 if A is positive semidefinite.

70. Let A be an $n \times n$ symmetric matrix. Prove that A is positive definite if and only if

$$\sum_{i,j} a_{ij} u_i u_j > 0$$

for all scalars $u_1, u_2, \ldots, u_n$, not all zero.

---

[16] This exercise is used in Section 6.7 (on page 439).

71. State and prove a result analogous to Exercise 70 if A is positive semidefinite.

72.[17] Prove that, for any matrix A, the matrices A^TA and AA^T are positive semidefinite.

73. Prove that, for any invertible matrix A, the matrices A^TA and AA^T are positive definite.

In Exercises 74–76, use either a calculator with matrix capabilities or computer software such as MATLAB to solve each problem.

74. Let

$$A = \begin{bmatrix} 4 & 0 & 2 & 0 & 2 \\ 0 & 4 & 0 & 2 & 0 \\ 2 & 0 & 4 & 0 & 2 \\ 0 & 2 & 0 & 4 & 0 \\ 2 & 0 & 2 & 0 & 4 \end{bmatrix}.$$

(a) Verify that A is symmetric.

(b) Find the eigenvalues of A.

(c) Find an orthonormal basis for $\mathcal{R}^5$ of eigenvectors of A.

(d) Use your answers to (b) and (c) to find a spectral decomposition of A.

(e) Compute A^6 by matrix multiplication.

(f) Use your answer to (d) to find a spectral decomposition of A^6.

(g) Use your answer to (f) to compute A^6.

75. Let

$$A = \begin{bmatrix} 56 & 62 & 96 & 24 & 3 \\ 62 & 61 & 94 & 25 & 1 \\ 96 & 94 & 167 & 33 & 1 \\ 24 & 25 & 33 & 9 & 1 \\ 3 & 1 & 1 & 1 & 2 \end{bmatrix}.$$

(a) Find an orthogonal matrix P and a diagonal matrix D such that $A = PDP^T$.

(b) Use your answer to (a) to find the associated spectral decomposition of A, arranged so that the coefficients (eigenvalues) of the orthogonal projection matrices are in order of decreasing magnitudes (absolute values). Include the orthogonal projection matrices.

76. Let A be the matrix in Exercise 75.

(a) Using the spectral decomposition for A, form an approximation A_1 of A, based on the largest eigenvalue.

(b) Compute the Frobenius norms of the error matrix $E_1 = A - A_1$ and the matrix A.

(c) Give the percentage of information lost when A_1 is used to approximate A.

SOLUTIONS TO THE PRACTICE PROBLEMS

1. Since A is symmetric, the matrices P and D exist. We compute the characteristic polynomial of A to be $-t(t-18)^2$. It can be shown that $\begin{bmatrix} -4 \\ 1 \\ 1 \end{bmatrix}$ is an eigenvector corresponding to the eigenvalue 0, and hence forms a basis for the corresponding eigenspace. Furthermore, it can be shown that

$$\left\{ \begin{bmatrix} 1 \\ 4 \\ 0 \end{bmatrix}, \begin{bmatrix} 1 \\ 0 \\ 4 \end{bmatrix} \right\}$$

is a basis for the eigenspace corresponding to the eigenvalue 18. Applying the Gram–Schmidt process to the vectors in this basis, we obtain an orthogonal basis

$$\left\{ \begin{bmatrix} 1 \\ 4 \\ 0 \end{bmatrix}, \begin{bmatrix} 16 \\ -4 \\ 68 \end{bmatrix} \right\}$$

for this eigenspace. Notice that the vectors in this basis are orthogonal to the chosen eigenvector corresponding to the eigenvalue 0. So the set

$$\left\{ \begin{bmatrix} -\frac{4}{3\sqrt{2}} \\ \frac{1}{3\sqrt{2}} \\ \frac{1}{3\sqrt{2}} \end{bmatrix}, \begin{bmatrix} \frac{1}{\sqrt{17}} \\ \frac{4}{\sqrt{17}} \\ 0 \end{bmatrix}, \begin{bmatrix} \frac{4}{3\sqrt{34}} \\ -\frac{1}{3\sqrt{34}} \\ \frac{17}{3\sqrt{34}} \end{bmatrix} \right\}$$

is an orthonormal basis of eigenvectors of A. It follows that one possible choice of the orthogonal matrix P and diagonal matrix D is

$$P = \begin{bmatrix} -\frac{4}{3\sqrt{2}} & \frac{1}{\sqrt{17}} & \frac{4}{3\sqrt{34}} \\ \frac{1}{3\sqrt{2}} & \frac{4}{\sqrt{17}} & -\frac{1}{3\sqrt{34}} \\ \frac{1}{3\sqrt{2}} & 0 & \frac{17}{3\sqrt{34}} \end{bmatrix}$$

and

$$D = \begin{bmatrix} 0 & 0 & 0 \\ 0 & 18 & 0 \\ 0 & 0 & 18 \end{bmatrix}.$$

2. (a) The entries of A are obtained from the coefficients of the quadratic form $a_{11} = -4$, $a_{22} = -11$, and

[17] This exercise is used in Section 6.7 (on page 439).

$a_{12} = a_{21} = \frac{1}{2}(24) = 12$. Thus

$$A = \begin{bmatrix} -4 & 12 \\ 12 & -11 \end{bmatrix}.$$

(b) It can be shown that A has the eigenvalues $\lambda_1 = 5$ and $\lambda_2 = -20$ with corresponding unit eigenvectors $\begin{bmatrix} 0.8 \\ 0.6 \end{bmatrix}$ and $\begin{bmatrix} -0.6 \\ 0.8 \end{bmatrix}$. Since the first of these lies in the first quadrant, we take

$$P = \begin{bmatrix} 0.8 & -0.6 \\ 0.6 & 0.8 \end{bmatrix},$$

which is the θ-rotation matrix in which $\cos\theta = 0.8$ and $\sin\theta = 0.6$. Since $0° < \theta < 90°$, it follows that

$$\theta = \cos^{-1}(0.8) \approx 36.9°.$$

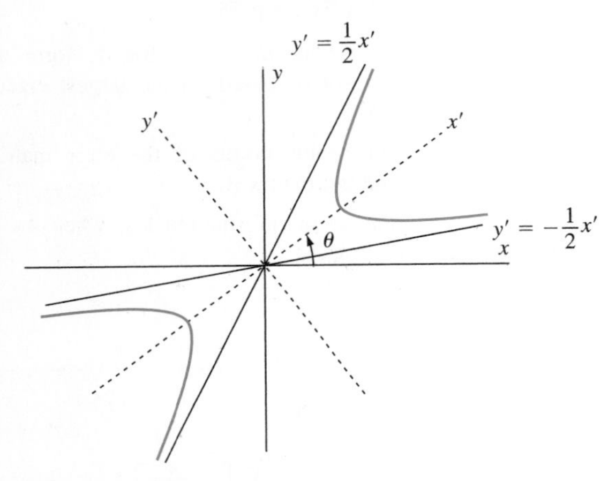

Figure 6.24 The graph of $-4x^2 + 24xy - 11y^2 = 20$

By equation (8), we have

$$-4x^2 + 24xy - 11y^2 = 5(x')^2 - 20(y')^2 = 20,$$

and so the original equation becomes

$$\frac{(x')^2}{4} - \frac{(y')^2}{1} = 1,$$

which is the equation of a hyperbola. For a sketch, see Figure 6.24, which includes the two asymptotes $y' = \pm\frac{1}{2}x'$.

3. First, observe (we omit the details) that $\lambda_1 = \lambda_2 = 3$ and $\lambda_3 = 6$ are the eigenvalues of A, with an orthonormal basis of $\mathcal{R}^3$ consisting of corresponding eigenvectors

$$\{\mathbf{u}_1, \mathbf{u}_2, \mathbf{u}_3\} = \left\{ \begin{bmatrix} \frac{1}{\sqrt{2}} \\ 0 \\ \frac{1}{\sqrt{2}} \end{bmatrix}, \begin{bmatrix} -\frac{1}{\sqrt{6}} \\ \frac{2}{\sqrt{6}} \\ \frac{1}{\sqrt{6}} \end{bmatrix} \begin{bmatrix} \frac{1}{\sqrt{3}} \\ \frac{1}{\sqrt{3}} \\ -\frac{1}{\sqrt{3}} \end{bmatrix} \right\}.$$

Thus a spectral decomposition of A is given by

$$A = \lambda_1\mathbf{u}_1\mathbf{u}_1^T + \lambda_2\mathbf{u}_2\mathbf{u}_2^T + \lambda_3\mathbf{u}_3\mathbf{u}_3^T$$

$$= 3\begin{bmatrix} \frac{1}{2} & 0 & \frac{1}{2} \\ 0 & 0 & 0 \\ \frac{1}{2} & 0 & \frac{1}{2} \end{bmatrix} + 3\begin{bmatrix} \frac{1}{6} & -\frac{2}{6} & -\frac{1}{6} \\ -\frac{2}{6} & \frac{4}{6} & \frac{2}{6} \\ -\frac{1}{6} & \frac{2}{6} & \frac{1}{6} \end{bmatrix}$$

$$+ 6\begin{bmatrix} \frac{1}{3} & \frac{1}{3} & -\frac{1}{3} \\ \frac{1}{3} & \frac{1}{3} & -\frac{1}{3} \\ -\frac{1}{3} & -\frac{1}{3} & \frac{1}{3} \end{bmatrix}.$$

6.7* SINGULAR VALUE DECOMPOSITION

We have seen that the easiest matrices to study are those that possess an orthonormal basis of eigenvectors. The basis of eigenvectors gives us complete insight into the way the matrix acts on vectors. If this basis is also an orthonormal set, then we have the added benefit of a set of mutually perpendicular coordinate axes that illuminate the geometric behavior of the matrix as it acts on vectors.

But alas, only the symmetric matrices enjoy all of these properties. Moreover, if the matrix is not square, then eigenvectors are not even defined.

In this section, we consider a generalization of an *orthonormal basis of eigenvectors* in the context of an arbitrary $m \times n$ matrix A of rank k. In this setting, we prove that there exist two orthonormal bases, one for $\mathcal{R}^n$ and one for $\mathcal{R}^m$, such that the product of A and each of the first k vectors in the first basis is a scalar multiple of the corresponding vector in the second basis. In the case that $m = n$ and the two bases

* This section can be omitted without loss of continuity.

are identical, this approach reduces to the usual situation of an orthonormal basis of eigenvectors, and A is, necessarily, a symmetric matrix.

With this in mind, we state the principal theorem of this section.

THEOREM 6.17

Let A be an $m \times n$ matrix of rank k. Then there exist orthonormal bases

$$\mathcal{B}_1 = \{\mathbf{v}_1, \mathbf{v}_2, \ldots, \mathbf{v}_n\} \quad \text{for } \mathcal{R}^n \quad \text{and} \quad \mathcal{B}_2 = \{\mathbf{u}_1, \mathbf{u}_2, \ldots, \mathbf{u}_m\} \quad \text{for } \mathcal{R}^m$$

and scalars

$$\sigma_1 \geq \sigma_2 \geq \cdots \geq \sigma_k > 0$$

such that

$$A\mathbf{v}_i = \begin{cases} \sigma_i \mathbf{u}_i & \text{if } 1 \leq i \leq k \\ \mathbf{0} & \text{if } i > k \end{cases} \tag{9}$$

and

$$A^T\mathbf{u}_i = \begin{cases} \sigma_i \mathbf{v}_i & \text{if } 1 \leq i \leq k \\ \mathbf{0} & \text{if } i > k. \end{cases} \tag{10}$$

PROOF By Exercise 72 of Section 6.6, A^TA is an $n \times n$ positive semidefinite matrix, and hence there is an orthonormal basis $\mathcal{B}_1 = \{\mathbf{v}_1, \mathbf{v}_2, \ldots, \mathbf{v}_n\}$ for $\mathcal{R}^n$ consisting of eigenvectors of A^TA with corresponding eigenvalues λ_i that are nonnegative (see Exercise 60 of Section 6.6). We order these eigenvalues and the vectors in $\mathcal{B}_1$ so that $\lambda_1 \geq \lambda_2 \geq \cdots \geq \lambda_n$. By Exercise 97 of Section 6.1, A^TA has rank k, and hence the first k eigenvalues are positive and the last $n - k$ are zero. For each $i = 1, 2, \ldots k$, let $\sigma_i = \sqrt{\lambda_i}$. Then $\sigma_1 \geq \sigma_2 \geq \cdots \geq \sigma_k > 0$.

Next, for each $i \leq k$, let $\mathbf{u}_i$ be the vector in $\mathcal{R}^m$ defined by $\mathbf{u}_i = \dfrac{1}{\sigma_i}A\mathbf{v}_i$. We show that $\{\mathbf{u}_1, \mathbf{u}_2, \ldots, \mathbf{u}_k\}$ is an orthonormal subset of $\mathcal{R}^m$. Consider any $\mathbf{u}_i$ and $\mathbf{u}_j$. Then

$$\begin{aligned}
\mathbf{u}_i \cdot \mathbf{u}_j &= \frac{1}{\sigma_i}A\mathbf{v}_i \cdot \frac{1}{\sigma_j}A\mathbf{v}_j \\
&= \frac{1}{\sigma_i\sigma_j}A\mathbf{v}_i \cdot A\mathbf{v}_j \\
&= \frac{1}{\sigma_i\sigma_j}\mathbf{v}_i \cdot A^TA\mathbf{v}_j \\
&= \frac{1}{\sigma_i\sigma_j}\mathbf{v}_i \cdot \lambda_j\mathbf{v}_j \\
&= \frac{\sigma_j^2}{\sigma_i\sigma_j}\mathbf{v}_i \cdot \mathbf{v}_j.
\end{aligned}$$

Thus

$$\mathbf{u}_i \cdot \mathbf{u}_j = \frac{\sigma_j}{\sigma_i}\mathbf{v}_i \cdot \mathbf{v}_j = \begin{cases} 0 & \text{if } i \neq j \\ 1 & \text{if } i = j, \end{cases}$$

and it follows that $\{\mathbf{u}_1, \mathbf{u}_2, \ldots, \mathbf{u}_k\}$ is an orthonormal set. By Exercise 56 of Section 6.2, this set extends to an orthonormal basis $\mathcal{B}_2 = \{\mathbf{u}_1, \mathbf{u}_2, \ldots, \mathbf{u}_m\}$ for $\mathcal{R}^m$.

Our final task is to verify equation (10). Consider any $\mathbf{u}_i$ in $\mathcal{B}_2$. First, suppose that $i \leq k$. Then

$$\begin{aligned} A^T\mathbf{u}_i &= A^T\left(\frac{1}{\sigma_i}A\mathbf{v}_i\right) \\ &= \frac{1}{\sigma_i}A^TA\mathbf{v}_i \\ &= \frac{1}{\sigma_i}\sigma_i^2\mathbf{v}_i \\ &= \sigma_i\mathbf{v}_i. \end{aligned}$$

Now suppose that $i > k$. We show that $A^T\mathbf{u}_i$ is orthogonal to every vector in $\mathcal{B}_1$. Since $\mathcal{B}_1$ is a basis for $\mathcal{R}^n$, it follows that $A^T\mathbf{u}_i = \mathbf{0}$. Consider any $\mathbf{v}_j$ in $\mathcal{B}_1$. If $j \leq k$, then

$$A^T\mathbf{u}_i \cdot \mathbf{v}_j = \mathbf{u}_i \cdot A\mathbf{v}_j = \mathbf{u}_i \cdot \sigma_j\mathbf{u}_j = \sigma_j\mathbf{u}_i \cdot \mathbf{u}_j = 0$$

because $i \neq j$. On the other hand, if $j > k$, then

$$A^T\mathbf{u}_i \cdot \mathbf{v}_j = \mathbf{u}_i \cdot A\mathbf{v}_j = \mathbf{u}_i \cdot \mathbf{0} = 0.$$

Thus $A^T\mathbf{u}_i$ is orthogonal to every vector in $\mathcal{B}_1$, and we conclude that $A^T\mathbf{u}_i = \mathbf{0}$. ■

In the proof of Theorem 6.17, the vectors $\mathbf{v}_i$ are chosen to be eigenvectors of A^TA. It can be shown (see Exercise 76) that if $\{\mathbf{v}_1, \mathbf{v}_2, \ldots, \mathbf{v}_n\}$ and $\{\mathbf{u}_1, \mathbf{u}_2, \ldots, \mathbf{u}_m\}$ are any orthonormal bases for $\mathcal{R}^n$ and $\mathcal{R}^m$, respectively, that satisfy equations (9) and (10), then each $\mathbf{v}_i$ is an eigenvector of A^TA corresponding to the eigenvalue σ_i^2 if $i \leq k$ and to the eigenvalue 0 if $i > k$. Furthermore, for $i = 1, 2, \ldots, k$, the vector $\mathbf{u}_i$ is an eigenvector of AA^T corresponding to the eigenvalue σ_i^2 and, for $i > k$, the vector $\mathbf{u}_i$ is an eigenvector of AA^T corresponding to the eigenvalue 0. It follows, therefore, that the σ_i's are the unique scalars satisfying equations (9) and (10). These scalars are called the **singular values** of the matrix A.

Although the singular values of a matrix are unique, the orthonormal bases $\mathcal{B}_1$ and $\mathcal{B}_2$ in Theorem 6.17 are not unique. Of course, this is also the case for bases consisting of eigenvectors of a matrix, even if the eigenvectors are orthonormal.

Example 1 Find the singular values of

$$A = \begin{bmatrix} 0 & 1 & 2 \\ 1 & 0 & 1 \end{bmatrix},$$

and orthonormal bases $\{\mathbf{v}_1, \mathbf{v}_2, \mathbf{v}_3\}$ for $\mathcal{R}^3$ and $\{\mathbf{u}_1, \mathbf{u}_2\}$ for $\mathcal{R}^2$ satisfying equations (9) and (10).

Solution The proof of Theorem 6.17 gives us the method for solving this problem. We first form the product

$$A^TA = \begin{bmatrix} 1 & 0 & 1 \\ 0 & 1 & 2 \\ 1 & 2 & 5 \end{bmatrix}.$$

Since A has rank 2, so does A^TA. In fact, it can be shown (we omit the details) that for

$$\mathbf{v}_1 = \frac{1}{\sqrt{30}}\begin{bmatrix} 1 \\ 2 \\ 5 \end{bmatrix}, \qquad \mathbf{v}_2 = \frac{1}{\sqrt{5}}\begin{bmatrix} 2 \\ -1 \\ 0 \end{bmatrix}, \qquad \text{and} \qquad \mathbf{v}_3 = \frac{1}{\sqrt{6}}\begin{bmatrix} 1 \\ 2 \\ -1 \end{bmatrix},$$

$\{\mathbf{v}_1, \mathbf{v}_2, \mathbf{v}_3\}$ is an orthonormal basis for $\mathcal{R}^3$ consisting of eigenvectors of A^TA with corresponding eigenvalues 6, 1, and 0. So $\sigma_1 = \sqrt{6}$, and $\sigma_2 = \sqrt{1} = 1$ are the singular values of A. Let

$$\mathbf{u}_1 = \frac{1}{\sigma_1}A\mathbf{v}_1 = \frac{1}{\sqrt{6}}\frac{1}{\sqrt{30}}\begin{bmatrix} 0 & 1 & 2 \\ 1 & 0 & 1 \end{bmatrix}\begin{bmatrix} 1 \\ 2 \\ 5 \end{bmatrix} = \frac{1}{6\sqrt{5}}\begin{bmatrix} 12 \\ 6 \end{bmatrix} = \frac{1}{\sqrt{5}}\begin{bmatrix} 2 \\ 1 \end{bmatrix},$$

and

$$\mathbf{u}_2 = \frac{1}{\sigma_2}A\mathbf{v}_2 = \frac{1}{\sqrt{5}}\begin{bmatrix} 0 & 1 & 2 \\ 1 & 0 & 1 \end{bmatrix}\begin{bmatrix} 2 \\ -1 \\ 0 \end{bmatrix} = \frac{1}{\sqrt{5}}\begin{bmatrix} -1 \\ 2 \end{bmatrix}.$$

Then $\{\mathbf{u}_1, \mathbf{u}_2\}$ is an orthonormal basis for $\mathcal{R}^2$. (See Figure 6.25.)

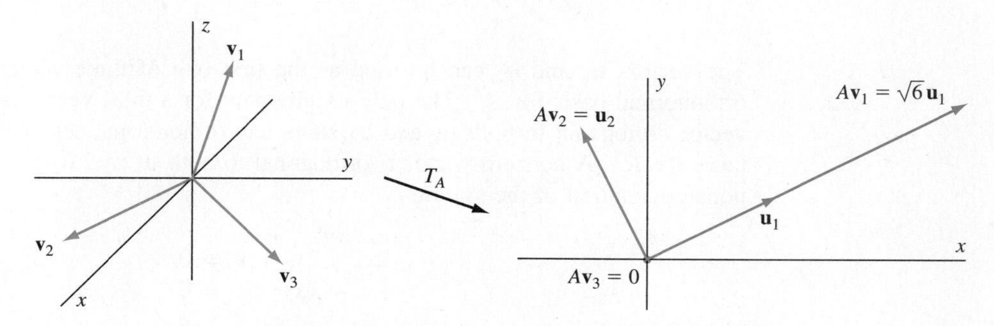

Figure 6.25 The relationships among A and the bases $\mathcal{B}_1$ and $\mathcal{B}_2$

Example 2 Find the singular values of

$$A = \begin{bmatrix} 1 & 3 & 2 & 1 \\ 3 & 1 & 2 & -1 \\ 1 & 1 & 1 & 0 \end{bmatrix},$$

and orthonormal bases $\{\mathbf{v}_1, \mathbf{v}_2, \mathbf{v}_3, \mathbf{v}_4\}$ for $\mathcal{R}^4$ and $\{\mathbf{u}_1, \mathbf{u}_2, \mathbf{u}_3\}$ for $\mathcal{R}^3$ satisfying equations (9) and (10).

Solution We first form the product

$$A^TA = \begin{bmatrix} 11 & 7 & 9 & -2 \\ 7 & 11 & 9 & 2 \\ 9 & 9 & 9 & 0 \\ -2 & 2 & 0 & 2 \end{bmatrix}.$$

It can be shown (we omit the details) that for

$$\mathbf{v}_1 = \frac{1}{\sqrt{3}}\begin{bmatrix} 1 \\ 1 \\ 1 \\ 0 \end{bmatrix}, \quad \mathbf{v}_2 = \frac{1}{\sqrt{3}}\begin{bmatrix} 1 \\ -1 \\ 0 \\ -1 \end{bmatrix}, \quad \mathbf{v}_3 = \frac{1}{\sqrt{3}}\begin{bmatrix} 1 \\ 0 \\ -1 \\ 1 \end{bmatrix}, \quad \text{and} \quad \mathbf{v}_4 = \frac{1}{\sqrt{3}}\begin{bmatrix} 0 \\ -1 \\ 1 \\ 1 \end{bmatrix},$$

$\{\mathbf{v}_1, \mathbf{v}_2, \mathbf{v}_3, \mathbf{v}_4\}$ is an orthonormal basis for $\mathcal{R}^4$ consisting of eigenvectors of A^TA with corresponding eigenvalues 27, 6, 0, and 0, respectively. So $\sigma_1 = \sqrt{27}$ and $\sigma_2 = \sqrt{6}$ are the singular values of A. Let

$$\mathbf{u}_1 = \frac{1}{\sigma_1}A\mathbf{v}_1 = \frac{1}{\sqrt{27}}\frac{1}{\sqrt{3}}\begin{bmatrix} 1 & 3 & 2 & 1 \\ 3 & 1 & 2 & -1 \\ 1 & 1 & 1 & 0 \end{bmatrix}\begin{bmatrix} 1 \\ 1 \\ 1 \\ 0 \end{bmatrix} = \frac{1}{9}\begin{bmatrix} 6 \\ 6 \\ 3 \end{bmatrix} = \frac{1}{3}\begin{bmatrix} 2 \\ 2 \\ 1 \end{bmatrix}$$

and

$$\mathbf{u}_2 = \frac{1}{\sigma_2}A\mathbf{v}_2 = \frac{1}{\sqrt{6}}\frac{1}{\sqrt{3}}\begin{bmatrix} 1 & 3 & 2 & 1 \\ 3 & 1 & 2 & -1 \\ 1 & 1 & 1 & 0 \end{bmatrix}\begin{bmatrix} 1 \\ -1 \\ 0 \\ -1 \end{bmatrix} = \frac{1}{3\sqrt{2}}\begin{bmatrix} -3 \\ 3 \\ 0 \end{bmatrix} = \frac{1}{\sqrt{2}}\begin{bmatrix} -1 \\ 1 \\ 0 \end{bmatrix}.$$

The vectors $\mathbf{u}_1$ and $\mathbf{u}_2$ can be used as the first two of three vectors in the desired orthonormal basis for $\mathcal{R}^3$. The only requirement for a third vector is that it be a unit vector orthogonal to both $\mathbf{u}_1$ and $\mathbf{u}_2$ since any orthonormal set of three vectors is a basis for $\mathcal{R}^3$. A nonzero vector is orthogonal to both $\mathbf{u}_1$ and $\mathbf{u}_2$ if and only if it is a nonzero solution of the system

$$\begin{aligned} 2x_1 + 2x_2 + x_3 &= 0 \\ -x_1 + x_2 \phantom{{}+x_3} &= 0. \end{aligned}$$

For example (we omit the details), the vector $\mathbf{w} = \begin{bmatrix} 1 \\ 1 \\ -4 \end{bmatrix}$ is a nonzero solution of this system. Therefore we let

$$\mathbf{u}_3 = \frac{1}{\|\mathbf{w}\|}\mathbf{w} = \frac{1}{\sqrt{18}}\begin{bmatrix} 1 \\ 1 \\ -4 \end{bmatrix} = \frac{1}{3\sqrt{2}}\begin{bmatrix} 1 \\ 1 \\ -4 \end{bmatrix}.$$

Thus we may choose $\{\mathbf{u}_1, \mathbf{u}_2, \mathbf{u}_3\}$ for our orthonormal basis for $\mathcal{R}^3$.

Practice Problem 1 ▶ Find the singular values of

$$A = \begin{bmatrix} -2 & -20 & 8 \\ 14 & -10 & 19 \end{bmatrix},$$

and orthonormal bases $\{\mathbf{v}_1, \mathbf{v}_2, \mathbf{v}_3\}$ for $\mathcal{R}^3$ and $\{\mathbf{u}_1, \mathbf{u}_2\}$ for $\mathcal{R}^2$ satisfying equations (9) and (10). ◀

Consider a linear transformation $T\colon \mathcal{R}^n \to \mathcal{R}^m$. Since the image of a vector in $\mathcal{R}^n$ is a vector in $\mathcal{R}^m$, geometric objects in $\mathcal{R}^n$ that can be constructed from vectors are transformed by T into objects in $\mathcal{R}^m$. The singular values of the standard matrix A of T can be used to describe how the shape of an object in $\mathcal{R}^n$ is affected by applying the transformation T. Consider, for instance, a vector $\mathbf{v}_i$ in an orthonormal basis $\mathcal{B}_1$ satisfying equations (9) and (10). The norm of the image of any vector $c\mathbf{v}_i$ parallel to $\mathbf{v}_i$ is σ_i times the norm of $c\mathbf{v}_i$ because

$$\|T(c\mathbf{v}_i)\| = \|Ac\mathbf{v}_i\| = \|c\sigma_i\mathbf{u}_i\| = |c|\sigma_i = \sigma_i\|c\mathbf{v}_i\|,$$

where $\mathbf{u}_i$ is the vector in $\mathcal{B}_2$ such that $A\mathbf{v}_i = \sigma_i\mathbf{u}_i$.

As a simple example of how singular values can be used to describe shape changes, we consider the image of the unit circle (the circle of radius 1 and center $\mathbf{0}$) under the matrix transformation T_A, where A is an invertible 2×2 matrix with distinct (nonzero) singular values.

Example 3

Let S be the unit circle in $\mathcal{R}^2$, and A be a 2×2 invertible matrix with the distinct singular values $\sigma_1 > \sigma_2 > 0$. Suppose $S' = T_A(S)$ is the image of S under the matrix transformation T_A. We describe S'. For this purpose, let $\mathcal{B}_1 = \{\mathbf{v}_1, \mathbf{v}_2\}$ and $\mathcal{B}_2 = \{\mathbf{u}_1, \mathbf{u}_2\}$ be orthonormal bases for $\mathcal{R}^2$ satisfying equations (9) and (10). For a vector $\mathbf{u}$ in $\mathcal{R}^2$, let $\mathbf{u} = x_1'\mathbf{u}_1 + x_2'\mathbf{u}_2$ for some scalars x_1' and x_2'. So

$$[\mathbf{u}]_{\mathcal{B}_2} = \begin{bmatrix} x_1' \\ x_2' \end{bmatrix}.$$

We wish to characterize S' by means of an equation in x_1' and x_2'.

For any vector $\mathbf{v} = x_1\mathbf{v}_1 + x_2\mathbf{v}_2$, the condition $\mathbf{u} = T_A(\mathbf{v})$ means that

$$x_1'\mathbf{u}_1 + x_2'\mathbf{u}_2 = T_A(x_1\mathbf{v}_1 + x_2\mathbf{v}_2) = x_1A\mathbf{v}_1 + x_2A\mathbf{v}_2 = x_1\sigma_1\mathbf{u}_1 + x_2\sigma_2\mathbf{u}_1,$$

and hence

$$x_1' = \sigma_1 x_1 \quad \text{and} \quad x_2' = \sigma_2 x_2.$$

Furthermore, $\mathbf{v}$ is in S if and only if $\|\mathbf{v}\|^2 = x_1^2 + x_2^2 = 1$. It follows that $\mathbf{u}$ is in S' if and only if $\mathbf{u} = T(\mathbf{v})$, where $\mathbf{v}$ is in S; that is,

$$\frac{(x_1')^2}{\sigma_1^2} + \frac{(x_2')^2}{\sigma_2^2} = x_1^2 + x_2^2 = 1.$$

This is the equation of an ellipse with the major and minor axes along the lines through the origin containing the vectors $\mathbf{u}_1$ and $\mathbf{u}_2$, respectively. (See Figure 6.26.)

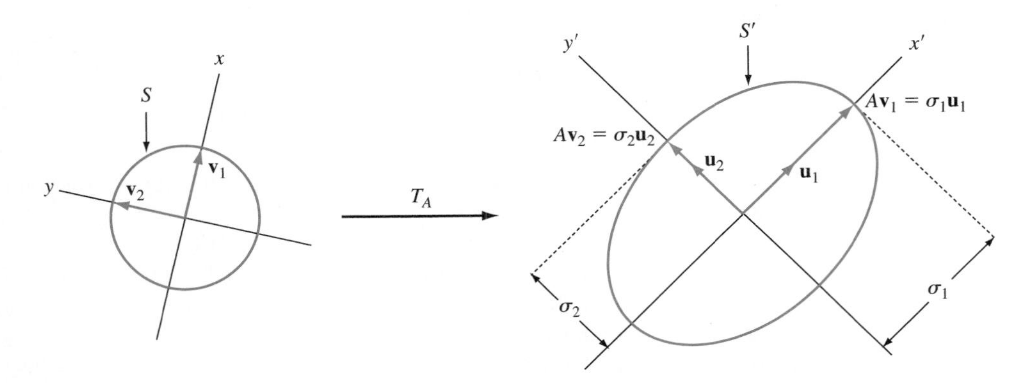

Figure 6.26 The image of the unit circle in $\mathcal{R}^2$ under T_A

THE SINGULAR VALUE DECOMPOSITION OF A MATRIX

Theorem 6.17 can be restated as a single matrix equation that has many useful applications. Using the notation of Theorem 6.17, let A be an $m \times n$ matrix of rank k with singular values $\sigma_1 \geq \sigma_2 \geq \cdots \geq \sigma_k > 0$, $\mathcal{B}_1 = \{\mathbf{v}_1, \mathbf{v}_2, \ldots, \mathbf{v}_n\}$ be an orthonormal basis for $\mathcal{R}^n$, and $\mathcal{B}_2 = \{\mathbf{u}_1, \mathbf{u}_2, \ldots, \mathbf{u}_m\}$ be an orthonormal basis for $\mathcal{R}^m$ such that equations (9) and (10) are satisfied. The $n \times n$ and $m \times m$ matrices V and U defined by

$$V = [\mathbf{v}_1 \ \ \mathbf{v}_2 \ \ \ldots \ \ \mathbf{v}_n] \quad \text{and} \quad U = [\mathbf{u}_1 \ \ \mathbf{u}_2 \ \ \ldots \ \ \mathbf{u}_m]$$

are orthogonal matrices because their columns form orthonormal bases. Let Σ be the $m \times n$ matrix whose (i,j)-entry, s_{ij}, is given by

$$\begin{cases} s_{ii} = \sigma_i & \text{for } i = 1, 2, \ldots, k \\ s_{ij} = 0 & \text{otherwise.} \end{cases}$$

So

$$\Sigma = \left[\begin{array}{cccc|cccc} \sigma_1 & 0 & \ldots & 0 & 0 & 0 & \ldots & 0 \\ 0 & \sigma_2 & \ldots & 0 & 0 & 0 & \ldots & 0 \\ \vdots & \vdots & \ddots & \vdots & \vdots & \vdots & & \vdots \\ 0 & 0 & & \sigma_k & 0 & 0 & \ldots & 0 \\ \hline 0 & 0 & \ldots & 0 & 0 & 0 & \ldots & 0 \\ \vdots & \vdots & & \vdots & \vdots & \vdots & & \vdots \\ 0 & 0 & \ldots & 0 & 0 & 0 & \ldots & 0 \end{array}\right]. \qquad (11)$$

By equation (9),

$$\begin{aligned} AV &= A[\mathbf{v}_1 \ \ \mathbf{v}_2 \ \ \ldots \ \ \mathbf{v}_n] \\ &= [A\mathbf{v}_1 \ \ A\mathbf{v}_2 \ \ \ldots \ \ A\mathbf{v}_n] \\ &= [\sigma_1\mathbf{u}_1 \ \ \sigma_2\mathbf{u}_2 \ \ \ldots \ \ \sigma_k\mathbf{u}_k \ \ \mathbf{0} \ \ \ldots \ \ \mathbf{0}] \end{aligned}$$

$$= [\mathbf{u}_1 \ \mathbf{u}_2 \ \dots \ \mathbf{u}_m] \left[\begin{array}{cccc|cccc} \sigma_1 & 0 & \dots & 0 & 0 & 0 & \dots & 0 \\ 0 & \sigma_2 & \dots & 0 & 0 & 0 & \dots & 0 \\ \vdots & \vdots & \ddots & \vdots & \vdots & \vdots & & \vdots \\ 0 & 0 & & \sigma_k & 0 & 0 & \dots & 0 \\ \hline 0 & 0 & \dots & 0 & 0 & 0 & \dots & 0 \\ \vdots & \vdots & & \vdots & \vdots & \vdots & & \vdots \\ 0 & 0 & \dots & 0 & 0 & 0 & \dots & 0 \end{array}\right]$$

$$= U\Sigma.$$

Thus $AV = U\Sigma$. Since V is an orthogonal matrix, we may multiply both sides of this last equation on the right by V^T to obtain

$$A = U\Sigma V^T.$$

In general, any factorization of an $m \times n$ matrix A into a product $A = U\Sigma V^T$, where U and V are orthogonal matrices and Σ is an $m \times n$ matrix of the form given in equation (11), is called a **singular value decomposition** of A.

We summarize the preceding discussion with the following result:

THEOREM 6.18

(Singular Value Decomposition) For any $m \times n$ matrix A of rank k, there exist $\sigma_1 \geq \sigma_2 \geq \cdots \geq \sigma_k > 0$, an $m \times m$ orthogonal matrix U, and an $n \times n$ orthogonal matrix V such that

$$A = U\Sigma V^T,$$

where Σ is the $m \times n$ matrix given in equation (11).

It can be proved that if $A = U\Sigma V^T$ is any singular value decomposition of an $m \times n$ matrix A, then the nonzero diagonal entries of Σ are the singular values of A, and the columns of V and the columns of U, which form orthonormal bases for $\mathcal{R}^n$ and $\mathcal{R}^m$, respectively, satisfy equations (9) and (10). (See Exercise 78.) For this reason, the columns of U and V in a singular value decomposition of a matrix A are sometimes referred to as *left* and *right singular vectors* of A, respectively.

Example 4 Find a singular value decomposition of the matrix

$$A = \begin{bmatrix} 0 & 1 & 2 \\ 1 & 0 & 1 \end{bmatrix}$$

in Example 1.

Solution We may use the results of Example 1 to obtain the decomposition. The columns of the matrix U are the vectors in $\mathcal{B}_2$, the columns of V are the vectors in

$\mathcal{B}_1$, and $\sigma_1 = \sqrt{6}$ and $\sigma_2 = 1$ are the singular values. Thus

$$A = U\Sigma V^T = \begin{bmatrix} \frac{2}{\sqrt{5}} & \frac{-1}{\sqrt{5}} \\ \frac{1}{\sqrt{5}} & \frac{2}{\sqrt{5}} \end{bmatrix} \begin{bmatrix} \sqrt{6} & 0 & 0 \\ 0 & 1 & 0 \end{bmatrix} \begin{bmatrix} \frac{1}{\sqrt{30}} & \frac{2}{\sqrt{5}} & \frac{1}{\sqrt{6}} \\ \frac{2}{\sqrt{30}} & \frac{-1}{\sqrt{5}} & \frac{2}{\sqrt{6}} \\ \frac{5}{\sqrt{30}} & 0 & \frac{-1}{\sqrt{6}} \end{bmatrix}^T.$$

Example 5 Find a singular value decomposition of the matrix

$$C = \begin{bmatrix} 0 & 1 \\ 1 & 0 \\ 2 & 1 \end{bmatrix}.$$

Solution Notice that C is the transpose of the matrix A in Examples 1 and 4. Using the singular value decomposition $A = U\Sigma V^T$, we have

$$C = A^T = (U\Sigma V^T)^T = (V^T)^T \Sigma^T U^T = V\Sigma^T U^T.$$

Observe that

$$\Sigma^T = \begin{bmatrix} \sqrt{6} & 0 \\ 0 & 1 \\ 0 & 0 \end{bmatrix}$$

is the matrix satisfying equation (11) for A^T. Therefore, since V^T and U^T are orthogonal matrices, $V\Sigma^T U^T$ is a singular value decomposition of $C = A^T$.

Practice Problem 2 ► Find a singular value decomposition of the matrix A in Practice Problem 1. ◄

There are efficient and accurate methods for finding a singular value decomposition $U\Sigma V^T$ of an $m \times n$ matrix A, and many practical applications of linear algebra use this decomposition. Because the matrices U and V^T are orthogonal, multiplication by U and V^T does not change the norms of vectors or the angles between them. Thus any roundoff errors that arise in calculations involving A are due solely to the matrix Σ. For this reason, calculations involving A are most reliable if a singular value decomposition of A is used.

ORTHOGONAL PROJECTIONS, SYSTEMS OF LINEAR EQUATIONS, AND THE PSEUDOINVERSE

Let A be an $m \times n$ matrix and $\mathbf{b}$ be in $\mathcal{R}^m$. We have seen that the system of linear equations $A\mathbf{x} = \mathbf{b}$ can be consistent or inconsistent.

In the case that the system is consistent, a vector $\mathbf{u}$ in $\mathcal{R}^n$ is a solution if and only if $\|A\mathbf{u} - \mathbf{b}\| = 0$.

In the case that the system is inconsistent, $\|A\mathbf{u} - \mathbf{b}\| > 0$ for every $\mathbf{u}$ in $\mathcal{R}^n$. However, it is often desirable to find a vector $\mathbf{z}$ in $\mathcal{R}^n$ that minimizes the distance between $A\mathbf{u}$ and $\mathbf{b}$, that is, a vector $\mathbf{z}$ such that

$$\|A\mathbf{z} - \mathbf{b}\| \leq \|A\mathbf{u} - \mathbf{b}\| \qquad \text{for all } \mathbf{u} \text{ in } \mathcal{R}^n.$$

This problem, called the *least-squares problem*, was encountered in Section 6.4, where we showed that $\|A\mathbf{z} - \mathbf{b}\|$ is a minimum if and only if

$$A\mathbf{z} = P_W\mathbf{b}, \tag{12}$$

where W is the column space of A and P_W is the orthogonal projection matrix for W.

The next theorem shows how we can use a singular value decomposition of A to compute P_W.

THEOREM 6.19

Let A be an $m \times n$ matrix of rank k having a singular value decomposition $A = U\Sigma V^T$, and let $W = \text{Col}\,A$. Let D be the $m \times m$ diagonal matrix whose first k diagonal entries are 1s and whose other entries are 0s. Then

$$P_W = UDU^T. \tag{13}$$

PROOF Let $P = UDU^T$. Observe that $P^2 = P^T = P$, and hence by Exercise 82 of Section 6.3, P is an orthogonal projection matrix for some subspace of $\mathcal{R}^m$. We must show that this subspace is, in fact, W. For this purpose, we modify the $m \times n$ matrix Σ to obtain a new $n \times m$ matrix $\Sigma^\dagger$ defined by

$$\Sigma^\dagger = \left[\begin{array}{cccc|cccc} \frac{1}{\sigma_1} & 0 & \dots & 0 & 0 & 0 & \dots & 0 \\ 0 & \frac{1}{\sigma_2} & \dots & 0 & 0 & 0 & \dots & 0 \\ \vdots & \vdots & \ddots & \vdots & \vdots & \vdots & & \vdots \\ 0 & 0 & & \frac{1}{\sigma_k} & 0 & 0 & \dots & 0 \\ \hline 0 & 0 & \dots & 0 & 0 & 0 & \dots & 0 \\ \vdots & \vdots & & \vdots & \vdots & \vdots & & \vdots \\ 0 & 0 & \dots & 0 & 0 & 0 & \dots & 0 \end{array}\right]. \tag{14}$$

Observe that $\Sigma\Sigma^\dagger = D$, and hence

$$A(V\Sigma^\dagger U^T) = U\Sigma V^T V\Sigma^\dagger U^T = U\Sigma\Sigma^\dagger U^T = UDU^T = P. \tag{15}$$

It follows that, for any vector $\mathbf{v}$ in $\mathcal{R}^n$, we have $P\mathbf{v} = A\mathbf{w}$, where $\mathbf{w} = (V\Sigma^\dagger U^T)\mathbf{v}$. Consequently, $P\mathbf{v}$ lies in W. Thus the column space of P is a subspace of W. Since D has rank k, P also has rank k, and hence the dimension of the column space of P is k. It follows that the column space of P is W by Theorem 4.9 in Section 4.3. We conclude that $P = P_W$. ■

We can use equation (15), whose right side we have shown to be P_W, to select the vector $\mathbf{z}$ in equation (12) that minimizes $\|A\mathbf{u} - \mathbf{b}\|$. Let

$$\mathbf{z} = V\Sigma^\dagger U^T\mathbf{b}. \tag{16}$$

Then, by equation (15), $A\mathbf{z} = A(V\Sigma^\dagger U^T)\mathbf{b} = P_W\mathbf{b}$. (See Figure 6.27.)

Besides the vector $\mathbf{z} = V\Sigma^\dagger U^T\mathbf{b}$, there may be other vectors in $\mathcal{R}^n$ that also minimize $\|A\mathbf{u} - \mathbf{b}\|$. However, we now show that among all such vectors, $\mathbf{z} = V\Sigma^\dagger U^T\mathbf{b}$

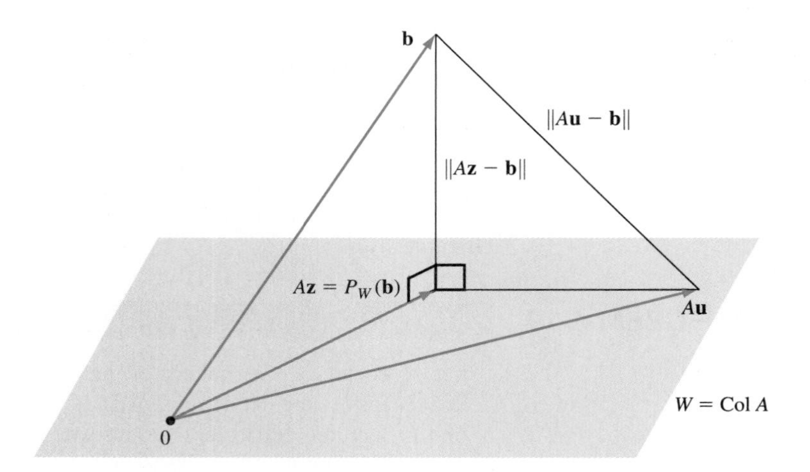

Figure 6.27 The vector $V\Sigma^{\dagger}U^{T}\mathbf{b}$ satisfies $A\mathbf{z} = P_W\mathbf{b}$.

is the unique vector of least norm. Suppose that $\mathbf{y}$ is any vector in $\mathcal{R}^n$, different from $\mathbf{z}$, that also minimizes $\|A\mathbf{u} - \mathbf{b}\|$. Then $A\mathbf{y} = P_W\mathbf{b}$ by equation (12). Let $\mathbf{w} = \mathbf{y} - \mathbf{z}$. Then $\mathbf{w} \neq \mathbf{0}$, but

$$A\mathbf{w} = A\mathbf{y} - A\mathbf{z} = P_W\mathbf{b} - P_W\mathbf{b} = \mathbf{0}.$$

Substituting the singular value decomposition of A, we have

$$U\Sigma V^T\mathbf{w} = \mathbf{0}.$$

Since U is invertible, it follows that

$$\Sigma V^T\mathbf{w} = \mathbf{0}.$$

This last equation tells us that the first k components of $V^T\mathbf{w}$ are zeros. Furthermore, since $\mathbf{z} = V\Sigma^{\dagger}U^T\mathbf{b}$, we have that $V^T\mathbf{z} = \Sigma^{\dagger}U^T\mathbf{b}$, and hence the last $n - k$ components of $V^T\mathbf{z}$ are zeros. It follows that $V^T\mathbf{w}$ and $V^T\mathbf{z}$ are orthogonal. Since V^T is an orthogonal matrix, it preserves dot products, and hence

$$\mathbf{z} \cdot \mathbf{w} = (V^T\mathbf{z}) \cdot (V^T\mathbf{w}) = 0.$$

Thus $\mathbf{z}$ and $\mathbf{w}$ are orthogonal. Since $\mathbf{y} = \mathbf{z} + \mathbf{w}$, we may apply the Pythagorean theorem to find that

$$\|\mathbf{y}\|^2 = \|\mathbf{z} + \mathbf{w}\|^2 = \|\mathbf{z}\|^2 + \|\mathbf{w}\|^2 > \|\mathbf{z}\|^2,$$

and hence $\|\mathbf{y}\| > \|\mathbf{z}\|$. It follows that $\mathbf{z} = V\Sigma^{\dagger}U^T\mathbf{b}$ is the vector of least norm that minimizes $\|A\mathbf{u} - \mathbf{b}\|$.

We summarize what we have learned in the following box:

Let A be an $m \times n$ matrix with a singular value decomposition $A = U\Sigma V^T$, $\mathbf{b}$ be a vector in $\mathcal{R}^m$, and $\mathbf{z} = V\Sigma^{\dagger}U^T\mathbf{b}$, where $\Sigma^{\dagger}$ is as in equation (14). Then the following statements are true:

(a) If the system $A\mathbf{x} = \mathbf{b}$ is consistent, then $\mathbf{z}$ is the unique solution of least norm.

(b) If the system $A\mathbf{x} = \mathbf{b}$ is inconsistent, then $\mathbf{z}$ is the unique vector of least norm such that

$$\|A\mathbf{z} - \mathbf{b}\| \leq \|A\mathbf{u} - \mathbf{b}\|$$

for all $\mathbf{u}$ in $\mathcal{R}^n$.

Example 6 Use a singular value decomposition to find the solution of least norm to the system

$$\begin{aligned} x_2 + 2x_3 &= 5 \\ x_1 \quad + x_3 &= 1. \end{aligned}$$

Solution Let A denote the coefficient matrix of this system, and let $\mathbf{b} = \begin{bmatrix} 5 \\ 1 \end{bmatrix}$. A singular value decomposition of A was computed in Example 4, where we obtained

$$\begin{bmatrix} 0 & 1 & 2 \\ 1 & 0 & 1 \end{bmatrix} = A = U\Sigma V^T = \begin{bmatrix} \frac{2}{\sqrt{5}} & \frac{-1}{\sqrt{5}} \\ \frac{1}{\sqrt{5}} & \frac{2}{\sqrt{5}} \end{bmatrix} \begin{bmatrix} \sqrt{6} & 0 & 0 \\ 0 & 1 & 0 \end{bmatrix} \begin{bmatrix} \frac{1}{\sqrt{30}} & \frac{2}{\sqrt{5}} & \frac{1}{\sqrt{6}} \\ \frac{2}{\sqrt{30}} & \frac{-1}{\sqrt{5}} & \frac{2}{\sqrt{6}} \\ \frac{5}{\sqrt{30}} & 0 & \frac{-1}{\sqrt{6}} \end{bmatrix}^T.$$

Let $\mathbf{z}$ denote the solution of least norm of the given system. Then

$$\begin{aligned} \mathbf{z} &= V\Sigma^{\dagger}U^T\mathbf{b} \\ &= \begin{bmatrix} \frac{1}{\sqrt{30}} & \frac{2}{\sqrt{5}} & \frac{1}{\sqrt{6}} \\ \frac{2}{\sqrt{30}} & \frac{-1}{\sqrt{5}} & \frac{2}{\sqrt{6}} \\ \frac{5}{\sqrt{30}} & 0 & \frac{-1}{\sqrt{6}} \end{bmatrix} \begin{bmatrix} \frac{1}{\sqrt{6}} & 0 \\ 0 & 1 \\ 0 & 0 \end{bmatrix} \begin{bmatrix} \frac{2}{\sqrt{5}} & \frac{-1}{\sqrt{5}} \\ \frac{1}{\sqrt{5}} & \frac{2}{\sqrt{5}} \end{bmatrix}^T \begin{bmatrix} 5 \\ 1 \end{bmatrix} \\ &= \frac{1}{6}\begin{bmatrix} -5 \\ 8 \\ 11 \end{bmatrix}. \end{aligned}$$

Practice Problem 3 ▶ Use your answer to Practice Problem 2 to find the solution of least norm to the system

$$\begin{aligned} -2x_1 - 20x_2 + 8x_3 &= 5 \\ 14x_1 - 10x_2 + 19x_3 &= -5. \end{aligned}$$

◀

Example 7 Let

$$A = \begin{bmatrix} 1 & 1 & 2 \\ 1 & -1 & 3 \\ 1 & 3 & 1 \end{bmatrix} \quad \text{and} \quad \mathbf{b} = \begin{bmatrix} 1 \\ 4 \\ -1 \end{bmatrix}.$$

It is easy to show that the equation $A\mathbf{x} = \mathbf{b}$ has no solution. Find a vector $\mathbf{z}$ in $\mathcal{R}^3$ such that

$$\|A\mathbf{z} - \mathbf{b}\| \leq \|A\mathbf{u} - \mathbf{b}\|$$

for all $\mathbf{u}$ in $\mathcal{R}^3$.

Solution First observe (we omit the details) that

$$\mathcal{B}_1 = \left\{ \frac{1}{\sqrt{6}}\begin{bmatrix}1\\1\\2\end{bmatrix}, \frac{1}{\sqrt{5}}\begin{bmatrix}0\\2\\-1\end{bmatrix}, \frac{1}{\sqrt{30}}\begin{bmatrix}-5\\1\\2\end{bmatrix} \right\}$$

is an orthonormal basis of eigenvectors of A^TA with corresponding eigenvalues 18, 10, and 0. So $\sigma_1 = \sqrt{18}$ and $\sigma_2 = \sqrt{10}$ are the singular values of A. Let $\mathbf{v}_1$ and $\mathbf{v}_2$ denote the first two vectors in $\mathcal{B}_1$, which correspond to the singular values σ_1 and σ_2, respectively, and let

$$\mathbf{u}_1 = \frac{1}{\sigma_1}A\mathbf{v}_1 = \frac{1}{\sqrt{18}}\frac{1}{\sqrt{6}}\begin{bmatrix}1 & 1 & 2\\1 & -1 & 3\\1 & 3 & 1\end{bmatrix}\begin{bmatrix}1\\1\\2\end{bmatrix} = \frac{1}{\sqrt{3}}\begin{bmatrix}1\\1\\1\end{bmatrix}$$

and

$$\mathbf{u}_2 = \frac{1}{\sigma_2}A\mathbf{v}_2 = \frac{1}{\sqrt{10}}\frac{1}{\sqrt{5}}\begin{bmatrix}1 & 1 & 2\\1 & -1 & 3\\1 & 3 & 1\end{bmatrix}\begin{bmatrix}0\\2\\-1\end{bmatrix} = \frac{1}{\sqrt{2}}\begin{bmatrix}0\\-1\\1\end{bmatrix}.$$

As in Example 2, we are able to extend $\{\mathbf{u}_1, \mathbf{u}_2\}$ to an orthonormal basis for $\mathcal{R}^3$ by adjoining the vector $\mathbf{u}_3 = \frac{1}{\sqrt{6}}\begin{bmatrix}2\\-1\\-1\end{bmatrix}$. This produces a set of left singular vectors $\mathcal{B}_2 = \{\mathbf{u}_1, \mathbf{u}_2, \mathbf{u}_3\}$ of A. Next, set

$$U = [\mathbf{u}_1\ \mathbf{u}_2\ \mathbf{u}_3], \qquad V = [\mathbf{v}_1\ \mathbf{v}_2\ \mathbf{v}_3],$$

and

$$\Sigma = \begin{bmatrix}\sigma_1 & 0 & 0\\0 & \sigma_2 & 0\\0 & 0 & 0\end{bmatrix} = \begin{bmatrix}\sqrt{18} & 0 & 0\\0 & \sqrt{10} & 0\\0 & 0 & 0\end{bmatrix}.$$

Then $A = U\Sigma V^T$ is a singular value decomposition of A. Hence

$$\begin{aligned}\mathbf{z} &= V\Sigma^\dagger U^T\mathbf{b}\\ &= \begin{bmatrix}\frac{1}{\sqrt{6}} & 0 & \frac{-5}{\sqrt{30}}\\ \frac{1}{\sqrt{6}} & \frac{2}{\sqrt{5}} & \frac{1}{\sqrt{30}}\\ \frac{2}{\sqrt{6}} & \frac{-1}{\sqrt{5}} & \frac{2}{\sqrt{30}}\end{bmatrix}\begin{bmatrix}\frac{1}{\sqrt{18}} & 0 & 0\\0 & \frac{1}{\sqrt{10}} & 0\\0 & 0 & 0\end{bmatrix}\begin{bmatrix}\frac{1}{\sqrt{3}} & 0 & \frac{2}{\sqrt{6}}\\ \frac{1}{\sqrt{3}} & \frac{-1}{\sqrt{2}} & \frac{-1}{\sqrt{6}}\\ \frac{1}{\sqrt{3}} & \frac{1}{\sqrt{2}} & \frac{-1}{\sqrt{6}}\end{bmatrix}^T\begin{bmatrix}1\\4\\-1\end{bmatrix}\\ &= \frac{1}{18}\begin{bmatrix}4\\-14\\17\end{bmatrix}\end{aligned}$$

is the vector of least norm that satisfies the condition $\|A\mathbf{z} - \mathbf{b}\| \le \|A\mathbf{u} - \mathbf{b}\|$ for all $\mathbf{u}$ in $\mathcal{R}^3$.

Practice Problem 4 ▶ Let

$$A = \begin{bmatrix}1 & 1 & 2\\1 & -1 & 3\\1 & 3 & 1\end{bmatrix} \quad \text{and} \quad \mathbf{b} = \begin{bmatrix}27\\36\\-18\end{bmatrix}.$$

Use the singular value decomposition in Example 7 to find a vector $\mathbf{z}$ in $\mathcal{R}^3$ such that $\|A\mathbf{z} - \mathbf{b}\| \leq \|A\mathbf{u} - \mathbf{b}\|$ for all $\mathbf{u}$ in $\mathcal{R}^3$. ◀

In the previous discussion, a singular value decomposition of the coefficient matrix $A = U\Sigma V^T$ of a system of linear equations $A\mathbf{x} = \mathbf{b}$ is used to find the solution of least norm or the vector of least norm that minimizes $\|A\mathbf{u} - \mathbf{b}\|$. The solution is the product of the matrix $V\Sigma^\dagger U^T$ and $\mathbf{b}$.

Although a singular value decomposition of a matrix $A = U\Sigma V^T$ is not unique, the matrix $V\Sigma^\dagger U^T$ is unique; that is, it is independent of the choice of a singular value decomposition of A. To see this, consider an $m \times n$ matrix A, and suppose that

$$A = U_1\Sigma V_1^T = U_2\Sigma V_2^T$$

are two singular value decompositions of A. Now consider any vector $\mathbf{b}$ in $\mathcal{R}^m$. Then we have seen that both $V_1\Sigma^\dagger U_1^T\mathbf{b}$ and $V_2\Sigma^\dagger U_2^T\mathbf{b}$ are the unique vector of least norm that minimizes $\|A\mathbf{u} - \mathbf{b}\|$. So

$$V_1\Sigma^\dagger U_1^T\mathbf{b} = V_2\Sigma^\dagger U_2^T\mathbf{b}.$$

Since $\mathbf{b}$ is an arbitrarily chosen vector in $\mathcal{R}^n$, it follows that $V_1\Sigma^\dagger U_1^T = V_2\Sigma^\dagger U_2^T$.

For a given matrix $A = U\Sigma V^T$, the matrix $V\Sigma^\dagger U^T$ is called the **pseudoinverse**, or **Moore–Penrose generalized inverse**, of A and is denoted by $A^\dagger$. Note that the pseudoinverse of the matrix Σ in equation (11) is the matrix $\Sigma^\dagger$ in equation (14). (See Exercise 81.) The terminology *pseudoinverse* is due to the fact that if A is invertible, then $A^\dagger = A^{-1}$, the ordinary inverse of A. (See Exercise 82.)

Replacing $U\Sigma V^T$ by $A^\dagger$, the pseudoinverse of A, we can restate equations (15) and (16) as follows:

Applications of the Pseudoinverse

For any $m \times n$ matrix A and any vector $\mathbf{b}$ in $\mathcal{R}^m$, the following statements are true:

1. The orthogonal projection matrix for Col A is $AA^\dagger$.
2. The unique vector of least norm that minimizes $\|A\mathbf{u} - \mathbf{b}\|$ for $\mathbf{u}$ in $\mathcal{R}^n$ is $A^\dagger\mathbf{b}$. Therefore, if $A\mathbf{x} = \mathbf{b}$ is consistent, $A^\dagger\mathbf{b}$ is the unique solution of least norm.

Example 8 Find the pseudoinverse of the matrix

$$A = \begin{bmatrix} 0 & 1 & 2 \\ 1 & 0 & 1 \end{bmatrix}$$

in Example 4. Then use the result to solve the problem posed in Example 6 again.

Solution From Example 4, a singular value decomposition of A is given by

$$A = U\Sigma V^T = \begin{bmatrix} \frac{2}{\sqrt{5}} & \frac{-1}{\sqrt{5}} \\ \frac{1}{\sqrt{5}} & \frac{2}{\sqrt{5}} \end{bmatrix} \begin{bmatrix} \sqrt{6} & 0 & 0 \\ 0 & 1 & 0 \end{bmatrix} \begin{bmatrix} \frac{1}{\sqrt{30}} & \frac{2}{\sqrt{5}} & \frac{1}{\sqrt{6}} \\ \frac{2}{\sqrt{30}} & \frac{-1}{\sqrt{5}} & \frac{2}{\sqrt{6}} \\ \frac{5}{\sqrt{30}} & 0 & \frac{-1}{\sqrt{6}} \end{bmatrix}^T.$$

Hence the pseudoinverse of A is

$$A^\dagger = V\Sigma^\dagger U^T = \begin{bmatrix} \frac{1}{\sqrt{30}} & \frac{2}{\sqrt{5}} & \frac{1}{\sqrt{6}} \\ \frac{2}{\sqrt{30}} & \frac{-1}{\sqrt{5}} & \frac{2}{\sqrt{6}} \\ \frac{5}{\sqrt{30}} & 0 & \frac{-1}{\sqrt{6}} \end{bmatrix} \begin{bmatrix} \frac{1}{\sqrt{6}} & 0 \\ 0 & 1 \\ 0 & 0 \end{bmatrix} \begin{bmatrix} \frac{2}{\sqrt{5}} & \frac{-1}{\sqrt{5}} \\ \frac{1}{\sqrt{5}} & \frac{2}{\sqrt{5}} \end{bmatrix}^T$$

$$= \begin{bmatrix} -\frac{1}{3} & \frac{5}{6} \\ \frac{1}{3} & -\frac{1}{3} \\ \frac{1}{3} & \frac{1}{6} \end{bmatrix}.$$

To solve the problem posed in Example 6, let $\mathbf{b} = \begin{bmatrix} 5 \\ 1 \end{bmatrix}$. The solution of least norm to the system of equations in Example 6 is

$$A^\dagger \mathbf{b} = \begin{bmatrix} -\frac{1}{3} & \frac{5}{6} \\ \frac{1}{3} & -\frac{1}{3} \\ \frac{1}{3} & \frac{1}{6} \end{bmatrix} \begin{bmatrix} 5 \\ 1 \end{bmatrix} = \frac{1}{6} \begin{bmatrix} -5 \\ 8 \\ 11 \end{bmatrix}.$$

EXERCISES

In Exercises 1–10, find a singular value decomposition for each matrix.

1. $\begin{bmatrix} 1 & 0 \\ 1 & 0 \end{bmatrix}$

2. $\begin{bmatrix} 1 & 1 \\ 0 & 0 \end{bmatrix}$

3. $\begin{bmatrix} 1 \\ 2 \\ 2 \end{bmatrix}$

4. $\begin{bmatrix} 1 \\ 1 \\ -1 \\ 1 \end{bmatrix}$

5. $\begin{bmatrix} 1 & 1 \\ 1 & -1 \\ 1 & 2 \end{bmatrix}$

6. $\begin{bmatrix} 1 & 2 \\ 3 & -1 \\ 1 & 0 \\ 1 & 1 \end{bmatrix}$

7. $\begin{bmatrix} 1 & 1 & 1 \\ 1 & -1 & -1 \end{bmatrix}$

8. $\begin{bmatrix} 1 & 0 & 0 & 0 \\ 0 & 2 & 0 & 0 \end{bmatrix}$

9. $\begin{bmatrix} 1 & 1 & 2 \\ 2 & 0 & -1 \\ 1 & -1 & 0 \end{bmatrix}$

10. $\begin{bmatrix} 1 & -1 & 3 \\ 1 & -1 & -1 \\ 2 & 1 & -1 \end{bmatrix}$

In Exercises 11–18, find a singular value decomposition of each matrix A. In each case, the characteristic polynomial of A^TA is given.

11. $A = \begin{bmatrix} 1 & -1 & 2 \\ -2 & 2 & -4 \end{bmatrix}$
$-t^2(t-30)$

12. $A = \begin{bmatrix} 1 & 0 & -2 \\ 2 & 0 & -4 \end{bmatrix}$
$-t^2(t-25)$

13. $A = \begin{bmatrix} 1 & -1 & 1 \\ -1 & 2 & 1 \end{bmatrix}$
$-t(t-2)(t-7)$

14. $A = \begin{bmatrix} 1 & 0 & -1 \\ -1 & 1 & 0 \end{bmatrix}$
$-t(t-1)(t-3)$

15. $A = \begin{bmatrix} 3 & 5 & 4 & 1 \\ 4 & 0 & 2 & -2 \\ 0 & 0 & 0 & 0 \end{bmatrix}, \quad t^2(t-60)(t-15)$

16. $A = \begin{bmatrix} 2 & -3 & 2 & -3 \\ 6 & 1 & 6 & 1 \\ 0 & 0 & 0 & 0 \end{bmatrix}, \quad t^2(t-80)(t-20)$

17. $A = \begin{bmatrix} 3 & 0 & 1 & 3 \\ 0 & 3 & 1 & 0 \\ 0 & -3 & -1 & 0 \end{bmatrix}, \quad t^2(t-18)(t-21)$

18. $A = \begin{bmatrix} -4 & 8 & 0 & -8 \\ 8 & -25 & 0 & 7 \\ 8 & -7 & 0 & 25 \end{bmatrix}, \quad t^2(t-324)(t-1296)$

In Exercises 19 and 20, sketch the image of the unit circle under the matrix transformation T_A induced by each matrix A.

19. $\begin{bmatrix} 2 & 1 \\ -2 & 1 \end{bmatrix}$

20. $\begin{bmatrix} 1 & 2 \\ 2 & 1 \end{bmatrix}$

In Exercises 21–28, find the unique solution of least norm to each system of linear equations. In one of the exercises, you can use an answer from Exercises 1–10.

21. $\begin{aligned} x_1 + x_2 &= 2 \\ 2x_1 + 2x_2 &= 4 \end{aligned}$

22. $\begin{aligned} x_1 - x_2 &= 2 \\ 2x_1 - 2x_2 &= 4 \end{aligned}$

23. $\begin{aligned} x_1 \quad\quad - x_3 &= 2 \\ -x_2 + x_3 &= 5 \end{aligned}$

24. $\begin{aligned} x_1 - x_2 + 2x_3 &= -1 \\ -x_1 + 2x_2 - 2x_3 &= 2 \end{aligned}$

25. $\begin{aligned} x_1 - 2x_2 + x_3 &= 3 \\ -x_1 + x_2 + 2x_3 &= -1 \end{aligned}$

26. $\begin{aligned} x_1 \quad + x_3 &= 3 \\ x_2 \quad &= 1 \end{aligned}$

27. $\begin{aligned} x_1 + x_2 + x_3 &= 5 \\ x_1 - x_2 - x_3 &= 1 \end{aligned}$

28. $\begin{aligned} x_1 + x_2 \quad &= 4 \\ x_1 - x_2 - x_3 &= 1 \end{aligned}$

In Exercises 29–36, the systems are inconsistent. For each system $A\mathbf{x} = \mathbf{b}$*, find the unique vector* $\mathbf{z}$ *of least norm such that* $\|A\mathbf{z} - \mathbf{b}\|$ *is a minimum. In a few of these exercises, you can use answers from Exercises 1–18.*

29. $\begin{aligned} x_1 + 2x_2 &= -1 \\ 2x_1 + 4x_2 &= 1 \end{aligned}$

30. $\begin{aligned} x_1 - x_2 + 2x_3 &= 3 \\ -2x_1 + 2x_2 - 4x_3 &= -1 \end{aligned}$

31. $\begin{aligned} x_1 + x_2 &= 3 \\ x_1 - x_2 &= 1 \\ x_1 + 2x_2 &= 2 \end{aligned}$

32. $\begin{aligned} x_1 + 2x_2 &= 4 \\ 3x_1 - x_2 &= 5 \\ x_1 \quad &= 1 \\ x_1 + x_2 &= 0 \end{aligned}$

33. $\begin{aligned} x_1 + x_2 - x_3 &= 4 \\ x_1 + x_2 + x_3 &= 6 \\ x_3 &= 3 \end{aligned}$

34. $\begin{aligned} 2x_1 + x_2 &= 1 \\ x_1 - x_2 &= -4 \\ x_1 + 2x_2 &= 0 \end{aligned}$

35. $\begin{aligned} x_1 - x_2 + x_3 &= 1 \\ -x_1 + 2x_2 + x_3 &= 1 \\ 2x_1 - x_2 + 4x_3 &= 0 \end{aligned}$

36. $\begin{aligned} x_1 \quad - x_3 &= -1 \\ -x_1 + x_2 \quad &= 2 \\ 3x_1 - x_2 - 2x_3 &= 1 \end{aligned}$

In Exercises 37–46, find the pseudoinverse of the given matrix. In most cases, you can use the results of Exercises 1–28.

37. $\begin{bmatrix} 1 \\ 2 \\ 2 \end{bmatrix}$

38. $\begin{bmatrix} 1 & -1 \\ -2 & 2 \end{bmatrix}$

39. $\begin{bmatrix} 1 & 0 & -1 \\ 0 & -1 & 1 \end{bmatrix}$

40. $\begin{bmatrix} 1 & 2 \\ 3 & -1 \\ 1 & 0 \\ 1 & 1 \end{bmatrix}$

41. $\begin{bmatrix} 1 & 1 \\ 1 & -1 \\ 1 & 2 \end{bmatrix}$

42. $\begin{bmatrix} 1 & 0 & 0 & 0 \\ 0 & 2 & 0 & 0 \end{bmatrix}$

43. $\begin{bmatrix} 1 & 1 & 1 \\ 1 & -1 & -1 \end{bmatrix}$

44. $\begin{bmatrix} 1 & -1 & 2 \\ -1 & 2 & -2 \end{bmatrix}$

45. $\begin{bmatrix} 1 & -1 & 1 \\ -1 & 2 & 1 \end{bmatrix}$

46. $\begin{bmatrix} 3 & 5 & 4 & 1 \\ 4 & 0 & 2 & -2 \\ 0 & 0 & 0 & 0 \end{bmatrix}$

In equation (13), a singular value decomposition of a matrix A is used to obtain the orthogonal projection on the subspace Col A*. Use this method in Exercises 47–54 to compute the orthogonal projection matrix* P_W *on the subspace* W*. You can use the results of Exercises 1–10.*

47. $W = \text{Span}\left\{ \begin{bmatrix} 1 \\ 2 \\ 2 \end{bmatrix} \right\}$

48. $W = \text{Span}\left\{ \begin{bmatrix} 1 \\ 1 \\ -1 \\ 1 \end{bmatrix} \right\}$

49. $W = \text{Span}\left\{ \begin{bmatrix} 1 \\ 0 \\ 1 \end{bmatrix}, \begin{bmatrix} 0 \\ -1 \\ 1 \end{bmatrix} \right\}$

50. $W = \text{Span}\left\{ \begin{bmatrix} 1 \\ 1 \\ -2 \end{bmatrix}, \begin{bmatrix} 1 \\ -1 \\ 1 \end{bmatrix} \right\}$

51. $W = \text{Span}\left\{ \begin{bmatrix} 1 \\ -2 \\ 3 \end{bmatrix}, \begin{bmatrix} 2 \\ 1 \\ -1 \end{bmatrix} \right\}$

52. $W = \text{Span}\left\{ \begin{bmatrix} 3 \\ -2 \\ 1 \end{bmatrix}, \begin{bmatrix} -2 \\ 1 \\ 2 \end{bmatrix} \right\}$

53. $W = \text{Span}\left\{ \begin{bmatrix} 1 \\ 1 \\ 1 \end{bmatrix}, \begin{bmatrix} 1 \\ -1 \\ 2 \end{bmatrix} \right\}$

54. $W = \text{Span}\left\{ \begin{bmatrix} 1 \\ 3 \\ 1 \\ 1 \end{bmatrix}, \begin{bmatrix} 2 \\ -1 \\ 0 \\ 1 \end{bmatrix} \right\}$

In Exercises 55–75, determine whether the statements are true or false. For the purpose of these exercises, we consider a particular $m \times n$ *matrix* A *and orthonormal bases* $\mathcal{B}_1$ *and* $\mathcal{B}_2$ *of* $\mathcal{R}^n$ *and* $\mathcal{R}^m$*, respectively, such that the singular values of* A *and these bases satisfy equations (9) and (10)*

55. If σ is a singular value of a matrix A, then σ is an eigenvalue of A^TA.
56. If the roles of $\mathcal{B}_1$ and $\mathcal{B}_2$ are reversed, then equations (9) and (10) are satisfied for the matrix A^T.
57. If a matrix is square, then $\mathcal{B}_1 = \mathcal{B}_2$.
58. Every matrix has the same singular values as its transpose.
59. A matrix has a pseudoinverse if and only if it is not invertible.
60. $\mathcal{B}_2$ is an orthonormal basis of eigenvectors of A^TA.
61. $\mathcal{B}_2$ is an orthonormal basis of eigenvectors of AA^T.
62. $\mathcal{B}_1$ is an orthonormal basis of eigenvectors of A^TA.
63. $\mathcal{B}_1$ is an orthonormal basis of eigenvectors of AA^T.
64. If matrix A has rank k, then A has k singular values.
65. Every matrix has a singular value decomposition.
66. Every matrix has a unique singular value decomposition.
67. In a singular value decomposition $U\Sigma V^T$ of A, each diagonal entry of Σ is a singular value of A.
68. Suppose that $A = U\Sigma V^T$ is a singular value decomposition, $\mathcal{A}_1$ is the set of columns of V, and $\mathcal{A}_2$ is the set of columns of U. Then equations (9) and (10) are satisfied if $\mathcal{A}_1$ replaces $\mathcal{B}_1$ and $\mathcal{A}_2$ replaces $\mathcal{B}_2$.

69. If $U\Sigma V^T$ is a singular value decomposition of A, then $V\Sigma U^T$ is a singular value decomposition of A^T.

70. If $U\Sigma V^T$ is a singular value decomposition of A having rank k, and if $W = \text{Col}\, A$, then $P_W = UDU^T$, where $d_{ii} = 1$ if $i = 1, 2, \cdots, k$ and $d_{ij} = 0$ otherwise.

71. If A is an $m \times n$ matrix and $U\Sigma V^T$ is a singular value decomposition of A, then a vector $\mathbf{u}$ in $\mathcal{R}^n$ that minimizes $\|A\mathbf{u} - \mathbf{b}\|$ is $V\Sigma^\dagger U^T\mathbf{b}$.

72. If A is an $m \times n$ matrix and $U\Sigma V^T$ is a singular value decomposition of A, then $V\Sigma^\dagger U^T\mathbf{b}$ is the unique vector $\mathbf{u}$ in $\mathcal{R}^n$ that minimizes $\|A\mathbf{u} - \mathbf{b}\|$.

73. If A is an $m \times n$ matrix and $U\Sigma V^T$ is a singular value decomposition of A, then $V\Sigma^\dagger U^T\mathbf{b}$ is the unique vector $\mathbf{u}$ in $\mathcal{R}^n$ with least norm that minimizes $\|A\mathbf{u} - \mathbf{b}\|$.

74. If $U\Sigma V^T$ is a singular value decomposition of A, then $A^\dagger = V\Sigma U^T$.

75. If A is an invertible matrix, then $A^\dagger = A^{-1}$.

---

76. Suppose that A is an $m \times n$ matrix of rank k with singular values $\sigma_1 \geq \sigma_2 \geq \cdots \geq \sigma_k > 0$ and bases $\mathcal{B}_1$ and $\mathcal{B}_2$ of $\mathcal{R}^n$ and $\mathcal{R}^m$, respectively, satisfying equations (9) and (10).
   (a) Prove that $\mathcal{B}_1$ is a basis consisting of eigenvectors of A^TA with corresponding eigenvalues $\sigma_1^2, \sigma_2^2, \ldots, \sigma_k^2, 0, \ldots, 0$.
   (b) Prove that $\mathcal{B}_2$ is a basis consisting of eigenvectors of AA^T with corresponding eigenvalues $\sigma_1^2, \sigma_2^2, \ldots, \sigma_k^2, 0, \ldots, 0$.
   (c) Prove that the singular values of A are unique.
   (d) Let $\mathcal{B}_1'$ and $\mathcal{B}_2'$ be the sets obtained from $\mathcal{B}_1$ and $\mathcal{B}_2$ by multiplying the first vector in each set by -1. Prove that, although $\mathcal{B}_1' \neq \mathcal{B}_1$ and $\mathcal{B}_2' \neq \mathcal{B}_2$, equations (9) and (10) are still satisfied if $\mathcal{B}_1'$ and $\mathcal{B}_2'$ replace $\mathcal{B}_1$ and $\mathcal{B}_2$, respectively.

77. Let A be an $m \times n$ matrix of rank m with singular values
$$\sigma_1 \geq \sigma_2 \geq \cdots \geq \sigma_m > 0.$$
   (a) Prove that $\sigma_m\|\mathbf{v}\| \leq \|A\mathbf{v}\| \leq \sigma_1\|\mathbf{v}\|$ for every vector $\mathbf{v}$ in $\mathcal{R}^n$.
   (b) Prove that there exist nonzero vectors $\mathbf{v}$ and $\mathbf{w}$ in $\mathcal{R}^n$ such that
$$\|A\mathbf{v}\| = \sigma_m\|\mathbf{v}\| \qquad \text{and} \qquad \|A\mathbf{w}\| = \sigma_1\|\mathbf{w}\|.$$

78. Let A be an $m \times n$ matrix with rank k, and suppose that $A = U\Sigma V^T$ is a singular value decomposition of A.
   (a) Prove that the nonzero diagonal entries of Σ are the singular values of A.
   (b) Let $\mathcal{B}_1$ and $\mathcal{B}_2$ be the orthonormal bases of $\mathcal{R}^n$ and $\mathcal{R}^m$ consisting of the columns of V and U, respectively. Prove that for these bases, equations (9) and (10) are satisfied.

79. Prove that the transpose of a singular value decomposition of A is a singular value decomposition of A^T, as illustrated in Example 5.

80. Prove that, for any matrix A, the matrices A^TA and AA^T have the same nonzero eigenvalues.

81. Prove that the pseudoinverse of the matrix Σ in equation (11) is the matrix $\Sigma^\dagger$ in equation (14).

82. Prove that if A is an invertible matrix, then $A^\dagger = A^{-1}$.

83. Prove that, for any matrix A, $(A^T)^\dagger = (A^\dagger)^T$.

84. Prove that if A is a symmetric matrix, then the singular values of A are the absolute values of the nonzero eigenvalues of A.

85. Let A be an $n \times n$ symmetric matrix of rank k, with singular values $\sigma_1 \geq \sigma_2 \geq \cdots \geq \sigma_k > 0$, and let Σ be the $n \times n$ matrix in equation (11). Prove that A is positive semidefinite if and only if there is an $n \times n$ orthogonal matrix V such that $V\Sigma V^T$ is a singular value decomposition of A.

86. Let Q be an $n \times n$ orthogonal matrix.
   (a) Determine the singular values of Q. Justify your answer.
   (b) Describe a singular value decomposition of Q.

87. Let A be an $n \times n$ matrix of rank n. Prove that A is an orthogonal matrix if and only if 1 is the only singular value of A.

88. Let A be an $m \times n$ matrix with a singular value decomposition $A = U\Sigma V^T$. Suppose that P and Q are orthogonal matrices of sizes $m \times m$ and $n \times n$, respectively, such that $P\Sigma = \Sigma Q$.
   (a) Prove that $(UP)\Sigma(VQ)^T$ is a singular value decomposition of A.
   (b) Use (a) to find an example of a matrix with two distinct singular value decompositions.
   (c) Prove the converse of (a): If $U_1\Sigma V_1^T$ is any singular value decomposition of A, then there exist orthogonal matrices P and Q of sizes $m \times m$ and $n \times n$, respectively, such that $P\Sigma = \Sigma Q$, $U_1 = UP$, and $V_1 = VQ$.

89. Prove that if $A = U\Sigma V^T$ is a singular value decomposition of an $m \times n$ matrix of rank k, then $\Sigma\Sigma^\dagger$ is the $m \times m$ diagonal matrix whose first k diagonal entries are 1s and whose last $m - k$ diagonal entries are 0s.

90. Prove that, for any matrix A, the product $A^\dagger A$ is the orthogonal projection matrix for Row A.

91. Let A be an $m \times n$ matrix with rank k and nonzero singular values $\sigma_1 \geq \sigma_2 \geq \cdots \geq \sigma_k$, and let $A = U\Sigma V^T$ be a singular value decomposition of A. Suppose that $U = [\mathbf{u}_1\ \mathbf{u}_2\ \cdots\ \mathbf{u}_m]$ and $V = [\mathbf{v}_1\ \mathbf{v}_2\ \cdots\ \mathbf{v}_n]$. For $1 \leq i \leq k$, let Q_i be the $m \times n$ matrix defined by $Q_i = \mathbf{u}_i\mathbf{v}_i^T$. Prove the following statements:

(a) $A = \sigma_1 Q_1 + \sigma_2 Q_2 + \cdots + \sigma_k Q_k$.

(b) For all i, rank $Q_i = 1$.

(c) For all i, $Q_i Q_i^T$ is the orthogonal projection matrix for the 1-dimensional subspace Span $\{\mathbf{u}_i\}$ of $\mathcal{R}^m$.

(d) For all i, $Q_i^T Q_i$ is the orthogonal projection matrix for the 1-dimensional subspace Span $\{\mathbf{v}_i\}$ of $\mathcal{R}^n$.

(e) For distinct i and j, $Q_i Q_j^T = O$ and $Q_i^T Q_j = O$.

In Exercises 92 and 93, use either a calculator with matrix capabilities or computer software such as MATLAB to find a singular value decomposition and the pseudoinverse of each matrix A.

92. $\begin{bmatrix} 2 & 0 & 1 & -1 \\ 1 & 3 & 1 & 2 \\ 1 & 1 & -1 & 1 \end{bmatrix}$

93. $\begin{bmatrix} 1 & 2 & 1 & 3 \\ 2 & -1 & 1 & 4 \\ -1 & 0 & 1 & 2 \end{bmatrix}$

SOLUTIONS TO THE PRACTICE PROBLEMS

1. First observe that

$$A^TA = \begin{bmatrix} 200 & -100 & 250 \\ -100 & 500 & -350 \\ 250 & -350 & 425 \end{bmatrix}.$$

Then

$$\mathcal{B}_1 = \{\mathbf{v}_1, \mathbf{v}_2, \mathbf{v}_3\} = \left\{ \frac{1}{3}\begin{bmatrix} 1 \\ -2 \\ 2 \end{bmatrix}, \frac{1}{3}\begin{bmatrix} 2 \\ 2 \\ 1 \end{bmatrix}, \frac{1}{3}\begin{bmatrix} 2 \\ -1 \\ -2 \end{bmatrix} \right\}$$

is an orthonormal basis for $\mathcal{R}^3$ consisting of eigenvectors of A^TA with corresponding eigenvalues $\lambda_1 = 900$, $\lambda_2 = 225$, and $\lambda_3 = 0$. Thus the singular values of A are $\sigma_1 = \sqrt{\lambda_1} = 30$ and $\sigma_2 = \sqrt{\lambda_2} = 15$.

Next, set

$$\mathbf{u}_1 = \frac{1}{\sigma_1} A\mathbf{v}_1 = \frac{1}{30} \cdot \frac{1}{3}\begin{bmatrix} 54 \\ 72 \end{bmatrix} = \frac{1}{5}\begin{bmatrix} 3 \\ 4 \end{bmatrix}$$

and

$$\mathbf{u}_2 = \frac{1}{\sigma_2} A\mathbf{v}_2 = \frac{1}{15} \cdot \frac{1}{3}\begin{bmatrix} -36 \\ 27 \end{bmatrix} = \frac{1}{5}\begin{bmatrix} -4 \\ 3 \end{bmatrix}.$$

Then $\mathcal{B}_2 = \{\mathbf{u}_1, \mathbf{u}_2\}$ is an orthonormal basis for $\mathcal{R}^2$ that satisfies equations (9) and (10).

2. Let $\mathbf{v}_1$, $\mathbf{v}_2$, $\mathbf{v}_3$, $\mathbf{u}_1$, and $\mathbf{u}_2$ be as in the solution to Practice Problem 1. Define

$$U = [\mathbf{u}_1\ \mathbf{u}_2] = \begin{bmatrix} \frac{3}{5} & -\frac{4}{5} \\ \frac{4}{5} & \frac{3}{5} \end{bmatrix},$$

$$V = [\mathbf{v}_1\ \mathbf{v}_2\ \mathbf{v}_3] = \begin{bmatrix} \frac{1}{3} & \frac{2}{3} & \frac{2}{3} \\ -\frac{2}{3} & \frac{2}{3} & -\frac{1}{3} \\ \frac{2}{3} & \frac{1}{3} & -\frac{2}{3} \end{bmatrix},$$

and

$$\Sigma = \begin{bmatrix} \sigma_1 & 0 & 0 \\ 0 & \sigma_2 & 0 \end{bmatrix} = \begin{bmatrix} 30 & 0 & 0 \\ 0 & 15 & 0 \end{bmatrix}.$$

Then $A = U\Sigma V^T$ is a singular value decomposition of A.

3. We want to find the solution of the system $A\mathbf{x} = \mathbf{b}$ with least norm, where A is the matrix in Practice Problem 1 and $\mathbf{b} = \begin{bmatrix} 5 \\ -5 \end{bmatrix}$. Using the matrices U and V in the solution of Practice Problem 2 and $\Sigma^\dagger = \begin{bmatrix} \frac{1}{30} & 0 \\ 0 & \frac{1}{15} \\ 0 & 0 \end{bmatrix}$, we find that the solution of least norm is

$$\mathbf{z} = V\Sigma^\dagger U^T\mathbf{b} = \frac{1}{90}\begin{bmatrix} 29 \\ 26 \\ 16 \end{bmatrix}.$$

4. Using the matrices V, U, and Σ in Example 7, we see that the desired vector is

$$\mathbf{z} = V\Sigma^\dagger U^T\mathbf{b} = \frac{1}{10}\begin{bmatrix} 25 \\ -83 \\ 104 \end{bmatrix}.$$

6.8* PRINCIPAL COMPONENT ANALYSIS

Consider a study concerned with health issues. We collect data on a large sample of people (subjects), using variables such as age, two cholesterol readings (high-density lipoprotein and low-density lipoprotein), two blood pressure readings (diastolic and systolic), weight, height, exercise habits, daily fat intake, and daily salt intake. It would be far more convenient if we could replace these ten variables with two or three new variables. How can this be done without losing a significant amount of information?

If two variables are very closely related to each other, say, height and weight, then we may be able to replace these two variables with one new variable. In this section, we use linear algebra to discover a smaller set of new variables. The method we use is called **principal component analysis** (PCA). It was developed by Pearson (1901) and Hotelling (1933). (See [11] and [7].)

There are a number of practical outcomes from the use of PCA.

(a) If we were doing a least-squares analysis or other type of statistical analysis for this or future groups, we could base it on fewer variables; this would add what statisticians call *power* to our analyses.

(b) Most often, PCA shows the groupings of variables—for example, quantitative scores versus verbal ones, or gross motor skills versus fine motor skills. This is in itself revealing to those in the particular area of interest.

(c) If the data are plotted, PCA finds lines, planes, and hyperplanes that approximate the data as well as possible in the least-squares sense. It also finds the "appropriate" rotation of axes to plot the data. Think of using the major and minor axes of an ellipse as the coordinate axes, if the data were elliptical.

(d) If we had a large amount of data to transmit, PCA's reduction to fewer variables could prove to be a useful tool, allowing us to submit a much smaller amount of data without losing much information.

To learn more about PCA, see [12] and [10].

We begin with some basic statistical concepts. Given a set of m observations, $x_1, x_2, \ldots, x_m$, a familiar measure of the *center* of these observations is the **(sample) mean**, $\overline{x}$ (read "x-bar"), defined as

$$\overline{x} = \frac{1}{m}(x_1 + x_2 + \cdots + x_m).$$

For example, if our data consist of 3, 8, 7, then $\overline{x} = \frac{1}{3}(3 + 8 + 7) = 6$. The mean, however, does not help us measure the *spread* or *variability* of the data. One reasonable approach to measuring variability is to average the squared deviations of the measurements from the mean. This leads to the definition of the **(sample) variance**, s^2 (we sometimes write $s_{\mathbf{x}}^2$ to emphasize that the observations can be expressed as the components of the vector $\mathbf{x}$), defined as

$$s_{\mathbf{x}}^2 = \frac{1}{m-1}\left[(x_1 - \overline{x})^2 + (x_2 - \overline{x})^2 + \cdots + (x_m - \overline{x})^2\right].$$

In most statistics books, the formula uses a denominator of $m - 1$ rather than m, because it can be shown that using $m - 1$ provides a more accurate approximation to the variance of the population from which the sample is drawn. (See [9].) For the same set of data, we have

$$s^2 = \frac{1}{2}\left[(3 - 6)^2 + (8 - 6)^2 + (7 - 6)^2\right] = 7.$$

* This section can be omitted without loss of continuity.

Notice that if the original measurements are heights measured in inches, then $\bar{x}$ is also given in inches, but s^2 is given in square inches. To keep the measure of spread in the same units as in the original data, we use the positive square root of the variance, namely, s, called the **standard deviation**. For the previous data, the standard deviation is $s = \sqrt{7}$.

It is very useful to represent the variance by matrix notation. Suppose we denote the measurements $x_1, x_2, \ldots, x_m$ by the vector $\mathbf{x} = \begin{bmatrix} x_1 \\ x_2 \\ \vdots \\ x_m \end{bmatrix}$ and introduce the vector $\bar{\mathbf{x}} = \begin{bmatrix} \bar{x} \\ \bar{x} \\ \vdots \\ \bar{x} \end{bmatrix}$. It follows easily that

$$s_{\mathbf{x}}^2 = \frac{1}{m-1}(\mathbf{x} - \bar{\mathbf{x}})^T(\mathbf{x} - \bar{\mathbf{x}}) = \frac{1}{m-1}(\mathbf{x} - \bar{\mathbf{x}}) \cdot (\mathbf{x} - \bar{\mathbf{x}}) = \frac{1}{m-1}\|\mathbf{x} - \bar{\mathbf{x}}\|^2.$$

Earlier, we suggested that if two variables were closely related, we may be able to replace both of them with one variable and not lose much information. One commonly used measure of the strength of the *linear* relationship between two variables is called the *(sample) covariance*. Specifically, let

$$\mathbf{x} = \begin{bmatrix} x_1 \\ x_2 \\ \vdots \\ x_m \end{bmatrix} \quad \text{and} \quad \mathbf{y} = \begin{bmatrix} y_1 \\ y_2 \\ \vdots \\ y_m \end{bmatrix}.$$

We define the **(sample) covariance** of $\mathbf{x}$ and $\mathbf{y}$ to be

$$\operatorname{cov}(\mathbf{x}, \mathbf{y}) = \frac{1}{m-1}\left[(x_1 - \bar{x})(y_1 - \bar{y}) + (x_2 - \bar{x})(y_2 - \bar{y}) + \cdots + (x_m - \bar{x})(y_m - \bar{y})\right],$$

or, using matrix notation,

$$\operatorname{cov}(\mathbf{x}, \mathbf{y}) = \frac{1}{m-1}(\mathbf{x} - \bar{\mathbf{x}})^T(\mathbf{y} - \bar{\mathbf{y}}) = \frac{1}{m-1}(\mathbf{x} - \bar{\mathbf{x}}) \cdot (\mathbf{y} - \bar{\mathbf{y}}).$$

One problem with using covariance, however, is that its size is affected by the units of measurement. For example, if $\mathbf{x}$ is measured in feet and $\mathbf{y}$ is measured in pounds, then the covariance is measured in foot-pounds. But if the units were given in inches and ounces, the covariance would be greatly increased. To avoid this problem, we use instead the quantity

$$\frac{\operatorname{cov}(\mathbf{x}, \mathbf{y})}{s_{\mathbf{x}} s_{\mathbf{y}}},$$

where $s_{\mathbf{x}}$ and $s_{\mathbf{y}}$ are the respective standard deviations of $\mathbf{x}$ and $\mathbf{y}$. This quantity is called the **(sample) correlation** between $\mathbf{x}$ and $\mathbf{y}$.[18] It is easily shown that correlation is a "unit-free" measurement; with a little more work (see Exercise 37), it can be

[18] When using the correlation between two variables, we assume that neither variable has a variance equal to zero.

proven that the correlation always lies between -1 and 1. In the extreme case that the relationship is perfectly linear—that is, all the points $(x_1, y_1), (x_2, y_2), \ldots, (x_m, y_m)$ lie on a line—the correlation is 1 if the line has a positive slope, and the correlation is -1 if the line has a negative slope. If there is little or no linear relationship between $\mathbf{x}$ and $\mathbf{y}$, the correlation is close to zero.

Practice Problem 1 ▶ Let $\mathbf{x} = \begin{bmatrix} 4 \\ -2 \\ 7 \end{bmatrix}$ and $\mathbf{y} = \begin{bmatrix} 3 \\ 4 \\ 5 \end{bmatrix}$. Compute the following quantities:

(a) The means $\overline{\mathbf{x}}$ and $\overline{\mathbf{y}}$.

(b) The variances $s_{\mathbf{x}}^2$ and $s_{\mathbf{y}}^2$.

(c) The covariance $\text{cov}(\mathbf{x}, \mathbf{y})$.

(d) The correlation between $\mathbf{x}$ and $\mathbf{y}$. ◀

Because of the properties of dot products, it is easy to see that $\text{cov}(\mathbf{x}, \mathbf{y}) = \text{cov}(\mathbf{y}, \mathbf{x})$; so the covariance of $\mathbf{x}$ and $\mathbf{y}$ is the same as the covariance of $\mathbf{y}$ and $\mathbf{x}$. The same symmetry holds for correlation.

In general, given n variables $\mathbf{x}_1, \mathbf{x}_2, \ldots, \mathbf{x}_n$, each considered as an $m \times 1$ vector, we define two $n \times n$ matrices. The **covariance matrix** is the $n \times n$ matrix whose (i,j)-entry is the covariance $\text{cov}(\mathbf{x}_i, \mathbf{x}_j)$. Using block notation, if we write $X = [\mathbf{x}_1 \; \mathbf{x}_2 \; \ldots \; \mathbf{x}_n]$ and $\overline{X} = [\overline{\mathbf{x}}_1 \; \overline{\mathbf{x}}_2 \; \ldots \; \overline{\mathbf{x}}_n]$, we may represent the covariance matrix as

$$C = \frac{1}{m-1}(X - \overline{X})^T (X - \overline{X}).$$

(See Exercise 32.) Note the similarity of this formula to the one for the variance $s_{\mathbf{x}}^2$ of a vector $\mathbf{x}$.

Likewise, the **correlation matrix** is defined to be the $n \times n$ matrix whose (i,j)-entry is the correlation between $\mathbf{x}_i$ and $\mathbf{x}_j$. To see the form of a correlation matrix, we need a little terminology. A variable $\mathbf{z}$ is a **scaled variable** if it has mean equal to zero and standard deviation equal to one. A variable $\mathbf{x}$ with a nonzero variance can be *scaled* or *standardized* by subtracting the vector $\overline{\mathbf{x}}$ and then dividing by the standard deviation $s_{\mathbf{x}}$. So $\dfrac{\mathbf{x} - \overline{\mathbf{x}}}{s_{\mathbf{x}}}$ is a scaled variable. (See Exercise 33.) Thus, given n variables $\mathbf{x}_1, \mathbf{x}_2, \ldots, \mathbf{x}_n$, we may scale them and produce the scaled variables $\mathbf{z}_1, \mathbf{z}_2, \ldots, \mathbf{z}_n$. This scaling is frequently performed in cases where the variables have widely varying units, in order to put all variables on an "equal footing."

Now if $Z = [\mathbf{z}_1 \; \mathbf{z}_2 \; \cdots \; \mathbf{z}_n]$, then the correlation matrix for the original variables can be represented as

$$C_0 = \frac{1}{m-1} Z^T Z.$$

(See Exercise 34.) It also follows that the correlation between $\mathbf{x}_i$ and $\mathbf{x}_j$ is the same as $\mathbf{z}_i$ and $\mathbf{z}_j$. Because of the symmetry mentioned earlier—and as is made clear through the preceding equations—it follows that both the covariance and correlation matrices are symmetric.

Example 1 The set of data in the table that follows was collected by one of the authors from an Honors Calculus class of 14 students. The four variables are ACT (a score from a national test, with range 1 to 36), FE (the final exam score with range 0 to 200), Qav (the mean of eight quiz scores, each with range 0 to 100), and Tav (the mean of three test scores, each with range 0 to 100). The scaled variables are given in the last four columns, with asterisks added to their names.

Student	ACT	FE	Qav	Tav	ACT*	FE*	Qav*	Tav*
1	33	181	95	89	1.27	0.94	1.3	0.95
2	31	169	81	89	0.8	0.48	0.29	0.95
3	21	176	65	68	−1.58	0.75	−0.88	−0.64
4	25	181	66	90	−0.63	0.94	−0.81	1.03
5	29	169	89	81	0.32	0.48	0.87	0.35
6	24	103	61	57	−0.86	−2.05	−1.17	−1.47
7	24	150	81	76	−0.86	−0.25	0.29	−0.03
8	29	147	86	76	0.32	−0.36	0.65	−0.03
9	36	181	98	102	1.98	0.94	1.52	1.94
10	26	163	72	70	−0.39	0.25	−0.37	−0.49
11	31	163	95	81	0.8	0.25	1.3	0.35
12	29	147	65	67	0.32	−0.36	−0.88	−0.71
13	23	160	62	68	−1.1	0.14	−1.1	−0.64
14	26	100	63	56	−0.39	−2.16	−1.02	−1.55

Using MATLAB to compute $\frac{1}{m-1}Z^TZ$, we obtain the 4×4 correlation matrix C_0.

The next table gives the entries of C_0, rounded to four decimal places.

	ACT*	FE*	Qav*	Tav*
ACT*	1	.3360	.8111	.7010
FE*	.3360	1	.4999	.7958
Qav*	.8111	.4999	1	.7487
Tav*	.7010	.7958	.7487	1

Observations:

1. Notice that each of the diagonal entries equals 1. This is because the correlation between any variable and itself is 1.
2. There is a very strong correlation of .8111 between the ACT score and the quiz average.
3. The weakest correlation of .3360 is between the ACT score and the final exam score.

Of course, this is a very small sample. If this were a much larger sample or if these results were replicated in other samples, then we might be able to draw general conclusions. For example, the test average correlates very highly with all other variables, including the variable ACT, a test given before a student enters college. So if one variable were to be used to represent these data, perhaps it should be Tav.

We are now ready to begin our discussion of principal component analysis. We start with two questions:

1. How do we find the new variables to replace the existing ones?
2. How do we measure how well these new variables capture the original data?

To illustrate the method, we use our original data set, and note that what we do is easily generalized to other data sets. We have four variables $\mathbf{x}_1$, $\mathbf{x}_2$, $\mathbf{x}_3$, $\mathbf{x}_4$, each of which is a 14×1 column vector. Because our variables have different scales, it is recommended to use the scaled variables instead. So we let $Z = \begin{bmatrix} \mathbf{z}_1 & \mathbf{z}_2 & \mathbf{z}_3 & \mathbf{z}_4 \end{bmatrix}$.

The 4×4 correlation matrix C_0 is computed as

$$C_0 = \begin{bmatrix} 1.0000 & 0.3360 & 0.8111 & 0.7010 \\ 0.3360 & 1.0000 & 0.4999 & 0.7958 \\ 0.8111 & 0.4999 & 1.0000 & 0.7487 \\ 0.7010 & 0.7958 & 0.7487 & 1.0000 \end{bmatrix}.$$

As we noted earlier, this matrix is necessarily symmetric. So by Theorem 6.15, there exists an orthonormal basis of eigenvectors $\mathbf{u}_1$, $\mathbf{u}_2$, $\mathbf{u}_3$, $\mathbf{u}_4$ of C_0 with corresponding eigenvalues λ_1, λ_2, λ_3, λ_4, where we assume $\lambda_1 \geq \lambda_2 \geq \lambda_3 \geq \lambda_4$. Using MATLAB, we obtain the orthogonal matrix P, shown next, whose columns are the eigenvectors $\mathbf{u}_1$, $\mathbf{u}_2$, $\mathbf{u}_3$, $\mathbf{u}_4$, and the diagonal matrix D of associated eigenvalues.[19]

$$P = \begin{bmatrix} 0.4856 & -0.5561 & 0.5128 & 0.4381 \\ 0.4378 & 0.7317 & -0.064 & 0.5185 \\ 0.5209 & -0.3275 & -0.7848 & -0.0744 \\ 0.5489 & 0.2192 & 0.3421 & -0.7305 \end{bmatrix}$$

$$D = \begin{bmatrix} 2.9654 & 0 & 0 & 0 \\ 0 & 0.7593 & 0 & 0 \\ 0 & 0 & 0.1844 & 0 \\ 0 & 0 & 0 & 0.0910 \end{bmatrix}$$

The first new variable is called the **first principal component**, and is defined to be the vector $\mathbf{y}_1 = Z\mathbf{u}_1$, where $\mathbf{u}_1$ is the eigenvector of C_0 with the largest eigenvalue (2.9654); and the **second principal component** is $\mathbf{y}_2 = Z\mathbf{u}_2$ because $\mathbf{u}_2$ has the second largest eigenvalue (0.7593). In our example, the first two principal components are

$$\begin{aligned} \mathbf{y}_1 &= .4856\mathbf{z}_1 + .4378\mathbf{z}_2 + .5209\mathbf{z}_3 + .5489\mathbf{z}_4 \\ \mathbf{y}_2 &= -.5561\mathbf{z}_1 + .7317\mathbf{z}_2 - .3275\mathbf{z}_3 + .2192\mathbf{z}_4. \end{aligned}$$

Or,

$$\begin{aligned} \mathbf{y}_1 &= .4856\text{ACT}^* + .4378\text{FE}^* + .5209\text{Qav}^* + .5489\text{Tav}^* \\ \mathbf{y}_2 &= -.5561\text{ACT}^* + .7317\text{FE}^* - .3275\text{Qav}^* + .2192\text{Tav}^*. \end{aligned}$$

The remaining principal components are defined similarly. The coefficients used in the definition of the principal components, given by the components of the eigenvectors, are called **loadings**. The magnitude of a variable's coefficient is related to its relative importance to the given principal component. We say more about this later. In this

[19] MATLAB often produces matrices with the columns in a different order than listed here. We follow the standard practice of statisticians and list the eigenvalues and eigenvectors so that the eigenvalues are in decreasing order. Also, we occasionally replace an eigenvector produced by MATLAB with a nonzero multiple of itself, which, of course, is still an eigenvector corresponding to the same eigenvalue.

example, it appears that the first principal component represents an average of the four variables in the sense that it gives a weighted sum of the four variables where the weights are approximately equal. The second principal component represents a "contrast" between the pair, FE* and Tav*, with positive loadings and the pair, ACT* and Qav*, with negative loadings.

One criterion used by statisticians when replacing original variables with fewer new variables is whether or not the new variables "explain" or "account for" a high percentage of the variance in the original data set. Intuitively, we understand that if variables are discarded which are highly correlated with the remaining ones, then the variability of what is left has not changed significantly. It can be shown (see Exercise 35) that the variance of a principal component is given by its associated eigenvalue. So, for example, the variance of $\mathbf{y}_1$ is 2.9654. Also, the **total variance**, defined as the sum of all the variances of the variables, is given by the sum of all the eigenvalues (which by Exercise 89 of Section 5.3 is trace(C_0)), namely, 4. So, using the language of statistics, we say that $\mathbf{y}_1$ *accounts for* $2.9654/4 = 74.14\%$ of the variance, while $\mathbf{y}_1$ and $\mathbf{y}_2$ together account for $(2.9654 + .7593)/4 = 93.12\%$ of the variance. It seems reasonable to use $\mathbf{y}_1$ and $\mathbf{y}_2$ as the new variables rather than all four of the original variables.

In the next table, we compute the (rounded) correlations between the scaled variables and the first two principal components. The strong correlation of 0.9452 between $\mathbf{y}_1$ and Tav* tells us that the test average contributes significantly to the first component. Likewise, the weak correlation of 0.1911 between $\mathbf{y}_2$ and Tav* tells us that the test average contributes very little to the second component. Notice also that the correlation between $\mathbf{y}_1$ and $\mathbf{y}_2$ in the table is (approximately) zero; that is, $\mathbf{y}_1$ and $\mathbf{y}_2$ are *uncorrelated*.

	ACT*	FE*	Qav*	Tav*	$\mathbf{y}_1$	$\mathbf{y}_2$
ACT*	1	0.3660	0.8111	0.7010	0.8362	−0.4846
FE*	0.3360	1	0.4999	0.7958	0.7539	0.6376
Qav*	0.8111	0.4999	1	0.7487	0.8969	−0.2854
Tav*	0.7010	0.7958	0.7487	1	0.9452	0.1911
$\mathbf{y}_1$	0.8362	0.7539	0.8969	0.9452	1	0.0000
$\mathbf{y}_2$	−0.4846	0.6376	−0.2854	0.1911	0.0000	1

Comments:

- Principal components are used to reduce large-dimensional data sets to data sets with a few dimensions that still retain most of the information in the original data. For example, if we needed to transmit most of the information in these data, we might decide to transmit only the first several principal components.
- Each principal component is a linear combination of the scaled variables.
- Any two principal components are uncorrelated. (See Exercise 36.)
- Usually, the first few principal components account for a large percentage of the total variance, so they are all that is needed for future analyses.
- The principal components are artificial variables and are not necessarily easy to interpret.
- We have neglected to make any assumptions about the underlying "statistical distributions" of the variables. This topic requires a much deeper background in mathematical statistics than we are assuming here.

Example 2

Suppose that a group of ten students takes four tests: two of the tests yield mathematics scores, Alg (algebra) and Trig (trigonometry); and two yield English scores, Englit (English literature) and Shakes (Shakespeare). The data are presented here in table form; the scaled variables are given in the last four columns.

Student	Alg	Trig	Englit	Shakes	Alg*	Trig*	Englit*	Shakes*
1	95	88	65	68	1.52	0.99	−1.02	−0.85
2	87	92	70	74	0.95	1.24	−0.58	−0.31
3	75	78	75	72	0.10	0.35	−0.15	−0.49
4	74	70	85	81	0.03	−0.16	0.72	0.31
5	46	51	92	95	−1.96	−1.37	1.33	1.56
6	62	55	88	90	−0.82	−1.11	0.98	1.11
7	82	91	85	90	0.60	1.18	0.72	1.11
8	68	52	55	60	−0.40	−1.30	−1.89	−1.56
9	82	78	80	75	0.60	0.35	0.29	−0.22
10	65	70	72	70	−0.61	−0.16	−0.41	−0.67

As in Example 1, we seek a smaller set of variables to describe the data. By using MATLAB, we obtain the 4×4 correlation matrix C_0. The next table gives the entries of C_0, rounded to four decimal places.

	Alg*	Trig*	Englit*	Shakes*
Alg*	1	0.871	−0.435	−0.450
Trig*	0.871	1	−0.148	−0.160
Englit*	−0.435	−0.148	1	0.942
Shakes*	−0.450	−0.160	0.942	1

Observations:

1. There are very high correlations of .871 between the algebra and trigonometry scores and .942 between the English literature and Shakespeare scores, but low negative correlations between any mathematics score and any English score.
2. If this pattern continues in other samples, then we might conclude that skills in trigonometry and English are unrelated.

As we noted earlier, this matrix is necessarily symmetric. So, as in Example 1, there is an orthonormal basis of eigenvectors $\mathbf{u}_1$, $\mathbf{u}_2$, $\mathbf{u}_3$, $\mathbf{u}_4$ of C_0 with corresponding eigenvalues λ_1, λ_2, λ_3, λ_4, where we again assume that $\lambda_1 \geq \lambda_2 \geq \lambda_3 \geq \lambda_4$. The orthogonal matrix P, whose columns are the eigenvectors $\mathbf{u}_1$, $\mathbf{u}_2$, $\mathbf{u}_3$, $\mathbf{u}_4$, and the diagonal matrix D of the associated eigenvalues are as follows:

$$P = \begin{bmatrix} -0.5382 & 0.4112 & 0.7312 & -0.0820 \\ -0.4129 & 0.6323 & -0.6525 & 0.0622 \\ 0.5169 & 0.4696 & 0.1936 & 0.6890 \\ 0.5222 & 0.4589 & 0.0459 & -0.7174 \end{bmatrix}$$

$$D = \begin{bmatrix} 2.5228 & 0 & 0 & 0 \\ 0 & 1.3404 & 0 & 0 \\ 0 & 0 & 0.0792 & 0 \\ 0 & 0 & 0 & 0.0577 \end{bmatrix}$$

Again, the first principal component is defined to be the vector $\mathbf{y}_1 = Z\mathbf{u}_1$, where $\mathbf{u}_1$ is the eigenvector of C_0 with the largest eigenvalue (2.5228); and the second principal component is $\mathbf{y}_2 = Z\mathbf{u}_2$, where $\mathbf{u}_2$ has the second largest eigenvalue (1.3404). The first two principal components are

$$\begin{aligned} \mathbf{y}_1 &= -0.5382\text{Alg}^* - 0.4129\text{Trig}^* + 0.5169\text{Englit}^* + 0.5222\text{Shakes}^* \\ \mathbf{y}_2 &= 0.4112\text{Alg}^* + 0.6323\text{Trig}^* + 0.4696\text{Englit}^* + 0.4589\text{Shakes}^*. \end{aligned}$$

As we saw before, the total variance, 4, is given by trace(D). The first two principal components together account for $(2.5228 + 1.3404)/4 = 96.58\%$ of the variance. It seems reasonable to use $\mathbf{y}_1$ and $\mathbf{y}_2$ as the new variables, rather than all four of the original variables.

Notice that $\mathbf{y}_1$ represents a contrast between the pair Englit* and Shakes*, with positive loadings, and the pair Alg* and Trig*, with negative loadings, while $\mathbf{y}_2$ gives a weighted sum of the four variables, where the weights are approximately equal.

If we compute the correlations between the scaled variables and the first two principal components, we obtain the next table (with rounding). Notice the very small correlation between the two principal components. Also notice that $\mathbf{y}_1$ has the strongest correlation with Alg*, which is reflected in the fact that its strongest loading is with Alg*. Indeed, for both principal components, observe that the signs of the loadings agree with the signs of the corresponding correlations.

	Alg*	Trig*	Englit*	Shakes*	$\mathbf{y}_1$	$\mathbf{y}_2$
Alg*	1	0.8708	−0.4351	−0.4491	−0.8549	0.4769
Trig*	0.8708	1	−0.1479	−0.1603	−0.6561	0.7322
Englit*	−0.4351	−0.1479	1	0.9418	0.8209	0.5431
Shakes*	−0.4491	−0.1603	0.9418	1	0.8289	0.5308
$\mathbf{y}_1$	−0.8549	−0.6561	0.8209	0.8209	1	−0.0011
$\mathbf{y}_2$	0.4769	0.7322	0.5431	0.5308	−0.0011	1

EXERCISES

In Exercises 1–8, $\mathbf{x} = \begin{bmatrix} 2 \\ -3 \\ 4 \end{bmatrix}$ *and* $\mathbf{y} = \begin{bmatrix} 4 \\ 2 \\ 3 \end{bmatrix}$. *Compute the following quantities:*

1. the mean of $\mathbf{x}$
2. the mean of $\mathbf{y}$
3. the variance of $\mathbf{x}$
4. the variance of $\mathbf{y}$
5. the covariance of $\mathbf{x}$ and $\mathbf{y}$
6. the correlation between $\mathbf{x}$ and $\mathbf{y}$
7. the covariance matrix, using the vectors $\mathbf{x}$ and $\mathbf{y}$
8. the correlation matrix, using the vectors $\mathbf{x}$ and $\mathbf{y}$

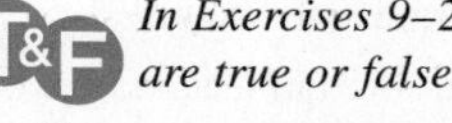

In Exercises 9–20, determine whether the statements are true or false.

9. The mean of m numbers is defined as the sum of the numbers divided by m.

10. The variance of m numbers is defined as the sum of the squared deviations of each of the numbers from the mean, divided by m.
11. The covariance is a number that is always nonnegative.
12. The correlation is a number that is always nonnegative.
13. If two variables have a linear relationship, then their correlation is one.
14. If the correlation between two variables is one, then the variables have a linear relationship.
15. The covariance is a number that always lies between -1 and 1.
16. The correlation is a number that always lies between -1 and 1.
17. A scaled variable always has mean 0 and standard deviation one.
18. PCA generally produces a smaller set of variables with little loss of information.
19. The variance of a principal component is given by an eigenvalue of the correlation matrix.
20. Distinct principal components are always uncorrelated.

In Exercises 21–25, $\mathbf{x}$, $\mathbf{y}$, *and* $\mathbf{z}$ *are* $m \times 1$ *vectors and* c *is a real number. Prove the following results:*

21. $\text{cov}(\mathbf{x}, \mathbf{y}) = \text{cov}(\mathbf{y}, \mathbf{x})$
22. (a) $\text{cov}(c\mathbf{x}, \mathbf{y}) = c \cdot \text{cov}(\mathbf{x}, \mathbf{y})$

    (b) $\text{cov}(\mathbf{x}, c\mathbf{y}) = c \cdot \text{cov}(\mathbf{x}, \mathbf{y})$
23. (a) $\text{cov}(\mathbf{x} + \mathbf{y}, \mathbf{z}) = \text{cov}(\mathbf{x}, \mathbf{z}) + \text{cov}(\mathbf{y}, \mathbf{z})$

    (b) $\text{cov}(\mathbf{x}, \mathbf{y} + \mathbf{z}) = \text{cov}(\mathbf{x}, \mathbf{y}) + \text{cov}(\mathbf{x}, \mathbf{z})$
24. $\text{cov}(\mathbf{x}, \mathbf{x}) = s_{\mathbf{x}}^2$
25. $\text{cov}(\mathbf{x}, \mathbf{x}) = 0$ if and only if all the components of $\mathbf{x}$ are equal.

In Exercises 26 and 27, $\mathbf{w}$ *and* $\mathbf{u}$ *are* $m \times 1$ *vectors, all of whose components are equal, and* $\mathbf{x}$ *and* $\mathbf{y}$ *are arbitrary* $m \times 1$ *vectors. Prove the following results:*

26. (a) $\text{cov}(\mathbf{w}, \mathbf{y}) = 0$

    (b) $\text{cov}(\mathbf{x}, \mathbf{u}) = 0$
27. (a) $\text{cov}(\mathbf{x} + \mathbf{w}, \mathbf{y}) = \text{cov}(\mathbf{x}, \mathbf{y})$

    (b) $\text{cov}(\mathbf{x}, \mathbf{y} + \mathbf{u}) = \text{cov}(\mathbf{x}, \mathbf{y})$

    (c) $\text{cov}(\mathbf{x} + \mathbf{w}, \mathbf{y} + \mathbf{u}) = \text{cov}(\mathbf{x}, \mathbf{y})$

In Exercises 28–30, $\mathbf{x}$ *and* $\mathbf{y}$ *are* $m \times 1$ *vectors with means* $\overline{\mathbf{x}}$ *and* $\overline{\mathbf{y}}$ *and variances* $s_{\mathbf{x}}^2$ *and* $s_{\mathbf{y}}^2$*, respectively, and* c *and* d *are real numbers. Prove the following results:*

28. (a) The mean of $c\mathbf{x}$ is $c\overline{\mathbf{x}}$.

    (b) The mean of $c\mathbf{x} + d\mathbf{y}$ is $c\overline{\mathbf{x}} + d\overline{\mathbf{y}}$.
29. The variance of $c\mathbf{x}$ is $c^2 s_{\mathbf{x}}^2$.
30. The variance of $\mathbf{x} \pm \mathbf{y}$ is $s_{\mathbf{x}}^2 + s_{\mathbf{y}}^2 \pm 2\text{cov}(\mathbf{x}, \mathbf{y})$.
31. Consider two variables, one measured in feet and the other in pounds. Show that their correlation is a "unit-free" measurement.
32. Using the notation developed in this section, prove that the covariance matrix may be represented as

$$C = \frac{1}{m-1}(X - \overline{X})^T(X - \overline{X}).$$

33. Let $\mathbf{x}$ be an $m \times 1$ vector such that $s_{\mathbf{x}} \neq 0$. Prove that $\frac{1}{s_{\mathbf{x}}}(\mathbf{x} - \overline{\mathbf{x}})$ is a scaled variable.
34. Suppose that $\mathbf{x}_1, \mathbf{x}_2, \ldots, \mathbf{x}_n$ are $m \times 1$ vectors. For each i, let $\mathbf{z}_i$ be the vector obtained by scaling $\mathbf{x}_i$, and let $Z = \begin{bmatrix} \mathbf{z}_1 & \mathbf{z}_2 & \cdots & \mathbf{z}_n \end{bmatrix}$. Prove that the correlation matrix C_0 for $X = \begin{bmatrix} \mathbf{x}_1 & \mathbf{x}_2 & \cdots & \mathbf{x}_n \end{bmatrix}$ can be represented as

$$C_0 = \frac{1}{m-1} Z^T Z.$$

35. Let Z be the $m \times n$ matrix in Exercise 34, and let $\mathbf{w}$ be any $n \times 1$ vector.

    (a) Show that the mean of $Z\mathbf{w}$ is 0.

    (b) Use (a) to show that the variance of a principal component is given by the associated eigenvalue of C_0.
36. Prove that any two principal components are uncorrelated.
37. Let $\mathbf{x}$ and $\mathbf{y}$ be $m \times 1$ vectors, each with nonzero variance, and let

$$\mathbf{x}^* = \frac{\mathbf{x} - \overline{\mathbf{x}}}{s_{\mathbf{x}}} \quad \text{and} \quad \mathbf{y}^* = \frac{\mathbf{y} - \overline{\mathbf{y}}}{s_{\mathbf{y}}}.$$

Suppose that r denotes the correlation between $\mathbf{x}$ and $\mathbf{y}$.

(a) Prove $r = \text{cov}(\mathbf{x}^*, \mathbf{y}^*)$.

(b) Use (a) to prove $|r| \leq 1$.

In the following exercise, use either a calculator with matrix capabilities or computer software such as MATLAB to solve the problem:

38. The data in the accompanying table were collected for a class of students enrolled in a general education mathematics course. The variables are PRE (the pre-final

Student	PRE	FE	ACTE	ACTM
1	96	100	24	25
2	76	90	18	16
3	79	87	24	20
4	86	90	21	24
5	80	71	18	23
6	77	54	18	18
7	72	78	19	23
8	59	71	21	13
9	64	78	22	16
10	61	67	21	13
11	57	79	16	22
12	36	50	17	20
13	42	55	22	15

examination average, with range 0 to 100), FE (the final examination results, with range 0 to 100), ACTE (an English score on a national test, with range from 1 to 36), and ACTM (a mathematics score on a national test, with range from 1 to 36). Students who withdrew from the class are not included.

(a) Extend the table to include the scaled variables PRE*, FE*, ACTE*, and ACTM*.

(b) Use your answer to (a) to compute the correlation matrix C_0.

(c) Which pair of variables shows the strongest correlation? What is this correlation?

(d) Which pair of variables shows the weakest correlation? What is this correlation?

(e) Find an orthogonal matrix P and a diagonal matrix D, whose diagonal entries are listed in decreasing order of magnitude, such that $C_0 = PDP^T$.

(f) Use your answer to (e) to express the first two principal components $\mathbf{y}_1$ and $\mathbf{y}_2$ as linear combinations of the scaled variables.

(g) What percentage of the variance is accounted for by $\mathbf{y}_1$?

(h) What percentage of the variance is accounted for by both $\mathbf{y}_1$ and $\mathbf{y}_2$?

SOLUTIONS TO THE PRACTICE PROBLEMS

1. (a) $\overline{\mathbf{x}} = \frac{1}{3}(4 - 2 + 7) = 3$ and $\overline{\mathbf{y}} = \frac{1}{3}(3 + 4 + 5) = 4$

(b) $s_{\mathbf{x}}^2 = \frac{1}{3-1}\left((4-3)^2 + (-2-3)^2 + (7-3)^2\right) = 21$

and

$$s_{\mathbf{y}}^2 = \frac{1}{3-1}\left((3-4)^2 + (4-4)^2 + (5-4)^2\right) = 1$$

(c) $\text{cov}(\mathbf{x}, \mathbf{y}) = \frac{1}{3-1}[(4-3)(3-4)$
$+ (-2-3)(4-4) + (7-3)(2-4)] = -\frac{9}{2}$

(d) The correlation between $\mathbf{x}$ and $\mathbf{y}$ is

$$\frac{(-9/2)}{\sqrt{21}\sqrt{1}} = -\frac{9}{2\sqrt{21}}.$$

6.9* ROTATIONS OF $\mathcal{R}^3$ AND COMPUTER GRAPHICS

In this section, we study rotations of $\mathcal{R}^3$ about a line that contains $\mathbf{0}$. These rotations can be described in terms of left multiplication by special 3×3 orthogonal matrices, just as rotations in the plane about the origin can be represented by left multiplication by 2×2 rotation matrices, which are orthogonal matrices. The most important lines about which rotations are performed are the x-, y-, and z-axes. We describe how to compute rotations about these axes and explain how they can be used for graphical representations of three-dimensional objects. We then consider rotations about arbitrary lines containing $\mathbf{0}$.

We begin with a description of rotations about the z-axis. For a given angle θ and a vector $\begin{bmatrix} x \\ y \\ z \end{bmatrix}$ in $\mathcal{R}^3$, let $\begin{bmatrix} x' \\ y' \end{bmatrix}$ be the result of rotating $\begin{bmatrix} x \\ y \end{bmatrix}$ by θ in the xy-plane. Then $\begin{bmatrix} x' \\ y' \\ z \end{bmatrix}$ is the rotation of $\begin{bmatrix} x \\ y \\ z \end{bmatrix}$ about the z-axis by θ. (See Figure 6.28.)

Hence we may use the rotation matrix A_θ introduced in Section 1.2 to obtain

$$\begin{bmatrix} x' \\ y' \end{bmatrix} = A_\theta \begin{bmatrix} x \\ y \end{bmatrix} = \begin{bmatrix} \cos\theta & -\sin\theta \\ \sin\theta & \cos\theta \end{bmatrix} \begin{bmatrix} x \\ y \end{bmatrix}.$$

It follows that

$$\begin{bmatrix} x' \\ y' \\ z \end{bmatrix} = \left[\begin{array}{cc|c} \cos\theta & -\sin\theta & 0 \\ \sin\theta & \cos\theta & 0 \\ \hline 0 & 0 & 1 \end{array}\right] \begin{bmatrix} x \\ y \\ z \end{bmatrix}.$$

* This section can be omitted without loss of continuity.

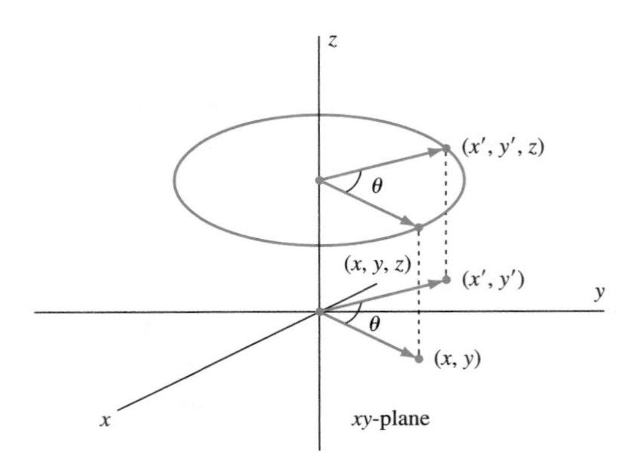

Figure 6.28 A rotation of $\mathcal{R}^3$ about the z-axis.

Let

$$R_\theta = \begin{bmatrix} \cos\theta & -\sin\theta & 0 \\ \sin\theta & \cos\theta & 0 \\ 0 & 0 & 1 \end{bmatrix}.$$

Then $R_\theta\left(\begin{bmatrix} x \\ y \\ z \end{bmatrix}\right) = \begin{bmatrix} x' \\ y' \\ z \end{bmatrix}$ produces a rotation of $\begin{bmatrix} x \\ y \\ z \end{bmatrix}$ about the z-axis by the angle θ. The matrix R_θ is an example of a 3×3 *rotation matrix*. Using arguments similar to the previous one, we can find the rotation matrices for rotating a vector by an angle θ about the other coordinate axes. Let P_θ and Q_θ be the matrices for rotations by an angle θ about the x-axis and the y-axis, respectively. Then

$$P_\theta = \begin{bmatrix} 1 & 0 & 0 \\ 0 & \cos\theta & -\sin\theta \\ 0 & \sin\theta & \cos\theta \end{bmatrix} \quad \text{and} \quad Q_\theta = \begin{bmatrix} \cos\theta & 0 & \sin\theta \\ 0 & 1 & 0 \\ -\sin\theta & 0 & \cos\theta \end{bmatrix}.$$

In each case, the positive direction of the rotation is counterclockwise when viewed from a position along the positive direction of the axis of rotation. Note that all of these rotation matrices are orthogonal matrices.

We can combine rotations by taking products of rotation matrices. For example, if a vector $\mathbf{v}$ is rotated about the z-axis by an angle θ and the result is then rotated about the y-axis by an angle ϕ, the final position of the rotated vector is given by $Q_\phi(R_\theta\mathbf{v}) = (Q_\phi R_\theta)\mathbf{v}$. One should be careful to note the order in which the rotations are made because $Q_\phi R_\theta$ is not necessarily equal to $R_\theta Q_\phi$. For example, let $\mathbf{v} = \mathbf{e}_1$, $\theta = 30^\circ$, and $\phi = 45^\circ$. Then

$$\begin{aligned} Q_\phi R_\theta \mathbf{v} &= \begin{bmatrix} \frac{1}{\sqrt{2}} & 0 & \frac{1}{\sqrt{2}} \\ 0 & 1 & 0 \\ -\frac{1}{\sqrt{2}} & 0 & \frac{1}{\sqrt{2}} \end{bmatrix} \begin{bmatrix} \frac{\sqrt{3}}{2} & -\frac{1}{2} & 0 \\ \frac{1}{2} & \frac{\sqrt{3}}{2} & 0 \\ 0 & 0 & 1 \end{bmatrix} \begin{bmatrix} 1 \\ 0 \\ 0 \end{bmatrix} \\ &= \begin{bmatrix} \frac{1}{\sqrt{2}} & 0 & \frac{1}{\sqrt{2}} \\ 0 & 1 & 0 \\ -\frac{1}{\sqrt{2}} & 0 & \frac{1}{\sqrt{2}} \end{bmatrix} \begin{bmatrix} \frac{\sqrt{3}}{2} \\ \frac{1}{2} \\ 0 \end{bmatrix} \\ &= \begin{bmatrix} \frac{\sqrt{3}}{2\sqrt{2}} \\ \frac{1}{2} \\ -\frac{\sqrt{3}}{2\sqrt{2}} \end{bmatrix}. \end{aligned}$$

On the other hand, a similar calculation shows that

$$R_\theta Q_\phi \mathbf{v} = \begin{bmatrix} \frac{\sqrt{3}}{2\sqrt{2}} \\ \frac{1}{2\sqrt{2}} \\ -\frac{1}{\sqrt{2}} \end{bmatrix},$$

which is not equal to $Q_\phi R_\theta \mathbf{v}$.

Practice Problem 1 ▶ Find the result of rotating the vector $\begin{bmatrix} 1 \\ -1 \\ 2 \end{bmatrix}$ by an angle of $60°$ about the y-axis followed by a rotation of $90°$ about the x-axis. ◀

Rotation matrices are used in computer graphics to present various orientations of the same three-dimensional shape. Although computers can store the information necessary to construct three-dimensional shapes, these shapes must be displayed on a two-dimensional surface such as the screen of a computer monitor or a sheet of paper. From a mathematical viewpoint, such a representation is projected on a plane. For example, the shape can be projected on the yz-plane by simply ignoring the first coordinates of the points that constitute the shape, and by plotting only the second and third coordinates.

To present different views, the shape can be rotated in various ways before each projection is made. To illustrate the results of these procedures, a simple program that creates three-dimensional shapes consisting of points connected by lines was written for a computer. The coordinates of these points (vertices) and the information about which of these points are connected by lines (edges) are used as data in the program. The program plots the projection of the resulting shape on the yz-plane and represents the results as a printout. Before making such a plot, the computer rotates the shape about any one or a combination of the three coordinate axes by multiplying the vertices of the shape by the appropriate rotation matrix.

In Figure 6.29, we use a crude rendering of a tower, originally oriented as shown with the coordinate axes superimposed. What we see here is the projection of the figure on the yz-plane without rotations. Notice that the x-axis is not visible because it is perpendicular to the plane of this page. For each subsequent figure, the tower is rotated about one or two axes before being projected. (See Figure 6.30.)

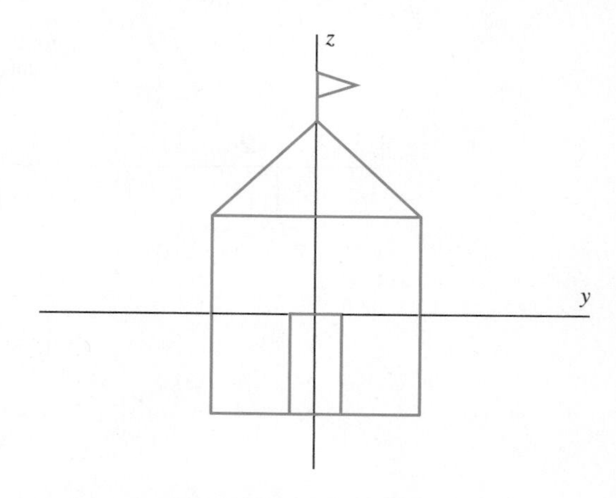

Figure 6.29 A front view of a tower

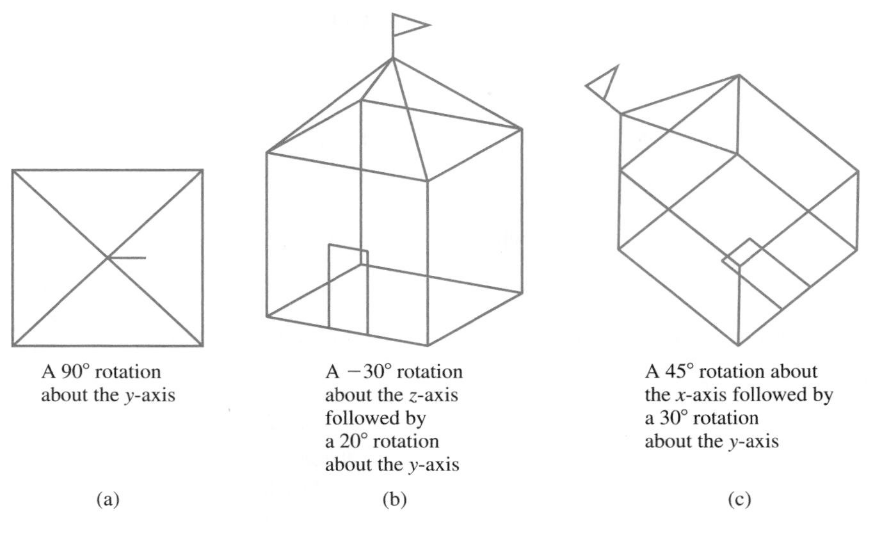

Figure 6.30

PERSPECTIVE

An object appears smaller when viewed from a greater distance. This effect, called **perspective**, is apparent when the object is viewed directly, as well as when it is viewed from photographic images. In both cases, light reflected from the object converges to a point, called a **focal point**, and then diverges along lines to a plane. This correspondence is called a **perspective projection**. For example, the plane on which the image is projected could be film in a camera. Figure 6.31 illustrates this phenomenon. In this figure, the focal point L is situated on the x-axis at $(a, 0, 0)$, and the plane on which any image is projected is perpendicular to the x-axis at $x = b$. Notice that the image of any point lies on the opposite side of the x-axis. This effect causes the image of an object to be reversed.

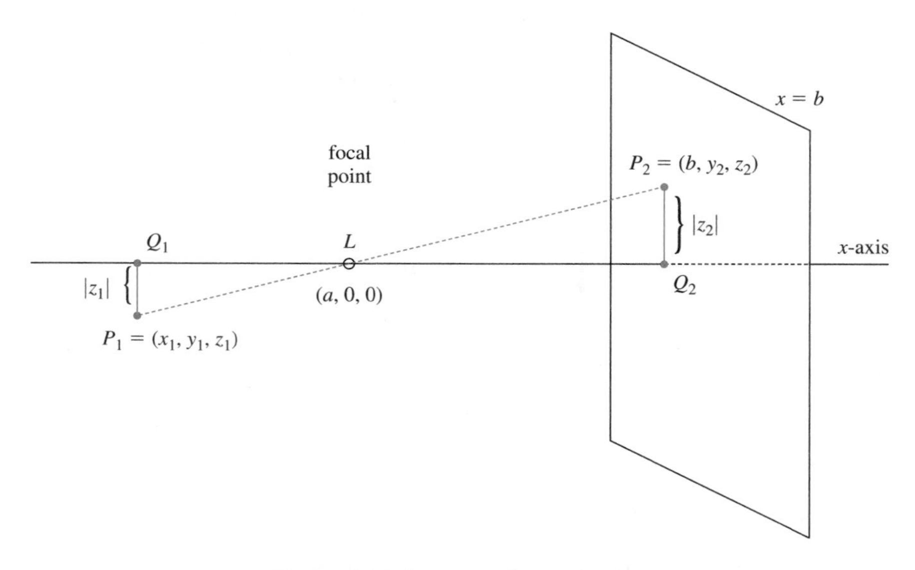

Figure 6.31 A perspective projection

We can use similar triangles to relate the location of a point with the location of its projected image. Consider an arbitrary point P_1 with coordinates (x_1, y_1, z_1). The point P_1 projects to the point $P_2 = (b, y_2, z_2)$ on the other side of the focal point. (See Figure 6.31.) Note that triangles P_1Q_1L and P_2Q_2L are similar, the length of P_1Q_1 is $|z_1|$, and the length of P_2Q_2 is $|z_2|$. It follows that

$$|z_2| = \frac{|z_1|(b-a)}{a-x_1}.$$

This equation tells us that the greater the distance P_1 is from the focal point, the smaller $|z_2|$ will be in comparison to $|z_1|$. That is, the larger the value of $a - x_1$ is, the smaller the size of the projected image will be. Since the image of P_1 is reversed, the signs of z_1 and z_2 are opposite, and hence

$$z_2 = \frac{-z_1(b-a)}{a-x_1}. \tag{17}$$

Similarly, we have that

$$y_2 = \frac{-y_1(b-a)}{a-x_1}. \tag{18}$$

The problem with equations (17) and (18) is that their application to the graphic projection of an actual object results in an image that is reversed. But if we simply replace y_2 by $-y_2$ and z_2 by $-z_2$, we invert the reversed image to obtain an image that is restored to its original orientation. Finally, we ignore the first coordinate, b, of the projected point, treating the plane $x = b$ as if it were the yz-plane. Thus we obtain a correspondence, called a *perspective projection*, that takes P_1, with coordinates (x_1, y_1, z_1), into P_2, with coordinates $\left(\frac{y_1(b-a)}{a-x_1}, \frac{z_1(b-a)}{a-x_1}\right)$. This correspondence enables us to create the illusion of perspective.

The difference between a computer graphic with perspective and one without perspective can be seen by comparing Figure 6.32 with Figures 6.29 and 6.30. In Figure 6.32, the graphic is drawn with the focal point located on the x-axis at $x = a = 100$ and the projected plane located at $x = b = 180$.

ROTATION MATRICES

In addition to rotations of $\mathcal{R}^3$ about the coordinate axes, a rotation of $\mathcal{R}^3$ about any line L that contains $\mathbf{0}$ can be produced by left multiplication by the appropriate orthogonal matrix. Such a matrix is called a **rotation matrix**, and the line L is called the **axis of rotation**. Notice that an axis of rotation is a 1-dimensional subspace of $\mathcal{R}^3$, and conversely, any 1-dimensional subspace of $\mathcal{R}^3$ is the axis of rotation for some rotation matrix.

In what follows, we discuss the problem of finding these more general rotation matrices. Let L be a 1-dimensional subspace of $\mathcal{R}^3$, and let θ be an angle. We wish to find the rotation matrix P so that left multiplication by P causes a rotation of θ about the axis L. Before examining this problem, we must first decide what is meant by a rotation by the angle θ. We want to adopt the convention, as is done in the xy-plane, that the rotation is counterclockwise if $\theta > 0$ and is clockwise if $\theta < 0$. However, what is clockwise and what is counterclockwise literally depends on one's point of view. Suppose we could physically transport ourselves to a point $\mathbf{p}$ on L, where $\mathbf{p} \neq \mathbf{0}$. From this vantage point, we can view the 2-dimensional subspace $L^\perp$, which is the plane perpendicular to L at $\mathbf{0}$, and observe the rotation of a vector $\mathbf{v}$ in $L^\perp$ in the counterclockwise direction to a vector $\mathbf{v}'$ in $L^\perp$. On the other hand, if

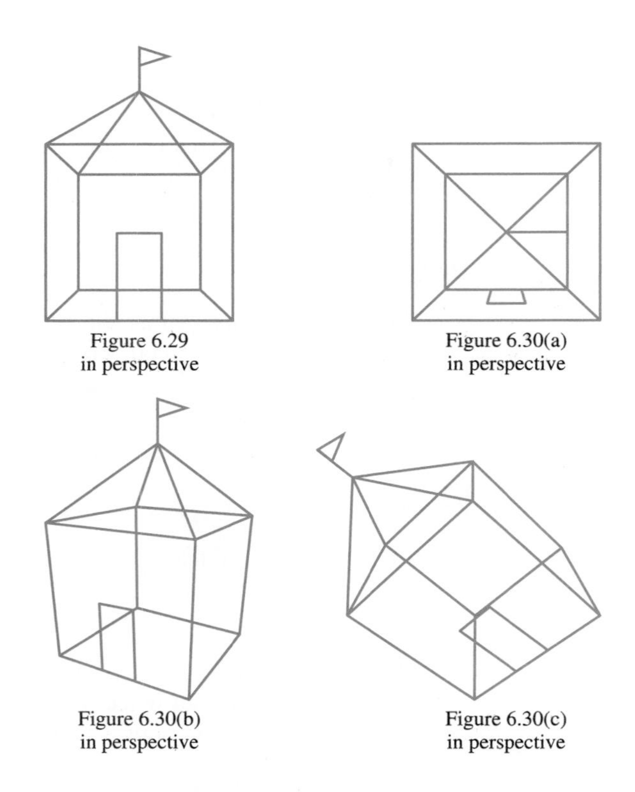

Figure 6.32 Views in perspective

we observe this same rotation from the opposite side of L, at the point $-\mathbf{p}$, then the same rotation is now seen to be in the clockwise direction. (See Figure 6.33.) Thus the direction of a rotation is affected by which side of $\mathbf{0}$ contains the vantage point in L. A side is determined by the unit vector in L that points in its direction. Since there are two sides, and these can be identified with the two unit vectors in L, the direction of rotation can be described unambiguously by choosing one of these two unit vectors. Such a choice is called an **orientation** of L.

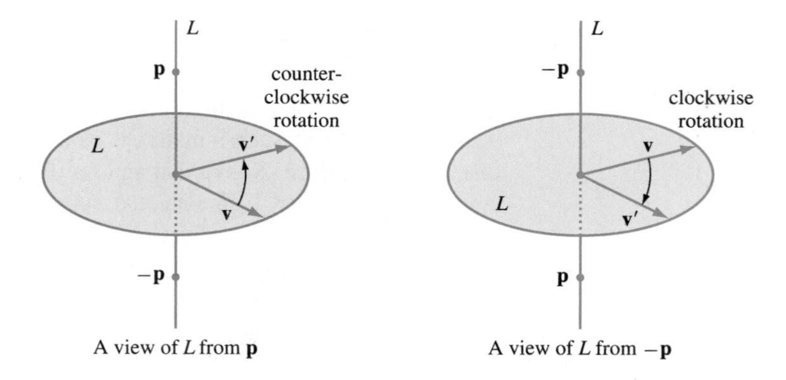

Figure 6.33 The direction of rotation from $\mathbf{p}$ to $-\mathbf{p}$

Choose a unit vector $\mathbf{v}_3$ in L, which determines an orientation of L and hence the direction of a counterclockwise rotation on $L^{\perp}$. Having chosen the axis L, the orientation $\mathbf{v}_3$, and the angle θ, we are ready to find the rotation matrix P. Since L is a 1-dimensional subspace of $\mathcal{R}^3$, its orthogonal complement, $L^{\perp}$, is a 2-dimensional subspace. Select an orthonormal basis $\{\mathbf{v}_1, \mathbf{v}_2\}$ for $L^\perp$ such that $\mathbf{v}_2$ is the result of rotating $\mathbf{v}_1$ counterclockwise in $L^\perp$ by 90° with respect to the chosen orientation of L. Let $\mathcal{B} = \{\mathbf{v}_1, \mathbf{v}_2, \mathbf{v}_3\}$. Then $\mathcal{B}$ is an orthonormal basis for $\mathcal{R}^3$, and hence the rotation matrix P can be obtained by finding $P\mathbf{v}_1$, $P\mathbf{v}_2$, and $P\mathbf{v}_3$, and then applying what we know about matrix representations. Since $P\mathbf{v}_1$ makes an angle θ with $\mathbf{v}_1$, we may apply Exercise 98 of Section 6.1 to obtain

$$P\mathbf{v}_1 \cdot \mathbf{v}_1 = \|P\mathbf{v}_1\| \, \|\mathbf{v}_1\| \cos\theta = \cos\theta.$$

Because $\mathbf{v}_2$ is obtained by rotating $\mathbf{v}_1$ by 90° in the counterclockwise direction, it follows that the angle between $P\mathbf{v}_1$ and $\mathbf{v}_2$ is $90^\circ - \theta$ if $\theta < 90^\circ$ and is $\theta - 90^\circ$ if $\theta > 90^\circ$. (See Figure 6.34.) In either case, $\cos(\theta - 90^\circ) = \cos(90^\circ - \theta) = \sin\theta$, and hence

$$P\mathbf{v}_1 \cdot \mathbf{v}_2 = \|P\mathbf{v}_1\| \, \|\mathbf{v}_2\| \cos(\pm(\theta - 90^\circ)) = (1)(1)\sin\theta = \sin\theta.$$

Therefore

$$P\mathbf{v}_1 = (P\mathbf{v}_1 \cdot \mathbf{v}_1)\mathbf{v}_1 + (P\mathbf{v}_1 \cdot \mathbf{v}_2)\mathbf{v}_2 = (\cos\theta)\mathbf{v}_1 + (\sin\theta)\mathbf{v}_2. \tag{19}$$

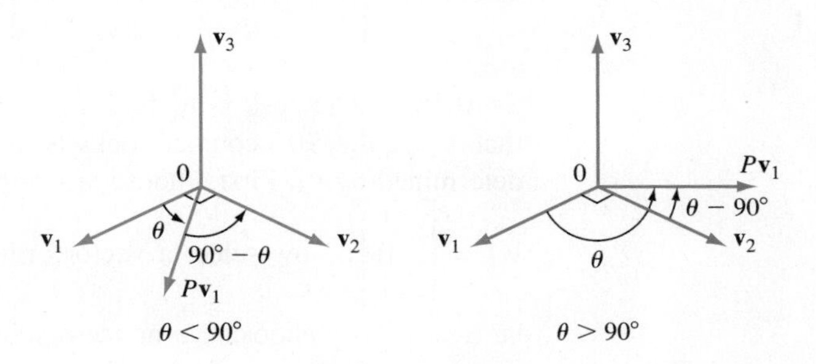

Figure 6.34 The angle between $P\mathbf{v}_1$ and $\mathbf{v}_2$

To find $P\mathbf{v}_2$, observe that $-\mathbf{v}_1$ can be obtained from $\mathbf{v}_2$ by a 90° counterclockwise rotation, and so we may apply the same arguments to the set $\{\mathbf{v}_2, -\mathbf{v}_1, \mathbf{v}_3\}$ to produce

$$P\mathbf{v}_2 = (\cos\theta)\mathbf{v}_2 + (\sin\theta)(-\mathbf{v}_1) = -(\sin\theta)(\mathbf{v}_1) + (\cos\theta)\mathbf{v}_2. \tag{20}$$

Finally, since $\mathbf{v}_3$ is in L and L remains unmoved by the rotation,

$$P\mathbf{v}_3 = \mathbf{v}_3. \tag{21}$$

We can now use equations (19), (20), and (21) to obtain the matrix representation of the matrix transformation T_P relative to $\mathcal{B}$:

$$[T_P]_{\mathcal{B}} = \begin{bmatrix} \cos\theta & -\sin\theta & 0 \\ \sin\theta & \cos\theta & 0 \\ 0 & 0 & 1 \end{bmatrix} = R_\theta \tag{22}$$

Let V be the 3×3 matrix $V = [\mathbf{v}_1 \; \mathbf{v}_2 \; \mathbf{v}_3]$. Then V is an orthogonal matrix because its columns are the vectors in $\mathcal{B}$, an orthonormal basis for $\mathcal{R}^3$. Furthermore,

$$V^{-1}PV = [T_P]_{\mathcal{B}} \tag{23}$$

since the columns of V are the vectors of $\mathcal{B}$. Combining equations (22) and (23), we obtain

$$P = VR_\theta V^{-1} = VR_\theta V^T. \tag{24}$$

We summarize this fact about rotation matrices in the following box:

> Any 3×3 rotation matrix has the form $VR_\theta V^T$ for some orthogonal matrix V and some angle θ.

Practice Problem 2 ▶ Let

$$W = \text{Span}\left\{\begin{bmatrix}1\\2\\3\end{bmatrix}, \begin{bmatrix}2\\3\\4\end{bmatrix}\right\}.$$

Suppose that R is a 3×3 rotation matrix such that for any vector $\mathbf{w}$ in W, the vector $R\mathbf{w}$ is also in W. Describe the axis of rotation. ◀

Example 1

Find the rotation matrix P that rotates $\mathcal{R}^3$ by an angle of 30° about the axis L containing $\begin{bmatrix}1\\1\\1\end{bmatrix}$, with the orientation determined by the unit vector $\mathbf{v}_3 = \dfrac{1}{\sqrt{3}}\begin{bmatrix}1\\1\\1\end{bmatrix}$.

Solution Our task is to find a pair of orthonormal vectors $\mathbf{v}_1$ and $\mathbf{v}_2$ in $L^\perp$ such that $\mathbf{v}_2$ is the 90° counterclockwise rotation of $\mathbf{v}_1$ with respect to the orientation determined by $\mathbf{v}_3$. First, choose any nonzero vector $\mathbf{w}_1$ orthogonal to $\mathbf{v}_3$; for example, $\mathbf{w}_1 = \begin{bmatrix}1\\0\\-1\end{bmatrix}$. Now select a vector orthogonal to both $\mathbf{v}_3$ and $\mathbf{w}_1$. Such a vector can be obtained by choosing a nonzero solution of the system of linear equations

$$\sqrt{3}\,\mathbf{v}_3 \cdot \mathbf{x} = \begin{bmatrix}1\\1\\1\end{bmatrix} \cdot \begin{bmatrix}x_1\\x_2\\x_3\end{bmatrix} = x_1 + x_2 + x_3 = 0$$

$$\mathbf{w}_1 \cdot \mathbf{x} = \begin{bmatrix}1\\0\\-1\end{bmatrix} \cdot \begin{bmatrix}x_1\\x_2\\x_3\end{bmatrix} = x_1 \qquad - x_3 = 0.$$

For example, $\mathbf{w}_2 = \begin{bmatrix}1\\-2\\1\end{bmatrix}$ is such a solution. Then $\mathbf{w}_1$ and $\mathbf{w}_2$ are orthogonal vectors in $L^\perp$. We use these vectors to find $\mathbf{v}_1$ and $\mathbf{v}_2$, but we must first consider a more difficult situation. For the orientation of L determined by $\mathbf{v}_3$, we must decide whether the 90° rotation from $\mathbf{w}_1$ to $\mathbf{w}_2$ is clockwise or counterclockwise. From Figure 6.35 we see that the 90° rotation from $\mathbf{w}_1$ to $\mathbf{w}_2$ is clockwise. There are two ways to correct this situation. One possibility is to reverse the order of $\mathbf{w}_1$ and $\mathbf{w}_2$ since the 90° rotation from $\mathbf{w}_2$ to $\mathbf{w}_1$ is counterclockwise. Or we can replace $\mathbf{w}_2$ by $-\mathbf{w}_2$ since the rotation from $\mathbf{w}_1$ to $-\mathbf{w}_2$ is also counterclockwise. Either way is acceptable. We arbitrarily select the first way. Finally, we replace the $\mathbf{w}_i$'s by unit vectors in the same direction. Thus we let

$$\mathbf{v}_1 = \frac{1}{\|\mathbf{w}_2\|}\mathbf{w}_2 = \frac{1}{\sqrt{6}}\begin{bmatrix}1\\-2\\1\end{bmatrix} \quad \text{and} \quad \mathbf{v}_2 = \frac{1}{\|\mathbf{w}_1\|}\mathbf{w}_1 = \frac{1}{\sqrt{2}}\begin{bmatrix}1\\0\\-1\end{bmatrix}.$$

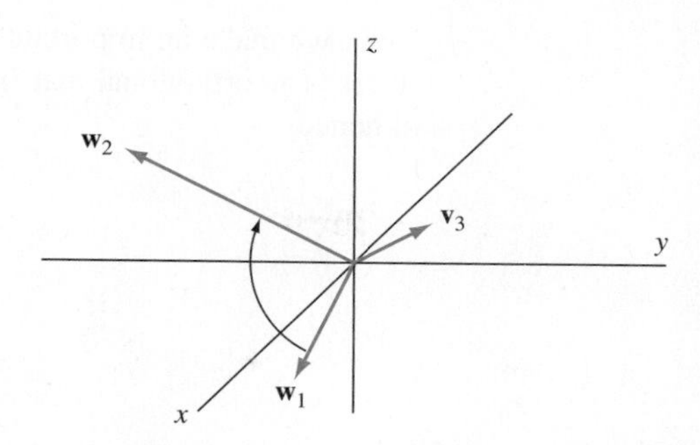

Figure 6.35 The 90° rotation from $\mathbf{w}_1$ to $\mathbf{w}_2$ with orientation determined by $\mathbf{v}_3$ is clockwise.

Then $\mathcal{B} = \{\mathbf{v}_1, \mathbf{v}_2, \mathbf{v}_3\}$ is the required orthonormal basis. Let

$$V = [\mathbf{v}_1\ \mathbf{v}_2\ \mathbf{v}_3] = \begin{bmatrix} \frac{1}{\sqrt{6}} & \frac{1}{\sqrt{2}} & \frac{1}{\sqrt{3}} \\ \frac{-2}{\sqrt{6}} & 0 & \frac{1}{\sqrt{3}} \\ \frac{1}{\sqrt{6}} & \frac{-1}{\sqrt{2}} & \frac{1}{\sqrt{3}} \end{bmatrix}.$$

Then, by equation (24),

$$\begin{aligned} P &= VR_{30^\circ}V^T \\ &= \begin{bmatrix} \frac{1}{\sqrt{6}} & \frac{1}{\sqrt{2}} & \frac{1}{\sqrt{3}} \\ \frac{-2}{\sqrt{6}} & 0 & \frac{1}{\sqrt{3}} \\ \frac{1}{\sqrt{6}} & \frac{-1}{\sqrt{2}} & \frac{1}{\sqrt{3}} \end{bmatrix} \begin{bmatrix} \cos 30^\circ & -\sin 30^\circ & 0 \\ \sin 30^\circ & \cos 30^\circ & 0 \\ 0 & 0 & 1 \end{bmatrix} \begin{bmatrix} \frac{1}{\sqrt{6}} & \frac{-2}{\sqrt{6}} & \frac{1}{\sqrt{6}} \\ \frac{1}{\sqrt{2}} & 0 & \frac{-1}{\sqrt{6}} \\ \frac{1}{\sqrt{3}} & \frac{1}{\sqrt{3}} & \frac{1}{\sqrt{3}} \end{bmatrix} \\ &= \begin{bmatrix} \frac{1}{\sqrt{6}} & \frac{1}{\sqrt{2}} & \frac{1}{\sqrt{3}} \\ \frac{-2}{\sqrt{6}} & 0 & \frac{1}{\sqrt{3}} \\ \frac{1}{\sqrt{6}} & \frac{-1}{\sqrt{2}} & \frac{1}{\sqrt{3}} \end{bmatrix} \begin{bmatrix} \frac{\sqrt{3}}{2} & \frac{-1}{2} & 0 \\ \frac{1}{2} & \frac{\sqrt{3}}{2} & 0 \\ 0 & 0 & 1 \end{bmatrix} \begin{bmatrix} \frac{1}{\sqrt{6}} & \frac{-2}{\sqrt{6}} & \frac{1}{\sqrt{6}} \\ \frac{1}{\sqrt{2}} & 0 & \frac{-1}{\sqrt{6}} \\ \frac{1}{\sqrt{3}} & \frac{1}{\sqrt{3}} & \frac{1}{\sqrt{3}} \end{bmatrix} \\ &= \frac{1}{3}\begin{bmatrix} 1+\sqrt{3} & 1-\sqrt{3} & 1 \\ 1 & 1+\sqrt{3} & 1-\sqrt{3} \\ 1-\sqrt{3} & 1 & 1+\sqrt{3} \end{bmatrix}. \end{aligned}$$

We now make an important observation about the eigenvectors of an arbitrary 3×3 rotation matrix P with axis of rotation L. For any vector $\mathbf{v}$ in L, $P\mathbf{v} = \mathbf{v}$, and hence 1 is an eigenvalue of P and L is contained in the eigenspace of P corresponding to the eigenvalue 1. If the angle of rotation of P is $\theta = 0^\circ$ (or, more generally, $\theta = 360n$ degrees for some integer n), then the matrix transformation T_P rotates every vector in $\mathcal{R}^3$ to itself, and hence $P = I_3$. In this case, the axis of rotation can be taken to be any 1-dimensional subspace of $\mathcal{R}^3$. With the exception of this rotation by 0°, P has a unique axis of rotation L. Furthermore, if $P \neq I_3$, then for any $\mathbf{v}$ not in L, we have $P\mathbf{v} \neq \mathbf{v}$ (see Exercise 68), and hence L is the eigenspace of P corresponding to the eigenvalue 1. Therefore the eigenspace of P corresponding to the eigenvalue 1 is 1-dimensional, except when $P = I_3$, in which case the eigenspace is 3-dimensional.

Next, we make an important observation about the determinant of P. By equation (24), there is an orthogonal matrix V such that $P = VR_\theta V^{-1}$. By Exercise 69, $\det R_\theta = 1$, and hence

$$\begin{aligned}\det P &= \det(VR_\theta V^{-1})\\ &= (\det V)(\det R_\theta)(\det V^{-1})\\ &= (\det V)(\det R_\theta)(\det V)^{-1}\\ &= \det R_\theta\\ &= 1.\end{aligned}$$

The converse is also true. We state the full result here, but defer the proof of the converse, which is more substantial, to the end of this section. Thus this condition regarding the determinant gives us a simple characterization of 3×3 rotation matrices. As an application, we give a simple proof that the transpose of a rotation matrix and the product of rotation matrices are rotation matrices.

THEOREM 6.20

Let P and Q be 3×3 orthogonal matrices.

(a) P is a rotation matrix if and only if $\det P = 1$.

(b) If P is a rotation matrix, then P^T is a rotation matrix.

(c) If P and Q are rotation matrices, then PQ is a rotation matrix.

PROOF The proof of (a) can be found on page 476.

(b) Suppose that P is a rotation matrix. Since P is an orthogonal matrix, P^T is an orthogonal matrix by Theorem 6.10(d) and $\det P^T = \det P = 1$ by (a). Thus P^T is a rotation matrix by (a).

(c) Suppose that P and Q are rotation matrices. Then $\det P = \det Q = 1$ by (a). Since P and Q are each orthogonal matrices, PQ is also an orthogonal matrix by Theorem 6.10(b). Furthermore,

$$\det(PQ) = (\det P)(\det Q) = 1 \cdot 1 = 1.$$

Therefore PQ is a rotation matrix by (a). ■

There is something unsatisfactory about our solution of Example 1: We relied on a figure to make the judgment that a certain 90° rotation is counterclockwise when viewed from a certain direction. In general, there is a problem with this approach, because accurately drawing and interpreting 3-dimensional figures depends on our ability to visualize space, and this can fail us. However, Theorem 6.20 gives us a way out of this predicament because it leads to a computational method for determining whether a certain rotation is counterclockwise. The following result describes this method:

THEOREM 6.21

Let $\{\mathbf{v}_1, \mathbf{v}_2, \mathbf{v}_3\}$ be an orthonormal basis for $\mathcal{R}^3$. The 90° rotation from $\mathbf{v}_1$ to $\mathbf{v}_2$ is counterclockwise as viewed from $\mathbf{v}_3$ if and only if $\det [\mathbf{v}_1\ \mathbf{v}_2\ \mathbf{v}_3] = 1$.

PROOF First, suppose that $\det [\mathbf{v}_1\ \mathbf{v}_2\ \mathbf{v}_3] = 1$. Since $V = [\mathbf{v}_1\ \mathbf{v}_2\ \mathbf{v}_3]$ is an orthogonal matrix, it is a rotation matrix by Theorem 6.20. Observe that the 90° rotation from $\mathbf{e}_1$ to $\mathbf{e}_2$ is counterclockwise as viewed from $\mathbf{e}_3$. Since V is a rotation, the relative positions of $\mathbf{e}_1$, $\mathbf{e}_2$, and $\mathbf{e}_3$ are the same as those of $V\mathbf{e}_1 = \mathbf{v}_1$, $V\mathbf{e}_2 = \mathbf{v}_2$, and $V\mathbf{e}_3 = \mathbf{v}_3$. Therefore the 90° rotation from $\mathbf{v}_1$ to $\mathbf{v}_2$ is counterclockwise as viewed from $\mathbf{v}_3$.

Now suppose that $\det [\mathbf{v}_1\ \mathbf{v}_2\ \mathbf{v}_3] \neq 1$. Since $[\mathbf{v}_1\ \mathbf{v}_2\ \mathbf{v}_3]$ is an orthogonal matrix, $\det [\mathbf{v}_1\ \mathbf{v}_2\ \mathbf{v}_3] = -1$ by Theorem 6.10. Therefore

$$\det [\mathbf{v}_2\ \mathbf{v}_1\ \mathbf{v}_3] = -\det [\mathbf{v}_1\ \mathbf{v}_2\ \mathbf{v}_3] = (-1)(-1) = 1.$$

By what was discussed previously, we can deduce that the 90° rotation from $\mathbf{v}_2$ to $\mathbf{v}_1$ as viewed from $\mathbf{v}_3$ is counterclockwise. It follows that the 90° rotation from $\mathbf{v}_1$ to $\mathbf{v}_2$ as viewed from $\mathbf{v}_3$ is clockwise. ■

Revisiting Example 1, we can apply Theorem 6.21 to verify that our choice of orthonormal vectors $\mathbf{v}_1$, $\mathbf{v}_2$, and $\mathbf{v}_3$ satisfies the requirement that the 90° rotation from $\mathbf{v}_1$ to $\mathbf{v}_2$ is counterclockwise as viewed from $\mathbf{v}_3$. In this case,

$$\det [\mathbf{v}_1\ \mathbf{v}_2\ \mathbf{v}_3] = \det \begin{bmatrix} \frac{1}{\sqrt{6}} & \frac{1}{\sqrt{2}} & \frac{1}{\sqrt{3}} \\ \frac{-2}{\sqrt{6}} & 0 & \frac{1}{\sqrt{3}} \\ \frac{1}{\sqrt{6}} & \frac{-1}{\sqrt{2}} & \frac{1}{\sqrt{3}} \end{bmatrix} = 1,$$

and hence the choice of orthonormal vectors $\mathbf{v}_1$, $\mathbf{v}_2$, and $\mathbf{v}_3$ made in Example 1 is acceptable.

Example 2 By Theorem 6.20(c), $P_\phi R_\theta$ is a rotation matrix for any angles ϕ and θ. Describe the axis of rotation for $P_\phi R_\theta$, where $\phi = 45^\circ$ and $\theta = 30^\circ$.

Solution The axis of rotation for $P_\phi R_\theta$ is Span $\{\mathbf{v}\}$, where $\mathbf{v}$ is an eigenvector of $P_\phi R_\theta$ corresponding to the eigenvalue 1. So $P_\phi R_\theta \mathbf{v} = \mathbf{v}$, and hence $R_\theta \mathbf{v} = P_\phi^{-1}\mathbf{v} = P_\phi^T \mathbf{v}$. Therefore $(R_\theta - P_\phi^T)\mathbf{v} = \mathbf{0}$. Conversely, any nonzero solution of the equation $(R_\theta - P_\phi^T)\mathbf{x} = \mathbf{0}$ is an eigenvector of $P_\phi R_\theta$ corresponding to the eigenvalue 1. Therefore we require a nonzero solution of the equation

$$(R_\theta - P_\phi^T)\begin{bmatrix} x_1 \\ x_2 \\ x_3 \end{bmatrix} = \frac{1}{2}\begin{bmatrix} \sqrt{3}-2 & -1 & 0 \\ 1 & \sqrt{3}-2 & -\sqrt{2} \\ 0 & \sqrt{2} & 2-\sqrt{2} \end{bmatrix}\begin{bmatrix} x_1 \\ x_2 \\ x_3 \end{bmatrix} = \begin{bmatrix} 0 \\ 0 \\ 0 \end{bmatrix}.$$

The vector

$$\mathbf{v} = \begin{bmatrix} 1-\sqrt{2} \\ (\sqrt{3}-2)(1-\sqrt{2}) \\ \sqrt{3}-2 \end{bmatrix}$$

is such a solution. Thus the axis of rotation for $P_\phi R_\theta$ is the subspace Span $\{\mathbf{v}\}$.

Example 3 Find the angle of rotation induced by the rotation matrix $P_\phi R_\theta$ of Example 2.

Solution Let α be the angle of rotation. Since no orientation of the axis of rotation is given, we assume that $\alpha > 0$. Choose a nonzero vector $\mathbf{w}$ in $L^\perp$, where L is the axis of rotation, and observe that α is the angle between $P_\phi R_\theta \mathbf{w}$ and $\mathbf{w}$. Any nonzero vector $\mathbf{w}$ orthogonal to the vector $\mathbf{v}$ in Example 2 suffices—for example,

$$\mathbf{w} = \begin{bmatrix} \sqrt{3}-2 \\ 0 \\ \sqrt{2}-1 \end{bmatrix}.$$

Since $P_\phi^T = P_\phi^{-1}$ is an orthogonal matrix, and orthogonal matrices preserve dot products and norms, it follows that

$$\cos\alpha = \frac{(P_\phi R_\theta \mathbf{w})\cdot\mathbf{w}}{\|P_\phi R_\theta \mathbf{w}\|\,\|\mathbf{w}\|} = \frac{(P_\phi^T P_\phi R_\theta \mathbf{w})\cdot(P_\phi^T \mathbf{w})}{\|P_\phi R_\theta \mathbf{w}\|\,\|\mathbf{w}\|} = \frac{(R_\theta \mathbf{w})\cdot(P_\phi^T \mathbf{w})}{\|\mathbf{w}\|^2}.$$

Thus

$$\begin{aligned}\cos\alpha &= \frac{\frac{1}{4}\begin{bmatrix} 3-2\sqrt{3} \\ \sqrt{3}-2 \\ 2\sqrt{2}-2 \end{bmatrix}\cdot\begin{bmatrix} 2\sqrt{3}-4 \\ 2-\sqrt{2} \\ 2-\sqrt{2} \end{bmatrix}}{10-4\sqrt{3}-2\sqrt{2}} \\ &= \frac{16\sqrt{3}-36-\sqrt{6}+8\sqrt{2}}{4(10-4\sqrt{3}-2\sqrt{2})} \\ &\approx 0.59275.\end{aligned}$$

Therefore

$$\alpha \approx \cos^{-1}(0.59275) \approx 53.65^\circ.$$

Finally, we complete the proof of Theorem 6.20, as promised earlier.

Proof of Theorem 6.20(a) Since the comments given immediately before the statement of Theorem 6.20 prove that the determinant of every 3×3 rotation matrix equals 1, only the converse needs to be proved.

Let P be a 3×3 orthogonal matrix such that $\det P = 1$. We first prove that 1 is an eigenvalue of P. We recall a few facts to prepare us for the calculation that follows. Since P is an orthogonal matrix, $P^{-1} = P^T$. Furthermore, $\det P^T = \det P$, and hence

$$\det A = \det P^T \det A = \det P^T A$$

for any 3×3 matrix A. Finally, $\det(-A) = -\det A$ for any 3×3 matrix A.

Let $f(t)$ be the characteristic polynomial of P. Then

$$\begin{aligned} f(1) &= \det(P - I_3) \\ &= (\det P^T)\det(P - I_3) \end{aligned}$$

$$\begin{aligned}
&= \det(P^T(P - I_3)) \\
&= \det(P^T P - P^T) \\
&= \det(I_3 - P^T) \\
&= \det(I_3 - P)^T \\
&= \det(I_3 - P) \\
&= \det(-(P - I_3)) \\
&= -\det(P - I_3) \\
&= -f(1),
\end{aligned}$$

and hence $2f(1) = 0$. Thus $f(1) = 0$, and it follows that 1 is an eigenvalue of P.

Let L be the eigenspace of P corresponding to the eigenvalue 1. We now establish that, for any $\mathbf{w}$ in $L^\perp$, $P\mathbf{w}$ is in $L^\perp$. To see this, consider any $\mathbf{v}$ in L. Then $P\mathbf{v} = \mathbf{v}$, and hence $P^T\mathbf{v} = P^{-1}\mathbf{v} = \mathbf{v}$. It now follows that

$$(P\mathbf{w}) \cdot \mathbf{v} = (P\mathbf{w})^T\mathbf{v} = (\mathbf{w}^T P^T)\mathbf{v} = \mathbf{w}^T(P^T\mathbf{v}) = \mathbf{w}^T\mathbf{v} = \mathbf{w} \cdot \mathbf{v} = 0.$$

Thus $P\mathbf{w}$ is in $L^\perp$.

If $\dim L = 3$, then $P = I_3$ is a rotation by 0°. So, suppose that $\dim L < 3$. We show that $\dim L = 1$. By way of contradiction, suppose that $\dim L = 2$. Then $\dim L^\perp = 1$. Select any nonzero vector $\mathbf{w}$ in $L^\perp$. Then $\{\mathbf{w}\}$ is a basis for $L^\perp$. Since $P\mathbf{w}$ is in $L^\perp$, there is a scalar λ such that $P\mathbf{w} = \lambda\mathbf{w}$. Thus $\mathbf{w}$ is an eigenvector of P corresponding to the eigenvalue λ. Since L is an eigenspace of P corresponding to the eigenvalue 1 and $\mathbf{w}$ is in $L^\perp$, $\lambda \neq 1$. By Exercise 49 of Section 6.5, $\lambda = \pm 1$, and hence $\lambda = -1$. Let $\{\mathbf{u}_1, \mathbf{u}_2\}$ be a basis for L. Then $\mathcal{S} = \{\mathbf{u}_1, \mathbf{u}_2, \mathbf{w}\}$ is a basis for $\mathcal{R}^3$. Let B be the 3×3 matrix $[\mathbf{u}_1\ \mathbf{u}_2\ \mathbf{w}]$. Then

$$B^{-1}PB = [T_P]_{\mathcal{S}} = \begin{bmatrix} 1 & 0 & 0 \\ 0 & 1 & 0 \\ 0 & 0 & -1 \end{bmatrix},$$

and hence

$$P = B \begin{bmatrix} 1 & 0 & 0 \\ 0 & 1 & 0 \\ 0 & 0 & -1 \end{bmatrix} B^{-1}.$$

Therefore

$$\begin{aligned}
\det P &= \det\left(B \begin{bmatrix} 1 & 0 & 0 \\ 0 & 1 & 0 \\ 0 & 0 & -1 \end{bmatrix} B^{-1}\right) \\
&= \det B \cdot \det \begin{bmatrix} 1 & 0 & 0 \\ 0 & 1 & 0 \\ 0 & 0 & -1 \end{bmatrix} \cdot \det(B^{-1}) \\
&= (\det B)(-1)(\det B)^{-1} = -1,
\end{aligned}$$

contrary to the assumption that $\det P = 1$. We conclude that $\dim L = 1$.

Thus $\dim L^\perp = 2$. Let $\{\mathbf{v}_1, \mathbf{v}_2\}$ be an orthonormal basis for $L^\perp$, and let $\mathbf{v}_3$ be a unit vector in L. Then $\mathcal{B} = \{\mathbf{v}_1, \mathbf{v}_2, \mathbf{v}_3\}$ is an orthonormal basis for $\mathcal{R}^3$, and

$P\mathbf{v}_3 = \mathbf{v}_3$. Furthermore, $P\mathbf{v}_1$ and $P\mathbf{v}_2$ are in $L^\perp$. Thus there are scalars a, b, c, and d such that

$$P\mathbf{v}_1 = a\mathbf{v}_1 + b\mathbf{v}_2 \qquad \text{and} \qquad P\mathbf{v}_2 = c\mathbf{v}_1 + d\mathbf{v}_2.$$

Let $V = [\mathbf{v}_1\ \mathbf{v}_2\ \mathbf{v}_3]$. Then V is an orthogonal matrix, and

$$[T_P]_{\mathcal{B}} = V^{-1}PV = \begin{bmatrix} a & c & 0 \\ b & d & 0 \\ 0 & 0 & 1 \end{bmatrix}.$$

Let $A = \begin{bmatrix} a & c \\ b & d \end{bmatrix}$. Comparing the columns of A with the first two columns of the orthogonal matrix $V^{-1}PV$, we see that the columns of A are orthonormal, and hence A is a 2×2 orthogonal matrix. Furthermore,

$$\begin{aligned} \det A &= \det \begin{bmatrix} a & c & 0 \\ b & d & 0 \\ 0 & 0 & 1 \end{bmatrix} \\ &= \det(V^{-1}PV) \\ &= \det(V^{-1}) \cdot \det P \cdot \det V \\ &= (\det V)^{-1} \det P \det V \\ &= \det P \\ &= 1, \end{aligned}$$

and hence A is a rotation matrix by Theorem 6.11. Therefore there is an angle θ such that

$$A = \begin{bmatrix} \cos\theta & -\sin\theta \\ \sin\theta & \cos\theta \end{bmatrix},$$

and thus

$$V^{-1}PV = \begin{bmatrix} \cos\theta & -\sin\theta & 0 \\ \sin\theta & \cos\theta & 0 \\ 0 & 0 & 1 \end{bmatrix} = R_\theta.$$

Hence $P = VR_\theta V^{-1}$. It follows by equation (24) that P is a rotation matrix. ■

EXERCISES

In Exercises 1–6, find the matrix M such that for each vector $\mathbf{v}$ in $\mathcal{R}^3$, $M\mathbf{v}$ is the result of the given sequence of rotations.

1. Each vector $\mathbf{v}$ is first rotated by 90° about the x-axis, and the result is then rotated 90° about the y-axis.
2. Each vector $\mathbf{v}$ is first rotated by 90° about the y-axis, and the result is then rotated 90° about the x-axis.
3. Each vector $\mathbf{v}$ is first rotated by 45° about the z-axis, and the result is then rotated 90° about the x-axis.
4. Each vector $\mathbf{v}$ is first rotated by 45° about the z-axis, and the result is then rotated 90° about the y-axis.
5. Each vector $\mathbf{v}$ is first rotated by 30° about the y-axis, and the result is then rotated 30° about the x-axis.
6. Each vector $\mathbf{v}$ is first rotated by 90° about the x-axis, and the result is then rotated 45° about the z-axis.

In Exercises 7–14, find the rotation matrix P that rotates $\mathcal{R}^3$ by the angle θ about the axis containing $\mathbf{v}$, with the orientation determined by each unit vector $\mathbf{u}$.

7. $\theta = 180°$, $\mathbf{v} = \begin{bmatrix} 1 \\ 0 \\ 1 \end{bmatrix}$, and $\mathbf{u} = \frac{1}{\sqrt{2}}\begin{bmatrix} 1 \\ 0 \\ 1 \end{bmatrix}$

8. $\theta = 90^\circ$, $\mathbf{v} = \begin{bmatrix} 1 \\ -1 \\ 1 \end{bmatrix}$, and $\mathbf{u} = \frac{1}{\sqrt{3}}\begin{bmatrix} 1 \\ -1 \\ 1 \end{bmatrix}$
9. $\theta = 45^\circ$, $\mathbf{v} = \begin{bmatrix} 1 \\ 1 \\ 0 \end{bmatrix}$, and $\mathbf{u} = \frac{-1}{\sqrt{2}}\begin{bmatrix} 1 \\ 1 \\ 0 \end{bmatrix}$
10. $\theta = 45^\circ$, $\mathbf{v} = \begin{bmatrix} 1 \\ 1 \\ 0 \end{bmatrix}$, and $\mathbf{u} = \frac{1}{\sqrt{2}}\begin{bmatrix} 1 \\ 1 \\ 0 \end{bmatrix}$
11. $\theta = 30^\circ$, $\mathbf{v} = \begin{bmatrix} 1 \\ -1 \\ 0 \end{bmatrix}$, and $\mathbf{u} = \frac{1}{\sqrt{2}}\begin{bmatrix} 1 \\ -1 \\ 0 \end{bmatrix}$
12. $\theta = 30^\circ$, $\mathbf{v} = \begin{bmatrix} 1 \\ -1 \\ 0 \end{bmatrix}$, and $\mathbf{u} = \frac{-1}{\sqrt{2}}\begin{bmatrix} 1 \\ -1 \\ 0 \end{bmatrix}$
13. $\theta = 60^\circ$, $\mathbf{v} = \begin{bmatrix} 1 \\ -1 \\ 1 \end{bmatrix}$, and $\mathbf{u} = \frac{1}{\sqrt{3}}\begin{bmatrix} 1 \\ -1 \\ 1 \end{bmatrix}$
14. $\theta = 45^\circ$, $\mathbf{v} = \begin{bmatrix} 1 \\ -1 \\ 1 \end{bmatrix}$, and $\mathbf{u} = \frac{1}{\sqrt{3}}\begin{bmatrix} 1 \\ -1 \\ 1 \end{bmatrix}$

In Exercises 15–22, a rotation matrix M is given. Find (a) a vector that forms a basis for each axis of rotation, and (b) the cosine of each angle of rotation.

15. The matrix M of Exercise 1.
16. The matrix M of Exercise 2.
17. The matrix M of Exercise 3.
18. The matrix M of Exercise 4.
19. The matrix M of Exercise 5.
20. The matrix M of Exercise 6.
21. $M = P_\theta Q_\phi$, where $\theta = 45^\circ$ and $\phi = 60^\circ$
22. $M = R_\theta P_\phi$, where $\theta = 30^\circ$ and $\phi = 45^\circ$

In Exercises 23–30, find the standard matrix of the reflection operator T_W of $\mathcal{R}^3$ about each subspace W.

23. $W = \text{Span}\left\{\begin{bmatrix} 1 \\ 2 \\ 3 \end{bmatrix}, \begin{bmatrix} 1 \\ 0 \\ -1 \end{bmatrix}\right\}$
24. $W = \text{Span}\left\{\begin{bmatrix} 1 \\ 1 \\ 1 \end{bmatrix}, \begin{bmatrix} 1 \\ 1 \\ -1 \end{bmatrix}\right\}$
25. $W = \{(x, y, z) : x + y + z = 0\}$
26. $W = \{(x, y, z) : x + 2y - z = 0\}$
27. $W = \{(x, y, z) : x + 2y - 2z = 0\}$
28. $W = \{(x, y, z) : x + y + 2z = 0\}$
29. $W = \{(x, y, z) : 3x - 4y + 5z = 0\}$
30. $W = \{(x, y, z) : x + 8y - 5z = 0\}$

In Exercises 31–38, find the standard matrix of the reflection operator T_W on $\mathcal{R}^3$ such that $T(\mathbf{v}) = -\mathbf{v}$ for each vector $\mathbf{v}$.

31. $\mathbf{v} = \begin{bmatrix} 1 \\ 2 \\ -1 \end{bmatrix}$
32. $\mathbf{v} = \begin{bmatrix} -1 \\ 1 \\ 1 \end{bmatrix}$
33. $\mathbf{v} = \begin{bmatrix} 1 \\ 0 \\ 2 \end{bmatrix}$
34. $\mathbf{v} = \begin{bmatrix} -1 \\ 2 \\ 3 \end{bmatrix}$
35. $\mathbf{v} = \begin{bmatrix} 3 \\ 4 \\ 5 \end{bmatrix}$
36. $\mathbf{v} = \begin{bmatrix} 3 \\ 0 \\ -1 \end{bmatrix}$
37. $\mathbf{v} = \begin{bmatrix} 2 \\ -1 \\ 2 \end{bmatrix}$
38. $\mathbf{v} = \begin{bmatrix} 1 \\ 1 \\ -2 \end{bmatrix}$

In Exercises 39–46,

(a) determine whether each orthogonal matrix is a rotation matrix, the standard matrix of a reflection operator, or neither of these;

(b) if the matrix is a rotation matrix, find a vector that forms a basis for the axis of rotation. If the matrix is the standard matrix of a reflection operator, find a basis for the 2-dimensional subspace about which $\mathcal{R}^3$ is reflected.

39. $\begin{bmatrix} 0 & 1 & 0 \\ -1 & 0 & 0 \\ 0 & 0 & -1 \end{bmatrix}$
40. $\begin{bmatrix} 0 & 0 & 1 \\ 0 & 1 & 0 \\ 1 & 0 & 0 \end{bmatrix}$
41. $\begin{bmatrix} 1 & 0 & 0 \\ 0 & -1 & 0 \\ 0 & 0 & -1 \end{bmatrix}$
42. $\begin{bmatrix} 0 & 0 & -1 \\ 0 & -1 & 0 \\ -1 & 0 & 0 \end{bmatrix}$
43. $\frac{1}{45}\begin{bmatrix} 35 & 28 & 4 \\ -20 & 29 & -28 \\ -20 & 20 & 35 \end{bmatrix}$
44. $\frac{1}{9}\begin{bmatrix} 1 & -4 & 8 \\ -4 & 7 & 4 \\ 8 & 4 & 1 \end{bmatrix}$
45. $\begin{bmatrix} \frac{1}{\sqrt{2}} & 0 & \frac{1}{\sqrt{2}} \\ 0 & 1 & 0 \\ \frac{1}{\sqrt{2}} & 0 & \frac{-1}{\sqrt{2}} \end{bmatrix}$
46. $\begin{bmatrix} \frac{1}{\sqrt{2}} & 0 & \frac{-1}{\sqrt{2}} \\ 0 & 1 & 0 \\ \frac{1}{\sqrt{2}} & 0 & \frac{1}{\sqrt{2}} \end{bmatrix}$

T&F *In Exercises 47–67, determine whether the statements are true or false.*

47. Every 3×3 orthogonal matrix is a rotation matrix.
48. For any 3×3 orthogonal matrix P, if $|\det P| = 1$, then P is a rotation matrix.
49. Every 3×3 orthogonal matrix has 1 as an eigenvalue.
50. Every 3×3 orthogonal matrix has -1 as an eigenvalue.
51. If P is a 3×3 rotation matrix and $P \neq I_3$, then 1 is an eigenvalue of P with multiplicity 1.
52. Every 3×3 orthogonal matrix is diagonalizable.
53. If P and Q are 3×3 rotation matrices, then PQ^T is a rotation matrix.
54. If P and Q are 3×3 rotation matrices, then PQ is a rotation matrix.
55. If P is a 3×3 rotation matrix, then P^T is a rotation matrix.
56. For any angles ϕ and θ, $Q_\phi R_\theta = R_\theta Q_\phi$.
57. The matrix that produces a rotation by the angle θ about the z-axis is
$$\begin{bmatrix} \cos\theta & -\sin\theta & 0 \\ \sin\theta & \cos\theta & 0 \\ 0 & 0 & 1 \end{bmatrix}.$$

58. The matrix that produces a rotation by the angle θ about the x-axis is
$$\begin{bmatrix} 1 & 0 & 0 \\ 0 & \cos\theta & -\sin\theta \\ 0 & \sin\theta & \cos\theta \end{bmatrix}.$$
59. The matrix that produces a rotation by the angle θ about the y-axis is
$$\begin{bmatrix} \cos\theta & 0 & -\sin\theta \\ 0 & 1 & 0 \\ \sin\theta & 0 & \cos\theta \end{bmatrix}.$$
60. A rotation about a line containing $\mathbf{0}$ can be produced by left multiplication by some orthogonal matrix.
61. An orientation of a line through the origin of $\mathcal{R}^3$ is determined by a unit vector that lies on the line.
62. Any 3×3 rotation matrix has the form $VR_\theta V^T$ for some orthogonal matrix V and some angle θ.
63. If $\{\mathbf{v}_1, \mathbf{v}_2, \mathbf{v}_3\}$ is an orthonormal basis for $\mathcal{R}^3$ and $\det\,[\mathbf{v}_1\ \mathbf{v}_2\ \mathbf{v}_3] = 1$, then the 90° rotation from $\mathbf{v}_1$ to $\mathbf{v}_2$ is clockwise.
64. If $\{\mathbf{v}_1, \mathbf{v}_2, \mathbf{v}_3\}$ is an orthonormal basis for $\mathcal{R}^3$ and the 90° rotation from $\mathbf{v}_1$ to $\mathbf{v}_2$ is clockwise, then
$$\det\,[\mathbf{v}_1\ \mathbf{v}_2\ \mathbf{v}_3] = 1.$$
65. If P is a 3×3 rotation matrix, then an eigenvector of P corresponding to eigenvalue -1 forms a basis for the axis of rotation for P.
66. Any nonzero solution of $(P_\theta - R_\phi)\mathbf{x} = \mathbf{0}$ forms a basis for the axis of rotation for $R_\phi P_\theta$.
67. For any vector $\mathbf{w}$ perpendicular to the axis of rotation for $R_\phi P_\theta$, the angle of rotation for $R_\phi P_\theta$ is the angle between $R_\phi P_\theta \mathbf{w}$ and $\mathbf{w}$.

---

68. Let P be a 3×3 rotation matrix with axis of rotation L, and suppose that $P \neq I_3$. Prove that for any vector $\mathbf{v}$ in $\mathcal{R}^3$, if $\mathbf{v}$ is not in L, then $P\mathbf{v} \neq \mathbf{v}$. *Hint:* For $\mathbf{v}$ not in L, let $\mathbf{v} = \mathbf{w} + \mathbf{z}$, where $\mathbf{w}$ is in L and $\mathbf{z}$ is in $L^\perp$. Now consider $P\mathbf{v} = P(\mathbf{w} + \mathbf{z})$.
69. Show by a direct computation that $\det P_\theta = \det Q_\theta = \det R_\theta = 1$ for any angle θ.
70. Prove that if P is a 3×3 orthogonal matrix, then P^2 is a rotation matrix.
71. Suppose that P is the 3×3 rotation matrix that rotates $\mathcal{R}^3$ by an angle θ about an axis L with orientation determined by the unit vector $\mathbf{v}$ in L. Prove that P^T rotates $\mathcal{R}^3$ by the angle $-\theta$ about L with the orientation determined by $\mathbf{v}$.
72. Let W be a 2-dimensional subspace of $\mathcal{R}^3$, and let B_W be the standard matrix of the reflection operator T_W. Prove that B_W is an orthogonal matrix, and hence T_W is an orthogonal operator.
73. Let T_W be the reflection of $\mathcal{R}^3$ about the 2-dimensional subspace W.
    (a) Prove that 1 is an eigenvalue of T_W and that W is the corresponding eigenspace.
    (b) Prove that -1 is an eigenvalue of T_W with corresponding eigenspace $W^\perp$.
74. Prove the converse of Exercise 73: Let T be an orthogonal operator on $\mathcal{R}^3$ with eigenvalues 1 and -1 having corresponding multiplicities 2 and 1, respectively. Then $T = T_W$ is the reflection operator of $\mathcal{R}^3$ about W, where W is the eigenspace of T corresponding to eigenvalue 1.
75. Let W be a 2-dimensional subspace of $\mathcal{R}^3$, and let T_W be the reflection operator of $\mathcal{R}^3$ about W with standard matrix B_W. Prove that $\det B_W = -1$.
76. Let B_W be the standard matrix of a reflection operator T_W about the 2-dimensional subspace W of $\mathcal{R}^3$. Prove the following statements:
    (a) $B_W^2 = I_3$.
    (b) B_W is a symmetric matrix.
77. Let B and C be standard matrices of reflection operators about 2-dimensional subspaces of $\mathcal{R}^3$. Prove that the product BC is a rotation matrix.
78. Let W_1 and W_2 be distinct 2-dimensional subspaces of $\mathcal{R}^3$, let B_1 and B_2 be the standard matrices of the reflection operators T_{W_1} and T_{W_2}, respectively, and consider the rotation matrix B_2B_1. Prove the following:
    (a) The axis of rotation of the rotation matrix B_2B_1 is the line determined by the intersection of the planes W_1 and W_2.
    (b) The axis of rotation of the rotation matrix B_2B_1 is the solution space of the system of linear equations $(B_1 - B_2)\mathbf{x} = \mathbf{0}$.
79. Let W_1, W_2, B_1, and B_2 be as in Exercise 78, and let $\mathbf{n}_1$ and $\mathbf{n}_2$ be unit vectors orthogonal to W_1 and W_2, respectively. Prove the following:
    (a) Both $\mathbf{n}_1$ and $\mathbf{n}_2$ are orthogonal to the axis of rotation of B_2B_1.
    (b) The rotation matrix B_2B_1 rotates both $\mathbf{n}_1$ and $\mathbf{n}_2$ about the axis of rotation by the angle θ, where
$$\cos\theta = -\mathbf{n}_1 \cdot (B_2\mathbf{n}_1) = -\mathbf{n}_2 \cdot (B_1\mathbf{n}_2).$$
80. Find a 3×3 orthogonal matrix C such that $\det C = -1$, but C is not the standard matrix of a reflection operator. *Hint:* Multiply the standard matrix of a reflection operator by a rotation matrix whose axis of rotation is the eigenspace of the reflection operator corresponding to the eigenvalue 1.
81. Suppose that $\{\mathbf{v}_1, \mathbf{v}_2\}$ is a basis for a 2-dimensional subspace W of $\mathcal{R}^3$. Let $\mathbf{v}_3$ be a nonzero vector that is orthogonal to both $\mathbf{v}_1$ and $\mathbf{v}_2$, and define
$$B = [\mathbf{v}_1\ \mathbf{v}_2\ \mathbf{v}_3] \quad \text{and} \quad C = [\mathbf{v}_1\ \mathbf{v}_2\ -\mathbf{v}_3].$$
Prove that CB^{-1} is the standard matrix of the reflection of $\mathcal{R}^3$ about W.

In Exercises 82 and 83, use either a calculator with matrix capabilities or computer software such as MATLAB to find the axis of rotation and the angle of rotation, to the nearest degree, for each rotation matrix.

82. $P_{22^\circ}Q_{16^\circ}$ 83. $R_{42^\circ}P_{23^\circ}$

SOLUTIONS TO THE PRACTICE PROBLEMS

1. For $\phi = 90^\circ$, we have

$$P_\phi = \begin{bmatrix} 1 & 0 & 0 \\ 0 & \cos 90^\circ & -\sin 90^\circ \\ 0 & \sin 90^\circ & \cos 90^\circ \end{bmatrix} = \begin{bmatrix} 1 & 0 & 0 \\ 0 & 0 & -1 \\ 0 & 1 & 0 \end{bmatrix},$$

and for $\theta = 60^\circ$, we have

$$Q_\theta = \begin{bmatrix} \cos 60^\circ & 0 & \cos 60^\circ \\ 0 & 1 & 0 \\ -\sin 60^\circ & 0 & \cos 60^\circ \end{bmatrix} = \begin{bmatrix} \frac{1}{2} & 0 & \frac{\sqrt{3}}{2} \\ 0 & 1 & 0 \\ -\frac{\sqrt{3}}{2} & 0 & \frac{1}{2} \end{bmatrix}.$$

Thus the desired vector is

$$P_\phi Q_\theta \begin{bmatrix} 1 \\ -1 \\ 2 \end{bmatrix} = \begin{bmatrix} 1 & 0 & 0 \\ 0 & 0 & -1 \\ 0 & 1 & 0 \end{bmatrix} \begin{bmatrix} \frac{1}{2} & 0 & \frac{\sqrt{3}}{2} \\ 0 & 1 & 0 \\ -\frac{\sqrt{3}}{2} & 0 & \frac{1}{2} \end{bmatrix} \begin{bmatrix} 1 \\ -1 \\ 2 \end{bmatrix} = \frac{1}{2} \begin{bmatrix} 1 + 2\sqrt{3} \\ -2 + \sqrt{3} \\ -2 \end{bmatrix}.$$

2. The axis of rotation is $W^\perp$, which is the solution space of the homogeneous system of linear equations

$$\begin{aligned} x_1 + 2x_2 + 3x_3 &= 0 \\ 2x_1 + 3x_2 + 4x_3 &= 0. \end{aligned}$$

The vector $\begin{bmatrix} 1 \\ -2 \\ 1 \end{bmatrix}$ forms a basis for this solution space, and hence the axis of rotation is the line through the origin containing this vector.

CHAPTER 6 REVIEW EXERCISES

In Exercises 1–19, determine whether the statements are true or false.

1. The norm of a vector in $\mathcal{R}^n$ is a scalar.
2. The dot product of two vectors in $\mathcal{R}^n$ is a scalar.
3. The dot product of any two vectors is defined.
4. If the endpoint of a vector lies on a given line, then the vector equals the orthogonal projection of the vector on that line.
5. The distance between two vectors in $\mathcal{R}^n$ is the norm of their difference.
6. The orthogonal complement of the row space of a matrix equals the null space of the matrix.
7. If W is a subspace of $\mathcal{R}^n$, then every vector in $\mathcal{R}^n$ can be written uniquely as a sum of a vector in W and a vector in $W^\perp$.
8. Every orthonormal basis of a subspace is also an orthogonal basis of the subspace.
9. A subspace and its orthogonal complement have the same dimension.
10. An orthogonal projection matrix is never invertible.
11. If $\mathbf{w}$ is the closest vector in a subspace W of $\mathcal{R}^n$ to a vector $\mathbf{v}$ in $\mathcal{R}^n$, then $\mathbf{w}$ is the orthogonal projection of $\mathbf{v}$ on W.
12. If $\mathbf{w}$ is the orthogonal projection of a vector $\mathbf{v}$ in $\mathcal{R}^n$ on a subspace W of $\mathcal{R}^n$, then $\mathbf{w}$ is orthogonal to $\mathbf{v}$.
13. For a given set of data plotted in the xy-plane, the least-squares line is the unique line in the plane that minimizes the sum of squared distances from the data points to the line.
14. If the columns of an $n \times n$ matrix P are orthogonal, then P is an orthogonal matrix.
15. If the determinant of the standard matrix of a linear operator on $\mathcal{R}^2$ equals one, then the linear operator is a rotation.
16. In $\mathcal{R}^2$, the composition of two reflections is a rotation.
17. In $\mathcal{R}^2$, the composition of two rotations is a rotation.
18. Every square matrix has a spectral decomposition.
19. If a matrix has a spectral decomposition, then the matrix must be symmetric.

In Exercises 20–23, two vectors $\mathbf{u}$ *and* $\mathbf{v}$ *are given. In each exercise,*

(a) compute the norm of each of the vectors;
(b) compute the distance d between the vectors;
(c) compute the dot product of the vectors;
(d) determine whether the vectors are orthogonal.

20. $\mathbf{u} = \begin{bmatrix} 2 \\ -3 \end{bmatrix}$ and $\mathbf{v} = \begin{bmatrix} 4 \\ 1 \end{bmatrix}$
21. $\mathbf{u} = \begin{bmatrix} 3 \\ -6 \end{bmatrix}$ and $\mathbf{v} = \begin{bmatrix} 4 \\ 2 \end{bmatrix}$
22. $\mathbf{u} = \begin{bmatrix} 2 \\ -1 \\ 3 \end{bmatrix}$ and $\mathbf{v} = \begin{bmatrix} 0 \\ 4 \\ 2 \end{bmatrix}$
23. $\mathbf{u} = \begin{bmatrix} 1 \\ -1 \\ 2 \end{bmatrix}$ and $\mathbf{v} = \begin{bmatrix} 2 \\ 4 \\ 1 \end{bmatrix}$

In Exercises 24 and 25, a vector $\mathbf{v}$ *and a line* $\mathcal{L}$ *in* $\mathcal{R}^2$ *are given. Compute the orthogonal projection* $\mathbf{w}$ *of* $\mathbf{v}$ *on* $\mathcal{L}$*, and use it to compute the distance d from the endpoint of* $\mathbf{v}$ *to* $\mathcal{L}$*.*

24. $\mathbf{v} = \begin{bmatrix} 3 \\ 5 \end{bmatrix}$ and $y = 4x$

25. $\mathbf{v} = \begin{bmatrix} 3 \\ 2 \end{bmatrix}$ and $y = -2x$

In Exercises 26–29, suppose that $\mathbf{u}$, $\mathbf{v}$, *and* $\mathbf{w}$ *are vectors in* $\mathcal{R}^n$ *such that* $\|\mathbf{u}\| = 3$, $\|\mathbf{v}\| = 4$, $\|\mathbf{w}\| = 2$, $\mathbf{u} \cdot \mathbf{v} = -2$, $\mathbf{u} \cdot \mathbf{w} = 5$, *and* $\mathbf{v} \cdot \mathbf{w} = -3$.

26. Compute $\|-2\mathbf{u}\|$.

27. Compute $(2\mathbf{u} + 3\mathbf{v}) \cdot \mathbf{w}$.

28. Compute $\|3\mathbf{u} - 2\mathbf{w}\|^2$.

29. Compute $\|\mathbf{u} - \mathbf{v} + 3\mathbf{w}\|^2$.

In Exercises 30 and 31, determine whether the given linearly independent set S is orthogonal. If the set is not orthogonal, apply the Gram–Schmidt process to S to find an orthogonal basis for the span of S.

30. $\left\{ \begin{bmatrix} 1 \\ 1 \\ 0 \end{bmatrix}, \begin{bmatrix} 2 \\ 0 \\ 1 \end{bmatrix}, \begin{bmatrix} 2 \\ 2 \\ 1 \end{bmatrix} \right\}$

31. $\left\{ \begin{bmatrix} 1 \\ 1 \\ -1 \\ 0 \end{bmatrix}, \begin{bmatrix} 0 \\ 0 \\ 1 \\ 1 \end{bmatrix}, \begin{bmatrix} 1 \\ 2 \\ 0 \\ 1 \end{bmatrix} \right\}$

In Exercises 32 and 33, find a basis for $S^\perp$.

32. $S = \left\{ \begin{bmatrix} 2 \\ -1 \\ 3 \end{bmatrix} \right\}$

33. $S = \left\{ \begin{bmatrix} 2 \\ 1 \\ -1 \\ 0 \end{bmatrix}, \begin{bmatrix} 3 \\ 4 \\ 2 \\ -2 \end{bmatrix} \right\}$

In Exercises 34 and 35, a vector $\mathbf{v}$ *in* $\mathcal{R}^n$ *and an orthonormal basis S for a subspace W of* $\mathcal{R}^n$ *are given. Use S to obtain the unique vectors* $\mathbf{w}$ *in W and* $\mathbf{z}$ *in* $W^\perp$ *such that* $\mathbf{v} = \mathbf{w} + \mathbf{z}$. *Use your answer to find the distance from* $\mathbf{v}$ *to W.*

34. $\mathbf{v} = \begin{bmatrix} 2 \\ 3 \end{bmatrix}$ and $S = \left\{ \frac{1}{\sqrt{5}} \begin{bmatrix} 2 \\ 1 \end{bmatrix} \right\}$

35. $\mathbf{v} = \begin{bmatrix} 1 \\ 2 \\ -3 \end{bmatrix}$ and $S = \left\{ \frac{1}{\sqrt{5}} \begin{bmatrix} 1 \\ 2 \\ 0 \end{bmatrix}, \frac{1}{\sqrt{14}} \begin{bmatrix} -2 \\ 1 \\ 3 \end{bmatrix} \right\}$

In Exercises 36–39, a subspace W and a vector $\mathbf{v}$ *are given. Find the orthogonal projection matrix* P_W, *and find the vector* $\mathbf{w}$ *in W that is closest to* $\mathbf{v}$.

36. $W = \text{Span}\left\{ \begin{bmatrix} 1 \\ -1 \\ 2 \end{bmatrix}, \begin{bmatrix} 1 \\ 0 \\ 1 \end{bmatrix} \right\}$ and $\mathbf{v} = \begin{bmatrix} 2 \\ -1 \\ 6 \end{bmatrix}$

37. $W = \text{Span}\left\{ \begin{bmatrix} 1 \\ 2 \\ 0 \\ -1 \end{bmatrix} \right\}$ and $\mathbf{v} = \begin{bmatrix} 2 \\ 1 \\ 3 \\ -8 \end{bmatrix}$

38. W is the solution set of

$$\begin{aligned} x_1 + 2x_2 - x_3 &= 0 \\ x_1 - x_2 - x_3 &= 0 \end{aligned} \quad \text{and} \quad \mathbf{v} = \begin{bmatrix} 2 \\ 1 \\ 4 \end{bmatrix}$$

39. W is the orthogonal complement of

$$\text{Span}\left\{ \begin{bmatrix} 1 \\ -1 \\ 0 \\ 0 \end{bmatrix}, \begin{bmatrix} 1 \\ 0 \\ 1 \\ 0 \end{bmatrix} \right\} \quad \text{and} \quad \mathbf{v} = \begin{bmatrix} 2 \\ -1 \\ 1 \\ 2 \end{bmatrix}$$

40. Find the equation of the least-squares line for the following data: (1, 4), (2, 6), (3, 10), (4, 12), (5, 13).

41. An object is moving away from point P at a constant speed v. At various times t, the distance d from the object to P was measured. The results are listed in the following table:

Time t (in seconds)	Distance d (in feet)
1	3.2
2	5.1
3	7.1
4	9.2
5	11.4

Assuming that d and t are related by the equation $d = vt + c$ for some constant c, use the method of least squares to estimate the speed of the object at time t and the distance between the object and P at time $t = 0$.

42. Use the method of least squares to find the best quadratic fit for the following data:

$$(1, 2), (2, 3), (3, 7), (4, 14), (5, 23)$$

In Exercises 43–46, determine whether the given matrix is orthogonal.

43. $\begin{bmatrix} 0.7 & 0.3 \\ -0.3 & 0.7 \end{bmatrix}$

44. $\frac{1}{13}\begin{bmatrix} 5 & -12 \\ 12 & -5 \end{bmatrix}$

45. $\frac{1}{\sqrt{2}}\begin{bmatrix} 1 & 0 & 1 \\ 0 & \sqrt{2} & 0 \\ 1 & 0 & -1 \end{bmatrix}$

46. $\frac{1}{\sqrt{6}}\begin{bmatrix} \sqrt{2} & -\sqrt{3} & 1 \\ \sqrt{2} & \sqrt{3} & -2 \\ \sqrt{2} & 0 & 1 \end{bmatrix}$

In Exercises 47–50, determine whether each orthogonal matrix is the standard matrix of a rotation or a reflection. If the operator is a rotation, determine the angle of rotation. If the operator is a reflection, determine the equation of the line of reflection.

47. $\frac{1}{2}\begin{bmatrix} 1 & \sqrt{3} \\ -\sqrt{3} & 1 \end{bmatrix}$

48. $\frac{1}{2}\begin{bmatrix} 1 & -\sqrt{3} \\ \sqrt{3} & 1 \end{bmatrix}$

49. $\frac{1}{2}\begin{bmatrix} 1 & \sqrt{3} \\ \sqrt{3} & -1 \end{bmatrix}$

50. $\frac{1}{5}\begin{bmatrix} -3 & 4 \\ 4 & 3 \end{bmatrix}$

51. Let $T\colon \mathcal{R}^3 \to \mathcal{R}^3$ be defined by

$$T\left(\begin{bmatrix} x_1 \\ x_2 \\ x_3 \end{bmatrix}\right) = \begin{bmatrix} -x_2 \\ x_3 \\ x_1 \end{bmatrix}.$$

Prove that T is an orthogonal operator.

52. Suppose that $T\colon \mathcal{R}^2 \to \mathcal{R}^2$ is an orthogonal operator. Let $U\colon \mathcal{R}^2 \to \mathcal{R}^2$ be defined by

$$U\left(\begin{bmatrix} x_1 \\ x_2 \end{bmatrix}\right) = \frac{1}{\sqrt{2}} T\left(\begin{bmatrix} x_1 + x_2 \\ -x_1 + x_2 \end{bmatrix}\right).$$

(a) Prove that U is an orthogonal operator.

(b) Suppose that T is a rotation. Is TU a rotation or a reflection?

(c) Suppose that T is a reflection. Is TU a rotation or a reflection?

In Exercises 53 and 54, a symmetric matrix A is given. Find an orthonormal basis of eigenvectors of A and their corresponding eigenvalues. Use this information to obtain a spectral decomposition of each matrix A.

53. $A = \begin{bmatrix} 2 & 3 \\ 3 & 2 \end{bmatrix}$

54. $A = \begin{bmatrix} 6 & 2 & 0 \\ 2 & 9 & 0 \\ 0 & 0 & -9 \end{bmatrix}$

In Exercises 55 and 56, the equation of a conic section is given in xy-coordinates. Find the appropriate angle of rotation so that each equation may be written in $x'y'$-coordinates with no $x'y'$-term. Give the new equation and identify the type of conic section.

55. $x^2 + 6xy + y^2 - 16 = 0$

56. $3x^2 - 4xy + 3y^2 - 9 = 0$

57. Let W be a subspace of $\mathcal{R}^n$, and let Q be an $n \times n$ orthogonal matrix. Prove that $Q^T P_W Q = P_Z$, where $Z = \{Q^T\mathbf{w}\colon \mathbf{w} \text{ is in } W\}$.

58. Prove that rank $P_W = \dim W$ for any subspace W of $\mathcal{R}^n$. *Hint:* Apply Exercise 68 of Section 6.3.

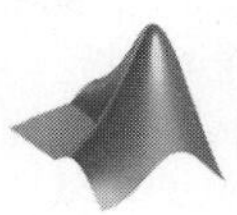

CHAPTER 6 MATLAB EXERCISES

For the following exercises, use MATLAB (or comparable software) or a calculator with matrix capabilities. The MATLAB functions in Tables D.1, D.2, D.3, D.4, and D.5 of Appendix D may be useful.

1. Let

$$\mathbf{u}_1 = \begin{bmatrix} 1 \\ 2 \\ -1 \\ 3 \\ 0 \\ 1 \end{bmatrix}, \quad \mathbf{u}_2 = \begin{bmatrix} 2 \\ -1 \\ -3 \\ -1 \\ 2 \\ -2 \end{bmatrix}, \quad \mathbf{u}_3 = \begin{bmatrix} 3 \\ -3 \\ 2 \\ -2 \\ 1 \\ 1 \end{bmatrix}, \quad \mathbf{u}_4 = \begin{bmatrix} -6 \\ 6 \\ -4 \\ 4 \\ -2 \\ -2 \end{bmatrix}.$$

(a) Compute $\mathbf{u}_1 \cdot \mathbf{u}_2$, $\|\mathbf{u}_1\|$, and $\|\mathbf{u}_2\|$.

(b) Compute $\mathbf{u}_3 \cdot \mathbf{u}_4$, $\|\mathbf{u}_3\|$, and $\|\mathbf{u}_4\|$.

(c) Verify the Cauchy–Schwarz inequality for $\mathbf{u}_1$ and $\mathbf{u}_2$.

(d) Verify the Cauchy–Schwarz inequality for $\mathbf{u}_3$ and $\mathbf{u}_4$.

(e) From your results, make a conjecture about when the Cauchy–Schwarz inequality is an equality. (See Exercise 88 of Section 6.1.)

2. Let W be the column space of an $n \times k$ matrix A.

(a) Prove that $W^\perp$ is the solution set of the equation $A^T\mathbf{x} = \mathbf{0}$.

(b) Use the MATLAB function null, described in Table D.2 of Appendix D, to obtain a basis for the subspace $W^\perp$ of $\mathcal{R}^5$, where W is the span of the columns of the matrix

$$A = \begin{bmatrix} 1 & 2 & 3 \\ -1 & 3 & 2 \\ 2 & 8 & 10 \\ 3 & -1 & 2 \\ 0 & 4 & 4 \end{bmatrix}.$$

3. Let

$$\mathcal{S} = \left\{ \begin{bmatrix} 1 \\ -1 \\ 2 \\ 3 \\ 5 \\ -4 \end{bmatrix}, \begin{bmatrix} 0 \\ 1 \\ -3 \\ 2 \\ -2 \\ 1 \end{bmatrix}, \begin{bmatrix} 2 \\ -2 \\ 1 \\ 4 \\ 0 \\ -1 \end{bmatrix}, \begin{bmatrix} 1 \\ 0 \\ -1 \\ 5 \\ 3 \\ -3 \end{bmatrix} \right\}$$

and W be the span of $\mathcal{S}$.

(a) Use the MATLAB function `orth` to find an orthonormal basis $\mathcal{B}$ for W.

(b) Use $\mathcal{B}$ to compute the orthogonal projection of each of the following vectors on W:

$$\text{(i)} \begin{bmatrix} 1 \\ -1 \\ 2 \\ 3 \\ 1 \\ -2 \end{bmatrix} \quad \text{(ii)} \begin{bmatrix} 1 \\ -2 \\ 2 \\ -1 \\ -3 \\ 2 \end{bmatrix} \quad \text{(iii)} \begin{bmatrix} -1 \\ -2 \\ -1 \\ 0 \\ 1 \\ 1 \end{bmatrix}$$

(c) Let M be the matrix whose columns are the vectors in $\mathcal{B}$, and let $P = MM^T$. For each vector $\mathbf{v}$ in (b), compute $P\mathbf{v}$ and compare the result with your answer in (b).

(d) State and prove a general result that justifies your observation in (c).

4. Let

$$A = \begin{bmatrix} 1.1 & 2.4 & -5.0 & 7.1 \\ 2.3 & 5.1 & -3.5 & 1.0 \\ 3.1 & 1.3 & -2.0 & 8.0 \\ 7.2 & -4.3 & 2.8 & 8.3 \\ 8.0 & -3.8 & 1.5 & 7.0 \end{bmatrix}.$$

In this exercise, we compute a QR factorization of A by the method described in Section 6.2 and use a MATLAB function to obtain an alternative QR factorization.

(a) Apply the MATLAB function gs, described in Table D.5 of Appendix D, to A to produce a matrix V whose columns form an orthogonal basis for Col A. (V is obtained by applying the Gram–Schmidt process to the columns of A.)

(b) Let V be the matrix obtained in (a). Compute $D = V^TV$. Observe that D is a diagonal matrix.[20] What are the values of the diagonal entries in relation to the orthogonal vectors obtained in (a)?

[20] Due to roundoff errors, the off-diagonal entries of D may have small nonzero values. For this reason, it is advisable to replace the off-diagonal entries of D with zeros before using D in subsequent calculations. One easy way to do this by using MATLAB is to enter the command $D =$ `diag(diag(D))`.

(c) For a matrix M whose entries are nonnegative, let M_s denote the matrix whose entries are the square roots of the corresponding entries of M.[21] Use the matrices V obtained in (a) and D obtained in (b) to compute the matrix $Q = V(D_s)^{-1}$. Verify that the columns of Q form an orthonormal set.

(d) Use the the matrix Q in (c) to compute the matrix $R = Q^T A$. Observe that R is an upper triangular matrix and $QR = A$, yielding a QR factorization of A.

(e) Use the MATLAB function $[Q\ R] =$ `qr`$(A, 0)$ to obtain another QR factorization of A and compare your results with those of (d).

(f) Prove that if a QR factorization of a matrix A is obtained by the method outlined in (a)–(d), then the diagonal entries of the upper triangular matrix R are positive.

5. Use the method in Exercise 4 to compute a QR factorization of the matrix

$$A = \begin{bmatrix} 1 & 3 & 2 & 2 \\ 4 & 2 & 1 & 1 \\ -1 & 1 & 5 & -1 \\ 2 & 0 & -3 & 0 \\ 1 & 5 & -4 & 4 \\ 1 & 1 & 2 & -2 \end{bmatrix}$$

in which the diagonal entries of R are positive.

6. Let

$$\mathcal{S} = \left\{ \begin{bmatrix} 1 \\ -1 \\ 2 \\ 3 \\ 5 \\ -4 \end{bmatrix}, \begin{bmatrix} 0 \\ 1 \\ -3 \\ 2 \\ -2 \\ 1 \end{bmatrix}, \begin{bmatrix} 2 \\ -2 \\ 1 \\ 4 \\ 0 \\ -1 \end{bmatrix}, \begin{bmatrix} 1 \\ 0 \\ -1 \\ 5 \\ 3 \\ -3 \end{bmatrix} \right\}$$

and W be the span of $\mathcal{S}$.

(a) Compute an orthonormal basis $\mathcal{B}_1$ for W.

(b) Use Exercise 61 in Section 6.3 to compute an orthonormal basis $\mathcal{B}_2$ for $W^\perp$.

(c) Let P be the matrix whose columns consist of the vectors in $\mathcal{B}_1$ followed by the vectors in $\mathcal{B}_2$. Compute PP^T and $P^T P$. Explain your results.

7. Let

$$\mathcal{S} = \left\{ \begin{bmatrix} 1 \\ 3 \\ 0 \\ -1 \\ 2 \\ 1 \end{bmatrix}, \begin{bmatrix} 0 \\ 1 \\ 3 \\ -2 \\ 1 \\ 1 \end{bmatrix}, \begin{bmatrix} -2 \\ -1 \\ 1 \\ 4 \\ 2 \\ 1 \end{bmatrix}, \begin{bmatrix} 1 \\ 1 \\ 2 \\ -1 \\ 3 \\ 0 \end{bmatrix}, \begin{bmatrix} -1 \\ 2 \\ 1 \\ 3 \\ 4 \\ 2 \end{bmatrix} \right\},$$

W be the span of $\mathcal{S}$, and P_W be the orthogonal projection matrix on W.

(a) Use the formula for P_W given in Section 6.4 to compute P_W.

(b) Use Exercise 75 of Section 6.4 to compute P_W. Compare your answer with the answer obtained in (a).

(c) Compute $P_W \mathbf{v}$ for each vector $\mathbf{v}$ in $\mathcal{S}$.

(d) Compute a basis $\mathcal{B}$ for $W^\perp$, and then compute $P_W \mathbf{v}$ for each vector $\mathbf{v}$ in $\mathcal{B}$. (You can use Exercise 61 in Section 6.3 for this purpose.)

8. Consider the data set of ordered pairs (x, y), where x is an integral multiple of 0.1 for $1 \leq x \leq 4$, and $y = \log(x)$.

(a) Plot the data.

(b) Use the method of least squares to find the equation of the least-squares line for the data, and then plot the line.

(c) Use the method of least squares to produce the best quadratic fit for the data, and then plot the result.

If possible, use software to display all three plots, each in a different color, so that the plots can be distinguished.

What follows is a description of how to create these plots with MATLAB. However, you should not use the instructions that follow unless you can give the explanation required in Exercise 9.

To plot the graph of the logarithm function, use the command `plot`(x, y), where x is the column vector consisting of the values $1, 1.1, \ldots, 3.9, 4$ and y is the column vector consisting of the corresponding logarithms. The result is a plot of the logarithm function in blue.

To plot the least-squares line by using these data, compute $a =$ `inv`$(C' * C) * C' * y$, where C is the matrix given on page 403. Then type the command `hold on` so that the plot in (a) remains, and type `plot`$(x, C * a,\ 'r')$ for the required plot, which is in red.

To plot the best quadratic fit for the data in green, imitate the process for plotting the least-squares line using the matrix C defined on page 405 and the command `plot`$(x, C * a,\ 'g')$.

9. With respect to the explanations for plotting the least-squares line and best quadratic fit given at the end of Exercise 8, explain the use of $C * a$ in the second argument of the `plot` function.

10. According to Theorem 6.12, the composition of a rotation and a reflection of $\mathcal{R}^2$ is a reflection. Let T be the rotation of $\mathcal{R}^2$ that rotates every vector by the angle 35°, and let U be the reflection of $\mathcal{R}^2$ about the line $y = 2.3x$.

(a) Find the equation of the line about which UT reflects $\mathcal{R}^2$.

(b) Find the equation of the line about which TU reflects $\mathcal{R}^2$.

11. Let

$$A = \begin{bmatrix} -2 & 4 & 2 & 1 & 0 \\ 4 & -2 & 0 & 1 & 2 \\ 2 & 0 & 0 & -3 & 6 \\ 1 & 1 & -3 & 9 & -3 \\ 0 & 2 & 6 & -3 & 0 \end{bmatrix}.$$

(a) Find an orthogonal matrix P and a diagonal matrix D such that $A = PDP^T$.

[21] In MATLAB, M_s can be obtained from M with the MATLAB command `sqrt`(M).

(b) Find an orthonormal basis of eigenvectors of A for $\mathcal{R}^5$. Indicate the corresponding eigenvalue for each eigenvector in your basis.

(c) Use your answer to (b) to find a spectral decomposition of A.

(d) Use the spectral decomposition obtained in (c) to form an approximation A_2 of A based on the two eigenvalues of largest magnitudes (absolute values).

(e) Compute the Frobenius norms of the error matrix $E_2 = A - A_2$ and the matrix A.

(f) Give the percentage of the information lost by using A_2 to approximate A.

12. Let

$$A = \begin{bmatrix} 1 & 1 & 2 & 1 & 3 & 2 \\ 1 & -1 & 4 & 1 & 1 & 2 \\ 1 & 0 & 3 & 1 & 2 & 2 \\ 0 & 1 & -1 & 0 & 1 & 0 \end{bmatrix}.$$

(a) Use the MATLAB command $[U,\ S,\ V] = \texttt{svd}(A)$ to obtain a singular value decomposition USV^T of A.

(b) Compare the columns of V obtained in (a) with the columns of Null A.

(c) Use the MATLAB function `orth` to obtain a matrix whose columns form an orthonormal basis for Col A. Compare the result with the columns of U.

(d) Make a conjecture relating the columns of U and V in a singular value decomposition of a matrix A to orthonormal bases for Null A and Col A.

(e) Prove your conjecture.

13. Use Exercise 90 of Section 6.7 to compute P_W, where W is the subspace of $\mathcal{R}^6$ in Exercise 7. Compare your result with your answers in Exercise 7.

14. Find the unique solution of least norm to the system of linear equations

$$\begin{aligned} x_1 - 2x_2 + 2x_3 + x_4 + 2x_5 &= 3 \\ x_1 + 3x_2 - x_3 - x_4 - x_5 &= -1 \\ 2x_1 + x_2 + x_3 - 2x_4 + 3x_5 &= 0. \end{aligned}$$

15. This problem applies singular value decompositions to extend spectral approximations of symmetric matrices to arbitrary matrices. The problem uses the results and notation of Exercise 91 of Section 6.7 and assumes that the reader is familiar with the subsection Spectral Approximations in Section 6.6.

Let

$$A = \begin{bmatrix} 56 & -33 & 25 & 78 & 9 \\ 28 & -76 & 134 & 32 & -44 \\ 17 & 83 & -55 & 65 & 25 \\ 36 & -57 & 39 & 18 & -1 \end{bmatrix}.$$

(a) Compute $\sigma_i Q_i$ for each i, and form the sum

$$\sigma_1 Q_1 + \sigma_2 Q_2 + \cdots + \sigma_k Q_k,$$

where k is the rank of A. Then compare this sum with A.

(b) Compute $A_2 = \sigma_1 Q_1 + \sigma_2 Q_2$ and $E_2 = A - A_2$, the corresponding error matrix.

(c) Compute the Frobenius norms of E_2 and A.

(d) The ratio of the norms in (c) is the portion of information lost by using A_2 as an approximation of A. Express this ratio as a percentage.

16. For each of the following rotations, find the axis of rotation and the angle of rotation to the nearest degree using the orientation so that the angle is positive:

(a) $P_{32^\circ} R_{21^\circ}$

(b) $R_{21^\circ} P_{32^\circ}$

17. Let W be the solution set of the equation

$$x + 2y - z = 0,$$

let T_W be the reflection of $\mathcal{R}^3$ about W as defined in the exercises for Section 6.9, and let A_W be the standard matrix of T_W.

(a) Compute A_W.

(b) Prove that $A_W Q_{23^\circ} A_W$ is a rotation matrix.

(c) For the rotation matrix in (b), find the axis of rotation and the angle of rotation to the nearest degree using the orientation that makes the angle positive.

(d) State and prove a conjecture based on your observation in (c).

Exercise 18 uses the graphics features of MATLAB. An imported function `grfig` is used to create simple 3-dimensional figures that can be rotated using the rotation matrices described in Section 6.9. This function uses an M-file in the folder of M-files that can be downloaded.

18. Consider a 3-dimensional figure defined by specifying vertices $\mathbf{v}_1, \mathbf{v}_2, \ldots, \mathbf{v}_n$ in $\mathcal{R}^3$, and line segments, called *edges*, connecting vertices. Let

$$V = \begin{bmatrix} \mathbf{v}_1, \mathbf{v}_2, \cdots, \mathbf{v}_n \end{bmatrix} = \begin{bmatrix} x_1 & x_2 & \cdots & x_n \\ y_1 & y_2 & \cdots & y_n \\ z_1 & z_2 & \cdots & z_n \end{bmatrix},$$

and let E be the $k \times 2$ matrix such that $\begin{bmatrix} i & j \end{bmatrix}$ is a row of E if and only if vertex $\mathbf{v}_i$ is connected to vertex $\mathbf{v}_j$. Then $\texttt{grfig}(V,E)$ produces a plot of the resulting figure as viewed from the point $(1,0,0)$ on the positive side of the x-axis. Repeated use of this command adds figures to the same window. To eliminate the earlier figures, simply close the figure window before creating the next figure.

Let M be a rotation matrix. The columns of the product MV are the rotations of the columns of V by M. Thus, if we set $C = MV$ and apply the command $\texttt{grfig}(C,E)$, the result is the rotation of the original figure according to the rotation specified by M. You should be aware that if M and N are rotation matrices, and the figure is rotated first by M and then by N, the product $C = NMV$ contains the final locations of the rotated vertices.

(a) Load `c6sMe18a.dat` and `c6sMe18b.dat`, and let $V =$ `c6sMe18a` and $E =$ `c6sMe18b`. When you type the command $\texttt{grfig}(C,E)$, a figure window should open containing Figure 6.29. Obtain the other graphics in Figure 6.30 in a similar manner. For example, to obtain Figure 6.30(b), use the following commands:

$$C = \texttt{Qdeg}(20) * \texttt{Rdeg}(-30) * V$$

$$\texttt{grfig}(C, E).$$

(To avoid superimposing this on the previous Figure 6.29, close the figure window before using the function `grfig`.)

To obtain, for example, Figure 6.30(b), type the command $C = \texttt{Qdeg}(20) * \texttt{Rdeg}(-30) * V$, and then $\texttt{grfig}(C, E)$. To avoid superimposing this on the previous figure, close the figure window before applying the function `grfig`.

Obtain the other graphics in Figure 6.30 in a similar manner.

(b) Use `grfig` to obtain the reflection of the tower in Figure 6.29 about the plane W in Exercise 17.

(c) Design your own figures, and rotate them in various ways by using the appropriate rotation matrices.

7 INTRODUCTION

Audio amplifiers lie at the heart of many modern electronic devices such as radio receivers, CD players, and television sets. The fundamental task of an audio amplifier is to take a weak signal and make it louder without making any other changes to it. This simple task is complicated by the fact that audio signals typically contain multiple frequencies. Even a simple note has a basic frequency plus harmonics at frequencies that are integral multiples of the basic frequency. All of these frequency components must be amplified by the same factor to avoid changing the waveform and hence the quality of the sound. For example, the part of the music at 440 Hz must be amplified by the same factor as the part at 880 Hz. An amplifier that multiplies the amplitudes of all frequencies by the same factor is called a *linear amplifier*. It produces a louder version of a piece of music with no distortion.

The sound waves that produce music can be analyzed in terms of an infinite set of orthogonal sine and cosine functions. The study of such functions is a branch of mathematics called *Fourier analysis*. It provides a way to describe how components at different frequencies come together to form a musical tone (see Section 7.5). It also offers a simple way to define linear amplification and to test the linearity of an amplifier.

Linear amplification of a Fourier series means that each term of the infinite series is multiplied by the same constant scale factor. The analysis of linearity uses a particular periodic signal called a *square wave*, which is illustrated by the first graph in the figure, and a display device called an oscilloscope. Electrically, the square wave alternates between fixed voltages $+V$ and $-V$ for time intervals of fixed length. The low frequency components of the Fourier series give the overall structure of the signal, and the high frequencies give the detail such as the corners of the square wave.

If the amplifier is linear, then the output for a square wave input is just another, taller square wave. If it is not linear, distortions occur. The second graph shows the output if the amplifier multiplies high frequencies more than low frequencies, so that there is "too much" high frequency detail. The third graph shows the output if the amplifier multiplies high frequencies less than low frequencies. Here, the overall structure of the square wave is visible, but with little detail.

CHAPTER

7 VECTOR SPACES

Up to this point in our development of linear algebra, we have accumulated a rich body of facts about vectors in $\mathcal{R}^n$ and linear transformations acting on these vectors. In this chapter, we consider other mathematical systems that share many of the formal properties of $\mathcal{R}^n$. For example, consider differentiable functions, which are encountered in the study of calculus. These functions can be added and multiplied by scalars to yield differentiable functions. The operations of differentiation and integration transform these functions in such a way that addition and multiplication by scalars are preserved, much as linear transformations preserve the corresponding operations on vectors. Because of these similarities, we can reformulate such notions as *linear combination, linear independence*, and *linear transformation* in the context of differentiable functions.

Transplanting these concepts to the context of *function* gives us a way to analyze functions with the tools that we have developed earlier. As a dramatic example of this, we see how to use the concept of *orthogonal projection* and the *closest vector property* to devise a method of approximating a given function by polynomials or by sines and cosines.

7.1 VECTOR SPACES AND THEIR SUBSPACES

The operations of addition and multiplication by scalars that are central to the study of vectors in $\mathcal{R}^n$ have their analogs in other mathematical systems. For example, real-valued functions defined on $\mathcal{R}$ can be added and multiplied by scalars to produce real-valued functions defined on $\mathcal{R}$. So again, we have the opportunity to start at the beginning and develop a theory of functions imitating, as much as possible, the definitions and theorems developed in the context of vectors in $\mathcal{R}^n$.

In fact, there are many mathematical systems in which addition and multiplication by scalars are defined. It is impractical to develop the formal properties of these operations for each such system as we have done for $\mathcal{R}^n$. For this reason, a general theory of *vector spaces* has been developed that applies to each of these systems. In this theory, a vector space is defined to be any mathematical system that satisfies certain prescribed axioms, and general theorems about vector spaces are then deduced from these axioms. Once it is shown that a particular mathematical system satisfies these axioms, it follows immediately that all of the theorems about vector spaces apply to that system.

We begin with the formal definition of vector space. The reader should compare the following axioms with Theorem 1.1:

Definition A (real) **vector space** is a set V on which two operations, called **vector addition** and **scalar multiplication**, are defined so that for any elements $\mathbf{u}$, $\mathbf{v}$, and $\mathbf{w}$

in V and any scalars a and b, the sum $\mathbf{u}+\mathbf{v}$ and the scalar multiple $a\mathbf{u}$ are unique elements of V, and such that the following axioms hold:

Axioms of a Vector Space

1. $\mathbf{u}+\mathbf{v}=\mathbf{v}+\mathbf{u}$. (commutative law of vector addition)
2. $(\mathbf{u}+\mathbf{v})+\mathbf{w}=\mathbf{u}+(\mathbf{v}+\mathbf{w})$. (associative law of vector addition)
3. There is an element $\mathbf{0}$ in V such that $\mathbf{u}+\mathbf{0}=\mathbf{u}$.
4. There is an element $-\mathbf{u}$ in V such that $\mathbf{u}+(-\mathbf{u})=\mathbf{0}$.
5. $1\mathbf{u}=\mathbf{u}$.
6. $(ab)\mathbf{u}=a(b\mathbf{u})$.
7. $a(\mathbf{u}+\mathbf{v})=a\mathbf{u}+a\mathbf{v}$.
8. $(a+b)\mathbf{u}=a\mathbf{u}+b\mathbf{u}$.

The elements of a vector space are called **vectors**. The vector $\mathbf{0}$ in axioms 3 and 4 is called the **zero vector**. We show in Theorem 7.2(c) that it is unique; that is, there cannot be distinct vectors in a vector space that both satisfy axiom 3. For any vector $\mathbf{u}$ in a vector space V, the vector $-\mathbf{u}$ in axiom 4 is called the **additive inverse** of $\mathbf{u}$. We show in Theorem 7.2(d) that the additive inverse of a vector in a vector space is unique. In view of the commutative law of vector addition (axiom 1), the zero vector must also satisfy $\mathbf{0}+\mathbf{u}=\mathbf{u}$ for every $\mathbf{u}$ in V. Likewise, the additive inverse of every vector $\mathbf{u}$ in V also satisfies $(-\mathbf{u})+\mathbf{u}=\mathbf{0}$.

By applying Theorem 1.1 to the set of $n\times 1$ matrices, we see that $\mathcal{R}^n$ is a vector space with the operations of addition and scalar multiplication defined in Chapter 1. Moreover, it can be shown that any subspace of $\mathcal{R}^n$ is also a vector space with the same operations. (See Exercise 96.) We are already familiar with these examples.

FUNCTION SPACES

Among the most important vector spaces are those consisting of functions. Such vector spaces are called **function spaces**. The area of modern mathematics called *functional analysis* is devoted to the study of function spaces.

For a given nonempty set S, let $\mathcal{F}(S)$ denote the set of all functions from S to $\mathcal{R}$. Recall that two functions f and g in $\mathcal{F}(S)$ are **equal** if $f(t)=g(t)$ for all t in S. (See Appendix B, page 554.) The **sum** $f+g$ of functions f and g in $\mathcal{F}(S)$, and the **scalar multiple** af of a function f in $\mathcal{F}(S)$ and a scalar a, are the functions in $\mathcal{F}(S)$ defined by

$$(f+g)(t)=f(t)+g(t) \qquad \text{and} \qquad (af)(t)=a(f(t))$$

for all t in S.

For example, suppose that S is the set $\mathcal{R}$ of real numbers and that f and g are in $\mathcal{F}(\mathcal{R})$ defined by $f(t)=t^2-t$ and $g(t)=2t+1$ for all t in $\mathcal{R}$. Then

$$(f+g)(t)=f(t)+g(t)=(t^2-t)+(2t+1)=t^2+t+1$$

for all t in $\mathcal{R}$. Also, the scalar multiple $3f$ is defined by

$$(3f)(t)=3f(t)=3(t^2-t)$$

for all t in $\mathcal{R}$.

Next we define the **zero function** $\mathbf{0}$ in $\mathcal{F}(S)$ by $\mathbf{0}(t)=0$ for all t in S. This function serves as the zero vector for axiom 3 in the definition of a vector space.

Finally, for any f in $\mathcal{F}(S)$, the function $-f$ in $\mathcal{F}(S)$ is defined by $(-f)(t) = -f(t)$ for all t in S. For example, if $S = \mathcal{R}$ and $f(t) = t - 1$, then

$$(-f)(t) = -f(t) = -(t-1) = 1 - t$$

for all t in $\mathcal{R}$. For any function f, the function $-f$ serves as the additive inverse of f for axiom 4 in the definition of a vector space.

In the context of $\mathcal{F}(S)$, each axiom is an equation involving functions.

THEOREM 7.1

With the operations previously defined, $\mathcal{F}(S)$ is a vector space.

PROOF To prove that $\mathcal{F}(S)$ is a vector space, we must verify the eight axioms of a vector space. We verify axioms 1, 3, and 7, leaving the verification of the other axioms as exercises.

Axiom 1 Let f and g be functions in $\mathcal{F}(S)$. Then, for any t in S,

$$\begin{aligned} (f+g)(t) &= f(t) + g(t) && \text{(definition of sum of functions)} \\ &= g(t) + f(t) && \text{(commutative law of addition for real numbers)} \\ &= (g+f)(t). && \text{(definition of sum of functions)} \end{aligned}$$

Therefore $f + g = g + f$, and axiom 1 is verified.

Axiom 3 Let f be any function in $\mathcal{F}(S)$. Then, for any t in S,

$$\begin{aligned} (f+\mathbf{0})(t) &= f(t) + \mathbf{0}(t) && \text{(definition of sum of functions)} \\ &= f(t) + 0 && \text{(definition of } \mathbf{0}\text{)} \\ &= f(t). \end{aligned}$$

Therefore $f + \mathbf{0} = f$, and hence axiom 3 is verified.

Axiom 7 Let f and g be functions in $\mathcal{F}(S)$, and let a be a scalar. Then, for any t in S,

$$\begin{aligned} [a(f+g)](t) &= a[(f+g)(t)] && \text{(definition of scalar multiplication)} \\ &= a[f(t) + g(t)] && \text{(definition of sum of functions)} \\ &= a[f(t)] + a[g(t)] && \text{(distributive law for real numbers)} \\ &= (af)(t) + (ag)(t) && \text{(definition of scalar multiplication)} \\ &= (af + ag)(t). && \text{(definition of sum of functions)} \end{aligned}$$

Therefore $a(f+g) = af + ag$, which verifies axiom 7. ■

OTHER EXAMPLES OF VECTOR SPACES

In what follows, we briefly consider three examples of vector spaces. In the first, the vectors are matrices; in the second, the vectors are linear transformations; and in the third, the vectors are polynomials. The notation and terminology introduced in these three examples are used in other examples throughout this chapter.

Example 1

For any given positive integers m and n, let $\mathcal{M}_{m\times n}$ denote the set of all $m \times n$ matrices. Then as a direct consequence of Theorem 1.1, $\mathcal{M}_{m\times n}$ is a vector space with the operations of matrix addition and multiplication of a matrix by a scalar. In this case, the $m \times n$ zero matrix plays the role of the zero vector.

Example 2

For any given positive integers m and n, let $\mathcal{L}(\mathcal{R}^n, \mathcal{R}^m)$ denote the set of all linear transformations from $\mathcal{R}^n$ to $\mathcal{R}^m$. Let T and U be in $\mathcal{L}(\mathcal{R}^n, \mathcal{R}^m)$ and c be a scalar. Define $(T+U)\colon \mathcal{R}^n \to \mathcal{R}^m$ and $cT\colon \mathcal{R}^n \to \mathcal{R}^m$ by

$$(T+U)(\mathbf{x}) = T(\mathbf{x}) + U(\mathbf{x}) \quad \text{and} \quad (cT)(\mathbf{x}) = cT(\mathbf{x})$$

for all $\mathbf{x}$ in $\mathcal{R}^n$. By Exercises 83 and 84 of Section 2.7, $T+U$ and cT are in $\mathcal{L}(\mathcal{R}^n, \mathcal{R}^m)$. It can be shown that $\mathcal{L}(\mathcal{R}^n, \mathcal{R}^m)$ is a vector space under these operations. The zero transformation T_0 plays the role of the zero vector, and for any transformation T, we have that $(-1)T$ is the additive inverse $-T$. The proof that $\mathcal{L}(\mathcal{R}^n, \mathcal{R}^m)$ is a vector space is very similar to the proof that $\mathcal{F}(S)$ is a vector space, because both of these are function spaces. For this reason, the proof is left as an exercise. (See Exercise 74.)

Example 3

Let $\mathcal{P}$ denote the set of all polynomials

$$p(x) = a_0 + a_1x + \cdots + a_nx^n,$$

where n is a nonnegative integer and $a_0, a_1, \ldots, a_n$ are real numbers. For each i, the scalar a_i is called the **coefficient** of x^i. We usually write x^i in place of $1x^i$, and $-a_ix^i$ in place of $(-a_i)x^i$. Furthermore, if $a_i = 0$, we often omit the term a_ix^i entirely. The unique polynomial $p(x)$ with only zero coefficients is called the **zero polynomial**. The **degree** of a nonzero polynomial $p(x)$ is defined to be the largest exponent of x that appears in the representation

$$p(x) = a_0 + a_1x + \cdots + a_nx^n$$

with a nonzero coefficient. The zero polynomial is not assigned a degree, and a constant polynomial has degree 0. Two nonzero polynomials $p(x)$ and $q(x)$ are called **equal** if they have the same degree and all their corresponding coefficients are equal. That is, if

$$p(x) = a_0 + a_1x + \cdots + a_nx^n \quad \text{and} \quad q(x) = b_0 + b_1x + \cdots + b_mx^m,$$

then $p(x)$ and $q(x)$ are equal if and only if $m = n$ and $a_i = b_i$ for $i = 0, 1, \ldots, n$. Notice that if two distinct polynomials $p(x)$ and $q(x)$ have different degrees, say, n and m, respectively, with $m < n$, then we may still represent $q(x)$ in the form

$$q(x) = b_0 + b_1x + \cdots + b_nx^n$$

by simply requiring that $b_i = 0$ for all $i > m$. With this in mind, for any polynomials

$$p(x) = a_0 + a_1x + \cdots + a_nx^n \quad \text{and} \quad q(x) = b_0 + b_1x + \cdots + b_nx^n$$

(not necessarily of the same degree), and any scalar a, we define the sum $p(x) + q(x)$ and the scalar multiple $ap(x)$ by

$$p(x) + q(x) = (a_0 + b_0) + (a_1 + b_1)x + \cdots + (a_n + b_n)x^n$$

and

$$ap(x) = (a \cdot a_0) + (a \cdot a_1)x + \cdots + (a \cdot a_n)x^n.$$

For example, $(1 - x + 2x^2) + (3 + 2x) = 4 + x + 2x^2$ and $4(3 + 2x) = 12 + 8x$. We can also define the additive inverse of a polynomial $p(x)$ by $-p(x) = (-1)p(x)$. With these definitions, it can be shown that $\mathcal{P}$ is a vector space with respect to the operations just defined. The zero polynomial serves the role of the zero vector. We leave the details to the exercises.

Practice Problem 1 ▶ In Example 3, show that $\mathcal{P}$ satisfies axiom 8. ◀

The following example illustrates a set with two operations that is not a vector space because it fails to satisfy at least one of the axioms:

Example 4 Let S be the set $\mathcal{R}^2$ with addition and scalar multiplication defined as

$$(a, b) \oplus (c, d) = (a + c, 0) \qquad \text{and} \qquad k \odot (a, b) = (ka, kb)$$

for all (a, b) and (c, d) in $\mathcal{R}^2$. Show that S is *not* a vector space.

Solution We show that S has no zero vector. Suppose that (z, w) is a zero vector for S. Then

$$(1, 1) \oplus (z, w) = (1 + z, 0) \neq (1, 1).$$

So (z, w) does not satisfy axiom 3, and hence S has no zero vector.

PROPERTIES OF VECTOR SPACES

The results that follow are deduced entirely from the axioms of a vector space. They are therefore valid for all examples of vector spaces.

THEOREM 7.2

Let V be a vector space. For any **u**, **v**, and **w** in V and any scalar a, the following statements are true:

(a) If $\mathbf{u} + \mathbf{v} = \mathbf{w} + \mathbf{v}$, then $\mathbf{u} = \mathbf{w}$. (right cancellation law)

(b) If $\mathbf{u} + \mathbf{v} = \mathbf{u} + \mathbf{w}$, then $\mathbf{v} = \mathbf{w}$. (left cancellation law)

(c) The zero vector $\mathbf{0}$ is unique; that is, it is the only vector in V that satisfies axiom 3.

(d) Each vector in V has exactly one additive inverse.

(e) $0\mathbf{v} = \mathbf{0}$.

(f) $a\mathbf{0} = \mathbf{0}$.

(g) $(-1)\mathbf{v} = -\mathbf{v}$.

(h) $(-a)\mathbf{v} = a(-\mathbf{v}) = -(a\mathbf{v})$.

PROOF We prove parts (a), (c), (e), and (g), leaving the proofs of parts (b), (d), (f), and (h) as exercises.

(a) Suppose that $\mathbf{u}+\mathbf{v}=\mathbf{w}+\mathbf{v}$. Then

$$\begin{aligned}
\mathbf{u} &= \mathbf{u}+\mathbf{0} && \text{(by axiom 3)}\\
&= \mathbf{u}+[\mathbf{v}+(-\mathbf{v})] && \text{(by axiom 4)}\\
&= (\mathbf{u}+\mathbf{v})+(-\mathbf{v}) && \text{(by axiom 2)}\\
&= (\mathbf{w}+\mathbf{v})+(-\mathbf{v}) &&\\
&= \mathbf{w}+[\mathbf{v}+(-\mathbf{v})] && \text{(by axiom 2)}\\
&= \mathbf{w}+\mathbf{0} && \text{(by axiom 4)}\\
&= \mathbf{w}. && \text{(by axiom 3)}
\end{aligned}$$

(c) Suppose that $\mathbf{0}'$ is also a zero vector for V. Then, by axioms 3 and 1,

$$\mathbf{0}'+\mathbf{0}'=\mathbf{0}'=\mathbf{0}'+\mathbf{0}=\mathbf{0}+\mathbf{0}',$$

and hence $\mathbf{0}'=\mathbf{0}$ by the right cancellation law. It follows that the zero vector is unique.

(e) For any vector $\mathbf{v}$,

$$\begin{aligned}
0\mathbf{v}+0\mathbf{v} &= (0+0)\mathbf{v} && \text{(by axiom 8)}\\
&= 0\mathbf{v} &&\\
&= \mathbf{0}+0\mathbf{v}. && \text{(by axioms 3 and 1)}
\end{aligned}$$

So $0\mathbf{v}+0\mathbf{v}=\mathbf{0}+0\mathbf{v}$, and hence $0\mathbf{v}=\mathbf{0}$ by the right cancellation law.

(g) For any vector $\mathbf{v}$,

$$\begin{aligned}
\mathbf{v}+(-1)\mathbf{v} &= (1)\mathbf{v}+(-1)\mathbf{v} && \text{(by axiom 5)}\\
&= [1+(-1)]\mathbf{v} && \text{(by axiom 8)}\\
&= 0\mathbf{v} &&\\
&= \mathbf{0}. && \text{(by (e))}
\end{aligned}$$

Therefore $(-1)\mathbf{v}$ is an additive inverse of $\mathbf{v}$. But by (d), additive inverses are unique, and hence $(-1)\mathbf{v}=-\mathbf{v}$. ■

SUBSPACES

As is the case for $\mathcal{R}^n$, vector spaces have *subspaces*.

Definition A subset W of a vector space V is called a **subspace** of V if W satisfies the following three properties:

1. The zero vector of V is in W.
2. Whenever $\mathbf{u}$ and $\mathbf{v}$ belong to W, then $\mathbf{u}+\mathbf{v}$ belongs to W. (In this case, we say that W is **closed under (vector) addition**.)
3. Whenever $\mathbf{u}$ belongs to W and c is a scalar, then $c\mathbf{u}$ belongs to W. (In this case, we say that W is **closed under scalar multiplication**.)

It is a simple matter to verify that if V is a vector space, then V is a subspace of itself. In fact, V is the largest subspace of V. Moreover, the set $\{\mathbf{0}\}$ is also a subspace of V. This subspace is called the **zero subspace** and is the smallest subspace of V. A subspace of a vector space other than $\{\mathbf{0}\}$ is called a **nonzero subspace**.

Example 5 Let S be a nonempty set, and let W be the subset of $\mathcal{F}(S)$ consisting of all functions f such that $f(s_0) = 0$ for some particular element s_0 in S. Clearly, the zero function lies in W. For any functions f and g in S, and any scalar a,

$$(f+g)(s_0) = f(s_0) + g(s_0) = 0 + 0 = 0,$$

and

$$(af)(s_0) = af(s_0) = a \cdot 0 = 0.$$

Hence $f + g$ and af are in W. We conclude that W is closed under the operations of $\mathcal{F}(S)$. Therefore W is a subspace of V.

Suppose that A is a square matrix. We define the **trace** of A, denoted trace(A), to be the sum of the diagonal entries of A. For any square matrices A and B of the same size, and any scalar c, we have

$$\begin{aligned} \text{trace}(A+B) &= \text{trace}(A) + \text{trace}(B) \\ \text{trace}(cA) &= c \cdot \text{trace}(A) \\ \text{trace}(A^T) &= \text{trace}(A). \end{aligned}$$

(See Exercise 82 of Section 1.1.)

Example 6 Let W be the set of all $n \times n$ matrices with trace equal to zero. Show that W is a subspace of $\mathcal{M}_{n\times n}$.

Solution Since the $n \times n$ zero matrix has trace equal to zero, it belongs to W. Suppose that A and B are matrices in W. Then

$$\text{trace}(A+B) = \text{trace}(A) + \text{trace}(B) = 0 + 0 = 0,$$

and, for any scalar c,

$$\text{trace}(cA) = c \cdot \text{trace}(A) = c \cdot 0 = 0.$$

Therefore $A + B$ and cA are in W. We conclude that W is closed under the operations of $\mathcal{M}_{n\times n}$. Therefore W is a subspace of $\mathcal{M}_{n\times n}$.

Practice Problem 2 ▶ Let W be the set of all 2×2 matrices of the form $\begin{bmatrix} a & a+b \\ b & 0 \end{bmatrix}$. Prove that W is a subspace of $\mathcal{M}_{2\times 2}$. ◀

If W is a subspace of a vector space, then W satisfies all of the axioms in the definition of *vector space* with the same operations as defined on V, and hence W is itself a vector space. (See Exercise 96.) This fact provides a simpler way to prove that certain sets are vector spaces, namely, by verifying that they are actually subspaces of a known vector space. The next two examples demonstrate this technique.

Example 7 Let $\mathsf{C}(\mathcal{R})$ denote the set of all continuous real-valued functions on $\mathcal{R}$. Then $\mathsf{C}(\mathcal{R})$ is a subset of $\mathcal{F}(\mathcal{R})$, the vector space of all real-valued functions defined on $\mathcal{R}$. Since the zero function is a continuous function, the sum of continuous functions is a continuous function, and any scalar multiple of a continuous function is a continuous function, it follows that $\mathsf{C}(\mathcal{R})$ is a subspace of $\mathcal{F}(\mathcal{R})$. In particular, $\mathsf{C}(\mathcal{R})$ is a vector space.

Example 8 Recall the vector space $\mathcal{P}$ of all polynomials considered in Example 3. Let n be a nonnegative integer, and let $\mathcal{P}_n$ denote the subset of $\mathcal{P}$ consisting of the zero polynomial and all polynomials of degree less than or equal to n. Since the sum of two polynomials of degree less than or equal to n is the zero polynomial or has degree less than or equal to n, and a scalar multiple of a polynomial of degree less than or equal to n is either the zero polynomial or a polynomial of degree less than or equal to n, it is clear that $\mathcal{P}_n$ is closed under both addition and scalar multiplication. Therefore $\mathcal{P}_n$ is a subspace of $\mathcal{P}$, and hence is a vector space.

LINEAR COMBINATIONS AND GENERATING SETS

As in Chapter 1, we can combine vectors in a vector space by taking *linear combinations* of other vectors. However, in contrast to subspaces of $\mathcal{R}^n$, there are important vector spaces that have no finite generating sets. Hence it is necessary to extend the definition of a linear combination to permit vectors from an infinite set.

Definition A vector $\mathbf{v}$ is a **linear combination** of the vectors in a (possibly infinite) subset S of a vector space V if there exist vectors $\mathbf{v}_1, \mathbf{v}_2, \ldots, \mathbf{v}_m$ in S and scalars $c_1, c_2, \ldots, c_m$ such that

$$\mathbf{v} = c_1\mathbf{v}_1 + c_2\mathbf{v}_2 + \cdots + c_m\mathbf{v}_m.$$

The scalars are called the **coefficients** of the linear combination.

We consider examples of linear combinations of vectors of both finite and infinite sets.

Example 9 In the vector space of 2×2 matrices,

$$\begin{bmatrix} -1 & 8 \\ 2 & -2 \end{bmatrix} = 2\begin{bmatrix} 1 & 3 \\ 1 & -1 \end{bmatrix} + (-1)\begin{bmatrix} 4 & 0 \\ 1 & 1 \end{bmatrix} + 1\begin{bmatrix} 1 & 2 \\ 1 & 1 \end{bmatrix}.$$

Hence $\begin{bmatrix} -1 & 8 \\ 2 & -2 \end{bmatrix}$ is a linear combination of the matrices

$$\begin{bmatrix} 1 & 3 \\ 1 & -1 \end{bmatrix}, \quad \begin{bmatrix} 4 & 0 \\ 1 & 1 \end{bmatrix}, \quad \text{and} \quad \begin{bmatrix} 1 & 2 \\ 1 & 1 \end{bmatrix},$$

with coefficients 2, -1, and 1.

Example 10 Let $S = \{1, x, x^2, x^3\}$, which is a subset of the vector space $\mathcal{P}$ of all polynomials. Then the polynomial $f(x) = 2 + 3x - x^2$ is a linear combination of the vectors in S because there are scalars, namely, 2, 3, and -1, such that

$$f(x) = (2)1 + (3)x + (-1)x^2.$$

In fact, the zero polynomial and any polynomial of degree less than or equal to 3 is a linear combination of the vectors in S. That is, the set of all linear combinations of the vectors in S is equal to $\mathcal{P}_3$, the subspace of $\mathcal{P}$ in Example 8.

Example 11 Let S be the set of real-valued functions given by

$$S = \{1, \sin t, \cos^2 t, \sin^2 t\},$$

which is a subset of $\mathcal{F}(R)$. Observe that the function $\cos 2t$ is a linear combination of the vectors in S because

$$\begin{aligned} \cos 2t &= \cos^2 t - \sin^2 t \\ &= (1)\cos^2 t + (-1)\sin^2 t. \end{aligned}$$

Example 12 Let

$$S = \{1, x, x^2, \ldots, x^n, \ldots\},$$

which is an infinite subset of $\mathcal{P}$. Then the polynomial $p(x) = 3 - 4x^2 + 5x^4$ is a linear combination of the vectors in S because it is a linear combination of a finite number of vectors in S, namely, 1, x^2, and x^4. In fact, any polynomial

$$p(x) = a_0 + a_1 x + \cdots + a_n x^n$$

is a linear combination of the vectors in S because it is a linear combination of $1, x, x^2, \ldots, x^n$.

Example 13 Determine if the polynomial x is a linear combination of the polynomials $1 - x^2$ and $1 + x + x^2$.

Solution Suppose that

$$\begin{aligned} x &= a(1 - x^2) + b(1 + x + x^2) \\ &= (a + b) + bx + (-a + b)x^2 \end{aligned}$$

for scalars a, b, and c. Then

$$\begin{aligned} a + b &= 0 \\ b &= 1 \\ -a + b &= 0. \end{aligned}$$

Since this system is inconsistent, we conclude that x is not a linear combination of the polynomials $1 - x^2$ and $1 + x + x^2$.

Now that we have extended the definition of *linear combination* to include infinite sets, we are ready to reintroduce the definition of *span*.

Definition The **span** of a nonempty subset S of a vector space V is the set of all linear combinations of vectors in S. This set is denoted by $\text{Span}\, S$.

By the remarks made in Example 10,

$$\text{Span}\,\{1, x, x^2, x^3\} = \mathcal{P}_3,$$

and by the remarks in Example 12,

$$\text{Span}\,\{1, x, \ldots, x^n, \ldots\} = \mathcal{P}.$$

Example 14 Describe the span of the subset

$$S = \left\{ \begin{bmatrix} 1 & 0 \\ 0 & -1 \end{bmatrix}, \begin{bmatrix} 0 & 1 \\ 0 & 0 \end{bmatrix}, \begin{bmatrix} 0 & 0 \\ 1 & 0 \end{bmatrix} \right\}.$$

Solution For any matrix A in $\text{Span}\, S$, there exist scalars a, b, and c, such that

$$A = a \begin{bmatrix} 1 & 0 \\ 0 & -1 \end{bmatrix} + b \begin{bmatrix} 0 & 1 \\ 0 & 0 \end{bmatrix} + c \begin{bmatrix} 0 & 0 \\ 1 & 0 \end{bmatrix} = \begin{bmatrix} a & b \\ c & -a \end{bmatrix}.$$

Therefore $\text{trace}(A) = a + (-a) = 0$. Conversely, suppose that $A = \begin{bmatrix} a & b \\ c & d \end{bmatrix}$ is a matrix in $\mathcal{M}_{2\times 2}$ such that $\text{trace}(A) = 0$. Then $a + d = 0$, and hence $d = -a$. It follows that $A = \begin{bmatrix} a & b \\ c & -a \end{bmatrix}$, which is in $\text{Span}\, S$ by the preceding calculation. Therefore $\text{Span}\, S$ is the subset of all 2×2 matrices with trace equal to zero. Since this set was proved to be a subspace of $\mathcal{M}_{2\times 2}$ in Example 6, $\text{Span}\, S$ is a subspace of $\mathcal{M}_{2\times 2}$.

In the previous examples, we have anticipated the next result, which is an extension of Theorem 4.1 (page 231) to vectors spaces. We omit the proof, which is similar to the proof of Theorem 4.1.

THEOREM 7.3

The span of a nonempty subset of a vector space V is a subspace of V.

This result gives us a convenient way to define vector spaces. For example, we let $\mathcal{T}[0, 2\pi]$ denote the subspace of $\mathcal{F}([0, 2\pi])$ defined by

$$\mathcal{T}[0, 2\pi] = \text{Span}\,\{1, \cos t, \sin t, \cos 2t, \sin 2t, \ldots, \cos nt, \sin nt, \ldots\}.$$

This vector space is called the **space of trigonometric polynomials** and is studied in Section 7.5.

Practice Problem 3 ▶ Determine whether the matrix $\begin{bmatrix} 1 & 2 \\ -1 & -3 \end{bmatrix}$ is in the span of the set

$$S = \left\{ \begin{bmatrix} 1 & -1 \\ 1 & 2 \end{bmatrix}, \begin{bmatrix} 0 & 1 \\ 1 & 2 \end{bmatrix}, \begin{bmatrix} 2 & 1 \\ 0 & -1 \end{bmatrix} \right\}.$$

◀

EXERCISES

In Exercises 1–9, determine whether each matrix is in the span of the set

$$\left\{ \begin{bmatrix} 1 & 2 & 1 \\ 0 & 0 & 0 \end{bmatrix}, \begin{bmatrix} 0 & 0 & 0 \\ 1 & 1 & 1 \end{bmatrix}, \begin{bmatrix} 1 & 0 & 1 \\ 1 & 2 & 3 \end{bmatrix} \right\}.$$

1. $\begin{bmatrix} 0 & 2 & 0 \\ 1 & 1 & 1 \end{bmatrix}$
2. $\begin{bmatrix} 1 & 2 & 1 \\ 1 & 1 & 1 \end{bmatrix}$
3. $\begin{bmatrix} 2 & 2 & 2 \\ 2 & 3 & 4 \end{bmatrix}$
4. $\begin{bmatrix} 2 & 2 & 2 \\ 2 & 2 & 2 \end{bmatrix}$
5. $\begin{bmatrix} 2 & 2 & 2 \\ 1 & 1 & 1 \end{bmatrix}$
6. $\begin{bmatrix} 1 & 10 & 1 \\ 3 & -1 & -5 \end{bmatrix}$
7. $\begin{bmatrix} 2 & 5 & 2 \\ 1 & -1 & 3 \end{bmatrix}$
8. $\begin{bmatrix} 1 & 3 & 6 \\ 3 & 5 & 7 \end{bmatrix}$
9. $\begin{bmatrix} -2 & -8 & -2 \\ 5 & 7 & 9 \end{bmatrix}$

In Exercises 10–15, determine whether each polynomial is in the span of the set

$$\{1 - x, 1 + x^2, 1 + x - x^3\}.$$

10. $-3 - x^2 + x^3$
11. $1 + x + x^2 + x^3$
12. $1 + x^2 + x^3$
13. $-2 + x + x^2 + x^3$
14. $2 - x^2 - 2x^3$
15. $1 - 2x - x^2$

In Exercises 16–21, determine whether each matrix is in the span of the set

$$\left\{ \begin{bmatrix} 1 & 0 \\ -1 & 0 \end{bmatrix}, \begin{bmatrix} 0 & 1 \\ 0 & 1 \end{bmatrix}, \begin{bmatrix} 1 & 1 \\ 0 & 0 \end{bmatrix} \right\}.$$

16. $\begin{bmatrix} 1 & 0 \\ 0 & 1 \end{bmatrix}$
17. $\begin{bmatrix} 1 & 2 \\ -3 & 4 \end{bmatrix}$
18. $\begin{bmatrix} 2 & -1 \\ -1 & -2 \end{bmatrix}$
19. $\begin{bmatrix} 2 & 1 \\ 0 & 1 \end{bmatrix}$
20. $\begin{bmatrix} 3 & 1 \\ -1 & 3 \end{bmatrix}$
21. $\begin{bmatrix} 1 & -2 \\ -3 & 0 \end{bmatrix}$

In Exercises 22–27, determine whether each polynomial is in the span of the set

$$\{1 + x, 1 + x + x^2, 1 + x + x^2 + x^3\}.$$

22. $3 + x - x^2 + 2x^3$
23. $3 + 3x + 2x^2 - x^3$
24. $4x^2 - 3x^3$
25. $1 + x$
26. x
27. $1 + 2x + 3x^2$

For Exercises 28–30, use the set

$$S = \{9 + 4x + 5x^2 - 3x^3, -3 - 5x - 2x^2 + x^3\}.$$

28. Prove that the polynomial $-6 + 12x - 2x^2 + 2x^3$ is a linear combination of the polynomials in S. Find the coefficients of the linear combination.
29. Prove that the polynomial $12 - 13x + 5x^2 - 4x^3$ is a linear combination of the polynomials in S. Find the coefficients of the linear combination.
30. Prove that the polynomial $8 + 7x - 2x^2 + 3x^3$ is *not* a linear combination of the polynomials in S.
31. Prove that Span $\{1 + x, 1 - x, 1 + x^2, 1 - x^2\} = \mathcal{P}_2$.
32. Prove that

$$\text{Span} \left\{ \begin{bmatrix} 0 & 1 \\ 1 & 1 \end{bmatrix}, \begin{bmatrix} 1 & 0 \\ 1 & 1 \end{bmatrix}, \begin{bmatrix} 1 & 1 \\ 0 & 1 \end{bmatrix}, \begin{bmatrix} 1 & 1 \\ 1 & 0 \end{bmatrix} \right\} = \mathcal{M}_{2\times 2}.$$

In Exercises 33–54, determine whether the statements are true or false.

33. Every vector space has a zero vector.
34. A vector space may have more than one zero vector.
35. In any vector space, $a\mathbf{v} = \mathbf{0}$ implies that $\mathbf{v} = \mathbf{0}$.
36. $\mathcal{R}^n$ is a vector space for every positive integer n.
37. Only polynomials of the same degree can be added.
38. The set of polynomials of degree n is a subspace of the vector space of all polynomials.
39. Two polynomials of the same degree are equal if and only if they have equal corresponding coefficients.
40. The set of all $m \times n$ matrices with the usual definitions of matrix addition and scalar multiplication is a vector space.
41. The zero vector of $\mathcal{F}(S)$ is the function that assigns 0 to every element of S.
42. Two functions in $\mathcal{F}(S)$ are equal if and only if they assign equal values to each element of S.
43. If V is a vector space and W is a subspace of V, then W is a vector space with the same operations that are defined on V.
44. The empty set is a subspace of every vector space.
45. If V is a nonzero vector space, then V contains a subspace other than itself.
46. If W is a subspace of vector space V, then the zero vector of W must equal the zero vector of V.
47. The set of continuous real-valued functions defined on a closed interval $[a, b]$ is a subspace of $\mathcal{F}([a, b])$, the vector space of real-valued functions defined on $[a, b]$.

48. The zero vector of $\mathcal{L}(\mathcal{R}^n, \mathcal{R}^m)$ is the zero transformation T_0.
49. In any vector space, addition of vectors is commutative; that is, $\mathbf{u} + \mathbf{v} = \mathbf{v} + \mathbf{u}$ for every pair of vectors $\mathbf{u}$ and $\mathbf{v}$.
50. In any vector space, addition of vectors is associative; that is, $(\mathbf{u} + \mathbf{v}) + \mathbf{w} = \mathbf{u} + (\mathbf{v} + \mathbf{w})$ for all vectors $\mathbf{u}$, $\mathbf{v}$, and $\mathbf{w}$.
51. In any vector space, $\mathbf{0} + \mathbf{0} = \mathbf{0}$.
52. In any vector space, if $\mathbf{u} + \mathbf{v} = \mathbf{v} + \mathbf{w}$, then $\mathbf{u} = \mathbf{w}$.
53. The zero vector is a linear combination of any nonempty set of vectors.
54. The span of any nonempty subset of a vector space is a subspace of the vector space.

---

55. Verify axiom 2 for $\mathcal{F}(S)$.
56. Verify axiom 4 for $\mathcal{F}(S)$.
57. Verify axiom 5 for $\mathcal{F}(S)$.
58. Verify axiom 6 for $\mathcal{F}(S)$.
59. Verify axiom 8 for $\mathcal{F}(S)$.

In Exercises 60–65, determine whether or not the set V is a subspace of the vector space $\mathcal{M}_{n\times n}$. Justify your answer.

60. V is the set of all $n \times n$ symmetric matrices.
61. V is the set of all $n \times n$ matrices with determinant equal to 0.
62. V is the set of all $n \times n$ matrices A such that $A^2 = A$.
63. Let B be a specific $n \times n$ matrix. V is the set of all $n \times n$ matrices A such that $AB = BA$.
64. V is the set of all 2×2 matrices of the form $\begin{bmatrix} a & 2a \\ 0 & b \end{bmatrix}$, and $n = 2$.
65. V is the set of all $n \times n$ skew-symmetric matrices. (See the definition given in Exercise 74 of Section 3.2.)

In Exercises 66–69, determine whether or not the set V is a subspace of the vector space $\mathcal{P}$. Justify your answer.

66. V is the subset of $\mathcal{P}$ consisting of the zero polynomial and all polynomials of the form $c_0 + c_1x + \cdots + c_mx^m$ with $c_k = 0$ if k is odd.
67. V is the subset of $\mathcal{P}$ consisting of the zero polynomial and all polynomials of the form $c_0 + c_1x + \cdots + c_mx^m$ with $c_k \neq 0$ if k is even.
68. V is the subset of $\mathcal{P}$ consisting of the zero polynomial and all polynomials of the form $c_0 + c_1x + \cdots + c_mx^m$ with $c_i \geq 0$ for all i.
69. V is the subset of $\mathcal{P}$ consisting of the zero polynomial and all polynomials of the form $c_0 + c_1x + \cdots + c_mx^m$ with $c_0 + c_1 = 0$.

In Exercises 70–72, determine whether or not the set V is a subspace of the vector space $\mathcal{F}(S)$, where S is a particular nonempty set. Justify your answer.

70. Let S' be a nonempty subset of S, and let V be the set of all functions f in $\mathcal{F}(S)$ such that $f(s) = 0$ for all s in S'.
71. Let $\{s_1, s_2, \ldots, s_n\}$ be a subset of S, and let V be the set of all functions f in $\mathcal{F}(S)$ such that
$$f(s_1) + f(s_2) + \cdots + f(s_n) = 0.$$
72. Let s_1 and s_2 be elements of S, and let V be the set of all functions f in $\mathcal{F}(S)$ such that $f(s_1) \bullet f(s_2) = 0$.
73. Show that the set of functions f in $\mathcal{F}(\mathcal{R})$ such that $f(1) = 2$ is not a vector space under the operations defined on $\mathcal{F}(\mathcal{R})$.

In Exercises 74–78, verify that the set V is a vector space with respect to the indicated operations.

74. $V = \mathcal{L}(\mathcal{R}^n, \mathcal{R}^m)$ in Example 2
75. $V = \mathcal{P}$ in Example 3
76. For a given nonempty set S and some positive integer n, let V denote the set of all functions from S to $\mathcal{R}^n$. For any functions f and g and any scalar c, define the sum $f + g$ and the product cf by
$$(f + g)(s) = f(s) + g(s) \quad \text{and} \quad (cf)(s) = cf(s)$$
for all s in S.
77. Let V be the set of all 2×2 matrices of the form $\begin{bmatrix} a & 2a \\ b & -b \end{bmatrix}$, where a and b are any real numbers. Addition and multiplication by scalars are defined in the usual way for matrices.
78. Let V be the set of all functions $f: \mathcal{R} \to \mathcal{R}$ for which $f(t) = 0$ whenever $t < 0$. Addition of functions and multiplication by scalars are defined as in $\mathcal{F}(\mathcal{R})$.
79. Prove Theorem 7.2(b).
80. Prove Theorem 7.2(d).
81. Prove Theorem 7.2(f).
82. Prove Theorem 7.2(h).
83. Use the axioms of a vector space to prove that
$$(a + b)(\mathbf{u} + \mathbf{v}) = a\mathbf{u} + a\mathbf{v} + b\mathbf{u} + b\mathbf{v}$$
for all scalars a and b and all vectors $\mathbf{u}$ and $\mathbf{v}$ in a vector space.
84. Prove that, for any vector $\mathbf{v}$ in a vector space, $-(-\mathbf{v}) = \mathbf{v}$.
85. Prove that, for any vectors $\mathbf{u}$ and $\mathbf{v}$ in a vector space, $-(\mathbf{u} + \mathbf{v}) = (-\mathbf{u}) + (-\mathbf{v})$.
86. Let $\mathbf{u}$ and $\mathbf{v}$ be vectors in a vector space, and suppose that $c\mathbf{u} = c\mathbf{v}$ for some scalar $c \neq 0$. Prove that $\mathbf{u} = \mathbf{v}$.
87. Prove that $(-c)(-\mathbf{v}) = c\mathbf{v}$ for any vector $\mathbf{v}$ in a vector space and any scalar c.
88. For a given nonzero vector $\mathbf{v}$ in $\mathcal{R}^n$, let V be the set of all linear operators T on $\mathcal{R}^n$ such that $T(\mathbf{v}) = \mathbf{0}$. Prove that V is a subspace of $\mathcal{L}(\mathcal{R}^n, \mathcal{R}^n)$.
89. Let W be the set of all differentiable functions from $\mathcal{R}$ to $\mathcal{R}$. Prove that W is a subspace of $\mathcal{F}(R)$.
90. Let S be the subset of the subspace W in Exercise 89 that consists of the functions f such that $f' = f$. Show that S is a subspace of W.
91. A function f in $\mathcal{F}(\mathcal{R})$ is called an **even function** if $f(t) = f(-t)$ for all t in $\mathcal{R}$ and is called an **odd function** if $f(-t) = -f(t)$ for all t in $\mathcal{R}$.
    (a) Show that the subset of all even functions is a subspace of $\mathcal{F}(\mathcal{R})$.
    (b) Show that the subset of all odd functions is a subspace of $\mathcal{F}(\mathcal{R})$.

92. Let V be the set of all continuous real-valued functions defined on the closed interval $[0, 1]$.
(a) Show that V is a subspace of $\mathcal{F}([0, 1])$.
(b) Let W be the subset of V defined by
$$W = \left\{ f \in V : \int_0^1 f(t)\,dt = 0 \right\}.$$
Prove that W is a subspace of V.

93. Suppose that W_1 and W_2 are subspaces of a vector space V. Prove that their intersection $W_1 \cap W_2$ is also a subspace of V.

94. Suppose that W_1 and W_2 are subspaces of a vector space V. Define
$$W = \{\mathbf{w}_1 + \mathbf{w}_2 : \mathbf{w}_1 \text{ is in } W_1 \text{ and } \mathbf{w}_2 \text{ is in } W_2\}.$$
Prove that W is a subspace of V.

95. Let W be a subset of a vector space V. Prove that W is a subspace of V if and only if the following conditions hold:
(i) $\mathbf{0}$ is in W.
(ii) $a\mathbf{w}_1 + \mathbf{w}_2$ is in W whenever $\mathbf{w}_1$ and $\mathbf{w}_2$ are in W and a is a scalar.

96. Suppose that W is a subspace of a vector space V. Prove that W satisfies the axioms in the definition of *vector space*, and hence W is itself a vector space.

97. Suppose that W is a subset of a vector space V such that W is a vector space with the same operations that are defined on V. Prove that W is a subspace of V.

SOLUTIONS TO THE PRACTICE PROBLEMS

1. Let $p(x) = a_0 + a_1x + \cdots + a_nx^n$ be a polynomial in $\mathcal{P}$, and let a and b be scalars. Then
$$\begin{aligned}(a+b)p(x) &= (a+b)(a_0 + a_1x + \cdots + a_nx^n)\\ &= (a+b)a_0 + (a+b)a_1x\\ &\quad + \cdots + (a+b)a_nx^n\\ &= (aa_0 + ba_0) + (aa_1 + ba_1)x\\ &\quad + \cdots + (aa_n + ba_n)x^n\\ &= (aa_0 + aa_1x + \cdots + aa_nx^n)\\ &\quad + (ba_0 + ba_1x + \cdots + ba_nx^n)\\ &= a(a_0 + a_1x + \cdots + a_nx^n)\\ &\quad + b(a_0 + a_1x + \cdots + a_nx^n)\\ &= ap(x) + bp(x).\end{aligned}$$

2. (i) Clearly, W is a subset of $\mathcal{M}_{2\times 2}$ that contains the 2×2 zero matrix.
(ii) Suppose that
$$A = \begin{bmatrix} a_1 & a_1 + b_1 \\ b_1 & 0 \end{bmatrix} \quad \text{and} \quad B = \begin{bmatrix} a_2 & a_2 + b_2 \\ b_2 & 0 \end{bmatrix}$$
are in W. Then
$$\begin{aligned}A + B &= \begin{bmatrix} a_1 & a_1 + b_1 \\ b_1 & 0 \end{bmatrix} + \begin{bmatrix} a_2 & a_2 + b_2 \\ b_2 & 0 \end{bmatrix}\\ &= \begin{bmatrix} a_1 + a_2 & (a_1 + b_1) + (a_2 + b_2) \\ b_1 + b_2 & 0 \end{bmatrix}\\ &= \begin{bmatrix} a_1 + a_2 & (a_1 + a_2) + (b_1 + b_2) \\ b_1 + b_2 & 0 \end{bmatrix},\end{aligned}$$
which is clearly in W. So W is closed under vector addition.
(iii) For the matrix A in (ii) and any scalar c, we have
$$\begin{aligned}cA = c\begin{bmatrix} a_1 & a_1 + b_1 \\ b_1 & 0 \end{bmatrix} &= \begin{bmatrix} ca_1 & c(a_1 + b_1) \\ cb_1 & c \cdot 0 \end{bmatrix}\\ &= \begin{bmatrix} ca_1 & ca_1 + cb_1 \\ cb_1 & 0 \end{bmatrix},\end{aligned}$$
which is clearly in W. So W is closed under scalar multiplication.

3. Suppose that
$$\begin{bmatrix} 1 & 2 \\ -1 & -3 \end{bmatrix} = a\begin{bmatrix} 1 & -1 \\ 1 & 2 \end{bmatrix} + b\begin{bmatrix} 0 & 1 \\ 1 & 2 \end{bmatrix} + c\begin{bmatrix} 2 & 1 \\ 0 & -1 \end{bmatrix}$$
for some scalars a, b, and c. Then
$$\begin{aligned} a \phantom{{}+b} + 2c &= 1\\ -a + b + c &= 2\\ a + b \phantom{{}+c} &= -1\\ 2a + 2b - c &= -3.\end{aligned}$$
Since this system has the solution $a = -1$, $b = 0$, $c = 1$, we conclude that the matrix is a linear combination of the matrices in S and hence lies in the span of S.

7.2 LINEAR TRANSFORMATIONS

In this section, we study *linear transformations*, those functions acting on vector spaces that preserve the operations of vector addition and scalar multiplication. The following definition extends the definition of linear transformation given in Section 2.7:

Definitions Let V and W be vector spaces. A mapping $T: V \to W$ is called a **linear transformation** (or simply, **linear**) if, for all vectors $\mathbf{u}$ and $\mathbf{v}$ in V and all scalars c, both of the following conditions hold:

(i) $T(\mathbf{u}+\mathbf{v}) = T(\mathbf{u}) + T(\mathbf{v})$. (In this case, we say that T **preserves vector addition**.)

(ii) $T(c\mathbf{u}) = cT(\mathbf{u})$. (In this case, we say that T **preserves scalar multiplication**.)

The vector spaces V and W are called the **domain** and **codomain** of T, respectively.

We gave examples of linear transformations from $\mathcal{R}^n$ to $\mathcal{R}^m$ in Chapter 2. In the following examples, we consider linear transformations that are defined on other vector spaces:

Example 1 Let $U: \mathcal{M}_{m\times n} \to \mathcal{M}_{n\times m}$ be the mapping defined by $U(A) = A^T$. The linearity of U is a consequence of Theorem 1.2 on page 7.

As in the case of $\mathcal{R}^n$, a linear transformation from a vector space to itself is called a **linear operator**.

Example 2 Let C^∞ denote the subset of $\mathcal{F}(\mathcal{R})$ consisting of those functions that have derivatives of all orders. That is, a function f in $\mathcal{F}(\mathcal{R})$ belongs to C^∞ if the nth derivative of f exists for every positive integer n. Theorems from calculus imply that C^∞ is a subspace of $\mathcal{F}(\mathcal{R})$. (See Exercise 58.) Consider the mapping $D: \mathsf{C}^\infty \to \mathsf{C}^\infty$ defined by $D(f) = f'$ for all f in C^∞. From elementary properties of the derivative, it follows that

$$D(f+g) = (f+g)' = f' + g' = D(f) + D(g)$$

and

$$D(cf) = (cf)' = cf' = cD(f)$$

for all functions f and g in C^∞ and for every scalar c. It follows that D is a linear operator on C^∞.

Example 2 shows that differentiation is a linear transformation. Integration provides another example of a linear transformation.

Example 3 Let $\mathsf{C}([a,b])$ denote the set of all continuous real-valued functions defined on the closed interval $[a,b]$. It can be shown that $\mathsf{C}([a,b])$ is a subspace of $\mathcal{F}([a,b])$, the set of real-valued functions defined on $[a,b]$. (See Exercise 59.) For each function f in $\mathsf{C}([a,b])$, the definite integral $\int_a^b f(t)\,dt$ exists, and so a mapping $T: \mathsf{C}([a,b]) \to \mathcal{R}$ is defined by

$$T(f) = \int_a^b f(t)\,dt.$$

The linearity of T follows from the elementary properties of definite integrals. For example, for any f and g in $C([a,b])$,

$$\begin{aligned} T(f+g) &= \int_a^b (f+g)(t)\,dt \\ &= \int_a^b [f(t)+g(t)]\,dt \\ &= \int_a^b f(t)\,dt + \int_a^b g(t)\,dt \\ &= T(f)+T(g). \end{aligned}$$

Similarly, $T(cf) = cT(f)$ for every scalar c. Thus T is a linear transformation.

Example 4 Let $T: \mathcal{P}_2 \to \mathcal{R}^3$ be defined by

$$T(f(x)) = \begin{bmatrix} f(0) \\ f(1) \\ 2f(1) \end{bmatrix}.$$

For example, $T(3+x-2x^2) = \begin{bmatrix} 3 \\ 2 \\ 4 \end{bmatrix}$. We show that T is linear. Consider any polynomials $f(x)$ and $g(x)$ in $\mathcal{P}_2$. Then

$$T(f(x)+g(x)) = \begin{bmatrix} f(0)+g(0) \\ f(1)+g(1) \\ 2[f(1)+g(1)] \end{bmatrix} = \begin{bmatrix} f(0) \\ f(1) \\ 2f(1) \end{bmatrix} + \begin{bmatrix} g(0) \\ g(1) \\ 2g(1) \end{bmatrix} = T(f(x)) + T(g(x)),$$

and hence T preserves vector addition. Furthermore, for any scalar c,

$$T(cf(x)) = \begin{bmatrix} cf(0) \\ cf(1) \\ 2cf(1) \end{bmatrix} = c\begin{bmatrix} f(0) \\ f(1) \\ 2f(1) \end{bmatrix} = cT(f(x)),$$

and so T preserves scalar multiplication. We conclude that T is linear.

Practice Problem 1 ▶ Let $T: \mathcal{P}_2 \to \mathcal{R}^3$ be the mapping defined by

$$T(f(x)) = \begin{bmatrix} f(0) \\ f'(0) \\ f''(0) \end{bmatrix}.$$

Prove that T is a linear transformation. ◀

ELEMENTARY PROPERTIES OF LINEAR TRANSFORMATIONS

We now generalize results about linear transformations from $\mathcal{R}^n$ to $\mathcal{R}^m$ to other vector spaces. In most cases, the proofs, with some notational changes, are identical to the corresponding results in Chapter 2.

The first result extends Theorem 2.8 to all vector spaces. Its proof is identical to the proof of Theorem 2.8, and hence we omit it.

THEOREM 7.4

Let V and W be vector spaces and $T: V \to W$ be a linear transformation. For any vectors $\mathbf{u}$ and $\mathbf{v}$ in V and any scalars a and b, the following statements are true:

(a) $T(\mathbf{0}) = \mathbf{0}$.
(b) $T(-\mathbf{u}) = -T(\mathbf{u})$.
(c) $T(\mathbf{u} - \mathbf{v}) = T(\mathbf{u}) - T(\mathbf{v})$.
(d) $T(a\mathbf{u} + b\mathbf{v}) = aT(\mathbf{u}) + bT(\mathbf{v})$.

As in Section 2.7, Theorem 7.4(d) extends to arbitrary linear combinations; that is, T *preserves linear combinations.*

Let $T: V \to W$ be a linear transformation. If $\mathbf{u}_1, \mathbf{u}_2, \ldots, \mathbf{u}_k$ are vectors in V and $a_1, a_2, \ldots, a_k$ are scalars, then

$$T(a_1\mathbf{u}_1 + a_2\mathbf{u}_2 + \cdots + a_k\mathbf{u}_k) = a_1T(\mathbf{u}_1) + a_2T(\mathbf{u}_2) + \cdots + a_kT(\mathbf{u}_k).$$

Given a linear transformation $T: V \to W$, where V and W are vector spaces, there are subspaces of V and W that are naturally associated with T.

Definitions Let $T: V \to W$ be a linear transformation, where V and W are vector spaces. The **null space** of T is the set of all vectors $\mathbf{v}$ in V such that $T(\mathbf{v}) = \mathbf{0}$. The **range** of T is the set of all images of T, that is, the set of all vectors $T(\mathbf{v})$ for $\mathbf{v}$ in V.

The null space and the range of a linear transformation $T: V \to W$ are subspaces of V and W, respectively. (See Exercises 56 and 57.)

Example 5

Let $U: \mathcal{M}_{m\times n} \to \mathcal{M}_{n\times m}$ be the linear transformation in Example 1 that is defined by $U(A) = A^T$. Describe the null space and the range of U.

Solution A matrix A in $\mathcal{M}_{m\times n}$ is in the null space of U if and only if $A^T = O$, where O is the $n \times m$ zero matrix. Clearly, $A^T = O$ if and only if A is the $m \times n$ zero matrix. Hence the null space of U contains only the $m \times n$ zero matrix.

For any matrix B in $\mathcal{M}_{n\times m}$, we have that $B = (B^T)^T = U(B^T)$, and hence B is in the range of U. We conclude that the range of U coincides with its codomain, $\mathcal{M}_{n\times m}$.

Example 6

Let $D: C^\infty \to C^\infty$ be the linear transformation in Example 2 defined by $D(f) = f'$. Describe the null space and the range of D.

Solution A function f in C^∞ is in the null space of D if and only if $D(f) = f' = \mathbf{0}$; that is, the derivative of f is the zero function. It follows that f is a constant function. So the null space of D is the subspace of constant functions.

Any function f in C^∞ has an antiderivative g in C^∞. Thus $f = g' = D(g)$ is in the range of D. Therefore the range of D is all of C^∞.

The following is an example of a linear transformation T in which the range of T does not coincide with its codomain:

Example 7 Let $T\colon \mathcal{P}_2 \to \mathcal{R}^3$ be the linear transformation in Example 4 defined by

$$T(f(x)) = \begin{bmatrix} f(0) \\ f(1) \\ 2f(1) \end{bmatrix}.$$

Describe the null space and the range of T.

Solution A polynomial $f(x) = a + bx + cx^2$ lies in the null space of T if and only if

$$T(f(x)) = \begin{bmatrix} f(0) \\ f(1) \\ 2f(1) \end{bmatrix} = \begin{bmatrix} 0 \\ 0 \\ 0 \end{bmatrix};$$

that is, $f(0) = a = 0$ and $f(1) = a + b + c = 0$. These two equations are satisfied if and only if $a = 0$ and $c = -b$; that is, $f(x) = bx - bx^2 = b(x - x^2)$ for some scalar b. Thus the null space of T is the span of the set $\{x - x^2\}$.

The image of an arbitrary polynomial $f(x) = a + bx + cx^2$ in $\mathcal{P}_2$ is

$$T(f(x)) = \begin{bmatrix} f(0) \\ f(1) \\ 2f(1) \end{bmatrix} = \begin{bmatrix} a \\ a+b+c \\ 2(a+b+c) \end{bmatrix} = a\begin{bmatrix} 1 \\ 1 \\ 2 \end{bmatrix} + b\begin{bmatrix} 0 \\ 1 \\ 2 \end{bmatrix} + c\begin{bmatrix} 0 \\ 1 \\ 2 \end{bmatrix}.$$

It follows that the range of T is the span of the set

$$\left\{ \begin{bmatrix} 1 \\ 1 \\ 2 \end{bmatrix}, \begin{bmatrix} 0 \\ 1 \\ 2 \end{bmatrix} \right\},$$

which is a 2-dimensional subspace of $\mathcal{R}^3$.

Recall that a linear transformation $T\colon V \to W$ is **onto** if its range equals W. Thus the linear transformations U and D in Examples 5 and 6, respectively, are onto. The transformation T is **one-to-one** if every pair of distinct vectors in V has distinct images in W.

The next result, which is an extension of Theorem 2.11 in Section 2.8, tells us how to use the null space of a linear transformation to determine whether the transformation is one-to-one. Its proof is identical to that given on page 182 and hence is omitted.

THEOREM 7.5

A linear transformation is one-to-one if and only if its null space contains only the zero vector.

The null space of the linear operator U in Example 5 contains only the zero vector of $\mathcal{M}_{m\times n}$. Thus U is one-to-one by Theorem 7.5. In contrast, the linear transformations in Examples 6 and 7 are not one-to-one because their null spaces contain vectors other than the zero vector.

Example 8 Let $T\colon \mathcal{P}_2 \to \mathcal{R}^3$ be the linear transformation defined by

$$T(f(x)) = \begin{bmatrix} f(0) \\ f'(0) \\ f''(0) \end{bmatrix}.$$

Use Theorem 7.5 to show that T is one-to-one.

Solution Let $f(x) = a + bx + cx^2$. Then $f'(x) = b + 2cx$ and $f''(x) = 2c$. If $f(x)$ is in the null space of T, then $f(0) = f'(0) = f''(0) = 0$. Hence $a = 0$, $b = 0$, and $2c = 0$. We conclude that $f(x)$ is the zero polynomial, and so it is the only polynomial in the null space of T. It follows that T is one-to-one by Theorem 7.5.

ISOMORPHISM

Linear transformations that are both one-to-one and onto play an important role in the next section.

Definitions Let V and W be vector spaces. A linear transformation $T\colon V \to W$ is called an **isomorphism** if it is both one-to-one and onto. In this case, we say that V is **isomorphic** to W.

We have already seen that the linear transformation U in Example 5 is one-to-one and onto, and hence it is an isomorphism.

Example 9 Show that the linear transformation T in Example 8 is an isomorphism.

Solution In Example 8, we showed that T is one-to-one, so it suffices to show that T is onto. Observe that for any vector $\begin{bmatrix} a \\ b \\ c \end{bmatrix}$ in $\mathcal{R}^3$,

$$T\left(a + bx + \frac{c}{2}x^2\right) = \begin{bmatrix} a \\ b \\ c \end{bmatrix}, \qquad (1)$$

and hence every vector in $\mathcal{R}^3$ is in the range of T. So T is onto, and therefore T is an isomorphism. So $\mathcal{P}_2$ is isomorphic to $\mathcal{R}^3$.

Practice Problem 2 ▶ Let $T\colon \mathcal{M}_{2\times 2} \to \mathcal{M}_{2\times 2}$ be defined by $T(A) = \begin{bmatrix} 1 & 1 \\ 1 & 2 \end{bmatrix} A$. Prove that T is an isomorphism. ◀

Because an isomorphism $T\colon V \to W$ is both one-to-one and onto, for every $\mathbf{w}$ in W there is a unique $\mathbf{v}$ in V such that $T(\mathbf{v}) = \mathbf{w}$. Thus the function T has an inverse $T^{-1}\colon W \to V$ defined by $T^{-1}(\mathbf{w}) = \mathbf{v}$. For this reason, an isomorphism can also be called an **invertible linear transformation**, or in the case that the isomorphism is an operator, an **invertible linear operator**.

The next result tells us that the inverse of an isomorphism is also an isomorphism.

THEOREM 7.6

Let V and W be vector spaces and $T\colon V \to W$ be an isomorphism. Then $T^{-1}\colon W \to V$ is linear, and hence is also an isomorphism.

PROOF We first show that T^{-1} preserves vector addition. Let $\mathbf{w}_1$ and $\mathbf{w}_2$ be vectors in W, and let $T^{-1}(\mathbf{w}_1) = \mathbf{v}_1$ and $T^{-1}(\mathbf{w}_2) = \mathbf{v}_2$. Then $T(\mathbf{v}_1) = \mathbf{w}_1$ and $T(\mathbf{v}_2) = \mathbf{w}_2$. Hence

$$\mathbf{w}_1 + \mathbf{w}_2 = T(\mathbf{v}_1) + T(\mathbf{v}_2) = T(\mathbf{v}_1 + \mathbf{v}_2),$$

from which it follows that

$$T^{-1}(\mathbf{w}_1 + \mathbf{w}_2) = \mathbf{v}_1 + \mathbf{v}_2 = T^{-1}(\mathbf{w}_1) + T^{-1}(\mathbf{w}_2).$$

To show that T^{-1} preserves scalar multiplication, let $\mathbf{w}$ be a vector in W, c be a scalar, and $\mathbf{v} = T^{-1}(\mathbf{w})$. Then $T(\mathbf{v}) = \mathbf{w}$, and hence

$$c\mathbf{w} = cT(\mathbf{v}) = T(c\mathbf{v}).$$

Therefore

$$T^{-1}(c\mathbf{w}) = c\mathbf{v} = cT^{-1}(\mathbf{w}). \qquad \blacksquare$$

We have seen that the linear transformation T in Example 8 is an isomorphism. Thus, by equation (1), the isomorphism T^{-1} satisfies

$$T^{-1}\left(\begin{bmatrix} a \\ b \\ c \end{bmatrix}\right) = a + bx + \frac{c}{2}x^2 \quad \text{for} \quad \begin{bmatrix} a \\ b \\ c \end{bmatrix} \text{ in } \mathcal{R}^3.$$

Theorem 7.6 shows that if $T\colon V \to W$ is an isomorphism, so is $T^{-1}\colon W \to V$. Applying Theorem 7.6 again, we see that $(T^{-1})^{-1}$ is also an isomorphism. In fact, it is easy to show that $(T^{-1})^{-1} = T$. (See Exercise 52.)

One consequence of Theorem 7.6 is that if a vector space V is isomorphic to a vector space W, then W is also isomorphic to V. For this reason, we simply say that V and W are **isomorphic**.

COMPOSITION OF LINEAR TRANSFORMATIONS

Let V, W, and Z be vector spaces and $T\colon V \to W$ and $U\colon W \to Z$ be linear transformations. As in Section 2.8, we can form the **composition** $UT\colon V \to Z$ of U and T defined by $UT(\mathbf{v}) = U(T(\mathbf{v}))$ for all $\mathbf{v}$ in V.

The following result tells us that UT is linear if both U and T are linear:

THEOREM 7.7

Let V, W, and Z be vector spaces and $T\colon V \to W$ and $U\colon W \to Z$ be linear transformations. Then the composition $UT\colon V \to Z$ is also a linear transformation.

PROOF Let $\mathbf{u}$ and $\mathbf{v}$ be vectors in V. Then

$$\begin{aligned} UT(\mathbf{u}+\mathbf{v}) &= U(T(\mathbf{u}+\mathbf{v})) \\ &= U(T(\mathbf{u})+T(\mathbf{v})) \\ &= U(T(\mathbf{u}))+U(T(\mathbf{v})) \\ &= UT(\mathbf{u})+UT(\mathbf{v}), \end{aligned}$$

and hence UT preserves vector addition. Similarly, UT preserves scalar multiplication. Therefore UT is linear. ■

Example 10 Let $T\colon \mathcal{P}_2 \to \mathcal{R}^3$ and $U\colon \mathcal{R}^3 \to \mathcal{M}_{2\times 2}$ be the functions defined by

$$T(f(x)) = \begin{bmatrix} f(0) \\ f(1) \\ 2f(1) \end{bmatrix} \quad \text{and} \quad U\left(\begin{bmatrix} s \\ t \\ u \end{bmatrix}\right) = \begin{bmatrix} s & t \\ t & u \end{bmatrix}.$$

Both T and U are linear transformations, and their composition UT is defined. For any polynomial $a + bx + cx^2$ in $\mathcal{P}_2$, UT satisfies

$$UT(a+bx+cx^2) = U\left(\begin{bmatrix} a \\ a+b+c \\ 2(a+b+c) \end{bmatrix}\right) = \begin{bmatrix} a & a+b+c \\ a+b+c & 2(a+b+c) \end{bmatrix}.$$

By Theorem 7.7, UT is linear.

Suppose that $T\colon V \to W$ and $U\colon W \to Z$ are isomorphisms. By Theorem 7.7, the composition UT is linear. Furthermore, since both U and T are one-to-one and onto, it follows that UT is also one-to-one and onto. Therefore UT is an isomorphism. The inverses of T, U, and UT are related by $(UT)^{-1} = T^{-1}U^{-1}$. (See Exercise 53.) We summarize these observations as follows:

> Let $T\colon V \to W$ and $U\colon W \to Z$ be isomorphisms.
>
> (a) $UT\colon V \to Z$ is an isomorphism.
>
> (b) $(UT)^{-1} = T^{-1}U^{-1}$.

Note that (b) in the preceding box extends to the composition of any finite number of isomorphisms. (See Exercise 54.)

EXERCISES

In Exercises 1–8, determine whether each linear transformation is one-to-one.

1. $T\colon \mathcal{M}_{2\times 2} \to \mathcal{M}_{2\times 2}$ is the linear transformation defined by $T(A) = A\begin{bmatrix} 1 & 2 \\ 3 & 4 \end{bmatrix}$.
2. $U\colon \mathcal{M}_{2\times 2} \to \mathcal{R}$ is the linear transformation defined by $U(A) = \text{trace}(A)$.
3. $T\colon \mathcal{M}_{2\times 2} \to \mathcal{R}^2$ is the linear transformation defined by $T(A) = A\mathbf{e}_1$.
4. $U\colon \mathcal{P}_2 \to \mathcal{R}^2$ is the linear transformation defined by $U(f(x)) = \begin{bmatrix} f(1) \\ f'(1) \end{bmatrix}$.
5. $T\colon \mathcal{P}_2 \to \mathcal{P}_2$ is the linear transformation defined by $T(f(x)) = xf'(x)$.

6. $T\colon \mathcal{P}_2 \to \mathcal{P}_2$ is the linear transformation defined by $T(f(x)) = f(x) + f'(x)$.

7. $U\colon \mathcal{R}^3 \to \mathcal{M}_{2\times 2}$ is the linear transformation in Example 10 defined by

$$U\left(\begin{bmatrix} s \\ t \\ u \end{bmatrix}\right) = \begin{bmatrix} s & t \\ t & u \end{bmatrix}.$$

8. $U\colon \mathcal{R}^2 \to \mathcal{R}$ is the linear transformation defined by $U(\mathbf{v}) = \det \begin{bmatrix} v_1 & 1 \\ v_2 & 3 \end{bmatrix}$.

In Exercises 9–16, determine whether each linear transformation is onto.

9. the linear transformation T in Exercise 1
10. the linear transformation U in Exercise 2
11. the linear transformation T in Exercise 3
12. the linear transformation U in Exercise 4
13. the linear transformation T in Exercise 5
14. the linear transformation T in Exercise 6
15. the linear transformation U in Exercise 7
16. the linear transformation U in Exercise 8

In Exercises 17–24, prove that each function is actually linear.

17. the function T in Exercise 1
18. the function U in Exercise 2
19. the function T in Exercise 3
20. the function U in Exercise 4
21. the function T in Exercise 5
22. the function T in Exercise 6
23. the function U in Exercise 7
24. the function T in Exercise 8

In Exercises 25–29, compute the expression determined by each composition of linear transformations.

25. $UT\left(\begin{bmatrix} a & b \\ c & d \end{bmatrix}\right)$, where U is the linear transformation in Exercise 2 and T is the linear transformation in Exercise 1

26. $UT(a + bx + cx^2)$, where U is the linear transformation in Exercise 4 and T is the linear transformation in Exercise 5

27. $UT(a + bx + cx^2)$, where U is the linear transformation in Exercise 4 and T is the linear transformation in Exercise 6

28. $UT\left(\begin{bmatrix} a & b \\ c & d \end{bmatrix}\right)$, where U is the linear transformation in Exercise 8 and T is the linear transformation in Exercise 3

29. $TU\left(\begin{bmatrix} s \\ t \\ u \end{bmatrix}\right)$, where $T\colon \mathcal{M}_{2\times 2} \to \mathcal{M}_{2\times 2}$ is the linear transformation defined by $T(A) = A^T$ and U is the linear transformation in Exercise 7

In Exercises 30–37, determine whether each transformation T is linear. If T is linear, determine if it is an isomorphism. Justify your conclusions.

30. $T\colon \mathcal{M}_{n\times n} \to \mathcal{R}$ defined by $T(A) = \det A$
31. $T\colon \mathcal{P} \to \mathcal{P}$ defined by $T(f(x)) = xf(x)$
32. $T\colon \mathcal{P}_2 \to \mathcal{R}^3$ defined by $T(f(x)) = \begin{bmatrix} f(0) \\ f(1) \\ f(2) \end{bmatrix}$
33. $T\colon \mathcal{P} \to \mathcal{P}$ defined by $T(f(x)) = (f(x))^2$
34. $T\colon \mathcal{M}_{2\times 2} \to \mathcal{M}_{2\times 2}$ defined by $T(A) = \begin{bmatrix} 1 & 1 \\ 1 & 1 \end{bmatrix} A$
35. $T\colon \mathcal{F}(\mathcal{R}) \to \mathcal{F}(\mathcal{R})$ defined by $T(f)(x) = f(x+1)$
36. $T\colon \mathcal{D}(\mathcal{R}) \to \mathcal{F}(\mathcal{R})$ defined by $T(f) = f'$, where f' is the derivative of f and $\mathcal{D}(\mathcal{R})$ is the set of functions in $\mathcal{F}(\mathcal{R})$ that are differentiable
37. $T\colon \mathcal{D}(\mathcal{R}) \to \mathcal{R}$ defined by $T(f) = \int_0^1 f(t)dt$, where $\mathcal{D}(\mathcal{R})$ is the set of functions in $\mathcal{F}(\mathcal{R})$ that are differentiable

38. Let $S = \{s_1, s_2, \ldots, s_n\}$ be a set consisting of n elements, and let $T\colon \mathcal{F}(S) \to \mathcal{R}^n$ be defined by

$$T(f) = \begin{bmatrix} f(s_1) \\ f(s_2) \\ \vdots \\ f(s_n) \end{bmatrix}.$$

Prove that T is an isomorphism.

In Exercises 39–48, determine whether the statements are true or false.

39. Every isomorphism is linear and one-to-one.
40. A linear transformation that is one-to-one is an isomorphism.
41. The vector spaces $\mathcal{M}_{m\times n}$ and $\mathcal{L}(\mathcal{R}^n, \mathcal{R}^m)$ are isomorphic.
42. The definite integral can be considered to be a linear transformation from $\mathsf{C}([a,b])$ to the real numbers.
43. The function $f(t) = \cos t$ belongs to C^∞.
44. The function $t^4 - 3t^2$ does not belong to C^∞.
45. Differentiation is a linear operator on C^∞.
46. The definite integral is a linear operator on $\mathsf{C}([a,b])$, the vector space of continuous real-valued functions defined on $[a,b]$.
47. The null space of every linear operator on V is a subspace of V.
48. The solution set of the differential equation $y'' + 4y = \sin 2t$ is a subspace of C^∞.

---

49. Let N denote the set of nonnegative integers, and let V be the subset of $\mathcal{F}(N)$ consisting of the functions that are zero except at finitely many elements of N.
   (a) Show that V is a subspace of $\mathcal{F}(N)$.
   (b) Show that V is isomorphic to $\mathcal{P}$. *Hint:* Choose $T\colon V \to \mathcal{P}$ to be the transformation that maps a function f to the polynomial having $f(i)$ as the coefficient of x^i.

50. Let V be the vector space of all 2×2 matrices with trace equal to 0. Prove that V is isomorphic to $\mathcal{P}_2$ by constructing an isomorphism from V to $\mathcal{P}_2$. Verify your answer.

51. Let V be the subset of $\mathcal{P}_4$ of polynomials of the form $ax^4 + bx^2 + c$, where a, b, and c are scalars.
    (a) Prove that V is a subspace of $\mathcal{P}_4$.
    (b) Prove that V is isomorphic to $\mathcal{P}_2$ by constructing an isomorphism from V to $\mathcal{P}_2$. Verify your answer.

52. Let V and W be vector spaces and $T: V \to W$ be an isomorphism. Prove that $(T^{-1})^{-1} = T$.

53. Let V, W, and Z be vector spaces and $T: V \to W$ and $U: W \to Z$ be isomorphisms. Prove that $UT: V \to Z$ is an isomorphism and $(UT)^{-1} = T^{-1}U^{-1}$.

54. Prove the following extension of Exercise 53: If $T_1, T_2, \ldots, T_k$ are isomorphisms such that the composition $T_1T_2\cdots T_k$ is defined, then $T_1T_2\cdots T_k$ is an isomorphism and
$$(T_1T_2\cdots T_k)^{-1} = (T_k)^{-1}(T_{k-1})^{-1}\cdots(T_1)^{-1}.$$

Definitions For vector spaces V and W, let $\mathcal{L}(V, W)$ denote the set of all linear transformations $T: V \to W$. For T and U in $\mathcal{L}(V, W)$ and any scalar c, define $T + U: V \to W$ and $cT: V \to W$ by
$$(T + U)(\mathbf{x}) = T(\mathbf{x}) + U(\mathbf{x}) \qquad \text{and} \qquad (cT)(\mathbf{x}) = cT(\mathbf{x})$$
for all $\mathbf{x}$ in V.

The preceding definitions are used in Exercise 55:

55. Let V and W be vector spaces, let T and U be in $\mathcal{L}(V, W)$, and let c be a scalar.
    (a) Prove that for any T and U in $\mathcal{L}(V, W)$, $T + U$ is a linear transformation.
    (b) Prove that for any T in $\mathcal{L}(V, W)$, cT is a linear transformation for any scalar c.
    (c) Prove that $\mathcal{L}(V, W)$ is a vector space with these operations.
    (d) Describe the zero vector of this vector space.

56. Let $T: V \to W$ be a linear transformation between vector spaces V and W. Prove that the null space of T is a subspace of V.

57. Let $T: V \to W$ be a linear transformation between vector spaces V and W. Prove that the range of T is a subspace of W.

58. Recall the set C^∞ in Example 2.
    (a) Prove that C^∞ is a subspace of $\mathcal{F}(\mathcal{R})$.
    (b) Let $T: \mathsf{C}^\infty \to \mathsf{C}^\infty$ be defined by $T(f)(t) = e^t f''(t)$ for all t in $\mathcal{R}$. Prove that T is linear.

59. Recall the set $\mathsf{C}([a, b])$ in Example 3.
    (a) Prove that $\mathsf{C}([a, b])$ is a subspace of $\mathcal{F}([a, b])$.
    (b) Let $T: \mathsf{C}([a, b]) \to \mathsf{C}([a, b])$ be defined by
$$T(f)(x) = \int_a^x f(t)dt \qquad \text{for } a \le x \le b.$$
    Prove that T is linear and one-to-one.

60. Review Examples 5 and 6 of Section 7.1, in which it is shown that certain subsets of vector spaces are subspaces. Give alternate proofs of these results using the fact that the null space of a linear transformation is a subspace of its domain.

SOLUTIONS TO THE PRACTICE PROBLEMS

1. Let
$$p(x) = a_0 + a_1x + a_2x^2 \text{ and } q(x) = b_0 + b_1x + b_2x^2.$$
Then
$$T(f(x) + g(x)) = T((a_0 + b_0) + (a_1 + b_1)x + (a_2 + b_2)x^2)$$
$$= \begin{bmatrix} a_0 + b_0 \\ a_1 + b_1 \\ 2(a_2 + b_2) \end{bmatrix} = \begin{bmatrix} a_0 \\ a_1 \\ 2a_2 \end{bmatrix} + \begin{bmatrix} b_0 \\ b_1 \\ 2b_2 \end{bmatrix}$$
$$= T(f(x)) + T(g(x)),$$
and for any scalar c,
$$T(cf(x)) = T(ca_0 + ca_1x + ca_2x^2) = \begin{bmatrix} ca_0 \\ ca_1 \\ 2ca_2 \end{bmatrix}$$
$$= c\begin{bmatrix} a_0 \\ a_1 \\ 2a_2 \end{bmatrix} = cT(f(x)).$$

2. Let $B = \begin{bmatrix} 1 & 1 \\ 1 & 2 \end{bmatrix}$.
    (a) For any matrices C and D in $\mathcal{M}_{2\times 2}$, and any scalar k, we have
$$T(C + D) = B(C + D) = BC + BD$$
$$= T(C) + T(D)$$
    and
$$T(kC) = B(kC) = k(BC) = kT(C).$$
    So T is linear.
    (b) Observe that B is invertible. If $T(A) = BA = O$, then $A = B^{-1}(BA) = B^{-1}O = O$, and hence T is one-to-one by Theorem 7.5. Now consider any matrix A in $\mathcal{M}_{2\times 2}$. Then $T(B^{-1}A) = B(B^{-1}A) = A$, and so T is onto. We conclude that T is an isomorphism.

7.3 BASIS AND DIMENSION

In this section, we develop the concepts of *basis* and *dimension* for vector spaces. Then we identify a special class of vector spaces, the *finite-dimensional* vector spaces, which are isomorphic to $\mathcal{R}^n$ for some n. We use these isomorphisms to transfer facts about $\mathcal{R}^n$ that we have learned in earlier chapters to all finite-dimensional vector spaces.

LINEAR DEPENDENCE AND LINEAR INDEPENDENCE

For general vector spaces, the concepts of *linear dependence* and *linear independence* are similar to the corresponding concepts for $\mathcal{R}^n$, except that we must allow for infinite sets. In fact, the definitions for finite sets are the same as for $\mathcal{R}^n$. (See page 75.)

Definitions An infinite subset S of a vector space V is **linearly dependent** if some finite subset of S is linearly dependent. An infinite set S is **linearly independent** if S is not linearly dependent, that is, if every finite subset of S is linearly independent.

Example 1 The subset $S = \{x^2 - 3x + 2, 3x^2 - 5x, 2x - 3\}$ of $\mathcal{P}_2$ is linearly dependent because

$$3(x^2 - 3x + 2) + (-1)(3x^2 - 5x) + 2(2x - 3) = \mathbf{0},$$

where $\mathbf{0}$ is the zero polynomial. As with linearly dependent subsets of $\mathcal{R}^n$, S is linearly dependent because we are able to represent the zero vector as a linear combination of the vectors in S with at least one nonzero coefficient.

Example 2 In Example 14 of Section 7.1, we noted that the set

$$S = \left\{ \begin{bmatrix} 1 & 0 \\ 0 & -1 \end{bmatrix}, \begin{bmatrix} 0 & 1 \\ 0 & 0 \end{bmatrix}, \begin{bmatrix} 0 & 0 \\ 1 & 0 \end{bmatrix} \right\}$$

is a generating set for the subspace of 2×2 matrices with trace equal to zero. We now show that this set is linearly independent. Consider any scalars a, b, and c such that

$$a \begin{bmatrix} 1 & 0 \\ 0 & -1 \end{bmatrix} + b \begin{bmatrix} 0 & 1 \\ 0 & 0 \end{bmatrix} + c \begin{bmatrix} 0 & 0 \\ 1 & 0 \end{bmatrix} = O,$$

where O is the zero matrix of $\mathcal{M}_{2\times 2}$. In Example 14 of Section 7.1, we observed that this linear combination equals $\begin{bmatrix} a & b \\ c & -a \end{bmatrix}$, and hence

$$\begin{bmatrix} a & b \\ c & -a \end{bmatrix} = \begin{bmatrix} 0 & 0 \\ 0 & 0 \end{bmatrix}.$$

By equating corresponding entries, we find that $a = 0$, $b = 0$, and $c = 0$. It follows that S is linearly independent.

Practice Problem 1 ► Determine if the set

$$S = \left\{ \begin{bmatrix} 1 & 1 \\ 1 & 0 \end{bmatrix}, \begin{bmatrix} 0 & 0 \\ 1 & 1 \end{bmatrix}, \begin{bmatrix} 0 & 2 \\ 0 & -1 \end{bmatrix}, \begin{bmatrix} 0 & 1 \\ 1 & 0 \end{bmatrix} \right\}$$

is a linearly independent subset of $\mathcal{M}_{2\times 2}$. ◄

Example 3 Let $S = \{e^t, e^{2t}, e^{3t}\}$. We show that S is a linearly independent subset of $\mathcal{F}(R)$. Consider any scalars a, b, and c such that

$$ae^t + be^{2t} + ce^{3t} = \mathbf{0},$$

where $\mathbf{0}$ is the zero function. By taking the first and second derivatives of both sides of this equation, we obtain the two equations

$$ae^t + 2be^{2t} + 3ce^{3t} = \mathbf{0}$$

and

$$ae^t + 4be^{2t} + 9ce^{3t} = \mathbf{0}.$$

Since the left sides of these equations equal the zero function, they must equal 0 for every real number t. So we may substitute $t = 0$ into these three equations to obtain the homogeneous system

$$\begin{aligned} a + b + c &= 0 \\ a + 2b + 3c &= 0 \\ a + 4b + 9c &= 0. \end{aligned}$$

It is easy to show that this system has only the zero solution $a = b = c = 0$, and hence S is linearly independent.

The next two examples involve infinite sets.

Example 4 The infinite subset $\{1, x, x^2, \ldots, x^n, \ldots\}$ of the vector space $\mathcal{P}$ is linearly independent. Observe that any linear combination of a nonempty finite subset of this set is not the zero polynomial unless all of the coefficients are zeros. So every nonempty finite subset is linearly independent.

Example 5 The infinite subset

$$\{1 + x, 1 - x, 1 + x^2, 1 - x^2, \ldots, 1 + x^n, 1 - x^n, \ldots\}$$

of $\mathcal{P}$ is linearly dependent because it contains a finite linearly dependent subset. One example of such a set is $\{1 + x, 1 - x, 1 + x^2, 1 - x^2\}$, because

$$1(1 + x) + 1(1 - x) + (-1)(1 + x^2) + (-1)(1 - x^2) = \mathbf{0}.$$

The next result tells us that isomorphisms preserve linear independence. We omit the proof. (See Exercise 66.)

THEOREM 7.8

Let V and W be vector spaces, $\{\mathbf{v}_1, \mathbf{v}_2, \ldots, \mathbf{v}_k\}$ be a linearly independent subset of V, and $T\colon V \to W$ be an isomorphism. Then $\{T(\mathbf{v}_1), T(\mathbf{v}_2), \ldots, T(\mathbf{v}_k)\}$, the set of images of $\mathbf{v}_1, \mathbf{v}_2, \ldots, \mathbf{v}_k$, is a linearly independent subset of W.

BASES FOR VECTOR SPACES

As in Chapter 4, we define a subset S of a vector space V to be a **basis** for V if S is both a linearly independent set and a generating set for V. Thus we see that the set S in Example 2 is a basis for the subspace of 2×2 matrices with trace equal to zero. In Example 4 of this section and Example 12 of Section 7.1, we saw that $\{1, x, x^2, \ldots, x^n, \ldots\}$ is a linearly independent generating set for $\mathcal{P}$; so it is a basis for $\mathcal{P}$. Thus, in contrast to the subspaces of $\mathcal{R}^n$, the vector space $\mathcal{P}$ has an infinite basis. The generating set for the vector space $\mathcal{T}[0, 2\pi]$ of trigonometric polynomials defined on page 498 is also linearly independent. (See Section 7.5.) This is another example of an infinite basis for a vector space.

In Chapter 4, it was shown that any two bases for the same subspace of $\mathcal{R}^n$ contain the same number of vectors. The preceding observations lead to three questions about bases of vector spaces. We list these questions together with their answers.

1. Is it possible for a vector space to have both an infinite and a finite basis? The answer is *no.*
2. If a vector space V has a finite basis, must any two bases for V have the same number of vectors (as is the case for subspaces of $\mathcal{R}^n$)? The answer is *yes.*
3. Does every vector space have a basis? If we use the *axiom of choice* from set theory, the answer is *yes.*

We justify the first two answers in this section. The justification for the third answer is beyond the scope of this book. (For a proof, see [4, pages 58–61].)

We begin with a study of vector spaces that have finite bases.[1] For this purpose, we revisit Theorem 4.10 on page 265. The next result is an extension of this theorem to general vector spaces that have finite bases. Its proof is identical to that of Theorem 4.10.

> Suppose that $\mathcal{B} = \{\mathbf{v}_1, \mathbf{v}_2, \ldots, \mathbf{v}_n\}$ is a finite basis for a vector space V. Then any vector $\mathbf{v}$ in V can be uniquely represented as a linear combination of the vectors in $\mathcal{B}$; that is, $\mathbf{v} = a_1\mathbf{v}_1 + a_2\mathbf{v}_2 + \cdots + a_n\mathbf{v}_n$ for unique scalars $a_1, a_2, \ldots, a_n$.

Consider a vector space V having a finite basis $\mathcal{B} = \{\mathbf{v}_1, \mathbf{v}_2, \ldots, \mathbf{v}_n\}$. Because of the uniqueness statement in the preceding box, we can define a mapping $\Phi_\mathcal{B}\colon V \to \mathcal{R}^n$ as follows: For any vector $\mathbf{v}$ in V, suppose that the unique representation of $\mathbf{v}$ as a linear combination of the vectors in $\mathcal{B}$ is given by

$$\mathbf{v} = a_1\mathbf{v}_1 + a_2\mathbf{v}_2 + \cdots + a_n\mathbf{v}_n.$$

Define $\Phi_\mathcal{B}$ by

$$\Phi_\mathcal{B}(\mathbf{v}) = \begin{bmatrix} a_1 \\ a_2 \\ \vdots \\ a_n \end{bmatrix}.$$

We call $\Phi_\mathcal{B}$ the **coordinate transformation** of V relative to $\mathcal{B}$. (See Figure 7.1.)

Then $\Phi_\mathcal{B}$ is one-to-one because the representation of a vector in V as a linear combination of the vectors of $\mathcal{B}$ is unique. Furthermore, any scalars $a_1, a_2, \ldots, a_n$ are the coefficients of some linear combination of the vectors of $\mathcal{B}$, and hence $\Phi_\mathcal{B}$ is onto.

---

[1] As in Chapter 4, we implicitly assume that the vectors in a given finite basis of a vector space are listed in a specific order; that is, the basis is an *ordered basis.*

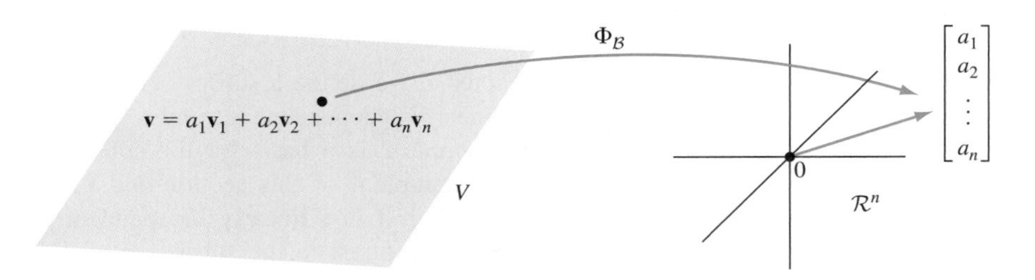

Figure 7.1 The coordinate transformation of V relative to $\mathcal{B}$

We leave as an exercise the straightforward proof that $\Phi_{\mathcal{B}}$ is linear. (See Exercise 73.) Thus we have the following result:

If V is a vector space with the basis $\mathcal{B} = \{\mathbf{v}_1, \mathbf{v}_2, \ldots, \mathbf{v}_n\}$, then the mapping $\Phi_{\mathcal{B}}\colon V \to \mathcal{R}^n$ defined by

$$\Phi_{\mathcal{B}}(a_1\mathbf{v}_1 + a_2\mathbf{v}_2 + \cdots + a_n\mathbf{v}_n) = \begin{bmatrix} a_1 \\ a_2 \\ \vdots \\ a_n \end{bmatrix}$$

is an isomorphism. Therefore if V has a basis of n vectors, then V is isomorphic to $\mathcal{R}^n$.

In view of this result, a vector space with a finite basis has the same vector space structure as $\mathcal{R}^n$. It follows that we can answer questions about the number of vectors in a basis for a vector space by comparing this vector space with $\mathcal{R}^n$.

THEOREM 7.9

Let V be a vector space with a finite basis. Then every basis for V is finite and contains the same number of vectors.

PROOF Let $\mathcal{B} = \{\mathbf{v}_1, \mathbf{v}_2, \ldots, \mathbf{v}_n\}$ be a finite basis for V and $\Phi_{\mathcal{B}}\colon V \to \mathcal{R}^n$ be the isomorphism defined previously. Suppose that some basis $\mathcal{A}$ for V contains more vectors than $\mathcal{B}$. Then there exists a subset $S = \{\mathbf{w}_1, \mathbf{w}_2, \ldots, \mathbf{w}_{n+1}\}$ of $\mathcal{A}$ consisting of $n+1$ distinct vectors. By Exercise 91 of Section 1.7, which applies to all vector spaces, S is linearly independent. So $\{\Phi_{\mathcal{A}}(\mathbf{w}_1), \Phi_{\mathcal{A}}(\mathbf{w}_2), \ldots, \Phi_{\mathcal{A}}(\mathbf{w}_{n+1})\}$ is a linearly independent subset of $\mathcal{R}^n$ by Theorem 7.8. But this is a contradiction because a linearly independent subset of $\mathcal{R}^n$ contains at most n vectors. It follows that $\mathcal{A}$ is finite and contains at most n vectors. Let m denote the number of vectors in $\mathcal{A}$. Then $m \leq n$. We can now apply the argument given earlier, but with the roles of $\mathcal{B}$ and $\mathcal{A}$ reversed, to deduce that $n \leq m$. Therefore $m = n$, and we conclude that any two bases for V contain the same number of vectors. ■

As a consequence of Theorem 7.9, vector spaces are of two types. The first type consists of the zero vector space and those vector spaces that have a finite basis. These vector spaces are called **finite-dimensional**. The second type of vector space, which is not finite-dimensional, is called **infinite-dimensional**. It can be shown that every infinite-dimensional vector space contains an infinite linearly independent set. (See Exercise 72.) In fact, every infinite-dimensional vector space contains an infinite basis. (See [4, page 61].)

The preceding boxed result shows that a finite-dimensional vector space V having a finite basis containing exactly n vectors is isomorphic to $\mathcal{R}^n$. Moreover, every basis for V must contain exactly n vectors. In such a case, we say that n is the **dimension** of V, and we denote it by $\dim V$. Furthermore, the dimension of a vector space is preserved under isomorphism. That is, if V and W are isomorphic vector spaces and V is finite-dimensional, then W is finite-dimensional and the two vector spaces have the same dimension. (See Exercise 66.) On the other hand, if one of the vector spaces is infinite-dimensional, then so is the other. (See Exercise 74.)

Because there is an isomorphism between an n-dimensional vector space and $\mathcal{R}^n$, this isomorphism can be used to transfer properties of linear dependence and linear independence from $\mathcal{R}^n$ to other n-dimensional vector spaces. The next box contains several properties that are analogous to results proved in Section 4.2 for $\mathcal{R}^n$. Their proofs are left as exercises. (See Exercises 69 and 70.)

Properties of Finite-Dimensional Vector Spaces

Let V be an n-dimensional vector space.

1. Any linearly independent subset of V contains at most n vectors.
2. Any linearly independent subset of V containing exactly n vectors is a basis for V.
3. Any generating set for V contains at least n vectors.
4. Any generating set for V containing exactly n vectors is a basis for V.

Example 6

Recall the vector space $\mathcal{P}_n$ in Example 8 of Section 7.1. The set $\mathcal{B} = \{1, x, x^2, \ldots, x^n\}$ is a linearly independent subset of $\mathcal{P}_n$. Furthermore, an arbitrary polynomial $p(x) = a_0 + a_1x + \cdots + a_nx^n$ of degree at most n can be expressed as a linear combination of the polynomials in $\mathcal{B}$. Therefore $\mathcal{B}$ is a generating set for $\mathcal{P}_n$, and we conclude that $\mathcal{B}$ is a basis for $\mathcal{P}_n$. Since $\mathcal{B}$ contains $n+1$ polynomials, $\mathcal{P}_n$ is finite-dimensional with dimension equal to $n+1$.

The mapping $\Phi_{\mathcal{B}}\colon \mathcal{P}_n \to \mathcal{R}^{n+1}$ defined by

$$\Phi_{\mathcal{B}}(a_0 + a_1x + \cdots + a_nx^n) = \begin{bmatrix} a_0 \\ a_1 \\ \vdots \\ a_n \end{bmatrix}$$

is an isomorphism.

Example 7

Recall $\mathcal{M}_{m\times n}$, the vector space of $m \times n$ matrices in Example 1 of Section 7.1. Let $m = n = 2$, and define

$$E_{11} = \begin{bmatrix} 1 & 0 \\ 0 & 0 \end{bmatrix}, \quad E_{12} = \begin{bmatrix} 0 & 1 \\ 0 & 0 \end{bmatrix}, \quad E_{21} = \begin{bmatrix} 0 & 0 \\ 1 & 0 \end{bmatrix}, \quad \text{and} \quad E_{22} = \begin{bmatrix} 0 & 0 \\ 0 & 1 \end{bmatrix}.$$

Then any matrix A in $\mathcal{M}_{2\times 2}$ can be written as a linear combination

$$\begin{aligned} A = \begin{bmatrix} a_{11} & a_{12} \\ a_{21} & a_{22} \end{bmatrix} &= a_{11}\begin{bmatrix} 1 & 0 \\ 0 & 0 \end{bmatrix} + a_{12}\begin{bmatrix} 0 & 1 \\ 0 & 0 \end{bmatrix} + a_{21}\begin{bmatrix} 0 & 0 \\ 1 & 0 \end{bmatrix} + a_{22}\begin{bmatrix} 0 & 0 \\ 0 & 1 \end{bmatrix} \\ &= a_{11}E_{11} + a_{12}E_{12} + a_{21}E_{21} + a_{22}E_{22} \end{aligned}$$

of the vectors in $S = \{E_{11}, E_{12}, E_{21}, E_{22}\}$. Furthermore, if O is written as a linear combination of the matrices in S, then all of the coefficients must be zeros, and so S is linearly independent. It follows that S is a basis for $\mathcal{M}_{2\times 2}$.

Example 7 can be extended to $\mathcal{M}_{m\times n}$ for any positive integers m and n. For each $1 \le i \le m$ and $1 \le j \le n$, let E_{ij} be the $m \times n$ matrix whose (i,j)-entry is 1, and whose other entries are 0. Then it can be shown that the set of all $m \times n$ matrices of the form E_{ij} constitutes a basis for $\mathcal{M}_{m\times n}$. Since there are mn such matrices, it follows that $\dim \mathcal{M}_{m\times n} = mn$.

Practice Problem 2 ▶ Use the boxed properties of finite-dimensional vector spaces on page 515 to determine whether the set S in Practice Problem 1 is a basis for $\mathcal{M}_{2\times 2}$. ◀

Example 8

Recall the vector space $\mathcal{L}(\mathcal{R}^n, \mathcal{R}^m)$ of linear transformations in Example 2 of Section 7.1. Let $U : \mathcal{M}_{m\times n} \to \mathcal{L}(\mathcal{R}^n, \mathcal{R}^m)$ be defined by $U(A) = T_A$, where T_A is the matrix transformation induced by A (defined in Section 2.7). Then for any $m \times n$ matrices A and B, we have

$$U(A+B) = T_{A+B} = T_A + T_B = U(A) + U(B).$$

Similarly, $U(cA) = c\,U(A)$ for any scalar c. By Theorem 2.9 on page 174, U is both one-to-one and onto. It follows that U is an isomorphism, and $\mathcal{M}_{m\times n}$ is isomorphic to $\mathcal{L}(\mathcal{R}^n, \mathcal{R}^m)$. Because the discussion immediately after Example 7 shows that $\mathcal{M}_{m\times n}$ has dimension mn, and because isomorphisms preserve dimension, it follows that $\mathcal{L}(\mathcal{R}^n, \mathcal{R}^m)$ is a finite-dimensional vector space of dimension mn.

Example 9

Let a_0, a_1, a_2, c_0, c_1, and c_2 be real numbers such that a_0, a_1, and a_2 are distinct. We show that there exists a unique polynomial $p(x)$ in $\mathcal{P}_2$ such that $p(a_i) = c_i$ for $i = 0, 1, 2$.

Let $p_0(x)$, $p_1(x)$, and $p_2(x)$ be the polynomials in $\mathcal{P}_2$ defined by

$$p_0(x) = \frac{(x-a_1)(x-a_2)}{(a_0-a_1)(a_0-a_2)}, \qquad p_1(x) = \frac{(x-a_0)(x-a_2)}{(a_1-a_0)(a_1-a_2)},$$

and

$$p_2(x) = \frac{(x-a_0)(x-a_1)}{(a_2-a_0)(a_2-a_1)}.$$

Observe that, for each i and j,

$$p_i(a_j) = \begin{cases} 0 & \text{if } i \ne j \\ 1 & \text{if } i = j. \end{cases}$$

Now set

$$p(x) = c_0p_0(x) + c_1p_1(x) + c_2p_2(x).$$

Then

$$\begin{aligned} p(a_0) &= c_0p_0(a_0) + c_1p_1(a_0) + c_2p_2(a_0) \\ &= c_0 \cdot 1 + c_1 \cdot 0 + c_2 \cdot 0 \\ &= c_0. \end{aligned}$$

Similarly, $p(a_1) = c_1$ and $p(a_2) = c_2$.

We now show that $\mathcal{A} = \{p_0(x), p_1(x), p_2(x)\}$ is a basis for $\mathcal{P}_2$. Suppose that

$$b_0p_0(x) + b_1p_1(x) + b_2p_2(x) = \mathbf{0}$$

for some scalars b_0, b_1, and b_2, where $\mathbf{0}$ is the zero polynomial. Substituting $x = a_i$ for $i = 0, 1, 2$ into this equation yields $b_i = 0$ for $i = 0, 1, 2$, and hence $\mathcal{A}$ is linearly independent. Since $\dim \mathcal{P}_2 = 3$ and $\mathcal{A}$ is a linearly independent subset of $\mathcal{P}_2$ consisting of three polynomials, it follows that $\mathcal{A}$ is a basis for $\mathcal{P}_2$ by the boxed result on page 515.

To show the uniqueness of $p(x)$, suppose that $q(x)$ is a polynomial in $\mathcal{P}_2$ such that $q(a_i) = c_i$ for $i = 0, 1, 2$. Since $\mathcal{A}$ is a basis for $\mathcal{P}_2$, there exist unique scalars d_0, d_1, and d_2 such that $q(x) = d_0p_0(x) + d_1p_1(x) + d_2p_2(x)$. Then

$$c_0 = q(a_0) = d_0p(a_0) + d_1p_1(a_0) + d_2p_2(a_0) = d_0 \cdot 1 + d_1 \cdot 0 + d_2 \cdot 0 = d_0.$$

Similarly, $c_1 = d_1$ and $c_2 = d_2$, and it follows that $q(x) = p(x)$.

To illustrate this method, we find a polynomial $p(x)$ in $\mathcal{P}_2$ such that $p(1) = 3$, $p(2) = 1$, and $p(4) = -1$. In the previous notation, we have $a_0 = 1$, $a_1 = 2$, and $a_2 = 4$ and $c_0 = 3$, $c_1 = 1$, and $c_2 = -1$. Then

$$p_0(x) = \frac{(x - a_1)(x - a_2)}{(a_0 - a_1)(a_0 - a_2)} = \frac{(x-2)(x-4)}{(1-2)(1-4)} = \frac{1}{3}(x^2 - 6x + 8)$$

$$p_1(x) = \frac{(x - a_0)(x - a_2)}{(a_1 - a_0)(a_1 - a_2)} = \frac{(x-1)(x-4)}{(2-1)(2-4)} = -\frac{1}{2}(x^2 - 5x + 4)$$

$$p_2(x) = \frac{(x - a_0)(x - a_1)}{(a_2 - a_0)(a_2 - a_1)} = \frac{(x-1)(x-2)}{(4-1)(4-2)} = \frac{1}{6}(x^2 - 3x + 2).$$

Thus

$$\begin{aligned} p(x) &= (3)p_0(x) + (1)p_1(x) + (-1)p_2(x) \\ &= \frac{3}{3}(x^2 - 6x + 8) - \frac{1}{2}(x^2 - 5x + 4) - \frac{1}{6}(x^2 - 3x + 2) \\ &= \frac{1}{3}x^2 - 3x + \frac{17}{3}. \end{aligned}$$

Example 9 extends to polynomials of any positive degree. In general, for any positive integer n and any distinct real numbers $a_0, a_1, \ldots, a_n$, define

$$p_i(x) = \frac{(x - a_0)(x - a_1)\cdots(x - a_{i-1})(x - a_{i+1})\cdots(x - a_n)}{(a_i - a_0)(a_i - a_1)\cdots(a_i - a_{i-1})(a_i - a_{i+1})\cdots(a_i - a_n)}$$

for all i. The set $\{p_0(x), p_1(x), \ldots, p_n(x)\}$ is a basis for $\mathcal{P}_n$. Using the same methods as in Example 9, we can show that for any real numbers $c_0, c_1, \ldots, c_n$,

$$p(x) = c_0p_0(x) + c_1p_1(x) + \cdots + c_np_n(x)$$

is the unique polynomial in $\mathcal{P}_n$ such that $p(a_i) = c_i$ for all i. The polynomials $p_i(x)$ are called the **Lagrange[2] interpolating polynomials** (associated with $a_0, a_1, \ldots, a_n$).

EXERCISES

In Exercises 1–8, determine whether each subset of $\mathcal{M}_{2\times 2}$ is linearly independent or linearly dependent.

1. $\left\{\begin{bmatrix}1 & 2\\3 & 1\end{bmatrix}, \begin{bmatrix}1 & -5\\-4 & 0\end{bmatrix}, \begin{bmatrix}3 & -1\\2 & 2\end{bmatrix}\right\}$
2. $\left\{\begin{bmatrix}1 & 2\\3 & 1\end{bmatrix}, \begin{bmatrix}1 & -1\\0 & 1\end{bmatrix}, \begin{bmatrix}1 & 0\\1 & 1\end{bmatrix}\right\}$
3. $\left\{\begin{bmatrix}1 & 2\\2 & 1\end{bmatrix}, \begin{bmatrix}4 & 3\\-1 & 0\end{bmatrix}, \begin{bmatrix}12 & 9\\-3 & 0\end{bmatrix}\right\}$
4. $\left\{\begin{bmatrix}1 & 2\\2 & 1\end{bmatrix}, \begin{bmatrix}1 & 3\\3 & 1\end{bmatrix}, \begin{bmatrix}1 & 2\\3 & 1\end{bmatrix}\right\}$
5. $\left\{\begin{bmatrix}1 & 0 & 1\\-1 & 2 & 1\end{bmatrix}, \begin{bmatrix}-1 & 1 & 2\\2 & -1 & 1\end{bmatrix}, \begin{bmatrix}-1 & 0 & 1\\1 & -1 & 0\end{bmatrix}\right\}$
6. $\left\{\begin{bmatrix}1 & 0 & 1\\-1 & 2 & 1\end{bmatrix}, \begin{bmatrix}-1 & 1 & 2\\2 & -1 & 1\end{bmatrix}, \begin{bmatrix}3 & 2 & 9\\-1 & 8 & 7\end{bmatrix}\right\}$
7. $\left\{\begin{bmatrix}1 & 0\\-2 & 1\end{bmatrix}, \begin{bmatrix}0 & -1\\1 & 1\end{bmatrix}, \begin{bmatrix}-1 & 2\\1 & 0\end{bmatrix}, \begin{bmatrix}2 & 1\\-4 & 4\end{bmatrix}\right\}$
8. $\left\{\begin{bmatrix}1 & 0\\-2 & 1\end{bmatrix}, \begin{bmatrix}0 & -1\\1 & 1\end{bmatrix}, \begin{bmatrix}-1 & 2\\1 & 0\end{bmatrix}, \begin{bmatrix}2 & 1\\2 & -2\end{bmatrix}\right\}$

In Exercises 9–16, determine whether each subset of $\mathcal{P}$ is linearly independent or linearly dependent.

9. $\{1+x, 1-x, 1+x+x^2, 1+x-x^2\}$
10. $\{x^2-2x+5, 2x^2-4x+10\}$
11. $\{x^2-2x+5, 2x^2-5x+10, x^2\}$
12. $\{x^3+4x^2-2x+3, x^3+6x^2-x+4, 3x^3+8x^2-8x+7\}$
13. $\{x^3+2x^2, -x^2+3x+1, x^3-x^2+2x-1\}$
14. $\{x^3-x, 2x^2+4, -2x^3+3x^2+2x+6\}$
15. $\{x^4-x^3+5x^2-8x+6, -x^4+x^3-5x^2+5x-3, x^4+3x^2-3x+5, 2x^4+3x^3+4x^2-x+1, x^3-x+2\}$
16. $\{x^4-x^3+5x^2-8x+6, -x^4+x^3-5x^2+5x-3, x^4+3x^2-3x+5, 2x^4+x^3+4x^2+8x\}$

In Exercises 17–24, determine whether each subset of $\mathcal{F}(\mathcal{R})$ is linearly independent or linearly dependent.

17. $\{t, t\sin t\}$
18. $\{t, t\sin t, e^{2t}\}$
19. $\{\sin t, \sin^2 t, \cos^2 t, 1\}$
20. $\{\sin t, e^{-t}, e^t\}$
21. $\{e^t, e^{2t}, \ldots, e^{nt}, \ldots\}$
22. $\{\cos^2 t, \sin^2 t, \cos 2t\}$
23. $\{t, \sin t, \cos t\}$
24. $\{1, t, t^2, \ldots\}$

In Exercises 25–30, use Lagrange interpolating polynomials to determine the polynomial in $\mathcal{P}_n$ whose graph passes through the given points.

25. $n=3$; $(0,1)$, $(1,0)$, and $(2,3)$
26. $n=3$; $(1,8)$, $(2,5)$, and $(3,-4)$
27. $n=3$; $(-1,-11)$, $(1,1)$, $(2,1)$
28. $n=3$; $(-2,-13)$, $(1,2)$, $(3,12)$
29. $n=4$; $(-1,5)$, $(0,2)$, $(1,-1)$, $(2,2)$
30. $n=4$; $(-2,1)$, $(-1,3)$, $(1,1)$, $(2,-15)$

In Exercises 31–48, determine whether the statements are true or false.

31. If a set is infinite, it cannot be linearly independent.
32. Every vector space has a finite basis.
33. The dimension of the vector space $\mathcal{P}_n$ equals n.
34. Every subspace of an infinite-dimensional vector space is infinite-dimensional.
35. It is possible for a vector space to have both an infinite basis and a finite basis.
36. If every finite subset of S is linearly independent, then S is linearly independent.
37. Every nonzero finite-dimensional vector space is isomorphic to $\mathcal{R}^n$ for some n.
38. If a subset of a vector space contains $\mathbf{0}$, then it is linearly dependent.
39. In $\mathcal{P}$, the set $\{x, x^3, x^5, \ldots\}$ is linearly dependent.
40. A basis for a vector space V is a linearly independent set that is also a generating set for V.
41. If $\mathcal{B} = \{\mathbf{v}_1, \mathbf{v}_2, \ldots, \mathbf{v}_n\}$ is a basis for a vector space V, then the mapping $\Psi\colon \mathcal{R}^n \to V$ defined by

$$\Psi\left(\begin{bmatrix}c_1\\c_2\\\vdots\\c_n\end{bmatrix}\right) = c_1\mathbf{v}_1 + c_2\mathbf{v}_2 + \cdots + c_n\mathbf{v}_n$$

is an isomorphism.

[2] Joseph Louis Lagrange (1736–1813) was one of the most important mathematicians and physical scientists of his time. Among his most significant accomplishments was the development of the *calculus of variations,* which he applied to problems in celestial mechanics. His 1788 treatise on analytical mechanics summarized the principal results in mechanics and demonstrated the importance of mathematics to mechanics. Lagrange also made important contributions to number theory, the theory of equations, and the foundations of calculus (by emphasizing functions and the use of Taylor series).

42. The dimension of $\mathcal{M}_{m \times n}$ is $m + n$.

43. If $T: V \to W$ is an isomorphism between vector spaces V and W and $\{\mathbf{v}_1, \mathbf{v}_2, \ldots, \mathbf{v}_k\}$ is a linearly independent subset of V, then $\{T(\mathbf{v}_1), T(\mathbf{v}_2), \ldots, T(\mathbf{v}_k)\}$ is a linearly independent subset of W.

44. If $T: V \to W$ is an isomorphism between finite-dimensional vector spaces V and W, then the dimensions of V and W are equal.

45. The dimension of $\mathcal{L}(\mathcal{R}^n, \mathcal{R}^m)$ is $m + n$.

46. The vector spaces $\mathcal{M}_{m \times n}$ and $\mathcal{L}(\mathcal{R}^n, \mathcal{R}^m)$ are isomorphic.

47. The Lagrange interpolating polynomials associated with $n + 1$ distinct real numbers form a basis for $\mathcal{P}_n$.

48. If a vector space contains a finite linearly dependent set, then the vector space is finite-dimensional.

---

49. Let N be the set of positive integers, and let f, g, and h be the functions in $\mathcal{F}(N)$ defined by $f(n) = n + 1$, $g(n) = 1$, and $h(n) = 2n - 1$. Determine if the set $\{f, g, h\}$ is linearly independent. Justify your answer.

50. Let N denote the set of nonnegative integers, and let V be the subset of $\mathcal{F}(N)$ consisting of the functions that are zero except at finitely many elements of N. By Exercise 49 of Section 7.2, V is a subspace of $\mathcal{F}(N)$. For each n in N, let $f_n: N \to \mathcal{R}$ be defined by

$$f_n(k) = \begin{cases} 0 & \text{for } k \neq n \\ 1 & \text{for } k = n. \end{cases}$$

Prove that $\{f_1, f_2, \ldots, f_n, \ldots\}$ is a basis for V.

In Exercises 51–58, find a basis for the subspace W of the vector space V.

51. Let W be the subspace of symmetric 3×3 matrices and $V = \mathcal{M}_{3\times 3}$.

52. Let W be the subspace of skew-symmetric 3×3 matrices and $V = \mathcal{M}_{3\times 3}$.

53. Let W be the subspace of 2×2 matrices with trace equal to 0 and $V = \mathcal{M}_{2\times 2}$.

54. Let W be the subspace of $V = \mathcal{P}_n$ consisting of polynomials $p(x)$ for which $p(0) = 0$.

55. Let W be the subspace of $V = \mathcal{P}_n$ consisting of polynomials $p(x)$ for which $p(1) = 0$.

56. Let $W = \{f \in \mathcal{D}(\mathcal{R}): f' = f\}$, where f' is the derivative of f and $\mathcal{D}(\mathcal{R})$ is the set of functions in $\mathcal{F}(\mathcal{R})$ that are differentiable, and let $V = \mathcal{F}(\mathcal{R})$.

57. Let $W = \{p(x) \in \mathcal{P}: p''(x) = 0\}$ and $V = \mathcal{P}$.

58. Let $W = \{p(x) \in \mathcal{P}: p(-x) = -p(x)\}$ and $V = \mathcal{P}$.

59. Let S be a subset of $\mathcal{P}_n$ consisting of exactly one polynomial of degree k for $k = 0, 1, \ldots, n$. Prove that S is a basis for $\mathcal{P}_n$.

The following definitions and notation apply to Exercises 60–65:

Definition An $n \times n$ matrix is called a **magic square of order n** if the sum of the entries in each row, the sum of the entries in each column, the sum of the diagonal entries, and the sum of the entries on the secondary diagonal are all equal. (The entries on the secondary diagonal are the $(1, n)$-entry, the $(2, n-1)$-entry, ..., the $(n, 1)$-entry.) This common value is called the **sum** of the magic square.

For example, the 3×3 matrix

$$\begin{bmatrix} 4 & 9 & 2 \\ 3 & 5 & 7 \\ 8 & 1 & 6 \end{bmatrix}$$

is a magic square of order 3 with sum equal to 15.

Let V_n denote the set of all magic squares of order n, and let W_n denote the subset of V_n consisting of magic squares with sum equal to 0.

60. (a) Show that V_n is a subspace of $\mathcal{M}_{n\times n}$.
    (b) Show that W_n is a subspace of V_n.

61. For each positive integer n, let C_n be the $n \times n$ matrix all of whose entries are equal to $1/n$.
    (a) Prove that C_n is in V_n.
    (b) Prove that, for any positive integer n and magic square A in V_n, if A has sum s, then there is a unique magic square B in W_n such that $A = B + sC_n$.

62. Prove that W_3 has dimension equal to 2.

63. Prove that V_3 has dimension equal to 3.

64. Use the result of Exercise 61 to prove that for any positive integer n, $\dim V_n = \dim W_n + 1$.

65. Prove that for any $n \geq 3$, $\dim W_n = n^2 - 2n - 1$, and hence V_n has dimension equal to $n^2 - 2n$ by Exercise 64. *Hint:* Identify $\mathcal{M}_{n\times n}$ with $\mathcal{R}^{n^2}$ and then analyze the description of W_n as the solution space of a system of homogeneous equations.

66. Let V and W be vector spaces and $T: V \to W$ be an isomorphism.
    (a) Prove that if $\{\mathbf{v}_1, \mathbf{v}_2, \ldots, \mathbf{v}_n\}$ is a linearly independent subset of V, then the set of images $\{T(\mathbf{v}_1), T(\mathbf{v}_2), \ldots, T(\mathbf{v}_n)\}$ is a linearly independent subset of W.
    (b) Prove that if $\{\mathbf{v}_1, \mathbf{v}_2, \ldots, \mathbf{v}_n\}$ is a basis for V, then the set of images $\{T(\mathbf{v}_1), T(\mathbf{v}_2), \ldots, T(\mathbf{v}_n)\}$ is a basis for W.
    (c) Prove that if V is finite-dimensional, then W is finite-dimensional and $\dim V = \dim W$.

67. Use Exercise 66 and Example 8 to find a basis for $\mathcal{L}(\mathcal{R}^n, \mathcal{R}^m)$.

In Exercises 68–71, use an isomorphism from V to $\mathcal{R}^n$ to prove the result.

68. Let n be a positive integer. Suppose that V is a vector space such that any subset of V consisting of more than n vectors is linearly dependent, and some linearly independent subset contains n vectors. Prove that any linearly independent subset of V consisting of n vectors is a basis for V, and hence $\dim V = n$.

69. Let V be a finite-dimensional vector space of dimension $n \geq 1$.

(a) Prove that any subset of V containing more than n vectors is linearly dependent.

(b) Prove that any linearly independent subset of V consisting of n vectors is a basis for V.

70. Let V be a vector space of dimension $n \geq 1$, and suppose that $\mathcal{S}$ is a finite generating set for V. Prove the following statements:

(a) $\mathcal{S}$ contains at least n vectors.

(b) If $\mathcal{S}$ consists of exactly n vectors, then $\mathcal{S}$ is a basis for V.

71. Let V be a finite-dimensional vector space and W be a subspace of V. Prove the following statements:

(a) W is finite-dimensional, and $\dim W \leq \dim V$.

(b) If $\dim W = \dim V$, then $W = V$.

72. Let V be an infinite-dimensional vector space. Prove that V contains an infinite linearly independent set. *Hint:* Choose a nonzero vector $\mathbf{v}_1$ in V. Next, choose a vector $\mathbf{v}_2$ not in the span of $\{\mathbf{v}_1\}$. Show that this process can be continued to obtain an infinite subset $\{\mathbf{v}_1, \mathbf{v}_2, \ldots, \mathbf{v}_n, \ldots\}$ of V such that for any n, $\mathbf{v}_{n+1}$ is not in the span of $\{\mathbf{v}_1, \mathbf{v}_2, \ldots, \mathbf{v}_n\}$. Now show that this infinite set is linearly independent.

73. Prove that if $\mathcal{B}$ is a basis for a vector space that contains exactly n vectors, then $\Phi_{\mathcal{B}}\colon V \to \mathcal{R}^n$ is a linear transformation.

74. Suppose that V and W are isomorphic vector spaces. Prove that if V is infinite-dimensional, then W is infinite-dimensional.

75. Let V and W be finite-dimensional vector spaces. Prove that $\dim \mathcal{L}(V, W) = (\dim V) \cdot (\dim W)$.

76. Let n be a positive integer. For $0 \leq i \leq n$, define $T_i\colon \mathcal{P}_n \to \mathcal{R}$ by $T_i(f(x)) = f(i)$. Prove that T_i is linear for all i and that $\{T_0, T_1, \ldots, T_n\}$ is a basis for $\mathcal{L}(\mathcal{P}_n, \mathcal{R})$. *Hint:* For each i, let $p_i(x)$ be the ith Lagrange interpolating polynomial associated with $0, 1, \ldots, n$. Show that for all i and j,

$$T_i(p_j(x)) = \begin{cases} 0 & \text{if } i \neq j \\ 1 & \text{if } i = j. \end{cases}$$

Use this to show that $\{T_0, T_1, \ldots, T_n\}$ is linearly independent. Now apply Exercises 75 and 69.

77. Apply Exercise 76 to prove that for any positive integer n and any scalars a and b, there exist unique scalars $c_0, c_1, \ldots, c_n$ such that

$$\int_a^b f(x)\,dx = c_0 f(0) + c_1 f(1) + \cdots + c_n f(n)$$

for every polynomial $f(x)$ in $\mathcal{P}_n$.

78. (a) Derive *Simpson's rule:* For any polynomial $f(x)$ in $\mathcal{P}_2$ and any scalars $a < b$,

$$\int_a^b f(x)\,dx = \frac{b-a}{6}\left[f(a) + 4f\left(\frac{a+b}{2}\right) + f(b)\right].$$

(b) Verify that Simpson's rule is valid for the polynomial x^3, and use this fact to justify that Simpson's rule is valid for every polynomial in $\mathcal{P}_3$.

In Exercises 79–83, use either a calculator with matrix capabilities or computer software such as MATLAB to solve each problem.

In Exercises 79–82, determine whether each set is linearly dependent. In the case that the set is linearly dependent, write some vector in the set as a linear combination of the others.

79. $\{1 + x - x^2 + 3x^3 - x^4, 2 + 5x - x^3 + x^4, 3x + 2x^2 + 7x^4, 4 - x^2 + x^3 - x^4\}$

80. $\{2 + 5x - 2x^2 + 3x^3 + x^4, 3 + 3x - x^2 + x^3 + x^4, 6 - 3x + 2x^2 - 5x^3 + x^4, 2 - x + x^2 + x^4\}$

81. $\left\{\begin{bmatrix} 0.97 & -1.12 \\ 1.82 & 2.13 \end{bmatrix}, \begin{bmatrix} 1.14 & 2.01 \\ 1.01 & 3.21 \end{bmatrix}, \begin{bmatrix} -0.63 & 7.38 \\ -3.44 & 0.03 \end{bmatrix}, \begin{bmatrix} 2.12 & -1.21 \\ 0.07 & -1.32 \end{bmatrix}\right\}$

82. $\left\{\begin{bmatrix} 1.23 & -0.41 \\ 2.57 & 3.13 \end{bmatrix}, \begin{bmatrix} 2.71 & 1.40 \\ -5.23 & 2.71 \end{bmatrix}, \begin{bmatrix} 3.13 & 1.10 \\ 2.12 & -1.11 \end{bmatrix}, \begin{bmatrix} 8.18 & 2.15 \\ -1.21 & 4.12 \end{bmatrix}\right\}$

83. In view of Exercise 77, find the scalars c_0, c_1, c_2, c_3, and c_4 such that

$$\int_0^1 f(x)\,dx = c_0 f(0) + c_1 f(1) + c_2 f(2) + c_3 f(3) + c_4 f(4)$$

for every polynomial $f(x)$ in $\mathcal{P}_4$. *Hint:* Apply this equation to 1, x, x^2, x^3, and x^4 to obtain a system of 5 linear equations in 5 variables.

SOLUTIONS TO THE PRACTICE PROBLEMS

1. Suppose that

$$c_1\begin{bmatrix} 1 & 1 \\ 1 & 0 \end{bmatrix} + c_2\begin{bmatrix} 0 & 0 \\ 1 & 1 \end{bmatrix} + c_3\begin{bmatrix} 0 & 2 \\ 0 & -1 \end{bmatrix} + c_4\begin{bmatrix} 0 & 1 \\ 1 & 0 \end{bmatrix} = \begin{bmatrix} 0 & 0 \\ 0 & 0 \end{bmatrix}$$

for scalars c_1, c_2, c_3, and c_4. This equation may be rewritten as

$$\begin{bmatrix} c_1 & c_1 + 2c_3 + c_4 \\ c_1 + c_2 + c_4 & c_2 - c_3 \end{bmatrix} = \begin{bmatrix} 0 & 0 \\ 0 & 0 \end{bmatrix}.$$

The last matrix equation is equivalent to the system

$$\begin{aligned} c_1 \qquad\qquad\qquad &= 0 \\ c_1 \qquad + 2c_3 + c_4 &= 0 \\ c_1 + c_2 \qquad\quad + c_4 &= 0 \\ c_2 - \quad c_3 \qquad &= 0. \end{aligned}$$

After we apply Gaussian elimination to this system, we discover that its only solution is $c_1 = c_2 = c_3 = c_4 = 0$. Therefore S is linearly independent.

2. From Example 7, we know that $\dim \mathcal{M}_{2\times 2} = 4$. By Practice Problem 1, the set S, which contains four matrices, is linearly independent. Therefore S is a basis for $\mathcal{M}_{2\times 2}$ by property 2 of finite-dimensional vector spaces on page 515.

7.4 MATRIX REPRESENTATIONS OF LINEAR OPERATORS

Given an n-dimensional vector space V with the basis $\mathcal{B}$, we wish to use $\mathcal{B}$ to convert problems concerning linear operators on V and vectors in V into ones involving $n \times n$ matrices and vectors in $\mathcal{R}^n$. We have already seen how to use the isomorphism $\Phi_{\mathcal{B}}\colon V \to \mathcal{R}^n$ to identify a vector in V with a vector in $\mathcal{R}^n$. In fact, we can use this isomorphism to extend the definition of *coordinate vector*, given in Section 4.4, to all finite-dimensional vector spaces.

COORDINATE VECTORS AND MATRIX REPRESENTATIONS

Definition Let V be a finite-dimensional vector space and $\mathcal{B}$ be a basis for V. For any vector $\mathbf{v}$ in V, the vector $\Phi_{\mathcal{B}}(\mathbf{v})$ is called the **coordinate vector of v relative to** $\mathcal{B}$ and is denoted by $[\mathbf{v}]_{\mathcal{B}}$.

Since $\Phi_{\mathcal{B}}(\mathbf{v}) = [\mathbf{v}]_{\mathcal{B}}$ and $\Phi_{\mathcal{B}}$ is linear, it follows that for any vectors $\mathbf{u}$ and $\mathbf{v}$ in V and any scalar c,

$$[\mathbf{u}+\mathbf{v}]_{\mathcal{B}} = \Phi_{\mathcal{B}}(\mathbf{u}+\mathbf{v}) = \Phi_{\mathcal{B}}(\mathbf{u}) + \Phi_{\mathcal{B}}(\mathbf{v}) = [\mathbf{u}]_{\mathcal{B}} + [\mathbf{v}]_{\mathcal{B}}$$

and

$$[c\mathbf{v}]_{\mathcal{B}} = \Phi_{\mathcal{B}}(c\mathbf{v}) = c\Phi_{\mathcal{B}}(\mathbf{v}) = c[\mathbf{v}]_{\mathcal{B}}.$$

Let T be a linear operator on an n-dimensional vector space V with the basis $\mathcal{B}$. We show how to represent T as an $n \times n$ matrix. Our strategy is to use $\Phi_{\mathcal{B}}$ and T to construct a linear operator on $\mathcal{R}^n$ and choose its standard matrix.

We can define a linear operator on $\mathcal{R}^n$ by using $\Phi_{\mathcal{B}}$ and T as follows: Starting at $\mathcal{R}^n$, apply $\Phi_{\mathcal{B}}^{-1}$, which maps $\mathcal{R}^n$ to V. Now apply T and then $\Phi_{\mathcal{B}}$ to return to $\mathcal{R}^n$. The result is the composition of linear transformations $\Phi_{\mathcal{B}} T \Phi_{\mathcal{B}}^{-1}$, which is a linear operator

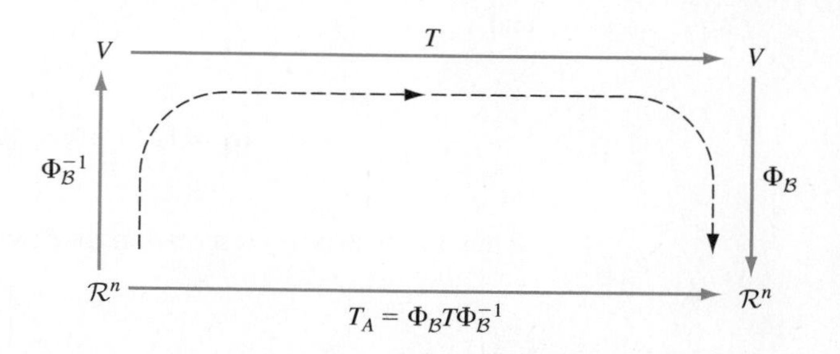

Figure 7.2 The linear operator T_A, where A is the standard matrix of $\Phi_{\mathcal{B}} T \Phi_{\mathcal{B}}^{-1}$

on $\mathcal{R}^n$. (See Figure 7.2.) This linear operator equals T_A, where A is the standard matrix of $\Phi_{\mathcal{B}} T \Phi_{\mathcal{B}}^{-1}$. Thus we make the following definition:

Definition Let T be a linear operator on an n-dimensional vector space V, and let $\mathcal{B}$ be a basis for V. The standard matrix of the linear operator $\Phi_{\mathcal{B}} T \Phi_{\mathcal{B}}^{-1}$ on $\mathcal{R}^n$ is called the **matrix representation of T with respect to $\mathcal{B}$** and is denoted by $[T]_{\mathcal{B}}$. So if $A = [T]_{\mathcal{B}}$, then $T_A = \Phi_{\mathcal{B}} T \Phi_{\mathcal{B}}^{-1}$.

For a linear operator T on a finite-dimensional vector space V with the basis $\mathcal{B} = \{\mathbf{v}_1, \mathbf{v}_2, \ldots, \mathbf{v}_n\}$, we now show how to compute the matrix representation $[T]_{\mathcal{B}}$. Observe that for each j, $\Phi_{\mathcal{B}}(\mathbf{v}_j) = \mathbf{e}_j$, the jth standard vector of $\mathcal{R}^n$, and hence $\Phi_{\mathcal{B}}^{-1}(\mathbf{e}_j) = \mathbf{v}_j$ for all j. Let $A = [T]_{\mathcal{B}}$. Then for each j, the jth column of A can be obtained as

$$\mathbf{a}_j = A\mathbf{e}_j = T_A(\mathbf{e}_j) = \Phi_{\mathcal{B}} T \Phi_{\mathcal{B}}^{-1}(\mathbf{e}_j) = \Phi_{\mathcal{B}} T(\mathbf{v}_j) = [T(\mathbf{v}_j)]_{\mathcal{B}}.$$

We summarize this result as follows:

The Matrix Representation of a Linear Operator

Let T be a linear operator on a finite-dimensional vector space V with the basis $\mathcal{B} = \{\mathbf{v}_1, \mathbf{v}_2, \ldots, \mathbf{v}_n\}$. Then $[T]_{\mathcal{B}}$ is the $n \times n$ matrix whose jth column is $[T(\mathbf{v}_j)]_{\mathcal{B}}$. Thus

$$[T]_{\mathcal{B}} = [\, [T(\mathbf{v}_1)]_{\mathcal{B}} \;\; [T(\mathbf{v}_2)]_{\mathcal{B}} \;\; \cdots \;\; [T(\mathbf{v}_n)]_{\mathcal{B}} \,].$$

Example 1 Let $T\colon \mathcal{P}_2 \to \mathcal{P}_2$ be defined by

$$T(p(x)) = p(0) + 3p(1)x + p(2)x^2.$$

For example, if $p(x) = 2 + x - 2x^2$, then $p(0) = 2$, $p(1) = 1$, and $p(2) = -4$. Therefore $T(p(x)) = 2 + 3x - 4x^2$. It can be shown that T is linear. Let $\mathcal{B} = \{1, x, x^2\}$, which is a basis for $\mathcal{P}_2$, and let $A = [T]_{\mathcal{B}}$. Then

$$\mathbf{a}_1 = [T(1)]_{\mathcal{B}} = [1 + 3x + x^2]_{\mathcal{B}} = \begin{bmatrix} 1 \\ 3 \\ 1 \end{bmatrix},$$

$$\mathbf{a}_2 = [T(x)]_{\mathcal{B}} = [0 + 3x + 2x^2]_{\mathcal{B}} = \begin{bmatrix} 0 \\ 3 \\ 2 \end{bmatrix},$$

and

$$\mathbf{a}_3 = [T(x^2)]_{\mathcal{B}} = [0 + 3x + 4x^2]_{\mathcal{B}} = \begin{bmatrix} 0 \\ 3 \\ 4 \end{bmatrix}.$$

Thus the matrix representation of T with respect to $\mathcal{B}$ is

$$A = \begin{bmatrix} 1 & 0 & 0 \\ 3 & 3 & 3 \\ 1 & 2 & 4 \end{bmatrix}.$$

Example 2

Let $\mathcal{B} = \{e^t \cos t,\ e^t \sin t\}$, a subset of C^∞, and let $V = \text{Span}\,\mathcal{B}$. It can be shown that $\mathcal{B}$ is linearly independent and hence is a basis for V. Let D be the linear operator on V defined by $D(f) = f'$ for all f in V. Then

$$D(e^t \cos t) = (1)e^t \cos t + (-1)e^t \sin t$$

and

$$D(e^t \sin t) = (1)e^t \cos t + (1)e^t \sin t.$$

Therefore the matrix representation of D with respect to $\mathcal{B}$ is

$$[D]_{\mathcal{B}} = \begin{bmatrix} 1 & 1 \\ -1 & 1 \end{bmatrix}.$$

Practice Problem 1 ▶ Define $T\colon \mathcal{P}_2 \to \mathcal{P}_2$ by $T(p(x)) = (x+1)p'(x) + p(x)$.

(a) Prove that T is linear.

(b) Determine the matrix representation of T with respect to the basis $\{1,\ x,\ x^2\}$ for $\mathcal{P}_2$. ◀

The next result enables us to express the image of any vector under a linear operator on a finite-dimensional vector space as a matrix-vector product. We use this description to acquire information about a linear operator by applying what we already know about matrices.

THEOREM 7.10

Let T be a linear operator on a finite-dimensional vector space V with the basis $\mathcal{B}$. Then, for any vector $\mathbf{v}$ in V,

$$[T(\mathbf{v})]_{\mathcal{B}} = [T]_{\mathcal{B}}[\mathbf{v}]_{\mathcal{B}}.$$

PROOF Setting $A = [T]_{\mathcal{B}}$, we have $T_A = \Phi_{\mathcal{B}} T \Phi_{\mathcal{B}}^{-1}$. Thus, for any vector $\mathbf{v}$ in V, we have

$$[T(\mathbf{v})]_{\mathcal{B}} = \Phi_{\mathcal{B}} T(\mathbf{v}) = \Phi_{\mathcal{B}} T \Phi_{\mathcal{B}}^{-1} \Phi_{\mathcal{B}}(\mathbf{v}) = T_A([\mathbf{v}]_{\mathcal{B}}) = A[\mathbf{v}]_{\mathcal{B}} = [T]_{\mathcal{B}}[\mathbf{v}]_{\mathcal{B}}.$$

■

Example 3

Let D be the linear operator on $\mathcal{P}_2$ defined by $D(p(x)) = p'(x)$, $\mathcal{B} = \{1, x, x^2\}$ (which is a basis for $\mathcal{P}_2$), and $A = [T]_{\mathcal{B}}$. Then

$$\mathbf{a}_1 = [D(1)]_{\mathcal{B}} = [\mathbf{0}]_{\mathcal{B}} = \begin{bmatrix} 0 \\ 0 \\ 0 \end{bmatrix}, \qquad \mathbf{a}_2 = [D(x)]_{\mathcal{B}} = [1]_{\mathcal{B}} = \begin{bmatrix} 1 \\ 0 \\ 0 \end{bmatrix},$$

and

$$\mathbf{a}_3 = [D(x^2)]_{\mathcal{B}} = [2x]_{\mathcal{B}} = \begin{bmatrix} 0 \\ 2 \\ 0 \end{bmatrix}.$$

Therefore

$$[T]_{\mathcal{B}} = [\mathbf{a}_1\ \mathbf{a}_2\ \mathbf{a}_3] = \begin{bmatrix} 0 & 1 & 0 \\ 0 & 0 & 2 \\ 0 & 0 & 0 \end{bmatrix}.$$

With this information, we can use Theorem 7.10 to compute the derivative of a polynomial in $\mathcal{P}_2$.

Consider the polynomial $p(x) = 5 - 4x + 3x^2$. Then

$$[p(x)]_{\mathcal{B}} = \begin{bmatrix} 5 \\ -4 \\ 3 \end{bmatrix}.$$

By Theorem 7.10,

$$\begin{aligned} [p'(x)]_{\mathcal{B}} = [D(p(x)]_{\mathcal{B}} &= [D]_{\mathcal{B}}[p(x)]_{\mathcal{B}} \\ &= \begin{bmatrix} 0 & 1 & 0 \\ 0 & 0 & 2 \\ 0 & 0 & 0 \end{bmatrix} \begin{bmatrix} 5 \\ -4 \\ 3 \end{bmatrix} \\ &= \begin{bmatrix} -4 \\ 6 \\ 0 \end{bmatrix}. \end{aligned}$$

This vector is the coordinate vector of the polynomial $-4 + 6x$, which is the derivative of $p(x)$. Thus we can compute the derivative of a polynomial by evaluating a matrix–vector product.

Practice Problem 2 ▶ Let $p(x) = 2 - 3x + 5x^2$. Use the linear transformation T in Practice Problem 1 to compute $T(p(x))$ in two ways—first by using the rule for T, and second by applying Theorem 7.10. ◀

MATRIX REPRESENTATIONS OF INVERTIBLE LINEAR OPERATORS

As in Theorem 2.13, we can use matrix representations to derive a test for invertibility of a linear operator on a finite-dimensional vector space and to obtain a method for computing the inverse of an invertible operator.

Let T be a linear operator on a finite-dimensional vector space V with the basis $\mathcal{B}$, and let $A = [T]_{\mathcal{B}}$.

First, suppose that T is invertible. Then $T_A = \Phi_{\mathcal{B}} T \Phi_{\mathcal{B}}^{-1}$ is a composition of isomorphisms and hence is invertible by the boxed result on page 508. Hence A is invertible by Theorem 2.13. Conversely, suppose that A is invertible. Then T_A is invertible by Theorem 2.13, and hence $T = \Phi_{\mathcal{B}}^{-1} T_A \Phi_{\mathcal{B}}$ is also invertible by the boxed result on page 508. Thus a linear operator is invertible if and only if any one of its matrix representations is invertible.

When T is invertible, we can obtain a simple relationship between the matrix representations of T and T^{-1} with respect to a basis $\mathcal{B}$. Let $C = [T^{-1}]_{\mathcal{B}}$. By definition, $T_C = \Phi_{\mathcal{B}} T^{-1} \Phi_{\mathcal{B}}^{-1}$. Furthermore,

$$T_{A^{-1}} = (T_A)^{-1} = (\Phi_{\mathcal{B}} T \Phi_{\mathcal{B}}^{-1})^{-1} = \Phi_{\mathcal{B}} T^{-1} \Phi_{\mathcal{B}}^{-1} = T_C.$$

Therefore $C = A^{-1}$; that is, the matrix representation of T^{-1} is the inverse of the matrix representation of T.

We summarize these results as follows:

The Matrix Representation of an Invertible Linear Operator

Let T be a linear operator on a finite-dimensional vector space V with the basis $\mathcal{B}$, and let $A = [T]_{\mathcal{B}}$.

(a) T is invertible if and only if A is invertible.

(b) If T is invertible, then $[T^{-1}]_{\mathcal{B}} = A^{-1}$.

Example 4 Find $[D^{-1}]_{\mathcal{B}}$ for the linear operator D and basis $\mathcal{B}$ in Example 2, and use it to find an antiderivative of $e^t \sin t$.

Solution In Example 2, we saw that

$$[D]_{\mathcal{B}} = \begin{bmatrix} 1 & 1 \\ -1 & 1 \end{bmatrix}.$$

Thus D is invertible because $[D]_{\mathcal{B}}$ is invertible, and

$$[D^{-1}]_{\mathcal{B}} = ([D]_{\mathcal{B}})^{-1} = \begin{bmatrix} 1 & 1 \\ -1 & 1 \end{bmatrix}^{-1} = \begin{bmatrix} \frac{1}{2} & -\frac{1}{2} \\ \frac{1}{2} & \frac{1}{2} \end{bmatrix}.$$

An antiderivative of $e^t \sin t$ equals $D^{-1}(e^t \sin t)$. Since

$$[e^t \sin t]_{\mathcal{B}} = \begin{bmatrix} 0 \\ 1 \end{bmatrix},$$

it follows that

$$[D^{-1}(e^t \sin t)]_{\mathcal{B}} = \begin{bmatrix} \frac{1}{2} & -\frac{1}{2} \\ \frac{1}{2} & \frac{1}{2} \end{bmatrix} \begin{bmatrix} 0 \\ 1 \end{bmatrix} = \begin{bmatrix} -\frac{1}{2} \\ \frac{1}{2} \end{bmatrix}.$$

Thus $D^{-1}(e^t \sin t) = (-\frac{1}{2})e^t \cos t + (\frac{1}{2})e^t \sin t$.

Practice Problem 3 ▶ Let T be the linear transformation in Practice Problem 1. Show that T is invertible and find a rule for its inverse. ◀

EIGENVALUES AND EIGENVECTORS

We now extend the definitions of *eigenvector*, *eigenvalue*, and *eigenspace* in Chapter 5 to general vector spaces.

Let T be a linear operator on a vector space V. A nonzero vector $\mathbf{v}$ in V is called an **eigenvector** of T if there is a scalar λ such that $T(\mathbf{v}) = \lambda\mathbf{v}$. The scalar λ is called the **eigenvalue** of T corresponding to $\mathbf{v}$. If λ is an eigenvalue of T, then the set of all vectors $\mathbf{v}$ in V such that $T(\mathbf{v}) = \lambda\mathbf{v}$ is called the **eigenspace** of T corresponding to λ. As in $\mathcal{R}^n$, this eigenspace is the subspace of V consisting of the zero vector and all the eigenvectors of T corresponding to λ. (See Exercise 49.)

Example 5 Let $D\colon \mathrm{C}^\infty \to \mathrm{C}^\infty$ be the linear operator in Example 2 of Section 7.2. Let λ be a scalar, and let f be the exponential function $f(t) = e^{\lambda t}$. Then

$$D(f)(t) = (e^{\lambda t})' = \lambda e^{\lambda t} = \lambda f(t) = (\lambda f)(t).$$

So $D(f) = \lambda f$, and therefore f is an eigenvector of D, and λ is the eigenvalue corresponding to f. Since λ was chosen arbitrarily, we see that every scalar is an eigenvalue of D. Therefore D has infinitely many eigenvalues, in contrast to linear operators on $\mathcal{R}^n$.

Example 6 Let D be the derivative operator on C^∞ in Example 2 of Section 7.2. Then $D^2 = DD$ is also a linear operator on C^∞, and $D^2(f) = f''$ for all f in C^∞. Show that the solution set of the differential equation

$$y'' + 4y = \mathbf{0}$$

coincides with the eigenspace of D^2 corresponding to the eigenvalue $\lambda = -4$.

Solution First, observe that the solutions of this differential equation lie in C^∞. For if f is a solution, then f must be twice differentiable, and $f'' = -4f$. Thus f'' is also twice differentiable, and so f has four derivatives. We can now differentiate both sides of this equation twice to find that $f'''' = -4f''$, from which we can infer that the fourth derivative of f is twice differentiable, and so f has six derivatives. Repetition of this argument leads to the conclusion that f has derivatives of any order.

Since $y'' = D^2y$, we can rewrite the given differential equation as

$$D^2y = -4y.$$

But this last equation shows that y is in the eigenspace of D^2 corresponding to the eigenvalue -4. Thus this eigenspace equals the solution set of the differential equation.

Note that the functions $\sin 2t$ and $\cos 2t$ are solutions of this differential equation. So they are also eigenvectors of D^2 corresponding to the eigenvalue $\lambda = -4$.

We have defined a square matrix A to be **symmetric** if $A^T = A$. We say that A is **skew-symmetric** if $A^T = -A$. Note that

$$\begin{bmatrix} 0 & 1 \\ -1 & 0 \end{bmatrix}$$

is a nonzero skew-symmetric matrix.

Example 7 Let $U: \mathcal{M}_{n\times n} \to \mathcal{M}_{n\times n}$ be the linear operator defined by $U(A) = A^T$. We saw in Section 7.2 that U is an isomorphism.

If A is a nonzero symmetric matrix, then $U(A) = A^T = A$. Thus A is an eigenvector of U with $\lambda = 1$ as the corresponding eigenvalue, and the corresponding eigenspace is the set of $n \times n$ symmetric matrices. In addition, if B is a nonzero skew-symmetric matrix, then $U(B) = B^T = -B$. So B is an eigenvector of U with $\lambda = -1$ as the corresponding eigenvalue, and the corresponding eigenspace is the set of all skew-symmetric $n \times n$ matrices. It can be shown that 1 and -1 are the only eigenvalues of U. (See Exercise 42.)

Finally, we apply Theorem 7.10 to analyze the eigenvalues and eigenvectors of a linear operator T on a finite-dimensional vector space V with the basis $\mathcal{B}$.

Suppose that $\mathbf{v}$ is an eigenvector of T with corresponding eigenvalue λ. Then $\mathbf{v} \neq \mathbf{0}$, and hence $[\mathbf{v}]_{\mathcal{B}} \neq \mathbf{0}$. So, by Theorem 7.10,

$$[T]_{\mathcal{B}}[\mathbf{v}]_{\mathcal{B}} = [T(\mathbf{v})]_{\mathcal{B}} = [\lambda\mathbf{v}]_{\mathcal{B}} = \lambda[\mathbf{v}]_{\mathcal{B}},$$

and therefore $[\mathbf{v}]_{\mathcal{B}}$ is an eigenvector of the matrix $[T]_{\mathcal{B}}$ with corresponding eigenvalue λ.

Conversely, suppose that $\mathbf{w}$ is an eigenvector in $\mathcal{R}^n$ of the matrix $[T]_{\mathcal{B}}$ with corresponding eigenvalue λ. Let $\mathbf{v} = \Phi_{\mathcal{B}}^{-1}(\mathbf{w})$, which is in V. Then $[\mathbf{v}]_{\mathcal{B}} = \Phi_{\mathcal{B}}(\mathbf{v}) = \mathbf{w}$. Applying Theorem 7.10, we have

$$\Phi_{\mathcal{B}}(T(\mathbf{v})) = [T(\mathbf{v})]_{\mathcal{B}} = [T]_{\mathcal{B}}[\mathbf{v}]_{\mathcal{B}} = \lambda[\mathbf{v}]_{\mathcal{B}} = [\lambda\mathbf{v}]_{\mathcal{B}} = \Phi_{\mathcal{B}}(\lambda\mathbf{v}).$$

Since $\Phi_{\mathcal{B}}$ is one-to-one, $T(\mathbf{v}) = \lambda\mathbf{v}$. It follows that $\mathbf{v}$ is an eigenvector of T with corresponding eigenvalue λ.

We summarize these results as follows, replacing $\Phi_{\mathcal{B}}(\mathbf{v})$ by $[\mathbf{v}]_{\mathcal{B}}$:

Eigenvalues and Eigenvectors of a Matrix Representation of a Linear Operator

Let T be a linear operator on a finite-dimensional vector space V with the basis $\mathcal{B}$, and let $A = [T]_{\mathcal{B}}$. Then a vector $\mathbf{v}$ in V is an eigenvector of T with corresponding eigenvalue λ if and only if $[\mathbf{v}]_{\mathcal{B}}$ is an eigenvector of A with corresponding eigenvalue λ.

Example 8

Let T and $\mathcal{B}$ be as in Example 1. To find the eigenvalues and eigenvectors of T, let $A = [T]_{\mathcal{B}}$. From Example 1, we have that

$$A = \begin{bmatrix} 1 & 0 & 0 \\ 3 & 3 & 3 \\ 1 & 2 & 4 \end{bmatrix}.$$

As in Chapter 5, we find that the characteristic polynomial of A is $-(t-1)^2(t-6)$, and so the eigenvalues of A, and hence of T, are 1 and 6. We now determine the eigenspaces of T.

Eigenspace corresponding to the eigenvalue 1: The vector

$$\begin{bmatrix} 0 \\ -3 \\ 2 \end{bmatrix}$$

forms a basis for the eigenspace of A corresponding to $\lambda = 1$. Since this vector is the coordinate vector of the polynomial $p(x) = -3x + 2x^2$, the eigenspace of T corresponding to the eigenvalue 1 equals Span $\{p(x)\}$.

Eigenspace corresponding to the eigenvalue 6: The vector

$$\begin{bmatrix} 0 \\ 1 \\ 1 \end{bmatrix}$$

forms a basis for the eigenspace of A corresponding to $\lambda = 6$. Since this vector is the coordinate vector of the polynomial $q(x) = x + x^2$, the eigenspace of T corresponding to the eigenvalue 6 equals Span $\{q(x)\}$.

Example 9

Let $U: \mathcal{M}_{2\times 2} \to \mathcal{M}_{2\times 2}$ be the linear operator defined by $U(A) = A^T$, which is a special case of Example 7. We saw in that example that 1 and -1 are eigenvalues of U. We now show that these are the only eigenvalues of U. Let

$$\mathcal{B} = \left\{ \begin{bmatrix} 1 & 0 \\ 0 & 0 \end{bmatrix}, \begin{bmatrix} 0 & 1 \\ 0 & 0 \end{bmatrix}, \begin{bmatrix} 0 & 0 \\ 1 & 0 \end{bmatrix}, \begin{bmatrix} 0 & 0 \\ 0 & 1 \end{bmatrix} \right\},$$

which is a basis for $\mathcal{M}_{2\times 2}$. Then

$$[U]_{\mathcal{B}} = \begin{bmatrix} 1 & 0 & 0 & 0 \\ 0 & 0 & 1 & 0 \\ 0 & 1 & 0 & 0 \\ 0 & 0 & 0 & 1 \end{bmatrix},$$

which has the characteristic polynomial $(t-1)^3(t+1)$. Therefore 1 and -1 are the only eigenvalues of A, and we conclude that 1 and -1 are the only eigenvalues of U.

Practice Problem 4 ▶ Let T be the linear transformation in Practice Problem 1.

(a) Find the eigenvalues of T.

(b) For each eigenvalue λ of T, describe the eigenspace corresponding to λ. ◀

EXERCISES

In Exercises 1–8, a vector space V, a basis $\mathcal{B}$ for V, and a vector **u** *in V are given. Determine the coordinate vector of* **u** *relative to $\mathcal{B}$.*

1. $V = \mathcal{M}_{2\times 2}$,

$$\mathcal{B} = \left\{ \begin{bmatrix} 1 & 0 \\ 0 & 0 \end{bmatrix}, \begin{bmatrix} 0 & 0 \\ 1 & 0 \end{bmatrix}, \begin{bmatrix} 0 & 0 \\ 0 & 1 \end{bmatrix}, \begin{bmatrix} 0 & 1 \\ 0 & 0 \end{bmatrix} \right\},$$

and **u** is the matrix $\begin{bmatrix} 1 & 2 \\ 3 & 4 \end{bmatrix}$.

2. $V = \mathcal{P}_2$, $\mathcal{B} = \{x^2, x, 1\}$, and **u** is the polynomial $2 + x - 3x^2$.

3. $V = \text{Span}\,\mathcal{B}$, where $\mathcal{B} = \{\cos^2 t, \sin^2 t, \sin t \cos t\}$, and **u** is the function $\sin 2t - \cos 2t$.

4. $V = \{\mathbf{u} \in \mathcal{R}^3 \colon u_1 - u_2 + 2u_3 = 0\}$,

$$\mathcal{B} = \left\{ \begin{bmatrix} 1 \\ 1 \\ 0 \end{bmatrix}, \begin{bmatrix} -2 \\ 0 \\ 1 \end{bmatrix} \right\}, \quad \text{and} \quad \mathbf{u} = \begin{bmatrix} 5 \\ -1 \\ -3 \end{bmatrix}.$$

5. $V = \{\mathbf{u} \in \mathcal{R}^4 \colon u_1 + u_2 - u_3 - u_4 = 0\}$,

$$\mathcal{B} = \left\{ \begin{bmatrix} -1 \\ 1 \\ 0 \\ 0 \end{bmatrix}, \begin{bmatrix} 1 \\ 0 \\ 1 \\ 0 \end{bmatrix}, \begin{bmatrix} 1 \\ 0 \\ 0 \\ 1 \end{bmatrix} \right\}, \quad \text{and} \quad \mathbf{u} = \begin{bmatrix} 6 \\ -3 \\ 2 \\ 1 \end{bmatrix}.$$

6. $V = \mathcal{P}_3$, $\mathcal{B} = \{x^3 - x^2, x^2 - x, x - 1, x^3 + 1\}$, and **u** is the polynomial $2x^3 - 5x^2 + 3x - 2$.

7. V is the vector space of all 2×2 matrices whose trace equals 0,

$$\mathcal{B} = \left\{ \begin{bmatrix} -1 & 0 \\ 0 & 1 \end{bmatrix}, \begin{bmatrix} 0 & 1 \\ 0 & 0 \end{bmatrix}, \begin{bmatrix} 0 & 0 \\ 1 & 0 \end{bmatrix} \right\},$$

and **u** is the matrix $\begin{bmatrix} 3 & -2 \\ 1 & -3 \end{bmatrix}$.

8. V is the vector space of all symmetric 2×2 matrices,

$$\mathcal{B} = \left\{ \begin{bmatrix} 1 & 0 \\ 0 & -1 \end{bmatrix}, \begin{bmatrix} 1 & 0 \\ 0 & 2 \end{bmatrix}, \begin{bmatrix} 1 & 1 \\ 1 & 1 \end{bmatrix} \right\},$$

and **u** is the matrix $\begin{bmatrix} 4 & -1 \\ -1 & 3 \end{bmatrix}$.

In Exercises 9–16, find the matrix representation $[T]_{\mathcal{B}}$, where T is a linear operator on the vector space V and $\mathcal{B}$ is a basis for V.

9. $\mathcal{B} = \{e^t, e^{2t}, e^{3t}\}$, $V = \text{Span}\,\mathcal{B}$, and $T = D$, the derivative operator.

10. $\mathcal{B} = \{e^t, te^t, t^2e^t\}$, $V = \text{Span}\,\mathcal{B}$, and $T = D$, the derivative operator.

11. $V = \mathcal{P}_2$, $\mathcal{B} = \{1, x, x^2\}$, and

$$T(p(x)) = p(0) + 3p(1)x + p(2)x^2.$$

12. $V = \mathcal{M}_{2\times 2}$,

$$\mathcal{B} = \left\{ \begin{bmatrix} 1 & 0 \\ 0 & 0 \end{bmatrix}, \begin{bmatrix} 0 & 1 \\ 0 & 0 \end{bmatrix}, \begin{bmatrix} 0 & 0 \\ 1 & 0 \end{bmatrix}, \begin{bmatrix} 0 & 0 \\ 0 & 1 \end{bmatrix} \right\},$$

and $T(A) = \begin{bmatrix} 1 & 2 \\ 3 & 2 \end{bmatrix} A$.

13. $V = \mathcal{P}_3$, $\mathcal{B} = \{1, x, x^2, x^3\}$, and $T(p(x)) = p'(x) - p''(x)$.

14. $V = \mathcal{P}_3$, $\mathcal{B} = \{1, x, x^2, x^3\}$, and

$$T(a + bx + cx^2 + dx^3) = d + cx + bx^2 + ax^3.$$

15. $V = \mathcal{M}_{2\times 2}$,

$$\mathcal{B} = \left\{ \begin{bmatrix} 1 & 0 \\ 0 & 0 \end{bmatrix}, \begin{bmatrix} 0 & 1 \\ 0 & 0 \end{bmatrix}, \begin{bmatrix} 0 & 0 \\ 1 & 0 \end{bmatrix}, \begin{bmatrix} 0 & 0 \\ 0 & 1 \end{bmatrix} \right\},$$

and $T(A) = A^T$.

16. V is the vector space of symmetric 2×2 matrices,

$$\mathcal{B} = \left\{ \begin{bmatrix} 1 & 0 \\ 0 & 0 \end{bmatrix}, \begin{bmatrix} 0 & 0 \\ 0 & 1 \end{bmatrix}, \begin{bmatrix} 0 & 1 \\ 1 & 0 \end{bmatrix} \right\},$$

and $T(A) = CAC^T$, where $C = \begin{bmatrix} 1 & 2 \\ -1 & -2 \end{bmatrix}$

17. Use the technique in Example 3 to find the derivatives of the following polynomials:

(a) $p(x) = 6 - 4x^2$

(b) $p(x) = 2 + 3x + 5x^2$

(c) $p(x) = x^3$

18. Use the technique in Example 4 to find an antiderivative of $e^t \cos t$.

19. Let $\mathcal{B} = \{e^t, te^t, t^2e^t\}$ be a basis for the subspace V of C^∞. Use the method in Example 4 to find antiderivatives of the following functions:

(a) te^t

(b) t^2e^t

(c) $3e^t - 4te^t + 2t^2e^t$

In Exercises 20–27, find the eigenvalues of T and a basis for each of the corresponding eigenspaces.

20. Let T be the linear operator in Exercise 10.

21. Let T be the linear operator in Exercise 9.

22. Let T be the linear operator in Exercise 12.

23. Let T be the linear operator in Exercise 11.

24. Let T be the linear operator in Exercise 14.

25. Let T be the linear operator in Exercise 13.

26. Let T be the linear operator in Exercise 16.

27. Let T be the linear operator in Exercise 15.

In Exercises 28–39, determine whether the statements are true or false.

28. Every linear operator has an eigenvalue.

29. Every linear operator can be represented by a matrix.

30. Every linear operator on a nonzero finite-dimensional vector space can be represented by a matrix.

31. For any positive integer n, taking the derivative of a polynomial in $\mathcal{P}_n$ can be accomplished by matrix multiplication.

32. The inverse of an invertible linear operator on a finite-dimensional vector space can be found by computing the inverse of a matrix.

33. It is possible for a matrix to be an eigenvector of a linear operator.

34. A linear operator on a vector space can have only a finite number of eigenvalues.

35. For the linear operator U on $\mathcal{M}_{n\times n}$ defined by $U(A) = A^T$, the eigenspace corresponding to the eigenvalue 1 is the set of skew-symmetric matrices.

36. If T is a linear operator on a vector space with the basis $\mathcal{B} = \{\mathbf{v}_1, \mathbf{v}_2, \ldots, \mathbf{v}_n\}$, then the matrix representation of T with respect to $\mathcal{B}$ is the matrix $[T(\mathbf{v}_1)\ T(\mathbf{v}_2)\ \cdots\ T(\mathbf{v}_n)]$.

37. If T is a linear operator on a vector space with the basis $\mathcal{B} = \{\mathbf{v}_1, \mathbf{v}_2, \ldots, \mathbf{v}_n\}$, then for any vector $\mathbf{v}$ in V, $T(\mathbf{v}) = [T]_\mathcal{B}\mathbf{v}$.

38. If T is a linear operator on a finite-dimensional vector space V with the basis $\mathcal{B}$, then a vector $\mathbf{v}$ in V is an eigenvector of T corresponding to eigenvalue λ if and only if $[\mathbf{v}]_\mathcal{B}$ is an eigenvector of $[T]_\mathcal{B}$ corresponding to the eigenvalue λ.

39. Let V be an n-dimensional vector space and $\mathcal{B}$ be a basis for V. Every linear operator on V is of the form $\Phi_\mathcal{B}^{-1} T_A \Phi_\mathcal{B}$ for some $n \times n$ matrix A.

---

40. Let T be the operator $D^2 + D$, where D is the derivative operator on C^∞.

(a) Show that 1 and e^{-t} lie in the null space of T.

(b) Show that for any real number a, the function e^{at} is an eigenvector of T corresponding to the eigenvalue $a^2 + a$.

41. Let D be the derivative operator on $\mathcal{P}_2$.

(a) Find the eigenvalues of D.

(b) Find a basis for each of the corresponding eigenspaces.

42. Let U be the linear operator on $\mathcal{M}_{n\times n}$ defined by $U(A) = A^T$. Prove that 1 and -1 are the only eigenvalues of A. *Hint:* Suppose that A is a nonzero $n \times n$ matrix and λ is a scalar such that $A^T = \lambda A$. Take the transpose of both sides of this equation, and show that $A = \lambda^2 A$.

43. Let P denote the set of positive integers, and let $E\colon \mathcal{F}(P) \to \mathcal{F}(P)$ be defined by $E(f)(n) = f(n+1)$.

(a) Prove that E is a linear operator on $\mathcal{F}(P)$.

(b) Since a sequence of real numbers is a function from P to $\mathcal{R}$, we may identify $\mathcal{F}(P)$ with the space of sequences. Recall the Fibonacci sequences defined in Section 5.5. Prove that a nonzero sequence f is a Fibonacci sequence if and only if f is an eigenvector of $E^2 - E$ with corresponding eigenvalue 1.

44. Let $\mathcal{B}$ be the basis for $\mathcal{M}_{2\times 2}$ given in Example 9, and let $T\colon \mathcal{M}_{2\times 2} \to \mathcal{M}_{2\times 2}$ be defined by

$$T\left(\begin{bmatrix} a & b \\ c & d \end{bmatrix}\right) = \begin{bmatrix} b & a+c \\ 0 & d \end{bmatrix}.$$

(a) Prove that T is linear.

(b) Determine the matrix representation $[T]_{\mathcal{B}}$.

(c) Find the eigenvalues of T.

(d) Find a basis for each eigenspace.

45. For a given matrix B in $\mathcal{M}_{2\times 2}$, let T be the function on $\mathcal{M}_{2\times 2}$ defined by $T(A) = (\text{trace}(A))B$.

(a) Prove that T is linear.

(b) Suppose that $\mathcal{B}$ is the basis for $\mathcal{M}_{2\times 2}$ given in Example 9, and that $B = \begin{bmatrix} 1 & 2 \\ 3 & 4 \end{bmatrix}$. Determine $[T]_{\mathcal{B}}$.

(c) Prove that if A is a nonzero matrix whose trace is zero, then A is an eigenvector of T.

(d) Prove that if A is an eigenvector of T with a corresponding nonzero eigenvalue, then A is a scalar multiple of B.

46. Let B be an $n \times n$ matrix and $T\colon \mathcal{M}_{n\times n} \to \mathcal{M}_{n\times n}$ be the function defined by $T(A) = BA$.

(a) Prove that T is linear.

(b) Prove that T is invertible if and only if B is invertible.

(c) Prove that a nonzero $n \times n$ matrix C is an eigenvector of T corresponding to the eigenvalue λ if and only if λ is an eigenvalue of B and each column of C lies in the eigenspace of B corresponding to λ.

47. Let $\mathcal{B} = \{\mathbf{v}_1, \mathbf{v}_2, \ldots, \mathbf{v}_n\}$ be a basis for a vector space V. Show that for any j, we have $[\mathbf{v}_j]_{\mathcal{B}} = \mathbf{e}_j$, where $\mathbf{e}_j$ is the jth standard vector in $\mathcal{R}^n$.

48. Let V be a finite-dimensional vector space with the basis $\mathcal{B}$. Prove that for any linear operators T and U on V,

$$[UT]_{\mathcal{B}} = [U]_{\mathcal{B}}[T]_{\mathcal{B}}.$$

Hint: Apply Theorem 7.10 to $(UT)\mathbf{v}$ and $U(T(\mathbf{v}))$, where $\mathbf{v}$ is an arbitrary vector in V.

49. Let T be a linear operator on a vector space V, and suppose that λ is an eigenvalue of T. Prove that the eigenspace of T corresponding to λ is a subspace of V and consists of the zero vector and the eigenvectors of T corresponding to λ.

The following definition of a diagonalizable linear operator is used in the next exercise. Compare this definition with the definition given in Section 5.4.

Definition A linear operator on a finite-dimensional vector space is **diagonalizable** if there is a basis for the vector space consisting of eigenvectors of the operator.

50. Let T be a linear operator on a finite-dimensional vector space V. Prove the following statements:

(a) T is diagonalizable if and only if there is a basis $\mathcal{B}$ for V such that $[T]_{\mathcal{B}}$ is a diagonal matrix.

(b) T is diagonalizable if and only if, for any basis $\mathcal{B}$, $[T]_{\mathcal{B}}$ is a diagonalizable matrix.

(c) Prove that the linear operator U in Example 9 is diagonalizable.

The following definition of a matrix representation of a linear transformation is used in Exercises 51–53:

Definition Let $T\colon V \to W$ be a linear transformation, where V and W are finite-dimensional vector spaces, and let $\mathcal{B} = \{\mathbf{b}_1, \mathbf{b}_2, \ldots, \mathbf{b}_n\}$ and $\mathcal{C}$ be (ordered) bases for V and W, respectively. The matrix

$$[\,[T(\mathbf{b}_1)]_{\mathcal{C}}\ [T(\mathbf{b}_2)]_{\mathcal{C}}\ \cdots\ [T(\mathbf{b}_n)]_{\mathcal{C}}\,]$$

is called the **matrix representation of T with respect to $\mathcal{B}$ and $\mathcal{C}$**. It is denoted by $[T]_{\mathcal{B}}^{\mathcal{C}}$.

51. Let $\mathbf{v} = \begin{bmatrix} 1 \\ 3 \end{bmatrix}$, and let $T\colon \mathcal{M}_{2\times 2} \to \mathcal{R}^2$ be defined by $T(A) = A\mathbf{v}$.

(a) Prove that T is a linear transformation.

(b) Let $\mathcal{B}$ be the basis for $\mathcal{M}_{2\times 2}$ given in Example 9, and let $\mathcal{C}$ be the standard basis for $\mathcal{R}^2$. Find $[T]_{\mathcal{B}}^{\mathcal{C}}$.

(c) Let $\mathcal{B}$ be the basis for $\mathcal{M}_{2\times 2}$ given in Example 9, and let

$$\mathcal{D} = \left\{\begin{bmatrix} 1 \\ 1 \end{bmatrix}, \begin{bmatrix} 1 \\ 2 \end{bmatrix}\right\},$$

which is a basis for $\mathcal{R}^2$. Find $[T]_{\mathcal{B}}^{\mathcal{D}}$.

52. Let $T\colon V \to W$ be a linear transformation, where V and W are finite-dimensional vector spaces, and let $\mathcal{B}$ and $\mathcal{C}$ be (ordered) bases for V and W, respectively. Prove the following results (parts (a) and (b) use the definitions given on page 510):

(a) $[sT]_{\mathcal{B}}^{\mathcal{C}} = s[T]_{\mathcal{B}}^{\mathcal{C}}$ for any scalar s.

(b) If $U\colon V \to W$ is linear, then

$$[T+U]_{\mathcal{B}}^{\mathcal{C}} = [T]_{\mathcal{B}}^{\mathcal{C}} + [U]_{\mathcal{B}}^{\mathcal{C}}.$$

(c) $[T(\mathbf{v})]_{\mathcal{C}} = [T]_{\mathcal{B}}^{\mathcal{C}}[\mathbf{v}]_{\mathcal{B}}$ for every vector $\mathbf{v}$ in V.

(d) Let $U\colon W \to Z$ be linear, where Z is a finite-dimensional vector space, and let $\mathcal{D}$ be an (ordered) basis for Z. Then

$$[UT]_{\mathcal{B}}^{\mathcal{D}} = [U]_{\mathcal{C}}^{\mathcal{D}}[T]_{\mathcal{B}}^{\mathcal{C}}.$$

53. Let $T\colon \mathcal{P}_2 \to \mathcal{R}^2$ be defined by $T(f(x)) = \begin{bmatrix} f(1) \\ f(2) \end{bmatrix}$, let $\mathcal{B} = \{1, x, x^2\}$, and let $\mathcal{C} = \{\mathbf{e}_1, \mathbf{e}_2\}$, the standard basis for $\mathcal{R}^2$.

(a) Prove that T is a linear transformation.

(b) Find $[T]_{\mathcal{B}}^{\mathcal{C}}$.

(c) Let $f(x) = a + bx + cx^2$, for scalars a, b, and c.

(i) Compute $T(f(x))$ directly from the definition of T. Then find $[T(f(x)]_{\mathcal{C}}$.

(ii) Find $[f(x)]_{\mathcal{B}}$, and then compute $[T]_{\mathcal{B}}^{\mathcal{C}}[f(x)]_{\mathcal{B}}$. Compare your results with your answer in (i).

In Exercises 54 and 55, use either a calculator with matrix capabilities or computer software such as MATLAB to solve each problem.

54. Let T be the linear operator on $\mathcal{P}_3$ defined by

$$T(f(x)) = f(x) + f'(x) + f''(x) + f(0) + f(2)x^2.$$

(a) Determine the eigenvalues of T.

(b) Find a basis for $\mathcal{P}_3$ consisting of eigenvectors of T.

(c) For $f(x) = a_0 + a_1x + a_2x^2 + a_3x^3$, find $T^{-1}(f(x))$.

55. Let T be the linear operator on $\mathcal{M}_{2\times 2}$ defined by

$$T(A) = \begin{bmatrix} 1 & 2 \\ 3 & 4 \end{bmatrix} A + 3A^T$$

for all A in $\mathcal{M}_{2\times 2}$.

(a) Determine the eigenvalues of T.

(b) Find a basis for $\mathcal{M}_{2\times 2}$ consisting of eigenvectors of T.

(c) For $A = \begin{bmatrix} a & b \\ c & d \end{bmatrix}$, find $T^{-1}(A)$.

SOLUTIONS TO THE PRACTICE PROBLEMS

1. (a) Let $q(x)$ and $r(x)$ be polynomials in $\mathcal{P}_2$, and let c be a scalar. Then

$$\begin{aligned} &T(q(x) + r(x)) \\ &= (x+1)(q(x) + r(x))' + (q(x) + r(x)) \\ &= (x+1)(q'(x) + r'(x)) + (q(x) + r(x)) \\ &= (x+1)q'(x) + (x+1)r'(x) + q(x) + r(x) \\ &= ((x+1)q'(x) + q(x)) + ((x+1)r'(x) + r(x)) \\ &= T(q(x)) + T(r(x)) \end{aligned}$$

and

$$\begin{aligned} T(cq(x)) &= (x+1)(cq(x))' + cq(x) \\ &= c((x+1)q'(x) + q(x)) \\ &= cT(q(x)). \end{aligned}$$

So T is linear.

(b) We have

$$\begin{aligned} T(1) &= (x+1)(0) + 1 = 1, \\ T(x) &= (x+1)(1) + x = 1 + 2x, \end{aligned}$$

and

$$T(x^2) = (x+1)(2x) + x^2 = 2x + 3x^2.$$

Let $\mathcal{B} = \{1, x, x^2\}$. Then

$$[T(1)]_{\mathcal{B}} = \begin{bmatrix} 1 \\ 0 \\ 0 \end{bmatrix}, \quad [T(x)]_{\mathcal{B}} = \begin{bmatrix} 1 \\ 2 \\ 0 \end{bmatrix}, \quad \text{and}$$

$$[T(x^2)]_{\mathcal{B}} = \begin{bmatrix} 0 \\ 2 \\ 3 \end{bmatrix}.$$

Therefore

$$[T]_{\mathcal{B}} = \begin{bmatrix} 1 & 1 & 0 \\ 0 & 2 & 2 \\ 0 & 0 & 3 \end{bmatrix}.$$

2. Using the rule for T, we have

$$\begin{aligned} T(p(x)) &= T(2 - 3x + 5x^2) \\ &= (x+1)(-3 + 10x) \\ &\quad + (2 - 3x + 5x^2) \\ &= -1 + 4x + 15x^2. \end{aligned}$$

Applying Theorem 7.10, we have

$$\begin{aligned} [T(2 - 3x + 5x^2)]_{\mathcal{B}} &= [T]_{\mathcal{B}}[2 - 3x + 5x^2]_{\mathcal{B}} \\ &= \begin{bmatrix} 1 & 1 & 0 \\ 0 & 2 & 2 \\ 0 & 0 & 3 \end{bmatrix} \begin{bmatrix} 2 \\ -3 \\ 5 \end{bmatrix} = \begin{bmatrix} -1 \\ 4 \\ 15 \end{bmatrix}, \end{aligned}$$

and hence $T(p(x)) = -1 + 4x + 15x^2$.

3. The operator T is invertible because $[T]_{\mathcal{B}}$ is an invertible matrix. To find the rule for T^{-1}, we can use the result that $[T^{-1}]_{\mathcal{B}} = [T]_{\mathcal{B}}^{-1}$. So

$$\begin{aligned} [T^{-1}(a + bx + cx^2)]_{\mathcal{B}} &= [T^{-1}]_{\mathcal{B}} \begin{bmatrix} a \\ b \\ c \end{bmatrix} \\ &= [T]_{\mathcal{B}}^{-1} \begin{bmatrix} a \\ b \\ c \end{bmatrix} \\ &= \begin{bmatrix} 1 & -\frac{1}{2} & \frac{1}{3} \\ 0 & \frac{1}{2} & -\frac{1}{3} \\ 0 & 0 & \frac{1}{3} \end{bmatrix} \begin{bmatrix} a \\ b \\ c \end{bmatrix} \\ &= \begin{bmatrix} a - \frac{1}{2}b + \frac{1}{3}c \\ \frac{1}{2}b - \frac{1}{3}c \\ \frac{1}{3}c \end{bmatrix}. \end{aligned}$$

Therefore

$$T^{-1}(a + bx + cx^2)$$
$$= (a - \tfrac{1}{2}b + \tfrac{1}{3}c) + (\tfrac{1}{2}b - \tfrac{1}{3}c)x + (\tfrac{1}{3}c)x^2.$$

4. (a) The eigenvalues of T are the same as the eigenvalues of $[T]_{\mathcal{B}}$. Since $[T]_{\mathcal{B}}$ is an upper triangular matrix, its eigenvalues are its diagonal entries, that is, 1, 2, and 3.

(b) *Eigenspace corresponding to the eigenvalue* 1: The vector

$$\begin{bmatrix} 1 \\ 0 \\ 0 \end{bmatrix},$$

which is the coordinate vector of the constant polynomial 1, forms a basis for the eigenspace of $[T]_{\mathcal{B}}$ corresponding to $\lambda = 1$. It follows that the multiples of 1 (that is, the constant polynomials), form the eigenspace of T corresponding to the eigenvalue 1.

Eigenspace corresponding to the eigenvalue 2: The vector

$$\begin{bmatrix} 1 \\ 1 \\ 0 \end{bmatrix},$$

which is the coordinate vector of the polynomial $1 + x$, forms a basis for the eigenspace of $[T]_{\mathcal{B}}$ corresponding to $\lambda = 2$. Thus the eigenspace of T corresponding to the eigenvalue 2 equals Span $\{1 + x\}$.

Eigenspace corresponding to the eigenvalue 3: The vector

$$\begin{bmatrix} 1 \\ 2 \\ 1 \end{bmatrix},$$

which is the coordinate vector of the polynomial $1 + 2x + x^2$, forms a basis for the eigenspace of $[T]_{\mathcal{B}}$ corresponding to $\lambda = 3$. Thus the eigenspace of T corresponding to the eigenvalue 3 equals

$$\text{Span}\,\{1 + 2x + x^2\}.$$

7.5 INNER PRODUCT SPACES

The dot product introduced in Chapter 6 provides a strong link between vectors and matrices and the geometry of $\mathcal{R}^n$. For example, we saw how the concept of dot product leads to deep results about symmetric matrices.

In certain vector spaces, especially function spaces, there are scalar-valued products, called *inner products*, that share the important formal properties of dot products. These inner products allow us to extend such concepts as *distance* and *orthogonality* to vector spaces.

Definitions An **inner product** on a vector space V is a real-valued function that assigns to any ordered pair of vectors $\mathbf{u}$ and $\mathbf{v}$ a scalar, denoted by $\langle \mathbf{u}, \mathbf{v} \rangle$, such that for any vectors $\mathbf{u}$, $\mathbf{v}$, and $\mathbf{w}$ in V and any scalar a, the following axioms hold:

Axioms of an Inner Product

1. $\langle \mathbf{u}, \mathbf{u} \rangle > 0$ if $\mathbf{u} \neq \mathbf{0}$
2. $\langle \mathbf{u}, \mathbf{v} \rangle = \langle \mathbf{v}, \mathbf{u} \rangle$
3. $\langle \mathbf{u} + \mathbf{v}, \mathbf{w} \rangle = \langle \mathbf{u}, \mathbf{w} \rangle + \langle \mathbf{v}, \mathbf{w} \rangle$
4. $\langle a\mathbf{u}, \mathbf{v} \rangle = a \langle \mathbf{u}, \mathbf{v} \rangle$

Suppose that $\langle \mathbf{u}, \mathbf{v} \rangle$ is an inner product on a vector space V. For any scalar $r > 0$, defining $\langle\langle \mathbf{u}, \mathbf{v} \rangle\rangle$ by $\langle\langle \mathbf{u}, \mathbf{v} \rangle\rangle = r \langle \mathbf{u}, \mathbf{v} \rangle$ gives another inner product on V. Thus there can be infinitely many different inner products on a vector space. A vector space endowed with a particular inner product is called an **inner product space**.

The dot product on $\mathcal{R}^n$ is an example of an inner product, where $\langle \mathbf{u}, \mathbf{v} \rangle = \mathbf{u} \cdot \mathbf{v}$ for $\mathbf{u}$ and $\mathbf{v}$ in $\mathcal{R}^n$. Notice that the axioms of an inner product are verified for the dot product in Theorem 6.1 on page 364.

Many facts about dot products are valid for inner products. Often, a proof of a result for inner product spaces requires little or no modification of the proof for the dot product on $\mathcal{R}^n$.

Example 1 presents a particularly important inner product on a function space.

Example 1 Let $\mathsf{C}([a,b])$ denote the vector space of continuous real-valued functions defined on the closed interval $[a,b]$, which was described in Example 3 of Section 7.2. For f and g in $\mathsf{C}([a,b])$, let

$$\langle f, g \rangle = \int_a^b f(t)g(t)\,dt.$$

This definition determines an inner product on $\mathsf{C}([a,b])$.

To verify axiom 1, let f be any nonzero function in $\mathsf{C}([a,b])$. Then f^2 is a nonnegative function that is continuous on $[a,b]$. Since f is nonzero, $f^2(t) > 0$ on some interval $[c,d]$, where $a < c < d < b$. Thus

$$\langle f, f \rangle = \int_a^b f^2(t)\,dt \geq \int_c^d f^2(t)\,dt > 0.$$

To verify axiom 2, let f and g be functions in $\mathsf{C}([a,b])$. Then

$$\langle f, g \rangle = \int_a^b f(t)g(t)\,dt = \int_a^b g(t)f(t)\,dt = \langle g, f \rangle\,.$$

We leave the verifications of axioms 3 and 4 as exercises.

Our next example introduces the *Frobenius inner product,* an important example of an inner product on $\mathcal{M}_{n\times n}$.

Example 2 For A and B in $\mathcal{M}_{n\times n}$, define

$$\langle A, B \rangle = \text{trace}(AB^T).$$

This definition determines an inner product, called the **Frobenius[3] inner product**, on $\mathcal{M}_{n\times n}$.

To verify axiom 1, let A be any nonzero matrix, and let $C = AA^T$. Then

$$\begin{aligned}\langle A, A \rangle = \text{trace}(AA^T) &= \text{trace}\, C \\ &= c_{11} + c_{22} + \cdots + c_{nn}.\end{aligned}$$

[3] Ferdinand Georg Frobenius (1849–1917) was a German mathematician best known for his work in group theory. His research combined results from the theory of algebraic equations, number theory, and geometry. His representation theory for finite groups made important contributions to quantum mechanics.

Furthermore, for each i,

$$c_{ii} = a_{i1}^2 + a_{i2}^2 + \cdots + a_{in}^2.$$

It follows that $\langle A, A \rangle$ is the sum of squares of all the entries of A. Since $A \neq O$, it follows that $a_{ij}^2 > 0$ for some i and j, and hence $\langle A, A \rangle > 0$.

To verify axiom 2, let A and B be matrices in V. Then

$$\begin{aligned} \langle A, B \rangle &= \operatorname{trace}(AB^T) \\ &= \operatorname{trace}(AB^T)^T \\ &= \operatorname{trace}(BA^T) \\ &= \langle B, A \rangle . \end{aligned}$$

We leave the verifications of axioms 3 and 4 as exercises.

It can be shown that the Frobenius inner product of two $n \times n$ matrices is simply the sum of the products of their corresponding entries. (See Exercises 73 and 74.) Thus the Frobenius inner product looks like an ordinary inner product in $\mathcal{R}^{n^2}$, except that the components are entries in a matrix. For example,

$$\left\langle \begin{bmatrix} 1 & 2 \\ 3 & 4 \end{bmatrix}, \begin{bmatrix} 5 & 6 \\ 7 & 8 \end{bmatrix} \right\rangle = 1 \cdot 5 + 2 \cdot 6 + 3 \cdot 7 + 4 \cdot 8 = 70.$$

As was done with the dot product on $\mathcal{R}^n$, we can define the *length* of a vector in an inner product space. For any vector $\mathbf{v}$ in an inner product space V, the **norm** (**length**) of $\mathbf{v}$, denoted by $\|\mathbf{v}\|$, is defined by

$$\|\mathbf{v}\| = \sqrt{\langle \mathbf{v}, \mathbf{v} \rangle}\,.$$

The **distance** between two vectors $\mathbf{u}$ and $\mathbf{v}$ in V is defined in the usual way as $\|\mathbf{u} - \mathbf{v}\|$.

The norm of a vector depends, of course, on the specific inner product used. To describe the norm defined in terms of a specific inner product, we may refer to the norm *induced* by that inner product. For instance, the norm induced by the Frobenius inner product on $\mathcal{M}_{n\times n}$ in Example 2 is given by

$$\|A\| = \sqrt{\langle A, A \rangle} = \sqrt{\operatorname{trace}(AA^T)}.$$

For obvious reasons, this norm is called the **Frobenius norm.**

Practice Problem 1 ► Let $A = \begin{bmatrix} 2 & 1 \\ 0 & 3 \end{bmatrix}$ and $B = \begin{bmatrix} 1 & 1 \\ 2 & 0 \end{bmatrix}$, matrices in $\mathcal{M}_{2\times 2}$. Use the Frobenius inner product to compute $\|A\|^2$, $\|B\|^2$, and $\langle A, B \rangle$. ◄

As stated earlier, many of the elementary properties of dot products are also valid for all inner products. In particular, all of the parts of Theorem 6.1 are valid for inner product spaces. For example, the analog of Theorem 6.1(d) for inner products follows from axioms 2 and 3 of inner products, since if $\mathbf{u}$, $\mathbf{v}$, and $\mathbf{w}$ are vectors in an inner product space, then

$$\begin{aligned} \langle \mathbf{u}, \mathbf{v} + \mathbf{w} \rangle &= \langle \mathbf{v} + \mathbf{w}, \mathbf{u} \rangle && \text{(by axiom 2)} \\ &= \langle \mathbf{v}, \mathbf{u} \rangle + \langle \mathbf{w}, \mathbf{u} \rangle && \text{(by axiom 3)} \\ &= \langle \mathbf{u}, \mathbf{v} \rangle + \langle \mathbf{u}, \mathbf{w} \rangle. && \text{(by axiom 2)} \end{aligned}$$

The Cauchy–Schwarz inequality (Theorem 6.3 on page 368) and the triangle inequality (Theorem 6.4 on page 369) are also valid for all inner product spaces because their proofs are based on the items of Theorem 6.1 that correspond to the axioms of an inner product. Thus, in Example 1, we can obtain an inequality about the integrals of functions in C([a, b]) by applying the Cauchy–Schwarz inequality and squaring both sides:

$$\left[\int_a^b f(t)g(t)\,dt\right]^2 \leq \left[\int_a^b f^2(t)\,dt\right]\left[\int_a^b g^2(t)\,dt\right]$$

Here f and g are continuous functions on the closed interval $[a, b]$.

Practice Problem 2 ▶ Let $f(t) = t$ and $g(t) = t^2$ be vectors in the inner product space C([0, 1]) in Example 1.

(a) Compute $\|f\|^2$, $\|g\|^2$, and $\langle f, g\rangle$.

(b) Verify that $|\langle f, g\rangle| \leq \|f\| \cdot \|g\|$. ◀

ORTHOGONALITY AND THE GRAM-SCHMIDT PROCESS

Let V be an inner product space. As in Chapter 6, two vectors $\mathbf{u}$ and $\mathbf{v}$ in V are called **orthogonal** if $\langle \mathbf{u}, \mathbf{v}\rangle = 0$, and a subset S of V is said to be **orthogonal** if any two distinct vectors in S are orthogonal. Again, a vector $\mathbf{u}$ in V is called a **unit vector** if $\|\mathbf{u}\| = 1$, and a subset S of V is called **orthonormal** if S is an orthogonal set and every vector in S is a unit vector. For any nonzero vector $\mathbf{v}$, the unit vector $\frac{1}{\|\mathbf{v}\|}\mathbf{v}$, which is a scalar multiple of $\mathbf{v}$, is called its **normalized vector**. If S is an orthogonal set of nonzero vectors, then replacing every vector in S by its normalized vector results in an orthonormal set whose span is the same as S.

It was shown in Section 6.2 that any finite orthogonal set of nonzero vectors is linearly independent (Theorem 6.5 on page 375). This result is valid for any inner product space, and the proof is identical. Furthermore, we can show that this result is also valid for infinite orthogonal sets. (See Exercise 76.)

Example 3 Let $f(t) = \sin 3t$ and $g(t) = \cos 2t$ be defined on the closed interval $[0, 2\pi]$. Then f and g are functions in the inner product space C([$0, 2\pi$]) of Example 1. Show that f and g are orthogonal.

Solution We apply the trigonometric identity

$$\sin\alpha\cos\beta = \frac{1}{2}\left[\sin(\alpha+\beta) + \sin(\alpha-\beta)\right]$$

with $\alpha = 3t$ and $\beta = 2t$ to obtain

$$\begin{aligned}\langle f, g\rangle &= \int_0^{2\pi} \sin 3t\cos 2t\,dt \\ &= \frac{1}{2}\int_0^{2\pi}[\sin 5t + \sin t]\,dt \\ &= \frac{1}{2}\left[-\frac{1}{5}\cos 5t - \cos t\right]\Bigg|_0^{2\pi} \\ &= 0.\end{aligned}$$

Hence f and g are orthogonal.

Example 4 Recall the vector space of trigonometric polynomials $\mathcal{T}[0, 2\pi]$ defined on page 498. This function space is defined as the span of

$$\mathcal{S} = \{1, \cos t, \sin t, \cos 2t, \sin 2t, \ldots, \cos nt, \sin nt, \ldots\},$$

a set of trigonometric functions defined on $[0, 2\pi]$. So $\mathcal{T}[0, 2\pi]$ is a subspace of $\mathrm{C}([0, 2\pi])$ and is an inner product space with the same inner product as Example 3.

To show any two distinct functions f and g in $\mathcal{S}$ are orthogonal, there are several cases to consider.

If $f(t) = 1$ and $g(t) = \cos nt$ for some positive integer n, then

$$\langle f, g \rangle = \int_0^{2\pi} \cos nt \, dt = \frac{1}{n} \sin nt \Big|_0^{2\pi} = 0.$$

In a similar manner, if $f(t) = \sin mt$ and $g(t) = 1$, then $\langle f, g \rangle = 0$.

If $f(t) = \sin mt$ and $g(t) = \cos nt$ for positive integers m and n, we can apply the trigonometric identity in Example 3 to find that $\langle f, g \rangle = 0$. The other two cases are treated in the exercises. (See Exercises 58 and 59.)

So $\mathcal{S}$ is an orthogonal set. Since $\mathcal{S}$ consists of nonzero functions, $\mathcal{S}$ is linearly independent and hence is a basis for $\mathcal{T}[0, 2\pi]$. It follows that $\mathcal{T}[0, 2\pi]$ is an infinite-dimensional vector space.

In Section 6.2, we saw that the Gram–Schmidt process (Theorem 6.6 on page 378) converts a linearly independent subset of $\mathcal{R}^n$ into an orthogonal set. It is also valid for any inner product space, and its justification is identical to the proof of Theorem 6.6. So we can use the Gram–Schmidt process to replace an arbitrary basis for a finite-dimensional inner product space with an orthogonal or an orthonormal basis. It follows that *every finite-dimensional inner product space has an orthonormal basis.*

For convenience, we restate Theorem 6.6 in the context of general vector spaces.

The Gram-Schmidt Process

Let $\{\mathbf{u}_1, \mathbf{u}_2, \ldots, \mathbf{u}_k\}$ be a basis for an inner product space V. Define

$$\mathbf{v}_1 = \mathbf{u}_1,$$

$$\mathbf{v}_2 = \mathbf{u}_2 - \frac{\mathbf{u}_2 \cdot \mathbf{v}_1}{\|\mathbf{v}_1\|^2}\mathbf{v}_1,$$

$$\mathbf{v}_3 = \mathbf{u}_3 - \frac{\mathbf{u}_3 \cdot \mathbf{v}_1}{\|\mathbf{v}_1\|^2}\mathbf{v}_1 - \frac{\mathbf{u}_3 \cdot \mathbf{v}_2}{\|\mathbf{v}_2\|^2}\mathbf{v}_2,$$

$$\vdots$$

$$\mathbf{v}_k = \mathbf{u}_k - \frac{\mathbf{u}_k \cdot \mathbf{v}_1}{\|\mathbf{v}_1\|^2}\mathbf{v}_1 - \frac{\mathbf{u}_k \cdot \mathbf{v}_2}{\|\mathbf{v}_2\|^2}\mathbf{v}_2 - \cdots - \frac{\mathbf{u}_k \cdot \mathbf{v}_{k-1}}{\|\mathbf{v}_{k-1}\|^2}\mathbf{v}_{k-1}.$$

Then $\{\mathbf{v}_1, \mathbf{v}_2, \ldots, \mathbf{v}_i\}$ is an orthogonal set of nonzero vectors such that

$$\text{Span}\,\{\mathbf{v}_1, \mathbf{v}_2, \ldots, \mathbf{v}_i\} = \text{Span}\,\{\mathbf{u}_1, \mathbf{u}_2, \ldots, \mathbf{u}_i\}$$

for each i. So $\{\mathbf{v}_1, \mathbf{v}_2, \ldots, \mathbf{v}_k\}$ is an orthogonal basis for V.

Example 5 Define an inner product on $\mathcal{P}_2$ by

$$\langle f(x), g(x)\rangle = \int_{-1}^{1} f(t)g(t)\,dt$$

for all polynomials $f(x)$ and $g(x)$ in $\mathcal{P}_2$. (It can be verified that this does indeed define an inner product on $\mathcal{P}_2$. See, for example, the argument in Example 1.) Use the Gram–Schmidt process to convert the basis $\{1, x, x^2\}$ into an orthogonal basis for $\mathcal{P}_2$. Then normalize the vectors of this orthogonal basis to obtain an orthonormal basis for $\mathcal{P}_2$.

Solution Using the notation of Theorem 6.6, we let $\mathbf{u}_1 = 1$, $\mathbf{u}_2 = x$, and $\mathbf{u}_3 = x^2$. Then

$$\mathbf{v}_1 = \mathbf{u}_1 = 1,$$

$$\mathbf{v}_2 = \mathbf{u}_2 - \frac{\langle \mathbf{u}_2, \mathbf{v}_1\rangle}{\|\mathbf{v}_1\|^2}\mathbf{v}_1 = x - \frac{\displaystyle\int_{-1}^{1} t \cdot 1\,dt}{\displaystyle\int_{-1}^{1} 1^2\,dt}(1) = x - 0 \cdot 1 = x,$$

and

$$\begin{aligned}
\mathbf{v}_3 &= \mathbf{u}_3 - \frac{\langle \mathbf{u}_3, \mathbf{v}_1\rangle}{\|\mathbf{v}_1\|^2}\mathbf{v}_1 - \frac{\langle \mathbf{u}_3, \mathbf{v}_2\rangle}{\|\mathbf{v}_2\|^2}\mathbf{v}_2 \\
&= x^2 - \frac{\displaystyle\int_{-1}^{1} t^2 \cdot 1\,dt}{\displaystyle\int_{-1}^{1} 1^2\,dt}(1) - \frac{\displaystyle\int_{-1}^{1} t^2 \cdot t\,dt}{\displaystyle\int_{-1}^{1} t^2\,dt}(x) \\
&= x^2 - \frac{\left(\frac{2}{3}\right)}{2} \cdot 1 - 0 \cdot x \\
&= x^2 - \frac{1}{3}.
\end{aligned}$$

Thus the set $\{1, x, x^2 - \frac{1}{3}\}$ is an orthogonal basis for $\mathcal{P}_2$.

Next we normalize each vector in this set to obtain an orthonormal basis for $\mathcal{P}_2$. Since

$$\|\mathbf{v}_1\| = \sqrt{\int_{-1}^{1} 1^2\,dx} = \sqrt{2}\,,$$

$$\|\mathbf{v}_2\| = \sqrt{\int_{-1}^{1} x^2\,dx} = \sqrt{\frac{2}{3}}\,,$$

and

$$\|\mathbf{v}_3\| = \sqrt{\int_{-1}^{1} \left(x^2 - \frac{1}{3}\right)^2 dx} = \sqrt{\frac{8}{45}}\,,$$

the desired orthonormal basis for $\mathcal{P}_2$ is

$$\left\{\frac{1}{\|\mathbf{v}_1\|}\mathbf{v}_1, \frac{1}{\|\mathbf{v}_2\|}\mathbf{v}_2, \frac{1}{\|\mathbf{v}_3\|}\mathbf{v}_3\right\} = \left\{\frac{1}{\sqrt{2}}, \sqrt{\frac{3}{2}}\,x, \sqrt{\frac{45}{8}}\left(x^2 - \frac{1}{3}\right)\right\}.$$

The method in this example extends to $\mathcal{P}_n$ for any positive integer n by using the same inner product and choosing $\{1, x, \ldots, x^n\}$ as the initial basis. As the Gram–Schmidt process is applied to polynomials of higher degree, the polynomials of lower degree remain unchanged. Thus we obtain an infinite sequence of polynomials $p_0(x)$, $p_1(x), \ldots, p_n(x), \ldots$ such that, for any n, the first $n+1$ polynomials in the sequence form an orthonormal basis for $\mathcal{P}_n$. These polynomials, called the **normalized Legendre[4] polynomials**, form an orthonormal basis for the infinite-dimensional vector space $\mathcal{P}$. They have applications to differential equations, statistics, and numerical analysis. In Example 5, we computed the first three normalized Legendre polynomials.

ORTHOGONAL PROJECTIONS AND LEAST-SQUARES APPROXIMATION

In an inner product space V, we define (as in Section 6.3) the **orthogonal complement** of a set S to be the set $S^\perp$ consisting of all vectors in V that are orthogonal to every vector in S. Just as in $\mathcal{R}^n$, it is easy to prove that $S^\perp$ is a subspace of V.

Suppose that V is an inner product space and that W is a finite-dimensional subspace of V with the *orthonormal* basis $\mathcal{B} = \{\mathbf{v}_1, \mathbf{v}_2, \ldots, \mathbf{v}_n\}$. Because the proof of Theorem 6.7 on page 392 applies directly to this context, we assume the result here. Thus, for any vector $\mathbf{v}$ in V, there exist unique vectors $\mathbf{w}$ in W and $\mathbf{z}$ in $W^\perp$ such that $\mathbf{v} = \mathbf{w} + \mathbf{z}$. Furthermore,

$$\mathbf{w} = \langle \mathbf{v}, \mathbf{v}_1 \rangle \mathbf{v}_1 + \langle \mathbf{v}, \mathbf{v}_2 \rangle \mathbf{v}_2 + \cdots + \langle \mathbf{v}, \mathbf{v}_n \rangle \mathbf{v}_n. \tag{2}$$

The vector $\mathbf{w}$ is called the **orthogonal projection** of $\mathbf{v}$ on W. In equation (2), the representation for $\mathbf{w}$ is independent of the choice of the orthonormal basis $\mathcal{B}$ because the orthogonal projection $\mathbf{w}$ is unique.

Of particular interest to us is the *closest vector property* of orthogonal projections, which is stated and justified on page 397 for subspaces of $\mathcal{R}^n$. The property easily extends to finite-dimensional subspaces of inner product spaces:

Closest Vector Property

Let W be a finite-dimensional subspace of an inner product space V and $\mathbf{u}$ be a vector in V. Among all vectors in W, the vector closest to $\mathbf{u}$ is the orthogonal projection of $\mathbf{u}$ on W.

If the inner product space V is a function space, then the closest vector property can be used to express the best approximation to a function in V as a linear combination of some specified finite set of functions. Here, "best" means nearest, as measured by the distance between two functions in V. In the examples that follow, we illustrate the use of an orthogonal projection to approximate a function. In this context, the orthogonal projection is called the **least-squares approximation** of the function.

Example 6 Find the least-squares approximation to the function $f(x) = \sqrt[3]{x}$ in $\mathrm{C}([-1,1])$ as a polynomial of degree less than or equal to 2.

Solution We may view the polynomials in $\mathcal{P}_2$ as functions and, restricting their domains to $[-1,1]$, we see that they form a finite-dimensional subspace of $\mathrm{C}([-1,1])$.

[4] Adrien Marie Legendre (1752–1833) was a French mathematician who taught at the École Militaire and the École Normale in Paris. He is best known for his research on elliptic functions, but he also produced important results in number theory such as the law of quadratic reciprocity. His paper *Nouvelles méthodes pour la détermination des orbites des comètes* contained the first mention of the method of least squares.

Then the required least-squares approximation to f is the orthogonal projection of f on $\mathcal{P}_2$. We apply equation (2) with $\mathbf{w} = f$ and with the $\mathbf{v}_i$'s as the vectors of the orthonormal basis for $\mathcal{P}_2$ obtained in Example 5. So we set

$$\mathbf{v}_1 = \frac{1}{\sqrt{2}}, \qquad \mathbf{v}_2 = \sqrt{\frac{3}{2}}\,x, \qquad \text{and} \qquad \mathbf{v}_3 = \sqrt{\frac{45}{8}}\left(x^2 - \frac{1}{3}\right)$$

and compute

$$\begin{aligned}
\mathbf{w} &= \langle \mathbf{v}, \mathbf{v}_1\rangle\, \mathbf{v}_1 + \langle \mathbf{v}, \mathbf{v}_2\rangle\, \mathbf{v}_2 + \langle \mathbf{v}, \mathbf{v}_3\rangle\, \mathbf{v}_3 \\
&= \left(\int_{-1}^{1} \sqrt[3]{x} \cdot \frac{1}{\sqrt{2}} dx\right) \frac{1}{\sqrt{2}} + \left(\int_{-1}^{1} \sqrt[3]{x} \cdot \sqrt{\frac{3}{2}}\, x\, dx\right) \sqrt{\frac{3}{2}}\, x \\
&\quad + \left(\int_{-1}^{1} \sqrt[3]{x} \cdot \sqrt{\frac{45}{8}}\left(x^2 - \frac{1}{3}\right) dx\right) \sqrt{\frac{45}{8}}\left(x^2 - \frac{1}{3}\right) \\
&= 0 \cdot \frac{1}{\sqrt{2}} + \frac{6}{7}\sqrt{\frac{3}{2}}\sqrt{\frac{3}{2}}\, x + 0 \cdot \sqrt{\frac{45}{8}}\left(x^2 - \frac{1}{3}\right) \\
&= \frac{9}{7}x.
\end{aligned}$$

Thus the function $g(x) = \frac{9}{7}x$, which is the orthogonal projection of $f(x) = \sqrt[3]{x}$ on $\mathcal{P}_2$ with respect to the inner product used here, is the least-squares approximation of f as a linear combination of 1, x, and x^2. This means that for any polynomial $p(x)$ of degree less than or equal to 2, if $p(x) \neq g(x)$, then

$$\|f - p\| > \|f - g\|.$$

(See Figure 7.3.)

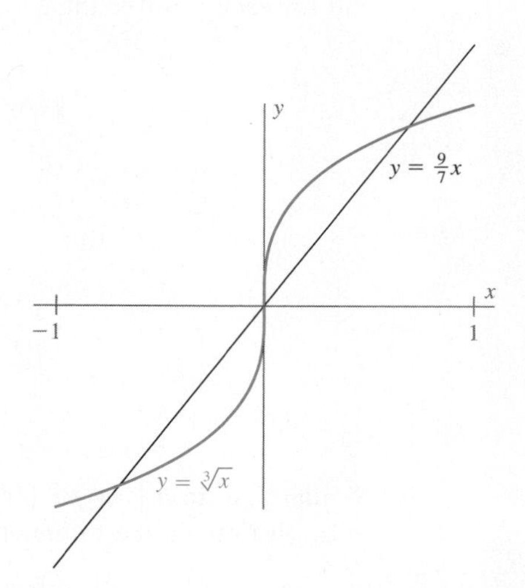

Figure 7.3 The orthogonal projection of $f(x) = \sqrt[3]{x}$ on $\mathcal{P}_2$

APPROXIMATION BY TRIGONOMETRIC POLYNOMIALS

A function $y = f(t)$ is called **periodic of period** p if $f(t + p) = f(t)$ for all t. Periodic functions are used to model phenomena that regularly repeat. An example of this is a vibration that generates a sound at a specific pitch or frequency. The frequency is the number of vibrations per unit of time, usually measured in seconds. In this case, we can define a function f of time t so that $f(t)$ is the relative pressure caused by the sound at time t at some particular location, such as the diaphragm of a microphone. For example, a musical instrument that plays a sustained note of middle C vibrates at the rate of 256 cycles per second. The length of one cycle, therefore, is $\frac{1}{256}$ of a second. So the function f associated with this sound has a period of $\frac{1}{256}$ seconds; that is, $f(t+\frac{1}{256})=f(t)$ for all t.

We can use orthogonal projections to obtain least-squares approximations of periodic functions by trigonometric polynomials. Suppose that $y=f(t)$ is a continuous periodic function. To simplify matters, we adjust the units of t so that f has period 2π. Then we may regard f and all trigonometric polynomials as continuous functions on $[0,2\pi]$, that is, as vectors in $\mathrm{C}([0,2\pi])$. The least-squares approximations of f of interest are the orthogonal projections of f on particular finite-dimensional subspaces of trigonometric polynomials.

For each positive integer n, let

$$\mathcal{S}_n=\{1,\cos t,\sin t,\cos 2t,\sin 2t,\ldots,\cos nt,\sin nt\}.$$

Then $\operatorname{Span}\mathcal{S}_n$ is a finite-dimensional subspace of trigonometric polynomials, which we denote by W_n. Furthermore, $\mathcal{S}_n$ is an orthogonal set, as we saw in Example 4. We can normalize each function in $\mathcal{S}_n$ to obtain an orthonormal basis for W_n, which can be used to compute the orthogonal projection of f on W_n. For this purpose, we compute the norms of the functions in $\mathcal{S}_n$ as

$$\|1\|=\sqrt{\int_0^{2\pi}1\,dt}=\sqrt{2\pi},$$

and for each positive integer k,

$$\begin{aligned}\|\cos kt\|&=\sqrt{\int_0^{2\pi}\cos^2 kt\,dt}\\&=\sqrt{\frac{1}{2}\int_0^{2\pi}(1+\cos 2kt)\,dt}\\&=\sqrt{\frac{1}{2}\left(t+\frac{1}{2k}\sin 2kt\right)\Big|_0^{2\pi}}\\&=\sqrt{\pi}.\end{aligned}$$

Similarly, $\|\sin kt\|=\sqrt{\pi}$ for every positive integer k. If we normalize each function in $\mathcal{S}_n$, we obtain the orthonormal basis

$$\mathcal{B}_n=\left\{\frac{1}{\sqrt{2\pi}},\frac{1}{\sqrt{\pi}}\cos t,\frac{1}{\sqrt{\pi}}\sin t,\frac{1}{\sqrt{\pi}}\cos 2t,\frac{1}{\sqrt{\pi}}\sin 2t,\ldots,\frac{1}{\sqrt{\pi}}\cos nt,\frac{1}{\sqrt{\pi}}\sin nt\right\}$$

for W_n. We can use $\mathcal{B}_n$ to compute the orthogonal projection of f on W_n to obtain its least-squares approximation, as described earlier in this section. The following example illustrates this approach:

Example 7 Sounds are detected by our ears or by a device such as a microphone that measures fluctuations in pressure as a function of time. The graph of such a function gives us

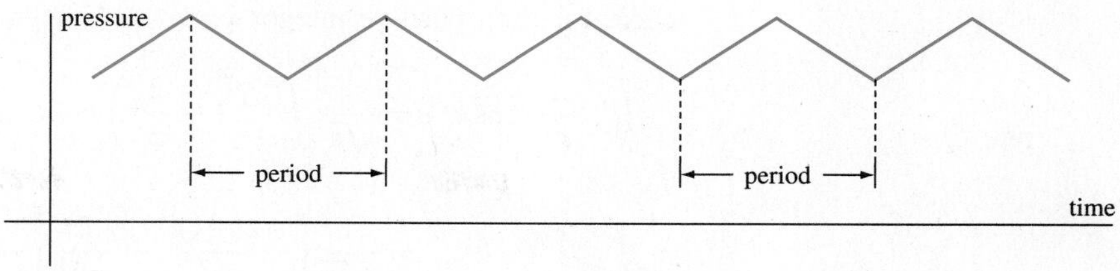

pressure

period period

time

Figure 7.4 A sawtooth tone

visual information about the sound. Consider a sound at a specific frequency whose graph is in the shape of sawteeth. (See Figure 7.4.) To simplify our computations, we adjust the units of time and relative pressure so that the function describing the relative pressure has period 2π and varies between 1 and -1. Furthermore, we select $t = 0$ at a maximum value of the relative pressure. Thus we obtain the function f in $\mathsf{C}([0, 2\pi])$ defined by

$$f(t) = \begin{cases} 1 - \dfrac{2}{\pi}t & \text{if } 0 \leq t \leq \pi \\[2ex] \dfrac{2}{\pi}t - 3 & \text{if } \pi \leq t \leq 2\pi. \end{cases}$$

(See Figure 7.5.)

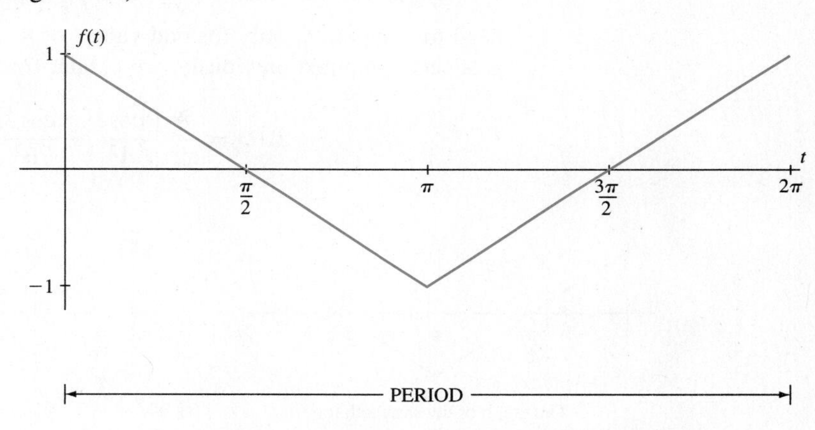

Figure 7.5 One period of the sawtooth tone

For each positive integer n, let f_n be the orthogonal projection of f on W_n. We can compute f_n by equation (2) in conjunction with the orthonormal basis $\mathcal{B}_n$:

$$\begin{aligned} f_n = \left\langle f, \tfrac{1}{\sqrt{2\pi}} \right\rangle \tfrac{1}{\sqrt{2\pi}} + \left\langle f, \tfrac{1}{\sqrt{\pi}} \cos t \right\rangle \tfrac{1}{\sqrt{\pi}} \cos t + \left\langle f, \tfrac{1}{\sqrt{\pi}} \sin t \right\rangle \tfrac{1}{\sqrt{\pi}} \sin t + \cdots \\ + \left\langle f, \tfrac{1}{\sqrt{\pi}} \cos nt \right\rangle \tfrac{1}{\sqrt{\pi}} \cos nt + \left\langle f, \tfrac{1}{\sqrt{\pi}} \sin nt \right\rangle \tfrac{1}{\sqrt{\pi}} \sin nt \end{aligned} \tag{3}$$

To find f_n, we must calculate the inner products in equation (3). First,

$$\begin{aligned} \left\langle f, \frac{1}{\sqrt{2\pi}} \right\rangle &= \frac{1}{\sqrt{2\pi}} \int_0^{\pi} \left(1 - \frac{2}{\pi}t\right) dt + \frac{1}{\sqrt{2\pi}} \int_{\pi}^{2\pi} \left(\frac{2}{\pi}t - 3\right) dt \\ &= 0 + 0 \\ &= 0. \end{aligned}$$

Next, for each positive integer k, we use integration by parts to compute

$$\begin{aligned}\left\langle f, \frac{1}{\sqrt{\pi}} \cos kt \right\rangle &= \frac{1}{\sqrt{\pi}} \int_0^{\pi} \left(1 - \frac{2}{\pi}t\right) \cos kt \, dt + \frac{1}{\sqrt{\pi}} \int_{\pi}^{2\pi} \left(\frac{2}{\pi}t - 3\right) \cos kt \, dt \\ &= \frac{1}{\sqrt{\pi}} \left[\frac{-2(-1)^k}{\pi k^2} + \frac{2}{\pi k^2}\right] + \frac{1}{\sqrt{\pi}} \left[\frac{2}{\pi k^2} - \frac{-2(-1)^k}{\pi k^2}\right] \\ &= \frac{4}{\pi\sqrt{\pi}k^2}(1 - (-1)^k) \\ &= \begin{cases} \dfrac{4}{\pi\sqrt{\pi}k^2} & \text{if } k \text{ is odd} \\ 0 & \text{if } k \text{ is even.} \end{cases}\end{aligned}$$

Finally, a similar calculation shows that

$$\left\langle f, \frac{1}{\sqrt{\pi}} \sin kt \right\rangle = 0 \qquad \text{for every positive integer } k.$$

In view of the fact that $\left\langle f, \frac{1}{\sqrt{\pi}} \cos kt \right\rangle = \left\langle f, \frac{1}{\sqrt{\pi}} \sin kt \right\rangle = 0$ for even integers k, we need to compute f_n only for odd values of n. Substituting into equation (3) the inner products computed previously, we obtain, for every odd positive integer n,

$$f_n(t) = \frac{8}{\pi^2} \left[\frac{\cos t}{1^2} + \frac{\cos 3t}{3^2} + \cdots + \frac{\cos nt}{n^2}\right].$$

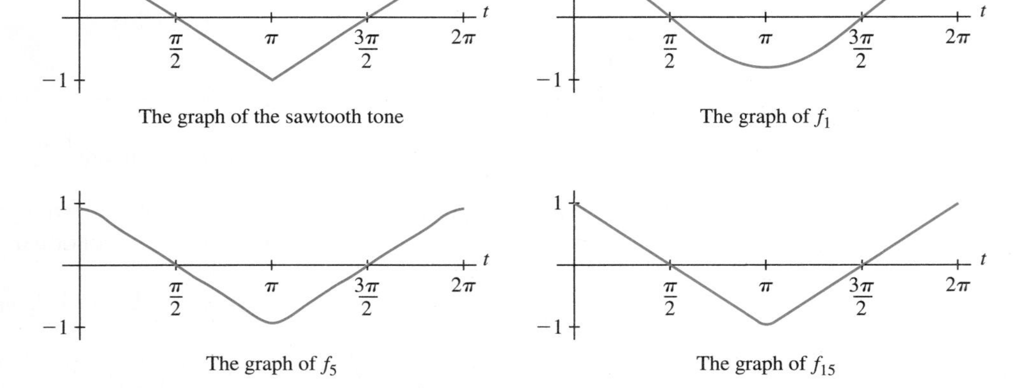

Figure 7.6 *f* and 3 least-squares approximations of *f*

Figure 7.6 allows us to compare the graphs of f with three least-squares approximations, f_1, f_5, and f_{15}, obtained by taking orthogonal projections of f on W_1, W_5, and W_{15}, respectively. Notice that as n increases, the graph of f_n more closely resembles the graph of f.

Simple electronic circuits can be designed to generate alternating currents described by functions of the form $\cos kt$ and $\sin kt$. These currents can be combined with a simple direct current (corresponding to a constant function) to produce a current that describes any chosen trigonometric polynomial. This current can be fed to an audio amplifier to produce an audible tone that approximates a given tone, such as the sawtooth tone of Example 7. Electronic devices called synthesizers do exactly this. Least-squares approximations of musical tones produced on different instruments such as violins and clarinets can be computed, and synthesizers can then use this information to produce sounds that convincingly mimic these instruments.

An area of mathematics called *Fourier*[5] *analysis* is concerned with the study of periodic functions, including many that are not continuous, and their approximations by trigonometric polynomials.

Practice Problem 3 ▶ Let W be the subset of $\mathcal{M}_{2\times 2}$ consisting of the 2×2 matrices A such that $\text{trace}(A) = 0$, and let $B = \begin{bmatrix} 1 & 2 \\ 3 & 5 \end{bmatrix}$. Find the matrix in W that is closest to B, where distance is defined by using the Frobenius inner product on $\mathcal{M}_{2\times 2}$. ◀

EXERCISES

In Exercises 1–8, use the inner product in Example 1 for $\mathsf{C}([1,2])$ *to compute each* $\langle f, g\rangle$.

1. $f(t) = t^3$ and $g(t) = 1$
2. $f(t) = 2t$ and $g(t) = t - 1$
3. $f(t) = t$ and $g(t) = t^2 + 1$
4. $f(t) = t^2$ and $g(t) = t^2$
5. $f(t) = t^3$ and $g(t) = t^2$
6. $f(t) = t^2$ and $g(t) = \dfrac{1}{t}$
7. $f(t) = t$ and $g(t) = e^t$
8. $f(t) = t^2$ and $g(t) = e^t$

In Exercises 9–16, use the Frobenius inner product for $\mathcal{M}_{2\times 2}$ *to compute each* $\langle A, B\rangle$.

9. $A = \begin{bmatrix} 5 & 0 \\ 0 & 5 \end{bmatrix}$ and $B = \begin{bmatrix} 1 & 2 \\ 3 & 4 \end{bmatrix}$
10. $A = \begin{bmatrix} 1 & 0 \\ 0 & 2 \end{bmatrix}$ and $B = \begin{bmatrix} 2 & 3 \\ 1 & 0 \end{bmatrix}$
11. $A = \begin{bmatrix} 1 & -1 \\ 2 & 3 \end{bmatrix}$ and $B = \begin{bmatrix} 2 & 4 \\ 1 & 0 \end{bmatrix}$
12. $A = \begin{bmatrix} 0 & 5 \\ -2 & 0 \end{bmatrix}$ and $B = \begin{bmatrix} 1 & 3 \\ 2 & 4 \end{bmatrix}$
13. $A = \begin{bmatrix} -1 & 2 \\ 0 & 4 \end{bmatrix}$ and $B = \begin{bmatrix} 3 & 2 \\ 1 & -1 \end{bmatrix}$
14. $A = \begin{bmatrix} 0 & -2 \\ 3 & 0 \end{bmatrix}$ and $B = \begin{bmatrix} 2 & -1 \\ 1 & 0 \end{bmatrix}$
15. $A = \begin{bmatrix} 3 & 2 \\ 1 & -1 \end{bmatrix}$ and $B = \begin{bmatrix} -1 & 2 \\ 0 & 4 \end{bmatrix}$
16. $A = \begin{bmatrix} 3 & -2 \\ -1 & 1 \end{bmatrix}$ and $B = \begin{bmatrix} 3 & -2 \\ -1 & 1 \end{bmatrix}$

In Exercises 17–24, use the inner product for $\mathcal{P}_2$ *in Example 5 to compute each* $\langle f(x), g(x)\rangle$.

17. $f(x) = 3$ and $g(x) = -x + 2$
18. $f(x) = x$ and $g(x) = 2x + 1$
19. $f(x) = x^2 - 2$ and $g(x) = 3x + 5$
20. $f(x) = x + 1$ and $g(x) = x - 1$
21. $f(x) = x^2 + 1$ and $g(x) = x$
22. $f(x) = x + 1$ and $g(x) = x^2$
23. $f(x) = x^2 + 1$ and $g(x) = x - 1$
24. $f(x) = x^2 - 1$ and $g(x) = x^2 + 2$

In Exercises 25–44, determine whether the statements are true or false.

25. The inner product of two vectors in an inner product space is a vector in the same inner product space.
26. An inner product is a real-valued function on the set of ordered pairs of vectors in a vector space.
27. An inner product on a vector space V is a linear operator on V.
28. There can be at most one inner product on a vector space.
29. Every nonzero finite-dimensional inner product space has an orthonormal basis.

[5] Jean Baptiste Joseph Fourier (1768–1830) was a French mathematician and professor of analysis at the École Polytechnique in Paris. His *Théorie analytique de la chaleur* (1822) was an important contribution to physics (to problems involving the radiation of heat) and to mathematics (to the development of what are now called Fourier or trigonometric series). In 1798, he was appointed governor of Lower Egypt by Napoleon.

30. Every orthogonal set in an inner product space is linearly independent.
31. Every orthonormal set in an inner product space is linearly independent.
32. It is possible to define an inner product on the set of $n \times n$ matrices.
33. The dot product is a special case of an inner product.
34. The definite integral can be used to define an inner product on $\mathcal{P}_2$.
35. The indefinite integral can be used to define an inner product on $\mathcal{P}_2$.
36. In an inner product space, the orthogonal projection of a vector $\mathbf{v}$ on a finite-dimensional subspace W is the vector in W that is closest to $\mathbf{v}$.
37. In an inner product space, $\langle \mathbf{v}, \mathbf{v} \rangle = 0$ if and only if $\mathbf{v} = \mathbf{0}$.
38. In an inner product space, the norm of a vector $\mathbf{v}$ equals $\langle \mathbf{v}, \mathbf{v} \rangle$.
39. In an inner product space, if $\langle \mathbf{u}, \mathbf{v} \rangle = \langle \mathbf{u}, \mathbf{w} \rangle$ for some vector $\mathbf{u}$, then $\mathbf{v} = \mathbf{w}$.
40. The Frobenius inner product on $\mathcal{M}_{m \times n}$ is defined by $\langle A, B \rangle = \text{trace}(AB)$.
41. In an inner product space, the distance between vectors $\mathbf{u}$ and $\mathbf{v}$ is defined to be $\|\mathbf{u} - \mathbf{v}\|$.
42. If W is a finite-dimensional subspace of an inner product space V, then every vector $\mathbf{v}$ in V can be written as $\mathbf{w} + \mathbf{z}$, where $\mathbf{w}$ is in W and $\mathbf{z}$ is in $W^{\perp}$.
43. The normalized Legendre polynomials are the polynomials obtained by applying the Gram–Schmidt process to $\{1, x, x^2, \ldots\}$.
44. If $\mathcal{B} = \{\mathbf{v}_1, \mathbf{v}_2, \ldots, \mathbf{v}_n\}$ is a basis for a subspace W of an inner product space V, then the orthogonal projection of $\mathbf{u}$ on W is the vector

$$\langle \mathbf{u}, \mathbf{v}_1 \rangle \mathbf{v}_1 + \langle \mathbf{u}, \mathbf{v}_2 \rangle \mathbf{v}_2 + \cdots + \langle \mathbf{u}, \mathbf{v}_n \rangle \mathbf{v}_n.$$

---

45. In Example 1, verify axioms 3 and 4 of the definition of inner product.
46. In Example 2, verify axioms 3 and 4 of the definition of inner product.
47. Let V be a finite-dimensional vector space and $\mathcal{B}$ be a basis for V. For $\mathbf{u}$ and $\mathbf{v}$ in V, define

$$\langle \mathbf{u}, \mathbf{v} \rangle = [\mathbf{u}]_{\mathcal{B}} \cdot [\mathbf{v}]_{\mathcal{B}}.$$

Prove that this rule defines an inner product on V.

48. Let A be an $n \times n$ invertible matrix. For $\mathbf{u}$ and $\mathbf{v}$ in $\mathcal{R}^n$, define

$$\langle \mathbf{u}, \mathbf{v} \rangle = (A\mathbf{u}) \cdot (A\mathbf{v}).$$

Prove that this rule defines an inner product on $\mathcal{R}^n$.

49. Let A be an $n \times n$ positive definite matrix (as defined in the exercises of Section 6.6). For $\mathbf{u}$ and $\mathbf{v}$ in $\mathcal{R}^n$, define

$$\langle \mathbf{u}, \mathbf{v} \rangle = (A\mathbf{u}) \cdot \mathbf{v}.$$

Prove that this rule defines an inner product on $\mathcal{R}^n$.

In Exercises 50–57, a vector space V and a rule are given. Determine whether the rule defines an inner product on V. Justify your answer.

50. $V = \mathcal{R}^n$ and $\langle \mathbf{u}, \mathbf{v} \rangle = |\mathbf{u} \cdot \mathbf{v}|$
51. $V = \mathcal{R}^n$ and $\langle \mathbf{u}, \mathbf{v} \rangle = 2(\mathbf{u} \cdot \mathbf{v})$
52. $V = \mathcal{R}^2$, $D = \begin{bmatrix} 3 & 0 \\ 0 & 2 \end{bmatrix}$, and $\langle \mathbf{u}, \mathbf{v} \rangle = (D\mathbf{u}) \cdot \mathbf{v}$
53. Let $V = \mathsf{C}([0, 2])$, and

$$\langle f, g \rangle = \int_0^1 f(t)g(t)\,dt$$

for all f and g in V. (Note that the limits of integration are not 0 and 2.)

54. $V = \mathcal{R}^n$ and $\langle \mathbf{u}, \mathbf{v} \rangle = -2(\mathbf{u} \cdot \mathbf{v})$
55. Let V be any vector space on which two inner products $\langle \mathbf{u}, \mathbf{v} \rangle_1$ and $\langle \mathbf{u}, \mathbf{v} \rangle_2$ are defined for $\mathbf{u}$ and $\mathbf{v}$ in V. Define $\langle \mathbf{u}, \mathbf{v} \rangle$ by

$$\langle \mathbf{u}, \mathbf{v} \rangle = \langle \mathbf{u}, \mathbf{v} \rangle_1 + \langle \mathbf{u}, \mathbf{v} \rangle_2 .$$

56. Let V be any vector space on which two inner products $\langle \mathbf{u}, \mathbf{v} \rangle_1$ and $\langle \mathbf{u}, \mathbf{v} \rangle_2$ are defined for $\mathbf{u}$ and $\mathbf{v}$ in V. Define $\langle \mathbf{u}, \mathbf{v} \rangle$ by

$$\langle \mathbf{u}, \mathbf{v} \rangle = \langle \mathbf{u}, \mathbf{v} \rangle_1 - \langle \mathbf{u}, \mathbf{v} \rangle_2 .$$

57. Let V be any vector space on which two inner products $\langle \mathbf{u}, \mathbf{v} \rangle_1$ and $\langle \mathbf{u}, \mathbf{v} \rangle_2$ are defined for $\mathbf{u}$ and $\mathbf{v}$ in V. Define $\langle \mathbf{u}, \mathbf{v} \rangle$ by

$$\langle \mathbf{u}, \mathbf{v} \rangle = a \langle \mathbf{u}, \mathbf{v} \rangle_1 + b \langle \mathbf{u}, \mathbf{v} \rangle_2 ,$$

where a and b are positive real numbers.

58. Use the inner product in Example 4 to prove that $\sin mt$ and $\sin nt$ are orthogonal for any two distinct integers m and n. *Hint:* Use the trigonometric identity

$$\sin a \sin b = \frac{\cos(a+b) - \cos(a-b)}{2}.$$

59. Use the inner product in Example 4 to prove that $\cos mt$ and $\cos nt$ are orthogonal for any two distinct integers m and n. *Hint:* Use the trigonometric identity

$$\cos a \cos b = \frac{\cos(a+b) + \cos(a-b)}{2}.$$

60. (a) Use the methods of Example 5 to obtain $p_3(x)$, the normalized Legendre polynomial of degree 3.
(b) Use the result of (a) to find the least-squares approximation of $\sqrt[3]{x}$ on $[-1, 1]$ as a polynomial of degree less than or equal to 3.
61. Find an orthogonal basis for the subspace $\mathsf{C}([0, 1])$ in Example 1 having the generating set $\{1, e^t, e^{-t}\}$.
62. Suppose that $\langle \mathbf{u}, \mathbf{v} \rangle$ is an inner product for a vector space V. For any scalar $r > 0$, define $\langle\langle \mathbf{u}, \mathbf{v} \rangle\rangle = r\langle \mathbf{u}, \mathbf{v} \rangle$.
(a) Prove that $\langle\langle \mathbf{u}, \mathbf{v} \rangle\rangle$ is an inner product on V.
(b) Why is $\langle\langle \mathbf{u}, \mathbf{v} \rangle\rangle$ not an inner product if $r \le 0$?

In Exercises 63–70, let $\mathbf{u}$, $\mathbf{v}$, and $\mathbf{w}$ be vectors in an inner product space V, and let c be a scalar.

63. Prove that $\|\mathbf{v}\| = 0$ if and only if $\mathbf{v} = \mathbf{0}$.
64. Prove that $\|c\mathbf{v}\| = |c|\|\mathbf{v}\|$.
65. Prove that $\langle \mathbf{0}, \mathbf{u} \rangle = \langle \mathbf{u}, \mathbf{0} \rangle = 0$.
66. Prove that $\langle \mathbf{u} - \mathbf{w}, \mathbf{v} \rangle = \langle \mathbf{u}, \mathbf{v} \rangle - \langle \mathbf{w}, \mathbf{v} \rangle$.

67. Prove that $\langle \mathbf{v}, \mathbf{u} - \mathbf{w} \rangle = \langle \mathbf{v}, \mathbf{u} \rangle - \langle \mathbf{v}, \mathbf{w} \rangle$.

68. Prove that $\langle \mathbf{u}, c\mathbf{v} \rangle = c \langle \mathbf{u}, \mathbf{v} \rangle$.

69. Prove that if $\langle \mathbf{u}, \mathbf{w} \rangle = 0$ for all $\mathbf{u}$ in V, then $\mathbf{w} = \mathbf{0}$.

70. Prove that if $\langle \mathbf{u}, \mathbf{v} \rangle = \langle \mathbf{u}, \mathbf{w} \rangle$ for all $\mathbf{u}$ in V, then $\mathbf{v} = \mathbf{w}$.

71. Let V be a finite-dimensional inner product space, and suppose that $\mathcal{B}$ is an orthonormal basis for V. Prove that, for any vectors $\mathbf{u}$ and $\mathbf{v}$ in V,

$$\langle \mathbf{u}, \mathbf{v} \rangle = [\mathbf{u}]_{\mathcal{B}} \cdot [\mathbf{v}]_{\mathcal{B}}.$$

72. Prove that if A is an $n \times n$ symmetric matrix and B is an $n \times n$ skew-symmetric matrix, then A and B are orthogonal with respect to the Frobenius inner product.

73. Prove that if A and B are 2×2 matrices, then the Frobenius inner product $\langle A, B \rangle$ can be computed as

$$\langle A, B \rangle = a_{11}b_{11} + a_{12}b_{12} + a_{21}b_{21} + a_{22}b_{22}.$$

74. Extend Exercise 73 to the general case. That is, prove that if A and B are $n \times n$ matrices, then the Frobenius inner product $\langle A, B \rangle$ can be computed as

$$\langle A, B \rangle = a_{11}b_{11} + a_{12}b_{12} + \cdots + a_{nn}b_{nn}.$$

75. Consider the inner product space $\mathcal{M}_{2\times 2}$ with the Frobenius inner product.
    (a) Find an orthonormal basis for the subspace of 2×2 symmetric matrices.
    (b) Use (a) to find the 2×2 symmetric matrix that is closest to

$$\begin{bmatrix} 1 & 2 \\ 4 & 8 \end{bmatrix}.$$

76. Prove that if $\mathcal{B}$ is an infinite orthogonal subset of nonzero vectors in an inner product space V, then $\mathcal{B}$ is a linearly independent subset of V.

77. Prove that if $\{\mathbf{u}, \mathbf{v}\}$ is a linearly dependent subset of an inner product space, then $\langle \mathbf{u}, \mathbf{v} \rangle^2 = \langle \mathbf{u}, \mathbf{u} \rangle \langle \mathbf{v}, \mathbf{v} \rangle$.

78. Prove the converse of Exercise 77: If $\mathbf{u}$ and $\mathbf{v}$ are vectors in an inner product space and $\langle \mathbf{u}, \mathbf{v} \rangle^2 = \langle \mathbf{u}, \mathbf{u} \rangle \langle \mathbf{v}, \mathbf{v} \rangle$, then $\{\mathbf{u}, \mathbf{v}\}$ is a linearly dependent set. *Hint:* Suppose that $\mathbf{u}$ and $\mathbf{v}$ are nonzero vectors. Show that

$$\left\| \mathbf{v} - \frac{\langle \mathbf{u}, \mathbf{v} \rangle}{\langle \mathbf{u}, \mathbf{u} \rangle} \mathbf{u} \right\| = 0.$$

79. Let V be an inner product space and $\mathbf{u}$ be a vector in V. Define $F_{\mathbf{u}}: V \to \mathcal{R}$ by

$$F_{\mathbf{u}}(\mathbf{v}) = \langle \mathbf{v}, \mathbf{u} \rangle$$

for all $\mathbf{v}$ in V. Prove that $F_{\mathbf{u}}$ is a linear transformation.

80. Prove the converse of Exercise 79 for finite-dimensional inner product spaces: If V is a finite-dimensional inner product space and $T: V \to \mathcal{R}$ is a linear transformation, then there exists a unique vector $\mathbf{u}$ in V such that $T = F_{\mathbf{u}}$. *Hint:* Let $\{\mathbf{v}_1, \mathbf{v}_2, \ldots, \mathbf{v}_n\}$ be an orthonormal basis for V, and let

$$\mathbf{u} = T(\mathbf{v}_1)\mathbf{v}_1 + T(\mathbf{v}_2)\mathbf{v}_2 + \cdots + T(\mathbf{v}_n)\mathbf{v}_n.$$

81. (a) Prove that B^TB is positive definite (as defined in the exercises of Section 6.6) for any invertible matrix B.
    (b) Use (a) and Exercise 71 to prove the converse of Exercise 49: For any inner product on $\mathcal{R}^n$, there exists a positive definite matrix A such that

$$\langle \mathbf{u}, \mathbf{v} \rangle = (A\mathbf{u}) \cdot \mathbf{v}$$

for all vectors $\mathbf{u}$ and $\mathbf{v}$ in $\mathcal{R}^n$.

The following definitions are used in Exercises 82 and 83:

Definition A linear transformation $T: V \to W$ is called a **linear isometry** if T is an isomorphism and $\langle T(\mathbf{u}), T(\mathbf{v}) \rangle = \langle \mathbf{u}, \mathbf{v} \rangle$ for every $\mathbf{u}$ and $\mathbf{v}$ in V. The inner product spaces V and W are called **isometric** if there exists a linear isometry from V to W.

82. Let V, W, and Z be inner product spaces. Prove the following statements:
    (a) V is isometric to itself.
    (b) If V is isometric to W, then W is isometric to V.
    (c) If V is isometric to W and W is isometric to Z, then V is isometric to Z.

83. Let V be an n-dimensional inner product space.
    (a) Prove that, for any orthonormal basis $\mathcal{B}$ of V, the linear transformation $\Phi_{\mathcal{B}}: V \to \mathcal{R}^n$ defined by $\Phi_{\mathcal{B}}(\mathbf{v}) = [\mathbf{v}]_{\mathcal{B}}$ is a linear isometry. Thus every n-dimensional inner product space is isometric to $\mathcal{R}^n$.
    (b) Consider the inner product space $\mathcal{M}_{n\times n}$ with the Frobenius inner product. For A in $\mathcal{M}_{n\times n}$, define

$$T(A) = \begin{bmatrix} a_{11} \\ a_{12} \\ \vdots \\ a_{1n} \\ \vdots \\ a_{n1} \\ \vdots \\ a_{nn} \end{bmatrix}.$$

Use (a) to prove that $T: \mathcal{M}_{n\times n} \to \mathcal{R}^{n^2}$ is a linear isometry.

84. Let $\{\mathbf{w}_1, \mathbf{w}_2, \ldots, \mathbf{w}_n\}$ be an orthonormal basis for a subspace W of an inner product space. Prove that, for any vector $\mathbf{v}$ in W,

$$\mathbf{v} = \langle \mathbf{v}, \mathbf{w}_1 \rangle \mathbf{w}_1 + \langle \mathbf{v}, \mathbf{w}_2 \rangle \mathbf{w}_2 + \cdots + \langle \mathbf{v}, \mathbf{w}_n \rangle \mathbf{w}_n.$$

85. Let $\{\mathbf{w}_1, \mathbf{w}_2, \ldots, \mathbf{w}_n\}$ be an orthonormal basis for a subspace W of an inner product space. Prove that, for any vectors $\mathbf{u}$ and $\mathbf{v}$ in W, we have

$$\begin{aligned}&\mathbf{u}+\mathbf{v}\\&= (\langle \mathbf{u}, \mathbf{w}_1\rangle + \langle \mathbf{v}, \mathbf{w}_1\rangle)\mathbf{w}_1 + \cdots + (\langle \mathbf{u}, \mathbf{w}_n\rangle + \langle \mathbf{v}, \mathbf{w}_n\rangle)\mathbf{w}_n.\end{aligned}$$

86. Let $\{\mathbf{w}_1, \mathbf{w}_2, \ldots, \mathbf{w}_n\}$ be an orthonormal basis for a subspace W of an inner product space. Prove that, for any vectors $\mathbf{u}$ and $\mathbf{v}$ in W, we have

$$\begin{aligned}\langle \mathbf{u}, \mathbf{v}\rangle = \langle \mathbf{u}, \mathbf{w}_1\rangle \langle \mathbf{v}, \mathbf{w}_1\rangle + \langle \mathbf{u}, \mathbf{w}_2\rangle \langle \mathbf{v}, \mathbf{w}_2\rangle + \cdots\\ + \langle \mathbf{u}, \mathbf{w}_n\rangle \langle \mathbf{v}, \mathbf{w}_n\rangle.\end{aligned}$$

87. Let W be the 1-dimensional subspace Span $\{I_n\}$ of $\mathcal{M}_{n\times n}$. That is, W is the set of all $n \times n$ scalar matrices. (See Exercise 85(c) of Section 2.4.) Prove that, for any $n \times n$ matrix A, the matrix in W that is nearest to A is $\left(\frac{\text{trace}(A)}{n}\right) I_n$, where distance is defined by using the Frobenius inner product on $\mathcal{M}_{n\times n}$.

In Exercise 88, use either a calculator with matrix capabilities or computer software such as MATLAB to solve the problem.

88. Let

$$A = \begin{bmatrix} 25 & 24 & 23 & 22 & 21\\ 20 & 19 & 18 & 17 & 16\\ 15 & 14 & 13 & 12 & 11\\ 10 & 9 & 8 & 7 & 6\\ 5 & 4 & 3 & 2 & 1\end{bmatrix}$$

and

$$B = \begin{bmatrix} 1 & 2 & 3 & 4 & 5\\ 6 & 7 & 8 & 9 & 10\\ 11 & 12 & 13 & 14 & 15\\ 16 & 17 & 18 & 19 & 20\\ 21 & 22 & 23 & 24 & 25\end{bmatrix}.$$

Find the matrix in the 1-dimensional subspace Span $\{A\}$ that is nearest to B, where distance is defined by using the Frobenius inner product on $\mathcal{M}_{5\times 5}$.

SOLUTIONS TO THE PRACTICE PROBLEMS

1. $$\begin{aligned}\|A\|^2 &= \text{trace}(AA^T) = \text{trace}\left(\begin{bmatrix}2&1\\0&3\end{bmatrix}\begin{bmatrix}2&1\\0&3\end{bmatrix}^T\right)\\ &= \text{trace}\left(\begin{bmatrix}2&1\\0&3\end{bmatrix}\begin{bmatrix}2&0\\1&3\end{bmatrix}\right) = \text{trace}\left(\begin{bmatrix}5&3\\3&9\end{bmatrix}\right)\\ &= 14\end{aligned}$$

$$\begin{aligned}\|B\|^2 &= \text{trace}(BB^T) = \text{trace}\left(\begin{bmatrix}1&1\\2&0\end{bmatrix}\begin{bmatrix}1&1\\2&0\end{bmatrix}^T\right)\\ &= \text{trace}\left(\begin{bmatrix}1&1\\2&0\end{bmatrix}\begin{bmatrix}1&2\\1&0\end{bmatrix}\right) = \text{trace}\left(\begin{bmatrix}2&2\\2&4\end{bmatrix}\right)\\ &= 6\end{aligned}$$

$$\begin{aligned}\langle A, B\rangle &= \text{trace}(AB^T) = \text{trace}\left(\begin{bmatrix}2&1\\0&3\end{bmatrix}\begin{bmatrix}1&1\\2&0\end{bmatrix}^T\right)\\ &= \text{trace}\left(\begin{bmatrix}2&1\\0&3\end{bmatrix}\begin{bmatrix}1&2\\1&0\end{bmatrix}\right) = \text{trace}\left(\begin{bmatrix}3&4\\3&0\end{bmatrix}\right)\\ &= 3\end{aligned}$$

2. (a) $$\begin{aligned}\|f\|^2 &= \langle f,f\rangle = \textstyle\int_0^1 f(t)\cdot f(t)\,dt = \int_0^1 t^2\,dt\\ &= \tfrac{1}{3}t^3\big|_0^1 = \tfrac{1}{3}\\ \|g\|^2 &= \langle g,g\rangle = \textstyle\int_0^1 g(t)\cdot g(t)\,dt = \int_0^1 t^5\,dt\\ &= \tfrac{1}{5}t^5\big|_0^1 = \tfrac{1}{5}\\ \langle f,g\rangle &= \textstyle\int_0^1 t\cdot t^2\,dt = \int_0^1 t^3\,dt = \tfrac{1}{4}\,t^4\big|_0^1 = \tfrac{1}{4}\end{aligned}$$

(b) Observe that $\frac{1}{4} \le \frac{1}{\sqrt{3}} \cdot \frac{1}{\sqrt{5}}$.

3. The set W is a subspace of $\mathcal{M}_{2\times 2}$ by Example 6 of Section 7.1, and hence the desired matrix A is the orthogonal projection of B on W. Since

$$\{A_1, A_2, A_3\} = \left\{\frac{1}{\sqrt{2}}\begin{bmatrix}1&0\\0&-1\end{bmatrix}, \begin{bmatrix}0&1\\0&0\end{bmatrix}, \begin{bmatrix}0&0\\1&0\end{bmatrix}\right\}$$

is an orthonormal basis for W, we can apply equation (2) to obtain the orthogonal projection. For this purpose, observe that

$$\begin{aligned}\langle B, A_1\rangle &= \text{trace}(BA_1^T) = \text{trace}\left(\begin{bmatrix}1&2\\3&5\end{bmatrix}\frac{1}{\sqrt{2}}\begin{bmatrix}1&0\\0&-1\end{bmatrix}^T\right)\\ &= -\frac{4}{\sqrt{2}}.\end{aligned}$$

Similarly,

$$\langle B, A_2\rangle = 2 \quad\text{and}\quad \langle B, A_3\rangle = 3.$$

Therefore

$$\begin{aligned}A &= \langle B, A_1\rangle A_1 + \langle B, A_2\rangle A_2 + \langle B, A_3\rangle A_3\\ &= -\frac{4}{\sqrt{2}}\left(\frac{1}{\sqrt{2}}\right)\begin{bmatrix}1&0\\0&-1\end{bmatrix} + 2\begin{bmatrix}0&1\\0&0\end{bmatrix} + 3\begin{bmatrix}0&0\\1&0\end{bmatrix}\\ &= \begin{bmatrix}-2&2\\3&2\end{bmatrix}.\end{aligned}$$

CHAPTER 7 REVIEW EXERCISES

T&F *In Exercises 1–7, determine whether the statements are true or false.*

1. Every subspace of a vector space is a subset of $\mathcal{R}^n$ for some integer n.
2. Every $m \times n$ matrix is a vector in the vector space $\mathcal{M}_{m\times n}$.
3. $\dim \mathcal{M}_{m\times n} = m + n$
4. A matrix representation of a linear operator on $\mathcal{M}_{m\times n}$ is an $m \times n$ matrix.
5. The Frobenius inner product of two matrices is a scalar.
6. Suppose that $\mathbf{u}$, $\mathbf{v}$, and $\mathbf{w}$ are vectors in an inner product space. If $\mathbf{u}$ is orthogonal to $\mathbf{v}$ and $\mathbf{v}$ is orthogonal to $\mathbf{w}$, then $\mathbf{u}$ is orthogonal to $\mathbf{w}$.
7. Suppose that $\mathbf{u}$, $\mathbf{v}$, and $\mathbf{w}$ are vectors in an inner product space. If $\mathbf{u}$ is orthogonal to both $\mathbf{v}$ and $\mathbf{w}$, then $\mathbf{u}$ is orthogonal to $\mathbf{v} + \mathbf{w}$.

In Exercises 8–11, determine whether each set V is a vector space with respect to the indicated operations. Justify your conclusions.

8. V is the set of all sequences $\{a_n\}$ of real numbers. For any sequences $\{a_n\}$ and $\{b_n\}$ in V and any scalar c, define the sum $\{a_n\} + \{b_n\}$ and the product $c\{a_n\}$ by
$$\{a_n\} + \{b_n\} = \{a_n + b_n\} \quad \text{and} \quad c\{a_n\} = \{ca_n\}.$$
9. V is the set of all real numbers with vector addition, $\oplus$, and scalar multiplication, $\odot$, defined by
$$a \oplus b = a + b + ab \quad \text{and} \quad c \odot a = ca,$$
where a and b are in V and c is any scalar.
10. V is the set of all 2×2 matrices with vector addition, $\oplus$, and scalar multiplication, $\odot$, defined by
$$A \oplus B = A + B \quad \text{and} \quad t \odot \begin{bmatrix} a & b \\ c & d \end{bmatrix} = \begin{bmatrix} ta & tb \\ c & d \end{bmatrix}$$
for all 2×2 matrices A and B and scalars t.
11. V is the set of all functions from $\mathcal{R}$ to $\mathcal{R}$ such that $f(x) > 0$ for all x in $\mathcal{R}$. Vector addition, $\oplus$, and scalar multiplication, $\odot$, are defined by
$$(f \oplus g)(x) = f(x)g(x) \quad \text{and} \quad (c \odot f)(x) = [f(x)]^c$$
for all f and g in V, x in $\mathcal{R}$, and scalars c.

In Exercises 12–15, determine whether each subset W is a subspace of the vector space V. Justify your conclusions.

12. $V = \mathcal{F}(\mathcal{R})$ and W is the set of all functions f in V such that $f(x) \geq 0$ for all x in $\mathcal{R}$.
13. $V = \mathcal{P}$ and W is the set consisting of the zero polynomial and all polynomials of even degree.
14. For a given nonzero vector $\mathbf{v}$ in $\mathcal{R}^n$, let W be the set of all $n \times n$ matrices A such that $\mathbf{v}$ is an eigenvector of A, and $V = \mathcal{M}_{n\times n}$.
15. For a given nonzero scalar λ, let W be the set of all $n \times n$ matrices A such that λ is an eigenvalue of A, and $V = \mathcal{M}_{n\times n}$.

In Exercises 16–19, determine whether each matrix is a linear combination of the matrices in the set
$$\left\{\begin{bmatrix} 1 & 2 \\ 1 & -1 \end{bmatrix}, \begin{bmatrix} 0 & 1 \\ 2 & 0 \end{bmatrix}, \begin{bmatrix} -1 & 3 \\ 1 & 1 \end{bmatrix}\right\}.$$

16. $\begin{bmatrix} 1 & 10 \\ 9 & -1 \end{bmatrix}$
17. $\begin{bmatrix} 2 & 8 \\ 1 & -5 \end{bmatrix}$
18. $\begin{bmatrix} 3 & 1 \\ -2 & -4 \end{bmatrix}$
19. $\begin{bmatrix} 4 & 1 \\ -2 & -4 \end{bmatrix}$

In Exercises 20 and 21, let $\mathcal{S}$ be the following subset of $\mathcal{P}$:
$$\mathcal{S} = \{x^3 - x^2 + x + 1, 3x^2 + x + 2, x - 1\}$$

20. Show that the polynomial $x^3 + 2x^2 + 5$ is a linear combination of the polynomials in $\mathcal{S}$.
21. Find a constant c so that $f(x) = 2x^3 + x^2 + 2x + c$ is a linear combination of the polynomials in $\mathcal{S}$.

In Exercises 22 and 23, find a basis for each subspace W of each vector space V. Then find the dimension of W.

22. $V = \mathcal{M}_{2\times 2}$ and
$$W = \left\{A \in V : \begin{bmatrix} 1 & 2 \\ 1 & 2 \end{bmatrix} A = \begin{bmatrix} 0 & 0 \\ 0 & 0 \end{bmatrix}\right\}$$
23. $V = \mathcal{P}_3$ and
$$W = \{f(x) \in V : f(0) + f'(0) + f''(0) = 0\}$$

In Exercises 24–27, determine whether each function T is linear. If T is linear, determine whether it is an isomorphism.

24. $T\colon \mathcal{R}^3 \to \mathcal{P}_2$ defined by
$$T\left(\begin{bmatrix} a \\ b \\ c \end{bmatrix}\right) = (a + b) + (a - b)x + cx^2$$
25. $T\colon \mathcal{M}_{2\times 2} \to \mathcal{R}$ defined by $T(A) = \operatorname{trace}(A^2)$
26. $T\colon \mathcal{R}^3 \to \mathcal{M}_{2\times 2}$ defined by
$$T\left(\begin{bmatrix} a \\ b \\ c \end{bmatrix}\right) = \begin{bmatrix} a & b \\ c & a + b + c \end{bmatrix}$$
27. $T\colon \mathcal{P}_2 \to \mathcal{R}^3$ defined by
$$T(f(x)) = \begin{bmatrix} f(0) \\ f'(0) \\ \int_0^1 f(t)\,dt \end{bmatrix}$$

In Exercises 28–31, a vector space V, a basis $\mathcal{B}$ for V, and a linear operator T on V are given. Find $[T]_{\mathcal{B}}$.

28. $V = \mathcal{P}_2$, $T(p(x)) = p(1) + 2p'(1)x - p''(1)x^2$ for all $p(x)$ in V, and $\mathcal{B} = \{1, x, x^2\}$
29. $V = \operatorname{Span} \mathcal{B}$, where $\mathcal{B} = \{e^{at}\cos bt, e^{at}\sin bt\}$ for some nonzero scalars a and b, and T is the derivative operator
30. $V = \operatorname{Span} \mathcal{B}$, where $\mathcal{B} = \{e^t \cos t, e^t \sin t\}$, and $T = D^2 + 2D$, where D is the derivative operator

31. $V = \mathcal{M}_{2\times 2}$,
$$\mathcal{B} = \left\{ \begin{bmatrix} 1 & 0 \\ 0 & 0 \end{bmatrix}, \begin{bmatrix} 0 & 1 \\ 0 & 0 \end{bmatrix}, \begin{bmatrix} 0 & 0 \\ 1 & 0 \end{bmatrix}, \begin{bmatrix} 0 & 0 \\ 0 & 1 \end{bmatrix} \right\},$$
and T is defined by $T(A) = 2A + A^T$ for all A in V.

32. Find an expression for $T^{-1}(a + bx + cx^2)$, where T is the linear operator in Exercise 28.

33. Find an expression for $T^{-1}(c_1e^{at}\cos bt + c_2e^{at}\sin bt)$, where T is the linear operator in Exercise 29.

34. Find an expression for $T^{-1}(c_1e^t\cos t + c_2e^t\sin t)$, where T is the linear operator in Exercise 30.

35. Find an expression for
$$T^{-1}\left(\begin{bmatrix} a & b \\ c & d \end{bmatrix}\right),$$
where T is the linear operator in Exercise 31.

36. Find the eigenvalues and a basis for each eigenspace of the linear operator in Exercise 28.

37. Find the eigenvalues and a basis for each eigenspace of the linear operator in Exercise 29.

38. Find the eigenvalues and a basis for each eigenspace of the linear operator in Exercise 30.

39. Find the eigenvalues and a basis for each eigenspace of the linear operator in Exercise 31.

In Exercises 40–43, $V = \mathcal{M}_{2\times 2}$ with the Frobenius inner product, and W is the subspace of V defined by
$$W = \left\{ A \in \mathcal{M}_{2\times 2} \colon \operatorname{trace}\left(\begin{bmatrix} 0 & 1 \\ 1 & 0 \end{bmatrix} A\right) = 0 \right\}.$$

40. Find $\langle A, B\rangle$ for $A = \begin{bmatrix} 1 & 2 \\ -1 & 3 \end{bmatrix}$ and $B = \begin{bmatrix} 2 & -1 \\ 1 & 1 \end{bmatrix}$.

41. Find a basis for the subspace of V consisting of all matrices that are orthogonal to $\begin{bmatrix} 1 & 3 \\ 4 & 2 \end{bmatrix}$.

42. Find an orthonormal basis for W.

43. Find the orthogonal projection of $\begin{bmatrix} 2 & 5 \\ 9 & -3 \end{bmatrix}$ on W.

In Exercises 44–47, let $V = \mathrm{C}([0,1])$ with the inner product defined by
$$\langle f, g\rangle = \int_0^1 f(t)g(t)\,dt,$$
and let W be the subspace of V consisting of all polynomial functions of degree less than or equal to 2 with domain restricted to $[0, 1]$.

44. Let f and g be the functions in V defined by $f(t) = \cos 2\pi t$ and $g(t) = \sin 2\pi t$. Prove that f and g are orthogonal.

45. Find an orthonormal basis for W.

46. Determine the orthogonal projection of the function $f(t) = t$ on W without doing any computations. Now compute the orthogonal projection of f to verify your answer.

47. Find the orthogonal projection of the function $f(t) = \sqrt{t}$ on W.

48. Prove that for any $n \times n$ matrix A, $\langle A, I_n\rangle = \operatorname{trace}(A)$, where the inner product is the Frobenius inner product.

Let $T\colon V_1 \to V_2$ be a linear transformation from a vector space V_1 to a vector space V_2. For any nonempty subset W of V_1, let $T(W)$ denote the set consisting of the images $T(\mathbf{w})$ for every $\mathbf{w}$ in W. Exercises 49 and 50 use this notation.

49. Prove that if W is a subspace of V_1, then $T(W)$ is a subspace of V_2.

50. Let V be an n-dimensional inner product space, and $T\colon V \to \mathcal{R}^n$ be a linear isometry. (See the definition in the exercises for Section 7.5.) Let W be a subspace of V and $Z = T(W)$, which is a subspace of $\mathcal{R}^n$. Prove that for any vector $\mathbf{v}$ in V, the orthogonal projection $\mathbf{p}$ of $\mathbf{v}$ on W is given by
$$\mathbf{p} = T^{-1}(P_Z T(\mathbf{v})),$$
where P_Z is the orthogonal projection matrix for Z.

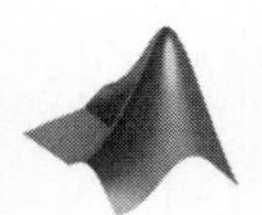

CHAPTER 7 MATLAB EXERCISES

For the following exercises, use MATLAB (or comparable software) or a calculator with matrix capabilities. The MATLAB functions in Tables D.1, D.2, D.3, D.4, and D.5 of Appendix D may be useful.

In Exercises 1 and 2, determine whether each set is linearly dependent. In the case that the set is linearly dependent, write some vector in the set as a linear combination of the others.

1. $\{1 + 2x + x^2 - x^3 + x^4, 2 + x + x^3 + x^4,$
$1 - x + x^2 + 2x^3 + 2x^4, 1 + 2x + 2x^2 - x^3 - 2x^4\}$

2. $\left\{ \begin{bmatrix} 1 & -1 \\ 3 & 1 \end{bmatrix}, \begin{bmatrix} 1 & 2 \\ 1 & 2 \end{bmatrix}, \begin{bmatrix} 0 & 1 \\ -1 & 1 \end{bmatrix}, \begin{bmatrix} 1 & -3 \\ 4 & 1 \end{bmatrix} \right\}$

3. (a) Use Exercise 76 of Section 7.3 to prove that there exist unique scalars $c_0, c_1, \ldots, c_n$ such that
$$f(-1) + f(-2) = c_0f(0) + c_1f(1) + \cdots + c_nf(n)$$
for every polynomial $f(x)$ in $\mathcal{P}_n$.

(b) Find the scalars c_0, c_1, c_2, c_3, and c_4 such that
$$\begin{aligned} &f(-1) + f(-2) \\ &\quad = c_0f(0) + c_1f(1) + c_2f(2) + c_3f(3) + c_4f(4) \end{aligned}$$
for every polynomial $f(x)$ in $\mathcal{P}_4$.

4. Let
$$\mathcal{B} = \{\cos t, \sin t, t\cos t, t\sin t, t^2\cos t, t^2\sin t\}$$
and $V = \operatorname{Span}\mathcal{B}$, which is a subspace of C^∞. Define $T\colon V \to V$ by
$$T(f) = f'' - 3f' + 2f$$

for all f in V.

(a) Prove that T is an invertible linear operator on V.

(b) Find $T^{-1}(t^2 \sin t)$ as a linear combination of the functions in $\mathcal{B}$.

Hint: It is easier to find $[D]_{\mathcal{B}}$ first (where D is the derivative operator) and then use this matrix to compute $[T]_{\mathcal{B}}$.

5. Let T be the linear operator on $\mathcal{M}_{2\times 3}$ defined by

$$T(A) = \begin{bmatrix} 1 & 3 \\ 1 & -1 \end{bmatrix} A \begin{bmatrix} 4 & -2 & 0 \\ 3 & -1 & 3 \\ -3 & 3 & 1 \end{bmatrix}$$

for all A in $\mathcal{M}_{2\times 3}$.

(a) Determine the eigenvalues of T.

(b) Find a basis for $\mathcal{M}_{2\times 3}$ consisting of eigenvectors of T. *Hint:* To avoid a messy answer, use the MATLAB function `null`(*A*, 'r') as explained in Appendix D.

The following exercise uses the definition preceding Exercise 49, as well as Exercises 49 and 50 of the Review Exercises for Chapter 7.

6. Apply Exercise 50 of the Review Exercises for Chapter 7 to find the orthogonal projection of any 3×3 matrix on the subspace of 3×3 magic squares. (See the definition in the exercises for Section 7.3.) In particular, let $T\colon \mathcal{M}_{3\times 3} \to \mathcal{R}^9$ be as in Exercise 83 of Section 7.5. Find a 9×9 matrix P such that, for any matrix A in $\mathcal{M}_{3\times 3}$, $T^{-1}(PT(A))$ is the orthogonal projection of A on the subspace of 3×3 magic squares.

Hints: Use Exercises 60 and 61 in Section 7.3 to perform the following steps:

(i) Let $Z = T(W_3)$. Find a 7×9 matrix B of rank 7 such that $Z = \text{Null}\, B$.

(ii) Apply Exercise 79 in Section 6.3 to obtain the orthogonal projection matrix P_Z.

(iii) Show that C_3 is orthogonal to every vector in W_3 and that $\|C_3\| = 1$.

(iv) Apply (i), (ii), (iii), Exercise 74 in Section 6.3, and Exercise 50 of the Review Exercises for Chapter 7 to obtain the desired matrix.

APPENDICES

APPENDIX A SETS

In the study of linear algebra, we often consider a collection of similar objects, for example, a collection of vectors or a collection of matrices. To describe such collections, the terminology and notation of set theory is useful.

For our purposes, a **set** may be considered to be a collection of objects for which it is possible to determine whether or not any given object is in the collection. The objects in a set are called its **elements**. For example, the collection of positive integers less than 7 is a set whose elements are the numbers 1, 2, 3, 4, 5, and 6. One way to specify the elements of a set is to list them between **set braces** {}. Thus the set described in the previous sentence is $\{1, 2, 3, 4, 5, 6\}$.

Two sets are called **equal** if they contain exactly the same elements. We denote the equality of sets by using an ordinary equals sign. Thus, if X denotes the set of integers whose absolute value is less than 3, then

$$X = \{-2, -1, 0, 1, 2\}.$$

Note that because the elements of a set are not ordered, we also have

$$X = \{0, 1, -1, 2, -2\}.$$

If the sets S and T are not equal, then we write $S \neq T$. For example, $X \neq \{0, 1, 2\}$.

To denote that an object x is an element of S, we write $x \in S$. On the other hand, if an object y is not an element of S, then we write $y \notin S$. So, for the set X in the preceding paragraph, we have $0 \in X$, but $3 \notin X$.

Example 1 Let P denote the set of presidents of the United States. Then

$$\text{Abraham Lincoln} \in P, \quad \text{but} \quad \text{Benjamin Franklin} \notin P.$$

An element is either in a set or not in the set; it cannot be an element of a set more than once. Thus, when the elements of a set are listed between set braces, the set is not different if an element appears more than once. For example,

$$\{1, 2\} = \{1, 1, 2, 2\} = \{1, 2, 2, 1, 1\}.$$

Similarly, in Example 1, a listing of the elements of P would include Franklin Roosevelt once, even though he was elected President of the United States four times.

If every element of S is also an element of a set T, then we call S a **subset** of T and write $S \subseteq T$. For example, let E denote the set of persons who were elected to be President of the United States. Then $E \subseteq P$. There are elements of P that are not in E; so $E \neq P$. For example, Gerald Ford became President after the resignation of Richard Nixon, but was never elected President. Therefore

$$\text{Gerald Ford} \in P, \quad \text{but} \quad \text{Gerald Ford} \notin E.$$

Note that two sets S and T are equal if and only if they are subsets of each other. This fact can be used to verify that they are equal; simply prove first that S is a subset of T, and then prove that T is a subset of S.

When a set contains a large number of elements, it may not be convenient to list all its elements. To describe such a set, we can identify an arbitrary element of the set in terms of one or more properties that characterize it. For example, the set of positive integers $1, 2, \ldots, 19$ can be written

$$\{x : x \text{ is an integer and } 0 < x < 20\}.$$

This notation is read "the set of all elements x such that x is an integer and $0 < x < 20$." Likewise, the subset E of P can be defined as

$$E = \{x \in P : x \text{ was elected to be President of the United States}\}.$$

This notation is read "the set of all elements x in P such that x was elected to be President of the United States."

Example 2 List two elements of the set

$$S = \{(x, y) : x \text{ and } y \text{ are real numbers and } xy > 0\}.$$

Solution The set S consists of the ordered pairs of real numbers for which the product of the coordinates is positive. Hence

$$(3, 7) \in S \quad \text{and} \quad (-5, -2) \in S$$

because $3, 7, -5$, and -2 are real numbers and $3 \cdot 7 = 21 > 0$ and $(-5) \cdot (-2) = 10 > 0$. On the other hand,

$$(3, -2) \notin S \quad \text{and} \quad (-5, 0) \notin S$$

because $3 \cdot (-2) = -6 < 0$ and $(-5) \cdot 0 = 0 \leq 0$.

For the set P in Example 1, the set

$$W = \{x \in P : x \text{ is a woman}\}$$

is a subset of P consisting of the women who have served as President of the United States. At the time of publication of this book, however, no women have served as President of the United States. Thus the set W contains no elements. Such a set is called the **empty set** and denoted by $\varnothing$.

In the study of mathematics, it is common to form new sets from existing sets. We now define two of the most useful ways to do this.

Definitions If S and T are sets, the **union** of S and T is the set consisting of all the elements that are in at least one of the sets S and T, and the **intersection** of S and T is the set consisting of all the elements that are in both of the sets S and T. The union and intersection of S and T are denoted by $S \cup T$ and $S \cap T$, respectively.

For example, if

$$X = \{1, 3, 5, 7, 9\}, \quad Y = \{5, 6, 7, 8, 9\}, \quad \text{and} \quad Z = \{2, 4, 6, 8\},$$

then

$$X \cup Y = \{1, 3, 5, 6, 7, 8, 9\} \quad \text{and} \quad X \cap Y = \{5, 7, 9\},$$

whereas

$$X \cup Z = \{1, 2, 3, 4, 5, 6, 7, 8, 9\} \quad \text{and} \quad X \cap Z = \varnothing.$$

If the intersection of two sets equals $\varnothing$, then the sets are called **disjoint**. Thus, in the preceding example, the sets X and Z are disjoint.

The definitions of union and intersection can be extended to apply to an infinite number of sets. In this case, the **union** of sets is the set consisting of all the elements that are in at least one of the sets, and the **intersection** of sets is the set consisting of all the elements that are in every set. For example, for each positive integer n, define

$$A_n = \{0, 1, 2, \dots, n\}$$

so that $A_5 = \{0, 1, 2, 3, 4, 5\}$. The union of the sets $A_1, A_2, A_3, \dots$ is the set of all nonnegative integers (because each nonnegative integer is in some set A_n), and the intersection of the sets $A_1, A_2, A_3, \dots$ is the set $\{0, 1\}$ (because 0 and 1 are the only integers that are elements of each set A_n).

APPENDIX B FUNCTIONS

In college algebra and calculus courses, the concept of *function* occurs frequently. In most cases, however, the input and output values of functions are real numbers. In the context of linear algebra, the functions that arise often have inputs and outputs that are *vectors*, and so a more general formulation of the concept of function is required. In this appendix, we provide a definition of a function in which the inputs and outputs are allowed to be elements from any sets whatsoever.

Definitions Let X and Y be sets. A **function** (or **mapping** or **transformation**) f from X to Y, denoted by $f : X \to Y$, is an assignment that associates to each element x in X a unique element $f(x)$ in Y. The element $f(x)$ is called the **image** of x (under f), X is called the **domain** of f, and Y is called the **codomain** of f.

Intuitively, we regard the elements of the domain as the *inputs* and the elements of the codomain as the possible *outputs* of the function. The functions that are considered in this book are those for which the outputs are related to the inputs by means of an algebraic equation. For example, the function $h : \mathcal{R} \to \mathcal{R}$ defined by $h(x) = x^2$ assigns to each real number x its square, so that the image of -2 is $h(-2) = (-2)^2 = 4$ and the image of 3 is $h(3) = 3^2 = 9$.

The **range** of a function $f : X \to Y$ is the set of all images $f(x)$ for x in X. Thus the range of the function h in the preceding paragraph is

$$\{y \in Y : y = h(x) \text{ for some } x \text{ in } \mathcal{R}\} = \{x^2 : x \text{ is a real number}\},$$

which is the set of all nonnegative real numbers. Notice that the range of a function is always a subset of its codomain.

Two functions f and g are said to be **equal** if they have the same domain and the same codomain, and for each element x of the domain, $f(x) = g(x)$; that is, their images are equal. If the functions f and g are equal, we write $f = g$.

If f and g are functions for which the domain of g equals the codomain of f, then g and f can be combined by the operation of **composition** to produce a new function $g \circ f$. The function that results from the composition of functions is called the *composition* of g and f. Specifically, if $f : X \to Y$ and $g : Y \to Z$ are functions, then the composition $g \circ f : X \to Z$ is defined by

$$(g \circ f)(x) = g(f(x)) \qquad \text{for every } x \text{ in } X.$$

As Figure B.1 suggests, we obtain the composition function $g \circ f$ by first applying f to an element x in X and then applying g to the image $f(x)$ (which is an element of Y) to produce the element $g(f(x))$ of Z.

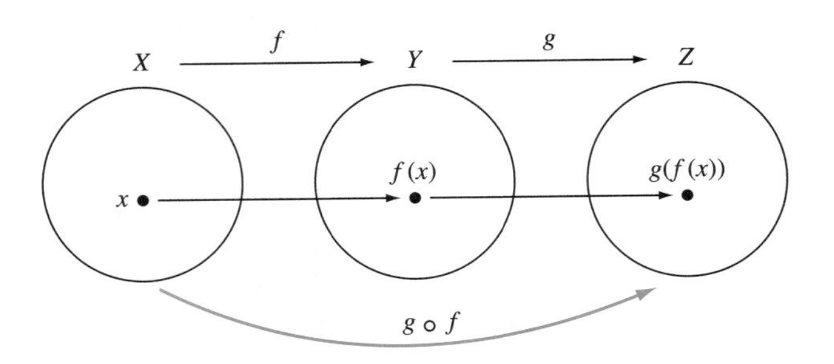

Figure B.1 The composition of functions

Example 1

Let $\mathcal{R}$ be the set of real numbers, and let $f : \mathcal{R} \to \mathcal{R}$ and $g : \mathcal{R} \to \mathcal{R}$ be the functions defined by

$$f(x) = 3x - 7 \qquad \text{and} \qquad g(x) = x^2 + 1.$$

Because the codomain of f equals the domain of g, the composition $g \circ f$ is defined. Its domain and codomain are both $\mathcal{R}$, and for every real number x,

$$(g \circ f)(x) = g(f(x)) = (3x - 7)^2 + 1 = 9x^2 - 42x + 50.$$

In this case, the composition $f \circ g$ is also defined, and for every real number x,

$$(f \circ g)(x) = 3(x^2 + 1) - 7 = 3x^2 - 4.$$

So $g \circ f \neq f \circ g$.

Example 2

Let X denote the set of positive real numbers, Y denote the set of real numbers, and $Z = \{x \in X : x > -2\}$. Define $f : X \to Y$ by $f(x) = \ln x$, and $g : Y \to Z$ by $g(x) = x^2 - 2$. Then the composition $g \circ f$ is defined, and

$$(g \circ f)(x) = (\ln x)^2 - 2.$$

In this case, the composition $f \circ g$ is not defined because the domain of f does not equal the codomain of g. (For instance, the expression $(f \circ g)(1) = \ln(-1)$ is undefined.)

As Example 1 illustrates, even if both the compositions $g \circ f$ and $f \circ g$ are defined, it is rarely true that $g \circ f = f \circ g$. However, there is an important property possessed by the composition of functions: The composition of functions is associative. That is, if $f: X \to Y$, $g: Y \to Z$, and $h: Z \to W$ are functions, then $(f \circ g) \circ h$ and $f \circ (g \circ h)$ are defined and $(f \circ g) \circ h = f \circ (g \circ h)$. To see why, note that both $(f \circ g) \circ h$ and $f \circ (g \circ h)$ have X as their domains and W as their codomains, and for every element x in X, we have

$$((f \circ g) \circ h)(x) = (f \circ g)(h(x)) = f(g(h(x)))$$

and

$$(f \circ (g \circ h))(x) = f((g \circ h)(x)) = f(g(h(x))).$$

Therefore

$$((f \circ g) \circ h)(x) = (f \circ (g \circ h))(x).$$

Because $(f \circ g) \circ h$ and $f \circ (g \circ h)$ have the same domain and codomain and are equal for every element in their common domain, $(f \circ g) \circ h = f \circ (g \circ h)$.

A function $f: X \to Y$ is called **invertible** if there exists a function $g: Y \to X$ such that

$$(g \circ f)(x) = x \text{ for every } x \text{ in } X \quad \text{and} \quad (f \circ g)(y) = y \text{ for every } y \text{ in } Y. \tag{1}$$

If such a function g exists, then the preceding conditions imply that

$$y = f(x) \quad \text{if and only if} \quad g(y) = x.$$

It follows that g is unique. If f is invertible, the unique function g satisfying (1) is called the **inverse** of f and is denoted by f^{-1}.

Let $f: X \to Y$ be a function. Then f is called **one-to-one** if every pair of distinct elements in X has distinct images in Y. The function f is called **onto** if its range is all of Y.

It is not difficult to show that (1) implies that f is one-to-one and onto. If x_1 and x_2 are distinct elements in X, then $x_1 = (g \circ f)(x_1) = g(f(x_1))$ and $x_2 = (g \circ f)(x_2) = g(f(x_2))$, and hence $f(x_1) \neq f(x_2)$. It follows that f is one-to-one. Furthermore, $(f \circ g)(y) = f(g(y)) = y$ for every y in Y, and hence f is onto. Thus, if f is invertible, then it is one-to-one and onto.

Conversely, suppose that f is one-to-one and onto. For each element y in Y, let $g(y)$ be the unique element x in X such that $f(x) = y$. This defines a function $g: Y \to X$, and it follows that f and g satisfy (1). Therefore f is invertible with inverse g.

Thus we have the following result:

THEOREM B.1

A function is invertible if and only if it is both one-to-one and onto.

Example 3 Let $\mathcal{R}$ denote the set of real numbers, and let $f: \mathcal{R} \to \mathcal{R}$ be defined by $f(x) = x^3 - 7$. Then f is invertible, and $f^{-1}(x) = \sqrt[3]{x+7}$ because

$$(f^{-1} \circ f)(x) = \sqrt[3]{(x^3 - 7) + 7} = \sqrt[3]{x^3} = x$$

and

$$(f \circ f^{-1})(x) = \left(\sqrt[3]{x+7}\right)^3 - 7 = (x+7) - 7 = x$$

for all x in $\mathcal{R}$.

APPENDIX C COMPLEX NUMBERS

Throughout this book, the word *scalar* is used almost interchangeably with the term *real number*. However, many of the results that are established can be reformulated to hold when scalars are allowed to be *complex numbers*.

Definition A **complex number** z is an expression of the form

$$z = a + bi,$$

where a and b are real numbers. The real numbers a and b are called the **real part** and the **imaginary part** of z, respectively. We denote the set of all complex numbers by $\mathcal{C}$.

Thus $z = 3 + (-2)i$ is a complex number, which can be written $3 - 2i$. The real part of z is 3, and the imaginary part is -2. When the imaginary part of a complex number is 0, we identify the number with its real part. Thus $4 + 0i$ is identified with the real number 4. In this way $\mathcal{R}$ may be regarded as a subset of $\mathcal{C}$.

Two complex numbers are called **equal** if their real parts are equal and their imaginary parts are equal. Thus two complex numbers $a + bi$ and $c + di$, where a, b, c, and d are real numbers, are equal if and only if $a = c$ and $b = d$.

The arithmetic operations on $\mathcal{R}$ can be extended to $\mathcal{C}$. The **sum** of two complex numbers $z = a + bi$ and $w = c + di$, where a, b, c, and d are real numbers, is defined by

$$z + w = (a + bi) + (c + di) = (a + c) + (b + d)i,$$

and their **product** is defined by

$$zw = (a + bi)(c + di) = (ac - bd) + (bc + ad)i.$$

Example 1 Compute the sum and product of $z = 2 + 3i$ and $w = 4 - 5i$.

Solution By definition,

$$z + w = (2 + 3i) + (4 - 5i) = (2 + 4) + [3 + (-5)]i = 6 + (-2)i = 6 - 2i,$$

and

$$zw = (2 + 3i)(4 - 5i) = [2(4) - 3(-5)] + [3(4) + 2(-5)]i = 23 + 2i.$$

Note that the complex number $i = 0 + 1i$ has the property that

$$i^2 = (0 + 1i)(0 + 1i) = [0(0) - 1(1)] + [1(0) + 0(1)]i = -1 + 0i = -1.$$

This provides an easy method for multiplying complex numbers: Multiply the numbers as though they were algebraic expressions, and then replace i^2 by -1. Thus the computation in Example 1 can be performed as follows:

$$\begin{aligned} zw &= (2+3i)(4-5i) \\ &= 8+(12-10)i-15i^2 \\ &= 8+2i-15(-1) \\ &= 23+2i \end{aligned}$$

The sum and product of complex numbers share many of the same properties as sums and products of real numbers. In particular, the following theorem can be proved:

THEOREM C.1

For all complex numbers x, y, and z, the following statements are true:

(a) $x+y=y+x$ and $xy=yx$. (commutativity of addition and multiplication)

(b) $x+(y+z)=(x+y)+z$ and $x(yz)=(xy)z$. (associativity of addition and multiplication)

(c) $0+x=x$. (0 is an identity element for addition)

(d) $1\cdot x=x$. (1 is an identity element for multiplication)

(e) $x+(-1)x=0$. (existence of additive inverses)

(f) If $x \neq 0$, there is a u in $\mathcal{C}$ such that $xu=1$. (existence of multiplicative inverses)

(g) $x(y+z)=xy+xz$. (distributive property of multiplication over addition)

The **difference** of complex numbers z and w is defined by $z-w=z+(-1)w$. Thus

$$(2+3i)-(4-5i)=(2+3i)+(-4+5i)=-2+8i.$$

Because of Theorem C.1(f), it is also possible to define division for complex numbers. In order to develop an efficient method for computing the quotient of complex numbers, we need the following concept:

Definition The **(complex) conjugate** of the complex number $z=a+bi$, where a and b are real numbers, is the complex number $a-bi$. It is denoted by $\overline{z}$.

Thus the conjugate of $z=4-3i$ is $\overline{z}=4-(-3)i=4+3i$. The following result lists some useful properties of conjugates:

THEOREM C.2

For all complex numbers z and w, the following statements are true:

(a) $\overline{\overline{z}}=z$

(b) $\overline{z+w}=\overline{z}+\overline{w}$

(c) $\overline{zw}=\overline{z}\cdot\overline{w}$

(d) z is a real number if and only if $z=\overline{z}$.

Complex numbers can be visualized as vectors in a plane with two axes, which are called the **real axis** and the **imaginary axis**. (See Figure C.1.) In this interpretation, the sum of complex numbers $z = a + bi$ and $w = c + di$, where a, b, c, and d are real numbers, corresponds to the sum of the vectors

$$\begin{bmatrix} a \\ b \end{bmatrix} \quad \text{and} \quad \begin{bmatrix} c \\ d \end{bmatrix}$$

in $\mathcal{R}^2$. The **absolute value** (or **modulus**) of z, denoted by $|z|$, corresponds to the length of a vector in $\mathcal{R}^2$ and is defined as the nonnegative real number

$$|z| = \sqrt{a^2 + b^2}.$$

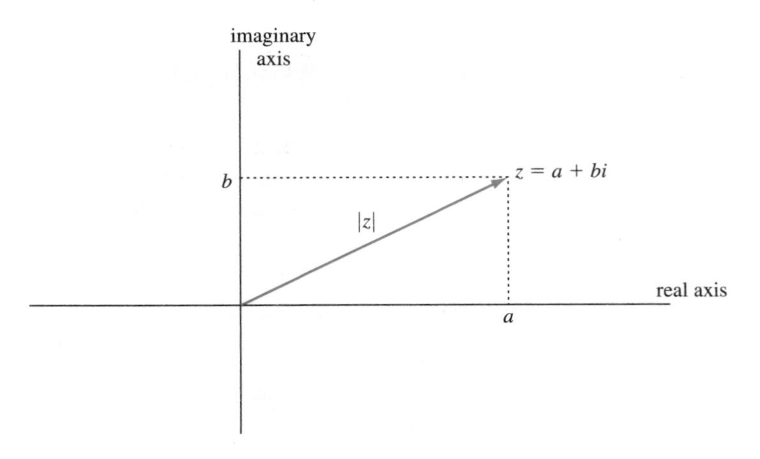

Figure C.1 The complex number $a + bi$

The following properties of absolute value are easy to verify:

THEOREM C.3

For all complex numbers z and w, the following statements are true:

(a) $|z| \geq 0$, and $|z| = 0$ if and only if $z = 0$.

(b) $z\bar{z} = |z|^2$

(c) $|zw| = |z||w|$

(d) $\left|\dfrac{z}{w}\right| = \dfrac{|z|}{|w|}$ if $w \neq 0$

(e) $|z + w| \leq |z| + |w|$

Note that Theorem C.3(b) tells us that the product of a complex number and its conjugate is a real number. This fact provides an easy method for evaluating the quotient of two complex numbers. Suppose that $z = a + bi$ and $w = c + di$, where a, b, c, and d are real numbers and $w \neq 0$. We wish to represent z/w in the form $r + si$, where r and s are real numbers. Since $w\bar{w} = |w|^2$ is real, we can calculate such a representation by multiplying the numerator and denominator of z/w by the conjugate of the denominator as follows:

$$\frac{z}{w} = \frac{z}{w}\frac{\bar{w}}{\bar{w}} = \frac{z\bar{w}}{|w|^2} = \frac{(a + bi)\cdot(c - di)}{c^2 + d^2} = \frac{ac + bd}{c^2 + d^2} + \frac{bc - ad}{c^2 + d^2}i$$

Example 2 Compute $\dfrac{9+8i}{2-i}$.

Solution Multiplying the numerator and denominator of the given expression by the conjugate of the denominator, we obtain

$$\frac{9+8i}{2-i} = \frac{9+8i}{2-i} \cdot \frac{2+i}{2+i} = \frac{10+25i}{5} = 2+5i.$$

When z is a complex number, it is possible to define the expression e^z in a manner that reduces to the familiar case when z is real. (Here, e is the base of the natural logarithm.) If $z = a + bi$, where a and b are real numbers, we define

$$e^z = e^{a+ib} = e^a(\cos b + i \sin b),$$

where b is in radians. This definition is called *Euler's formula.*[1] Notice that if $b = 0$ so that z is real, this expression reduces to e^a. Moreover, with this definition, all of the familiar properties of exponents are preserved. For example, the equations

$$e^z e^w = e^{z+w} \quad \text{and} \quad \frac{e^z}{e^w} = e^{z-w}$$

are true for all complex numbers z and w.

Recall that there are polynomials with real coefficients that have no real roots—for example, $t^2 + 1$. The principal reason for the importance of the complex number system is the following result, due to Gauss,[2] which shows that such an occurrence is impossible in C:

The Fundamental Theorem of Algebra

Any polynomial of positive degree with complex coefficients has a (complex) root.

An important consequence of this result is that every polynomial of positive degree with complex coefficients can be factored into a product of linear factors. For example, the polynomial $t^3 - 2t^2 + t - 2$ can be factored as

$$t^3 - 2t^2 + t - 2 = (t-2)(t^2+1) = (t-2)(t+i)(t-i).$$

This fact is useful in our discussion of eigenvalues in Chapter 5.

We conclude Appendix C by proving Theorem 6.15. This requires a preliminary result.

[1] Leonhard Euler (1707–1783), a Swiss mathematician, wrote more than 500 books and papers during his lifetime. He is responsible for much of our present-day mathematical notation, including the symbols e (for the base of the natural logarithm) and i (for the complex number whose square is -1).

[2] The German mathematician Karl Friedrich Gauss (1777–1885) is regarded by many as the greatest mathematician of all time. Although the fundamental theorem of algebra had been previously stated by others, Gauss gave the first successful proof in his doctoral thesis at the University of Helmstädt.

THEOREM C.4

Every eigenvalue of a symmetric matrix having real entries is real.

PROOF Let A be a symmetric matrix having real entries and $\mathbf{v}$ be an eigenvector of A with corresponding eigenvalue λ. As mentioned on page 309, if λ is an eigenvalue of A, then so is $\overline{\lambda}$, and an eigenvector of A corresponding to $\overline{\lambda}$ is the vector

$$\mathbf{w} = \begin{bmatrix} \overline{v_1} \\ \overline{v_2} \\ \vdots \\ \overline{v_n} \end{bmatrix},$$

whose components are the complex conjugates of the components of $\mathbf{v}$. Thus $A\mathbf{v} = \lambda\mathbf{v}$ and $A\mathbf{w} = \overline{\lambda}\mathbf{w}$. Note that

$$\begin{aligned} \mathbf{v}^T\mathbf{w} &= [v_1 \ v_2 \ \cdots \ v_n] \begin{bmatrix} \overline{v_1} \\ \overline{v_2} \\ \vdots \\ \overline{v_n} \end{bmatrix} \\ &= v_1\overline{v_1} + v_2\overline{v_2} + \cdots + v_n\overline{v_n} \\ &= |v_1|^2 + |v_2|^2 + \cdots + |v_n|^2 > 0 \end{aligned}$$

because $\mathbf{v} \neq \mathbf{0}$. We compute the scalar $\mathbf{v}^T A\mathbf{w}$ in two ways:

$$\mathbf{v}^T A\mathbf{w} = \mathbf{v}^T(A\mathbf{w}) = \mathbf{v}^T\overline{\lambda}\mathbf{w} = \overline{\lambda}\mathbf{v}^T\mathbf{w}$$

and

$$\mathbf{v}^T A\mathbf{w} = (\mathbf{v}^T A)\mathbf{w} = (\mathbf{v}^T A^T)\mathbf{w} = (A\mathbf{v})^T\mathbf{w} = (\lambda\mathbf{v})^T\mathbf{w} = \lambda\mathbf{v}^T\mathbf{w}$$

Since these values are equal and $\mathbf{v}^T\mathbf{w} \neq 0$, it follows that $\overline{\lambda} = \lambda$; that is, λ is a real number. ■

THEOREM 6.15

An $n \times n$ matrix A is symmetric if and only if there is an orthonormal basis for $\mathcal{R}^n$ consisting of eigenvectors of A. In this case, there exists an orthogonal matrix P and a diagonal matrix D such that $P^T AP = D$.

PROOF It was proved on page 425 that if there is an orthonormal basis for $\mathcal{R}^n$ consisting of eigenvectors of A, then A is a symmetric matrix.

To prove the converse, assume that A is a symmetric $n \times n$ matrix. By the fundamental theorem of algebra, A has an eigenvalue λ, and λ is real by Theorem C.4. Let $\mathbf{v}_1$ be an eigenvector of A corresponding to λ such that $\mathbf{v}_1$ is a unit vector. Extend $\{\mathbf{v}_1\}$ to a basis for $\mathcal{R}^n$, and then apply the Gram–Schmidt process to make this basis into an orthonormal basis $\{\mathbf{v}_1, \mathbf{v}_2, \ldots, \mathbf{v}_n\}$ for $\mathcal{R}^n$. Let $Q = [\mathbf{v}_1 \ \mathbf{v}_2 \ \cdots \ \mathbf{v}_n]$. Then Q is an orthogonal matrix, and

$$(Q^T AQ)^T = Q^T A^T (Q^T)^T = Q^T AQ$$

because A is symmetric. So Q^TAQ is symmetric. Furthermore, because $Q\mathbf{e}_1 = \mathbf{v}_1$, we have $\mathbf{e}_1 = Q^{-1}\mathbf{v}_1 = Q^T\mathbf{v}_1$ by Theorem 6.9. Thus the first column of Q^TAQ is

$$(Q^TAQ)\mathbf{e}_1 = (Q^TA)(Q\mathbf{e}_1) = (Q^TA)\mathbf{v}_1 = Q^T(A\mathbf{v}_1) = Q^T(\lambda\mathbf{v}_1) = \lambda(Q^T\mathbf{v}_1) = \lambda\mathbf{e}_1.$$

So Q^TAQ has the form

$$\begin{bmatrix} \lambda & O \\ O & B \end{bmatrix},$$

where B is a symmetric $(n-1)\times(n-1)$ matrix. Repeating this argument $n-1$ times, we see that there is an orthogonal $(n-1)\times(n-1)$ matrix R and a diagonal $(n-1)\times(n-1)$ matrix E such that $R^TBR = E$. Since the $n \times n$ matrix

$$S = \begin{bmatrix} 1 & O \\ O & R \end{bmatrix}$$

is orthogonal, $P = QS$ is orthogonal by Theorem 6.10. Moreover,

$$\begin{aligned} P^TAP &= (QS)^TA(QS) = (S^TQ^T)A(QS) = S^T(Q^TAQ)S \\ &= \begin{bmatrix} 1 & O \\ O & R^T \end{bmatrix}\begin{bmatrix} \lambda & O \\ O & B \end{bmatrix}\begin{bmatrix} 1 & O \\ O & R \end{bmatrix} = \begin{bmatrix} \lambda & O \\ O & R^TBR \end{bmatrix} = \begin{bmatrix} \lambda & O \\ O & E \end{bmatrix}. \end{aligned}$$

Thus there exists an orthogonal matrix P such that P^TAP equals the diagonal matrix

$$D = \begin{bmatrix} \lambda & O \\ O & E \end{bmatrix},$$

completing the proof. ■

APPENDIX D MATLAB

At the end of many of the exercise sets, as well as at the end of each chapter, you are asked to *use a calculator with matrix capabilities or computer software such as MATLAB*, to solve numerical problems. This appendix provides a rudimentary introduction to MATLAB that is adequate for solving these problems. In addition to describing operations and functions that are native to MATLAB, we describe useful functions that are not native, but are available as M-files from our website.

ENTERING AND STORING DATA

To solve numerical problems, we must enter, store, and manipulate numerical data by means of various operations and functions. Each of these activities requires typing a statement after a prompt. After entering a statement, the *enter* key (for an MS Windows operating system) or the *return* or *enter* key (for a Macintosh operating system) is pressed to execute the statement.

Suppose we want to enter and store the scalar 5 as the variable c. The statement

$c = 5$

returns the following:

$c =$

5

Suppose we want to enter and store the matrix $\begin{bmatrix} 1 & 2 & 3 \\ 4 & 5 & 6 \end{bmatrix}$ as the variable A. The statement

$$A = [1\ 2\ 3; 4\ 5\ 6]$$

returns the following:

$$A = \begin{matrix} 1 & 2 & 3 \\ 4 & 5 & 6 \end{matrix}$$

Notice that we enclose the entries in brackets, separating the entries in a row by spaces and the rows by a semicolon. Extra spaces are ignored. Until we quit MATLAB or type a new matrix A, this 2×3 matrix is used in all future computations involving A. For example, the statement

$$c * A$$

returns the following:

$$\text{ans} = \begin{matrix} 5 & 10 & 15 \\ 20 & 25 & 30 \end{matrix}$$

The syntax for matrix arithmetic is given in Table D.1.

In addition to entering data by these methods, you can load data from special data files that are available from our website. This saves the work required to enter large amounts of data from the exercises in this textbook. We explain how to use these files at the end of this appendix.

DISPLAYING DATA

The way that numerical values are displayed in MATLAB depends on the *format*. Enter, for example, the statement $c = 7/9$. If the result is displayed as 0.7778, MATLAB is in the *short* format. Enter the statement `format long` to change the format, and then enter c. MATLAB now displays the value to 14 places after the decimal. (Of course, neither of these formats gives the exact value of c.)

In cases where it is known that numerical values of scalars and matrix entries are rational numbers, it is often desirable to have their precise values represented as ratios of integers. For example, we may want rational solutions of systems of linear equations whose coefficients and constant terms are rational numbers. For this purpose, there is a format that requires MATLAB to represent numbers as ratios of integers. Enter the statement `format rat` to change to this rational format. Now enter c. The resulting display is $7/9$. To restore the original format, type `format short`.

Although the various MATLAB formats display a number to different accuracies, the actual value stored by MATLAB is not affected by the chosen format.

VARIABLES

As we have seen, variables are used as names for scalars and matrices that are stored for subsequent use. A *variable* is a string of characters chosen from upper- and lower-case letters, digits, and the underscore symbol _ , with the requirement that the first character of the string is a letter. Upper- and lower-case letters are considered distinct. For example, a and A are distinct variables. The strings cD_2 and $a33$ are other examples of variables. Avoid using i, j, pi, and ans as variables because these are reserved for special purposes. For example, pi initially stores the value of π. So using pi as a variable may affect the values of certain functions that use π in their computations. Similarly, both i and j store the complex number whose square is -1. Thus, for example, any complex number $a + bi$ can be entered in a statement as $a + b * i$.

A new variable is defined by setting it equal to a scalar, a matrix, or a valid expression using previously defined variables and arithmetic data. For example, typing the statement $B = 3 * A$ creates a new matrix equal to the scalar product of 3 and A (see Table D.1), which is stored in the variable B. A variable that has been assigned a value can be reassigned a new value by setting it equal to a new expression. Note that this new expression may use the variable that is to be redefined. For example, typing the statement $A = 2 * A$ stores in variable A the matrix equal to 2 times the former value of A.

OPERATIONS AND FUNCTIONS ON DATA

Operations can be performed on variables that contain data. For example, if two $m \times n$ matrices are stored in the variables A and B, typing the statement $A + B$ displays the sum of these matrices. Typing $C = A + B$, in addition to displaying the sum, stores it in the variable C. Scalars can be summed in a similar manner.

Tables D.1, D.2, D.3, and D.4, which are not comprehensive, list various operations and functions that can be performed on scalars and matrices. These are useful in solving numerical problems in this text that require the use of a calculator or a computer.

Table D.1

Scalar Operation	Matrix Operation	Description
$a + b$	$A + B$	addition
$a - b$	$A - B$	subtraction
$a * b$	$A * B$	multiplication
a/b		division
sqrt(a)	sqrt(A)	square root of a or the matrix of square roots of the entries of A
a^n	A^n	a or A raised to the power n
	A'	the matrix whose (i, j)-entry is the complex conjugate of a_{ji}. Thus, if A has only real entries, $A' = A^T$.
	$a * A$	the product of a scalar and a matrix

For example, if A and B are variables denoting square matrices of the same size, and B is invertible, then typing $A*$inv(B) causes the matrix AB^{-1} to be displayed. If $C = A*$inv(B) is typed, then the matrix AB^{-1} is stored as variable C.

In some exercises, you are asked to experiment with random matrices. These can be generated by the MATLAB function rand(m, n) found in Table D.2.

In Table D.3, the variables P, D, Q, R, U, S, and V are created (if they have not yet been defined) and given the values described in the table.

Table D.2

Function	Description
`det`(A)	the determinant of A
`inv`(A)	the inverse of A
`norm`($\mathbf{v}$)	the norm of the $1 \times n$ or $n \times 1$ vector $\mathbf{v}$
`null`(A)	a matrix whose columns form an orthonormal basis for Null A
`null`(A, 'r')	a matrix whose columns form a basis for Null A computed by the reduced row echelon form. Therefore its entries are rational if the entries of A are rational.
`orth`(A)	a matrix whose columns form an orthonormal basis for Col A
`pinv`(A)	the pseudoinverse of A
`rand`(m, n)	an $m \times n$ matrix with entries randomly selected between 0 and 1
`rank`(A)	the rank of A
`rref`(A)	the reduced row echelon form of A
`trace`(A)	the trace of A

Table D.3

Function	Description
$[P\ D] =$ `eig`(A)	D is a diagonal matrix whose diagonal entries are the eigenvalues of A with repetition according to multiplicity, and the columns of P are eigenvectors of unit length corresponding to the eigenvalues in the diagonal of D. If A is diagonalizable, then $PDP^{-1} = A$. (See Section 5.3.) If A is symmetric, then the columns of P are orthonormal, and hence P is an orthogonal matrix and $PDP^T = A$. (See Sections 6.5 and 6.6.) (Note that MATLAB considers any root of the characteristic polynomial of A, real or complex, as an eigenvalue of A. See Section 5.2, page 307.)
$[Q\ R] =$ `qr`(A, 0)	where $A = QR$ is a QR factorization of A. The columns of Q form an orthonormal basis for Col A.
$[U\ S\ V] =$ `svd`(A)	where $A = USV^T$ is a singular value decomposition of A

Table D.4

Matrix Operation	Description
`eye`(n)	the $n \times n$ identity matrix
`zeros`(m, n)	the $m \times n$ zero matrix
`ones`(m, n)	the $m \times n$ matrix whose entries are all 1
$A(i, j)$	the (i, j)-entry of A (This operation can be used to return the (i, j)-entry of A or to change it.)
$A(:, j)$	the jth column of A
$A(i, :)$	the ith row of A
$A(:, [c_1\ c_2\ \ldots\ c_k])$	the matrix whose columns are those with the column numbers $c_1, c_2, \ldots, c_k$ of A, in the same order
$A([r_1\ r_2\ \ldots\ r_k], :)$	the matrix whose rows are those with the row numbers $r_1, r_2, \ldots, r_k$ of A, in the same order
$[A\ B]$	the matrix consisting of the columns of A followed by the columns of B, provided that A and B have the same number of rows
$[A; B]$	the matrix consisting of the rows of A followed by the rows of B, provided that A and B have the same number of columns
`diag`(A)	an $n \times n$ diagonal matrix whose diagonal entries are the entries of an $n \times 1$ or $1 \times n$ matrix A
$[m{:}\ n]$	the row matrix whose entries are the consecutive integers from m to n

SYMBOLIC VARIABLES

In many cases where numerical values of scalars are not rational numbers, exact values are available if expressed *symbolically*. For example, the exact values of the solutions of the equation $x^2 = 2$ are irrational numbers, and hence cannot be displayed precisely by decimal representations or ratios of integers. However, the exact solutions, expressed symbolically, are $\sqrt{2}$ and $-\sqrt{2}$, or in the notation of MATLAB, `sqrt`(2) and $-$`sqrt`(2). For this purpose, MATLAB allows for the use of **symbolic variables**, which store exact values (when possible) in symbolic form with the `sym` function. Rather than abstractly describing how to use this function, we illustrate its use in the following example:

Example 1

Let $A = \begin{bmatrix} 0 & 1 \\ 2 & 0 \end{bmatrix}$. Store this matrix as the variable A, and enter the statement $[P\ D] =$ `eig`(A), as shown earlier, to obtain a diagonal matrix D whose diagonal entries are the eigenvalues of A, and a matrix P whose columns are corresponding eigenvectors. Assuming that MATLAB is in the short format, P and D are displayed as

$$P = \begin{bmatrix} 0.5775 & -0.5774 \\ 0.8165 & 0.8165 \end{bmatrix} \quad \text{and} \quad D = \begin{bmatrix} 1.4142 & 0 \\ 0 & -1.4142 \end{bmatrix}.$$

From experience, we could guess that the diagonal entries of D are decimal approximations of $\pm\sqrt{2}$. However, it is not obvious what the entries of P approximate. To gain more insight, we obtain symbolic representations of the entries of P and D by entering the following statements that use the sym function:

$$P1 = \texttt{sym}(P) \quad \text{and} \quad D1 = \texttt{sym}(D)$$

The values of $P1$ and $D1$ are now displayed as

$$P1 = \begin{bmatrix} \texttt{sqrt(1/3)}, & -\texttt{sqrt(1/3)} \\ \texttt{sqrt(2/3)}, & \texttt{sqrt(2/3)} \end{bmatrix} \quad \text{and} \quad D1 = \begin{bmatrix} \texttt{sqrt(2)}, & 0 \\ 0, & -\texttt{sqrt(2)} \end{bmatrix}.$$

Although the columns of $P1$ are eigenvectors of A and the entries of $P1$ are given precise symbolic values, there are other eigenvectors with simpler entries that can replace these. For example, multiplying $P1$ by $\sqrt{3}$ would produce another matrix whose columns are eigenvectors of A, but whose entries are less complicated:

$$\sqrt{3}P1 = \begin{bmatrix} 1 & -1 \\ \sqrt{2} & \sqrt{2} \end{bmatrix}$$

Now try the following experiment: Enter sqrt(3) $* P1$ and view the result. Then enter sym(sqrt(3) $* P$) and view the result. Compare the two displays. Both are symbolic representations of $\sqrt{3}P1$, but the first display, which uses only symbolic manipulations, is rather ugly. In contrast, the second display uses numerical computations followed by conversions to symbolic form. The moral here is that the algorithms that MATLAB uses do not always produce an answer in simplest form. However, once the answer is displayed in symbolic form, it is usually clear how to simplify it.

The next examples use operations and functions in Tables D.1, D.2, and D.4 for computations. The first example uses material found in Section 1.2.

Example 2

Let $A = \begin{bmatrix} .85 & .03 \\ .15 & .97 \end{bmatrix}$ and $\mathbf{p} = \begin{bmatrix} 500 \\ 700 \end{bmatrix}$ be the stochastic matrix and population vector given in Example 3 of Section 1.2. To use MATLAB to compute the population distributions for several years, first store the matrix and population vector as variables A and p, respectively. Now enter

$$p = A * p$$

to display next year's population distribution, storing the result as variable p. Enter $p = A * p$ again to produce the population distribution for the following year, also storing this result in the variable p. Each time the statement $p = A * p$ is entered, the vector stored in p is multiplied by A, and the result is displayed and stored in the variable p. Continuing this process, we obtain a list of population distributions for consecutive years starting with the initial population.

To find the population distribution n years after the initial population, the statement $p = A * p$ can be entered a total of n times. In Section 2.1, the operation of matrix multiplication is defined. This operation can also be used to compute the population distribution for a future year in one step, namely, by entering the statement $(A^\wedge n) * p$. (See Example 6 of Section 2.1.)

The following example requires the content of Section 2.4:

Example 3 Suppose that A is the 3×4 matrix

$$A = \begin{bmatrix} 1 & 1 & 3 & 4 \\ 4 & 5 & 6 & 8 \\ 2 & 3 & 0 & 0 \end{bmatrix}.$$

Use MATLAB to obtain a 3×3 invertible matrix P such that PA is the reduced row echelon form of A.

Solution We first form the 3×7 matrix $[A\ I_3]$.

The statement

$B = [A$ `eye(3)`$]$

returns the following:

```
B =

     1     1     3     4     1     0     0
     4     5     6     8     0     1     0
     2     3     0     0     0     0     1
```

Now we seek the reduced row echelon C of B. The statement

$C =$ `rref`(B)

returns the following:

```
C =

    1.0000         0    9.0000   12.0000         0    1.5000   -2.5000
         0    1.0000   -6.0000   -8.0000         0   -1.0000    2.0000
         0         0         0         0    1.0000   -0.5000    0.5000
```

Depending on the chosen format (as described on page 562), the matrix C may be displayed differently than shown here. Notice that the first 4 columns of C constitute the reduced row echelon form of A, and the last 3 columns of C constitute the required matrix P. Can you justify these claims? Finally, we want to create the matrix P. The statement

$P = C(:, [5\ 6\ 7])$

returns the following:

```
P =

         0    1.5000   -2.5000
         0   -1.0000    2.0000
    1.0000   -0.5000    0.5000
```

IMPORTED FUNCTIONS

Table D.5 contains a list of functions that are not native to MATLAB. These functions are computed by using small programs stored as text files, called *M-files*, that can be downloaded from the website given at the end of the Preface. The M-files have the

Table D.5 Imported functions

Function	Description
grfig(V, E)	produces a graphic plot with vertices specified by the columns of V and edges specified by the rows of E. For a complete description, see Exercise 18 of the MATLAB exercises at the end of Chapter 6.
rotdeg(t)	the 2×2 rotation matrix of t degrees (see A_θ defined in Section 1.2)
Pdeg(t)	the 3×3 rotation matrix of t degrees about the x-axis (see P_θ defined in Section 6.9)
Qdeg(t)	the 3×3 rotation matrix of t degrees about the y-axis (see Q_θ defined in Section 6.9)
Rdeg(t)	the 3×3 rotation matrix of t degrees about the z-axis (see R_θ defined in Section 6.9)
pvtcol(A)	the matrix whose columns are the pivot columns of A listed in the same order
gs(A)	a matrix whose columns form an orthogonal basis for the column space of A (obtained by applying the Gram–Schmidt process to the columns of A), provided that the columns of A are linearly independent
[L U] = elu(A)	the matrices L and U in the LU decomposition of A (see Section 2.6)
[L U P] = elu2(A)	the matrices L, U, and P, where P is a permutation matrix such that PA has an LU decomposition with matrices L and U (see Section 2.6)
cpoly(A)	the $1 \times (n+1)$ matrix whose entries are the coefficients of the characteristic polynomial of the $n \times n$ matrix A, starting with the coefficient of the nth degree term

same names as their corresponding functions and end with .m. For instance, the file cpoly.m contains the function cpoly. On your computer, move the downloaded M-files to the folder containing the MATLAB application. In Table D.5, A is a matrix, $\mathbf{b}$ is a column vector, c and t are scalars, and i and j are positive integers. These letters may be any variables, or they may be actual values.

You can then use these functions as you would any other MATLAB function, such as a function from Table D.1, D.2, D.3, or D.4. For example, entering the statement

$$C = \texttt{pvtcol}(A)$$

creates a matrix whose columns are the pivot columns of A and stores it in the variable C.

There are functions native to MATLAB, namely, lu and poly, that resemble the functions elu and cpoly in Table D.5. Although the function lu factors a square matrix into a product of a lower and upper triangular matrix when possible, the resulting lower triangular matrix does not necessarily have 1s as its diagonal entries and multipliers as its subdiagonal entries. The function poly computes $\det(tI_n - A)$ for the characteristic polynomial of a matrix A, rather than $\det(A - tI_n)$ as defined in Chapter 5.

DATA FILES

A collection of data files is available for downloading. These files contain the data for the exercises that require the use of MATLAB. To obtain these files, go to the website given at the end of the Preface and choose the link that reads "Download MATLAB datafiles." The instructions enable you to download these files to your computer.

A data file that is used in Chapter P, Section Q, Exercise R is named

cPsQeRa.dat

For example, the matrix used in Exercise 97 of Chapter 1, Section 4 is contained in the file named c1s4e97a.dat. (For the MATLAB exercises in the Chapter Review, use M for the section number. So c3sMe1a.dat is the name of the file containing the matrix used in Exercise 1 of the Chapter 3 MATLAB Exercises.) If several data files are needed for Chapter P, Section Q, Exercise R, they are denoted cPsQeRa.dat, cPsQeRb.dat, cPsQeRc.dat, and so forth. For example, the data needed for Exercise 79 of Chapter 1, Section 6 is contained in two files, which are named c1s6e79a.dat and c1s6e79b.dat.

To obtain the data in the file named *filename.dat*, type the command

```
load filename.dat
```

The data is then stored as a variable named *filename* (without the suffix `.dat`).

For example, the data for Exercise 91 in Chapter 1, Section 2, is contained in the following files:

c1s2e91a.dat, c1s2e91b.dat, c1s2e91c.dat, and c1s3e91d.dat

Type the four commands

```
load c1s2e91a.dat
load c1s2e91b.dat
load c1s2e91c.dat
load c1s2e91d.dat
```

Now the variables c1s2e91a, c1s2e91b, c1s2e91c, and c1s2e91d contain the data used in this exercise. By displaying these four variables, you will see that they contain the matrices A and B and the vectors $\mathbf{u}$ and $\mathbf{v}$, respectively. So the commands

$$A = \text{c1s2e91a}$$
$$B = \text{c1s2e91b}$$
$$u = \text{c1s2e91c}$$
$$v = \text{c1s2e91d}$$

store the data as the variables A, B, u, and v. Thus, to compute the matrix-vector product $A\mathbf{u}$, you can now enter the command `A*u`.

ADDITIONAL EXAMPLES

In conclusion, we give several examples illustrating the use of MATLAB throughout the text.

The following example requires familiarity with the material in Section 1.7:

Example 4 Determine whether the given set

$$\mathcal{S} = \left\{ \begin{bmatrix} 1 \\ 2 \\ -1 \\ 1 \\ 0 \\ 1 \end{bmatrix}, \begin{bmatrix} 2 \\ 1 \\ 0 \\ 1 \\ 1 \\ -2 \end{bmatrix}, \begin{bmatrix} 3 \\ -1 \\ 1 \\ 2 \\ 1 \\ 0 \end{bmatrix}, \begin{bmatrix} 3 \\ 2 \\ -1 \\ 3 \\ 0 \\ 4 \end{bmatrix} \right\}$$

is linearly dependent or linearly independent.

Solution Let A be the matrix whose columns are the vectors in $\mathcal{S}$ in the same order. Compute the reduced row echelon form of A by entering `rref(A)`, which returns the matrix

$$\begin{bmatrix} 1 & 0 & 0 & 2 \\ 0 & 1 & 0 & -1 \\ 0 & 0 & 1 & 1 \\ 0 & 0 & 0 & 0 \\ 0 & 0 & 0 & 0 \\ 0 & 0 & 0 & 0 \end{bmatrix}.$$

Because one of the columns of this matrix is not a standard vector, the columns of A are linearly dependent by Theorem 1.8. Therefore $\mathcal{S}$ is linearly dependent.

The following example requires familiarity with the material in Section 2.1:

Example 5 There are three supermarkets in town, numbered 1, 2, and 3. If a person shops at market j this week, the probability that the person will shop at market i during the next week is a_{ij}, the (i,j)-entry of the matrix

$$A = \begin{bmatrix} .1 & .2 & .4 \\ .3 & .5 & .2 \\ .6 & .3 & .4 \end{bmatrix}.$$

Suppose that the numbers of people shopping this week at markets 1, 2, and 3 are 350, 200, and 150, respectively. Determine the approximate numbers of people shopping at the various markets next week and 10 weeks from now.

Solution Let

$$\mathbf{p} = \begin{bmatrix} 350 \\ 200 \\ 150 \end{bmatrix}.$$

To answer the two questions, we must compute $A\mathbf{p}$ for next week's distribution, and $A^{10}\mathbf{p}$ for the distribution of shoppers 10 weeks from now. After storing the values for A and $\mathbf{p}$ in the variables A and p, enter the expression $A * p$ to return the 3×1 vector

$$\begin{bmatrix} 135 \\ 235 \\ 330 \end{bmatrix},$$

which tells us that next week the numbers of people who will be shopping at stores 1, 2, and 3, are 135, 235, and 330, respectively. These numbers are based on probabilities, and hence may not be exact. Now enter the expression $A \wedge 10 * p$ to obtain the 3×1 vector

$$\begin{bmatrix} 180.6454 \\ 225.8064 \\ 293.5481 \end{bmatrix},$$

which tells us that in ten weeks, the approximate numbers of people who will be shopping at stores 1, 2, and 3 are 181, 226, and 294, respectively.

The following example requires familiarity with the material in Section 2.3:

Example 6 Determine the pivot columns of the matrix

$$A = \begin{bmatrix} 1 & 1 & 1 & 1 & 3 & 2 & 1 \\ 2 & -1 & 5 & 2 & 3 & 1 & 2 \\ -1 & 2 & -4 & 0 & 1 & 0 & -2 \\ 3 & 4 & 2 & -1 & 6 & 1 & 4 \\ 2 & 1 & 3 & 3 & 6 & -1 & 5 \end{bmatrix},$$

and write each nonpivot column of A as a linear combination of the pivot columns of A.

Solution After entering A, compute the reduced row echelon form of A by typing $R =$ `rref(A)`, which returns the matrix

$$R = \begin{bmatrix} 1 & 0 & 2 & 0 & 1 & 0 & 0 \\ 0 & 1 & -1 & 0 & 1 & 0 & 0 \\ 0 & 0 & 0 & 1 & 1 & 0 & 0 \\ 0 & 0 & 0 & 0 & 0 & 1 & 0 \\ 0 & 0 & 0 & 0 & 0 & 0 & 1 \end{bmatrix},$$

stored in R. Since $\mathbf{r}_1$, $\mathbf{r}_2$, $\mathbf{r}_4$, $\mathbf{r}_6$, and $\mathbf{r}_7$ are distinct standard vectors, they are the pivot columns of R, and hence $\mathbf{a}_1$, $\mathbf{a}_2$, $\mathbf{a}_4$, $\mathbf{a}_6$, and $\mathbf{a}_7$ are the pivot columns of A. Clearly, we have $\mathbf{r}_3 = 2\mathbf{r}_1 - \mathbf{r}_2$ and $\mathbf{r}_5 = \mathbf{r}_1 + \mathbf{r}_2 + \mathbf{r}_4$. Therefore, by the column correspondence property,

$$\mathbf{a}_3 = 2\mathbf{a}_1 - \mathbf{a}_2 \quad \text{and} \quad \mathbf{a}_5 = \mathbf{a}_1 + \mathbf{a}_2 + \mathbf{a}_4.$$

Returning to Example 4, we can apply the column correspondence property to the reduced row echelon form of the matrix A in that example to conclude that $\mathbf{v}_4 = 2\mathbf{v}_1 + (-1)\mathbf{v}_2 + \mathbf{v}_3$, where $\mathbf{v}_1$, $\mathbf{v}_2$, $\mathbf{v}_3$, and $\mathbf{v}_4$ are the vectors in $\mathcal{S}$.

The following example requires familiarity with the material in Section 2.8:

Example 7 Let $T\colon \mathcal{R}^5 \to \mathcal{R}^5$ be the linear transformation defined by

$$T(\mathbf{x}) = T\left(\begin{bmatrix} x_1 \\ x_2 \\ x_3 \\ x_4 \\ x_5 \end{bmatrix}\right) = \begin{bmatrix} x_1 + 2x_2 + x_3 + 3x_5 \\ 2x_1 + x_2 - 5x_3 + x_4 + 2x_5 \\ 3x_1 - 2x_2 + 2x_3 - 5x_4 + 6x_5 \\ x_1 + x_3 - 2x_4 + x_5 \\ x_2 + 3x_3 - 2x_4 - x_5 \end{bmatrix}.$$

Show that T is invertible and determine the rule for $T^{-1}(\mathbf{x})$.

Solution First, observe that the standard matrix of T is

$$A = \begin{bmatrix} 1 & 2 & 1 & 0 & 3 \\ 2 & 1 & -5 & 1 & 2 \\ 3 & -2 & 2 & -5 & 6 \\ 1 & 0 & 1 & -2 & 1 \\ 0 & 1 & 3 & -2 & -1 \end{bmatrix}.$$

After storing A, enter the statement `rank(A)` to return the value 5, which indicates that A, and hence T, is invertible. Enter the statement `inv(A)` to compute A^{-1}, which is the standard matrix of T^{-1}. The returned matrix is

$$\begin{bmatrix} 2.5000 & -4.6000 & -6.4000 & 26.9000 & -13.2000 \\ -0.5000 & 1.4000 & 1.6000 & -7.1000 & 3.8000 \\ 1.0000 & -1.8000 & -2.2000 & 9.2000 & -4.6000 \\ 1.5000 & -2.6000 & -3.4000 & 13.9000 & -7.2000 \\ -0.5000 & 1.2000 & 1.8000 & -7.3000 & 3.4000 \end{bmatrix}.$$

Using this matrix, we can write the rule for $T^{-1}(\mathbf{x})$ (dropping the unnecessary zeros in the decimal expansions) as

$$T^{-1}(\mathbf{x}) = A^{-1}\begin{bmatrix} x_1 \\ x_2 \\ x_3 \\ x_4 \\ x_5 \end{bmatrix} = \begin{bmatrix} 2.5x_1 - 4.6x_2 - 6.4x_3 + 26.9x_4 - 13.2x_5 \\ -0.5x_1 + 1.4x_2 + 1.6x_3 - 7.1x_4 + 3.8x_5 \\ 1.0x_1 - 1.8x_2 - 2.2x_3 + 9.2x_4 - 4.6x_5 \\ 1.5x_1 - 2.6x_2 - 3.4x_3 + 13.9x_4 - 7.2x_5 \\ -0.5x_1 + 1.2x_2 + 1.8x_3 - 7.3x_4 + 3.4x_5 \end{bmatrix}.$$

The following example requires familiarity with the material in Section 3.1.

Example 8 Compute the determinant of the matrix

$$\begin{bmatrix} 1.1 & 3.1 & -4.2 & 3.7 \\ 5.1 & 2.5 & -3.3 & -2.4 \\ 4.0 & -0.6 & 0.9 & 3.1 \\ 1.2 & 2.4 & -2.5 & 3.1 \end{bmatrix}.$$

Solution After storing the matrix as A, enter the statement `det(A)` to compute the determinant of A. The returned value (displayed in `format short`) is 89.4424.

The following example requires familiarity with the material in Section 4.2:

Example 9 Find a basis for the column space and the null space of the matrix

$$A = \begin{bmatrix} 1 & -1 & 1 & 3 & 3 & 4 \\ 2 & 1 & 5 & -1 & 2 & 1 \\ 1 & 1 & 3 & 1 & 3 & 0 \\ 3 & -2 & 4 & 2 & 3 & 1 \\ 0 & 1 & 1 & -3 & -2 & 2 \end{bmatrix}.$$

Solution After storing A, use the imported function `pvtcol`, described in Table D.5, to produce a matrix whose columns are the pivot columns of A. Entering the statement `pvtcol(A)` returns the following matrix:

$$\begin{bmatrix} 1 & -1 & 3 & 4 \\ 2 & 1 & -1 & 1 \\ 1 & 1 & 1 & 0 \\ 3 & -2 & 2 & 1 \\ 0 & 1 & -3 & 2 \end{bmatrix}$$

The columns of this matrix form a basis for the column space of A by Theorem 2.4. We can use the function `null`, described in Table D.2, to obtain a matrix whose columns form an orthonormal basis for the the null space of A. Entering the statement `null(A)` returns the following matrix:

$$\begin{bmatrix} -0.1515 & -0.8023 \\ -0.3865 & -0.3425 \\ -0.2350 & 0.4598 \\ -0.6215 & 0.1173 \\ 0.6215 & -0.1173 \\ -0.0000 & 0.0000 \end{bmatrix}$$

As an alternative, we can use the function `null(A, 'r')`, which also returns a matrix whose columns form a basis for the null space of A. Although these columns are usually not orthogonal, they tend to have simpler entries. Entering the statement `null(A, 'r')` returns the following matrix:

$$\begin{bmatrix} -2 & -1 \\ -1 & -1 \\ 1 & 0 \\ 0 & -1 \\ 0 & 1 \\ 0 & 0 \end{bmatrix}$$

In Chapter 5, you are often required to find a basis for an eigenspace of a matrix A. Although we have seen that columns of P returned by the statement `[P D] = eig(A)` can be used to produce such a basis, the components of the basis vectors found in this way are usually irrational numbers, and hence their decimal representations are approximate. In order to obtain a "friendlier" basis for the eigenspace of a matrix A with rational entries corresponding to a rational eigenvalue λ, use the function null(A, 'r') applied to the matrix $(A - \lambda I)$. So, if A is an $n \times n$ matrix and the eigenvalue λ is stored in the variable c, then the statement

$$\texttt{null}\ (A - c * \texttt{eye}(n), \text{'r'})$$

returns a matrix whose columns form such a basis.

Example 10 Given that the matrix

$$A = \begin{bmatrix} 3 & -5 & 10 & 3 & -7 \\ 5 & -12 & 19 & 6 & -15 \\ 3 & -7 & 10 & 3 & -7 \\ -2 & 6 & -10 & -2 & 8 \\ -1 & 4 & -7 & -2 & 7 \end{bmatrix}$$

has the eigenvalue $\lambda = 2$, find a basis for the corresponding eigenspace. As explained earlier, the MATLAB statement null($A - 2 *$ eye(5), 'r') yields the basis whose columns are given by the matrix

$$\begin{bmatrix} -3 & 2 \\ -2 & 1 \\ -1 & 1 \\ 1 & 0 \\ 0 & 1 \end{bmatrix}.$$

The following example requires familiarity with the material in Section 5.3:

Example 11 Suppose that A is a 4×4 matrix and

$$\mathcal{B} = \left\{ \begin{bmatrix} 1 \\ -1 \\ 2 \\ 4 \end{bmatrix}, \begin{bmatrix} 2 \\ 1 \\ -1 \\ 1 \end{bmatrix}, \begin{bmatrix} -2 \\ 3 \\ 1 \\ 1 \end{bmatrix}, \begin{bmatrix} 0 \\ 2 \\ -3 \\ 1 \end{bmatrix} \right\}$$

is a basis for $\mathcal{R}^4$ consisting of eigenvectors of A with corresponding eigenvalues (listed in the order of the vectors in $\mathcal{B}$) of $2, -1, 1, 3$. Find A.

Solution Let P denote the 4×4 matrix whose columns are the vectors in $\mathcal{B}$, in the given order, and let D be the diagonal matrix whose diagonal entries are the eigenvalues, listed in the same order. Then $A = PDP^{-1}$.

Store the matrix P, and enter the statement $D =$ diag([2 −1 1 3]) to obtain the 4×4 diagonal matrix D whose diagonal entries are the eigenvalues of A. Now enter $A = P * D*$inv(P) to compute A. The returned matrix is

$$\begin{bmatrix} -2.1250 & -1.8750 & -2.3750 & 1.7500 \\ -2.2750 & -0.1250 & -2.4250 & 1.2500 \\ 2.7500 & 1.2500 & 4.2500 & -1.5000 \\ -2.1000 & -1.5000 & -1.7000 & 3.0000 \end{bmatrix}.$$

The following example requires familiarity with the material in Section 6.3:

Example 12 Let

$$\mathcal{B} = \left\{ \begin{bmatrix} 1 \\ 2 \\ 1 \\ -1 \\ 3 \end{bmatrix}, \begin{bmatrix} 1 \\ 0 \\ 1 \\ 1 \\ -2 \end{bmatrix}, \begin{bmatrix} 2 \\ 1 \\ -3 \\ 1 \\ 1 \end{bmatrix} \right\} \quad \text{and} \quad \mathbf{v} = \begin{bmatrix} 1 \\ -3 \\ 4 \\ 2 \\ 1 \end{bmatrix}.$$

Find the vector in $W = \operatorname{Span} \mathcal{B}$ that is closest to $\mathbf{v}$, and then find the distance from $\mathbf{v}$ to W.

Solution The vector in W that is closest to $\mathbf{v}$ is the orthogonal projection of $\mathbf{v}$ onto W. Let C be the 5×3 matrix whose columns are the vectors in $\mathcal{B}$. After storing C and $\mathbf{v}$, enter the statement $\texttt{rank}(C)$. The returned value is 3, indicating that $\mathcal{B}$ is a basis for W. Therefore we can apply Theorem 6.8 of Section 6.3 to obtain the orthogonal projection $\mathbf{w} = C(C^TC)^{-1}C^T\mathbf{v}$ of $\mathbf{v}$ onto W. Enter the statement

$$w = C * \texttt{inv}(C' * C) * C' * v$$

to compute $\mathbf{w}$. The returned vector (represented to 4 places after the decimal) is

$$\mathbf{w} = \begin{bmatrix} -0.0189 \\ 0.1321 \\ 2.9434 \\ -0.1509 \\ -1.1132 \end{bmatrix}.$$

The distance from $\mathbf{v}$ to W equals the distance between $\mathbf{v}$ and $\mathbf{w}$, that is, $\|\mathbf{v} - \mathbf{w}\|$. To compute this distance, enter the statement $\texttt{norm}(v - w)$. The returned value (represented to 4 places after the decimal) is 4.5887.

APPENDIX E THE UNIQUENESS OF THE REDUCED ROW ECHELON FORM

In this appendix, we delve more deeply into the column correspondence property and its consequences. In addition to a formal proof of this property, we give proofs of Theorem 2.4 and the uniqueness part of Theorem 1.4, which asserts that a given matrix can have only one reduced row echelon form. This last result is important because certain properties of a matrix A, such as *rank*, *nullity*, and *pivot column*, are defined in terms of the reduced row echelon form of A.

THEOREM E.1

(Column Correspondence Property) Let A be a matrix with reduced row echelon form R. Then the following statements are true:

(a) If column j of A is a linear combination of other columns of A, then column j of R is a linear combination of the corresponding columns of R with the same coefficients.

(b) If column j of R is a linear combination of other columns of R, then column j of A is a linear combination of the corresponding columns of A with the same coefficients.

PROOF (a) By Theorem 2.3, there is an invertible matrix P such that $PA = R$. Hence $P\mathbf{a}_i = \mathbf{r}_i$ for all i. Suppose that column j of A is a linear combination of other columns of A. Then there are scalars $c_1, c_2, \ldots, c_k$ such that

$$\mathbf{a}_j = c_1\mathbf{a}_1 + c_2\mathbf{a}_2 + \cdots + c_k\mathbf{a}_k.$$

Therefore

$$\begin{aligned}\mathbf{r}_j = P\mathbf{a}_j &= P(c_1\mathbf{a}_1 + c_2\mathbf{a}_2 + \cdots + c_k\mathbf{a}_k)\\ &= c_1P\mathbf{a}_1 + c_2P\mathbf{a}_2 + \cdots + c_kP\mathbf{a}_k\\ &= c_1\mathbf{r}_1 + c_2\mathbf{r}_2 + \cdots + c_k\mathbf{r}_k.\end{aligned}$$

The proof of (b) is similar, applying the equations $\mathbf{r}_i = P^{-1}\mathbf{a}_i$. ■

Because of the column correspondence property, conditions involving linear dependence, linear independence, and linear combinations of columns of the reduced row echelon form R of a matrix A can be directly translated into the corresponding conditions for the columns of A. Of particular importance are the relationships between the pivot columns and non-pivot columns of a matrix.

To help us understand these relationships, consider the matrix

$$R = [\mathbf{r}_1\ \mathbf{r}_2\ \cdots\ \mathbf{r}_7] = \begin{bmatrix} 0 & 1 & 2 & 0 & 3 & 0 & 2\\ 0 & 0 & 0 & 1 & 4 & 0 & 5\\ 0 & 0 & 0 & 0 & 0 & 1 & 3\\ 0 & 0 & 0 & 0 & 0 & 0 & 0 \end{bmatrix},$$

which is in reduced row echelon form. Notice that $\mathbf{r}_2$, $\mathbf{r}_4$, and $\mathbf{r}_6$ are the pivot columns of R. Notice also that these columns are the first three standard vectors of $\mathcal{R}^4$, and hence they are linearly independent. Also, the first pivot column $\mathbf{r}_2$ is the first nonzero column of R, and no pivot column is a linear combination of the columns to its left. Except for the first column, which is the zero vector, any column of R that is not a pivot column is a linear combination of the preceding pivot columns. For example, the fifth column, which is not a pivot column, can be written as a linear combination of the preceding pivot columns, of which there are two. That is, $\mathbf{r}_5 = 3\mathbf{r}_2 + 4\mathbf{r}_4$. Notice that the coefficients in this linear combination, 3 and 4, are the first two entries of $\mathbf{r}_5$. Furthermore, the rest of the entries of $\mathbf{r}_5$ are zeros. As a consequence, every column of R is a linear combination of the pivot columns of R.

These properties are obviously true for any matrix in reduced row echelon form. We summarize them here.

Properties of a Matrix in Reduced Row Echelon Form

Let R be an $m \times n$ matrix in reduced row echelon form. Then the following statements are true:

(a) A column of R is a pivot column if and only if it is nonzero and not a linear combination of the preceding columns of R.

(b) The jth pivot column of R is $\mathbf{e}_j$, the jth standard vector of $\mathcal{R}^m$, and hence the pivot columns of R are linearly independent.

(c) Suppose that $\mathbf{r}_j$ is not a pivot column of R, and there are k pivot columns of R preceding it. Then $\mathbf{r}_j$ is a linear combination of the k preceding pivot columns, and the coefficients of the linear combination are the first k entries of $\mathbf{r}_j$. Furthermore, the other entries of $\mathbf{r}_j$ are zeros.

The following result lists two of these properties, translated by means of the column correspondence property, into the corresponding property for any matrix A:

THEOREM 2.4

The following statements are true for any matrix A:

(a) The pivot columns of A are linearly independent.

(b) Each nonpivot column of A is a linear combination of the previous pivot columns of A, where the coefficients of the linear combination are the entries of the corresponding column of the reduced row echelon form of A.

The following result completes the proof of Theorem 1.4:

THEOREM 1.4

(Uniqueness of the Reduced Row Echelon Form) The reduced row echelon form of a matrix is unique.

PROOF In what follows, we refer to the previous box.

Let A be a matrix, and let R be a reduced row echelon form of A. By property (a), a column of R is a pivot column of R if and only if it is nonzero and it is not a linear combination of the preceding pivot columns of R. These two conditions can be combined with the column correspondence property to produce a test for the pivot columns of A exclusively in terms of the columns of A: *A column of A is a pivot column if and only if it is nonzero and it is not a linear combination of the previous columns of A.* Thus the positions of the pivot columns of R are uniquely determined by the columns of A. Furthermore, since the jth pivot column of R is the jth standard vector of $\mathcal{R}^m$, the pivot columns of R are completely determined by the columns of A.

We show that the other columns of R are also determined by the columns of A. Suppose that $\mathbf{r}_j$ is not a pivot column of R. If $\mathbf{r}_j = \mathbf{0}$, then $\mathbf{a}_j = \mathbf{0}$ by the column correspondence property. (See Exercise 85 of Section 2.3.) Now suppose that $\mathbf{r}_j \neq \mathbf{0}$. Then by property (c), $\mathbf{r}_j$ is a linear combination of the preceding pivot columns of R, which are linearly independent. Furthermore, the coefficients in this linear combination are the beginning entries of $\mathbf{r}_j$, one for each of the preceding pivot columns, while the other entries of $\mathbf{r}_j$ are zeros. By the column correspondence property, $\mathbf{a}_j$ is a linear combination of the preceding pivot columns of A, which are linearly independent, with the same corresponding coefficients. Because of the linear independence of the pivot columns of A, these coefficients are unique, and completely determined by the columns of A. Thus $\mathbf{r}_j$ is completely determined by A. We conclude that R is unique. ■

BIBLIOGRAPHY

[1] Abbot, Stephen D. and Matt Richey. "Take a Walk on the Boardwalk." *The College Mathematics Journal*, 28 (1997), pp. 162–171.

[2] Ash, Robert and Richard Bishop. "Monopoly as a Markov Process." *Mathematics Magazine*, 45 (1972), pp. 26–29.

[3] Bourne, Larry S. "Physical Adjustment Process and Land Use Succession: A Conceptual Review and Central City Example." *Economic Geography,* 47 (1971), pp. 1–15.

[4] Friedberg, Stephen H., Arnold J. Insel, and Lawrence E. Spence. *Linear Algebra*, 4th ed. Prentice-Hall, Inc., 2003.

[5] Gabriel, K. R. and J. Neumann. "A Markov Chain Model for Daily Rainfall Occurrence at Tel Aviv." *Quarterly Journal of the Royal Meteorological Society,* 88 (1962), pp. 90–95.

[6] Hampton, P. "Regional Economic Development in New Zealand." *Journal of Regional Science,* 8 (1968), pp. 41–51.

[7] Hotelling, H. "Analysis of Statistical Variables into Principal Components." *Journal of Educational Psychology*, 24 (1933), pp. 417–441, 498–520.

[8] Hunter, Albert. "Community Change: A Stochastic Analysis of Chicago's Local Communities, 1930–1960." *American Journal of Sociology,* vol. 79 (January 1974), pp. 923–947.

[9] Larsen, R. and L. Marx. *An Introduction to Mathematical Statistics and Its Applications*, 3rd ed. Prentice-Hall, Inc., 2001.

[10] Morrison, Donald F. *Multivariate Statistical Methods*, 4th ed. Pacific Grove, CA: Brooks/Cole, 2005.

[11] Pearson, K. "On Lines and Planes of Closest Fit to Systems of Points in Space." *Philosophical Magazine*, 6(2) (1901), 559–572.

[12] Vogt, W. Paul. *Quantitative Research Methods for Professionals in Education and Other Fields*. Allyn & Bacon, 2005.

ANSWERS TO SELECTED EXERCISES

Chapter 1

Section 1.1

1. $\begin{bmatrix} 8 & -4 & 20 \\ 12 & 16 & 4 \end{bmatrix}$

3. $\begin{bmatrix} 6 & -4 & 24 \\ 8 & 10 & -4 \end{bmatrix}$

5. $\begin{bmatrix} 2 & 4 \\ 0 & 6 \\ -4 & 8 \end{bmatrix}$

7. $\begin{bmatrix} 3 & -1 & 3 \\ 5 & 7 & 5 \end{bmatrix}$

9. $\begin{bmatrix} 2 & 3 \\ -1 & 4 \\ 5 & 1 \end{bmatrix}$

11. $\begin{bmatrix} -1 & -2 \\ 0 & -3 \\ 2 & -4 \end{bmatrix}$

13. $\begin{bmatrix} -3 & 1 & -2 & -4 \\ -1 & -5 & 6 & 2 \end{bmatrix}$

15. $\begin{bmatrix} -6 & 2 & -4 & -8 \\ -2 & -10 & 12 & 4 \end{bmatrix}$

17. not possible

19. $\begin{bmatrix} 7 & 1 \\ -3 & 0 \\ 3 & -3 \\ 4 & -4 \end{bmatrix}$

21. not possible

23. $\begin{bmatrix} -7 & -1 \\ 3 & 0 \\ -3 & 3 \\ -4 & 4 \end{bmatrix}$

25. -2

27. $\begin{bmatrix} 3 \\ 0 \\ 2\pi \end{bmatrix}$

29. $\begin{bmatrix} 2 \\ 2e \end{bmatrix}$

31. $[2 \ \ -3 \ \ 0.4]$

33. $\begin{bmatrix} 150 \\ 150\sqrt{3} \\ 10 \end{bmatrix}$ mph

35. (a) $\begin{bmatrix} 150\sqrt{2}+50 \\ 150\sqrt{2} \end{bmatrix}$ mph

(b) $50\sqrt{37+6\sqrt{2}} \approx 337.21$ mph

37. T	38. T	39. T	40. F	41. F
42. T	43. F	44. F	45. T	46. F
47. T	48. T	49. T	50. F	51. T
52. T	53. T	54. T	55. T	56. T

71. $\begin{bmatrix} 2 & 5 \\ 5 & 8 \end{bmatrix}$ and $\begin{bmatrix} 2 & 5 & 6 \\ 5 & 7 & 8 \\ 6 & 8 & 4 \end{bmatrix}$

77. No. Consider $\begin{bmatrix} 2 & 5 & 6 \\ 5 & 7 & 8 \\ 6 & 8 & 4 \end{bmatrix}$ and $\begin{bmatrix} 2 & 6 \\ 5 & 8 \end{bmatrix}$.

79. They must equal 0.

Section 1.2

1. $\begin{bmatrix} 12 \\ 14 \end{bmatrix}$

3. $\begin{bmatrix} 9 \\ 0 \\ 10 \end{bmatrix}$

5. $\begin{bmatrix} a \\ b \end{bmatrix}$

7. $\begin{bmatrix} 22 \\ 5 \end{bmatrix}$

9. $\begin{bmatrix} sa \\ tb \\ uc \end{bmatrix}$

11. $\begin{bmatrix} 2 \\ -6 \\ 10 \end{bmatrix}$

13. $\begin{bmatrix} -1 \\ 6 \end{bmatrix}$

15. $\begin{bmatrix} 21 \\ 13 \end{bmatrix}$

17. $\frac{1}{2}\begin{bmatrix} \sqrt{2} & -\sqrt{2} \\ \sqrt{2} & \sqrt{2} \end{bmatrix}, \ \frac{1}{2}\begin{bmatrix} -\sqrt{2} \\ \sqrt{2} \end{bmatrix}$

19. $\frac{1}{2}\begin{bmatrix} 1 & -\sqrt{3} \\ \sqrt{3} & 1 \end{bmatrix}, \ \frac{1}{2}\begin{bmatrix} 3-\sqrt{3} \\ 3\sqrt{3}+1 \end{bmatrix}$

21. $\frac{1}{2}\begin{bmatrix} -\sqrt{3} & 1 \\ -1 & -\sqrt{3} \end{bmatrix}, \ \frac{1}{2}\begin{bmatrix} \sqrt{3}-3 \\ 3\sqrt{3}+1 \end{bmatrix}$

23. $\begin{bmatrix} 3 \\ 2 \end{bmatrix}$

25. $\frac{1}{2}\begin{bmatrix} 3-\sqrt{3} \\ 3\sqrt{3}+1 \end{bmatrix}$

27. $\frac{1}{2}\begin{bmatrix} 3 \\ -3\sqrt{3} \end{bmatrix}$

29. $\begin{bmatrix} 1 \\ 1 \end{bmatrix} = (1)\begin{bmatrix} 1 \\ 0 \end{bmatrix} + (1)\begin{bmatrix} 0 \\ 1 \end{bmatrix}$

31. not possible

33. not possible

35. $\begin{bmatrix} -1 \\ 11 \end{bmatrix} = 3\begin{bmatrix} 1 \\ 3 \end{bmatrix} - 2\begin{bmatrix} 2 \\ -1 \end{bmatrix}$

37. $\begin{bmatrix} 3 \\ 8 \end{bmatrix} = 7\begin{bmatrix} 1 \\ 2 \end{bmatrix} - 2\begin{bmatrix} 2 \\ 3 \end{bmatrix} + 0\begin{bmatrix} -2 \\ -5 \end{bmatrix}$

39. not possible

41. $\begin{bmatrix} 3 \\ -2 \\ 1 \end{bmatrix} = 0\begin{bmatrix} 2 \\ -1 \\ 2 \end{bmatrix} + 1\begin{bmatrix} 3 \\ -2 \\ 1 \end{bmatrix} + 0\begin{bmatrix} -4 \\ 1 \\ 3 \end{bmatrix}$

43. $\begin{bmatrix} -4 \\ -5 \\ -6 \end{bmatrix} = -4\begin{bmatrix} 1 \\ 0 \\ 0 \end{bmatrix} - 5\begin{bmatrix} 0 \\ 1 \\ 0 \end{bmatrix} - 6\begin{bmatrix} 0 \\ 0 \\ 1 \end{bmatrix}$

45. T	46. F	47. T	48. T	49. T
50. F	51. F	52. F	53. T	54. F
55. F	56. T	57. F	58. T	59. F
60. T	61. F	62. F	63. T	64. T

69. (a) 349,000 in the city and 351,000 in the suburbs
(b) 307,180 in the city and 392,820 in the suburbs

73. $B = \begin{bmatrix} 1 & 0 \\ 0 & -1 \end{bmatrix}$

91. (a) $\begin{bmatrix} 24.6 \\ 45.0 \\ 26.0 \\ -41.4 \end{bmatrix}$ (b) $\begin{bmatrix} 134.1 \\ 44.4 \\ 7.6 \\ 104.8 \end{bmatrix}$

(c) $\begin{bmatrix} 128.4 \\ 80.6 \\ 63.5 \\ 25.8 \end{bmatrix}$ (d) $\begin{bmatrix} 653.09 \\ 399.77 \\ 528.23 \\ -394.52 \end{bmatrix}$

Section 1.3

1. (a) $\begin{bmatrix} 0 & -1 & 2 \\ 1 & 3 & 0 \end{bmatrix}$ (b) $\begin{bmatrix} 0 & -1 & 2 & 0 \\ 1 & 3 & 0 & -1 \end{bmatrix}$

3. (a) $\begin{bmatrix} 1 & 2 \\ -1 & 3 \\ -3 & 4 \end{bmatrix}$ (b) $\begin{bmatrix} 1 & 2 & 3 \\ -1 & 3 & 2 \\ -3 & 4 & 1 \end{bmatrix}$

5. (a) $\begin{bmatrix} 0 & 2 & -3 \\ -1 & 1 & 2 \\ 2 & 0 & 1 \end{bmatrix}$ (b) $\begin{bmatrix} 0 & 2 & -3 & 4 \\ -1 & 1 & 2 & -6 \\ 2 & 0 & 1 & 0 \end{bmatrix}$

7. $\begin{bmatrix} 0 & 2 & -4 & 4 & 2 \\ -2 & 6 & 3 & -1 & 1 \\ 1 & -1 & 0 & 2 & -3 \end{bmatrix}$

9. $\begin{bmatrix} 1 & -1 & 0 & 2 & -3 \\ 0 & 4 & 3 & 3 & -5 \\ 0 & 2 & -4 & 4 & 2 \end{bmatrix}$

11. $\begin{bmatrix} 1 & -1 & 0 & 2 & -3 \\ -2 & 6 & 3 & -1 & 1 \\ 0 & 1 & -2 & 2 & 1 \end{bmatrix}$

13. $\begin{bmatrix} 1 & -1 & 0 & 2 & -3 \\ -2 & 6 & 3 & -1 & 1 \\ -8 & 26 & 8 & 0 & 6 \end{bmatrix}$

15. $\begin{bmatrix} -2 & 4 & 0 \\ -1 & 1 & -1 \\ 2 & -4 & 6 \\ -3 & 2 & 1 \end{bmatrix}$ 17. $\begin{bmatrix} 1 & -2 & 0 \\ -1 & 1 & -1 \\ 0 & 0 & 6 \\ -3 & 2 & 1 \end{bmatrix}$

19. $\begin{bmatrix} 1 & -2 & 0 \\ 2 & -4 & 6 \\ -1 & 1 & -1 \\ -3 & 2 & 1 \end{bmatrix}$ 21. $\begin{bmatrix} 1 & -2 & 0 \\ -1 & 1 & -1 \\ 2 & -4 & 6 \\ -1 & 0 & 3 \end{bmatrix}$

23. yes 25. no 27. no 29. yes
31. yes 33. yes 35. no 37. no

39. $\begin{aligned} x_1 &= 2 + x_2 \\ x_2 &\ \text{free} \end{aligned}$ 41. $\begin{aligned} x_1 &= 6 + 2x_2 \\ x_2 &\ \text{free} \end{aligned}$

43. not consistent 45. $\begin{aligned} x_1 &= 4 + 2x_2 \\ x_2 &\ \text{free} \\ x_3 &= 3 \end{aligned}$

47. $\begin{aligned} x_1 &= 3x_4 \\ x_2 &= 4x_4 \\ x_3 &= -5x_4 \\ x_4 &\ \text{free} \end{aligned}$, $\begin{bmatrix} x_1 \\ x_2 \\ x_3 \\ x_4 \end{bmatrix} = x_4 \begin{bmatrix} 3 \\ 4 \\ -5 \\ 1 \end{bmatrix}$

49. $\begin{aligned} x_1 &\ \text{free} \\ x_2 &= -3 \\ x_3 &= -4 \\ x_4 &= 5 \end{aligned}$, $\begin{bmatrix} x_1 \\ x_2 \\ x_3 \\ x_4 \end{bmatrix} = x_1 \begin{bmatrix} 1 \\ 0 \\ 0 \\ 0 \end{bmatrix} + \begin{bmatrix} 0 \\ -3 \\ -4 \\ 5 \end{bmatrix}$

51. $\begin{aligned} x_1 &= 6 - 3x_2 + 2x_4 \\ x_2 &\ \text{free} \\ x_3 &= 7 - 4x_4 \\ x_4 &\ \text{free} \end{aligned}$, $\begin{bmatrix} x_1 \\ x_2 \\ x_3 \\ x_4 \end{bmatrix} = \begin{bmatrix} 6 \\ 0 \\ 7 \\ 0 \end{bmatrix} + x_2 \begin{bmatrix} -3 \\ 1 \\ 0 \\ 0 \end{bmatrix} + x_4 \begin{bmatrix} 2 \\ 0 \\ -4 \\ 1 \end{bmatrix}$

53. not consistent 55. $n - k$

57. F 58. F 59. T 60. F 61. T
62. T 63. F 64. T 65. T 66. F
67. T 68. T 69. F 70. T 71. T
72. T 73. F 74. T 75. F 76. T
81. 7

Section 1.4

1. $\begin{aligned} x_1 &= -2 - 3x_2 \\ x_2 &\ \text{free} \end{aligned}$ 3. $\begin{aligned} x_1 &= 4 \\ x_2 &= 5 \end{aligned}$

5. not consistent 7. $\begin{aligned} x_1 &= -1 + 2x_2 \\ x_2 &\ \text{free} \\ x_3 &= 2 \end{aligned}$

9. $\begin{aligned} x_1 &= 1 + 2x_3 \\ x_2 &= -2 - x_3 \\ x_3 &\ \text{free} \\ x_4 &= -3 \end{aligned}$ 11. $\begin{aligned} x_1 &= -4 - 3x_2 + x_4 \\ x_2 &\ \text{free} \\ x_3 &= 3 - 2x_4 \\ x_4 &\ \text{free} \end{aligned}$

13. not consistent 15. $\begin{aligned} x_1 &= -2 + x_5 \\ x_2 &\ \text{free} \\ x_3 &= 3 - 3x_5 \\ x_4 &= -1 - 2x_5 \\ x_5 &\ \text{free} \end{aligned}$

17. -12 19. $r \neq 0$ 21. no r
23. $r = 3$
25. no r
27. (a) $r = 2$, $s \neq 15$ (b) $r \neq 2$ (c) $r = 2$, $s = 15$
29. (a) $r = -8$, $s \neq -2$ (b) $r \neq -8$ (c) $r = -8$, $s = -2$
31. (a) $r = \frac{5}{2}$, $s \neq -6$ (b) $r \neq \frac{5}{2}$ (c) $r = \frac{5}{2}$, $s = -6$
33. (a) $r = 3$, $s \neq \frac{2}{3}$ (b) $r \neq 3$ (c) $r = 3$, $s = \frac{2}{3}$

35. 3, 1 37. 2, 3 39. 3, 1 41. 2, 3

43. (a) 10, 20, and 25 days, respectively (b) no

45. (a) 15 units (b) no

47. $2x^2 - 5x + 7$ 49. $4x^2 - 7x + 2$ 51. It is $\mathbf{e}_3$.

53. T 54. F 55. T 56. T 57. T

58. T 59. F 60. F 61. T 62. T

63. T 64. F 65. F 66. T 67. T

68. F 69. T 70. T 71. F 72. T

73. the $m \times n$ zero matrix

75. 4 77. 3

79. the minimum of m and n

81. no 93. no

95. $\begin{aligned} x_1 &= 2.32 + 0.32x_5 \\ x_2 &= -6.44 + 0.56x_5 \\ x_3 &= 0.72 - 0.28x_5 \\ x_4 &= 5.92 + 0.92x_5 \\ x_5 & \text{ free} \end{aligned}$

97. 3, 2 99. 4, 1

Section 1.5

1. T 2. T 3. F 4. F 5. T 6. T

7. \$11 million 9. services 11. entertainment

13. \$16.1 million of agriculture, \$17.8 million of manufacturing, \$18 million of services, and \$10.1 million of entertainment

15. \$13.9 million of agriculture, \$22.2 million of manufacturing, \$12 million of services, and \$9.9 million of entertainment.

17. (a) \$15.5 million of transportation, \$1.5 million of food, and \$9 million of oil
    (b) \$128 million of transportation, \$160 million of food, and \$128 million of oil

19. (a) $\begin{bmatrix} .1 & .4 \\ .3 & .2 \end{bmatrix}$
    (b) \$34 million of electricity and \$22 million of oil
    (c) \$128 million of electricity and \$138 million of oil

21. (a) \$49 million of finance, \$10 million of goods, and \$18 million of services
    (b) \$75 million of finance, \$125 million of goods, and \$100 million of services
    (c) \$75 million of finance, \$104 million of goods, and \$114 million of services

25. $I_1 = 9, I_2 = 4, I_3 = 5$

27. $I_1 = 21, I_2 = 18, I_3 = 3$

29. $I_1 = I_4 = 12.5, I_2 = I_6 = 7.5, I_3 = I_5 = 5$

Section 1.6

1. yes 3. no 5. yes 7. no

9. no 11. yes 13. yes 15. no

17. 3 19. −6 21. no 23. yes

25. yes 27. no 29. yes 31. no

33. no 35. yes

37. $\left\{\begin{bmatrix} 1 \\ 3 \end{bmatrix}, \begin{bmatrix} 0 \\ 1 \end{bmatrix}\right\}$ 39. $\left\{\begin{bmatrix} 1 \\ 0 \\ -1 \end{bmatrix}, \begin{bmatrix} 0 \\ 1 \\ 0 \end{bmatrix}\right\}$

41. $\left\{\begin{bmatrix} 1 \\ -2 \\ 1 \end{bmatrix}\right\}$ 43. $\left\{\begin{bmatrix} -1 \\ 0 \\ 1 \end{bmatrix}, \begin{bmatrix} 0 \\ 1 \\ 2 \end{bmatrix}\right\}$

45. T 46. T 47. T 48. F 49. T

50. T 51. T 52. F 53. F 54. F

55. T 56. T 57. T 58. T 59. T

60. T 61. T 62. T 63. T 64. T

65. (a) 2 (b) infinitely many

73. no 79. yes 81. no

Section 1.7

1. yes 3. yes 5. no

7. yes 9. no 11. yes

13. $\left\{\begin{bmatrix} 1 \\ -2 \\ 3 \end{bmatrix}\right\}$ 15. $\left\{\begin{bmatrix} -3 \\ 2 \\ 0 \end{bmatrix}, \begin{bmatrix} 1 \\ 6 \\ 0 \end{bmatrix}\right\}$

17. $\left\{\begin{bmatrix} 2 \\ -3 \\ 5 \end{bmatrix}, \begin{bmatrix} 1 \\ 0 \\ 2 \end{bmatrix}\right\}$ 19. $\left\{\begin{bmatrix} 4 \\ 3 \end{bmatrix}, \begin{bmatrix} -2 \\ 5 \end{bmatrix}\right\}$

21. $\left\{\begin{bmatrix} -2 \\ 0 \\ 3 \end{bmatrix}, \begin{bmatrix} 0 \\ 4 \\ 0 \end{bmatrix}\right\}$

23. no 25. yes 27. yes 29. no

31. $-3\begin{bmatrix} -1 \\ 1 \\ 2 \end{bmatrix} = \begin{bmatrix} 3 \\ -3 \\ -6 \end{bmatrix}$

33. $5\begin{bmatrix} 0 \\ 1 \\ 1 \end{bmatrix} + 4\begin{bmatrix} 1 \\ 0 \\ -1 \end{bmatrix} = \begin{bmatrix} 4 \\ 5 \\ 1 \end{bmatrix}$

35. $1\begin{bmatrix} 1 \\ -1 \end{bmatrix} + 5\begin{bmatrix} 0 \\ 1 \end{bmatrix} + 0\begin{bmatrix} 3 \\ -2 \end{bmatrix} = \begin{bmatrix} 1 \\ 4 \end{bmatrix}$

37. $5\begin{bmatrix} 1 \\ 2 \\ -1 \end{bmatrix} - 3\begin{bmatrix} 0 \\ 1 \\ -1 \end{bmatrix} + 3\begin{bmatrix} -1 \\ -2 \\ 0 \end{bmatrix} = \begin{bmatrix} 2 \\ 1 \\ -2 \end{bmatrix}$

39. all real numbers 41. −2

43. every real number
45. every real number
47. $r = 4$
49. no r

51. $\begin{bmatrix} x_1 \\ x_2 \\ x_3 \end{bmatrix} = x_2 \begin{bmatrix} 4 \\ 1 \\ 0 \end{bmatrix} + x_3 \begin{bmatrix} -2 \\ 0 \\ 1 \end{bmatrix}$

53. $\begin{bmatrix} x_1 \\ x_2 \\ x_3 \\ x_4 \end{bmatrix} = x_2 \begin{bmatrix} -3 \\ 1 \\ 0 \\ 0 \end{bmatrix} + x_4 \begin{bmatrix} -2 \\ 0 \\ 6 \\ 1 \end{bmatrix}$

55. $\begin{bmatrix} x_1 \\ x_2 \\ x_3 \\ x_4 \end{bmatrix} = x_3 \begin{bmatrix} -4 \\ 3 \\ 1 \\ 0 \end{bmatrix} + x_4 \begin{bmatrix} 2 \\ -5 \\ 0 \\ 1 \end{bmatrix}$

57. $\begin{bmatrix} x_1 \\ x_2 \\ x_3 \\ x_4 \\ x_5 \\ x_6 \end{bmatrix} = x_2 \begin{bmatrix} 0 \\ 1 \\ 0 \\ 0 \\ 0 \\ 0 \end{bmatrix} + x_4 \begin{bmatrix} -1 \\ 0 \\ 2 \\ 1 \\ 0 \\ 0 \end{bmatrix} + x_6 \begin{bmatrix} -3 \\ 0 \\ -1 \\ 0 \\ 0 \\ 1 \end{bmatrix}$

59. $\begin{bmatrix} x_1 \\ x_2 \\ x_3 \\ x_4 \end{bmatrix} = x_3 \begin{bmatrix} 0 \\ 0 \\ 1 \\ 0 \end{bmatrix} + x_4 \begin{bmatrix} 2 \\ -3 \\ 0 \\ 1 \end{bmatrix}$

61. $\begin{bmatrix} x_1 \\ x_2 \\ x_3 \\ x_4 \\ x_5 \\ x_6 \end{bmatrix} = x_2 \begin{bmatrix} -2 \\ 1 \\ 0 \\ 0 \\ 0 \\ 0 \end{bmatrix} + x_3 \begin{bmatrix} 1 \\ 0 \\ 1 \\ 0 \\ 0 \\ 0 \end{bmatrix} + x_5 \begin{bmatrix} -2 \\ 0 \\ 0 \\ -4 \\ 1 \\ 0 \end{bmatrix} + x_6 \begin{bmatrix} 1 \\ 0 \\ 0 \\ -3 \\ 0 \\ 1 \end{bmatrix}$

63. T 64. F 65. F 66. T 67. T
68. T 69. F 70. T 71. F 72. F
73. F 74. T 75. T 76. T 77. F
78. T 79. F 80. T 81. T 82. T

83. $A = \begin{bmatrix} 1 & 0 \\ 0 & 1 \end{bmatrix}$

101. The set is linearly dependent, and $\mathbf{v}_5 = 2\mathbf{v}_1 - \mathbf{v}_3 + \mathbf{v}_4$, where $\mathbf{v}_j$ is the jth vector in the set.

103. The set is linearly independent.

Chapter 1 Review Exercises

1. F 2. T 3. T 4. T 5. T
6. T 7. T 8. F 9. F 10. T
11. T 12. T 13. F 14. T 15. T
16. F 17. F

19. (a) There is at most one solution.
(b) There is at least one solution.

21. $\begin{bmatrix} 3 & 2 \\ -2 & 7 \\ 4 & 3 \end{bmatrix}$

23. undefined because A has 2 columns and D^T has 3 rows

25. $\begin{bmatrix} 3 \\ 3 \end{bmatrix}$

27. undefined because C^T and D don't have the same number of columns

29. The components are the average values of sales for all stores during January of last year for produce, meats, dairy, and processed foods, respectively.

31. $\begin{bmatrix} 0 \\ -4 \\ 3 \\ -2 \end{bmatrix}$
33. $\frac{1}{2} \begin{bmatrix} 2\sqrt{3} - 1 \\ -2 - \sqrt{3} \end{bmatrix}$

35. $\mathbf{v} = (-1) \begin{bmatrix} -1 \\ 5 \\ 2 \end{bmatrix} + 3 \begin{bmatrix} 1 \\ 3 \\ 4 \end{bmatrix} + 1 \begin{bmatrix} 1 \\ -1 \\ 1 \end{bmatrix}$

37. $\mathbf{v}$ is not in the span of $\mathcal{S}$.

39. $x_1 = 1 - 2x_2 + x_3$
x_2 free
x_3 free

41. inconsistent

43. $x_1 = 7 - 5x_3 - 4x_4$
$x_2 = -5 + 3x_3 + 3x_4$
x_3 free
x_4 free

45. The rank is 1, and the nullity is 4.

47. The rank is 3, and the nullity is 2.

49. 20 of the first pack, 10 of the second pack, 40 of the third pack

51. yes 53. no 55. yes
57. yes 59. no

61. linearly independent
63. linearly dependent

65. $\begin{bmatrix} 3 \\ 3 \\ 8 \end{bmatrix} = 2 \begin{bmatrix} 1 \\ 2 \\ 3 \end{bmatrix} + 1 \begin{bmatrix} 1 \\ -1 \\ 2 \end{bmatrix}$

67. $\begin{bmatrix} 1 \\ -1 \\ 1 \\ -1 \end{bmatrix} = 2 \begin{bmatrix} 1 \\ 0 \\ 1 \\ 0 \end{bmatrix} + (-1) \begin{bmatrix} 1 \\ 1 \\ 1 \\ 1 \end{bmatrix}$

69. $\begin{bmatrix} x_1 \\ x_2 \\ x_3 \end{bmatrix} = x_3 \begin{bmatrix} -3 \\ 2 \\ 1 \end{bmatrix}$

71. $\begin{bmatrix} x_1 \\ x_2 \\ x_3 \\ x_4 \end{bmatrix} = x_4 \begin{bmatrix} -2 \\ 5 \\ 0 \\ 1 \end{bmatrix}$

Chapter 1 MATLAB Exercises

1. (a) $\begin{bmatrix} 3.38 \\ 8.86 \\ 16.11 \\ 32.32 \\ 15.13 \end{bmatrix}$ (b) $\begin{bmatrix} 13.45 \\ -4.30 \\ -1.89 \\ 7.78 \\ 10.69 \end{bmatrix}$ (c) $\begin{bmatrix} 20.18 \\ -11.79 \\ 7.71 \\ 8.52 \\ 0.28 \end{bmatrix}$

2. (a) $\begin{bmatrix} -0.3 & 8.5 & -12.3 & 3.9 \\ 27.5 & -9.0 & -22.3 & -2.7 \\ -11.6 & 4.9 & 16.2 & -2.1 \\ 8.0 & 12.7 & 34.2 & -24.7 \end{bmatrix}$

(b) $\begin{bmatrix} -7.1 & 20.5 & -13.3 & 6.9 \\ 10.5 & -30.0 & -22.1 & -14.3 \\ -7.0 & -31.7 & 16.4 & 27.3 \\ -14.6 & 19.3 & -9.6 & -23.9 \end{bmatrix}$

(c) $\begin{bmatrix} 1.30 & 4.1 & -2.75 & 3.15 \\ 4.10 & 2.4 & 1.90 & 1.50 \\ -2.75 & 1.9 & 3.20 & 4.65 \\ 3.15 & 1.5 & 4.65 & -5.10 \end{bmatrix}$

(d) $\begin{bmatrix} 0.00 & -2.00 & -.55 & .95 \\ 2.00 & 0.00 & -3.20 & -4.60 \\ .55 & 3.20 & 0.00 & -2.55 \\ -.95 & 4.60 & 2.55 & 0.00 \end{bmatrix}$

(e) $P^T = P$, $Q^T = -Q$, $P + Q = A$

(f) $\begin{bmatrix} 17.67 \\ -15.87 \\ -9.83 \\ -44.27 \end{bmatrix}$ (g) $\begin{bmatrix} -143.166 \\ -154.174 \\ -191.844 \\ -202.945 \end{bmatrix}$

(h) $\begin{bmatrix} -64.634 \\ 93.927 \\ -356.424 \\ -240.642 \end{bmatrix}$ (i) $\begin{bmatrix} -3.30 \\ 6.94 \\ 3.50 \\ 19.70 \end{bmatrix}$, $\mathbf{w} = \begin{bmatrix} 3.5 \\ -1.2 \\ 4.1 \\ 2.0 \end{bmatrix}$

(j) $M\mathbf{u} = B(A\mathbf{u})$ for every $\mathbf{u}$ in $\mathcal{R}^4$

3. (a) $\begin{bmatrix} -0.0864 \\ 3.1611 \end{bmatrix}$ (b) $\begin{bmatrix} -1.6553 \\ 2.6944 \end{bmatrix}$

(c) $\begin{bmatrix} -1.6553 \\ 2.6944 \end{bmatrix}$ (d) $\begin{bmatrix} 1.0000 \\ 3.0000 \end{bmatrix}$

4. (b) $\begin{bmatrix} 1 & 0 & 2.0000 & 0 & .1569 & 9.2140 \\ 0 & 1 & 1.0000 & 0 & .8819 & -.5997 \\ 0 & 0 & & 0 & 1 & -.2727 & -3.2730 \\ 0 & 0 & & 0 & 0 & 0 & 0 \end{bmatrix}$

5. Answers are given correct to 4 places after the decimal point.

(a) $\begin{bmatrix} x_1 \\ x_2 \\ x_3 \\ x_4 \\ x_5 \\ x_6 \end{bmatrix} = \begin{bmatrix} -8.2142 \\ -0.4003 \\ 0.0000 \\ 3.2727 \\ 0.0000 \\ 0.0000 \end{bmatrix} + x_3 \begin{bmatrix} -2.0000 \\ -1.0000 \\ 1.0000 \\ 0.0000 \\ 0.0000 \\ 0.0000 \end{bmatrix}$

$+ x_5 \begin{bmatrix} -0.1569 \\ -0.8819 \\ 0.0000 \\ 0.2727 \\ 1.0000 \\ 0.0000 \end{bmatrix} + x_6 \begin{bmatrix} -9.2142 \\ 0.5997 \\ 0.0000 \\ 3.2727 \\ 0.0000 \\ 1.0000 \end{bmatrix}$

(b) inconsistent

(c) $\begin{bmatrix} x_1 \\ x_2 \\ x_3 \\ x_4 \\ x_5 \\ x_6 \end{bmatrix} = \begin{bmatrix} -9.0573 \\ 1.4815 \\ 0.0000 \\ 4.0000 \\ 0.0000 \\ 0.0000 \end{bmatrix} + x_3 \begin{bmatrix} -2.0000 \\ -1.0000 \\ 1.0000 \\ 0.0000 \\ 0.0000 \\ 0.0000 \end{bmatrix}$

$+ x_5 \begin{bmatrix} -0.1569 \\ -0.8819 \\ 0.0000 \\ 0.2727 \\ 1.0000 \\ 0.0000 \end{bmatrix} + x_6 \begin{bmatrix} -9.2142 \\ 0.5997 \\ 0.0000 \\ 3.2727 \\ 0.0000 \\ 1.0000 \end{bmatrix}$

(d) inconsistent

6. The gross production for each of the respective sectors is \$264.2745 billion, \$265.7580 billion, \$327.9525 billion, \$226.1281 billion, and \$260.6357 billion.

7. (a) $\begin{bmatrix} 0 \\ 1 \\ 1 \\ 2 \\ 2 \\ 1 \end{bmatrix} = \begin{bmatrix} 1 \\ 2 \\ -1 \\ 3 \\ 2 \\ 1 \end{bmatrix} + \begin{bmatrix} 1 \\ 0 \\ 1 \\ 1 \\ 0 \\ 1 \end{bmatrix} - \begin{bmatrix} 2 \\ 1 \\ -1 \\ 2 \\ 0 \\ 1 \end{bmatrix}$

(b) linearly independent

8. Let $\mathbf{v}_1, \mathbf{v}_2, \ldots, \mathbf{v}_5$ denote the vectors in $\mathcal{S}_1$ in the order listed in Exercise 7(a).

(a) no

(b) yes, $2\mathbf{v}_1 - \mathbf{v}_2 + 0\mathbf{v}_3 + \mathbf{v}_4 + 0\mathbf{v}_5$

(c) yes, $2\mathbf{v}_1 - \mathbf{v}_2 + 0\mathbf{v}_3 + \mathbf{v}_4 + 0\mathbf{v}_5$

(d) no

Chapter 2

Section 2.1

1. AB is defined and has size 2×2.

3. undefined

5. $C\mathbf{y} = \begin{bmatrix} 22 \\ -18 \end{bmatrix}$

7. $\mathbf{xz} = \begin{bmatrix} 14 & -2 \\ 21 & -3 \end{bmatrix}$

9. $AC\mathbf{x}$ is undefined.

11. $AB = \begin{bmatrix} 5 & 0 \\ 25 & 20 \end{bmatrix}$

13. $BC = \begin{bmatrix} 29 & 56 & 23 \\ 7 & 8 & 9 \end{bmatrix}$

15. CB^T is undefined.

17. $A^3 = \begin{bmatrix} -35 & -30 \\ 45 & 10 \end{bmatrix}$

19. C^2 is undefined.

25. -2

27. 24

29. $\begin{bmatrix} -4 \\ -9 \\ -2 \end{bmatrix}$

31. $\begin{bmatrix} 7 \\ 16 \end{bmatrix}$

33. F 34. F 35. F 36. T 37. F
38. F 39. T 40. T 41. F 42. F
43. T 44. F 45. F 46. T 47. F
48. T 49. T 50. T

51. (a) $B = \begin{bmatrix} .70 & .95 \\ .30 & .05 \end{bmatrix}$

53. (a)

		Today: Hot Lunch	Today: Bag Lunch
Next Day	Hot Lunch	.3	.4
Next Day	Bag Lunch	.7	.6

$$A = \begin{bmatrix} .3 & .4 \\ .7 & .6 \end{bmatrix}$$

(b) $A^3 \begin{bmatrix} u_1 \\ u_2 \end{bmatrix} = \begin{bmatrix} 109.1 \\ 190.9 \end{bmatrix}$. Approximately 109 students will buy hot lunches and 191 students will bring bag lunches 3 school days from today.

(c) $A^{100} \begin{bmatrix} u_1 \\ u_2 \end{bmatrix} = \begin{bmatrix} 109.0909 \\ 190.9091 \end{bmatrix}$ (rounded to 4 places after the decimal)

63. $A = \begin{bmatrix} 1 & 0 \\ 1 & 0 \end{bmatrix}$, $B = \begin{bmatrix} 0 & 0 \\ 1 & 1 \end{bmatrix}$

69. (a), (b), and (c) have the same answer, namely,
$\begin{bmatrix} -1 & 0 \\ 0 & -1 \end{bmatrix}$.

71. (b) The population of the city is 205,688. The population of the suburbs is 994,332.
(c) The population of the city is 200,015. The population of the suburbs is 999,985.

Section 2.2

1. F 2. F 3. F 4. T 5. T
6. F 7. T 8. F 9. F

11. (a) all of them
(b) 0 from the first and 1 from the second
(c) $\begin{bmatrix} a \\ b \end{bmatrix}$ in an even number of years after the current year and $\begin{bmatrix} b \\ a \end{bmatrix}$ in an odd number of years after the current year

13. (a) $\begin{bmatrix} 0 & 2 & 1 \\ q & 0 & 0 \\ 0 & .5 & 0 \end{bmatrix}$
(b) The population grows without bound.
(c) The population approaches $\mathbf{0}$.
(d) $q = .4$, $\begin{bmatrix} 400 \\ 160 \\ 80 \end{bmatrix}$
(e) Over time, it approaches $\begin{bmatrix} 450 \\ 180 \\ 90 \end{bmatrix}$.
(f) $q = .4$
(g) $\mathbf{x} = x_3 \begin{bmatrix} 5 \\ 2 \\ 1 \end{bmatrix}$. The stable distributions have this form.

15. (a) $\begin{bmatrix} 0 & 2 & b \\ .2 & 0 & 0 \\ 0 & .5 & 0 \end{bmatrix}$
(b) The population approaches $\mathbf{0}$.
(c) The population grows without bound.
(d) $b = 6$, $\begin{bmatrix} 1600 \\ 320 \\ 160 \end{bmatrix}$
(e) Over time, it approaches $\begin{bmatrix} 1500 \\ 300 \\ 150 \end{bmatrix}$.
(f) $b = 6$
(g) $\mathbf{q} = c \begin{bmatrix} 10 \\ 2 \\ 1 \end{bmatrix}$ where $c = \frac{1}{26}(p_1 + 5p_2 + 6p_3)$.

17. $\begin{bmatrix} 0.644 & 0.628 \\ 0.356 & 0.372 \end{bmatrix}$

19. (a) There are no nonstop flights from any of the cities 1, 2, and 3 to the cities 4 and 5, and vice versa.

(b) $A^2 = \begin{bmatrix} B^2 & O_1 \\ O_2 & C^2 \end{bmatrix}$, $A^3 = \begin{bmatrix} B^3 & O_1 \\ O_2 & C^3 \end{bmatrix}$, and

$$A^k = \begin{bmatrix} B^k & O_1 \\ O_2 & C^k \end{bmatrix}.$$

(c) There are no flights with any number of layovers from any of the cities 1, 2, and 3 to the cities 4 and 5, and vice versa.

21. (a) 1 and 2, 1 and 4, 2 and 3, 3 and 4

(c) $\begin{bmatrix} 0 & 1 & 0 & 1 \\ 1 & 0 & 1 & 0 \\ 0 & 1 & 0 & 1 \\ 1 & 0 & 1 & 0 \end{bmatrix}$, yes

23. (c) Students 1 and 2 have 1 common course preference, and students 1 and 9 have 3 common course preferences.

(d) For each i, the ith diagonal entry of AA^T represents the number of courses preferred by student i.

25. (a)

k	Sun	Noble	Honored	MMQ
1	100	300	500	7700
2	100	400	800	7300
3	100	500	1200	6800

(b)

k	Sun	Noble	Honored	MMQ
9	100	1100	5700	1700
10	100	1200	6800	500
11	100	1300	8000	−800

Section 2.3

1. no 3. yes 5. yes 7. no

9. $\begin{bmatrix} 1 & 2 & 1 \\ 2 & 0 & 1 \\ 3 & 1 & -1 \end{bmatrix}$ 11. $\begin{bmatrix} 3 & 7 & 2 \\ 4 & 4 & -4 \\ 0 & 7 & 6 \end{bmatrix}$

13. $\begin{bmatrix} 5 & 7 & 3 \\ -3 & -4 & -1 \\ 12 & 7 & 12 \end{bmatrix}$ 15. $\begin{bmatrix} 1 & 0 \\ -1 & 1 \end{bmatrix}$

17. $\begin{bmatrix} 1 & 0 & 0 \\ 2 & 1 & 0 \\ 0 & 0 & 1 \end{bmatrix}$ 19. $\begin{bmatrix} 1 & 0 & 0 & 0 \\ 0 & .25 & 0 & 0 \\ 0 & 0 & 1 & 0 \\ 0 & 0 & 0 & 1 \end{bmatrix}$

21. $\begin{bmatrix} 1 & 0 & 0 & 0 \\ 0 & 0 & 0 & 1 \\ 0 & 0 & 1 & 0 \\ 0 & 1 & 0 & 0 \end{bmatrix}$ 23. $\begin{bmatrix} -1 & 0 \\ 0 & 1 \end{bmatrix}$

25. $\begin{bmatrix} 0 & 1 \\ 1 & 0 \end{bmatrix}$ 27. $\begin{bmatrix} 0 & 1 \\ 1 & 0 \end{bmatrix}$

29. $\begin{bmatrix} 1 & 0 & 0 \\ 0 & 1 & 0 \\ 0 & -5 & 1 \end{bmatrix}$ 31. $\begin{bmatrix} 1 & 0 & 0 \\ 0 & 0 & 1 \\ 0 & 1 & 0 \end{bmatrix}$

33. F 34. T 35. T 36. F 37. T
38. T 39. F 40. T 41. T 42. T
43. F 44. F 45. T 46. F 47. T
48. T 49. T 50. T 51. F 52. T

67. $\begin{bmatrix} 3 & 2 & 7 \\ -1 & 5 & 9 \end{bmatrix}$

69. $\begin{bmatrix} -1 & 1 & 1 & 4 & 13 \\ 2 & -2 & -1 & 1 & 3 \\ -1 & 1 & 0 & 3 & 8 \end{bmatrix}$

71. $\begin{bmatrix} 1 & 2 & 3 & 13 \\ 2 & 4 & 5 & 23 \end{bmatrix}$

73. $\begin{bmatrix} 1 & -1 & 1 & 1 & 1 & -1 \\ 0 & 0 & 2 & 6 & 1 & -7 \\ 1 & -1 & 0 & 2 & 1 & 3 \end{bmatrix}$

75. $\mathbf{a}_2 = -2\mathbf{a}_1 + 0\mathbf{a}_3$

77. $\mathbf{a}_4 = 2\mathbf{a}_1 - 3\mathbf{a}_3$

79. $\mathbf{b}_3 = \mathbf{b}_1 + (-1)\mathbf{b}_2 + 0\mathbf{b}_5$

81. $\mathbf{b}_5 = 0\mathbf{b}_1 + 0\mathbf{b}_2 + \mathbf{b}_5$

83. $R = \begin{bmatrix} 1 & 2 & 0 & -1 \\ 0 & 0 & 1 & 1 \\ 0 & 0 & 0 & 0 \end{bmatrix}$

87. Every nonzero column is a standard vector.

95. (a) $A^{-1} = \begin{bmatrix} -7 & 2 & 3 & -2 \\ 5 & -1 & -2 & 1 \\ 1 & 0 & 0 & 1 \\ -3 & 1 & 1 & -1 \end{bmatrix}$

(b) $B^{-1} = \begin{bmatrix} 3 & 2 & -7 & -2 \\ -2 & -1 & 5 & 1 \\ 0 & 0 & 1 & 1 \\ 1 & 1 & -3 & -1 \end{bmatrix}$ and

$$C^{-1} = \begin{bmatrix} -7 & -2 & 3 & 2 \\ 5 & 1 & -2 & -1 \\ 1 & 1 & 0 & 0 \\ -3 & -1 & 1 & 1 \end{bmatrix}$$

(c) B^{-1} can be obtained by interchanging columns 1 and 3 of A^{-1}, and C^{-1} can be obtained by interchanging columns 2 and 4 of A^{-1}.

(d) B^{-1} can be obtained by interchanging columns i and j of A^{-1}.

97. (b) $(A^2)^{-1} = (A^{-1})^2 = \begin{bmatrix} 113 & -22 & -10 & -13 \\ -62 & 13 & 6 & 6 \\ -22 & 4 & 3 & 2 \\ 7 & -2 & -1 & 0 \end{bmatrix}$

99. $A^{-1} = \begin{bmatrix} 10 & -2 & -1 & -1 \\ -6 & 1 & -1 & 2 \\ -2 & 0 & 1 & 0 \\ 1 & 0 & 1 & -1 \end{bmatrix}$ For each i, the solution of $A\mathbf{x} = \mathbf{e}_i$ is the ith column of A^{-1}.

Section 2.4

1. $\begin{bmatrix} -2 & 3 \\ 1 & -1 \end{bmatrix}$

3. not invertible

5. $\begin{bmatrix} 5 & -3 \\ -3 & 2 \end{bmatrix}$

7. not invertible

9. $\frac{1}{3}\begin{bmatrix} -7 & 2 & 3 \\ -6 & 0 & 3 \\ 8 & -1 & -3 \end{bmatrix}$

11. not invertible

13. $\begin{bmatrix} -1 & -5 & 3 \\ 1 & 2 & -1 \\ 1 & 4 & -2 \end{bmatrix}$

15. not invertible

17. $\frac{1}{3}\begin{bmatrix} 1 & 1 & 1 & -2 \\ 1 & 1 & -2 & 1 \\ 1 & -2 & 1 & 1 \\ -2 & 1 & 1 & 1 \end{bmatrix}$

19. $A^{-1}B = \begin{bmatrix} -1 & 3 & -4 \\ 1 & -2 & 3 \end{bmatrix}$

21. $A^{-1}B = \begin{bmatrix} -1 & -4 & 7 & -7 \\ 2 & 6 & -6 & 10 \end{bmatrix}$

23. $A^{-1}B = \begin{bmatrix} 1.0 & -0.5 & 1.5 & 1.0 \\ 6.0 & 12.5 & -11.5 & 12.0 \\ -2.0 & -5.5 & 5.5 & -5.0 \end{bmatrix}$

25. $A^{-1}B = \begin{bmatrix} -5 & -1 & -6 \\ -1 & 1 & 0 \\ 4 & 1 & 3 \\ 3 & 1 & 2 \end{bmatrix}$

27. $R = \begin{bmatrix} 1 & 0 & -1 \\ 0 & 1 & -3 \end{bmatrix}$, $P = \begin{bmatrix} -1 & -1 \\ -2 & -1 \end{bmatrix}$

29. $R = \begin{bmatrix} 1 & 0 & -2 & -1 \\ 0 & 1 & 1 & -1 \\ 0 & 0 & 0 & 0 \end{bmatrix}$

One possibility is $P = \begin{bmatrix} -1 & 0 & 0 \\ 0 & 1 & 0 \\ 2 & -3 & 1 \end{bmatrix}$.

31. $R = \begin{bmatrix} 1 & 0 & 0 & 0 \\ 0 & 1 & 0 & 0 \\ 0 & 0 & 1 & 0 \\ 0 & 0 & 0 & 1 \end{bmatrix}$, $P = \begin{bmatrix} -4 & -15 & -8 & 1 \\ 1 & 4 & 2 & 0 \\ 1 & 3 & 2 & 0 \\ -4 & -13 & -7 & 1 \end{bmatrix}$

33. $R = \begin{bmatrix} 1 & 0 & 0 & 5 & 2.5 \\ 0 & 1 & 0 & -4 & -1.5 \\ 0 & 0 & 1 & -3 & -1.5 \\ 0 & 0 & 0 & 0 & 0 \end{bmatrix}$,

$P = \frac{1}{6}\begin{bmatrix} 0 & -5 & 4 & 1 \\ 0 & 7 & -2 & 1 \\ 0 & 1 & -2 & 1 \\ 6 & 4 & -2 & -2 \end{bmatrix}$

35. T 36. F 37. T 38. T 39. T
40. T 41. T 42. T 43. T 44. T
45. T 46. T 47. T 48. F 49. T
50. F 51. F 52. T 53. T 54. T

57. (a) $\begin{bmatrix} -1 & -3 \\ 2 & 5 \end{bmatrix}\begin{bmatrix} x_1 \\ x_2 \end{bmatrix} = \begin{bmatrix} -6 \\ 4 \end{bmatrix}$

(b) $A^{-1} = \begin{bmatrix} 5 & 3 \\ -2 & -1 \end{bmatrix}$

(c) $\begin{bmatrix} x_1 \\ x_2 \end{bmatrix} = A^{-1}\mathbf{b} = \begin{bmatrix} -18 \\ 8 \end{bmatrix}$

59. (a) $\begin{bmatrix} -1 & 0 & 1 \\ 1 & 2 & -2 \\ 2 & -1 & 1 \end{bmatrix}\begin{bmatrix} x_1 \\ x_2 \\ x_3 \end{bmatrix} = \begin{bmatrix} -4 \\ 3 \\ 1 \end{bmatrix}$

(b) $A^{-1} = \frac{1}{5}\begin{bmatrix} 0 & 1 & 2 \\ 5 & 3 & 1 \\ 5 & 1 & 2 \end{bmatrix}$

(c) $\begin{bmatrix} x_1 \\ x_2 \\ x_3 \end{bmatrix} = A^{-1}\mathbf{b} = \begin{bmatrix} 1 \\ -2 \\ -3 \end{bmatrix}$

61. (a) $\begin{bmatrix} 2 & 3 & -4 \\ -1 & -1 & 2 \\ 0 & -1 & 1 \end{bmatrix}\begin{bmatrix} x_1 \\ x_2 \\ x_3 \end{bmatrix} = \begin{bmatrix} -6 \\ 5 \\ 3 \end{bmatrix}$

(b) $A^{-1} = \begin{bmatrix} 1 & 1 & 2 \\ 1 & 2 & 0 \\ 1 & 2 & 1 \end{bmatrix}$

(c) $\begin{bmatrix} x_1 \\ x_2 \\ x_3 \end{bmatrix} = A^{-1}\mathbf{b} = \begin{bmatrix} 5 \\ 4 \\ 7 \end{bmatrix}$

63. (a) $\begin{bmatrix} 1 & -2 & -1 & 1 \\ 1 & 1 & 0 & -1 \\ -1 & -1 & 1 & 1 \\ -3 & 1 & 2 & 0 \end{bmatrix}\begin{bmatrix} x_1 \\ x_2 \\ x_3 \\ x_4 \end{bmatrix} = \begin{bmatrix} 4 \\ -2 \\ 1 \\ 1 \end{bmatrix}$

(b) $A^{-1} = \begin{bmatrix} -1 & 0 & 1 & -1 \\ -3 & -2 & 1 & -2 \\ 0 & 1 & 1 & 0 \\ -4 & -3 & 2 & -3 \end{bmatrix}$

(c) $\begin{bmatrix} x_1 \\ x_2 \\ x_3 \\ x_4 \end{bmatrix} = A^{-1}\mathbf{b} = \begin{bmatrix} -2 \\ -5 \\ -1 \\ -5 \end{bmatrix}$

67. (b) $A^{-1} = A^{k-1}$

75. (a) $\begin{aligned} x_1 &= -3 + x_3 \\ x_2 &= 4 - 2x_3 \\ x_3 &\text{ free} \end{aligned}$ (b) No, A is not invertible.

77. \$2 million of electricity and \$4.5 million of oil

79. \$12.5 million of finance, \$15 million of goods, and \$65 million of services

89. The reduced row echelon form of A is I_4.

91. $\operatorname{rank} A = 4$

Section 2.5

1. $[-4 \mid 2]$ 3. $\begin{bmatrix} -2 \\ 7 \end{bmatrix}$

5. $\left[\begin{array}{cc|cc} -2 & 4 & 6 & 0 \\ -1 & 8 & 8 & 2 \\ \hline 11 & 8 & -8 & 10 \\ 3 & 6 & 1 & 4 \end{array}\right]$

7. $\left[\begin{array}{c|ccc} -2 & 4 & 6 & 0 \\ \hline -1 & 8 & 8 & 2 \\ 11 & 8 & -8 & 10 \\ 3 & 6 & 1 & 4 \end{array}\right]$

9. $\left[\begin{array}{cc} 3 & 6 \\ 9 & 12 \\ \hline 2 & 4 \\ 6 & 8 \end{array}\right]$ 11. $\left[\begin{array}{cc|cc} 1 & 1 & 2 & 1 \\ \hline 1 & 0 & 1 & -1 \\ 0 & 1 & -1 & 1 \end{array}\right]$

13. $[16 \quad -4]$ 15. $[16 \quad 9 \quad 24]$

17. $[-2 \quad -3 \quad 1]$ 19. $[-12 \quad -3 \quad 2]$

29. T 30. T 31. F 32. T 33. F

34. F 35. $2I_n$

37. $\begin{bmatrix} O & AC \\ BD & O \end{bmatrix}$

39. $\begin{bmatrix} A^TA + C^TC & A^TB + C^TD \\ B^TA + D^TC & B^TB + D^TD \end{bmatrix}$

49. $\begin{bmatrix} A^k & A^{k-1}B \\ O & O \end{bmatrix}$

51. $\begin{bmatrix} A & O \\ I_n & B \end{bmatrix}^{-1} = \begin{bmatrix} A^{-1} & O \\ -B^{-1}A^{-1} & B^{-1} \end{bmatrix}$

53. (c) $A^k = \begin{bmatrix} B^k & * \\ 0 & D^k \end{bmatrix}$, where * represents some 2×2 matrix.

Section 2.6

1. $L = \begin{bmatrix} 1 & 0 & 0 \\ 3 & 1 & 0 \\ -1 & 1 & 1 \end{bmatrix}$, $U = \begin{bmatrix} 2 & 3 & 4 \\ 0 & -1 & -2 \\ 0 & 0 & 3 \end{bmatrix}$

3. $L = \begin{bmatrix} 1 & 0 & 0 \\ 2 & 1 & 0 \\ -3 & 1 & 1 \end{bmatrix}$, $U = \begin{bmatrix} 1 & -1 & 2 & 1 \\ 0 & -1 & 1 & 2 \\ 0 & 0 & 1 & 1 \end{bmatrix}$

5. $L = \begin{bmatrix} 1 & 0 & 0 \\ -1 & 1 & 0 \\ 2 & 1 & 1 \end{bmatrix}$, $U = \begin{bmatrix} 1 & -1 & 2 & 1 & 3 \\ 0 & 1 & 2 & -1 & 1 \\ 0 & 0 & 1 & -2 & -6 \end{bmatrix}$

7. $L = \begin{bmatrix} 1 & 0 & 0 & 0 \\ 2 & 1 & 0 & 0 \\ -1 & -1 & 1 & 0 \\ 0 & -1 & 0 & 1 \end{bmatrix}$,

$U = \begin{bmatrix} 1 & 0 & -3 & -1 & -2 & 1 \\ 0 & -1 & -2 & 1 & -1 & -2 \\ 0 & 0 & 0 & 1 & 1 & 1 \\ 0 & 0 & 0 & 2 & 2 & 2 \end{bmatrix}$

9. $\begin{bmatrix} x_1 \\ x_2 \\ x_3 \end{bmatrix} = \begin{bmatrix} 2 \\ -1 \\ 0 \end{bmatrix}$

11. $\begin{bmatrix} x_1 \\ x_2 \\ x_3 \\ x_4 \end{bmatrix} = \begin{bmatrix} -7 \\ -4 \\ 2 \\ 0 \end{bmatrix} + x_4 \begin{bmatrix} 2 \\ 1 \\ -1 \\ 1 \end{bmatrix}$

13. $\begin{bmatrix} x_1 \\ x_2 \\ x_3 \\ x_4 \\ x_5 \end{bmatrix} = \begin{bmatrix} -3 \\ 3 \\ 1 \\ 0 \\ 0 \end{bmatrix} + x_4 \begin{bmatrix} -8 \\ -3 \\ 2 \\ 1 \\ 0 \end{bmatrix} + x_5 \begin{bmatrix} -28 \\ -13 \\ 6 \\ 0 \\ 1 \end{bmatrix}$

15. $\begin{bmatrix} x_1 \\ x_2 \\ x_3 \\ x_4 \\ x_5 \\ x_6 \end{bmatrix} = \begin{bmatrix} 3 \\ -4 \\ 0 \\ 2 \\ 0 \\ 0 \end{bmatrix} + x_3 \begin{bmatrix} 3 \\ -2 \\ 1 \\ 0 \\ 0 \\ 0 \end{bmatrix} + x_5 \begin{bmatrix} 1 \\ -2 \\ 0 \\ -1 \\ 1 \\ 0 \end{bmatrix} + x_6 \begin{bmatrix} -2 \\ -3 \\ 0 \\ -1 \\ 0 \\ 1 \end{bmatrix}$

17. $P = \begin{bmatrix} 1 & 0 & 0 \\ 0 & 0 & 1 \\ 0 & 1 & 0 \end{bmatrix}$, $L = \begin{bmatrix} 1 & 0 & 0 \\ -1 & 1 & 0 \\ 2 & 0 & 1 \end{bmatrix}$, and

$U = \begin{bmatrix} 1 & -1 & 3 \\ 0 & 1 & 2 \\ 0 & 0 & -1 \end{bmatrix}$

19. $P = \begin{bmatrix} 1 & 0 & 0 \\ 0 & 0 & 1 \\ 0 & 1 & 0 \end{bmatrix}$, $L = \begin{bmatrix} 1 & 0 & 0 \\ -1 & 1 & 0 \\ 2 & 0 & 1 \end{bmatrix}$, and

$U = \begin{bmatrix} 1 & 1 & -2 & -1 \\ 0 & -1 & -3 & 0 \\ 0 & 0 & 1 & 1 \end{bmatrix}$

21. $P = \begin{bmatrix} 0 & 1 & 0 & 0 \\ 1 & 0 & 0 & 0 \\ 0 & 0 & 1 & 0 \\ 0 & 0 & 0 & 1 \end{bmatrix}$, $L = \begin{bmatrix} 1 & 0 & 0 & 0 \\ 0 & 1 & 0 & 0 \\ -2 & 0 & 1 & 0 \\ -1 & -1 & -1 & 0 \end{bmatrix}$,

and $U = \begin{bmatrix} -1 & 2 & -1 \\ 0 & 1 & -2 \\ 0 & 0 & 1 \\ 0 & 0 & 0 \end{bmatrix}$

23. $P = \begin{bmatrix} 1 & 0 & 0 & 0 \\ 0 & 0 & 0 & 1 \\ 0 & 1 & 0 & 0 \\ 0 & 0 & 1 & 0 \end{bmatrix}$, $L = \begin{bmatrix} 1 & 0 & 0 & 0 \\ 2 & 1 & 0 & 0 \\ 2 & 0 & 1 & 0 \\ 3 & -4 & 0 & 1 \end{bmatrix}$, and

$U = \begin{bmatrix} 1 & 2 & 1 & -1 \\ 0 & 1 & 1 & 2 \\ 0 & 0 & -1 & 3 \\ 0 & 0 & 0 & 9 \end{bmatrix}$

25. $\begin{bmatrix} x_1 \\ x_2 \\ x_3 \end{bmatrix} = \begin{bmatrix} -2 \\ 1 \\ 3 \end{bmatrix}$

27. $\begin{bmatrix} x_1 \\ x_2 \\ x_3 \\ x_4 \end{bmatrix} = \begin{bmatrix} 16 \\ -9 \\ 3 \\ 0 \end{bmatrix} + x_4 \begin{bmatrix} -4 \\ 3 \\ -1 \\ 1 \end{bmatrix}$

29. $\begin{bmatrix} x_1 \\ x_2 \\ x_3 \end{bmatrix} = \begin{bmatrix} 5 \\ 2 \\ 1 \end{bmatrix}$

31. $\begin{bmatrix} x_1 \\ x_2 \\ x_3 \\ x_4 \end{bmatrix} = \begin{bmatrix} -3 \\ 2 \\ 1 \\ -1 \end{bmatrix}$

33. F 34. T 35. F 36. F 37. F

38. T 39. F 40. T 41. T

49. $m(2n-1)p$

51. $L = \begin{bmatrix} 1 & 0 & 0 & 0 & 0 \\ -1 & 1 & 0 & 0 & 0 \\ 2 & 3 & 1 & 0 & 0 \\ 3 & -3 & 2 & 1 & 0 \\ 2 & 0 & 1 & -1 & 1 \end{bmatrix}$ and

$U = \begin{bmatrix} 2 & -1 & 3 & 2 & 1 \\ 0 & 1 & 2 & 3 & 5 \\ 0 & 0 & 3 & -1 & 2 \\ 0 & 0 & 0 & 1 & 8 \\ 0 & 0 & 0 & 0 & 13 \end{bmatrix}$

53. $P = \begin{bmatrix} 0 & 1 & 0 & 0 & 0 \\ 1 & 0 & 0 & 0 & 0 \\ 0 & 0 & 1 & 0 & 0 \\ 0 & 0 & 0 & 1 & 0 \\ 0 & 0 & 0 & 0 & 1 \end{bmatrix}$,

$L = \begin{bmatrix} 1.0 & 0 & 0 & 0 & 0 \\ 0.0 & 1 & 0 & 0 & 0 \\ 0.5 & 2 & 1 & 0 & 0 \\ -0.5 & -1 & -3 & 1 & 0 \\ 1.5 & 7 & 9 & -9 & 1 \end{bmatrix}$, and

$U = \begin{bmatrix} 2 & -2 & -1.0 & 3.0 & 4 \\ 0 & 1 & 2.0 & -1.0 & 1 \\ 0 & 0 & -1.5 & -0.5 & -2 \\ 0 & 0 & 0.0 & -1.0 & -2 \\ 0 & 0 & 0.0 & 0.0 & -9 \end{bmatrix}$

Section 2.7

1. The domain is $\mathcal{R}^3$, and the codomain is $\mathcal{R}^2$.
3. The domain is $\mathcal{R}^2$, and the codomain is $\mathcal{R}^3$.
5. The domain is $\mathcal{R}^3$, and the codomain is $\mathcal{R}^3$.

7. $\begin{bmatrix} 11 \\ 8 \end{bmatrix}$ 9. $\begin{bmatrix} 8 \\ -6 \\ 11 \end{bmatrix}$ 11. $\begin{bmatrix} 6 \\ -7 \\ 6 \end{bmatrix}$

13. $\begin{bmatrix} 5 \\ 22 \end{bmatrix}$ 15. $\begin{bmatrix} -1 \\ 6 \\ 17 \end{bmatrix}$ 17. $\begin{bmatrix} -3 \\ -9 \\ 2 \end{bmatrix}$

19. $T_{(A+C^T)}\left(\begin{bmatrix} 2 \\ 1 \\ 1 \end{bmatrix}\right) = T_A\left(\begin{bmatrix} 2 \\ 1 \\ 1 \end{bmatrix}\right) +$

$T_{C^T}\left(\begin{bmatrix} 2 \\ 1 \\ 1 \end{bmatrix}\right) = \begin{bmatrix} 8 \\ 9 \end{bmatrix}$

21. $n = 3,\quad m = 2$ 23. $n = 2,\quad m = 4$

25. $\begin{bmatrix} 0 & 1 \\ 1 & 1 \end{bmatrix}$ 27. $\begin{bmatrix} 1 & 1 & 1 \\ 2 & 0 & 0 \end{bmatrix}$

29. $\begin{bmatrix} 1 & -1 \\ 2 & -3 \\ 0 & 0 \\ 0 & 1 \end{bmatrix}$ 31. $\begin{bmatrix} 1 & -1 \\ 0 & 0 \\ 3 & 0 \\ 0 & 1 \end{bmatrix}$

33. $\begin{bmatrix} 1 & 0 & 0 \\ 0 & 1 & 0 \\ 0 & 0 & 1 \end{bmatrix}$

35. F 36. T 37. F 38. T 39. F

40. T 41. F 42. T 43. F 44. F

45. F 46. T 47. T 48. T 49. T
50. T 51. T 52. F 53. F 54. T
55. They are equal.

57. $\begin{bmatrix} 4 \\ -8 \\ 12 \end{bmatrix}$ and $\begin{bmatrix} -1 \\ 2 \\ -3 \end{bmatrix}$

59. $\begin{bmatrix} -16 \\ 12 \\ 4 \end{bmatrix}$ and $\begin{bmatrix} 20 \\ -15 \\ -5 \end{bmatrix}$

61. $\begin{bmatrix} 16 \\ 2 \\ 0 \end{bmatrix}$ 63. $\begin{bmatrix} -4 \\ 3 \end{bmatrix}$

65. $T\left(\begin{bmatrix} x_1 \\ x_2 \end{bmatrix}\right) = \begin{bmatrix} 2x_1 + 4x_2 \\ 3x_1 + x_2 \end{bmatrix}$

67. $T\left(\begin{bmatrix} x_1 \\ x_2 \\ x_3 \end{bmatrix}\right) = \begin{bmatrix} -x_1 + 3x_2 \\ -x_2 - 3x_3 \\ 2x_1 + 2x_3 \end{bmatrix}$

69. $T\left(\begin{bmatrix} x_1 \\ x_2 \end{bmatrix}\right) = \begin{bmatrix} 12x_1 + 5x_2 \\ 3x_1 + x_2 \end{bmatrix}$

71. $T\left(\begin{bmatrix} x_1 \\ x_2 \\ x_3 \end{bmatrix}\right) = \begin{bmatrix} x_1 + 3x_2 - x_3 \\ 2x_1 + 3x_2 + x_3 \\ 2x_1 + 3x_2 + 2x_3 \end{bmatrix}$

73. linear 75. not linear
77. linear 79. not linear

89. (b) $\begin{bmatrix} 1 & 0 \\ 0 & 0 \end{bmatrix}$

91. $T = T_A$ for $A = \begin{bmatrix} -1 & 0 \\ 0 & 1 \end{bmatrix}$ (b) $\mathcal{R}^2$

93. (b) $\mathcal{R}^n$ 97. Both are $\mathbf{v}$.

103. The given vector is in the range of T.

Section 2.8

1. $\left\{\begin{bmatrix} 2 \\ 4 \end{bmatrix}, \begin{bmatrix} 3 \\ 5 \end{bmatrix}\right\}$ 3. $\left\{\begin{bmatrix} 0 \\ 2 \\ 1 \end{bmatrix}, \begin{bmatrix} 3 \\ -1 \\ 1 \end{bmatrix}\right\}$

5. $\left\{\begin{bmatrix} 2 \\ 2 \\ 4 \end{bmatrix}, \begin{bmatrix} 1 \\ 2 \\ 1 \end{bmatrix}, \begin{bmatrix} 1 \\ 3 \\ 0 \end{bmatrix}\right\}$ 7. $\left\{\begin{bmatrix} 1 \\ 0 \end{bmatrix}\right\}$

9. $\left\{\begin{bmatrix} 1 \\ 0 \\ 0 \end{bmatrix}, \begin{bmatrix} 0 \\ 1 \\ 0 \end{bmatrix}\right\}$ 11. $\left\{\begin{bmatrix} 0 \\ 0 \end{bmatrix}\right\}$

13. $\{\mathbf{0}\}$, one-to-one 15. $\left\{\begin{bmatrix} 0 \\ -1 \\ 1 \end{bmatrix}\right\}$, not one-to-one

17. $\left\{\begin{bmatrix} 1 \\ -1 \\ 1 \end{bmatrix}\right\}$, not one-to-one 19. $\{\mathbf{0}\}$, one-to-one

21. $\{\mathbf{e}_2\}$, not one-to-one

23. $\left\{\begin{bmatrix} 1 \\ -3 \\ 1 \\ 0 \end{bmatrix}, \begin{bmatrix} 3 \\ -5 \\ 0 \\ 1 \end{bmatrix}\right\}$, not one-to-one

25. $\begin{bmatrix} 2 & 3 \\ 4 & 5 \end{bmatrix}$, one-to-one 27. $\begin{bmatrix} 0 & 3 \\ 2 & -1 \\ 1 & 1 \end{bmatrix}$, one-to-one

29. $\begin{bmatrix} 1 & -1 & 0 \\ 0 & 1 & -1 \\ 1 & 0 & -1 \end{bmatrix}$, not one-to-one

31. $\begin{bmatrix} 1 & 2 & 2 & 1 & 8 \\ 1 & 2 & 1 & 0 & 6 \\ 1 & 1 & 1 & 2 & 5 \\ 3 & 2 & 0 & 5 & 8 \end{bmatrix}$, not one-to-one

33. $\begin{bmatrix} 2 & 3 \\ 4 & 5 \end{bmatrix}$, onto

35. $\begin{bmatrix} 0 & 3 \\ 2 & -1 \\ 1 & 1 \end{bmatrix}$, not onto

37. $\begin{bmatrix} 0 & 1 & -2 \\ 1 & 0 & -1 \\ -1 & 2 & -3 \end{bmatrix}$, not one-to-one

39. $\begin{bmatrix} 1 & -2 & 2 & -1 \\ -1 & 1 & 3 & 2 \\ 1 & -1 & -6 & -1 \\ 1 & -2 & 5 & -5 \end{bmatrix}$, onto

41. T 42. F 43. F 44. T 45. T
46. T 47. F 48. T 49. F 50. T
51. F 52. F 53. T 54. F 55. F
56. T 57. T 58. F 59. T 60. T

61. (a) $\{\mathbf{0}\}$ (b) yes
(c) $\mathcal{R}^2$ (d) yes

63. (a) Span $\{\mathbf{e}_1\}$ (b) no
(c) Span $\{\mathbf{e}_2\}$ (d) no

65. (a) Span $\{\mathbf{e}_3\}$ (b) no
(c) Span $\{\mathbf{e}_1, \mathbf{e}_2\}$ (d) no

67. (a) one-to-one (b) onto

69. The domain and codomain are $\mathcal{R}^2$. The rule is
$UT\left(\begin{bmatrix} x_1 \\ x_2 \end{bmatrix}\right) = \begin{bmatrix} 16x_1 + 4x_2 \\ 4x_1 - 8x_2 \end{bmatrix}$.

71. $A = \begin{bmatrix} 1 & 1 \\ 1 & -3 \\ 4 & 0 \end{bmatrix}$ and $B = \begin{bmatrix} 1 & -1 & 4 \\ 1 & 3 & 0 \end{bmatrix}$

73. The domain and codomain are $\mathcal{R}^3$. The rule is

$$TU\left(\begin{bmatrix} x_1 \\ x_2 \\ x_3 \end{bmatrix}\right) = \begin{bmatrix} 2x_1 + 2x_2 + 4x_3 \\ -2x_1 - 10x_2 + 4x_3 \\ 4x_1 - 4x_2 + 16x_3 \end{bmatrix}.$$

75. $\begin{bmatrix} 2 & 2 & 4 \\ -2 & -10 & 4 \\ 4 & -4 & 16 \end{bmatrix}$ 77. $\begin{bmatrix} -1 & 5 \\ 15 & -5 \end{bmatrix}$

79. $\begin{bmatrix} -1 & 5 \\ 15 & -5 \end{bmatrix}$ 81. $\begin{bmatrix} 2 & 9 \\ 6 & -8 \end{bmatrix}$

83. $T^{-1}\left(\begin{bmatrix} x_1 \\ x_2 \end{bmatrix}\right) = \begin{bmatrix} \frac{1}{3}x_1 + \frac{1}{3}x_2 \\ -\frac{1}{3}x_1 + \frac{2}{3}x_2 \end{bmatrix}$

85. $T^{-1}\left(\begin{bmatrix} x_1 \\ x_2 \\ x_3 \end{bmatrix}\right) = \begin{bmatrix} 2x_1 + x_2 - x_3 \\ -9x_1 - 2x_2 + 5x_3 \\ 4x_1 + x_2 - 2x_3 \end{bmatrix}$

87. $T^{-1}\left(\begin{bmatrix} x_1 \\ x_2 \\ x_3 \end{bmatrix}\right) = \begin{bmatrix} x_1 - 2x_2 + x_3 \\ -x_1 + x_2 - x_3 \\ 2x_1 - 7x_2 + 3x_3 \end{bmatrix}$

89. $T^{-1}\left(\begin{bmatrix} x_1 \\ x_2 \\ x_3 \\ x_4 \end{bmatrix}\right) = \frac{1}{2}\begin{bmatrix} x_1 - 3x_2 - 6x_3 + 3x_4 \\ 3x_1 - 2x_2 - 3x_3 + 3x_4 \\ -3x_1 + 3x_2 + 4x_3 - 3x_4 \\ -3x_1 + 6x_2 + 9x_3 - 5x_4 \end{bmatrix}$

91. yes

99. (a) $A = \begin{bmatrix} 1 & 3 & -2 & 1 \\ 3 & 0 & 4 & 1 \\ 2 & -1 & 0 & 2 \\ 0 & 0 & 1 & 1 \end{bmatrix}$ and

$B = \begin{bmatrix} 0 & 1 & 0 & -3 \\ 2 & 0 & 1 & -1 \\ 1 & -2 & 0 & 4 \\ 0 & 5 & 1 & 0 \end{bmatrix}$

(b) $AB = \begin{bmatrix} 4 & 10 & 4 & -14 \\ 4 & 0 & 1 & 7 \\ -2 & 12 & 1 & -5 \\ 1 & 3 & 1 & 4 \end{bmatrix}$

(c) $TU\left(\begin{bmatrix} x_1 \\ x_2 \\ x_3 \\ x_4 \end{bmatrix}\right) = \begin{bmatrix} 4x_1 + 10x_2 + 4x_3 - 14x_4 \\ 4x_1 + x_3 + 7x_4 \\ -2x_1 + 12x_2 + x_3 - 5x_4 \\ x_1 + 3x_2 + x_3 + 4x_4 \end{bmatrix}$

Chapter 2 Review Exercises

1. T	2. F	3. F	4. T	5. F
6. F	7. T	8. T	9. T	10. F
11. T	12. F	13. T	14. F	15. T
16. T	17. F	18. T	19. F	20. F
21. T				

23. (a) BA is defined if and only if $q = m$. (b) $p \times n$

25. $\begin{bmatrix} 64 & -4 \\ 32 & -2 \end{bmatrix}$ 27. $\begin{bmatrix} 2 \\ 29 \\ 4 \end{bmatrix}$

29. incompatible dimensions

31. $\frac{1}{6}\begin{bmatrix} 5 & 10 \\ 2 & 4 \end{bmatrix}$ 33. $\begin{bmatrix} 30 \\ 42 \end{bmatrix}$

35. incompatible dimensions

37. $\begin{bmatrix} 1 \\ 3 \end{bmatrix} + \begin{bmatrix} 7 \\ 4 \end{bmatrix} = \begin{bmatrix} 8 \\ 7 \end{bmatrix}$ 39. $\frac{1}{50}\begin{bmatrix} 22 & 14 & -2 \\ -42 & -2 & 11 \\ -5 & -10 & 5 \end{bmatrix}$

43. $\begin{bmatrix} -2 \\ 7 \end{bmatrix}$ 45. $\begin{bmatrix} 3 & 6 & 2 & 2 & 2 \\ 5 & 10 & 0 & -1 & -11 \\ 2 & 4 & -1 & 3 & -4 \end{bmatrix}$

47. The codomain is $\mathcal{R}^3$, and the range is the span of the columns of B.

49. $\begin{bmatrix} 20 \\ -2 \\ 2 \end{bmatrix}$ 51. $\begin{bmatrix} 2 & 0 & -1 \\ 4 & 0 & 0 \end{bmatrix}$

53. The standard matrix is $\begin{bmatrix} 4 & 1 \\ 3 & 2 \end{bmatrix}$.

55. linear 57. linear

59. $\left\{\begin{bmatrix} 1 \\ 0 \end{bmatrix}, \begin{bmatrix} 2 \\ 1 \end{bmatrix}, \begin{bmatrix} 0 \\ -1 \end{bmatrix}\right\}$

61. $\left\{\begin{bmatrix} -2 \\ 1 \\ 1 \end{bmatrix}\right\}$ T is not one-to-one.

63. $\begin{bmatrix} 1 & 1 \\ 0 & 0 \\ 2 & -1 \end{bmatrix}$ The columns are linearly independent, so T is one-to-one.

65. $\begin{bmatrix} 3 & -1 \\ 0 & 1 \\ 1 & 1 \end{bmatrix}$ The rank is 2, so T is not onto.

67. $\begin{bmatrix} 5 & -1 & 4 \\ 1 & 1 & -1 \\ 3 & 1 & 0 \end{bmatrix}$ 69. $\begin{bmatrix} 5 & -1 & 4 \\ 1 & 1 & -1 \\ 3 & 1 & 0 \end{bmatrix}$

71. $\begin{bmatrix} 7 & -1 \\ 2 & -1 \end{bmatrix}$ 73. $T^{-1}\left(\begin{bmatrix} x_1 \\ x_2 \end{bmatrix}\right) = \frac{1}{5}\begin{bmatrix} 3x_1 - 2x_2 \\ x_1 + x_2 \end{bmatrix}$

Chapter 2 MATLAB Exercises

1. (a) $AD = \begin{bmatrix} 4 & 10 & 9 \\ 1 & 2 & 9 \\ 5 & 8 & 15 \\ 5 & 8 & -8 \\ -4 & -8 & 1 \end{bmatrix}$

(b) $DB = \begin{bmatrix} 6 & -2 & 5 & 11 & 9 \\ -3 & -1 & 10 & 7 & -3 \\ -3 & 1 & 2 & -1 & -3 \\ 2 & -2 & 7 & 9 & 3 \\ 0 & -1 & 10 & 10 & 2 \end{bmatrix}$

(c), (d) $(AB^T)C = A(B^TC) =$

$$\begin{bmatrix} 38 & -22 & 14 & 38 & 57 \\ 10 & -4 & 4 & 10 & 11 \\ -12 & -9 & -11 & -12 & 12 \\ 9 & -5 & 4 & 9 & 14 \\ 28 & 10 & 20 & 28 & -9 \end{bmatrix}$$

(e) $D(B - 2C) = \begin{bmatrix} -2 & 10 & 5 & 3 & -17 \\ -31 & -7 & -8 & -21 & -1 \\ -11 & -5 & -4 & -9 & 7 \\ -14 & 2 & -1 & -7 & -11 \\ -26 & -1 & -4 & -16 & -6 \end{bmatrix}$

(f) $\begin{bmatrix} 11 \\ 8 \\ 20 \\ -3 \\ -9 \end{bmatrix}$ (g), (h) $C(A\mathbf{v}) = (CA)\mathbf{v} = \begin{bmatrix} 1 \\ -18 \\ 81 \end{bmatrix}$

(i) $A^3 = \begin{bmatrix} 23 & 14 & 9 & -7 & 46 \\ 2 & 11 & 6 & -2 & 10 \\ 21 & 26 & -8 & -17 & 11 \\ -6 & 18 & 53 & 24 & -36 \\ -33 & -6 & 35 & 25 & -12 \end{bmatrix}$

2. (a) The entries of the following matrices are rounded to four places after the decimal:

$$A^{10} = \begin{bmatrix} 0.2056 & 0.2837 & 0.2240 & 0.1380 & 0.0589 & 0 \\ 0.1375 & 0.2056 & 0.1749 & 0.1101 & 0.0471 & 0 \\ 0.1414 & 0.1767 & 0.1584 & 0.1083 & 0.0475 & 0 \\ 0.1266 & 0.1616 & 0.1149 & 0.0793 & 0.0356 & 0 \\ 0.0356 & 0.0543 & 0.0420 & 0.0208 & 0.0081 & 0 \\ 0.0027 & 0.0051 & 0.0051 & 0.0036 & 0.0016 & 0 \end{bmatrix}$$

$$A^{100} = \begin{bmatrix} 0.0045 & 0.0062 & 0.0051 & 0.0033 & 0.0014 & 0 \\ 0.0033 & 0.0045 & 0.0037 & 0.0024 & 0.0010 & 0 \\ 0.0031 & 0.0043 & 0.0035 & 0.0023 & 0.0010 & 0 \\ 0.0026 & 0.0036 & 0.0029 & 0.0019 & 0.0008 & 0 \\ 0.0008 & 0.0011 & 0.0009 & 0.0006 & 0.0003 & 0 \\ 0.0001 & 0.0001 & 0.0001 & 0.0001 & 0.0000 & 0 \end{bmatrix}$$

$A^{500} =$

$$\frac{1}{10^9}\begin{bmatrix} 0.2126 & 0.2912 & 0.2393 & 0.1539 & 0.0665 & 0 \\ 0.1552 & 0.2126 & 0.1747 & 0.1124 & 0.0486 & 0 \\ 0.1457 & 0.1996 & 0.1640 & 0.1055 & 0.0456 & 0 \\ 0.1216 & 0.1665 & 0.1369 & 0.0880 & 0.0381 & 0 \\ 0.0381 & 0.0521 & 0.0428 & 0.0275 & 0.0119 & 0 \\ 0.0040 & 0.0054 & 0.0045 & 0.0029 & 0.0012 & 0 \end{bmatrix}$$

The colony will disappear.

(b) (ii) The entries of the following vectors are rounded to four places after the decimal:

$$\mathbf{x}_1 = \begin{bmatrix} 4.7900 \\ 3.2700 \\ 4.0800 \\ 3.4400 \\ 0.7200 \\ 0.1800 \end{bmatrix}, \quad \mathbf{x}_2 = \begin{bmatrix} 5.3010 \\ 4.4530 \\ 5.0430 \\ 3.2640 \\ 1.0320 \\ 0.0720 \end{bmatrix},$$

$$\mathbf{x}_3 = \begin{bmatrix} 6.1254 \\ 4.8107 \\ 6.1077 \\ 4.0344 \\ 0.9792 \\ 0.1032 \end{bmatrix}, \quad \mathbf{x}_4 = \begin{bmatrix} 7.2115 \\ 5.3878 \\ 6.4296 \\ 4.8862 \\ 1.2103 \\ 0.0979 \end{bmatrix},$$

and

$$\mathbf{x}_5 = \begin{bmatrix} 8.1259 \\ 6.1480 \\ 6.9490 \\ 5.1437 \\ 1.4658 \\ 0.1210 \end{bmatrix}$$

(iv) Assuming that $\mathbf{x}_{n+1} = \mathbf{x}_n$ we have that $\mathbf{x}_n = A\mathbf{x}_n + \mathbf{b}$, and hence

$$(I_6 - A)\mathbf{x}_n = \mathbf{x}_n - A\mathbf{x}_n = \mathbf{b}.$$

Therefore $\mathbf{x}_n = (I_6 - A)^{-1}\mathbf{b}$.

Using the given $\mathbf{b}$, we obtain (with entries rounded to four places after the decimal)

$$\mathbf{x}_n = \begin{bmatrix} 28.1412 \\ 20.7988 \\ 20.8189 \\ 16.6551 \\ 4.9965 \\ 0.4997 \end{bmatrix}.$$

3. The eight airports divide up into two sets: $\{1,2,6,8\}$ and $\{3,4,5,7\}$.

4. (a) $R = \begin{bmatrix} 1 & 2 & 0 & -1 & 0 & 0 & 0 \\ 0 & 0 & 1 & 1 & 0 & 0 & 1 \\ 0 & 0 & 0 & 0 & 1 & 0 & -1 \\ 0 & 0 & 0 & 0 & 0 & 1 & -1 \\ 0 & 0 & 0 & 0 & 0 & 0 & 0 \end{bmatrix}$

(b) $S = \begin{bmatrix} -2 & 1 & 0 \\ 1 & 0 & 0 \\ 0 & -1 & -1 \\ 0 & 1 & 0 \\ 0 & 0 & 1 \\ 0 & 0 & 1 \\ 0 & 0 & 1 \end{bmatrix}$

5. (b) $P = \begin{bmatrix} 0.0 & -0.8 & -2.2 & -1.8 & 1.0 \\ 0.0 & -0.8 & -1.2 & -1.8 & 1.0 \\ 0.0 & 0.4 & 1.6 & 2.4 & -1.0 \\ 0.0 & 1.0 & 2.0 & 2.0 & -1.0 \\ 1.0 & 0.0 & -1.0 & -1.0 & 0.0 \end{bmatrix}$

6. (a) $M = \begin{bmatrix} 1 & 2 & 1 & 2 \\ 2 & 0 & -1 & 3 \\ -1 & 1 & 0 & 0 \\ 2 & 1 & 1 & 2 \\ 4 & 4 & 1 & 6 \end{bmatrix}$

(b) $S = \begin{bmatrix} 1 & 0 & 1 & 0 & 1 & 0 \\ 0 & 1 & 1 & 0 & 0 & 0 \\ 0 & 0 & 0 & 1 & 2 & 0 \\ 0 & 0 & 0 & 0 & 0 & 1 \end{bmatrix}$

7. $A^{-1}B = \begin{bmatrix} 6 & -4 & 3 & 19 & 5 & -2 & -5 \\ -1 & 2 & -4 & -1 & 4 & -3 & -2 \\ -2 & 0 & 2 & 6 & -1 & 6 & 3 \\ 0 & 1 & -3 & -8 & 2 & -3 & 1 \\ -1 & 0 & 2 & -6 & -5 & 2 & 2 \end{bmatrix}$

8. (a) $L = \frac{1}{3}\begin{bmatrix} 3 & 0 & 0 & 0 \\ 6 & 3 & 0 & 0 \\ 3 & 1 & 3 & 0 \\ 6 & 1 & 3 & 3 \end{bmatrix}$ and

$$U = \frac{1}{3}\begin{bmatrix} 3 & -3 & 6 & 0 & -6 & 12 \\ 0 & 9 & -9 & -6 & 15 & -15 \\ 0 & 0 & -6 & 11 & -8 & -1 \\ 0 & 0 & 0 & -6 & 21 & -9 \end{bmatrix}$$

(b) $\mathbf{x} = \begin{bmatrix} \frac{17}{4} \\ -\frac{5}{4} \\ -\frac{1}{4} \\ \frac{3}{2} \\ 0 \\ 0 \end{bmatrix} + x_5 \begin{bmatrix} -\frac{29}{12} \\ \frac{23}{4} \\ \frac{61}{12} \\ \frac{7}{2} \\ 1 \\ 0 \end{bmatrix} + x_6 \begin{bmatrix} -\frac{5}{12} \\ -\frac{9}{4} \\ -\frac{35}{12} \\ -\frac{3}{2} \\ 0 \\ 1 \end{bmatrix}$

9. (a) $A = \begin{bmatrix} 1 & 2 & 0 & 1 & -3 & -2 \\ 0 & 1 & 0 & -1 & 0 & 0 \\ 1 & 0 & 1 & 0 & 0 & 3 \\ 2 & 4 & 0 & 3 & -6 & -4 \\ 3 & 2 & 2 & 1 & -2 & -4 \\ 4 & 4 & 2 & 2 & -5 & 3 \end{bmatrix}$

(b) $A^{-1} = \frac{1}{9}$

$$\times \begin{bmatrix} 10 & -18 & -54 & -27 & 1 & 26 \\ -18 & 9 & 0 & 9 & 0 & 0 \\ -7 & 18 & 63 & 27 & 2 & -29 \\ -18 & 0 & 0 & 9 & 0 & 0 \\ -17 & 0 & -18 & 0 & 1 & 8 \\ -1 & 0 & 0 & 0 & -1 & 1 \end{bmatrix}$$

(c) $T^{-1}\left(\begin{bmatrix} x_1 \\ x_2 \\ x_3 \\ x_4 \\ x_5 \\ x_6 \end{bmatrix}\right) =$

$$\begin{bmatrix} \frac{10}{9}x_1 + -2x_2 + -6x_3 + -3x_4 + \frac{1}{9}x_5 + \frac{26}{9}x_6 \\ -2x_1 + x_2 + x_4 \\ -\frac{7}{9}x_1 + 2x_2 + 7x_3 + 3x_4 + \frac{2}{9}x_5 - \frac{29}{9}x_6 \\ -2x_1 + x_4 \\ -\frac{17}{9}x_1 - 2x_3 + \frac{1}{9}x_5 + \frac{8}{9}x_6 \\ -\frac{1}{9}x_1 - \frac{1}{9}x_5 + \frac{1}{9}x_6 \end{bmatrix}$$

10. (a) $B = \begin{bmatrix} 1 & 0 & 2 & 0 & 0 & 1 \\ 2 & -1 & 0 & 1 & 0 & 0 \\ 0 & 3 & 0 & 0 & -1 & 0 \\ 2 & 1 & -1 & 0 & 0 & 1 \end{bmatrix}$

(b) The standard matrix of UT is

$$BA = \begin{bmatrix} 7 & 6 & 4 & 3 & -8 & 7 \\ 4 & 7 & 0 & 6 & -12 & -8 \\ -3 & 1 & -2 & -4 & 2 & 4 \\ 5 & 9 & 1 & 3 & -11 & -4 \end{bmatrix}.$$

(c) $UT\left(\begin{bmatrix} x_1 \\ x_2 \\ x_3 \\ x_4 \\ x_5 \\ x_6 \end{bmatrix}\right) =$

$$\begin{bmatrix} 7x_1 + 6x_2 + 4x_3 + 3x_4 - 8x_5 + 7x_6 \\ 4x_1 + 7x_2 + 6x_4 - 12x_5 - 8x_6 \\ -3x_1 + x_2 - 2x_3 - 4x_4 + 2x_5 + 4x_6 \\ 5x_1 + 9x_2 + x_3 + 3x_4 - 11x_5 - 4x_6 \end{bmatrix}$$

(d) The standard matrix of UT^{-1} is

$$BA^{-1} = \begin{bmatrix} -\frac{5}{9} & 2 & 8 & 3 & \frac{4}{9} & -\frac{31}{9} \\ \frac{20}{9} & -5 & -12 & -6 & \frac{2}{9} & \frac{52}{9} \\ -\frac{37}{9} & 3 & 2 & 3 & -\frac{1}{9} & -\frac{8}{9} \\ \frac{8}{9} & -5 & -19 & -8 & -\frac{1}{9} & \frac{82}{9} \end{bmatrix},$$

and hence

$UT^{-1}\left(\begin{bmatrix} x_1 \\ x_2 \\ x_3 \\ x_4 \\ x_5 \\ x_6 \end{bmatrix}\right) =$

$$\begin{bmatrix} -\frac{5}{9}x_1 + 2x_2 + 8x_3 + 3x_4 + \frac{4}{9}x_5 - \frac{31}{9}x_6 \\ \frac{20}{9}x_1 - 5x_2 - 12x_3 - 6x_4 + \frac{2}{9}x_5 + \frac{52}{9}x_6 \\ -\frac{37}{9}x_1 + 3x_2 + 2x_3 + 3x_4 - \frac{1}{9}x_5 - \frac{8}{9}x_6 \\ \frac{8}{9}x_1 - 5x_2 - 19x_3 - 8x_4 - \frac{1}{9}x_5 + \frac{82}{9}x_6 \end{bmatrix}.$$

Chapter 3

Section 3.1

1. 0 3. −25 5. 0 7. 2
9. 16 11. −30 13. 19 15. −2
17. 20 19. 2 21. 60 23. 180
25. −147 27. −24 29. 31 31. 0
33. 22 35. 22 37. 2 39. −9
41. ±4 43. no c
45. F 46. F 47. F 48. F 49. T
50. F 51. T 52. T 53. F 54. T
55. F 56. T 57. T 58. F 59. T
60. T 61. F 62. T 63. F 64. F
67. 2 79. $\frac{1}{2}|\det[\mathbf{u}\ \mathbf{v}]|$
81. (c) no 83. (c) yes

Section 3.2

1. −9 3. 19 5. 12 7. −2
9. −2 11. −60 13. −15 15. 30
17. −20 19. −3 21. 18 23. −95
25. −8 27. −6 and 2 29. 5 31. −14
33. −5 and 3 35. −1 37. $\frac{1}{2}$
39. F 40. T 41. F 42. T 43. F
44. T 45. F 46. F 47. T 48. F
49. T 50. T 51. F 52. F 53. T
54. T 55. T 56. F 57. T 58. F

59. $\begin{bmatrix} x_1 \\ x_2 \end{bmatrix} = \begin{bmatrix} -15.0 \\ 10.5 \end{bmatrix}$ 61. $\begin{bmatrix} x_1 \\ x_2 \end{bmatrix} = \begin{bmatrix} 11 \\ -6 \end{bmatrix}$

63. $\begin{bmatrix} x_1 \\ x_2 \\ x_3 \end{bmatrix} = \begin{bmatrix} 2 \\ 3 \\ -2 \end{bmatrix}$ 65. $\begin{bmatrix} x_1 \\ x_2 \\ x_3 \end{bmatrix} = \begin{bmatrix} -0.4 \\ 1.8 \\ -2.4 \end{bmatrix}$

67. Take $k = 2$ and $A = I_2$.

83. (a)

$$A \longrightarrow \begin{bmatrix} 2.4 & 3.0 & -6 & -9 \\ 0.0 & -3.0 & -2 & -5 \\ -4.8 & 6.3 & 4 & -2 \\ 9.6 & 1.5 & 5 & 9 \end{bmatrix}$$

$$\longrightarrow \begin{bmatrix} 2.4 & 3.0 & -6 & 9 \\ 0.0 & -3.0 & -2 & -5 \\ 0.0 & 12.3 & -8 & 16 \\ 0.0 & -10.5 & 29 & -27 \end{bmatrix}$$

$$\longrightarrow \begin{bmatrix} 2.4 & 3 & -6.0 & 9.0 \\ 0.0 & -3 & -2.0 & -5.0 \\ 0.0 & 0 & -16.2 & -4.5 \\ 0.0 & 0 & 36.0 & -9.5 \end{bmatrix}$$

$$\longrightarrow \begin{bmatrix} 2.4 & 3 & -6.0 & 9.0 \\ 0.0 & -3 & -2.0 & -5.0 \\ 0.0 & 0 & -16.2 & -4.5 \\ 0.0 & 0 & 0.0 & -19.5 \end{bmatrix}$$

(b) 2274.48

85. $\begin{bmatrix} 13 & -8 & -3 & 6 \\ -10 & -16 & -10 & -12 \\ -17 & 8 & -1 & 2 \\ -12 & 0 & -12 & -8 \end{bmatrix}$

Chapter 3 Review Exercises

1. F 2. F 3. T 4. F 5. T 6. F
7. F 8. T 9. F 10. F 11. F
13. 5 15. −3
17. $2(-3) + 1(-1) + 3(1)$ 19. $1(7) + (-1)5 + 2(-3)$
21. 0 23. 3 25. −3 and 4
27. −3 29. 25 31. $x_1 = 2.1$, $x_2 = 0.8$
33. 5 35. 40 37. 5
39. 20 41. $\det B = 0$ or $\det B = 1$

Chapter 3 MATLAB Exercises

1. Matrix A can be transformed into an upper triangular matrix U by means of only row addition operations. The diagonal entries of U (rounded to 4 places after the decimal point) are $-0.8000, -30.4375, 1.7865, -0.3488, -1.0967$, and 0.3749. Thus

$$\det A = (-1)^0(-0.8000)(-30.4375)(1.7865)(-0.3488)(-1.0967)(0.3749) = 6.2400.$$

2. The following sequence of elementary row operations transforms A into an upper triangular matrix: $\mathbf{r}_1 \leftrightarrow \mathbf{r}_2$, $-2\mathbf{r}_1 + \mathbf{r}_3 \to \mathbf{r}_3$, $-2\mathbf{r}_1 + \mathbf{r}_4 \to \mathbf{r}_4$, $2\mathbf{r}_1 + \mathbf{r}_5 \to \mathbf{r}_5$, $-\mathbf{r}_1 + \mathbf{r}_6 \to \mathbf{r}_6$, $-\mathbf{r}_2 + \mathbf{r}_4 \to \mathbf{r}_4$, $4\mathbf{r}_2 + \mathbf{r}_5 \to \mathbf{r}_5$, $-2\mathbf{r}_2 + \mathbf{r}_6 \to \mathbf{r}_6$, $\mathbf{r}_3 \leftrightarrow \mathbf{r}_6$, $17\mathbf{r}_3 + \mathbf{r}_5 \to \mathbf{r}_5$, $\mathbf{r}_4 \leftrightarrow \mathbf{r}_6$, $\mathbf{r}_5 \leftrightarrow \mathbf{r}_6$, and $-\frac{152}{9}\mathbf{r}_5 + \mathbf{r}_6 \to \mathbf{r}_6$. (Other sequences are possible.) This matrix is

$$\begin{bmatrix} 1 & 1 & 2 & -2 & 1 & 2 \\ 0 & 1 & 2 & -2 & 3 & 1 \\ 0 & 0 & -1 & 1 & -10 & 2 \\ 0 & 0 & 0 & -1 & 4 & -1 \\ 0 & 0 & 0 & 0 & -9 & 0 \\ 0 & 0 & 0 & 0 & 0 & 46 \end{bmatrix}.$$

Thus

$$\det A = (-1)^4(1)(1)(-1)(-1)(-9)(46) = -414.$$

3. (a) $\det\begin{bmatrix}\mathbf{v}\\ A\end{bmatrix} = 2$ and $\det\begin{bmatrix}\mathbf{w}\\ A\end{bmatrix} = -10$.
   (b) $\det\begin{bmatrix}\mathbf{v}+\mathbf{w}\\ A\end{bmatrix} = \det\begin{bmatrix}\mathbf{v}\\ A\end{bmatrix} + \det\begin{bmatrix}\mathbf{w}\\ A\end{bmatrix} = -8$.
   (c) $\det\begin{bmatrix}3\mathbf{v}-2\mathbf{w}\\ A\end{bmatrix} = 3\det\begin{bmatrix}\mathbf{v}\\ A\end{bmatrix} - 2\det\begin{bmatrix}\mathbf{w}\\ A\end{bmatrix} = 26$.
   (d) Any such function is a linear transformation.
   (f) Any such function is a linear transformation.
   (g) Any such function is a linear transformation.

Chapter 4

Section 4.1

1. $\{\mathbf{e}_2\}$

3. $\left\{\begin{bmatrix}4\\-1\end{bmatrix}\right\}$

5. $\left\{\begin{bmatrix}-1\\2\\1\end{bmatrix}, \begin{bmatrix}1\\-1\\3\end{bmatrix}\right\}$

7. $\left\{\begin{bmatrix}-1\\0\\0\\3\end{bmatrix}, \begin{bmatrix}1\\4\\0\\0\end{bmatrix}, \begin{bmatrix}0\\-3\\0\\-1\end{bmatrix}\right\}$

9. $\left\{\begin{bmatrix}0\\3\\1\\-1\end{bmatrix}, \begin{bmatrix}2\\1\\-4\\2\end{bmatrix}, \begin{bmatrix}-5\\-2\\3\\0\end{bmatrix}\right\}$

11. yes 13. no 15. yes 17. yes
19. no 21. yes 23. yes 25. yes

27. $\left\{\begin{bmatrix}7\\5\\1\end{bmatrix}\right\}$

29. $\left\{\begin{bmatrix}2\\-1\\1\\0\end{bmatrix}, \begin{bmatrix}-1\\-3\\0\\1\end{bmatrix}\right\}$

31. $\left\{\begin{bmatrix}-5\\3\\1\\0\end{bmatrix}, \begin{bmatrix}3\\-4\\0\\1\end{bmatrix}\right\}$

33. $\left\{\begin{bmatrix}3\\1\\0\\0\\0\\0\end{bmatrix}, \begin{bmatrix}-1\\0\\-2\\1\\0\\0\end{bmatrix}, \begin{bmatrix}-2\\0\\-3\\0\\-2\\1\end{bmatrix}\right\}$

35. $\{1, 2, -1\}$, $\left\{\begin{bmatrix}-2\\1\\0\end{bmatrix}, \begin{bmatrix}1\\0\\1\end{bmatrix}\right\}$

37. $\left\{\begin{bmatrix}1\\1\\1\\0\end{bmatrix}, \begin{bmatrix}1\\-1\\0\\1\end{bmatrix}\right\}$, $\left\{\begin{bmatrix}0\\0\end{bmatrix}\right\}$

39. $\left\{\begin{bmatrix}1\\0\\2\end{bmatrix}, \begin{bmatrix}1\\0\\0\end{bmatrix}, \begin{bmatrix}-1\\0\\-1\end{bmatrix}\right\}$, $\left\{\begin{bmatrix}1\\1\\2\end{bmatrix}\right\}$

41. $\left\{\begin{bmatrix}1\\-1\\2\\0\end{bmatrix}, \begin{bmatrix}-1\\2\\-1\\2\end{bmatrix}, \begin{bmatrix}-5\\7\\-8\\4\end{bmatrix}\right\}$, $\left\{\begin{bmatrix}3\\-2\\1\end{bmatrix}\right\}$

43. T 44. F 45. F 46. T 47. T
48. F 49. F 50. F 51. T 52. T
53. T 54. T 55. F 56. T 57. T
58. T 59. T 60. T 61. T 62. T

63. $\left\{\begin{bmatrix}-1\\1\end{bmatrix}, \begin{bmatrix}1\\-2\end{bmatrix}\right\}$

65. $\left\{\begin{bmatrix}1\\3\\0\end{bmatrix}, \begin{bmatrix}1\\2\\-1\end{bmatrix}, \begin{bmatrix}0\\1\\1\end{bmatrix}, \begin{bmatrix}2\\6\\-1\end{bmatrix}\right\}$

67. $\left\{\begin{bmatrix}1\\-2\end{bmatrix}, \begin{bmatrix}-3\\4\end{bmatrix}\right\}$

69. $\left\{\begin{bmatrix}-2\\4\\5\\-1\end{bmatrix}, \begin{bmatrix}-1\\1\\2\\0\end{bmatrix}, \begin{bmatrix}3\\-4\\-5\\1\end{bmatrix}\right\}$

71. $\mathcal{R}^n$, the zero subspace of $\mathcal{R}^m$, the zero subspace of $\mathcal{R}^n$

73. no 75. $\begin{bmatrix}1 & -1\\ -1 & 1\end{bmatrix}$

81. $\begin{bmatrix}1\\0\end{bmatrix}$ and $\begin{bmatrix}0\\1\end{bmatrix}$ are in the set, but $\begin{bmatrix}1\\0\end{bmatrix}+\begin{bmatrix}0\\1\end{bmatrix}$ is not.

83. $\begin{bmatrix}0\\0\\0\end{bmatrix}$ is not in the set.

85. $\begin{bmatrix}1\\0\\-1\end{bmatrix}$ is in the set, but $(-2)\begin{bmatrix}1\\0\\-1\end{bmatrix}$ is not.

87. $\begin{bmatrix}6\\2\\3\end{bmatrix}$ is in the set, but $(-1)\begin{bmatrix}6\\2\\3\end{bmatrix}$ is not.

101. (a) yes (b) no 103. (a) yes (b) no

Section 4.2

1. (a) $\left\{\begin{bmatrix}1\\-1\end{bmatrix}\right\}$ (b) $\left\{\begin{bmatrix}3\\1\\0\\0\end{bmatrix},\begin{bmatrix}-4\\0\\1\\0\end{bmatrix},\begin{bmatrix}2\\0\\0\\1\end{bmatrix}\right\}$

3. (a) $\left\{\begin{bmatrix}1\\-1\\-1\end{bmatrix},\begin{bmatrix}2\\-1\\0\end{bmatrix}\right\}$ (b) $\left\{\begin{bmatrix}2\\-3\\1\end{bmatrix}\right\}$

5. (a) $\left\{\begin{bmatrix}1\\-1\\2\end{bmatrix},\begin{bmatrix}0\\1\\3\end{bmatrix}\right\}$ (b) $\left\{\begin{bmatrix}2\\1\\0\\0\end{bmatrix},\begin{bmatrix}-2\\0\\1\\1\end{bmatrix}\right\}$

7. (a) $\left\{\begin{bmatrix}-1\\2\\1\\0\end{bmatrix},\begin{bmatrix}1\\0\\-1\\1\end{bmatrix},\begin{bmatrix}2\\-5\\-1\\-2\end{bmatrix}\right\}$ (b) $\left\{\begin{bmatrix}-4\\-4\\-1\\1\end{bmatrix}\right\}$

9. (a) $\left\{\begin{bmatrix}1\\2\\1\end{bmatrix},\begin{bmatrix}2\\3\\2\end{bmatrix},\begin{bmatrix}1\\3\\4\end{bmatrix}\right\}$
(b) The null space of T is $\{\mathbf{0}\}$.

11. (a) $\left\{\begin{bmatrix}1\\2\\1\end{bmatrix},\begin{bmatrix}-2\\-5\\-3\end{bmatrix}\right\}$ (b) $\left\{\begin{bmatrix}-3\\-1\\1\\0\end{bmatrix},\begin{bmatrix}1\\1\\0\\1\end{bmatrix}\right\}$

13. (a) $\left\{\begin{bmatrix}1\\2\\0\\3\end{bmatrix},\begin{bmatrix}1\\1\\0\\1\end{bmatrix}\right\}$ (b) $\left\{\begin{bmatrix}1\\-3\\1\\0\end{bmatrix},\begin{bmatrix}-1\\2\\0\\1\end{bmatrix}\right\}$

15. (a) $\left\{\begin{bmatrix}1\\3\\7\end{bmatrix},\begin{bmatrix}2\\1\\4\end{bmatrix}\right\}$ (b) $\left\{\begin{bmatrix}1\\-2\\1\\0\\0\end{bmatrix},\begin{bmatrix}0\\0\\0\\1\\0\end{bmatrix},\begin{bmatrix}2\\-3\\0\\0\\1\end{bmatrix}\right\}$

17. $\left\{\begin{bmatrix}1\\-2\end{bmatrix}\right\}$ 19. $\left\{\begin{bmatrix}5\\2\\0\\0\end{bmatrix},\begin{bmatrix}-3\\0\\0\\-4\end{bmatrix}\right\}$

21. $\left\{\begin{bmatrix}3\\1\\0\end{bmatrix},\begin{bmatrix}-5\\0\\1\end{bmatrix}\right\}$ 23. $\left\{\begin{bmatrix}2\\1\\0\\0\end{bmatrix},\begin{bmatrix}-3\\0\\1\\0\end{bmatrix},\begin{bmatrix}4\\0\\0\\1\end{bmatrix}\right\}$

25. $\left\{\begin{bmatrix}1\\2\\1\end{bmatrix},\begin{bmatrix}2\\1\\3\end{bmatrix}\right\}$ 27. $\left\{\begin{bmatrix}1\\-1\\3\end{bmatrix},\begin{bmatrix}0\\-1\\1\end{bmatrix},\begin{bmatrix}1\\-2\\0\end{bmatrix}\right\}$

29. $\left\{\begin{bmatrix}1\\0\\-1\\2\end{bmatrix},\begin{bmatrix}1\\1\\-2\\1\end{bmatrix},\begin{bmatrix}0\\1\\-1\\2\end{bmatrix}\right\}$

31. $\left\{\begin{bmatrix}-2\\4\\5\\-1\end{bmatrix},\begin{bmatrix}3\\-4\\-5\\1\end{bmatrix},\begin{bmatrix}1\\5\\4\\-2\end{bmatrix}\right\}$

33. F 34. T 35. F 36. T 37. T
38. T 39. T 40. F 41. F 42. T
43. F 44. T 45. T 46. T 47. T
48. T 49. T 50. F 51. T 52. T

53. Because $\dim \mathcal{R}^4 = 4$, a generating set for $\mathcal{R}^4$ must contain at least 4 vectors.

55. A basis for $\mathcal{R}^3$ must contain exactly 3 vectors.

57. A subset of $\mathcal{R}^2$ containing more than 2 vectors is linearly dependent.

67. 1 69. $n-2$

79. $\left\{\begin{bmatrix}2\\3\\0\end{bmatrix},\begin{bmatrix}1\\0\\0\end{bmatrix},\begin{bmatrix}0\\0\\1\end{bmatrix}\right\}$ 81. $\left\{\begin{bmatrix}0\\2\\1\\0\end{bmatrix},\begin{bmatrix}1\\1\\0\\0\end{bmatrix},\begin{bmatrix}-1\\0\\0\\1\end{bmatrix}\right\}$

83. (c) No, $\mathcal{S}$ is not a subset of V.

85. (a) $\left\{\begin{bmatrix}0.1\\0.7\\-0.5\end{bmatrix},\begin{bmatrix}0.2\\0.9\\0.5\end{bmatrix},\begin{bmatrix}0.5\\-0.5\\-0.5\end{bmatrix}\right\}$
(b) $\left\{\begin{bmatrix}1.2\\-2.3\\1.0\\0.0\\0.0\end{bmatrix},\begin{bmatrix}-1.4\\2.9\\0.0\\-0.7\\1.0\end{bmatrix}\right\}$

Section 4.3

1. (a) 2 (b) 2 (c) 2 (d) 1
3. (a) 3 (b) 2 (c) 3 (d) 0
5. (a) 1 (b) 3 (c) 1 (d) 0
7. (a) 2 (b) 1 (c) 2 (d) 0
9. (a) 2 (b) 2 (c) 2 (d) 1
11. (a) 2 (b) 1 (c) 2 (d) 2
13. 1 15. 2
17. $\{[1\ \ 0\ \ 3], [0\ \ 1\ \ 2]\}$

19. $\{[1\ \ 0\ \ 0\ \ 1], [0\ \ 1\ \ 1\ \ -1]\}$

21. $\{[1\ \ 0\ \ 0\ \ -3\ \ 1\ \ 3], [0\ \ 1\ \ 0\ \ 2\ \ -1\ \ -2], [0\ \ 0\ \ 1\ \ 0\ \ 0\ \ -1]\}$

23. $\{[1\ \ 0\ \ 0\ \ 1\ \ 0], [0\ \ 1\ \ 0\ \ -1\ \ 0], [0\ \ 0\ \ 1\ \ 0\ \ 0], [0\ \ 0\ \ 0\ \ 0\ \ 1]\}$

25. $\{[1\ \ -1\ \ 1], [0\ \ 1\ \ 2]\}$

27. $\{[-1\ \ 1\ \ 1\ \ -2], [2\ \ -1\ \ -1\ \ 3]\}$

29. $\{[1\ \ 0\ \ -1\ \ -3\ \ 1\ \ 4], [2\ \ -1\ \ -1\ \ -8\ \ 3\ \ 9], [0\ \ 1\ \ 1\ \ 2\ \ -1\ \ -3]\}$

31. $\{[1\ \ 0\ \ -1\ \ 1\ \ 3], [2\ \ -1\ \ -1\ \ 3\ \ -8], [0\ \ 1\ \ -1\ \ -1\ \ 2], [-1\ \ 1\ \ 1\ \ -2\ \ 5]\}$

33. (a) 2 (b) 0 one-to-one and onto

35. (a) 1 (b) 2 neither one-to-one nor onto

37. (a) 2 (b) 0 one-to-one, not onto

39. (a) 2 (b) 1 onto, not one-to-one

41. F	42. T	43. T	44. F	45. F
46. T	47. T	48. F	49. T	50. F
51. F	52. F	53. T	54. T	55. F
56. F	57. T	58. T	59. T	60. T

69. (a) $\left\{\begin{bmatrix}1\\0\\6\\0\end{bmatrix}, \begin{bmatrix}0\\1\\-4\\1\end{bmatrix}\right\}, \left\{\begin{bmatrix}-6\\4\\1\\0\end{bmatrix}, \begin{bmatrix}0\\-1\\0\\1\end{bmatrix}\right\}$

79. Take $V = \text{Span}\,\{\mathbf{e}_1, \mathbf{e}_2\}$ and $W = \text{Span}\,\{\mathbf{e}_4, \mathbf{e}_5\}$.

85. (a) $\begin{bmatrix}1 & 2 & 0 & 0\\-1 & 1 & 0 & 0\\1 & 0 & 0 & 0\\0 & 1 & 0 & 0\end{bmatrix}$

87. (a) No, the first vector in $\mathcal{A}_1$ is not in W.
(b) yes
(c) $[\mathbf{e}_1\ \mathbf{e}_2\ \mathbf{e}_3]$, $[\mathbf{e}_1\ \mathbf{e}_2\ \mathbf{e}_3]$, $[\mathbf{e}_1\ \mathbf{e}_2\ \mathbf{e}_3]$,
$\begin{bmatrix}1 & 0 & 0 & -.4 & -.2\\0 & 1 & 0 & .8 & .4\\0 & 0 & 1 & -.2 & -.6\end{bmatrix}$,
$\begin{bmatrix}1 & 0 & 0 & -.4 & -.2\\0 & 1 & 0 & .8 & .4\\0 & 0 & 1 & -.2 & -.5\end{bmatrix}$,
$\begin{bmatrix}1 & 0 & 0 & -.4 & -.2\\0 & 1 & 0 & .8 & .4\\0 & 0 & 1 & -.2 & -.6\end{bmatrix}$

Section 4.4

1. $\begin{bmatrix}1\\2\end{bmatrix}$ 3. $\begin{bmatrix}-5\\11\end{bmatrix}$ 5. $\begin{bmatrix}-3\\8\end{bmatrix}$ 7. $\begin{bmatrix}4\\5\\4\end{bmatrix}$

9. $\begin{bmatrix}-7\\-3\\2\end{bmatrix}$ 11. (b) $\begin{bmatrix}5\\-3\end{bmatrix}$ 13. (b) $\begin{bmatrix}3\\0\\-1\end{bmatrix}$

15. $\begin{bmatrix}-5\\-1\end{bmatrix}$ 17. $\begin{bmatrix}7\\2\end{bmatrix}$ 19. $\begin{bmatrix}0\\-1\\3\end{bmatrix}$ 21. $\begin{bmatrix}-5\\1\\2\end{bmatrix}$

23. $(a + 2b)\mathbf{b}_1 + (a + 3b)\mathbf{b}_2 = \mathbf{u}$

25. $(-5a - 3b)\mathbf{b}_1 + (-3a - 2b)\mathbf{b}_2 = \mathbf{u}$

27. $(-4a - 3b + 2c)\mathbf{b}_1 + (-2a - b + c)\mathbf{b}_2 + (3a + 2b - c)\mathbf{b}_3 = \mathbf{u}$

29. $(-a - b + 2c)\mathbf{b}_1 + b\mathbf{b}_2 + (-a - b + c)\mathbf{b}_3 = \mathbf{u}$

31. F	32. T	33. T	34. T	35. T
36. T	37. T	38. T	39. T	40. T
41. T	42. T	43. T	44. F	45. T
46. T	47. T	48. F	49. T	50. T

51. (b) $\begin{bmatrix}-3 & 2\\2 & -1\end{bmatrix}$ (c) $A = B^{-1}$

53. (b) $\begin{bmatrix}1 & 0 & 1\\1 & 1 & 3\\0 & -1 & -1\end{bmatrix}$ (c) $A = B^{-1}$

55. $\begin{aligned}x' &= \frac{\sqrt{3}}{2}x + \frac{1}{2}y\\ y' &= -\frac{1}{2}x + \frac{\sqrt{3}}{2}y\end{aligned}$

57. $\begin{aligned}x' &= -\frac{\sqrt{2}}{2}x + \frac{\sqrt{2}}{2}y\\ y' &= -\frac{\sqrt{2}}{2}x - \frac{\sqrt{2}}{2}y\end{aligned}$

59. $\begin{aligned}x' &= -5x - 3y\\ y' &= -2x - y\end{aligned}$

61. $\begin{aligned}x' &= -x - y\\ y' &= -2x - y\end{aligned}$

63. $\begin{aligned}x' &= -x + y + 2z\\ y' &= 2x - y - 2z\\ z' &= x - y - z\end{aligned}$

65. $\begin{aligned}x' &= x - y + z\\ y' &= -3x + 4y - 2z\\ z' &= x - 2y + z\end{aligned}$

67. $\begin{aligned}x &= \frac{1}{2}x' - \frac{\sqrt{3}}{2}y'\\ y &= \frac{\sqrt{3}}{2}x' + \frac{1}{2}y'\end{aligned}$

69. $\begin{aligned}x &= -\frac{\sqrt{2}}{2}x' - \frac{\sqrt{2}}{2}y'\\ y &= \frac{\sqrt{2}}{2}x' - \frac{\sqrt{2}}{2}y'\end{aligned}$

71. $\begin{aligned}x &= x' + 3y'\\ y &= 2x' + 4y'\end{aligned}$

73. $\begin{aligned}x &= -x' + 3y'\\ y &= 3x' + 5y'\end{aligned}$

75. $\begin{aligned}x &= x' - y'\\ y &= 3x' + y' - z'\\ z &= y' + z'\end{aligned}$

77. $\begin{aligned}x &= x' - y' - z'\\ y &= -x' + 3y' + z'\\ z &= x' + 2y' + z'\end{aligned}$

79. $73x^2 + 18\sqrt{3}xy + 91y^2 = 1600$

81. $8x^2 - 34xy + 8y^2 = 225$

83. $-23x^2 - 26\sqrt{3}xy + 3y^2 = 144$

85. $-11x^2 + 50\sqrt{3}xy + 39y^2 = 576$

87. $2(x')^2 - 5(y')^2 = 10$

89. $4(x')^2 + 3(y')^2 = 12$

91. $4(x')^2 - 3(y')^2 = 60$

93. $5(x')^2 + 2(y')^2 = 10$

95. $\begin{bmatrix} \frac{a_1}{c_1} \\ \frac{a_2}{c_2} \\ \vdots \\ \frac{a_n}{c_n} \end{bmatrix}$ 97. $\begin{bmatrix} a_1 \\ a_2 - a_1 \\ \vdots \\ a_n - a_1 \end{bmatrix}$ 99. no

109. (b) $\begin{bmatrix} 29 \\ 44 \\ -52 \\ 33 \\ 39 \end{bmatrix}$ 111. $\begin{bmatrix} 0 \\ 2 \\ -2 \\ 2 \\ 1 \end{bmatrix}$

Section 4.5

1. $\begin{bmatrix} 1 & 1 \\ 3 & 0 \end{bmatrix}$ 3. $\begin{bmatrix} 1 & 2 \\ 1 & 1 \end{bmatrix}$

5. $\begin{bmatrix} 10 & 19 & 16 \\ -5 & -8 & -8 \\ 2 & 2 & 3 \end{bmatrix}$ 7. $\begin{bmatrix} 0 & -19 & 28 \\ 3 & 34 & -47 \\ 3 & 23 & -31 \end{bmatrix}$

9. $\begin{bmatrix} -10 & -12 & -9 & 1 \\ 20 & 26 & 20 & -7 \\ -10 & -15 & -12 & 7 \\ 7 & 7 & 5 & 1 \end{bmatrix}$

11. $\begin{bmatrix} 10 & -19 \\ 3 & -4 \end{bmatrix}$ 13. $\begin{bmatrix} 45 & 25 \\ -79 & -44 \end{bmatrix}$

15. $\begin{bmatrix} 2 & 5 & 10 \\ -6 & 1 & -7 \\ 2 & -2 & 0 \end{bmatrix}$ 17. $\begin{bmatrix} -1 & -1 & 0 \\ 1 & 3 & -1 \\ -1 & 0 & 1 \end{bmatrix}$

19. F 20. T 21. T 22. F 23. T
24. F 25. F 26. T 27. F 28. F
29. F 30. T 31. F 32. F 33. T
34. T 35. T 36. T 37. F 38. T

39. $\begin{bmatrix} 1 & -3 \\ 4 & 0 \end{bmatrix}$ 41. $\begin{bmatrix} 3 & 2 \\ -5 & 4 \end{bmatrix}$

43. $\begin{bmatrix} 0 & 2 & 3 \\ -5 & 0 & 0 \\ 4 & -7 & 1 \end{bmatrix}$ 45. $\begin{bmatrix} 1 & 0 & -3 & 0 \\ -1 & 2 & 0 & 4 \\ 1 & 0 & 5 & -1 \\ -1 & -1 & 0 & 3 \end{bmatrix}$

47. (a) $\begin{bmatrix} 0 & 3 \\ 1 & 0 \end{bmatrix}$ (b) $\begin{bmatrix} -1 & 2 \\ 1 & 1 \end{bmatrix}$

(c) $T\left(\begin{bmatrix} x_1 \\ x_2 \end{bmatrix}\right) = \begin{bmatrix} -x_1 + 2x_2 \\ x_1 + x_2 \end{bmatrix}$

49. (a) $\begin{bmatrix} 3 & 2 \\ -1 & 0 \end{bmatrix}$ (b) $\begin{bmatrix} -8 & -6 \\ 15 & 11 \end{bmatrix}$

(c) $\begin{bmatrix} -8x_1 - 6x_2 \\ 15x_1 + 11x_2 \end{bmatrix}$

51. (a) $\begin{bmatrix} 0 & 0 & 1 \\ -1 & 0 & 2 \\ 0 & 2 & 0 \end{bmatrix}$ (b) $\begin{bmatrix} -1 & 2 & 1 \\ 0 & 2 & -1 \\ 1 & 0 & -1 \end{bmatrix}$

(c) $T\left(\begin{bmatrix} x_1 \\ x_2 \\ x_3 \end{bmatrix}\right) = \begin{bmatrix} -x_1 + 2x_2 + x_3 \\ 2x_2 - x_3 \\ x_1 - x_3 \end{bmatrix}$

53. (a) $\begin{bmatrix} 0 & -1 & 2 \\ 3 & 0 & 5 \\ -2 & 4 & 0 \end{bmatrix}$ (b) $\begin{bmatrix} 2 & -7 & -1 \\ -8 & -8 & 11 \\ -4 & -9 & 6 \end{bmatrix}$

(b) $\begin{bmatrix} 2x_1 - 7x_2 - x_3 \\ -8x_1 - 8x_2 + 11x_3 \\ -4x_1 - 9x_2 + 6x_3 \end{bmatrix}$

55. $9\mathbf{b}_1 + 12\mathbf{b}_2$ 57. $-3\mathbf{b}_1 - 17\mathbf{b}_2$

59. $-2\mathbf{b}_1 - 10\mathbf{b}_2 + 15\mathbf{b}_3$ 61. $8\mathbf{b}_1 + 5\mathbf{b}_2 - 16\mathbf{b}_3 - \mathbf{b}_4$

63. I_n 65. $T\left(\begin{bmatrix} x_1 \\ x_2 \end{bmatrix}\right) = \begin{bmatrix} .8x_1 + .6x_2 \\ .6x_1 - .8x_2 \end{bmatrix}$

67. $T\left(\begin{bmatrix} x_1 \\ x_2 \end{bmatrix}\right) = \begin{bmatrix} -.6x_1 - .8x_2 \\ -.8x_1 + .6x_2 \end{bmatrix}$

69. $U\left(\begin{bmatrix} x_1 \\ x_2 \end{bmatrix}\right) = \begin{bmatrix} .5x_1 + .5x_2 \\ .5x_1 + .5x_2 \end{bmatrix}$

71. $U\left(\begin{bmatrix} x_1 \\ x_2 \end{bmatrix}\right) = \begin{bmatrix} .1x_1 - .3x_2 \\ -.3x_1 + .9x_2 \end{bmatrix}$

73. (a) $T_W\left(\begin{bmatrix} -2 \\ 1 \\ 0 \end{bmatrix}\right) = \begin{bmatrix} -2 \\ 1 \\ 0 \end{bmatrix}$,

$T_W\left(\begin{bmatrix} 3 \\ 0 \\ 1 \end{bmatrix}\right) = \begin{bmatrix} 3 \\ 0 \\ 1 \end{bmatrix}$, and

$T_W\left(\begin{bmatrix} 1 \\ 2 \\ -3 \end{bmatrix}\right) = \begin{bmatrix} -1 \\ -2 \\ 3 \end{bmatrix}$

(c) $\begin{bmatrix} 1 & 0 & 0 \\ 0 & 1 & 0 \\ 0 & 0 & -1 \end{bmatrix}$

(d) $\frac{1}{7}\begin{bmatrix} 6 & -2 & 3 \\ -2 & 3 & 6 \\ 3 & 6 & -2 \end{bmatrix}$

(e) $T\left(\begin{bmatrix} x_1 \\ x_2 \\ x_3 \end{bmatrix}\right) = \frac{1}{7}\begin{bmatrix} 6x_1 - 2x_2 + 3x_3 \\ -2x_1 + 3x_2 + 6x_3 \\ 3x_1 + 6x_2 - 2x_3 \end{bmatrix}$

75. $T\left(\begin{bmatrix} x_1 \\ x_2 \\ x_3 \end{bmatrix}\right) = \frac{1}{13}\begin{bmatrix} 12x_1 + 4x_2 - 3x_3 \\ 4x_1 - 3x_2 + 12x_3 \\ -3x_1 + 12x_2 + 4x_3 \end{bmatrix}$

77. $T\left(\begin{bmatrix} x_1 \\ x_2 \\ x_3 \end{bmatrix}\right) = \frac{1}{41}\begin{bmatrix} 39x_1 - 12x_2 + 4x_3 \\ -12x_1 - 31x_2 + 24x_3 \\ 4x_1 + 24x_2 + 33x_3 \end{bmatrix}$

79. $T\left(\begin{bmatrix} x_1 \\ x_2 \\ x_3 \end{bmatrix}\right) = \frac{1}{21}\begin{bmatrix} 19x_1 + 4x_2 + 8x_3 \\ 4x_1 + 13x_2 - 16x_3 \\ 8x_1 - 16x_2 - 11x_3 \end{bmatrix}$

81. (a) $U_W\left(\begin{bmatrix} -2 \\ 1 \\ 0 \end{bmatrix}\right) = \begin{bmatrix} -2 \\ 1 \\ 0 \end{bmatrix}$,

$U_W\left(\begin{bmatrix} 3 \\ 0 \\ 1 \end{bmatrix}\right) = \begin{bmatrix} 3 \\ 0 \\ 1 \end{bmatrix}$, and

$U_W\left(\begin{bmatrix} 1 \\ 2 \\ -3 \end{bmatrix}\right) = \begin{bmatrix} 0 \\ 0 \\ 0 \end{bmatrix}$

(b) $\begin{bmatrix} 1 & 0 & 0 \\ 0 & 1 & 0 \\ 0 & 0 & 0 \end{bmatrix}$

(c) $\frac{1}{14}\begin{bmatrix} 13 & -2 & 3 \\ -2 & 10 & 6 \\ 3 & 6 & 5 \end{bmatrix}$

(d) $U\left(\begin{bmatrix} x_1 \\ x_2 \\ x_3 \end{bmatrix}\right) = \frac{1}{14}\begin{bmatrix} 13x_1 - 2x_2 + 3x_3 \\ -2x_1 + 10x_2 + 6x_3 \\ 3x_1 + 6x_2 + 5x_3 \end{bmatrix}$

83. $U\left(\begin{bmatrix} x_1 \\ x_2 \\ x_3 \end{bmatrix}\right) = \frac{1}{30}\begin{bmatrix} 29x_1 + 2x_2 - 5x_3 \\ 2x_1 + 26x_2 + 10x_3 \\ -5x_1 + 10x_2 + 5x_3 \end{bmatrix}$

85. $U\left(\begin{bmatrix} x_1 \\ x_2 \\ x_3 \end{bmatrix}\right) = \frac{1}{35}\begin{bmatrix} 34x_1 + 3x_2 + 5x_3 \\ 3x_1 + 26x_2 - 15x_3 \\ 5x_1 - 15x_2 + 10x_3 \end{bmatrix}$

87. $U\left(\begin{bmatrix} x_1 \\ x_2 \\ x_3 \end{bmatrix}\right) = \frac{1}{75}\begin{bmatrix} 74x_1 + 5x_2 - 7x_3 \\ 5x_1 + 50x_2 + 35x_3 \\ -7x_1 + 35x_2 + 26x_3 \end{bmatrix}$

101. (b) $\begin{bmatrix} -2 & 14 & -16 \\ 2 & -8 & 10 \end{bmatrix}$

103. (a) $\begin{bmatrix} 11 & 5 & 13 & 1 \\ -2 & 0 & -5 & -3 \\ -8 & -3 & -9 & 0 \\ 6 & 1 & 8 & 1 \end{bmatrix}$, $\begin{bmatrix} -5 & 10 & -38 & -31 \\ 2 & -3 & 9 & 6 \\ 6 & -10 & 27 & 17 \\ -4 & 7 & -25 & -19 \end{bmatrix}$, $\begin{bmatrix} 43 & 58 & -21 & -66 \\ -8 & -11 & 8 & 17 \\ -28 & -34 & 21 & 53 \\ 28 & 36 & -14 & -44 \end{bmatrix}$

105. (a) $\begin{bmatrix} 0 & 0 & 0 & 1 \\ 1 & 0 & 0 & 0 \\ 0 & 1 & 0 & 0 \\ 0 & 0 & 1 & 0 \end{bmatrix}$,

$T\left(\begin{bmatrix} x_1 \\ x_2 \\ x_3 \\ x_4 \end{bmatrix}\right) = \begin{bmatrix} 8x_1 - 4x_2 + 3x_3 + x_4 \\ -11x_1 + 7x_2 - 4x_3 - 2x_4 \\ -35x_1 + 20x_2 - 13x_3 - 5x_4 \\ -9x_1 + 4x_2 - 3x_3 - 2x_4 \end{bmatrix}$

107. $[T^{-1}]_{\mathcal{B}} = ([T]_{\mathcal{B}})^{-1}$

Chapter 4 Review Exercises

1. T	2. T	3. F	4. F	5. F
6. T	7. T	8. F	9. F	10. F
11. T	12. T	13. T	14. T	15. T
16. T	17. F	18. T	19. F	20. F
21. T	22. T	23. F	24. T	25. T

27. (a) There are at most k vectors in a linearly independent subset of V.

(c) There are at least k vectors in a generating set for V.

29. No, $\begin{bmatrix} -1 \\ 0 \\ 1 \\ 0 \end{bmatrix}$ and $\begin{bmatrix} 1 \\ 0 \\ 1 \\ 0 \end{bmatrix}$ are in the set, but their sum is not.

31. (a) $\left\{\begin{bmatrix} -3 \\ 2 \\ 1 \end{bmatrix}\right\}$ (b) $\left\{\begin{bmatrix} 1 \\ -1 \\ 2 \\ 1 \end{bmatrix}, \begin{bmatrix} 2 \\ -1 \\ 1 \\ 4 \end{bmatrix}\right\}$

(c) $\left\{\begin{bmatrix} 1 \\ 0 \\ 3 \end{bmatrix}, \begin{bmatrix} 0 \\ 1 \\ -2 \end{bmatrix}\right\}$

33. (a) $\left\{\begin{bmatrix} 0 \\ -1 \\ 1 \\ 2 \end{bmatrix}, \begin{bmatrix} 1 \\ 3 \\ -4 \\ -1 \end{bmatrix}, \begin{bmatrix} -2 \\ 1 \\ 1 \\ 3 \end{bmatrix}\right\}$

(b) The null space of T is $\{\mathbf{0}\}$.

35. The given set is a linearly independent subset of the null space that contains 2 vectors.

37. (b) $\begin{bmatrix} -1 \\ -2 \\ 5 \end{bmatrix}$ (c) $\begin{bmatrix} 1 \\ -8 \\ -6 \end{bmatrix}$

39. (a) $\begin{bmatrix} -17 & 1 \\ -10 & 1 \end{bmatrix}$

(b) $\begin{bmatrix} -7 & -5 \\ -14 & -9 \end{bmatrix}$

(c) $T\left(\begin{bmatrix} x_1 \\ x_2 \end{bmatrix}\right) = \begin{bmatrix} -7x_1 - 5x_2 \\ -14x_1 - 9x_2 \end{bmatrix}$

41. $T\left(\begin{bmatrix} x_1 \\ x_2 \\ x_3 \end{bmatrix}\right) = \begin{bmatrix} x_1 + 6x_2 - 5x_3 \\ -4x_1 + 4x_2 + 5x_3 \\ -x_1 + 3x_2 + x_3 \end{bmatrix}$

43. $21x^2 - 10\sqrt{3}xy + 31y^2 = 144$

45. $50(x')^2 + 8(y')^2 = 200$

47. $T\left(\begin{bmatrix} x_1 \\ x_2 \end{bmatrix}\right) = \dfrac{1}{13}\begin{bmatrix} -5x_1 - 12x_2 \\ -12x_1 + 5x_2 \end{bmatrix}$

53. $\left\{\begin{bmatrix} 1 \\ -1 \\ 2 \end{bmatrix}, \begin{bmatrix} -2 \\ -1 \\ 1 \end{bmatrix}\right\}$

Chapter 4 MATLAB Exercises

1. (a) yes (b) no (c) no (d) yes
2. (a) yes (b) yes (c) no (d) yes

3. (a) $\left\{\begin{bmatrix} 1.2 \\ -1.1 \\ 2.3 \\ -1.2 \\ 1.1 \\ 0.1 \end{bmatrix}, \begin{bmatrix} 2.3 \\ 3.2 \\ 1.1 \\ 1.4 \\ -4.1 \\ -2.1 \end{bmatrix}, \begin{bmatrix} 1.2 \\ -3.1 \\ 2.1 \\ -1.4 \\ 5.1 \\ 1.2 \end{bmatrix}\right\}$

(b) $\left\{\begin{bmatrix} 1.2 \\ -1.1 \\ 2.3 \\ -1.2 \\ 1.1 \\ 0.1 \end{bmatrix}, \begin{bmatrix} 2.3 \\ 3.2 \\ 1.1 \\ 1.4 \\ -4.1 \\ -2.1 \end{bmatrix}, \begin{bmatrix} 1.2 \\ -3.1 \\ 2.1 \\ -1.4 \\ 5.1 \\ 1.2 \end{bmatrix}, \begin{bmatrix} 1 \\ 0 \\ 0 \\ 0 \\ 0 \\ 0 \end{bmatrix}, \mathbf{e}_1, \mathbf{e}_2, \mathbf{e}_3\right\}$

(c) $\left\{\begin{bmatrix} -1 \\ -1 \\ -1 \\ 1 \\ 0 \end{bmatrix}, \begin{bmatrix} 0 \\ 2 \\ 1 \\ 0 \\ 1 \end{bmatrix}\right\}$

(d) $\left\{\begin{bmatrix} 1.2 \\ 2.3 \\ 1.2 \\ 4.7 \\ -5.8 \end{bmatrix}, \begin{bmatrix} -1.1 \\ 3.2 \\ -3.1 \\ -1.0 \\ -3.3 \end{bmatrix}, \begin{bmatrix} 2.3 \\ 1.1 \\ 2.1 \\ 5.5 \\ -4.3 \end{bmatrix}\right\}$

4. (a) $\left\{\begin{bmatrix} 1.3 \\ 2.2 \\ -1.2 \\ 4.0 \\ 1.7 \\ -3.1 \end{bmatrix}, \begin{bmatrix} 2.1 \\ -1.4 \\ 1.3 \\ 2.7 \\ 4.1 \\ 1.0 \end{bmatrix}, \begin{bmatrix} 2.9 \\ -3.0 \\ 3.8 \\ 1.4 \\ 6.5 \\ 5.1 \end{bmatrix}\right\}$

(b) $\left\{\begin{bmatrix} 1.3 \\ 2.2 \\ -1.2 \\ 4.0 \\ 1.7 \\ -3.1 \end{bmatrix}, \begin{bmatrix} 2.1 \\ -1.4 \\ 1.3 \\ 2.7 \\ 4.1 \\ 1.0 \end{bmatrix}, \begin{bmatrix} 2.9 \\ -3.0 \\ 3.8 \\ 1.4 \\ 6.5 \\ 5.1 \end{bmatrix}, \mathbf{e}_1, \mathbf{e}_3, \mathbf{e}_4\right\}$

(c) $\left\{\begin{bmatrix} -2 \\ 1 \\ 1 \\ 0 \end{bmatrix}\right\}$

(d) $\left\{\begin{bmatrix} 1 \\ 0 \\ 2 \\ 0 \end{bmatrix}, \begin{bmatrix} 0 \\ 1 \\ -1 \\ 0 \end{bmatrix}, \begin{bmatrix} 0 \\ 0 \\ 0 \\ 1 \end{bmatrix}\right\}$

5. For simplicity, let $\mathbf{b}_i, 1 \le i \le 6$, denote the vectors in $\mathcal{B}$.

(a) $\mathcal{B}$ is a linearly independent set of 6 vectors from $\mathcal{R}^6$.

(b) (i) $2\mathbf{b}_1 - \mathbf{b}_2 - 3\mathbf{b}_3 + 2\mathbf{b}_5 - \mathbf{b}_6$
(ii) $\mathbf{b}_1 - \mathbf{b}_2 + \mathbf{b}_3 + 2\mathbf{b}_4 - 3\mathbf{b}_5 + \mathbf{b}_6$
(iii) $-3\mathbf{b}_2 + \mathbf{b}_3 + 2\mathbf{b}_4 - 4\mathbf{b}_5$

(c) (i) $\begin{bmatrix} 2 \\ -1 \\ -3 \\ 0 \\ 2 \\ -1 \end{bmatrix}$ (ii) $\begin{bmatrix} 1 \\ -1 \\ 1 \\ 2 \\ -3 \\ 1 \end{bmatrix}$ (iii) $\begin{bmatrix} 0 \\ -3 \\ 1 \\ 2 \\ -4 \\ 0 \end{bmatrix}$

6. $\begin{bmatrix} -47.6 & 0.6 & 3.4 & -44.6 & 23.5 \\ -30.9 & 1.4 & 2.1 & -28.9 & 12.5 \\ 22.2 & -0.2 & -1.8 & 21.2 & -10.5 \\ 0.7 & -1.2 & 1.7 & -0.3 & 0.0 \\ -38.5 & -1.0 & 4.5 & -38.5 & 21.5 \end{bmatrix}$

7. $\begin{bmatrix} -1 & 2 & 1 & 0 & -1 & -2 \\ -8 & -4 & -9 & 3 & 4 & -10 \\ 6 & 1 & 6 & 0 & -1 & 7 \\ 1 & 1 & -1 & -1 & -2 & 2 \end{bmatrix}$

9. $A = \begin{bmatrix} 1.00 & 0.00 & 0.00 & 0.75 & -0.50 \\ 0.00 & 0.00 & 0.00 & 0.00 & 0.00 \\ 0.00 & 0.00 & 1.00 & -0.75 & 0.50 \\ 0.00 & 0.00 & 0.00 & 0.00 & 0.00 \\ 0.00 & 0.00 & 0.00 & 0.00 & 0.00 \end{bmatrix}$

10. (b) $\left\{ \begin{bmatrix} -3 \\ 6 \\ 2 \\ 0 \end{bmatrix}, \begin{bmatrix} 0 \\ 1 \\ 0 \\ 1 \end{bmatrix} \right\}$

Chapter 5

Section 5.1

1. 6 3. 3 5. -2 7. -3

9. -4 11. 2 13. $\left\{ \begin{bmatrix} -1 \\ 1 \end{bmatrix} \right\}$ 15. $\left\{ \begin{bmatrix} -3 \\ 1 \end{bmatrix} \right\}$

17. $\left\{ \begin{bmatrix} -1 \\ 1 \\ 0 \end{bmatrix} \right\}$ 19. $\left\{ \begin{bmatrix} -2 \\ -1 \\ 1 \end{bmatrix} \right\}$

21. $\left\{ \begin{bmatrix} -1 \\ 3 \\ 0 \end{bmatrix}, \begin{bmatrix} 2 \\ 0 \\ 3 \end{bmatrix} \right\}$ 23. $\left\{ \begin{bmatrix} 1 \\ 1 \\ 0 \end{bmatrix}, \begin{bmatrix} 1 \\ 0 \\ 1 \end{bmatrix} \right\}$

25. 6 27. 4 29. -3 31. 5

33. $\left\{ \begin{bmatrix} 2 \\ 3 \end{bmatrix} \right\}$ 35. $\left\{ \begin{bmatrix} -2 \\ 3 \end{bmatrix} \right\}$

37. $\left\{ \begin{bmatrix} -1 \\ 1 \\ 0 \end{bmatrix}, \begin{bmatrix} -3 \\ 0 \\ 1 \end{bmatrix} \right\}$ 39. $\left\{ \begin{bmatrix} 1 \\ -2 \\ 2 \end{bmatrix} \right\}$

41. F 42. F 43. T 44. T 45. T
46. F 47. T 48. T 49. F 50. T
51. F 52. T 53. T 54. T 55. F
56. T 57. F 58. T 59. F 60. T

61. The only eigenvalue is 1; its eigenspace is $\mathcal{R}^n$.

65. Null A

71. Either $\mathbf{v} = \mathbf{0}$ or $\mathbf{v}$ is an eigenvector of A.

77. no

81. yes, $\begin{bmatrix} -1 \\ 1 \\ -2 \\ 1 \end{bmatrix}, \begin{bmatrix} 2 \\ 0 \\ 3 \\ 3 \end{bmatrix}, \begin{bmatrix} 1 \\ -1 \\ 2 \\ 0 \end{bmatrix}, \begin{bmatrix} 0 \\ -1 \\ 0 \\ 1 \end{bmatrix}$

Section 5.2

1. 5, $\left\{ \begin{bmatrix} -3 \\ 2 \end{bmatrix} \right\}$, 6, $\left\{ \begin{bmatrix} -1 \\ 1 \end{bmatrix} \right\}$

3. 0, $\left\{ \begin{bmatrix} 3 \\ 5 \end{bmatrix} \right\}$, -1, $\left\{ \begin{bmatrix} 2 \\ 3 \end{bmatrix} \right\}$

5. -3, $\left\{ \begin{bmatrix} 1 \\ 1 \\ 1 \end{bmatrix} \right\}$, 2, $\left\{ \begin{bmatrix} 1 \\ 0 \\ 1 \end{bmatrix} \right\}$

7. 6, $\left\{ \begin{bmatrix} 1 \\ -1 \\ 1 \end{bmatrix} \right\}$, -2, $\left\{ \begin{bmatrix} 1 \\ 2 \\ 0 \end{bmatrix}, \begin{bmatrix} 1 \\ 0 \\ 2 \end{bmatrix} \right\}$

9. -3, $\left\{ \begin{bmatrix} -1 \\ 1 \\ 1 \end{bmatrix} \right\}$, -2, $\left\{ \begin{bmatrix} -1 \\ 1 \\ 0 \end{bmatrix} \right\}$, 1, $\left\{ \begin{bmatrix} 1 \\ 0 \\ 1 \end{bmatrix} \right\}$

11. 3, $\left\{ \begin{bmatrix} 1 \\ 1 \\ 0 \\ 0 \end{bmatrix} \right\}$, 4, $\left\{ \begin{bmatrix} 0 \\ 1 \\ 0 \\ 1 \end{bmatrix} \right\}$, -1, $\left\{ \begin{bmatrix} 0 \\ 1 \\ 1 \\ 0 \end{bmatrix}, \begin{bmatrix} -1 \\ 1 \\ 0 \\ 1 \end{bmatrix} \right\}$

13. -4, $\left\{ \begin{bmatrix} -3 \\ 5 \end{bmatrix} \right\}$, 1, $\left\{ \begin{bmatrix} 1 \\ 0 \end{bmatrix} \right\}$

15. 3, $\left\{ \begin{bmatrix} -2 \\ 3 \end{bmatrix} \right\}$, 5, $\left\{ \begin{bmatrix} -1 \\ 2 \end{bmatrix} \right\}$

17. -3, $\left\{ \begin{bmatrix} 1 \\ 0 \\ 1 \end{bmatrix} \right\}$, 1, $\left\{ \begin{bmatrix} 1 \\ 0 \\ 2 \end{bmatrix} \right\}$

19. -1, $\left\{ \begin{bmatrix} 0 \\ 0 \\ 1 \end{bmatrix} \right\}$, 5, $\left\{ \begin{bmatrix} 0 \\ -3 \\ 1 \end{bmatrix} \right\}$

21. -6, $\left\{ \begin{bmatrix} -1 \\ 1 \\ 1 \end{bmatrix} \right\}$, -2, $\left\{ \begin{bmatrix} 1 \\ 1 \\ 1 \end{bmatrix} \right\}$, 4, $\left\{ \begin{bmatrix} 0 \\ 1 \\ 0 \end{bmatrix} \right\}$

23. -1, $\left\{ \begin{bmatrix} 1 \\ 0 \\ 0 \\ 0 \end{bmatrix} \right\}$, 1, $\left\{ \begin{bmatrix} -1 \\ 1 \\ 0 \\ 0 \end{bmatrix} \right\}$, -2, $\left\{ \begin{bmatrix} -1 \\ -2 \\ 3 \\ 0 \end{bmatrix} \right\}$,
2, $\left\{ \begin{bmatrix} 7 \\ -2 \\ -1 \\ 4 \end{bmatrix} \right\}$

25. 5, $\left\{ \begin{bmatrix} 1 \\ 1 \end{bmatrix} \right\}$, 7, $\left\{ \begin{bmatrix} 3 \\ 4 \end{bmatrix} \right\}$

27. 2, $\left\{ \begin{bmatrix} -2 \\ 1 \end{bmatrix} \right\}$, 6, $\left\{ \begin{bmatrix} -3 \\ 2 \end{bmatrix} \right\}$

29. -2, $\left\{ \begin{bmatrix} 1 \\ 1 \\ 0 \end{bmatrix} \right\}$, 4, $\left\{ \begin{bmatrix} 1 \\ 0 \\ 1 \end{bmatrix} \right\}$

31. 1, $\left\{ \begin{bmatrix} 1 \\ 2 \\ 3 \end{bmatrix} \right\}$, 2, $\left\{ \begin{bmatrix} -2 \\ 1 \\ 0 \end{bmatrix}, \begin{bmatrix} 2 \\ 0 \\ 1 \end{bmatrix} \right\}$

33. $-3, \left\{\begin{bmatrix}1\\1\end{bmatrix}\right\}, -2, \left\{\begin{bmatrix}1\\2\end{bmatrix}\right\}$

35. $5, \left\{\begin{bmatrix}2\\5\end{bmatrix}\right\}, 4, \left\{\begin{bmatrix}1\\2\end{bmatrix}\right\}$

37. $-3, \left\{\begin{bmatrix}1\\1\\0\end{bmatrix}\right\}, 2, \left\{\begin{bmatrix}0\\0\\1\end{bmatrix}\right\}$

39. $-3, \left\{\begin{bmatrix}1\\2\\3\end{bmatrix}\right\}, 1, \left\{\begin{bmatrix}0\\1\\0\end{bmatrix}, \begin{bmatrix}0\\0\\1\end{bmatrix}\right\}$

45. $1-2i, \left\{\begin{bmatrix}1\\-2\end{bmatrix}\right\}, 1+2i, \left\{\begin{bmatrix}-1\\3\end{bmatrix}\right\}$

47. $8-12i, \left\{\begin{bmatrix}1+4i\\1\end{bmatrix}\right\}, 8+12i, \left\{\begin{bmatrix}1-4i\\1\end{bmatrix}\right\}$

49. $2i, \left\{\begin{bmatrix}1\\0\\0\end{bmatrix}\right\}, 4, \left\{\begin{bmatrix}i\\2\\0\end{bmatrix}\right\}, 1, \left\{\begin{bmatrix}2\\1\\i\end{bmatrix}\right\}$

51. $i, \left\{\begin{bmatrix}1\\0\\0\end{bmatrix}\right\}, 1, \left\{\begin{bmatrix}0\\1\\0\end{bmatrix}\right\}, 2, \left\{\begin{bmatrix}1\\1\\2\end{bmatrix}\right\}$

53. F 54. T 55. T 56. F 57. F
58. F 59. F 60. T 61. F 62. T
63. F 64. F 65. T 66. F 67. T
68. F 69. T 70. T 71. F 72. T

73. c is not an eigenvalue of A.

77. (a) $(t-5)^3(t+9)$
(b) $(t-5)^3(t+9)$, $(t-5)^2(t+9)^2$, $(t-5)(t+9)^3$
(c) $(t-5)^2(t+9)^2$, $(t-5)^3(t+9)$

81. (a) $\left\{\begin{bmatrix}-1\\1\end{bmatrix}\right\}, \left\{\begin{bmatrix}-2\\1\end{bmatrix}\right\},$
(b) $\left\{\begin{bmatrix}-1\\1\end{bmatrix}\right\}, \left\{\begin{bmatrix}-2\\1\end{bmatrix}\right\}$
(c) $\left\{\begin{bmatrix}-1\\1\end{bmatrix}\right\}, \left\{\begin{bmatrix}-2\\1\end{bmatrix}\right\}$
(d) $\mathbf{v}$ is an eigenvector of B if and only if $\mathbf{v}$ is an eigenvector of cB.
(e) λ is an eigenvalue of B if and only if $c\lambda$ is an eigenvalue of cB.

83. (a) $(t-6)(t-7)$
(b) The characteristic polynomials of B and B^T are equal.
(c) The eigenvalues of B and B^T are the same.
(d) no

87. $-t^3 + \frac{23}{15}t^2 - \frac{127}{720}t + \frac{1}{2160}$

89. $\begin{bmatrix}0&0&0&5\\1&0&0&-7\\0&1&0&-23\\0&0&1&11\end{bmatrix}$

91. (a) $3, \begin{bmatrix}1\\1\end{bmatrix}, -0.5, \begin{bmatrix}1\\2\end{bmatrix}$
(b) $\frac{1}{3}, \begin{bmatrix}1\\1\end{bmatrix}, -2, \begin{bmatrix}1\\2\end{bmatrix}$

Section 5.3

1. $P = \begin{bmatrix}-2&-3\\1&1\end{bmatrix}, D = \begin{bmatrix}4&0\\0&5\end{bmatrix}$

3. The eigenspace corresponding to 2 is 1-dimensional.

5. $P = \begin{bmatrix}0&-2&-1\\1&3&1\\1&2&1\end{bmatrix}, D = \begin{bmatrix}-5&0&0\\0&2&0\\0&0&3\end{bmatrix}$

7. There is only one real eigenvalue, and its multiplicity is one.

9. $P = \begin{bmatrix}-1&-1&1\\4&1&0\\2&0&1\end{bmatrix}, D = \begin{bmatrix}5&0&0\\0&3&0\\0&0&3\end{bmatrix}$

11. $P = \begin{bmatrix}0&-1&-1&1\\0&1&0&0\\1&0&1&0\\0&0&0&1\end{bmatrix}, D = \begin{bmatrix}4&0&0&0\\0&-1&0&0\\0&0&-1&0\\0&0&0&-1\end{bmatrix}$

13. The eigenspace corresponding to 1 is 1-dimensional.

15. $P = \begin{bmatrix}-2&-3\\1&2\end{bmatrix}, D = \begin{bmatrix}3&0\\0&2\end{bmatrix}$

17. $P = \begin{bmatrix}1&-1&-1\\0&1&1\\0&0&5\end{bmatrix}, D = \begin{bmatrix}-1&0&0\\0&-3&0\\0&0&2\end{bmatrix}$

19. The eigenspace corresponding to 0 is 1-dimensional.

21. $P = \begin{bmatrix}-i&i\\1&1\end{bmatrix}, D = \begin{bmatrix}2-i&0\\0&2+i\end{bmatrix}$

23. $P = \begin{bmatrix}-2i&2i\\1&1\end{bmatrix}, D = \begin{bmatrix}1-3i&0\\0&1+i\end{bmatrix}$

25. $P = \begin{bmatrix}-1&-1&1\\2+i&2-i&-1\\1&1&1\end{bmatrix}, D = \begin{bmatrix}1+i&0&0\\0&1-i&0\\0&0&2\end{bmatrix}$

27. $P = \begin{bmatrix}1&0&0\\0&1&i\\0&0&1\end{bmatrix}, D = \begin{bmatrix}2i&0&0\\0&1&0\\0&0&0\end{bmatrix}$

29. F 30. T 31. T 32. T 33. F

34. F 35. F 36. T 37. F 38. F
39. T 40. F 41. F 42. T 43. T
44. F 45. F 46. T 47. F 48. F

51. (a) the eigenspace corresponding to -1 is 2-dimensional
(b) the eigenspace corresponding to -1 is 1-dimensional

53. (a) the eigenspace corresponding to -3 is 4-dimensional
(b) the eigenspace corresponding to -3 is not 4-dimensional

55. (a) $-(t-4)^2(t-5)(t-8)^2$
(b) There is insufficient information because the dimensions of W_1 and W_3 are not given. Therefore the multiplicities of the eigenvalues 4 and 8 are not determined.
(c) $-(t-4)(t-5)^2(t-8)^2$

57. $\begin{bmatrix} -4^k+2\cdot 3^k & 2\cdot 4^k-2\cdot 3^k \\ -4^k+3^k & 2\cdot 4^k-3^k \end{bmatrix}$

59. $\begin{bmatrix} -2\cdot 2^k+3\cdot 3^k & -6\cdot 2^k+6\cdot 3^k \\ 2^k-3^k & 3\cdot 2^k-2\cdot 3^k \end{bmatrix}$

61. $\begin{bmatrix} -5^k+2 & -2\cdot 5^k+2 & 0 \\ 5^k-1 & 2\cdot 5^k-1 & 0 \\ 0 & 0 & 5^k \end{bmatrix}$

63. 3 65. all c 67. no c

69. -2 and -1 71. 2

73. $\begin{bmatrix} -7 & 4 \\ -12 & 9 \end{bmatrix}$ 75. $\begin{bmatrix} -1 & 5 & -4 \\ 0 & -2 & 0 \\ 2 & -5 & 5 \end{bmatrix}$

77. $\begin{bmatrix} 0 & 0 \\ 0 & 1 \end{bmatrix}$ and $\begin{bmatrix} 0 & -1 \\ 0 & -1 \end{bmatrix}$ 89. (c) $\lambda_1\lambda_2\cdots\lambda_n$

91. $P=\begin{bmatrix} -3 & -1 & -8 & -1 \\ -1 & -1 & -1 & -2 \\ 2 & 0 & 3 & 0 \\ 0 & 2 & 0 & 3 \end{bmatrix}$,

$D=\begin{bmatrix} -1 & 0 & 0 & 0 \\ 0 & -1 & 0 & 0 \\ 0 & 0 & -2 & 0 \\ 0 & 0 & 0 & -2 \end{bmatrix}$

93. The eigenspace corresponding to 1 has dimension 2.

Section 5.4

1. $\begin{bmatrix} 0 & 0 & 2 \\ 0 & 3 & 0 \\ 4 & 0 & 0 \end{bmatrix}$, no 3. $\begin{bmatrix} 2 & 1 & 0 \\ 0 & 2 & 0 \\ 0 & 0 & 1 \end{bmatrix}$, no

5. $\begin{bmatrix} 0 & 0 & 3 \\ 0 & -2 & 0 \\ -4 & 0 & 0 \end{bmatrix}$, no 7. $\begin{bmatrix} 2 & 0 & 0 \\ 0 & -1 & 0 \\ 0 & 0 & -3 \end{bmatrix}$, yes

9. There are no real eigenvalues.

11. $\left\{\begin{bmatrix} 1 \\ 2 \end{bmatrix}, \begin{bmatrix} 1 \\ 1 \end{bmatrix}\right\}$ 13. $\left\{\begin{bmatrix} 1 \\ -1 \\ 1 \end{bmatrix}, \begin{bmatrix} 0 \\ -1 \\ 1 \end{bmatrix}, \begin{bmatrix} 0 \\ 0 \\ 1 \end{bmatrix}\right\}$

15. The eigenspace corresponding to -1 is 1-dimensional.

17. $\left\{\begin{bmatrix} 1 \\ 1 \\ 0 \end{bmatrix}, \begin{bmatrix} 0 \\ 1 \\ 1 \end{bmatrix}, \begin{bmatrix} 1 \\ 0 \\ 0 \end{bmatrix}\right\}$

19. $\left\{\begin{bmatrix} 1 \\ 0 \\ 1 \\ 0 \end{bmatrix}, \begin{bmatrix} -1 \\ 2 \\ 0 \\ 0 \end{bmatrix}, \begin{bmatrix} 1 \\ 0 \\ 2 \\ 0 \end{bmatrix}, \begin{bmatrix} -1 \\ 0 \\ 0 \\ 2 \end{bmatrix}\right\}$

21. T has no real eigenvalues.

23. $\left\{\begin{bmatrix} 1 \\ 1 \end{bmatrix}, \begin{bmatrix} -3 \\ 4 \end{bmatrix}\right\}$ 25. $\left\{\begin{bmatrix} 0 \\ 1 \\ 0 \end{bmatrix}, \begin{bmatrix} -1 \\ 0 \\ 1 \end{bmatrix}, \begin{bmatrix} 0 \\ 1 \\ 1 \end{bmatrix}\right\}$

27. The eigenspace corresponding to 1 is 2-dimensional.

29. F 30. F 31. T 32. F 33. F
34. T 35. T 36. T 37. T 38. F
39. T 40. F 41. F 42. F 43. T
44. F 45. T 46. T 47. T 48. F

49. $c=7$ 51. all scalars c 53. -3 and -1

55. all scalars c 57. no scalars c

59. $T_W\left(\begin{bmatrix} x_1 \\ x_2 \\ x_3 \end{bmatrix}\right) = \frac{1}{3}\begin{bmatrix} x_1-2x_2-2x_3 \\ -2x_1+x_2-2x_3 \\ -2x_1-2x_2+x_3 \end{bmatrix}$

61. $T_W\left(\begin{bmatrix} x_1 \\ x_2 \\ x_3 \end{bmatrix}\right) = \frac{1}{3}\begin{bmatrix} 2x_1-2x_2+x_3 \\ -2x_1-x_2+2x_3 \\ x_1+2x_2+2x_3 \end{bmatrix}$

63. $T_W\left(\begin{bmatrix} x_1 \\ x_2 \\ x_3 \end{bmatrix}\right) = \frac{1}{90}\begin{bmatrix} 88x_1-16x_2+10x_3 \\ -16x_1-38x_2+80x_3 \\ 10x_1+80x_2+40x_3 \end{bmatrix}$

65. $2x+2y+z=0$

67. $U_W\left(\begin{bmatrix} x_1 \\ x_2 \\ x_3 \end{bmatrix}\right) = \frac{1}{3}\begin{bmatrix} 2x_2-x_2-x_3 \\ -x_1+2x_2-x_3 \\ -x_1-x_2+2x_3 \end{bmatrix}$

69. $U_W\left(\begin{bmatrix} x_1 \\ x_2 \\ x_3 \end{bmatrix}\right) = \frac{1}{6}\begin{bmatrix} 5x_1-2x_2+x_3 \\ -2x_1+2x_2+2x_3 \\ x_1+2x_2+5x_3 \end{bmatrix}$

71. $U_W\left(\begin{bmatrix}x_1\\x_2\\x_3\end{bmatrix}\right)=\frac{1}{90}\begin{bmatrix}89x_1-8x_2+5x_3\\-8x_1+26x_2+40x_3\\5x_1+40x_2+65x_3\end{bmatrix}$

73. $U_W\left(\begin{bmatrix}x_1\\x_2\\x_3\end{bmatrix}\right)=\frac{1}{9}\begin{bmatrix}5x_1-4x_2-2x_3\\-4x_1+5x_2-2x_3\\-2x_1-2x_2+8x_3\end{bmatrix}$

85. $\left\{\begin{bmatrix}2\\-2\\-4\\3\\0\end{bmatrix},\begin{bmatrix}-1\\1\\2\\0\\3\end{bmatrix},\begin{bmatrix}-1\\1\\-3\\2\\0\end{bmatrix},\begin{bmatrix}1\\1\\3\\0\\2\end{bmatrix},\begin{bmatrix}1\\0\\1\\0\\0\end{bmatrix}\right\}$

Section 5.5

1. F 2. F 3. T 4. F 5. T
6. T 7. T 8. F 9. F 10. T
11. F 12. T

13. no 15. yes 17. no 19. yes

21. $\begin{bmatrix}.75\\.25\end{bmatrix}$ 23. $\begin{bmatrix}.25\\.25\\.50\end{bmatrix}$ 25. $\frac{1}{6}\begin{bmatrix}1\\3\\2\end{bmatrix}$

27. $\frac{1}{29}\begin{bmatrix}3\\4\\10\\12\end{bmatrix}$

29. (a) $\begin{bmatrix}.25 & .5\\.75 & .5\end{bmatrix}$
(b) .375
(c) .6

31. (a) $\begin{bmatrix}.7 & .1 & .1\\.1 & .6 & .1\\.2 & .3 & .8\end{bmatrix}$
(b) .6
(c) .33
(d) .25 buy brand A, .20 buy brand B, and .55 buy brand C.

35. (a) $\begin{bmatrix}\frac{117}{195} & \frac{16}{123}\\\frac{78}{195} & \frac{107}{123}\end{bmatrix}$
(b) $\frac{107}{123}$
(c) about .809
(d) about .676
(e) about .245

37. (a) .05 (b) .1 (c) .3 (d) $\begin{bmatrix}.6\\.3\\.1\end{bmatrix}$

39. (a) $\mathbf{u}$
(b) 1 is an eigenvalue of A^T.
(d) 1 is an eigenvalue of A.

45. $\begin{aligned}y_1&=-ae^{-3t}+2be^{4t}\\y_2&=3ae^{-3t}+be^{4t}\end{aligned}$

47. $\begin{aligned}y_1&=-2ae^{-4t}-be^{-2t}\\y_2&=3ae^{-4t}+be^{-2t}\end{aligned}$

49. $\begin{aligned}y_1&=-ce^{2t}\\y_2&=-ae^{-t}+be^{2t}\\y_3&=ae^{-t}+ce^{2t}\end{aligned}$

51. $\begin{aligned}y_1&=ae^{-2t}+be^{-t}\\y_2&=-ae^{-2t}-ce^{2t}\\y_3&=2ae^{-2t}+2be^{-t}+ce^{2t}\end{aligned}$

53. $\begin{aligned}y_1&=10ae^{-t}+5e^{3t}\\y_2&=-20e^{-t}+10e^{3t}\end{aligned}$

55. $\begin{aligned}y_1&=-3e^{4t}+5e^{6t}\\y_2&=6e^{4t}-5e^{6t}\end{aligned}$

57. $\begin{aligned}y_1&=4e^{-t}+5e^{t}-9e^{2t}\\y_2&=5e^{t}-3e^{2t}\\y_3&=4e^{-t}-3e^{2t}\end{aligned}$

59. $\begin{aligned}y_1&=6e^{-t}-4e^{t}-6e^{-2t}\\y_2&=6e^{-t}-8e^{t}-3e^{-2t}\\y_3&=-6e^{-t}+12e^{t}-3e^{-2t}\end{aligned}$

61. $y=ae^{3t}+be^{-t}$

63. $y=3e^{-t}-2e^{t}+e^{2t}$

65. $y=e^{-t}(c\cos\sqrt{3}t+d\sin\sqrt{3}t)$

67. (a) $y_1=100e^{-2t}+800e^{t},\ y_2=100e^{-2t}+200e^{t}$
(b) 2188 and 557 at time 1, 5913 and 1480 at time 2, and 16069 and 4017 at time 3
(c) .25, no

71. $r_n=8(-3)^n$

73. $r_n=.6(-1)^n+.4(4)^n$, $r_6=1639$

75. $r_n=3(-3)^n+5(2^n)$, $r_6=2507$

77. $r_n=6(-1)^n+3(2^n)$, $r_6=198$

79. (a) $r_0=1,\ r_1=2,\ r_2=7,\ r_3=20$
(b) $r_n=2r_{n-1}+3r_{n-2}$
(c) $r_n=\left(\frac{3}{4}\right)3^n+\left(\frac{1}{4}\right)(-1)^n$

81. $\begin{bmatrix}r_n\\r_{n+1}\\r_{n+2}\end{bmatrix}=\begin{bmatrix}0 & 1 & 0\\0 & 0 & 1\\5 & -2 & 4\end{bmatrix}\begin{bmatrix}r_{n-1}\\r_n\\r_{n+1}\end{bmatrix}$

89.
$$\begin{aligned} y_1 &= -6e^{-0.8t} - 2e^{-0.1t} + e^{0.3t} + 8e^t \\ y_2 &= \qquad\qquad -2e^{-0.1t} + 2e^{0.3t} - 4e^t \\ y_3 &= 6e^{-0.8t} \qquad\qquad\qquad\qquad - 4e^t \\ y_4 &= -6e^{-0.8t} + 4e^{-0.1t} - 3e^{0.3t} + 8e^t. \end{aligned}$$

91. (a)
$$\begin{bmatrix} 0.0150 & 0.0150 & 0.1850 & 0.1 & 0.0150 & 0.0150 & 0.2275 & 0.0150 & 0.0150 & 0.1 \\ 0.2275 & 0.0150 & 0.0150 & 0.1 & 0.4400 & 0.0150 & 0.0150 & 0.0150 & 0.0150 & 0.1 \\ 0.0150 & 0.0150 & 0.0150 & 0.1 & 0.0150 & 0.0150 & 0.0150 & 0.0150 & 0.0150 & 0.1 \\ 0.0150 & 0.4400 & 0.1850 & 0.1 & 0.0150 & 0.0150 & 0.0150 & 0.4400 & 0.0150 & 0.1 \\ 0.0150 & 0.0150 & 0.1850 & 0.1 & 0.0150 & 0.4400 & 0.2275 & 0.0150 & 0.4400 & 0.1 \\ 0.2275 & 0.0150 & 0.0150 & 0.1 & 0.0150 & 0.0150 & 0.2275 & 0.0150 & 0.0150 & 0.1 \\ 0.0150 & 0.4400 & 0.0150 & 0.1 & 0.0150 & 0.0150 & 0.0150 & 0.0150 & 0.0150 & 0.1 \\ 0.2275 & 0.0150 & 0.1850 & 0.1 & 0.0150 & 0.0150 & 0.0150 & 0.4400 & 0.0150 & 0.1 \\ 0.2275 & 0.0150 & 0.0150 & 0.1 & 0.0150 & 0.4400 & 0.2275 & 0.0150 & 0.0150 & 0.1 \\ 0.0150 & 0.0150 & 0.1850 & 0.1 & 0.4400 & 0.0150 & 0.0150 & 0.0150 & 0.4400 & 0.1 \end{bmatrix}$$

(b) $\mathbf{v}^T = [0.0643 \;\; 0.1114 \;\; 0.0392 \;\; 0.1372 \;\; 0.1377 \;\; 0.0712 \;\; 0.0865 \;\; 0.1035 \;\; 0.1015 \;\; 0.1475]^T$, which results in the rankings 10, 5, 4, 2, 8, 9, 7, 6, 1, 3.

Chapter 5 Review Exercises

1. T 2. F 3. T 4. T 5. T
6. T 7. F 8. F 9. F 10. F
11. T 12. T 13. F 14. T 15. T
16. T 17. T

19. $\left\{\begin{bmatrix}-3\\2\end{bmatrix}\right\}$ for 1 and $\left\{\begin{bmatrix}-2\\1\end{bmatrix}\right\}$ for 2

21. $\left\{\begin{bmatrix}-1\\1\\0\end{bmatrix}\right\}$ for -2 and $\left\{\begin{bmatrix}0\\1\\0\end{bmatrix}\right\}$ for -1

23. $P = \begin{bmatrix}2 & 1\\1 & 3\end{bmatrix}$ and $D = \begin{bmatrix}2 & 0\\0 & 7\end{bmatrix}$

25. The eigenspace corresponding to -1 has dimension 1.

27. $\left\{\begin{bmatrix}-2\\1\end{bmatrix}, \begin{bmatrix}-1\\4\end{bmatrix}\right\}$

29. $\left\{\begin{bmatrix}0\\0\\1\end{bmatrix}, \begin{bmatrix}1\\1\\0\end{bmatrix}, \begin{bmatrix}-1\\0\\1\end{bmatrix},\right\}$

31. none

33. -2 and 2

35. $\begin{bmatrix}(-1)^{k+1} + 2^{k+1} & 2(-1)^k - 2^{k+1}\\ (-1)^{k+1} + 2^k & 2(-1)^k - 2^k\end{bmatrix}$

37. $\left\{\begin{bmatrix}-1\\0\\2\end{bmatrix}, \begin{bmatrix}-1\\1\\0\end{bmatrix}, \begin{bmatrix}-1\\0\\1\end{bmatrix},\right\}$

Chapter 5 MATLAB Exercises

1. (a) $P = \begin{bmatrix} 1.00 & 0.80 & 0.75 & 1.00 & 1.00 \\ -0.50 & -0.40 & -0.50 & 1.00 & -1.00 \\ 0.00 & -0.20 & -0.25 & 0.00 & -0.50 \\ 0.50 & 0.40 & 0.50 & 0.00 & 0.00 \\ 1.00 & 1.00 & 1.00 & 1.00 & 1.00 \end{bmatrix}$

$$D = \begin{bmatrix} 3 & 0 & 0 & 0 & 0 \\ 0 & 1 & 0 & 0 & 0 \\ 0 & 0 & 0 & 0 & 0 \\ 0 & 0 & 0 & -1 & 0 \\ 0 & 0 & 0 & 0 & 2 \end{bmatrix}$$

(b) rank $(A - \frac{1}{2}I_4) = 3$, but eigenvalue $\frac{1}{2}$ has multiplicity 2

(c) $P = \begin{bmatrix} -1.25 & -1.00 & -0.50 & -1.00 \\ -0.25 & -0.50 & 0.50 & 0.00 \\ 0.75 & 0.50 & 1.00 & 0.00 \\ 1.00 & 1.00 & 0.00 & 1.00 \end{bmatrix}$

$$D = \begin{bmatrix} -1 & 0 & 0 & 0 \\ 0 & 2 & 0 & 0 \\ 0 & 0 & 1 & 0 \\ 0 & 0 & 0 & 1 \end{bmatrix}$$

(d) rank $(A - 0I_5) = 4$, but eigenvalue 0 has multiplicity 2

2. The eigenvalues of B and C are the same. Let J be the matrix obtained from I_n by interchanging columns i and j. If $\mathbf{v}$ is an eigenvector of B with corresponding eigenvalue λ, then $J\mathbf{v}$ is an eigenvector of C with corresponding eigenvalue λ. Similarly, if $\mathbf{w}$ is an eigenvector of C with corresponding eigenvalue λ, then $J^{-1}\mathbf{w}$ is an eigenvector of B with corresponding eigenvalue λ.

3. $\begin{bmatrix} -9 & 20 & -5 & -8 & 2 \\ -5 & 11 & -2 & -4 & 1 \\ 0 & 0 & 2 & 0 & 0 \\ -10 & 16 & -4 & -5 & 2 \\ -27 & 60 & -15 & -24 & 6 \end{bmatrix}$

4. (a)
$$\begin{bmatrix} 0.0150 & 0.2983 & 0.0150 & 0.0150 & 0.2275 & 0.0150 & 0.2275 & 0.1 & 0.0150 & 0.4400 \\ 0.2275 & 0.0150 & 0.0150 & 0.0150 & 0.0150 & 0.0150 & 0.0150 & 0.1 & 0.4400 & 0.0150 \\ 0.0150 & 0.0150 & 0.0150 & 0.0150 & 0.0150 & 0.0150 & 0.2275 & 0.1 & 0.0150 & 0.4400 \\ 0.0150 & 0.0150 & 0.0150 & 0.0150 & 0.2275 & 0.2983 & 0.0150 & 0.1 & 0.0150 & 0.0150 \\ 0.2275 & 0.0150 & 0.0150 & 0.4400 & 0.0150 & 0.0150 & 0.2275 & 0.1 & 0.4400 & 0.0150 \\ 0.0150 & 0.0150 & 0.0150 & 0.4400 & 0.0150 & 0.0150 & 0.2275 & 0.1 & 0.0150 & 0.0150 \\ 0.2275 & 0.0150 & 0.4400 & 0.0150 & 0.2275 & 0.2983 & 0.0150 & 0.1 & 0.0150 & 0.0150 \\ 0.0150 & 0.2983 & 0.0150 & 0.0150 & 0.0150 & 0.2983 & 0.0150 & 0.1 & 0.0150 & 0.0150 \\ 0.0150 & 0.2983 & 0.0150 & 0.0150 & 0.2275 & 0.0150 & 0.0150 & 0.1 & 0.0150 & 0.0150 \\ 0.2275 & 0.0150 & 0.4400 & 0.0150 & 0.0150 & 0.0150 & 0.0150 & 0.1 & 0.0150 & 0.0150 \end{bmatrix}$$

(b) $\begin{bmatrix}0.1442\\0.0835\\0.0895\\0.0756\\0.1463\\0.0835\\0.1442\\0.0681\\0.0756\\0.0895\end{bmatrix}$, which results in the rankings 5, 1, 7, 3, 10, 2, 6, 4, 9, 8.

5. (a) A basis does not exist because the sum of the multiplicities of the eigenvalues of the standard matrix of T is not 4.

(b) $\left\{\begin{bmatrix}-1\\-1\\0\\1\\0\end{bmatrix}, \begin{bmatrix}0\\-1\\-1\\0\\1\end{bmatrix}, \begin{bmatrix}11\\10\\-3\\-13\\3\end{bmatrix}, \begin{bmatrix}15\\8\\-4\\-15\\1\end{bmatrix}, \begin{bmatrix}5\\10\\0\\-7\\1\end{bmatrix}\right\}$

6. (b) $T\left(\begin{bmatrix}x_1\\x_2\\x_3\\x_4\end{bmatrix}\right) =$
$\begin{bmatrix}1.5x_1 - 3.5x_2 + 1.0x_3 + 0.5x_4\\ -3.0x_1 + 3.6x_2 - 0.2x_3 + 1.0x_4\\ -16.5x_1 + 22.3x_2 - 3.6x_3 + 5.5x_4\\ 4.5x_1 - 8.3x_2 + 1.6x_3 + 1.5x_4\end{bmatrix}.$

(c) T is not diagonalizable.

7. (a) $T\left(\begin{bmatrix}x_1\\x_2\\x_3\\x_4\end{bmatrix}\right) =$
$\begin{bmatrix}11.5x_1 - 13.7x_2 + 3.4x_3 - 4.5x_4\\ 5.5x_1 - 5.9x_2 + 1.8x_3 - 2.5x_4\\ -6.0x_1 + 10.8x_2 - 1.6x_3\\ 5.0x_1 - 5.6x_2 + 1.2x_3 - 3.0x_4\end{bmatrix}$

(b) Answers are given correct to 4 places after the decimal point.

$\left\{\begin{bmatrix}3.0000\\2.0000\\1.0000\\1.0000\end{bmatrix}, \begin{bmatrix}2.0000\\1.0000\\-2.0000\\1.0000\end{bmatrix}, \begin{bmatrix}-2.4142\\-8.2426\\-24.7279\\1.0000\end{bmatrix}, \begin{bmatrix}0.4142\\0.2426\\0.7279\\1.0000\end{bmatrix}\right\}$

8. Answers are given correct to 4 places after the decimal point.

(b) $\begin{bmatrix}.2344\\.1934\\.1732\\.2325\\.1665\end{bmatrix},$

(c) $\begin{bmatrix}5.3\\5.2\\5.1\\6.1\\3.3\end{bmatrix}, \begin{bmatrix}5.8611\\4.8351\\4.3299\\5.8114\\4.1626\end{bmatrix}, \begin{bmatrix}5.8610\\4.8351\\4.3299\\5.8114\\4.1625\end{bmatrix}$

(d) $A^{100}\mathbf{p} \approx 25\mathbf{v}$

9. $r_n = (0.2)3^n - 2^n - (0.2)(-2)^n + 4 + 2(-1)^n$

Chapter 6

Section 6.1

1. $\|\mathbf{u}\| = \sqrt{34}$, $\|\mathbf{v}\| = \sqrt{20}$, and $d = \sqrt{58}$
3. $\|\mathbf{u}\| = \sqrt{2}$, $\|\mathbf{v}\| = \sqrt{5}$, and $d = \sqrt{5}$
5. $\|\mathbf{u}\| = \sqrt{11}$, $\|\mathbf{v}\| = \sqrt{5}$, and $d = \sqrt{14}$
7. $\|\mathbf{u}\| = \sqrt{7}$, $\|\mathbf{v}\| = \sqrt{15}$, and $d = \sqrt{26}$

9. 0, yes 11. 1, no 13. 0, yes 15. −2, no

17. $\|\mathbf{u}\|^2 = 20$, $\|\mathbf{v}\|^2 = 45$, $\|\mathbf{u}+\mathbf{v}\|^2 = 65$
19. $\|\mathbf{u}\|^2 = 13$, $\|\mathbf{v}\|^2 = 0$, $\|\mathbf{u}+\mathbf{v}\|^2 = 13$
21. $\|\mathbf{u}\|^2 = 14$, $\|\mathbf{v}\|^2 = 3$, $\|\mathbf{u}+\mathbf{v}\|^2 = 17$
23. $\|\mathbf{u}\|^2 = 14$, $\|\mathbf{v}\|^2 = 138$, $\|\mathbf{u}+\mathbf{v}\|^2 = 152$
25. $\|\mathbf{u}\| = \sqrt{13}$, $\|\mathbf{v}\| = \sqrt{44}$, $\|\mathbf{u}+\mathbf{v}\| = \sqrt{13}$
27. $\|\mathbf{u}\| = \sqrt{20}$, $\|\mathbf{v}\| = \sqrt{10}$, $\|\mathbf{u}+\mathbf{v}\| = \sqrt{50}$
29. $\|\mathbf{u}\| = \sqrt{21}$, $\|\mathbf{v}\| = \sqrt{11}$, $\|\mathbf{u}+\mathbf{v}\| = \sqrt{34}$
31. $\|\mathbf{u}\| = \sqrt{14}$, $\|\mathbf{v}\| = \sqrt{17}$, $\|\mathbf{u}+\mathbf{v}\| = \sqrt{53}$
33. $\|\mathbf{u}\| = \sqrt{13}$, $\|\mathbf{v}\| = \sqrt{34}$, $\mathbf{u}\cdot\mathbf{v} = -1$
35. $\|\mathbf{u}\| = \sqrt{17}$, $\|\mathbf{v}\| = 2$, $\mathbf{u}\cdot\mathbf{v} = -2$
37. $\|\mathbf{u}\| = \sqrt{41}$, $\|\mathbf{v}\| = \sqrt{18}$, $\mathbf{u}\cdot\mathbf{v} = 0$
39. $\|\mathbf{u}\| = \sqrt{21}$, $\|\mathbf{v}\| = \sqrt{6}$, $\mathbf{u}\cdot\mathbf{v} = 5$
41. $\mathbf{w} = \begin{bmatrix}5\\0\end{bmatrix}$ and $d = 0$
43. $\mathbf{w} = \frac{1}{2}\begin{bmatrix}-1\\1\end{bmatrix}$ and $d = \frac{7\sqrt{2}}{2}$
45. $\mathbf{w} = \begin{bmatrix}0.7\\2.1\end{bmatrix}$ and $d = 1.1\sqrt{10}$

49. −3 51. 11 53. 441 55. 21
57. 7 59. 49

61. T	62. F	63. F	64. F	65. F
66. T	67. T	68. T	69. T	70. F
71. F	72. T	73. T	74. T	75. T
76. F	77. T	78. F	79. T	80. T

99. 135° 101. 180° 103. 60° 105. 150°

121. $\overline{v} = \frac{26}{3} \approx 8.6667$ and $v^* = \frac{244}{26} \approx 9.3846$
123. $\overline{v} = v^* = 22$

Section 6.2

1. no 3. no 5. no 7. yes

9. (a) $\left\{\begin{bmatrix}1\\1\\1\end{bmatrix}, \begin{bmatrix}3\\-3\\0\end{bmatrix}\right\}$

(b) $\left\{\frac{1}{\sqrt{3}}\begin{bmatrix}1\\1\\1\end{bmatrix}, \frac{1}{\sqrt{2}}\begin{bmatrix}1\\-1\\0\end{bmatrix}\right\}$

11. (a) $\left\{\begin{bmatrix}1\\-2\\-1\end{bmatrix}, \begin{bmatrix}9\\3\\3\end{bmatrix}\right\}$

(b) $\left\{\frac{1}{\sqrt{6}}\begin{bmatrix}1\\-2\\-1\end{bmatrix}, \frac{1}{\sqrt{11}}\begin{bmatrix}3\\1\\1\end{bmatrix}\right\}$

13. (a) $\left\{ \begin{bmatrix} 0 \\ 1 \\ 1 \\ 1 \end{bmatrix}, \frac{1}{3}\begin{bmatrix} 3 \\ -2 \\ 1 \\ 1 \end{bmatrix}, \frac{1}{5}\begin{bmatrix} 3 \\ 3 \\ -4 \\ 1 \end{bmatrix} \right\}$

(b) $\left\{ \frac{1}{\sqrt{3}}\begin{bmatrix} 0 \\ 1 \\ 1 \\ 1 \end{bmatrix}, \frac{1}{\sqrt{15}}\begin{bmatrix} 3 \\ -2 \\ 1 \\ 1 \end{bmatrix}, \frac{1}{\sqrt{35}}\begin{bmatrix} 3 \\ 3 \\ -4 \\ 1 \end{bmatrix} \right\}$

15. (a) $\left\{ \begin{bmatrix} 1 \\ 0 \\ -1 \\ 1 \end{bmatrix}, \begin{bmatrix} 1 \\ 1 \\ 0 \\ -1 \end{bmatrix}, \begin{bmatrix} 2 \\ -1 \\ 3 \\ 1 \end{bmatrix} \right\}$

(b) $\left\{ \frac{1}{\sqrt{3}}\begin{bmatrix} 1 \\ 0 \\ -1 \\ 1 \end{bmatrix}, \frac{1}{\sqrt{3}}\begin{bmatrix} 1 \\ 1 \\ 0 \\ -1 \end{bmatrix}, \frac{1}{\sqrt{15}}\begin{bmatrix} 2 \\ -1 \\ 3 \\ 1 \end{bmatrix} \right\}$

17. $\mathbf{v} = 2\begin{bmatrix} 2 \\ 1 \end{bmatrix} + 3\begin{bmatrix} -1 \\ 2 \end{bmatrix}$

19. $\mathbf{v} = (-1)\begin{bmatrix} -1 \\ 3 \\ -2 \end{bmatrix} + (-2)\begin{bmatrix} -1 \\ 1 \\ 2 \end{bmatrix} + 4\begin{bmatrix} 1 \\ 1 \\ 1 \end{bmatrix}$

21. $\mathbf{v} = \frac{5}{2}\begin{bmatrix} 1 \\ 0 \\ 1 \end{bmatrix} + \frac{3}{6}\begin{bmatrix} 1 \\ 2 \\ -1 \end{bmatrix} + 0\begin{bmatrix} 1 \\ -1 \\ -1 \end{bmatrix}$

23. $\mathbf{v} = (-4)\begin{bmatrix} 1 \\ -1 \\ -1 \\ 1 \end{bmatrix} + 2\begin{bmatrix} 2 \\ 1 \\ 1 \\ 0 \end{bmatrix} + (-1)\begin{bmatrix} -1 \\ 1 \\ 1 \\ 3 \end{bmatrix}$

25. $Q = \begin{bmatrix} \frac{1}{\sqrt{3}} & \frac{1}{\sqrt{2}} \\ \frac{1}{\sqrt{3}} & -\frac{1}{\sqrt{2}} \\ \frac{1}{\sqrt{3}} & 0 \end{bmatrix}$ and $R = \begin{bmatrix} \sqrt{3} & 2\sqrt{3} \\ 0 & 3\sqrt{2} \end{bmatrix}$

27. $Q = \begin{bmatrix} \frac{1}{\sqrt{6}} & \frac{3}{\sqrt{11}} \\ -\frac{2}{\sqrt{6}} & \frac{1}{\sqrt{11}} \\ -\frac{1}{\sqrt{6}} & \frac{1}{\sqrt{11}} \end{bmatrix}$ and $R = \begin{bmatrix} \sqrt{6} & -2\sqrt{6} \\ 0 & 3\sqrt{11} \end{bmatrix}$

29. $Q = \begin{bmatrix} 0 & \frac{3}{\sqrt{15}} & \frac{3}{\sqrt{35}} \\ \frac{1}{\sqrt{3}} & -\frac{2}{\sqrt{15}} & \frac{3}{\sqrt{35}} \\ \frac{1}{\sqrt{3}} & \frac{1}{\sqrt{15}} & -\frac{4}{\sqrt{35}} \\ \frac{1}{\sqrt{3}} & \frac{1}{\sqrt{15}} & \frac{1}{\sqrt{35}} \end{bmatrix}$ and

$$R = \begin{bmatrix} \sqrt{3} & \frac{2}{\sqrt{3}} & \frac{2}{\sqrt{3}} \\ 0 & \frac{\sqrt{15}}{3} & \frac{2}{\sqrt{15}} \\ 0 & 0 & \frac{7}{\sqrt{35}} \end{bmatrix}$$

31. $Q = \begin{bmatrix} \frac{1}{\sqrt{3}} & \frac{1}{\sqrt{3}} & \frac{2}{\sqrt{15}} \\ 0 & \frac{1}{\sqrt{3}} & -\frac{1}{\sqrt{15}} \\ -\frac{1}{\sqrt{3}} & 0 & \frac{3}{\sqrt{15}} \\ \frac{1}{\sqrt{3}} & -\frac{1}{\sqrt{3}} & \frac{1}{\sqrt{15}} \end{bmatrix}$ and

$$R = \begin{bmatrix} \sqrt{3} & \sqrt{3} & 2\sqrt{3} \\ 0 & \sqrt{3} & -\frac{2}{\sqrt{3}} \\ 0 & 0 & \frac{5}{\sqrt{15}} \end{bmatrix}$$

33. $\begin{bmatrix} 2 \\ -1 \end{bmatrix}$ 35. $\begin{bmatrix} 3 \\ -2 \end{bmatrix}$ 37. $\begin{bmatrix} -2 \\ 1 \\ 3 \end{bmatrix}$ 39. $\begin{bmatrix} 2 \\ -4 \\ 3 \end{bmatrix}$

41. F 42. T 43. T 44. T 45. T
46. T 47. T 48. T 49. F 50. F
51. T 52. F

67. (a) $\operatorname{rank} A = 3$

(b) $Q = \begin{bmatrix} -0.3172 & -0.4413 & -0.5587 \\ 0.2633 & -0.4490 & -0.2951 \\ 0.7182 & -0.4040 & -0.0570 \\ -0.5386 & -0.5875 & 0.3130 \\ -0.1556 & 0.3087 & -0.7068 \end{bmatrix}$ and

$$R = \begin{bmatrix} -16.7096 & 6.4460 & 6.3700 \\ 0.0000 & -20.7198 & -3.7958 \\ 0.0000 & 0.0000 & -15.6523 \end{bmatrix}$$

Section 6.3

1. $\left\{ \begin{bmatrix} 1 \\ 1 \\ 0 \end{bmatrix}, \begin{bmatrix} -2 \\ 0 \\ 1 \end{bmatrix} \right\}$ 3. $\left\{ \begin{bmatrix} 1 \\ 1 \\ -1 \end{bmatrix} \right\}$

5. $\left\{ \begin{bmatrix} -5 \\ -2 \\ 1 \\ 0 \end{bmatrix}, \begin{bmatrix} -3 \\ -1 \\ 0 \\ 1 \end{bmatrix} \right\}$ 7. $\left\{ \begin{bmatrix} 1 \\ -2 \\ 1 \\ 0 \end{bmatrix}, \begin{bmatrix} -3 \\ 1 \\ 0 \\ 1 \end{bmatrix} \right\}$

9. (a) $\mathbf{w} = \begin{bmatrix} -1 \\ 1 \end{bmatrix}$ and $\mathbf{z} = \begin{bmatrix} 2 \\ 2 \end{bmatrix}$

(b) $\begin{bmatrix} -1 \\ 1 \end{bmatrix}$ (c) $\sqrt{8}$

11. (a) $\mathbf{w} = \mathbf{u}$ and $\mathbf{z} = \mathbf{0}$

(b) $\begin{bmatrix} 1 \\ 4 \\ -1 \end{bmatrix}$ (c) 0

13. (a) $\mathbf{w} = \begin{bmatrix} 2 \\ 2 \\ 3 \\ 1 \end{bmatrix}$ and $\mathbf{z} = \begin{bmatrix} 0 \\ 2 \\ -2 \\ 2 \end{bmatrix}$

(b) $\begin{bmatrix} 2 \\ 2 \\ 3 \\ 1 \end{bmatrix}$ (c) $\sqrt{12}$

15. (a) $\mathbf{w} = \begin{bmatrix} 3 \\ 4 \\ -2 \\ 3 \end{bmatrix}$ and $\mathbf{z} = \begin{bmatrix} -3 \\ 1 \\ -1 \\ 1 \end{bmatrix}$

(b) $\begin{bmatrix} 3 \\ 4 \\ -2 \\ 3 \end{bmatrix}$ (c) $\sqrt{12}$

17. (a) $P_W = \frac{1}{25}\begin{bmatrix} 9 & -12 \\ -12 & 16 \end{bmatrix}$

(b) $\mathbf{w} = \begin{bmatrix} -6 \\ 8 \end{bmatrix}$, $\mathbf{z} = \begin{bmatrix} -4 \\ -3 \end{bmatrix}$, (c) 5

19. (a) $P_W = \frac{1}{6}\begin{bmatrix} 1 & -2 & -1 \\ -2 & 4 & 2 \\ -1 & 2 & 1 \end{bmatrix}$

(b) $\mathbf{w} = \frac{1}{3}\begin{bmatrix} -1 \\ 2 \\ 1 \end{bmatrix}$, $\mathbf{z} = \frac{4}{3}\begin{bmatrix} 1 \\ 1 \\ -1 \end{bmatrix}$ (c) $\frac{4}{\sqrt{3}}$

21. (a) $P_W = \frac{1}{33}\begin{bmatrix} 22 & 11 & 0 & 11 \\ 11 & 19 & 9 & -8 \\ 0 & 9 & 6 & -9 \\ 11 & -8 & -9 & 19 \end{bmatrix}$

(b) $\mathbf{w} = \begin{bmatrix} 3 \\ 0 \\ -1 \\ 3 \end{bmatrix}$, $\mathbf{z} = \begin{bmatrix} -2 \\ 1 \\ 3 \\ 3 \end{bmatrix}$ (c) $\sqrt{23}$

23. (a) $P_W = \frac{1}{12}\begin{bmatrix} 11 & 1 & -3 & 1 \\ 1 & 11 & 3 & -1 \\ -3 & 3 & 3 & 3 \\ 1 & -1 & 3 & 11 \end{bmatrix}$

(b) $\mathbf{w} = \begin{bmatrix} 3 \\ -1 \\ 0 \\ 4 \end{bmatrix}$, $\mathbf{z} = \begin{bmatrix} -1 \\ 1 \\ -3 \\ 1 \end{bmatrix}$ (c) $\sqrt{12}$

25. (a) $P_W = \frac{1}{6}\begin{bmatrix} 5 & -2 & 1 \\ -2 & 2 & 2 \\ 1 & 2 & 5 \end{bmatrix}$

(b) $\mathbf{w} = \begin{bmatrix} 2 \\ -1 \\ 0 \end{bmatrix}$, $\mathbf{z} = \begin{bmatrix} 1 \\ 2 \\ -1 \end{bmatrix}$ (c) $\sqrt{6}$

27. (a) $P_W = \frac{1}{42}\begin{bmatrix} 25 & -20 & 5 \\ -20 & 16 & -4 \\ 5 & -4 & 1 \end{bmatrix}$

(b) $\mathbf{w} = \begin{bmatrix} 5 \\ -4 \\ 1 \end{bmatrix}$, $\mathbf{z} = \begin{bmatrix} 3 \\ 4 \\ 1 \end{bmatrix}$ (c) $\sqrt{26}$

29. (a) $P_W = \frac{1}{11}\begin{bmatrix} 6 & -2 & -1 & -5 \\ -2 & 8 & 4 & -2 \\ -1 & 4 & 2 & -1 \\ -5 & -2 & -1 & 6 \end{bmatrix}$

(b) $\mathbf{w} = \begin{bmatrix} 0 \\ 4 \\ 2 \\ -2 \end{bmatrix}$, $\mathbf{z} = \begin{bmatrix} 1 \\ 1 \\ -1 \\ 1 \end{bmatrix}$ (c) 2

31. (a) $P_W = \begin{bmatrix} 1 & 0 \\ 0 & 1 \end{bmatrix}$

(b) $\mathbf{w} = \begin{bmatrix} 2 \\ 3 \end{bmatrix}$, $\mathbf{z} = \begin{bmatrix} 0 \\ 0 \end{bmatrix}$ (c) 0

33. F	34. F	35. T	36. F	37. T
38. T	39. T	40. T	41. T	42. F
43. F	44. T	45. T	46. F	47. T
48. F	49. F	50. T	51. F	52. T
53. T	54. F	55. T	56. T	

79. (b) $\frac{1}{6}\left[\begin{array}{ccc|c} 4 & 2 & 0 & -2 \\ 2 & 3 & 2 & 1 \\ 0 & 2 & 2 & 2 \\ -2 & 1 & 2 & 3 \end{array}\right]$

81. $\frac{1}{3}\begin{bmatrix} 2 & -1 & 0 & 1 \\ -1 & 1 & -1 & 0 \\ 0 & -1 & 2 & -1 \\ 1 & 0 & -1 & 1 \end{bmatrix}$

85. (a) There is no unique answer. Using Q in the MATLAB command $[Q\ R] = \texttt{qr}(A, 0)$ (see Table D.3 in Appendix D), where A is the matrix whose columns are the vectors in $\mathcal{S}$, we obtain

$$\left\{ \begin{bmatrix} 0 \\ 0.2914 \\ -0.8742 \\ 0 \\ 0.3885 \end{bmatrix}, \begin{bmatrix} 0.7808 \\ -0.5828 \\ -0.1059 \\ 0 \\ 0.1989 \end{bmatrix}, \begin{bmatrix} -0.0994 \\ -0.3243 \\ -0.4677 \\ 0.1082 \\ -0.8090 \end{bmatrix}, \begin{bmatrix} -0.1017 \\ -0.1360 \\ -0.0589 \\ -0.9832 \\ -0.0304 \end{bmatrix} \right\}.$$

(b) $\mathbf{w} = \begin{bmatrix} -6.3817 \\ 6.8925 \\ 7.2135 \\ 1.3687 \\ 2.3111 \end{bmatrix}$

(c) $\|\mathbf{u} - \mathbf{w}\| = 4.3033$

87. $P_W =$

$$\begin{bmatrix} 0.6298 & -0.4090 & -0.0302 & 0.0893 & 0.2388 \\ -0.4090 & 0.5482 & -0.0334 & 0.0986 & 0.2638 \\ -0.0302 & -0.0334 & 0.9975 & 0.0073 & 0.0195 \\ 0.0893 & 0.0986 & 0.0073 & 0.9785 & -0.0576 \\ 0.2388 & 0.2638 & 0.0195 & -0.0576 & 0.8460 \end{bmatrix}$$

Section 6.4

1. $y = 13.5 + x$ 3. $y = 3.2 + 1.6x$

5. $y = 44 - 3x$ 7. $y = 9.6 + 2.6x$

9. $y = -6.35 + 2.1x$, the estimates of k and L are 2.1 and 3.02, respectively.

11. $3 - x + x^2$ 13. $2 + 0.5x + 0.5x^2$

15. $-1 + x^2 + x^3$

17. $\frac{1}{3}\begin{bmatrix} 4 \\ -2 \\ 0 \end{bmatrix} + x_3 \begin{bmatrix} -1 \\ 1 \\ 1 \end{bmatrix}$

19. $\frac{1}{19}\begin{bmatrix} -35 \\ -50 \\ 31 \\ 0 \end{bmatrix} + x_4 \begin{bmatrix} 0 \\ 1 \\ -1 \\ 1 \end{bmatrix}$

21. $\frac{1}{3}\begin{bmatrix} 7 \\ -1 \\ 8 \end{bmatrix}$ 23. $\begin{bmatrix} 1 \\ -1 \\ 1 \\ 0 \end{bmatrix}$

25. $\frac{1}{3}\begin{bmatrix} 2 \\ 0 \\ 2 \end{bmatrix}$ 27. $\frac{1}{19}\begin{bmatrix} -35 \\ -23 \\ 4 \\ 27 \end{bmatrix}$

28. F 29. T 30. F 31. F 32. T

39. $y = 1.42 + 0.49x + 0.38x^2 + 0.73x^3$

41. $a = 3.0$ and $b = -2.0$

Section 6.5

1. no 3. no 5. yes 7. no

9. a reflection, $y = (\sqrt{2} - 1)x$

11. a rotation, $\theta = 30°$

13. a reflection, $y = \frac{2}{3}x$

15. a rotation, $\theta = 270°$

17. T 18. F 19. F 20. T 21. T

22. T 23. F 24. F 25. T 26. F

27. F 28. T 29. T 30. T 31. F

32. F 33. F 34. T 35. T 36. T

37. One possibility is to let $T = T_A$ for

$$A = \frac{1}{35}\begin{bmatrix} 10 & 33 & 6 \\ -30 & 6 & 17 \\ 15 & -10 & 30 \end{bmatrix}.$$

39. (b) The only eigenvalue is $\lambda = 1$, and the corresponding eigenspace is Span $\{\mathbf{e}_3\}$.

41. One possibility is to let $T = T_A$ for

$$A = \frac{1}{21}\begin{bmatrix} 20 & 4 & -5 \\ -5 & 20 & -4 \\ 4 & 5 & 20 \end{bmatrix}.$$

61. $Q = \begin{bmatrix} 1 & 0 \\ 0 & -1 \end{bmatrix}$ and $\mathbf{b} = \begin{bmatrix} 1 \\ 4 \end{bmatrix}$

63. $Q = \begin{bmatrix} .8 & -.6 \\ .6 & .8 \end{bmatrix}$ and $\mathbf{b} = \begin{bmatrix} 0 \\ 3 \end{bmatrix}$

67. (a) $A_2 = \begin{bmatrix} 0 & -1 \\ -1 & 0 \end{bmatrix}$, $A_3 = \frac{1}{3}\begin{bmatrix} 1 & -2 & -2 \\ -2 & 1 & -2 \\ -2 & -2 & 1 \end{bmatrix}$,

and A_6 is the 6×6 matrix whose diagonal entries are $\frac{2}{3}$ and whose off diagonal entries are $-\frac{1}{3}$.

71. $\begin{bmatrix} 0.7833 & 0.6217 \\ 0.6217 & -0.7833 \end{bmatrix}$ (rounded to 4 places after the decimal)

73. 231°

Section 6.6

1. (a) $\begin{bmatrix} 2 & -7 \\ -7 & 50 \end{bmatrix}$

(b) about 8.1°

(c) $x = \frac{7}{\sqrt{50}}x' - \frac{1}{\sqrt{50}}y'$
$y = \frac{1}{\sqrt{50}}x' + \frac{7}{\sqrt{50}}y'$

(d) $(x')^2 + 51(y')^2 = 255$

(e) an ellipse

3. (a) $\begin{bmatrix} 1 & -6 \\ -6 & -4 \end{bmatrix}$

(b) about 56.3°

(c) $x = \frac{2}{\sqrt{13}}x' - \frac{3}{\sqrt{13}}y'$
$y = \frac{3}{\sqrt{13}}x' + \frac{2}{\sqrt{13}}y'$

(d) $-8(x')^2 + 5(y')^2 = 40$

(e) a hyperbola

5. (a) $\begin{bmatrix} 5 & 2 \\ 2 & 5 \end{bmatrix}$

(b) $45°$

(c) $x = \frac{1}{\sqrt{2}}x' - \frac{1}{\sqrt{2}}y'$
$y = \frac{1}{\sqrt{2}}x' + \frac{1}{\sqrt{2}}y'$

(d) $7(x')^2 + 3(y')^2 = 9$

(e) an ellipse

7. (a) $\begin{bmatrix} 1 & 2 \\ 2 & 1 \end{bmatrix}$

(b) $45°$

(c) $x = \frac{1}{\sqrt{2}}x' - \frac{1}{\sqrt{2}}y'$
$y = \frac{1}{\sqrt{2}}x' + \frac{1}{\sqrt{2}}y'$

(d) $3(x')^2 - (y')^2 = 7$

(e) a hyperbola

9. (a) $\begin{bmatrix} 2 & -6 \\ -6 & -7 \end{bmatrix}$

(b) about $63.4°$

(c) $x = \frac{1}{\sqrt{5}}x' - \frac{2}{\sqrt{5}}y'$
$y = \frac{2}{\sqrt{5}}x' + \frac{1}{\sqrt{5}}y'$

(d) $-10(x')^2 + 5(y')^2 = 200$

(e) a hyperbola

11. (a) $\begin{bmatrix} 1 & 1 \\ 1 & 1 \end{bmatrix}$

(b) $45°$

(c) $x = \frac{1}{\sqrt{2}}x' - \frac{1}{\sqrt{2}}y'$
$y = \frac{1}{\sqrt{2}}x' + \frac{1}{\sqrt{2}}y'$

(d) $2\sqrt{2}(x')^2 + 9x' - 7y' = 0$

(e) a parabola

13. $\left\{ \frac{1}{\sqrt{2}}\begin{bmatrix} 1 \\ -1 \end{bmatrix}, \frac{1}{\sqrt{2}}\begin{bmatrix} 1 \\ 1 \end{bmatrix} \right\}$, 2 and 4,

$$A = 2\begin{bmatrix} 0.5 & -0.5 \\ -0.5 & 0.5 \end{bmatrix} + 4\begin{bmatrix} 0.5 & 0.5 \\ 0.5 & 0.5 \end{bmatrix}$$

15. $\left\{ \frac{1}{\sqrt{2}}\begin{bmatrix} 1 \\ 1 \end{bmatrix}, \frac{1}{\sqrt{2}}\begin{bmatrix} 1 \\ -1 \end{bmatrix} \right\}$, 3 and -1,

$$A = 3\begin{bmatrix} 0.5 & 0.5 \\ 0.5 & 0.5 \end{bmatrix} + (-1)\begin{bmatrix} 0.5 & -0.5 \\ -0.5 & 0.5 \end{bmatrix}$$

17. $\left\{ \frac{1}{3}\begin{bmatrix} -1 \\ -2 \\ 2 \end{bmatrix}, \frac{1}{3}\begin{bmatrix} 2 \\ 1 \\ 2 \end{bmatrix}, \frac{1}{3}\begin{bmatrix} -2 \\ 2 \\ 1 \end{bmatrix} \right\}$, 3, 6, and 0,

$$A = 3\begin{bmatrix} \frac{1}{9} & \frac{2}{9} & -\frac{2}{9} \\ \frac{2}{9} & \frac{4}{9} & -\frac{4}{9} \\ -\frac{2}{9} & -\frac{4}{9} & \frac{2}{9} \end{bmatrix} + 6\begin{bmatrix} \frac{4}{9} & \frac{2}{9} & \frac{4}{9} \\ \frac{2}{9} & \frac{1}{9} & \frac{2}{9} \\ \frac{4}{9} & \frac{2}{9} & \frac{4}{9} \end{bmatrix} +$$
$$0\begin{bmatrix} \frac{4}{9} & -\frac{4}{9} & -\frac{2}{9} \\ -\frac{4}{9} & \frac{4}{9} & \frac{2}{9} \\ -\frac{2}{9} & \frac{2}{9} & \frac{1}{9} \end{bmatrix}$$

19. $\left\{ \begin{bmatrix} 1 \\ 0 \\ 0 \end{bmatrix}, \frac{1}{\sqrt{5}}\begin{bmatrix} 0 \\ -2 \\ 1 \end{bmatrix}, \frac{1}{\sqrt{5}}\begin{bmatrix} 0 \\ 1 \\ 2 \end{bmatrix} \right\}$ -1, -1, and 4,

$$A = (-1)\begin{bmatrix} 1 & 0 & 0 \\ 0 & 0 & 0 \\ 0 & 0 & 0 \end{bmatrix} + (-1)\begin{bmatrix} 0 & 0 & 0 \\ 0 & .8 & -.4 \\ 0 & -.4 & .2 \end{bmatrix}$$
$$+ 4\begin{bmatrix} 0 & 0 & 0 \\ 0 & .2 & .4 \\ 0 & .4 & .8 \end{bmatrix}$$

21. T	22. F	23. T	24. F	25. F
26. T	27. T	28. F	29. T	30. T
31. T	32. F	33. T	34. F	35. F
36. F	37. T	38. T	39. F	40. F

41. $2\begin{bmatrix} 1 & 0 \\ 0 & 0 \end{bmatrix} + 2\begin{bmatrix} 0 & 0 \\ 0 & 1 \end{bmatrix}$ and

$2\begin{bmatrix} .5 & .5 \\ .5 & .5 \end{bmatrix} + 2\begin{bmatrix} .5 & -.5 \\ -.5 & .5 \end{bmatrix}$

Section 6.7

1. $\begin{bmatrix} \frac{1}{\sqrt{2}} & -\frac{1}{\sqrt{2}} \\ \frac{1}{\sqrt{2}} & \frac{1}{\sqrt{2}} \end{bmatrix}\begin{bmatrix} \sqrt{2} & 0 \\ 0 & 0 \end{bmatrix}\begin{bmatrix} 1 & 0 \\ 0 & 1 \end{bmatrix}^T$

3. $\begin{bmatrix} \frac{1}{3} & \frac{2}{\sqrt{5}} & \frac{2}{3\sqrt{5}} \\ \frac{2}{3} & \frac{-1}{\sqrt{5}} & \frac{4}{3\sqrt{5}} \\ \frac{2}{3} & 0 & \frac{-5}{3\sqrt{5}} \end{bmatrix}\begin{bmatrix} 3 \\ 0 \\ 0 \end{bmatrix}[1]^T$

5. $\begin{bmatrix} \frac{3}{\sqrt{35}} & \frac{1}{\sqrt{10}} & \frac{-3}{\sqrt{14}} \\ \frac{-1}{\sqrt{35}} & \frac{3}{\sqrt{10}} & \frac{1}{\sqrt{14}} \\ \frac{5}{\sqrt{35}} & 0 & \frac{2}{\sqrt{14}} \end{bmatrix}\begin{bmatrix} \sqrt{7} & 0 \\ 0 & \sqrt{2} \\ 0 & 0 \end{bmatrix}\begin{bmatrix} \frac{1}{\sqrt{5}} & \frac{2}{\sqrt{5}} \\ \frac{2}{\sqrt{5}} & \frac{-1}{\sqrt{5}} \end{bmatrix}^T$

7. $\begin{bmatrix} \frac{1}{\sqrt{2}} & \frac{1}{\sqrt{2}} \\ \frac{-1}{\sqrt{2}} & \frac{1}{\sqrt{2}} \end{bmatrix} \begin{bmatrix} 2 & 0 & 0 \\ 0 & \sqrt{2} & 0 \end{bmatrix} \begin{bmatrix} 0 & 1 & 0 \\ \frac{1}{\sqrt{2}} & 0 & \frac{1}{\sqrt{2}} \\ \frac{1}{\sqrt{2}} & 0 & \frac{-1}{\sqrt{2}} \end{bmatrix}^T$

9. $\begin{bmatrix} \frac{1}{\sqrt{6}} & \frac{5}{\sqrt{30}} & 0 \\ \frac{2}{\sqrt{6}} & \frac{-2}{\sqrt{30}} & \frac{1}{\sqrt{5}} \\ \frac{1}{\sqrt{6}} & \frac{-1}{\sqrt{30}} & \frac{-2}{\sqrt{5}} \end{bmatrix} \begin{bmatrix} \sqrt{6} & 0 & 0 \\ 0 & \sqrt{6} & 0 \\ 0 & 0 & 1 \end{bmatrix} \begin{bmatrix} 1 & 0 & 0 \\ 0 & \frac{1}{\sqrt{5}} & \frac{2}{\sqrt{5}} \\ 0 & \frac{2}{\sqrt{5}} & \frac{-1}{\sqrt{5}} \end{bmatrix}^T$

11. $\begin{bmatrix} \frac{1}{\sqrt{5}} & \frac{2}{\sqrt{5}} \\ \frac{-2}{\sqrt{5}} & \frac{1}{\sqrt{5}} \end{bmatrix} \begin{bmatrix} \sqrt{30} & 0 & 0 \\ 0 & 0 & 0 \end{bmatrix} \begin{bmatrix} \frac{1}{\sqrt{6}} & \frac{1}{\sqrt{3}} & \frac{1}{\sqrt{2}} \\ \frac{-1}{\sqrt{6}} & \frac{-1}{\sqrt{3}} & \frac{1}{\sqrt{2}} \\ \frac{2}{\sqrt{6}} & \frac{-1}{\sqrt{3}} & 0 \end{bmatrix}^T$

13. $\begin{bmatrix} \frac{1}{\sqrt{5}} & \frac{2}{\sqrt{5}} \\ \frac{-2}{\sqrt{5}} & \frac{1}{\sqrt{5}} \end{bmatrix} \begin{bmatrix} \sqrt{7} & 0 & 0 \\ 0 & \sqrt{2} & 0 \end{bmatrix} \begin{bmatrix} \frac{3}{\sqrt{35}} & \frac{1}{\sqrt{10}} & \frac{3}{\sqrt{14}} \\ \frac{-5}{\sqrt{35}} & 0 & \frac{2}{\sqrt{14}} \\ \frac{-1}{\sqrt{35}} & \frac{3}{\sqrt{10}} & \frac{-1}{\sqrt{14}} \end{bmatrix}^T$

15. The singular value decomposition is $U\Sigma V^T$, where

$$U = \begin{bmatrix} \frac{2}{\sqrt{5}} & \frac{1}{\sqrt{5}} & 0 \\ \frac{1}{\sqrt{5}} & \frac{-2}{\sqrt{5}} & 0 \\ 0 & 0 & 1 \end{bmatrix},\ \Sigma = \begin{bmatrix} \sqrt{60} & 0 & 0 & 0 \\ 0 & \sqrt{15} & 0 & 0 \\ 0 & 0 & 0 & 0 \end{bmatrix},$$

and $V = \begin{bmatrix} \frac{1}{\sqrt{3}} & \frac{-1}{\sqrt{3}} & \frac{1}{\sqrt{6}} & \frac{1}{\sqrt{6}} \\ \frac{1}{\sqrt{3}} & \frac{1}{\sqrt{3}} & \frac{1}{\sqrt{6}} & \frac{-1}{\sqrt{6}} \\ \frac{1}{\sqrt{3}} & 0 & \frac{-2}{\sqrt{6}} & 0 \\ 0 & \frac{1}{\sqrt{3}} & 0 & \frac{2}{\sqrt{6}} \end{bmatrix}$.

17. The singular value decomposition is $U\Sigma V^T$, where

$$U = \begin{bmatrix} \frac{1}{\sqrt{3}} & \frac{2}{\sqrt{6}} & 0 \\ \frac{1}{\sqrt{3}} & \frac{-1}{\sqrt{6}} & \frac{1}{\sqrt{2}} \\ \frac{-1}{\sqrt{3}} & \frac{1}{\sqrt{6}} & \frac{1}{\sqrt{2}} \end{bmatrix},\ \Sigma = \begin{bmatrix} \sqrt{21} & 0 & 0 & 0 \\ 0 & \sqrt{18} & 0 & 0 \\ 0 & 0 & 0 & 0 \end{bmatrix},$$

and $V = \begin{bmatrix} \frac{1}{\sqrt{7}} & \frac{1}{\sqrt{3}} & \frac{1}{\sqrt{11}} & \frac{1}{\sqrt{2}} \\ \frac{2}{\sqrt{7}} & \frac{-1}{\sqrt{3}} & \frac{1}{\sqrt{11}} & 0 \\ \frac{1}{\sqrt{7}} & 0 & \frac{-3}{\sqrt{11}} & 0 \\ \frac{1}{\sqrt{7}} & \frac{1}{\sqrt{3}} & 0 & \frac{-1}{\sqrt{2}} \end{bmatrix}$.

19. In the accompanying figure, $\mathbf{u}_1 = \frac{1}{\sqrt{2}}\begin{bmatrix} 1 \\ -1 \end{bmatrix}$, $\mathbf{u}_2 = \frac{1}{\sqrt{2}}\begin{bmatrix} 1 \\ 1 \end{bmatrix}$, $OP = 2\sqrt{2}$, and $OQ = \sqrt{2}$.

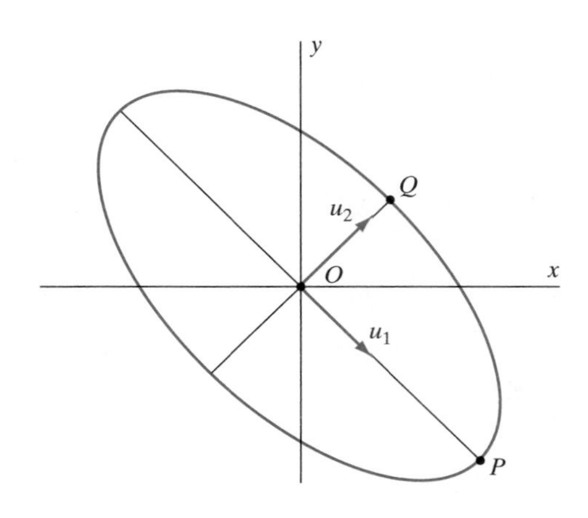

Figure for Exercise 19

21. $\begin{bmatrix} 1 \\ 1 \end{bmatrix}$ 23. $\begin{bmatrix} 3 \\ -4 \\ 1 \end{bmatrix}$ 25. $\frac{1}{35}\begin{bmatrix} 20 \\ -37 \\ 11 \end{bmatrix}$

27. $\begin{bmatrix} 3 \\ 1 \\ 1 \end{bmatrix}$ 29. $\begin{bmatrix} 0.04 \\ 0.08 \end{bmatrix}$ 31. $\frac{1}{7}\begin{bmatrix} 12 \\ 3 \end{bmatrix}$

33. $\frac{5}{6}\begin{bmatrix} 3 \\ 3 \\ 2 \end{bmatrix}$ 35. $\frac{1}{154}\begin{bmatrix} -11 \\ 30 \\ 27 \end{bmatrix}$ 37. $\frac{1}{9}[1\ 2\ 2]$

39. $\frac{1}{3}\begin{bmatrix} 2 & 1 \\ -1 & -2 \\ -1 & 1 \end{bmatrix}$ 41. $\frac{1}{14}\begin{bmatrix} 4 & 8 & 2 \\ 1 & -5 & 4 \end{bmatrix}$

43. $\frac{1}{4}\begin{bmatrix} 2 & 2 \\ 1 & -1 \\ 1 & -1 \end{bmatrix}$ 45. $\frac{1}{14}\begin{bmatrix} 4 & -1 \\ -2 & 4 \\ 8 & 5 \end{bmatrix}$

47. $\frac{1}{9}\begin{bmatrix} 1 & 2 & 2 \\ 2 & 4 & 4 \\ 2 & 4 & 4 \end{bmatrix}$ 49. $\frac{1}{3}\begin{bmatrix} 2 & 1 & 1 \\ 1 & 2 & -1 \\ 1 & -1 & 2 \end{bmatrix}$

51. $\frac{1}{75}\begin{bmatrix} 74 & 7 & 5 \\ 7 & 26 & -35 \\ 5 & -35 & 50 \end{bmatrix}$ 53. $\frac{1}{14}\begin{bmatrix} 5 & 3 & 6 \\ 3 & 13 & -2 \\ 6 & -2 & 10 \end{bmatrix}$

55. F 56. T 57. F 58. T 59. F
60. F 61. T 62. T 63. F 64. T
65. T 66. F 67. F 68. T 69. F
70. T 71. T 72. F 73. T 74. F
75. T

93. (rounded to 4 places after the decimal)

$$U = \begin{bmatrix} 0.5836 & 0.7289 & -0.3579 \\ 0.7531 & -0.6507 & -0.0970 \\ 0.3036 & 0.2129 & 0.9287 \end{bmatrix},$$

$$\Sigma = \begin{bmatrix} 5.9073 & 0 & 0 & 0 \\ 0 & 2.2688 & 0 & 0 \\ 0 & 0 & 1.7194 & 0 \end{bmatrix},$$

$$V = \begin{bmatrix} 0.3024 & -0.3462 & -0.8612 & -0.2170 \\ 0.0701 & 0.9293 & -0.3599 & 0.0434 \\ 0.2777 & 0.1283 & 0.2755 & -0.9113 \\ 0.9091 & 0.0043 & 0.2300 & 0.3472 \end{bmatrix},$$

and

$$A^\dagger = \begin{bmatrix} 0.0979 & 0.1864 & -0.4821 \\ 0.3804 & -0.2373 & -0.1036 \\ 0.0113 & -0.0169 & 0.1751 \\ 0.0433 & 0.1017 & 0.1714 \end{bmatrix}$$

Section 6.8

1. $\overline{\mathbf{x}} = 1$ 3. $s_{\mathbf{x}}^2 = 13$

5. $\text{cov}(\mathbf{x}, \mathbf{y}) = \frac{5}{2}$ 7. $\begin{bmatrix} 13 & \frac{5}{2} \\ \frac{5}{2} & 1 \end{bmatrix}$

9. T 10. F 11. F 12. F 13. F
14. T 15. F 16. T 17. T 18. T
14. T 20. T

Section 6.9

1. $\begin{bmatrix} 0 & 1 & 0 \\ 0 & 0 & -1 \\ -1 & 0 & 0 \end{bmatrix}$ 3. $\frac{1}{\sqrt{2}}\begin{bmatrix} 1 & -1 & 0 \\ 0 & 0 & -\sqrt{2} \\ 1 & 1 & 0 \end{bmatrix}$

5. $\frac{1}{4}\begin{bmatrix} 2\sqrt{3} & 0 & 2 \\ 1 & 2\sqrt{3} & -\sqrt{3} \\ -\sqrt{3} & 2 & 3 \end{bmatrix}$

7. $\begin{bmatrix} 0 & 0 & 1 \\ 0 & -1 & 0 \\ 1 & 0 & 0 \end{bmatrix}$

9. $\frac{1}{2\sqrt{2}}\begin{bmatrix} \sqrt{2}+1 & \sqrt{2}-1 & -\sqrt{2} \\ \sqrt{2}-1 & \sqrt{2}+1 & \sqrt{2} \\ \sqrt{2} & -\sqrt{2} & 2 \end{bmatrix}$

11. $\frac{1}{4}\begin{bmatrix} \sqrt{3}+2 & \sqrt{3}-2 & -\sqrt{2} \\ \sqrt{3}-2 & \sqrt{3}+2 & -\sqrt{2} \\ \sqrt{2} & \sqrt{2} & 2\sqrt{3} \end{bmatrix}$

13. $\frac{1}{3}\begin{bmatrix} 2 & -2 & -1 \\ 1 & 2 & -2 \\ 2 & 1 & 2 \end{bmatrix}$

15. (a) $\begin{bmatrix} -1 \\ -1 \\ 1 \end{bmatrix}$ (b) $-\frac{1}{2}$

17. (a) $\begin{bmatrix} \sqrt{2}+1 \\ -1 \\ 1 \end{bmatrix}$ (b) $\frac{1-\sqrt{2}}{2\sqrt{2}}$

19. (a) $\begin{bmatrix} 1 \\ 1 \\ 2-\sqrt{3} \end{bmatrix}$ (b) $\frac{4\sqrt{3}-1}{8}$

21. (a) $\begin{bmatrix} \sqrt{3} \\ \sqrt{2}+1 \\ 1 \end{bmatrix}$ (b) $\frac{3\sqrt{2}-2}{8}$

23. $\frac{1}{3}\begin{bmatrix} 2 & 2 & -1 \\ 2 & -1 & 2 \\ -1 & 2 & 2 \end{bmatrix}$ 25. $\frac{1}{3}\begin{bmatrix} 1 & -2 & -2 \\ -2 & 1 & -2 \\ -2 & -2 & 1 \end{bmatrix}$

27. $\frac{1}{9}\begin{bmatrix} 7 & -4 & 4 \\ -4 & 1 & 8 \\ 4 & 8 & 1 \end{bmatrix}$ 29. $\frac{1}{25}\begin{bmatrix} 16 & 12 & -15 \\ 12 & 9 & 20 \\ -15 & 20 & 0 \end{bmatrix}$

31. $\frac{1}{3}\begin{bmatrix} 2 & -2 & 1 \\ -2 & -1 & 2 \\ 1 & 2 & 2 \end{bmatrix}$ 33. $\frac{1}{5}\begin{bmatrix} 3 & 0 & -4 \\ 0 & 5 & 0 \\ -4 & 0 & 3 \end{bmatrix}$

35. $\frac{1}{25}\begin{bmatrix} 16 & -12 & -15 \\ -12 & 9 & -20 \\ -15 & -20 & 0 \end{bmatrix}$ 37. $\frac{1}{9}\begin{bmatrix} 1 & 4 & -8 \\ 4 & 7 & 4 \\ -8 & 4 & 1 \end{bmatrix}$

39. (a) neither

41. (a) a rotation (b) $\mathbf{e}_1$

43. (a) a rotation (b) $\begin{bmatrix} 2 \\ 1 \\ -2 \end{bmatrix}$

45. (a) a reflection (b) $\left\{ \begin{bmatrix} 1 \\ 0 \\ \sqrt{2}-1 \end{bmatrix}, \begin{bmatrix} 0 \\ 1 \\ 0 \end{bmatrix} \right\}$

47. F 48. F 49. F 50. F 51. T
52. F 53. T 54. T 55. T 56. F
57. T 58. T 59. F 60. T 61. T
62. T 63. F 64. F 65. F 66. F
67. T

79. $\begin{bmatrix} 0 & -1 & 0 \\ 1 & 0 & 0 \\ 0 & 0 & -1 \end{bmatrix}$

83. (rounded to 4 places after the decimal)

Span $\left\{ \begin{bmatrix} .4609 \\ .1769 \\ .8696 \end{bmatrix} \right\}$, 48°

Chapter 6 Review Exercises

1. T 2. T 3. F 4. T 5. T
6. T 7. T 8. T 9. F 10. F

11. T 12. F 13. F 14. F 15. F
16. T 17. T 18. F 19. T

21. (a) $\|\mathbf{u}\| = \sqrt{45}$, $\|\mathbf{v}\| = \sqrt{20}$
(b) $d = \sqrt{65}$
(c) $\mathbf{u} \cdot \mathbf{v} = 0$
(d) $\mathbf{u}$ and $\mathbf{v}$ are orthogonal.

23. (a) $\|\mathbf{u}\| = \sqrt{6}$, $\|\mathbf{v}\| = \sqrt{21}$
(b) $d = \sqrt{27}$
(c) $\mathbf{u} \cdot \mathbf{v} = 0$
(d) $\mathbf{u}$ and $\mathbf{v}$ are orthogonal.

25. $\mathbf{w} = \frac{1}{5}\begin{bmatrix} -1 \\ 2 \end{bmatrix}$, $d = 3.5777$

27. 1 29. 113

31. $\left\{ \begin{bmatrix} 1 \\ 1 \\ -1 \\ 0 \end{bmatrix}, \frac{1}{3}\begin{bmatrix} 1 \\ 1 \\ 2 \\ 3 \end{bmatrix}, \frac{1}{5}\begin{bmatrix} -2 \\ 3 \\ 1 \\ -1 \end{bmatrix} \right\}$

33. $\left\{ \begin{bmatrix} 6 \\ -7 \\ 5 \\ 0 \end{bmatrix}, \begin{bmatrix} -2 \\ 4 \\ 0 \\ 5 \end{bmatrix} \right\}$

35. $\mathbf{w} = \frac{1}{14}\begin{bmatrix} 32 \\ 19 \\ -27 \end{bmatrix}$ and $\mathbf{z} = \frac{1}{14}\begin{bmatrix} -18 \\ 9 \\ -15 \end{bmatrix}$

37. $P_W = \frac{1}{6}\begin{bmatrix} 1 & 2 & 0 & -1 \\ 2 & 4 & 0 & -2 \\ 0 & 0 & 0 & 0 \\ -1 & -2 & 0 & 1 \end{bmatrix}$ and

$\mathbf{w} = \begin{bmatrix} 2 \\ 4 \\ 0 \\ -2 \end{bmatrix}$

39. $P_W = \frac{1}{3}\begin{bmatrix} 1 & 1 & -1 & 0 \\ 1 & 1 & -1 & 0 \\ -1 & -1 & 1 & 0 \\ 0 & 0 & 0 & 3 \end{bmatrix}$ and $\mathbf{w} = \begin{bmatrix} 0 \\ 0 \\ 0 \\ 2 \end{bmatrix}$

41. $v \approx 2.05$ and $c \approx 1.05$

43. no 45. yes

47. a rotation, $\theta = -60^\circ$

49. a reflection, $y = \frac{1}{\sqrt{3}}x$

53. $\left\{ \frac{1}{\sqrt{2}}\begin{bmatrix} -1 \\ 1 \end{bmatrix}, \frac{1}{\sqrt{2}}\begin{bmatrix} 1 \\ 1 \end{bmatrix} \right\}$, -1 and 5,

$A = (-1)\begin{bmatrix} 0.5 & -0.5 \\ -0.5 & 0.5 \end{bmatrix} + 5\begin{bmatrix} 0.5 & 0.5 \\ 0.5 & 0.5 \end{bmatrix}$

55. 45°, $\frac{(x')^2}{4} - \frac{(y')^2}{8} = 1$, a hyperbola

Chapter 6 MATLAB Exercises

1. (a) $\mathbf{u}_1 \cdot \mathbf{u}_2 = -2$, $\|\mathbf{u}_1\| = 4$, $\|\mathbf{u}_2\| = \sqrt{23} \approx 4.7958$
(b) $\mathbf{u}_3 \cdot \mathbf{u}_4 = -56$, $\|\mathbf{u}_3\| = \sqrt{28} \approx 5.2915$, $\|\mathbf{u}_4\| = \sqrt{112}$
(c) $|\mathbf{u}_1 \cdot \mathbf{u}_2| = 2 \leq 4\sqrt{23} = \|\mathbf{u}_1\| \cdot \|\mathbf{u}_2\|$
(d) $|\mathbf{u}_3 \cdot \mathbf{u}_4| = 56 = \sqrt{28} \cdot \sqrt{112} = \|\mathbf{u}_3\| \cdot \|\mathbf{u}_4\|$

2. (b) $\left\{ \begin{bmatrix} -2.8 \\ -0.8 \\ 1.0 \\ 0.0 \\ 0.0 \end{bmatrix}, \begin{bmatrix} -1.6 \\ 1.4 \\ 0.0 \\ 1.0 \\ 0.0 \end{bmatrix}, \begin{bmatrix} -0.8 \\ -0.8 \\ 0.0 \\ 0.0 \\ 1.0 \end{bmatrix} \right\}$

3. Answers are given correct to 4 places after the decimal point.

(a) $\left\{ \begin{bmatrix} -0.1994 \\ 0.1481 \\ -0.1361 \\ -0.6282 \\ -0.5316 \\ 0.4924 \end{bmatrix}, \begin{bmatrix} 0.1153 \\ 0.0919 \\ -0.5766 \\ 0.6366 \\ -0.4565 \\ 0.1790 \end{bmatrix}, \begin{bmatrix} 0.3639 \\ -0.5693 \\ 0.5469 \\ 0.1493 \\ -0.4271 \\ 0.1992 \end{bmatrix} \right\}$

(b) (i) $\begin{bmatrix} 1.3980 \\ -1.5378 \\ 1.4692 \\ 2.7504 \\ 1.4490 \\ -1.6574 \end{bmatrix}$ (ii) $\begin{bmatrix} 1 \\ -2 \\ 2 \\ -1 \\ -3 \\ 2 \end{bmatrix}$ (iii) $\begin{bmatrix} 0 \\ 0 \\ 0 \\ 0 \\ 0 \end{bmatrix}$

(c) They are the same.

(d) If M is a matrix whose columns form an orthonormal basis for a subspace W of R^n, then $P_w = MM^T$; that is, MM^T is the orthogonal projection matrix for W.

4. Answers are given correct to 4 places after the decimal point.

(a) $V = \begin{bmatrix} 1.1000 & 2.7581 & -2.6745 & -0.3438 \\ 2.3000 & 5.8488 & 1.4345 & -1.0069 \\ 3.1000 & 2.3093 & -0.2578 & 3.1109 \\ 7.2000 & -1.9558 & 0.4004 & 1.5733 \\ 8.0000 & -1.1954 & -0.3051 & -2.2847 \end{bmatrix}$

(b) $D = \begin{bmatrix} 131.9500 & 0.0000 & 0.0000 & 0.0000 \\ 0.0000 & 52.4032 & 0.0000 & 0.0000 \\ 0.0000 & 0.0000 & 9.5306 & 0.0000 \\ 0.0000 & 0.0000 & 0.0000 & 18.5046 \end{bmatrix}$

(c) $Q = \begin{bmatrix} 0.0958 & 0.3810 & -0.8663 & -0.0799 \\ 0.2002 & 0.8080 & 0.4647 & -0.2341 \\ 0.2699 & 0.3190 & -0.0835 & 0.7232 \\ 0.6268 & -0.2702 & 0.1297 & 0.3657 \\ 0.6964 & -0.1651 & -0.0988 & -0.5311 \end{bmatrix}$

(d) $R = \begin{bmatrix} 11.4869 & -3.7399 & 1.0804 & 13.1166 \\ 0.0000 & 7.2390 & -6.3751 & 2.6668 \\ 0.0000 & 0.0000 & 3.0872 & -5.9697 \\ 0.0000 & 0.0000 & 0.0000 & 4.3017 \end{bmatrix}$

(e) In this case, we have

$$Q = \begin{bmatrix} -0.0958 & -0.3810 & 0.8663 & -0.0799 \\ -0.2002 & -0.8080 & -0.4647 & -0.2341 \\ -0.2699 & -0.3190 & 0.0835 & 0.7232 \\ -0.6268 & 0.2702 & -0.1297 & 0.3657 \\ -0.6964 & 0.1651 & 0.0988 & -0.5311 \end{bmatrix}$$

$$R = \begin{bmatrix} -11.4869 & 3.7399 & -1.0804 & -13.1166 \\ 0.0000 & -7.2390 & 6.3751 & -2.6668 \\ 0.0000 & 0.0000 & -3.0872 & 5.9697 \\ 0.0000 & 0.0000 & 0.0000 & 4.3017 \end{bmatrix}$$

5. Answers are given correct to 4 places after the decimal point.

$$Q = \begin{bmatrix} 0.2041 & 0.4308 & 0.3072 & 0.3579 \\ 0.8165 & -0.1231 & 0.2861 & 0.2566 \\ -0.2041 & 0.3077 & 0.6264 & -0.1235 \\ 0.4082 & -0.2462 & -0.3222 & -0.2728 \\ 0.2041 & 0.8001 & -0.4849 & -0.1253 \\ 0.2041 & 0.0615 & 0.3042 & -0.8371 \end{bmatrix}$$

$$R = \begin{bmatrix} 4.8990 & 3.2660 & -1.4289 & 1.8371 \\ 0.0000 & 5.4160 & -0.0615 & 3.5081 \\ 0.0000 & 0.0000 & 7.5468 & -2.2737 \\ 0.0000 & 0.0000 & 0.0000 & 2.2690 \end{bmatrix}$$

6. Answers are given correct to 4 places after the decimal point.

(a) $\mathcal{B}_1 = \left\{ \begin{bmatrix} -0.1194 \\ 0.1481 \\ -0.1361 \\ -0.6282 \\ -0.5316 \\ 0.4924 \end{bmatrix}, \begin{bmatrix} 0.1153 \\ 0.0919 \\ -0.5766 \\ 0.6366 \\ -0.4565 \\ 0.1790 \end{bmatrix}, \begin{bmatrix} 0.3639 \\ -0.5693 \\ 0.5469 \\ 0.1493 \\ -0.4271 \\ 0.1992 \end{bmatrix} \right\}$

(b) $\mathcal{B}_2 = \left\{ \begin{bmatrix} 0.8986 \\ 0.3169 \\ -0.1250 \\ -0.2518 \\ 0.1096 \\ 0.0311 \end{bmatrix}, \begin{bmatrix} -0.0808 \\ 0.6205 \\ 0.5183 \\ 0.3372 \\ 0.1022 \\ 0.4644 \end{bmatrix}, \begin{bmatrix} 0.0214 \\ -0.4000 \\ -0.2562 \\ 0.0246 \\ 0.5514 \\ 0.6850 \end{bmatrix} \right\}$

(c) $PP^T = P^TP = I_6$

7. Answers are given correct to 4 places after the decimal point.

(a) $P_W =$

$$\begin{bmatrix} 0.3913 & 0.0730 & -0.1763 & -0.2716 & 0.2056 & -0.2929 \\ 0.0730 & 0.7180 & -0.1688 & -0.1481 & 0.1328 & 0.3593 \\ -0.1763 & -0.1688 & 0.8170 & -0.2042 & 0.1690 & 0.1405 \\ -0.2716 & -0.1481 & -0.2042 & 0.7594 & 0.1958 & 0.0836 \\ 0.2056 & 0.1328 & 0.1690 & 0.1958 & 0.8398 & -0.0879 \\ -0.2929 & 0.3593 & 0.1405 & 0.0836 & -0.0879 & 0.4744 \end{bmatrix}$$

(b) same as (a)

(c) $P_W\mathbf{v} = \mathbf{v}$ for all $\mathbf{v}$ in $\mathcal{S}$.

(d) $\left\{ \begin{bmatrix} -1.75 \\ -0.50 \\ -1.00 \\ -1.25 \\ 1.00 \\ 0.00 \end{bmatrix}, \begin{bmatrix} 0.85 \\ -0.60 \\ -0.10 \\ 0.05 \\ 0.00 \\ 1.00 \end{bmatrix} \right\}$ In each case, $P_W\mathbf{v} = \mathbf{0}$.

8. Answers are given correct to 4 places after the decimal point.

(b) $y = 0.5404 + 0.4091x$

(c) $y = 0.2981 + 0.7279x - 0.0797x^2$

9. In the case of the least squares line, the ith entry of $C * a$ is $a_0 + a_1x_i$, where x_i is the second entry of the ith row of C. Similarly, for the best quadratic fit, the ith entry of $C * a$ is $a_0 + a_1x_i + a_2x_i^2$.

10. Answers are given correct to 4 places after the decimal point.

(a) $y = 1.1504x$

(b) $y = 9.5167x$

11. Answers are given correct to 4 places after the decimal point.

(a) $P =$

$$\begin{bmatrix} -0.5000 & -0.5477 & -0.5000 & -0.4472 & 0.0000 \\ 0.5000 & -0.5477 & 0.5000 & -0.4472 & 0.0000 \\ -0.5000 & 0.3651 & 0.5000 & -0.4472 & 0.4082 \\ 0.0000 & 0.3651 & 0.0000 & -0.4472 & -0.8165 \\ 0.5000 & 0.3651 & -0.5000 & -0.4472 & 0.4082 \end{bmatrix}$$

$$D = \begin{bmatrix} -4 & 0 & 0 & 0 & 0 \\ 0 & 0 & 0 & 0 & 0 \\ 0 & 0 & -8 & 0 & 0 \\ 0 & 0 & 0 & 5 & 0 \\ 0 & 0 & 0 & 0 & 12 \end{bmatrix}$$

(b) The columns of P form an orthonormal basis, and the diagonal entries of D (in the same order) are the corresponding eigenvalues.

(c) $A = -4 \begin{bmatrix} 0.25 & -0.25 & 0.25 & 0.00 & -0.25 \\ -0.25 & 0.25 & -0.25 & 0.00 & 0.25 \\ 0.25 & -0.25 & 0.25 & 0.00 & -0.25 \\ 0.00 & 0.00 & 0.00 & 0.00 & 0.00 \\ -0.25 & 0.25 & -0.25 & 0.00 & 0.25 \end{bmatrix}$

$$+0\begin{bmatrix} 0.3000 & 0.3000 & -0.2000 & -0.2000 & -0.2000 \\ 0.3000 & 0.3000 & -0.2000 & -0.2000 & -0.2000 \\ -0.2000 & -0.2000 & 0.1333 & 0.1333 & 0.1333 \\ -0.2000 & -0.2000 & 0.1333 & 0.1333 & 0.1333 \\ -0.2000 & -0.2000 & 0.1333 & 0.1333 & 0.1333 \end{bmatrix}$$

$$-8\begin{bmatrix} 0.25 & -0.25 & -0.25 & 0.00 & 0.25 \\ -0.2500 & 0.2500 & 0.2500 & 0.00 & -0.25 \\ -0.2500 & 0.2500 & 0.2500 & 0.00 & -0.25 \\ 0.00 & 0.00 & 0.00 & 0.00 & 0.00 \\ 0.2500 & -0.2500 & -0.2500 & 0.00 & 0.25 \end{bmatrix}$$

$$+5\begin{bmatrix} 0.2 & 0.2 & 0.2 & 0.2 & 0.2 \\ 0.2 & 0.2 & 0.2 & 0.2 & 0.2 \\ 0.2 & 0.2 & 0.2 & 0.2 & 0.2 \\ 0.2 & 0.2 & 0.2 & 0.2 & 0.2 \\ 0.2 & 0.2 & 0.2 & 0.2 & 0.2 \end{bmatrix}$$

$$+12\begin{bmatrix} 0.0000 & 0.0000 & 0.0000 & 0.0000 & 0.0000 \\ 0.0000 & 0.0000 & 0.0000 & 0.0000 & 0.0000 \\ 0.0000 & 0.0000 & 0.1667 & -0.3333 & 0.1667 \\ 0.0000 & 0.0000 & -0.3333 & 0.6667 & -0.3333 \\ 0.0000 & 0.0000 & 0.1667 & -0.3333 & 0.1667 \end{bmatrix}$$

(d) $A_2 = \begin{bmatrix} -2 & 2 & 2 & 0 & -2 \\ 2 & -2 & -2 & 0 & 2 \\ 2 & -2 & 0 & -4 & 4 \\ 0 & 0 & -4 & 8 & -4 \\ -2 & 2 & 4 & -4 & 0 \end{bmatrix}$

(e) $\|E_2\| = 6.4031$ $\|A\| = 15.7797$

(f) 40.58%

12. Answers are given correct to 4 places after the decimal point.

(a) $U = \begin{bmatrix} -0.5404 & 0.6121 & 0.2941 & 0.4968 \\ -0.6121 & -0.5404 & -0.4968 & 0.2941 \\ -0.5762 & 0.0359 & 0.2028 & -0.7909 \\ 0.0359 & 0.5762 & -0.7909 & -0.2028 \end{bmatrix}$

$S = \begin{bmatrix} 7.5622 & 0 & 0 & 0 & 0 & 0 \\ 0 & 2.9687 & 0 & 0 & 0 & 0 \\ 0 & 0 & 0 & 0 & 0 & 0 \\ 0 & 0 & 0 & 0 & 0 & 0 \end{bmatrix}$

$V =$
$$\begin{bmatrix} -0.2286 & 0.0363 & -0.8821 & -0.4036 & 0.0507 & -0.0528 \\ 0.0142 & 0.5823 & 0.3581 & -0.7024 & 0.1371 & -0.1429 \\ -0.7000 & -0.4735 & 0.2658 & -0.1473 & 0.3045 & -0.3172 \\ -0.2286 & 0.0363 & 0.0538 & -0.0529 & -0.9050 & -0.3488 \\ -0.4430 & 0.6548 & -0.0922 & 0.5550 & 0.1674 & -0.1744 \\ -0.4572 & 0.0725 & 0.1076 & -0.1058 & -0.1970 & 0.8510 \end{bmatrix}$$

(b) The last 4 columns of V are the columns of Null A.

(c) the first 2 columns of U are the columns of `orth(A)`.

(d) Let $A = USV^T$ is a singular value decomposition of an $m \times n$ matrix A with k (not necessarily distinct) singular values. Then the first k columns of U form an orthonormal basis for Col A, and the last $n - k$ columns of V form an orthonormal basis for Null A.

13. Answers are given correct to 4 places after the decimal point.

$P_w =$
$$\begin{bmatrix} 0.3913 & 0.0730 & -0.1763 & -0.2716 & 0.2056 & -0.2929 \\ 0.0730 & 0.7180 & -0.1688 & -0.1481 & 0.1328 & 0.3593 \\ -0.1763 & -0.1688 & 0.8170 & -0.2042 & 0.1690 & 0.1405 \\ -0.2716 & -0.1481 & -0.2042 & 0.7594 & 0.1958 & 0.0836 \\ 0.2056 & 0.1328 & 0.1690 & 0.1958 & 0.8398 & -0.0879 \\ -0.2929 & 0.3593 & 0.1405 & 0.0836 & -0.0879 & 0.4744 \end{bmatrix}$$

14. Answers are given correct to 4 places after the decimal point.

$$\begin{bmatrix} 0.7550 \\ -0.0861 \\ 0.6556 \\ 0.9205 \\ -0.0795 \end{bmatrix}$$

15. Answers are given correct to 4 places after the decimal point.

(a) $A =$
$$\sigma_1\begin{bmatrix} 0.0618 & -0.1724 & 0.2088 & 0.0391 & -0.0597 \\ 0.1709 & -0.4769 & 0.5774 & 0.1082 & -0.1651 \\ -0.0867 & 0.2420 & -0.2930 & -0.0549 & 0.0838 \\ 0.0767 & -0.2141 & 0.2592 & 0.0486 & -0.0741 \end{bmatrix}$$

$$+\sigma_2\begin{bmatrix} 0.3163 & 0.1374 & -0.0541 & 0.5775 & 0.1199 \\ 0.0239 & 0.0104 & -0.0041 & 0.0436 & 0.0090 \\ 0.3291 & 0.1429 & -0.0563 & 0.6008 & 0.1247 \\ 0.0640 & 0.0278 & -0.0109 & 0.1168 & 0.0243 \end{bmatrix}$$

$$+\sigma_3\begin{bmatrix} 0.1316 & -0.2635 & -0.2041 & -0.0587 & 0.1452 \\ -0.2036 & 0.4076 & 0.3157 & 0.0908 & -0.2246 \\ -0.1470 & 0.2943 & 0.2280 & 0.0656 & -0.1622 \\ 0.1813 & -0.3630 & -0.2811 & -0.0808 & 0.2000 \end{bmatrix}$$

$$+\sigma_4\begin{bmatrix} -0.3747 & -0.2063 & -0.1482 & 0.2684 & -0.1345 \\ 0.0334 & 0.0184 & 0.0132 & -0.0239 & 0.0120 \\ 0.2571 & 0.1415 & 0.1016 & -0.1841 & 0.0922 \\ 0.5180 & 0.2852 & 0.2048 & -0.3710 & 0.1859 \end{bmatrix}$$

where $\sigma_1 = 205.2916$, $\sigma_2 = 123.3731$, $\sigma_3 = 50.3040$, and $\sigma_4 = 6.2391$

(b) $A_2 =$
$$\begin{bmatrix} 51.7157 & -18.4559 & 36.1913 & 79.2783 & 2.5334 \\ 38.0344 & -96.6198 & 118.0373 & 27.5824 & -32.7751 \\ 22.7926 & 67.3103 & -67.1013 & 62.8508 & 32.5841 \\ 23.6467 & -40.5194 & 51.8636 & 24.3814 & -12.2222 \end{bmatrix}$$

$E_2 =$
$$\begin{bmatrix} 4.2843 & -14.5441 & -11.1913 & -1.2783 & 6.4666 \\ -10.0344 & 20.6198 & 15.9627 & 4.4176 & -11.2249 \\ -5.7926 & 15.6897 & 12.1013 & 2.1492 & -7.5841 \\ 12.3533 & -16.4806 & -12.8636 & -6.3814 & 11.2222 \end{bmatrix}$$

(c) $\|E_2\| = 50.6894$ $\|A\| = 244.8163$

(d) $\dfrac{\|E_2\|}{\|A\|} = 0.2071$

16. Vector entries are given correct to 4 places after the decimal point.

(a) $\begin{bmatrix} 0.8298 \\ -0.1538 \\ 0.5364 \end{bmatrix}$, $\theta = 38^\circ$ (b) $\begin{bmatrix} 0.8298 \\ 0.1538 \\ 0.5364 \end{bmatrix}$, $\theta = 38^\circ$

17. (a) We use the rational format in MATLAB to obtain

$$A_W = \begin{bmatrix} 2/3 & -2/3 & 1/3 \\ -2/3 & -1/3 & 2/3 \\ 1/3 & 2/3 & 2/3 \end{bmatrix}.$$

(c) $\mathbf{v} = \begin{bmatrix} -2 \\ -1 \\ 2 \end{bmatrix}$ is a vector that lies on the axis of rotation, and the angle of rotation is 23°.

18. (b) Let $C = A_W V$, where A_W is the matrix in 17(a) and V is obtained in (a). Then apply the MATLAB command grfig(C,E), where E is obtained in (a).

Chapter 7

Section 7.1

1. no 3. yes 5. no 7. no
9. yes 11. no 13. yes 15. yes
17. yes 19. no 21. yes 23. yes
25. yes 27. no
29. The coefficients are 3 and 5.
33. T 34. F 35. F 36. T 37. F
38. F 39. T 40. T 41. T 42. T
43. T 44. F 45. T 46. T 47. T
48. T 49. T 50. T 51. T 52. T
53. T 54. T
61. no 63. yes 65. yes 67. no
69. yes 71. yes

Section 7.2

1. yes 3. no 5. no 7. yes
9. yes 11. yes 13. no 15. no
25. $a + 3b + 2c + 4d$
27. $\begin{bmatrix} a + 2b + 3c \\ b + 4c \end{bmatrix}$ 29. $\begin{bmatrix} s & t \\ t & u \end{bmatrix}$
31. linear, not an isomorphism
33. not linear
35. linear, an isomorphism
37. linear, not an isomorphism
39. T 40. F 41. T 42. T 43. T
44. F 45. T 46. F 47. T 48. F
55. (d) the zero transformation T_0

Section 7.3

1. linearly dependent 3. linearly dependent
5. linearly independent 7. linearly independent
9. linearly dependent 11. linearly independent
13. linearly independent 15. linearly independent
17. linearly independent 19. linearly dependent
21. linearly independent 23. linearly independent
25. $2x^2 - 3x + 1$ 27. $-2x^2 + 6x - 3$
29. $x^3 - 4x + 2$
31. F 32. F 33. F 34. F 35. F
36. T 37. T 38. T 39. F 40. T
41. T 42. F 43. T 44. T 45. F
46. T 47. T 48. F
49. linearly dependent

51. $$\left\{ \begin{bmatrix} 1 & 0 & 0 \\ 0 & 0 & 0 \\ 0 & 0 & 0 \end{bmatrix}, \begin{bmatrix} 0 & 0 & 0 \\ 0 & 1 & 0 \\ 0 & 0 & 0 \end{bmatrix}, \begin{bmatrix} 0 & 0 & 0 \\ 0 & 0 & 0 \\ 0 & 0 & 1 \end{bmatrix}, \begin{bmatrix} 0 & 1 & 0 \\ 1 & 0 & 0 \\ 0 & 0 & 0 \end{bmatrix}, \begin{bmatrix} 0 & 0 & 1 \\ 0 & 0 & 0 \\ 1 & 0 & 0 \end{bmatrix}, \begin{bmatrix} 0 & 0 & 0 \\ 0 & 0 & 1 \\ 0 & 1 & 0 \end{bmatrix} \right\}$$

53. $$\left\{ \begin{bmatrix} 1 & 0 \\ 0 & -1 \end{bmatrix}, \begin{bmatrix} 0 & 1 \\ 0 & 0 \end{bmatrix}, \begin{bmatrix} 0 & 0 \\ 1 & 0 \end{bmatrix} \right\}$$

55. $\{1 - x^n, x - x^n, \ldots, x^{n-1} - x^n\}$ 57. $\{1, x\}$
79. The set is linearly independent.
81. The set is linearly dependent, and $M_3 = (-3)M_1 + 2M_2$, where M_j is the jth matrix in the set.
83. (rounded to 4 places after the decimal) $c_0 = 0.3486$, $c_1 = 0.8972$, $c_2 = -0.3667$, $c_3 = 0.1472$, $c_4 = -0.0264$

Section 7.4

1. $[A]_{\mathcal{B}} = \begin{bmatrix} 1 \\ 3 \\ 4 \\ 2 \end{bmatrix}$

3. $[\sin 2t - \cos 2t]_{\mathcal{B}} = [2 \sin t \cos t - \cos^2 t + \sin^2 t]_{\mathcal{B}}$
$= \begin{bmatrix} -1 \\ 1 \\ 2 \end{bmatrix}$

5. $\begin{bmatrix} -3 \\ 2 \\ 1 \end{bmatrix}$ 7. $\begin{bmatrix} -3 \\ -2 \\ 1 \end{bmatrix}$ 9. $\begin{bmatrix} 1 & 0 & 0 \\ 0 & 2 & 0 \\ 0 & 0 & 3 \end{bmatrix}$

11. $\begin{bmatrix} 1 & 0 & 0 \\ 3 & 3 & 3 \\ 1 & 2 & 4 \end{bmatrix}$ 13. $\begin{bmatrix} 0 & 1 & -2 & 0 \\ 0 & 0 & 2 & -6 \\ 0 & 0 & 0 & 3 \\ 0 & 0 & 0 & 0 \end{bmatrix}$

15. $\begin{bmatrix} 1 & 0 & 0 & 0 \\ 0 & 0 & 1 & 0 \\ 0 & 1 & 0 & 0 \\ 0 & 0 & 0 & 1 \end{bmatrix}$

17. (a) $-8x$
(b) $3 + 10x$
(c) $3x^2$

19. (a) $-e^t + te^t$
(b) $2e^t - 2te^t + t^2e^t$
(c) $11e^t - 8te^t + 2t^2e^t$

21. 1, 2, 3, $\{e^t\}$, $\{e^{2t}\}$, $\{e^{3t}\}$

23. 1, 6, $\{3x - 2x^2\}$, $\{x + x^2\}$ 25. 0, $\{1\}$

27. $1, -1, \left\{\begin{bmatrix} 1 & 0 \\ 0 & 0 \end{bmatrix}, \begin{bmatrix} 0 & 1 \\ 1 & 0 \end{bmatrix}, \begin{bmatrix} 0 & 0 \\ 0 & 1 \end{bmatrix}\right\}, \left\{\begin{bmatrix} 0 & 1 \\ -1 & 0 \end{bmatrix}\right\}$

28. F 29. F 30. T 31. T 32. T
33. T 34. F 35. F 36. F 37. F
38. T 39. T

41. (a) $\lambda = 0$ (b) $\{1\}$

45. (b) $\begin{bmatrix} 1 & 0 & 0 & 1 \\ 2 & 0 & 0 & 2 \\ 3 & 0 & 0 & 3 \\ 4 & 0 & 0 & 4 \end{bmatrix}$

51. (b) $\begin{bmatrix} 1 & 3 & 0 & 0 \\ 0 & 0 & 1 & 3 \end{bmatrix}$
(c) $\begin{bmatrix} 2 & 6 & -1 & -3 \\ -1 & -3 & 1 & 3 \end{bmatrix}$

53. (b) $\begin{bmatrix} 1 & 1 & 1 \\ 1 & 2 & 4 \end{bmatrix}$
(c) $[T(f(x))]_{\mathcal{C}} = [T]_{\mathcal{B}}^{\mathcal{C}}[f(x)]_{\mathcal{B}} = \begin{bmatrix} a + b + c \\ a + 2b + 4c \end{bmatrix}$

55. (rounded to 4 places after the decimal)
(a) $-1.6533, 2.6277, 6.6533, 8.3723$
(b) $\left\{\begin{bmatrix} -0.1827 & -0.7905 \\ 0.5164 & 0.2740 \end{bmatrix}, \begin{bmatrix} 0.6799 & -0.4655 \\ -0.4655 & 0.3201 \end{bmatrix}, \begin{bmatrix} 0.4454 & 0.0772 \\ 0.5909 & -0.6681 \end{bmatrix}, \begin{bmatrix} 0.1730 & 0.3783 \\ 0.3783 & 0.8270 \end{bmatrix}\right\}$
(c) $\begin{bmatrix} 0.2438a - 0.1736b & -0.2603a - 0.2893b \\ 0.0124a + 0.3471b & -0.1116a + 0.1240b \end{bmatrix} + \begin{bmatrix} 0.0083c + 0.0496d & 0.3471c + 0.0826d \\ 0.0165c - 0.0992d & -0.1488c + 0.1074d \end{bmatrix}$

Section 7.5

1. $\frac{15}{4}$ 3. $\frac{21}{4}$ 5. $\frac{21}{2}$ 7. e^2
9. 25 11. 0 13. -3 15. -3
17. 12 19. $-\frac{50}{3}$ 21. 0 23. $-\frac{8}{3}$
25. F 26. T 27. F 28. F 29. T
30. F 31. T 32. T 33. T 34. T
35. F 36. T 37. T 38. F 39. F
40. F 41. T 42. T 43. F 44. F
51. yes 53. no 55. yes 57. yes

61. $\left\{1, e^t - e + 1, e^{-t} + \frac{e^2 - 2e - 1}{e(e-3)} - \frac{2(e^2 - 3e + 1)}{e(e-3)(e-1)}e^t\right\}$

75. (a) $\left\{\begin{bmatrix} 1 & 0 \\ 0 & 0 \end{bmatrix}, \frac{1}{\sqrt{2}}\begin{bmatrix} 0 & 1 \\ 1 & 0 \end{bmatrix}, \begin{bmatrix} 0 & 0 \\ 0 & 1 \end{bmatrix}\right\}$ (b) $\begin{bmatrix} 1 & 3 \\ 3 & 8 \end{bmatrix}$

Chapter 7 Review Exercises

1. F 2. T 3. F 4. F 5. T
6. F 7. T
9. no 11. yes 13. no 15. no
17. no 19. yes
21. $c = 5$
23. $\{-1 + x, -2 + x^2, x^3\}$, $\dim W = 3$
25. not linear
27. linear, an isomorphism

29. $\begin{bmatrix} a & b \\ -b & a \end{bmatrix}$ 31. $\begin{bmatrix} 3 & 0 & 0 & 0 \\ 0 & 2 & 1 & 0 \\ 0 & 1 & 2 & 0 \\ 0 & 0 & 0 & 3 \end{bmatrix}$

33. $\frac{1}{a^2 + b^2}(ac_1 - bc_2)e^{at}\cos bt + \frac{1}{a^2 + b^2}(bc_1 + ac_2)e^{at}\sin bt$

35. $\frac{1}{3}\begin{bmatrix} a & 2b - c \\ -b + 2c & d \end{bmatrix}$

37. T has no (real) eigenvalues.

39. 3 and 1, with corresponding bases $\left\{\begin{bmatrix} 1 & 0 \\ 0 & 0 \end{bmatrix}, \begin{bmatrix} 0 & 1 \\ 1 & 0 \end{bmatrix}, \begin{bmatrix} 0 & 0 \\ 0 & 1 \end{bmatrix}\right\}$ and $\left\{\begin{bmatrix} 0 & 1 \\ -1 & 0 \end{bmatrix}\right\}$, respectively

41. $\left\{\begin{bmatrix} -3 & 1 \\ 0 & 0 \end{bmatrix}, \begin{bmatrix} -4 & 0 \\ 1 & 0 \end{bmatrix}, \begin{bmatrix} -2 & 0 \\ 0 & 1 \end{bmatrix}\right\}$

43. $\begin{bmatrix} 2 & -2 \\ 2 & -3 \end{bmatrix}$

45. $\left\{1, \sqrt{3}(2x - 1), \sqrt{5}(6x^2 - 6x + 1)\right\}$

47. $\frac{6}{35} + \frac{48}{35}x - \frac{4}{7}x^2$

Chapter 7 MATLAB Exercises

1. The set is linearly independent.
2. The set is linearly dependent.

$$\begin{bmatrix} 1 & -3 \\ 4 & 1 \end{bmatrix} = 2\begin{bmatrix} 1 & -1 \\ 3 & 1 \end{bmatrix} + (-1)\begin{bmatrix} 1 & 2 \\ 1 & 2 \end{bmatrix}$$
$$+ (1)\begin{bmatrix} 0 & 1 \\ -1 & 1 \end{bmatrix}$$

3. (b) $c_0 = 20,\ c_1 = -50,\ c_2 = 55,\ c_3 = -29,\ c_4 = 6$
4. (b) $T^{-1}(t^2 \sin t) = 0.324 \cos t - 0.532 \sin t$
$$+\, 0.680t \cos t - 0.240t \sin t$$
$$+\, 0.300t^2 \cos t + 0.100t^s \sin t.$$
5. (a) $8, 4, -4, -8$

(b) $\left\{\begin{bmatrix} -3 & 3 & 3 \\ -1 & 1 & 1 \end{bmatrix}, \begin{bmatrix} 3 & 0 & 3 \\ 0 & 1 & 0 \end{bmatrix}, \begin{bmatrix} 2 & 1 & 2 \\ 1 & 0 & 1 \end{bmatrix},\right.$
$\left.\begin{bmatrix} -3 & 2 & -3 \\ 0 & 1 & 0 \end{bmatrix}, \begin{bmatrix} 0 & -1 & 0 \\ 1 & 0 & 1 \end{bmatrix}, \begin{bmatrix} 1 & -1 & -1 \\ -1 & 1 & 1 \end{bmatrix}\right\}$

Note that the first matrix in the basis has corresponding eigenvalue 8, the second and third have corresponding eigenvalue 4, the fourth and fifth have corresponding eigenvalue −4, and the sixth has corresponding eigenvalue −8.

6. $$P = \frac{1}{18}\begin{bmatrix} 5 & -1 & 2 & -1 & 2 & 5 & 2 & 5 & -1 \\ -1 & 8 & -1 & 2 & 2 & 2 & 5 & -4 & 5 \\ 2 & -1 & 5 & 5 & 2 & -1 & -1 & 5 & 2 \\ -1 & 2 & 5 & 8 & 2 & -4 & -1 & 2 & 5 \\ 2 & 2 & 2 & 2 & 2 & 2 & 2 & 2 & 2 \\ 5 & 2 & -1 & -4 & 2 & 8 & 5 & 2 & -1 \\ 2 & 5 & -1 & -1 & 2 & 5 & 5 & -1 & 2 \\ 5 & -4 & 5 & 2 & 2 & 2 & -1 & 8 & -1 \\ -1 & 5 & 2 & 5 & 2 & -1 & 2 & -1 & 5 \end{bmatrix}$$

INDEX

Q

R

S

T

U